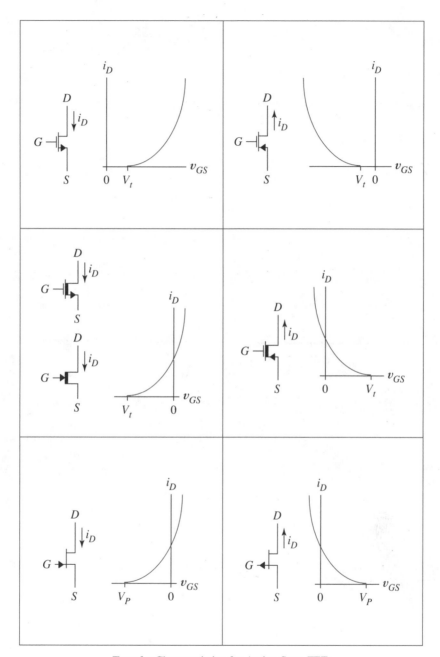

Transfer Characteristics for Active-State FETs

ELECTRONIC CIRCUITS

ELECTRONIC CIRCUITS

Analysis, Simulation, and Design

Norbert R. Malik

Department
of Electrical
and Computer Engineering

University of Iowa

Prentice Hall
Englewood Cliffs, New Jersey 07632

Library of Congress Cataloging-in-Publication Data

MALIK, NORBERT R.
    Electronic circuits : analysis, simulation, and design / Norbert
  R. Malik.   p.   cm.
    Includes bibliographical references and index.
    ISBN 0-02-374910-5
    1. Electronic circuits.   I. Title.
TK7867.M3244 1995
621.381'078—dc20

94-26013
CIP

Acquisitions Editor: Alan Apt
Editorial production and supervision: Gretchen K. Chenenko
Production Coordinators:
    Aliza Greenblatt/Bayani Mendoza de Leon
Cover Designer: Gryphon Three
Buyer: Lori Bulwin
Illustrations: Academy ArtWorks, Inc.

© 1995 by Prentice-Hall, Inc.
A Simon & Schuster Company
Englewood Cliffs, New Jersey 07632

Printed in the United States of America

10 9 8 7 6 5 4 3 2 1

0-02-374910-5

Prentice-Hall International (UK) Limited, *London*
Prentice-Hall of Australia Pty. Limited, *Sydney*
Prentice-Hall Canada Inc., *Toronto*
Prentice-Hall Hispanoamericana, S.A., *Mexico*
Prentice-Hall of India Private Limited, *New Delhi*
Prentice-Hall of Japan, Inc., *Tokyo*
Simon & Schuster Asia Pte. Ltd., *Singapore*
Editora Prentice-Hall do Brasil, Ltda., *Rio de Janeiro*

Preface

This text is primarily intended for a two-semester electronics sequence for majors in electrical engineering. The text covers a multitude of topics in reasonable depth; therefore, the instructor can use selective omission to tailor the courses to meet individual program needs. The text organization also serves the special needs of computer engineers by allowing readers to advance rapidly to digital topics, omitting many details of linear applications.

The text assumes a prerequisite course in circuit theory that includes sinusoidal steady-state and transient analysis of first-order RC circuits. A concurrent course in systems or circuit theory that includes Laplace transforms and transfer functions is helpful when studying the second half of the book. Chapters 13 and 14 assume familiarity with ideal logic gates.

The chapter sequence begins with instrumentation-oriented topics such as operational amplifiers and waveshaping, an ordering that gives students an early return on their study investment and facilitates a concurrent laboratory where interesting concepts and simple designs can immediately be explored. Early emphasis on operational amplifiers also promotes top-down design by encouraging students, from the very beginning, to visualize electronic circuits in terms of functional modules.

Beginning in Chapter 1, the book emphasizes the interrelationships between graphical, mathematical, and circuit representations of devices. In the author's experience such emphasis, continued throughout the text, helps students better integrate and retain the material. Chapter 1 also sets the stage for using computer simulation to support the study of electronics as well as its application.

SPICE modeling is fully integrated into the text, not relegated to an appendix. This integration establishes a learning environment in which students take for granted the powerful numerical capabilities of computers as they learn the subject. The text continually focuses students' attention on the importance of using algebraic hand analysis and simple models to develop understanding and then emphasizes, by carefully selected examples, how to use SPICE to extend this understanding beyond the limits of the simple models. This approach enables meaningful treatment of important topics that are traditionally omitted or covered only superficially because of their mathematical difficulty:

v

charge storage delays, sensitivity, and distortion analysis are three examples. SPICE discussions and examples are very basic and do not favor any particular version of SPICE. Text examples were developed using the student version of PSPICE™, and class tested by students using a network implementation of SPICE II.

The text develops new concepts in traditional fashion, by analysis; however, ideas are then summarized or re-posed from the perspective of the designer, thereby adding new meaning to what would otherwise be simply a collection of analysis equations. SPICE is used to complement this practical design perspective. For example, SPICE numerical sensitivities are used to identify those circuit and device parameters most critical in meeting specifications. To discover how to improve the circuit, the designer then works out algebraic sensitivity expressions for only the most critical parameters. Practical design ideas of this kind permeate the development of virtually all topics rather than being relegated to a few special examples and problems. Examples and exercises emphasize critical analysis of SPICE results in terms of theoretical principles.

The accompanying flow chart shows the essential prerequisites for each chapter. Chapters 4 and 5, which introduce BJTs and FETs, exclusively stress large-signal models and their relationship to the characteristic curves. Small-signal models are briefly introduced in Chapter 6 and then formally defined in Chapter 7. This approach avoids the confusion usually associated with trying to learn too many models before the models can be adequately reinforced by applications. The initial emphasis on large-signal models also equips students to either branch directly to digital circuits in Chapter 13 or, more conventionally, to continue into Chapter 6 where the models are applied to bias circuit analysis and design.

Any chapter sequence consistent with the flow chart can be taught with few difficulties. For example, operational amplifiers can be deferred until after transistor circuits, a preference for some instructors. Three years of class testing involved several variations on the following basic theme: a first course covering most of Chapters 1–5 plus the bias circuit design portion of Chapter 6 and a second course beginning with sensitivity in Chapter 6, and including the more important topics from Chapters 7–10, 12, and 13.

I acknowledge many debts to contemporary authors, especially Sedra and Smith, Gray and Meyer, and Hodges and Jackson, whose lucid and creatively written textbooks have strongly infuenced my thinking, teaching, and presumably, this text. From an earlier work by Angelo I first learned what a pleasure it could be to teach from a text that students are able to read. Also greatly appreciated are the support of the Department of Electrical and Computer Engineering, especially for granting leave at a critical time, and the nurturing environment and excellent facilities of the University of Iowa Center for Advanced Studies. Many students and teaching assistants contributed to the project, both in the classroom and behind the scenes. Especially noteworthy were the efforts of Terry Shie, Tom Cross, Geetani Edirisooriya, Brian Sobeks, and Alissa Chan who helped formulate and check the solutions manual.

In a more personal vein, I credit Rosemary Malik and Lumir Samek with vital early guidance, Bill Streib and Margaret McDowell for classroom inspiration, and my wife Margaret for unfailing support.

Norbert R. Malik

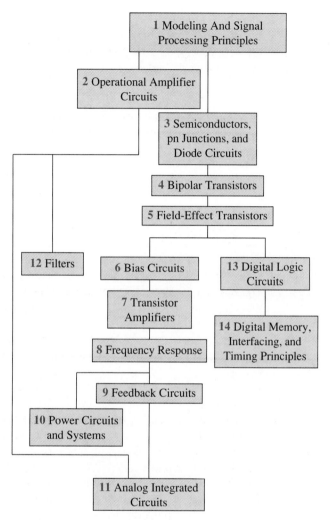

Chapter Prerequisite Structure

Contents

CHAPTER

1

Modeling and Signal Processing Principles

The subject of electronics differs significantly from the prerequisite circuit theory course. In circuit theory, voltages and currents were either elementary time functions such as sinusoids or exponentials, or were dc quantities having no special meaning or significance. In *electronics,* applied voltages and currents are usually *signals* that contain information to be *processed* in some way by the circuit. In *analog* systems, the information is encoded as changes in voltage or current waveforms, and processing might involve making the signal larger or smaller, removing noise, changing its shape, determining its peak or average value, or combining it with another signal. In digital systems, sequences of binary numbers are the signals; high and low values of a current or voltage, respectively, represent the binary ones and zeros. Digital electronic circuits perform arithmetic and other sophisticated information processing operations on the binary numbers using interconnections of special-purpose electronic circuits called logic gates.

We can identify electronic circuits by the presence of special devices that modify signal shape or amplify signal power. By this criterion, electronics originated in 1896 with wireless signal transmission by Marconi in Italy and Popov in Russia. Early radio receivers detected signals using either point contact devices, called cat whiskers, or vacuum diodes to change the signal waveshape. *Modern* electronics dates to 1907 when Lee DeForest added a control grid to the vacuum diode, an 1883 discovery of Thomas Edison. DeForest's discovery, that a tiny input signal applied to the control grid could control a large output signal, is the basis of amplification. In 1947 a team at Bell Telephone Laboratories, William Schockley, John Bardeen, and Walter Brattain, developed the point contact transistor, a solid-state nonvacuum device that performed the same function. This discovery earned a Nobel Prize for the inventors and led to advances in miniaturization, reliability, operating speed, and cost of electronic circuits. The integrated circuit (IC), independently invented in 1959 by Jack Kilby at Texas Instruments and Robert Noyce at Fairchild Semiconductor, led to similar revolutionary advances. The IC concept enabled fabrication of circuits consisting of thousands of transistors plus associated resistors and capacitors, all at the same time on a single semiconductor chip. The IC also eliminated the need for circuit-level assembly and led to great cost reductions.

1

Associated with signal processing is a sense of directed information flow through the circuit. One side of the circuit, by convention the left-hand side, functions as the *input,* where the signal enters in the form of a voltage or current. The other side of the circuit, the right-hand side, functions as *output,* where the processed signal leaves the circuit and becomes available for observation, for performing some useful function, or for processing by another circuit.

Sources connected to the input sometimes represent *input transducers,* devices that convert a physical quantity of interest such as sound, temperature, pressure, flow rate, or acceleration into an electrical signal encoded in voltage or current fluctuations. We sometimes refer to these as alternating current (ac) signals because they change with time. In some cases, however, the signal may remain constant for long intervals (direct current or dc) and still be a signal in the sense of containing information. Examples are temperature, light intensity, mechanical strain, and pressure.

Diagrams of electronic circuits often show a special resistor at the output, designated as the *load* or *load resistor,* across or through which the output signal is developed. This load resistor sometimes represents an *output transducer* that converts electrical energy into another physical form, for example, a sonar source that converts electrical energy into underwater pressure waves. Or instead, the load resistor might represent the input of another signal processing circuit, or the load might simply *be* a resistor.

Some electronic circuits serve as sources of signals that are used for various purposes such as timing and laboratory testing of other circuits. These sources are examples of electronic circuits that have outputs but require no input signals.

Another idea introduced in electronics is *biasing.* Devices are able to amplify signals only by virtue of converting dc power into signal power. Therefore, in addition to the signals, we must apply dc voltages and currents to our electronic circuits. Consequently, in most electronic circuits, signal voltages and currents are superimposed upon dc *bias* levels. Figure 1.1a represents such a circuit. A dc *power supply* V_{PP} with average current I_P provides dc power $V_{PP}I_P$ to the circuit. Signal voltage v_i and signal current i_i contain information to be processed, and R_L is the load resistor, the destination of the processed signal. Suppose that the signal v_i is off until $t = 0$, turned on for T seconds, and then turned off again. The output current and voltage waveforms might then take the form of Figs. 1.1b and c. Before $t = 0$ and after $t = T$, only a dc biasing voltage V_L and current I_L appear at the output. When the input signal is present, however, v_L and i_L both contain signal components *superimposed* upon the underlying dc levels V_L and I_L.

Most digital circuits do not directly employ biasing; however, their ability to refresh signal levels that have deteriorated during transmission depend upon power-amplifying devices that operate in conjunction with power supplies. The concept of signals combined with dc levels is important in designing digital interface circuits.

To distinguish between the different kinds of voltages and currents, the electronics community observes the following notational conventions. Uppercase characters with uppercase subscripts such as V_{PP}, I_P, V_L, and I_L describe dc voltages and currents. Lowercase characters with lowercase subscripts such as v_i, i_i, v_l, and i_l denote signals. Lowercase characters with uppercase subscripts such as v_L and i_L denote *total* voltages and currents, which might include both dc bias values and signal components. Thus in Fig.

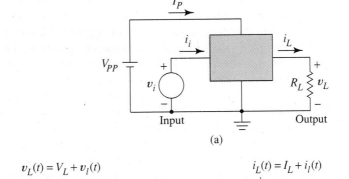

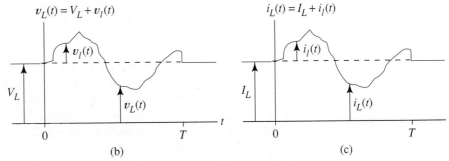

Figure 1.1 Notation for electronic circuits: (a) dc, signal, and total currents and voltages in a circuit; (b) components of a typical voltage waveform; (c) components of a typical current waveform.

1.1b, for $t < 0$ and for $t > T$, $i_L = I_L$. For $0 \leq t \leq T$, $i_L = i_l + I_L$. These distinctions become particularly important with the introduction of small-signal analysis in Chapter 7.

1.1
Human-Computer Synergism

Once, much of an engineer's time was occupied with tedious calculations. Today engineers have been largely supplanted in such tasks by computers because of the computer's superior speed, precision, and memory. Nevertheless, computers still remain inferior in most tasks that involve "judgment" or "understanding." Recognizing that humans and computers have abilities that are largely complementary, it is important that we engineers perfect those traits that best enable us to work in synergism rather than in competition with computers. Thus we should concentrate on remembering basic principles and how to apply them, while concurrently learning to utilize computers to extend our understanding into areas where we are limited by our more modest computational abilities.

Electronic devices and circuits are complex, yet their fundamental operation is readily understood through models that sacrifice accuracy for simplicity. These models help us visualize and make rough mathematical predictions of how the physical circuit will perform. Meanwhile, powerful computer simulation packages are available to aid us in our analysis and design tasks. These programs use sophisticated models that make mathematical predictions far closer to the actual performance than do the simple models humans use directly. The computer, however, cannot interpret the results of its predictions nor (usually) improve upon the original design. Here human understanding and intuition, honed by experience with simple models, are required to evaluate the results and suggest improvements.

In this chapter we learn some elementary modeling concepts and use them to describe simple amplifiers. In subsequent chapters we apply these same concepts to a wide variety of devices and circuits. The models provide the mind with an efficient and relatively effortless coding mechanism for collecting, storing, relating, and retrieving many important facts and ideas without much need for memorizing. Further, they impart a quality of understanding that enables us to bring previous experience to bear in learning new concepts and in analyzing new and unfamiliar circuits.

A computer program called SPICE (Simulation Program, Integrated Circuit Emphasis) helps us apply the modeling ideas to complex circuits. Simulations also make us more effective in the laboratory because preliminary design decisions are easily tested, verified, and refined by simulation before a prototype circuit is actually constructed.

SPICE performs dc, ac, or transient analysis on any circuit that we describe to the SPICE program. The information SPICE requires is exactly what we would need for hand analysis. We provide a circuit diagram by entering a statement for each element: resistor, inductor, capacitor, or source. This statement gives a node number for each connection point and, for passive elements and dc sources, the appropriate numerical value. For time-varying sources, we use SPICE conventions to indicate the nature of the function. Special control statements indicate the desired type of analysis and the numerical output values we wish to examine. As each new electronic device is introduced in subsequent chapters, we describe the corresponding SPICE model so we can gradually grow accustomed to having this powerful computational tool to complement our efforts.

SPICE results—*numbers*—differ significantly from the results of hand analysis—*algebraic equations*. The equations reveal how the individual components contribute to overall circuit performance, and thereby give an insight that is not provided by numerical results alone. This insight is most important to designers who often need to improve on the circuit being analyzed.

1.2
Volt-Ampere and Transfer Characteristics

We use three different device descriptions to analyze and design electronic circuits and to visualize how they operate: (1) mathematical equations, (2) circuit models, and (3) graphical device curves. The mathematical equations give precise quantitative descriptions of the individual devices, the circuit models permit systematic analysis of interconnected devices using the formalisms of circuit theory, and the graphical representations facilitate use of human pattern-recognition capabilities to make associations with idealized devices. Although they represent different viewpoints, these descriptions are closely related. A clear understanding of exactly how they are related makes electronics a much easier subject. The key concept is a simple plot of current versus voltage for a given device. By convention such plots are widely known as *volt-ampere curves*, although ampere-volt curve would be a more appropriate name.

1.2.1 VOLT-AMPERE CURVES

We begin with some familiar *one-port* circuit elements, that is, elements having only a single pair of nodes called a *port*.

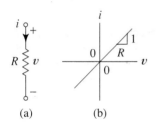

Figure 1.2 Resistor: (a) schematic symbol; (b) volt-ampere curve.

Resistor. To introduce the idea of a volt-ampere curve, consider first the resistor. Associated with this device is the schematic symbol of Fig. 1.2a. Whenever we insert a resistor into a circuit, we add two quantities, v and i, to the set of unknown voltages and currents; we also introduce a circuit constraint. For the resistor this constraint is Ohm's law, Eq. (1.1), which characterizes the resistor by relating v and i.

$$i = \left(\frac{1}{R}\right)v \tag{1.1}$$

The volt-ampere curve for the resistor is a plot of Eq. (1.1) in a current versus voltage coordinate system, the straight-line graph of Fig. 1.2b, where the reciprocal of the slope is the resistance R. The volt-ampere curve is simply the graphical representation of the equation that characterizes the circuit element, or of the constraint that the element places on the circuit.

The schematic symbol, the equation, and the volt-ampere curve are equivalent representations of the same object, and one must be able to translate between these representations with ease. For example, for any device having a straight line through the origin for its volt-ampere curve, we can always write an equation of the form of Eq. (1.1) to describe this device mathematically, and we can always represent it by the schematic symbol of Fig. 1.2a.

Independent Voltage Source. An abstract view of an independent voltage source starts with the schematic symbol of Fig. 1.3a. When we add this component to a circuit, we introduce new quantities, v and i, just as we do for a resistor. For the voltage source, however, there is no equation to relate v and i. Instead, the device is characterized by

$$v = V_{SS} \tag{1.2}$$

and

$$i = ? \tag{1.3}$$

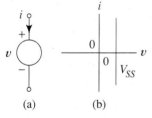

Figure 1.3 Independent voltage source: (a) schematic symbol; (b) volt-ampere curve.

Equation (1.2) states that the voltage can take on only the prescribed value V_{SS}. Equation (1.3) emphasizes that the device's current is not constrained in any way by the source definition. This means that the current can have *any* value whatsoever—positive, negative, or zero.

Figure 1.3 shows the constraint more clearly. The volt-ampere curve is a line of constant voltage located at $v = V_{SS}$. The amount of current depends entirely on the other circuit components that are connected to the source, not on the source itself. In fact, the *only* way to determine the current in an ideal voltage source is to apply Kirchhoff's current law (KCL) at one of its nodes. The ability of the ideal voltage source to deliver unlimited power, vi, to an external circuit is implicit in its definition, because i can be arbitrarily large while v is fixed.

So far, we have characterized a dc voltage source. For an ac source we replace Eq. (1.2) by a given time function $v = v_s(t)$. This means that the vertical volt-ampere curve in Fig. 1.2 shifts to the right and left with time as required by $v_s(t)$. All other observations about the voltage source remain valid.

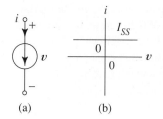

Figure 1.4 Independent current source: (a) schematic symbol; (b) volt-ampere curve.

Independent Current Source. The ideal current source of Fig. 1.4a also introduces two element quantities and one circuit constraint. Mathematically,

$$i = I_{SS} \tag{1.4}$$

and

$$v = ? \tag{1.5}$$

The constraint is that the current have the known value, I_{SS}. The voltage is unspecified and can only be determined by applying Kirchhoff's voltage law (KVL) to some loop containing the source. Figure 1.4b, a constant current (horizontal) line intersecting the current axis at value I_{SS}, shows this graphically. The independent current source is also capable of delivering infinite power. For an ac current source the constant current line shifts up and down in some prescribed fashion.

Short and Open Circuits. Hardly meriting the term "circuit element," short circuits and open circuits nevertheless provide circuit constraints similar to those of real circuit elements, and it is useful to study their abstract descriptions. The short circuit of Fig. 1.5a adds to the circuit a voltage, $v = 0$, and also a current, i, that can have any value. This constraint is graphed in Fig. 1.5b. When we compare Fig. 1.5b with Fig. 1.3b, we notice that the short circuit is an independent voltage source with $V_{ss} = 0$, for the two have the same volt-ampere curves. Often, in superposition arguments, it is necessary to *turn off* voltage sources. From the volt-ampere curve we see that turning off a voltage source involves sliding the constant voltage line of Fig. 1.3b to the origin, that is, setting the specified voltage to zero. Thus turning off a voltage source is equivalent to replacing the voltage source with a short circuit in the circuit diagram.

A similar comparison of Figs. 1.5c and d with Fig. 1.4 shows that an open circuit in a circuit diagram is equivalent to an independent current source that has been turned off or set to zero.

There is another useful way to view short and open circuits. Figures 1.5 and 1.2 show that a short circuit is the limiting case of a resistor as the resistance approaches zero, and an open circuit is the limiting case of a resistor as its resistance approaches infinity.

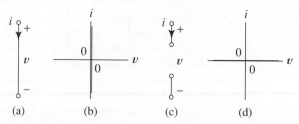

Figure 1.5 Degenerate circuit elements: (a) short circuit; (b) volt-ampere curve for short circuit; (c) open circuit; (d) volt-ampere curve for open circuit.

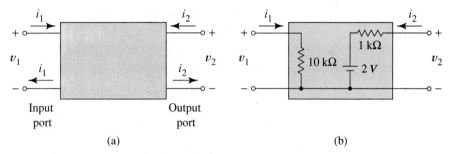

Figure 1.6   Two-port circuit: (a) general notion and notation; (b) example.

1.2.2   INPUT, OUTPUT, AND TRANSFER CHARACTERISTICS FOR TWO-PORT DEVICES

Many electronic devices are classified as *two-ports,* circuits that communicate with the outside world exclusively though an input *port* and an output *port,* as in Fig. 1.6a. The word "port" implies a node pair in which the current that enters one port node also leaves the other port node. Inside the two-port box is either an electronic device or some interconnection of circuit elements, such as those in Fig. 1.6b.

An interesting two-port example is the voltage-controlled voltage source (VCVS) of Fig. 1.7a. A two-port adds *four* quantities to the circuit unknowns, input variables v_1 and i_1, and output variables v_2 and i_2. It also introduces *two* circuit constraints. For the VCVS these are

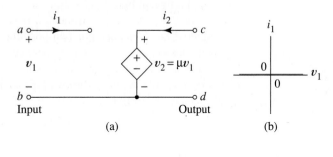

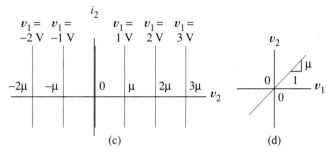

Figure 1.7   Voltage-controlled voltage source: (a) schematic symbol; (b) input characteristic; (c) output characteristic; (d) transfer characteristic.

$$i_1 = 0 \tag{1.6}$$

and

$$v_2 = \mu \, v_1 \tag{1.7}$$

In Eq. (1.7), input voltage, v_1, controls the dependent voltage, v_2, constraining v_2 to be always μ times v_1. We call the factor that relates a two-port output quantity to an input quantity the *transmittance* of the two-port. The transmittance of the VCVS is μ. Notice that the remaining quantities, v_1 and i_2, are unspecified.

A graphical description for a two-port requires *two* volt-ampere curves, an *input characteristic* to relate the input variables and an *output characteristic* to relate output variables. For the VCVS, the input characteristic is the volt-ampere curve of an open circuit as in Fig. 1.7b. The graphical description of the output port is the curve family shown in Fig. 1.7c. Because the dependent source is a voltage source, its output is described by a constant voltage line. Since the source is dependent, however, the exact location of this line is determined by the *controlling variable*, v_1. It is customary to suggest this control graphically by showing a *family* of lines, one line for each of a number of representative values of the controlling variable, as in Fig. 1.7c. The value of parameter μ determines the spacing between the lines. Thus, when $v_1 = 1$ volt, the constant voltage line is positioned at $v_2 = \mu$ volts, when $v_1 = 2$ volts, the line is at $2\,\mu$ volts, and so forth, for both positive and negative values of v_1.

In addition to input and output characteristics, a two-port always has a third graph called its *transfer characteristic* (TC), which relates output variable to input variable. For the VCVS this is the plot of Eq. (1.7) shown in Fig. 1.7d, where the slope of the curve is the transmittance. The transfer characteristic is important for it shows exactly how the two-port *transfers* information from input to output.

Figure 1.8a shows another two-port, the current-controlled current source (CCCS). Its input characteristic is the volt-ampere curve of a short circuit. Its output characteristic is a family of constant current lines, with the value of the input current used to select a particular curve from the infinity of possible output curves. These curves imply the mathematical description:

$$v_1 = 0 \tag{1.8}$$

and

$$i_2 = \beta i_1 \tag{1.9}$$

The transfer characteristic of Fig. 1.8d is a plot of Eq. (1.9) with the current ratio from Eq. (1.9) for its slope.

There are two other kinds of dependent sources, the voltage-controlled current source (VCCS) and the current-controlled voltage source (CCVS), both shown in Fig. 1.9.

The following examples illustrate how these simple modeling ideas provide a degree of understanding of unfamiliar devices.

Exercise 1.1 For the two-port of Fig. 1.6b write the equation of (a) the input characteristic, and (b) the output characteristic.

Ans. (a) $i_1 = 10^{-4} \, v_1$, $i_2 = (v_2 - 2)/10^3$

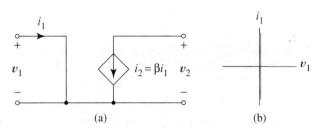

(a)   (b)

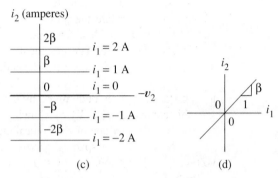

(c)   (d)

Figure 1.8 Current-controlled current source: (a) schematic; (b) input characteristic; (c) output characteristic; (d) transfer characteristic.

EXAMPLE 1.1 Figure 1.10a is the schematic representation of a two-port device called a bipolar transistor. Measured characteristics, shown in Figs. 1.10b and c, apply only to first quadrant voltage and current values. Draw a linear circuit model for the transistor, and indicate any necessary restrictions on the use of this model.

Solution. We first recognize Fig.1.10b as the volt-ampere curve of an independent voltage source of 0.7 V. A dc voltage source of 0.7 V connected between nodes B and E would give the same input characteristic. To restrict operation to the first quadrant, we specify the input current range $i_B \geq 0$.

   We next recognize the output characteristics, constant-current lines labeled with values of input current, as the description of a CCCS, connected between nodes C and E and controlled by i_B. From the line spacing, the transmittance must be $\beta = 400$. Since $i_B \geq 0$ already precludes operation in the third and fourth quadrants (see Fig. 1.10c), we

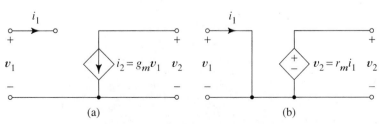

(a)   (b)

Figure 1.9 (a) Voltage-controlled current source; (b) current-controlled voltage source.

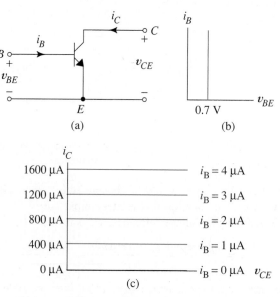

(a)

(b)

(c)

Figure 1.10 Two-port device modeling: (a) transistor schematic; (b) measured input characteristic; (c) measured output characteristic.

need only the additional requirement $v_{CE} \geq 0$ to restrict operation to the first quadrant. Figure 1.11 shows the circuit that represents the transistor when $i_B \geq 0$ and $v_{CE} \geq 0$. ❑

Exercise 1.2 Describe in words the input and output characteristics of Fig. 1.9a if $g_m = 2$ mS.

Ans. The input characteristic is the horizontal line, $i_1 = 0$. Output characteristics are a family of evenly spaced horizontal lines, each labeled with a value of v_1. For $v_1 = \pm 1$ V, ± 2 V, . . . , the lines are spaced at 2 mA intervals.

EXAMPLE 1.2 Figure 1.12a is the schematic of De Forest's vacuum triode, and Figure 1.12b is a circuit that represents the triode under certain conditions. Sketch the first-quadrant output characteristics of the triode for $v_G = 0$ V, -1 V, -2V, and -3 V. Assume μ is a positive-valued parameter.

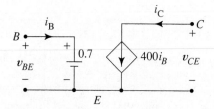

Figure 1.11 Circuit model for two-port device of Figure 1.10a.

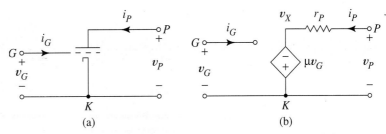

Figure 1.12   Schematic and circuit model for vacuum triode.

Solution. Without r_P, the curves would be those of the VCVS, Fig. 1.7c. To include the effect of r_P, we write an equation that relates output current, i_P, to output voltage, v_P. If we use node K for node-voltage reference, we can write

$$i_P = \frac{v_P - v_X}{r_P} = \frac{v_P - (-\mu v_G)}{r_P} = \frac{v_P + \mu v_G}{r_P}$$

an equation that relates output variables i_P and v_P using input variable v_G as a controlling variable. When $v_G = 0$, this is the equation of a resistor, thus the curve labeled $v_G = 0$ in Fig. 1.13. Substituting, in turn, $v_G = -1$, -2, and -3 into the equation gives three additional straight-line curves of the same slope, but shifted to the right by increments of μ volts. ❏

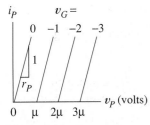

Figure 1.13   Output characteristic of vacuum triode.

In the limit as r_P approaches zero, the circuit of Fig. 1.12b and the characteristic of Fig. 1.13 both approach the VCVS. We conclude that a small resistance added in series with a dependent voltage source causes its output characteristics to depart slightly from the vertical lines. Conversely, a family of nearly vertical lines leads to a circuit consisting of a dependent voltage source and series resistor. Problem 1.6 shows that placing a large resistor r_o in parallel with the current source in Fig. 1.8a causes the lines in Fig. 1.8c to have nonzero slope $1/r_o$. Examples 1.1 and 1.2 show the versatility and general applicability of our methods.

The transfer curves for the four dependent sources show straight-line input/output behavior that extends to infinity in each direction. One implication is that the device can accommodate any voltage or current values, no matter how large. Output characteristics also suggest devices that can deliver infinite power to an external circuit. Unfortunately, practical devices are more limited in their capabilities. Later in the chapter we compare real transfer and output curves with ideal device curves to better understand the practical limitations of real devices.

1.2.3   DEVICE MODELING, THE BIG PICTURE

To take stock of our progress in device modeling, refer to Fig. 1.14. At left-center is a physical device, such as a transistor, and we wish to predict what happens if we place this device in a circuit. There is usually a mathematical description based upon physical principles that we can use in computer simulations, but this description is often too complicated to give us the understanding we require in the initial stages of analysis or de-

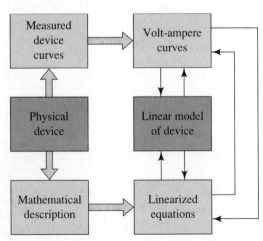

Figure 1.14   Role of modeling and volt-ampere curves in describing electronic devices.

sign. There is also graphical data in the form of measured device curves, also rather complex. Our objective is a simple linear circuit model, perhaps valid only under some restricted operating conditions, that helps us understand the device and visualize how it behaves in a circuit. There are two practical ways to obtain such a model. One is to follow the lower path in the figure. Here we simplify the mathematical description to a set of *approximate linear equations* and then derive a circuit model that satisfies these equations. An alternative (upper path) is to *approximate the measured device curves* by simplified volt-ampere curves; this also leads us to a circuit model. As we have already seen, once we have one of the three linear descriptions on the right we have them all, for we can easily convert from one to the other.

Up to now we have mastered only the right-hand side of Fig. 1.14. In subsequent chapters, as we study mathematical and graphical descriptions of various devices, we rely on the techniques of this chapter to relate these descriptions to simple ideas from our circuit theory course. These modeling techniques are very general and very powerful, applying equally to historic devices such as vacuum tubes, modern devices such as field-effect transistors, and devices yet to be invented.

In the next section we show why we need amplifiers, and we use our basic modeling ideas to study some of their practical properties.

1.3
Ideal Amplifiers

One characteristic of any amplifier is that it enables an input signal to control an output signal of greater power. This section describes an ideal amplifier that embodies the essence of this idea in a simple model. Once we understand the ideal, we then incorporate additional features that make it less ideal and more typical of real amplifiers.

To review the meaning of signal power, suppose that voltage $v(t)$ in Fig. 1.15a is developed across a resistor R_L. The *instantaneous power* dissipated in R_L at time t is

$$P_i(t) = \frac{v^2(t)}{R_L}$$

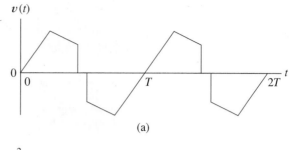

(a)

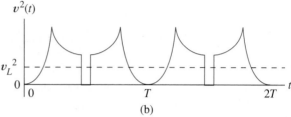

(b)

Figure 1.15 rms value of a time-varying voltage: (a) voltage waveform; (b) average squared value of $v(t)$.

Restricting our discussion to voltages like $v(t)$ that are *periodic* with *period T* (repeat themselves every T seconds), the *average power* in R_L is

$$P_{ave} = \frac{1}{T}\int_0^T P_i(\tau)d\tau = \frac{1}{T}\int_0^T \frac{v^2(\tau)}{R_L}d\tau = \frac{\left[\sqrt{\frac{1}{T}\int_0^T v^2(\tau)d\tau}\right]^2}{R_L}$$

If we define the *root-mean-square* (rms) value of $v(t)$ by

$$v_L = \sqrt{\frac{1}{T}\int_0^T v^2(\tau)d\tau} \tag{1.10}$$

then we can calculate the average power in $v(t)$ using

$$P_{ave} = \frac{v_L^2}{R_L}$$

Figure 1.15b shows the squared voltage $v^2(t)$; the dashed line is its average. The rms value, v_L, is the square root of this *average squared value*, a number associated with any periodic signal that enables us to find average power using a familiar equation from dc circuits. A similar derivation for the rms value of a time-varying periodic current $i(t)$ gives

$$i_{Lrms} = \sqrt{\frac{1}{T}\int_0^T i^2(\tau)d\tau}$$

Unless otherwise specified, the term "power" hereafter always means average power. A useful by-product of this discussion is the concept of the rms value of a time-varying *signal* current or voltage.

1.3.1 THE NEED FOR AN AMPLIFIER

Figure 1.16a exemplifies the need for an amplifier. A *given signal source* with open-circuit voltage $v_i(t)$ and given internal resistance R_i is available. To achieve some useful objective, the signal must drive a *given* load R_L. For example, suppose the source is a

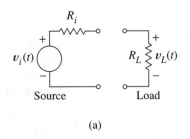

(a)

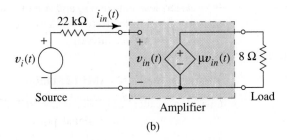

(b)

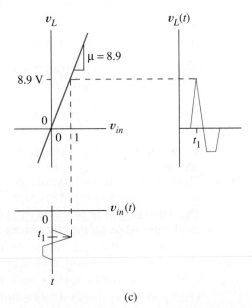

(c)

Figure 1.16 (a) The amplifier problem: given source and load; (b) coupling source to load with an ideal amplifier; (c) amplification visualized as projection from the transfer characteristic.

tape play head with open-circuit voltage of 100 mV rms and internal resistance of R_i = 22 kΩ, the load a 100 mW speaker of 8 Ω, and the object to produce audible music. Since the speaker must produce pressure waves in the air sufficiently intense to stimulate the ear, it is no surprise that the load requires a specified quantity of signal power. To produce music at the 100 mW level, the speaker's rms voltage must satisfy

$$\frac{V_L^2}{8} = 100 \text{ mW}$$

therefore an output voltage of V_L = 894 mV rms is required. Clearly the given tape head is incapable of producing 100 mW because the rms value of $v_i(t)$, V_I, is less than the required 894 mV. The situation is actually even worse than this because a direct connection between the tape head and the speaker produces the voltage division:

$$V_L = \left(\frac{8}{8 + 22000} \right) V_I = (0.00036 \times 100) = 0.036 \text{ mV}$$

It is easier, less expensive, and a more flexible approach to amplify the existing signal instead of trying to design a transducer (improved tape head) that can provide the required output power directly.

1.3.2   THE IDEAL AMPLIFIER

One kind of ideal amplifier is the VCVS. Figure 1.16b helps us discover how this amplifier increases output signal power. Our circuit analysis involves rms signal values. The amplifier input current $i_{in}(t)$ is zero because of the open circuit. This makes the internal voltage drop across the tape head resistance $i_{in}(t)[22 \text{ k}] = 0$. The amplifier input voltage is

$$v_{in}(t) = v_i(t) - i_{in}(t)[22 \text{ k}] = v_i(t)$$

The dependent source produces $\mu v_i(t)$ volts across the load resistor.

The voltage gain, A_v, of an amplifier is defined as

$$A_v = \frac{\text{output signal voltage}}{\text{input signal voltage}} \tag{1.11}$$

In the notation of Fig. 1.16b $A_v = \mu$. To produce the required 100 mW in the speaker (894 mV output voltage), the amplifier must have a voltage gain that satisfies

$$A_v V_{in} = \mu 100 = V_{out} = 894$$

Thus voltage gain of $\mu \geq 8.94$ ensures adequate sound power.

Figure 1.16c uses the VCVS transfer characteristic to show how the ideal amplifier processes the input $v_i(t)$. The graphical construction arises from the idea of composition of functions in mathematics. Recall that if a function $x = f(y)$ is known, and if y itself is a function of another variable such as t, then $x(t)$ can be found from the substitution $x(t) = f(y(t))$. For each t we first find the number $y(t)$; we then find the x that corresponds to this y. The transfer characteristic in Figure 1.16c, a graph of the straight-line function $v_{out} = f(v_{in})$, takes the place of $x = f(y)$. To discover how v_L varies with t, we first sketch $v_{in}(t)$ below the transfer curve. A coordinate system for $v_L(t)$ is also set up at the right. At each instant, such as $t = t_1$, we find the value of $v_{in}(t_1)$ graphically. Projecting this value up to the transfer characteristic using the vertical dashed line is equivalent to calculating $v_L(t_1) = f(v_{in}(t_1))$. Projecting this value horizontally to time t_1 completes the composition of functions operation. By doing this for every t we establish the complete output waveform for any input waveform. (Usually, we can sketch the general waveshape using mostly imagination, that is, without actually projecting point by point.) The Fig. 1.16c example shows that the output signal has exactly the same shape as the input signal, but is 8.9 times as large. An amplifier with a higher voltage gain would have a steeper transfer characteristic and would produce a larger output signal for the same input. In terms of a v_L versus v_{in} transfer characteristic, the voltage gain definition of Eq. (1.11) becomes the slope

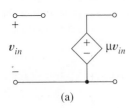

$$A_v = \frac{dv_L}{dv_{in}}$$

Projecting from a curve is an important procedure that we use often because it gives insight into complicated problems without requiring much effort.

Another useful gain parameter for an amplifier is the power gain, A_p, defined by

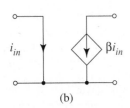

$$A_p = \frac{\text{output signal power}}{\text{input signal power}} \tag{1.12}$$

We saw that when $\mu = 8.9$, the amplifier of Fig. 1.16b delivers 100 mW to the speaker but draws no input power from the source because its input current is zero. Thus the power gain of the ideal amplifier is infinite. In later chapters we learn that many real amplifiers draw negligible input power and therefore have approximately infinite power gain.

When an amplifier increases signal power, it is always within the context of Fig. 1.1a. There is a dc source, V_{PP}, that supplies power to the amplifier by means of a constant voltage and a pulsating dc current. The amplifier *converts* some of this dc power into signal power, thus providing power gain for the signal without violating the principle of conservation of energy. By describing only signals, our equivalent circuits tacitly conceal these details of power transfer.

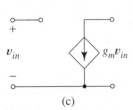

So far, "amplifier" has meant a *voltage amplifier* in which an input voltage controls an output voltage, the VCVS of Fig. 1.17a serving as its model. We can control an output signal with an input signal in three other ways. One way is to use a *current amplifier* modeled by a CCCS as in Fig. 1.17b. The transfer characteristic for this amplifier is a plot of output current versus input current. For this circuit gain means current gain, β, the slope of the transfer characteristic. Physical devices called bipolar transistors function as current amplifiers.

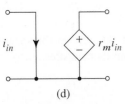

Figure 1.17 Four types of amplifiers: (a) voltage amplifier (VCVS); (b) current amplifier (CCCS), (c) transconductance amplifier (VCCS); (d) transresistance amplifier (CCVS).

Another possibility is a *transconductance amplifier,* in which input voltage controls output current as in the VCCS in Fig. 1.17c. Its name nicely describes its function, since conductance is dimensionally the ratio of current to voltage and since trans means across. That is, control passes across from an *input* voltage to an *output* current. The transfer characteristic for a transconductance amplifier is a plot of output current versus

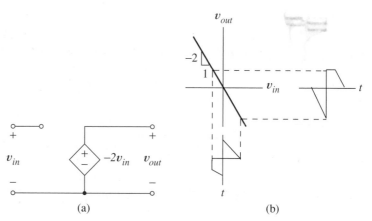

Figure 1.18 Inverting amplifier: (a) a VCVS with negative gain; (b) effect of negative gain on output signal.

input voltage; gain is g_m, and the slope of the curve has the dimension of conductance. Devices called field-effect transistors are inherently transconductance amplifiers.

The transresistance amplifier, modeled by the CCVS, completes the set. Gain in this case is the *transresistance* r_m in Fig. 1.17d. We conclude that the volt-ampere characteristics of the four dependent sources are idealized descriptions of four basic amplifiers.

So far we have treated gain as a positive number; however, negative gain is also commonly encountered. Negative gain simply means that the transfer curve has negative slope. Figure 1.18 shows that the result of negative gain is an *inverted* output signal, that is, a signal reflected about the time axis in addition to being amplified. Such incidental signal inversions present no information processing problem.

The tape-head/speaker example captured the essence of many practical amplifier applications. Instead of a tape head, the signal source might be an electrocardiogram lead system, a strain gauge, or a receiving antenna; instead of music, the desired output might be a strip-chart recording, magnetic memory tracks on a computer disk, or the rotational position of a motor that winds elevator cables. There is always a *given* signal source and a *given* load, each with given parameter values, that must somehow work together. An amplifier provides the power required by the load while preserving the essence of the signal. The next section addresses practical complications that arise in implementing these basic amplifier concepts.

1.4
Input, Output, and Interstage Loading

1.4.1 EFFECTS OF INPUT AND OUTPUT RESISTANCE

To make a more realistic amplifier model, we add internal resistances to the dependent source as in Fig. 1.19a. These are an output resistance, R_{out}, and an input resistance, R_{in}. Amplifier descriptions commonly list values for μ, R_{in}, and R_{out}, and we must understand the significance of these parameters before we can design, order, or even intelligently use a real amplifier.

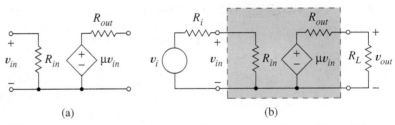

(a)   (b)

Figure 1.19 Realistic amplifier model: (a) amplifier equivalent circuit; (b) amplifier used to couple given source to given load.

Figure 1.19b is an equivalent circuit in which a signal source, v_i and R_i, is connected to load R_L by a *nonideal amplifier*. Because of R_i and R_{in}, the voltage developed at the amplifier input is only

$$v_{in} = \frac{R_{in}}{R_{in} + R_i} v_i \tag{1.13}$$

This voltage division at the amplifier input is *input loading*. Input voltage v_{in} is internally amplified by μ; however, the entire amplified voltage does not appear across the load resistor because of output *loading*. Thus

$$v_{out} = (\mu v_{in}) \frac{R_L}{R_L + R_{out}} \tag{1.14}$$

From Eqs. (1.13) and (1.14) it follows that the voltage gain of the nonideal amplifier is

$$A_v = \frac{v_{out}}{v_{in}} = \mu \frac{R_L}{R_L + R_{out}} \tag{1.15}$$

Notice that μ is not actually the amplifier gain, but only an upper limit to its gain. Since maximum gain occurs as $R_L \to \infty$, we call μ the *open-circuit voltage gain*.

A second voltage gain definition, which includes loading at the input, is often useful. We define this new gain, A_v', by

$$A_v' = \frac{v_{out}}{v_i} \tag{1.16}$$

Substituting v_{in} from Eq. (1.13) into Eq. (1.14) gives

$$A_v' = \frac{R_{in}}{R_{in} + R_i} \mu \frac{R_L}{R_L + R_{out}} \tag{1.17}$$

This definition shows that when amplifier requirements involve both v_i and v_{out}, we must consider not only μ, but also R_{in} and R_{out} in our choice. It further shows that we make best use of a given amplifier's inherent amplifying capabilities if the load is such that $R_L \gg R_{out}$ and the source satisfies $R_i \ll R_{in}$. Only then does the voltage gain approach the gain μ of the ideal amplifier. Having two gain expressions is no problem provided we take care to indicate which is being used at a particular time.

Since the amplifier of Fig. 1.19b has nonzero input current, its power gain is not infinite. We could derive a general expression for the power gain for Fig. 1.19b; however, it is easier to rely on the definition of Eq. (1.12), and use the circuit model when power gain is needed.

Another amplifier gain quantity is current gain,

$$A_i = \frac{\text{amplifier output current}}{\text{amplifier input current}} \tag{1.18}$$

In the following exercise and example, we apply the various definitions and modeling concepts.

Exercise 1.3 A signal source with 5 mV rms open-circuit voltage and 10 kΩ internal resistance is connected to a 100 Ω load using a nonideal amplifier. The amplifier has open-circuit voltage gain of 200, 90 kΩ input resistance, and 100 Ω output resistance. Compute the numerical values for the amplifier output voltage and the voltage gain from amplifier input to output.

Ans. $V_{out} = 450$ mV, $A_v = 100$.

EXAMPLE 1.3 For the amplifier of Exercise 1.3 find (a) the current gain, (b) the power gain, and (c) the voltage that would appear across the load resistor if the source were attached directly to the load without using an amplifier.

Solution. (a) The first step is to summarize the known information using a circuit diagram like Fig. 1.20a. By using the exercise answers and continuing to work with rms values, we find:

$$i_{in} = \frac{v_{in}}{R_{in}} = \frac{4.5 \times 10^{-3}}{90k} = 5 \times 10^{-8} \text{ A}$$

$$i_{out} = \frac{v_{out}}{R_L} = \frac{4.5 \times 10^{-1}}{100} = 4.5 \text{ mA}$$

Now Eq. (1.18) gives

$$A_i = \frac{4.5 \times 10^{-3}}{5 \times 10^{-8}} = 9 \times 10^4$$

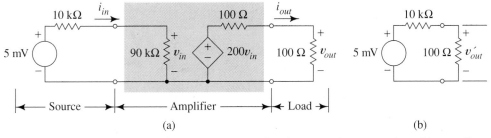

Figure 1.20 Circuit diagrams for Example 1.3. (a) Load connected to source through an amplifier; (b) load connected directly to the source.

(b) The amplifier's output power goes into R_L. Therefore,

$$P_{out} = \frac{v_{out}^2}{R_L} = \frac{0.45^2}{100} = 2.03 \text{ mW}$$

The signal power entering the amplifier is

$$P_{in} = v_{in} i_{in} = \frac{(4.5 \times 10^{-3})^2}{9 \times 10^4} = 2.25 \times 10^{-10} \text{ W}$$

Thus power gain is

$$A_p = \frac{2.03 \times 10^{-3}}{2.25 \times 10^{-10}} = 9.02 \times 10^6$$

(c) Figure 1.20b shows that without the amplifier the 5 mV open-circuit voltage would be reduced by voltage division to

$$V'_{out} = \left(\frac{100}{10,100} \right) \times 5 = 0.0495 \text{ mV} \qquad \square$$

The voltage amplifier may be the most widely used amplifier model; however, special circumstances sometimes favor one of the other ideal amplifiers of Fig. 1.17. The available signal source dictates the appropriate amplifier input circuit: a signal voltage favors an open-circuit (voltage-controlled) input; a signal current favors the short-circuit (current-controlled) input. Figure 1.21a shows the source as a signal voltage in series with a small source impedance Z_i. This impedance may be a function or frequency, is often highly variable, and is possibly nonlinear (Z_i value changes with current). If appreciable current flows through Z_i, then amplifier input v_{in} can differ significantly from $v_i(t)$, meaning the amplifier signal is distorted. The open-circuit inputs of the voltage and transconductance amplifiers ensure that $v_{in} = v_i(t)$. A real amplifier with input resistance $R_{in} >> |Z_i|$ is next best.

Although we tend to think of signals as voltages, some sources inherently provide information as a current. One example is a photodetector, which converts light intensity into electrical current. A nonideal current source, which has high internal impedance, is best visualized as in Fig. 1.21b, where Z_i might be frequency dependent, highly variable, and nonlinear (a function of voltage). An ideal amplifier with zero input resistance is

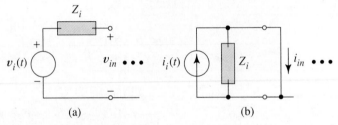

(a)   (b)

Figure 1.21 Signal sources and ideal amplifier input circuits: (a) voltage source favors an open circuit; (b) current source favors a short circuit.

best because the voltage across Z_i is zero; thus $i_{in} = i_i(t)$, and Z_i has no effect whatsoever on the input signal. A real amplifier with $R_{in} << |Z_i|$ is almost as good.

The amplifier's output requirements suggest the best amplifier output circuit. Most often, output information is to be encoded as voltage, so a voltage or transresistance amplifier is best. However, occasionally information must take the form of current. A prosaic example is a video display terminal that uses the magnetic field produced by a coil of wire to deflect an electron beam. Because the magnetic deflection force is proportional to coil current, deflection information should be encoded as current. Current and transconductance amplifiers are best in such applications.

1.4.2 CASCADED AMPLIFIER STAGES

If one amplifier has insufficient gain to satisfy specifications, two amplifiers can be *cascaded*, as in Fig. 1.22. From the figure, the gain of the first amplifier is

$$A_{v_1} = \frac{v_2}{v_1} = \mu_1 \frac{R_{in2}}{R_{in2} + R_{out1}} \tag{1.19}$$

The gain of the second amplifier is

$$\frac{v_{out}}{v_2} = A_{v2} = \mu_2 \frac{R_L}{R_L + R_{out2}} \tag{1.20}$$

The gain of the two-stage amplifier is the product of these individual gains; that is,

$$\frac{v_{out}}{v_1} = A_v = \frac{v_2}{v_1} \times \frac{v_{out}}{v_2} = A_{v1} \times A_{v2} \tag{1.21}$$

that becomes

$$A_v = \mu_1 \frac{R_{in2}}{R_{in2} + R_{out1}} \mu_2 \frac{R_L}{R_L + R_{out2}} \tag{1.22}$$

after substituting Eqs. (1.19) and (1.20). We recognize the last factor as an output loading term. The second factor is called an *interstage loading factor* because it comes from loading at the interface between amplifier stages. The principle of multiplying the individual stage gains (including *interstage loading effects*) to obtain the overall gain of a succession of cascaded stages as in Eq. (1.21) extends to three or more cascaded amplifiers.

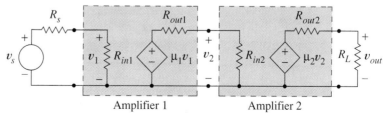

Figure 1.22   Two-stage amplifier formed by connecting two amplifiers in cascade.

Notice the simplifying role played by the amplifier models in the preceding developments. If we concentrate on definitions and models, we can readily derive any equations we might need using simple circuit analysis, and thereby avoid memorizing. This pattern persists throughout the book.

1.4.3 DECIBEL UNITS

We defined voltage, current, and power gain as simple ratios of output to input quantities. Gains and other ratios are sometimes more conveniently expressed in terms of logarithmic units called *decibels* (dB), especially when dealing with large gain variations or cascaded amplifier stages. The definition of voltage *gain in decibels*, A_{vdB}, is

$$A_{vdB} = 20 \times \log|A_v| = 20 \times \log|v_{out}/v_{in}| \qquad (1.23)$$

where log denotes the base 10 logarithm. The absolute value signs are necessary because gain is sometimes negative. Current gain in decibels is given by

$$A_{idB} = 20 \times \log|A_i| = 20 \times \log|i_{out}/i_{in}| \qquad (1.24)$$

Power gain has a different decibel definition, namely,

$$A_{pdB} = 10 \times \log|A_p| = 10 \times \log|P_{out}/P_{in}| \qquad (1.25)$$

Problem 1.29 explores how A_{pdB} is related to A_{vdB} and A_{idB}.

Exercise 1.4   An amplifier has rms input and output voltages of 35 mV and 6 V, respectively. Its input and load resistances are 10 kΩ and 20 kΩ, respectively. Calculate the voltage gain, current gain and power gain of this amplifier in decibels.

Ans. $A_{vdB} = 44.7$ dB, $A_{idB} = 38.7$ dB, $A_{pdB} = 41.7$ dB.

To convert backward from decibels to ratios, we use the inverse of the definition. For example, an amplifier with a power gain of 23 dB, from Eq. (1.25), has a power gain ratio of

$$A_P = 10^{23/10} = \frac{P_{out}}{P_{in}} = 199.5$$

Decibels, named for Alexander Graham Bell, inventor of the telephone, were originally formulated as measures conveniently related to the response of the human ear. Since the ear perceives an increase in speaker output power from 0.1 watt to 1 watt (0.9 watt increase) to be the same as an increase from 1 watt to 10 watts (9 watt increase), it makes sense to describe both as 10 dB power increases. Also, with decibel scales we can conveniently graph collections of numbers that range from the very large to the very small. For example, a curve of amplifier gain that varies from 1 to 10^5 with variations in some parameter can be conveniently plotted on a decibel scale that ranges from 0 to 100 dB.

TABLE 1.1 Decibel Gains and Gain Ratios

Decibel Gain		Voltage or Current Ratio
+20	--------	×10
+6	--------	×2
+3	--------	×1.414
+1	--------	×1.12
+0	--------	×1.0
−1	--------	×0.892
−3	--------	×0.707
−6	--------	×0.5
−20	--------	×0.1

Another useful feature is that the decibel gain of a multistage amplifier is the sum of the decibel gains of the individual stages. If a three-stage amplifier has a voltage gain given by

$$A_v = A_{v1}A_{v2}A_{v3}$$

then, in decibels, we have

$$A_{vdB} = 20\,[\log|A_{v_1}| + \log|A_{v_2}| + \log|A_{v_3}|]$$
$$= A_{v1dB} + A_{v2dB} + A_{v3dB}$$

Table 1.1 shows a few important decibel values and the corresponding ratios. Zero dB corresponds to unity gain, positive dB values correspond to gain greater than one, and negative values correspond to fractional gains. Plus and minus signs in the first column remind us that decibel gains are additive; multiplication signs in the second column remind us that gain ratios are multiplicative. By remembering a few numbers from Table 1.1 we can make quick mental decibel conversions. For a voltage gain of 47 dB, we mentally compute $47 = 20 + 20 + 6 + 1$. We then reason that the gain ratio is $10 \times 10 \times 2 \times 1.12$, or 224. For a gain of 57 dB, think $57 = 20 + 20 + 20 - 3$, which makes the gain ratio $10 \times 10 \times 10 \times 0.707 = 707$.

1.5 _____

Difference Amplifiers

We next introduce a new kind of amplifier, known variously as the *difference amplifier,* the *differential amplifier,* or the *diff amp.* Its special properties make it a versatile, off-the-shelf, packaged device, especially useful for processing low-amplitude signals in noisy environments. The differential amplifier is also a building block or subcircuit used within high-quality integrated circuit amplifiers, linear and nonlinear signal processing circuits, and even certain logic gates and digital interfacing circuits. A working knowledge of the difference amplifier is therefore essential, both for using commercially available amplifiers, and for understanding the sophisticated integrated circuits described in subsequent chapters.

1.5.1 BASIC DEFINITIONS

The difference amplifier, Fig.1.23, has a ground reference, and three nodes associated with signal processing. The node marked with a plus sign is the *noninverting input;* a minus sign identifies the *inverting input.* Because information entry involves these *two* ungrounded nodes, we say this amplifier has a *double-ended input.* The remaining node is the output node. Since there is only one ungrounded output node, this amplifier has a *single-ended output.* All amplifiers in preceding sections were *single ended* because they had single-ended inputs and outputs. Some authors use a triangular schematic symbol for the difference amplifier; however, in this book we reserve the triangular symbol for the operational amplifier, a special difference amplifier having exceptionally high voltage gain.

Signals enter the difference amplifier as a *pair* of node voltages v_a and v_b. Information encoded in *two* node voltages in this manner is called a *double-ended signal.* We describe the double-ended signal in terms of two components, a *difference-mode* component defined by

$$v_d = v_a - v_b \tag{1.26}$$

and a *common-mode* component defined by

$$v_c = \frac{v_a + v_b}{2} \tag{1.27}$$

In terms of these components the difference amplifier of Fig. 1.23 performs the operation:

$$v_o = A_d v_d + A_c v_c \tag{1.28}$$

where A_d is the *difference-mode gain* and A_c *is the common-mode gain.* Ordinarily, the difference-mode component is useful information to be amplified; the common-mode component is an undesired element such as noise. The difference amplifier is therefore designed so that A_d is large, usually much greater than one, and A_c is small, often less than one.

When a signal happens to be combined with excessive additive noise the single-ended amplifier is not very useful, for it amplifies signal and noise equally. However, we

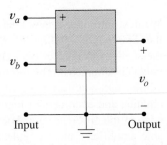

Figure 1.23   Difference amplifier schematic symbol.

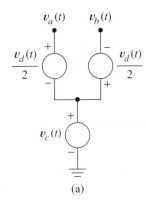

(a)

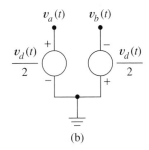

(b)

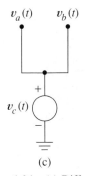

(c)

Figure 1.24 (a) Difference-and common-mode components of an arbitrary input signal; (b) pure difference-mode signal; (c) pure common-mode signal.

can often configure a difference amplifier so that the signal enters the amplifier mostly as the difference-mode component while the noise enters primarily *in the common mode.* Then the amplifier reduces the noise as it amplifies the signal. These ideas gradually become more clear after closely examining difference and common-mode components.

1.5.2 DOUBLE-ENDED SIGNALS AS A SUPERPOSITION OF DIFFERENCE- AND COMMON-MODE COMPONENTS

Equations (1.26) and (1.27) imply that we can visualize any *arbitrary* pair of node voltages as a *superposition* of common- and difference-mode components. To aid this visualization, we first give a short derivation. Let $v_a(t)$ and $v_b(t)$ be arbitrary time functions. Multiplying Eq. (1.26) by 1/2 and adding to Eq. (1.27) gives

$$v_a(t) = v_c(t) + \frac{v_d(t)}{2} \tag{1.29}$$

Subtracting one-half of Eq. (1.26) from (1.27) gives

$$v_b(t) = v_c(t) - \frac{v_d(t)}{2} \tag{1.30}$$

The circuit of Fig. 1.24a is an interpretation of Eqs. (1.29) and (1.30). The former expresses node voltage $v_a(t)$ as the sum of voltages from the two series sources connected between ground and node a in Fig. 1.24a. Equation (1.30) is a similar statement about $v_b(t)$. Using superposition, we can now isolate the difference-mode component by turning off the common-mode source of Fig. 1.24a. Doing this gives Fig. 1.24b, which shows that the essence of a difference-mode signal is a pair of node voltages of equal amplitude and opposite polarity, an important subtlety that is not apparent in Eq.(1.26). We hereafter refer to the excitation scheme of Fig. 1.24b as a *pure difference-mode signal.*

To capture the essence of the common-mode component, we turn off both difference-mode sources in Fig. 1.24a giving Fig. 1.24c. We conclude that a common-mode component (a *pure common-mode signal*) is a voltage applied with identical phase to both nodes.

The superposition idea expressed in Fig. 1.24 not only shows the basic nature of common- and difference-mode signals, but it also provides an important conceptual and analytical tool. To find the response of any linear circuit to a double-ended signal voltage, we can separately compute the responses to the difference- and common-mode components and then sum the results. We now recognize Eq. (1.28) as a simple statement of superposition for the difference amplifier.

Exercise 1.5 Find the common- and difference-mode components when the input node voltages are $v_a = +5$ V and $v_b = -2$ V.

Ans. $v_d = +7$ V, $v_c = +1.5$ V.

EXAMPLE 1.4   Find the common- and difference-mode components when the input node voltages are $v_a(t) = 2 + 3 \cos 30t + 8 \cos 16t$ and $v_b(t) = 6 - 4 \cos 30t + 8 \cos 16t$.

Solution. $v_d(t) = (2 + 3 \cos 30t + 8 \cos 16t) - (6 - 4 \cos 30t + 8 \cos 16t)$

$$= -4 + 7 \cos 30t$$

$$v_c(t) = 0.5[(2 + 3 \cos 30t + 8 \cos 16t) + (6 - 4 \cos 30t + 8 \cos 16t)]$$

$$= 4 - 0.5 \cos 30t + 8 \cos 16t \qquad ❑$$

EXAMPLE 1.5   The voltages applied to a difference amplifier are

$$v_a(t) = 0.010 \cos(2\pi\, 400t) + 0.20 \cos(2\pi\, 60t)$$

and

$$v_b(t) = -0.010 \cos(2\pi\, 400t) + 0.20 \cos(2\pi\, 60t)$$

Find $v_o(t)$ if $A_d = 100$ and $A_c = 0.5$.

Solution. The difference- and common-mode components are

$$v_d(t) = 0.02 \cos(2\pi\, 400t)$$

and

$$v_c(t) = 0.20 \cos(2\pi\, 60t)$$

Using Eq.(1.28),

$$v_o(t) = 2 \cos(2\pi\, 400t) + 0.10 \cos(2\pi\, 60t) \qquad ❑$$

Exercise 1.6   Find $v_o(t)$ for a difference amplifier having difference-mode gain of 2 and common-mode gain of 0.001 if $v_a(t)$ and $v_b(t)$ are the functions in Example 1.4.

Ans. $v_o(t) \approx -7.996 + 13.9995 \cos 30t + 0.008 \cos 16t$.

1.5.3   COMMON MODE REJECTION

In Example 1.5 the amplitude of the 60 Hz common-mode noise was ten times that of the 400 Hz difference-mode "signal" at the input; however, at the output the noise was only 1/20 the signal. Not only was the signal amplified by 100, but the noise was also reduced relative to the signal, demonstrating the difference amplifier's ability to discriminate in favor of the desired component.

A *figure of merit* is a number that measures some special capability of a device. Figures of merit are useful as specifications, and form a basis for comparing similar devices. We describe the difference amplifier's ability to reduce the common-mode component relative to the difference-mode component by a figure of merit called the *common-mode rejection ratio* (CMRR),

$$\text{CMRR} = \frac{A_d}{A_c} \qquad (1.31)$$

In decibels CMRR is

$$\text{CMRR}_{dB} = 20 \log \left| \frac{A_d}{A_c} \right| \qquad (1.32)$$

Difference amplifier specifications typically include A_d and CMRR_{dB}; we find A_c from Eq. (1.32) if needed.

To better appreciate how CMRR relates to noise reduction, we write Eq. (1.31) as

$$\text{CMRR} = \frac{A_d}{A_c} = \frac{v_{od}/v_{id}}{v_{oc}/v_{ic}} \times 1 = \left(\frac{v_{od}/v_{id}}{v_{oc}/v_{ic}} \times \frac{v_{id}/v_{oc}}{v_{id}/v_{oc}} \right) = \frac{v_{od}/v_{oc}}{v_{id}/v_{ic}}$$

where subscripts i and o denote amplifier input and output and d and c denote difference- and common-mode components. Solving for v_{od}/v_{oc} gives

$$\frac{v_{od}}{v_{oc}} = \text{CMRR} \left(\frac{v_{id}}{v_{ic}} \right)$$

This equation shows that the output difference- to common-mode voltage ratio is the input ratio multiplied by the CMRR. If we are able to configure our amplifier so that difference- and common-mode components are exactly signal and noise, respectively, then CMRR is the factor of improvement in signal-to-noise amplitude ratio.

Using Difference Amplifier with Single-Ended Input. We can use difference amplifiers as single-ended amplifiers by applying signals at one input with the other input grounded. Suppose, for example, that $v_a(t) = v_s(t)$, and $v_b(t) = 0$. Equations (1.26) and (1.27) then give

$$v_d(t) = v_s(t) \quad \text{and} \quad v_c(t) = 0.5 v_s(t)$$

From Eq. (1.28),

$$v_o(t) = \left(A_d + \frac{A_c}{2} \right) v_s(t) \approx A_d v_s(t)$$

where the approximation assumes high CMRR. Of course the amplifier loses its ability to reject common-mode signals when used in this manner.

Section 1.4 described single-ended amplifiers in terms of simple circuit models to show how input and output loading affect amplification. We now apply these same techniques to the difference amplifier.

1.5.4 IDEALIZED DIFFERENCE AMPLIFIER MODEL

Figure 1.25a shows a simple circuit model of Fig. 1.23 and Eq. (1.28). When the input is a pure difference-mode signal, $v_c(t) = 0$, the transfer characteristic is the straight line of Fig. 1.25b. This transfer characteristic resembles the single-ended amplifier characteristic of Fig. 1.16c; however the independent variable is *difference-mode input voltage* and the slope is *difference-mode gain*.

(a)

(b)

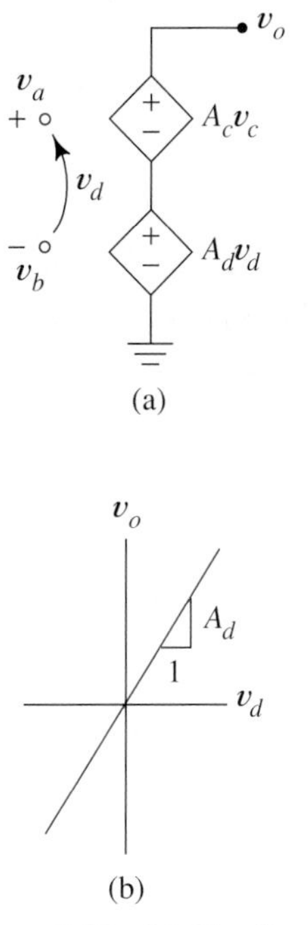

Figure 1.25 (a) Circuit model for idealized difference amplifier; (b) transfer characteristic for pure difference-mode input signals.

Sources of Double-Ended Signals. Many modern signal sources have double-ended outputs specifically designed for use with difference amplifiers. Instead of introducing such a source directly, we use a roundabout approach to *discover* the idea.

Figure 1.26a shows a voltage divider that consists of a fixed and a variable resistor. The latter represents a *transducer* that encodes a signal as changes in resistance, ΔR. An

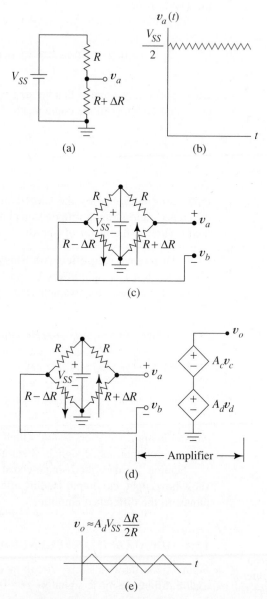

Figure 1.26 Difference amplifier application: (a) voltage divider; (b) divider output containing signal plus undesired dc level; (c) Wheatstone bridge; (d) bridge providing difference amplifier input; (e) amplifier output when CMRR is high.

example of such a transducer is a strain gauge attached to a beam that is being mechanically deflected. When the beam is in its rest position, ΔR is zero. When the beam bends downward, the gauge, fastened to the top of the beam, stretches, producing positive ΔR. When the beam bends upward, ΔR becomes negative. Thus ΔR, which can either be constant or a function of time, indicates the extent of the beam deflection.

From the voltage divider,

$$v_a = \frac{R + \Delta R}{2R + \Delta R} V_{SS}$$

Factoring R from numerator and $2R$ from denominator gives

$$v_a = V_{SS} \frac{R}{2R} \left[\frac{1 + (\Delta R/R)}{1 + (\Delta R/2R)} \right]$$

where $\Delta R/R$ is usually a very small fraction. Performing long division within the brackets gives

$$v_a = \frac{V_{SS}}{2} \left[1 + \frac{\Delta R}{R} - \frac{\Delta R}{2R} - \frac{1}{2} \left(\frac{\Delta R}{R} \right)^2 + \frac{1}{4} \left(\frac{\Delta R}{R} \right)^2 + \cdots \right] \approx \frac{V_{SS}}{2} \left(1 + \frac{\Delta R}{2R} \right)$$

where the first three terms of the series give a good approximation because $\Delta R/R$ is so small. Thus

$$v_a = \frac{V_{SS}}{2} + \frac{\Delta R}{4R} V_{SS} \qquad (1.33)$$

Figure 1.26b shows how the voltage at node a might look on an oscilloscope when ΔR changes with time. The information, encoded as small fractional changes in R, is superimposed upon a large dc level, $V_{SS}/2$. A single-ended amplifier is unsuitable in this situation, for it would amplify the dc term along with the information. In fact, with the dc component included, the total voltage might be larger than the amplifier allows. Clearly the dc term, $V_{SS}/2$, of Eq. (1.32) is an undesired component that the divider provides in addition to the signal.

Figure 1.26c shows a *Wheatstone bridge*, which produces output information in double-ended format. The bridge circuit consists of two back-to-back dividers, each with a variable resistor. In a beam measurement, the second strain gauge is attached to the bottom of the beam. When the beam is at rest, ΔR is zero for both gauges. When the beam bends, one gauge stretches as the other relaxes producing changes ΔR of complementary sign. In Fig. 1.26c, v_b differs from v_a only in the algebraic sign of ΔR. The mathematical development just given, including this sign change, gives

$$v_b = \frac{V_{SS}}{2} - \frac{\Delta R}{4R} V_{SS}$$

The bridge gives an open-circuit difference-mode output voltage of

$$v_d = v_a - v_b = \frac{V_{SS}}{2} \frac{\Delta R}{R}$$

showing v_d to be directly proportional to signal ΔR. The common-mode output component of the double-ended output voltage is the troublesome dc term.

$$v_c = \frac{V_{SS}}{2}$$

In Fig. 1.26d the bridge provides input for a difference amplifier. The difference amplifier output is

$$v_o = A_d \frac{V_{SS}}{2} \frac{\Delta R}{R} + A_c \frac{V_{SS}}{2}$$

which might resemble Fig. 1.26e for large CMRR. Notice that the difference amplifier has also converted the *double-ended signal* into a *single-ended signal*. If the signal of Fig. 1.26e is still too small we can use a single-ended circuit for additional amplification because the dc component in the output is now small.

The two *ground references* in Fig. 1.26d imply the same *potential*. That is, the circuit is unchanged electrically by connecting the two ground points in the diagram. Multiple ground symbols are commonly used in circuit diagrams to avoid some of the clutter that would otherwise occur.

1.5.5 FORCING INPUT NOISE INTO THE COMMON MODE

The bridge discussion showed us how to use the difference amplifier with a source of double-ended signals. We also use diff amps to reduce noise associated with single-ended signals. The ability of a difference amplifier to reduce common-mode noise is especially useful for signals of small amplitude—in biological signal processing, for example. In this application small amounts of noise, nearly always present in the environment, may actually be larger than the signal, making some sort of noise reduction mandatory. We next discuss three common noise situations of this kind.

Ground-Loop Noise. Figure 1.27a shows a single-ended amplifier about to be connected to a source, v_s. The dashed source and resistance, v_n and R_n, represent a source of noise present in all real measuring systems, a potential difference between the amplifier and the signal generator chassis grounds—often 60 Hz power line *hum*. Although this noise source is quite invisible in the laboratory, we can detect it with a voltmeter. When we connect the amplifier input to the source with two connecting wires (called *leads*), we expect to short out the noise. However, real leads always have resistance, R_w, as in Fig. 1.27b. Notice that the resistance of the bottom lead combines with the noise source to produce a circulating noise current, i_n. This *ground-loop* current produces a Thevenin equivalent noise voltage v'_n in series with v_s in the amplifier input as in Fig.1.27c. Both are amplified equally, and both appear in the amplifier output. If $v_s >> v'_n$ the added noise is no problem. However, in *low-level measurements*, v'_n can be quite troublesome.

Figure 1.27d shows how to connect a difference amplifier using three wires of resistance R_w. In Fig. 1.27e the ground loop is replaced by its Thevenin equivalent, the

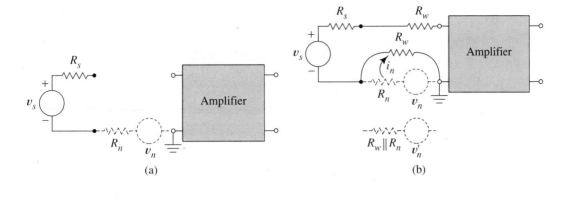

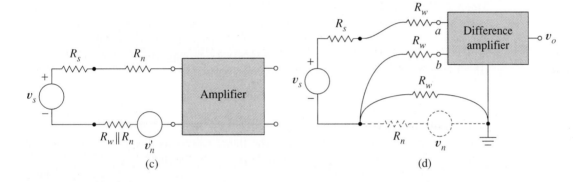

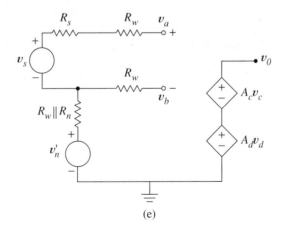

Figure 1.27 Use of difference amplifier to reduce ground-loop noise: (a) invisible noise source between signal source and amplifier; (b) ground-loop noise; (c) ground-loop noise as an input component; (d) input circuit using difference amplifier; (e) equivalent input circuit when amplifier input resistance is high.

difference amplifier by the circuit model of Fig.1.25a. Since the currents in all resistors are zero,

$$v_a = v'_n + v_s \quad \text{and} \quad v_b = v'_n$$

Then, by definition, $v_d = v_s$ and $v_c = v'_n + 0.5\ v_s$. Thus, by virtue of the input wiring, ground-loop noise enters the amplifier in the common mode while signal enters in the difference mode. The output voltage is

$$v_o = A_d v_s + A_c(v'_n + 0.5 v_s) = (A_d + 0.5\ A_c)v_s + A_c v'_n$$

showing that the signal-to-noise amplitude ratio is greatly improved by the amplifier's ability to reject common-mode signals. Notice that no harm is done when some signal enters in the common mode, as in this example. In fact output signal amplitude is slightly increased.

Magnetically Coupled Noise. Figure 1.28a shows a single-ended amplifier connected to the source by the usual two leads. This time, noise appears in the form of time-varying magnetic flux $\phi(t)$, created by a time-varying current i_n in some neighboring circuit. In physics we learn Faraday's law: "When time-varying magnetic flux $\phi(t)$ links with a circuit loop it induces an electromotive force or voltage

$$v_n = \frac{d\phi}{dt}$$

along the conductors forming the loop." The figure represents this noise as the lumped source v_n. Other than reducing the physical area, A, of the input loop as much as possible to reduce the number of flux linkages, there is little we can do to solve this noise problem with a single-ended amplifier.

Figure 1.28b shows how to connect the difference amplifier to the source with three wires: a *twisted pair* of input *leads* and a ground lead. Because the tightly twisted leads allow little linkage with the magnetic flux, there is little noise voltage induced in the loop formed by the pair. Instead, most magnetic flux links with the large open loop; therefore, most of the induced noise voltage v_n enters the circuit in the common mode while the signal again enters in the difference mode as shown by the equivalent circuit of Fig.1.28c.

Capacitively Coupled Noise. Another principle from elementary physics is that any two conductors separated by a dielectric constitute a capacitor. Thus capacitively coupled noise enters a single-ended amplifier as in Figure 1.28d, where noise current i_n depends upon the time derivative of the difference between noise voltage v_n and the voltage of the ungrounded amplifier input node. By forming the Thevenin equivalent of the two sources and their internal impedances, it is easy to show that noise also adds to the input signal in this case.

Figures 1.28e and f show how a twisted pair enclosed in a conducting shield, with shield connected only at the source end, forces capacitively coupled noise to enter the amplifier in the common mode. The electric field lines that define the distributed capacitance must terminate on the grounded shield because the twisted pair is inaccessible. The result is that v_x, noise capacitor C_n, and wire resistance R_W combine into a Thevenin

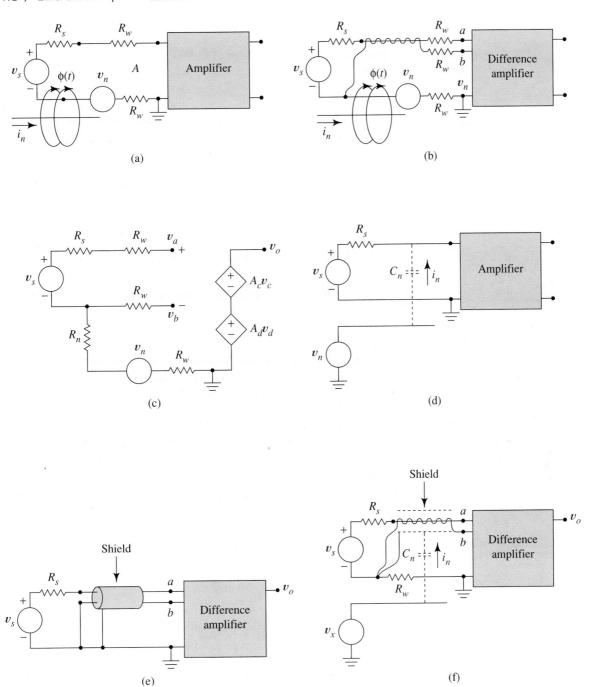

Figure 1.28 (a) Single-ended amplifier with magnetically induced noise; (b) difference amplifier configured to reduce magnetically induced noise; (c) induced noise entering amplifier in the common mode; (d) capacitively coupled input noise; (e) shielded pair connection that reduces all three kinds of noise; (f) entrance of capacitively coupled noise into diff amp input circuit.

equivalent noise source and impedance, giving an equivalent circuit resembling Fig. 1.28c. The configuration of Fig. 1.28e is the standard for low-level measurements because this connection forces all three kinds of noise: ground loop, inductively coupled, and capacitively coupled, to enter the amplifier in the common mode, while the signal enters in the difference mode. This connection is frequently used because all three sources of noise are simultaneously present in most circuits.

In measurement systems inductive and capacitively coupled noises are often 60 Hz power supply hum. More generally, however, a signal in a circuit can be inadvertently coupled to a physically adjacent circuit by both these mechanisms, where it then constitutes noise. When both circuits are integral to system operation, such noise is not easily avoided, and often requires that great care be given to practical implementation matters such as circuit layout and shielding. It is obvious that rapidly changing signals are much more troublesome than slowly varying signals.

1.5.6 MORE COMPLETE MODEL FOR THE DIFFERENCE AMPLIFIER

As with single-ended amplifiers, we incorporate input and output resistances into our circuit models to predict what happens when realistic sources and loads are attached to difference amplifiers. The equivalent circuit of Fig. 1.29a includes the input and output resistance of the difference amplifier. To see how this equivalent works, we apply pure difference- and pure common-mode excitations in turn to the input nodes.

Figure 1.29b is Fig. 1.29a with difference-mode excitation. Because of the symmetry of the sources and the input circuit the two loop currents are identical, making $v_x = R_{cx} (i_1 - i_2)$ always zero. Therefore R_{cx} is an open circuit to difference-mode signals. We conclude that a difference-mode signal source sees only the *difference-mode input resistance R_d* of Fig. 1.29c. Also, from Fig. 1.29b,

$$v_c = \frac{v_a + v_b}{2} = \frac{0.5\, v_d + (-0.5\, v_d)}{2} = 0$$

effectively turning off the dependent source controlled by v_c. Thus the general circuit diagram of Fig. 1.29a reduces to the simplified circuit of Fig. 1.29c for pure difference-mode excitation.

With common-mode excitation, as in Fig. 1.29d, the common-mode source sees *common-mode input resistance R_{cm}*, given by

$$R_{cm} = \frac{R_d}{4} + R_{cx} \tag{1.34}$$

Also, we see that $v_d = v_a - v_b = 0$, which turns off the difference-mode source in the output circuit. We conclude that the simplified model of Fig. 1.29e is valid for pure common-mode signals. Input resistances R_d and R_{cm} are specified for difference amplifiers, and we use Eq. (1.34) if we want to use the equivalent circuits of Figs. 1.29a or 1.29e.

For many problems we can decompose the excitation into difference- and common-mode components and then use superposition. This approach allows us to avoid the cir-

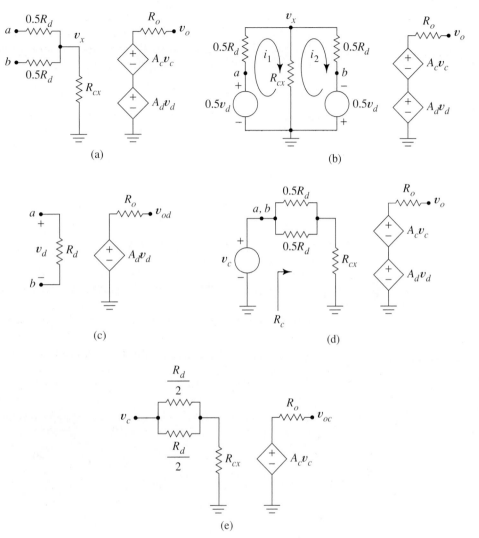

Figure 1.29 Derivation of special difference amplifier models: (a) complete difference amplifier model; (b) complete model with pure difference-mode excitation; (c) simplified model for pure difference-mode signals; (d) complete model with pure common-mode excitation; (e) simplified model for pure common-mode signals.

cuit of Fig. 1.29a entirely and use, instead, the simpler circuits, 1.29c and e. The following example demonstrates this approach and also shows output loading.

Exercise 1.7 The specifications for the difference amplifier in Fig. 1.30a are listed in the box. Draw an equivalent circuit for the amplifier alone, excluding signal sources and load.

Ans. Figure 1.30b.

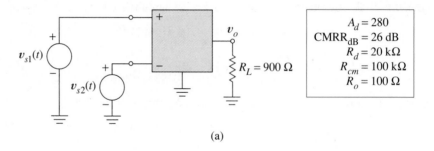

(a)

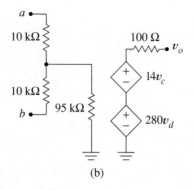

(b)

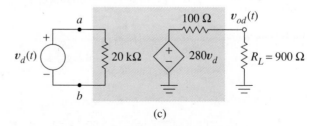

(c)

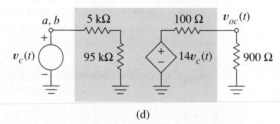

(d)

Figure 1.30 (a) Difference amplifier with resistive load; (b) equivalent circuit for difference amplifier; (c) circuit for computing difference-mode output component; (d) circuit for computing common-mode component.

EXAMPLE 1.6   Compute $v_o(t)$ for the difference amplifier of Fig. 1.30a when

$$v_{s1}(t) = 0.03 \, \sin(2\pi 30t) + 0.081 \, \sin(2\pi 60t)$$

and

$$v_{s2}(t) = -0.04 \, \sin(2\pi 30t) + 0.080 \, \sin(2\pi 60t)$$

Solution.   The difference- and common-mode components of the excitation are

$$v_d(t) = v_{s1}(t) - v_{s2}(t) = +0.07 \, \sin(2\pi 30t) + 0.001 \, \sin(2\pi 60t)$$

and

$$v_c(t) = 0.5 \, [v_{s1}(t) + v_{s2}(t)] = -0.005 \, \sin(2\pi 30t) + 0.081 \, \sin(2\pi 60t)$$

To compute the difference-mode component of $v_o(t)$, we reduce the general circuit of Fig. 1.30b to the special difference-mode circuit, Fig. 1.30c. Because of output loading, the difference-mode component of the output voltage is

$$v_{od}(t) = \frac{900}{900 + 100} \, 280 \, v_d(t) = 252 v_d(t)$$

which gives

$$v_{od}(t) = 17.6 \, \sin(2\pi 30t) + 0.252 \, \sin(2\pi 60t)$$

To find the common-mode component we reduce the general circuit of Fig. 1.30b to the special common-mode circuit of Fig. 1.30d. From this

$$v_{oc}(t) = \frac{900}{900 + 100} \, 14 \, v_c(t) = -0.063 \, \sin(2\pi 30t) + 1.021 \, \sin(2\pi 60t)$$

Superimposing the results gives the total output

$$v_o(t) = 17.5 \, \sin(2\pi 30t) + 1.27 \, \sin(2\pi 60t) \qquad\qquad ❏$$

1.5.7   INPUT LOADING

When the double-ended signal source we connect to an amplifier has internal resistance, input loading might reduce the signal amplitude. Figure 1.31a shows a source with internal resistances R_s connected to the general input circuit of a difference amplifier. Common- and difference-mode components, v_{sc} and v_{sd} describe the source voltage.

Because resistances, R_s, are equal, turning off the common-mode component results in $v_x = 0$ by the argument we used for Fig. 1.29b. Thus Fig. 1.31b applies to difference-mode analysis. A voltage division, $R_d/(R_d + 2R_s)$, reduces v_{sd} to the difference v_d that actually controls the dependent source.

When we turn off the difference-mode sources in Fig. 1.31a to find the common-mode response, since the source resistors R_s are identical, $v_a = v_b$, and the input circuit reduces to Fig. 1.31c. Because $v_a = v_b$ by circuit symmetry, one could connect a short circuit between nodes a and b without changing any currents or voltages. Doing so makes analysis easy, giving

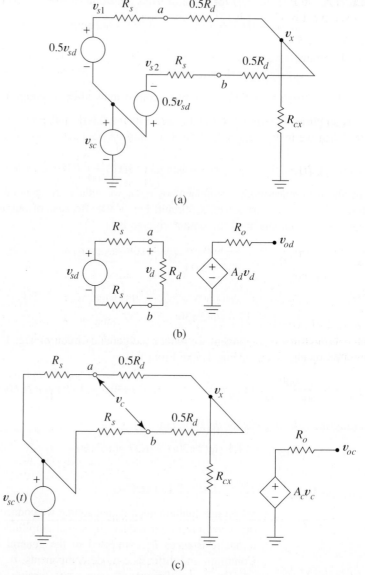

Figure 1.31 Input loading: (a) difference-mode source connected to general difference ampli-fier input; (b) equivalent circuit for calculating difference-mode output component; (c) circuit for calculating common-mode output.

$$v_c = \frac{(R_{cx} + 0.25\,R_d)}{(R_{cx} + 0.25\,R_d) + 0.5R_s}\,v_{sc} \approx v_{sc} \qquad (1.35)$$

where the approximation assumes that R_{cx} is large compared with both $0.25R_d$ and $0.5R_s$.

We next see how Example 1.6 changes when there is input loading.

EXAMPLE 1.7   In Fig. 1.32a $v_{s1}(t)$ and $v_{s2}(t)$ and the amplifier specifications are the same as in Example 1.6. Compute $v_o(t)$ for this circuit.

Solution. The two 5 kΩ source resistances caution us to look for input loading. To compute the difference-mode component of $v_o(t)$, we use Fig. 1.31b where, from Example 1.6,

$$v_{sd}(t) = +0.07\ \sin(2\pi30t) + 0.001\ \sin(2\pi60t)$$

However, it is amplifier input, $v_d(t)$, not $v_{sd}(t)$, that is amplified by the difference-mode gain of 280. When we include input loading, gain, and output loading we obtain

$$v_{od}(t) = \left[\frac{20\ k}{20\ k + 5\ k + 5\ k}\ v_{sd}(t) \right] 280 \left(\frac{900}{900 + 100} \right)$$

$$= (0.667 \times 280 \times 0.9)v_{sd}(t) = 168\ v_{sd}(t)$$

Therefore

$$v_{od}(t) = 11.8\ \sin(2\pi30t) + 0.168\ \sin(2\pi60t)$$

To compute the common-mode component we use Fig. 1.32c, where

$$v_{sc}(t) = -0.005\ \sin(2\pi30t) + 0.081\ \sin(2\pi60t)$$

from our work in Example 1.6. Node voltage v_c, to which the amplifier responds, is related to v_{sc} through a voltage division. When we include input loading, open-circuit gain, and output loading,

$$v_{oc}(t) = \left[\frac{(95\ k + 5\ k)}{(95\ k + 5\ k) + 2.5\ k}\ v_{sc}(t) \right] 14 \left(\frac{900}{900 + 100} \right)$$

$$= (0.976 \times 14 \times 0.9)\ v_{sc}(t) = 12.3 v_{sc}(t)$$

Therefore

$$v_{oc}(t) = -0.062\ \sin(2\pi30t) + 0.996\ \sin(2\pi60t)$$

The combined difference- and common-mode response is

$$v_o(t) = 11.7\ \sin(2\pi30t) + 1.16\ \sin(2\pi60t)$$

compared with

$$v_o(t) = 17.5\ \sin(2\pi30t) + 1.27\ \sin(2\pi60t)$$

from Example 1.6 when source resistances were zero.   ❑

> **Exercise 1.8**   Find $v_d(t)$, $v_c(t)$, and $v_o(t)$ for the circuit of Fig. 1.33.
>
> **Ans.** $v_d(t) = 0.04\ \cos(2\pi65t)$
> $v_c(t) = 4.2\ \cos(2\pi60t)$
> $v_o(t) = 2.67\ \cos(2\pi65t) + 0.028\ \cos(2\pi60t).$
>
> If R_{cx} is sufficiently large, which is almost always true, we need not worry too much about source resistances R_s. Problem 1.39 helps us see that even with unequal source

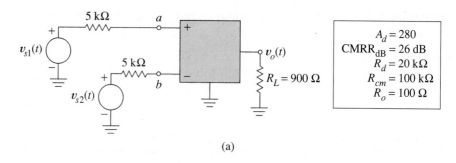

(a)

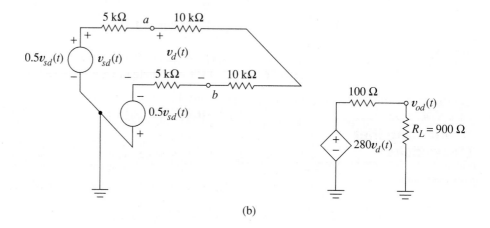

(b)

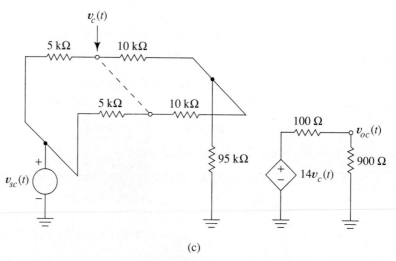

(c)

Figure 1.32 Circuits for calculating output voltage components for Example 1.7: (a) original circuit; (b) circuit for difference-mode component; (c) circuit for common-mode component.

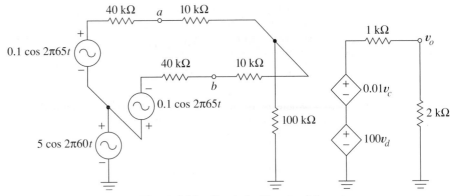

Figure 1.33 Circuit for Exercise 1.8.

resistances, $R_{s1} \neq R_{s2}$, the amplifier equivalent of Fig. 1.31b still applies to the difference-mode signal, albeit with a slight change in gain.

1.6

Other Amplifier Limitations

1.6.1 GENERAL DEFINITIONS OF INPUT AND OUTPUT RESISTANCE

We next examine several *second-order effects*, additional reasons why real amplifiers differ from the ideal. The new concepts help us better understand what we see in laboratory and they provide a foundation for more detailed studies in later chapters. There we learn how internal amplifier components such as transistors inevitably give rise to these effects.

Input and output resistance were simple concepts in Sec.1.4 because we examined only *unilateral* circuits. In these circuits, signals pass from input to output, but not from output to input. Many circuits we study in later chapters, however, are *bilateral*. This means they contain *feedback* paths through which changes made at the output affect voltages and currents at the input. A consequence is that we cannot usually find the input or output resistance of a bilateral circuit by simply inspecting the circuit diagram. To extend our analysis techniques to such circuits, we give general definitions of input and output resistance that reduce to the simple ideas of Sec. 1.4 for unilateral circuits.

Figure 1.34a defines input resistance, R_{in}, for a two-port. We must turn off all independent sources within the two-port; however, dependent sources remain active. If the

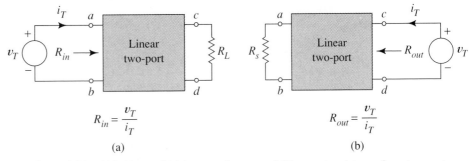

$$R_{in} = \frac{v_T}{i_T}$$

$$R_{out} = \frac{v_T}{i_T}$$

(a)   (b)

Figure 1.34 Definitions of (a) input resistance and (b) output resistance for a two-port.

circuit is to be used with a load resistor R_L attached to the output port, we connect this resistor as shown. Given these conditions, input resistance is

$$R_{in} = \frac{v_T}{i_T} \qquad (1.36)$$

where v_T is the voltage of a test generator applied to the defining input nodes and i_T is the calculated or measured test current that flows in response to v_T. For sinusoidal signals R_{in} generalizes to a complex *input impedance*, Z_{in}, the ratio of the test voltage phasor to test current phasor at some prescribed frequency.

Figure 1.34b defines output resistance of a two-port. We attach any input signal source (voltage or current) to be used with the two-port to the input terminals and then turn it off, a process that always leaves the source resistance R_s connected across the input nodes. Within the two-port, we turn off independent sources, however dependent sources remain active. Under these conditions the output resistance R_{out}, is

$$R_{out} = \frac{v_T}{i_T}$$

Output resistance generalizes to *output impedance* for sinusoidal signals.

Both R_{in} and R_{out} are the Thevenin resistances of the appropriately terminated two-port. As in Thevenin's theorem, one can alternatively use the ratio of open-circuit voltage and short-circuit current for the resistance or impedance.

When we rewrite the input and output resistance equations in the forms

$$i_T = (1/R_{in})v_T \quad \text{and} \quad i_T = (1/R_{out})v_T$$

the definitions become statements about the proportionality of response i_T to excitation v_T. For any circuit consisting of linear resistors and dependent sources, the volt-ampere curves of R_{in} and R_{out} must be straight lines through the origin; the test-voltage technique is simply a mechanism for determining another point on the line to determine its slope. From this line of reasoning we conclude that equivalent definitions of input and output resistance involve applying an arbitrary test current and measuring or computing the resulting voltage. This procedure is more convenient for some circuits.

EXAMPLE 1.8 (a) Find the input resistance of the two-port of Fig. 1.35a, including the effect of the external load, R_L.
(b) Find the output resistance when the source shown in the figure is to be used with the amplifier.
(c) Find the numerical values of R_{in} and R_{out} when $R_L = 10$ kΩ, $R_S = 100$ Ω, $R_F = 20$ kΩ, and $g_m = 2 \times 10^{-3}$ S.
(d) Use R_{in} to find input voltage v_1 if $v_s = 2$ V, $R_S = 10$ kΩ, and all other values are the same as in part (c).

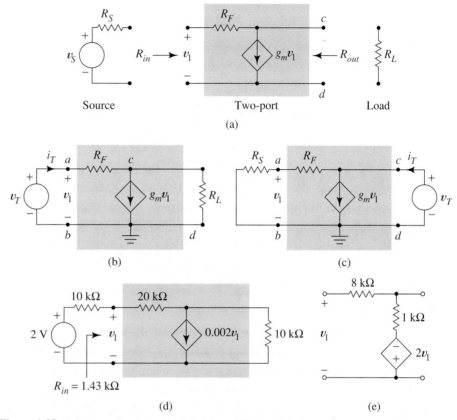

Figure 1.35 Diagrams for Example 1.8: (a) amplifier and signal source; (b) circuit for calculating R_{in}; (c) circuit for calculating R_{out}; (d) using R_{in} to calculate v_1; (e) circuit for Exercise 1.9.

Solution. (a) In Fig. 1.35a, as is often the case, the input nodes of the two-port are *implied* by an arrow labeled R_{in}, instead of being directly specified. Attaching the load and applying a test generator gives Fig. 1.35b. Applying KCL at node c gives

$$i_T = g_m v_T + \frac{v_T - i_T R_F}{R_L}$$

Solving for the ratio v_T / i_T gives

$$R_{in} = \frac{v_T}{i_T} = \frac{R_L + R_F}{1 + g_m R_L}$$

Notice that R_{in} is a function of R_L.
(b) After we attach the signal source and turn off the signal voltage, as required by the definition, we obtain Fig. 1.35c. KCL at node c gives

$$i_T = g_m v_1 + \frac{v_T}{R_S + R_F}$$

Since one term involves neither i_T nor v_T we go back to the circuit in search of a way to express v_1 in terms of either i_T or v_T. By using the voltage divider we can write

$$v_1 = \frac{R_S}{R_S + R_F} \, v_T$$

After we replace v_1 in the first equation we find

$$R_{out} = \frac{v_T}{i_T} = \frac{R_S + R_F}{1 + g_m R_S}$$

Notice that R_{out} is a function of R_S.

(c) We substitute the given numbers into the general expressions to find numerical values $R_{in} = 1.43 \text{ k}\Omega$ and $R_{out} = 16.8 \text{ k}\Omega$.

(d) This question gets to the heart of what input resistance *means*. Once we know the numerical value of R_{in}, we can use it to compute input loading for any R_S. Figure 1.35d shows that voltage division at the input reduces signal $v_s = 2$ V to

$$v_1 = \frac{1.43 \text{ k}}{1.43 \text{ k} + 10 \text{ k}} \, 2 = 0.250 \text{ V}$$

that is, R_{in} causes the usual voltage division at the input.   ❏

Exercise 1.9   Find the input resistance for the two-port of Fig. 1.35e. The load resistance is infinite.

Ans. $R_{in} = 3 \text{ k}\Omega$.

The appearance of g_m in the final expressions in Example 1.8a and b showed that the dependent source influenced the values of R_{in} and R_{out}. Sometimes a dependent source lowers an input (or output) resistance, sometimes it raises it over what it would otherwise be, say, for $g_m = 0$. We conclude that the effect of the feedback path and the dependent source cannot be ignored in the analysis, and the definitions of R_{in} and R_{out} must be formally applied in all but the simplest circuits.

1.6.2 OFFSET VOLTAGE

Figure 1.36a illustrates a transfer characteristic problem called *offset voltage*. The transfer characteristic is displaced, giving a nonzero output voltage for $v_{in} = 0$. The amount and direction of the offset are unpredictable, and it often *drifts* with temperature. Amplifier specifications describe offset by listing a typical value for *input offset voltage* V_{OS}, defined as the magnitude of input voltage needed to force v_{out} to zero.

The effect of offset is to add a dc component to the amplified signal. Offset voltages are especially troublesome when the input signal varies slowly. Examples are temperature and atmospheric pressure, which remain nearly constant over extended intervals. In such cases, there is no way to tell whether the observed output voltage is the signal, an offset, or both. After learning about amplifier saturation, our next topic, we will see that the offset can also be troublesome with rapidly varying signals.

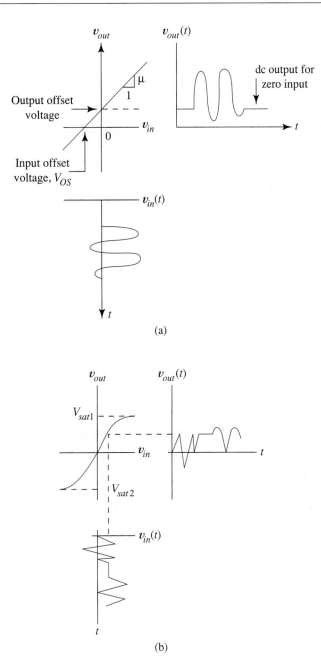

Figure 1.36 Examples of nonideal transfer characteristics: (a) offset voltage; (b) curvature of the transfer characteristic.

1.6.3 NONLINEAR DISTORTION

Figure 1.36b illustrates another amplifier problem, curvature of the transfer characteristic. It is common for the transfer characteristic to become increasingly horizontal for large $|v_{in}|$. We say that the transfer characteristic *saturates* for large input signals. With saturation, the output voltage cannot exceed an upper limit V_{sat1} nor can it fall below a lower limit V_{sat2} no matter how large the input signal. The curve of Fig. 1.36b approximates a straight line only for values of v_{in} that are not too large in magnitude. Small signals, which project from this linear part of the curve, are amplified correctly; however, large input voltages are flattened at their extremities. When the output signal is not simply an amplified version of the input waveform, we say the signal is *distorted* by the amplifier. Distortion produced by projecting from a nonlinear transfer characteristic is appropriately named *nonlinear distortion*. We shall encounter other kinds of distortion later in the chapter.

In audio amplifiers nonlinear distortion adds harsh, unpleasant sounds to music, and reduces the intelligibility of speech. Both effects occur because the nonlinear transfer characteristic produces new frequencies in the output that were not present in the input signal. Since all real amplifiers have nonlinear transfer curves, we must simply be careful to use signal amplitudes that keep operation in the linear (straight-line) region of the transfer characteristic. Practically speaking, it is the saturation of the gain curve and our desire to avoid distortion rather than the amplifier's output resistance that often limits the output power of an amplifier. Notice that a dc offset in a preceding amplifier stage, combined with the signal, may be large enough to drive the next amplifier into its saturation region as in Fig. 1.36b.

1.6.4 FREQUENCY RESPONSE

So far, all amplifier properties we have studied are related to the *static* behavior of the amplifier, that is, the response of the amplifier to dc signals or signals that are slow relative to the amplifier's inherent speed limitations. To understand an amplifier's response to rapidly changing signals, we need to know something of amplifier *dynamics*.

Real amplifiers always contain capacitances, some inherent in the amplifier components, some unavoidably associated with fabrication or circuit wiring, and some intentionally added to serve some useful purpose. Because capacitor voltages cannot change instantaneously, intuition suggests that internal capacitances might limit an amplifier's response time. Since capacitors block dc, we suspect that capacitors might also limit the amplifier's ability to respond to slowly changing signals.

The most important concept associated with these dynamic effects is the amplifier's *frequency response*, for it embodies the cumulative effect of the many capacitances in one simple pair of diagrams. Frequency response is one important consideration in selecting an amplifier that has dynamic capabilities suitable for a given application.

We can obtain the frequency response of an amplifier experimentally by applying a variable-frequency sinusoidal input signal, and measuring the gain over a range of frequencies. To ensure that operation is linear, we take care that the signal amplitude is small enough to avoid nonlinear distortion. Since the response of any linear circuit to a sinusoidal input is a sinusoid of the same frequency (possibly differing in magnitude and phase), we can conveniently represent input and output signals by phasors. At any frequency, ω, the amplifier gain is a complex number, $A(j\omega)$, the ratio of the output phasor to the input phasor at radian frequency ω. The frequency response is the collection of these complex numbers, one for each frequency. The gain magnitude, $|A(j\omega)|$, is the ratio of output to input amplitude at radian frequency ω, and the angle, $\Theta(\omega)$, is the phase of the output sine wave relative to the input. We summarize this information conveniently in the form of separate graphs of the magnitude, $|A(j\omega)|$, and phase, $\Theta(\omega)$. Sometimes the magnitude graph shows gain in decibels, sometimes as a ratio.

Figure 1.37 shows frequency response curves typical of an audio or video amplifier. The magnitude curve has a broad central region, called *midrange,* in which gain is more or less constant and independent of frequency. The two radian frequencies, ω_L and ω_H, where the gain is 3 dB below the midrange value, are called the *lower and upper −3 dB frequencies* or, alternately, the *lower and upper half-power frequencies.* (A −3 dB change in voltage ratio equals a −6 dB change in squared voltage ratio, or a power ratio of 1/2.) The midrange region is the region of useful amplification. At frequencies above ω_H the circuit capacitances cannot charge and discharge rapidly enough to keep up with the rapidly changing input signal, so output signal amplitude, and consequently gain, decreases. Below ω_L the high impedances of certain capacitors impede signal flow, and the output amplitude decreases. An important figure of merit is called the *bandwidth;,*

$$\omega_B = \omega_H - \omega_L \tag{1.37}$$

This is the width of the useful region in rad/s. In cases where $\omega_L = 0$ or where $\omega_L \ll \omega_H$, bandwidth is ω_H. The second curve in Fig. 1.37 indicates one possible amplifier phase characteristic. Often phase shift is also nearly constant over passband frequencies as in Fig.1.37b.

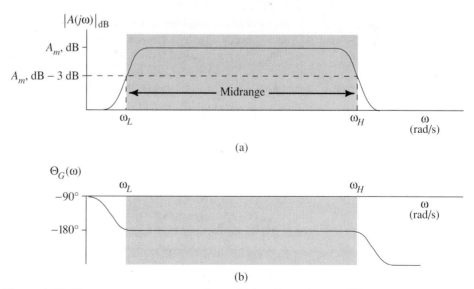

(a)

(b)

Figure 1.37   Frequency response curves for a wideband inverting amplifier: (a) magnitude of gain versus radian frequency; (b) phase shift of gain versus radian frequency.

1.6.5 EFFECT OF FREQUENCY RESPONSE ON INPUT SIGNALS

It is easy, initially, to underestimate the importance of frequency response, for we rarely amplify simple sine waves. To relate the amplifier's frequency response to more important signals, the *Fourier series* concept is of central importance. Most curricula address this topic in detail in another course. For our purposes, we need only understand the main ideas.

Jean Baptiste Joseph Fourier (1768–1830) was a French mathematician, interested in heat transfer associated with boring cannon barrels. In the course of his studies he discovered that *any* signal, *v(t)*, that repeats itself every *T* seconds, that is, any *periodic signal* of *period T*, is equal to a certain sum of sinusoidal signals with frequencies that are integer multiples of a *fundamental frequency*, $\omega_0 = 2\pi(1/T)$ rad/s. This sum is the Fourier series

$$v(t) = \sum_{n=0}^{\infty} A_n \cos(n\omega_0 t + \phi_n)$$

where A_n and ϕ_n are, respectively, the amplitude and phase of the cosine at the *n*th *harmonic* frequency, $n\omega_0$. A detailed discussion would include formulas for computing numerical A_n and ϕ_n values for a given signal; here we need only know that such calculations can be done. Engineers often approximate the infinite series for practical signals by a finite series that includes only those terms large enough to make important contributions to the sum.

The Fourier series enables us to visualize a periodic signal, such as a single note from a clarinet, as the *superposition* of some number *M* of sinusoidal signals, each having its own amplitude and phase. A voltage source that reproduces this signal is equivalent to a collection of voltage sources in series, one for each Fourier frequency.

An ideal linear amplifier scales up *each* of these sinusoidal *components* by the *same* gain factor, producing an output Fourier series that is an amplitude-scaled version of the input Fourier series. As each sinusoid passes through a real amplifier, however, its amplitude is multiplied by the gain magnitude, $|A(j\omega)|$, that corresponds to its specific frequency, and $\Theta(j\omega)$ is added to its phase angle. The dependence of gain on frequency can result in a distorted output signal. We explore this next.

1.6.6 AMPLITUDE DISTORTION

In Fig. 1.38a a *line spectrum* represents a given periodic input signal. The spectrum is a diagram that uses a line at each input frequency to show the amplitude, A_n, of the Fourier cosine at that frequency. (Phase is not shown.) The same figure shows an amplifier frequency response that is appropriate for this signal. Since all important Fourier components fall into the midrange region where gain is uniform, they are amplified equally.

The amplifier's frequency response is not appropriate for the signal spectrum of Fig. 1.38b, however, because cosines at frequencies below ω_L and above ω_H are amplified significantly less than midrange cosines. Clearly the output Fourier series will not be a scaled-up replica of the input series. We call this *amplitude distortion*. To avoid amplitude distortion, the amplifier needs lower ω_L and higher ω_H for this signal. In Chapter 2

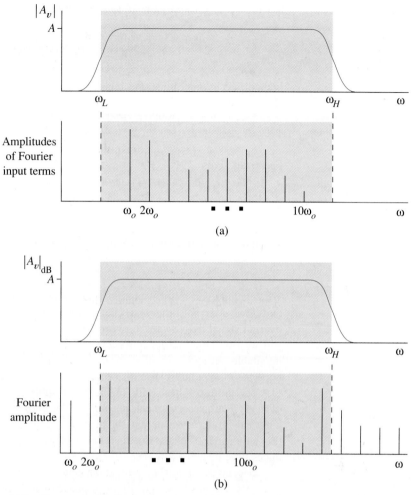

Figure 1.38 Frequency amplitude distortion concept: (a) magnitude gain curve and signal spectrum for correct amplification; (b) signal spectrum that results in amplitude distortion.

and elsewhere we shall sometimes deliberately tailor a circuit's bandwidth to fit exactly a particular signal spectrum. It is a good idea to reduce gain at frequencies where there is no signal to reduce noise, a concept called *filtering*.

1.6.7 PHASE DISTORTION

When an amplifier has phase shift, the signal is delayed as well as amplified. The easiest case to understand is when the amplifier's phase shift is a linear function of frequency. Then the complex gain is

$$A\,(j\omega) = A\,\underline{/\Theta}(\omega) = A\,\underline{/-k}\,\omega = A\,e^{-jk\omega}$$

where A is a constant gain. In this amplifier, each term in the input Fourier series

$$v_i(t) = \sum_{n=0}^{\infty} V_n \cos(n\omega_0 t + \phi_n)$$

is modified in magnitude and phase by the complex gain, giving the output signal

$$v_o(t) = \sum_{n=0}^{N} AV_n \cos(n\omega_0 t + \phi_n - kn\omega_0)$$

Factoring gives

$$v_o(t) = A \sum_{n=0}^{N} V_n \cos[n\omega_0(t-k) + \phi_n]$$

When we compare output and input expressions we find that

$$v_o(t) = Av_i\,(t-k)$$

We conclude that a linear phase curve with slope of k rad/s introduces a delay of k seconds, but does not otherwise change the signal shape. If the phase curve is not linear over all signal frequencies, however, the output signal is *not* shaped like the input signal, and we say the amplifier introduces *phase distortion*. Since some phase shift is usually inevitable, an approximation to the ideal linear phase characteristic is the best one can hope for in avoiding phase distortion.

The uniform phase shift in Fig.1.37b subtracts approximately 180° from the phase of *every* Fourier component that falls in the midrange region. This inverts *every* component and produces an output signal that is an inverted version of the input. Notice that this uniform 180° phase shift is a linear phase characteristic of slope $k = 0$, which introduces time delay of $k = 0$ s.

Depending upon the application, phase distortion may or may not cause problems. Since the ear has little sensitivity to phase distortion, phase shift is often ignored in audio amplifiers. Nevertheless, for most signals, video signals or pulses in digital communication systems for example, the shape of the waveform is important and phase distortion must be minimized.

The general term *frequency distortion* applies to any undesired changes in the signal that result from inadequacies in the frequency response curve and includes both amplitude or phase distortion. Notice that frequency distortion is quite different from nonlinear distortion. The former is *incorrect processing of certain input frequencies by a linear circuit;* the latter involves a *nonlinear circuit* producing *new* frequencies for the output that were *not present in the input.* A large sine wave applied to the curve of Fig. 1.36b, for example, gives a periodic output waveform that is not a sine wave. The nonlinear circuit thus produces harmonics of the input signal.

We introduced the Fourier series to illustrate in a relatively simple fashion the importance of the amplifier's frequency response. Advanced courses use another signal representation, the Fourier transform, to show that a nonperiodic input signal also occupies a specific *band* of frequencies. Even though music is generally far from periodic,

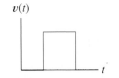

$v(t)$

Figure 1.39 Pulse signal.

the Fourier transform implies that in a stereo amplifier ω_L must be low enough to amplify the low frequencies of the organ pedals and ω_H must be high enough to process the higher harmonics of the violins. More generally, to avoid frequency distortion an amplifier requires bandwidth suitable for its signal.

1.6.8 EFFECT OF FREQUENCY LIMITATIONS ON PULSE RESPONSE

Some important signals, video signals for example, consist of successive dc levels (picture brightness levels). We can view such signals as successions of *pulse* waveforms having individual amplitudes. Figure 1.39 shows a pulse. To see how the amplifier's frequency response affects its ability to process pulselike signals, we examine some time-domain ideas. One important result is an intuitive notion of bandwidth as a measure of the amplifier's information processing speed. The main point is that frequency response and transient response are closely related.

High-Frequency Response and Rise Time. To relate frequency response to transient response we review some properties of the *RC* circuit of Fig. 1.40a. If the input is a sinusoidal voltage of frequency ω, the ratio of output to input voltage phasors is

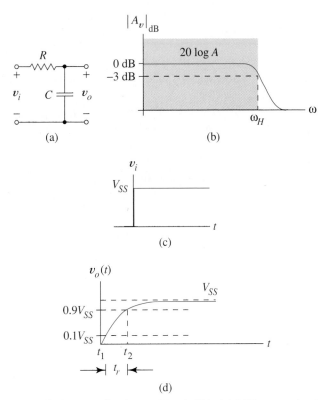

Figure 1.40 Relationship between rise time and bandwidth: (a) *RC* lowpass circuit; (b) frequency response of lowpass circuit; (c) step input to *RC* circuit; (d) output response to step input.

$$\frac{V_o}{V_i} = A(j\omega) = \frac{1/j\omega C}{(1/j\omega C) + R} = \frac{1}{1 + j\omega RC} = \frac{1}{1 + j(\omega/\omega_H)} \tag{1.38}$$

where V_o and V_i are the output and input phasors and $\omega_H = 1/RC$. The complex-valued voltage gain, $A(j\omega)$, is a function of frequency ω. Figure 1.40b is the amplitude frequency response curve for the circuit, a plot of $20 \log |A(j\omega)|$ versus frequency. This RC circuit is called a *lowpass filter* because it passes sine waves of low frequency, multiplying them by 1 (0 dB), but *attenuates* high frequencies, scaling them by values $|A(j\omega)| < 1$. Circuit bandwidth is ω_H.

If we apply the voltage step of Fig. 1.40c to the input with the capacitor initially uncharged, solving the first-order differential equation that describes this circuit gives

$$v_o(t) = V_{SS}(1 - e^{-t/RC}) = V_{SS}(1 - e^{-\omega_H t}) \tag{1.39}$$

Figure 1.40d is a sketch of this function. Notice that the capacitance that reduces amplification at frequencies higher than ω_H also prevents the output voltage from responding instantaneously to sudden changes in input amplitude. To quantify this idea, we define the *rise time*, t_r, of the circuit by $t_r = t_2 - t_1$, where t_1 is the time when the output reaches 10% of final value and t_2 the time when it reaches 90% of final value.

When we apply the definitions of t_1 and t_2 from Fig. 1.40d to Eq. (1.39), we obtain

$$0.1 \, V_{SS} = V_{SS}(1 - e^{-\omega_H t_1}) \tag{1.40}$$

and

$$0.9 \, V_{SS} = V_{SS}(1 - e^{-\omega_H t_2}) \tag{1.41}$$

Solving for t_1 and t_2 leads to the rise time expression

$$t_r = t_2 - t_1 = \frac{2.2}{\omega_H} = \frac{0.35}{f_H} \tag{1.42}$$

where $f_H = \omega_H/2\pi$ is the bandwidth in hertz. Equation (1.42) shows that the rise time is inversely proportional to bandwidth for the lowpass RC circuit. It is evident that the rise time and bandwidth contain the same information about the circuit, but in different form.

The gain of a simple amplifier of finite bandwidth is $A_m A(j\omega)$, where $A(j\omega)$ is given by Eq. (1.38) and A_m is a constant. For such an amplifier, Eq. (1.42) is exactly true. Amplifiers with more complicated gain functions often have one time constant that dominates the high-frequency behavior. When this is true, Eq. (1.42) serves as a useful *approximation*. Rise time is an important figure of merit that allows us to compare amplifiers in terms of their response speeds.

We can view the finite rise time in Fig. 1.40d as a distortion of the *leading edge* of the amplifier's step or pulse response caused by the finite value of ω_H. We associate another kind of time-domain distortion with a nonzero lower half-power frequency, ω_L. This we examine next.

Low-Frequency Limitation: Sag. The *highpass RC* circuit of Fig. 1.41a has complex gain given by

$$\frac{V_o}{V_i} = A(j\omega) = \frac{j(\omega/\omega_L)}{j(\omega/\omega_L) + 1} \tag{1.43}$$

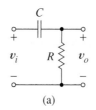

(a)

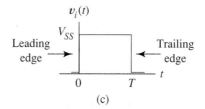

(b)

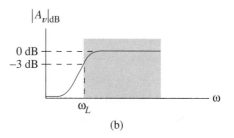

(c)

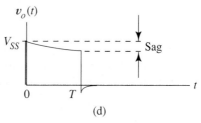

(d)

Figure 1.41 Relationship between low-frequency response and pulse sag: (a) highpass *RC* circuit; (b) frequency response of highpass circuit; (c) input pulse to *RC* circuit; (d) response of highpass circuit to input pulse.

where V_o and V_i are output and input phasors and $\omega_L = 1/RC$ is the lower half-power frequency. Figure 1.41b shows the frequency response of this circuit. As a *highpass filter*, it passes high-frequency signals ($\omega \gg \omega_L$), multiplying them by one, but reduces the amplitudes of low-frequency signals. Now suppose the input is the pulse of Fig. 1.41c. Since the circuit cannot distinguish a pulse from a step input until the *trailing edge* arrives at $t = T$, the initial response to the pulse is the same as the response to an input step of amplitude, V_{SS}, namely

$$v_o(t) = V_{SS}e^{-t/\tau} = V_{SS}e^{-\omega_L t}$$

By the time the end of the pulse arrives, there is a *sag* in the response relative to its maximum value as shown in Fig. 1.41d. The amount of sag is $V_{SS} - v_o(T)$. Most often we refer to *fractional sag,*

$$\text{fractional sag} = \frac{\text{sag}}{V_{SS}} = \frac{V_{SS}(1 - e^{-\omega_L T})}{V_{SS}} = 1 - [1 - \omega_L T + (\omega_L T)^2 + \cdots]$$

where the exponential term is represented by its infinite series expansion. For small $\omega_L T$, the series is approximated by its first two terms, giving

$$\text{fractional sag} = \omega_L T \tag{1.44}$$

Thus improved low-frequency response, lower ω_L, gives less sag in the pulse response. Simple amplifiers with low frequency limitations have gain expressions $A_m A(j\omega)$, where $A(j\omega)$ is given by Eq. (1.43) and A_m is a constant gain factor. Thus Eq. (1.44) applies exactly to such amplifiers and is often a useful approximation for more complex amplifiers.

When amplifiers have *both* low and high frequency limitations, widely separated, as in Fig. 1.37, the upper 3 dB frequency leads to a finite rise time *approximated* by Eq. (1.42), and the lower 3 dB frequency leads to sag *approximated* by Eq. (1.44). We sometimes use these equations to estimate the limits of the frequency response curve from time-domain measurements.

1.6.9 IMPORTANCE OF BANDWIDTH

Now let us use Eqs. (1.42) and (1.44) to relate an amplifier's bandwidth to its information processing speed. For an intuitive example, imagine a television set producing a black and white picture. An electron beam sweeps across the screen from left to right, moving at constant speed and using changes in intensity caused by an amplified *video signal* to construct a picture line by line. To produce high-resolution transitions between black and white (sharp edges in the picture), there must be large changes in the video signal amplitude that take only small fractions of a sweep time. Thus faster video amplifier output response facilitates more finely detailed screen constructions. This makes it possible to display more information during the fixed time used to produce a picture frame. We conclude from Eq. (1.42) that bandwidth provides a fundamental limitation on the information processing rate of the amplifier. Equation (1.44) implies that the video amplifier's low-frequency response must be adequate to *sustain* required gray levels during intervals when the input signal is constant.

1.7
Summary

The principles of this chapter form a foundation of basic ideas to which we often refer in the remainder of the text. Central to electronics are special devices that are characterized by measured graphical characteristics, theoretical mathematical descriptions, or both. By learning to visualize familiar devices in terms of volt-ampere curves, we equip ourselves to understand graphical descriptions of unfamiliar components and, further, to correlate their graphical and mathematical descriptions. Once we understand a new device, we are next interested in predicting how it operates when interconnected with other devices. To this end we integrate schematic device representations with the graphical and mathematical descriptions. Such integration leads to simple equivalent circuits that represent the devices when they are used as components in a network.

There are four fundamental amplifying structures, each based upon one of the four dependent sources. Real amplifiers differ from simple dependent sources in having finite, nonzero, input, and/or output resistance. These resistances cause loading effects, voltage or current divisions at the amplifier input, at its output, and between amplifier stages. Equivalent circuits consisting of dependent sources and resistors serve as circuit models for real amplifiers. To increase amplification, we can use the output of one amplifier to provide input to another, a process called cascading amplifier stages. When loading is included, gains of cascaded amplifiers multiply. Logarithmic gain units called decibels are useful for many purposes including dealing with gain of cascaded amplifier stages.

Amplifier inputs and outputs can be either single or double ended. When an amplifier has a double-ended input, we call it a difference amplifier because its principal function is to amplify the difference of the voltages applied to its two input nodes. Arbitrary double-ended signals, such as the input to a difference amplifier, can be decomposed into difference- and common-mode components. A second function of a difference amplifier is to reduce the common-mode input component relative to the difference-mode component. CMRR is a figure of merit indicative of the amplifier's ability to accomplish this goal. Difference amplifiers also have input and output resistances that cause loading effects.

We studied several important second-order effects that limit any amplifier's performance. We learned that we must use input and output resistance definitions that include internal feedback within the circuit from output back to the input. Other second-order effects are offset and curvature of the transfer characteristic that cause, respectively, a dc output component that is unrelated to the input signal and nonlinear distortion. Internal capacitances limit the gain of an amplifier at high frequencies and sometimes at low frequencies as well and also introduces phase shift or signal delay. Bandwidth is a figure of merit that gives us an idea of the high-frequency limits of an amplifier. If the frequency response of an amplifier is inadequate for a particular signal, there will be frequency distortion—either amplitude distortion or phase distortion, or both.

REFERENCES _____

1. ANGELO, E. J. *Electronics, BJTs, FETs, and Microcircuits,* McGraw-Hill, New York, 1969.
2. Editors of *Electronics Magazine,* "An age of innovation: The world of electronics 1930–2000." McGraw-Hill, New York, 1981.
3. IRWIN, DAVID J. *Basic Engineering Circuit Analysis,* 2nd ed., Macmillan, New York, 1987.
4. NILSSON, J. W. *Electric Circuits,* Addison-Wesley, Reading, MA, 1986.
5. SEARS, F. W., M. W. ZEMANSKY, and H. D. YOUNG. *University Physics.* Addison-Wesley, Reading, MA, 1977.

PROBLEMS _____

Section 1.2

1.1 Sketch the volt-ampere curve for a 5 V battery using the voltage and current sign conventions of Fig. 1.3a. Show the points on the curve where the source (a) delivers 10 watts of power, (b) absorbs 2 watts of power, (c) neither delivers nor absorbs power.

1.2 For each component, fill in the letter for the matching volt-ampere curve of Fig. P1.2. Then fill in the letter for the matching equation listed.

Name	Curve	Equation
resistor	___	___ (u) $v = -6$
current source	___	___ (v) $v = 0$
short circuit	___	___ (w) $i = -2v$
voltage source	___	___ (x) $i = 0$
open circuit	___	___ (y) $i = 3v$
negative resistor	___	___ (z) $i = 16$

1.3 Sketch input, output, and transfer characteristics for a VCCS with transmittance of $g_m = 2 \times 10^{-3}$ S. Label the curves.

1.4 Work P1.3 for a CCVS with $r_m = 2 \times 10^3 \ \Omega$.

1.5 Sketch volt-ampere curves for each nonideal source in P1.5.
Hint: Write an equation for each.

1.6 Place a resistor r_o in parallel with the current source in Fig. 1.8a. Then write the equation for the output characteristics and sketch them.

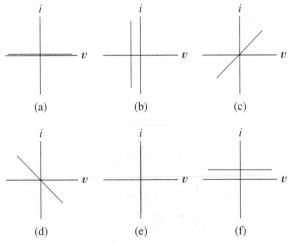

Figure P1.2

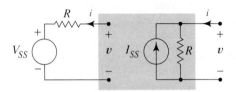

Figure P1.5

1.7 A device called an ideal diode is described by the expressions: $v = 0$ when $i \geq 0$ and $i = 0$ when $v < 0$.
(a) Sketch the volt-ampere curve for the diode.
(b) Sketch the schematic diagram of a simple one-port device that could replace the diode when $i > 0$.
(c) Sketch the schematic diagram of a simple one-port device that could replace the diode when $v < 0$.

1.8 P1.8 is the output characteristic of some device.
(a) What simple one-port model best describes the output circuit of this device when $i_1 = 0$?
(b) When $i_1 = 3$ mA and $v_2 > 5$ V?
(c) When operation is on the straight line indicated by *?
(d) Draw and label the diagram of a linear two-port circuit whose output characteristic is identical to the given one in the region ($v_2 \geq 5$, $0 \leq i_2 < 50$ mA, $0 \leq i_1 < 5$ mA). Assume that the input voltage v_1 of this two-port is always zero.

1.9 A two-port device has the transfer characteristic $i_2 = 2 \times 10^{-3}v_1$. The input current in mA is always four times the input voltage in volts.
(a) Sketch input and output characteristics.
(b) Draw an equivalent circuit to represent this two-port.

1.10 A two-port device has the transfer characteristic $v_2 = 2v_1$. The input current i_1 is always 2 mA.
(a) Sketch input and output characteristics.

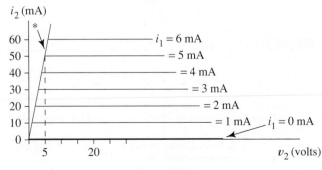

Figure P1.8

(b) Draw an equivalent circuit to represent this two-port. Observe the two-port sign convention of Fig. 1.6a.
(c) Use your equivalent circuit to find the output current when a 1 kΩ resistor is connected across the input port and a 5 kΩ resistor is also connected across the output port.

1.11 The current-source and the battery-resistor combination of P1.11 are to be connected together at nodes a and b. Notice that once the connection is made, i and v are the *same* for both devices.
(a) Sketch the volt-ampere curve for the battery-resistor combination alone.
Hint: Write an equation.
(b) On the *same i* versus v coordinate system sketch the volt-ampere curves for both the current-source and the battery-resistor combination. Then show on your curves the resulting numerical value of v.
(c) Redraw the sketch of part (b). Then add a *dashed* line to show the new volt-ampere curve for the current source if I_{SS} increases to 4 mA. Indicate the new value of v with an arrow labeled c.
(d) Add to the part (c) diagram a *dotted* line for $I_{SS} = 1$ mA. Use an arrow labeled d to show the new current-source voltage.
(e) Redraw the sketch of part (b). Then show how v changes for increases and decreases in V_{CC} when R_L remains constant.
(f) Carefully describe in words how this graphical solution for the current-source voltage would change if the dc source, V_{CC} in part (b), were replaced with a sinusoidal source, $2 \sin 5t$.
(g) Starting with another new sketch of part (b), show how v changes for R_L larger and smaller than 1 kΩ but with the same V_{CC}.

1.12 P1.12 shows the input and output characteristics of a two-port.
(a) Draw a circuit model of the two-port.
(b) Sketch the transfer characteristic of the two-port.

1.13 Sketch and label those output characteristic curves for the device of P1.13 that correspond to $v_1 = -1, 0, 1,$ and 2.

1.14 For P1.14 sketch and label the output characteristic curves that correspond to $v_1 = 0, 1,$ and 2V.

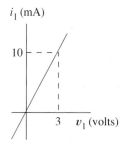

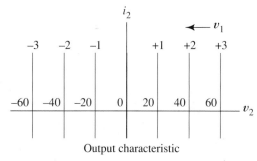

Figure P1.12

1.15 How much power can a source of V_{SS} volts with internal resistance of R ohms deliver to an external load?
Hint: Maximum power transfer theorem.

1.16 P1.16 shows a mystery device and its volt-ampere characteristic.
(a) For each separate region, A through D, describe as completely as possible the simple circuit element that the mystery device resembles.
(b) What is the value of i when an ideal voltage source of 2 V is connected in parallel with the mystery device and oriented from node b to node a?
(c) What is the value of i when an ideal voltage source of 6 V is connected in parallel with the mystery device and oriented from node b to node a?

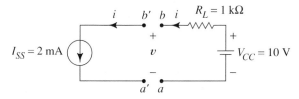

Figure P1.11

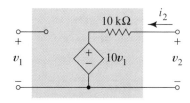

Figure P1.13

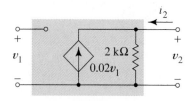

Figure P1.14

(d) What is the value of v when an ideal current source of 2.5 A is connected in parallel with the mystery device and oriented from node b to node a?

1.17 (a) Write an expression for the voltage developed across R_L in P1.17 if the given source is directly connected to the given load.
(b) Find an expression for the voltage developed across R_L if the given source is connected to the given load using a CCCS of $\beta = 1$.
(c) For which case is the output voltage higher? Explain.

1.18 The device of P1.18a has input volt-ampere curve of P1.18b and output characteristic of P1.18c.
(a) Draw a circuit model that represents the device when it operates in the first quadrants of the input and output characteristics.
(b) Use your circuit model to find the value of i_x in P1.18d.

1.19 We can visualize a time-varying source as a moving volt-ampere curve. Carefully describe in words the volt-ampere curves of
(a) $v(t) = 5 \sin 10t$ (b) $i(t) = 0.2 \cos 5t$ (c) $R(t) = 100e^{-t}$
First describe each curve at $t = 0$. Then for $t > 0$.

Section 1.3

1.20 Find the voltage gain needed if an ideal voltage amplifier connects a 2 millivolt (rms) signal source having 200 Ω source resistance to a 50 Ω load that requires 1/2 watt of signal power.

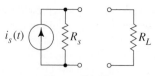

Figure P1.17

1.21 A 5 mV rms signal must be amplified so that 1 watt of signal power can be delivered to a 100 Ω resistor. How many ideal voltage amplifiers must be cascaded if each has voltage gain of 8?

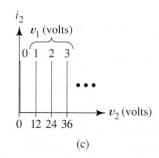

(a)

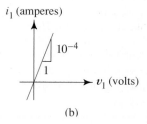

(b)

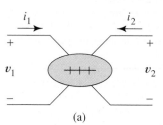

(c)

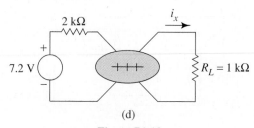

(d)

Figure P1.18

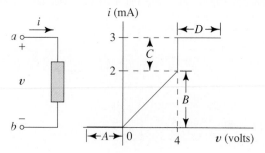

Figure P1.16

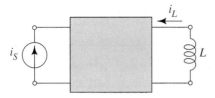

Figure P1.27

1.22 A signal source with open-circuit voltage of 3 mV rms and internal resistance of 10 kΩ is to be used with a 5 kΩ load resistor. Calculate the load voltage and load power if
(a) the source is connected directly to the load,
(b) an ideal voltage amplifier with gain of 1 connects the source to the load,
(c) an ideal amplifier with voltage gain of 10 connects the source to the load.

1.23 If we use a CCCS with transmittance β to deliver 100 mW of signal power to an 8 Ω load from a voltage source of 0.1 V rms and 1 kΩ resistance, find the minimum permissible value of β.

1.24 Find the transresistance gain needed to solve Problem 1.20 using a CCVS.

1.25 Show how to make a VCVS out of two other dependent sources from Fig. 1.17. The answer is a diagram and an equation.

1.26 Show how to make a CCCS using two other kinds of dependent sources from Fig. 1.17. The answer is a diagram and an equation.

1.27 The box in P1.27 must contain a two-port device that makes the inductor current larger than the signal current by a factor of 80, but which does not violate either of Kirchhoff's laws. Sketch input, output, and transfer characteristics for the device.

Section 1.4

1.28 A signal source, $v_i = 3$ mV and $R_i = 4$ kΩ, is attached to a 10 Ω load resistor through two identical cascaded amplifier

stages. Amplifier specifications are $R_{in} = 6$ kΩ, $R_{out} = 100$ Ω, open-circuit gain = 80.
(a) Find the voltage gain, v_L/v_i, where v_L is the voltage across the 10 Ω load.
(b) Find the power gain of the two-stage amplifier.
(c) For each amplifier stage, calculate the input and output power, and determine how much power must be added to the amplifier stage from an external source (power supply) to satisfy conservation of energy.

1.29 An amplifier develops an output voltage of v_2 volts rms across a load resistor of R_2 ohms. At the amplifier input, a voltage of v_1 volt rms is developed across an input resistance of R_1 ohm.
(a) Write an expression for the decibel power gain in terms of v_1, v_2, R_1, and R_2.
(b) Use the expression of part (a) to relate the decibel power gain to the voltage gain $A_v = v_2/v_1$.
(c) Use the results of part (b) to prove that in the special case $R_1 = R_2$, the decibel power gain and decibel voltage gain definitions give the same numerical values.
(d) Write an expression for A_{PdB} in terms of R_1, R_2, and input and output currents i_1 and i_2, respectively. Starting from this expression, find a relationship between A_{pdB} and A_{idB}.

1.30 For the two-stage amplifier of P1.30, calculate (a) the voltage gain from v_i to v_L, (b) the current gain, i_L/i_1, and (c) the power gain. Define input power as signal power that enters the first stage.

1.31 P1.31 shows a two-stage amplifier with an adjustable voltage divider placed between the stages to control the size of the output signal. We can think of the voltage divider as an *amplifier* with negative dB gain.
(a) How are the db gains of A_1 and A_2 and the voltage divider, A_{DIV}, related to the overall dB gain,

$$A_{TdB} = 20 \log|v_L/v_s|?$$

(b) With $R_1 = 0$ and $R_2 = 10$ kΩ, the voltage gain from v_s to v_L is 70 dB. Find the values of R_1 and R_2 so that the voltage divider reduces the overall gain to 18 dB by increasing R_1

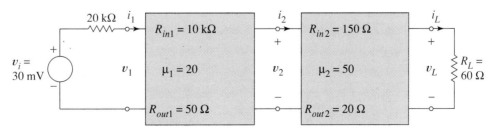

Figure P1.30

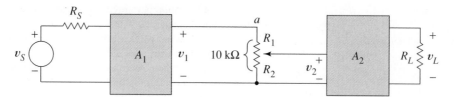

Figure P1.31

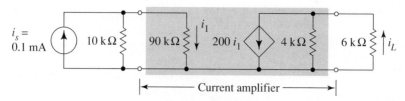

Current amplifier

Figure P1.32

while keeping $R_1 + R_2 = 10$ kΩ. Assume that the input resistance of the second stage $>>10$ kΩ.

1.32 Find the output current, current gain, and power gain for the current amplifier of Fig. P1.32.

1.33 Design a nonideal amplifier that satisfies all the following conditions: (1) delivers 1 W of signal power into a 1 kΩ load resistor, (2) has output resistance of at least 100 Ω, (3) has input resistance of at least 2 kΩ, (4) does all of the above when the signal source is a current of 0.1 mA rms having internal resistance of 50 kΩ. Any correct answer is acceptable.
Hint: Develop an equivalent circuit by working backward from the load.

Section 1.5

1.34 A difference amplifier has infinite input resistance (difference and common mode), zero output resistance, and the pa-

rameters $A_d = 75$, CMRR = 40 dB. Find the output voltage when

(a) $v_a = 2.3$ mV and $v_b = 1.6$ mV.
(b) $v_a(t) = 0.01 \sin(1000t) + 0.015 \sin(2000t)$
    $v_b(t) = -0.012 \sin(1000t) + 0.0151 \sin(2000t)$.

1.35 The output of a difference amplifier with $A_d = 20$ and $A_c = 0.5$ is $v_o(t) = 16 \sin(1000t) + 0.1 \sin(100t)$. Assume that the 1000 rad/s signal component entered the amplifier in the difference mode and the 100 rad/s signal entered in the common mode. Compute the node voltage at the noninverting input and at the inverting input.

1.36 Assume the input resistance of the difference amplifier in Fig. 1.27d is a resistor R_d connected between nodes a and b. (i.e., $R_{cx} = \infty$) Find the difference- and common-mode components of the input voltage in terms of v_s and v_n when $R_S = 600$ Ω, $R_w = 1$ Ω, $R_d = 10$ kΩ, and $R_n = 0.5$ Ω.

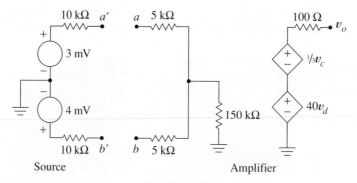

Figure P1.38

1.37 In Fig. 1.28d assume v_s and v_n are sinusoids of frequency ω_s and ω_n and amplitude V_S and V_N, respectively. Find a phasor expression for each component of the amplifier input voltage if the input impedance of the amplifier is infinite.

1.38 P1.38 shows the model for a difference amplifier and a double-ended dc source.
(a) What are the values of A_d, CMRR, R_d, and R_c?
(b) Find the difference-mode component of v_o.
(c) Find the common-mode component of v_o.
(d) Find the value of v_o.
(e) Find the new value of v_o if the amplifier output is connected to ground through an 800 Ω load resistor.

1.39 Redraw Fig. 1.31a with unequal resistors R_{s1} and R_{s2} replacing resistors R_s.
(a) Use superposition to find expressions for v_a and for v_b, each in terms of v_{sd} and v_{sc}. Assume R_{cx} is so large that it can be omitted.
(b) Use your results to show that $v_a - v_b$ is proportional to v_{sd}.

1.40 P1.40 shows a signal v_s and a noise v_n connected to a 100 Ω load through a differential amplifier. Let us examine what happens to signal and to noise individually as they pass through the amplifier.
(a) Redraw the circuit with amplifier replaced by its complete equivalent circuit.
(b) With v_n off, use circuit analysis to find v_a, v_b, v_d, and v_c, each in terms of v_s. Finally, find v_o in terms of v_s.
(c) With v_s turned off, find v_a, v_b, v_d, and v_c, each in terms of v_n. Find v_o in terms of v_n.
(d) Combine answers from (b) and (c) to find the total v_o.

1.41 A difference amplifier becomes a single-ended amplifier when we ground one input and attach a signal source $v_s(t)$ to the other input. Use the equivalent circuit of Fig. 1.29a to find
(a) the input resistance, and
(b) the open-circuit output voltage, $v_o(t)$, for this signal connection.

1.42 In Chapter 7 we study differential amplifiers that have double-ended outputs as well as double-ended inputs. Figure P1.42a represents such an amplifier and P1.42b is its equivalent circuit, where v_d is still the difference-mode input component defined by Eq. (1.26).
(a) Find $v_x(t)$ for the circuit of Fig. P1.42c.
(b) Find $v_o(t)$ for the circuit of Fig. P1.40d.

1.43 P1.43 shows the equivalent circuit for a difference amplifier with two output nodes instead of one. Difference- and com-

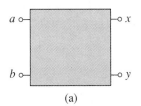

(a)

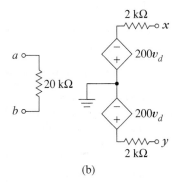

(b)

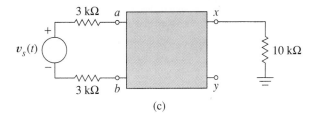

(c)

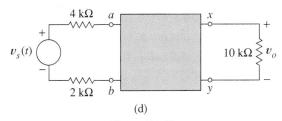

(d)

Figure P1.42

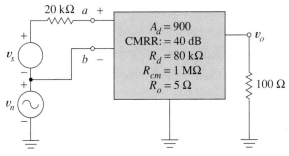

Figure P1.40

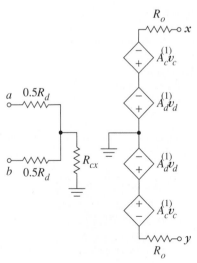

Figure P1.43

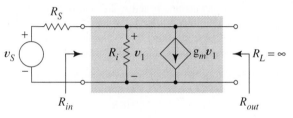

Figure P1.46

1.44 Draw the equivalent circuit for a difference amplifier with single-ended output that satisfies all the following conditions (1) develops 4 V across a load resistor of 2 kΩ when a pure difference-mode signal of 50 mV is applied at the input, (2) has output resistance less than 50 Ω and CMRR of 63 dB, (3) presents 10 MΩ input resistance to a common-mode input signal, (4) develops 50 mV between its input nodes when a difference-mode source having output resistance $R_s = 20$ kΩ and total difference-mode voltage of $v_s = 150$ mV is applied at the input.

Section 1.6

1.45 Find the output resistance of the two-port of Fig. 1.35e when a current-source signal generator with 2 kΩ internal resistance is connected to the input.

1.46 Find the R_{in} and R_{out} for the two-port of P1.46.

1.47 Find the output resistance of the circuit of P1.46 using a current source as the test generator. Use numerical values $R_s = R_i = 10$ kΩ, $g_m = 0.002$ S.

1.48 The dependent source in the input circuit makes the amplifier in P1.48 bilateral. Find R_{in} and R_{out} for the source and load resistance values given.

1.49 An amplifier with voltage gain of 400 has an input offset voltage of 0.5 mV.

mon-mode input components, v_d and v_c, are still defined by Eqs. (1.26) and (1.27). The relative dependent-source polarities are important.

(a) Suppose we apply a pure difference-mode signal to this circuit. Draw a simplified equivalent circuit that describes this special case. Use words to classify the resulting output signal.

(b) Draw a simplified equivalent circuit that describes the special case of pure common-mode excitation. Use words to classify the resulting output signal.

(c) Suppose for a general input signal, we use node voltage v_x as the output signal. Express v_x in terms of v_d and v_c. What is the CMRR for this case?

(d) For a general input signal, we use $v_o = v_y - v_x$ as output signal. Express v_o in terms of v_d and v_c. What is the CMRR for this case?

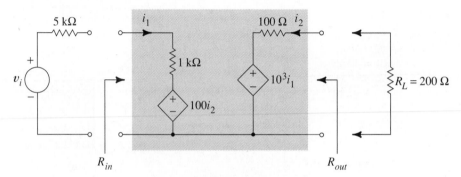

Figure P1.48

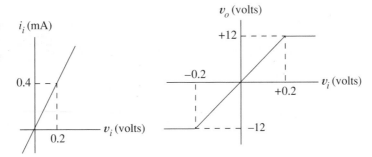

Figure P1.50

(a) Sketch the transfer characteristic.
(b) Find the value of output voltage when the input voltage is zero.
(c) Write an expression for the output voltage if the input voltage is $10^{-3} \sin 50t$.
Hint: First project from the transfer characteristic.

1.50 An amplifier is described by the input and transfer characteristics in Fig. P1.50.
(a) What is the amplifier gain for small signals?
(b) Draw a circuit model that describes the amplifier for $-0.2 \text{ V} < v_i < 0.2 \text{ V}$.
(c) Draw a circuit model that describes the amplifier for $v_i > 0.2 \text{ V}$.
Hint: The output is constant for this input range.
(d) Draw a circuit model that describes the amplifier for $v_i < -0.2 \text{ V}$.

1.51 The input voltage to the amplifier of P1.50 is

$$v_i = -0.15 + A \sin \omega t$$

where the sinusoid's amplitude is the information of interest and 0.15 is an offset that arose in a preceding amplifier stage.

(a) Find the maximum amplitude A so that the information bearing signal is not distorted.
(b) If there were no offset, how large could amplitude A be before distortion begins?

1.52 An amplifier has a nonlinear transfer characteristic given by $v_o = 80v_i - 10v_i^3$.
(a) Carefully sketch and label values on the transfer characteristic for $-V_x < v_i < V_x$ where V_x is that positive value of v_i at which the slope of the curve is zero.
(b) What is the output voltage range for this amplifier when the input voltage is constrained to the range $-V_x < v_i < V_x$?
(c) Find the largest value, V_y, of v_i such that the size of the cubic term does not exceed 10% of the size of the linear term in the transfer characteristic equation.
(d) If the input signal amplitude is strictly limited to the range defined in part (c), draw a *linear* VCVS model that approximates the behavior of this nonlinear amplifier without excessive errors for $-V_y < v_i < V_y$.

1.53 P1.53 gives the frequency response curves for the voltage gain of an amplifier. Write the Fourier series for the output voltage for

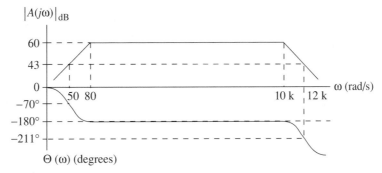

Figure P1.53

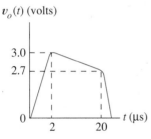

Figure P1.56

(a) input $v_i (t) = 0.002$ sin(90t + 10°) + 0.015 sin(270t + 16°) + 0.001 sin(9000t − 12°).
(b) input $v_i (t) = 0.017$ sin(50t + 8°) + 0.020 sin(250t − 90°) + 0.005 sin(12,000t + 15°).

1.54 A linear amplifier is required to process pulses of 10 µs. duration. The sag must not exceed 3% of the pulse amplitude and the rise time must not exceed 0.5 µs. Estimate the upper and lower half-power frequencies needed.

1.55 To amplify speech the frequency response must be approximately flat (within 3 dB) from 30 Hz to 3000 Hz. Estimate the rise time and fractional sag if a pulse of 1 ms duration is applied to the input of such an amplifier.

1.56 Figure P1.56 shows the response of an amplifier to an ideal pulse of 20 µs duration. The amplifier has a frequency response similar to Fig. 1.37. Estimate the −3 dB frequencies of the amplifier.

1.57 Design a two-stage amplifier in the form of Fig. P1.57. Use rms quantities to represent signals.

The second stage has voltage gain of one and must deliver at least 20 watts of signal power to the load. Common-mode output noise must not exceed 0.1 W.

The open-circuit difference-mode gain of the input stage is less than 200; the output difference-mode signal power must be less than 100 mW. You may assume that R_{cx} is infinite. Otherwise, for each amplifier stage, specify nonzero values for output resistance and finite values for input resistance.

The differential signal source consists of a signal v_s with rms voltage of 100 mV. There is also a common-mode noise v_c having rms value of 200 mV.

Draw a complete circuit diagram for your amplifier including values for all circuit elements. This problem requires you to make some arbitrary parameter choices; however, any design that meets specifications is acceptable. Show that your amplifier meets or exceeds all specifications.

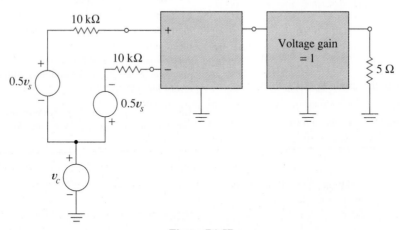

Figure P1.57

CHAPTER
2

Operational Amplifier Circuits

In this chapter we introduce active devices called operational amplifiers (op amps) and study many circuits that employ them. Using op amps, we will learn to design circuits that perform useful *operations* such as generating sine waves or square waves; amplifying, combining, integrating, or differentiating signals; removing noise, turning ac into dc, changing the shape of a waveform, producing an output change when an input signal reaches a certain level, or providing constant current or voltage. From op amp circuits we develop valuable intuition about how electronic circuits work in general, even circuits that do not contain op amps. Also, our experience with op amp circuits conditions us to visualize design tasks as successions of elementary operations to be performed by simple circuits. In later chapters we study the op amp's internal construction and then use this additional background to explain its external behavior.

2.1
The Operational Amplifier

An operational amplifier is a 25¢ integrated circuit that we depict schematically as in Fig. 2.1a. Signals enter through *noninverting* and *inverting* input nodes, denoted respectively by plus and minus signs, and exit through a single output node. To the physical op amp circuit we must also connect power supplies as in Fig. 2.1b. The internal op amp construction is such that the connection point of these supplies is the circuit ground; this point serves as the ground reference for node analysis. By convention, op amp circuit diagrams omit the power supplies as in Fig. 2.1a, and sometimes even leave the ground node to the reader's imagination.

Although its last name is "amplifier," we can avoid some initial confusion if we think of the op amp, like the resistor or capacitor, as a building block that we can use to construct circuits. The op amp is, however, an *active* building block that adds signal power to our circuits. Included among the many circuits that utilize this building block are—amplifiers!

Infinite-Gain Op Amp. The operational amplifier is essentially a difference amplifier in which R_{cx} is nearly infinite and A_c and R_o nearly zero. Thus for starters, we visualize the op amp as in Fig. 2.1c, where a *grounded* VCVS provides an output voltage proportional to the difference-mode component of the input voltage. The term "operational am-

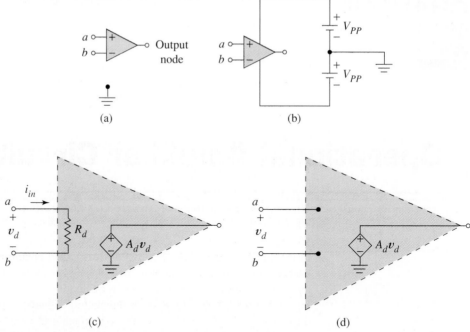

Figure 2.1 Operational amplifier: (a) schematic; (b) schematic showing supply connections; (c) as an idealized difference amplifier; (d) further idealized.

plifier" implies still more, however—exceptionally high difference-mode gain A_d. This gain is so high, in fact, that we can simplify our analysis of circuits that contain op amps by assuming A_d to be infinite. This assumption imposes two powerful constraints upon the op amp's input *signals:*

$$1. \quad v_a = v_b \qquad (2.1)$$

and

$$2. \quad i_{in} = 0 \qquad (2.2)$$

We call this equation pair *the infinite-gain assumption.* Next we give a rationale for this assumption, comment on its significance, and learn how to apply it to a variety of different circuits.

We apply the infinite-gain assumption only to op amp circuits that employ *negative feedback.* This means that the op amp's output and inverting input nodes are interconnected through the external circuit in such a way that for any reasonable value of v_o, high gain forces the difference $v_d = v_a - v_b$ to be small compared to other voltages in the circuit. Because

$$v_o = A_d(v_a - v_b) \qquad (2.3)$$

Eq. (2.1) is a consequence of infinite gain. In effect, we assume that A_d in Eq. (2.3) approaches infinity in such a manner that v_o retains its true value while $v_a - v_b$ is driven to zero.

The second consequence of infinite gain has to do with the input signal current

$$i_{in} = \frac{v_a - v_b}{R_d}$$

in Fig. 2.1c. Although R_d is often so large that i_{in} would be negligible compared to other circuit currents anyway, the same limiting process that forces $v_a = v_b$ simultaneously implies Eq. (2.2), even when R_d is not exceedingly large. (Many prefer to visualize the infinite-gain op amp as the limiting case of Fig. 2.1d as $A_d \rightarrow \infty$.)

Taken together, the infinite-gain constraints are quite remarkable in that they over-constrain the op amp by providing twice the *usual* amount of input information. Equations (2.1) and (2.2) imply that the input characteristic of an infinite gain op amp (in an i_{in} versus v_d coordinate system) is exactly the single point at the origin! When we apply the infinite-gain approximation to circuits we tacitly ignore any voltage constraint at the output node. Doing this, we always end up with the proper number of circuit equations, and it all turns out fine. This unusual constraint makes analysis of infinite-gain op amp circuits at first seem a bit peculiar (because it is). Once mastered, however, the principle is easy, and even fun, to apply.

A hypothetical one-port called a *nullator* is defined by its volt-ampere curve, the *single point (i, v) = (0, 0)*. A companion device, the *norator,* has the *entire i versus v plane* for its volt-ampere curve. That is, the norator is entirely unconstrained. Nullators and norators can be used in pairs to model many useful electronic objects, including infinite-gain op amps, and they provide a theoretical framework to justify analysis using Eqs. (2.1) and (2.2). (Interested readers may find Sanjit Mitra's chapter on nullators and norators a rewarding dessert to complement the main course of this text. Mitra's text is included in the references at the end of the chapter.)

We shall examine the limiting process for A_d in Sec. 2.5.2, define more clearly the role of R_d in Sec. 2.5.3, and justify the infinite-gain approximation using negative feedback principles in Chapter 9. In the meantime, we learn how to use Eqs. (2.1) and (2.2) to obtain simple equations that closely approximate the essential operation of most op amp circuits.

The negative feedback op amp circuits that lend themselves to infinite-gain analysis logically divide into two broad categories: *memoryless* and *memory circuits.* (See Chua's text in the end-of-chapter references.) A memoryless op amp circuit contains only resistors and op amps, and its operation is characterized by an algebraic equation. A memory circuit contains at least one essential energy-storage element, usually a capacitor, and is characterized by a time differential equation. There are also useful op amp circuits that require different analysis methods because they have no negative feedback. We consider these in Sec. 2.6.

2.2
Memoryless Op Amp Circuits

We now describe a sizable collection of circuits based upon the operational amplifier. Each circuit, important in its own right, also serves as an example of how to apply the infinite gain approximation. Notice in *each* case how the key to simple analysis is the application of Eqs. (2.1) and (2.2) to the op amp input nodes.

2.2.1 INVERTING AMPLIFIER

Figure 2.2a shows the *inverting amplifier.* We shall now use infinite-gain principles to find the voltage gain, input resistance, and output resistance of this circuit.

First, we find the voltage gain, v_{out}/v_{in}. Infinite op amp gain requires the node voltage at the inverting node to be identical to the voltage of the grounded noninverting node, namely, zero. We say there is a *virtual ground* at the inverting node. Using this virtual ground, the *amplifier* input current is

$$i_1 = \frac{v_{in} - 0}{R_1} \qquad (2.4)$$

(Equation (2.4) employs a convention widely used in electronics. Even though a voltage source is not explicitly shown, we treat v_{in} as if an independent voltage source were connected between the input node and ground.) Using the virtual ground idea again in the same diagram gives

$$i_2 = \frac{0 - v_{out}}{R_2} \qquad (2.5)$$

Infinite gain also requires that $i_1 = i_2$. Thus we equate Eqs. (2.4) and (2.5), giving

$$A_v = \frac{v_{out}}{v_{in}} = -\frac{R_2}{R_1} \qquad (2.6)$$

For $R_2 > R_1$ the circuit amplifies; the minus sign explains why the circuit is called an *inverting* amplifier. Setting $R_2 = R_1$ gives a circuit that simply inverts the signal, a useful property in itself.

Although an infinite-gain op amp always has zero *input* signal current, its *output* current is not zero. In Fig. 2.2a the signal current i_2, which satisfies Eq. (2.5), flows through R_2 and then to ground through the op amp output node. (See Fig. 2.1d.)

We next apply the input resistance definition to the inverting amplifier. If we use v_{in} and i_1 in Fig. 2.2a for the test voltage and current in Eq. (1.36), we see from Eq. (2.4) that the input resistance of the inverting amplifier circuit is R_1.

The inverting amplifier circuit provides a simple and practical way to design an amplifier like Fig. 1.19a. Since the inverting amplifier has $R_{in} = R_1$ and $\mu = -R_2/R_1$, it is

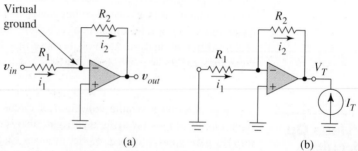

(a) (b)

Figure 2.2 (a) Inverting amplifier; (b) circuit for finding output resistance.

easy to design an amplifier with prescribed input resistance and open-circuit gain. Example 2.1 illustrates the idea.

EXAMPLE 2.1 Design an inverting amplifier that has voltage gain of -40 and input resistance of 5 kΩ

Solution. To achieve the specified input resistance, $R_1 = R_{in} = 5$ kΩ. For a gain of -40, Eq. (2.6) requires

$$R_2 = 40\,R_1 = 200 \text{ k}\Omega \qquad\qquad \square$$

The values computed for R_1 and R_2 are the *nominal values* for the design; as a practical matter the values of resistors actually selected for the circuit can deviate from nominal because of manufacturing tolerances. According to Eq. (2.6) the amplifier would have highest gain when R_2 happens to be higher and R_1 lower than nominal. For resistors that can deviate from nominal by as much as 1%, gain can be as high as

$$A_v = -\frac{202 \text{ k}\Omega}{4.95 \text{ k}\Omega} = -40.8$$

Exercise 2.1 Find the range of gain values for the Example 2.1 design if resistors vary by $\pm10\%$.

Ans. $-48.9 \le A_v \le -32.7$.

It is fairly easy to show that an inverting amplifier with gain uncertainty of approximately $2x\%$ results from resistor tolerances of $x\%$, for small x. (See Problem 2.4.)

We use Fig. 2.2b to find the output resistance of the inverting amplifier; I_T is the test generator. Because infinite gain establishes a virtual ground at the inverting node,

$$V_T = 0 - i_2 R_2$$

Infinite gain also implies

$$i_1 = \frac{-0}{R_1} = 0 \quad \text{and} \quad i_2 = i_1 = 0$$

Thus $V_T = 0$. Test current I_T, however, need not be zero because nonzero current *can* flow into the output node of the op amp. Thus

$$R_{out} = \frac{V_T}{I_T} = \frac{0}{I_T} = 0$$

Collecting our results, we conclude that the circuit of Fig. 2.2a is a practical implementation of Fig. 1.19a in which $R_{in} = R_1$, $\mu = -R_2/R_1$, and $R_{out} = 0$.

We proved that the inverting amplifier has *no output loading*. In fact, we can similarly prove that *zero output resistance is a feature common to almost all negative feedback op amp circuits*. (Exceptions are a few circuits in which the output nodes of the op amp and the circuit do not coincide.) Because giving separate proofs for all cases would introduce no new principles, we hereafter simply assume zero output resistance.

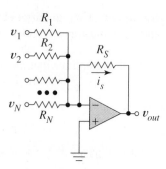

Figure 2.3   Summing amplifier.

2.2.2   SUMMING AMPLIFIER

In some signal processing situations we require a voltage that is the sum of two or more existing voltages. For example, we may wish to combine signals or add a dc component to a given signal. The *summing amplifier* of Fig. 2.3 performs this function.

When we apply Kirchhoff's current law (KCL) at the noninverting input node, because of the virtual ground and op amp input current of zero, we obtain

$$\frac{v_1 - 0}{R_1} + \frac{v_2 - 0}{R_2} + \cdots + \frac{v_N - 0}{R_N} = i_s = \frac{0 - v_{out}}{R_s}$$

Solving for v_{out} gives

$$v_{out} = -\left(\frac{R_S}{R_1} v_1 + \frac{R_S}{R_2} v_2 + \cdots + \frac{R_S}{R_N} v_N \right) \tag{2.7}$$

which reduces to the desired summing operation (plus an inversion) if all resistors are identical. Nonequal resistors give additional flexibility by allowing the designer to attach individual weights to the input voltages if needed.

Because of the virtual ground, each input voltage source looks into the circuit and sees only a resistor to ground; thus there is no interaction between the various input voltages as the combined output is produced. This point is perhaps fully appreciated only after one has tried to find some other way to combine voltages (using an all-resistor network, for example).

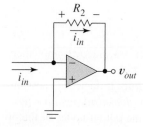

Figure 2.4   Current-to-voltage converter.

2.2.3   CURRENT-TO-VOLTAGE CONVERTER

Chapter 1 mentioned that information is sometimes available as a current because it originates in a device of high output resistance such as a photodetector. The converter circuit of Fig. 2.4 is a practical form of the CCVS that transfers such information into the more useful form of a node voltage.

To analyze, we note that the inverting input node is at virtual-ground voltage of zero; i_{in} takes on whatever value the excitation circuit (not shown) establishes. Because of infinite gain, i_{in} flows through R_2, and Kirchhoff's voltage law (KVL) gives

$$v_{out} = 0 - i_{in}R_2 = -R_2 i_{in} \tag{2.8}$$

Thus any information carried by i_{in} appears at the output as a node voltage, and we can use R_2 as a scaling parameter. Since input current is arbitrary and infinite gain constrains the input voltage to be zero, the input resistance of the circuit is zero. (To apply the input resistance definition formally, we must use a test current because a nonzero test voltage applied to the virtual ground would violate Kirchhoff's voltage law.)

2.2.4 VOLTAGE-TO-CURRENT CONVERTER

We noted in Chapter 1 that some applications favor information encoded as current. Figure 2.5 shows a *voltage-to-current converter,* which produces an output current i_L that is proportional to v_{in}, *regardless of the particular value of* Z_L. That is, the output current is voltage controlled.

To analyze the circuit, first notice that Eq. (2.4) also applies to Fig. 2.5. Because of infinite gain, $i_L = i_1$. Combining these equations gives

$$i_L = \frac{v_{in}}{R_1} \qquad (2.9)$$

Even if Z_L changes in value, its current is still given by Eq. (2.9). In effect, Z_L *sees* a VCCS, a device with a horizontal volt-ampere curve that follows the input voltage.

For an application example consider the design of a dc voltmeter based upon an analog meter movement. The latter is an electromagnetic system that produces in an indicator needle deflection that is proportional to meter current. A problem is that the internal impedance, Z_L, of such a meter varies widely from one instrument to another. When the meter movement occupies the Z_L position in Fig. 2.5, however, needle deflection is proportional to v_{in} for *any* Z_L. This permits us to work out the initial design, and even replace the meter movement, without any need for calibration. Without transforming information to current, it would be necessary to custom design a circuit for a particular Z_L.

By the same reasoning used for the inverting amplifier, the input resistance of the converter is R_1. Problem 2.9 shows that the output resistance, as defined by the two nodes of Z_L, is infinite, exactly what we need. A limitation of this particular circuit is that both ends of Z_L must be ungrounded. Later we will consider a more complex current-to-voltage converter that does not have this restriction.

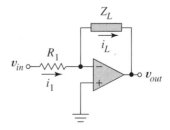

Figure 2.5 Voltage-to-current converter.

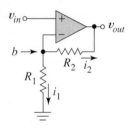

Figure 2.6 Noninverting amplifier.

2.2.5 NONINVERTING AMPLIFIER

So far, all our op amp circuits have been second cousins to the inverting amplifier. The noninverting amplifier of Fig. 2.6 has a somewhat different structure; however, applying the same infinite-gain principles reveals its properties.

In Fig. 2.6 infinite gain requires zero op amp input current; therefore,

$$i_2 = -i_1$$

Infinite gain also requires the voltage at node b to be an exact copy of v_{in} giving,

$$i_1 = \frac{v_{in}}{R_1}$$

By Ohm's law,

$$i_2 = \frac{(v_{in} - v_{out})}{R_2}$$

Substituting for i_1 and i_2 in the first equation and solving for v_{out} gives

$$A_v = \frac{v_{out}}{v_{in}} = 1 + \frac{R_2}{R_1} \tag{2.10}$$

Unlike Eq. (2.6) for the inverting amplifier, this gain expression has no minus sign, hence *noninverting amplifier.*

Inverting and noninverting amplifiers both have zero output resistance but differ in input resistance. When we apply a test voltage between ground and the input node in Fig. 2.6, the test current is zero; otherwise the infinite-gain assumption would be contradicted. Therefore the input resistance of the noninverting amplifier is infinite. Combining this, the gain of Eq. (2.10), and zero output resistance, we conclude that the noninverting amplifier provides a practical way to obtain the VCVS of Fig. 1.7a, with $\mu = (1 + R_2/R_1)$.

2.2.6 FOLLOWER CIRCUIT

Figure 2.7 shows a simple and useful operational amplifier connection called a *follower* or follower circuit. As usual, v_{in} and v_{out} are node voltages relative to the op amp's ground reference. The infinite-gain approximation implies that $v_{out} = v_{in}$; that is, the output voltage *follows* the input voltage. By visualizing first a test voltage applied to the input and then a test current applied to the output, we see that infinite gain also implies infinite input resistance and zero output resistance. We conclude that the follower circuit is a practical way to construct a VCVS with gain of one.

This follower is frequently used to eliminate loading (provide *isolation*) between circuits. We call an amplifier used for this purpose a *buffer* or *buffer amplifier.* To demonstrate how important isolation can be in practical circuits, we next examine a sample-and-hold circuit that employs followers for isolation.

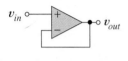

Figure 2.7 Follower circuit.

2.2.7   SAMPLE-AND-HOLD CIRCUIT

Figure 2.8a is the ideal sample-and-hold circuit, a capacitor and an electronic switch. In response to a clock, $\phi(t)$, the switch closes for an instant and the capacitor instantly charges to voltage v_s using infinite current from the ideal voltage source. When the switch opens, the isolated capacitor stores *(holds)* this voltage *sample* until the next sample time, when the switch again closes. Figure 2.8b shows input, clock, and output waveforms. The sample-and-hold operation is often the first processing step when we convert an analog voltage into a digital signal. This is because the analog-to-digital conversion circuit, which converts analog voltages like $v_s(t)$ into a sequence of binary numbers, requires an input voltage like $v_o(t)$ that does not change during the conversion process.

Figure 2.8c shows the sample-and-hold environment made more realistic by including source and load resistances; these resistances introduce some practical problems. First, because of R_S, the capacitor charges exponentially, not instantly. Second, if the

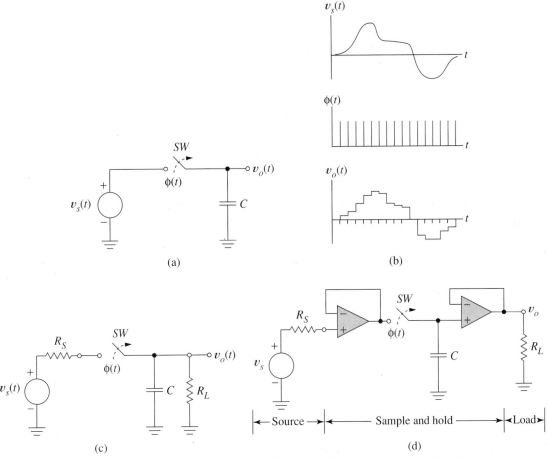

Figure 2.8   Sample-and-hold circuit: (a) ideal; (b) input and output waveforms; (c) with realistic source and load; (d) employing followers for isolation.

switch reopens before the capacitor is completely charged, the IR drop across R_S reduces the stored voltage. Finally, the capacitor discharges through R_L causing $v_o(t)$ to sag between sampling instants. Figure 2.8d shows how we can add followers to achieve near-ideal performance in the presence of R_S and R_L. Because of its infinite input resistance, the first follower reproduces v_s across the capacitor without creating an IR drop in R_S. This follower provides the large output current required to charge the capacitor quickly because it has zero output resistance. The second follower replicates the stored voltage across R_L but does not allow the capacitor to discharge between samples. (The reader can create a visual aid for this discussion, if needed, by redrawing Fig. 2.8d with each follower replaced by Fig. 1.17a where $\mu = 1$.)

2.2.8 DIFFERENCE AMPLIFIERS

The *instrumentation amplifier* circuit of Fig. 2.9a affords us one easy way to realize the difference amplifier of Fig. 1.23. With properly matched resistors, the circuit amplifies the difference-mode component and removes the common-mode component of input voltage pair v_a, v_b.

To derive a gain expression, we first use superposition to find the individual contributions of v_a and v_b to v_{out}. Setting $v_b = 0$ in Fig. 2.9a grounds one end of R_1. This gives a noninverting amplifier like Fig. 2.6 with v^+ its input voltage; a voltage divider provides v^+. Because the current flowing into the noninverting input node must be zero,

$$v^+ = \left(\frac{R_4}{R_3 + R_4} \right) v_a \tag{2.11}$$

This voltage is amplified by the gain of Eq. (2.10) giving

$$v_{out,a} = \left(1 + \frac{R_2}{R_1} \right) \frac{R_4}{R_3 + R_4} v_a = \frac{R_4}{R_1} \left(\frac{R_1 + R_2}{R_3 + R_4} \right) v_a \tag{2.12}$$

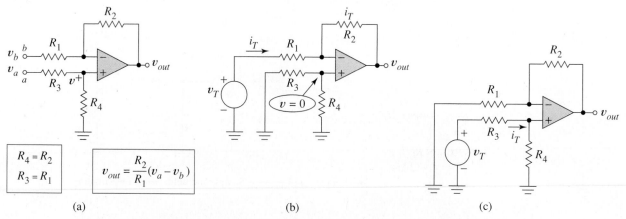

(a)   (b)   (c)

Figure 2.9 Instrumentation amplifier: (a) basic circuit; (b) finding input resistance at inverting input; (c) finding input resistance at noninverting input; (d) signal source for Exercise 2.2; (e) buffers added to remove input loading.

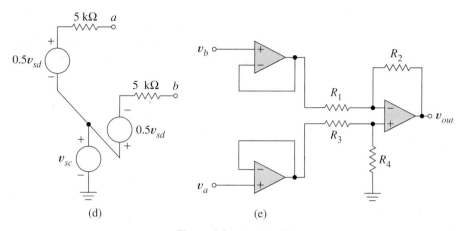

(d)   (e)

Figure 2.9   (continued)

To find the second superposition term, we turn off v_a. Now v_b enters a circuit that resembles the inverting amplifier of Fig. 2.2, except that $R_3 \| R_4$ connects the noninverting node to ground. However, $R_3 \| R_4$ has no effect on the inverting amplifier. Because of infinite gain, the current in $R_3 \| R_4$ and, therefore, the voltage drop across it, must be zero. We conclude that when $v_a = 0$, the noninverting node is at ground potential and there is a virtual ground at the inverting node of Fig. 2.9a just as in Fig. 2.2. Thus Eq. (2.6) applies and

$$v_{out,b} = -\left(\frac{R_2}{R_1}\right) v_b \tag{2.13}$$

Superimposing Eqs. (2.12) and (2.13) gives

$$v_{out} = \frac{R_4}{R_1}\left(\frac{R_1 + R_2}{R_3 + R_4}\right) v_a - \left(\frac{R_2}{R_1}\right) v_b \tag{2.14}$$

When we design we make $R_4 = R_2$ and $R_3 = R_1$, so Eq. (2.14) becomes

$$v_{out} = \frac{R_2}{R_1}(v_a - v_b) = \frac{R_2}{R_1}(v_d) \tag{2.15}$$

In view of Eq. (1.28), Eq. (2.15) describes a difference amplifier with common-mode gain of zero and difference-mode gain of R_2/R_1.

If we do not satisfy the design conditions exactly, performance is less than ideal. By substituting Eqs. (1.29) and (1.30) for v_a and v_b in Eq. (2.14), we discover that a resistor mismatch adds a nonzero common-mode component to the output voltage. (See Problem 2.25.) In a discrete design we can replace R_3 by a variable resistor called a *trim pot* and then adjust the pot to minimize output voltage in the presence of a common-mode test signal. In integrated circuit designs an equivalent operation is to laser-trim resistors to final values.

Another problem with the difference amplifier of Fig. 2.9a is that the input resistances between each input node and ground differ. Example 2.2 shows that the input resistances at the inverting and noninverting *difference amplifier* input nodes are R_1 and $R_3 + R_4$, respectively. Exercise 2.2 shows how unsymmetric input loading can cause a common-mode noise to enter the amplifier in the difference mode and be amplified along with the signal.

EXAMPLE 2.2   Find the resistance seen between each input node and ground in Fig. 2.9a.

Solution. Figure 2.9b defines input resistance between node b and ground, with the other half of the symmetrical excitation turned off. The current through $R_3 \| R_4$ is forced to zero by the infinite gain, producing a virtual ground at the op amp input nodes. Therefore, the input resistance at the inverting node is $v_T / i_T = R_1$.

In Fig. 2.9c there is *not* a virtual ground at the op amp input nodes because current can flow in R_1; however, the op amp input current must be zero. Therefore, source v_T sees $R_3 + R_4$.   ❑

Exercise 2.2   The signal source of Fig. 2.9d is connected to the input nodes of an instrumentation amplifier having $R_1 = R_3 = 5 \text{ k}\Omega$, $R_2 = R_4 = 50 \text{ k}\Omega$.
(a) Find the component of v_{out} that comes from the difference-mode input component v_{sd} alone.
Hint: First use the input resistances found in Example 2.2 to find v_a, v_b, and v_d.
(b) Find the component of v_{out} that comes from v_{sc} alone. The hint of part (a) also applies here.
(c) What is the CMRR of the circuit when input loading is included?

Ans. $7.08v_{sd}$, $4.17v_{sc}$, 1.70.

We can solve the loading problem caused by unequal input resistances by adding follower circuits at the input nodes as in Fig. 2.9e. This circuit closely approximates the ideal diff amp model of Fig. 1.25a with $A_d = R_2/R_1$ and $A_c = 0$.

Fig. 2.10a shows a difference amplifier with *double-ended input and output*. To verify this we use superposition. With a pure difference-mode signal applied, infinite

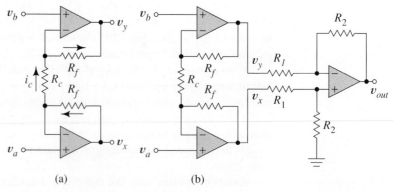

(a)   (b)

Figure 2.10   Difference amplifiers: (a) with double-ended output; (b) with single-ended output.

gain in both operational amplifiers forces the difference voltage to be developed across R_c, giving current

$$i_c = \frac{v_a - v_b}{R_c} = \frac{v_d}{R_c}$$

Because the op amps have infinite gain, this current has nowhere to go except through resistors R_f. The currents produce a voltage difference $v_x - v_y$ between the output nodes. Working out the details gives

$$v_x - v_y = i_c(2R_f + R_c) = \left(\frac{2R_f}{R_c} + 1\right)v_d$$

Thus v_d is amplified, and appears as the difference between the two output-node voltages. A common-mode input voltage implies $v_a = v_b$; therefore, $i_c = 0$ and $v_x - v_y = 0$. In summary, the double-ended output voltage, $v_x - v_y$, contains no common-mode component but only an amplified difference-mode component. Clearly the input resistance of the circuit is infinite to both common- and difference-mode signals, and both output nodes have zero output resistance. Best of all, unlike the circuit of Fig. 2.9a, there are no critical resistor requirements. The circuit of Fig. 2.10b combines the common-mode rejection of Fig. 2.10a with the single-ended output of Fig. 2.9a. This circuit gives performance close to

$$A_d = \left(\frac{2R_f}{R_c} + 1\right)\frac{R_2}{R_1}, \qquad A_c = 0$$

Because the common-mode signal is not amplified by the first stage, a slight common-mode gain in the second stage caused by imperfectly matched resistors is not harmful. (In Sec. 2.5, we learn that the operational amplifiers themselves have nonzero common-mode gain, a second-order effect we have ignored in analyzing these difference amplifier circuits.)

2.2.9 CURRENT SOURCE

The voltage-to-current converter of Fig. 2.5 applies a voltage-controlled current to the load impedance; however, there is no way in which Z_L can be grounded. The current source of Fig. 2.11 does not have this limitation.

We first use the infinite-gain approximation to derive a necessary design equation. The voltage v at the inverting op amp input equals the voltage at the noninverting node, giving $v = i_L Z_L$. The input current is

$$i_1 = \frac{v_{in} - i_L Z_L}{R_1} = i_2 = \frac{i_L Z_L - v_x}{R_2} \tag{2.16}$$

Applying KCL at the noninverting input gives

$$\frac{v_x - i_L Z_L}{R_3} = i_L + \frac{i_L Z_L}{R_4} \tag{2.17}$$

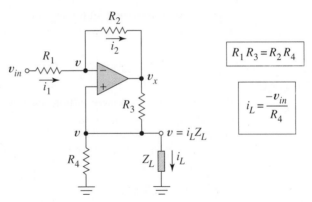

Figure 2.11 Constant current source.

Solving (2.16) for $(v_x - i_L Z_L)$ and substituting this in the numerator of (2.17) leads to

$$\frac{-R_2(v_{in} - i_L Z_L)}{R_1 R_3} = i_L + \frac{i_L Z_L}{R_4} \qquad (2.18)$$

Solving (2.18) for i_L gives

$$i_L\left(1 + \frac{Z_L}{R_4} - \frac{R_2 Z_L}{R_1 R_3}\right) = -\frac{R_2 v_{in}}{R_1 R_3} \qquad (2.19)$$

To have a current source, we must design the circuit so that load current is independent of Z_L. We do this by selecting components that make the coefficient of Z_L zero in Eq. (2.19). This takes

$$\frac{1}{R_4} = \frac{R_2}{R_1 R_3}$$

Substituting this into Eq. (2.19) gives

$$i_L = -\frac{v_{in}}{R_4} \qquad (2.20)$$

which describes the overall operation of the circuit.

The input resistance of this current source is finite, and is in fact a function of Z_L. To see this, notice that when v_{in} is constant i_L is also constant because of Eq. (2.20). A change in Z_L in Fig. 2.11 causes a change in v and, therefore, in i_1. If we add a follower in series with R_1, the result is a practical VCCS realization.

Anyone keeping track has noticed that we now have an op amp *realization* for each ideal amplifier except the CCCS. Since op amp circuits have negligible output resistance, we can make a CCCS by cascading a current-to-voltage circuit with a current source. This complete collection of dependent-source building blocks means that if we can invent a new circuit based on (grounded) ideal amplifiers, we can use op amp circuits to implement the idea. The reader interested in clever inventions is directed to the text by Chua in the references. This book describes a dazzling array of circuits, many realized using infinite-gain op amps, that do everything from rotating volt-ampere characteristics to converting nonlinear capacitors into nonlinear inductors.

2.3
Op Amp Circuits with Memory

The common feature of the op amp circuits in this section is that their operating principles depend upon at least one energy storage element. We describe such circuits, called *memory circuits,* either by differential equations or by phasor equations related to sinusoidal steady-state operation. For most, the filter being a good example, the phasor description is the more appropriate. We use the integrator to demonstrate both differential-equation and phasor descriptions.

2.3.1 INTEGRATOR

The integrating circuit of Fig. 2.12 produces an output voltage that is proportional to the integral of the input voltage. We confirm this by the following. Because of infinite gain,

$$\frac{v_i(t) - 0}{R} = i_1(t) = i_2(t) = C\frac{d[0 - v_o(t)]}{dt} \tag{2.21}$$

Integrating both sides and solving for $v_o(t)$ gives

$$v_o(t) = -\frac{1}{RC}\int v_i(t)\,dt \tag{2.22}$$

By using a specific starting time, t_o, and an initial capacitor voltage, we can write this as the definite integral

$$v_o(t) = -\frac{1}{RC}\int_{t_o}^{t} v_i(u)\,du + v_o(t_o)$$

One of many uses of the integrator is to convert a square wave into a triangular wave. Figure 2.12b shows the waveforms if the capacitor is initially uncharged. For the

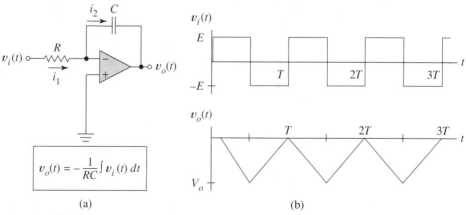

(a)   (b)

Figure 2.12 Integrator: (a) circuit; (b) waveform with square-wave input.

first $T/2$ s the circuit integrates E. At the end of this interval, the integral is the area under the first half-cycle of the input square wave, $ET/2$. Therefore, at $t = T/2$ s the output voltage is $-1/RC$ times this area, or $V_o = -ET/2RC$. This establishes the peak-to-peak amplitude; integrating successive positive and negative constants produces the periodic output. Because this square to triangular conversion works over a wide range of square-wave frequencies, the circuit is useful in laboratory signal generators that produce multiple output waveforms. Real op amps, as contrasted with the ideal ones we are now studying, have second-order effects that present a practical upper limit to useful operating frequencies.

Using a circuit to intentionally change a waveform in some prescribed fashion is an operation we call *wave shaping*. Square-to-triangular conversion is our first example. Chapter 3 contains many others.

2.3.2 ANALYSIS USING COMPLEX IMPEDANCE

Sinusoidal steady-state analysis using complex impedances is an alternative to writing differential or integral equations for memory circuits. Because analysis with complex impedances produces algebraic expressions analogous to dc circuit equations, a useful shortcut is to replace resistances with impedances in previously derived gain expressions. For example, we can view the integrator of Fig. 2.12a as an inverting amplifier with the capacitor replacing R_2. Therefore Eq. (2.6) applies to Fig. 2.12a if we replace R_2 with $1/j\omega C$, and R_1 with R. The ratio of phasors, V_o/V_i, is then

$$\frac{V_o}{V_i} = -\frac{Z_C}{R} = -\frac{1}{j\omega RC} \tag{2.23}$$

which is equivalent to Eq. (2.22) for sinusoidal steady-state analysis.

2.3.3 DIFFERENTIAL, NONINVERTING, AND SUMMING INTEGRATORS

The *differential integrator* of Fig. 2.13a integrates the difference-mode component of a double-ended input voltage. To verify this description, we first notice the structure is that of the instrumentation amplifier of Fig. 2.9a, but with R_2 and R_4 replaced by $1/j\omega C$ and R_1 and R_3 by R. From Eq. (2.15) we deduce that

$$v_{out} = \frac{1}{j\omega RC}(V_2 - V_1)$$

which is the expected result.

To describe the related *noninverting integrator* of Fig. 2.13b, we set $V_1 = 0$ and $V_2 = V_i$ in the preceding equation. A different approach is to view Fig. 2.13b as an RC circuit followed by a noninverting *amplifier* in which R_2 is replaced by an impedance Z_C.

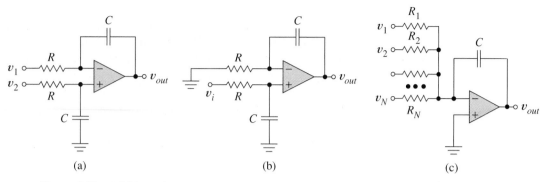

Figure 2.13 Additional integrating circuits: (a) differential integrator; (b) noninverting integrator; (c) summing integrator.

Figure 2.13c shows a *summing integrator.* Because it structurally resembles the summing amplifier of Fig. 2.5, we can use Eq. (2.7) to establish that

$$V_{out} = -\frac{1}{j\omega C}\left(\frac{1}{R_1}V_1 + \frac{1}{R_2}V_2 + \cdots + \frac{1}{R_N}V_N\right)$$

This means the output is the negative integral of the weighted sum of the input voltages.

The discussion that follows illustrates another analysis approach, applying phasor techniques directly to the op amp circuit.

2.3.4 FIRST-ORDER ACTIVE FILTERS

Figure 2.14a shows a first-order, lowpass, active filter circuit. It is called *first order* because it contains only one energy-storage element, and is therefore described by a first-order differential equation. It is a *lowpass filter* because it passes low frequencies while *attenuating* high frequencies. *Active means* it contains an amplifying device, the op amp. Between the op amp's inverting input and output nodes is the impedance

$$Z_2 = \frac{R_2}{j\omega R_2 C_2 + 1}$$

Because of the virtual ground, the output phasor is $V_o = -I_2 Z_2$. Infinite gain implies $I_2 = I_1 = V_i/R_1$. Combining these equations gives

$$\frac{V_o}{V_i} = \frac{-I_2 Z_2}{I_2 R_1} = \frac{-Z_2}{R_1}$$

Substituting for Z_2 gives the final expression

$$\frac{V_o}{V_i} = \frac{-Z_2}{R_1} = \frac{-R_2/R_1}{1 + j\omega R_2 C_2} = \frac{-R_2/R_1}{1 + j(\omega/\omega_H)} \tag{2.24}$$

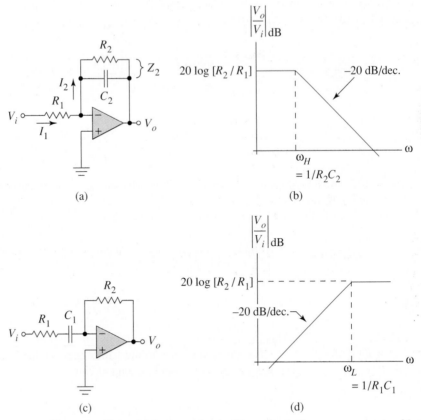

Figure 2.14 First-order filters: (a) lowpass circuit; (b) amplitude frequency response of lowpass circuit; (c) highpass circuit; (d) frequency response of highpass circuit.

where $\omega_H = 1/R_2C_2$ is the filter's *cut-off frequency*. Since the magnitude of the denominator is $\sqrt{2}$ when $\omega = \omega_H$, the cut-off frequency is the -3 dB frequency or bandwidth of the filter. Figure 2.14b shows how the magnitude of Eq. (2.24) in dB changes with radian frequency. Below ω_H, gain is reasonably constant. For frequencies much greater than ω_H, the gain decreases by 20 dB/decade. That is, gain drops by one-tenth for each factor of ten increase in ω.

The lowpass filter is useful for separating a signal from additive noise or an undesired second signal. It is necessary that all important frequencies (see Sec. 1.6.5) in the desired signal lie below ω_H and that all frequencies in the noise or undesired signal lie well above ω_H. We then visualize all these *superimposed* frequencies as passing through the filter simultaneously. The filter amplifies each desired signal frequency by approximately $-R_2/R_1$ while reducing amplitudes of all other input frequencies.

EXAMPLE 2.3 A signal $s(t)$ in which all important frequencies lie well below 4 kHz is available, but only in the form

$$v_i(t) = s(t) + 0.2 \cos (2\pi\ 10^5 t)$$

That is, $s(t)$ is combined with an undesired additive 100 kHz sinusoid.
(a) Design a lowpass filter with gain of -10 to reduce the noise.
(b) Find the filter output voltage.

Solution. (a) Since we must find the three component values in Fig. 2.14a and have only two given conditions, gain and cut-off frequency, we can start with any convenient value for R_1, say, 10 kΩ. Then to satisfy the gain condition, $R_2 = 100$ kΩ. For cut-off frequency of 4 kHz,

$$R_2 C_2 = \frac{1}{2\pi 4000} = 3.98 \times 10^{-5}$$

giving

$$C_2 = \frac{3.98 \times 10^{-5}}{(100\ k)} = 398\ \text{pF}$$

(b) Because the lowpass filter is linear, we can find its response by superposition. Since all frequency components of $s(t)$ are below 4 kHz, $s(t)$ is simply amplified by -10. The filter multiplies the noise phasor, $0.2\underline{/0°}$, by the complex gain computed from Eq. (2.24), which is

$$\frac{-10}{1 + j[(2\pi\ 10^5)/(2\pi\ 4000)]} = \frac{-10}{1 + j25} \approx 0.4\underline{/90°}$$

The complete output is

$$v_o(t) = -10s(t) + 0.08\ \cos\ (2\pi 10^5 t + 90°) \qquad \qquad ❏$$

Exercise 2.3   Design an integrator that processes the voltage

$$v_i\ (t) = 0.1\ \cos\ (2\pi\ 10^2 t) + 0.3\ \cos\ (2\pi\ 10^3 t) + 0.2\ \cos\ (2\pi\ 2 \times 10^4 t)$$

without changing the amplitude of the first component. Use Eq. (2.23) to find the output voltage. Assume steady-state operation.

Ans.   R = 10 kΩ,
C = 0.159 μF,
$v_o(t) = 0.1\ \cos\ (2\pi\ 10^2 t + 90°) + 0.03\ \cos$
$(2\pi\ 10^3 t + 90°) + 0.001\ \cos\ (2\pi\ 2 \times 10^4 t + 90°).$

The active filter of Fig. 2.14a is superior to the *RC* circuit of Fig. 1.40a in three ways; it has voltage gain, zero output impedance, and input impedance that is independent of frequency. The active circuit not only filters but also amplifies. Because output resistance is zero, Eq. (2.24) applies even when a load impedance is connected between the output node and ground. In the passive filter, any load impedance becomes part of the filter and changes the gain function. (Also, we can cascade active filters like Fig. 2.14a without loading effects. The filter gain functions simply multiply, increasing both low-frequency gain and high-frequency attenuation.) Because the passive filter has input impedance

$$Z_{in} = R + \frac{1}{j\omega C}$$

input loading varies with frequency. Source resistance R_S and filter resistance R both appear in the expression for ω_H, and filter bandwidth depends upon the particular source to be used. In the active filter, input resistance, R_1, is independent of frequency, bandwidth depends only upon R_2, and input loading affects only gain, not ω_H.

Figure 2.14c shows a first-order *highpass filter*. By using complex impedances, it is easy to show that the circuit's filter function is

$$\frac{V_o}{V_i} = \left(-\frac{R_2}{R_1}\right)\frac{j\omega R_1 C_1}{1 + j\omega R_1 C_1} = -\frac{R_2}{R_1}\frac{j(\omega/\omega_L)}{1 + j(\omega/\omega_L)} \tag{2.25}$$

As $\omega \to \infty$ the gain approaches $-R_2/R_1$, and as $\omega \to 0$, gain approaches zero. The overall frequency response is that of Fig. 2.14d. This filter removes low frequency noise from signals consisting of frequencies higher than ω_L. Because the op amp output node provides the circuit output, this filter has zero output resistance. Its input impedance is a complex function of frequency. To avoid frequency-dependent input loading, we could add a follower in series with the input.

2.3.5 SECOND-ORDER FILTER/OSCILLATOR

Figure 2.15a shows a second-order lowpass filter that uses a noninverting amplifier as a building block. Figure 2.15b shows this subcircuit as a VCVS of gain A. Because of the dependent source,

$$V_x = \frac{V_o}{A}$$

thus we can write Kirchhoff's current law at node y in the form

$$\frac{V_i - V_y}{R} = j\omega C\frac{V_o}{A} + (V_y - V_o)j\omega C$$

and at node x as

$$\frac{V_y - V_o/A}{R} = j\omega C\frac{V_o}{A}$$

Multiplying both equations by R, replacing RC everywhere by $1/\omega_o$, and rearranging gives

$$\left(j\frac{\omega}{\omega_o A} - j\frac{\omega}{\omega_o}\right)V_o + \left(j\frac{\omega}{\omega_o} + 1\right)V_y = V_i$$

$$\left(j\frac{\omega}{\omega_o A} + \frac{1}{A}\right)V_o \qquad -V_y = 0$$

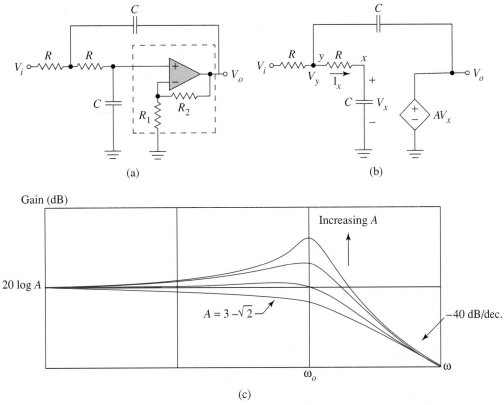

Figure 2.15 Second-order lowpass filter/oscillator: (a) circuit; (b) model including VCVS of gain A; (c) frequency response for different values of A.

Solving for V_o leads to the complex gain function

$$\frac{V_o}{V_i} = \frac{A\omega_o^2}{-\omega^2 + j(3 - A)\omega_o\omega + \omega_o^2} \qquad (2.26)$$

where $A = (1 + R_2/R_1)$ and $\omega_o = 1/RC$.

Figure 2.15c is the frequency response, showing voltage gain in decibels. If we select A to satisfy $3 - A = \sqrt{2}$, the filter has -3 dB frequency at ω_o. For this value of A, frequency response has the general form of the first-order filter, but at high frequencies, the gain drops off by 40 dB per decade instead of 20 dB/dec. This means a greater reduction in the undesired signal or noise than would be possible using a first-order filter of the same bandwidth. For values of $0 < 3 - A < \sqrt{2}$ the frequency response shows a *peaking* effect. Filters with frequency response peaks are useful as subcircuits of more complicated filters described in Chapter 12.

For a certain value of A the circuit becomes a sinusoidal *oscillator*, that is, a *generator* of a sinusoidal signal. With internal gain A of precisely three, the imaginary term in the denominator in Eq. (2.26) disappears, and gain becomes infinite at the single frequency $\omega = \omega_o$. Physically, for $A = 3$ the amplifier replaces exactly the energy dissipated in the resistors, and the circuit generates its own signal *(oscillates)* of frequency ω_o, even when the input voltage is zero.

2.3.6 DIFFERENTIATOR

Figure 2.16a shows a *differentiator circuit*. Because of the virtual ground,

$$i_1(t) = C\frac{d[v_i(t) - 0]}{dt}$$

and $v_o(t) = -Ri_2(t)$. Since $i_2 = i_1$, these equations imply

$$v_o = -RC\frac{dv_i(t)}{dt}$$

To find a phasor representation, we think of Fig. 2.16a as an "inverting amplifier" with $R_2 = R$ and with $R_1 = 1/j\omega C$; thus

$$\frac{V_o}{V_i} = -j\omega RC \tag{2.27}$$

which corresponds to differentiation for sinusoidal signals.

One common use of differentiators is to detect and emphasize rapid transitions in signals. For example, Fig. 2.16b shows how the leading and trailing edges of an input pulse are identified in the differentiator output. Because of the minus sign in the differential equation, negative output spikes occur for positive slopes and conversely. The output signal can be used to synchronize some other event with the onset or end of the pulse. In this application the designer must select a differentiator time constant RC that is short compared to the pulse width.

Differentiators are sometimes troublesome because they also differentiate the random noise that usually coexists with the input signal, creating a poor output signal to noise ratio. Equation (2.27) shows that the circuit has very high gain for high-frequency noise. This problem is alleviated by placing a capacitor in parallel with the resistor. For proper choice of capacitance, this modified circuit differentiates low-frequency signals but has constant high-frequency gain.

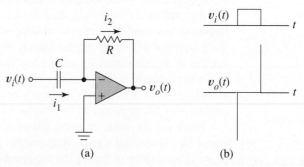

(a)   (b)

Figure 2.16 Differentiator: (a) circuit; (b) example input and output waveforms.

2.4 _____

Simulations of Infinite-Gain Op Amp Circuits

SPICE computer simulations of infinite-gain op amp circuits are useful for confirming that a design is correct and for testing our intuition. In SPICE we handle resistors, capacitors, inductors, and dc sources in a straightforward fashion; operational amplifiers require a special approach.

For a SPICE simulation we first number the circuit nodes, denoting the ground reference by zero. Each resistor, capacitor, inductor, and source requires a separate line of code, its *element line*. The first entry on a resistor's element line is a unique name beginning with R but otherwise arbitrary. Then comes the resistor's node numbers and, last, its resistance in ohms. One or more spaces separate these entries. Element lines for capacitors, inductors, voltage, and current sources follow the same format; however, the element names begin with $C, L, V,$ and $I,$ respectively; we specify values in farads, henries, volts, and amperes. For a voltage source, SPICE interprets the listed value as the voltage at the first-listed node relative to the second. For a current source the value is the current in amperes that moves through the source from the first listed node to the second. An optional identification "DC" precedes the value on a voltage- or current-source element line to indicate explicitly a direct current source. Appendix A shows some simple examples.

Special control statements allow us to sweep an input voltage or current over a specified range of values. Other control statements direct SPICE to print or plot a node voltage, the difference between two node voltages, or the current that flows through a voltage source, for each value of the swept variable. Because the only currents available as SPICE output are source currents, we sometimes insert sources of zero volts into the circuit to serve as "ammeters" at points where we wish to observe currents. Examples distributed through the text gradually introduce the more important control concepts.

SPICE lacks a special code format for the operational amplifier, so we use a circuit model, the dependent-source model of Fig. 2.1d. The first entry on the VCVS element line is a name beginning with "E." The nodes of the dependent source are listed next, followed by the nodes that define the controlling voltage. The sixth entry on the element line is the dependent-source transmittance, with algebraic sign consistent with the usual conventions for reference directions. SPICE does not allow infinite gain; however, if we use a very large value for A_d such as 10^8, the resulting numerical values for circuit voltages and currents are, for all practical purposes, the same as we would obtain using infinite gain. Because SPICE requires a path to ground from every node, it is sometimes necessary to include a large resistance, say, 10^9 Ω, between the two input nodes of the op amp to prevent an error message. Then we model the op amp like Fig. 2.1c. As our studies advance to the point where we are concerned with real rather than idealized op amps, we use variations of these basic models to study deviations from ideal behavior.

The following examples illustrate how to use SPICE to explore infinite-gain op amp circuits.

EXAMPLE 2.4 Use SPICE to verify that the circuit of Fig. 2.11 is a current source and satisfies Eq. (2.20).

Solution. Because SPICE requires numerical values, we must design a specific circuit. To satisfy the design equation we set R_1 and R_4 to 10 kΩ and R_2 and R_3 to 20 kΩ, and let the input voltage be 5 V. With these values, Eq. (2.20) predicts a constant output current of 0.5 mA. To verify this, we examine the SPICE plot of the volt-ampere characteristic, seen by Z_L, namely, i_L versus v_L.

Figure 2.17a shows the equivalent circuit, with the op amp modeled as an ideal VCVS. Voltage source V_T replaces Z_L to provide the independent variable for our volt-ampere plot. Figure 2.17b lists the code.

SPICE requires that the first line of code be a title such as Example 2.4. The next eight lines of code are element lines as discussed earlier. SPICE always requires ".END" to be the last line of code. Immediately preceding ".END" are two lines of special control code. The ".DC" control line requests that dc source V_T, nominally zero volts, be swept over a range of −5 to +5 volts with output computed at 0.5 V increments. The ".PLOT" line requests a plot of the dc current $I(VZ)$ in ammeter-source V_Z versus voltage V_T for each value of V_T. (We need the ammeter for proper current polarity because SPICE defines the current in a voltage source as the current that flows into its positive reference node.)

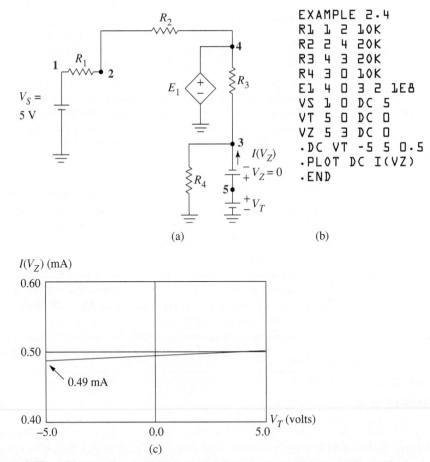

(a) (b)

(c)

Figure 2.17 SPICE model of current source: (a) circuit; (b) code; (c) partial answer to Exercise 2.5.

The SPICE output plot is the expected horizontal line that characterizes a constant current source of 0.5 mA. Changing VS to other values, positive and negative, gives similar results as Eq. (2.20) predicts.   ❏

Once SPICE code is written, it is always tempting to explore other aspects of a design because so little additional effort is required. For example, we might suspect that a mismatch in resistor values in Fig. 2.17a would lead to an imperfect current source, that is, would produce nonzero slope in the output volt-ampere characteristic. It would also be easy to explore the variation of input resistance with loading. The following exercise demonstrates the first point.

Exercise 2.4 Find the code change that increases R_1 by 10%. Use SPICE to determine the effect of this change on the output volt-ampere curve. If change does indeed cause finite output resistance r_o, estimate its value.

Ans. The R_1 element line becomes "R1 1 2 10.1K", see Fig. 2.17c, $R_{out} = 1$ MΩ.

The preceding example and exercise involved SPICE dc analysis. We next demonstrate applying SPICE ac and transient analysis to infinite-gain op amp circuits.

EXAMPLE 2.5 For a differentiator with $1/RC = 2\pi \times 10^3$ rad/s, use SPICE ac analysis to confirm Eq. (2.27), and use SPICE transient analysis to verify Fig. 2.16.

Solution. Figure 2.18a shows a differentiator with $C = 0.01$ μF, and R chosen to give the prescribed time constant. Figure 2.18b shows the SPICE code for the ac analysis. As before, a VCVS of high-gain models the infinite-gain op amp. The *VS* line describes the source as an ac phasor of unit amplitude and zero phase (by default). The .AC line asks SPICE to do ac computations, specifically 20 analyses for each decade, as the source frequency changes logarithmically from 10 to 10^5 Hz. The .PLOT line requests a plot of ac data; VDB(2) is SPICE code for 20 times the log of the voltage at node 2. Since the input voltage is one at all frequencies, the plot shows numerically the decibel gain of the circuit. We could add to the output a plot of the phase of the node 2 voltage by adding the statement VP(2) to the ".PLOT" line.

Figure 2.18c, the plot of SPICE output data, shows the linear variation in gain predicted by Eq. (2.27). Notice that the gain is 0 dB at $\omega = 2\pi f = 1/RC = 2\pi 10^3$, that is, at $f = 10^3$ Hz. One decade above and below 1 kHz, the gain is +20 and −20 dB, respectively.

For the transient analysis, we replace the three code lines preceding ".END" by

```
VS 1 0 PULSE(0 1E-3 1E-3 0 0 6E-3 10E-3)

.TRAN .03E-3 10E-3

.PLOT TRAN V(1) V(2)
```

The first changes the source waveform into a pulse. The ordered parameters in the argument of PULSE() specify that the source should change from 0 to 1 mV after a 1 ms delay. The input pulse is to have rise and fall times of zero, and pulse width is 6 ms. The last entry requests that the pulse repeat itself every 10 ms. The .TRAN line requests a

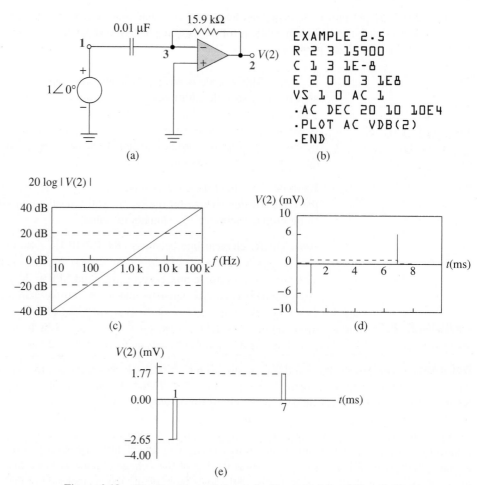

Figure 2.18 Circuit, code, and output for Example 2.5 and Exercise 2.5.

transient analysis, with output every 0.03 ms, for 10 ms. This implies about 330 output points, more than enough data to define the shape of the curve. The .PLOT line asks for graphs of input and output voltage.

The dashed curve of Fig. 2.18d is the 1 mV input pulse; the solid spikes are the differentiator output. A real pulse has a finite rate of increase and decrease at its beginning and end, and the differentiator output amplitude is determined by this rate and by the RC multiplier. In the simulation, the 1 mV change in voltage and the time between computations, 0.03 ms, gave an effective derivative of

$$\frac{\Delta V}{\Delta t} = \frac{1 \text{ mV}}{0.03 \text{ ms}} = 33.3 \text{ V/s}$$

and resulted in spike amplitude of $RC \times 33.3 = 5.3$ mV. ❑

Exercise 2.5 Estimate the new amplitudes of the output voltage spikes if the rise time and the fall times of the pulse in Example 2.5 are changed, respectively, to 0.06 and 0.09 ms. Use SPICE to check your prediction.

Ans. −2.65 mV, 1.77 mV, Fig. 2.18d.

2.5
Second-Order Effects in Operational Amplifiers

2.5.1 INTRODUCTION

So far we have viewed the operational amplifier as an idealized device of infinite gain having no static or dynamic limitations whatsoever. To better understand the practical limitations of circuits that employ op amps, we now explore various *second-order effects,* deviations from ideal operation that are characterized by numerical parameters. We obtain parameter values either from the manufacturer's data sheet or by direct measurement. Table 2.1 lists parameter values for two op amps: the μA741, a general-purpose, low-frequency device, and the HA2544, which was designed for processing high-speed signals in video applications.

We explore the effects of these parameters, one by one, introducing simple additions to our VCVS model to demonstrate how second-order effects limit op amp circuits. The modified op amp models provide a general basis for hand analysis in simple cases and for computer simulation of complex circuits. Since there are many combinations of specific circuits, inputs, and parameters, it is important (as always) to master the *concepts* and be able to apply them to any situation we encounter. We learn the physical origins of these second-order effects in later chapters.

2.5.2 OPEN-LOOP GAIN

We begin our study of second-order effects by critically examining our key assumption, infinite gain. Real operational amplifiers have only finite difference-mode gain, usually

TABLE 2.1 Data Sheet Information for Two Op Amps

Static Parameters	μA741	HA2544
Open-loop gain	2×10^5	6×10^3
Input resistance	2 MΩ	90 kΩ
Output resistance	75 Ω	20 Ω
Short-circuit current	25 mA	40 mA
Saturation limits: p-p output voltage (with ±15 V supplies)	28 V	22 V
Input offset voltage	5 mV	6 mV
Input bias current	80 nA	7 mA
Input offset current	20 nA	0.2 mA
Common mode rejection ratio	90 dB	89 dB
Dynamic Parameters		
Unity gain frequency	10^6 Hz	45×10^6 Hz
Slew rate	0.5 V/μs	150 V/μs

called *open-loop gain* on data sheets. "Open loop" implies that A_d is the *gain of the op amp itself,* measured in the absence of any external feedback loop. To determine the effect of finite gain on any op amp circuit, we must include the dependent source explicitly in our analysis.

Figure 2.19 shows how to determine the effect of finite gain in an inverting amplifier. Simply replace the op amp with its VCVS model. Although the A_d is finite, i_1 equals i_2 because we still assume infinite input resistance. Node voltage v_b, however, is not zero. From KCL,

$$\frac{v_i - v_b}{R_1} = \frac{v_b - A_d(-v_b)}{R_2}$$

After we multiply by R_1 and collect coefficients of v_b

$$v_i = \left(1 + \frac{R_1}{R_2} + \frac{R_1 A_d}{R_2}\right)v_b \qquad (2.28)$$

Because $v_o = -A_d v_b$, we can substitute

$$v_b = -\frac{v_o}{A_d}$$

into Eq. (2.28). Doing so and solving for circuit gain gives

$$\frac{v_o}{v_i} = \frac{-A_d}{(R_1/R_2)(1 + A_d) + 1} \qquad (2.29)$$

Notice that circuit gain approaches $-R_2/R_1$ as $A_d \to \infty$.

Obviously finite-gain analysis requires more effort than infinite-gain analysis. Detailed analysis is the method of choice, however, when we need to derive *quantitative criteria* for practical circuit design. For example, the denominator of (2.29) suggests that a *good* inverting amplifier design (with circuit gain independent of A_d) requires the inequality

$$\frac{R_1}{R_2} A_d \gg 1 + \frac{R_1}{R_2} \quad \text{or} \quad A_d \gg 1 + \frac{R_2}{R_1}$$

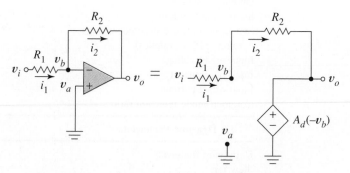

Figure 2.19 Inverting amplifier using finite-gain op amp.

not just $A_d \gg 1$. In practical engineering terms, such an inequality usually translates to

$$A_d \geq 10\left(1 + \frac{R_2}{R_1}\right)$$

This inequality ensures good designs because A_d, unlike the resistor ratio, is subject to large variations during manufacturing. The infinite-gain assumption is nearly always valid for an op amp like the 741 where $A_d \approx 2 \times 10^5$; however, $R_2/R_1 = 599$ would be marginal for the HA2544.

Similar limiting processes occur in the other infinite-gain circuits we have examined; Problem 2.49 gives the finite-gain expression for the noninverting amplifier. We next examine input resistance in the operational amplifier.

2.5.3 INPUT RESISTANCE

Figure 2.20 shows an inverting amplifier constructed with an op amp that has finite input resistance R_d as well as finite gain. When we apply KCL at the junction of the three resistors we obtain

$$\frac{v_i - v_b}{R_1} = \frac{v_b}{R_d} + \frac{v_b - A_d(-v_b)}{R_2} \tag{2.30}$$

To eliminate v_b, we substitute $v_b = -v_o/A_d$. This gives

$$\frac{v_i}{R_1} = -\left[\frac{1}{R_1} + \frac{1}{R_d} + \frac{1 + A_d}{R_2}\right]\frac{v_o}{A_d}$$

The voltage gain is

$$\frac{v_o}{v_i} = \frac{-A_d}{(R_1/R_2)(1 + A_d) + 1 + (R_1/R_d)} \tag{2.31}$$

When we compare Eq. (2.31) with Eq. (2.29) we see that finite R_d reduces voltage gain of the circuit by making the denominator larger. Of course Eq. (2.31) reduces to Eq. (2.29) as R_d approaches infinity. Problem 2.54 suggests using Eq. (2.31) to determine a lower bound for R_d.

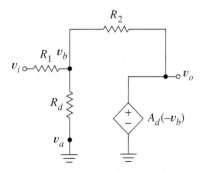

Figure 2.20   Inverting amplifier using op amp of input resistance R_d and open loop-gain A_d.

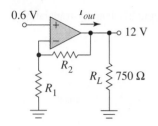

Figure 2.21 Amplifier output current.

2.5.4 OUTPUT RESISTANCE

We can determine the effect of op amp output resistance on any circuit by adding a resistor, r_{out}, in series with the dependent source in the op amp model. Problem 2.53 uses this method to show that the noninverting amplifier of Fig. 2.6 has output resistance given by

$$R_{out} = \frac{r_{out}}{1 + r_{out}/(R_1 + R_2) + A_d R_1/(R_1 + R_2)}$$

Notice that $R_{out} \rightarrow 0$ as $A_d \rightarrow \infty$ as we previously stated. The effect of feedback on R_{out} is explored at some length in a general feedback amplifier context in Chapter 9.

2.5.5 MAXIMUM OUTPUT CURRENT

An operational amplifier usually contains internal short-circuit protection to limit the current that leaves its output terminal. If output current ever exceeds this *short-circuit* value, the op amp protects itself from internal damage instead of continuing to provide amplification. Notice that the circuit need not be short-circuited to exceed this current limit. The following example illustrates how an op amp's short-circuit current rating places a lower bound on resistor values in a circuit design.

Exercise 2.6   Design a noninverting amplifier with gain of 20 using a model 741 op amp. The maximum input voltage is 0.6 V. In your design use the smallest possible resistor values so that the rated short-circuit current of 25 mA is not exceeded.

Ans. $R_1 = 24 \; \Omega, R_2 = 456 \; \Omega$.

EXAMPLE 2.6   Redesign the amplifier of Exercise 2.6 so that the current maximum is not exceeded when a 750 Ω load resistor is connected between the output node and ground.

Solution. Figure 2.21 helps us visualize the problem. At the output node it is necessary that

$$I_{out} = \frac{12}{750} + \frac{12}{R_1 + R_2} \leq I_{sc} = 25 \text{ mA}$$

Simplifying the equation and substituting the gain condition gives

$$R_1 + 19 \, R_1 \geq 1.33 \text{ k}\Omega$$

Thus $R_1 = 66.5 \; \Omega$ and $R_2 = 19R_1 = 1.26 \text{ k}\Omega$. The effect of the external load is to absorb some of the available I_{sc} *resource* compared to the open-circuit case, thereby making larger resistances necessary in the amplifier. ❏

This example shows that the short-circuit current places a lower bound on practical resistor values in op amp circuits. Resistors much larger than minimum are common, however, because they dissipate less power for the same voltage levels. An upper limit to available resistor values is about 20 MΩ. When we examine offset currents we will discover another upper limit for resistor sizes. Other resistor size constraints come up at various points in the text, especially in Chapter 8.

2.5.6 SATURATION LIMITS

The output voltage of an op amp is limited by *saturation* of its transfer characteristic, an idea previously illustrated for amplifiers in general in Fig. 1.36b, for an op amp in Fig. 2.22. The transfer characteristic of an op amp usually saturates when output voltage is slightly less than the power supply voltages. For example, Table 2.1 shows that with ± 15 V supplies, the 741 output is limited to peak-to peak voltage of 28 V.

Figure 2.22 is the *open-loop gain characteristic* (transfer characteristic) of the 741 op amp. The immense slope, $A_d = 2 \times 10^5$, is difficult to grasp intuitively. Because $|v_o|$ is limited to 14 V or less for linear operation, input voltage must lie within the range

$$-V_X \leq v_d \leq V_X$$

where $V_X = 14/2 \times 10^5 = 70$ μV. This helps us understand how "infinite gain" forces the difference between the input node voltages to approach zero. The input difference voltage for a real op amp, while not *exactly* zero, is most assuredly small compared to the other voltages in the circuit.

The *zero input current* approximation is also intuitively evident from considering the meaning of high gain. With v_d confined to such small values, the op amp input cur-

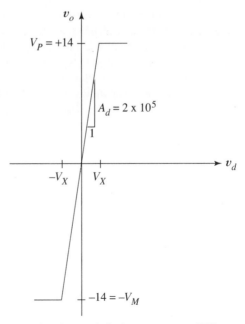

Figure 2.22   Open-loop transfer characteristic for an op amp. (Different scales used for v_o and v_d.)

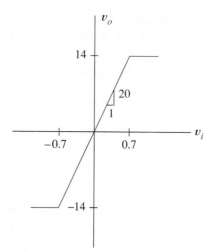

Figure 2.23 Transfer characteristic of a noninverting amplifier of gain 20.

rent is negligible compared to other circuit currents. Even for an op amp of relatively low input resistance, say, 10 kΩ, $|v_d| \leq 70$ μV implies

$$|i_{in}| = \left| \frac{v_d}{R_d} \right| \leq 7 \times 10^{-9} \text{ A}$$

This argument is consistent with the unimportance of the term R_d in Eq. (2.31) when A_d is large.

In all op amp circuits we have studied, linear operation ceases when the output voltage reaches one of the saturation limits. For example, Fig. 2.23 shows the effect of op amp saturation on the transfer characteristic of a noninverting amplifier with gain of 20 that employs a μA741. Notice that the independent variable in this figure is *the circuit's* input voltage v_i, not the op amp input voltage v_d.

Figure 2.24a shows the curve of Fig. 2.22 when the same voltage scale is used for both axes. This curve depicts the op amp as a device that produces constant output voltage of either V_P or $-V_M$, depending upon whether v_d is positive or negative. The two-

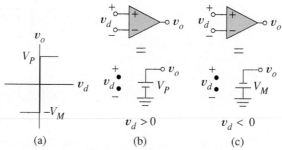

Figure 2.24 Large-signal open-loop op amp models: (a) transfer characteristic; (b) model for $v_d > 0$; (c) model for $v_d < 0$.

state, *large signal,* op amp model of Figs. 2.24b and c expresses this idea in terms of two equivalent circuits. These are useful for analyzing op amp circuits that have no negative feedback to keep operation on the steep part of the transfer curve, that is, for *open-loop* operation. In Section 2.6 we study some nonlinear op amp circuits in which saturation is central to correct circuit operation, not an undesired practical limitation as in linear circuits.

2.5.7 INPUT OFFSET VOLTAGE

In Sec. 1.6.2 we learned that the transfer characteristic of an amplifier sometimes fails to pass through the origin. The *input offset voltages* in Table 2.1 quantify this shortcoming. As previously noted, offset voltage is unpredictable in polarity, and drifts with temperature. Thus the published offset voltage is a *typical* value that allows us to estimate whether offset is likely to be a problem in a particular circuit.

We can visualize the output offset as the result of a dc input *offset voltage source* attached in series with the input nodes of an ideal amplifier. Figure 2.25(a) shows the idea. Since the sign of V_{OS} is unknown, the polarity of the added source is arbitrary.

It is customary to assume infinite gain when calculating the effects of offset voltage, a procedure justified in Fig. 2.25b. Since A_d is the *slope* of the curve, we assume for infinite-gain analysis that gain becomes infinite in such a way that the input offset remains unchanged. Thus V_{OS} is *not* forced to zero in Fig. 2.25a when using the infinite-gain approximation even though v_x is. Consequently, we can determine the effect of input offset voltage in *any op amp circuit* by adding an external source V_{OS} and then using superposition and infinite-gain analysis in the usual fashion.

For example, to find the effect of input offset voltage on the inverting amplifier of Fig. 2.26a we turn off v_s, as in Fig. 2.26b, and then analyze. With luck, we might notice that Fig. 2.26b is a noninverting amplifier with input voltage $-V_{OS}$, and immediately arrive at Eq. (2.32), the correct answer. For unlucky days or more complicated circuits,

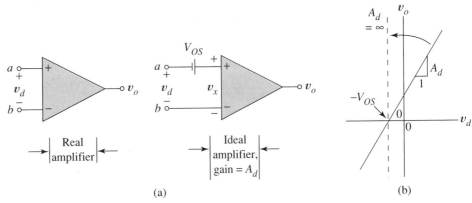

(a)   (b)

Figure 2.25 Offset voltage: (a) viewed as consequence of an input offset source applied to an op amp with no offset; (b) unaffected by infinite-gain assumption.

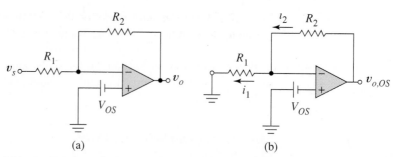

Figure 2.26 Including input offset voltage: (a) inverting amplifier; (b) circuit to compute output component caused by offset.

we should be able to work out answers from basic principles. Infinite gain implies that the inverting node has voltage $-V_{OS}$. Therefore

$$i_1 = \frac{-V_{OS}}{R_1} = i_2 = \frac{V_{o,OS} - (-V_{OS})}{R_2}$$

where $V_{o,OS}$ is the output offset voltage of the *circuit*. Solving gives

$$V_{o,OS} = -\left(1 + \frac{R_2}{R_1}\right)V_{OS} \tag{2.32}$$

Superposition in Fig. 2.26a gives

$$v_o(t) = -\frac{R_2}{R_1}v_s(t) - \left(1 + \frac{R_2}{R_1}\right)V_{OS}$$

the undesired result we first saw in Fig. 1.36a. Notice that an inverting amplifier with gain of -20 using an op amp with input offset of only 5 mV, has a dc *output* offset of -105 mV.

We observed in Chapter 1 that if the input signal is slowly varying, an output offset that drifts with temperature is easily mistaken for a signal variation. If the output is to be amplified by another circuit, the offset is amplified along with the signal and can drive the next amplifier into saturation. Offsets can cause problems in circuits other than inverting amplifiers. The preceding analysis only demonstrates how to carry out an offset voltage investigation.

Exercise 2.7 Derive an expression for the output offset component of a noninverting amplifier in terms of the op amp input offset V_{OS}.

Ans. Eq. (2.32).

2.5.8 INPUT BIAS CURRENT AND INPUT OFFSET CURRENT

Many op amps *require* a dc path from each input node to ground to carry small dc *bias currents* required by internal components. If we ground both inputs, dc currents flow as in Fig. 2.27a. Op amp data sheets list *input bias current*, I_B, and *input offset current*, I_{OS}. These are defined by

$$I_B = \frac{I_{B1} + I_{B2}}{2}, \qquad I_{OS} = I_{B1} - I_{B2} \qquad (2.33)$$

(Offset current is a positive number on data sheets; although the current can actually be either positive or negative.) As with the defining equations for common- and difference-mode voltages, we can solve Eqs. (2.33) for I_{B1} and I_{B2}, and then draw a circuit to explain the results. This gives

$$I_{B1} = I_B + 0.5I_{OS} \qquad I_{B2} = I_B - 0.5I_{OS}$$

and Fig. 2.27b. This figure, in a modeling sense, "explains" bias and offsets by attributing them to fictitious internal current sources. More important, it suggests how we can compute the effects of bias and offset on circuit performance. By superpostion, we can isolate the effect of the bias currents by turning off $0.5I_{OS}$ as in Fig. 2.27c, and we can find the effect of I_{os} by turning off currents I_B as in Fig. 2.27d. Because dc bias and offset currents are *independent of signals,* they are *not* forced to zero as the (signal) gain

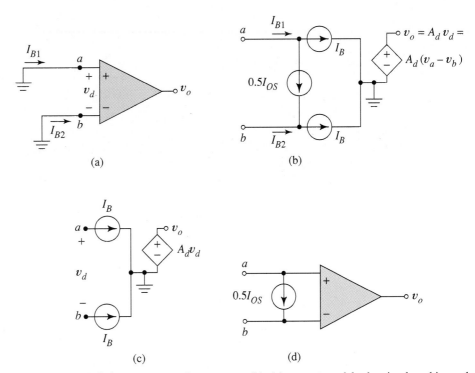

Figure 2.27 (a) dc input currents of an op amp; (b), (c) op amp models showing how bias and offset currents are modeled; (d) op amp with offset current viewed as ideal op amp with external offset source.

of an op amp approaches infinity; only input *signal current* must be zero. Therefore, we can treat the op amp as an infinite-gain device, and bias and offset as external dc currents *superimposed* upon the signals.

Figure 2.28a uses Fig. 2.27c to study bias currents in a noninverting amplifier. The op amp has infinite gain. With the bias sources off, the circuit amplifies v_s in the usual fashion. Turning off v_s grounds the noninverting input node, and infinite gain forces the voltage at node b to zero. With both nodes of R_1 at ground potential, all I_B (in the lower source) flows through R_2, producing dc output voltage $V_{o,IB} = I_B R_2$. The total output is then

$$v_o(t) = \left(1 + \frac{R_2}{R_1}\right) v_s + I_B R_2$$

Good designers anticipate this offset and add a resistor R_x in series with the noninverting input as in Fig. 2.28b. Proper choice of R_x causes I_B to produce equal dc node voltages at inverting and noninverting nodes, thereby canceling offset caused by I_B. In

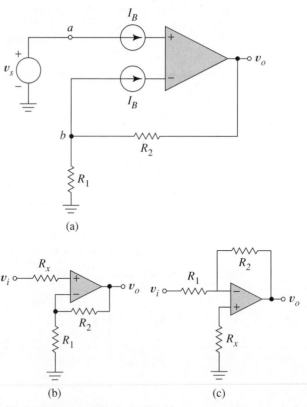

(a)

(b)   (c)

Figure 2.28 (a) Noninverting amplifier including bias currents; (b) noninverting amplifier with resistor added to cancel out effect of bias currents; (c) inverting amp with effect of bias current canceled.

Problem 2.61 we find that this takes $R_x = R_1 \| R_2$. To the uninitiated, R_x appears to be a redundant component since R_x has no effect on the signal. Similar analysis for the inverting amplifier shows that a resistor $R_x = R_1 \| R_2$, inserted as in Fig. 2.28c, cancels the bias current offset in this circuit. (A person more clever than energetic would omit the second analysis, observing that with $v_i = 0$ the circuits of Figs. 2.28b and c are identical.) "dc resistance from each op amp input pin to ground must be the same" is an easy-to-remember summary for both circuits.

Figure 2.29 shows a noninverting amplifier with an input offset current. When v_s is off and gain infinite, both nodes of R_1 are at ground potential and offset current flows only through R_2. The dc output component caused by offset current alone, $v_{o,OS}$, is $-(0.5I_{OS})R_2$.

At the end of Sec. 2.5.5 we said that 20 MΩ or so is a practical upper limit for discrete resistors. Using $R_2 = 20$ MΩ in a 741-based noninverting amplifier ($I_{OS} = 20$ nA) gives output offset of only 0.2 V. If we replace the 741 with an HA2544 op amp ($I_{OS} = 0.2$ mA), however, we instead compute an output offset of 2000 V! Clearly the offset current causes a saturated amplifier. We conclude that it is good practice to check I_{OS} when we use large resistors.

Op amp IC chips ordinarily have external nodes provided for correcting input offset voltage and current, and the manufacturer gives specific directions for *nulling* offsets using special external components. The procedure used is peculiar to each op amp. As a rule, drift of the offset values with temperature still causes problems in these devices. Some op amps have *built-in bias current compensation* and thus require no external dc path for bias currents.

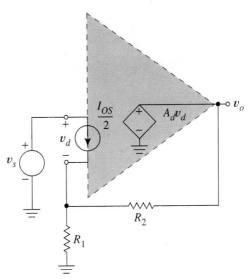

Figure 2.29 Circuit for calculating effect of input offset current on performance of noninverting amplifier circuit.

2.5.9 COMMON-MODE REJECTION

Like other difference amplifiers, op amps have large but finite CMRR as suggested in Table 2.1. A real op amp thus resembles Fig. 1.25a, and external components affect both common- and difference-mode output signals.

It is interesting to think about what happens to the common-mode component of the input as A_d approaches infinity. Because v_d is forced to zero, the voltage common to the input nodes *is* v_c. In circuits like the inverting amplifier, which have a virtual ground at the op amp input, v_c is zero; and circuit operation is not affected by the common-mode gain. When input nodes are not at ground potential, however, as in a noninverting amplifier, a common-mode voltage *is* developed at the output. Usually this causes negligible change in circuit operation. For the difference amplifier circuit, however, the op amp's CMRR is an important consideration because the nonzero common-mode gain of the *op amp* results in a common-mode output for the *circuit*. That is, a common-mode term is added to Eq. (2.15), even when the resistors are exactly balanced.

To predict consequences of input offset voltage, we added an independent voltage source in series with the input of an ideal op amp. A similar approach works for common-mode gain. For the op amp, as for any difference amplifier, Eq. (1.28) describes the output voltage. Factoring out A_d gives

$$v_o = A_d \left(v_d + \frac{A_c}{A_d} v_c \right) = A_d \left(v_d + \frac{1}{\text{CMRR}} v_c \right) = A_d v_x \qquad (2.34)$$

The equivalence of Figs. 2.30a and b is a creative interpretation of Eq. (2.34). The finite CMRR op amp is equivalent to an infinite CMRR op amp that has a VCVS in series with its input. The VCVS is controlled by v_c and has voltage gain of 1/CMRR. This model is easy to implement with SPICE.

For hand calculations the approximate circuit of Fig. 2.30c is simpler. Because CMRR is large for an op amp, the dependent voltage in Fig. 2.30b actually contributes little to the value of v_c. Thus we can *estimate* v_c from Fig. 2.30a. Once we have this estimate, we use the *independent source* (1/CMRR)v_c in Fig. 2.30c, superposition, and infinite-gain analysis to determine the common-mode component in the output. An example that compares hand and computer analysis of CMRR in op amps follows shortly.

2.5.10 SPICE MODELS OF STATIC SECOND-ORDER EFFECTS

Second-order effects are important because they potentially degrade circuit performance; however, they make op amp circuits more difficult to analyze by hand. Nowadays SPICE software installations often include library models for frequently used devices such as particular op amps. A sophisticated op amp model makes realistic simulation of a given circuit quite easy; however, simpler "home-made" models are more versatile in allowing us to study individual second-order effects in isolation. With the exception of saturation, we can simulate all the second-order effects studied thus far by simply adding resistors and independent and dependent sources to the basic VCVS op amp model. After we learn about diodes in the next chapter, we will be able to simulate saturation as well. The next example demonstrates using SPICE analysis to investigate second-order effects.

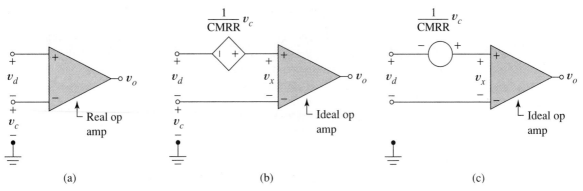

Figure 2.30 Finite CMRR: (a) real op amp with finite CMRR; (b) model that uses an op amp with infinite CMRR and an external VCVS; (c) real op amp approximated by ideal op amp and independent source.

EXAMPLE 2.7   (a) Use hand analysis to estimate $v_{out}(t)$ for Fig. 2.31a. The operational amplifier has $CMRR_{dB} = 90$. Input voltages are

$$v_a(t) = 0.010 \sin 2\pi 400t + 0.20 \sin 2\pi 60t$$

and

$$v_b(t) = -0.010 \sin 2\pi 400t + 0.20 \sin 2\pi 60t$$

(b) Use SPICE and the model of Fig. 2.30b to check the 60 Hz output component estimated in part (a).
(c) Use SPICE to determine the common-mode gain of the circuit.

Solution.   (a) First we find the difference-mode output component using Eq. (2.15). The difference-mode gain of the amplifier is 10. Therefore,

$$v_{o,dm}(t) = 10 \left[v_a(t) - v_b(t) \right] = 0.200 \sin 2\pi 400t$$

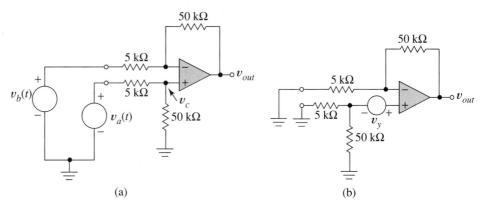

Figure 2.31   Finding common-mode output signal: (a) original amplifier with common-mode input v_c; (b) circuit to find common-mode output component using superposition; (c) circuit for SPICE simulation; (d) SPICE code.

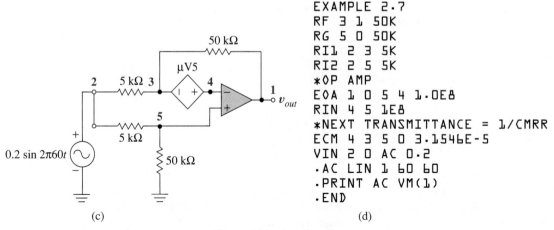

```
EXAMPLE 2.7
RF 3 1 50K
RG 5 0 50K
RI1 2 3 5K
RI2 2 5 5K
*OP AMP
EOA 1 0 5 4 1.0E8
RIN 4 5 1E8
*NEXT TRANSMITTANCE = 1/CMRR
ECM 4 3 5 0 3.1546E-5
VIN 2 0 AC 0.2
.AC LIN 1 60 60
.PRINT AC VM(1)
.END
```

(c)　　　　　　　　　　　　　　　　(d)

Figure 2.31　(continued)

Next we find the common-mode component. From the voltage divider in Fig. 2.31a, the common-mode voltage at the op amp input nodes is

$$v_c(t) = \left(\frac{50}{55}\right)v_a(t)$$

Since 90 dB corresponds to CMRR = 3.16×10^4, the independent source of Fig. 2.31b develops

$$v_y = \frac{1}{3.16 \times 10^4}v_c(t) = 2.87 \times 10^{-5}v_a(t)$$

or

$$v_y(t) = 2.88 \times 10^{-7} \sin 2\pi 400t + 5.76 \times 10^{-6} \sin 2\pi 60t$$

Because Figure 2.31b is a noninverting amplifier with difference-mode gain of 11,

$$v_{out,cm}(t) = 11v_y(t) = 3.16 \times 10^{-6} \sin 2400t + 6.34 \times 10^{-5} \sin 2\pi 60t$$

Superimposing difference- and common-mode results gives

$$v_{out}(t) = v_{out,dm}(t) + v_{out,cm}(t) = 0.200 \sin 2\pi 400t + 6.34 \times 10^{-5} \sin 2\pi 60t$$

In spite of perfectly matched resistors, there is a common-mode output component because the op amp common-mode gain is nonzero; nevertheless the amplifier is still quite effective in suppressing the 60 Hz common-mode noise relative to the 400 Hz signal.

(b) The 60 Hz component is a pure common-mode component, so we use the circuit of Fig. 2.31c where

$$\mu = \frac{1}{\text{CMRR}} = 3.16 \times 10^{-5}$$

Figure 2.31d is the SPICE code. Source EOA, a VCVS, is the op amp, with 10^8 approximating infinite gain. Lines with asterisks are comment statements, ignored by SPICE but useful for documentation. In order that node 4 have the required SPICE minimum of two components attached, we included input resistance of 100 MΩ in the op amp model. Dependent-source ECM introduces the op amp's nonzero common-mode gain into the circuit. SPICE output showed the voltage at node 1 to be 6.309×10^{-5} V, which verifies the hand analysis.

(c) To determine the common-mode gain of the *circuit*, we change the amplitude of VIN to one in the SPICE listing and print the output voltage. The result of this analysis is $V(1) = A_c = 3.155 \times 10^{-4}$ V. Since $A_d = 10$, the difference amplifier circuit has CMRR = 90 dB, the same as that of the op amp itself.   ❏

The second-order effects we have studied are all *static limitations;* that is, they describe the op amp when signals change slowly with time. When signals change rapidly, as we learned in Sec. 1.6.4, it is also necessary to consider dynamic limitations imposed by the capacitances within the amplifier. In op amps these limitations are two, frequency response and slewing.

2.5.11 FREQUENCY RESPONSE OF AN OPERATIONAL AMPLIFIER

When we excite the open-loop op amp using a difference-mode sine wave, gain A_d becomes a complex function $A(\omega)$, the ratio of output voltage phasor to input voltage phasor. For a large class of op amps, this function takes the form

$$A(\omega) = \frac{A_d}{1 + j(\omega/\omega_H)} = A_d\left(\frac{\omega_H}{j\omega + \omega_H}\right) \tag{2.35}$$

where ω_H is a constant called the half-power frequency or *open-loop bandwidth*. Notice that $A(\omega)$ has the same functional form as the *RC* circuit gain function, Eq. (1.38), except that the op amp gain approaches A_d rather than one as ω approaches zero.

The magnitude frequency response curve of Eq. (2.35) in dB is Fig. 2.32. Specification sheets do not list ω_H, but instead give the *unity gain frequency* ω_τ, the frequency where the op amp's gain is zero dB. We can relate ω_τ to ω_H and A_d by the following argument. When $\omega >> \omega_H$, Eq. (2.35) is approximated by

$$A(\omega) = \frac{A_d\omega_H}{j\omega} \tag{2.36}$$

From Eq. (2.36) ω_τ satisfies

$$|A(\omega_\tau)| = \frac{A_d\omega_H}{\omega_\tau} = 1$$

thus

$$\omega_\tau = A_d \omega_H \tag{2.37}$$

For the 741 op amp of Table 2.1, $\omega_\tau = 2\pi \times 10^6$ rad/s. Since $A_d = 2 \times 10^5$, we infer that the bandwidth ω_H of the 741 is only 31.4 rad/s and $f_H = 5$ Hz, a very small bandwidth indeed! By contrast, the HA2544 has $f_H = 7500$ Hz.

2.5.12 GAIN-BANDWIDTH PRODUCT

An important figure of merit for an amplifier, including the op amp, is its *gain-bandwidth product,* the product of maximum gain and bandwidth. In terms of Eq. (2.35) and Fig. 2.32 this is $A_d \omega_H = \omega_\tau$. For this reason, a commonly used alias for unity-gain frequency, ω_τ, is *gain-bandwidth product.* We learned in Chapter 1 that bandwidth is indicative of a device's information processing speed. The gain-bandwidth product combines speed and amplifying ability into one useful parameter.

The equations we derived for op amp circuits in previous sections all assume op amps of very high voltage gain; however, Fig. 2.32 shows that for signal frequencies significantly greater than ω_H, the op amp's gain will be too small for the infinite-gain approximation to be valid. Therefore we expect the op amp's finite bandwidth to produce deviations from ideal performance when we apply high frequencies or abrupt changes to an op amp circuit. To determine the exact effect of the op amp's frequency limitations, we represent the op amp by a VCVS having complex transmittance, for example, Eq. (2.35), when we analyze the circuit. To simplify the algebra we can first use the constant, A_d, for the transmittance, and then replace A_d by Eq. (2.35) in the final expression. To demonstrate this technique we will determine the frequency response of a noninverting amplifier.

Gain × Bandwidth for Noninverting Amplifier. Problem 2.49 establishes that the gain of the noninverting amplifier is

$$\frac{V_{out}}{V_{in}} = \frac{A_d}{1 + A_d R_1/(R_1 + R_2)}$$

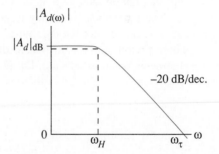

Figure 2.32   Frequency response of an internally compensated op amp.

To find the frequency response we replace A_d by Eq. (2.35), giving

$$\frac{V_{out}}{V_{in}} = A_{NI}(\omega) = \frac{A_d\omega_H/(j\,\omega + \omega_H)}{1 + [A_d\omega_H/(j\,\omega + \omega_H)][R_1/(R_2 + R_1)]}$$

$$= A_d \frac{\omega_H}{j\,\omega + \omega_H + A_d\,\omega_H R_1/(R_2 + R_1)}$$

$$= A_d \frac{\omega_H}{j\,\omega + \omega_H[1 + A_d R_1/(R_2 + R_1)]}$$

$$= \frac{A_d}{1 + A_d R_1/(R_2 + R_1)} \frac{\omega_H[1 + A_d R_1/(R_2 + R_1)]}{j\omega + \omega_H[1 + A_d R_1/(R_2 + R_1)]}$$

The final expression has the form

$$A_{NI}(\omega) = A_v \frac{\omega_B}{j\omega + \omega_B}$$

where

$$A_v = \frac{A_d}{1 + A_d R_1/(R_2 + R_1)}$$

is the low-frequency gain of the noninverting circuit and

$$\omega_B = \omega_H[1 + A_d R_1/(R_2 + R_1)] \tag{2.38}$$

is its bandwidth. These expressions show that

$$A_v\omega_B = A_d\omega_H = \omega_\tau$$

that is, *for the noninverting amplifier,* low-frequency gain times bandwidth equals *exactly* the gain-bandwidth product of the *op amp* itself! To express this conclusion in more familiar notation, we assume A_d so large in the gain expression that $A_v \approx \dfrac{R_2 + R_1}{R_1}$. This gives

$$A_v\omega_B \approx \left(1 + \frac{R_2}{R_1}\right)\omega_B = \omega_\tau \tag{2.39}$$

Each resistor ratio we consider in selecting gain therefore corresponds to some particular bandwidth ω_B. Since the product is invariant, lower gain gives higher bandwidth and conversely. An extreme case is the follower with gain of one. In this case all of the gain is traded for bandwidth, giving in a -3 dB bandwidth of ω_τ.

Gain × Bandwidth for Identical Cascaded Amplifiers. When a design task specifies *both* gain and bandwidth, a single-stage design requires an op amp with ω_τ equal to or greater than the required product. If an amplifier design has sufficient bandwidth, excess gain is easily reduced by a voltage divider.

If we have no op amp with adequate ω_τ for a particular design, we can use cascaded amplifiers. Consider a cascade of n identical amplifiers, each having low-frequency gain $A_L > 0$ and bandwidth ω_B. Because gains of cascaded amplifiers multiply, the overall gain of the n-stage cascade is

$$A_n(\omega) = \left[\frac{A_L}{1 + j(\omega/\omega_B)} \right]^n = \frac{A_L^n}{[1 + j(\omega/\omega_B)]^n}$$

By definition, the -3 dB bandwidth is that frequency, ω_{Bn}, at which the magnitude $|A_n(\omega)|$ is $1/\sqrt{2}$ times its low-frequency value; that is,

$$|A_n(\omega_{Bn})| = \frac{A_L^n}{\left[1 + (\omega_{Bn}/\omega_B)^2 \right]^{n/2}} = \frac{A_L^n}{\sqrt{2}}$$

Equating denominators gives

$$\omega_{Bn} = \omega_B \sqrt{2^{1/n} - 1} \tag{2.40}$$

We conclude that the gain-bandwidth product of n cascaded stages is

$$A_N \omega_{Bn} = A_L^n \omega_B \sqrt{2^{1/n} - 1} = A_L \omega_B [A_L^{n-1} \sqrt{2^{1/n} - 1}\,]$$

which is larger than the single-stage product $A_L \omega_B$. For cascaded stages, gain-bandwidth product increases with n because A_L^n increases faster with n than the square root expression decreases. We can easily use Eq. (2.40) iteratively to converge on a design that satisfies given specifications as illustrated in the following example.

EXAMPLE 2.8   Design a noninverting audio amplifier with gain of $A_v = 5000$ and bandwidth of $\omega_B = 20,000$ rad/s using a 741 op amp. See Eq. (2.39).

Solution. Required (gain) × (bandwidth) is

$$A_v \omega_B = 5000(20,000) = 10^8 \text{ rad/s}$$

Since the 741 has $\omega_\tau = 2\pi \times 10^6$, the design conditions cannot be satisfied by a single noninverting amplifier. (Had the 741 not been specified, we could use the HA2544, for Table 2.1 shows this device has $\omega_\tau = 2.8 \times 10^8$.)

Now consider $n = 2$ identical noninverting amplifiers in cascade. To meet specifications, we make $A_L = (5000)^{1/2} = 70.7$. According to Eq. (2.39) the bandwidth of each stage is

$$\omega_B = \frac{2\pi \times 10^6}{70.7} = 8.8 \times 10^4 \text{ rad/s}$$

From Eq. (2.40), the bandwidth of the two-stage amplifier is

$$\omega_{B2} = (8.8 \times 10^4)(2^{1/2} - 1)^{1/2} = 56.6 \times 10^3 \text{ rad/s}$$

which is more bandwidth than needed; however, exceeding minimum specifications in either gain or bandwidth is usually acceptable.

To meet the bandwidth specification exactly, we could instead have worked backward from Eq. (2.40) with $n = 2$ and $\omega_{B2} = 20,000$ rad/s. Then

$$\omega_B = \frac{20,000}{(2^{1/2} - 1)^{1/2}} = 31.1 \times 10^3 \text{ rad/s}$$

We then construct two stages, each of gain

$$A_v = \frac{\omega_\tau}{\omega_B} = \frac{2\pi \times 10^6}{31.1 \times 10^3} = 202$$

Since the gain of the cascade, $202^2 = 4.08 \times 10^4$, exceeds specifications we use a voltage divider to reduce the gain to 5000. ❏

Gain × Bandwidth for Inverting Amplifier. For the inverting amplifier, the gain-bandwidth product is *not* constant, so this circuit does not allow us to trade off gain directly against bandwidth. Problem 2.71 shows that the complex gain of the inverting amplifier is

$$A_I(\omega) \approx \frac{(-R_2/R_1)}{1 + j(\omega/\omega_B)} \tag{2.41}$$

with ω_B again given by Eq. (2.38). Therefore, for this circuit

$$|\text{gain}| \times (\text{bandwidth}) = \frac{R_2}{R_1}\omega_B \approx \frac{R_2}{R_1}\omega_H\left(A_d \frac{R_1}{R_2 + R_1}\right) \approx \frac{R_2}{R_2 + R_1}\omega_\tau \tag{2.42}$$

a function of resistor values and, therefore, of gain. Because the gain-bandwidth product changes with R_2/R_1, the inverting amplifier does not allow us to directly trade gain for bandwidth as does the noninverting circuit. Furthermore, *the gain bandwidth product of the inverting amplifier is always less than ω_τ.* For example, when $R_2/R_1 = 1$ for a unity-gain inverter, the product is $\omega_\tau/2$, only one-half the bandwidth of the follower circuit that uses the same op amp.

Exercise 2.8 Find the bandwidth of an inverting amplifier of gain -5 that employs the HA2544.

Ans. 7.5 MHz.

Our discussions of bandwidth for inverting and noninverting amplifiers have been tacitly limited to *internally compensated* op amps, a simple but important special case described by Eq. (2.35). We study more general frequency response concepts in Chapters 8 and 9.

The effect of the op amp's frequency limitations on other circuits, filters for example, can be difficult to determine analytically. For such circuits an effective approach is to work out the initial design using infinite-gain equations and then follow up with computer simulations that include the frequency limitations of the op amps. We next describe a simple SPICE op amp model that is useful for this purpose.

2.5.13   SPICE LINEAR DYNAMIC OP AMP MODEL

Figure 2.33a shows a linear, dynamic, SPICE op amp model. The internal RC-VCVS circuit provides the frequency response function of Eq. (2.35). Any values for R and C that satisfy

$$\frac{1}{RC} = \omega_H = \frac{\omega_\tau}{A_d}$$

are satisfactory. A second VCVS, with transmittance of one ensures that components connected to the output pin do not affect the op amp's frequency response; resistors R_{in} and R_{out} provide realistic loading effects at input and output. The next example shows why the designer must be mindful of the frequency response of the op amp when he or she works with equations based on ideal op amps.

EXAMPLE 2.9   (a) Design a first-order, lowpass filter with low-frequency gain of -10 and a -3 dB frequency of 1 kHz. Then redesign for a -3 dB frequency of 50 kHz. Assume an ideal op amp in both designs.
(b) Use SPICE to test both designs using a nonideal op amp described by

$$\text{open-loop gain} = 10^4$$

$$R_{in} = 50 \text{ k}\Omega$$

$$R_{out} = 200 \ \Omega$$

$$\omega_\tau = 2\pi \times 10^5 \text{ rad/s}$$

Solution. (a) Equation (2.24), with $\omega_H = 1/R_2C_2$, describes the filter of Fig. 2.14a. The partially arbitrary choice, $R_1 = 2$ kΩ and $R_2 = 20$ kΩ, satisfies the gain requirement. For a 1 kHz cut-off frequency we need

$$\frac{1}{R_2C_2} = \omega_H = 2\pi \times 10^3 \text{ rad/s}$$

thus $C_2 = 8000$ pF. Because cut-off frequency for the second design is 50 times that of the first,

$$C_2 = \frac{8000 \text{ pF}}{50} = 159 \text{ pF}$$

for the second design. Figure 2.33b shows the design.
(b) To predict the filter responses for an op amp with the given specifications, we replace the op amp by the model of Fig. 2.33a, and obtain Fig. 2.33c. For the op amp to have the correct bandwidth we need

$$\frac{1}{RC} = \frac{\omega_\tau}{A_d} = \frac{2\pi 10^5}{10^4}$$

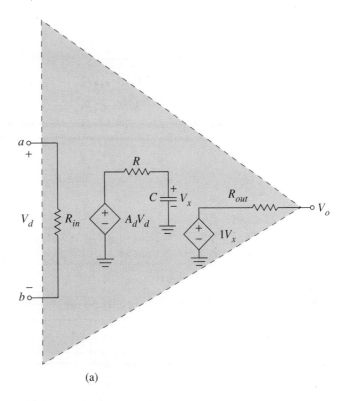

(a)

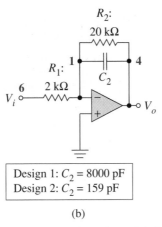

(b)

Figure 2.33 Use of op amp dynamic model: (a) model including first-order frequency response effects; (b) filter schematic; (c) filter equivalent circuit including op amp model; (d) SPICE code; (e) simulation results for Example 2.9.

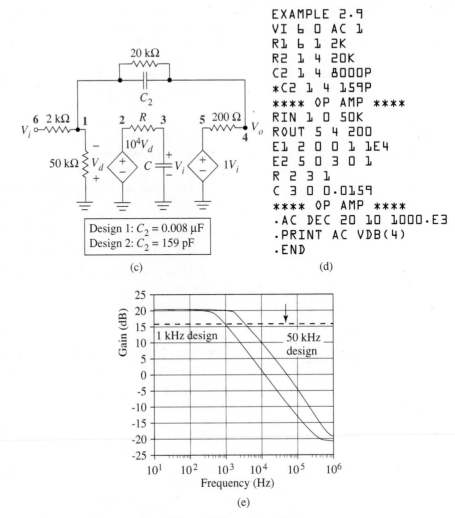

Design 1: $C_2 = 0.008\ \mu\text{F}$
Design 2: $C_2 = 159\ \text{pF}$

(c)

```
EXAMPLE 2.9
VI 6 0 AC 1
R1 6 1 2K
R2 1 4 20K
C2 1 4 8000P
*C2 1 4 159P
**** OP AMP ****
RIN 1 0 50K
ROUT 5 4 200
E1 2 0 0 1 1E4
E2 5 0 3 0 1
R 2 3 1
C 3 0 0.0159
**** OP AMP ****
.AC DEC 20 10 1000.E3
.PRINT AC VDB(4)
.END
```

(d)

(e)

Figure 2.33 (continued)

After we arbitrarily choose $R = 1\ \Omega$, this equation gives

$$C = \frac{10^4}{2\pi \times 10^5} = 0.0159\ \text{F}$$

Figure 2.33d shows the SPICE code, with the op amp description set off by comments. Notice that the second value for the filter capacitor C_2 is included as a comment. After the first SPICE run, we move the asterisk to the first C_2 line and then make a second run to simulate the second filter. With this strategy we can change a circuit parameter while retaining the original information for later use and for documentation.

Figure 2.33d shows computed SPICE results for both designs. The 1 kHz design matches specifications; however, because of the limited frequency response of the op amp, the cut-off frequency of the second filter is only 7.28 kHz instead of the design value of 50 kHz marked by the arrow. This confirms our suspicion that equations like Eq. (2.24), which assume infinite op amp gain at all frequencies, lose their validity at higher frequencies where op amp dynamic limitations become important.

The simulations also reveal a more subtle, real-world feature of op amp circuits. Notice that at high frequencies the 1 kHz filter curve "bottoms out" at about -20 dB instead of continuing to drop as predicted by Eq. (2.24). At frequencies so high that the gain of the op amp becomes negligible, the active filter degenerates to a passive RC circuit, where the op amp's only contributions are its input and output resistances. In Fig. 2.33c, as both capacitors become short circuits the circuit degenerates to the voltage divider

$$\frac{V_o}{V_i} = \frac{200 \,\|\, 50\text{ k}}{200 \,\|\, 50\text{ k} + 2\text{ k}} \approx \frac{200}{2000} = -20 \text{ dB}$$

The reader is urged to extend the frequency response another decade to verify that the 50 kHz filter design also bottoms out at -20 dB. ❏

SPICE Subcircuit for an Op Amp. With SPICE we can define an arbitrary *subcircuit* with a single block of code, and then refer to this same code for each occurrence of the subcircuit. Subcircuits reduce the amount of code we must write, organize a simulation in a manner that reduces errors, and enable us to change subcircuit parameters throughout the circuit by changing only the subcircuit code. Subcircuits are useful in simulating circuits that contain several op amps.

Figure 2.34a shows a subcircuit description for the 741 op amp of Table 2.1. The first line names the subcircuit and indicates that it communicates with the *outside world* by means of ordered nodes, 1, 2, and 6. Code details show that these *external* nodes are, respectively, the noninverting, inverting, and output nodes. Statement .ENDS marks the end of the subcircuit description.

Figure 2.34b demonstrates using 741 subcircuits to find the frequency response of a two-stage amplifier. The SPICE code refers to each 741 by an individual element statement. We identify each subcircuit by a name beginning with X, list in order its connection nodes, and refer to the subcircuit name. Node numbers internal to a subcircuit definition are independent of external node numbering in a SPICE simulation. (Notice repeated use of node numbers 1–5 in the example.) We give the subcircuit description only once, no matter how many times the op amp appears in a circuit, and the subcircuit definition can occur anywhere in the circuit code.

Exercise 2.9 Use SPICE to determine how the differentiator frequency response of Fig. 2.18c changes when we replace the ideal op amp model by the μA741 subcircuit of Fig. 2.34a. From the frequency response, determine a practical upper limit to the input frequencies we should use with this differentiator.

Ans. Figure 2.35, about 10 kHz.

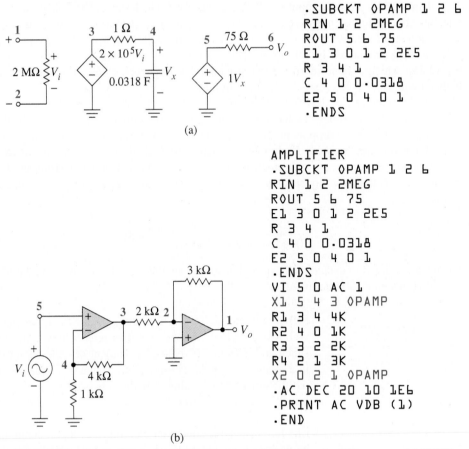

```
.SUBCKT OPAMP 1 2 6
RIN 1 2 2MEG
ROUT 5 6 75
E1 3 0 1 2 2E5
R 3 4 1
C 4 0 0.0318
E2 5 0 4 0 1
.ENDS
```

(a)

```
AMPLIFIER
.SUBCKT OPAMP 1 2 6
RIN 1 2 2MEG
ROUT 5 6 75
E1 3 0 1 2 2E5
R 3 4 1
C 4 0 0.0318
E2 5 0 4 0 1
.ENDS
VI 5 0 AC 1
X1 5 4 3 OPAMP
R1 3 4 4K
R2 4 0 1K
R3 3 2 2K
R4 2 1 3K
X2 0 2 1 OPAMP
.AC DEC 20 10 1E6
.PRINT AC VDB (1)
.END
```

(b)

Figure 2.34 Subcircuit for an op amp: (a) subcircuit definition for 741 op amp; (b) SPICE code for a circuit containing two op amps.

Frequently computer simulations lead to theoretical investigations that improve our understanding of the circuit. Our explanation of the -20 dB asymptote in Example 2.9 was one example. Another is Problem 2.73, which shows how to develop some differentiator theory to explain Fig. 2.35.

2.5.14 SLEWING

A second-order effect called *slewing* usually provides the dominant dynamic limitation for an op amp circuit that produces rapid voltage transitions. In contrast to finite bandwidth, a linear, dynamic limitation of the op amp, slewing, is a nonlinear, dynamic limitation.

From the viewpoint of linear theory, the follower of Fig. 2.36 is a first-order circuit with bandwidth ω_τ. Thus, when we apply a pulse smaller than the op amp's saturation

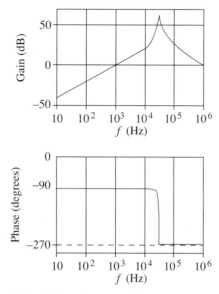

Figure 2.35 Differentiator amplitude and phase curves.

limit, we expect $v_o(t)$ to increase exponentially with time constant $1/\omega_\tau$ and to then decrease exponentially to zero at the end of the pulse as in Fig. 2.36b. When we examine this scenario more carefully, the possibility of trouble within the op amp is immediately evident. Because the op amp's internal capacitance requires $v_o(0^+) = 0$ at $t = 0^+$, there are $v_d = V_F$ *volts* developed *between the input pins of the op amp* at the leading edge of the pulse, and again at the trailing edge. But if $v_d = V_F$ exceeds the few microvolts corresponding to the high-gain portion of Fig. 2.22, the transistors inside the op amp protest by behaving nonlinearly until the input voltage becomes more reasonable. (We discuss these internal details in Chapter 9.) The external manifestation of these internal nonlinearities is that the op amp's output voltage increases or decreases, not exponentially, but *linearly* at a fixed rate as in Fig. 2.36c. This *slew rate, sr,* is predetermined by the internal structure of the operational amplifier. The pulse experiment of Fig. 2.36 is a convenient way to measure an unknown slew rate.

The practical consequence is that whenever linear theory predicts a rate of change for v_o that exceeds *sr,* the output voltage, instead, changes at the slew rate, and we say the op amp *slews.* Since the derivative of the exponential in Fig. 2.36b at the switching instants is $V_F\omega_\tau$ volt per second, the follower slews for any pulse such that $V_F > sr/\omega_\tau$. From Table 2.1, for example, we see that the 741 slews for $V_F > 0.08$ V, the 2544 for $V_F > 0.5$ V. Square-wave generators, sample-and-holds, and switched capacitor filters are examples of circuits we study in which slewing might impose the major limitation on the circuit operation.

Slewing can also limit circuits operating with smooth waveforms such as sinusoids. Suppose, for example, that an op amp circuit is to produce the waveform $v_o(t) = V_M \sin \omega_t$. Since the maximum rate of change of this sine wave is ωV_M, the op amp is able

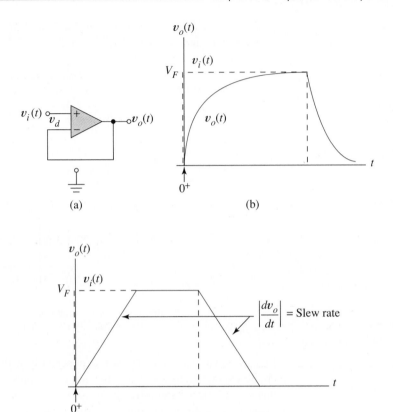

Figure 2.36 Comparison of linear and nonlinear dynamic constraints: (a) follower circuit; (b) waveforms for op amp predicted by linear theory; (c) waveforms for op amp dominated by slewing.

to produce this waveform only if $sr > \omega V_M$. If the product of the voltage and frequency exceeds the slew rate, the output waveform is distorted, more closely resembling a triangular wave than a sine wave.

Op amp data sheets sometimes specify a *power bandwidth,* ω_P, defined by

$$\omega_P = \frac{sr}{V_M} \text{ rad/s}$$

where sr is the slew rate and V_M the maximum voltage the op amp can develop at its output node, that is, the saturation limit. The op amp can deliver this full-power output sinusoid of amplitude V_M at all frequencies up to ω_P without *slewing distortion.*

Exercise 2.10 Find the power bandwidths in kHz for the 741 and the 2544 of Table 2.1 assuming 15 V supplies.

Ans. 5.68 kHz, 2.17 MHz.

The following example demonstrates how to incorporate certain op amp second-order effects into practical design problems.

EXAMPLE 2.10 Figure 2.37a shows the sample-and-hold circuit of Sec. 2.2.7, with bias currents of the second op amp shown explicitly. Figure 2.37b shows the switch timing. The switch must be closed long enough so that C can charge completely even when OA_1 slews. The interval between switch closures, $T - \tau$, must be short enough that the input bias current of the second op amp does not reduce the stored voltage appreciably.

Suppose the slew rate of the op amp is 5 V/μs and v_s is confined to

$$0 \le v_s \le 10 \text{ V}$$

(a) Find the minimum closure time τ so that C can charge completely for the worst case input voltage.
(b) Find a value for C such that the maximum capacitor charging current is only three quarters of the 25 mA short-circuit current of the op amp.
(c) If the input bias current is 0.2 μA, find the minimum T so that the capacitor voltage decreases by no more than 10 mV before the switch closes again. Use the capacitance found in part (b).

Solution. (a) The worst case is when v_s equals 10 V after the capacitor has previously been charged to 0 V, say, during a sudden signal transition. The voltage change of 10 V then requires

$$\tau = \frac{10}{sr} = \frac{10}{5 \times 10^6} = 2 \text{ μs}$$

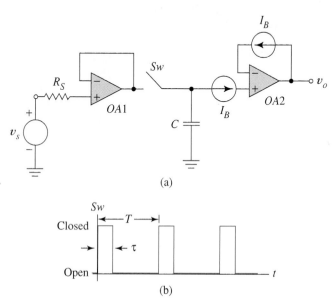

(a)

(b)

Figure 2.37 (a) Sample-and-hold circuit; (b) switch timing function.

(b) The maximum charging rate is $sr = 5 \times 10^6$ V/s. Since the maximum allowed current is $0.75 \times 25\text{mA} = 18.8$ mA,

$$18.8 \times 10^{-3} \geq C \frac{dv}{dt} = C \, sr$$

therefore, $C \leq 3760$ pF.

(c) With Sw open, the capacitor discharges at the rate

$$\frac{dv}{dt} = \frac{I_B}{C} = \frac{0.2 \times 10^{-6}}{3760 \times 10^{-12}} = 532 \text{ V/s}$$

Therefore

$$T - \tau = \frac{10 \text{ mV}}{53.2 \text{ V/s}} = 188 \text{ } \mu\text{s}$$

and $T < 190$ μs is required. ❏

2.6 _____

Circuits Without Negative Feedback

The circuits of this section differ from previous op amp circuits in having either no signal feedback or in having *positive* rather than negative feedback. In such circuits the op amp is best viewed as a two-state device defined by Fig. 2.24. Principal applications of these circuits are in nonlinear signal processing.

2.6.1 BASIC COMPARATOR CIRCUIT

The *comparator* circuit of Fig. 2.38a continuously compares signal $v_i(t)$ with reference voltage V_r and produces a binary output, one value for $v_i > V_r$ and another for $v_i < V_r$.

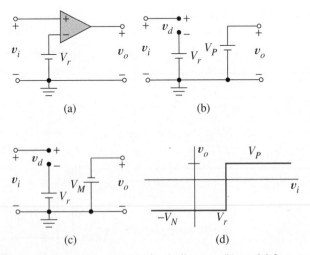

Figure 2.38 Basic op amp comparator: (a) circuit diagram; (b) model for $v_i > V_r$; (c) model for $v_i < V_r$; (d) transfer characteristic.

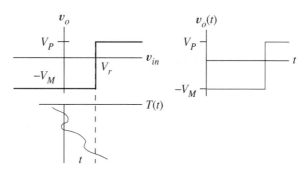

Figure 2.39 Use of a comparator in a temperature warning system.

Because there is no negative feedback, operation is *not* confined to the steep portion of the op amp gain curve of Fig. 2.22. Because

$$v_d = v_i - V_r,$$

whenever $v_i > V_r$, the circuit is equivalent to Fig. 2.38b; when $v_i < V_r$, the equivalent is Fig. 2.38c. These inequalities define the transfer characteristic, Fig. 2.38d.

The comparator extracts binary information from its analog input signal. Figure 2.39 suggests one of many applications. Input voltage, v_{in}, represents temperature T. So long as temperature remains below some level specified by V_r, the output voltage remains at $-V_M$. Should the temperature ever become too high, however, the comparator output switches to V_P, a change that we can use to turn on an alarm. Since V_r can be positive, negative, or zero, the comparator is a versatile building block. In some applications high slew rate is important to ensure rapid transitions. Comparator applications we encounter later in the text include analog-to-digital converters, event timers, and waveform generators.

2.6.2 SCHMITT TRIGGER CIRCUIT

The final circuit of this section, the Schmitt trigger of Fig. 2.40a, is an example of a *bistable,* a circuit that has two stable states. (Chapter 14 discusses a number of related circuits.) A vacuum tube version of the circuit was first described in 1938 by the biomedical engineer Otto Schmitt, for whom the circuit is named. (This work is included in the references at the end of the chapter.) The Schmitt trigger differs from the structurally similar inverting amplifier in that R_2 feeds output information back to the *noninverting* op amp input, our first example of positive rather than negative feedback.

To find the transfer characteristic, assume for a start that v_i takes on some very negative value. We then *guess* that op amp input v_d is also negative, causing op amp output to go to negative saturation. Next we redraw the circuit, replacing the op amp with the model of Fig. 2.24c to give Fig. 2.40b. To check our guess, we now calculate v_d. By superposition

$$v_d = \frac{R_2}{R_2 + R_1} v_i + \frac{R_1}{R_1 + R_2}(-V_M) \qquad (2.43)$$

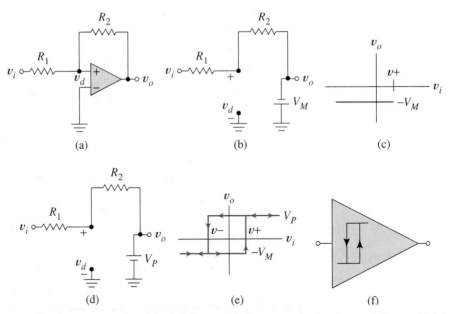

Figure 2.40 (a) Schmitt trigger circuit; (b) Schmitt trigger model for very negative v_i; (c) initial segment of the transfer characteristic; (d) model after v_i passes switching threshold; (e) entire transfer characteristic; (f) special Schmitt trigger schematic.

Since v_d is negative for any negative v_i, our guess was correct. In fact, Eq. (2.43) shows that v_i must take on some positive value before the model becomes *invalid*. To find this critical v_i, we set v_d to zero in Eq. (2.43) and solve for v_i. The result is

$$v_i = \left(\frac{R_1}{R_2}\right) V_M = v^+ \tag{2.44}$$

We have found so far that the transfer curve has the shape of Fig. 2.40c.

Once v_i exceeds v^+, *the op amp changes state,* meaning the circuit model changes to that of Fig. 2.40d. On the transfer curve there is a discontinuity as v_o changes to positive saturation. By superposition

$$v_d = \frac{R_2}{R_2 + R_1} v_i + \frac{R_1}{R_1 + R_2} V_P \tag{2.45}$$

which verifies that the current model is correct for all positive v_i; it also suggests something rather special! According to Eq. (2.45), v_i must now take on a *negative* value before the sign of v_d and, therefore, the state of the op amp changes again. Setting v_d in Eq. (2.45) to zero and solving for v_i gives

$$v_i = -\left(\frac{R_1}{R_2}\right) V_P = v^- \tag{2.46}$$

Figure 2.40e superimposes this result on Fig. 2.40c. For the Schmitt trigger, v_o is not a *function* of v_i as in most transfer characteristics because it is double valued. The arrows show the path the ordered pair (v_o, v_i) traverses during a conceptual experiment in which v_i advances from a very negative value to a very positive value and then back again. In a sense the circuit *remembers* its previous state. For example at $v_i = 0$, if v_i has *most recently been* more positive than v^+, the output is V_P; if v_i *has most recently been* more negative than v^-, then the output is $-V_M$. This *memory effect*, implied by a transfer curve like Fig. 2.40e, is called *hysteresis*. (A more prosaic example of hysteresis is the curve of magnetic flux density versus magnetomotive force for a ferromagnetic material.) The Schmitt trigger, called a *comparator with hysteresis*, is so often used as a subcircuit that it has its own schematic representation, Fig. 2.40f.

Figure 2.41 shows an application of the Schmitt trigger. The input is a two-valued digital signal, corrupted with high-amplitude, additive noise. By using Eqs. (2.44) and (2.46), the designer has set v^+ and v^- to fit the expected worst case noise levels. From Eqs. (2.44) and (2.46), the width of the hysteresis region is

$$v^+ - v^- = \frac{R_1}{R_2}(V_M + V_P) = H \tag{2.47}$$

The designer uses parameters R_1 and R_2 to define the width H of the hysteresis region. The arrows show times t_1 and t_2 when the input crosses a threshold and causes the output to switch to the other level. The circuit gives a *clean*, binary-valued output signal that we can visualize by projecting instantaneous values from the hysteresis curve. Notice that, because of the hysteresis, signal plus noise for the positive digital value can be more negative than signal plus noise for the negative digital signal.

By adding a positive reference V_R between inverting node and ground in Fig. 2.40a the designer can translate the transfer characteristic of Fig. 2.40e to the right by

$$\frac{R_1 + R_2}{R_2} V_R$$

while the hysteresis width remains unchanged. A negative reference voltage shifts the curve to the left. Because of its positive feedback the Schmitt trigger changes state very

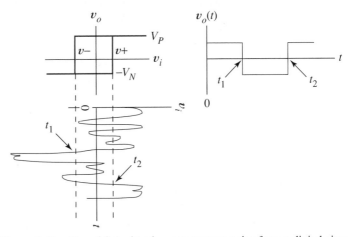

Figure 2.41   Use of Schmitt trigger to remove noise from a digital signal.

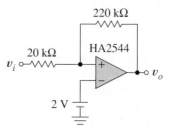

Figure 2.42 Schmitt trigger designed in Example 2.11.

rapidly, making it useful for interfaces between slowly changing analog signals and high-speed digital circuits. The latter sometimes malfunction if input voltage transitions are too slow.

EXAMPLE 2.11   Design a Schmitt trigger circuit that has a hysteresis width of 2 V, centered at $v_i = +2.18$ V. Use the HA2554 op amp of Table 2.1.

Solution.   For $H = 2$ V, Eq. (2.47) requires

$$\frac{R_1}{R_2}(11 + 11) = 2$$

We can satisfy this using $R_1 = 20$ kΩ and $R_2 = 220$ kΩ. Centering takes $V_R = 2$V. Figure 2.42 is the final design.   ❏

Exercise 2.11   What is the worst case effect of ± 5% resistor tolerances on the hysteresis width in Example 2.11?

Ans. 2.21 V ≥ H ≥ 1.81.

2.7 _____
Summary

This chapter describes many useful circuits that utilize the operational amplifier as the active device. Each circuit performs an important signal processing function, and we can easily combine these functional units into complex signal processing systems owing to negligible output resistances.

Most op amp circuits use negative feedback to make the circuit's function more or less independent of the parameters of the op amp. Inverting and noninverting amplifiers, summing circuits, difference amplifiers, and current sources are the more important negative feedback circuits that do not depend upon energy-storage elements for their operation. Memory circuits such as integrators, differentiators, and filters also employ negative feedback. We can readily derive the basic operating principles of op amp feedback circuits, familiar and unfamiliar, by using the infinite-gain approximation. This requires us to assume that op amp input voltage and current signals are simultaneously zero. Because op amp feedback circuits are so numerous, it is imperative that we master infinite-gain analysis so that we can apply it to any problem we might encounter.

Operational amplifiers have a number of internal second-order effects that represent deviations from ideal operation. These are important because they determine the practical operating limits of circuits that employ op amps, and therefore present us with constraints when we design circuits. Op amp data sheets provide numerical parameters that describe these second-order effects. Static effects include finite gain and input resistance, nonzero output resistance, bias currents, common-mode gain, input offset voltage and current, and limits on output signal voltages. Dynamic second-order effects are finite bandwidth, phase shift, and slewing. To include second-order effects in our analyses, we add additional circuit components to our basic VCVS op amp circuit model. As our circuit models become more realistic, they become more complicated, and SPICE simulations perform an increasingly important role in developing and supporting our intuition.

Op amp circuits that do not employ negative feedback use the op amp as a comparator—not an amplifier—and infinite-gain analysis does not apply. When the op amp's input voltage is positive, the output is a dc voltage, the positive saturation value; when input voltage is negative, the output is the negative saturation voltage. For such operation we say the op amp operates in *open-loop* fashion (even though the circuit might employ a positive feedback loop). Various types of comparators including Schmitt triggers and relaxation oscillators use the op amp as an open-loop device. The next chapter describes some other nonlinear waveshaping circuits that use open-loop op amps.

REFERENCES _____

1. CHUA, L.O. *Introduction to Nonlinear Network Theory,* McGraw-Hill, New York, 1969, p. 153.
2. JUNG, W.G. *ic Op Amp Cookbook,* Howard Sams, Indianapolis, 1974.
3. MAURO, R. *Engineering Electronics,* Prentice Hall, Englewood Cliffs, NJ, 1989.
4. MITRA, S.K. *Analysis and Synthesis of Linear Active Networks,* John Wiley, New York, 1969.
5. SEDRA, A., and K. SMITH. *Microelectronic Circuits,* 3rd Ed. Saunders College Publishing, Philadelphia, 1993.
6. SCHMITT, O.H. "A thermionic trigger," *J. Sci. Instrum.,* Vol. 15, no. 1, January 1938, pp. 24–26.
7. TUINENGA, P. *SPICE: A Guide to Circuit Simulation & Analysis Using PSPICE,* Prentice Hall, Englewood Cliffs NJ, 1988.

PROBLEMS _____

SPECIAL DIRECTIONS FOR SPICE HOMEWORK PROBLEMS: *Do not hand in lengthy spice printout for homework.* Instead, abstract the useful information from the SPICE output file as in the SPICE examples in the text. Include your SPICE code and a circuit diagram with nodes numbered to agree with the code. Cite relevant numerical values from the SPICE output file and discuss when appropriate. Make sketches of any relevant curves, and label appropriate points. Make small tables to present numerical data if useful for clarity.

Sections 2.2.1–2.2.4

2.1 Design an inverting amplifier that has voltage gain of -20 and $R_{in} = 1.5$ kΩ.

2.2 Design an inverting amplifier with gain of -30. The op amp output current should be 0.5 mA when $v_{out} = -2$ V.

2.3 To measure the difference-mode gain of a difference amplifier in the laboratory, we need a pure difference-mode signal like the one in Fig. 1.24b. All that is available is a sinusoidal signal generator of voltage v_s. Design an op amp circuit to provide the missing phase-inverted signal, and draw a circuit diagram of the resulting signal source. Express v_d in terms of v_s for your circuit.

2.4 Starting with Eq. (2.6), find an expression for $A_v \pm \Delta A_v$ by replacing R_2 by $R_2 \pm \Delta R_2$ and R_1 by $R_1 \pm \Delta R_1$. Show that for equal fractional resistor variations, $\Delta R/R$, the fractional gain variations $\Delta A_v/A_v \approx 2(\Delta R/R)$.
Hint: Factor and then use long division, discarding terms that are small.

2.5 Use the infinite-gain approximation to find v_o/v_i for the circuit of Fig. P2.5
Hint: Start with the current in R_3.

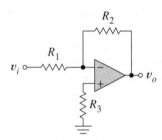

Figure P2.5

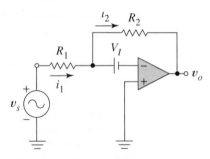

Figure P2.7

2.6   Figure P2.6 shows an inverting amplifier designed for gain of −4. It is connected to a given source with $R_s = 5$ kΩ and a load of $R_L = 5$ kΩ.
(a) Find the amplifier input voltage v_i in terms of v_s.
Hint: Input loading.
(b) Find v_o in terms of v_s.
(c) Find i_L in terms of v_s.

2.7   In Fig. P2.7, V_I is a dc voltage. The op amp has infinite gain.
(a) Express i_1 as a function of v_s and V_I.
(b) Express i_2 as a function of v_s and V_I.
(c) Express v_o as a function of v_s and V_I.

2.8   Sketch the input and output volt-ampere curves for the inverting amplifier of Fig. P2.8 Label each curve with suitable quantitative information.

2.9   Prove that in the voltage-to-current converter of Fig. 2.5 Z_L sees an infinite output resistance.
Hint: Apply the definition of Fig. 1.34b after replacing Z_L by a test voltage and turning off voltage source v_{in}.

2.10   Figure P2.10a shows the schematic symbol for a phototransistor, and Fig. P2.10b its output characteristics. This device converts light intensity, I, in W/m² into output current, i_s.

(a) Design a current-to-voltage converter to process the phototransistor output. Your circuit must provide output voltages between 0 and +10 V for light intensity between 0 and 400 W/m², respectively. Draw a circuit diagram for the system, using a dependent source controlled by light intensity I to represent the phototransistor.
(b) Real phototransistor characteristics have nonzero slope. Redraw the circuit diagram of part (a) with the phototransistor replaced by a circuit model that is appropriate if the curves leave the current axis with a slope of 1/40 mA per volt. How does the output voltage of the current-to-voltage circuit change if this *real* transistor is used instead of the ideal one? Explain.

2.11   Voltage sources that produce 0.2 sin 20t and 9 V dc, respectively, are available. Two *signal conditioning* circuits are needed to accept the available voltages as inputs and produce output node voltages of
(a) $v_o(t) = -4 - 4 \sin 20t$,
(b) $v_o(t) = -4 + 4 \sin 20t$.
Draw diagrams of the two circuits showing component values. More than one op amp may be used in each circuit if necessary.

2.12   In P2.12, find v_o/i_s for
(a) $R_{in} = 20$ kΩ,
(b) $R_{in} = 0$.

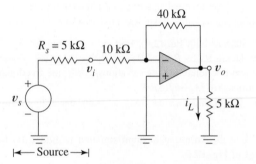

Figure P2.6

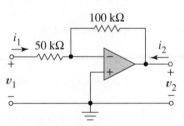

Figure P2.8

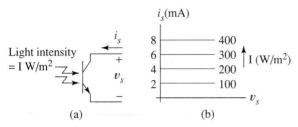

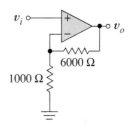

Figure P2.10 Phototransistor: (a) schematic; (b) output characteristics.

Figure P2.14

Sections 2.2.5–2.2.7

2.13 Design a noninverting amplifier with gain of 20. Maximum op amp output current of 0.5 mA should occur for maximum output voltage of 8 V.

2.14 Find the maximum and minimum values of voltage gain for the circuit of Fig. P2.14 if 5% resistors are used.

2.15 Replace the inverting amplifier in Fig. P2.6 with a noninverting amplifier with gain of +4.
(a) Find v_i in terms of v_s.
(b) Find v_o in terms of v_s.
(c) Find i_L in terms of v_s.
(d) Do you expect $|v_o|$ to be higher for this design than for the original circuit? Explain.

2.16 Replace each amplifier in Fig. P2.16 by a voltage amplifier like Fig. 1.19a with numerical values for μ, R_{in}, and R_{out}. Use these circuits to find v_1, v_2, v_3, and v_4.

2.17 P2.17 shows a *noninverting summing amplifier.* Prove that for an infinite-gain op amp, the circuit is described by

$$v_o = \frac{1}{N}\left(1 + \frac{R_2}{R_1}\right)\sum_{i=1}^{N} v_i$$

Hint: Define v_x as the voltage of the noninverting node.

2.18 (a) Find the input resistance for the circuit of Fig. P2.18.
(b) Use infinite gain to derive an expression for the voltage gain, v_o/v_i.

2.19 Prove that the output resistance of a noninverting amplifier (with infinite-gain op amp) is zero. Apply the definition of R_{out}, and explain your reasoning at each step.

2.20 For the circuit of Fig. P2.20 find the value of v_o if
(a) $v_i = 5$ V
(b) $v_i = 0$ V.
(c) Write an equation that relates v_o to v_i.

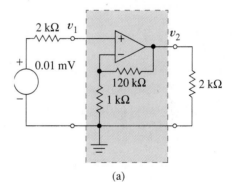

(a)

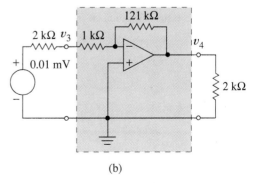

(b)

Figure P2.16

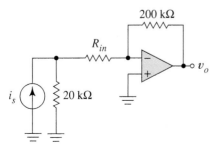

Figure P2.12

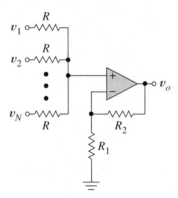

Figure P2.17

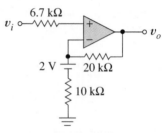

Figure P2.20

(a) Find numerical values for the two coefficients in Eq. (2.14).
(b) Resolve v_a and v_b into their common- and difference-mode components. Then find the numerical values of A_d and A_c and the CMRR for the amplifier.

2.25 (a) By resolving v_a and v_b of Eq. (2.14) into difference- and common-mode components, show that in general the diff amp of Fig. 2.9a has difference- and common-mode gains given by

$$A_d = \frac{R_4(R_1 + R_2) + R_2(R_3 + R_4)}{2R_1(R_3 + R_4)}$$

and

$$A_c = \frac{R_4R_1 - R_2R_3}{R_1(R_3 + R_4)}$$

(b) Suppose R_2 and R_1 have correct nominal values of 20 kΩ and 1 kΩ, respectively, but R_4 is 10% high and R_3 is 1% low. Compute the numerical values of A_d, A_c, and CMRR.

2.26 Rework Exercise 2.2 after changing the 5 k source resistors to 50 Ω.

2.27 The op amp in Fig. P2.27 has infinite gain.
(a) Find the current in the dc source.
(b) Find v_{out}.
(c) Find the common-mode input voltage.

2.21 The op amp in Fig. P2.21 has infinite gain.
(a) Find the value of the node voltage v_o for the given circuit.
(b) Add a 1 kΩ resistor from the node labeled v_o to ground. Then find the current i_o that flows *out of the op amp* with this resistor attached

2.22 Both circuits within dashed boxes in Fig. P2.22 have voltage gain of one. Calculate v_o for both circuits and explain why the answers differ.

Sections 2.2.8–2.2.9

2.23 The difference amplifier of Fig. P2.23 has an infinite-gain op amp so its gain is

$$v_o = 4(v_a - v_b)$$

The rated short-circuit current of the op amp is 25 mA. Show that for $v_a = 5$ V and $v_b = 2$ V (simultaneously), the rated short-circuit current is not exceeded.

2.24 In the laboratory we construct an instrumentation amplifier like Fig. 2.9a with a nominal gain of 10. R_2 and R_1 have exactly the desired values of 100 kΩ and 10 kΩ, respectively; however, R_4 is 10% larger than nominal and R_3 is 10% short.

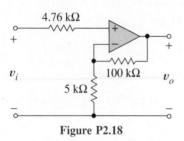

Figure P2.18

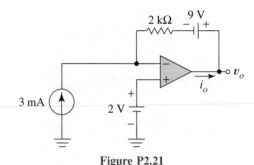

Figure P2.21

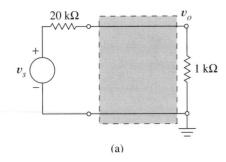

(a)

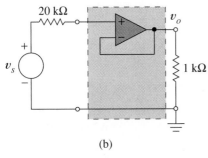

(b)

Figure P2.22

2.28 Design an amplifier like Fig. 2.10b that has difference-mode gain of 2000.
(a) Draw the schematic showing the specific resistor values you have chosen.
(b) Calculate the common- and difference-mode components of the input and output node voltages of the first amplifier stage when $v_a = +3$ mV and $v_b = -2.5$ mV.
(c) Starting with input values of part (b), label each node on your diagram with its node voltage. Use infinite gain analysis, and carry four significant figures

2.29 Design a VCCS that has transconductance of -0.3 mA/V using infinite-gain op amps. In your design make all resistor values as close to each other as possible.

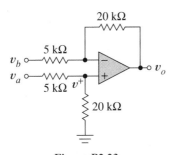

Figure P2.23

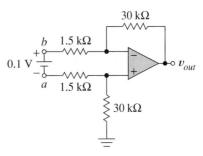

Figure P2.27

(a) Draw the circuit diagram showing all component values.
(b) Find the magnitude and polarity of each op amp output voltage and current for this current source when $v_{in} = 0.4$V and when Z_L is a 10 kΩ resistor.
(c) Find the magnitude and polarity of each op amp output current for this current source when $v_{in} = 0.4$V and when Z_L is a 1 kΩ resistor.
(d) Repeat Part (c) for $v_{in} = -0.4$ V.

2.30 Using infinite-gain op amps, design a CCCS with current gain of 20. Use four identical 1 kΩ resistors in your design.

2.31 Prove that the output resistance of the current source in Fig. P2.31 is given by

$$R_{out} = \frac{R_1 R_3 R_4}{R_1 R_3 - R_2 R_4}$$

Find the value of R_{out} when $R_1 = R_3 = 20.2$ kΩ and $R_2 = R_4 = 19.8$ kΩ.

2.32 The op amp in P2.32 has infinite gain.
(a) Find the input resistance between terminals xx and sketch the input characteristic. Analysis is simplest if you recognize a

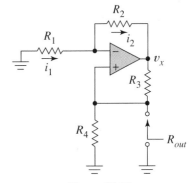

Figure P2.31

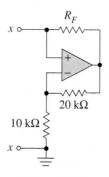

Figure P2.32

familiar subcircuit; however, applying infinite gain principles directly also works.

This class of circuit can be used to cancel out energy losses in a tuned circuit, producing a *negative-resistance oscillator*.

2.33 The op amp in Fig. P2.33 has infinite gain.
(a) Express v_x in terms of v_s.
(b) Use KCL at the v_x node to find an expression for v_o in terms of v_x.
(c) Find an expression for v_o/v_s.

Sections 2.3.1–2.3.5

2.34 Find and sketch the response of the integrator of Fig. 2.12 when $RC = 10^{-3}$ s and
(a) the input voltage is the constant $v_i = 0.2$ V.
(b) the input, $v_i(t)$, is a pulse of amplitude 1 V and duration 2 ms, and the initial condition is $v_o(0^+) = 0$ V.
(c) input is $v_i(t) = 0.2 \sin 20t$ V (assume steady-state operation).

2.35 Derive the gain expression for Fig. 2.13b by viewing the circuit as an RC filter followed by a noninverting "amplifier" in which $1/j\omega C$ replaces R_2 and R replaces R_1. In your derivation

look for a reason why the two RC products must be closely matched for the circuit to function as an integrator.

2.36 Figure P2.36 shows a *synthetic inductor* constructed from infinite-gain op amps, resistors and a capacitor. Use sinusoidal steady-state analysis and the infinite-gain approximation to find the input impedance, V_{in}/I_{in}.
Hints: (1) Justify writing $V_5 = V_{in}$. (2) Write equations that relate I_5 to I_4, I_4 to I_3, . . . , I_1 to I_{in}. (3) Starting with V_{in}/I_{in}, use substitutions until the ratio involves only resistor and capacitor values.

Notice that this circuit can realize inductance of practically any value for a given capacitor because the values of R_1, R_3, R_4, and R_5 are available for scaling. This and related circuits are important because it is not practical to fabricate inductors in integrated circuits.

2.37 Two filters are designed to have the same half-power frequency, $\omega_H = 1/R_2C_2$. Each is then connected to a source of internal resistance R_S as in Fig. P2.37. Find and compare the actual filter functions V_o/V_S. What conclusion can we make from this comparison?

2.38 Design a filter having the *bandpass characteristic* of Fig. P2.38 by using a cascade of two filter circuits. Design so that the smallest capacitor is 1000 pF, and the largest 1 µFd, and distribute the 20 dB gain equally between the filters. Give the final schematic with component values.

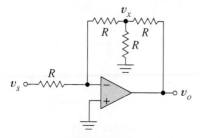

Figure P2.33

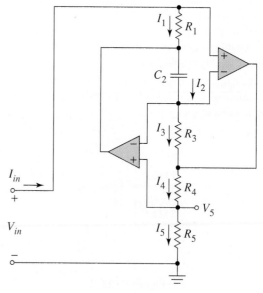

Figure P2.36

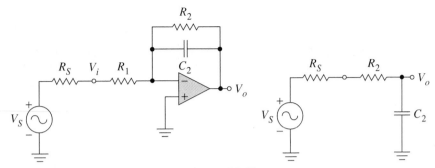

Figure P2.37

2.39 Derive Eq. (2.25).

2.40 Rework Example 2.3 using the second-order filter of Fig. 2.15a with $A = 3 - \sqrt{2}$. Use $C = 0.001 \ \mu\text{F}$ for the capacitance and $R_1 = 2 \ \text{k}\Omega$ in the amplifier.

2.41 Design an 800 Hz sinusoidal oscillator based upon the circuit of Fig. 2.15a.

2.42 The waveform of Fig. P2.42 is input for the differentiator of Fig. 2.16a.
(a) Sketch the general shape of the output waveform.
(b) Find RC if the output voltage must always be in the range

$$+12 \geq v_o \geq -12 \ \text{V}$$

(c) Make a labeled sketch of $v_o(t)$ for the design of part (b).

2.43 The differentiator of Fig. 2.16a uses $RC = 10^{-4}$ s. Find $v_o(t)$ when $v_i(t) = 0.04 \ \sin \ 25t + 0.015 \ \cos \ 200t$. Assume steady-state operation.

2.44 Place a capacitor C_1 in parallel with R in the differentiator of Fig. 2.16a. Find and sketch the magnitude of gain as a function of ω for the circuit.

2.45 Use cascaded op amp circuits to design a two-port that satisfies all the following conditions: input resistance is zero, output is a dependent current source connected to ground, and

$$i_{out} = 20 \int i_{in}(t)dt$$

Section 2.4

2.46 (a) Use SPICE transient analysis to plot the response of the differentiator of Fig. 2.16a to a 1 volt pulse of 20 ms duration. Use $R = 10 \ \text{k}\Omega$ and $C = 0.01 \ \mu\text{F}$.
(b) Repeat for $R = 10 \ \text{k}\Omega$ and $C = 2 \ \mu\text{F}$.
(c) What do you conclude from comparing results of parts (a) and (b)?

2.47 Construct a SPICE model of the difference amplifier of Fig. 2.9a with $R_2 = R_4 = 10 \ \text{k}\Omega$ and $R_3 = R_1 = 1 \ \text{k}\Omega$.
(a) Apply a dc common-mode signal to the inputs and find the dc output voltage using SPICE. What is the common-mode gain?
(b) Repeat part (a) after changing R_4 to 9 kΩ.
(c) Use the expression for A_c given in Problem 2.25 to check your part (b) result.

2.48 Use SPICE to plot the frequency response of the filter designed in Example 2.3. Use your results to verify that the design has the desired gain and -3dB frequency.

Sections 2.5.2–2.5.5

2.49 Show that when the op amp has finite gain, A_d, the non-inverting amplifier gain is

$$\frac{v_{out}}{v_{in}} = \frac{A_d}{1 + A_d R_1/(R_1 + R_2)}$$

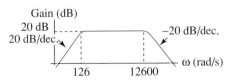

Figure P2.38

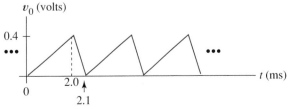

Figure P2.42

2.50 When the op amp has infinite gain, the integrator is described by Eq. (2.23).
(a) Prove that for finite gain A_d the expression is

$$\frac{V_{out}}{V_{in}} = \frac{-1}{1/A_d + j\omega RC(1 + 1/A_d)}$$

(b) Find an inequality that describes the range of ω over which Eq. (2.23) is a good approximation, even when the gain is finite.

2.51 Equation (2.27) describes the differentiator when the op amp has infinite gain.
(a) Prove that for finite gain A_d the expression becomes

$$\frac{V_o}{V_i} = \frac{-j\omega RC}{(1 + 1/A_d) + j\omega\, RC/A_d}$$

(b) Find an inequality that describes the range of ω over which Eq. (2.27) is a good approximation, even when the gain is finite.

2.52 An inverting amplifier using an infinite-gain op amp has circuit gain of -500. Find the actual circuit gain and the % error in circuit gain if the op amp has
(a) $A_d = 10^4$.
(b) $A_d = 10^5$.

2.53 Figure P 2.53 is the equivalent circuit for a noninverting amplifier in which the op amp has gain A_d and output resistance r_{out}. Show that the output resistance of this circuit is

$$R_{out} = \frac{r_{out}}{1 + r_{out}/(R_1 + R_2) + A_d R_1/(R_1 + R_2)}$$

Hint: The circuit is driven by an ideal voltage source of v_i volts.

2.54 (a) From Eq. (2.31), estimate the smallest R_d for which the infinite-gain answer, $v_o/v_i \approx -R_2/R_1$ is still a good approximation. The answer involves A_d, R_1, and R_2.

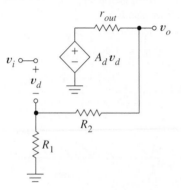

Figure P2.53

Hint: Begin by writing a "much greater than" inequality. Then use the engineering approximation: *much greater means at least a factor of 10.*
(b) For an inverting amplifier using $R_2 = 20$ k, $R_1 = 1$ k and an op amp having $A_d = 10^5$, what is the smallest value of R_d that cannot be ignored when gain is calculated using the infinite gain approximation?

2.55 Make a SPICE model of the filter designed in Example 2.3. Use ac analysis to plot the frequency response of the filter for op amp gains of
(a) $A_d = 10^8$ and $R_{in} = \infty$.
(b) $A_d = 10^2$ and $R_{in} = 100$ kΩ.
Examine the curves carefully and identify any differences in gain and/or cut-off frequency between the SPICE curves and curves expected from the example.

2.56 A 5 V dc source is needed in laboratory but only a $+15$ V supply is available.
(a) Show how one might use a voltage divider and an op amp buffer to produce the desired source.
(b) Why was the buffer needed in part (a)?
(c) If a 741 op amp is used for the buffer, what is the maximum dc current and power the 5 V source can deliver? (In Chapter 10 we learn of *power op amps* that are designed to deliver high output power.)

Sections 2.5.6–2.5.9

2.57 An inverting amplifier of gain -80 uses an op amp that saturates at ± 10 volts.
(a) Find the maximum allowable input voltage if operation is to remain linear.
(b) Find the smallest design values of R_2 and R_1 if the op amp also has a short-circuit current of 5 mA.
(c) For the values of R_2 and R_1 of part (b), calculate the power dissipation in each resistor when the output voltage is 10 V.
(d) For resistors 100 times the values of part (b), compute the op amp output current and the power dissipation in each resistor when $v_o = 10$V.

2.58 An integrator employs an infinite-gain op amp. Relate output voltage, v_o, to the op amp's input offset current when $v_i = 0$.

2.59 When the op amp in Fig. 2.28c has input bias current I_B, show that $R_x = R_1 \| R_2$ results in $v_o = 0$ when $v_i = 0$.

2.60 The infinite-gain op amp in Fig. P2.60a has input offset voltage, V_{OS}.
(a) Show that v_o increases at the rate $(1/RC)V_{OS}$.
(b) If $RC = 0.01$ s and $V_{OS} = 5$ mV, how long does it take for the output to reach the op amp's 12 V saturation limit?

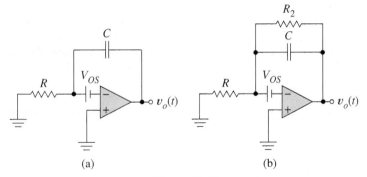

Figure P2.60

(c) After the op amp saturates, how much voltage is dropped across the capacitor?

To prevent such offset-induced saturation we sometimes place a large resistor R_2 in parallel with C as in Fig. P2.60b. As usual, the capacitor charges only until equilibrium ($Cdv/dt = 0$) is achieved.

(d) Verify that for $V_{OS} = 5$ mV and $R_2 = 100 \times R$ the final value of v_o is less than the op amp saturation value.

Hint: With the signal source turned off, the circuit at equilibrium resembles Fig. 2.26b.

2.61 Figures 2.28b and 2.28c show resistor R_x added to eliminate the effect of input bias current for the noninverting and inverting amplifiers, respectively. Notice that when $v_i = 0$, the circuits are identical. Thus we need to study only one circuit.

Assuming infinite gain, find an expression for v_o in terms of I_B, R_2, R_1, and R_x for Fig. 2.28c when $v_i = 0$. Find the value of R_x that makes the output offset zero.

2.62 (a) Show that the output component of the differentiator due to input bias current is RI_B.
(b) Show that the output of the integrator due to input bias current is $dv_o/dt = I_B/C$.
(c) How long does it take an integrator to saturate when $C = 0.001$ F, $I_B = 80$ nA, and $|V_{saturation}| = 14$ V?

2.63 Find the magnitude of the *output* offset voltage for the differentiator when the input offset voltage of the op amp is 5 mV.

2.64 The infinite-gain op amps in the two-stage amplifier of Fig. P2.64 saturate at ±10 V. If each op amp has input offset voltage $V_{OS,1} = V_{OS,2} = 8$ mV, with polarities as shown.
(a) Find the component of v_o due to each input offset voltage.
(b) Sketch the transfer characteristic, v_o/v_s, of the two-stage amplifier showing gain, total output offset, and saturation values.

(c) Suppose the input signal v_s is expected to produce ±8 volt variations in v_o when offset is disregarded. How large are the v_s variations? Compare them to $V_{OS,1}$.
(d) Will the amplifier saturate with such a combination of signal and offset? Show the input signal of part (c) on the transfer characteristic of part (b).

2.65 Use the model of Fig. 2.30c to explore the effect of CMRR on the gain of a noninverting amplifier.

Sections 2.5.10–2.5.14

2.66 (a) Draw the circuit diagram for a SPICE op amp model that simultaneously incorporates *all* of the following parameters: difference-mode gain, input resistance, output resistance, input offset voltage, input offset current, input bias currents.
(b) Write the SPICE code for your model appropriate for the HA2544 of Table 2.1.

2.67 Design a noninverting amplifier with gain of 200. Find the rise time if the op amp used is
(a) the 741 described in Table 2.1.
(b) the 2544.

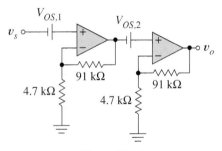

Figure P2.64

2.68 Design an inverting amplifier with gain of -70. Find the rise time if the op amp used is
(a) the 741 described in Table 2.1.
(b) the 2544.

2.69 An amplifier with gain of 800 and bandwidth of 1 MHz is required. Explore one-stage, two-stage, three-stage, and four-stage designs using identical noninverting stages. Do this by constructing a table showing stage gain and bandwidth in Hz versus number of stages for both the 741 and the 2544.

2.70 (a) Design a first-order filter using an infinite-gain op amp, two resistors, and a capacitor. Your filter should have an input resistance of 10 kΩ, a cut-off frequency of 5×10^5 Hz, and low-frequency gain of 10 dB.
(b) Make a SPICE model of your filter using a VCVS with gain of 10^8 for the infinite-gain op amp. Obtain the amplitude response in dB to verify correct operation of your design.
(c) Replace the op amp by a model that includes the bandwidth and gain of a 741 op amp. Compute the amplitude response of the filter using this model.

2.71 Equation (2.29) gives the gain of an inverting amplifier circuit with op amp of gain A_d.
(a) Find the frequency response expression for the inverting amplifier if the op amp has complex gain given by Eq. (2.35).
(b) Show that the gain expression of part (a) can be written in the form

$$A_{inv}(\omega) = \frac{-A_d R_2/(R_1 + A_d R_1 + R_2)}{1 + j(\omega/\omega_B)} \approx \frac{(-R_2/R_1)}{1 + j(\omega/\omega_B)}$$

where

$$\omega_B = \omega_H \left[1 + A_d \frac{R_1}{R_2 + R_1} \right]$$

2.72 An op amp has the following parameters:

input resistance $= 40$ kΩ output resistance $= 150\ \Omega$

gain $= 20{,}000$ input offset current $= 0.2\ \mu$A

unity gain frequency $= 5 \times 10^6$ rad/s.

(a) Construct a SPICE model for this op amp.
(b) Design an inverting amplifier with nominal gain of -20.
(c) Use SPICE to plot the amplitude frequency response of the amplifier when the op amp of part (a) is used. Compare the frequency response of the simulation with the one expected.

2.73 Exercise 2.9 suggested that the differentiator has magnitude and phase curves like Fig. 2.35 when we include op amp dynamics.
(a) Derive an expression for gain as a function of radian frequency for this differentiator using the following procedure.

Observe that replacing R_1 by $1/j\omega C$ and R_2 by R in the inverting amplifier gives the differentiator. Begin with the frequency response of the inverting amplifier given in Problem 2.71b (the complete expression). Your final gain expression should involve A_d, RC, and ω_H.
(b) Assume that the peak of the frequency response curve occurs near that frequency that makes the real part of the denominator go to zero in your frequency response expression. Determine this frequency using $A_d = 2 \times 10^5$, $RC = 2\pi \times 10^3$, and $\omega_\tau = 2\pi \times 10^6$, the same values employed in the simulation.
(c) Use the theoretical expression to confirm that at high frequencies the phase shift of the differentiator approaches $-270°$.

2.74 An inverting amplifier using a 741 op amp has gain of -80.
(a) Find its bandwidth, ω_B.
(b) Find the maximum amplitude *input* sinusoid of frequency ω_B that the amplifier can handle without slewing.
(c) repeat parts (a) and (b) for the HA2544.

2.75 (a) Estimate the rise time t_r of the output of a noninverting amplifier of gain 20 constructed using an 741 op amp.
(b) If the amplifier slews from 0.1 to 0.9 of the final value of an output pulse in t_r second, where t_r is the answer to part (a), how large is the output pulse?

2.76 The circuit of Fig. 2.12 integrates a variable-frequency square wave of 2 V amplitude. Find the minimum slew rate acceptable for the op amp in terms of the square-wave frequency, f. Notice that the op amp's slew rate does not limit the frequency of the square wave that the circuit can integrate.

Section 2.6

2.77 The comparator of Fig. 2.38a uses an op amp with ± 15 V saturation limits. If its slew rate is 150 V/μs, how fast are its output transitions?

2.78 An electronic thermometer produces an output voltage in millivolts given by

$$v_s(T) = 30 + 3T$$

where T is the temperature in °C.
(a) Sketch the thermometer characteristic.
    A comparator like Fig. 2.38a with a reference voltage of 5 V is to change its output state whenever the temperature exceeds 50° C.
(b) Draw a system diagram showing how an op amp circuit can be used to interface $v_s(T)$ to the comparator input. Include component values on your diagram.

(c) If the op amps have input offset voltages of ± 5 mV, the comparator might change state at an incorrect temperature. Use the more critical of the offset voltages to estimate the upper and lower ends of the temperature range where the comparator might change state.

2.79 A triangular wave of 10 V peak value provides input to the Schmitt trigger of Fig. 2.40a. Sketch the input and output waveforms if $V_P = V_M = 10$ V, $R_1 = 2.4$ kΩ, and $R_2 = 4.7$ kΩ. Your sketch should show the relative locations of the zero crossings of the two waveforms.

2.80 Add to Fig. 2.40a a reference voltage V_R that makes the inverting input positive relative to ground.
(a) Using this modified circuit, derive new versions of Eqs. (2.43)–(2.46) that are suitable for this modified circuit. Draw equivalent circuits to assist your reasoning.
(b) Sketch the transfer characteristic for $V_P = V_M = 8$ V, $V_R = +4.0$ V, $R_1 = 5$ kΩ, $R_2 = 10$ kΩ.

2.81 The *relaxation oscillator* of Fig. P2.81 uses the op amp as a two-state circuit to generate periodic oscillations at the output. The op amp operates at its saturation limits, in this case ± 14 V.
(a) Sketch the equivalent circuit assuming $v_o = -14$ V. At $t = 0$ the capacitor is charged to $v_B \approx -3.24$ V. Calculate v_x and tell why the op amp must soon change state.
(b) Draw a new diagram showing the new equivalent circuit. Use the ending capacitor voltage from part (a) as the initial condition. How long will the op amp remain in the assumed state?
(c) Review the preceding parts of the problem. Then try to sketch $v_o(t)$ and $v_B(t)$ for all t. Find the frequency of the waveform.

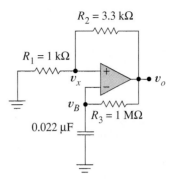

Figure P2.81

2.82 Use the two-state, open-loop op amp model of Fig. 2.24 to show that the Schmitt trigger circuit of Fig. P2.82 has the transfer characteristic indicated. Find equations for v^+ and v^- in terms of the resistor values and the saturation limits, V_P and V_M.

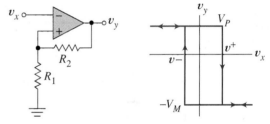

Figure P2.82

CHAPTER
3

Semiconductors, pn Junctions, and Diode Circuits

In this chapter we review electrical conduction in two familiar classes of materials, insulators and conductors. We then study special materials called semiconductors that have electrical properties intermediate to conductors and insulators. We focus special attention on junctions between different semiconductor types because these junctions are critical to the operation of many important solid-state devices. The simplest of these devices is the junction itself, the *junction diode*. Equipped with theoretical background, we then construct simple circuit models to represent the diode. Finally, we use the models to analyze and design interesting and useful circuits.

When electronic circuits were interconnections of discrete, independently fabricated devices, we lost little by delaying solid-state studies until a later course; we simply introduced devices in terms of their measured volt-ampere curves. Today, however, most electronic circuits are *integrated*. That is, all circuit components are fabricated at the same time on the same silicon *chip*. By relating the physical characteristics of all the chip components, this common fabrication process presents unique constraints and opportunities to the circuit designer. To appreciate better the constraints and utilize the opportunities, a brief introduction to solid-state theory is useful. This theory provides the mathematical description we need to access the lower modeling path in Fig. 1.14 and, in the process, provides us with physical insight into the devices. To avoid long, distracting explanations, we present solid-state concepts in a highly simplified and intuitive fashion. (Many excellent textbooks cover this material more rigorously. Streetman and Mattson, included among the end-of-chapter references, are good examples.)

This chapter also introduces important concepts related to nonlinear circuit operation. If linear, dependent sources were available, we would use them to construct the amplifiers of the preceding chapters; however, the only devices we have are semiconductor building blocks called *transistors*. These have output characteristics like dependent current sources—but only in certain regions of the i versus v plane. Diode models play a key role in helping us understand and analyze such non-linear behavior. We also learn how to use diode circuits for nonlinear waveform processing.

134

3.1 _____

Conduction in Insulators and Metals

3.1.1 INSULATORS AND CONDUCTORS

Consider an insulating solid such as diamond at absolute zero temperature. Lacking energy, the diamond's atoms are bound together in a stationary crystal lattice of regular geometry. As we increase the temperature, atoms begin to vibrate about their original lattice positions, and at progressively higher temperatures, this random motion becomes increasingly violent. The valence electrons share in this motion, but, because of strong atomic forces, they remain tightly bound to their parent atoms. Because there are no electrons free from atomic forces and able to move through the lattice, an electric field established in the insulator by means of an external voltage source produces no current. The essence of an <u>insulator</u>, then, is that all valence electrons remain tightly bound to parent atoms, even at high temperatures and in the presence of electric fields.

At the other extreme is a <u>conductor</u> such as silver, gold, or copper in which valence electrons are secured to parent atoms only by weak atomic forces. At absolute zero the valence electrons are bound as in the insulator. Because the atomic forces are weak, however, at a slightly elevated temperature, *all* valence electrons are shaken loose from their parent atoms and are then able to move freely through the lattice. The resulting concentration of *free electrons,* about 10^{22} per cm^3, is an enormous number. The result is a randomly vibrating latticework of immobile positive ions immersed in a sea of highly mobile free electrons.

3.1.2 DRIFT IN AN ELECTRIC FIELD

Because of its thermal energy, the typical free electron moves at high speed until it collides with a lattice ion, then changes direction. Because its motion is random, there is no directional bias, and the average displacement is zero. The solid trajectories in Fig. 3.1 depict in two dimensions successive three-dimensional random displacements, starting at point 0 and ending at point 6. For comparison, the blue lines show electron motion in the presence of an electric field **E**. Since field intensity is force per unit charge, the field acts on the electron, imparting a directional bias to each trajectory. Because of its negative charge, the electron's net displacement is opposite to the field direction. Although the net displacement or *drift* is slight compared to the sum of the individual displacements, it explains the important process of electrical conduction.

The average displacement per unit time caused by the electric field is the drift velocity, v_d, of the electron. Provided that the field intensity is not too large, the average drift velocity, $\langle v_d \rangle$, of free electrons varies in direct proportion to the electric field intensity **E**. Quantitatively,

$$\langle v_d \rangle = \mu_n \mathbf{E} \tag{3.1}$$

where the proportionality constant μ_n is called the *electron mobility.* Increased temperatures cause more violent lattice motion, more frequent collisions of drifting electrons with lattice ions, and shorter distances between the collisions. The result is a lower average drift velocity for the same electric field. Thus electron mobility *decreases with temperature.*

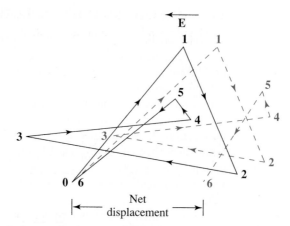

Figure 3.1 Typical electron displacements with no electric field (solid lines) and with an electric field (dashed lines). Starting point is point 0.

3.1.3 RESISTANCE

These simple physical ideas help us understand resistors. Figure 3.2 shows a conducting solid, of length L and uniform cross sectional area A, to which an external voltage is applied. This voltage establishes within the material a uniform electric field of intensity $|\mathbf{E}| = V/L$ volts per meter as shown, where $|\ |$ denotes the magnitude of vector $\mathbf{E}$. To account for average behavior, we visualize all free electrons, including those explicitly shown in an imaginary test volume, as drifting together toward the left at drift velocity

$$|\langle v_d \rangle| = \mu_n \frac{V}{L} \tag{3.2}$$

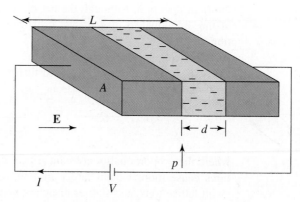

Figure 3.2 Drift of free electrons in a resistor.

The external source simultaneously removes electrons from the positive edge of the material and supplies them at the negative edge to support an external current of I amperes. This current equals the number of coulombs of negative charge that drift past observation point p each second. From Fig. 3.2 we see that the charge Q in coulombs within the imaginary volume is

$$Q = qnAd \qquad (3.3)$$

where $q = 1.6 \times 10^{-19}$ C is the electron charge and n is the free-electron *concentration* in electrons per unit volume. The time, Δt, required for this charge to drift past the observation point is

$$\Delta t = \frac{d}{\langle v_d \rangle} \qquad (3.4)$$

Dividing Eq. (3.3) by Eq. (3.4) and using $|\langle v_d \rangle|$ from Eq. (3.2) gives

$$I = (qn\mu_n) \times \left(\frac{A}{L}\right) \times V$$

or

$$V = \left[\frac{1}{qn\mu_n}\left(\frac{L}{A}\right)\right]I \qquad (3.5)$$

Since the ratio of V to I is the resistance R of the material, Eq. (3.5) shows that R depends upon both spatial dimensions and material properties. To separate the geometrical and material contributions, we express resistance R in the form

$$R = \frac{1}{\sigma} \times \frac{L}{A} \qquad (3.6)$$

where σ, the *conductivity,* is a parameter that characterizes the particular conducting material from which we fabricate the resistor. From Eqs. (3.5) and (3.6),

$$\sigma = qn\mu_n \text{ siemens/cm} \qquad (3.7)$$

To produce appreciable voltage drops in a small physical space, resistors utilize the lower-mobility conducting materials such as nichrome, carbon, carbon film, and metal film.

Equation (3.5) is related to a more fundamental form of Ohm's law. Substituting σ from Eq. (3.7) and dividing both sides by L gives

$$\frac{I}{A} = \sigma\frac{V}{L}$$

The left side is the magnitude of the *current density* **J**, the current per unit area at a point in the material resulting from the field **E**. In vector notation the Ohm's law relationship is

$$\mathbf{J} = \sigma \mathbf{E} \tag{3.8}$$

Thus current density is directly proportional to field intensity at any point, and conductivity is the proportionality constant.

3.1.4 TEMPERATURE COEFFICIENT OF RESISTANCE

We now show how the resistance of a conducting material varies with temperature. Eventually these concepts will help us understand how to design electronic circuits that operate properly over wide temperature ranges.

In a conductor the free-electron concentration, n, in Eq. (3.7) is a constant because *all* valence electrons contribute to the drift current; however, the mobility μ_n decreases with temperature because of more frequent collisions. We therefore expect σ to decrease, and the resistance in Eq. (3.6) to increase with temperature. This is consistent with measurements of metallic resistors, which typically show a linear variation as in Fig. 3.3. For T close to some reference temperature, T_R, this straight-line variation takes the form

$$R(T) = R(T_R)[1 + \alpha(T - T_R)] \tag{3.9}$$

where $R(T)$ and $R(T_R)$ are the resistances at temperatures T and T_R, respectively. The constant α, called the *temperature coefficient of resistance,* is the fractional change in R per unit change in temperature,

$$\alpha = \frac{[R(T) - R(T_R)]/R(T_R)}{T - T_R} = \frac{\Delta R/R}{\Delta T} \tag{3.10}$$

(We use similar temperature coefficients to describe nonmetallic resistors, capacitors, reference voltages, and other quantities where temperature variation is a matter of concern.)

With Eqs. (3.9) and (3.10) we can deal quantitatively with temperature-induced changes in resistance, as we see in the following.

Exercise 3.1 At 18°C a resistor has resistance of 20 kΩ. If $\alpha = 0.0039/°C$, what will be the resistance at 50°C?

Ans. $R = 22.5$ kΩ.

Figure 3.3 Variation of resistance with temperature for a metal resistor.

EXAMPLE 3.1   Find a temperature range centered at 18°C such that the resistor of Exercise 3.1 varies by no more than ±5% from its nominal value of 20 kΩ.

Solution.   A 5% variation means $\Delta R/R = 0.05$. From Eq. (3.10),

$$\alpha = 0.0039 = \frac{0.05}{\Delta T}$$

so $\Delta T = 12.8$°C. Therefore, the resistor will be within the specified tolerances for

$$(18°C - 12.8°C) < T < (18°C + 12.8°C)$$

or

$$5.2°C < T < 30.8°C \qquad \qquad ❏$$

3.2
Conduction in Intrinsic Semiconductors

3.2.1   INTRINSIC (PURE) SEMICONDUCTORS

Semiconductors are materials such as silicon, germanium, or gallium arsenide that have conduction properties intermediate to insulators and metals. We distinguish between pure or *intrinsic* semiconductors and *extrinsic* or *doped* semiconductors, the subject of the next section. Because of its continuing importance, the focus is primarily on silicon; however, our general results apply to the other semiconductors as well.

3.2.2   FREE ELECTRONS AND HOLES

Figure 3.4a depicts a valence 4 semiconductor atom as an ion and four associated valence electrons. In the solid state, many such atoms in close proximity form a regular crystal lattice in which each valence electron is shared in a covalent bond with one of the atom's four nearest neighbors. Figure 3.4b is a two-dimensional representation of this three-dimensional lattice structure at 0 K.

The covalent bonds that secure valence electrons to parent atoms in semiconductors are much stronger than metallic bonds but considerably weaker than bonds in insulators. Therefore, as the lattice acquires thermal energy and the temperature increases, *some* covalent bonds break. Electrons pull loose from the attraction of their parent atoms and become *conduction electrons,* free to drift in response to an electric field. Figure 3.4c shows one such conduction electron, which has drifted to the right under the influence of the field. Unlike conductors where virtually *all* valence electrons are available for conduction at low temperatures, *only a small fraction of the bonds are broken in a semiconductor.* At higher temperatures, covalent bonds break at a higher rate, providing additional free electrons.

In a conductor, the products of each broken bond are a mobile conduction electron and a positive ion, the latter *locked in the lattice.* In a semiconductor, however, when a bond is broken and the electron drifts away, there remains a net positive charge called a *hole that is also mobile.* In Fig. 3.4d an arrow shows a hole associated with the ionized central atom; the electron that formerly occupied the vacant site has drifted completely out of the picture. Figure 3.4e shows the mechanism by which the hole moves. The electron in a neighboring covalent bond, by chance, approaches so close to the vacated lat-

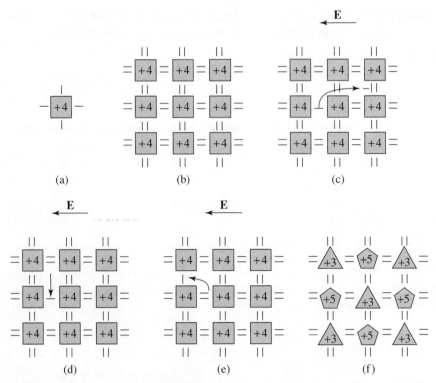

Figure 3.4   Intrinsic semiconductor: (a) atom of valence 4 semiconductor; (b) atomic lattice structure at 0 K; (c) thermal generation of a free electron; (d) hole remaining after free electron has drifted away; (e) movement of the hole under influence of the electric field; (f) compound semiconductor (gallium arsenide) with valence 3 and valence 5 atoms at 0 K.

tice site that it is captured by the local atomic forces, leaves its parent atom, and forms a covalent bond with the neighboring atom. In effect the hole has moved to an *adjacent* location in the lattice.

Not only is the hole mobile, but it drifts. An electric field makes involvement by electrons "downfield" from a vacated site more probable because electrons are pulled "upfield." This encourages hole motion in the direction of the field as in Figs. 3.4d and e. Since electrons actually change positions, treating holes as mobile, positively charged particles at first seems contrived and unnecessary. Quantum mechanical studies show, however, that hole motion is mathematically equivalent to drift of a positively charged particle having mass slightly greater than that of an electron. (From the more restricted nature of hole motion, it seems logical that hole mobility should be lower than electron mobility.) Thus, for all practical purposes, whenever a covalent bond is broken in a semiconductor, a *hole-electron pair* is generated, and *both* "mobile, charged particles" participate in conduction.

Since free electrons and holes in intrinsic materials always occur in pairs, the electron concentration, n, and the hole concentration, p, are always equal. The standard notation is

$$n = p = n_i \text{ carriers/cm}^3$$

where n_i, the *intrinsic carrier concentration,* is characteristic of the particular semiconductor material at a particular temperature.

There are also *compound semiconductor materials* that consist of atoms of more than one kind. An important example is gallium arsenide, which contains equal numbers of valence 3 gallium atoms and valence 5 arsenic atoms. Figure 3.4f shows the gallium arsenide crystal structure. As in valence 4 semiconductors, each atom shares electrons in covalent bonds with its four nearest neighbors; however, in the GaAs lattice each neighbor is an atom of the complementary type. Thermal generation of hole-electron pairs occurs in compound materials just as in valence 4 materials. The principal difference is quantitative, with n_i having a much lower and μ_n a much higher value in gallium arsenide than in either silicon or germanium.

3.2.3   RECOMBINATION OF ELECTRONS AND HOLES

Along with thermal generation of hole-electron pairs in semiconductors, a *recombination* process simultaneously occurs. When a conduction electron happens upon a hole, the electron is sometimes recaptured, reestablishing a shared covalent bond at the lattice site—the free electron and hole spontaneously disappear. The rate of recombination is proportional to the concentration of hole-electron pairs, an idea that makes sense physically, since higher concentrations make chance encounters of holes and electrons more likely. The generation and recombination processes coexist in a state of dynamic balance that is described next.

3.2.4   THERMAL EQUILIBRIUM AND $n_i(T)$

In intrinsic material, a *thermal equilibrium* balances generation and recombination of charge carriers, and establishes the *intrinsic carrier concentration, n_i,* at any temperature. The fluid analog of Fig. 3.5a shows the idea. Inflow at a rate determined by the faucet setting represents generation of new hole-electron pairs at a rate determined by the lattice temperature; outflow represents disappearance of hole-electron pairs by recombination. The outflow rate is proportional to fluid volume, just as recombination is proportional to carrier concentration. Steady-state fluid volume represents the equilibrium carrier concentration n_i. Equilibrium implies identical inflow and outflow rates—constant fluid volume in the container—fixed n_i in the semiconductor. Increasing the faucet opening upsets the equilibrium, temporarily causing inflow to exceed outflow. Fluid volume then increases until outflow matches the increased inflow at a higher fluid volume. Partially closing the faucet similarly produces a new equilibrium with a smaller fluid volume. By analogy, an increase or decrease in lattice temperature results in a new equilibrium involving larger or smaller n_i. The first row of Table 3.1 lists room temperature intrinsic carrier concentrations for commonly used semiconductor materials.

The rate of generation of new hole-electron pairs is a sensitive function of temperature. For silicon, the intrinsic concentration is

$$n_i(T) = 3.88 \times 10^{16} T^{3/2} e^{-6957/T} \text{cm}^{-3} \qquad (3.11)$$

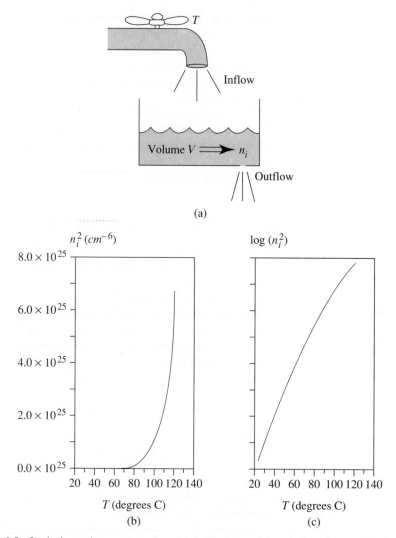

Figure 3.5 Intrinsic carrier concentration: (a) fluid analog; (b) variation of squared value with temperature; (c) log of squared value.

where T is absolute temperature. Germanium and gallium arsenide give similar expressions with different constants.

As we study solid-state devices in subsequent chapters, we repeatedly find the square, $n_i^2(T)$, associated with important parameter values. Indeed, this factor is our major adversary in trying to design circuits that work over large ranges of temperature. We need not know the specific equation; however, it is useful to know and remember that n_i^2 approximates an exponentially increasing function of T over a wide range of temperature, as we see in Figs. 3.5b and c.

TABLE 3.1   Room Temperature Values for Semiconductor Constants

	Silicon	Germanium	Gallium Arsenide	Units
n_i	1.45×10^{10}	2.4×10^{13}	1.8×10^{6}	cm^{-3}
μ_n	1500	3900	8500	cm^2/V-s
μ_p	480	1900	400	cm^2/V-s

3.2.5   CONDUCTIVITY OF INTRINSIC SILICON

In a semiconductor, holes and electrons respond to an electric field by drifting in opposite directions. This gives *two* additive current components to equate to the external current, not just an electron current as in a metal. The average drift velocity of each charge carrier is related to the field intensity by its own mobility, μ_n for electrons and μ_p for holes. By adding an imaginary volume of holes moving toward the right in Fig. 3.2 we could show that the conductivity of a semiconductor *in general* is

$$\sigma = q(n\mu_n + p\mu_p) \tag{3.12}$$

where p is the hole concentration. In *intrinsic* semiconductors, hole and electron concentrations are both equal to n_i, and Eq. (3.12) reduces to

$$\sigma_i = qn_i(\mu_n + \mu_p) \tag{3.13}$$

Typical values for electron and hole mobilities at room temperature, 27°C, are listed in Table 3.1. Because of its high electron mobility, gallium arsenide is playing an increasingly important role in high-speed devices.

Equations Eq. (3.6) and Eq. (3.8) both apply to drift in semiconductor materials as well as metals, if we use Eq. (3.12) for conductivity as in the following example.

EXAMPLE 3.2   Figure 3.6 shows a cylinder of intrinsic silicon. Calculate its resistance at room temperature.

Solution. The electron and hole concentrations are $n = p = n_i$ carriers per cm^3. We use values from Table 3.1 in Eq. (3.13) to find

$$\sigma_i = 1.6 \times 10^{-19} \times 1.45 \times 10^{10} \times (1500 + 480) = 4.59 \times 10^{-6} \text{ S/cm}$$

The length of the resistive path is 8 cm and the area through which the current flows is $A = \pi(1)^2/4 = 7.85 \times 10^{-1}$ cm^2. Using these values in Eq. (3.6), gives $R = 2.22$ MΩ. ❏

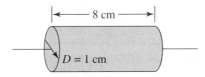

Figure 3.6   Resistor constructed from intrinsic silicon.

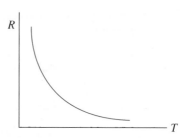

Figure 3.7   Change in intrinsic semiconductor resistance with temperature.

In intrinsic semiconductors, μ_n and μ_p in Eq.(3.13) both decrease slightly with temperature because of more frequent lattice collisions; however, this is *more than offset* by the large increases in n and p caused by thermal generation of new hole-electron pairs. A special resistor constructed from intrinsic semiconductor material, called a *thermistor*, is useful as a sensitive temperature-sensing transducer because its resistance decreases dramatically with temperature as in Fig. 3.7; that is, it has a *negative* temperature coefficient. This behavior is compared with the modest *increase* of Fig. 3.3 for a metallic resistor.

> **Exercise 3.2**   In Example 3.2 we found that $R = 2.22$ MΩ at 27°C for the thermistor of Fig. 3.6. Find R at 35°C. Assume that changes in the mobilities are negligible.
>
> *Ans.* $R = 990$ kΩ.

In the preceding example and exercise we found that an 8°C rise in temperature produced a 55% drop in thermistor resistance. We can make a rough estimate of the thermistor's temperature coefficient by using these numbers in Eq. (3.10). This gives

$$\alpha = \frac{\Delta R/R}{\Delta T} = \frac{(9.9 \times 10^5 - 2.22 \times 10^6)/2.22 \times 10^6}{35 - 27} = -0.069$$

or $-69,000$ *parts per million* (ppm) per degree C. Notice that α itself is a function of temperature, not a constant.

3.3 _____

Doped Semiconductors

To produce semiconductors with conductivity precisely controlled and relatively constant over a wide temperature range, we add carefully measured amounts of certain impurities to the material. There are two types of impurities, donor and acceptor, and two corresponding classes of doped semiconductors, n-type and p-type.

3.3.1   n-TYPE SEMICONDUCTORS

We produce an *n-type semiconductor* by introducing a small concentration of a valence 5 impurity, such as antimony or phosphorus, into the lattice structure. Figure 3.8a shows one of the widely separated lattice sites in which such an impurity is located. At 0 K, four of the impurity atom's valence electrons share covalent bonds with electrons from neighboring silicon atoms, but the fifth is unshared, and therefore bound only weakly to its parent atom. Only a slight amount of thermal energy is sufficient to shake loose *all*

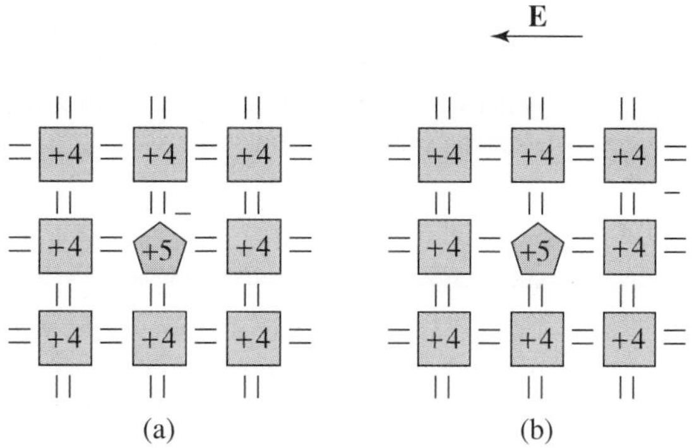

Figure 3.8 Donor-type semiconductor: (a) at $T = 0$ K; (b) at $T > 0$ K.

these unshared *donor* electrons. Once freed from their original bonds, these function much like free electrons in a metal. The impurity atom in Fig. 3.8b, for example, has been *ionized* in this fashion, and the electron has drifted to the right under the influence of the electric field.

Valence 5 impurities are called *donor* or *n-type* impurities because they *donate* free electrons. Notice that there are no mobile holes generated when impurity atoms donate electrons in n-type material. Left behind when the valence electrons drift away are *immobile positive ions,* locked in the lattice, and themselves unable to drift in response to an applied electric field. There is obviously a close similarity between n-type semiconductors and conductors; however, the free-electron concentration, n, in the semiconductor is much lower (hence *semi*conductor) and its value is closely controlled during the doping process.

In doped silicon there are also *ordinary* hole-electron pairs continually being generated and recombining just as in intrinsic materials; however, at moderate temperatures the donor electrons dominate the electrical behavior by virtue of their greatly superior numbers. For example, the donor concentration, N_d, might be of the order of 10^{19} atoms/cm^3, compared with $n_i = 1.5 \times 10^{10}$ electrons/cm^3 available in intrinsic silicon at 27°C. Since this makes 6.67×10^8 donated electrons for every intrinsic electron, conductivity of doped material is much higher than that of intrinsic material. On the other hand, the conductivity of the doped semiconductor is much lower than that of a conductor. A monovalent metal would have about 10^{22} free electrons/cm^3, about one thousand times the number in the doped impurity!

3.3.2 CARRIER CONCENTRATIONS IN n-TYPE SILICON

The presence of donor electrons profoundly alters the equilibrium that establishes n_i^2. The *law of mass action,* which applies to all semiconductors, intrinsic or doped, requires that the product of hole- and free-electron concentrations be always constant. Because the law applies to intrinsic materials, we know

$$np = n_i^2 \tag{3.14}$$

In n-type material, some electrons come from donor atoms and some from thermally generated hole-electron pairs in the silicon; however, holes arise only from thermally generated hole-electron *pairs*. Therefore, when the donor concentration is N_d, the free-electron concentration is

$$n = N_d + p \qquad (3.15)$$

We can find the *exact* values for n and p at any temperature by simultaneously solving Eqs. (3.14) and Eq. (3.15) for n and p, respectively. Figure 3.9 shows the resulting electron concentration. For heavily doped material there is a wide *extrinsic* temperature range over which the concentration of donated electrons is much greater than the intrinsic concentration. At sufficiently high temperatures the material *goes intrinsic*. That is, the thermally generated carrier pairs outnumber the donated carriers. In the important *extrinsic region,* however, the donated electrons dominate the electrical behavior of the material. This observation leads to the useful donor-material approximation:

$$n \approx N_d \qquad (3.16)$$

Substituting n from Eq. (3.16) into Eq. (3.14) gives an approximation for the hole concentration appropriate for extrinsic temperatures:

$$p \approx \frac{n_i^2}{N_d} = \left(\frac{n_i}{N_d}\right)n_i \qquad (3.17)$$

Because $N_d \gg n_i$ at room temperature, Eq. (3.17) shows that the hole concentration is much lower than n_i in n-type materials. Because $n \gg p$, we eall electrons *majority carriers* and holes *minority carriers* in n-type materials.

In the extrinsic temperature range, the approximation

$$\sigma \approx q\mu_n N_d \qquad (3.18)$$

describes the conductivity of n-type semiconductors. This follows from Eq. (3.12) by substituting Eqs. (3.16) and (3.17) and then noting that hole concentration is negligible over extrinsic temperatures.

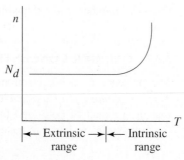

Figure 3.9 Free-electron concentration in n-type silicon as a function of temperature.

Figure 3.10 p-Type silicon impurity: (a) at absolute zero; (b) at an elevated temperature.

3.3.3   p-TYPE SEMICONDUCTORS

We produce another kind of doped semiconductor, p-type, by introducing a small concentration of *acceptor* impurities of valence 3 such as boron or indium. Figure 3.10a shows how all three valence electrons from each impurity atom join in covalent bonds with neighboring silicon atoms at absolute zero. At each impurity site this leaves one valence electron from a silicon atom without a partner to share its bond. This electrical asymmetry represents a *low-energy* location capable of capturing and holding any electron that happens to come too close. When the lattice acquires thermal energy, it is nearly certain that an electron from an adjacent silicon atom will be captured (accepted) at a given impurity site as shown by the arrow in Fig. 3.10b. This creates an *immobile negative lattice ion* and leaves a single electron in a covalent bond *originally shared by two silicon electrons*. We recognize that a hole has been formed, which is now free to drift through the lattice. In the material as a whole, raising the temperature slightly above absolute zero results in N_a holes per cm³, each free to drift with an applied electric field. For each of these holes there is a negative ion locked in the crystal lattice.

3.3.4   CARRIER CONCENTRATIONS IN p-TYPE SILICON

Arguments parallel to those for donor-type impurities show that the total number of holes satisfies the exact equations (3.14) and

$$p = N_a + n \tag{3.19}$$

Close examination of these equations at low and moderate temperatures leads to the useful approximations

$$p \approx N_a \tag{3.20}$$

and

$$n \approx \frac{n_i^2}{N_a} = \left(\frac{n_i}{N_a}\right) n_i \tag{3.21}$$

Therefore, in p-type materials the free-electron concentration is much lower than the intrinsic value at moderate temperatures. Because $p \gg n$, holes are the *majority carriers* and electrons the *minority carriers* in p-type materials. The equation

$$\sigma = q\mu_p N_a \tag{3.22}$$

approximates the electrical conductivity of p-type material at moderate temperatures.

It is also possible to create n- or p-type compound semiconductors. In gallium arsenide, for example, we can do this by selectively replacing a few gallium or arsenic atoms with silicon. The result is n- or p-type material, respectively, with silicon atoms for impurities.

3.3.5 COMPENSATION

While fabricating doped semiconductors, we can change n-type material into p type, and conversely, by a process, called *compensation*. To change n-type into p-type semiconductors, for example, we add donor impurities of concentration $N_a > N_d$. The properties of this *compensated* material are the same as ordinary p-type material with donor concentration of $N_a - N_d$. Similarly, we can add donor impurities to acceptor material to produce n-type material with an effective impurity concentration of $N_d - N_a$. All the approximate equations for carrier concentrations and conductivity of doped materials apply when these effective doping concentrations are used.

Although not drawn to scale, Fig. 3.11 summarizes the differences in conductivities of three kinds of materials. At extrinsic temperatures, where most solid-state devices operate, doped semiconductor conductivities are much lower than are those of conductors because of lower concentrations of charge carriers, but they are much higher than are those of intrinsic material. In doped materials, as in conductors, the number of charge carriers is constant; thus conductivities have small negative temperature coefficients caused by variations in mobility with temperature. This means that resistors made from doped semiconductors, like those made from conductors, have positive temperature coefficients. At intrinsic temperatures, doped semiconductors begin to resemble intrinsic materials; however, doping concentrations are kept high to avoid this problem.

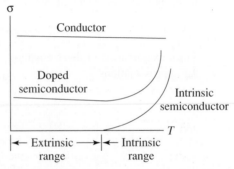

Figure 3.11 Conductivities of conductor, and of doped and intrinsic semiconductor.

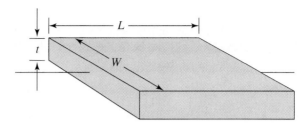

Figure 3.12 Integrated circuit resistor fabricated from doped material.

3.3.6 INTEGRATED CIRCUIT RESISTORS

In Chapter 4 we outline how integrated circuits are fabricated. There we see that the least expensive integrated circuit resistors are simply thin slabs of doped semiconductor material as in Fig. 3.12. All resistors on the same integrated circuit chip are fabricated at the same time, resulting in identical thickness t and conductivity σ. This means that in the same integrated circuit, we can obtain different resistor values only by variations in the geometrical dimensions L and W. If we replace area, A, in Eq. (3.6) by the product Wt from Fig. 3.12, resistance becomes

$$R = \frac{1}{\sigma t} \times \frac{L}{W} = R_s \times \frac{L}{W} \qquad (3.23)$$

which defines the *sheet resistance, R_s*. Notice that R_s contains those parameters common to all resistors on the IC chip. Technically, R_s has units of ohms; however, we use *ohms per square* to emphasize that sheet resistance is the resistance of a square of the semiconductor of any size as viewed from above.

To appreciate the convenience of sheet resistance, note that to produce a 6 kΩ resistor from a slab of 600 Ω per square semiconductor, we need only to lay out a surface area with a length-to-width ratio of 10 to 1. This is useful because the length-to-width ratio is controlled directly in the fabrication process. Integrated resistors typically satisfy Eq. (3.9) with temperature coefficient $\alpha \approx 1000$ parts per million. For practical sheet resistance values, large resistances require long, narrow geometries and excessive space on the integrated circuit chip. It is so hard to reconcile such geometry with other design and layout requirements that IC designers simply learn to work with circuits with relatively small resistor values.

We discover another important design principle from Eq. (3.23). Because integrated circuits are so small, resistors physically close on the chip have nearly identical temperatures. Since temperature variation in an integrated resistor is embodied in R_s (through σ), it follows that in a *ratio* of resistances, each of the form of Eq. (3.23), R_s cancels, and the ratio is virtually invariant over a wide temperature range. Since a ratio of IC resistances depends only upon geometry, it turns out that we can establish its value with a precision approaching 1% during fabrication. We conclude that good IC design involves making overall circuit performance depend, insofar as possible, upon resistor ratios. Many of the op amp circuits of Chapter 2 have overall performance determined by resistor ratios (e.g., inverting noninverting amplifiers, summing circuits, difference amplifiers) and therefore provide excellent bases for integrated circuit designs.

EXAMPLE 3.3   A semiconductor resistor R is made from a $W \times L = 2 \ \mu m$ by $20 \ \mu m$ section of semiconductor with $900 \ \Omega$ per square sheet resistance. Find the maximum and minimum values of R if R_s varies by $\pm 20\%$. Assume that W and L are constants.

Solution.   (a) From Eq. (3.23), the nominal resistance is

$$R = R_s \left(\frac{L}{W} \right) = 900 \left(\frac{20}{2} \right) = 9000 \ \Omega$$

Including 20% variations in R_s gives

$$R \pm \Delta R = 9000(1 \pm 0.2) = 9000 \pm 1800 \ \Omega$$

Therefore,

$$7200 \ \Omega \leq R \leq 10{,}800 \ \Omega \qquad\qquad ❑$$

The preceding example suggests the kind of variations we expect if we compare corresponding 9 kΩ resistors in a pair of ICs fabricated at different times. In the following exercise, we estimate variability of resistance *on a given chip* that results from quality control limitations on fabrication geometry.

Exercise 3.3   Find maximum and minimum values for the resistor in Example 3.3 if R_s is constant but W and L each vary by $\pm 1\%$.

Ans. $8822 \ \Omega \leq R \leq 9182 \ \Omega$.

3.3.7 MINORITY CARRIER LIFETIME

We already know of the thermal equilibrium that regulates hole and electron concentrations in semiconductors. Whenever we disturb this equilibrium, we encounter an important time constant called the minority carrier "lifetime." A classical experiment sheds some "light" on the dynamics of reestablishing the equilibrium.

Figure 3.13a shows a thin section of n-type semiconductor that constitutes resistor R in a voltage divider. The material is briefly illuminated by a pulse of light of intensity

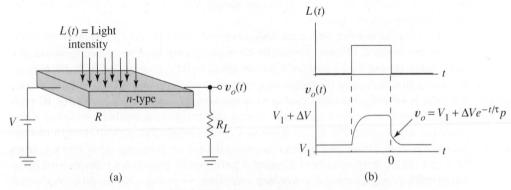

Figure 3.13   Experiment to measure carrier lifetime: (a) physical setup; (b) waveforms.

$L(t)$ as shown in Fig. 3.13b. Photons of light energy upset the equilibrium by continuously creating new hole-electron pairs. If $R >> R_L$, the voltage divider implies

$$v_o(t) = \frac{R_L}{R_L + R}V \approx \frac{R_L}{R}V = R_L V \frac{A}{L}q[\mu_p p(t) + \mu_n n(t)]$$

During the pulse, carrier concentrations increase, lowering R and raising the output voltage above its equilibrium value V_1 by ΔV volts, where ΔV is attributed to the photogenerated hole-electron pairs. When the pulse ends, as the *excess* carriers generated by the light recombine to reestablish the original equilibrium, an exponential decay toward the equilibrium condition occurs as in Fig. 3.13b. The time constant, τ_p, called the *minority carrier lifetime*, is a function of the material and is indicative of the time it takes for 68% of the excess carriers to recombine. A similar experiment using p-type material establishes the existence of a minority lifetime, τ_n, for electrons. In silicon, minority lifetimes range from 0.01 μs to 1 μs. We will see several important situations in which an exponential recombination process is characterized by a carrier *lifetime* when we investigate speed limitations of electronic devices.

3.4
Diffusion of Holes and Electrons

Mobile holes and electrons produce electrical currents by two distinct mechanisms: drift, which was already introduced, and diffusion, which is the subject of this section. Both processes are important in solid-state devices.

Grammar school teachers demonstrate diffusion by pouring ink into clear water. Random thermal motion of the ink molecules causes them to spread or *diffuse* in all directions until they are uniformly distributed throughout the solution. Essential to this or any diffusion process are three requirements: *mobile particles, random particle motion, and a nonuniform spatial distribution of the particles*. We have already seen that holes and electrons in a solid semiconductor satisfy the first two requirements.

3.4.1 DIFFUSION CURRENTS

If the holes and/or electrons somehow acquire a nonuniform spatial distribution, then, by virtue of their thermal energy, they diffuse like the ink molecules until their concentration is again uniform. Because holes and electrons are charged particles, however, this directed motion constitutes an electrical current called a *diffusion current*. Even when spatially concentrated, the holes and free electrons within materials remain in rather low concentrations ($\approx 10^{19}$ carriers among 10^{22} atoms in each cm^3) and are therefore widely spaced and separated from one another by lattice atoms. Therefore, electrical forces of repulsion or attraction between carriers are negligible.

In a diffusion process the number of particles that cross a unit area per unit time in the direction perpendicular to the area is proportional to the negative *concentration gradient* of the particles, where the proportionality coefficient is called the *diffusion constant*. The negative gradient means that net particle movement is always from the region of higher concentration to the region of lower concentration. In simple geometries in which the concentration varies with only one spatial variable, the gradient reduces to a simple spatial derivative. For holes with a concentration that varies only in the x direction, the case most useful in simple semiconductor geometries, the diffusion current density is

$$J_p = -qD_p \frac{dp}{dx} \text{ amperes/cm}^2 \tag{3.24}$$

where D_p, the *diffusion constant* for holes, is a property of the material. Similarly for electrons, the diffusion current density is

$$J_n = qD_n \frac{dn}{dx} \text{ amperes/cm}^2 \tag{3.25}$$

where D_n is the diffusion constant for electrons. The minus sign is missing in Eq. (3.25) because of the negative charge on the electrons and our use of conventional currents. Both D_p and D_n are functions of temperature.

Since diffusion constants and mobilities both relate to flows of charged particles through a lattice possessing thermal energy, it should not be a complete surprise to learn that they are related. Because a justification is beyond the scope of this book, we simply cite an important result known as the *Einstein relationship*,

$$\frac{D_p}{\mu_p} = \frac{D_n}{\mu_n} = V_T = \frac{kT}{q} \tag{3.26}$$

where k is Boltzmann's constant, T denotes absolute temperature, and q is the electron charge. The first equality shows how the diffusion constants and mobilities for holes and electrons relate to each other. The second shows that each ratio equals V_T, a quantity we call the *thermal voltage*. Since Boltzmann's constant is 1.38×10^{-23} J/K, the numerical value of V_T is about 1/40 V or 25 mV *at room temperature*—a number worth remembering for we shall use it often.

Exercise 3.4   In a material of 4.5 cm^2 cross section, the electron concentration at some instant is $n(x) = 10^6 e^{-x/0.001}$ cm^{-3}, when x is measured in cm perpendicular to the cross-sectional area. The concentration is uniform in the y and z directions. If the diffusion constant is $D_n = 20$ cm^2/s, determine the conventional electron current at $x = 0.005$ cm.

Ans. $I_n(0.005) = -97$ pA.

3.4.2   DIFFUSION LENGTH

One way to establish diffusion currents is to illuminate the end of a long semiconductor specimen. The classical experiment of Fig. 3.14a uses n-type material to measure a hole diffusion current indirectly. When the light is off, holes in the n-type silicon have a *uniform* concentration of $p_{n0} \approx n_i^2/N_d$, and no diffusion current exists. When we turn on the light, photogeneration of hole-electron pairs upsets the equilibrium near the illuminated surface; far away near the shaded edge, carrier concentrations are unchanged. Holes diffuse to the right because their concentrations are not uniform; however, many recombine

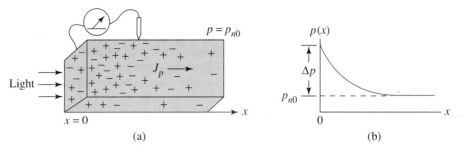

Figure 3.14 (a) Nonuniform hole concentration created by illumination of n-type silicon; (b) steady-state concentration profile of holes defining diffusion length L_p.

because concentrations exceed equilibrium values. The resulting steady-state hole concentration, $p(x)$, determined by measuring conductivity with a movable probe, is

$$p(x) = P_{n0} + \Delta p e^{-x/L_p} \tag{3.27}$$

where Δ_p is the excess minority concentration at the illuminated surface as shown in Fig. 3.14b. L_p is a "space constant" we call the *diffusion length for holes*. We can now compute the diffusion current density for holes at any point x by substituting Eq. (3.27) into Eq. (3.24).

The preceding discussion highlighted hole diffusion; however, the light also causes excess electrons to diffuse away from the lighted surface. This second current, conventionally from right to left, cancels the hole current at every x, making the net current zero, consistent with the absence of a current path outside the material. A diffusion length for electrons, L_n, is measured using illuminated p-type material. In silicon, hole and electron diffusion lengths are of the order of 1 to 100 μ, with 10 μ being typical.

EXAMPLE 3.4 For the material in Fig. 3.14, $n_i = 1.45 \times 10^{10}$ cm^{-3} and $N_d = 10^{16}$ atoms/cm^3,
(a) Compute the hole concentration before the light is turned on.
(b) If the light causes the hole concentration at the illuminated surface to increase to 11 times its equilibrium value, find the hole current density as a function of x. Assume that the diffusion length for holes is $L_P = 12 \times 10^{-4}$ cm and the diffusion constant is $D_P = 24$ cm^2/s.
(c) Compute the values of hole current at $x = 0$ and at $x = 10^{-3}$ cm if the cross-sectional area is 1.5 cm^2.

Solution. (a) Without the light, the hole concentration is the equilibrium value

$$p_{n0} = \frac{n_i^2}{N_d} = \frac{(1.45 \times 10^{10})^2}{10^{16}} = 2.1 \times 10^4 \text{ cm}^{-3}$$

(b) At the illuminated surface, $p_{n0} + \Delta_p = 11 p_{n0}$, or

$$\Delta P = 10 \, p_{n0} = 2.1 \times 10^5 \text{ cm}^{-3}$$

From Eq. (3.27)

$$p(x) = 2.1 \times 10^4 + 2.1 \times 10^5 e^{-x/0.0012}$$

From Eq. (3.24)

$$J_p(x) = -1.6 \times 10^{-19}(24)\frac{-2.1 \times 10^5}{0.0012}e^{-x/0.0012} = 6.72 \times 10^{-10}e^{-x/0.0012}$$

(c)
$$I_p(0) = 1.5 \times 6.72 \times 10^{-10}e^{-0/0.0012} = 1008 \text{ pA}$$

$$I_p(0.001) = 10^{-9}e^{-0.001/0.0012} = 435 \text{ pA} \qquad ❑$$

3.5 _____
The pn Junction in Equilibrium

The operating principles of many electronic devices critically depend upon the electrical activity at an interface between n- and p-type materials, a region only a few microns in width called a *pn junction*. A complete understanding of the junction requires principles of quantum mechanics that are beyond the scope of this text; however, this section's intuitive description gives a physical foundation that is adequate for most basic circuit analysis and design tasks. The text by Mattson, cited in the end-of-chapter references, gives a more complete mathematical description and excellent graphical aids to comprehension.

3.5.1 DIFFUSION CURRENT

An imaginary experiment introduces some important features of the pn junction. Imagine a single-crystal silicon specimen in which one-half is suddenly doped with p-type impurities and the other half with n-type impurities. Holes from the p-side initially diffuse into the n-type material where the hole concentration is low, and electrons diffuse into the p-type material, each mobile carrier leaving behind an immobile lattice ion of polarity opposite to its own. The diffusing holes and electrons, having become minority carriers, eventually recombine with other charge carriers. The result of this diffusion is a region virtually depleted of mobile carriers as shown in Fig. 3.15a, the *depletion* or *space charge region*. A continuing diffusion of charged particles across the depletion region in this fashion constitutes additive hole and electron components of the *diffusion current* I_{dif}.

The diffusion current is automatically limited by an electric field associated with the uncovered lattice ions. As a hole diffuses from the p-side to the n-side, it loses kinetic energy because of the retarding force of the electric field. The potential difference between the n- and p-sides that results from this electric field is called the *barrier voltage* V_{J0}, and is shown in Fig. 3.15b. V_{J0} is the work done or kinetic energy lost per unit charge by each hole crossing the depletion region. An analogous *barrier potential* of identical size opposes electrons diffusing in the opposite direction. Because majority carriers of widely varying thermal energies are always available, the barrier voltage functions as an *energy filter*, allowing only those carriers whose thermal energy exceeds the minimal value, V_{J0}, to climb the potential hill and contribute to I_{dif}.

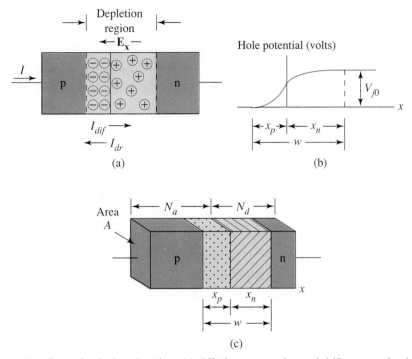

Figure 3.15 Open-circuited pn junction: (a) diffusion current, I_{dif}, and drift current, I_{dr}, in the depletion region; (b) potential hill seen by holes; (c) geometry of the depletion region.

The depletion region does not generally extend equal distances into the p- and n-materials. Figure 3.15c shows a depletion region of width $w = x_p + x_n$, where x_p denotes the p-material width and x_n the n-material width. Because each ion was abandoned by a mobile carrier that recombined with a carrier of opposite sign, the number of negative ions equals the number of positive ions. For a junction of cross-sectional area A, equating the numbers of positive and negative ions gives $N_a x_p A = N_d x_n A$, or

$$x_p = \frac{N_d}{N_a} x_n \qquad (3.28)$$

This shows that *the depletion region extends further into the more lightly doped material.*

3.5.2 DRIFT CURRENT

Opposed to the diffusion current is a drift current I_{dr}, also indicated in Fig. 3.15a. The drift current consists of thermally generated holes and electrons, continually being created throughout the material, that somehow find themselves in the depletion region. Each such hole (electron) is forced by the field into the p- (n-) material. Notice that hole and electron components of I_{dr} are both opposite in direction to corresponding components of I_{dif}.

3.5.3 BARRIER POTENTIAL AND CHARGE CONCENTRATIONS

Drift and diffusion are ongoing processes regulated by an equilibrium established in the vicinity of the junction. When there is no external current path, the current $I = I_{dif} - I_{dr}$ in Fig. 3.15a must be zero to satisfy Kirchhoff's current law. This condition establishes the junction equilibrium. The following development leads to a mathematical expression for the barrier potential when the junction is in equilibrium, and introduces some important new ideas about operation of the junction.

Solid and dashed lines in Fig. 3.16 show, respectively, the concentrations of holes and electrons at equilibrium. Because our key result depends only on carrier concentrations *outside* the depletion region, we select the x-axis scale so as to shrink the depletion region width w down to the bold vertical line. To simplify the discussion, we consider only the hole components of I_{dif} and I_{dr}.

If we assume uniform impurity concentrations in the y and z directions, from Eq. (3.24) the diffusion current is

$$I_{dif}(x) = AJ_P(x) = A\left[-qD_P \frac{dp(x)}{dx} \right]$$

From Eq. (3.8) the drift current is

$$I_{dr}(x) = A\sigma E_x = Aq\mu_p p(x)E_x$$

Equating diffusion current and drift current gives

$$-D_p \frac{dp(x)}{dx} = \mu_p p(x)E_x$$

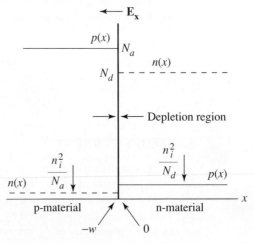

Hole and electron concentrations

Figure 3.16 Hole and electron concentrations in an open-circuited pn junction.

Dividing by $\mu_p p(x)$, substituting V_T for D_p/μ_p from Eq. (3.26), and integrating across the depletion region gives

$$-V_T\int_{-w}^{0}\frac{1}{p(x)}\left[\frac{dp(x)}{dx}\right]dx = -V_T\int_{p(-w)}^{p(0)}\frac{dp(x)}{p(x)} = \int_{-w}^{0}E_x dx \qquad (3.29)$$

The integral on the right involves summing field intensity (force per unit charge) times distance across the depletion region. This totals the work in joules done by each coulomb of charge in crossing the depletion region, or the kinetic energy it loses in climbing the barrier. This is precisely the definition of the barrier potential, V_{J0} of Fig. 3.15. Completing the integration and substituting concentration values from Fig. 3.16 at the integration limits gives

$$-V_T\ln\left[\frac{p(0)}{p(-w)}\right] = -V_T\ln\left(\frac{n_i^2/N_d}{N_a}\right) = V_{J0}$$

Dividing by $-V_T$ and taking the antilog gives

$$\frac{n_i^2}{N_d} = N_a e^{-V_{J0}/V_T} \qquad (3.30)$$

which is equivalent to

$$V_{J0} = V_T\ln\left(\frac{N_a N_d}{n_i^2}\right) \qquad (3.31)$$

Exercise 3.5 For silicon at room temperature, find the barrier voltage, V_{J0}, if
(a) $N_a = N_d = 10^{19}$ cm^{-3}.
(b) $N_a = 10^{17}$ and $N_d = 10^{19}$ cm^{-3}.
(c) For part (b), determine the relative extents of the depletion region on the two sides of the junction.

Ans. 1.02 V, 0.902 V, $x_p = 100\ x_n$.

3.6
The Junction Diode

We next introduce a nonlinear one-port device called the *junction diode*—essentially the pn junction itself. Knowing the general operating principles of the diode enables us to understand a multitude of nonlinear circuits used for diverse purposes such as changing waveform shapes, detecting radio signals, and converting ac to dc. By also examining some theoretical aspects of diode operation, we prepare ourselves for understanding more complex devices—transistors, that function as controlled switches in digital circuits and as dependent sources in analog circuits.

Figure 3.17a shows the diode's physical structure and schematic symbol. We call the p-side of the diode the *anode* and the n-side the *cathode*. The schematic also defines reference directions for positive diode current i_D and voltage v_D. These same reference directions apply to the diode's volt-ampere curve, Fig. 3.17b, and to the *diode equation*

$$i_D = I_s(e^{v_D/V_T} - 1) \qquad (3.32)$$

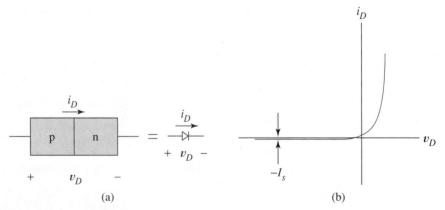

Figure 3.17 Junction diode: (a) physical structure and schematic symbol; (b) volt-ampere characteristic.

where $V_T = kT/q$ is the *thermal voltage* and I_s is called the *reverse saturation current*. We now explain the volt-ampere curve and the diode equation in terms of the pn junction theory. A qualitative discussion comes first, followed by a quantitative derivation.

3.6.1 VOLT-AMPERE CURVE AND DIODE EQUATION

Figure 3.18a shows the pn junction with its depletion region and barrier potential V_{J0}. So we can control the diode current with an external voltage, metal conductors are bonded to either side of the silicon. Contact potentials develop at the silicon-metal interfaces, a fraction γ of V_{J0} on the p-side, and the remaining $(1 - \gamma)V_{J0}$ on the n-side, with polarities as shown. Taken together, the contact potentials exactly cancel out the barrier potential, making it impossible to directly measure V_{J0}.

In Fig. 3.18b we apply an external voltage v_D to the diode; V_J denotes the barrier potential with this *bias* voltage present. From Kirchhoff's voltage law

$$V_J = \gamma V_{J0} - v_D + (1 - \gamma)V_{J0} = V_{J0} - v_D$$

For $v_D = 0$, this reduces to $V_J = V_{J0}$, the value we found for the equilibrium case of Fig. 3.15. When $v_D = 0$ then $i_D = 0$, which is consistent with this equilibrium condition.

For $v_D > 0$, the external source upsets the equilibrium by moving electrons from the p-material to the n-material through the outside circuit. This adds electrons to the n-side and holes to the p-side, neutralizing lattice ions on both sides of the junction as in Fig. 3.18b. This narrows the depletion region and thus reduces the barrier potential. Lowering V_J below V_{J0} greatly increases the diffusion current, for many additional carriers now possess sufficient energy to climb the lower potential hill, while drift current changes little. This explains physically the increased current for $v_D > 0$ in Eq. (3.32). When we make the anode so positive relative to the cathode that appreciable forward current results, we say the diode is *forward biased*. For a forward-biased diode, we approximate Eq. (3.32) by

$$i_D = I_s e^{v_D/V_T} \tag{3.33}$$

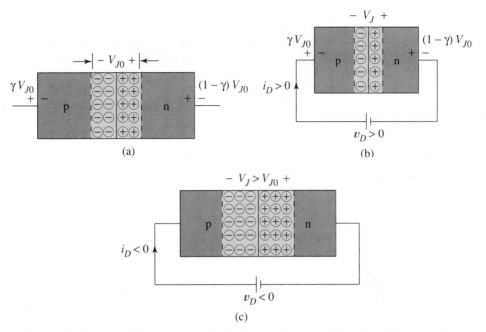

Figure 3.18 Junction diode: (a) open circuited; (b) with forward bias; (c) with reverse bias.

For $v_D < 0$, the battery moves electrons from n- to p-material, creating additional ions on both sides of the junction as in Fig. 3.18c. This *increases* the barrier potential above its equilibrium value, reducing the diffusion current; drift current, I_{dr}, again is hardly affected. If the barrier is made high enough by a large *reverse bias* voltage, the diffusion current becomes negligible and only the negative drift current flows through the diode. When the diode is strongly reverse biased, Eq. (3.32) reduces to $i_D = -I_s$.

We next justify the mathematical form of the diode equation. The result gives insight into how diode parameters are controlled during fabrication and prepares us for later studies of diode and transistor dynamic effects.

3.6.2 MINORITY CHARGE PROFILES

The diode equation arises from small concentrations of *excess minority charge* that exist within a few diffusion lengths of the depletion region. By excess we mean over and above the concentrations that exist when the diode is in equilibrium. When forward bias lowers the barrier potential, we say that additional holes from the p-side are *injected* into the n-material (where they become minority carriers); electrons are concurrently injected into the p-material. The resulting diffusion currents are just like those in the experiment described in Fig. 3.14. The only difference is that excess minority carriers are introduced by injection rather than by photons of light. On each side of the junction the new minority carriers diffuse away, declining in number by recombination and establishing exponential concentration profiles characterized by diffusion lengths L_p and L_n. This adds tiny *fillets* to the thermal equilibrium *minority* charge profiles of Fig. 3.16 as we see in

Fig. 3.19a. Again we select the x-axis scale to emphasize concentrations outside of the depletion region.

Assuming for the time being that hole current is dominant, we now derive the diode equation. Applied voltage v_D upsets the junction equilibrium; however, if v_D is not too large, *near-equilibrium* conditions apply. For such *low-level injection* we can equate diffusion and drift components of hole current and integrate across the depletion region as we did in Eq. (3.29). For this case, however, integrating the field E_x across the depletion region gives V_J instead of V_{J0}. Thus

$$-V_T \int_{p(-w)}^{p(0)} \frac{dp(x)}{p(x)} = \int_{-w}^{0} E_x \, dx = V_J = V_{J0} - v_D$$

which leads to

$$p(0) = p(-w)e^{-(V_{J0} - v_D)/V_T} = N_a e^{-V_{J0}/V_T} e^{v_D/V_T}$$

after we integrate, substitute $p(-w) = N_a$, and factor. Equation (3.30) allows us to rename the coefficient of the exponential giving

$$p(0) = \frac{n_i^2}{N_d} e^{v_D/V_T} \qquad (3.34)$$

This equation shows how the external voltage, v_D, controls the minority hole concentration, $p(0)$, at the edge of the depletion region in Fig. 3.19a.

Since the *excess* hole concentration in the n-material, $p(x) - n_i^2/N_d$ in Fig. 3.19a, decays with *space constant* L_p, we can write

$$p(x) - \frac{n_i^2}{N_d} = \left[p(0) - \frac{n_i^2}{N_d} \right] e^{-x/L_p}, \qquad x \geq 0$$

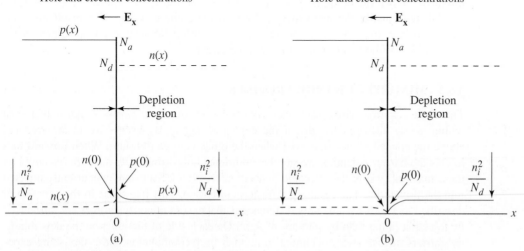

Figure 3.19 Minority stored charge concentrations for (a) forward-biased diode; (b) reverse-biased diode.

Substituting for $p(0)$ using Eq. (3.34) and solving for $p(x)$ shows that hole concentration in the n-material is

$$p(x) = \frac{n_i^2}{N_d}[e^{v_D/V_T} - 1]e^{-x/L_p} + \frac{n_i^2}{N_d}, \qquad x \geq 0 \tag{3.35}$$

Since Eq. (3.35) gives the hole concentration in the n-material as a function of x, we can use it to compute the diffusion current. By substituting $p(x)$ into Eq. (3.24) and multiplying by A to obtain current, we find that the hole component of diffusion current just outside the depletion region is

$$i_{D,p}(x) = \frac{qAn_i^2 D_p}{N_d L_p}(e^{v_D/V_T} - 1)e^{-x/L_p} \tag{3.36}$$

Current decreases with x because the flow of minority holes gradually diminishes by recombination. A stream of new electrons, supplied by the external source, flows into the n-material (from the right in Fig. 3.18b) to support this recombination. The sum of these two currents at every point x is constant. That is, hole current decreases and electron current simultaneously increases with increasing x. The easiest way to find the external diode current is to evaluate Eq. (3.36) at $x = 0$, where the electron component of the diode current is zero. This gives

$$i_D(x) = \frac{qAn_i^2 D_p}{N_d L_p}(e^{v_D/V_T} - 1) \tag{3.37}$$

an equation that describes many diodes. Let us now use the details we have already established to generalize a little.

Figure 3.19a shows that electrons are also injected into the p-material, a fact ignored in our derivation. Recombining as they diffuse away, these form a concentration profile analogous to that of the holes. They also form an electron current that adds to the hole current. Combining this additive electron current to the hole current of Eq. (3.37) gives the more complete diode equation

$$i_D(x) = I_s(e^{v_D/V_T} - 1) = An_i^2 q\left[\frac{D_P}{N_d L_P} + \frac{D_N}{N_a L_N}\right](e^{v_D/V_T} - 1) \tag{3.38}$$

In deriving Eq. (3.38) we tacitly assumed forward bias; however, it turns out that for negative bias, the minority charge distributions takes on the shapes shown in Fig. 3.19b, and the equation remains valid.

From Eq. (3.38) we learn that the diode's *reverse saturation current*, I_s, is related to physical parameters by

$$I_s = An_i^2 q\left[\frac{D_P}{N_d L_P} + \frac{D_N}{N_a L_N}\right] \tag{3.39}$$

This equation reveals some important facts about the diode. First, I_s is directly proportional to the diode area A, a parameter we can easily control during fabrication. This enables us to fabricate diodes having larger or smaller values of current for the same applied voltage. Second, notice that the highly temperature-sensitive term n_i^2 (recall Fig.

3.5) is a factor of I_s. Finally, since the two terms within brackets correspond to hole and electron currents, respectively, we see that if the p-material is doped much more heavily than the n-material, that is, if $N_a >> N_d$, then the first (hole) current dominates the operation of the diode and Eq. (3.38) reduces to Eq. (3.37). We will use this when we study diode dynamics, and again when we introduce bipolar transistors.

We made two key assumptions in deriving the diode equation. First, we ignored thermally generated holes and electrons in the depletion region itself. This leads to errors in Eq. (3.32) for low v_D. Second, we assumed *low-level injection*; that is, we assumed v_D was small enough that $p(0)$ and $n(0)$ in Fig. 3.19 were both small compared to the majority doping levels on their respective sides of the depletion region. When there is *high-level injection,* this is no longer true and Eq. (3.32) becomes inaccurate. A factor η, the *emission coefficient,* sometimes appears in the diode equation to correct these cases, giving

$$i_D(x) = I_s(e^{v_D/\eta V_T} - 1) \tag{3.40}$$

We choose the emission coefficient to give good agreement with measurements in the region of interest; useful values fall in the range $1 \leq \eta \leq 2$. Hereafter we assume $\eta = 1$ unless there is some special reason to do otherwise.

Having explained the diode's nonlinear volt-ampere characteristic, we next learn how to use the diode for nonlinear processing of signal and power waveforms.

3.7
Large-Signal Diode Models

Because the diode's mathematical description is nonlinear, we employ simplified models to help us understand and estimate how the diode operates in circuits. Whenever needed, computer programs that implement the nonlinear diode equation are available to back up our intuitive analyses and rough estimates with more accurate calculations.

The models of this chapter are called *large-signal models* because they are useful in predicting the behavior of circuits when the voltages and currents of interest are large. We use them to analyze and design digital logic circuits, power supplies, waveshaping circuits, and bias circuits. Most important conceptually is the ideal diode model.

3.7.1 IDEAL DIODE CHARACTERISTIC

The *ideal diode* elegantly characterizes the *essence* of diode operation, high forward conduction and negligible reverse conduction. Defined by

$$v_D = 0 \text{ when } i_D \geq 0$$

and

$$i_D = 0 \text{ when } v_D \leq 0$$

the ideal diode has the volt-ampere curve of Fig. 3.20a. When the diode conducts current, we say the diode is ON or in its *ON state.* When it blocks current, we say the diode is OFF or in its *OFF state.* The ideal diode emulates these states, respectively, with short and open circuits as Fig. 3.20b suggests. When $i_D = 0$ and $v_D = 0$, operation is at the *break point* of Fig. 3.20a, where the short and open circuit *both* correctly represent the ideal diode.

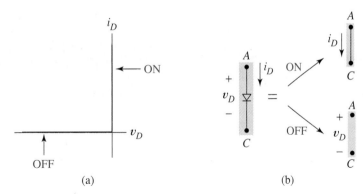

Figure 3.20 (a) Volt-ampere curve of ideal diode; (b) circuit models for ideal diode.

The ideal diode model leads to simple equivalent circuits that are easy to analyze. For $i_D > 0$ the model is somewhat inaccurate because it neglects the forward voltage drop across the real diode. Nevertheless, the model is invaluable for helping us understand unfamiliar diode circuits, and for making rough estimates of diode voltage and current.

3.7.2 dc ANALYSIS OF CIRCUITS CONTAINING IDEAL DIODES

A dc analysis using ideal diodes is a good way to begin studying an unfamiliar diode circuit. Once we understand how the circuit works for a single value of dc input voltage, it is usually easy to deduce how it works for other dc values or even for an ac input. The problem with a diode is that we do not initially know whether it is ON or OFF. The following procedure is usually effective.

1. Make an educated guess about the state of every diode.
2. Redraw the circuit diagram, replacing each ON diode with a short circuit and each OFF diode by an open circuit.
3. By circuit analysis, determine the current in each short circuit that represents an ON diode and the voltage across each open circuit that represents an OFF diode.
4. Check the validity of the assumption for each diode. If there is a contradiction—a negative current in an ON diode or a positive voltage across an OFF diode—anywhere in the circuit, return to step 1 with an improved guess.
5. When no contradictions exist, the voltages and currents we compute from the circuit approximate the true values.

The following examples illustrate the procedure.

EXAMPLE 3.5   The diode in Fig. 3.21a is ideal. Find V_A.

Solution. Figure 3.21a introduces a kind of circuit diagram widely used in electronics. Arrows indicate dc voltage constraints relative to a zero-volt ground reference. This ground reference is implied but often omitted from the diagram. Figure 3.21b is the more conventional notation for the same circuit. The compact notation conveniently

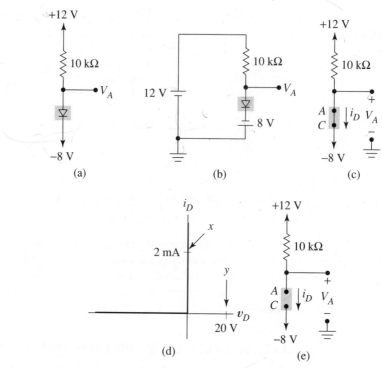

Figure 3.21 Ideal diode models of a diode circuit: (a) original circuit; (b) more familiar notation for the circuit; (c) diode assumed ON; (d) diode operating points; (e) diode assumed OFF.

shows how components are connected to *power rails,* conducting strips that route dc power to the various components on an integrated circuit chip or printed circuit board, and the notation lends itself nicely to the node analysis used extensively in electronics. We use these conventions without further comment throughout the remainder of the book. Now we return to the circuit analysis.

We expect current to flow from the +12 V supply rail, down through the diode *in the positive* reference direction, and into the −8 V supply; therefore, we *guess* that the diode is ON. Figure 3.21c shows the resulting circuit model. From this equivalent circuit the current is

$$i_D = \frac{12 - (-8)}{10\,\text{k}} = 2\,\text{mA}$$

Point x in Fig. 3.21d shows the meaning of our 2 mA result and confirms that the initial assumption was correct. From Fig. 3.21c, $V_A = -8$ V.

Although the solution is complete, let us see what would have happened had we guessed incorrectly. Figure 3.21e shows the circuit that follows from assuming that the diode is OFF. Since the resistor current is zero, $V_A = +12$ V and $V_C = -8$ V. The model therefore predicts a diode voltage of $v_D = V_{AC} = 12 - (-8) = 20$ V. But our OFF assumption requires that anode be negative relative to the cathode, hence a contradiction.

Point y in Fig. 3.21d helps us visualize what has happened. The open-circuit *model* predicts an operating point that is not on the ideal diode curve, a point at which the volt-ampere curves for the open circuit and the ideal diode are not coincident. We conclude that the OFF assumption is incorrect and rework the problem assuming the diode ON. Obviously, the way to minimize labor is to guess correctly the first time. This comes with problem-solving experience. ❑

Exercise 3.6 Find V_A if the $+12$ and -8 V sources in Fig. 3.21a are interchanged.

Ans. -8 V.

EXAMPLE 3.6 The diode in Fig. 3.22a is ideal. Find V_C.

Solution. Assume that the diode is ON and draw Fig. 3.22b. Apply Thevenin's theorem to the left of the diode and draw Fig. 3.22c. It is now clear that i_D is negative, meaning our original assumption was incorrect.

Assuming the diode OFF gives Fig. 3.22d. Because zero current flows in the circuit, $v_A = 5$ V and $v_C = 6$ V; therefore, $v_D = v_A - v_C = -1$ V. From the diagram, $V_C = +6$ V. ❑

Exercise 3.7 Find V_C if we reverse the orientation of the diode in Fig. 3.22a.

Ans. $V_C = 5.33$ V.

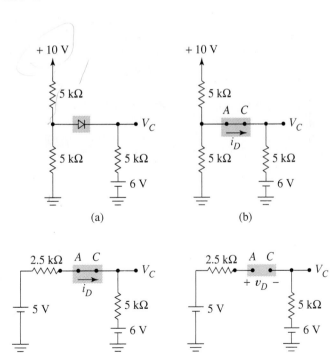

Figure 3.22 Circuit for Example 3.6: (a) original circuit; (b) circuit with diode assumed ON; (c) after applying Thevenin's theorem; (d) circuit with diode OFF.

EXAMPLE 3.7   Find I_i in Fig. 3.23a when (a) $V_i = +12$ V, (b) when $V_i = -6$ V. Assume ideal diodes.

Solution.   (a) When $V_i > 0$, we expect a positive I_i to divide, with part flowing through the 5 kΩ resistor and part through D_1, so we guess D_1 is ON. This current would continue through the 2 kΩ resistor, with D_2 turning OFF, since reverse diode current is impossible.

Figure 3.23b shows our circuit model. Since D_1 shorts out the 5 kΩ resistor, $I_i = 12/2$ k $= 6$ mA. For D_1, $i_{D1} = +6$ mA because the voltage across the 5 k resistor is zero. D_2 is OFF because $v_{D2} = -12$ V. Thus the original assumptions are verified.

(b) When $V_i = -6$V in Fig. 3.23a, we expect negative I_i. This suggests D_1 OFF and D_2 ON, giving Fig. 3.23c. From this circuit, $I_i = -6/5$ mA. For D_1, $v_{D1} = -6$ V and for D_2, $i_{D2} = +6/5$ mA. Since i_{D2} is positive and v_{D1} is negative, there is no contradiction.   ❏

Exercise 3.8   Find I_i and v_{D2} in Fig. 3.23a if $V_i = 5$ V.

Ans.   2.5 mA, −5 V.

These examples show how the ideal diode model reduces the circuit analysis problem to a familiar one that does not involve nonlinear diode equations. Also, the short or open circuits that replace the diodes often lead to simple circuits. We retain the advan-

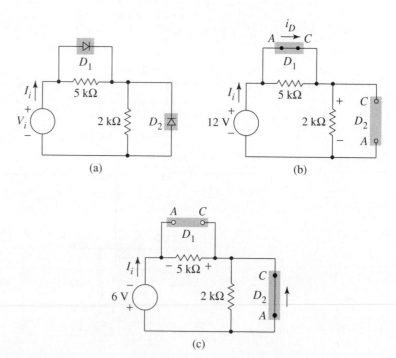

Figure 3.23   (a) Resistor-diode circuit; (b) equivalent assumed for $V_i = +12$ V; (c) equivalent assumed for $V_i = -6$ V.

tage of linear circuit analysis and achieve greater accuracy by using the more complicated models described next.

3.7.3 OTHER DIODE MODELS

Offset Model. In Fig. 3.24a, a constant voltage line approximates the diode for $i_D \geq 0$,
providing a voltage *offset* like that of a real diode. An offset voltage of 0.7 V is usually
a good approximation for a silicon pn junction carrying moderate forward current at
room temperature. (For germanium diodes, 0.25 V is more appropriate; for gallium arsenide, 1.2 V.) For $v_D \leq 0.7$V, an open circuit approximates the diode. Figure 3.24b
shows the corresponding circuit models.

We modify the procedure for analyzing ideal diode circuits slightly to fit the offset
model. ON diodes are replaced by batteries of 0.7 V (for silicon) and OFF diodes by
open circuits. Verifying an ON assumption requires that *battery current* i_D be positive.
Verifying an OFF assumption requires showing that $v_D \leq 0.7$ V, not $v_D \leq 0$ V as for the
ideal diode.

Example 3.8 repeats Example 3.5 but uses a more accurate model.

EXAMPLE 3.8 Find V_A in Fig. 3.25a using the offset model.

Solution. Figure 3.25b shows the equivalent circuit assuming the diode is ON. From
this circuit, $V_A = -8 + 0.7 = -7.3$ V, and

$$i_D = \frac{12 - (-7.3)}{10 \text{ k}} = +1.93 \text{ mA}$$

confirming that the diode is ON. Figure 3.25c shows the operating point on the idealized
diode curve. The offset did not make a great deal of difference in the answer compared
with the ideal diode estimate of Fig. 3.21d because the offset is rather small compared to
the 20 V placed across the diode-resistor combination. ❏

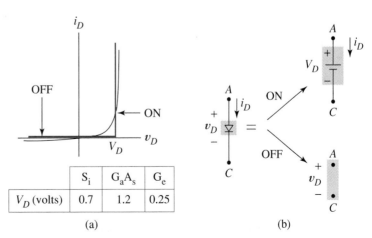

	S$_i$	G$_a$A$_s$	G$_e$
V_D (volts)	0.7	1.2	0.25

(a)   (b)

Figure 3.24 Offset model for silicon diode: (a) volt-ampere curve; (b) circuit models.

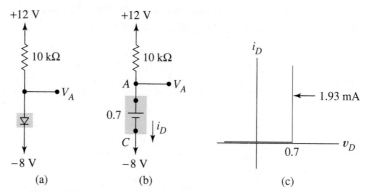

Figure 3.25 Analysis of Example 3.7 using offset diode model.

Exercise 3.9 Use the offset model to find the diode current if the (silicon) diode in Fig. 3.22a is reversed in its orientation.

Ans. 40 μA.

EXAMPLE 3.9 Assume that D_1 and D_2 in Fig. 3.23a are both idealized GaAs diodes with 1.2 V offset voltages. Find the value of I_i when (a) $V_i = 12$ V, (b) $V_i = 1.7$ V. (c) Find the smallest V_i for which the diode states of part (b) are invalid.

Solution. (a) Assuming D_1 ON and D_2 OFF gives Fig. 3.26a for $V_i = 12$ V. From the diagram,

$$I_i = \frac{12 - 1.2}{2\text{ k}} = +5.4\text{ mA}$$

The equivalent circuit also shows that

$$v_{D2} = -(12 - 1.2) = -10.8\text{ V}$$

As this corresponds to point w in Fig. 3.26b, D_2 is OFF as assumed.

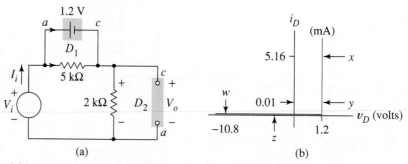

Figure 3.26 (a) Equivalent circuit for Fig. 3.23 when diodes are G_aA_s with D_1 ON and D_2 OFF; (b) operating points on diode curves.

To verify that D_1 is ON, we must show that $i_{D1} \geq 0$. From Fig. 3.26a, the current in the 5 kΩ resistor is $1.2/5k = 0.24$ mA. Since $I_i = 5.4$ mA,

$$i_{D1} = 5.4 - 0.24 = 5.16 \text{ mA}$$

The operating point for D_1 is point x in Fig. 3.26b, verifying that D_1 is ON. We conclude that

$$I_i = 5.4 \text{ mA}$$

(b) For $V_i = 1.7$ V we still expect $I_i > 0$; therefore, Fig. 3.26a is again the starting point. Analysis gives

$$I_i = \frac{1.7 - 1.2}{2 \text{ k}} = +0.25 \text{ mA}$$

The current in D_1 is

$$i_{D1} = I_i - \frac{1.2}{5 \text{ k}} = 0.01 \text{ mA}$$

placing the new operating point of D_1 at point y in Fig. 3.26b. The voltage across D_2 is

$$V_o = 1.7 - 1.2 = 0.5 \text{ V}$$

Since this corresponds to point z in Fig. 3.26b, the D_2 assumption is verified. Since both assumptions are correct, $I_i = 0.25$ mA.

(c) Notice that in part (b), reducing the source voltage caused both diode operating points to move closer to the *break point* at 1.2 V (from x to y and from w to z). As V_i is further reduced, either D_2 changes state or D_1 changes state. (Also possible but less likely is that both change state for exactly the same value of V_i.)

Let us *assume* that for some value of V_i, D_2 begins to turn ON while D_1 stays ON. For D_2 to operate at its break point, its cathode must be 1.2 V negative relative to its anode, or $V_o = -1.2$ V. To reach this point, V_i in Fig. 3.26a must be reduced to $-1.2 + 1.2 = 0$.

Next, we examine the other possibility. Assume D_1 begins turn OFF in Fig. 3.26a while D_2 remains OFF. D_1 turns OFF when its current reaches zero. This assumption gives

$$i_{D1} = I_i - \frac{1.2}{5 \text{ k}} = \frac{V_i - 1.2}{2 \text{ k}} - \frac{1.2}{5 \text{ k}} = 0$$

Solving gives $V_i = 1.68$ V.

Comparing the two results shows that as V_i is reduced from 1.7 V, the first important event is that D_1 turns OFF. This happens when $V_i = 1.68$ V, and the equivalent circuit becomes a simple voltage divider. ❏

> **Exercise 3.10** In Fig. 3.23, D_2 is reversed in polarity. Both diodes are made of germanium with 0.25 V offsets, and both are initially OFF. As V_i starts from zero and becomes

more positive, first one diode turns on, then the other. Determine the order in which the diodes turn on, and the two critical values of V_i.

Ans. D_1 turns on at 0.35 V, D_2 at 0.5 V.

Cut-in Voltage. The offset model provides reasonable estimates for a forward-biased diode at high currents; however, Fig. 3.24a shows that this model gives a large voltage error near $i_D = 0$. Sometimes a critical point in the operation of a diode circuit is the voltage at which *appreciable current begins to flow.* We call this the diode's *cut-in voltage, V_γ.* To formalize this idea, let $I_{0.7}$ denote the current in a silicon diode when its voltage is 0.7 V, and define V_γ as the voltage at which the diode current is 0.1% of $I_{0.7}$. From Eq. (3.33)

$$I_{0.7} \approx I_s e^{0.7/V_T}$$

By definition, V_γ satisfies

$$0.001 I_{0.7} \approx I_s e^{V_\gamma/V_T}$$

From the ratio

$$\frac{0.001 I_{0.7}}{I_{0.7}} = e^{(V_\gamma - 0.7)/V_T}$$

or

$$V_\gamma = 0.7 + V_T \; \ell n(0.001) \approx 0.5 \text{ volt}$$

This value is used consistently hereafter, with only rare exceptions for special cases.

The model of Fig. 3.27 approximates the ON diode by a series resistor r_f and a source V_γ. We choose the resistor value to give a good fit to the true curve over the intended range of operation. This model provides improved accuracy but further increases the complexity of hand analysis, a serious shortcoming when computer simulations are readily available. The principal merit of this model today is in the conceptual value it provides, in particular, the concept of a voltage V_γ where significant forward current begins to flow.

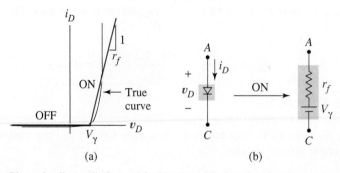

Figure 3.27 Piecewise linear diode model: (a) approximation to volt-ampere characteristic; (b) circuit model.

Beginners should not worry unduly about which diode model to use. The only rule is to use the simplest model that improves your understanding of the problem at hand. If additional precision is needed, a second analysis with a more complex model is usually straightforward. Finally, computer simulation can always be used to obtain accurate numerical values.

The next discussion introduces diode breakdown, an important event that is not included in the equations, theory, or models developed so far.

3.7.4 JUNCTION BREAKDOWN

If we apply sufficient negative voltage to a pn junction, its characteristic breaks sharply downward at almost *constant voltage* $-V_z$, as in Fig. 3.28a. V_z is called the *breakdown voltage* of the diode. The *breakdown region,* in which the diode functions as a negative voltage source, is delineated by a minimum current, I_{min}, and a maximum current, I_{max}. The former is defined by the knee of the curve, the latter by the maximum power the diode can dissipate without overheating. The diode overheats if its average power dissipation exceeds $P_{max} = V_z I_{max}$.

The schematic symbol in Fig. 3.28b denotes a diode intentionally operated in breakdown. Because the current-voltage product is positive everywhere on the diode curve, even in breakdown, the diode never delivers power to a circuit. This restricts its use as a voltage source to applications in which the source absorbs power. Examples are providing a reference voltage, compensating for temperature changes in other elements, and setting limits in waveshaping circuits.

There are two diode breakdown mechanisms, zener and avalanche. *Zener breakdown* occurs in heavily doped junctions that have very thin depletion regions, when the electric field becomes so intense that it breaks covalent bonds. This provides large numbers of additional carriers to contribute to the reverse current I_s. In *avalanche breakdown,* electrons, accelerated to high velocities by the field, break bonds by collisions thereby creating new hole-electron pairs that, in turn, break still more bonds, and so

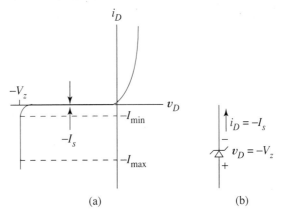

(a)   (b)

Figure 3.28   (a) Diode characteristic showing breakdown region; (b) schematic symbol for a breakdown diode.

forth—all these new carriers adding to I_s. When the breakdown voltage is less than 5.7 volts, the zener mechanism is usually responsible; for higher voltages, avalanche is probably taking place. We call diodes especially designed for breakdown operation *zener diodes,* regardless of the physical mechanism actually responsible for breakdown, and V_z is the *zener voltage.*

Zener Diode Models. Including breakdown, the diode is a *three-state device* approximated by the idealized characteristic of Fig. 3.29a. An open circuit represents the silicon diode for $-V_z \le v_D \le 0.7$ V; and voltage sources of 0.7 V and $-V_z$ replace the diode for $i_D > 0$ and $i_D < 0$, respectively. Interesting problems occasionally arise in digital logic circuits that require this three-state description.

In most applications, the zener schematic symbol indicates breakdown operation, and a circuit model needs only to approximate the diode in that region. Figures 3.29b and c show the most popular breakdown models. In both, the positive battery terminal is at the cathode because v_D is negative.

Zener Voltage Regulator. Figure 3.30 shows a zener diode in a voltage regulator circuit. The function of the diode is to maintain a constant voltage across load resistor R_L in spite of variations in R_L and V_{BB}. A key consideration in regulator design is to ensure that zener current remains within the limits $-I_{min}$ and $-I_{max}$ in Fig. 3.28a. The following example explores variations in R_L.

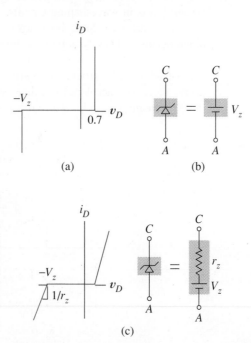

Figure 3.29   (a) Stylized diode characteristic suggesting a three-state model; (b) simple model for zener operation; (c) more complex zener model with a resistor accounting for the slope of the zener region.

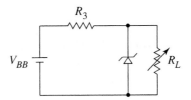

Figure 3.30 Zener voltage regulator circuit.

EXAMPLE 3.10 Use the zener model of Fig. 3.29b to find maximum and minimum values for R_L in the regulator circuit of Fig. 3.30 when $R_s = 48.7 \ \Omega$ and $V_{BB} = 15$ V. The 5 V zener has a minimum reverse current of 10 mA, and its maximum power dissipation is 1 W.

Solution. First, draw the equivalent circuit for the regulator, Fig. 3.31. For zener operation we need $I_z \geq 10$ mA. Also, since the maximum power dissipation of the diode is 1 W, it is necessary that

$$5 \, I_z \leq 1 \text{ watt}$$

or

$$I_z \leq 200 \text{ mA}$$

Notice from the equivalent circuit that I_o is constant at

$$I_o = \frac{15 - 5}{48.7} = 205 \text{ mA}$$

Thus changes in R_L are accompanied by changes in I_z. From KCL at node x,

$$I_o = 205 \text{ mA} = I_z + \frac{5}{R_L}$$

From this we deduce that the smallest R_L corresponds to the smallest I_z and conversely. Thus $R_{L,min}$ satisfies

$$205 \text{ mA} = 10 \text{ mA} + \frac{5}{R_{L,min}}$$

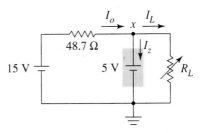

Figure 3.31 Equivalent circuit for zener voltage regulator of Fig. 3.30.

and $R_{L,max}$ satisfies

$$205 \text{ mA} = 200 \text{ mA} + \frac{5}{R_{L,max}}$$

We conclude that the regulator operates correctly for

$$25.6 \text{ }\Omega \le R_L \le 1 \text{ k}\Omega. \qquad\qquad\qquad ❑$$

Exercise 3.11 In Fig. 3.30, $R_L = 500 \text{ }\Omega$ and $R_S = 10 \text{ }\Omega$. For a 10 V zener with current limits of 20 mA and 2 mA, use the constant voltage zener model to determine the permissible limits of V_{BB}.

Ans. $10.2 \text{ V} \le V_{BB} \le 10.4 \text{ V}$.

The internal resistance, r_z, of the zener (Fig. 3.29c) causes the operation of the regulator to deviate from the idealized behavior described earlier. Because $r_z \| 48.7$ appears in the Thevenin equivalent circuit seen by R_L, zener resistance causes the output voltage to drop slightly with increases in load current.

3.7.5 VARIATION WITH TEMPERATURE

Figure 3.32a shows how offset voltage, reverse current, and breakdown voltage change with temperature for a silicon diode. Offset, V_D, decreases at 2 mV/°C. This means the battery voltage in the circuit model changes according to

$$V_D(T) = V_D(T_R) - 0.002 \, (T - T_R) \qquad\qquad (3.41)$$

where T_R is a reference temperature. From this we can estimate the temperature coefficient, α_D, of the diode offset voltage. We begin with the definition

$$\alpha_D = \frac{\Delta V_D / V_D}{\Delta T} = \frac{[V_D(T) - V_D(T_R)]/V_D(T_R)}{T - T_R} \qquad (3.42)$$

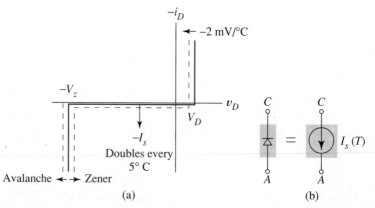

Figure 3.32 (a) Change in diode characteristics with temperature; (b) model for reverse-biased diode that includes temperature variation.

If we use room temperature for T_R, then $V_D(T_R) = 0.7$ V. The -2 mV/°C variation gives

$$\alpha_D = -\frac{0.002}{0.7} \approx -0.0029/\text{C}° = 2900 \text{ ppm/C}°$$

The reverse saturation current of the diode doubles for every 5°C increase in temperature. When this current might be important, for a diode operating over a wide temperature range for example, we replace the open circuit that models the OFF diode by an independent current source of value $i_D = -I_s$ that changes according to

$$I_S(T) = I_S(T_R) \times 2^{(T - T_R)/5} \tag{3.43}$$

Figure 3.32b shows the model.

Temperature coefficients for V_z in the range $\pm0.1\%$ per °C are typical. The sign of the temperature variation depends upon the physical breakdown mechanism. For $V_z < 5.7$ V (zener mechanism) the temperature coefficient is negative; for $V_z > 5.7$ V (avalanche) it is positive. A zener having the special value 5.7 V is favored as a reference voltage because its temperature coefficient is close to zero.

3.8
Static SPICE Model for the Diode

SPICE contains a built-in mathematical model for the diode, which reduces to Eq. (3.40) for dc analysis. SPICE also includes breakdown region calculations as a user option. The table of Fig. 3.33a shows SPICE notation and the automatically assigned *(default)* values for some of the diode parameters we have studied. Since $BV = \infty$, the *default diode* has no breakdown.

Figure 3.33c shows the SPICE code for the diodes in Fig. 3.33b. Each diode has an element line starting with a name that begins with "D." Node numbers appear next in anode-to-cathode order. An arbitrarily chosen *class name* (for example, KAY) appears next. The role of the class name is to link the diode to some particular .MODEL code line. The .MODEL line specifies the class and the type of device, and lists its parameter values.

Notation		Default
Text	SPICE	value
I_s	IS	1.0E–14 A
η	N	1
V_z	BV	∞ V
I_{min}	IBV	1.0E–3 A

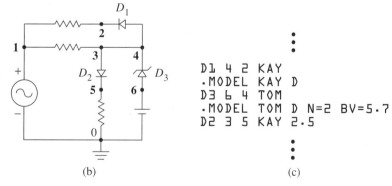

(a)   (b)   (c)

Figure 3.33 SPICE code for a diode: (a) parameter notation and default values; (b) circuit containing diodes; (c) coding for default diode (D_1), diode with nondefault values (D_3), default diode except for area (D_2).

The first code line in Fig. 3.33c states that D_1 belongs to a *class* of circuit elements named KAY. The .MODEL line that follows describes all devices that belong to class KAY. After the class name a D denotes that all KAYs are diodes. (Devices other than diodes also use .MODEL lines.) Since there is no other code on the .MODEL line, D_1 is a *default* diode whose parameter values are those listed in the last column of Fig. 3.22a.

D_3 belongs to a second class, named TOM. Notice that the .MODEL line for this class *overrides* the default values of the emission coefficient and zener voltage by listing replacement values, but retains the default value for I_s. (Parameter statements may be written in any order.) If there were other diodes identical to D_3 in the circuit, each would have an individual element line to describe its connections, but all would share the same .MODEL description by using class name TOM.

Because the scaling of reverse saturation current I_s by area A [see Eq (3.39)] is an important design technique, it is useful to be able to assign different relative areas to otherwise identical diodes. SPICE employs an optional element-line AREA factor for this purpose. This number, added on at the end of the element line, tells SPICE the number of parallel diodes of the type described in the .MODEL line it should use for the present diode description. For example, in Fig. 3.33c the element line of D_2 associates this diode with class KAY; however, the 2.5 at the end indicates that the area of D_2 is 2.5 times that of D_1. When an area is indicated on the element line, SPICE appropriately scales all pertinent diode parameters, including I_s.

SPICE assumes that all input data apply to devices at room temperature, 27°C, and always performs a dc analysis for this temperature. If we include an additional line of code specifying some other temperature in degrees Celsius, for example .TEMP = 50, SPICE calculates parameter values for this temperature and performs a second dc analysis.

The ready availability of nonlinear SPICE models extends our problem-solving ability, as the following examples suggest.

EXAMPLE 3.11   In Fig. 3.34a, D_2 has reverse saturation current of 3×10^{-15} A. D_1 is identical except that its area is 9.5 times the area of D_2. Find the node voltages V_a and V_b and the diode current at $T = 27°C$ and at $T = 100°C$.

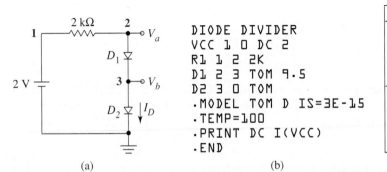

```
DIODE DIVIDER
VCC 1 0 DC 2
R1 1 2 2K
D1 2 3 TOM 9.5
D2 3 0 TOM
.MODEL TOM D IS=3E-15
.TEMP=100
.PRINT DC I(VCC)
.END
```

T	Answers
27°C	$V_a = 1.2626$ V $V_b = 0.6604$ V $I_D = 3.687 \times 10^{-4}$ A
100°C	$V_a = 1.0069$ V $V_b = 0.5397$ V $I_D = 4.965 \times 10^{-4}$ A

(a)	(b)	(c)

Figure 3.34   Example using SPICE diode model: (a) circuit; (b) SPICE code; (c) analysis results for 27°C and for 100°C.

Solution. The SPICE code is Fig. 3.34b, the answers obtained from the SPICE print-out are Fig. 3.34c.

Because of different junction areas, the 27°C data show the two diode drops are not identical even though diode currents are the same. Because of the low current, both drops are smaller than the 0.7 V we employ in our simple linear models.

In the second answer set, the diode voltages both decreased with increased temperature (as suggested by Fig. 3.32), but by a little less than our 2mV/°C rule of thumb suggests. SPICE results usually better reflect the actual performance of a physical circuit than our estimates. ❏

Exercise 3.12 In Fig. 3.34c, assume the 27°C values for V_a and V_b are correct. Compute the values we expect at 100°C if the diode drops increase by -2 mV/°C.

Ans. $V_b = 0.5144$ V, $V_a = 0.9706$ V.

Example 3.11 exemplifies problems that are conceptually simple but not easily solved by hand analysis. To solve this problem without simulation software, we would have to estimate the current, compute corresponding diode voltages by Eq. (3.32) using individual I_s values, recompute the current from the new voltages, and continue computing recursively in this way until the solutions converged. SPICE does such a recursion, but in a manner transparent to the user.

EXAMPLE 3.12 Figure 3.35a shows the voltage regulator of Example 3.10 with a dc current source replacing the load resistor. Use SPICE to plot load voltage versus load current, I_L, as load current ranges from zero to 300 mA. Explain the results.

Solution. Figures 3.35a–c show the simulation circuit, the SPICE code, and the simulation results. In Example 3.10 we found that the 15 V supply delivers a constant current of 205 mA. When the load current is zero, all this supply current flows through the diode. As I_L increases, the diode current diminishes. When load current equals supply current minus minimum diode current, $205 - 2 = 203$ mA, we expect the diode operating point to come out of the breakdown region and enter the constant current region of Fig. 3.28a. ❏

Exercise 3.13 Estimate the value of load current, I_L, for which the curve of Fig. 3.35c would begin to drop off if the 15 V source were replaced by a 10 V source. Use a SPICE simulation to check your result.

Ans. 101 mA, change source line to VBB 2 0 DC 10, Fig. 3.35d.

3.9
Nonlinear Waveshaping Circuits

We next study a circuit class that exploits diode nonlinearities to change the shapes of time-varying waveforms. Recall that a linear circuit responds to an input sine wave by producing an output sine wave of the *same frequency,* possibly differing in magnitude and/or phase. By contrast, the output of a *waveshaping* circuit contains *new frequencies,* each a multiple of the frequency of the input sine wave. *In an amplifier* output harmonics are not desired, and we call them distortion. In waveshaping, generating harmonics is the point. Sinusoids are not the only useful inputs for waveshaping circuits; it is both feasible and useful to shape general periodic and even nonperiodic input signals.

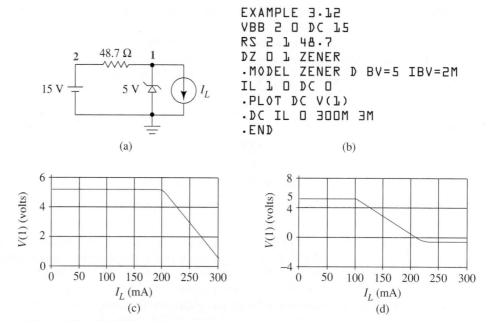

```
EXAMPLE 3.12
VBB 2 0 DC 15
RS 2 1 48.7
DZ 0 1 ZENER
.MODEL ZENER D BV=5 IBV=2M
IL 1 0 DC 0
.PLOT DC V(1)
.DC IL 0 300M 3M
.END
```

Figure 3.35 SPICE simulation of voltage regulator: (a) circuit; (b) SPICE code; (c) output voltage plot; (d) output voltage for Exercise 3.13.

We study a waveshaping circuit by comparing its time-varying input and output waveforms. Often we begin with a simple input signal such as a sine wave, and use circuit analysis with ideal diodes to determine the nature of the output waveform. To generalize such an analysis to any input signal, we use a nonlinear _static transfer characteristic,_ a simple point-by-point plot of output variable versus input variable. Once we know the transfer characteristic, we use the waveform projection technique of Fig. 1.16 to understand how the circuit processes _any_ input waveform. The transfer characteristic enables us to _classify_ a waveshaping circuit according to the character of its waveform processing.

The ideal diode is the most effective tool for exploring an unfamiliar waveshaping circuit; however, transfer characteristics based on ideal diode analysis may not be sufficiently accurate in some situations. In such cases, a second analysis using diode offset models is usually straightforward; computer simulations allow even more accurate determinations by including the exact diode nonlinearities.

While waveshaping circuits made up only of diodes, resistors, dc sources, and sometimes capacitors are widely used, operational amplifiers introduce additional benefits. Op amps can compensate for diode deficiencies, provide low-resistance outputs for convenience in cascading circuits, and elegantly combine outputs of nonlinear internal circuits to produce the desired overall nonlinear operation. Once we master circuits of resistors, sources, capacitors, and diodes, we examine circuits that also include op amps.

Limiter/Clipper Circuit. Figure 3.36a shows a circuit known as a limiter or a clipper. When $v_i > 0$, forward diode current flows, and the diode becomes the short circuit of

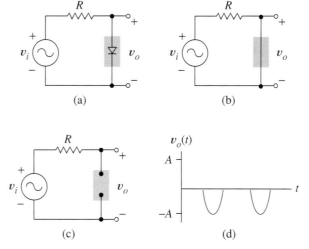

Figure 3.36   Limiter/clipper circuit: (a) circuit schematic; (b) circuit model for $v_i(t) \geq 0$; (c) circuit model for $v_i(t) \leq 0$; (d) output waveform for sinusoidal input.

Fig. 3.36b. When v_i is negative, the open circuit of Fig. 3.36c results. From these figures we conclude that

$$v_o = 0, \qquad \text{for } v_i \geq 0 \qquad (3.44)$$

$$v_o = v_i, \qquad \text{for } v_i \leq 0 \qquad (3.45)$$

These equations imply that if $v_i(t)$ is the time function

$$v_i(t) = A \sin \omega t$$

then $v_o(t)$ is the waveform of Fig. 3.36d.

To better visualize the waveshaping operation performed by the circuit, we sketch its static transfer characteristic, the graphical representation of Eqs. (3.44) and (3.45), shown as Fig. 3.37a. The sharp corners that appear in transfer characteristics at the point where diodes change state are called *break points*.

Figure 3.37b shows how the circuit processes a sine wave. Notice that the output amplitude is *limited* to the range $-\infty < v_o \leq 0$, a result the circuit achieves by *clipping* off the positive part of the input waveform. It should be obvious that *any* input waveform applied to the circuit will emerge with its positive values clipped off. In terms of the Fourier series concept of Sec. 1.6.5, the limiter produces a periodic output waveform having fundamental frequency equal to the frequency of the input sinusoid. The output waveform also contains many harmonic frequencies that were not present in the input waveform.

Reversing the sense of the diode in Fig. 3.36a causes the circuit to clip off the *negative* part of the waveform, a result easily confirmed by applying our original reasoning to the modified circuit.

Figure 3.38a shows a clipper that includes a source V_x. If the diode is ideal, it conducts current only when $v_i \geq V_x$. Then $v_o = V_x$ as in Figure 3.38b. When $v_i \leq V_x$, the

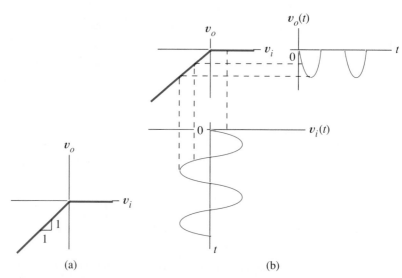

Figure 3.37 Limiter/clipper: (a) transfer characteristic; (b) waveshaping for sinusoidal input signal.

diode opens as in Fig. 3.38c, and the output is $v_o = v_i$. The transfer characteristic described by these inequalities, Fig. 3.38d, shows that for a time-varying wave, $v_i(t)$, the circuit clips off that part of the waveform more positive than V_x and reproduces the remainder. This result, valid for positive, negative, or zero V_x, is modified in an interesting way by reversing the diode polarity. Several arrangements of the circuit provide different but related clipping functions. Problem 3.43 explores the possibilities.

Figure 3.39a shows a more complicated limiter. When there are two diodes it is a little harder to see how to begin an ideal diode analysis. What often works is to select a particular value of v_i, such as zero, for a starting point. Once we draw the equivalent circuit, the next step becomes apparent.

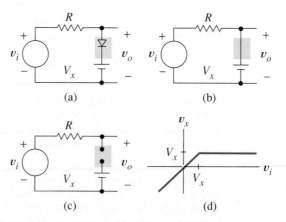

Figure 3.38 Limiter including an independent source: (a) circuit schematic; (b) with ideal diode conducting; (c) with ideal diode OFF; (d) transfer characteristic.

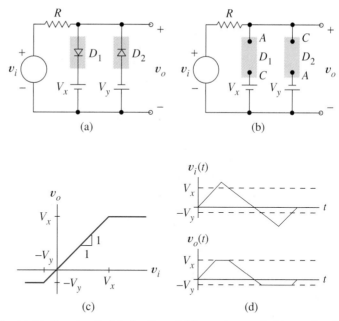

Figure 3.39 (a) Limiter circuit; (b) circuit model for v_i close to 0; (c) transfer characteristic; (d) input and output waveforms.

In Fig. 3.39a assume that $v_i = 0$. Since V_x and V_y both try to force diode currents in the negative diode direction, we guess that both diodes are OFF and draw Fig. 3.39b. When $v_i = 0$ in this circuit, no current flows in R, each anode is negative relative to its cathode, and we see that our guess was correct for $v_i = 0$. Furthermore, the transfer characteristic is $v_o = v_i$ whenever Fig. 3.39b is valid.

Examining Fig. 3.39b more carefully, we notice that D_1 and D_2 both remain OFF as v_i increases to V_x. For $v_i \geq V_x$, D_1 turns ON producing a short circuit that makes $v_o = V_x$. Once D_1 conducts, more positive values of v_i increase the current in R and D_1, but do not affect v_o. Thus the output is constant at V_x and D_2 stays OFF.

As v_i becomes negative, D_2 does not turn on until $v_i = -V_y$. For $v_i \leq -V_y$, D_2 becomes a short circuit and v_o equals $-V_y$. The transfer characteristic of Fig. 3.39c summarizes these results.

Figure 3.39d shows a triangular input $v_i(t)$ and the resulting output $v_o(t)$. Those portions of the input waveform above V_x and below $-V_y$ are clipped off by the nonlinear circuit, limiting the output waveform to the range $-V_y \leq v_o \leq V_x$. Notice that the values of V_x and $-V_y$ are arbitrary design parameters because they depend upon the sources used in the circuit.

Since any waveform that lies between the two thresholds emerges without change, this limiter can be used to protect circuits from excessive voltage amplitudes.

Effect of Diode Offset Voltage. Now let us determine the practical effect of diode offset voltage in a clipping circuit. If we use the voltage offset model of Fig. 3.24 to analyze the clipping circuit of Fig. 3.36a, it takes $v_i > 0.7$ V to turn on the diode. The equivalent circuit for the ON condition is Fig. 3.38b, with $V_x = 0.7$ V. That is, we can regard V_x as the battery in the offset model. When $v_i < 0.7$ V, the diode would open, giving a circuit like that in Fig. 3.38c, but without V_x. Thus the transfer characteristic is that of Fig. 3.38d, with $V_x = 0.7$ V. Using a diode with offset in a clipper like Fig. 3.38a, effectively adds 0.7 V to V_x when the diode conducts, causing the circuit to clip at $V_x + 0.7$ V rather than at V_x. In the limiter of Fig. 3.39, diode voltage offsets simply increase V_x and V_y by 0.7 V.

Actual diodes, having exponential volt-ampere characteristics, give transfer characteristics similar to the idealized ones, but with smooth transitions replacing the abrupt breakpoints.

Half-Wave Rectifier. The *half-wave rectifier* circuit of Fig. 3.40a converts ac voltage into pulsating dc. If the diode is ideal,

$$v_o = v_i \text{ when } v_i \geq 0$$

and

$$v_o = 0 \text{ when } v_i < 0$$

These equations define the transfer characteristic of Fig. 3.40b. The projection shows that a sinusoidal input results in a pulsating dc output waveform called a *half-wave rectified sine wave*. Because the object is to change ac into dc, the average value or *dc component* is the part of the output that is of interest. The remainder is a superposition of sine waves of harmonic frequencies that can be removed with a filter.

To quantify the rectifier's success in producing dc, we compute the *average* of $v_o(t)$. By definition, this is the area under $v_o(t)$ over the period, T, divided by the period. For $v_i(t) = V_M \sin \omega_o t$, the average is

$$V_{dc} = \left[\frac{1}{T} \int_0^{T/2} V_M \sin \omega_o t \, dt + \int_{T/2}^{T} 0 dt \right] = \frac{V_M}{\pi} \qquad (3.46)$$

where period T is related to ω_o by $\omega_o = 2\pi f = 2\pi/T$.

To determine the effect of diode offset on rectifier output, we use the offset model. When the diode conducts, this gives Fig. 3.40c. The diode now conducts only for $v_i \geq 0.7$ V, giving $v_o = v_i - 0.7$ as in Fig. 3.40d. Only that part of the input waveform larger than 0.7 V appears at the output. More specifically, diode conduction begins at time t_1 and ends at t_2 where

$$V_M \sin \omega_o t_1 = 0.7 \quad \text{and} \quad V_M \sin \omega_o t_2 = 0.7$$

Thus t_1 and t_2 are distinct principal value solutions of

$$t = \frac{1}{\omega_o} \sin^{-1} \left(\frac{0.7}{V_M} \right)$$

Since current conduction begins at angle $\omega_o t_1$ and ends at $\omega_o t_2$, the difference $\omega_o t_2 - \omega_o t_1$ is called the *conduction angle*. Since the area under $v_o(t)$ in Fig. 3.40d is smaller than in Fig. 3.40b, the dc output component is lower when the rectifier uses a real diode. To find the average for any V_M we replace the limits on the first integral of Eq. (3.46) by t_1 and t_2.

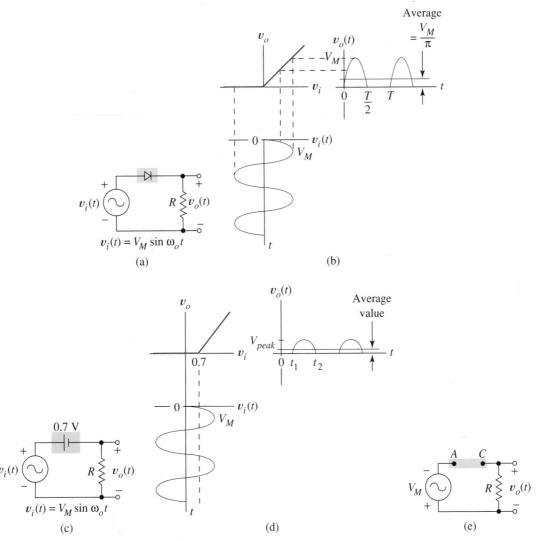

Figure 3.40 Half-wave rectifier: (a) circuit; (b) transfer characteristic and waveform processing; (c) offset model assuming the diode conducts; (d) transfer characteristic and waveforms when diode offset voltage is included; (e) diagram used to determine PIV required of the diode.

In some applications, $V_M >> 0.7$ V, and the error involved in ignoring the reduced conduction angle is negligible.

Diodes used in practical rectifier circuits must satisfy a specification called the *peak-inverse-voltage* (PIV) rating to ensure that the diode does not break down when the input voltage swings negative. To examine PIV, notice from Fig. 3.40e that when $v_i(t)$ reaches its negative peak, no current flows in R and $v_A - v_C = v_D = -V_M$ volts. Thus we must select a diode such that its rated PIV $> V_M$ to prevent breakdown.

Diodes selected for rectifiers must also meet a *peak current* specification. Notice from Fig. 3.40a that when the diode is ON, its peak current is V_M/R. If we include offset, the peak current for silicon is

$$i_{peak} = \frac{V_M - 0.7}{R}$$

In Chapter 10, we find that the rectifier is a key subcircuit in the power supplies that provide the dc power required by electronic circuits; however, the rectifier also has important signal processing applications. An automatic volume control circuit in a radio uses a rectifier to produce a dc voltage related to signal strength. When atmospheric conditions cause the signal to fade, the reduced dc component is used to increase, automatically, the gain of an amplifier that processes the signal. When the signal is too strong, the circuit automatically reduces gain to prevent amplifier saturation. The rectifier also removes information from high-frequency carrier signals, and is central to the operation of the inexpensive analog ac measuring instrument we discuss next.

Analog ac Ammeter. A dc meter movement produces an indicator needle deflection that is directly proportional to meter current. When the meter current is the pulsating dc output of a rectifier, inertia prevents the indicator needle from following the instantaneous variations in the signal. Instead, the observed deflection is proportional to the average current. Equation (3.46) shows that with an ideal diode the average component of the rectifier output is proportional to the amplitude of the input sinusoid, V_M; thus needle deflection is proportional to the amplitude of the ac signal, giving an ac voltmeter.

EXAMPLE 3.13 We need an analog meter to measure sinusoidal voltages with peak values ranging from 0 to 400 V. The circuit is Fig. 3.40a with $R = R_x + r_m$, where r_m is the 10,000 Ω resistance of an available meter movement and R_x is an unknown series resistor. The diode is ideal. Determine the value of R_x if the 400 V sinusoid is to result in a peak dc meter current of 2 mA.

Solution. When we apply a sinusoid of $V_M = 400$ V, from Eq. (3.46) the dc voltage across R is $400/\pi = 127.3$ V. The 2 mA dc meter current comes from

$$2 \times 10^{-3} = \frac{127.3}{R_x + 10,000}$$

Solving gives $R_x = 53.7$ kΩ.   ❏

Exercise 3.14 *Compute* the dc current in the meter of Example 3.13 when $v_i(t) = 50 \sin \omega t$.

Ans. $I_{meter} = 0.250$ mA.

Diode offset severely limits the usefulness of rectifiers in signal processing applications, because with offset there is no longer proportionality between input and average output amplitude. We will soon learn to use op amps to compensate for diode offsets in such applications.

Battery Charger. Figure 3.41a is a simple battery charger circuit that charges V_{BB} from the ac source

$$v_i(t) = V_M \sin \omega t = V_M \sin \theta \tag{3.47}$$

The ideal diode conducts when $v_i \geq V_{BB}$ and turns OFF for $v_i \leq V_{BB}$, giving a battery current of

$$i = \frac{v_i - V_{BB}}{R}, \qquad \text{for } v_i \geq V_{BB}$$

and

$$i = 0, \qquad \text{for } v_i \leq V_{BB} \tag{3.48}$$

The transfer characteristic of the circuit, with battery current i the output variable, is the plot of Eqs. (3.48), sketched in Fig. 3.41b. This figure also shows graphically how the nonlinear transfer characteristic shapes the battery current. In each input cycle of 2π radians, the diode conducts current only for $\theta_C = \theta_2 - \theta_1$ radians. It is the average, I_{DC}, of the current waveform that recharges the battery.

In a design problem, a given value for the charging current I_{DC} is a typical starting point. Thus we require a design equation that relates I_{DC} to the circuit parameters R and V_M. From Fig. 3.41b, average battery current is

$$I_{DC} = \frac{1}{2\pi} \int_{\theta_1}^{\theta_2} \frac{V_M \sin \theta - V_{BB}}{R} d\theta \tag{3.49}$$

where we obtain the integrand by substituting v_i from Eq. (3.47) into Eq. (3.48). The integration limits are those angles θ_i that satisfy

$$V_M \sin \theta_i = V_{BB} \tag{3.50}$$

The following example shows how these circuit equations relate to a design problem.

EXAMPLE 3.14 Design a battery charger to deliver $I_{DC} = 4$ A to a 12 V battery. Design implies finding the diode specifications, PIV and peak current, in addition to R and V_M. Assume that our equations, based upon an ideal diode, are sufficiently accurate.

Solution. The available design equations, Eq. (3.49) and (3.50), do not indicate a direct way to proceed to a solution. In such cases, we begin with an educated choice. Once the design is completed and we see the consequences, we can redesign, if necessary, using hindsight to improve on our initial choice.

If we knew θ_1 and θ_2, we could use (3.50) to find V_M, since $V_{BB} = 12$ V. Then, with known values for V_M and I_{DC}, we could solve Eq. (3.49) for R. Figure 3.41b shows that θ_1 and θ_2 are symmetric relative to $\pi/2$, or 90°. So we bravely but arbitrarily select $\theta_1 = 20°$ and $\theta_2 = 160°$. With $V_{BB} = 12$ V, Eq. (3.50) is

$$V_M \sin 20° = 12$$

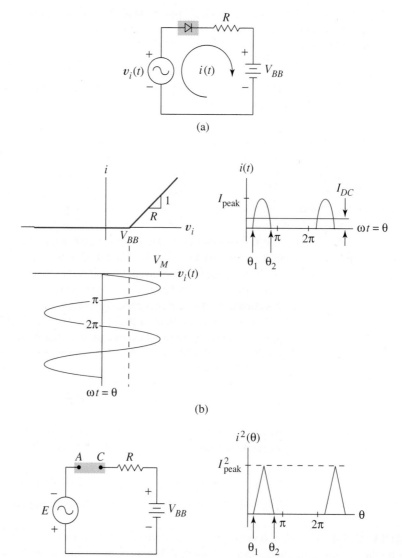

Figure 3.41 Battery charging circuit: (a) schematic; (b) transfer characteristic and waveforms; (c) equivalent for determining PIV; (d) sketch used to estimate power dissipation in R.

which gives $V_M = 35.1$ V. For a 4 A charging current, Eq. (3.49) requires

$$R = \frac{1}{4 \times 2\pi} \int_{0.349}^{2.79} (35.1 \sin \theta - 12) d\theta$$

where the angles are given in radians. Integrating gives

$$R = \frac{1}{8\pi}(-35.1 \cos \theta - 12\,\theta) \Big|_{0.349}^{2.790} = 1.46 \; \Omega$$

This is a smaller value than we ordinarily encounter in electronic circuits, but not unreasonable for a power application. We next find the requirements for the rectifier diode.

In Fig. 3.41b we see that the peak diode current occurs when v_i is maximum, at 35.1 V. With the ideal diode a short circuit,

$$I_{peak} = \frac{35.1 - 12}{1.46} = 15.8 \text{ A}$$

Figure 3.41c shows that in this circuit the diode must withstand a reverse voltage of $V_M + V_{BB} = 35.1 + 12 = 47.1$ V. A prudent designer adds another 20% or so to both PIV and I_{peak} ratings as a safety factor. This completes the initial design.

Now let us examine the arbitrary decision we made about conduction angles. Because the current in Fig. 3.41b must be 4 A, choosing a smaller conduction interval increases the peak current, since the area under the current waveform must be the same. But with $V_{BB} = 12$V in Eq. (3.50), this also implies a larger value of V_M, increasing the PIV rating required of the diode. Roughly speaking, smaller conduction intervals require more expensive diodes. The original conduction angle appears to be a satisfactory compromise. ❏

A good engineer learns to be careful, *actively searching* for possible problems in a design even though they may not be covered by the specifications. For example, it is a good idea to be concerned about power ratings of resistors in circuits that have high currents. To determine the power that resistor R must dissipate in the preceding example, we need to find the rms value, I_{rms}, of the resistor current and then use $I_{rms}^2 R$. To make a rough estimate while avoiding some integration, approximate the squared current pulses from Fig. 3.41b by triangles of height $(15.8)^2$ A^2 having bases of $(160° - 20°)$ as in Fig. 3.41d. The average squared value of i for this waveform is

$$I_{rms}^2 = \frac{\frac{1}{2}(160° - 20°)(15.8)^2}{360°} = 48.5 \text{ A}^2$$

or, $I_{rms} = 6.96$ A. This gives estimated average power dissipation in the 1.46 Ω resistor of

$$P_R = 48.5 \times 1.46 = 70.8 \text{ W!}$$

Overlooking such a detail might lead to embarrassment, as an underrated resistor can vaporize when the circuit is turned on.

In design there are often many possible approaches and many acceptable solutions. For example, since V_M was not specified, we could have begun by setting V_M to 170 V, the peak value of a 120 V rms power line voltage. Equation (3.50) would then have established the integration limits and led to R. Rectifier circuits for battery chargers that operate from ac line voltage are discussed after we introduce another important power component, the transformer.

Exercise 3.15 Determine R, I_{peak}, PIV, and resistor power rating if $V_M = 170$ V in the design problem of Example 3.14.

Ans. 12.1 Ω, 13.1 A, 182 V, 494 W.

Transformer. The *transformer* of Fig. 3.42a often provides the input voltage for rectifier circuits. The transformer consists of two windings of wire, a *primary* winding of N_1 turns and a *secondary* of N_2 turns, both wound around a common core of magnetic material. Magnetic flux ϕ, created in the core by ac currents i_1 and i_2, transfers power magnetically between the primary and secondary sides even though there is no direct electrical connection. When N_1 and N_2 differ, v_1 and v_2 differ in the same ratio. This makes it possible to transform ac voltage to lower levels before converting to dc, a feature that enables us to produce low dc voltages efficiently from the 120 V power lines in applications such as battery chargers, calculators, answering machines, and electric razors.

The *electrical isolation* a transformer imposes between the primary and secondary circuits is useful in reducing the possibility of electrical shock. A person touching circuit components in a low-voltage secondary circuit is never exposed to the higher voltages that might exist in the primary circuit. The transformer similarly isolates delicate electronic devices from possible damage through accidental connection to high voltages. (We see another isolation method, *optical isolation,* later in the chapter.) Transformers also have applications in high-frequency communication circuits, where the core is air rather than ferromagnetic material. An idealized model, the *ideal transformer,* captures the essence of the transformer.

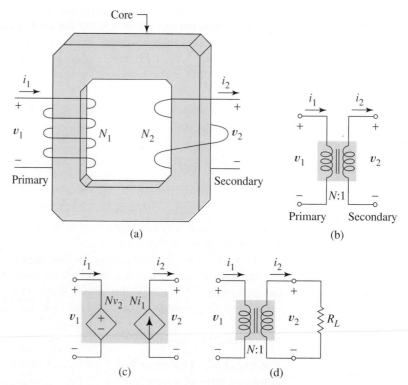

Figure 3.42 Transformer: (a) physical construction; (b) ideal transformer schematic; (c) SPICE model for ideal transformer; (d) circuit showing resistance transformation.

Ideal Transformer. The ideal transformer, schematically represented as in Fig. 3.42b, has a single parameter called the *turns ratio, N,* and two equations:

$$v_1 = Nv_2 \tag{3.51}$$

and

$$i_2 = Ni_1 \tag{3.52}$$

When electrical power moves from primary to secondary, if N is greater than one, voltage is stepped down and current is stepped up, both by the factor N. This implies that at every instant, input power $v_1(t)i_1(t)$ exactly equals the output power, $v_2(t)i_2(t)$, making the ideal transformer a lossless device. The primary receives power at high voltage and low current and then delivers the power at low voltage and high current. Figure 3.42c is a convenient SPICE model of an ideal transformer, which uses dependent sources to implement Eqs. (3.51) and (3.52).

Another feature, especially useful in high-frequency electronics, is the transformer's ability to transform resistances. From Fig. 3.42d and the transformer equations,

$$\frac{v_1}{i_1} = \frac{Nv_2}{(1/N)i_2} = N^2 \frac{v_2}{i_2} = N^2 R_L$$

We conclude that at the primary side, the circuit sees an input resistance of

$$R_{in} = N^2 R_L$$

The same kind of analysis shows that a resistor R_s attached at the primary transforms into

$$R_{out} = \frac{1}{N^2} R_s$$

Real Transformers. Real transformers also contain parasitic inductance, resistance, and capacitance in both primary and secondary circuits. Circuits that employ transformers with iron cores must have average currents of zero in both primary and secondary circuits to prevent *saturation* of the magnetic circuit, a nonlinear phenomenon that degrades performance and results in high power dissipation. In the rectifier circuits that follow, the dc currents in the transformers are zero.

Full-Wave Rectifier. Figure 3.43a shows a bridge-type *full-wave rectifier.* It is called a *bridge rectifier* because its structure is similar to the bridge circuit of Fig. 1.26c. We learn why it is called full wave by finding its output waveform.

When $v_i(t) \geq 0$, assume D_1 and D_2 are both ON while D_3 and D_4 are OFF. This gives Fig. 3.43b, where arrows show the direction of conventional current, and confirm that D_1 and D_2 are ON. The cathodes of D_3 and D_4 are both positive relative to their anodes, consistent with assuming D_3 and D_4 OFF. We conclude that

$$v_o(t) = v_i(t), \qquad \text{for } v_i(t) \geq 0$$

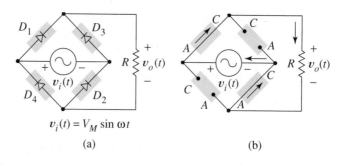

$$v_i(t) = V_M \sin \omega t$$

(a)   (b)

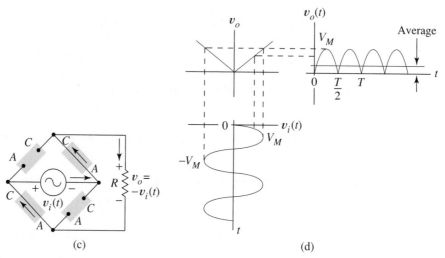

(c)   (d)

Figure 3.43 Bridge-type full-wave rectifier: (a) circuit diagram; (b) circuit when $v_i(t) > 0$; (c) circuit when $v_i(t) < 0$; (d) output waveform.

For $v_i(t) \leq 0$, assume that D_3 and D_4 conduct while D_1 and D_2 open as in Fig. 3.43c. Conventional current flows as indicated by the arrows because $v_i(t) < 0$. We conclude that

$$v_o(t) = -v_i(t), \qquad \text{for } v_i(t) \leq 0$$

Combining the two input/output equations gives

$$v_o = |v_i(t)|$$

the *absolute value* transfer characteristic of Fig. 3.43d. This figure also shows the input and output waveforms for a sinusoidal input. Comparing the output with the half-rectified wave, we find the area twice as large for the same period. Thus the dc component of the output waveform is

$$V_{DC} = \frac{2V_M}{\pi} \tag{3.53}$$

It is easy to see from Figs. 3.43b and c that the PIV rating for each diode is V_M.

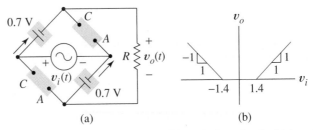

(a)   (b)

Figure 3.44   Effects of diode offset voltages on: (a) equivalent circuit; (b) transfer characteristic.

Although the ideal full-wave rectifier produces twice as much dc as the half-wave circuit, it has some disadvantages. The most obvious is the requirement for four diodes rather than one. Less obvious is the effect of diode offsets when real diodes are used. Notice in Fig. 3.44a that when real diodes conduct, both offsets subtract from the source voltage. For silicon diodes,

$$v_o(t) = v_i(t) - 1.4$$

when $v_i(t) \geq 1.4$ V, causing D_1 and D_2 to conduct. Similarly,

$$v_o(t) = -v_i(t) - 1.4$$

for $v_i(t) \leq -1.4$, when D_3 and D_4 conduct. The resulting transfer characteristic, Fig. 3.44b, has an input *dead zone* of width 2.8 V for which output is zero. This gives shorter conduction angles and a reduction in the dc output component compared to the theoretical value. By contrast, the half-wave rectifier introduces only a single diode offset when it conducts.

Notice in Figs. 3.43b and c that the *source current* alternates in direction, resulting in an average of zero. This means we can derive $v_i(t)$ from a real transformer with iron core. Figure 3.45a shows how to use a transformer to increase or reduce ac voltage levels and provide electrical isolation between input and output circuits. The primary voltage of peak value, V_P, is converted into a secondary voltage source of peak value $V_M = V_p/N$. The output equivalent circuit of Fig. 3.45b is the same as Fig. 3.43a.

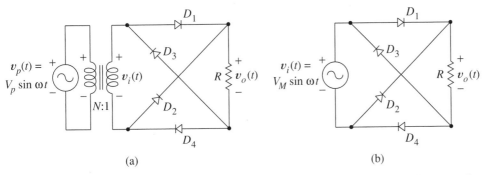

(a)   (b)

Figure 3.45   Bridge rectifier with input transformer: (a) complete circuit; (b) equivalent secondary circuit.

EXAMPLE 3.15 Design a bridge rectifier with transformer input that produces a dc output voltage of 9 V from the 120 V (rms) power line voltage. Select the ideal transformer and diodes such that the peak diode current does not exceed 20 mA. Specify the value of R and the minimum PIV rating of the diode as part of your design.

Solution. We assume ideal diodes for the initial design, and then examine consequences of this assumption after the design is complete. Starting with the 9 V dc output voltage, the excitation of the bridge is given from Eq. (3.53),

$$9 = \frac{2V_M}{\pi}$$

which gives $V_M = 14.1$ V for the peak voltage at the transformer secondary. The peak value of the primary voltage is $\sqrt{2} \times 120 = 170$ V; therefore, the transformer needs a turns ratio of $N = 170/14.1 = 12.1$. Peak current flows through the diodes when $v_i = V_M = 14.1$ V. To limit currents to 20 mA, choose

$$R = 14.1/20\text{mA} = 705 \ \Omega.$$

Larger values of R, which give smaller diode currents, are also acceptable. The PIV requirement is $V_M = 14.1$ V. ❏

The next exercise shows that the ideal diode assumption in the preceding example led to a conservative design.

Exercise 3.16 Find the actual peak current and PIV ratings needed when we replace the ideal diodes in Example 3.15 with silicon diodes.

Ans. 18 mA, 12.7 V.

Two-Diode Full-Wave Rectifier. Figure 3.46a shows a full-wave rectifier that uses only two diodes. The label *CT* denotes that the secondary windings are *center tapped.* A center-tapped secondary produces two voltages, each one-half of the full secondary voltage. Figure 3.46b, an equivalent for the secondary circuit, shows the phase relationship of these voltages. The equation for each source is

$$v_i(t) = 0.5\frac{V_p}{N} \sin \omega t \tag{3.54}$$

When primary voltage is positive, D_1 is ON and D_2 is OFF, giving Fig. 3.46c, where

$$v_o(t) = v_i(t), \quad \text{for } v_i(t) \geq 0$$

With negative primary voltage, sources $v_i(t)$ in Fig. 3.46b go negative, giving Fig. 3.46d, and

$$v_o(t) = -v_i(t), \quad \text{for } v_i(t) \leq 0$$

Thus the transfer characteristic and output voltage waveforms are those of 3.43d.

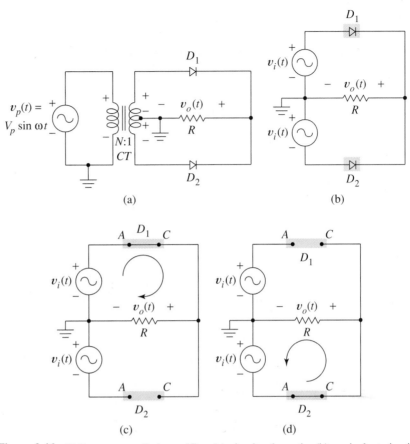

Figure 3.46 Full-wave two-diode rectifier: (a) circuit schematic; (b) equivalent circuit; (c) equivalent when $v_i(t) \geq 0$; (d) equivalent when $v_i(t) \leq 0$.

The PIV rating required by this circuit differs from that of the bridge rectifier. In Fig. 3.46c, we see that the maximum reverse voltage across D_2 occurs at the instant that $v_i(t)$ is maximum. At that instant

$$v_i = v_o = V_M$$

thus D_2 must withstand $2V_M$ compared with V_M for the bridge rectifier. Because operation is symmetric, the PIV rating for D_1 is also $2V_M$. The effect of diode offset on the transfer characteristic also differs from that of the bridge rectifier.

In Chapter 10 we show how to combine rectifiers with filters to produce the constant dc output voltage required of power supplies.

Clamping Circuit. Figure 3.47a shows a *clamping circuit*. Provided that the positive peak of $v_i(t)$ exceeds V_{BB}, this circuit clamps the positive peak of any periodic input waveform to V_{BB}.

Assume the diode is ideal. Let $v_i(t)$ be any periodic waveform, for example, Fig. 3.47b, and assume C is initially uncharged. Because $v_i(0)$ and the voltage across the ca

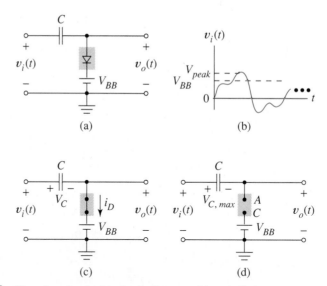

Figure 3.47 Clamping circuit: (a) circuit diagram; (b) periodic input voltage; (c) equivalent during transient when C charges; (d) equivalent once C is charged.

pacitor are both zero, the diode is OFF, for its anode is at zero volts and its cathode at V_{BB}. Since the capacitor cannot charge because of the diode open circuit, the node voltages at the two sides of the capacitor rise together as $v_i(t)$ grows more positive. Assuming $V_{peak} > V_{BB}$, the diode turns ON when

$$v_i = V_{BB}$$

giving the equivalent circuit of Fig. 3.48c. As $v_i(t)$ becomes still more positive, diode current i_D charges C, and capacitor voltage is

$$V_C = v_i(t) - V_{BB}$$

During this initial transient, the input voltage eventually reaches V_{peak}, while the output voltage remains at V_{BB}. When the node voltage on the left side of C begins to decrease, the capacitor cannot discharge, for this would require a negative diode current. Instead, the diode opens, giving Fig. 3.47d. Charge stored during the initial transient is trapped by the open circuit, and the diode never conducts again. From Fig. 3.47c, the capacitor voltage is

$$V_{C,max} = V_{peak} - V_{BB}$$

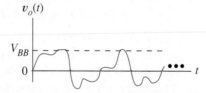

Figure 3.48 Output waveform of clamping circuit.

Once the transient is completed, the output voltage is

$$v_o(t) = v_i(t) - V_{C,max} = v_i(t) - V_{peak} + V_{BB} \qquad (3.55)$$

This equation shows how to obtain $v_o(t)$ from $v_i(t)$ once the transient is over. First, we must shift $v_o(t)$ downward by the amount of its peak value, and then shift it upward by V_{BB}. The overall result is Fig. 3.48. The derivation shows that the circuit clamps the positive peak of $v_i(t)$ to V_{BB} *automatically* for *any* periodic input waveform and *any* V_{BB}, positive, negative, or zero, provided only that the diode conducts at some time to charge C.

As a practical matter, notice that connecting a load resistor across the output terminals of Fig. 3.47d provides a discharge path for the capacitor and alters the operation of the circuit. If the load resistor is large, the capacitor discharges briefly during each cycle, slightly distorting the output waveform. This is one more of many situations where the infinite input resistance of the follower circuit of Chapter 2 is useful. A follower, inserted between the clamper output and load resistor, reproduces the ideal clamping circuit output voltage across the resistor without modification. But because the follower is essentially a VCVS, its output provides the current required by a resistive load without discharging the capacitor.

As with the diode clipping circuits, we can accomplish various wonders with the clamping circuit by reversing the sense of the diode and changing the magnitude and polarity of V_{BB}. Generally speaking, one can clamp the (positive, negative) peak of any periodic input waveform to V_{BB} volts, where V_{BB} is positive, negative, or zero, provided only that the diode conducts at some point to charge the capacitor. Further exploration is left to the reader, with Problem 3.52 provided for guidance.

ac to dc Converter and Voltage Doubler. The circuit of Fig. 3.49a converts ac to nonpulsating dc. We recognize this converter as the clamping circuit of Fig. 3.47a with $V_{BB} = 0$ and with output taken across the capacitor instead of the diode. After an initial transient, the dc output equals the positive peak value of the circuit's input voltage. In Chapter 10 we add a load resistor to the converter and call the result *a half-wave rectifier with capacitor filter,* an important circuit in simple power supplies. In a peak-reading analog meter, an op amp voltage-to-current circuit transforms the converter's dc output into a meter deflection proportional to the peak value of *any periodic input signal.*

Figure 3.49b is a related circuit, the *voltage doubler,* which produces a dc output voltage equal to twice the positive peak value of $v_i(t)$. To understand the doubler, consider the input transient with both capacitors initially uncharged and a *sinusoidal* input of amplitude V_M. As v_i swings positive, ideal diodes D_1 and then D_2 turn OFF and ON, respectively, giving Fig. 3.49c. The capacitors both initially charge to $0.5V_M$ with the indicated polarities. As v_i decreases and then swings to its negative peak, D_2 opens, D_1 closes, and the input capacitor charges to V_M as in Fig. 3.49d. During the next few cycles, the diodes act as switches, transferring charge from input to output capacitance, and then recharging the input capacitance. The following exercise explores this charge transfer. The sequence continues until the circuit reaches the final state shown in Fig. 3.49e.

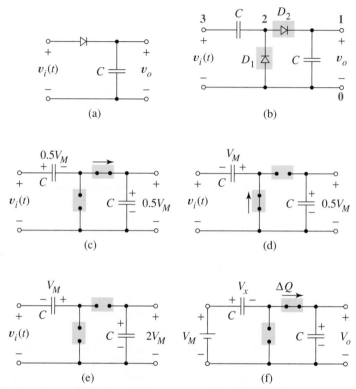

Figure 3.49 ac to dc converters: (a) basic converter; (b) voltage doubler circuit; (c) doubler during initial transient; (d) equivalent circuit showing how input capacitances charge during negative input cycle; (e) equivalent circuit after initial transients; (f) circuit showing how output voltage changes at the end of an input cycle.

Exercise 3.17   Figure 3.49f helps us understand how the output voltage changes at the end of the first input cycle. Initial values of V_o and V_x are those of Fig. 3.49d. The diodes then change states, and the input voltage goes to V_M, giving Fig. 3.49f. Charge ΔQ moves from the input to the output capacitor to satisfy Kirchhoff's voltage law, increasing V_o and V_x by the same ΔV because the capacitances are equal. Find the new value of V_o after the first, second, and third input cycles. This might take a little algebra.

Ans. $1.25V_M$, $1.625V_M$, $1.813V_M$.

EXAMPLE 3.16   Use SPICE transient analysis to plot the output voltage of Fig. 3.49b when the input is a 60 Hz sine wave of 10 V amplitude.

Solution. The SPICE code of Fig. 3.50 employs default diodes. In transient analysis we use a special SIN function to describe a sinusoidal source. The V_1 element line describes a sinusoidal source with zero time delay, 10 V amplitude, and 60 Hz frequency. The ".TRAN" statement specifies a run time lasting for 0.1s, six-cycles of the input waveform, and 100 output values. Figure 3.50b shows an output waveform that agrees

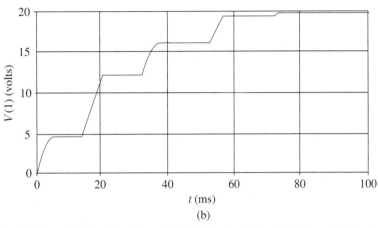

```
EXAMPLE3.16
VI 3 0 SIN(0 10 60)
C1 3 2 1000U
C2 1 0 1000U
D1 0 2 RECT
D2 2 1 RECT
.MODEL RECT D
.TRAN 0.001 0.1
.PLOT TRAN V(1)
.END
```

(a)

(b)

Figure 3.50 Voltage doubler simulation: (a) SPICE code of Fig. 3.49b circuit; (b) output voltage.

rather well with the answers to Exercise 3.17. (We expect diode cut-in voltages to make small contributions to the output voltage that we did not predict.) ❏

3.10
Waveshaping Circuits That Use Op Amps

The waveshaping circuits that follow use infinite-gain op amps as well as diodes and resistors. Most of the circuits use negative feedback around the op amp, justifying use of infinite gain analysis. We note the special contributions of the operational amplifiers as they arise in specific circuits.

Center Clipping Circuit. Figure 3.51a shows an op amp *center clipping circuit*. To simplify, we visualize the diode as a three-state device like that of Fig. 3.29a. For positive v_i, current i is positive, causing diode breakdown, as in Fig. 3.51b, where V_z is the breakdown voltage of the diode. Using the virtual ground at the op amp input, we see that

$$v_o = -V_z$$

For negative v_i, current is negative giving Fig. 3.51c and the equation

$$v_o = V_D$$

These equations give Fig. 3.51d.

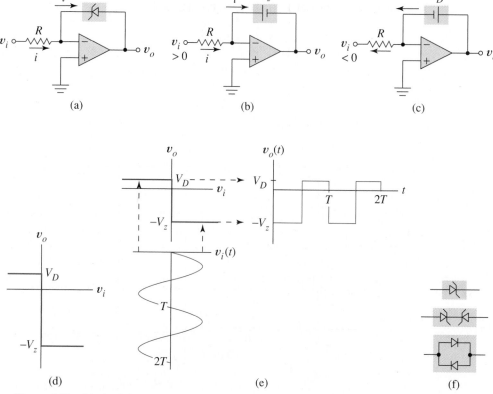

Figure 3.51   Limiter/clipper circuit using an op amp: (a) circuit schematic; (b) equivalent when $v_i \geq 0$; (c) equivalent when $v_i \leq 0$; (d) transfer characteristic; (e) producing a center-clipped waveform; (f) alternative feedback elements giving other clipping functions.

One use of this transfer characteristic is to convert a sine wave into a square wave of the same fundamental frequency, as shown in Fig. 3.51e. (Because the clipping transfer characteristic is not symmetric relative to the time axis, the square wave is offset by a dc term.) The center clipper is also used to extract information about the instantaneous algebraic sign of an analog waveform, because the binary or two-level output of the center clipper changes value whenever the input voltage crosses zero.

Three variations of the basic circuit result from replacing the zener diode by one of the feedback elements in Fig. 3.51f. Finding and sketching the transfer characteristics of each of these limiters is left for the interested reader.

Exercise 3.18   Design a diode limiter like Fig. 3.51a that turns an input sine wave of 10 V peak value into a square wave of 4 V peak-to-peak value (plus a dc component). Specify the zener voltage and peak zener power dissipation for your circuit.

Ans. $R = 1$ kΩ (arbitrary), 3.3 V, 33 mW.

Limiter/Clipper Circuits. The limiter circuit of Fig. 3.52a is an inverting amplifier with back-to-back zener diodes in parallel with feedback resistor R_F. Assume that for small $|v_i|$ both diodes are OFF. Figure 3.52b shows that this assumption gives an inverting amplifier. It also confirms our OFF assumption, for unless $|v_o|$ is at least $V_Z + V_D$ volts above virtual ground, the diodes cannot conduct. When the amplitude of v_o is sufficiently large, however, one diode conducts in the forward direction and the other breaks down. Figure 3.52c shows the equivalent circuit for v_i large and positive. Considering both polarities of input voltage gives Fig. 3.52d, where $V_L = V_z + V_D$.

One advantage of the op amp limiters over diode-resistor limiters is that load resistance does not affect the transfer characteristic. For example, the limiter of Fig. 3.39b becomes a voltage divider when a load R_L is attached to the output. Another advantage is that the op amp's power gain makes it possible to have a slope greater than one in the voltage transfer characteristic.

Precision Rectifiers. We noticed earlier in the chapter that diode offset voltages lead to offsets in half- and full-wave rectifier transfer characteristics. In the *precision rectifiers* that follow, operational amplifiers automatically compensate for diode offsets, producing transfer characteristics just like those of ideal diode circuits. This feature is of critical importance in those signal processing circuits that require a dc output voltage proportional to the amplitude of the input signal.

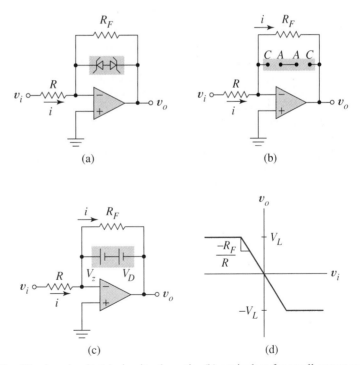

Figure 3.52 Clipping circuit: (a) circuit schematic; (b) equivalent for small output voltages; (c) circuit for large v_i; (d) transfer characteristic.

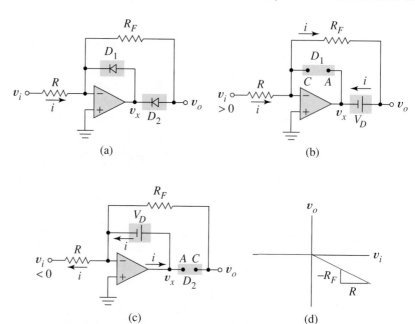

Figure 3.53 Precision rectifier: (a) circuit; (b) equivalent when $v_i \geq 0$; (c) equivalent when $v_i \leq 0$; (d) transfer characteristic.

Precision Half-Wave Rectifier. Figure 3.53a shows a *precision half-wave rectifier.* Assuming an infinite-gain op amp, let us reason out circuit operation when v_i is positive. Since i can neither enter the op amp nor flow in the reverse direction through D_1, assume it flows through R_F and D_2. The result, using offset diode models, is Fig. 3.53b. Routine infinite gain analysis gives

$$i = \frac{v_i - 0}{R} = \frac{0 - v_o}{R_F}$$

From which

$$v_o = \frac{-R_F}{R} v_i, \qquad \text{for } v_i \geq 0 \qquad (3.56)$$

When $v_i < 0$, D_2 turns OFF and D_1 ON, giving Fig. 3.53c (where feedback through D_1 justifies infinite-gain analysis). The virtual ground requires that

$$v_o = 0, \qquad \text{for } v_i \leq 0 \qquad (3.57)$$

Figure 3.53d, a plot of Eqs. (3.56) and (3.57), shows that the precision rectifier gives the ideal half-wave transfer characteristic (plus a harmless phase inversion) *even though real diodes are employed.* The secret of the circuit is that the op amp output, v_x, takes on whatever value necessary to ensure that the transfer characteristic is ideal. From Fig. 3.53b we see that

$$v_x = v_o - V_D, \qquad \text{for } v_i \geq 0$$

and from Fig. 3.53c

$$v_x = V_D, \qquad \text{for } v_i \le 0$$

The op amp also supplies voltage gain if needed. Unlike most op amp circuits we have seen, the precision rectifier does not have zero output resistance because the circuit output, v_o, does not coincide with op amp output, v_x. A follower at the output corrects this problem, should it be important in a particular application.

Precision Full-Wave Rectifier. In Fig. 3.54 we see the precision half-wave rectifier and a summing amplifier used as subcircuits in the *precision full-wave rectifier.* The summing circuit gives

$$v_o = -(v_i + 2v_R) \tag{3.58}$$

Since v_R is the output of the precision half-wave circuit, Eq. (3.56) gives

$$v_R = -v_i, \qquad \text{for } v_i \ge 0$$

and, from Eq. (3.57),

$$v_R = 0, \qquad \text{for } v_i \le 0$$

Substituting these v_R values into Eq. (3.58) gives

$$v_o = v_i, \qquad \text{for } v_i \ge 0$$

and

$$v_o = -v_i, \qquad \text{for } v_i \le 0$$

But the last two equations describe the *ideal diode* full-wave rectifier (even though the circuit employs real diodes). Figure 3.54b shows waveforms for v_i and $2v_R$ in Eq. (3.58) when $v_i(t)$ is a sine wave. This sketch suggests that resistor values are rather critical because the circuit must correctly match amplitudes of the two halves of the output waveform.

Comparator. We first encountered the comparator, a circuit having two analog input waveforms, $v_a(t)$ and $v_b(t)$ and a digital or binary output $v_o(t)$, in Chapter 2. Now that we know a little about op amp dynamics, it is important to recognize that most comparator

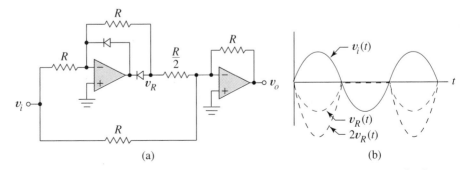

(a)   (b)

Figure 3.54 Precision full-wave rectifier: (a) circuit; (b) sketch showing how the circuit combines $v_R(t)$ and $v_i(t)$.

applications require a rapid output transition at the instant when the two input wave-forms cross. Large capacitors, built into some op amps to ensure stability, cause poor rise times and low slew rates. Consequently, we use very fast op amps in the more critical comparator applications.

Sometimes we add a limiter circuit to the op amp comparator, as in Fig. 3.55a. The limiter makes the output levels independent of the power supply, and the circuit switches faster because output voltage transitions are not so large. A limiter also allows us to customize output levels to fit the particular requirements of a given digital logic family.

Figure 3.55b shows a comparator that develops its output at the output node of an infinite-gain op amp. Because of the virtual ground at the input of the second op amp,

$$i(t) = \frac{1}{R}[v_a(t) - v_b(t)]$$

When $v_a(t) > v_b(t)$, then $i(t) > 0$, and

$$v_o(t) = -(V_z + V_D)$$

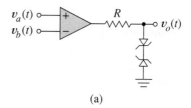

(a)

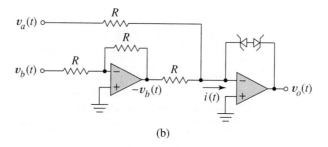

(b)

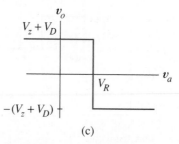

(c)

Figure 3.55 Comparator circuits: (a) op amp with output limiter; (b) comparator with low output resistance; (c) transfer characteristic for circuit of (b) when $v_b(t) = V_R$.

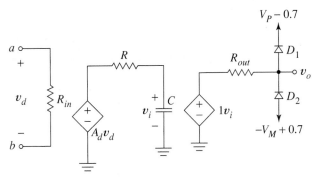

Figure 3.56 SPICE model for op amp with saturation limits V_P and $-V_M$.

When $v_a(t) > v_b(t)$, then $i(t) < 0$ and

$$v_o(t) = V_z + V_D$$

Sometimes a constant reference voltage, V_R, is used for one comparator input instead of a time function. In this case, the transfer function from the variable voltage input to the output is useful. Using

$$v_b(t) = V_R$$

in Fig. 3.55b, for example, gives Figure 3.55c.

SPICE Model for a Saturating Op Amp. Figure 3.56 shows our dynamic op amp model with diode limiting added at the output. Values for V_P and V_M match the saturation limits listed on the op amp data sheet. The diodes are OFF for $-V_M < v_o < V_P$; however, should v_o try to take on a value beyond these limits, a diode conducts, clamping the output voltage. The result is a nonlinear transfer characteristic resembling Fig. 2.22.

EXAMPLE 3.17 Construct a *static* SPICE subcircuit for the HA2544 op amp that includes saturation of the open-loop gain curve. Use this subcircuit and SPICE to find the transfer characteristic of the comparator of Fig. 2.38, when the comparator uses the HA2544 and $V_r = 2$ V.

Solution. Figure 3.57a shows subcircuit and comparator diagrams, and Fig. 3.57b the SPICE code. Figure 3.57c shows the output. For better modeling of the ± 11 V saturation limits, supply voltages should be changed to ± 10 V, since the current that flows through R_{out} within the subcircuit is very high. ❏

3.11 _____

The Diode as a Switch

This short section introduces the most elegant, versatile, and important device in modern electronics—the electronic switch. The simple switch underlies all the logic circuits we use for decision making, forms the basis for digital information storage, and enables us to construct interfaces by which computers control peripheral devices and peripheral devices control computers. The switch also forms the basis for many information processing circuits—modulators and detectors that transfer information to and from carrier sig-

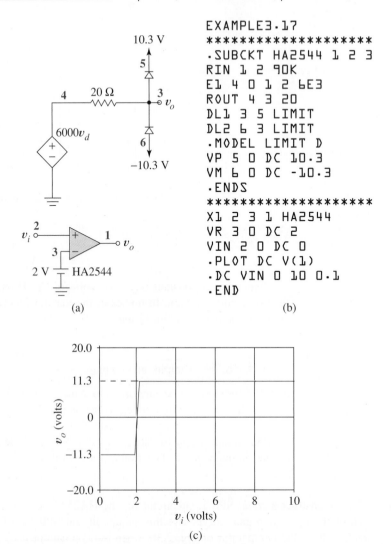

```
EXAMPLE3.17
********************
.SUBCKT HA2544 1 2 3
RIN 1 2 90K
E1 4 0 1 2 6E3
ROUT 4 3 20
DL1 3 5 LIMIT
DL2 6 3 LIMIT
.MODEL LIMIT D
VP 5 0 DC 10.3
VM 6 0 DC -10.3
.ENDS
********************
X1 2 3 1 HA2544
VR 3 0 DC 2
VIN 2 0 DC 0
.PLOT DC V(1)
.DC VIN 0 10 0.1
.END
```

(a)   (b)

(c)

Figure 3.57 Comparator transfer characteristic: (a) SPICE subcircuit and comparator diagrams; (b) SPICE code for the HA2544 subcircuit and for comparator; (c) transfer characteristic.

nals—switched-capacitor and charge-control devices that filter signals. In this section we encounter the electronic switch in its simplest form, the junction diode.

3.11.1 IDEAL SWITCH

The ideal switch, Fig. 3.58a, is a two-state device—open circuit or short circuit. Two volt-ampere curves, Fig. 3.58b, describe the switch in terms of switch current and voltage, i_{sw} and v_{sw}. In Fig. 3.58c the switch controls a simple on/off load such as a display lamp, relay, or an alarm, represented by resistance R. When closed, the switch places

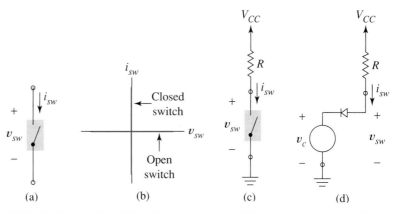

Figure 3.58 (a) Ideal switch; (b) volt-ampere curves of ideal switch; (c) a switching application; (d) application using a diode switch.

supply voltage V_{CC} across the load, current flows through the load, and the load device turns ON. When the switch opens, load current is zero, and the load turns OFF.

3.11.2 DIODE SWITCH

The diode is useful in this kind of application because its ON state resembles a closed switch and its OFF state an open switch. With the diode we also encounter, for the first time, the idea of a controlled, or electronic switch.

In Fig. 3.58d, we use an ideal diode as a switch and provide a source, v_c, to turn the switch ON or OFF. The control voltage takes on only two values, zero and V_{CC}. When $v_c = 0$ the diode turns ON, connecting R to V_{CC} exactly like the closed switch in Fig. 3.58c. When $v_c = V_{CC}$ volts (or more), the diode turns OFF and i_{sw} is zero, just as in Fig. 3.58c when the switch is open. Notice that the control source, v_c, *never delivers power to the circuit*. When v_c carries current, its voltage is zero; when its voltage is high, its current is zero. The diode thus enables us to use a low-power *signal* to control a device R that, itself, might consume a considerable amount of power—power provided by V_{CC}.

A real diode switches just like the ideal diode, except there is a small voltage drop V_D across the closed switch. Also, to open the switch, v_c must exceed $V_{CC} - V_D$ instead of V_{CC}. We examine these practical details further using our diode models in Problem 3.68.

An extremely important practical concern is the speed of the transitions between OFF and ON states of electronic switches, as this limits the operating speed of our electronics. For diode switches, and for transistor switches we study later, switching speeds depend upon the dynamic properties of the pn junction, the topic of the next section.

3.12
Dynamic Properties of the pn Junction

We begin by distinguishing between static and dynamic properties of an electronic device, using the diode as the simplest example. The diode's familiar volt-ampere curve is a *static characteristic*. For a one-port, the word "static" implies that operation is described by an *algebraic* equation, $i_D = f(v_D)$, that relates corresponding *dc* voltage and current values; for the diode, this is Eq. (3.32). A device's static characteristic also relates time-varying voltages and currents, but only if the time variations are not "too fast." Electronic devices always contain small internal *parasitic capacitances* that modify device behavior for fast signals. Therefore, in general, these devices are described by differential equations that happen to reduce to algebraic equations for sufficiently slow signals. Thus "static" means that the device variables are changing at such low rates that the time derivatives in the differential equation are small enough to ignore. In this context, the word "equilibrium" refers to a static system in the sense that all derivatives are zero (or negligible)—a *dynamic equilibrium*—in contrast to the *thermal equilibrium* of the unbiased pn junction. In summary, a device's *dynamic properties* have to do with its internal capacitances, and "fast" or "slow" signals for a particular device are defined relative to time constants associated with these capacitances.

Diode dynamics are important because they limit the validity of the theory we have studied for diodes, both as switches and as key components in waveshaping circuits. All previous descriptions of diode circuits assumed static diode operation. When we test the circuits in the laboratory, we find they work as predicted over a range of frequencies; however, we can always experimentally find some high-input frequency above which the output waveform is noticeably less than ideal. At this frequency the internal diode capacitance begins to adversely affect circuit operation, for the diode is no longer adequately described by its static volt-ampere curve. After a quick review of the basic idea of capacitance we introduce diode parasitics.

We define a *linear capacitor* by its characteristic curve, the linear charge versus voltage curve of Fig. 3.59. When operation moves from point *a* to point *b*, the voltage change ΔV must be accompanied by a change in stored charge ΔQ. *Large-signal capacitance* is defined by $C = \Delta Q/\Delta V$. Thus large capacitances require that incremental increases in voltage, ΔV, be accompanied by relatively large accumulations of charge ΔQ. Furthermore, large C requires that *rapid* changes in voltage (large $\Delta V/\Delta t$) be accompanied by a large charging current

$$\frac{\Delta Q}{\Delta t} = C\frac{\Delta V}{\Delta t}$$

In pn junctions there are two kinds of parasitic capacitance, diffusion capacitance and depletion capacitance, both characterized by *nonlinear* charge versus voltage curves.

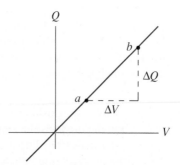

Figure 3.59   Characteristic curve of a linear capacitor.

3.12.1 DIFFUSION CAPACITANCE

Figure 3.60 is an enlarged view of the minority concentration profiles of Fig. 3.19a. The *diffusion capacitance* of a pn junction (or diode) has to do with the excess (above the equilibrium level) minority charges, Q_p and Q_n. In a diode of cross-sectional area A, each tiny volume, $A\,dx$, on the n-side of the junction contains

$$\left[p(x) - \frac{n_i^2}{N_d}\right]A\,dx$$

excess holes, each with a charge of q coulombs. The total excess charge stored near the junction in holes is therefore

$$Q_p = \int_{0^+}^{\infty} q\left[p(x) - \frac{n_i^2}{N_d}\right]A\,dx$$

where the infinite integral is a good approximation of the true integral if the p-side of the diode is longer than four or five diffusion lengths. Substituting $p(x)$ from Eq. (3.35) and integrating gives

$$Q_p = Aqn_i^2\frac{L_p}{N_d}(e^{v_D/V_T} - 1) \tag{3.59}$$

Since diode voltage v_D is the independent variable and Q_p is the dependent variable, Eq. (3.59) describes a nonlinear capacitor.

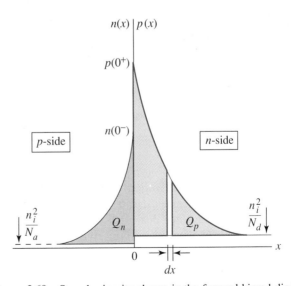

Figure 3.60 Stored minority charge in the forward-biased diode.

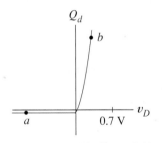

Figure 3.61 Nonlinear Q-V curve for diode diffusion capacitance.

Performing the same operations to find the excess charge Q_n stored in electrons in the p-material gives

$$Q_n = Aqn_i^2 \frac{L_n}{N_a}(e^{v_D/V_T} - 1) \tag{3.60}$$

If we denote the total minority stored charge by $Q_d = Q_p + Q_n$, Eqs. (3.59) and (3.60) together give an equation of the form

$$Q_d = K(e^{v_D/V_T} - 1) \tag{3.61}$$

where K is a constant. Plotting Eq. (3.61) gives the "diodelike" *diffusion capacitance curve* of Fig. 3.61.

A nonlinear capacitor such as Fig. 3.61 shares with the linear capacitor the fundamental property that a change in stored charge ΔQ must accompany any change in voltage ΔV. From the shape of Fig. 3.61 we infer that diffusion capacitance is of little importance when the diode remains reverse biased, since changes in negative voltage require negligible charge movement. In switching applications, however, where the diode changes between reverse and forward bias (points a and b), diffusion capacitance and the delay it creates are matters of concern.

3.12.2 DEPLETION CAPACITANCE

Depletion capacitance (alias transition, barrier, or space-charge capacitance) has to do with the charge stored in immobile ions in the depletion region. The basic idea was illustrated in Fig. 3.18 for zero, positive, and negative applied voltages. Because the amount of bound charge in the depletion region changes with voltage, it is no surprise that there is a capacitive effect. A straightforward derivation (see the text by Mattson in the references listed at the end of the chapter) shows that the depletion charge, Q_{dep}, varies with diode voltage as in Fig. 3.62a. Depletion capacitance dominates the dynamic behavior of the diode when the junction is reverse biased. While less important than diffusion capacitance for the forward-biased diode, depletion capacitance still contributes to switching delays during transitions between points such as a and b.

For nonlinear capacitors, a second definition, *small-signal* or *incremental capacitance*, is widely used. Incremental capacitance is *defined at a point* on the Q versus V curve as the slope of the curve at that point; that is,

$$C = \frac{dQ}{dV}$$

This definition prompts us to use the tangent to the curve to estimate the change in stored charge dQ required for a small change dV in voltage. (Figure 3.59 shows that incremental capacitance and "capacitance" are the same for the linear capacitor.) The incremental depletion capacitance of the pn junction is especially interesting and useful. Computing dQ_{dep}/dv_D from the theoretical expression for Fig. 3.62a gives the result

$$C_{dep} = \frac{C_{j0}}{(1 - v_D/V_{j0})^m} \tag{3.62}$$

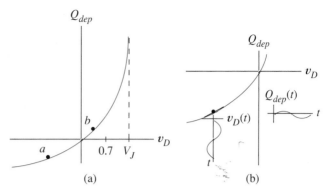

Figure 3.62 Depletion capacitance: (a) characteristic curve; (b) idea of incremental capacitance.

where V_{j0} is the built-in barrier potential, C_{j0} is the *zero-voltage capacitance,* and m is a constant called the *grading coefficient.* For the junctions we have illustrated, where the doping concentration changes abruptly from N_a to N_d (abrupt junctions), $m \approx 0.5$. For junctions characterized by a gradual transition in doping concentration, called *graded junctions, m* ≈ 0.33.

What makes incremental depletion capacitance interesting is that as long as the pn junction is kept reverse biased, the dc diode current is negligible, and for *small signals* the junction functions as a voltage-controlled capacitor as illustrated in Fig. 3.62b. Since the slope differs at different points on the curve, the small-signal capacitance can be changed by changing the dc operating point. Junctions especially designed for operation in this manner, known as *varactor diodes,* find use in automatic tuning of radios, frequency modulators, temperature compensation of crystal oscillators, and many other applications.

The following intuitive development shows that the diode's diffusion and depletion capacitances and its static diode equation all originate in a single differential equation. For the *static case,* when derivatives are zero, this equation reduces to the diode equation and predicts the familiar diode volt-ampere curve.

3.12.3 DYNAMIC CIRCUIT MODEL FOR THE DIODE

We first rewrite the static diode equation, Eq. (3.38), in terms of excess minority stored charge by substituting Eqs. (3.59) and (3.60). This gives

$$i_D = Q_p \frac{D_p}{L_p^2} + Q_n \frac{D_n}{L_n^2} \qquad (3.63)$$

Solid-state books show that diffusion lengths and minority carrier lifetimes are related by the equations

$$\tau_p = \frac{1}{D_p} L_p^2 \quad \text{and} \quad \tau_n = \frac{1}{D_n} L_n^2$$

Substituting these into Eq. (3.63) gives the elegant expression

$$i_D = \frac{Q_p}{\tau_p} + \frac{Q_n}{\tau_n} \tag{3.64}$$

called the *static charge control equation* for the pn junction.

For the special case where acceptor doping is much greater than donor doping, the minority charge concentrations more closely resemble Fig. 3.63 than Fig. 3.60. Then hole motion dominates the junction behavior, and Eq. (3.64) reduces to the yet simpler form

$$i_D = \frac{Q_p}{\tau_p} \tag{3.65}$$

Now we consider the total time-varying diode current. At any instant, t, if v_D is changing, Figs. 3.61 and 3.62a show that stored charges $Q_d = Q_p$ and Q_{dep} are also changing. Then diode current $i_D(t)$ must provide for not only the ordinary dc charge flow through the junction, but also for changes in stored charge. Adding current terms to Eq. (3.65) to account for the changing stored charge gives

$$i_D(t) = \frac{Q_p}{\tau_p} + \frac{dQ_p}{dt} + \frac{dQ_{dep}}{dt} \tag{3.66}$$

Notice that this differential equation reduces to the static diode equation, Eq. (3.65), when the derivatives are sufficiently small.

Figure 3.64a is a KCL interpretation of Eq. (3.66), showing a circuit element for each term. The diode symbol corresponds to the first term in Eq. (3.66), the static diode function. Two nonlinear large-signal capacitors account for the derivative terms, C_d for the diffusion capacitance and C_{dep} for depletion capacitance. The arrows on the capaci-

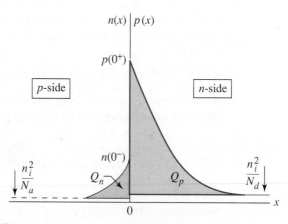

Figure 3.63 Excess minority charge when p-material is doped much more heavily than n-material.

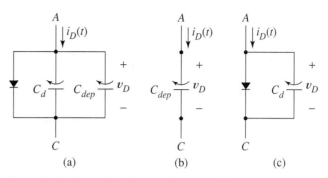

Figure 3.64 Dynamic diode models: (a) complete model for any v_D; (b) approximate model for $v_D < 0$; (c) approximate model for $v_D > 0$.

tors denote that the capacitors are nonlinear. For slow signals, when the derivative terms in Eq. (3.66) are negligible, the capacitors are effectively open circuits, and the circuit reduces to the ordinary static diode.

For a reverse-biased diode, the effects of diffusion capacitance and diode conduction are both small, and we can approximate the diode by the depletion capacitance model of Fig. 3.64b [third term of Eq. (3.66)]. When forward-biased, we represent the diode by Fig. 3.64c, Eq. (3.66) with the last term omitted.

The following development shows how these circuit models contribute to our understanding of diode switching.

3.12.4 DYNAMIC SWITCHING

To understand how internal capacitances limit the diode's switching speed, consider the circuit of Fig. 3.65a, where $v_i(t)$ is the test signal of Fig. 3.65b. If the diode had no dynamic parameters, diode voltage would instantly jump from its OFF value of $-V_{NN}$ volts to its ON value of 0.7 V at $t = 0$, and then instantly jump back to $-V_{NN}$ at $t = T$ as in Fig. 3.65b. This is not the behavior of a real diode.

Assume that p-side doping is much greater than n-side doping so that Eq. (3.66) applies. Consider first the interval $t < 0$. Assuming equilibrium, both derivative terms in Eq. (3.66) are zero, leaving only the algebraic diode equation. The tiny drop $I_s R$ caused by the diode reverse saturation current is negligible, so the source reverse biases the diode at $v_D \approx -V_{NN}$ volts. From Eq. (3.65),

$$i_D(0^-) = \frac{Q_p}{\tau_p} = I_s(e^{-V_{NN}/V_T} - 1) \approx -I_s$$

where 0^- denotes the instant just before $t = 0$. This establishes operating point a on each diode capacitance in Fig. 3.65c. From the circuit approximation of Fig. 3.64b, the diode is essentially a nonlinear depletion capacitor, charged to $v_D = -V_{NN}$ volts for $t < 0$.

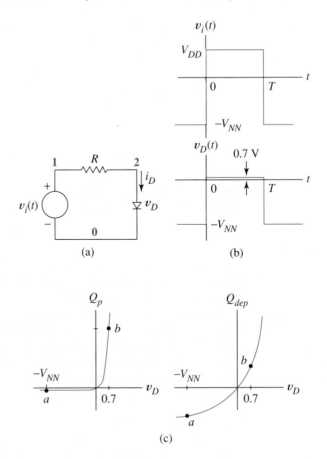

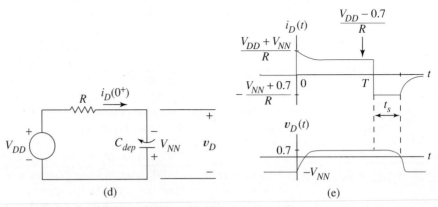

Figure 3.65 Diode switching transients: (a) given circuit; (b) waveform of voltage source and output voltage if capacitances are ignored; (c) diode capacitance curves showing initial and final equilibrium points; (d) circuit with diode replaced by its dynamic OFF circuit model; (e) waveforms of diode current and voltage during switching; (f) circuit with diode replaced by its dynamic ON model; (g) circuit model during storage time.

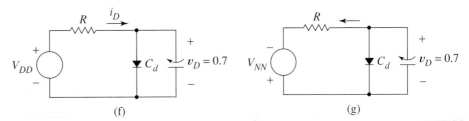

(f) (g)

Figure 3.65 (continued)

Now consider the instant $t = 0^+$ just after $v_i(t)$ changes to V_{DD}. The diode voltage cannot change instantly because of the depletion capacitance. Figure 3.65d shows that at this instant diode current suddenly jumps to

$$i_D(0^+) = \frac{V_{DD} + V_{NN}}{R}$$

The depletion capacitance begins to charge as large, positive diode current reduces the depletion region width. The resulting current and voltage waveforms of Figure 3.65e, just after $t = 0$, *resemble* exponential RC functions; however, the capacitor is nonlinear. When $v_D(t)$ crosses cut-in, the diode begins to conduct, and both capacitors charge toward their new equilibrium points, b in Fig. 3.65c. For this new equilibrium, the circuit model of Fig. 3.65f shows that the diode current is

$$i_D(T) = \frac{V_{DD} - 0.7}{R}$$

Equation (3.66) shows that the excess minority charge stored near the depletion region for this new equilibrium is

$$Q_P = \tau_p \frac{V_{DD} - 0.7}{R} \qquad (3.67)$$

which corresponds to point b on the diffusion capacitance curve of Fig. 3.65c.

At $t = T$, $v_i(t)$ switches back to its original value, $-V_{NN}$, again upsetting an equilibrium. Because of diffusion capacitance, v_D initially remains unchanged at 0.7 V, and a *negative diode current*

$$i_D = -\frac{V_{NN} + 0.7}{R}$$

flows as in Fig. 3.65g, discharging the diode capacitances. Because the Q_p curve of Fig. 3.65c is nearly vertical, diode voltage cannot change much until the excess minority charge has been removed (by the negative diode current and by recombination). Thus diode voltage and the negative diode current remain relatively constant for an interval t_s called the *storage time*, as shown in Fig. 3.65e. Once the operating point on the Q_p curve passes $v_D = 0$, the curve is nearly flat, and operation quickly drops to point a on both Q-V curves as the depletion region charge is restored. This is reflected in Fig. 3.65e as exponential-like transitions of $i_D(t)$ and $v_D(t)$ to their original equilibrium values.

In summary, the delay in switching from OFF to ON relates mainly to discharging the depletion capacitance, and is relatively short. The ON to OFF delay, mainly due to discharging diffusion capacitance, can be quite long. Until the minority charge has been removed, the diode voltage remains positive. Only after excess charge is removed from the base do the diode voltage and current drop to the equilibrium values we associate with an OFF diode.

Switching has the same general form for diodes characterized by injection of both holes and electrons, and for diodes dominated by electrons. Operating speeds of electronic computers are limited primarily by switching delays of this type, an observation that underscores the importance of understanding pn junction dynamics.

3.13
Dynamic SPICE Model for the Diode

The preceding discussion was somewhat qualitative because the nonlinear capacitances make diode switching dynamics difficult to handle mathematically. Fortunately, the diode's dynamic nonlinearities are modeled in SPICE. This makes it easy to obtain realistic *numerical solutions* by simulation.

Figure 3.66a is a SPICE equivalent circuit for the dynamic diode. Figure 3.66b lists five new parameters required to model dynamic behavior. Resistance *RS* adds the ohmic drop caused by diode current flowing through the doped semiconductor material external to the junction. Parameters CJO, M, and VJ define the depletion capacitance as given in Eq. (3.62). A general *transit time parameter* TT replaces the hole recombination time, τ_p, we used in the last section. Transit time is appropriate when both holes and electrons contribute to the diode action, but reduces to τ_p or τ_n when holes or electrons individually dominate the diode behavior. It also enables simulations when the diode dimensions are much less than the diffusion length, a geometry that changes some details of our derivations but not the general concepts.

The default values of the two key *dynamic* diode parameters, CJO and TT, are both zero. This means that *the default diode model in SPICE is a static model*. To simulate

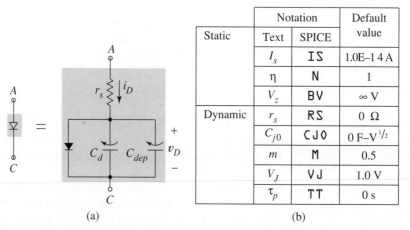

		Notation		Default
	Static	Text	SPICE	value
		I_s	IS	1.0E–14 A
		η	N	1
		V_z	BV	∞ V
	Dynamic	r_s	RS	0 Ω
		C_{j0}	CJO	0 F–V$^{1/2}$
		m	M	0.5
		V_J	VJ	1.0 V
		τ_p	TT	0 s

(a)   (b)

Figure 3.66 Dynamic SPICE model for the diode: (a) circuit model; (b) table of SPICE parameters and notation.

dynamic effects, these default values of zero must be overridden by nonzero values in a .MODEL statement.

The next example simulates the switching experiment of Fig. 3.65.

EXAMPLE 3.18 Use SPICE to find the current and voltage waveforms depicted in Fig. 3.65e and the value of the storage time. Use $V_{NN} = -10$ V, $V_{DD} = +10$ V, $R = 10$ kΩ, and T = 20 ns. The diode parameters are RS = 5 Ω, CJ0 = 1 pF, TT = 10 ns.

Solution. Figure 3.67a is the SPICE code for the circuit, including ammeter VM to monitor positive diode current; Figs. 3.67b and c show the results. From the current plot, the storage time is about 7 ns. To determine the value more precisely, we could examine a printout of current values from another SPICE run that uses smaller output intervals.   ❏

We gave basic equations and sketches related to switching dynamics, but notice that nowhere were equations of $i_D(t)$ and $v_D(t)$ given. To produce the data in Example 3.18, the computer numerically solved a very complicated nonlinear differential equation. Even with such powerful tools at our disposal, however, it is clear that we must *understand* the theoretical concepts. Notice that Figs. 3.67b and c give neither an understanding of the *causes* for the waveforms, nor any insight into *how to change them*. By con-

```
EXAMPLE 3.18
VS 1 0 PULSE(-10 +10 0 0 0 20N)
R 1 3 10K
VM 3 2 DC 0
D1 2 0 MAX
.MODEL MAX D RS=5 CJ0=1.0P TT=10.0N
.PLOT TRAN I(VM) V(2)
.TRAN 1N 50N
.END
```

(a)

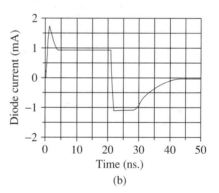

(b)

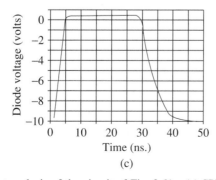

(c)

Figure 3.67   Transient analysis of the circuit of Fig. 3.61a: (a) SPICE code; (b) and (c) SPICE plots of diode current and voltage.

trast, our intuitive development suggests that if we wish to reduce the storage time we should increase V_{NN}. This would increase the magnitude of the negative diode current, which, in turn, would remove the stored charge faster than the 1.1 millicoulomb/s rate we see in Fig. 3.67b.

Exercise 3.19   From Fig. 3.67b
(a) estimate the minority stored charge. Then
(b) estimate how large V_{NN} must be to reduce storage time to 3 ns.

Ans. 7.7 pC, 25 V.

3.14
Special Kinds of Diodes

3.14.1 SCHOTTKY DIODE

A *Schottky diode* is a junction that has an aluminum anode and a lightly doped n-type semiconductor cathode as in Fig. 3.68. A depletion region at the metal-semiconductor interface gives a volt-ampere curve like Fig. 3.68b, which resembles that of a silicon pn junction. One important difference is that the reverse saturation current is significantly higher, say, 2×10^{-11} A compared to 1.0×10^{-14} A. This results in a forward voltage of the order of 0.4 V rather than 0.7 V as in silicon. Another difference is that for forward bias, electrons rather than holes are injected into the n-material from the metal anode; therefore there is no stored minority charge, no diffusion capacitance, and no storage time. The result is a diode that switches much faster than a pn junction. Low forward voltage and the fast switching make Schottky diodes important in switching power supplies and in the logic circuits of Chapter 13.

The Schottky junction is called a *rectifying junction,* to distinguish it from the ordinary, *ohmic,* or *nonrectifying* metal-semiconductor interfaces we have taken for granted until now. Since it is the low doping concentration that produces the special Schottky effect, highly doped material, denoted by n^+, is used for *ohmic contacts* between semiconductor materials and metal contacts as in Fig. 3.68a.

3.14.2 LIGHT-EMITTING DIODES AND PHOTODIODES

Light-emitting diodes (LEDs) are junctions constructed of special semiconductor materials such as gallium arsenide-phosphide. In these devices, the injected minority carriers that result from forward biasing give up energy in the form of radiated light when they

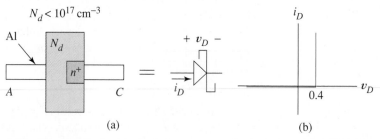

(a)   (b)

Figure 3.68   Schottky diode: (a) physical structure and schematic symbol; (b) typical volt-ampere curve.

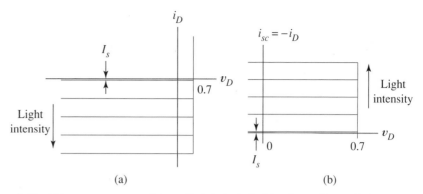

Figure 3.69 Volt-ampere curves of photodiode/solar cell: (a) using ordinary diode current convention; (b) using solar cell current convention.

recombine. Wavelengths ranging from ultraviolet through the visible and into the infrared bands are obtainable by use of different semiconductors and doping impurities. LEDs are widely used as display devices, for transferring information into optical fibers, and for optical isolation as discussed shortly.

A photodiode is a junction in which photons of energy in incident light break covalent bonds, adding the drift of these new carriers to the existing reverse saturation current. Figure 3.69a shows the volt-ampere curve of such a diode as a function of increasing incident light intensity. In the third and fourth quadrants the device functions as a (dependent) current source controlled by light.

Operation in the third quadrant represents passive conversion of light intensity *information* into electrical information, the device functioning as an optical to electrical transducer. Applications include light meters and communication systems that receive information coded in the form of light. The obvious circuit model is a light-intensity-controlled current source.

Diodes designed for fourth quadrant operation are called *photovoltaic cells* or *solar cells*. Points on the curves in the fourth quadrant correspond to voltages and currents of opposite sign, implying a device that delivers power to an external circuit. Figure 3.69b shows the form of the characteristic curves published for solar cells. By redefining the positive reference current, these curves emphasize that the cell is an active device for voltages not exceeding 0.7 V. Series connections of many solar cells produce large dc voltages. Parallel connections of solar cells of large area generate high output currents. Thus series-parallel interconnections can convert large amounts of solar power into electrical power at convenient voltage levels. Solar cell efficiency, electrical output power divided by solar input power, is typically of the order of 10–15%.

3.14.3 OPTICAL ISOLATION

Figure 3.70a shows how the light-emitting diode and the photodiode are combined in a useful device called an optical isolator. The signal $v_s(t)$ is transferred to the load by means of light, with no physical connection whatsoever between the input and output circuits. The optical isolator is useful in many applications, including computer inter-

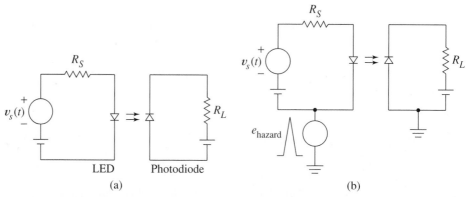

Figure 3.70 Optical isolation: (a) basic circuit; (b) protecting R_L from voltage hazard.

faces and biomedical instrumentation. In these applications, we want to transmit information while protecting delicate equipment or a human subject from dangers imposed by high voltages in the input circuit. Figure 3.70b illustrates the principle. Notice that the hazard voltage that exists in the circuit on the left is not transferred to the load circuit on the right. Transformers also give isolation; however, the optical isolator operates at frequencies down to dc and is smaller, lighter, and less expensive.

3.15
Summary

The materials studies that began the chapter have prepared us well for later studies of integrated circuits. We now understand the basic drift process in semiconductors, how drift of holes and electrons explains electrical resistance, and how and why resistance varies with temperature. In terms of these basic ideas we understand semiconductor resistors, resistor temperature coefficients, and the reasons why resistor ratios are important in modern circuit design.

We used details of the drift and diffusion processes for holes and electrons to describe our first true electronic device, the junction diode. We then applied the basic modeling techniques of Chapter 1, in the form of ideal and offset diode models, to determine how diodes function as components in a variety of circuits. We found that analysis of diode circuits depends upon viewing the diode as a three-state device that, at a particular time, operates in its ON, OFF, or breakdown state. For each state, we replace the diode by a simple equivalent that approximates the diode's volt-ampere characteristic and we then solve the resulting linear circuit to check our state assumption. When assumptions prove to be correct we can estimate diode current or voltage. We also showed how diode volt-ampere characteristics and diode model parameters change with temperature.

With these modeling and analysis techniques, we can describe many useful diode circuits such as limiters, half- and full-wave rectifiers, clamping circuits, voltage doublers, and voltage regulators. These perform a variety of useful functions such as turning ac into dc, shifting waveform levels, modifying the shapes of time-varying periodic waveforms, and producing a dc output that is relatively independent of changes in loading and source voltage. We also learned how the diode can be used as a switch to control power in a load.

A brief introduction to the idea of nonlinear capacitance led to descriptions of the two sources of parasitic capacitance within the pn junction: depletion capacitance, associated with the layers of bound ions in the depletion region, and diffusion capacitance, associated with storing excess minority charge carriers just outside the depletion region. When signals change slowly in diode circuits, these capacitances are inactive, and the diode satisfies its static equation. For rapid transitions or fast signals, however, a more basic diode description is revealed—a differential equation that is nonlinear when posed in terms of diode current or voltage. Because of diode capacitances, delays occur in diode switches, and waveform processing predicated upon static diode theory deteriorates at high frequencies.

SPICE models for diodes were introduced. By understanding how SPICE parameters relate to the diode parameters we use in circuit analysis, we can use simulation to study with increased accuracy and precision both static and dynamic operation of diodes. With SPICE we can formulate more realistic circuit descriptions, relying on computation to verify and reinforce our understanding of diode circuits.

REFERENCES _____

1. BANZHAF, W. *Computer-Aided Circuit Analysis Using SPICE,* Prentice Hall, Englewood Cliffs, NJ, 1989.
2. BURNS, S.G., and P.R. BOND. *Principles of Electronic Circuits,* West Publishing, St. Paul, MN, 1987.
3. DEBOO, G., and C. BURROUS. *Integrated Circuits and Semiconductor Devices: Theory and Application,* McGraw-Hill, New York, 1971.
4. GRAY, P.R., and R.G. MEYER. *Analysis and Design of Analog Integrated Circuits,* John Wiley, New York, 1984.
5. HODGES, D.A., and H.C. JACKSON. *Analysis and Design of Digital Integrated Circuits,* McGraw-Hill, New York, 1988.
6. IRWIN, J. DAVID. *Basic Engineering Circuit Analysis,* 2nd ed., Macmillan, New York, 1987.
7. MATTSON, R.H. *Basic Junction Devices and Circuits,* John Wiley, New York, 1963.
8. NILSSON, J.W. *Electric Circuits,* Addison-Wesley, Reading MA, 1986.
9. SCHILLING, D., and C. BELOVE. *Electronic Circuits, Discrete and Integrated,* McGraw-Hill, New York, 1989.
10. STREETMAN, B. *Solid State Electronic Devices,* Prentice Hall, Englewood Cliffs, NJ, 1980.
11. TOBEY, G., J. GRAEME, and L. HUELSMAN. *Operational Amplifiers,* McGraw-Hill, New York, 1971.

PROBLEMS _____

SPECIAL DIRECTIONS FOR SPICE HOMEWORK PROBLEMS: Do not hand in lengthy SPICE printout for homework. Instead, abstract the useful information from the SPICE output file as in the SPICE examples. Include your SPICE code and a circuit diagram with nodes numbered to agree with the code. Cite relevant numerical values from the SPICE output file and discuss them when appropriate. Make sketches of any relevant curves, and label appropriate points. Make small tables to present numerical data if useful for clarity.

Section 3.1

3.1 The following data apply to solid aluminum at room temperature:

Conductivity: 3.54×10^5 (ohm-cm)$^{-1}$
Number of valence electrons/atom: 3
Density: 2.7 gm/cm^3
Atomic weight: 26.98 gm/mole

(a) Recall Avogadro's number, 6.02×10^{23} atoms/mole. Compute the number of atoms per cubic centimeter in solid aluminum and the density of free electrons at room temperature.

(b) Find the electron mobility in solid aluminum.

(c) Compute the average drift velocity of electrons in solid aluminum in an electric field of 0.2 volts/cm.

3.2 The resistor in Fig. P3.2 is constructed of solid copper having conductivity of 5.65×10^5 (ohm-cm)$^{-1}$. Find its resistance.

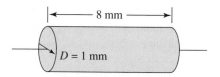

Figure P3.2

3.3 The following data have been obtained from measurements:

Temperature	Resistance
40°C	1.300 Ω
45°C	1.215 Ω

(a) Find the temperature coefficient of resistance.
(b) What is the resistance at 43°C?
(c) Compute the ratio of the conductivity of the resistor material at 43°C to the conductivity at 38°C.

Section 3.2

3.4 Using Eq. (3.11), plot $n_i^2(T)$ for a few temperatures between 273 K and 370 K (using a computer routine or programmable calculator).

3.5 If Fig. P3.2 is a cylinder of intrinsic silicon,
(a) Calculate its resistance at room temperature.
(b) Compute the average drift velocities of holes and electrons when 5 volts is applied to this resistor.
(c) Repeat parts (a) and (b) assuming the cylinder is made of intrinsic gallium arsenide.

3.6 Figure P3.6 shows a silicon thermistor. Assuming room temperature,
(a) Compute the current density inside the material.

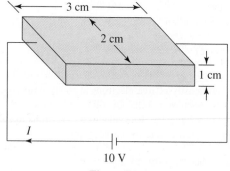

Figure P3.6

(b) Find the individual hole and electron components of current density.
(c) Find the external current, I.

3.7 Repeat Problem 3.6 when the temperature is 120°C.

3.8 A thermistor has the shape of a cylinder with length L and area A.
(a) Write an expression for the resistance $R(T)$ of the thermistor, using Eq. (3.11) to show explicitly its variation with temperature.
(b) Using the answer to part (a), write an expression for the ratio, $R(T)/R(T_R)$, where T_R is some reference temperature T_R.
(c) Use the result of part (b) to express $R(T)$ as the product of $R(T_R)$ and a simplified expression that involves $1/T$ and $1/T_R$. This is a form of the thermistor equation in common use.

Section 3.3

3.9 The intrinsic concentration, n_i, of an n-type semiconductor is 1.5×10^{10} cm^{-3} at room temperature.
(a) Estimate the concentrations of free electrons and holes in this material at room temperature if $N_d = 10^{14}$ cm^{-3}.
(b) Find the room temperature conductivity of the doped material if the electron and hole mobilities are 1500 and 480 cm^2/Vs, respectively.

3.10 For a piece of n-type semiconductor the intrinsic concentration is 1.5×10^{10} cm^{-3} and the doping concentration is 10^{14} cm^{-3}. Estimate the number of free electrons and holes if the material dimensions are 1 μm by 10 μm by 5 μm.

3.11 The intrinsic concentration of acceptor-type silicon has values of 1.5×10^{10} cm^{-3} and 10^{13} cm^{-3} at 300 K and 405 K, respectively, and the doping concentration is 2×10^{15} cm^{-3}.
(a) Estimate the electron and hole concentrations at both temperatures.
(b) Use the exact equations to determine the concentrations at both temperatures.

3.12 Some p-type silicon is fabricated using $N_A = 10^{14}$ cm^{-3}.
(a) Estimate its conductivity at room temperature.
(b) n-type impurities are now added to the p-material of part (a). Determine the required concentration N_D of the new impurities so that the compensated material has room temperature conductivity equal to 1% of its part (a) value.

3.13 In the sketches of Fig. P3.13, "−" denotes a free electron and "+" indicates a hole. Match the letter for each of these material samples with the best description from the list.
1. intrinsic silicon 2. donor-type silicon 3. acceptor-type silicon 4. conductor 5. insulator

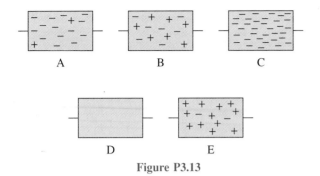

Figure P3.13

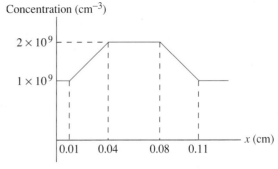

Figure P3.17

3.14 The integrated circuit resistor of Fig. 3.12 has a sheet resistance of 200 Ω/square.
(a) If one micron is the smallest feasible dimension, sketch and dimension an 800 Ω resistor of minimum area.
(b) If the sheet resistance varies by $\pm 20\%$ relative to the design value from one IC batch to the next, assign tolerances (maximum and minimum values) to the resistor of part (a).

3.15 Two IC resistors on the same IC chip, $R_1 = 5$ kΩ and $R_2 = 1$ kΩ, use sheet resistance of 600 Ω/square. ICs made on day 1 ended up with sheet resistances 20% higher than intended; day 2 ICs ended up with sheet resistances 20% lower.

If the masking tolerances (that is, tolerances on W and L) were $\pm 1\%$ for the fabrication process on both days, find
(a) Maximum and minimum values for R_1 and R_2 for the chip made on day 1.
(b) Maximum and minimum values for R_1 and R_2 for the chip made on day 2.
(c) Maximum and minimum values for the ratio R_1/R_2 for the chip made on day 1.
(d) Maximum and minimum values for the ratio R_1/R_2 for the chip made on day 2.

3.16 For the experiment in Fig. 3.13,
(a) Sketch the voltage waveform $v_m(t)$ that appears across the material.
(b) Sketch the hole concentration $p(t)$ as a function of time.
(c) Sketch the electron concentration $n(t)$ as a function of time.

Section 3.4

3.17 Figure P3.17 shows a nonuniform charge distribution in a material at some instant. Concentration is uniform in the y and z directions.
(a) Sketch the corresponding diffusion current density if the charge consists of holes and if $D_p = 20$ cm^2/s.

(b) Sketch the corresponding diffusion current density if the charge consists of electrons and if $D_n = 60$ cm^2/s.

3.18 Sketch the electron concentration and the electron current density for Exercise 3.4.

3.19 For the material in Fig. 3.14, $n_i = 2.4 \times 10^{13}$ cm^{-3}, $N_d = 10^{16}$ atoms/cm^3, $D_p = 10$ cm^2/s, and $L_p = 5 \times 10^{-3}$ cm. When the light is turned on, the hole concentration at the surface becomes 1000 times its equilibrium value. Find the hole current density as a function of x.

3.20 Electron and hole concentrations in a semiconductor are

$$n(x) = N_d + p_{n0} + \Delta p e^{-x/L_p}$$

and

$$p(x) = p_{n0} + \Delta p e^{-x/L_p}$$

Find an expression for the conductivity, σ, of the material.

Section 3.5

3.21 (a) In a pn junction, what kind of carriers make up the diffusion current, holes, electrons, or both?
(b) In the depletion region there are positive and negative ions. Where did the mobile charge carriers go, and by what process did they move?

3.22 A silicon pn junction is fabricated using $N_d = 10^{16}$ and $N_a = 10^{19}$ impurity atoms/cm^3.
(a) Make an unscaled sketch like Fig. 3.16. Label your sketch with the four appropriate numerical values. Assume room temperature.
(b) Find the barrier potential in volts at room temperature.
(c) Find the barrier potential in volts at 70°C.
Hint: See Eqs. (3.11) and (3.26).

3.23 Equation (3.31) is the seventh in a step-by step derivation of V_{J0} based upon holes. Demonstrate your understanding of the derivation by working out the analogous one based on electrons. To make a good start, use the correct form for the diffusion equation when you find the diffusion current.

Section 3.6

3.24 A diode has $I_s = 3.0 \times 10^{-14}$ amperes. Compute room temperature values of i_D for $v_D =$
(a) 0.8 V,
(b) 0.6 V,
(c) −0.2 V,
(d) −10 V.

3.25 At room temperature the forward current of a silicon diode is found to be 1 mA when 0.7 V is applied.
(a) Find I_s.
(b) What I_s value would be necessary for 1 mA at 0.4 V?

3.26 A junction is constructed with $N_a >> N_d$ so that the diode current is essentially hole current. Equation (3.36) describes how the hole current in the n-material decreases with x as holes recombine. A stream of electrons supplied by the external source flows in from the right in Fig. 3.18b to support this recombination. This electron current is such that at any point x, the hole and electron current components add up to the same value. At $x = 0$, the electron current is zero. Write an expression for the electron current, $i_{D,n}(x)$.

3.27 Diodes D_1 and D_2 are identical except that D_2 has cross-sectional area four times that of D_1.
(a) If the current of D_1 is 12 mA when its voltage is 0.62 V, what is the current of D_2 for 0.62 V?
(b) If D_1 and D_2 are in series and carry a positive current, find a relation between v_{D2} and v_{D1}.
Hint: Assume the forward current of each diode is much greater than the reverse saturation current.
(c) If D_1 and D_2 are in parallel and their currents sum to 4 mA, find the current in each diode and the voltage across the diodes.
Hint: See part (a) for needed information.

3.28 Equation (3.37) assumes that only the hole current in the diode is appreciable. Starting with the electron concentration profile in the p-material in Fig. 3.19a, derive the electron current counterpart of Eq. (3.37). Use Eq. (3.38) to check your answer.

Section 3.7

3.29 The diodes in Fig. P3.29 are ideal.
(a) Assume both diodes are OFF. Draw an equivalent circuit for this condition.

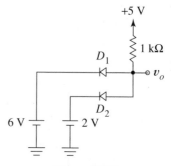

Figure P3.29

(b) Which diode gives a contradiction and what, exactly, is the contradiction? Be quantitative.
(c) Make a new assumption about the diode states, redraw the equivalent circuit, and check your new assumptions. What is the value of v_o in Fig. P3.29?

3.30 Find v_o, i_1, and i_2 for Fig. P3.30 if the diodes are ideal and
(a) $v_1 = 5$ V and $v_2 = 0.2$ V.
(b) $v_1 = 0.1$ V and $v_2 = 0.3$ V.
(c) $v_1 = 5.1$ V and $v_2 = 5.3$ V.

3.31 Solve Problem 3.30 using the silicon diode offset model.

3.32 Assuming ideal diodes in Fig. P3.32, find i_1 and i_2
(a) when $v_1 = 0.2$ V.
Hint: A KVL violation can also lead us to reexamine our diode assumptions.
(b) Find i_1 and i_2 when $v_1 = -9$ V.

3.33 Solve Problem 3.32 using the gallium arsenide diode offset model.

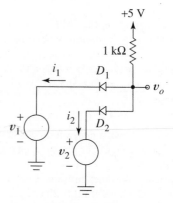

Figure P3.30

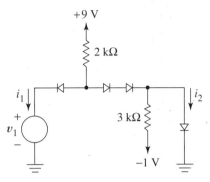

Figure P3.32

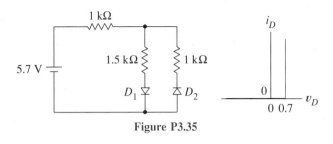

Figure P3.35

3.34 The diodes in Figure P3.34 are ideal. Assume both are OFF for $v_i = 0$ volts.
(a) Draw the equivalent circuit for these conditions.
(b) Is the assumption about D_1 correct? Give quantitative evidence.
Assume hereafter that D_1 is ON and D_2 is OFF for $v_i = 0$ volts.
(c) Draw the equivalent circuit for these assumptions.
(d) Is the assumption about D_1 correct? Be quantitative.
(e) Is the assumption about D_2 correct? Be quantitative.

3.35 The volt-ampere curve in Fig. P3.35 describes both diodes. D_1 is ON and D_2 is OFF.
(a) On a copy of the volt-ampere curve, mark the operating point of D_1 with an "x," and give the numerical value.
(b) Add to your sketch a "y" to mark the operating point of D_2. Show the numerical value.

3.36 Use the zener model of Fig. 3.29b to find the maximum and minimum values for V_{BB} in the regulator circuit of P3.36. The 5 V zener has a minimum reverse current of 10 mA, and its maximum power dissipation is 1 W.

3.37 In Fig. P3.36 V_{BB} is fixed at 15 V.
(a) Draw the Thevenin equivalent of the circuit to the left of the load resistor if the zener has internal resistance of 0.1 Ω.
(b) Use the result of part (a) to sketch the output voltage of the Thevenin equivalent versus output current.

(c) From part (b), what is the effect of zener resistance on the operation of the regulator when R_L varies?

3.38 Draw the equivalent circuit for the zener voltage regulator of Fig. 3.30 using the zener model of Fig. 3.29c. For values $V_{BB} = 9$ V, $R_s = 10$ Ω, $V_z = 6.8$ V, and $r_z = 0.1$ Ω, write an equation for output voltage as a function of R_L. Sketch the equation. This is called the *voltage regulation* curve for the regulator circuit.

3.39 The diode in Fig. P3.39 is silicon and its voltage varies linearly with temperature at a rate of -2 mV/°C. Write an expression for v_o as a function of temperature T, and sketch $v_o(T)$ versus T.

3.40 At 27°C, the diode in Fig. P3.40 is OFF. If $I_s = 2 \times 10^{-14}$ A at 27°C and varies according to Eq. (3.43), find the temperature at which the diode begins to turn ON (anode voltage exceeds cathode voltage by the cut-in value of 0.5 V). *Hint*: Refer to Fig. 3.32b.

Section 3.8

3.41 Figure P3.41 is an op amp model that includes the saturation of the open-loop transfer characteristic. If the diodes are ideal, both are open circuits until v_o exceeds one of the saturation limits. With real diodes, we can modify the voltage sources in the model to allow for the diode drops.

Make such a SPICE model for the HA2544 operational amplifier of Table 2.1 that includes saturation. Select dc voltage values to allow for drops of about 0.7 V when the diodes conduct, and use the default diode model.

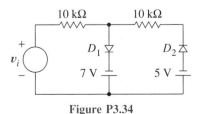

Figure P3.34

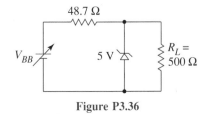

Figure P3.36

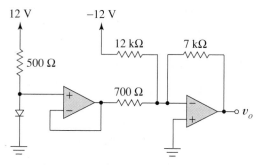

Figure P3.39

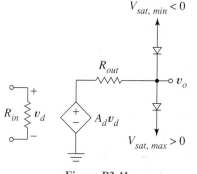

Figure P3.41

(a) Use dc analysis to plot v_o versus v_d over a voltage range that clearly shows both knees of the saturation curve.
(b) Add feedback resistors to your op amp model to make an inverting amplifier with gain of -20. Use SPICE transient analysis to plot the output waveforms for 1000 Hz sinusoidal input voltages of peak value 10 mV and 1 V.

3.42 The operational amplifier in Fig. P3.42 has infinite gain, and the diode has a volt-ampere curve like Fig. 3.29a with $V_Z = 5$ V.
(a) Sketch v_o versus v_i for the circuit.
Hint: Find equations relating v_o and v_i for $v_i > 0$ and for $v_i < 0$.
(b) Make a SPICE model of the circuit using a high-gain VCVS for the op amp. Use the default diode model, except include the 5 V zener voltage. Use dc analysis to produce the plot corresponding to the sketch of part (a).
(c) For $v_i = +2$ V, use SPICE to determine how much current the op amp supplies.
(d) Add a 1 kΩ resistor from the output node to ground and again find the current the op amp supplies when $v_i = +2$ V.

Section 3.9

3.43 Sketch the transfer characteristic for each ideal diode circuit of Fig. P3.43.

3.44 In Figure P3.44 the diode characteristic is shown on the right.
(a) Write an equation for v_o in terms of v_i when the diode is off.
(b) There is some input voltage value, $v_i = E$, where the diode just begins to turn on. Find E.
(c) Write an equation for v_o in terms of v_i when the diode is on.
(d) Sketch the transfer characteristic.

3.45 Reverse the diode polarity in Fig. P3.44. Then
(a) Write an equation for v_o in terms of v_i when the diode is off.
(b) There is some negative input voltage value, $v_i = -E$, where the diode just begins to turn on. Find E.
(c) Write an equation for v_o in terms of v_i when the diode is on.
(d) Sketch the transfer characteristic.

3.46 Both diodes in Fig. P3.46 have the given volt-ampere curve.

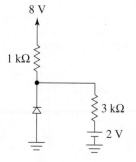

Figure P3.40

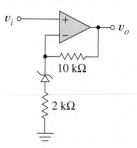

Figure P3.42

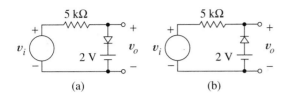

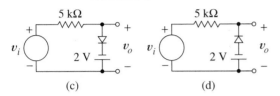

Figure P3.43

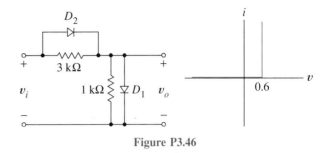

Figure P3.46

(a) When $v_i = 0$, both diodes are OFF with zero current. Redraw the circuit, replacing the diodes with appropriate models. Then write an equation that relates v_o to v_i.
(b) As v_i becomes more positive, which diode turns ON first? For what value of v_i?
(c) Redraw the circuit, changing the model for the diode that turns ON first. Write an equation relating v_o to v_i for this circuit.
(d) For what value of v_i will the second diode turn ON?
(e) Sketch the transfer characteristic of the circuit, v_o versus v_i.

3.47 (a) Design a half-wave rectifier like Fig. 3.40a so that the dc component of the output is 9.3 V and the peak current delivered by the diode is 0.1 A. Assume the diode is ideal. Find the PIV rating for the diode.
(b) Replace the ideal diode in your design with a real diode having an offset of 0.7 V. For the modified circuit, find the dc component of the output voltage, the peak diode current, and the actual PIV requirement.

3.48 The transformer and the diode in Fig. P3.48 are both ideal.
(a) Sketch and label the current waveforms $i_2(t)$ and $i_1(t)$.
(b) Part (a) is a good exercise on ideal diodes and ideal trans-

formers, but the circuit is not practical. What important restriction on real transformers is violated?

3.49 Design a bridge-type full-wave rectifier like Fig. 3.43a to give a dc output voltage component of 10 V. Use ideal diodes. Give the PIV rating required of the diodes and specify V_M. Find the value of load resistor that gives a peak diode current of 20 mA.

3.50 In the bridge rectifier of Fig. 3.43a, $v_i(t) = 10 \sin \omega t$. The diodes are silicon.
(a) Sketch the output waveform including the diode offset voltages.
(b) Compute the dc component of the output voltage, including diode offset, and compare it with Eq. (3.53).

3.51 Design a two-diode full wave rectifier like Fig. 3.46a that fits the specifications given in Problem 3.49. Find the transformer turns ratio required when 120 V line voltage is used.

3.52 Figure P3.52 shows a clamping circuit and its input voltage waveform.
(a) Carefully sketch the output waveform if the diode is ideal.
(b) Sketch the output waveform if the diode has an offset of 0.6 V.
(c) Repeat part (a) after reversing the diode orientation.
(d) Repeat part (b) after reversing the diode orientation.

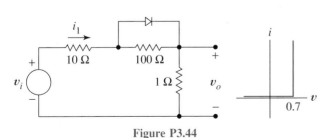

Figure P3.44

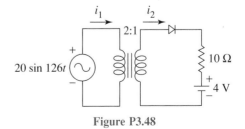

Figure P3.48

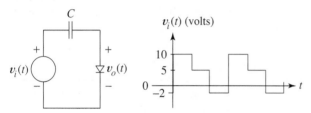

Figure P3.52

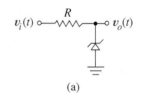

(a)

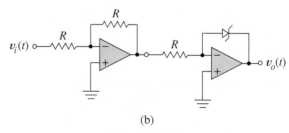

(b)

Figure P3.57

(e) Sketch $v_o(t)$ if the waveform of Fig. P3.52 provides input to the clamping circuit of Fig. 3.47a. The diode has offset of 0.6 V and $V_{BB} = 2$ V.

3.53 Draw the schematic of the peak-reading meter mentioned in the discussion of the ac to dc converter. Design the circuit so that a periodic input of 80 V peak value gives a full-scale meter current of 10 mA and so that the charged capacitor sees a resistance of at least 100 kΩ.

Hint: You may need to add something extra to the circuit to satisfy specifications.

3.54 Use SPICE to find the peak current ratings for the two diodes in Example 3.16. Add 20% to the values suggested by the simulation.

3.55 Use SPICE to find the output voltage waveform when a 100 Ω load resistor is attached to the output of the voltage doubler of Example 3.16.

Section 3.10

3.56 Sketch *each* transfer characteristic that results from replacing the zener in Fig. 3.51a by one of the feedback elements of Fig. 3.51f.

3.57 The zener diodes in Fig. P3.57a and b have $V_{on} = 0.7$ V and $V_z = 4$ V.
(a) Sketch the transfer characteristics of the circuits.
(b) Repeat part (a) after attaching a load resistor of R ohms between each output node and ground.
(c) What do you conclude from part (b)?

3.58 In Fig. P3.58, the volt-ampere characteristic defines the diode.
(a) Write an equation that relates v_o to v_i when $v_i > 0$.
(b) Write an equation that relates v_o to v_i when $v_i < 0$.
(c) Sketch the transfer characteristic.
(d) Use your transfer characteristics to sketch $v_o(t)$ when $v_i(t) = 2 \sin \omega t$.

3.59 Follow the instructions of Problem 3.58 but use the circuit and diode characteristic of Fig. P3.59.

3.60 In Fig. P3.60, both diodes are described by the given volt-ampere curve.
(a) Write two v_o versus v_i equations for the circuit, one for $v_i \geq 0$ and one for $v_i \leq 0$.
(b) Sketch the transfer characteristic for the circuit. Label slopes and critical values.

3.61 If $v_i = 8 \sin \omega t$ for the precision half-wave rectifier of Fig. 3.53, find the PIV and peak current ratings required of D_1 and of D_2. Assume each has an offset of 0.7 V.

3.62 Sketch the transfer characteristic of the circuit formed by reversing the orientation of both diodes in Fig. 3.53a.

3.63 (a) Find the input resistances of the rectifier circuits of Figs. 3.53a and 3.54.
(b) Sketch the input volt-ampere characteristic, that is, i_{in} versus v_i, for the rectifier of Fig. 3.40a.

3.64 (a) Use the precision full-wave rectifier like that of Fig. 3.54 in a system that produces a dc output current of 0–10 mA in a 100 Ω load resistor when the input is a sine wave of

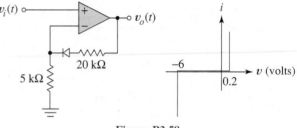

Figure P3.58

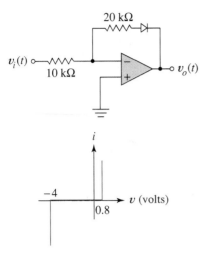

Figure P3.59

0–30 V peak value. The input resistance must be infinite and the output resistance zero. Use as many infinite gain op amps as you need.

(b) Design another system that differs from that of part (a) in the following ways. The output resistance of the new system must be infinite, and the 0–10 mA output current must be established in an arbitrary load impedance having one end attached to system ground.

3.65 The diodes in the comparator of Fig. 3.55a have volt-ampere characteristics like those in Fig. P3.58.

(a) Sketch the transfer characteristic of the comparator, v_o versus v_d, where $v_d = v_a - v_b$.

(b) Sketch the output time waveform if $v_d(t) = v_a(t) - v_b(t)$ is a square wave that instantaneously switches between -1 V and $+1$ V and the op amp has no dynamic limitations.

(c) Repeat part (b) except this time assume the op amp has a slew rate of 5 V/µs and saturation limits of ±13 V.

Hint: Draw an equivalent circuit model for the entire circuit, including op amp, valid for $v_d = -1$ V and a second equivalent

valid for $v_d = +1$ V. Then think about how the op amp acts during the transition.

3.66 In Fig. P3.66, $i_1(t)$ and $i_2(t)$ contain information about two physical events. Because of the complex internal impedances, Z_1 and Z_2, however, node voltages $v_1(t)$ and $v_2(t)$ are unusable because they contain the information in distorted form.

Design a current comparator that attaches to the given circuits and compares $i_1(t)$ and $i_2(t)$. The circuit should produce $v_o = +8$ V when $i_1 > i_2$ and $v_o = -8$ V when $i_1 < i_2$. The output should in no way be related to Z_1 or Z_2. Base your design on the *concept* of Fig. 3.55b. (The most elegant solution uses only two op amps.) Specify the diode breakdown voltages and the values of any resistors needed in your circuit. Draw a diagram showing the circuits of P3.66 connected to your comparator.

3.67 (a) Use SPICE to plot the transfer characteristic of Fig. P3.60. For the diode use a zener voltage of $V_z = 4$ V on the .MODEL line but use default values for the other diode parameters.

(b) Use SPICE transient analysis to find the output voltage when input is a sine wave of 12 V peak value and frequency of 60 Hz.

Section 3.11

3.68 In Fig. 3.58d, R requires a minimum ON current of 50 mA. Source v_C takes on only the values 0.2 volt and V_{CC} volts. Find the minimum value of V_{CC} required if $R = 200\ \Omega$ and the diode is

(a) ideal (offset = 0)

(b) silicon (offset = 0.7 V)

(c) gallium arsenide (offset = 1.2 V)

(d) germanium (offset = 0.25 V)

(e) Quantitatively analyze the power in the circuit for (b) when the switch is closed. What is the origin of the power? Where does it go?

3.69 Figure P3.69 shows a diode switching circuit that employs a fixed source I and a controlling current i_C. The output is

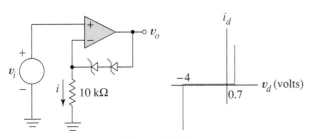

Figure P3.60

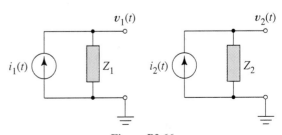

Figure P3.66

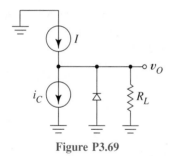

Figure P3.69

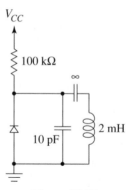

Figure P3.72

a two-valued voltage v_O. Describe how the circuit works if the diode is ideal and i_C takes on only the two values, $2I$ and 0, each value for half of the time. Include in your description the average power requirements of the two current sources.

Section 3.12

3.70 If $K = 5 \times 10^{-12}$ coulomb in Eq. (3.61),
(a) Compute how much stored charge must be moved to change the diode voltage from 0.2 V to 0.6 V,
(b) Use the definition $C = \Delta Q/\Delta V$ to find the large-signal capacitance corresponding to the two voltages of part (a).
(c) If the voltage change in part (a) must occur in 3 ns, estimate the average current needed to store the depletion region charge.

3.71 The depletion capacitance of Eq. (3.62) has $m = 0.33$, $C_{j0} = 10$ pF, and $V_{j0} = 1.0$ volt.
(a) Sketch C_{dep} as a function of voltage.
(b) On the sketch of part (a), show two regions where operation as a voltage-controlled capacitor might be impractical, and state your reasons.
(c) Find the value of diode voltage that produces a 4 pF capacitor.

3.72 Figure P3.72 shows a resonant circuit consisting of an inductor and a parallel capacitor. The capacitance seen by the inductor is the 10 pF capacitor in parallel with the depletion capacitance of the reverse-biased pn junction. The infinite capacitor is a short circuit to ac signals but allows V_{CC} to develop dc voltage across the diode. The resonant frequency of an inductor L and a parallel capacitor C is given by $\omega_o = (LC)^{-0.5}$.
If the diode has a graded junction, a barrier potential of 1 V, and zero bias capacitance of 3 pF, find the resonant frequency of the circuit for $V_{CC} = 2$ V and for $V_{CC} = 10$ V.

3.73 We saw in Chapter 1 that the bandwidth of the lowpass filter of Fig. P3.73a is $\omega_B = 1/RC$. To control the bandwidth with voltage, we use the depletion capacitance of a reverse-biased pn junction to replace C as in Fig. P3.73b.

When the small signal v_s is turned off, by superposition V_{CC} biases the diode through a voltage divider, producing the desired small-signal capacitance. When V_{CC} is turned off, because the 100 k resistor is so large, the diode sees a Thevenin equivalent approximated by v_s in series with 10 k. Therefore, signal v_s is filtered by an RC circuit like Fig. P3.73a. Find the value of V_{CC} required for a filter bandwidth of
(a) 8 MHz.
(b) 10 MHz.

3.74 For the Q versus V characteristic of Eq. (3.61), if $K = 2 \times 10^{-11}$ coulomb and $V_T = 0.9$ volt, compute the *small-signal* capacitance, dQ_d/dv_D, when
(a) $v_D = 0.2$ volt.
(b) $v_D = 0.5$ volt.
(c) $v_D = -3$ volts.

3.75 (a) Assume the circuit of Fig. P3.75 is in equilibrium when the applied voltage is -5 V. What is the diode current?
(b) What is the numerical value of the diode current *immediately after* switching to 3 V?

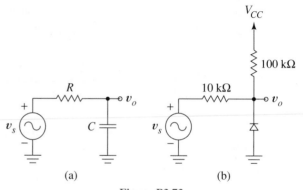

(a)   (b)

Figure P3.73

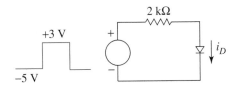

Figure P3.75

(c) Estimate the value that the diode current eventually reaches if the 3 V pulse is "long."
(d) What is the value of the diode current *immediately after* the voltage switches back to -5 V?

3.76 The circuit of Fig. 3.65a has $R = 10$ k, $V_{NN} = 5$ V, and $V_{DD} = 10$ V. In Fig. 3.65e the diode current, $i_D(t)$, is seen to reach an equilibrium value of $(V_{DD} - 0.7)/R$ just before the switching event that occurs at $t = T$.
(a) If $\tau_p = 10^{-11}$ s, find the amount of charge stored in the depletion region at this time.
(b) Assuming the $i_D(t)$ curve to be horizontal, estimate the storage time τ_s.
Hint: Ignore recombination of stored charge within the diode.
(c) Sketch the diode current and voltage waveforms and label known values.
(d) Estimate the value of V_{NN} that would halve the storage time.

Section 3.13

3.77 The diode switching circuit of Fig. 3.58d uses $V_{CC} = 5$ V and $R = 5$ kΩ. The diode has SPICE default values except for $C_{j0} = 2$ pF and $\tau_p = 12$ ns. Use a SPICE transient analysis to find the resistor current waveform when v_c is a 5 volt pulse of 500 ns duration.

3.78 The diode of Fig. 3.36a is described by the .MODEL statement in Fig. 3.67a.
(a) Use SPICE dc analysis to plot the transfer characteristic for the shaping circuit over the range $-5 \le v_i \le +5$ V.

(b) Use SPICE transient analysis to plot two cycles of $v_o(t)$ when the input is a sine wave of 5 V peak value. Starting at 1 kHz, increase the frequency of the input signal in several SPICE runs until the output waveform no longer has the shape it had at 1 kHz. Sketch such an "incorrect" output waveform and indicate the frequency at which it occurred.

Section 3.14

3.79 Use SPICE to plot diode static volt-ampere curves for
(a) a pn junction diode with $I_S = 1.0 \times 10^{-14}$ A.
(b) a Schottky diode with $I_S = 2 \times 10^{-11}$ A.

3.80 The Schottky diode has depletion capacitance but no diffusion capacitance.
(a) Describe explicitly the two parameter changes that must be made to convert a SPICE dynamic pn junction diode model into a Schottky diode model.
(b) Work Problem 3.77 for a Schottky diode following your own advice from part (a).

3.81 A solar cell has output characteristics that resemble Fig. 3.69b. The cell has active area of 0.5 cm^2 and 14% efficiency.
(a) The manufacturer specifies that the "dark current" of this device is 10 μA. Draw the curve corresponding to this condition and label it dark current.
(b) Sketch and label the curves corresponding to light intensities of 100, 200, and 300 mW/cm^2.
Hint: The cell delivers maximum power at maximum output voltage.
(c) If illumination of the cell is 250 mW/cm^2, how much power does the solar cell deliver to an external constant voltage source of $+0.2$ V connected from cathode to anode?
(d) Draw an array of solar cells, each represented by a diode symbol, that is capable of delivering 1.5 amperes at 3 V. Indicate which terminal of the array is positive and also indicate the positive current direction.

CHAPTER 4

Bipolar Transistors

4.1
Physical Principles

Bipolar junction transistors, or BJTs, are versatile, three-terminal, solid-state devices, central to the operation of many signal processing and switching circuits. In analog applications, BJTs perform a variety of functions. For example, a single operational amplifier might contain 30 or 40 transistors, used for amplifying signals, providing reference voltages and dc currents for biasing, reducing common-mode gain, providing high signal power to the external load without overheating, and protecting the op amp from destructive accidental overloads. In digital circuits, BJTs perform some of these same functions, but are most important as current-controlled switches. These electronic switches make possible high-speed decision making and, in combination with capacitance, provide digital memory.

BJTs are called *bipolar* because their operation depends upon the flow of both electrons and holes. According to physical structure, bipolar transistors fall into two categories, npn and pnp. Figure 4.1a shows the composition of the npn transistor. A *single-crystal* semiconductor is alternately doped n, p, and n, with a p-type *base,* sandwiched between n-type *emitter* and *collector* materials. The two pn junctions, base emitter and base collector, operate exactly like the isolated junctions we studied in Chapter 3 and satisfy the same equations. However, the transistor features one new principle; because they exist within the same crystal lattice, the junctions can *communicate.*

Figure 4.1a shows the two expected depletion regions with their bound ions, and also the electron potential diagram for the case when no external voltage is applied. It takes work for emitter or collector electrons to diffuse into the base against the electric fields established by the lattice ions. A similar barrier potential controls movement of holes out of the base region. As in isolated junctions, these barriers allow passage only of those charge carriers with kinetic energy exceeding the barrier potential.

Figure 4.1b shows the schematic symbol for the npn transistor. An arrowhead, always directed from the p-material to the n-material, identifies the emitter. Amplifier applications require external *biasing* voltages with polarities as shown in Figs. 4.1c and d. (We investigate polarities used for transistor switches later.) The battery in the emitter-base circuit is oriented so as to force current in the direction of the arrowhead; the base-

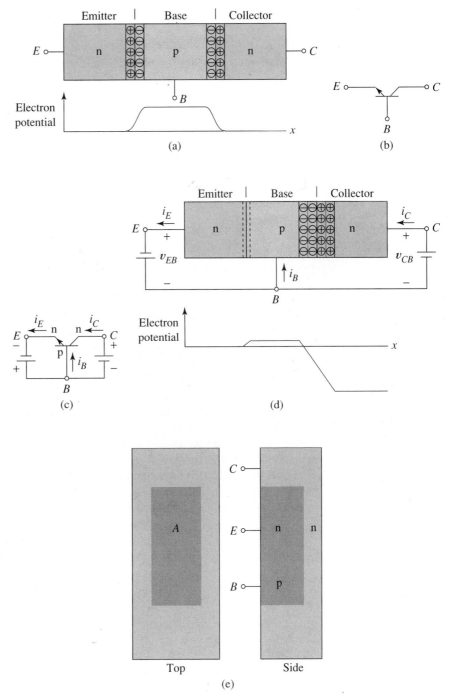

Figure 4.1 npn transistor: (a) stylized physical structure and electron potential diagram; (b) schematic symbol; (c) external biasing polarities for forward active operation; (d) depletion region widths and electron potential during forward active operation; (e) actual physical geometry.

to-collector biasing is of opposite polarity. When the base-emitter junction is forward biased and the base-collector junction reverse biased, as in the diagram, we say the transistor is in its *forward active operating mode* or *forward active state.*

As Fig. 4.1d shows, forward active biasing reduces the width of the base-emitter depletion region and increases the width of the base-collector depletion region relative to Fig. 4.1a values. The result of such biasing is the change in electron potentials shown in the two figures. Lowering the base-emitter barrier causes electrons to be continuously *injected* from the emitter into the p-type base, where they become minority carriers. These minority electrons diffuse across the base, with a large fraction α_t, the *transport factor,* reaching the base-collector depletion region and then falling down the steep potential hill. This *collection* of injected electrons is a desired event, part of the interjunction communication that characterizes the transistor.

Since the injected electrons are minority carriers while in the base, some recombine with base region holes, and thus fail to reach the collector. To minimize such recombination and make α_t as close as possible to unity, we make the base very narrow, only a fraction of a diffusion length. The side view in Fig. 4.1e shows how the actual physical transistor geometry employs a narrow base to help ensure high α_t. The collected electrons add directly to the tiny reverse saturation current of the reverse-biased base-collector junction, the two together becoming the total collector current.

The external emitter current i_E in Figs. 4.1c and d continually supplies new electrons to the emitter to replace those injected into the base, and the external collector current i_C removes electrons to make room for those that have been collected. A small base current i_B removes electrons from the base, producing new holes to replace those lost through recombination with injected electrons.

Figure 4.2 shows the internal electron flow and its relation to external currents. It also shows a secondary injection process simultaneously taking place at the emitter-base junction. The reduced barrier potential at this junction also causes *holes* to be injected from the *base into the emitter* as indicated by the blue arrow. This hole current is undesirable in the BJT because it adds to the conventional currents in the base and emitter leads without adding to the communication between the junctions. To make the fraction of injected carriers that are electrons, the *injection efficiency* γ, as close as possible to unity, we dope the emitter much more heavily than the base, that is, $N_{d,emitter} \gg N_{a,base}$. This results in a base-emitter diode equation dominated by electron current, the second term in Eq. (3.38). Because the base-collector junction is reverse biased, the collector current also contains a reverse saturation component indicated by the dashed line in Fig. 4.2. This is usually negligible compared to $\alpha_F i_E$.

In summary, the important current components of Fig. 4.2 are related in the following manner. Of the total emitter current, i_E, only a fraction, γi_E, consists of electrons. And of this, only the fraction α_t reaches the collector and contributes to collector current. Thus

$$i_C = \alpha_t \gamma i_E$$

or

$$i_C = \alpha_F i_E$$

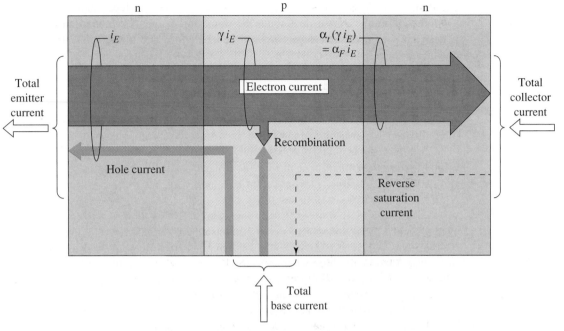

Figure 4.2 Internal charge flow and external currents of a bipolar transistor.

where $\alpha_F = \alpha_t \gamma$, the *forward alpha,* is an important parameter that describes the internal forward communication in the transistor. Values of α_F in the range 0.990 to 0.997 are common in transistors designed for analog signal processing applications.

4.2 _____
Ebers–Moll Model

So far, we have characterized the transistor as a pair of oppositely biased pn junctions that possess special "communications" embodied in the parameter α_F. Investigators Ebers and Moll introduced a more general transistor description, the *Ebers–Moll model,* which extends these introductory ideas to *arbitrary biasing conditions*. Through this generalization we come to understand the limits of forward active operation and, in the process, discover how to operate the transistor as a switch.

The Ebers–Moll equivalent circuit of Fig. 4.3 generalizes our introductory description, by showing the transistor as a pair of back-to-back diodes, with two dependent sources added to account for special interactions that can arise in the single-crystal con-

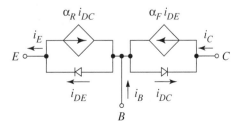

Figure 4.3 Ebers–Moll equivalent circuit.

figuration. Both diodes follow the theory and equations we developed in the last chapter, so we already know a great deal about the transistor. To generalize our understanding, we need only to justify the dependent source labeled α_R and then explore the implications of the interconnection.

Applying KCL at the collector gives

$$i_C = \alpha_F i_{DE} - i_{DC}$$

where i_{DC} is the ordinary diode current in the collector-base junction and $\alpha_F i_{DE}$ represents the special transistor action. These are exactly the components of i_C we saw in Fig. 4.2, where we assumed the base-collector junction was reverse biased; however, Fig. 4.3 assumes nothing about the junction biasing. After substituting Eq. (3.32)-type expressions for i_{DE} and i_{DC}, this equation becomes

$$i_C = \alpha_F I_{ES}(e^{v_{BE}/V_T} - 1) - I_{CS}(e^{v_{BC}/V_T} - 1) \tag{4.1}$$

where I_{ES} and I_{CS} are the reverse saturation currents of the respective junctions.

The Ebers–Moll model is _so_ general it even describes running the transistor "backward," that is, forward biasing the collector-base junction and reverse biasing the emitter-base junction. In this _reverse-active_ operating mode, electrons injected from the collector diffuse across the base and are collected at the emitter; α_R is the _reverse alpha_ of the transistor. Like α_F, α_R is the product of an injection efficiency and a transport factor, these being appropriately defined for reverse operation.

KCL at the emitter in Fig. 4.3 gives

$$i_E = I_{ES}(e^{v_{BE}/V_T} - 1) - \alpha_R I_{CS}(e^{v_{BC}/V_T} - 1) \tag{4.2}$$

Because the transistor geometry is optimized for large forward transport factor (Fig. 4.1e), the reverse transport factor is small. The collector is also doped lightly relative to the base in order to limit the expansion of the depletion region into the base. This gives a small reverse injection ratio. For these reasons the reverse alpha, α_R, is considerably less than one, ordinarily in the range 0.05–0.5, unlike the forward alpha, α_F, which is very close to one.

More complete solid-state discussions describe a _reciprocity law_

$$\alpha_F I_{ES} = \alpha_R I_{CS} = I_S$$

that relates the two reverse saturation currents and defines the quantity I_S. Using reciprocity to simplify Eqs. (4.1) and (4.2) gives the final form of the _Ebers–Moll equations_

$$i_C = I_S(e^{v_{BE}/V_T} - 1) - \frac{I_S}{\alpha_R}(e^{v_{BC}/V_T} - 1) \tag{4.3}$$

$$i_E = \frac{I_S}{\alpha_F}(e^{v_{BE}/V_T} - 1) - I_S(e^{v_{BC}/V_T} - 1) \tag{4.4}$$

Because $\alpha_F \approx 1$, I_S is essentially the reverse saturation current of the base-emitter junction, that is, the same as I_S in Eq. (3.39). Therefore I_S is directly proportional to the area A of the base-emitter junction, shown in the top view of Fig. 4.1e. In integrated circuits, this area is an important design parameter. Equations (4.3) and (4.4) show that transistors with larger area, that is, larger I_S, develop higher currents for the same junction voltages. The proportionality of I_S to n_i^2 in Eq. (3.39) and the appearance of V_T in the exponents in Eqs. (4.3) and (4.4) suggest strong temperature dependence in the transistor equations.

The circuit model, Fig. 4.3, and equations Eqs. (4.3) and (4.4) constitute a large-signal, nonlinear, static description that explains many aspects of transistor operation in both communication and switching applications. The equations are useful analysis tools when computers perform the computations; however, they lend themselves neither to simple hand calculations nor to a clear understanding of the transistor. To deal with these concerns we derive, from the Ebers–Moll equations, simplified transistor models that depict the transistor as a four-state device. Though somewhat artificial, this viewpoint is logical because most of the complexity of a transistor originates from two internal two-state devices, the pn junctions.

Transistor States. The four BJT *states* or *operating modes* correspond to the four possible ways we can bias the transistor junctions. Using external sources we can either forward or reverse bias each junction independently, giving rise to the state definition of Table 4.1. A critical feature of these definitions is that forward bias means appreciable current flows through a junction, and reverse bias means that negligible current flows. Thus it is the cut-in voltage $V_\gamma \approx 0.5$ V rather than zero volts that best distinguishes forward from reverse bias for silicon transistors. Figure 4.4 associates the state definitions with four regions in a v_{BE} versus v_{BC} plane, those delineated by dashed lines.

In *forward-active* operation, the transistor functions as a CCCS, providing the power gain needed for amplifiers and other linear applications. The states of special importance in computer and interfacing circuits are *cut-off* and *saturation* in which the transistor approximates, respectively, an open switch and a closed switch. The fourth mode, *reverse active,* is occasionally seen but not widely used.

4.3 _____
Forward Active State

At the beginning of the chapter we assumed forward active operation to introduce a new idea—interaction between two pn junctions. Now we revisit forward active operation, this time in a broader context. Starting with the Ebers–Moll description, we derive approximate characteristic curves and circuit models for this operating mode. Our principal result will be an understanding of how the BJT functions as a current-controlled current source.

TABLE 4.1 Definitions of Transistor States

Transistor State or Operating Mode	Junction Biasing	
	Base Emitter	Base Collector
Forward active	Forward $(v_{BE} \geq V_\gamma)$	Reverse $(v_{BC} \leq V_\gamma)$
Reverse active	Reverse $(v_{BE} \leq V_\gamma)$	Forward $(v_{BC} \geq V_\gamma)$
Cut-off	Reverse $(v_{BE} \leq V_\gamma)$	Reverse $(v_{BC} \leq V_\gamma)$
Saturation	Forward $(v_{BE} \geq V_\gamma)$	Forward $(v_{BC} \geq V_\gamma)$

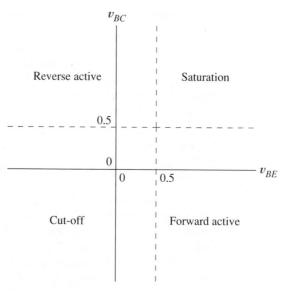

Figure 4.4 Transistor states defined by junction biasing.

4.3.1 COMMON-EMITTER CONFIGURATION

Most often the transistor operates in the *common-emitter configuration*. This means the emitter is common to input and output circuits as in Fig. 4.5a. The source in the base circuit forward biases the base-emitter junction directly. A source of higher voltage in the collector circuit reverse biases the base-collector junction.

Equations. For strong forward active operation, assume that v_{BE} is so far above cut-in in Eqs. (4.3) and (4.4) that both exponential terms involving v_{BE} are much greater than one. Also assume that v_{BC} is so much less than cut-in that both exponentials involving v_{BC} are much less than one. These assumptions give the approximate equations

$$i_C = I_S e^{v_{BE}/V_T} + \frac{I_S}{\alpha_R}$$

and

$$i_E = \frac{I_S}{\alpha_F} e^{v_{BE}/V_T} + I_S$$

Because in each equation the second term is much smaller than the first, an excellent approximation is

$$i_C = I_S e^{v_{BE}/V_T} \tag{4.5}$$

and

$$i_E = \frac{I_S}{\alpha_F} e^{v_{BE}/V_T} \tag{4.6}$$

An implication, now obvious, is

$$i_C = \alpha_F i_E \tag{4.7}$$

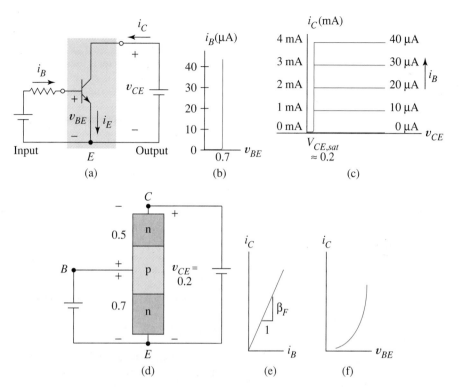

Figure 4.5 (a) BJT in common-emitter configuration; (b) common-emitter input characteristic; (c) output characteristics; (d) voltages at the limit of forward active biasing; (e) and (f) transfer characteristics.

which describes the physical process of Fig. 4.2, except for tiny reverse saturation currents in the junctions.

Although Eqs. (4.5)–(4.7) are always valid for forward active operation, equations that involve only the input and output variables in Fig. 4.5a are more useful. To derive such equations, we use Kirchhoff's current law,

$$i_E = i_B + i_C \tag{4.8}$$

to eliminate variable i_E in favor of input current i_B. Doing this in Eq. (4.7) and solving for i_C gives

$$i_C = \frac{\alpha_F}{1 - \alpha_F} i_B$$

After defining the transistor *forward beta* β_F, as

$$\beta_F = \frac{\alpha_F}{1 - \alpha_F} \tag{4.9}$$

we write

$$i_C = \beta_F i_B \tag{4.10}$$

By contrast with α_F, a number close to one and hard to measure, β_F typically falls in the range of 100 to 300 and is easy to measure. Thus β_F is the parameter more often used in transistor circuit analysis and design.

We next derive an input description. Substituting Eq. (4.10) into Eq. (4.8) gives

$$i_E = i_B + \beta_F i_B$$

or

$$i_E = (\beta_F + 1)i_B \tag{4.11}$$

We now replace i_E in Eq. (4.6) with Eq. (4.11) and solve for i_B. This gives the common-emitter *input equation*

$$i_B = \frac{I_S}{\alpha_F(\beta_F + 1)} e^{v_{BE}/V_T} \tag{4.12}$$

Characteristic Curves. The input characteristic of the common-emitter transistor is a plot of Eq. (4.12). To better relate this curve to circuit models and practical operating ranges, we sometimes sketch it in stylized form, as in Fig. 4.5b.

The output characteristics of the transistor, graphical representations of Eq. (4.10), are the family of curves shown in stylized form for $\beta_F = 100$ as Fig. 4.5c. When actually measured, the curves of constant base current do not continue into the second quadrant, but instead appear to converge into a single vertical line called the *collector-emitter saturation voltage*, $V_{CE,sat}$. Figure 4.5d shows that this occurs because our assumption of a reverse-biased base-collector junction becomes invalid. With v_{BE} at 0.7 V and with v_{BC} biased just at cut-in ($V\gamma \approx 0.5V$), v_{CE} is about +0.2V. Thus, for $v_{CE} = V_{CE,sat} \approx$ 0.2 V, the base-collector junction is no longer reverse biased and the forward active equations reach the limits of their validity—the saturation region, which we examine later.

The forward active BJT has two transfer characteristics. When the input variable is base current, we have Fig. 4.5e, a sketch of Eq. (4.10). It is sometimes useful to use voltage as the transistor's input variable, for example, when a transistor operates at low currents. For this case we use the transfer characteristic of Fig. 4.5f, a sketch of Eq. (4.5). This exponential description allows us to visualize and compute more accurately input voltages for small currents instead of simply assuming $V_{BE} = 0.7$ as prompted by Fig. 4.5b.

Large-Signal Equivalent Circuit. We learned in Chapter 1 that circuit models, or *equivalent circuits,* help us understand how electronic devices interact with other circuit elements. The *large-signal transistor models* developed in this section are used for studying digital logic circuits and for analyzing and designing dc biasing circuits that are required for amplifiers.

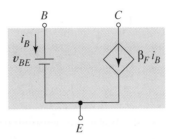

Figure 4.6 Large-signal circuit model for the common-emitter transistor.

Common-Emitter Model. From the nearly constant voltage in Fig. 4.5b and the current-controlled curves of Fig. 4.5c, we infer the *common-emitter model* of Fig. 4.6. For silicon transistors V_{BE} is 0.7 V. When the BJT schematic appears in a circuit diagram, we mentally visualize this model. This helps us recognize the circuit constraints of the transistor: the base must be V_{BE} volts positive relative to the emitter, and β_F times the base current flows into the collector *independent of* v_{CE} (because of the current source), provided that $v_{CE} \geq V_{CE,sat}$. Notice from the diagram that Kirchhoff's current law implies Eq. (4.11). Also sometimes useful is

$$i_E = \frac{(\beta_F + 1)}{\beta_F} i_C \qquad (4.13)$$

which is easy to derive.

4.3.2 MINORITY CHARGE STORAGE

We can improve our overall understanding of forward active operation by examining minority charge storage near the junctions. From studying the forward-biased pn junction, we might expect the minority charge concentration in the base to be an exponential function like $n(x)$ of Fig. 3.60. Instead it is approximately linear as in Fig. 4.7. This is because base width W is much smaller than an electron diffusion length, and because the electric field at the collector junction sweeps away electrons as fast as they arrive, pinning down the concentration at the collector end of the base to zero.

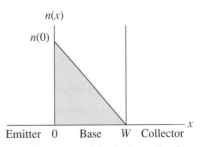

Figure 4.7 Excess minority charge stored in the base of a forward active npn transistor.

Diode theory, specifically Eq. (3.34), for injected holes, tells us that the *electron concentration* at the emitter end of the base should be controlled by the base-emitter junction voltage $v_D = v_{BE}$; that is,

$$n(0) = \frac{n_i^2}{N_a} e^{v_{BE}/V_T} \tag{4.14}$$

where n replaces p and N_a replaces N_d in Eq. (3.34) because we are describing injected electrons rather than holes. We next show that the slope, $-n(0)/W$, of this concentration profile is proportional to collector current.

Using Eq. (4.14), the slope is

$$|\text{slope}| = \frac{n(0)}{W} = \frac{n_i^2}{WN_a} e^{v_{BE}/V_T} \tag{4.15}$$

Next we substitute the exponential factor from Eq. (4.5), giving

$$|\text{slope}| = \frac{n_i^2}{WN_a I_S} i_C \tag{4.16}$$

This development is consistent with visualizing the transistor as a voltage-controlled device, by the following reasoning. Base-emitter voltage in Eq. (4.15) is the independent variable that controls $n(0)$, and thus the slope of the minority charge distribution in Fig. 4.7. Since electrons in the base flow by diffusion, changing this slope controls the collector current according to Eq. (4.16). As in the diode, we again characterize the essential operation of an important device in terms of a tiny concentration of minority charge next to a junction. We shall eventually see that this minority charge concentration curve also has a great deal to do with transistor dynamics.

4.4
Cut-Off, Saturation, and Reverse Active States

We derived approximate equations, characteristic curves, and circuit models for the forward active state by applying our state definitions directly to the Ebers–Moll equations. We could also examine the other transistor states in this fashion; however, we leave this for homework problems. Instead we use transistor characteristic curves, the circuit of Fig. 4.3, and our intuition to explore the remaining three states, an approach that provides a good overall perspective of the transistor.

4.4.1 TRANSISTOR STATES AND THEIR REGIONS OF OPERATION

Figure 4.8b shows in stylized form the output characteristics of a common-emitter transistor; each operating mode corresponds to operation in a particular region. First, we must understand that each base current value, for example, $i_B = I_{B1}$, labels a *continuous curve* that extends from the first quadrant into the third quadrant. The regions in the first and third quadrants where the lines of constant base current appear to merge into a single curve correspond to *saturation*. If we expand the v_{CE} scale in the saturation region, we can distinguish individual curves; however, the scale of Fig. 4.8b is more appropriate for purposes of simple modeling. At this scale the first quadrant saturation region resembles a constant voltage line at the *saturation voltage* $V_{CE,sat}$. For a silicon transistor,

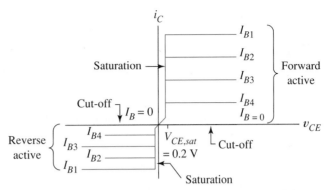

Figure 4.8 Regions of operation for the four transistor states in terms of the output characteristic curves.

$V_{CE,sat} \approx 0.2$ V. The voltage axis, labeled $I_B = 0$, is the *cut-off region*. Forward active operation occurs in the first quadrant region bounded by saturation and cut-off, *reverse active* in the third quadrant region bounded by saturation and cut-off. Because forward active operation is frequent and reverse active operation rather rare, hereafter the term *active* implies forward active operation. Also, the symbol β hereafter denotes β_F unless otherwise indicated.

4.4.2 CUT-OFF

When both junctions are reverse biased, the diodes in Fig. 4.3 approximate open circuits. Since diode currents control the dependent sources, these sources also turn off. The result is the *cut-off model* of Fig. 4.9a, which is adequate for most purposes. This circuit is consistent with operation in the cut-off region of Fig. 4.8: $i_C = i_B = 0$.

Figure 4.9b, a cut-off model of improved accuracy, is useful for investigating cut-off transistors at high temperatures. The temperature-sensitive dc current, $I_{CB0,}$ is measured by open circuiting the emitter lead and measuring the collector-to-base current that results from reverse biasing the collector-base junction. This value is often given on the manufacturer's data sheet for the transistor. I_{CB0} is quite small, typically of the order of picoamperes.

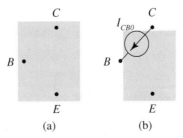

Figure 4.9 Large-signal cut-off models: (a) simple model; (b) model for high temperatures.

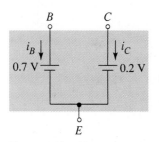

Figure 4.10 Circuit model for a saturated transistor.

4.4.3 SATURATION

Now consider the first quadrant saturation region in Fig. 4.8. Because the curves merge into a constant v_{CE} line, a battery of $V_{CE,sat} \approx 0.2$ V connects the collector and emitter in the saturation model of Fig. 4.10. Since the base-emitter junction is forward biased, the saturation model employs a dc input source of $V_{BE} = 0.7$ V between base and emitter. In saturation the transistor does *not* function as a CCCS, and $i_C = \beta_F i_B$ usually does *not* apply. In fact, we shall establish that a necessary condition for a transistor to be saturated is

$$\beta i_B \geq i_C \tag{4.17}$$

4.4.4 REVERSE ACTIVE OPERATION

In reverse active operation, the roles of emitter and collector are reversed compared to forward active. Thus the equivalent in Fig. 4.11a shows the 0.7 V source connected between base and collector and the dependent source directed from emitter to collector. The emitter current is $\beta_R i_B$ where

$$\beta_R = \frac{\alpha_R}{1 - \alpha_R}$$

defines the *reverse beta* of the transistor. Comparing Figs. 4.11a and b we see that our new model redefines the reference direction of the emitter current. In terms of the current reference directions we have been using, $\beta_R i_B = -i_E$. By Kirchhoff's current law in Fig. 4.11b

$$i_C = i_E - i_B = -(\beta_R + 1)i_B$$

This equation describes the third quadrant curves in Fig. 4.8. Because $\beta_R + 1 << \beta_F$, the third quadrant curves are more closely spaced than the first quadrant curves.

The load line, introduced in the next section, helps us understand how to bias the transistor for any of these operating modes using resistors and voltage sources.

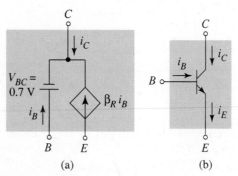

(a)   (b)

Figure 4.11 Model for BJT in reverse active operation.

4.5

Load Line

The *load line* is a graphical aid we use to find device currents and voltages when the device is described by its characteristic curves. Even when the device curves are not available (often), the load line is still a powerful conceptual aid for visualizing device operation.

4.5.1 LOAD LINE CONSTRAINT

Consider the problem of finding input current and voltage, i_B and v_{BE}, in Fig. 4.12a when V_{BB} and R_B are known. The input quantities must simultaneously satisfy two constraints. As coordinates of a point on the transistor input characteristic, i_B and v_{BE} must lie somewhere on that nonlinear curve. They must also satisfy a linear constraint imposed by the external circuit. From Fig. 4.12a this is

$$i_B = \frac{V_{BB} - v_{BE}}{R_B} \qquad (4.18)$$

On the i_B versus v_{BE} coordinate system that we use for the input characteristic, Eq. (4.18) describes a straight line of slope $-1/R_B$ called the *input load line*. It is easy to construct this load line because it always passes through the points

$$(v_{BE}, i_B) = (V_{BB}, 0)$$

and

$$(v_{BE}, i_B) = (0, V_{BB}/R_B)$$

and has slope of $-1/R_B$, as we see in Fig. 4.12b. The intersection of the load line and the transistor's input characteristic represents the simultaneous solution of the device and circuit constraints. The point of intersection is called the *quiescent operating point*, or *Q-point*. For the load line in the figure, the base current is 20 μA and the base-emitter voltage is 0.7 V.

We can change the Q-point by changing R_B. If we hold V_{BB} constant, we change the slope of the load line but not its voltage axis intercept. In Fig. 4.12c we see that a smaller resistor might increase the base current to 30 μA; a larger resistor might give $i_B = 10$ μA. For sufficiently large R_B, v_{BE} drops below 0.7 V. We can also move the Q-point by changing V_{BB} and leaving resistance fixed. This slides the load line to the right or left without a change in slope.

Now consider finding the output quantities, i_C and v_{CE} in Fig. 4.12a. Once base current is fixed at $i_B = I_B$, the output variables are constrained to lie somewhere on *the single transistor output curve labeled I_B*, for example, the bold $i_B = 20$ μA curve in Fig 4.12d. Collector current and voltage must also satisfy the collector *circuit constraint* of Fig. 4.12a, namely,

$$i_C = \frac{V_{CC} - v_{CE}}{R_C}$$

To find the point that simultaneously satisfies both constraints, we construct a load line on the output characteristic and find its intersection with the $i_B = I_B$ curve. The output

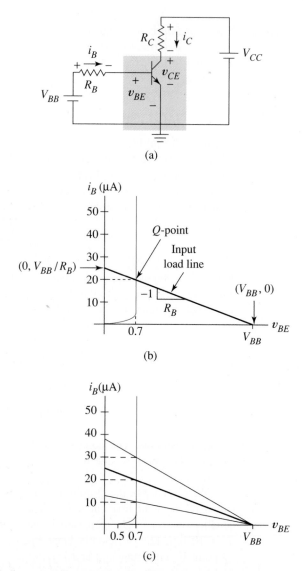

Figure 4.12 Finding operating point by using load lines: (a) transistor circuit; (b) input circuit load line and operating point; (c) changes in operating point with changes in R_B; (d) output load line and operating point.

load line always passes through the points $(v_{CE}, i_C) = (V_{CC}, 0)$ and $(v_{CE}, i_C) = (0, V_{CC}/R_C)$, and its slope is always $-1/R_C$. The output circuit load line is shown in Fig. 4.12d, for $V_{CC} = 8$ V and $R_C = 2$ kΩ. If the base current is set at 20 μA by the input biasing circuit, we see that the quiescent operating point is $(v_{CE}, i_C) = (4$ V, 2 mA). Since the voltage coordinate of the output circuit Q-point is $V_{CE} = 4$ V, we can also *graphically* find the voltage drop across R_C, $V_{CC} - v_{CE} = 8 - 4 = 4$ V.

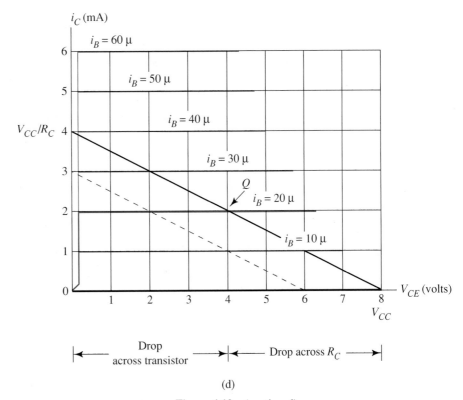

(d)

Figure 4.12   (continued)

If the base current is now raised to 30 μA, by reducing the resistance in series with the base, the Q-point moves to the intersection of the load line and the $i_B = 30$ μA curve, $(V_{CE}, I_C) = (2$ V, 3 mA$)$. At this new Q-point, there would be $8 - 2 = 6$ volts across R_C and 2 V across the transistor. Since R_C is in series with the transistor's collector, the resistor current also equals $i_C = 3$ mA.

We can change the output circuit Q-point by changing V_{CC} and/or R_C. For example, Fig. 4.12d shows that with i_B fixed at 20 μA, lowering V_{CC} to 6 V shifts the load line to the dashed location, making $v_{CE} = 2$ V.

Exercise 4.1   (a) Find the base current for the transistor in Fig. 4.12a if $V_{BB} = 5$ V and $R_B = 430$ kΩ.
Hint: Assume the transistor operates on the vertical portion of Fig. 4.12c. Then check your assumption.
(b) Given the base circuit biasing of part (a), use a load line and Fig.4.12d to find the Q-point if $V_{CC} = 6$ V and $R_C = 1.2$ kΩ.

Ans. $I_B = 10$ μA, $(V_{CE}, I_C) = (4.8$ V, 1 mA$)$.

In Sec. 4.9 we learn that transistor output characteristics are not always horizontal and evenly spaced as we like to assume. Problem 4.13 demonstrates that load line techniques work for these more realistic cases.

We next use load lines to show how the output Q-point can be positioned at any desired first quadrant location. The development also adds greatly to our understanding of transistor saturation.

4.5.2 LOAD LINE STUDY OF SATURATION

To understand how a transistor saturates in a circuit, we consider the effect of gradually increasing V_{BB} in Fig. 4.13a from 0 to 8 volts. When $V_{BB} = 0$, load line 1 of Fig. 4.13b shows $i_B = 0$. On the output load line of Fig. 4.13c, for $i_B = 0$ the Q-point is at (8V, 0 mA)—the transistor is cut-off. As V_{BB} increases from 0 to $V\gamma$ in Fig. 4.13b, i_B remains at zero, the transistor remains cut-off, and the Q-point in Fig. 4.13c does not change.

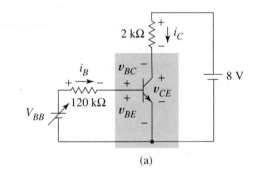

(a)

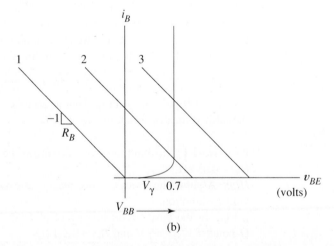

(b)

Figure 4.13 Use of external sources to establish a transistor operating point: (a) circuit diagram; (b) input load lines; (c) output characteristics and load line; (d) graphical meaning of a saturated transistor.

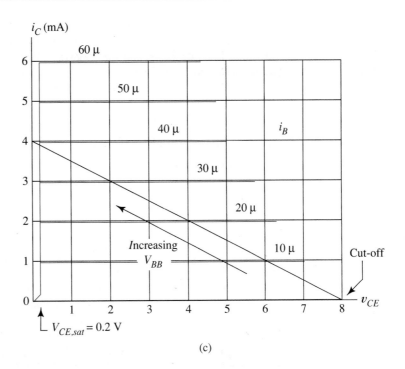

(c)

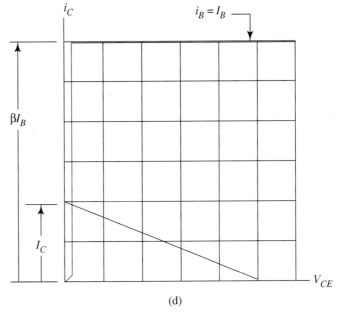

(d)

Figure 4.13 (continued)

For $V_{BB} > V\gamma$, the base current increases with V_{BB} (see 2 in Fig. 4.13b), and therefore the Q-point moves up the load line in Fig.4.13c into the active region. Once the input load line moves beyond the 2 position in Fig. 4.13b, base current increases with V_{BB} according to

$$i_B = \frac{V_{BB} - 0.7}{120 \text{ k}}$$

the value we would predict using our equivalent circuit.

When the base current reaches 39 μA in Fig. 4.13c, the transistor arrives at saturation. Since the base-emitter voltage drop is 0.7 V, we predict from the equation above that this *edge of saturation* value of base current, $I_{B,EOS}$, occurs when

$$i_{B,EOS} = 39 \text{ μA} = \frac{V_{BB,EOS} - 0.7}{120 \text{ k}}$$

or $V_{BB,EOS} = 5.38$ V. *At this one point,* the active and saturation equivalent circuits are *both* correct. That is, $i_C = \beta_F i_B$ from Fig. 4.6, and $v_{CE} = 0.2$ V from Fig. 4.10 are *both true.* For *further* increases in V_{BB}, *base current continues to increase* (see Fig. 4.13b), but the *collector current remains constant* at its edge-of-saturation value, approximately 3.9 mA in Fig. 4.13c. Because the curves for $i_B \geq 39$ μA all descend together along the vertical saturation line, the Q-point remains at the intersection of the load line and the line $v_{CE} = V_{CE,sat}$. This is the reason that Eq. (4.17) is necessary for a transistor to be saturated.

When there is a need for a saturated transistor, it is usually appropriate to drive the transistor *deeply* into saturation, meaning $\beta_F i_B >> i_C$. Figure 4.13d shows the idea graphically. For example, when $V_{BB} = 8$V in Fig. 4.13a,

$$I_B = \frac{8 - 0.7}{120 \text{ k}} = 60.8 \text{ μA}$$

For this base current, Fig. 4.13c shows that $\beta_F I_B \approx 6.0$ mA while $I_C \approx 3.9$ mA. A quantitative measure of the degree of saturation of a saturated transistor is the *forced beta,* defined *for the saturated transistor* by

$$\beta_{forced} = \frac{i_C}{i_B} \bigg|_{\text{saturated transistor}}$$

The equality $\beta_{forced} = \beta_F$ applies to a transistor just at the edge of saturation. $\beta_{forced} < \beta_F$ implies a transistor driven into the saturation region. When the base current in Fig. 4.13c is 60 μA, for example, $\beta_{forced} = 3.9$ mA/60 μA = 65. From the spacing of the characteristic curves we see that $\beta_F = 100$, thus $\beta_{forced} < \beta_F$ when $I_B = 60$ μA.

Since saturated transistors often operate with very high base currents, some authors use a value of 0.75 or 0.8 V for the base-emitter voltage in the saturation model of Fig. 4.14. Also, some authors use $V_{CE,sat}$ values less that 0.2 for a deeply saturated transistor. In this book we consistently use $V_{BE} = 0.7$ V and $V_{CE,sat} = 0.2$ V for a saturated transistor unless otherwise indicated.

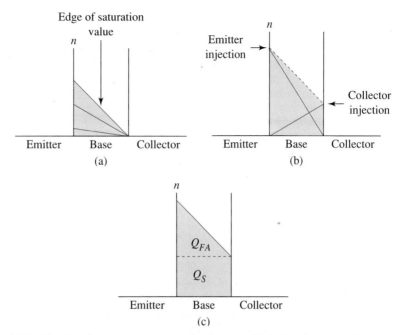

Figure 4.14 Minority electron concentration in the base: (a) during forward active operation; (b) as transistor is driven into saturation; (c) for a saturated transistor.

Minority Charge Storage for the Saturated Transistor. Figure 4.14 shows how minority charge concentration builds up in the base as V_{BB} in Fig. 4.13 increases. When the transistor is active, increases in v_{BE} cause a higher emitter-end electron concentration, greater slope, and higher collector current as in Fig.4.14a—until the edge of saturation value is reached. Once the transistor enters saturation, the excess minority stored charge profile in the base changes radically. At saturation the collector-emitter voltage *clamps* or becomes constant at approximately 0.2 V. Since

$$v_{BE} = v_{BC} + v_{CE} = v_{BC} + 0.2$$

additional increments of v_{BE} translate into *equal* increments of v_{BC} in the saturated transistor, causing electrons to be injected into the base from *both* collector and emitter. The resultant minority charge concentration (dashed line) in Fig. 4.14b is the superposition of the individual concentrations created as identical increases in v_{BE} and v_{BC} increase the heights of both solid triangles. As we drive the transistor more deeply into saturation, the minority charge concentration increases while the slope of the charge concentration, and therefore collector current, remain constant.

For a saturated transistor, it is useful to view the minority charge concentration as the superposition of rectangular and triangular concentration profiles as in Fig. 4.14c. We can compute the total minority charge Q_T stored in the base from

$$Q_T = qA \int n(x)dx = Q_{FA} + Q_S$$

where q and A denote the electron charge and junction area. This equation expresses Q_T as the sum of Q_{FA}, caused by forward active operation at the edge of saturation, and Q_S, the extra charge added as the transistor is driven into saturation. In Fig. 4.14c each part of the concentration curve is labeled with the minority charge it contains in the three-dimensional transistor base. To bring the transistor out of saturation, it is first necessary to remove Q_S coulombs of charge from the base. Only after this is done can we restore active operation. This important dynamic effect is discussed in Sec. 4.11.

4.6
pnp Transistor

Figures 4.15a and b show the physical structure and the schematic symbol for a *pnp transistor*. For forward active operation we forward bias the emitter-base junction, injecting holes into the base. The fraction α_t which diffuse across the base without recombining are collected at a reverse-biased CB junction. In fact, operation is the same as for the npn transistor, except that the roles of electrons and holes are interchanged. Consequently, all currents and voltages have algebraic signs opposite those of the npn transistor. The reference directions in Fig. 4.15a are the actual directions of the conventional dc currents. In the pnp schematic of Fig. 4.15b the emitter arrow points in the p-to-n direction relative to the physical transistor. The battery, which forward biases the emitter-base junction, forces conventional emitter current in the arrow direction.

A development that exactly mirrors the details worked out for the npn transistor shows that the Ebers–Moll model of Fig. 4.15c leads to a mathematical description, characteristic curves, and circuit models. Only the results are summarized here. Figure 4.16a shows the polarity of the biasing sources required for forward active operation in the common-emitter configuration. Figures 4.16b and c show, respectively the

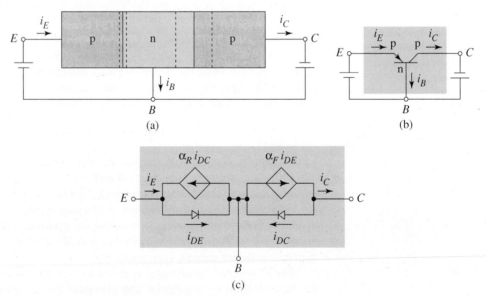

Figure 4.15 pnp transistor: (a) physical structure biased for forward active operation; (b) schematic symbol; (c) Ebers–Moll model.

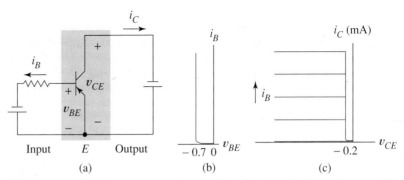

Figure 4.16 pnp transistor in common-emitter configuration: (a) biasing and current directions; (b) input characteristic; (c) output characteristic.

input and output characteristics. In this text, we use second quadrant characteristics for pnp devices in the interest of using identical equations wherever possible for both transistor classes. Others prefer to reflect each curve into the first quadrant by reversing the subscript order, then using v_{EB} and v_{EC} as the independent variables.

Figure 4.17 shows the forward active, cut-off, and saturation equivalent circuits for the pnp transistor. The cut-off equivalents for pnp and npn transistors are identical; active and saturation circuit models differ only in the source orientations.

4.7

Q-Point Analysis

Circuit analysis means finding the voltages and currents in a given circuit when component values are known. A complication in circuits that contain BJTs is that each transistor may be in any of four possible states. When transistor characteristic curves are available and the circuit is rather simple, we can use the load line method of Section 4.5 for circuit analysis. For more general circuits we use transistor equivalent circuits. We first learn to analyze a circuit when the state of the BJT is known. Then we examine the more general analysis problem.

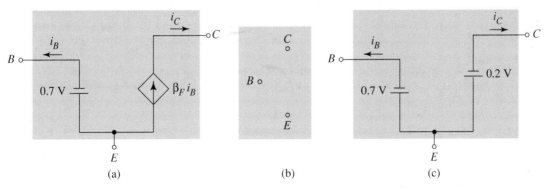

Figure 4.17 pnp large-signal models: (a) forward active; (b) cut-off; (c) saturation.

4.7.1 ANALYSIS WHEN TRANSISTORS ARE ACTIVE

In amplifiers and other linear circuits, transistors operate in their forward active states. Knowing this, we can replace each transistor with the model of Fig. 4.6 or 4.17a, and then analyze the resulting equivalent circuit. The objective of the circuit analysis usually includes finding the Q-point coordinates, (V_{CE}, I_C), for each transistor. Another technique, infinite beta analysis, is less accurate, but useful in special cases.

Analysis Using Large-Signal Circuit Model. The following examples and exercises illustrate analysis using the forward active model.

EXAMPLE 4.1   The transistor in Fig. 4.18a is biased for forward active operation. Find the dc operating point, (V_{CE}, I_C), of the transistor.

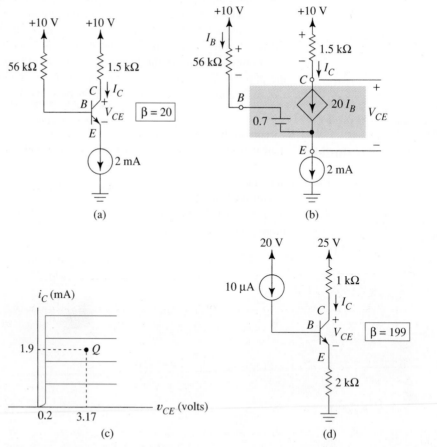

Figure 4.18   (a) Simple amplifier circuit; (b) equivalent circuit; (c) transistor operating point; (d) circuit for Exercise 4.2.

Solution. We first replace the transistor with its forward active model, Fig. 4.6, as in Fig. 4.18b. Applying KCL at the emitter node gives

$$I_B + 20I_B = 2 \text{ mA}$$

Thus, $I_B = 0.095$ mA and $I_C = 20 \times I_B = 1.90$ mA, the first Q-point coordinate. In terms of node voltages, the second coordinate is

$$V_{CE} = V_C - V_E$$

Because the voltage difference between the collector node and the 10 V supply node is the drop in the 1.5 k resistor,

$$V_C = 10 - 1.5 \text{ k}I_C = 7.15 \text{ V}$$

To find V_E, start at the supply and subtract drops along the path through the base circuit. The result is

$$V_E = 10 - 56 \text{ k}I_B - 0.7 = 3.98 \text{ V}$$

(When summing voltage drops between node pairs in this fashion, always avoid paths that contain current sources, because current source voltages are unknown.) From V_C and V_E,

$$V_{CE} = 7.15 - 3.98 = 3.17 \text{ V}$$

Visualizing our results in terms of the output characteristics as in Fig. 4.18c confirms that this operating point is well within the forward active region as assumed. ❏

> **Exercise 4.2** Find the Q-point coordinates and the base voltage for the transistor in Fig. 4.18d.
>
> *Ans.* $(V_{CE}, I_C) = (19 \text{ V}, 1.99 \text{ mA})$, $V_B = 4.7$ V.

EXAMPLE 4.2 Find the transistor Q-point in Fig. 4.19a.

Solution. Figure 4.19b is equivalent to Fig. 4.19a because the two voltage sources place exactly the same constraint on the circuit; that is, they maintain the node voltages at the tops of the 47 kΩ and 39 kΩ resistors at +15 V relative to ground. This equivalence is worth remembering.

Next, draw the Thevenin equivalent of the input circuit and replace the transistor with its large-signal model as in Fig. 4.19c. Applying Kirchhoff's voltage law to the left-hand loop gives

$$8.87 - 27.8 \text{ k} \times I_B - 0.7 - 68 \text{ k}(101 \ I_B) = 0$$

from which

$$I_B = 1.18 \ \mu\text{A}$$

and

$$I_C = 100 \times I_B = 0.118 \text{ mA}$$

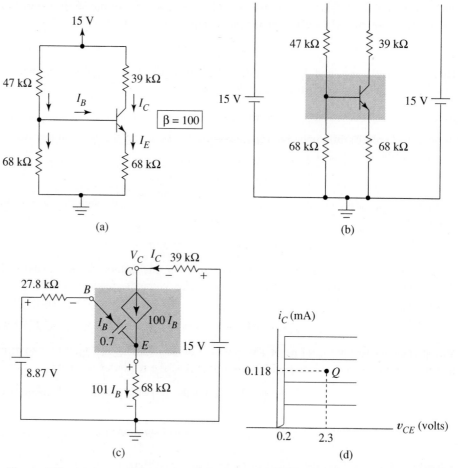

Figure 4.19 (a) An amplifier biasing circuit; (b) amplifier with equivalent power sources; (c) equivalent circuit; (d) transistor operating point.

To find the node voltage at the collector, add voltage rises along the right-hand path. This gives

$$V_C = 15 - 39\,\text{k}(0.118\ \text{mA}) = 10.4\ \text{V}$$

Also,

$$V_E = 68\,\text{k} \times 101(1.18\ \mu\text{A}) = 8.1\ \text{V}$$

giving

$$V_{CE} = 10.4 - 8.1 = 2.3\ \text{V}$$

From the sketch of Fig. 4.19d, the transistor is biased for active operation. ❏

Exercise 4.3 Does the transistor in Fig. 4.19a remain in the active region if the 68 kΩ emitter resistor is changed to 50 kΩ?

Ans. Yes. Q-point changes to (0.59V, 0.161 mA), which is still in the active region.

EXAMPLE 4.3 The amplifier of Fig. 4.20a contains a pnp transistor. Find the Q-point.

Solution. Using the pnp model of Fig. 4.17a gives Fig. 4.20b. A KVL equation along the path that avoids the currents source gives

$$-9 + 1 \text{ k}(31 I_B) + 51 \text{ k} I_B + 0.7 + 2.4 \text{ k}(31 I_B) = 0$$

Solving gives $I_B = 53.1$ μA and $I_C = 1.59$ mA. The emitter node voltage is

$$V_E = 0 - 2.4 \text{ k}(31 I_B) = -3.95 \text{ V}$$

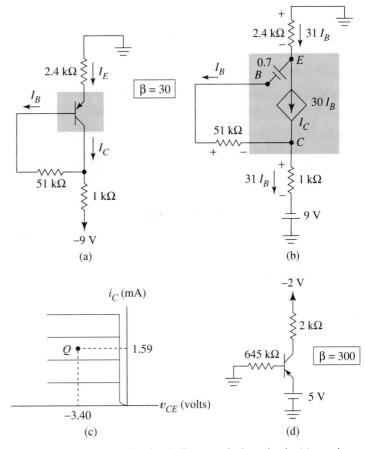

Figure 4.20 (a) Amplifier biasing circuit; (b) linear equivalent circuit; (c) transistor operating point; (d) circuit for Exercise 4.4.

and the collector node voltage is

$$V_C = -9 + 1\,\text{k}(31)53.1 \times 10^{-6} = -7.35 \text{ V}$$

Therefore, $V_{CE} = -7.35 - (-3.95) = -3.40$ V. Figure 4.20c shows the operating point on the output characteristics and confirms active operation. ❏

> **Exercise 4.4** Find V_{CE} for the transistor in Fig. 4.20d.
>
> *Ans.* $V_{CE} = -3$ V.
>
> **Infinite Beta Analysis.** In Chapter 2 we used an infinite gain assumption to analyze circuits containing high-gain op amps. A similar technique helps us *estimate* BJT operating points, especially for those integrated circuits that are too complicated to analyze by hand. This method, *infinite beta analysis,* makes three assumptions about each transistor:
>
> **1.** $V_{BE} = 0.7$ V for npn transistors, -0.7 for pnp.
> **2.** $I_B = 0$.
> **3.** $I_C = I_E$.
>
> Assumptions 2 and 3 imply $\alpha_F = 1$ in Eq. (4.9), hence infinite beta analysis. It is most efficient to work directly on the *given schematic,* adding to the diagram the value of each node voltage and current as it is found.
>
> Example 4.4 demonstrates the technique. It also demonstrates using node and branch voltage variables and single and double subscript notation in ways that are helpful, both in understanding circuit operation, and in efficient hand calculations.

EXAMPLE 4.4 Analyze the circuit of Fig. 4.21a assuming infinite beta transistors.

Solution. Since the base current of Q_1 is zero by assumption 2, the 18 and 27 kΩ resistors form a voltage divider. The result is a base voltage of

$$V_{B1} = \frac{27\,\text{k}}{27\,\text{k} + 18\,\text{k}} 24 = 14.4 \text{ V}$$

for Q_1 as in Fig. 4.21b. In this diagram, blue characters denote quantities added to the diagram as they are computed.

The emitter of Q_1 is 0.7 V below the base by assumption 1; therefore, $V_{E1} = 14.4 - 0.7 = 13.7$ V. Since the base current of Q_2 is zero,

$$I_{E1} = 1 \text{ mA}$$

By assumption 3, this is also the collector current of Q_1. At the collector of Q_1

$$V_{C1} = 24 - 5.6\,\text{k}I_{C1} = 18.4 \text{ V}$$

Using node voltages already added to the diagram gives

$$V_{CE1} = V_{C1} - V_{E1} = 18.4 - 13.7 = 4.7 \text{ V}$$

To summarize, the Q-point of Q_1 is

$$(V_{CE1}, I_{C1}) = (4.7 \text{ V}, 1 \text{ mA})$$

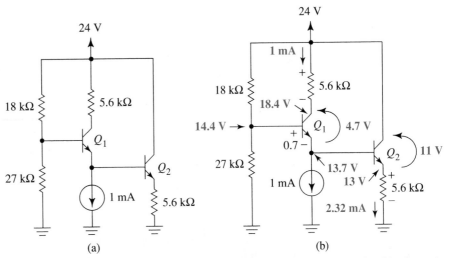

Figure 4.21 Infinite beta analysis of a biasing circuit: (a) circuit schematic; (b) schematic showing voltage and current values added during analysis.

Since the emitter of Q_2 is 0.7 V below V_{E1},

$$V_{E2} = V_{E1} - 0.7 = 13.7 - 0.7 = 13 \text{ V}$$

The emitter current of Q_2 flows through the 5.6 kΩ resistor; thus

$$I_{E2} = \frac{13.0}{5.6 \text{ k}} = 2.32 \text{ mA} = I_{C2}$$

For Q_2,

$$V_{CE2} = V_{C2} - V_{E2} = 24 - 13 = 11 \text{ V}$$

so its Q-point is

$$(V_{CE2}, I_{C2}) = (11.0 \text{ V}, 2.32 \text{ mA})$$

A SPICE simulation using β = 100 for both transistors gives operating points:

$$Q_1: (5.05\text{V}, 1.05 \text{ mA})$$

$$Q_2: (11.74 \text{ V}, 2.09 \text{ mA})$$

We conclude that our infinite beta results were reasonable estimates considering the complexity of the circuit and the slight effort required. ❑

Exercise 4.5 Use infinite beta analysis to estimate the transistor Q-point in Fig. 4.19a.

Ans. $(V_{CE}, I_C) = (2.13 \text{ V}, 0.120 \text{ mA})$ compared with the more accurate result, (2.3V, 0.118 mA), calculated in Example 4.2 using the large-signal model.

Cautions About Infinite Beta Analysis. Infinite beta analysis tacitly assumes a circuit that was designed to be relatively independent of β. If the circuit is poorly designed or if β is small, the results can be quite inaccurate. For these reasons, analysis using the transistor circuit model remains our method of choice; we use infinite beta analysis only for circuits too complex for ordinary hand analysis, for example, large integrated circuits, and then we entertain a healthy skepticism until our answers are verified by simulation.

The assumed base-emitter drop of 0.7 V sometimes leads to errors in infinite beta analysis as well as in analysis with equivalent circuits. We discuss this next.

Analysis of Circuits with Low Current Levels. Whenever we approximate the BJT's exponential input characteristic by the constant $V_{BE} = 0.7$ V, we tacitly assume an emitter current of the order of milliamperes. This assumption is reasonable in most circuits, but fails for circuits that operate at low current levels. For low currents a better transistor representation is Eq. (4.5), which represents the transistor as a *nonlinear VCCS*.

Doing hand analysis with Eq. (4.5) can be tedious. Usually we must estimate V_{BE}, find the corresponding i_C by circuit analysis, then use i_C and Eq. (4.5) to update $V_{BE}, \ldots$, continuing iteratively in this fashion until successive V_{BE} values converge to the correct solution. Circuits that require us to do this are more appropriate for SPICE analysis, where the iterations are transparent to the user. Needless to say, it is important that we try to recognize such problems when they arise and not blindly use $V_{BE} = 0.7$ V when it is inappropriate. Since low current levels usually appear in somewhat specialized circumstances, we postpone any further discussion of this point until the occasion arises later in the book.

All examples in Sec. 4.7.1 involved transistors in the forward active state; however, the general procedure of replacing each transistor by an equivalent circuit also works for transistors in cut-off, saturation, and reverse active operation. The need to analyze a circuit containing non forward-active transistors usually arises in circuits where transistor states are *unknown*. This adds a new element of uncertainty to the analysis problem, which we address next.

4.7.2 ANALYSIS WHEN TRANSISTOR STATES ARE UNKNOWN

In digital logic circuits, transistors change states during normal operation. We handled state uncertainties in diode circuits by assuming each diode either ON or OFF, replacing each with a suitable equivalent, and then verifying or disproving the assumed states by linear analysis. A similar procedure applies to transistor circuits. Because reverse active operation is rare, we emphasize the other three states.

Analyzing Circuits Containing Three-State Transistors

1. Make an educated guess about the state of each transistor.
2. In the circuit diagram, replace each transistor by the circuit model for its assumed state.
3. Analyze the resulting circuit to find *test* variables associated with each model.
4. Examine the test variables, looking for contradictions to the assumed state.
5. If there is a contradiction, make a new guess based upon the calculated information and return to step 2.
6. When no contradictions remain, the voltages and currents calculated from the equivalent circuit approximate those of the actual circuit.

We need not be unduly concerned about correctly guessing transistor states. Although the general problem is very complex, available information usually leads to valid initial guesses in practical problems. As in diode circuit analysis, incorrect guesses simply result in a little extra work, and correct guesses become more frequent with problem-solving experience.

Validity Tests for Assumed Transistor States. The guidelines of Fig. 4.22 help us in our search for contradictions. Since the reverse active state occurs rarely, the guidelines assume first quadrant operation where reverse active operation cannot occur.

Forward active operation. We replace each forward active transistor by the model of Fig. 4.22a, and find the direction of its base current. Figure 4.8 shows that base current must be positive to avoid cut-off; to avoid saturation, V_{CE} must be more positive than $V_{CE,sat}$.

Cut-off operation. The cut-off model of Fig. 4.22b forces all transistor currents in the equivalent circuit to be zero. As in using the OFF model for the diode, we must test the open-circuit voltage, in this case V_{BE}, to look for a contradiction.

Saturation. To be saturated, as in Fig. 4.22c, the transistor must first be ON; that is, its base current must be nonnegative. If the transistor is ON we must also verify that the base current is large enough for saturation, as illustrated in Fig. 4.13d. Because voltage

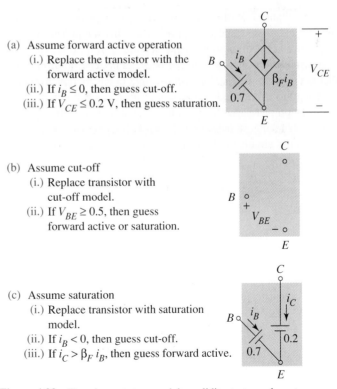

(a) Assume forward active operation
    (i.) Replace the transistor with the forward active model.
    (ii.) If $i_B \leq 0$, then guess cut-off.
    (iii.) If $V_{CE} \leq 0.2$ V, then guess saturation.

(b) Assume cut-off
    (i.) Replace transistor with cut-off model.
    (ii.) If $V_{BE} \geq 0.5$, then guess forward active or saturation.

(c) Assume saturation
    (i.) Replace transistor with saturation model.
    (ii.) If $i_B < 0$, then guess cut-off.
    (iii.) If $i_C > \beta_F i_B$, then guess forward active.

Figure 4.22 Transistor states, models, validity tests, and next guesses.

sources in the model *force* the base node to be positive relative to both emitter and collector nodes, we must test *currents* to ensure the validity of the model.

Similar models, test variables, and tests apply to pnp transistors. Some examples now demonstrate how to apply this trial-and-test procedure to given circuits.

EXAMPLE 4.5 Find the collector current in Fig. 4.23a.

Solution. Since there is a positive voltage in series with the base and a negative voltage in series with the emitter, the transistor is probably ON. Try forward active operation. Replacing the transistor with its forward active model gives Fig. 4.23b. Kirchhoff's voltage law applied to the base circuit gives

$$1 - (5 \text{ k})(I_B) - 0.7 = -10$$

Therefore, $I_B = 2.06$ mA. Since the base current is positive, all that remains is to check V_{CE}.

$$V_C = 5 - (8 \text{ k})(10)(2.06 \text{ mA}) = -159.8 \text{ V}$$

Since $V_E = -10$ V, this gives

$$V_{CE} = -159.8 - (-10) = -149.8 \text{ V}$$

Because V_{CE} is less than 0.2 V, the forward active assumption is not valid.

The sketch of Fig. 4.23c shows exactly what has happened. The forward active *model* incorrectly predicted a solution in the second quadrant, where the CCCS model and the true transistor characteristics are not coincident. The possibility of incorrect solutions like this is the price we pay for using simple linear models to represent nonlinear devices. A good way to avoid modeling errors is to develop the habit of visualizing the computed operating point relative to the characteristic curves. Both Fig 4.23c and the guideline in Fig. 4.22a suggest saturation should be our next guess.

Figure 4.23d is the circuit implied by saturation. The base current is still 2.06 mA. Collector current is

$$I_C = \frac{5 - (-10 + 0.2)}{8 \text{ k}} = 1.85 \text{ mA}$$

The inequality

$$\beta_F I_B = 20.6 \text{ mA} > 1.85 \text{ mA} = I_C.$$

confirms that the transistor is saturated. Figure 4.23e shows the correct operating point. ❏

Exercise 4.6 Prove that the transistor in Fig. 4.23a remains saturated when the base resistor is changed to 50 kΩ. Then find the base resistance R_B that brings the transistor to the verge of active operation.

Ans. $\beta_F I_B = 2.06$ mA > 1.85 mA $= I_C$; $R_B = 55.7$ kΩ.

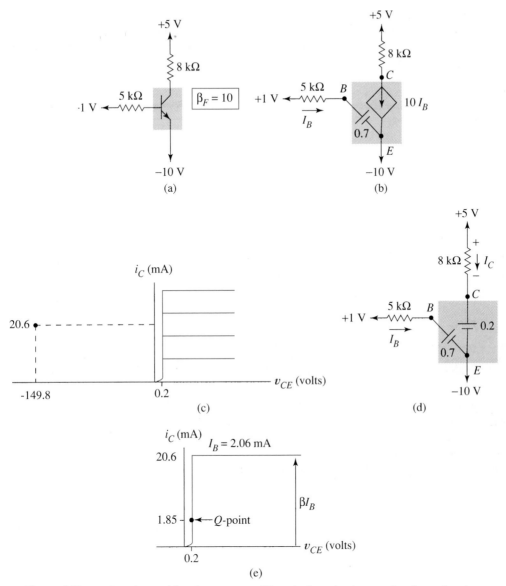

Figure 4.23 (a) Transistor with unknown state; (b) equivalent circuit assuming forward active state; (c) incorrect operating point from assuming forward active operation; (d) equivalent circuit assuming saturation; (e) correct Q-point.

EXAMPLE 4.6 Find the state of the transistor in Fig. 4.24a.

Solution. Because of the large positive voltage in the base circuit, we assume that the transistor is saturated. However, when we replace the transistor with its saturation model and apply Thevenin's theorem, we obtain Fig. 4.24b. In this circuit it is obvious that I_B is negative, contradicting the saturation hypothesis and suggesting cut-off.

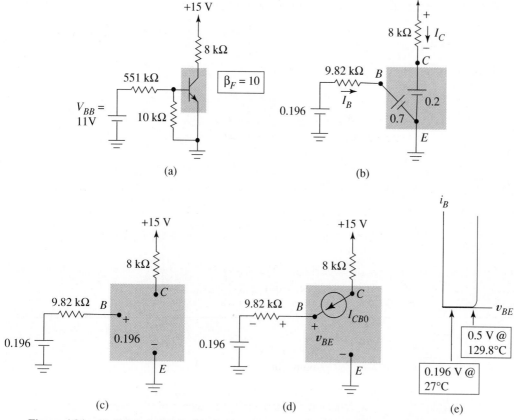

Figure 4.24 (a) Original circuit; (b) circuit assuming saturated transistor; (c) circuit with cut-off transistor; (d) cut-off transistor including leakage current; (e) input operating points at 27°C and at 129.8°C.

Figure 4.24c shows the circuit with the cut-off model replacing the transistor. Notice that $V_{BE} = 0.196$ V, a value considerably lower than the cut-in voltage. Thus the transistor is in cut-off. ❑

> **Exercise 4.7** If we slowly increase V_{BB} in Fig. 4.24a to values higher than 11 V, the transistor eventually begins to leave cut-off. Find the value of V_{BB} where this occurs.
>
> *Ans.* $V_{BB} = 28.5$ V.

EXAMPLE 4.7 Find the temperature at which the transistor of Fig. 4.24a will just begin to come out of cut-off if its leakage current is $I_{CB0} = 25$ nA at 27°C and doubles for each 10°C increase in temperature.

Solution. For this type of problem we use the cut-off model of Fig. 4.9b to replace the transistor as in Fig. 4.24d. When we include the drop in the base resistor produced by I_{CBO} at 27°C we obtain

$$V_{BE} = 0.196 + (9.82 \text{ k})(25 \times 10^{-9}) \approx 0.196$$

To bring V_{BE} to the cut-in point, temperature must increase until $I_{CBO}(T)$ satisfies

$$0.196 + (9.82 \text{ k})I_{CBO}(T) = V_\gamma = 0.5 \text{ V}$$

This gives $I_{CBO}(T) = 3.1 \times 10^{-5}$ A. Since this value results from doubling the 27°C value of I_{CBO} for each 10°C increase in temperature,

$$3.1 \times 10^{-5} = (25 \times 10^{-9})(2)^{(T-27)/10}$$

When we divide by 25×10^{-9} and then take the base-10 logarithm, we find

$$\log\left(\frac{3.1 \times 10^{-5}}{25 \times 10^{-9}}\right) = \frac{T-27}{10}\log(2)$$

Solving gives $T = 129.8$°C.

Figure 4.24e shows how V_{BE} shifts as temperature increases. For simplicity, this example assumes the input characteristic itself does not change with temperature. (In fact, the input characteristic of a BJT, like the characteristic of a diode, shifts to the left as the temperature increases.) ❏

EXAMPLE 4.8 One of the transistors in Fig. 4.25a is cut-off and the other is forward active. Verify cut-off for the former and find the Q-point for the latter.

Solution. Except for the voltage sources applied to the bases, the circuit is symmetrical about an axis through the center of the diagram. Since the node voltage at the base of

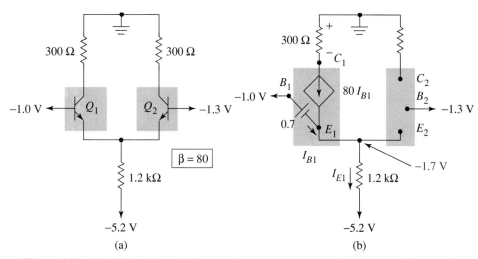

Figure 4.25 (a) Original circuit; (b) equivalent circuit assuming Q_1 active and Q_2 cut-off.

Q_1 is more positive than that of Q_2, we guess that Q_1 is active and Q_2 is OFF and draw Fig. 4.25b. The emitter node voltage in this diagram is -1.7 V. Since the voltages at both nodes of the 1.2 kΩ resistor are known,

$$I_{E1} = \frac{-1.7 - (-5.2)}{1.2 \text{ k}} = 2.92 \text{ mA}$$

The diagram shows that $I_{E1} = (80 + 1)I_{B1}$, therefore $I_{B1} = 36$ μA. Since the base current is positive, we have established that Q_1 is ON; however it could be saturated. Since

$$I_{C1} = 80I_{B1} = 2.88 \text{ mA}$$

$$V_{C1} = 0 - (300)(2.88 \times 10^{-3}) = -0.864 \text{ V}$$

therefore

$$V_{CE1} = -0.864 - (-1.7) = 0.836$$

Since $V_{CE} > 0.2$ V, Q_1 is active, not saturated.

To verify that Q_2 is cut-off, notice from Fig. 4.25b that

$V_{BE2} = -1.3 - (-1.7) = 0.4$ V, which is less than $V_\gamma = 0.5$ V, while

$V_{BC2} = -1.3 - 0 = -1.3$ V

Since both junctions are reverse biased, the transistor is cut-off. ❏

The circuit models we have been studying provide a good general understanding of how circuits work under dc conditions and allow us to make reasonably good estimates without much effort. The SPICE analysis discussed in the next section enables us to obtain more precise numerical results and to investigate effects of nonlinearities that we ignore when using our linear models, for example, the curvature of the BJT input characteristic.

4.8
SPICE Static BJT Model

To SPICE, the transistor is a nonlinear, active circuit like Fig. 4.3 that satisfies a generalized form of Eqs. (4.3) and (4.4). SPICE dc analysis employs iterative numerical calculations to find transistor operating points rather than assuming operating states and using linear circuit analysis as we do. No initial guess about the transistor states is necessary in SPICE analysis, and SPICE output never directly refers to the transistor state.

Figure 4.26a shows the SPICE code required for *default BJTs*. We use an element statement to assign each transistor a unique name beginning with Q, and we list its node numbers in collector-base-emitter order. On the element line we also assign a class name. This name associates the transistor with a .MODEL line that provides the parameter values for all transistors belonging to the class. For a default transistor the .MODEL line merely specifies whether the transistor is npn or pnp.

In any simulation of a circuit containing BJTs, SPICE does a dc analysis to find the operating point of each transistor, and it computes numerical values for all node voltages. If we wish to investigate output data in terms of transistor states, we must determine the junction voltages from SPICE and then apply the definitions of Table 4.1.

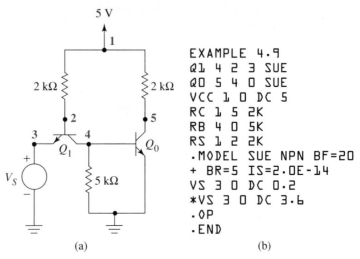

Notation		Default value
Text	SPICE	
β_F	BF	100
β_R	BR	1
I_S	IS	1.0E − 16

(b)

Figure 4.26 (a) SPICE code for default transistors; (b) SPICE notation and values of default parameters.

Figure 4.26b shows the SPICE default values for BJT parameters we have previously discussed. Values of β rather than α are used in SPICE input code since they are more readily available. To override any default parameter value, we add an appropriate statement on the .MODEL code line as illustrated in the example that follows.

EXAMPLE 4.9 Transistor parameters for Fig. 4.27a are

$$\beta_F = 20, \quad \beta_R = 5, \quad \text{and } I_S = 2.0 \times 10^{-14} \text{ A}$$

Use SPICE to determine the state of each transistor for (a) $VS = 0.2$ V, (b) $VS = 3.6$ V.

Solution. Figure 4.27b shows the SPICE code for the circuit. The value for VS for part (b), preceded by an asterisk to label it as a comment, is ignored in the first SPICE run. To override default values on the .MODEL line we simply indicate parameter values we wish to use. The "+" sign indicates continuation of the .MODEL code to a second

```
EXAMPLE 4.9
Q1 4 2 3 SUE
Q0 5 4 0 SUE
VCC 1 0 DC 5
RC 1 5 2K
RB 4 0 5K
RS 1 2 2K
.MODEL SUE NPN BF=20
+ BR=5 IS=2.0E-14
VS 3 0 DC 0.2
*VS 3 0 DC 3.6
.OP
.END
```

(a)   (b)

Figure 4.27 (a) Given BJT circuit; (b) SPICE code.

line. The special statement, .OP, asks for a listing of operating point information in the output file. For transistor circuits, this listing includes junction voltages, collector currents, and a wealth of other useful information that helps us detect errors and verify correct operation of a simulation. In this example we use the junction voltages to determine the transistor state. In the absence of a specific request for dc, ac, or transient analysis, SPICE automatically performs a dc analysis.

(a) The SPICE output file contained the following information after the first run.

	V_{BE}	V_{BC}
Q_1	0.695 V	0.691 V
Q_0	0.205 V	−4.8 V

From the state definitions of Table 4.1, we conclude that Q_1 is saturated. and Q_0 is cut off.

(b) After changing the input voltage to 3.6 V by moving the asterisk to the preceding statement, the junction voltages were

	V_{BE}	V_{BC}
Q_1	−2.17 V	0.694 V
Q_0	0.739 V	0.733 V

We conclude that Q_1 is reverse active and Q_0 is saturated.

The observant reader may have noticed that in part (b) the SPICE results for Q_0 gave

$$V_{CE,sat} = V_{CB} + V_{BE} = -0.733 + 0.739 = 0.006 \text{ V}$$

a value considerably less than the value of 0.2 V we normally assume for a saturated transistor. This is partly a result of inaccurate SPICE modeling algorithms and partly the result of second-order effects in real transistors that were not included in this simulation. The paper by J.R. Hines, cited in the references at the end of the chapter, discusses the problem with SPICE saturation calculations. ❏

Because scaling the transistor's base-emitter junction area, A, in Eq. (3.39), is so important in IC circuit design, SPICE facilitates assigning different relative areas to otherwise identical transistors. A relative area factor, optionally placed at the end of a transistor's element line, specifies the junction area of this particular transistor relative to the area of the reference transistor described in the .MODEL line. This is the same technique we use for diode areas, for example, D_1 in Fig. 3.34. When we explicitly indicate an area on the element line, SPICE appropriately scales all pertinent transistor parameters that depend on area.

Exercise 4.8 Find the new values for V_{BE} and V_{BC} in Example 4.9b if the area of Q_0 is five time the area of Q_1.

Ans. Q_1: $V_{BE} = -2.2$ V, $V_{BC} = 0.694$ V; Q_0: $V_{BE} = 0.697$ V, $V_{BC} = 0.691$ V.

Example 4.9 showed that the states of both transistors in Fig. 4.27a depend on the input voltage V_S. SPICE enables one to pursue an interesting problem like this in much more detail with little additional effort. For example, we could use SPICE to find the transistor currents or to plot a transfer characteristic for this highly nonlinear circuit.

EXAMPLE 4.10 Taking node 5 as the output and node 3 as the input, use SPICE to plot the transfer characteristic, v_{out} versus v_{in} for Fig. 4.27a.

Solution. SPICE code is Fig. 4.28a. We describe source VS, the independent variable in our plot, as a dc source of zero volts. The .DC line specifies that VS vary from 0.2 V to 3.6 V in voltage intervals of 0.17 V. The .PLOT line asks for a plot of output, $V(5)$, versus input voltage VS.

The initial plot showed that most of the change in the output occurred in the input range $0.5 \text{ V} \leq VS \leq 0.7 \text{ V}$. To explore this region more carefully, the .DC statement was replaced by the line marked with an asterisk in the code. This second SPICE run gave Fig. 4.28b. ❑

```
EXAMPLE 4.10
QI 4 2 3 SUE
Q0 5 4 0 SUE
VCC 1 0 DC 5
RC 1 5 2K
RB 4 0 5K
RS 1 2 2K
.MODEL SUE NPN BF=20 BR=5
IS=2.0E-14
VS 3 0 DC 0
.DC VS 0.2 3.6 0.17
*.DC VS 0.5 0.7 0.01
.PLOT DC V(5)
.END
```

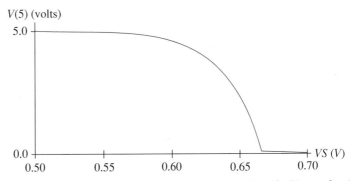

Figure 4.28 (a) Code for plotting $V(5)$ versus $V(3)$ for Example 4.10; (b) transfer characteristic.

The smooth change of output voltage from high to low values is a detail our idealized linear transistor models do not reveal, because actual transistors make smooth transitions *between* states that our models cannot describe. On the other hand, a study of this same circuit with simplified models in Chapter 13 gives a rough, piecewise linear approximation to this curve that also imparts a quality of understanding that SPICE does not give. We learn from our models that for low input voltages the output is 5 V because Q_0 is in cut-off. As input increases, an increasing fraction of the base current of Q_1 is diverted toward the base of Q_0 and flows through the 5 k resistor, increasing v_{BE} of Q_0. Finally Q_0 turns on and eventually saturates, producing a low output voltage. Meanwhile Q_1 changes to reverse active operation requiring V_S to deliver current to the circuit.

After studying some important second-order transistor effects in the next section, we introduce additional features of SPICE models.

4.9
Second-Order Effects

In this section we learn of the more important *second-order static effects* in bipolar transistors. This new knowledge reveals some important limitations of transistors themselves and some of the shortcomings of our idealized models. We continue to focus on simplified models and characteristic curves.

4.9.1 PARAMETER VARIATIONS WITH TEMPERATURE AND Q-POINT

Input and output characteristics of the BJT change with temperature as do the equivalent circuit parameters that describe them. The common-emitter input characteristic translates to the left with increased temperature as indicated in Figure 4.29a. As in the diode, $V_T = kT/q$ in the exponent of Eq. (4.12) causes the curve to shift by -2 mV/°C. Of course, this means that V_{BE} in Fig. 4.6 decreases by 2 mV for every °C increase in temperature.

The output characteristics increase in separation and translate upward with increasing temperature as in Fig. 4.29b. Since these output characteristics come from Eq. (4.10), we see that the increase in curve separation reflects increases in β with temperature. Minority carrier lifetime in the base increases with temperature, increasing the transport factor α_t and bringing α closer to one. Because β is such a sensitive function

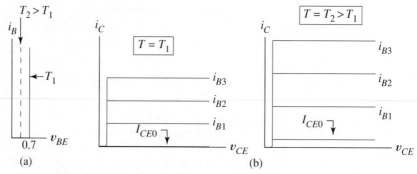

Figure 4.29 Transistor characteristics at two temperatures: (a) input; (b) output.

of α, the result is a large increase in β, of the order of 7000 ppm, as temperature increases. An empirical relationship that predicts variations in β over a rather wide temperature range is

$$\beta(T) = \beta(T_R)\left(\frac{T}{T_R}\right)^{XTB} \tag{4.19}$$

where T and T_R are temperatures in degrees Kelvin and *XTB* is an appropriately chosen constant called the *temperature exponent*. Choosing $XTB = 1.7$, for example, predicts that β approximately doubles from 27°C to 175°C.

To explain the upward translation, we need to notice the two terms we dropped just before we wrote Eq. (4.5). Had we not made this approximation, Eq. (4.10) would have the form

$$i_C = \beta_F i_B + I_{CE0}$$

where I_{CE0} is a small dc saturation current that doubles for every 5°C increase in temperature.

Of the three temperature-sensitive transistor parameters, V_{BE}, with its rather modest and linear variation of 2900 ppm/°C, actually proves to be the most troublesome in practical circuits. Even at elevated temperatures I_{CE0} is usually too small to play a major role in silicon transistors, and it is relatively easy to design circuits that work well for any high value of β, just as we can design op amp circuits that work well for any high value of op amp gain. Many IC designs use the V_{BE} drops of matched transistors to cancel each other over wide temperature ranges.

Figure 4.30 shows that β is also a function of dc collector current. At low values of I_C, recombination within the base-emitter depletion region itself becomes important, lowering the value of beta. At high values of I_C, a tacit assumption that the electron concentrations in the emitter were not disturbed by the injection process is no longer valid, resulting in a reduction in β. Since the curves are generally broad and flat at moderate values of I_C, we usually assume that β is independent of operating point as a first approximation.

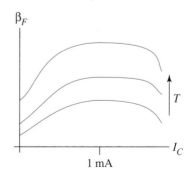

Figure 4.30 Variation of β with temperature and Q-point.

4.9.2 BREAKDOWN VOLTAGES

In Chapter 3 we learned that breakdown occurs when we apply large reverse bias to a pn junction. The polarity of the collector-base voltage in Fig. 4.1d suggests that breakdown of the collector-base junction might occur during forward active operation. This indeed happens—at a *collector-base breakdown voltage* denoted by BV_{CB0}. As the subscripts suggest, we measure BV_{CB0} between collector and base with emitter open circuited.

Collector-base breakdown is directly apparent when we operate the transistor in its *common-base* orientation, Fig. 4.1d. For this orientation, output current is i_C, output voltage is v_{CB}, and input current is i_E. Measured common-base output characteristics resemble Fig. 4.31a; the transistor operates as a CCCS with current gain α_F, as Eq. (4.7) suggests. As v_{CB} approaches BV_{CB0}, the collector current for each value of emitter current becomes high, and the transistor ceases to function as a CCCS. Thus collector-base breakdown sets an upper limit to the useful operating region on the common-base characteristics.

Collector-base breakdown also occurs in the common-emitter configuration; however, because of the active nature of the transistor it happens at a collector-emitter voltage, BV_{CE0}, significantly lower than BV_{CB0}. There is also some spreading of the characteristic curves for voltages just below breakdown, as shown in Fig. 4.31b. It is BV_{CE0} that establishes the upper limit of useful output voltage for the BJT. BV_{CE0} is measured between collector and emitter with the base open circuited. Values of 40 or 50 V for BV_{CE0} are typical for transistors designed for amplifier applications. This gives constant current curves that remain reasonably flat for voltages we usually encounter using power supplies of 15 V or so.

4.9.3 BASE WIDTH MODULATION

When the collector-base junction is reverse biased for forward active operation, the effective base width narrows. This *base width modulation* makes actual input and output characteristic curves differ in still other ways from those predicted by the Ebers–Moll equations.

The shading of Fig. 4.32a suggests how the two depletion regions extend into the p-type base material during forward active operation. The base-emitter depletion region is narrow, the reverse-biased base-collector junction has a wider depletion region. For increased collector-emitter voltage in Fig. 4.32b, base-collector depletion extends still further into the base. Since injected minority carriers need only diffuse to the edge of the

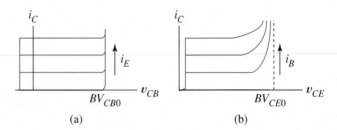

Figure 4.31 Collector breakdown voltage: (a) common-base; (b) common-emitter.

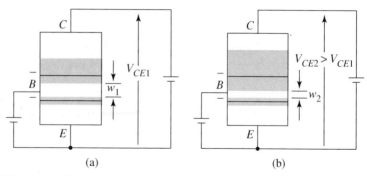

Figure 4.32 Base-width modulation: (a) effective base width, w_1, with moderate collector-base biasing; (b) effective base width w_2 with large collector-base bias voltage.

collector-base depletion region before being collected (see Fig. 4.1d), the *effective* base width decreases from w_1 to w_2 as v_{CE} increases to V_{CE2}. It is possible for v_{CE} to become so large that the base-collector depletion region expands all the way through the base, resulting in high currents that can damage the transistor. This irreversible *punch-through* breakdown is different from the avalanche breakdown discussed in Sec. 4.9.2.

A more important practical consequence of base-width modulation is a change in the common-emitter output characteristics. With a shorter distance for the injected carriers to travel, the transport factor increases, bringing $\alpha = \alpha_F$ closer to one as v_{CE} increases. The change in the common-base characteristics, with spacing determined by α, is hardly noticeable because the percent change in α is small. However, because β in Eq. (4.9) is such a sensitive function of α, the separation of the common-emitter output characteristics increases noticeably with v_{CE} when base width modulation is severe, as in Fig. 4.33. It was Early who noticed that the measured common-emitter output curves all appear to extrapolate backward to the same point, $-V_A$, on the voltage axis. Today V_A is known as the *Early voltage*. Figure 4.33 uses dashed lines to show this *Early effect*.

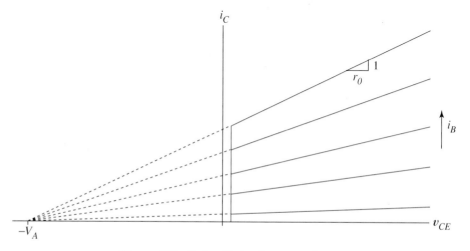

Figure 4.33 Early effect and output resistance.

We can generalize the idea of transistor current gain to include dependence on collector-emitter voltage by adding an *Early effect correction factor* to the transistor beta. Equation (4.10), so modified, becomes

$$i_C = f(v_{CE}, i_B) = \left[\left(1 + \frac{v_{CE}}{V_A} \right) \beta \right] i_B \tag{4.20}$$

which describes i_C as a function of two variables. High-quality integrated npn transistors have Early voltages of the order of 100 to 120 V; so for typical v_{CE} values of 15 V or so the Early effect correction is small, and the output characteristics appear to be nearly horizontal. On the other hand, integrated pnp transistors might have Early voltage as low as 50 V, which gives a very pronounced slope to the output characteristics. For simplicity, we often ignore the Early effect in hand estimates of Q-point, but include the Early effect correction in computer simulations for improved accuracy.

Output Resistance. By producing nonzero slope in the output characteristics, the Early effect gives rise to a transistor *output resistance*, r_o, defined by

$$\frac{1}{r_o} = \frac{\partial i_C}{\partial v_{CE}} \bigg|_{Q\text{-point}}$$

Taking the partial derivative of Eq. (4.20) and evaluating it at the dc operating point gives

$$\frac{1}{r_o} = \frac{1}{V_A} \beta I_B$$

where I_B is the dc base current at the Q-point. When $V_{CE} \ll V_A$ in Eq. (4.20),

$$I_C = \left(1 + \frac{V_{CE}}{V_A} \right) \beta I_B \approx \beta I_B$$

Substituting this into the preceding equation and taking the reciprocal gives the important result

$$r_o = \frac{V_A}{I_C} \tag{4.21}$$

By using Eq. (4.21) to compute the numerical value of r_o, we can include the nonzero slope of the output characteristics in our transistor equivalent circuit. We simply add a resistor r_o between collector and emitter nodes.

Most often r_o is so large that it contributes little to numerical accuracy. In derivations it makes the algebra more tedious, and often results in theoretical expressions too complicated to provide much insight. On the other hand, we will notice that certain circuits (for example, those with active loads) and certain analyses (for example, output resistance calculations) critically depend upon r_o. The best approach is to exclude r_o in hand calculations except where it obviously serves some useful purpose. We observe this rule throughout the remainder of the text.

Internal Feedback. Another consequence of base width modulation is *internal feedback,* in which a portion of the output signal voltage is fed back through the transistor

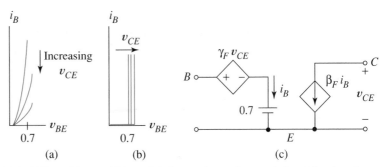

Figure 4.34 Effect of base width modulation on input characteristics: (a) reduced recombination with increased v_{CE} lowers base current; (b) stylized view of input characteristics; (c) circuit model of internal feedback.

into the input circuit. If we hold base-emitter bias constant in Fig. 4.32 while increasing v_{CE} from V_{CE1} to V_{CE2}, the base current becomes smaller because recombination in the base is reduced, and fewer holes in the base need to be replaced. Because of this, *the input "characteristic" of a real transistor is actually a family of input characteristics selected by v_{CE},* as in Figs. 4.34a and b. The stylized input curves of Fig. 4.34b suggest that we might modify the equivalent circuit of Fig. 4.6 to include internal feedback by adding a dependent source controlled by v_{CE} in series with the source in the input circuit. This is done in Fig. 4.34c, where γ_F is a *reverse voltage-gain* parameter determined from the spacing of the input curves. In the general linear two-port theory of Appendix B, there is always a parameter subscripted "12" that is associated with internal feedback of this kind. Because finite output resistance and internal feedback are both undesirable, the transistor designer minimizes base width modulation by doping the collector much more lightly than the base so that the expansion of the depletion region occurs mostly in the collector. A consequence is that the input curves are so closely spaced (γ_F is so small) that we usually represent them by the single vertical line at 0.7 V. This gives little error for most applications; however, for high-frequency operation this internal feedback is quite important and cannot be ignored.

4.9.4 BASE AND COLLECTOR RESISTANCE

All components in an integrated circuit are fabricated on a single piece of semiconductor material called the *substrate.* A consequence is that external connections to the base, emitter, and collector are located at the top of the IC as shown in Fig. 4.35. The transistor in this figure is a *vertical transistor* because current flows vertically through the *active region* beneath the emitter material. Three parasitic resistances are important in the bipolar transistor. We denote the first, the *base spreading resistance,* by r_b. The ohmic resistance of the base current path from the junction to the surface is high because the base width, very narrow as in Fig. 4.1e to give high α_F, results in a small cross-sectional area for current flow. A typical value for r_b is 100 Ω. This parameter is important in high-current transistors; and it often significantly influences amplifier frequency response. Next in importance is the ohmic resistance of the collector, r_c in Fig. 4.35. Because N_d in the collector is relatively low to reduce base width modulation, Eq. (3.18)

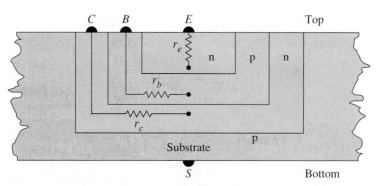

Figure 4.35   Ohmic resistances in an integrated transistor.

shows that the collector has a low conductivity. The result is a collector resistance r_c of the order of 10–100 Ω that causes an *IR* drop when current leaves the collector. A special highly conducting *buried layer,* mentioned in Sec. 4.13, helps reduce r_c. For a saturated transistor, the Ebers–Moll equations predict a numerically smaller value of $V_{CE,sat}$ than the 0.2 V we use in our linear model. The 0.2 V offset is a realistic value for the typical high collector currents that flow in saturated transistors when the effect of r_c is included. Of lesser importance than the other two is r_e, the ohmic resistance in the highly doped emitter material. A typical value is 1 Ω.

4.9.5   STATIC SPICE MODEL INCLUDING SECOND-ORDER EFFECTS

The static SPICE BJT model of Fig. 4.36a includes most of the second-order effects discussed in this section. Parameters r_c, r_b, and r_e model the parasitic resistances. Figure 4.36b shows the expanded set of SPICE parameters and their default values.

SPICE includes temperature variations in I_S and in V_{BE} by default, as for diodes. If we override the default value of zero for the temperature exponent, XTB, SPICE also models changes in forward beta with temperature according to Eq. (4.19). SPICE assumes that all parameter values are correct at a room temperature of 27°C and makes appropriate corrections when analyses at other temperatures are requested by using a .TEMP statement.

Finite values for forward and/or reverse Early voltages, VAF and VAR, in the .MODEL line lead to SPICE simulations that include the output resistances r_{oc} and r_{oe}, respectively. These are automatically computed by SPICE at the *Q*-point it determines by dc analysis.

EXAMPLE 4.11   The circuit of Fig. 4.37a was designed using the large-signal model of Fig. 4.6 with $V_{BE} = 0.7$ V and $\beta = 300$. For simplicity, Early voltage was ignored. The object of the design was to establish the operating point $(V_{CE}, I_C) = (1.5$ V, 1 mA) so that the transistor would function as a CCCS. Use SPICE to check the design at the centigrade temperatures: −40, −20, 0, 27, 50, 70, 100, and 125.

Solution.   To check the design, we write the SPICE code of Fig. 4.37b. The SPICE model iteratively determines more accurate values for V_{BE} than we could achieve by

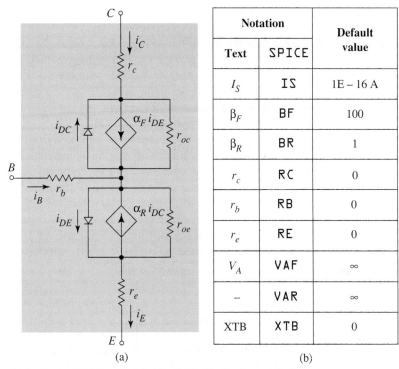

Notation		Default value
Text	**SPICE**	
I_S	IS	1E – 16 A
β_F	BF	100
β_R	BR	1
r_c	RC	0
r_b	RB	0
r_e	RE	0
V_A	VAF	∞
–	VAR	∞
XTB	XTB	0

(a)   (b)

Figure 4.36 Static SPICE model for the BJT: (a) circuit model; (b) parameter names and default values.

simple hand analysis. Because of the $VA = 90$ statement on the .MODEL line, the simulation also incorporates the Early-effect correction into the Q-point calculations. Finally, the .TEMP line asks SPICE to compute the Q-point at the specified temperatures. SPICE Q-point calculations include temperature-induced variations in β, V_{BE}, and I_S.

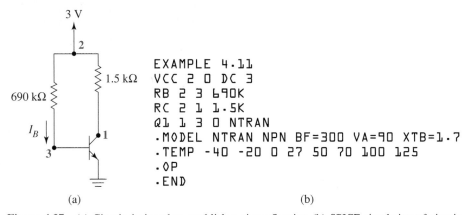

```
EXAMPLE 4.11
VCC 2 0 DC 3
RB 2 3 690K
RC 2 1 1.5K
Q1 1 3 0 NTRAN
.MODEL NTRAN NPN BF=300 VA=90 XTB=1.7
.TEMP -40 -20 0 27 50 70 100 125
.OP
.END
```

(a)   (b)

Figure 4.37 (a) Circuit designed to establish a given Q-point; (b) SPICE simulation of circuit performance over a range of temperatures.

Figure 4.38a shows the output values of V_{BE} and β, and Fig. 4.38b shows the Q-point variations. First, examine the room temperature results. At 27°C, the SPICE printout indicated β = 302.5 rather than the 300 in the SPICE code. This slight increase resulted from curve separation due to Early effect (see Fig. 4.33). Also, SPICE iteratively calculated a room temperature value of V_{BE} of 0.77 V. Since this exceeded the 0.7 V used in the design, we expect from Fig. 4.37a a simulation value of base current lower than the design value. This explains why the collector current in Fig. 4.38b is slightly lower than the 1 mA design value at 27°C even though β > 300.

For lower temperatures, β decreased and V_{BE} increased as expected from Fig. 4.29, and the Q-point moved along the load line toward cut-off. As the temperature increased, the Q-point moved along the load line toward saturation, that is, lower v_{CE}. These data demonstrate what can happen to the Q-point in a poorly designed circuit. In the early years of transistors, temperature variations like these were troublesome; however, improved transistors and modern design techniques now result in excellent circuit performance over wide variations in temperature. In Chapter 6 we will learn to design circuits that minimize this tendency of the Q-point to drift with changes in temperature.   ❏

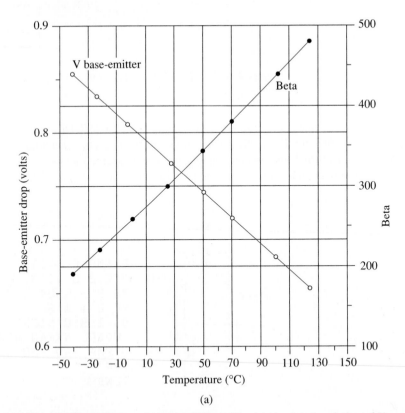

(a)

Figure 4.38 SPICE simulation of circuit of Fig. 4.37a. (a) simulated variation of V_{BE} and β with temperature; (b) Q-point variation with temperature.

Our simple models play a vital role, even after more accurate SPICE simulations, in helping us quickly confirm that the results "make sense" in terms of our intuition.

Exercise 4.9 Use simple hand analysis and the $T = 125°C$ parameter values from Fig. 4.38a to estimate the Q-point.

Ans. $I_C = 1.67$ mA, $V_{CE} = 0.497$ V.

This introductory discussion omitted many SPICE features related to BJT modeling. For example, SPICE includes both the variations of β with I_C as in Fig. 4.30 and the internal feedback depicted in Fig. 4.34, however, default parameter values render these features unavailable to the beginner. We leave further exploration to the curiosity of the reader, supplementary instruction, or an advanced course.

4.9.6 PARASITIC CAPACITANCES

We know that diodes possess parasitic internal capacitances, nonlinear Q versus V relationships that are transparent to slow signals but important when rapid changes in voltage or current are imposed upon the device. Since the BJT contains two pn junctions, we expect it to have similar dynamic limitations. Associated with *each junction* are deple-

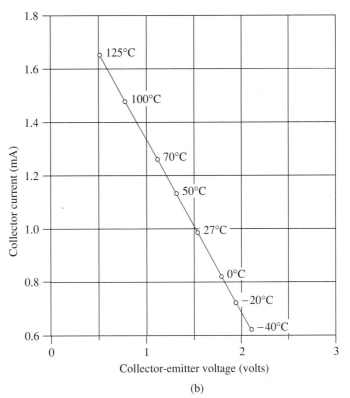

(b)

Figure 4.38 (continued)

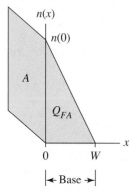

Figure 4.39 Excess stored minority charge in a forward active transistor.

tion and diffusion capacitances, which limit high-frequency performance. In forward active transistors, depletion capacitance is dominant at the reverse-biased collector-base junction. At the forward-biased base-emitter junction, diffusion capacitance and depletion capacitance are both important. In integrated transistors, fabricated on a substrate as in Fig. 4.35, there is also depletion capacitance between collector and substrate.

The transistor depletion capacitances are exactly those we described in diode theory. Thus the parameter values in Eq. (3.62) and the capacitance characteristic of Fig. 3.62, within the transistor, depend upon the individual junction grading coefficients and junction geometry.

The diffusion capacitance for a transistor differs slightly from that of an isolated diode because of the narrow base. For the forward active transistor we saw that the minority charge concentration profile in the base is triangular as in Fig. 4.7. For a base-emitter junction of cross-sectional area A, we use Fig. 4.39 to calculate that the quantity of minority charge, Q_{FA}, stored in the base in coulombs is

$$Q_{FA} = qA\frac{1}{2}n(0)W$$

Substituting $n(0)$ from Eq. (4.14) gives the charge-voltage curve that characterizes the diffusion capacitance of the forward active transistor

$$Q_{FA}(v_{BE}) = \frac{qAW}{2}\frac{n_i^2}{N_a}e^{v_{BE}/V_T} \qquad (4.22)$$

This particular collection of stored charge represents electrons in transit from emitter to collector. On the average, these injected electrons take τ_T seconds to traverse the base, where τ_T is called the *transit time*. For integrated npn transistors, the transit time is of the order of 1 ns, for integrated pnp transistors $\tau_T \approx 30$ ns. Since this electron flow *constitutes* the collector current, we have

$$I_C = \frac{Q_{FA}}{\tau_T} \text{ coulombs/s} \qquad (4.23)$$

an important equation in transistor dynamics.

This discussion briefly reviewed those capacitances most important for the forward active operating mode. In the next section we take a more general view that applies to all four transistor states.

4.10
Dynamic Model of the Transistor

Figure 4.40 introduces a *dynamic BJT model*. We recognize it as the Ebers–Moll model of Fig. 4.3 augmented by the nonlinear diffusion (subscript *diff*) and depletion (subscript *dep*) capacitances explained in Sec. 3.12 and introduced in Fig. 3.64a. Resistors represent ohmic drops in the semiconductor material. There is also a collector-substrate depletion capacitance for integrated transistors. Each diffusion capacitance is a nonlinear Q versus V relation of the form of Fig. 3.61 caused by storage of minority charge in the vicinity of that particular junction. Each depletion capacitance is a nonlinear Q versus V relation of the form of Fig. 3.62 caused by the separation of positive and negative ions

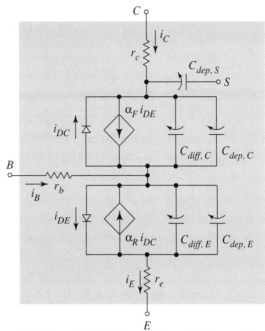

Figure 4.40 Dynamic BJT model including nonlinear diffusion and depletion capacitances.

in the depletion region of that junction. As in the diode, diffusion capacitance is dominant for a forward-biased junction, and depletion capacitance is more important for a reverse-biased junction, although both are present regardless of the junction biasing. Dynamic computer simulations can include all five nonlinear capacitances. For hand calculations, usually only the dominant capacitances are considered, and they are sometimes lumped into a single linear capacitor to simplify calculations.

In amplifiers, these internal capacitances reduce the high-frequency gain as discussed in Sec.1.6.4. Chapter 8 addresses this issue in detail. In digital circuits, the capacitances introduce switching delays similar to the delays we saw in diodes in Sec. 3.12.4. The following development introduces transistor switching delays.

4.11
The Transistor Switch

Many transistor applications such as digital logic gates, interfacing circuits, power supplies, and communication circuits use the BJT as a switch operated by a control signal. In such applications the transistor operates as a two-state device, with saturation corresponding to a closed switch and cut-off to an open switch. Advantages of transistor switches are speed, versatility, and convenience.

In this section we learn the static and dynamic principles of BJT switches. Chapter 5 introduces field-effect transistors that are also used as switches. Chapters 10 and 12 show how switches are used in power supplies and filters, respectively. We study detailed applications of switches in logic gates, digital memory elements, and interfacing circuits in Chapters 13 and 14.

4.11.1 STATIC SWITCH OPERATION

Figure 4.41a shows a switch connected in series with a load resistor and a power source. Figure 4.41b shows the corresponding characteristic curves, short circuit and open circuit, with a load line superimposed. In Figure 4.41c the transistor output circuit replaces the switch, with collector current becoming switch current and collector-emitter voltage becoming switch voltage; v_C is a signal that controls the switch. Figure 4.41d is the counterpart of Fig. 4.41b for the transistor switch. Control voltage v_C determines the state of the switch by taking on two values, a high value that produces a base current $i_B = I_B$ sufficiently large to saturate the transistor, and a value less than cut-in that makes $i_B = 0$. Out of all possible transistor output curves, only the two shown in Fig. 4.41d are relevant to switch operation. Comparing Figs. 4.41b and d shows that for R_L sufficiently large and for $V_{CE,sat}$ sufficiently small, transistor switching closely resembles ideal switching. When v_C is low, the load current is zero, and all of V_{CC} is dropped across the transistor. When v_C is high, a current of approximately V_{CC}/R_L flows through R_L, and nearly all of V_{CC} drops across R_L. The slight voltage drop across the switch, $V_{CE,sat}$, is usually not a problem. In the next chapter we study a field-effect transistor switch that does not have this *voltage offset*.

Exercise 4.10 In Fig. 4.41c find the minimum control voltage necessary to close the switch if $V_{CC} = 9$ V, $R_L = 800$ Ω, $R_B = 1$ kΩ, $V_{CE,sat} = 0.2$ V, and $\beta = 25$. Also find the load current when the switch is closed.

Ans. 1.14 V, 11 mA.

EXAMPLE 4.12 Use SPICE to simulate the switching circuit of Fig. 4.41c for $V_{CC} = 9$ V, $R_L = 800$ Ω, $R_B = 1$ kΩ, and $\beta = 25$. Control voltage, $v_C(t)$, is a 5 volt pulse of 0.5 μs duration. Make a SPICE plot of the switch voltage.

Solution. Figure 4.41e shows the code, and Fig. 4.41f the output waveform. The latter shows that the switch indeed opens and closes in response to its control voltage. Notice that a 5 V pulse switches a 9 V supply. Compared to the minimum control voltage of 1.14 V that we found in Exercise 4.10, this circuit uses 5 V to drive the transistor *deeply* into saturation for purposes of reliability. This design switches properly in spite of reasonable variations in resistance, beta, and supply voltage, as the reader can easily verify with SPICE. ❏

The simulation in Example 4.12 ignored something very important—transistor dynamics. In transistor switching, as in diode switching, we must be concerned not only with the switch conditions at equilibrium, but also with the nature, duration, and causes of the delays associated with the changes in the switch states, our next topic.

4.11.2 DYNAMIC SWITCHING

Transistor switching delays result from the nonlinear junction capacitances. Perhaps the best way to appreciate the rather complicated interaction between the four nonlinear capacitances and the transistor states is to examine a switching example in detail.

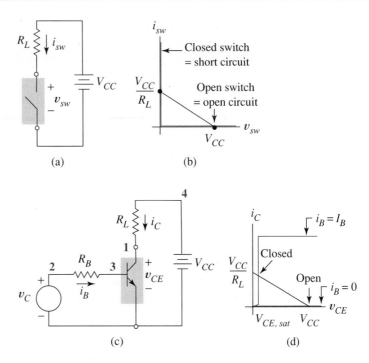

(a)   (b)

(c)   (d)

```
EXAMPLE4.12
VCC 4 0 DC 9
RL 4 1 800
RB 2 3 1K
QSW 1 3 0 SWITCH
.MODEL SWITCH NPN BF=25
VC 2 0 PULSE(0 5 0.5E-6 0 0 0.5E-6 2E-6)
.TRAN 0.02E-6 2E-6
.PLOT TRAN V(1)
.END
```

(e)

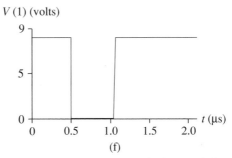

(f)

Figure 4.41 (a) Ideal switch; (b) volt-ampere curves for ideal switch; (c) transistor switch and control voltage v_c; (d) volt-ampere curves for transistor switch with load line superimposed; (e) SPICE code for Example 4.12; (f) simulated switch voltage.

Figure 4.42a shows a simple transistor switch. The top diagram of Fig. 4.42b shows that v_C is a 14 V pulse superimposed upon a dc level of -5 V. From the static switch discussion of the last section, we understand that the intention is to close the switch at $t = 0$ for an interval of T seconds, thereby connecting the 9 V supply to the 2 kΩ resistor by saturating the transistor, and then to reopen the switch at $t = T$ by cutting off the transistor. The actual output voltage, v_o in Fig. 4.42b, however, does not change instantaneously between cut-off and saturation values, but instead changes gradually, and then only after certain time delays. The diagrams in Fig. 4.43 use simplified models to describe the sequence of events internal to the transistor. We ignore substrate capacitance by assuming the transistor is discrete.

Figure 4.43a shows the equilibrium condition prior to $t = 0$. Depletion capacitances associated with its two reverse-biased pn junctions represent the cut-off transistor. Each capacitor is charged to the voltage of its junction, and all currents are zero.

At $t = 0$, the control voltage changes to 9 V as in Fig. 4.43b. The capacitor voltages cannot change instantaneously. The transistor initially stays in cut-off, as the capacitors charge toward new equilibrium values with current directions shown in the figure. It takes the *delay time, t_D,* shown in Fig. 4.42b, before the base-emitter capacitance charges to the cut-in value and the transistor turns on.

As the transistor begins active operation, Fig. 4.43c, base current gradually settles to the final ON value of

$$I_B = \frac{9 - 0.7}{8.3 \text{ k}} = 1 \text{ mA}$$

shown in Fig. 4.42b. Because of the high current delivered by the dependent current source, the base-collector capacitance quickly discharges, output voltage drops to 0.2 V,

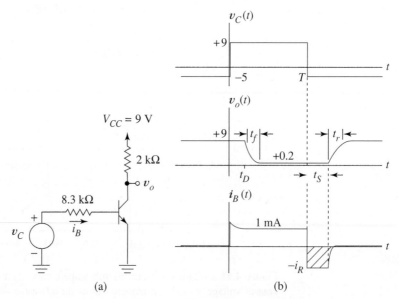

(a)   (b)

Figure 4.42 Switching waveforms for a BJT switch: (a) switching circuit; (b) waveforms.

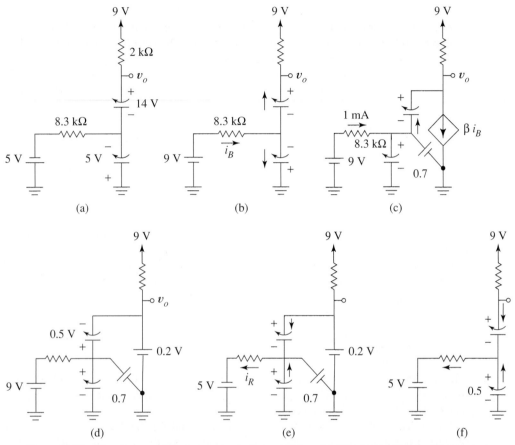

Figure 4.43 Transistor models that apply during switching: (a) initial cut-off state; (b) transient cut-off state after input voltage changes; (c) forward-active transient state; (d) saturation equilibrium state; (e) saturation state before transistor cuts off; (f) cut-off transistor with capacitors prepared to decay to cut-off equilibrium.

and the transistor enters saturation as in Fig. 4.43d. This is the equilibrium corresponding to a closed switch, and the currents all assume the values we would predict by static circuit analysis using the saturation model. Now is the time to recall that the charge stored in the base of the saturated transistor is as shown in Fig. 4.14c.

At $t = T$, the control voltage suddenly changes back to $-5V$ as in Figs. 4.42b and 4.43e. Because the base-emitter diffusion capacitance is charged to $+0.7$ V, v_{BE} cannot change instantaneously, and so the transistor cannot cut off. Instead, it remains in saturation until the excess minority charge in the base, Q_S from Fig. 4.14c, is cleared out. There are two processes simultaneously removing this charge, recombination inside the transistor and base current, $-i_R$, shown in Figs. 4.42b and Fig. 4.43e. (When the external current is the dominant charge removal process, the cross-hatched area under the reverse current curve in Fig. 4.42b approximates Q_S.) The result is a delay called the *storage time, t_s,* shown in Fig. 4.42b. During this interval the base current remains negative, and v_{BE} and v_o change little. Only after charge Q_S is removed does the transistor move through the forward active state to cut-off, Fig. 4.43f. The depletion capacitances now charge to their original equilibrium values. As in the diode switching transient of Fig. 3.65, the storage time, t_s, is usually the dominant switching delay. We can reduce storage time by providing a large reverse current to remove the excess charge quickly.

The output voltage waveform also shows the 10–90% rise time t_r we defined using Fig. 1.40. This part of the transient is mostly related to charging the depletion capacitances of the cut-off transistor to their final equilibrium values. There is also a 90–10% fall time, t_f. Because these particular rise and fall times both involve *nonlinear* capacitors, they are not simple *RC* exponential functions.

Figures 4.42 and 4.43 and the preceding discussion provide an important *qualitative* understanding of the switching process. However, since the transistors make gradual rather than abrupt state changes and since the capacitances are actually nonlinear and four in number, we need simulations to predict with accuracy the true delays expected of a given device. The dynamic SPICE model of the BJT described in the next section serves this purpose.

4.12
Dynamic SPICE BJT Model

The SPICE parameters of Fig. 4.36 account for most of the important static features of the transistor. To simulate dynamic behavior, we also need parameters to characterize the diffusion and depletion capacitances of Fig. 4.40. To describe the nonlinear voltage-controlled depletion capacitances, $C_{dep,E}$ and $C_{dep,C}$, we must specify zero voltage capacitances, CJE and CJC, barrier potentials, VJC and VJE, and grading coefficients, MJC and MJE. These all have the same meaning as they did for the diode in Sec. 3.12. For integrated transistors we also need to give parameters for the collector-substrate capacitance, $C_{dep,S}$—CJS, VJS, and MJS.

For the diffusion capacitances, $C_{diff,E}$ and $C_{diff,C}$, we have forward and reverse lifetimes: TF for emitter-base capacitance and TR for collector-base capacitance. The table of Fig. 4.44a updates our list of BJT SPICE parameters, including default values. All key dynamic parameters have default values of 0; thus *the default transistor model is a static model.* Figure 4.44b shows representative values for a low-power integrated npn transistor designed for linear applications such as amplification.

SPICE uses the model of Fig. 4.40 for all types of analysis. During a simulation, it first computes the dc operating point of every transistor. From this information, the Q-point values of the depletion and diffusion capacitances and the Early voltage correction factors are calculated. For sinusoidal analysis, SPICE then uses linear analysis, treating depletion and diffusion capacitances as constants. For transient analysis, SPICE updates capacitance values as required throughout the simulation as the operating point changes.

Figure 4.45a shows a digital logic circuit in which the transistor is a switch. Conceptually, operation is simple. When input VIN is close to zero, the transistor cuts off, bringing v_o close to VCC; when VIN is close to 4 V, the transistor saturates, and v_o is close to 0.2V. However, internal capacitances of transistor and diodes cause time delays

SPICE parameters				Typical IC values			
Static		**Dynamic**		IS	1E-16A	CJE	1.0 pF
	Default value		**Default value**	BF	200	VJE	0.7 V
				BR	2	MJE	0.33
IS	1E-16A	CJE	0	RC	200 Ω	CJC	0.3 pF
BF	100	VJE	0.785 V	RB	200 Ω	VJC	0.55 V
BR	1	MJE	0.33	RE	2 Ω	MJC	0.5
RC	0	CJC	0	VAF	130 V	CJS	3 pF
RB	0	VJC	0.75 V	VAR	50 V	VJS	0.52 V
RE	0	MJC	0.33	XTB	1.7	MJS	0.5 V
VAF	∞	CJS	0			TF	0.35 ns
VAR	∞	VJS	0.75 V			TR	10 ns
XTB	0	MJS	0				
		TF	0				
		TR	0				

(a)   (b)

Figure 4.44 SPICE parameters for dynamic transistor model: (a) parameters and default values; (b) representative values for integrated transistor.

and waveform shapes that are difficult to calculate by hand analysis. In the following example we use SPICE to examine the details.

EXAMPLE 4.13 Use SPICE to plot input voltage, base current, and output voltage for Fig. 4.45a when VIN is a 4 V pulse of 60 ns duration.

Solution. Figure 4.45b shows code for finding the output voltage and base current waveforms when VBB = 0 V. VIN is a 4 V pulse of 60 ns duration, delayed by 10 ns so that the transient associated with each pulse edge is easy to see. Source VM is an ammeter that allows us to observe the base current, an artifice necessary because SPICE prints or plots currents only when associated with voltage sources. The .MODEL lines show that the simulation includes three internal transistor capacitances and six diode capacitances, all nonlinear.

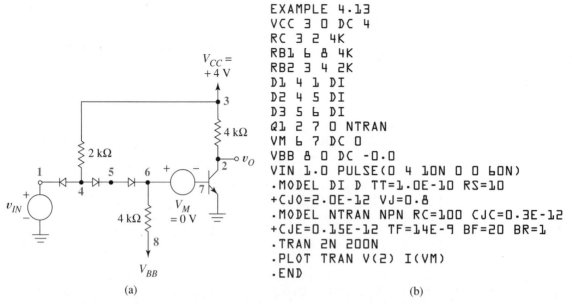

```
EXAMPLE 4.13
VCC 3 0 DC 4
RC 3 2 4K
RB1 6 8 4K
RB2 3 4 2K
D1 4 1 DI
D2 4 5 DI
D3 5 6 DI
Q1 2 7 0 NTRAN
VM 6 7 DC 0
VBB 8 0 DC -0.0
VIN 1.0 PULSE(0 4 10N 0 0 60N)
.MODEL DI D TT=1.0E-10 RS=10
+CJO=2.0E-12 VJ=0.8
.MODEL NTRAN NPN RC=100 CJC=0.3E-12
+CJE=0.15E-12 TF=14E-9 BF=20 BR=1
.TRAN 2N 200N
.PLOT TRAN V(2) I(VM)
.END
```

(a)   (b)

Figure 4.45 Example logic gate: (a) schematic diagram with nodes numbered for SPICE analysis; (b) SPICE code for analysis when VBB = 0.

Figure 4.46a compares input and output waveforms. We see the expected delays in the output voltage as it changes between cut-off and saturation values. The delay associated with leaving saturation is clearly dominant. Certainly the curves themselves provide us with little insight into the *causes* of the delays or *how* to reduce them. Forearmed with theory, however, we would expect a negative value of VBB to accelerate the exit from saturation by removing stored charge.

Figure 4.46b shows the results of a second SPICE simulation that uses VBB = −2 V to verify this conjecture. As expected, the negative base current transient at the end of the input pulse becomes larger in magnitude but shorter in duration. □

A less desirable change in the second run was that the high-to-low transition of output voltage took longer for VBB = −2 V. After some thought, we might explain this by changes associated with the base-emitter depletion capacitance of the cut-off transistor. This capacitance is charged to a more negative voltage prior to the input pulse when VBB = −2 V than for 0 V. Also, notice in Fig. 4.45a that the current that flows through the 2 kΩ resistor to charge the depletion capacitance and turn on the transistor is diminished by current that flows to VBB. This is consistent with the reduced positive base-current in Fig. 4.46b when the transistor is on. The point is to show how qualitative theoretical reasoning complements quantitative simulations.

Exercise 4.11 From the curves in Figs. 4.46, make rough estimates of delay time, storage time, rise time, and fall time for the gate circuit of the preceding example. Because

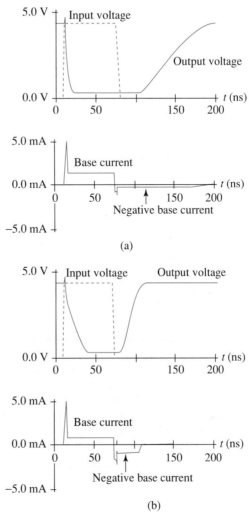

Figure 4.46 (a) Input and output waveforms when VBB = 0 V; (b) input and output voltage and current waveforms when VBB = −2.0 V.

the data are so imprecise, use 0–100% for rise and fall times instead of the usual 10–90%.

Ans. When $V_{BB} = 0$, $t_D = 0$, $t_S \approx 35$ ns, $t_r = 75$ ns, $t_f = 15$ ns. When $V_{BB} = -2$ V, $t_D = 0$, $t_S \approx 5$ ns, $t_r = 34$ ns, $t_f = 28$ ns.

4.13

Integrated Circuit Fabrication

Because integrated circuit design and fabrication are closely related, an understanding of modern electronics requires a basic understanding of how the IC is made. Integrated circuits favor particular devices and circuit structures over others, and IC designs exploit special features closely related to fabrication such as availability of circuit parameters that track over wide temperature ranges. Furthermore, fabrication techniques cause interrelationships between component parameter values everywhere on an IC chip. This section clarifies these ideas by describing one simplified process for producing bipolar ICs. For clarity and brevity we present only those fundamentals necessary for an introductory understanding. Courses in VLSI (very-large-scale integrated circuit) design, many excellent textbooks, and a variety of trade magazines are available to those interested in learning the latest techniques in this rapidly changing area.

Historical Perspective. The number of circuit elements contained in a single IC has increased rapidly over the years. The first historical landmark, called _small-scale integration_ or SSI, produced IC chips containing 1 to 10 logic gates, each the size of Fig. 4.45a. Improvements in fabrication procedures then led to _medium-scale integration,_ MSI, involving 10–100 gates/chip, and _large-scale integration,_ LSI, yielding 100–10,000 gates/per chip. Today _very-large-scale integration,_ VLSI, circuits include more than 10^5 gates/chip. As IC technology improved, the physical size of the circuit components diminished, until now the smallest component dimension on a chip is less than a micron. By comparison a red blood cell has a diameter of about 7 microns.

Each stage of IC development led to increased sophistication of the basic functional units available to circuit designers. Before ICs, circuit design involved interconnecting individual transistors, resistors, and diodes. SSI offered the high-quality logic gate as a design element. With MSI came still larger circuits such as registers and counters for digital design and operational amplifiers for analog design. With LSI came a novel marketing problem. A technology capable of making highly sophisticated products such as digital wristwatches and calculators was available; however, the process was inherently expensive, and it relied upon large sales volumes for reasonable unit prices. The problem was to identify other products able to command the necessary sales volumes. The microprocessor IC chip provided the ideal solution, for basic designs could be mass produced at low cost, and then individually tailored by the user for specific applications by customized software. VLSI is now extending this trend toward more complex functional elements, making realities of heretofore impractical ideas such as parallel processing, automatic error detection and correction, and fault-tolerant operation. Programmable logic arrays (PLAs) and application-specific integrated circuits (ASICs) are recent methods of adapting inexpensive multipurpose IC hardware elements to specific applications while reducing software development expenses through factory programming.

The advantages of ICs are legend:

1. The small physical size and weight are useful in themselves.
2. Unit cost decreases with the miniaturization, as more ICs are produced in each batch
3. Smaller size gives reduced device capacitances and higher speed.
4. Greater circuit complexity is possible because more components can be used per circuit.
5. Reliability is higher because device interconnections are formed during the manufacturing process rather than by individuals with soldering irons as pre-IC days.
6. Availability of closely matched circuit parameters now make circuit structures feasible that were previously impractical.
7. Devices physically close on the chip have nearly identical temperatures; thus parameters track with temperature.

ICs also have some disadvantages:

1. Circuit testing is difficult because the limited number of external connecting pins provide limited access to circuit nodes. Because complete testing is so difficult, component failures can sometimes go undetected.
2. Increased circuit complexity trades off against reliability. A circuit with many components has many possible failure points, which reduce yields.
3. With thousands of devices operating on a single IC chip, there are sometimes problems in disposing of excess heat.

IC Fabrication Fundamentals. We now describe one simple IC fabrication process. The scales of the simplified diagrams show concepts rather than accurately depicting the relative feature sizes.

An integrated circuit is sometimes termed *monolithic* (single stone) to emphasize that *all* circuit elements are fabricated together *at the same time* on a single semiconductor crystal. Figure 4.47a shows the first step in IC fabrication. A *seed crystal* of silicon is placed into molten p-type silicon and slowly rotated and raised. The result is a cylinder of *single-crystal,* acceptor-type solid silicon, about 1 m long and 15 cm in diameter. This is sliced into *wafers,* 0.5 mm in thickness, as shown in Fig. 4.47b, which are then polished and cleaned with an etching solution to produce a very smooth surface. Each wafer is the future home of 1000 or so identical integrated circuits, or *IC chips,* each 1 to 5 mm on an edge, as in Fig. 4.47c. The p-type silicon of the wafer, called the *substrate* or *body,* provides a rigid structural base for the IC components. A batch of several wafers is subjected to a sequence of fabrication steps that *simultaneously form every circuit component on every integrated circuit on every wafer.*

The wafers are first heated to 1000°C in a vacuum chamber in the presence of oxygen. This coats the top of each wafer with a layer of silicon dioxide, SiO_2 (glass), which

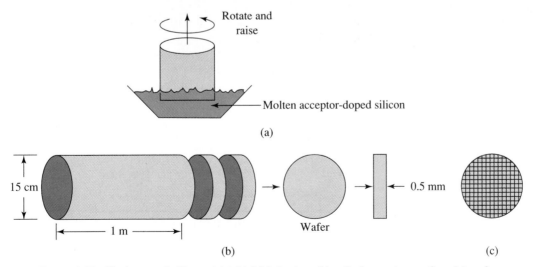

Figure 4.47 Single-crystal silicon: (a) initial fabrication; (b) cylinder cut into wafers; (c) wafer as future home of 1000 ICs.

protects the silicon from contamination by impurities. Silicon dioxide is also an electrical insulator and dielectric, two features that prove useful during IC fabrication.

Photolithography. The *photolithographic process* is a succession of photographic masking, developing, and etching steps that make selected regions of silicon accessible for processing. To introduce photolithography, we describe how it is used to create the *buried layer,* the first step in fabricating an integrated npn transistor. This buried layer of highly conducting material is needed to give transistors with low collector resistance, r_c. (See Figs. 4.35a and 4.40.)

Buried Layer Formation. The oxide that covers the substrate is coated with photoresist, and then covered with a photographic mask. The latter establishes the location of each transistor by an opaque region as in Fig. 4.48a. Ultraviolet light then hardens the photoresist in those areas unprotected by the mask. The mask is then removed, the un-

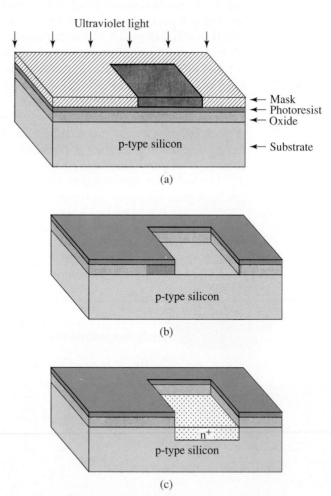

Figure 4.48 Photolithography: (a) mask defines target location; (b) photoresist and oxide removed over unprotected region; (c) impurities inserted by ion implant or diffusion.

hardened photoresist dissolved by developing solution, and the exposed oxide removed by chemical etching, giving Fig. 4.48b. Each selected region of the silicon is now accessible from above.

The *buried layer* is now formed by introducing a high concentration of donor impurities such as phosphorous or antimony into the substrate in those regions selected by the masking. The p-material is thereby converted into n^+-material by compensation. (n-Type material with impurity concentration $>10^{18}$ atoms/cm^2 is denoted as n^+.)

There are two techniques available for introducing impurities, diffusion, and ion implant. Diffusion involves heating the wafer to 1000°C in an atmosphere containing the impurities, and then allowing these impurities to diffuse down into the crystal. Ion implant is a room temperature process in which an electric field accelerates and focuses impurity ions into a narrow beam. The beam scans the wafer surface, bombarding it with impurities that penetrate the surface to a depth controlled by the accelerating field. The temperature is briefly elevated, and impurity atoms replace silicon atoms in the crystal lattice. Ion implant is used most often for the buried layer. After the buried layer is created, resulting in Fig. 4.48c, the remainder of the photoresist and oxide are removed in preparation for the next processing step, epitaxial crystal growth.

Epitaxial Growth. The wafer is heated to 1200°C in an atmosphere of silicon chloride and phosphine gas to provide silicon and donor-type impurity atoms, respectively. During the ensuing *epitaxial growth* process, a layer of n-doped silicon forms upon the p-type substrate, extending the *single-crystal* structure as in Fig. 4.49. Crystal growth proceeds until the epitaxial layer has a thickness of about 1 micron. A protective layer of SiO$_2$ is then formed on the surface, completing the first stage of fabrication. The epitaxial layer provides the collector material for each npn transistor in *every* integrated circuit. There may also be pnp transistors that use it for their bases. Notice that the doping concentration, N_d, established in this step is *common to all devices* fabricated from the material.

Oxide Isolation. The purpose of the next sequence of fabrication steps is to locate each component site and electrically isolate it by surrounding it on four sides with a silicon dioxide insulator. Before applying the photoresist and isolation mask, the surface oxide is coated with silicon nitrite, Si$_3$N$_4$ (an oxide growth inhibitor), as in Fig. 4.50a. The second mask shades all regions between future component sites. In this figure, the two unprotected areas above the buried layer locate two parts of a future npn transistor; the other unprotected area marks a location reserved for a resistor. After hardening the ex-

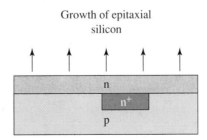

Figure 4.49 Epitaxial growth of single-crystal n-type silicon.

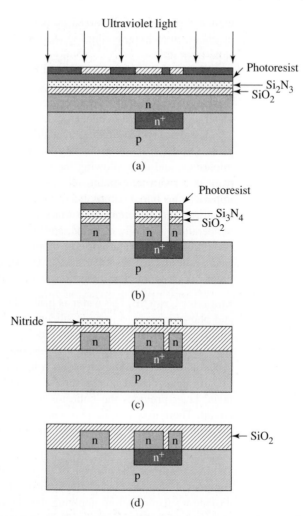

Figure 4.50 Oxide isolation: (a) hardening photoresist over protected areas; (b) removal of selected epitaxial silicon; (c) forming insulating oxide; (d) end result of the isolation process.

posed photoresist with light, then photoresist, nitrite, oxide, and n-type silicon are all removed. This produces isolated projections of protected silicon at component sites as in Fig. 4.50b. The hardened photoresist is then removed and a *thick oxide* formed. Because the silicon nitrite inhibits oxide growth at the component sites, the result resembles Fig. 4.50c, which, after removal of the nitrite, gives Fig. 4.50d. Now all components are electrically isolated on all sides except the bottom, where they share a common substrate. When we use ICs, we connect the p-type substrate to the most negative point in the circuit, thereby electrically isolating the components from the substrate by reverse-biased pn junctions.

Base and Emitter Diffusion. The next two steps are called base and emitter diffusion even though they are sometimes performed by ion implant. A new mask identifies loca-

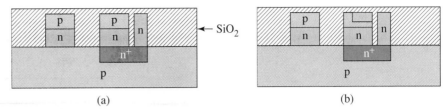

Figure 4.51 Base and emitter diffusion steps: (a) result of base diffusion; (b) result of emitter diffusion.

tions of transistor bases as well as sites for resistors and other components. After the protective oxide is etched away, p-type impurities diffuse from above into exposed n-material, and a new protective oxide is installed over the surface. Figure 4.51a shows the end result. The p-type material at both sites has conductivity appropriate for the base of a high-quality npn transistor. The p-type material at the resistor site also has the physical dimensions of the finished resistor. The final dimension of the base, its base width or thickness, depends upon the next fabrication step.

The next masking operation defines the emitters of npn transistors. Diffusion of n^+ dopants completes the npn transistors, as we see in Fig. 4.51b. Like the collectors, bases and emitters of *all transistors of all chips* on the wafer are fabricated simultaneously. Thus impurity concentrations and penetration depths are identical for *all chips and wafers in the same batch.* From one batch to another, however, it is difficult to reproduce exactly the same impurity conditions. Mask dimensions, on the other hand, are closely controlled. The original large masks, computer constructed, are photographically reduced to actual IC size. Thus the surface dimensions are well matched even across batches.

Metalization. Once all components are fabricated, they are next interconnected using two *metalization* steps. A mask defines locations where device connections are to be made, and the oxide is selectively etched away as shown in Fig. 4.52a. A thin layer of n^+ impurities (not shown in the diagrams) is made in the exposed contact regions to ensure ohmic rather than rectifying contacts. Aluminum is then vacuum deposited over the entire wafer surface, making all the necessary contacts to the silicon as in Fig. 4.52b. A final masking step protects the interconnection metal, and the undesired aluminum is removed with etching solution. Removing the photoresist completes the basic IC fabrication process, resulting in Fig. 4.52c. This narrative omits fabrication steps associated with providing external contacts to the ICs, testing, identifying defective chips, and separation of individual chips; because these important tasks are unrelated to circuit understanding and design.

Fabricating Resistors, Diodes, and Capacitors. The resistor between terminals X and Y in Fig. 4.52c is called a *base-diffused resistor* because it is formed during the base fabrication step. It is simply a segment of p-type semiconductor material with resistance given by Eq. (3.23). In this equation, L and W are established by a photographic mask; however, the value of the sheet resistance cannot be independently specified for the resistor. This is because the thickness, t, of the material and the impurity concentration (a factor in σ), must both be appropriate for forming bases of npn transistors. This limits the available sheet resistance to about 200 Ω/square, a rather low value that makes it

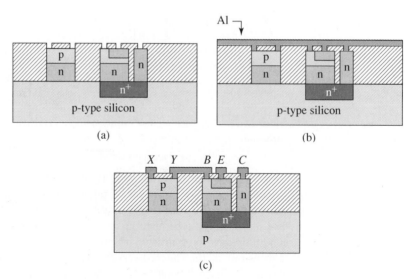

Figure 4.52   Metalization: (a) after masking; (b) after deposition of Al; (c) after removal of undesired Al.

impractical to fabricate resistors much larger than 50 kΩ by this process. *Emitter-diffused resistors* can be fabricated during the emitter formation; however, because the emitter is more heavily doped than the base for high injection efficiency, sheet resistances are even lower than those of base-diffused resistors. Other schemes are sometimes used for making IC resistors of higher sheet resistance, but these require extra fabrication steps that add to the expense. The general conclusion is that large resistor values are not readily available in IC circuit designs.

Since sheet resistance depends on the concentration of dopants in the atmosphere *at the time of fabrication,* resistor *values* are subject to rather wide variations. This means circuit designs that require precise control of resistor *values* are impractical for ICs. However, from Eq. (3.23), we see that *resistor ratios* on the same IC chip are determined by the relative *W/L* values, and these are precisely controlled by the photographic masks. A practical consequence is that resistor *ratios* can be held to tolerances of the order of 1%. For this reason, IC designs favor circuits whose operation is determined by resistor ratios.

IC diodes are usually base-collector junctions of transistors, with the emitter either left open or connected to the base during the metalization step.

Figure 4.53a shows one way to make an integrated capacitor. A thin layer of SiO_2 is the dielectric of a parallel-plate capacitor, with one plate made during metalization, and with highly doped p^+-material forming the other conductor. Diffusion of the p^+ impurities and fabrication of the thin dielectric introduce extra fabrication steps, adding to the expense. The depletion capacitance of a reverse-biased diode is a less expensive capacitor because it does not require extra fabrication steps.

pnp Transistors. A *lateral* pnp transistor, Fig. 4.53b, can be included in the IC without special fabrication steps. It is called a *lateral transistor* because holes injected from the

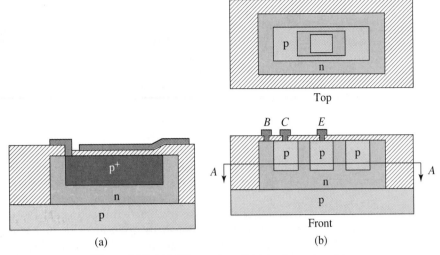

Figure 4.53 (a) IC capacitor; (b) lateral pnp transistor.

p-type emitter flow laterally through the n-type base to the ring-shaped collector. Emitter and collector are fabricated during the base-diffusion step, and the base is made of epitaxial n-type material. Because the geometry is not particularly favorable to efficiently collecting the emitted holes and preventing recombination in the base, the transport factor is lower than in npn transistors. Because the doping concentrations are selected to optimize npn transistors, lateral pnp transistors have higher doping concentrations in the collector than in the base, causing most of the collector-base depletion region to extend into the base. This gives rather large bases, with adverse effects on injection efficiency, base width modulation and depletion capacitance. Because of such factors, lateral pnp transistors tend to have low β, low output resistance, low breakdown voltage, and large internal capacitances compared to npn devices. High-quality pnp transistors can be produced in ICs by using extra diffusion steps that optimize doping concentrations for these transistors.

Also of interest is IC fabrication of other devices such as vertical pnp transistors, zener diodes, pinch resistors, thin film resistors. Many authors such as Gray and Meyer (listed in the references at the end of the chapter) address these subjects.

4.14 _____
Summary

The bipolar junction transistor is a solid-state device consisting of two pn junctions fabricated in close proximity on single-crystal semiconductor. A large-signal, nonlinear, static, Ebers–Moll model, embodied in an equivalent circuit and a pair of simultaneous equations, helps us understand the transistor in terms of the theory of pn junctions.

For small-signal, linear applications we use external sources to bias the transistor in its forward active state. For this state, the Ebers–Moll model predicts the characteristic curves and linear equations of a current-controlled current source. This leads to a simple linear equivalent circuit that we use to predict how the transistor interacts with other circuit elements. To do the circuit analysis, we replace the transistor with its equivalent circuit.

The transistor has two other states of major importance: saturation, in which the transistor resembles a closed switch, and cut-off, where the transistor represents an open switch. A third state, reverse active, is of little practical importance. Each of the four states corresponds to operation in a particular region of the transistor output characteristics, and each has a circuit model, related in an obvious way to the characteristic curves.

In some important problems the state of the transistor is unknown. In these cases we must first assume a state for each transistor, and then use its equivalent circuit to check our assumption. An essential part of using any equivalent circuit is learning to watch for and recognize a contradiction that proves the equivalent invalid.

An important aid in developing understanding and intuition about transistor circuits is a graphical tool, the load line. This is a simple volt-ampere plot that shows the constraint placed upon the transistor's input or output variables by the linear circuit in which the transistor is imbedded. A key example used the load line to give an intuitive appreciation for how a transistor saturates and for the meaning of the inequality, $\beta I_B \geq I_C$, that partially describes the saturated transistor.

A number of static second-order effects describe differences between real transistors and the idealized devices. These include temperature variations of parameters, output resistance, junction breakdown, internal feedback, and parasitic resistances. There are also dynamic transistor limitations embodied in the depletion and diffusion capacitances of the junctions and between the collector and substrate. Except for minor differences, these are the same nonlinear capacitances that we previously encountered in diodes.

Sophisticated SPICE transistor models enable us to simulate both static and dynamic behavior of transistors, including all of the major nonlinearities. We generally use our simplified and more intuitive transistor concepts to design, and then follow up with accurate computer simulations that include the second-order effects we consciously ignored to make our initial design work tractable. We examine and evaluate the simulation results in terms of our simple conceptual ideas of what should have happened, and then redesign, if necessary, again turning to simple models but under the guidance of simulation results.

A discussion of one basic integrated circuit fabrication process gave us some appreciation for the circuit design opportunities and constraints imposed by the fabrication process itself. In IC designs, we can employ transistors with closely matched parameter values that track with temperature. Adding additional transistors to serve some useful purpose in a circuit adds little to the cost. Resistor ratios that are temperature invariant and realized to 1% accuracy are readily available. On the other hand, parameter values of all circuit components on a chip are interdependent, for all components are fabricated at the same time. Some parameters, such as absolute resistor values, cannot be precisely controlled; and inexpensive pnp transistors have parameter values greatly inferior to npn parameters. Special components that require additional fabrication steps, for example, resistors of large ohmic value or high-quality pnp transistors, add significantly to circuit cost.

REFERENCES _____

1. ANTOGNETTI, D.A., and G. MASOBRIO (ed.). *Semiconductor Device Modeling with SPICE*, McGraw-Hill, New York, 1988.
2. BANZHAF, W. *Computer-Aided Circuit Analysis Using SPICE*, Prentice Hall, Englewood Cliffs, NJ, 1989.
3. BURNS, S.G., and P.R. BOND. *Principles of Electronic Circuits*, West Publishing, St. Paul, MN, 1987, p. 134.
4. GRAY, P.R., and R.G. MEYER. *Analysis and Design of Analog Integrated Circuits*, John Wiley, New York, 1984.
5. HINES, J.R. "A SPICE model for saturating NPNs," *Design Automation*, pp. 57–59, April 1992.
6. HODGES, D.A., and H.G. JACKSON. *Analysis and Design of Digital Integrated Circuits*, 2nd ed., McGraw-Hill, New York, 1988, p. 166.
7. HORENSTEIN, M.N. *Microelectronic Circuits and Devices*, Prentice Hall, Englewood Cliffs, NJ, 1990.
8. SEDRA, A.S., and K.C. SMITH. *Microelectronic Circuits*, 3rd. ed., Saunders, Philadelphia, 1991.
9. STREETMAN, B.G. *Solid State Electronic Devices*, 2nd ed., Prentice Hall, Englewood Cliffs, NJ, 1980, p. 271.
10. TUINENGA, P.W. *SPICE A Guide to Circuit Simulation & Analysis Using PSpice*, Prentice Hall, Englewood Cliffs, NJ, 1988.

PROBLEMS _____

SPECIAL DIRECTIONS FOR SPICE HOMEWORK PROBLEMS: *Do not hand in lengthy SPICE printout for homework.* Instead, abstract the useful information from the SPICE output file as in the SPICE examples. Include your SPICE code and a circuit diagram with nodes numbered to agree with the code. Cite relevant numerical values from the SPICE output file and discuss when appropriate. Make sketches of any relevant curves, and label appropriate points. Make small tables to present numerical data if useful for clarity.

Section 4.1

4.1 In Fig. 4.1c, $i_C = \alpha_F i_E$. Using this expression and KCL,
(a) express i_C in terms of i_B and α_F,
(b) express i_E in terms of i_B and α_F.
(c) Write numerical expressions for parts (a) and (b) using $\alpha_F = 0.995$.

4.2 A BJT design is modified by decreasing the base width. Would this change the transport factor or the injection efficiency? In which direction?

4.3 In Fig. 4.2, the emitter current is approximated as the current of a forward-biased diode having an electron component and a hole component; that is, $i_E = i_{En} + i_{Ep}$. Use the diode current expression of Eq. (3.38) to find separate expressions for i_{En} and i_{Ep}.

Section 4.2

4.4 In Fig.4.3, suppose the base-emitter diode is forward biased and the base-collector diode is reverse biased. Draw the simplified equivalent circuit that results from using the offset diode models. If you do this thoughtfully, your equivalent circuit will have only two components and you will be able to explain why.

4.5 A pnp transistor has p- and n-impurities interchanged, resulting in Fig. 4.15c instead of Fig. 4.3. Use KCL at emitter and collector nodes of Fig. 4.15c to find the pnp counterparts of Eqs. (4.1) and (4.2). In your equations use v_{EB} and v_{CB} to represent the voltages applied to the junctions.

Section 4.3

4.6 The curves of Figs. 4.5b and c describe the transistor in Fig. P4.6. Find the numerical values of i_B, i_C, and v_{CB}.
Hint: Replace the transistor in the circuit diagram with the model of Fig. 4.6.

4.7 Use the common-emitter equation Eq. (4.10) and Kirchhoff's current law for the transistor to solve for
(a) i_C as a function of i_E and β_F,
(b) i_B as a function of i_E and β_F.

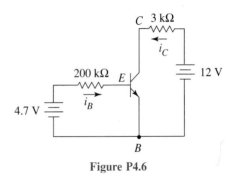

Figure P4.6

Section 4.4

4.8 (a) Reduce the Ebers–Moll equations, Eqs. (4.3) and (4.4), to simplified expressions that represent a cut-off transistor by assuming both v_{BE} and v_{BC} are $<< V_\gamma$. From the simplified equations derive an expression for base current.
(b) Rewrite your equations for the case $\alpha_F = 1$.
(c) Draw a circuit model of your equations from part (b) and compare it with Fig. 4.9b. Give a theoretical expression for I_{CBO}.

4.9 Reduce the Ebers–Moll equations, (4.3) and (4.4), to simpler forms suitable for a reverse-active transistor by assuming $v_{BE} << V_\gamma$ and $v_{BC} >> V_\gamma$.

4.10 Reduce Eqs. (4.3) and (4.4), to simpler forms suitable for a saturated transistor by assuming that v_{BE} and v_{BC} are both $>> V_\gamma$.

4.11 Sketch the first and third quadrant common-emitter output characteristics for a transistor having the parameters, $\beta_F = 10$, $\beta_R = 1$, $V_{CE,sat} = 0.1$ V. Use 1 mA increments of base current for representative curves.

4.12 Figure P4.12 shows BJT input and output characteristics. From the curves draw four equivalent circuits, one for each operating mode.

Section 4.5

4.13 Suppose the transistor in Fig. 4.13a has the output characteristics of Fig. P4.13. Use a load line to find the collector current and collector-emitter voltage if the base current is:
(a) $i_B = 0.04$ mA,
(b) $i_B = 0.1$ mA,
(c) $i_B = 0.085$ mA.

4.14 Figure P4.14 shows a transistor circuit and the transistor's output characteristic. Assume $V_{BE} = 0.7$ V.
(a) Use a load line to find values for I_C and V_{CE}.
(b) Graphically find the new value of V_{CE} if we change V_{CC} to $+2$ V.
(c) Graphically find the new value of V_{CE} if we change V_{CC} to -3 V.

4.15 Figures P4.15a–c show a device, its input characteristic, and its output characteristics. Use load lines to find the value of R that makes $i_2 = 20$ mA in Fig. P4.15d if $R_1 = 1.6$ kΩ, $V_{XX} = 4$ V, and $V_{YY} = 48$ V.

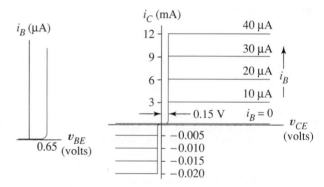

Figure P4.12

4.16 Figures P4.15a–c show a device, its input characteristic, and its output characteristics. We wish to set the output circuit operating point at $v_2 = 12$ V and $i_2 = 20$ mA.
(a) If $R = 1.2$ kΩ, what V_{YY} value do we need?
(b) If $V_{XX} = 2$ V, use the input characteristic to estimate R_1 for the values of part (a).

4.17 (a) In the circuit of Fig. P4.17, find the minimum value of V_S required to saturate the transistor.
(b) What value of V_S does it take to saturate the transistor with forced beta of 50?

4.18 The transistor in Fig. P4.18 has $\beta = 10$.
(a) Sketch its output characteristics.
(b) On the output characteristics, sketch the load line for the 5 V source and the inductor.
Hint: An inductor is a short circuit to dc.
(c) On the load line of part (b) indicate the Q-point when $v_I = 5$ V.

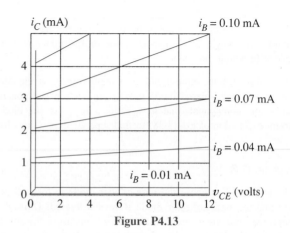

Figure P4.13

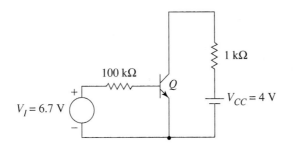

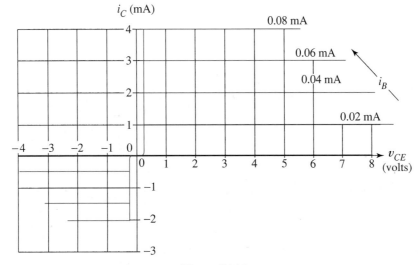

Figure P4.14

Section 4.6

4.19 In Fig. P4.19 assume $V_{BE} = -0.7$ V. Use the output circuit load line to help find the value of V_S that
(a) biases the transistor at $V_{CE} = -4$ V,

(b) puts the transistor at the edge of saturation,
(c) gives a Q-point with forced beta = 100.

4.20 For the transistor of Fig. P4.19,

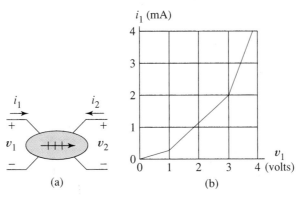

(a)

(b)

Figure P4.15

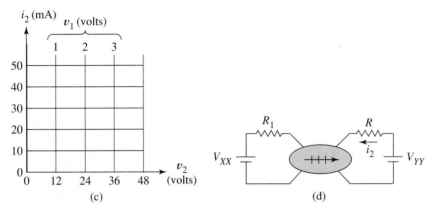

(c)   (d)

Figure P4.15 (continued)

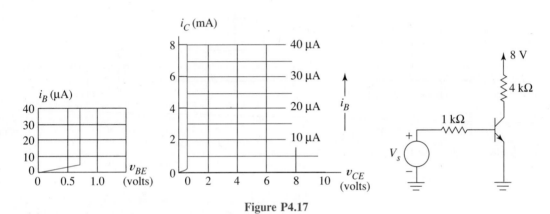

Figure P4.17

(a) draw the large-signal forward active model, assuming the transistor is made of silicon.

(b) Replace the transistor with its model from part (a); set V_S at a value that gives a base current of 5 μA. Then use the circuit to find the value of V_{CE}.

Section 4.7

4.21 For both transistors in Fig. P4.21, β = 99.

(a) Use forward active models to find the collector current and the collector, emitter, and base node voltage of each transistor.

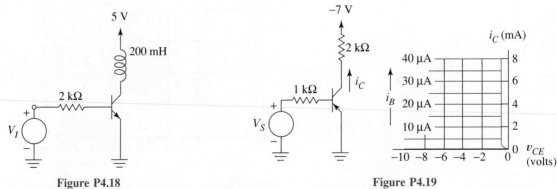

Figure P4.18   **Figure P4.19**

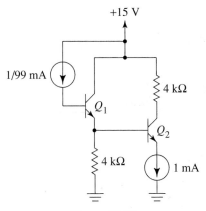

+15 V

1/99 mA

Q_1

4 kΩ

Q_2

4 kΩ

1 mA

Figure P4.21

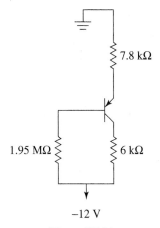

7.8 kΩ

1.95 MΩ

6 kΩ

−12 V

Figure P4.24

(b) Show each Q-point on an i_C versus v_{CE} coordinate system.

(c) Find the voltage across each independent current source.

4.22 The silicon transistor in Fig. P4.22 has $\beta = 20$ and is in the forward active mode.

(a) Use the large-signal model to find the node voltages at base and collector.

(b) Find the quiescent operating point of the transistor, (V_{CE}, I_C).

(c) Show the Q-point location on an i_C versus v_{CE} coordinate system.

(d) To the coordinate system used for part (c), add the new Q-point location if beta is changed to 100.

4.23 Redraw Fig. 4.19d. Add to this sketch the new Q-point of the transistor in Fig. 4.19a if we change the power supply to 12 V and simultaneously replace the transistor with one for which the transistor $\beta = 20$. Assume forward active.

4.24 Redraw Fig. P4.24, adding reference directions for each transistor current. Use the large-signal transistor model to find the Q-point of the transistor, and show it on an i_C versus v_{CE} coordinate system. The transistor is made of silicon with $\beta = 40$.

4.25 For the circuit of Fig. P4.25, use infinite beta analysis to find

(a) the node voltage at the base of Q_1,

(b) V_{CE} for Q_1,

(c) I_C for Q_1,

(d) the Q-point of Q_2.

4.26 Use infinite beta analysis to find the operating point of each transistor in Fig. P4.26. Show each Q-point on an output characteristic coordinate system.

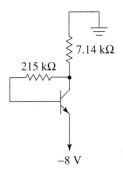

7.14 kΩ

215 kΩ

−8 V

Figure P4.22

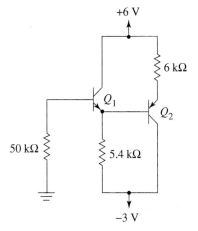

+6 V

6 kΩ

Q_1

Q_2

50 kΩ

5.4 kΩ

−3 V

Figure P4.25

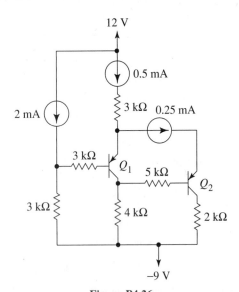

Figure P4.26

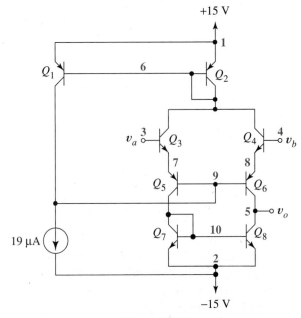

Figure P4.28

4.27 Assume Q_1 and Q_2 in Fig. P4.27 are both biased in the forward active region. Use infinite beta analysis to check this assumption by finding I_C and V_{CE} for each transistor.

4.28 Use infinite beta analysis to estimate node voltages 6–10 in Fig. P4.28 when nodes 3 and 4 are grounded.

4.29 (a) Using infinite beta analysis, derive a linear equation for v_o in terms of v_i for Fig. P4.29.
(b) As v_i takes on larger and larger values, the collector voltage decreases, the emitter voltage increases, and the transistor eventually comes into saturation. Write an equation that relates v_i and v_o when the transistor just reaches saturation.

(c) Simultaneously solve the equations of parts (a) and (b) to find the value of v_i that will saturate the transistor.

4.30 Assume that the transistor in Fig. P4.30 is forward active. Use a circuit model to check this assumption and, if incorrect, state precisely the contradiction.

4.31 Assume that the transistor in Fig. P4.30 is cut off. Use a circuit model to check this assumption and, if incorrect, state precisely the contradiction.

4.32 Assume that the transistor in Fig. P4.30 is saturated.
(a) Use the saturation model to find the values of collector current and base current.

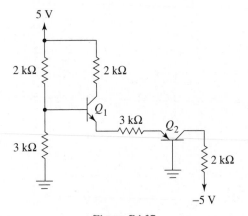

Figure P4.27

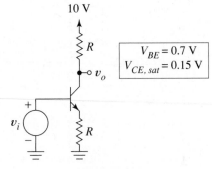

Figure P4.29

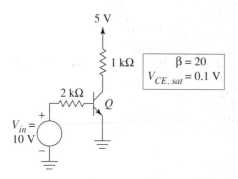

Figure P4.30

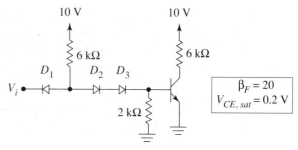

Figure P4.36

(b) Use the results of part (a) to confirm that the transistor is saturated.
(c) For what value of input voltage will the transistor just begin to come out of saturation?

4.33 Assume the transistor in Fig. P4.33 is cut off and the diode is ON. Use models to test both assumptions.
Clearly state any contradictions you find.

4.34 Assume the transistor in Fig. P4.33 is forward active and the diode is ON. Use models to test both assumptions. Clearly state any contradictions you find.

4.35 In Fig. P4.33, assume that the transistor is saturated and the diode is ON with a drop of 0.7 V.
(a) Compute the value of collector current.
(b) Verify the assumption that the transistor is saturated.

4.36 In Fig. P4.36, let $V_i = 10$ V. Assume the following states apply for this condition: D_1 OFF, D_2 and D_3 ON, transistor saturated. If ON diodes have an offset of 0.7 V,
(a) draw a circuit model showing these assumptions. Use your model to verify the assumed state of each device.
(b) Now let $V_i = 0.2$ V. Assume D_1 is ON, D_2 and D_3 OFF, and the transistor in cut-off. Verify the assumed state of each device.

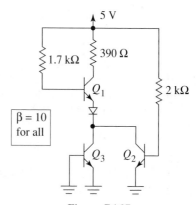

Figure P4.37

4.37 In Fig. P4.37 assume Q_1 and Q_2 are saturated and Q_3 is in cut-off.
(a) Estimate the diode current.
(b) Verify that Q_1 is saturated.

4.38 When $V_i = 5$ V in Fig. P4.38, assume Q_1 is reverse active and Q_2 is saturated. Verify these assumptions and compute the forced β for Q_2.

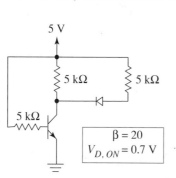

Figure P4.33

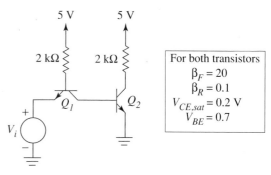

Figure P4.38

4.39 In Fig. P4.38 verify that Q_1 is saturated and Q_2 is cut off when $V_i = 0.2$ V.

Section 4.8

4.40 Use SPICE to find the transistor Q-point in the amplifier analyzed in Example 4.3. Repeat for $\beta = 150$ and for $\beta = 300$.

4.41 (a) Use SPICE to plot the collector voltage of Q_2 as a function of V_i for the circuit of Fig. P4.38 as V_i ranges from 0 to 5 V.
(b) From the SPICE data, estimate the values of V_i at which Q_2 changes from cut-off to active mode and from active mode to saturation.

4.42 In Fig. P4.42, when $v_i = 0$, Q cuts off, directing the 10 mA current into diode D and producing $v_o = 0.7$ V = one diode drop. If v_i becomes sufficiently positive, Q turns on—and for even higher v_i, saturates. This diverts the 10 mA current into Q's collector, and v_o drops to $V_{CE.sat}$ of the transistor.
(a) Use SPICE to obtain a dc transfer characteristic for the circuit when $\beta_F = 35$, $\beta_R = 0.3$, and $I_S = 3 \times 10^{-15}$A for both transistor and diode.
(b) Repeat part (a) with the current source changed to 40 mA.

4.43 Sketch transistor output characteristics in a way that shows how β_F varies with I_C. (See Fig. 4.30.) Assume temperature is constant.

Section 4.9

4.44 Figure P4.44 gives values for β and V_{BE} at $T = 27°C$. Assume the transistor is forward active.
(a) Use the transistor equivalent circuit to find the Q-point at 27°C.
(b) Use the temperature variation rules of thumb of Sec. 4.9.1 to calculate new model parameters for 60°C. Use these new values to find the Q-point at 60°C. Does the transistor remain active?

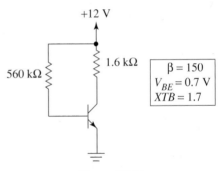

Figure P4.44

(c) Now calculate model parameters for $-25°C$, and use these to find the Q-point at $-25°C$. Does the transistor remain forward active?

4.45 The cut-in voltage for the transistor of Fig. P4.45 is 0.5 V and $I_{CB0} = 2.5 \times 10^{-8}$ A at 27°C.
(a) Use the model of Fig. 4.9b to verify that the transistor is cut off at 27°C.
(b) If I_{CB0} doubles for each 5°C increase in temperature, find the temperature at which the transistor begins to turn on.

4.46 Ordinarily transistor output resistance has little to do with establishing the transistor Q-point and is safely ignored for simplicity; however, if the Early voltage is small and the dc collector current high, output resistance can affect the Q-point.
(a) Ignoring output resistance, find the transistor Q-point in Fig. P4.46.
(b) Use I_C from part (a) to estimate r_o. Then repeat the Q-point analysis with r_o added to the transistor model. (Because collector current depends upon r_o, and r_o depends upon collector current, this is another case where recursion is needed to converge to a true solution. Do not bother with additional iterative calculations on this problem.)

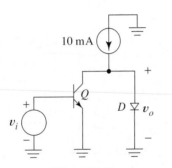

Figure P4.42

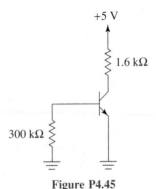

Figure P4.45

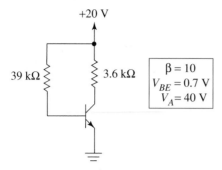

Figure P4.46

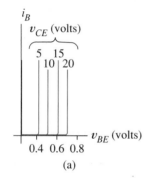

(a)

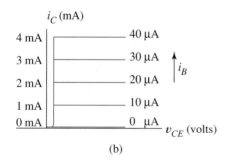

(b)

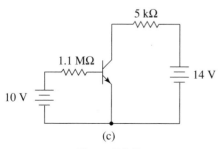

(c)

Figure P4.47

4.47 The input and output characteristics of Figs. P4.47a and b describe the transistor of Fig. P4.47c.

(a) Draw a large-signal circuit model for the transistor that applies to the first quadrants (where the graphical information is valid).

(b) Use your model to determine the BJT's Q-point.

4.48 Make a SPICE model of the circuit of Fig. P4.48. Use simulation results to find the state of each transistor for

(a) $v_i = 0$ V

(b) $v_i = 0.5$ V

(c) $v_i = 0.8$ V

4.49 Use SPICE to plot the transfer characteristic for the circuit of Fig. P4.48 for temperatures of $-45°C$, $27°C$, and $+75°C$. Use $XTB = 1.5$ for the temperature exponent of β.

Sections 4.10–4.12

4.50 Figure 4.40 models transistor dynamics for all operating modes.

(a) Draw a *simplified* model for the cut-off transistor. The nature of the biasing eliminates some of the static components and some of the dynamic components as well.

Hint: We examined static components when we specialized the Ebers–Moll model for our original cut-off equivalent circuit, the relative importance of depletion and diffusion capacitance when we introduced diode parasitics.

(b) Draw a simplified dynamic model for the forward active transistor.

(c) Draw a simplified dynamic model for the saturated transistor.

4.51 In Fig. 4.41c, $R_L = 200$ Ω. When the transistor switch is closed, the minimum permissible collector current is 2 mA. If $R_B = 20$ kΩ, $V_{CE,sat} = 0.2$ V, and $\beta = 20$, find nominal values

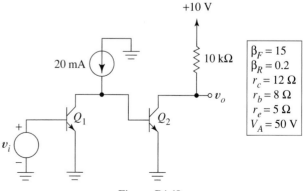

Figure P4.48

for V_{CC} and for v_C so that load current exceeds the minimum by 20%, even if v_C should happen to be 20% low.

4.52 The transistor in Fig. P4.52 is saturated when $v_{in} = 3.6$ V and cut off when $v_{in} = 0$ V.
(a) Sketch the general shape of the output waveform compared to the input waveform, when the ideal 3.6 V pulse shown is applied to the input. Assume the pulse is long enough for transients to die out.
(b) Draw the equivalent circuit that applies when $v_{in} = 0$ V. Then write an inequality involving v_{in}, R_1, and R_2 that must be satisfied for the transistor to remain off.
(c) Draw an equivalent circuit that applies when $v_{in} = 3.6$ V. Then write an inequality involving v_{in}, R_1, and R_2 that must be satisfied for the circuit to be valid. Notice that v_{in} is a *node voltage*.

4.53 In Fig. P4.53, $I_i(t)$ is initially zero.
(a) Replace the transistor by a simple cut-off model like that used in Fig.4.43a. Denote the two capacitors as C_E and C_C. Find the equilibrium voltage across each capacitor.
(b) At $t = 0$, $I_i(t)$ switches to its ON value of 1 mA. Compute the value of the base current at the instant after the current source turns on.
(c) Redraw the circuit with transistor replaced with a simple forward active model like that used in Fig. 4.43c. Assume this is the final steady-state circuit, where capacitor voltages are no longer changing. Calculate the capacitor voltages and the col-

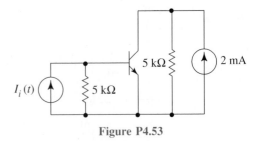

Figure P4.53

lector-emitter voltage. Is there any problem with this answer?
(d) Redraw the circuit with transistor replaced with a simple saturation model like that used in Fig. 4.43d. Find the base and collector currents.

4.54 In Fig. P4.54 the transistor switch is to be closed for $v_i = 0$ and open for $v_i = 5$ V. For the transistor, $\beta_F = 30$ and $V_{CE,sat} = -0.2$ V. Find the value of R_B so that when the switch is closed, the forced beta is 10.

4.55 Make a SPICE model of the circuit of Fig. P4.48 using transistor and diode parameters from Fig. 4.45b.
(a) Let v_i change instantaneously from 0 to 0.8 V at $t = 0$. Plot $v_o(t)$ and the base currents of Q_1 and Q_2. Determine how long it takes the circuit to reach equilibrium.
(b) Make $v_i(t)$ a pulse of 0.8 V, with width sufficient for the transistor to reach equilibrium. Compare the waveforms Q_1, Q_2, and $v_o(t)$ with $v_i(t)$.

4.56 Use SPICE to simulate the circuit of Fig. P4.48 with $R_B = 4$ kΩ in series with the base and the signal source. Transistor parameters are $BF = 35$, $BR = 0.5$, $CJC = 3E-12$, $CJE = 0.3E-12$, $TF = 40E-9$, and $RC = 250$.
(a) Let v_i change instantaneously from 0 to 5 V at $t = 0$. Plot $v_o(t)$ and the base current. Determine how long it takes the circuit to reach equilibrium.
(b) Make $v_i(t)$ a pulse of 5 V, with pulse width sufficient for the transistor to reach equilibrium. Plot v_o and base current.

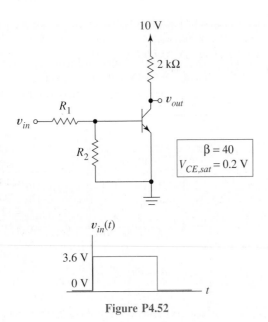

Figure P4.52

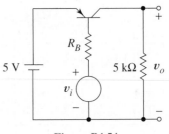

Figure P4.54

Field-Effect Transistors

This chapter introduces the *field-effect transistor* (FET), a solid-state device in which an electric field controls the flow of charge carriers through a conducting channel. Like BJTs, field-effect transistors function either as dependent current sources or controlled switches. FETs have lower noise than BJTs and often lead to simpler circuits because they have infinite input resistance. Furthermore, they occupy less space on an integrated circuit chip. An important disadvantage of the FET compared to the BJT is lower transconductance, g_m. In contrast with bipolar transistors, FETs are *unipolar* because they employ only one kind of charge to carry current.

Figure 5.1 classifies the most commonly used transistors. Field-effect transistors divide by physical structure into two classes: insulated gate and junction gate devices. The former, usually called *metal oxide semiconductor FETs* (MOSFETs), subdivide into *enhancement* and *depletion* types. The junction gate category consists of *metal semiconductor FETs* (MESFETS) and *junction FETs* (JFETs). Also, most FETs are available in both n-channel and p-channel forms. Fortunately, strong similarities between the equations and characteristic curves for the various FETs make the learning task easier than it might first appear.

5.1
n-Channel MOSFET

Figure 5.2a shows the physical structure of the enhancement-type n-channel MOSFET. Wells of highly doped n-type silicon called *source* and *drain* are diffused or implanted into a p-type *substrate*. A conducting *gate* is insulated from the silicon by a thin layer of SiO_2. We always connect the substrate, alternately called the *body,* to the most negative point in the circuit to ensure that drain and source wells remain reverse biased. The MOSFET is physically symmetric, with source and drain ultimately defined by the current direction.

In operation, positive charge is placed on the gate by an external source. This source attracts electrons from the wells into the region just beneath the oxide, creating a conducting *channel* between source and drain. This *enhancement* of the channel region with negative charge carriers gives the device its name. In all n-channel devices, the physical current in the channel is an *electron flow;* thus we regard one terminal as

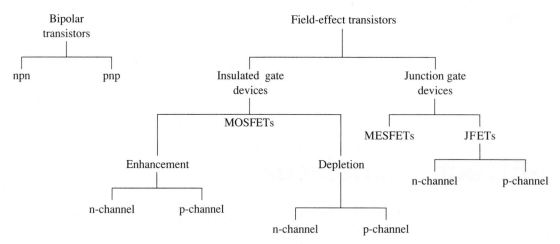

Figure 5.1   Basic transistor classes.

the *source of electrons* that flow through the channel to the drain when external voltage is applied.

In early MOSFETs the gate was always made of metal, deposited during the metalization step of the IC fabrication process; hence the name ***metal oxide semiconductor*** denoted the layers of material. Today, gates are more often constructed of highly doped silicon that is vacuum deposited over the oxide. This makes the more general term *insulated gate FET* (IGFET) a more appropriate generic name. However, bowing to common usage, we employ the term MOSFET, regardless of the gate material.

Figure 5.2b shows two schematic symbols for the n-channel MOSFET. Both symbolically suggest the electrical isolation of the gate. In the first, an arrow marks the source (of electrons) but shows the direction of conventional current. The second symbol shows explicitly how the substrate or body is connected in a circuit, and also better reflects the symmetry of drain and source. The arrowhead that denotes the body has p-to-n orientation, suggesting the channel of electrons that exists next to the oxide during operation.

Figure 5.2c shows biasing polarities and reference directions for the n-channel MOSFET in its *common-source* configuration. A dc source in the drain circuit establishes drain current i_D, which flows in the direction of the source arrow. A positive voltage applied to the gate relative to the source controls this current in a manner we explain next.

5.1.1 MOSFET OPERATING PRINCIPLES

Channel Formation. Because n-type drain and source are separated from p-type substrate by depletion regions, an electron attempting to leave a well is repelled by negative ions in the p-material and retarded by positive ions in the n-material. This means *work is required for an electron to leave either well*. Figure 5.3a shows this energy barrier for $v_{GS} = 0$ as a potential energy *hill* that separates the electron wells. To reach the p-material, electrons from the source or drain must possess energy corresponding to the *energy*

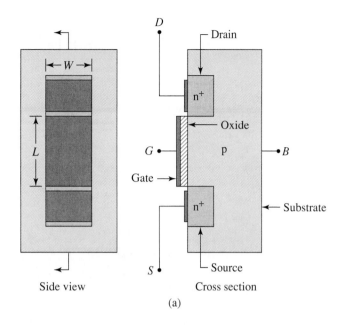

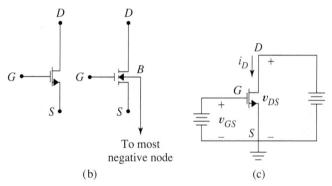

Figure 5.2 n-Channel enhancement MOSFET: (a) physical structure; (b) schematic symbols; (c) biasing polarity for active operation.

depth of the well, but because of the high doping concentrations of the well material, the number of electrons that possess this energy is negligible. Notice that an electron on the flat part of the energy diagram can move about freely because the p-type material under the oxide is free of electric fields; thus an electron placed here experiences no electrical forces.

The gate and body of the MOSFET are a parallel-plate capacitor with an oxide dielectric. If we make the gate-to-body voltage positive ($v_{GS} > 0$) as in Fig. 5.3b, we charge this capacitor. The positive charge on the gate then repels the mobile holes from the region just beneath the oxide, and the lines of electric force terminate on negative impurity ions left behind. The electric field thus creates a narrow *depletion region*, relatively free of mobile charges, extending from source to drain near $x = 0$. A test electron

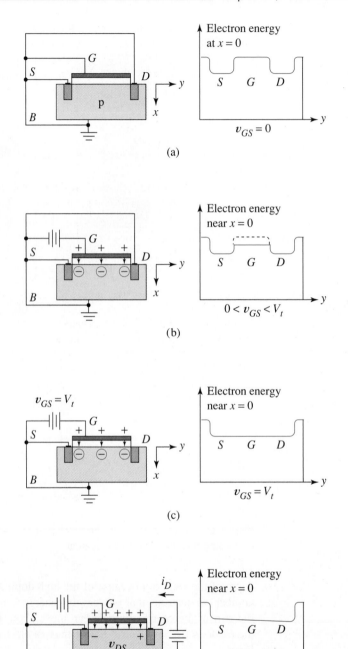

Figure 5.3 n-MOSFET channel formation: (a) transistor and electron energy when $v_{GS} = 0$; (b) lowered surface potential under gate for positive v_{GS}; (c) electron energy when channel appears; (d) energy "ramp" with v_{DS} slightly positive.

in this region is attracted toward the oxide by the positive charge on the gate. Displacing it away from the oxide in the $+x$ direction therefore requires energy. This indicates that making v_{GS} positive lowers the electron potential energy just beneath the oxide near $x = 0$ as in Fig. 5.3b. Technically speaking, we lower the *surface potential* in the p-material just under the oxide by making $v_{GS} > 0$.

Still larger values of v_{GS} further lower the surface potential until v_{GS} reaches a special value called the *threshold voltage, V_t.* Figure 5.3c suggests that V_t is precisely that voltage that lowers the surface potential to the level of the electron wells. Once $v_{GS} = V_t$, electrons are free to move in the y direction between source and drain without an energy expenditure.

Making v_{GS} larger than V_t does not lower the surface potential further, but instead attracts additional conduction electrons into the channel. This increases the channel conductivity. Once the p-material beneath the oxide is populated with free electrons instead of its indigenous holes, we say an *inversion* has taken place, meaning that p-material has been changed locally into n-material, a kind of electrostatic alchemy! Figure 5.3d also shows that making drain positive relative to source creates a longitudinal voltage along the channel that lowers the electron potential at the drain end. This causes electrons to drift through the channel from source to drain, the physical basis for the external drain current i_D.

Controlling the Channel Current. The family of linear volt-ampere curves of Fig. 5.4a describes the operation of the channel for small drain-source voltage using the circuit of Fig. 5.4b. Making v_{GS} more positive than the transistor's threshold voltage provides additional channel electrons, reducing channel resistance and increasing the slope of the volt-ampere curve. The result is a voltage-controlled linear resistor. Notice that each resistor curve passes through the origin and into the third quadrant.

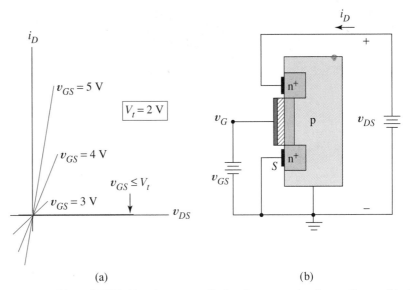

(a) (b)

Figure 5.4 n-Channel FET: (a) voltage-controlled resistor operation for small v_{DS}; (b) channel geometry when $|v_{DS}|$ is small.

For large v_{DS}, the voltage at the drain end of the channel becomes so large that the channel pinches off as in Fig. 5.5a. We already saw in Fig. 5.3c that the gate-to-channel voltage must be at least V_t volts for a channel to exist. In Fig. 5.5a this condition continues to exist at the source end of the channel regardless of v_D; however, because

$$v_{GD} = v_{GS} - v_{DS}$$

increasingly positive values of v_{DS} reduce the gate-to-channel voltage *at the drain end*. First, the density of channel electrons near the drain begins to diminish with increases in v_{DS}. This reduces the average channel conductivity, making the slope of each characteristic curve in Fig. 5.5b begin to decrease with v_{DS}. That is, successive voltage increases Δv_{DS} are accompanied by smaller current increases Δi_D. What is happening is that large, positive drain-source voltages are "undoing" the inversion at the drain end of the channel. Eventually, a critical value of v_{DS} is reached, for which the channel depletion region disappears at the drain end, and we say the channel is *pinched off*. In terms of node voltages, this is the value of v_{DS} that satisfies

$$v_{GS} - v_{DS} = V_t \qquad (5.1)$$

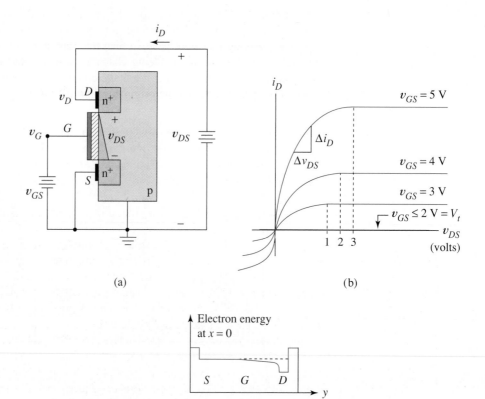

(a)

(b)

(c)

Figure 5.5 Channel pinch-off for large v_{DS}: (a) channel geometry; (b) output characteristic curves; (c) energy diagram as channel pinches off.

We rewrite this more compactly as

$$v_{DG} = -V_t$$

For still higher values of $v_{DG} = v_{DS} - v_{GS}$, that is, for

$$v_{DG} \geq -V_t \tag{5.2}$$

the channel remains pinched off. Once the channel pinches off, the voltage drop across the channel remains approximately constant, giving the constant current curves in Fig. 5.5b. Further increases in v_{DS}, (and, therefore, v_{DG}) widen the depletion region between drain and channel instead of increasing the drift of carriers in the channel. That is, the electron potential diagram begins to resemble Fig. 5.5c for large v_{DS}. For now increases in v_{DS} increase the depth of the potential well at the drain rather than increasing the slope of the energy curve in the channel. We conclude that Eq. (5.2) gives the condition for which each of the output curves of Fig. 5.5b becomes horizontal. Equation (5.2) can be written as

$$v_{DS} \geq v_{GS} - V_t$$

Substituting the values of v_{GS} that label each curve in Fig. 5.5b into this equation gives the range of v_{DS} over which that output curve is horizontal. Figure 5.5b is then the family of output characteristics for the n-channel MOSFET. Because of the insulator between the gate and the rest of the device, gate current is always zero, and the *input characteristic* is that of an open circuit.

5.1.2   THE MOSFET AS A THREE-STATE DEVICE

It is useful to view the MOSFET as a *three-state device* that has an equation, a circuit model, and a particular region of the output characteristics associated with each state. In its *active* state, the FET is used in amplifiers and other linear applications as a VCCS. There are also *cut-off* and *ohmic* states used in digital and switching circuits, where the FET acts as a controlled switch. In its ohmic state the FET sometimes functions as a voltage-controlled resistor, an application that has no BJT counterpart.

The diagram of Fig. 5.6a uses inequalities to define the three states, with each state name designated by blue characters. The first possibility is that the MOSFET is cut off. This means v_{GS} is so low that there is no channel. In Fig. 5.6b, cut-off corresponds to the single curve $i_D = 0$, reminiscent of cut-off in a bipolar transistor.

When the MOSFET is not cut off, a channel exists and we say the device is ON. A second inequality determines the state of this ON transistor. If the MOSFET is ON and $v_{DG} \geq -V_t$, the device is in its *active state*. Active operation corresponds to the region on and to the right of the dashed curve in Fig. 5.6b, where the MOSFET approximates a voltage-controlled current source. If the MOSFET is ON but $v_{DG} \leq -V_t$, the transistor is in its *ohmic* state, the region on and to the left of the dashed curve. Here the MOSFET operates as a nonlinear resistor controlled by v_{GS}. In the region near the origin, for v_{DS} sufficiently small, this nonlinear resistor reduces to a linear, controlled resistor. As in bipolar transistors, there is a smooth transition between states, which is reflected in the inequalities used to define them. This means that along the boundaries between states the transistor is correctly described by the equations and models of both states.

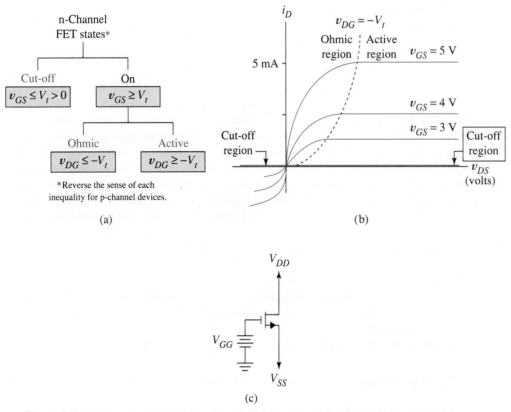

Figure 5.6 Three-state model for n-channel FET: (a) state definitions; (b) operating regions corresponding to the states; (c) example circuit for determining MOSFET states.

The reader should know that there is some variation in the names assigned to the transistor states. Some authors call the active-state region of the output characteristics the *saturation* region; others call it the *pinch-off* region. The ohmic region is sometimes called the *triode* region. For convenience the BJT and FET state definitions are also given inside the front cover.

The following example and exercise involve making deductions about transistor states.

EXAMPLE 5.1 The transistor of Fig. 5.6c has $V_t = +1$ V.
(a) If $V_{GG} = 7$ V, find the values of V_{SS} and V_{DD} for which the transistor is active.
(b) If $V_{GG} = 7$ V, find the values of V_{SS} and V_{DD} for which the transistor is ohmic.

Solution. (a) For the transistor to be active, it must first be ON. From Fig. 5.6a, this requires $V_{GS} = V_{GG} - V_{SS} \geq V_t = 1$ V. Therefore, $7 - V_{SS} \geq 1$, or $V_{SS} \leq 6$ V. To be active, it is *also* necessary that $V_{DG} = V_{DD} - V_{GG} \geq -V_t$; therefore,

$$V_{DD} \geq 7 - 1 = 6 \text{ V}$$

(b) To be ON, we found in part (a) that $V_{SS} \leq 6$ V. To be ohmic, also requires

$$V_{DG} = V_{DD} - V_{GG} \leq -V_t$$

therefore, $V_{DD} \leq 6$ V.   ❏

Exercise 5.1   In Fig. 5.6c, $V_t = 1$ V, $V_{SS} = -8$ V, and $V_{DD} = 3$ V. Find the range of values of V_{GG} for which the transistor is active.

Ans. -7 V $\leq V_{GG} \leq 4$ V.

We next give equations and large-signal models for each state and show how the equations relate to the MOSFET characteristic curves.

Cut-off State. In cut-off the transistor has no channel. Therefore, whenever $v_{GS} \leq V_t$

$$i_D = 0 \tag{5.3}$$

and the cut-off circuit of Fig. 5.7a is a good model. Notice that Eq. (5.3) mathematically describes the cut-off region in Fig. 5.6b.

Ohmic State. When

$$v_{GS} \geq V_t$$

and

$$v_{DG} \leq -V_t$$

the output equation is

$$i_D = \frac{k}{2}[2(v_{GS} - V_t)v_{DS} - v_{DS}^2] \tag{5.4}$$

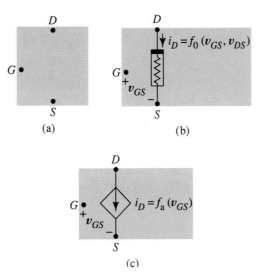

Figure 5.7   Large-signal models for FETs: (a) cut-off; (b) ohmic; (c) active.

where threshold voltage V_t and *transconductance parameter* k are two constants that characterize a particular MOSFET. To understand this equation, think of v_{DS} as the independent variable, i_D as the dependent variable, and v_{GS} is a control parameter. The v_{DS} terms show that Eq. (5.4) is the equation of an inverted parabola that is offset from the origin. Different values of parameter v_{GS} give a *family* of parabolas with different offsets. This is exactly the family of output characteristics in the ohmic region of Fig. 5.6b. In future problem solving it will be important to understand that only the "left" branch of each parabola has physical meaning for the transistor.

Figure 5.7b is a circuit model for the ohmic MOSFET, a voltage-controlled, nonlinear resistor, in which the function $f_0(v_{GS}, v_{DS})$ is Eq. (5.4).

Voltage-Controlled Linear Resistor. If v_{DS} in Eq. (5.4) is kept so small that the squared term is negligible, drain current and drain-source voltage satisfy

$$i_D = \frac{k}{2}[2(v_{GS} - V_t)v_{DS}] = \frac{1}{R_{NMOS}} v_{DS}$$

which describes linear volt-ampere curves like those of Fig. 5.4a. The resistance of this controlled resistor is

$$R_{NMOS} = \frac{1}{k(v_{GS} - V_t)} \tag{5.5}$$

We next derive a condition necessary for Eq. (5.5) to be valid.

In practical engineering terms, saying that the squared term in Eq. (5.4) is "negligible" means we assume it is at least an order of magnitude smaller than the other term in the sum. Stated mathematically,

$$|v_{DS}^2| << |2(v_{GS} - V_t)v_{DS}|$$

means

$$|v_{DS}^2| \leq (1/10) |2(v_{GS} - V_t)v_{DS}|$$

or

$$|v_{DS}| \leq 0.2 |v_{GS} - V_t| \tag{5.6}$$

This constraint between magnitudes of the control voltage, v_{GS}, and the resistor voltage, v_{DS}, applies whenever we use Eq. (5.5).

Active State. Because the channel is pinched off for active operation, we replace v_{DS} in Eq. (5.4) with the pinch-off condition

$$v_{DS} = v_{GS} - V_t$$

from Eq. (5.1). This gives the active-region equation

$$i_D = \frac{k}{2}(v_{GS} - V_t)^2 \tag{5.7}$$

The active-state circuit model is then the nonlinear VCCS of Fig. 5.7c, where function $f_a(v_{GS})$ is Eq. (5.7). In its active state the MOSFET is a voltage-controlled current source, but for general signals, this dependent source is nonlinear.

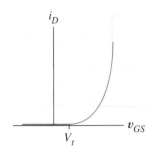

Figure 5.8 Transfer characteristic for n-channel MOSFET.

An important curve for an active MOSFET is its transfer characteristic. Figure 5.8 shows the transfer characteristic for the n-channel MOSFET. For $v_{GS} \leq V_t$ the transistor is cut off and satisfies Eq. (5.3). For $v_{GS} \geq V_t$ the transfer curve is a plot of Eq. (5.7), which shows graphically the nonlinearity of the VCVS of Fig. 5.7c. This curve also predicts the nonuniform spacing of the active-region curves of Fig. 5.6b for equal increments in v_{GS}.

Solid-state theory for the MOSFET shows that k in Eqs. (5.4), (5.5), and (5.7) can be written as

$$k = \frac{W}{L} \mu_n C_{ox} \qquad (5.8)$$

where C_{ox} is the capacitance per unit area of the oxide under the gate, and μ_n is the electron mobility. All MOSFETs on an IC chip have identical $\mu_n C_{ox}$, and therefore differ only in W and L, the width and length of the channel area beneath the oxide in Fig. 5.2a. Because the photographic masks define W/L, the ratio is easily and precisely controlled during IC fabrication, and emerges as a key design parameter. Since the drain current in each transistor state is proportional to k, we can fabricate devices, otherwise identical, that carry different currents for the same applied voltages by assigning different W/L values during mask design.

5.1.3 DEPLETION MOSFET

Figure 5.9a shows another kind of transistor, the n-channel *depletion* MOSFET. The difference between a depletion and an enhancement MOSFET is the thin channel of n-type impurities deposited just under the gate oxide during fabrication, usually by ion implant. Because of this, a channel already exists for $v_{GS} = 0$. Positive v_{GS} further increases channel conductivity by drawing additional electrons from the wells to reduce the channel resistance. Negative v_{GS} causes *depletion* of the carrier concentration in the channel, thereby increasing channel resistance. If we make v_{GS} sufficiently negative, that is, $v_{GS} \leq V_t$, where V_t is a *negative* threshold voltage, the channel electrons are all repelled into the wells, and the conducting channel ceases to exist.

For small v_{DS}, the device operates as a voltage-controlled, linear resistor. For large v_{DS} the channel pinches off at the drain end, producing voltage-controlled curves of constant current. The salient effect of the pre-formed channel is that the depletion MOSFET behaves exactly like an enhancement MOSFET with a *negative* threshold voltage. *The equations, state definitions, and models of the two devices are otherwise identical.* Figure 5.9b shows the schematic symbol for the depletion MOSFET. The dark bar connecting drain and source suggests the fabricated channel.

Figure 5.9d shows the output characteristics for the depletion transistor. They differ from the enhancement MOSFET curves only in the v_{GS} labels. The transfer characteristic, Fig. 5.9c, is a plot of Eq. (5.7). Since the principal difference between enhancement and depletion MOSFETs is the algebraic sign of V_t, it follows that Fig. 5.9c is a left-shifted version of the MOSFET transfer curve of Fig. 5.8. The transfer and output characteristics are juxtaposed in Figs. 5.9c and d to show how the current axes of the two curves relate.

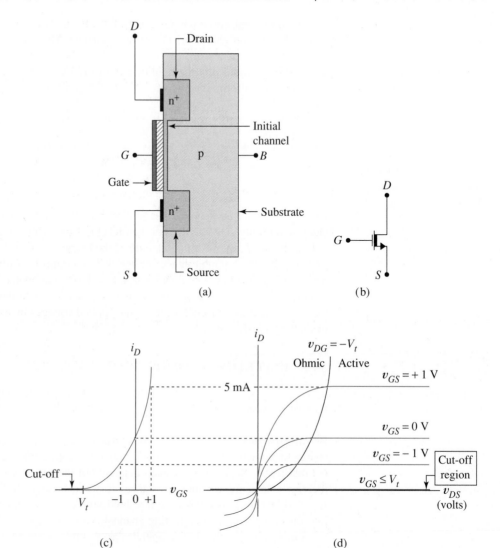

Figure 5.9 n-Channel depletion FET: (a) physical structure; (b) schematic; (c) transfer characteristic; (d) output characteristics.

5.2
MOSFET Operating Point Analysis

There are two large-signal analysis problems involving MOSFETs: finding the Q-point when the transistor state is known, and finding the Q-point when the state is unknown. As in BJT analysis, we replace transistors with circuit models and then analyze. There are some differences in detail because MOS and BJT transistor models differ. As in BJT circuits, the known-state problem usually involves a circuit in which the transistor is active.

5.2.1 ANALYSIS FOR ACTIVE-STATE TRANSISTORS

Both enhancement and depletion MOSFETs satisfy Eq. (5.7) when active; therefore, the solution details differ only in the signs of v_{GS} and V_t. Beginners should draw an equivalent circuit, because the transistor models make circuit constraints more obvious. With increased problem-solving experience, we gradually come to visualize the constraints in the original schematic, and eventually learn to solve simple problems without drawing equivalents. Another good habit is to label each voltage and current on the circuit diagram as it is found, for this often gives visual hints about what to do next. It is also good practice to make a simple sketch of the output or transfer characteristic at the end of the analysis to verify that the solution is sensible.

Analysis problems for active MOSFETs involve solving two simultaneous equations, one a quadratic, in two unknowns. Because only one branch of the transfer function parabola, Eq. (5.7), has physical meaning, we invariably obtain two mathematical solutions, only one of which is physically meaningful. If we formulate our equations so that V_{GS} (not I_D) is the unknown, we can easily recognize the correct solution as the one for which the transistor is turned on. The following example and exercise help clarify these points.

EXAMPLE 5.2 Find the Q-point of the active transistor in Fig. 5.10a, and show it on a sketch.

Solution. Since the transistor is active, draw Fig. 5.10b. To determine the Q-point we need to determine unknowns I_D and V_{DS}, *where we use uppercase because this is a dc analysis problem.* Because the transistor is active,

$$I_D = \frac{0.5 \times 10^{-4}}{2}(V_{GS} - 2)^2$$

Since this equation has two unknowns, we seek a second expression for V_{GS}. The diagram shows that $V_S = -10$ V. Also,

$$V_G = V_D = 10 - 160\,\text{k}\,I_D$$

Thus

$$V_{GS} = V_G - V_S = (10 - 160\,\text{k}\,I_D) - (-10) = 20 - 160\,\text{k}\,I_D$$

a second equation in the same unknowns. The simplest strategy is to solve the second equation for I_D and then equate it to the first. This gives

$$\frac{20 - V_{GS}}{160\,\text{k}} = \frac{0.5 \times 10^{-4}}{2}(V_{GS} - 2)^2$$

Simplifying gives

$$V_{GS}^2 - 3.75\,V_{GS} - 1 = 0$$

which has solutions $V_{GS} = +4$ V and -0.25 V.

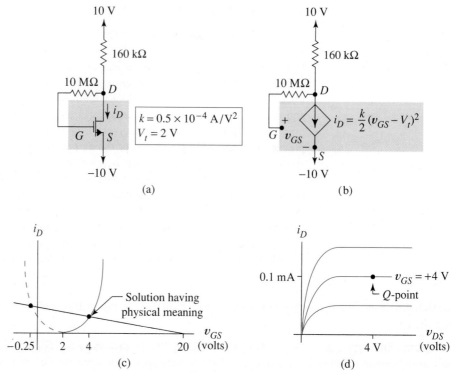

Figure 5.10   Example 5.3: (a) given circuit; (b) large-signal equivalent circuit; (c) meaning of two solutions to circuit equations; (d) location of Q-point.

Figure 5.10c shows graphically the two drain-current expressions we equated: a load line and the MOSFET transfer characteristic. Only the solution $V_{GS} = 4$ V is consistent with operation of an active transistor. The intersection with the dashed branch of the parabola is a mathematically correct solution, but it has no physical meaning because the transistor has no channel for $V_{GS} < 2$ V.

From Fig. 5.10b, one Q-point coordinate is $V_{DS} = V_{GS} = +4$ V. Substituting $V_{GS} = 4$ V into either drain-current equation gives $I_D = 0.1$ mA. Figure 5.10d shows the Q-point on the output characteristics.   ❑

Exercise 5.2   Find the new Q-point in Fig. 5.10a if we replace the transistor with one of $k = 2 \times 10^{-4}$ A/V^2 and $V_t = 2$ V.

Ans. $(V_{DS}, I_D) = (3.03$ V, 106 μA$)$.

The next example involves a *four-resistor biasing circuit,* a popular amplifier circuit because it establishes a Q-point that depends little on transistor parameter values. We learn how to *design* such circuits in Chapter 6.

EXAMPLE 5.3   Find the Q-point of the transistor in Fig. 5.11a.

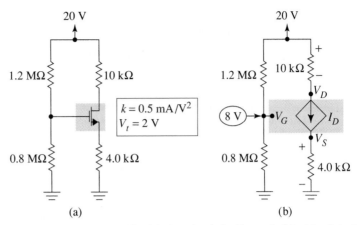

Figure 5.11 Four-resistor amplifier biasing circuit for Example 5.3: (a) original circuit; (b) equivalent circuit for active transistor.

Solution. Since the transistor is active, draw Fig. 5.11b, where

$$I_D = \frac{0.5 \times 10^{-3}}{2} (V_{GS} - 2)^2$$

Because the 0.8 MΩ and 1.2 MΩ resistors form a voltage divider, the gate voltage is

$$V_G = \frac{0.8 \text{ M}}{0.8 \text{ M} + 1.2 \text{ M}} \, 20 = 8\text{V}$$

Since $V_S = 4 \text{ k} \times I_D$, the gate-source voltage is

$$V_{GS} = 8 - 4 \text{ k} \times I_D$$

Equating the two expressions for I_D gives

$$\frac{8 - V_{GS}}{4 \text{ k}} = \frac{0.5 \times 10^{-3}}{2} (V_{GS} - 2)^2$$

or

$$V_{GS}^2 - 3V_{GS} - 4 = 0$$

which has solutions $V_{GS} = -1$ V and $+4$ V. Of these, only the second satisfies the condition that the transistor be ON, $v_{GS} \geq V_t = 2$ V. Substituting $V_{GS} = 4$ V gives $I_D = 1$ mA.

To find V_{DS} we use KVL, with the voltage across the current source, V_{DS}, as the unknown.

$$4 \text{ k}(1 \text{ mA}) + V_{DS} + 10 \text{ k}(1 \text{ mA}) = 20 \text{ V}$$

This gives $V_{DS} = 6.0$ V. In conclusion, the Q-point is

$$(V_{DS}, I_D) = (6 \text{ V}, 1 \text{ mA})$$ ❑

Exercise 5.3 Find the new Q-point if the 10 kΩ resistor in Example 5.3 is replaced by a 5 kΩ resistor.

Ans. $(V_{DS}, I_D) = (11 \text{ V}, 1 \text{ mA})$.

Exercise 5.4 Find the Q-point of the depletion MOSFET in Fig. 5.12a.

Ans. $(V_{DS}, I_D) = (3 \text{ V}, 2 \text{ mA})$.

EXAMPLE 5.4 Find the dc drop across the resistor, and also the operating point of the active-state transistor in Fig. 5.12b.

Solution. The equivalent circuit is Fig. 5.12c. Since $v_{GS} = 0$, Eq. (5.7) gives

$$I_D = \frac{22 \times 10^{-6}}{2} (0 + 3)^2 = 99 \text{ } \mu\text{A}$$

The dc drop across in the resistor is

$$51 \text{ k}(99 \times 10^{-6}) = 5.05 \text{ V}$$

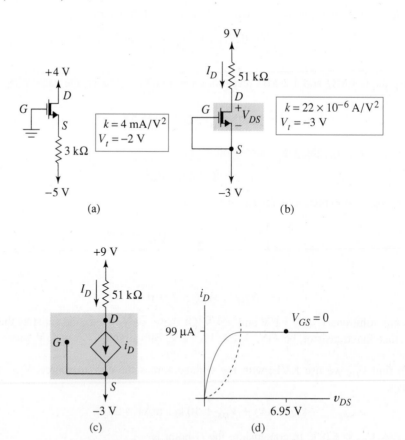

Figure 5.12 (a) Circuit for Exercise 5.4; (b) circuit for Example 5.4; (c) equivalent for Example 5.4; (d) Q-point for Example 5.4.

This means the node voltage at the drain is 5.05 V below the 9 V supply, or $V_D = 3.95$ V. Thus

$$V_{DS} = 3.95 - (-3) = 6.95 \text{ V}$$

Figure 5.12d shows the Q-point, (6.95 V, 99 µA)   ❑

The depletion transistor with gate connected to source as in Fig. 5.12b makes a simple constant current source;. The output characteristic that corresponds to $V_{GS} = 0$ (Fig. 5.12d) is a constant current curve provided that v_{DS} is sufficiently large to prevent ohmic-region operation. Sometimes the key observation in an analysis is to recognize this current source connection, as the next example shows. This example also illustrates some new circuit ideas that are useful in later chapters.

EXAMPLE 5.5 In Fig. 5.13a the enhancement transistors are identical, with $k = 0.5 \times 10^{-4}$ A/V^2 and $V_t = 3$ V. The depletion MOSFET has $k = 0.1$ mA/V^2 and $V_t = -2$ V. Find the Q-point of each transistor, assuming all transistors are active.

Solution. Since M_3 is active, Eq. (5.7) gives

$$I_{D3} = \frac{10^{-4}}{2} [0 - (-2)]^2 = 0.2 \text{ mA}$$

Next, we use this in the equivalent circuit of 5.13b, where KCL requires that $I_{D1} + I_{D2} = 0.2$ mA. Because M_1 and M_2 are identical, and the circuit symmetric, $I_{D1} = I_{D2} = 0.1$ mA.

If M_1 and M_2 are active, their drain currents satisfy Eq. (5.7). Therefore, for each,

$$0.1 \times 10^{-3} = 0.25 \times 10^{-4} (V_{GS1} - 3)^2$$

The solutions are $V_{GS1} = 1$ V and 5 V. For M_1 and M_2 to have channels, V_{GS1} must exceed $V_t = 3$ V, therefore only $V_{GS1} = 5$ V has physical meaning.

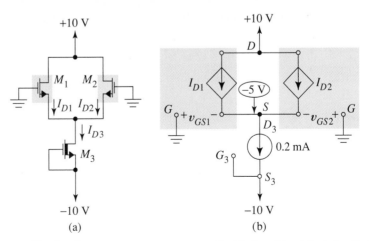

Figure 5.13   Circuit using a constant current source for biasing: (a) original circuit; (b) equivalent used for analysis when transistors are active.

In Fig. 5.13b $V_{GS1} = 5$ V implies that the source nodes of M_1 and M_2 are 5 V below ground, since their gates are at zero volts. We now add the value $V_S = -5$ V to Fig. 5.13b (indicated by a circle) for future use. Since the drains of M_1 and M_2 are at 10 V, V_{DS} for each is $10 - (-5) = 15$ V. Thus the enhancement transistors are each biased at (15 V, 0.1 mA). ❏

> **Exercise 5.5** Use the results in Example 5.5 and the state definitions of Fig. 5.6a to verify that all three transistors operate in their active states as assumed.
>
> **Ans.** $V_{DG1} = V_{DG2} = 10$ V satisfies $V_{DG} \geq -3$V, $V_{DG3} = +5$ V satisfies $V_{DG3} \geq -(-2)$ V.

5.2.2 ANALYSIS WHEN TRANSISTOR STATES ARE UNKNOWN

To analyze a MOSFET circuit when transistor states are unknown, we must find the transistor voltages and then use the state definitions of Fig. 5.6a to determine the transistor states. Once we know the states, we can compute transistor currents. In all but the simplest problems, we use the general procedure previously applied to diodes and BJTs.

1. Make an educated guess about the state of each transistor.
2. Replace each transistor with the appropriate model from Fig. 5.7; then analyze the resulting circuit.
3. Use the analysis results and state definitions to confirm each transistor state.
4. If a contradiction occurs, make a new guess and repeat the analysis.

We now apply these ideas to a variety of circuits.

EXAMPLE 5.6 Find the Q-point, (V_{DS}, I_D), of the transistor in Fig. 5.14a if $V_t = 1$ V and $k = 2 \times 10^{-3}$ A/V^2.

Solution. Because V_G and V_S are given, we can solve this problem without circuit models. Because $V_{GS} = 2$ V, the transistor is ON; therefore, the Q-point is somewhere on the single output characteristic of Fig. 5.14b. Since the active equation is the simpler, *assume* active. This gives

$$I_D = 10^{-3} (2 - 1)^2 = 1 \text{ mA}$$

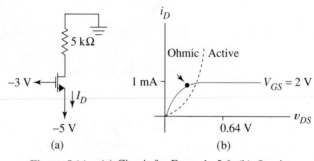

Figure 5.14 (a) Circuit for Example 5.6; (b) Q-point.

Now we test our assumption by comparing V_{DG} with $-V_t$. From Fig. 5.14a, if $I_D = 1$ mA, then $V_D = -5$ V, and $V_{DG} = -5 - (-3) = -2$ V. Since this is less than $-V_t = -1$ V, the transistor is in its ohmic state.

Recomputing the current using Eq. (5.4) gives

$$I_D = 10^{-3} [2(2-1)V_{DS} - V_{DS}^2]$$

From the schematic, $V_D = -5 kI_D$ and $V_S = -5$; thus

$$V_{DS} = -5 kI_D + 5$$

Solving for I_D and equating current expressions gives

$$\frac{5 - V_{DS}}{5 k} = 10^{-3} (2V_{DS} - V_{DS}^2)$$

Solving this polynomial equation gives the solutions: $V_{DS} = 1.56$ V and 0.64 V.

The transistor state is not in doubt; however, we must test the inequality $V_{DG} \le -1$ to see which value of V_{DS} has physical meaning. Since

$$V_{DG} = V_{DS} - V_{GS} = V_{DS} - 2$$

the two solutions correspond, respectively, to $V_{DG} = -0.44$ V and -1.36 V. Since only the second is consistent with ohmic operation, we conclude that $V_{DS} = 0.64$ V. We now substitute V_{DS} into a drain-current equation; for example,

$$\frac{5 - 0.64}{5 k} = I_D = 0.872 \text{ mA}$$

We conclude that the Q-point is (0.64 V, 0.872 mA), as indicated by the arrow in Fig. 5.14b. ❏

Exercise 5.6 We wish to change the 5 kΩ resistor in Fig. 5.14a to a value that will bring the transistor to the edge of the active region in Fig. 5.14b. Find a new value for this resistor.

Ans. $R_D = 4$ kΩ.

EXAMPLE 5.7 Find V_X in Fig. 5.15a. Transistor parameters are $k_1 = 0.5$ mA/V^2, $V_{t1} = -2$ V, $k_2 = 2/9$ mA/V^2, and $V_{t2} = 3$ V.

Solution. *Assume* both transistors are active, and draw Fig. 5.15b. Because $V_{GS1} = 0$ for M_1,

$$I_{D1} = 0.25 [0 - (-2)]^2 \text{ mA} = 1 \text{ mA}$$

The transistors are in series, so $I_{D2} = I_{D1}$. But this implies

$$I_{D2} = 1 \text{ mA} = (1/9)(V_{GS2} - 3)^2 \text{ mA}$$

which gives the solutions $V_{GS2} = 0, +6$ V. Only the latter exceeds $V_{t2} = 3$ V and is, therefore, consistent with M_2 being ON.

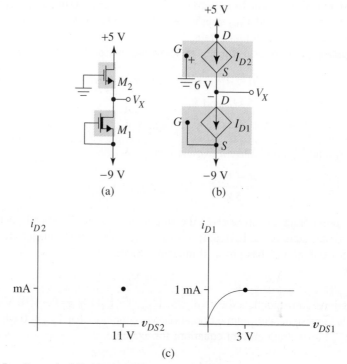

Figure 5.15   Example 5.7: (a) given circuit; (b) equivalent when both transistors are active; (c) Q-point coordinates of problem solution.

To verify that M_2 is active, notice that

$$V_{DG2} = 5 - 0 = 5 \text{ V}$$

which satisfies $V_{DG2} \geq -V_{t2} = -3\text{V}.$

Now we check the active assumption for M_1. Since we already found $V_{GS2} = 6$ V, we see from Fig. 5.15b that $V_X = -6$ V. For M_1 this means

$$V_{DG1} = -6 - (-9) = 3 \text{ V}$$

This satisfies $V_{DG1} \geq -V_{t1} = 2$ V. We conclude that both transistors are active as assumed and $V_X = -6$ V. To summarize, M_2 is biased at $(V_{DS}, I_D) = (11 \text{ V}, 1 \text{ mA})$ and M_1 is biased at $= (3 \text{ V}, 1 \text{ mA})$, as illustrated in Fig. 5.15c.   ❏

Exercise 5.7   Find how V_X changes in Example 5.7 if k_2 increases to 8/9 mA/V^2 but all other parameters remain the same.

Ans. -4.5 V.

Exercise 5.8   Find how much power each transistor in Fig. 5.15 is dissipating, and how much power is being delivered by each supply when the Q-points are those of Fig. 5.15c.

Ans. M_1 dissipates 3 mW, M_2 dissipates 11 mW, the 5 V supply delivers 5 mW, the -9 V supply delivers 9 mW.

EXAMPLE 5.8 Find the Q-point and power dissipation for each transistor in Fig. 5.16a.

Solution. Some circuits constrain transistor node voltages in ways that help us determine the transistor states. This is such a circuit. Because $V_{GS1} > V_{t1,}$ it is clear that M_1 is ON. For M_2, we see that $V_{DG2} = 0$, which satisfies $V_{DG2} \geq -V_{t2}$. Therefore, M_2 is either OFF ($V_{GS2} \leq V_{t2}$) or active. *Assume* both transistors are active, and draw Fig. 5.16b. If M_1 is active, then

$$I_{D1} = \frac{9 \times 10^{-3}}{2}(2-1)^2 = 4.5 \text{ mA}$$

Because $I_{D2} = I_{D1}$ and M_2 is assumed active

$$I_{D2} = \frac{4 \times 10^{-3}}{2}(V_{GS2} - 2)^2 = 4.5 \text{ mA}$$

Solving gives $V_{GS2} = 3.5$ V and 0.5 V, of which $V_{GS2} = 3.5$ V exceeds V_{t2}. We tentatively conclude that M_2 is ON, active, and biased at (3.5 V, 4.5 mA).

Next we check the state of M_1. From the diagram, $V_{DS1} + V_{DS2} = 9$ V; therefore

$$V_{DS1} = 9 - 3.5 = 5.5 \text{ V}$$

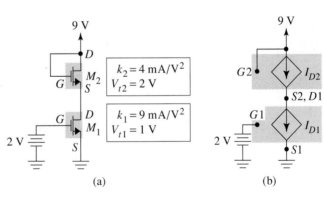

(a)   (b)

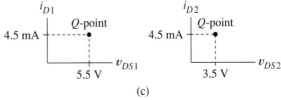

(c)

Figure 5.16 Example 5.8: (a) given circuit; (b) equivalent circuit for both transistors active; (c) Q-point for each transistor.

Thus

$$V_{DG1} = V_{DS1} - V_{GS1} = 5.5 - 2 = 3.5 \text{ V}$$

Because this satisfies $V_{DG1} \geq -V_{t1} = -1$ V, we conclude that both transistors are active, and M_1 is biased at (5.5 V, 4.5 mA). Figure 5.16c shows the two Q-points.

For M_1, the power dissipation is $I_{D1} \times V_{DS1} = 5.5(4.5 \times 10^{-3}) = 24.8$ mW; for M_2, the power is $3.5(4.5 \times 10^{-3}) = 15.8$ mW. ❏

EXAMPLE 5.9 Repeat the last example after replacing the 2 V source at the gate of M_1 with an 8 V source.

Solution. As in the last example we know that M_1 is ON and M_2 is either OFF or active; again we assume both are active. With the 8 V for V_{GS1},

$$I_{D1} = \frac{9 \times 10^{-3}}{2} (8 - 1)^2 = 220.5 \text{ mA}$$

Because drain currents are identical,

$$I_{D2} = 220.5 \text{ mA} = \frac{4 \times 10^{-3}}{2} (V_{GS} - 2)^2$$

This gives $V_{GS2} = 12.5$ and -8.5 V. Because $V_{GS2} = V_{DS2} = 12.5$ V is consistent with our assumptions, we conclude that the drain of M_1 is 12.5 V below the 9 V supply, or

$$V_{D1} = 9 - 12.5 \text{ V} = -3.5 \text{ V}.$$

But this gives $V_{DG1} = V_{D1} - V_{G1} = -3.5 - 8 = -11.5$ V, which contradicts the assumption that M_1 is active.

Because M_2 is OFF or active, we now assume M_2 active and M_1 ohmic. Equating drain currents, using Eq. (5.4) for M_1 and Eq. (5.7) for M_2, gives

$$I_D = \frac{4 \times 10^{-3}}{2} (V_{GS2} - 2)^2 = \frac{9 \times 10^{-3}}{2} [2(8 - 1)V_{DS1} - V_{DS1}^2] \qquad (5.9)$$

From the diagram $V_{GS2} = 9 - V_{DS1}$. Making this substitution gives

$$I_D = \frac{4 \times 10^{-3}}{2} (9 - V_{DS1} - 2)^2 = \frac{9 \times 10^{-3}}{2} [2(8 - 1)V_{DS1} - V_{DS1}^2]$$

or

$$13 V_{DS1}^2 - 182 V_{DS1} + 196 = 0$$

Solving gives $V_{DS1} = 1.18$ V and 12.8 V. The corresponding values of V_{DG1} are, respectively,

$$1.18 - 8 = -6.82 \text{ V}$$

and

$$12.8 - 8 = 4.8 \text{ V}$$

Only the former satisfies

$$V_{DG1} \le -V_{t1} = -1 \text{ V}$$

as required by our assumption of ohmic operation for M_1. Thus $V_{DS1} = 1.18$ V.
    To find the current, substitute $V_{DS1} = 1.18$ V into Eq. (5.9). This gives $I_D = 67.7$ mA. For M_2,

$$V_{DS2} = 9 - 1.18 = 7.82 \text{ V}$$

Thus M_1 is biased at (1.18 V, 67.7 mA) and M_2 at (7.82 V, 67.7 mA). The power dissipated in M_1 is $67.7 \times 1.18 = 79.9$ mW. Power in M_2 is $67.7 \times 7.82 = 529.4$ mW. ❏

> **Exercise 5.9**   In Fig. 5.16, change the 2 V source to zero and assume M_2 is active. With these assumptions, find V_{DS1}.
>
> *Ans.* $V_{DS1} = 7$ V.

5.3

FET Resistors and Nonlinear Load Lines

We learned in Chapter 4 that large diffused resistors occupy excessive chip space in integrated circuits. A practical way to obtain large resistances without extra fabrication steps is to use field-effect transistors as nonlinear resistors. Both enhancement and depletion transistors serve this purpose. The load lines we used to analyze bipolar transistors in Sec. 4.5 are equally useful for FETs. Here we carry the load line idea farther by generalizing it to include nonlinear resistors.

5.3.1   ENHANCEMENT-TYPE NONLINEAR RESISTOR

The gate-to-drain connection of Fig. 5.17a makes the enhancement transistor a nonlinear resistor; we deduce its volt-ampere description by the following reasoning. For $v_R = v_{GS} \le V_t$, the transistor is OFF for it has no channel; thus $i_R = 0$. When $v_R \ge V_t$, the transistor is not only ON, but active, because "$v_{DG} = 0$" satisfies the inequality $v_{DG} \ge -V_t$. Thus

$$i_R = 0, \qquad \text{for } v_R \le V_t$$

and

$$i_R = \frac{k}{2}(v_R - V_t)^2, \qquad \text{for } v_R \ge V_t$$

Figure 5.17b shows the corresponding volt-ampere curve. Though quasi-quadratic rather than exponential, the curve suggests the rectifying property of the diode; therefore, we call Fig. 5.17a a *diode-connected transistor*.
    Figure 5.17c shows a variation that requires an external source. Since $v_{GS} = v_R + V_t$ volts, whenever $v_R \le 0$, the transistor is OFF and whenever $v_R \ge 0$ the transistor is ON. Also, because $v_{DG} = -V_t$, whenever the transistor is ON, it is active. Substituting $v_{GS} = v_R + V_t$ into Eq. (5.7) gives the resistor equation

$$i_R = \frac{k}{2}v_R^2$$

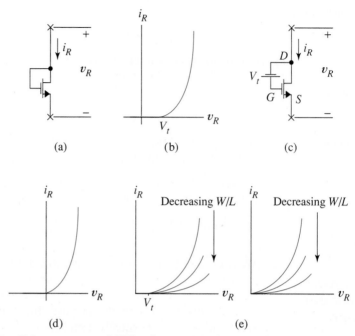

Figure 5.17 Enhancement-type MOS resistors: (a) basic connection; (b) volt-ampere curve for basic connection; (c) modified resistor connection; (d) volt-ampere curve for modified connection; (e) as W/L geometry is modified.

and the volt-ampere curve of Fig. 5.17d. We see that the external source shifts Fig. 5.17b to the left by V_t volts. By using other source values we can obtain any of a family of volt-ampere curves fitting between Figs. 5.17b and d; however, the two shown in the figure are used most frequently.

Equation (5.8) shows that the steepness of the resistor curves can be controlled by W/L as in Fig. 5.17e, with long, narrow channels giving lower currents for given voltages, thus higher resistances.

5.3.2 DEPLETION-TYPE NONLINEAR RESISTOR

Figure 5.18a shows a depletion-type resistor. Its volt-ampere characteristic is the "$v_{GS} = 0$ member" of the output characteristic family, shown in Fig. 5.18b. When $v_R = v_{DG} \leq -V_t$ (a positive number for a depletion transistor) the device is ohmic; for larger v_R the device is active. (In Example 5.5 we viewed Fig. 5.18a as an independent current source, an alternate interpretation that is correct for active-region operation.)

5.3.3 LOAD LINES FOR NONLINEAR RESISTORS

By showing graphically the constraints imposed upon transistor voltage and current by the external circuit, a load line helps us visualize in a very general way how the constrained transistor operates. We next learn how to sketch the load line for *any*

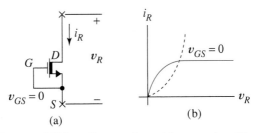

Figure 5.18 Depletion-type MOS nonlinear resistor: (a) connection; (b) volt-ampere characteristic.

resistor, linear or nonlinear. The first step is to examine the general idea of load line construction.

Figure 5.19a shows a common load line scenario. A resistor described by a function $i_R = f(v_R)$ is in series with a transistor output port and a supply V_{DD}. The constraint imposed by the external components upon the transistor, the load line, is to be superimposed upon the transistor's output characteristics, with both plotted on an i_D versus v_{DS} coordinate system.

The diagram shows that two constraints follow from the series connection, $i_D = i_R$ and $v_R = V_{DD} - v_{DS}$. These constitute a simple *change in variables* from resistor vari-

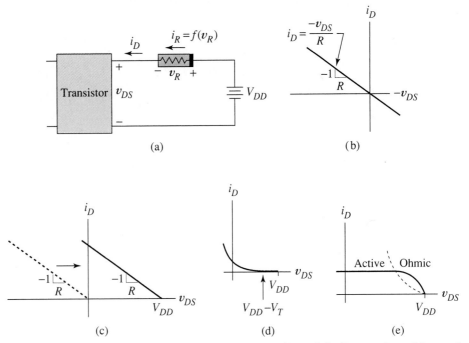

Figure 5.19 Load lines: (a) load line scenario; (b) $i_D = f(-v_{DS})$ for linear resistor; (c) second operation in change of variables for linear resistor; (d) load line for enhancement MOS resistor; (e) load line for depletion transistor.

ables (v_R, i_R) to transistor variables (v_{DS}, i_D). For a *linear* resistor R, the change in variables gives the load line equation by the process

$$i_D = i_R = f(v_R) = \frac{v_R}{R} = \frac{(V_{DD} - v_{DS})}{R} \qquad (5.10)$$

The substitution, $v_R = V_{DD} - v_{DS}$ in Eq. (5.10) consists of two elementary variable transformations. First, $i_D = f(-v_{DS})$ reflects the resistor's volt-ampere curve about the current axis, as sketched for a linear resistor in Fig. 5.14b. The second transformation, replacing $-v_{DS}$ by $V_{DD} - v_{DS}$, translates this reflected curve to the right by V_{DD} volts, as in Fig. 5.19c.

For a nonlinear resistor, $i_D = f(v_R)$ is not a straight line; however, because the change in variables is the same, the function is still reflected about the current axis and then translated to the right by V_{DD} volts. Figures 5.19d and e show the resulting load lines for the resistors of Figs. 5.17b and 5.18b. Nonlinear load lines rarely need to be plotted precisely. Their principal use is in making sketches that enhance our understanding of device and circuit operation. We see this in the next example.

EXAMPLE 5.10   In Examples 5.8 and 5.9 we used state definitions and circuit analysis to determine transistor Q-points in Fig. 5.20a (called Fig. 5.16a in the ex-

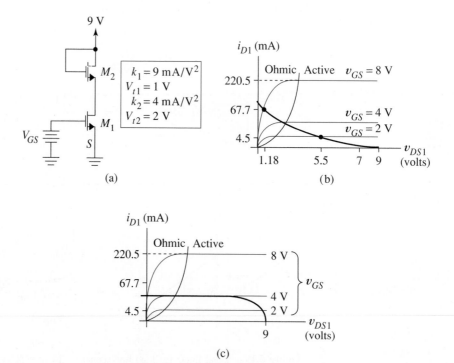

Figure 5.20   Nonlinear load lines: (a) given circuit; (b) nonlinear load line for Fig. 5.20a; (c) nonlinear load line when M_2 is a depletion transistor.

amples) for $V_{GS} = 2$ V and 8 V, respectively. Use a load line to help visualize the meaning of the analysis used in the examples.

Solution. We now recognize M_2 as a nonlinear load resistor with a volt-ampere curve like Fig. 5.17b. Figure 5.20b shows the corresponding nonlinear load line superimposed upon the output characteristics of M_1.

In the first example, M_1 was described by the curve $V_{GS} = 2$ V in Fig. 5.20b, which intersects the load line at $v_{DS} = 5.5$ V and $i_D = 4.5$ mA. At this Q-point $9 - 5.5 = 3.5$ V is the voltage across M_2. Clearly M_1 and M_2 are both active. As we increase V_{GS}, the operating point moves up the nonlinear load line, with M_1 changing to ohmic operation when $V_{GS} \approx 3.8$ V.

For $V_{GS} = 8$V, M_1 is driven far into the ohmic region, with the Q-point at the intersection of the $V_{GS} = 8$ V curve and the load line, that is, at 1.18 V and 67.7 mA. Figure 5.20b also shows how we initially obtained the incorrect value $I_D = 220.5$ mA in Example 5.9 by incorrectly assuming M_1 was still active with $V_{GS} = 8$ V.   ❏

Even a crude sketch of the nonlinear load line gives a perspective that we lack when we use equations and state models alone. The load line can also save time by helping us make correct initial assumptions about transistor states.

Exercise 5.10   When we replace M_2 in Fig. 5.20a by a depletion-type nonlinear resistor, the new load line is that of Fig. 5.20c. From the load line, determine the states of both transistors when $V_{GS} = 0$, when $V_{GS} = 2$ V and $V_{GS} = 5$ V.

Ans. M_1 cut-off and M_2 ohmic, M_1 active and M_2 ohmic, M_1 ohmic and M_2 active.

5.3.4 MOS VOLTAGE DIVIDER

MOS enhancement resistors make voltage dividers that occupy little chip space and draw low currents, for example, the divider in Fig. 5.21a that establishes node voltage V_X. The total voltage applied to the series combination, in this case 16 V, must be sufficient to keep both transistors ON. Given this, because of the gate-drain connections both

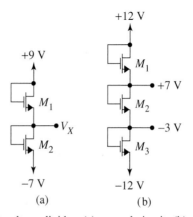

Figure 5.21   All MOS voltage divider: (a) general circuit; (b) circuit for Example 5.11.

transistors are active; their currents are identical because they are in series. Equating currents of M_1 and M_2 and expressing V_{GS} for each in terms of circuit node voltages gives

$$\frac{k_1}{2}[(9 - V_X - V_{t1}]^2 = \frac{k_2}{2}[(V_X + 7) - V_{t2}]^2$$

If the threshold voltages are known, for example, $V_{t1} = 2$ V and $V_{t2} = 1$ V, V_X is established by the values of the transconductance parameters; that is,

$$k_1(7 - V_X)^2 = k_2(V_X + 6)^2$$

Taking the square root of both sides and solving for V_X gives

$$V_X = \frac{7\sqrt{k_1/k_2} - 6}{\sqrt{k_1/k} + 1} = \frac{7\lambda - 6}{\lambda + 1} \tag{5.11}$$

This expression has two interesting features. First, V_X is a function of the ratio k_1/k_2, which, according to Eq. (5.8), is determined in an IC by the the masking geometry; that is,

$$\frac{k_1}{k_2} = \frac{W_1 L_2}{L_1 W_2}$$

Second, V_X is a continuous function of λ that varies from $V_X = -6$ V for $\lambda = 0$ to $V_X = +7$ for $\lambda = \infty$. We conclude that V_X can be set to any (reasonable) prescribed value within this range by fixing the relative dimensions of the two transistors. A final observation is that since V_X is set by the *ratio* of k_1 to k_2, we can use one of these parameters to set the current that flows through the divider, and therefore the total power dissipation of the divider. Suppose, for example, we select $\lambda = 6/7$ to place the output voltage near ground potential. This condition requires $k_2 = 1.361k_1$. Choosing $k_1 = 10^{-6}$ A/V^2 according to Eq. (5.7) gives a divider current of

$$i_D = 0.5 \times 10^{-6} (9 - 0 - 2)^2 = 25.5 \ \mu A$$

and a total power dissipation of

$$P_D = 25.5 \times 10^{-6} \times [9 - (-7)] = 408 \ \mu W$$

A comparable voltage divider using linear resistors would require a total series equivalent resistance of 16 V/25.5 μA = 627 kΩ, a value that would occupy considerably more chip space. To design a voltage divider we apply these principles to given specifications as in the next example.

Exercise 5.11 Find k_1 and k_2 for the divider of Fig. 5.21a so that $V_X = -3$ V and circuit power dissipation is 200 μW. Use threshold voltages, $V_{t1} = 2$ V and $V_{t2} = 1$ V.

Ans. $k_1 = 2.5 \times 10^{-7}$ A/V^2, $k_2 = 2.78 \times 10^{-6}$ A/V^2.

EXAMPLE 5.11 Design an MOS voltage divider that uses ± 12 V supplies to establish node voltages of $+7$ V and -3 V. The power dissipation of the divider must not exceed 1.2 mW.

Solution. Figure 5.21b shows the circuit. To limit the number of variables to those needed to satisfy the given requirements, specify $V_t = 2$ V for all transistors. Since a single dc current I_D flows from the $+12$ to the -12 V supply, the power dissipated is $24 \times I_D$ watt; therefore, 1.2 mW dissipation requires a current of $I_D = 0.05$ mA in each transistor. Since the drops across M_1, M_2, and M_3 are to be 5 V, 10 V, and 9 V, respectively, M_2 has the highest resistance and M_1 the lowest. Use the given current to fix the value of k_3, that is, the middle-sized transistor—a somewhat arbitrary choice. Using Eq. (5.7),

$$0.05 \times 10^{-3} = \frac{k_3}{2} [-3 - (-12) - 2]^2$$

which gives $k_3 = 2.04 \times 10^{-6}$ A/V^2.

Equating the currents in the three devices gives

$$\frac{2.04 \times 10^{-6}}{2} (9 - 2)^2 = \frac{k_1}{2} (5 - 2)^2 = \frac{k_2}{2} (10 - 2)^2$$

Solving gives $k_1 = (49/9)k_3$ and $k_2 = (49/64) k_3$, expressions that translate into W/L specifications for designing the IC masks.   ❏

We would be remiss not to mention *body effect,* a second-order transistor property (Sec. 5.11) that makes divider design somewhat more complicated than this development suggests; nevertheless, the salient features, low power dissipation and small chip space, are retained.

5.4
p-Channel MOSFET

p-Channel enhancement and depletion MOSFETs physically resemble the corresponding n-channel devices of Figs 5.2a and 5.9a, except that p- and n-doped materials are interchanged. Consequently, current directions, voltage polarities, and algebraic signs of the threshold voltages are all reversed. Also, in p-channel devices the n-type substrate must always be connected to the most *positive* node in the circuit.

For a channel to exist, the gate of the p-type enhancement MOSFET must be *negative* enough to lower the hole potential and create a channel. That is, the gate must be made negative relative to the source by an amount at least equal to a threshold voltage V_t, which is itself *negative*. The p-channel depletion device has a positive threshold voltage. Because both have insulated gates, their input characteristics are open circuits.

Figure 5.22 shows schematic symbols, biasing polarities, and reference directions for p-channel enhancement and depletion devices. Notice that the arrows in both schematics are consistent with sourcing holes, and orientations of drain-source supplies cause conventional current to flow in the directions of the arrows on the schematics.

Equations (5.3), (5.4) and (5.7) *apply without change to p-channel devices, provided we observe the reference current conventions of Fig. 5.22.* The output and transfer characteristics of the devices lie in the second quadrant as in Figs. 5.23 and 5.24. (Some prefer to invert the subscript order in the equations so that output characteristics plot in the first quadrant.)

The state definitions for p-channel MOSFETS are those of Fig. 5.6a *except that the sense of every inequality is reversed,* a statement consistent with Figs. 5.23 and 5.24.

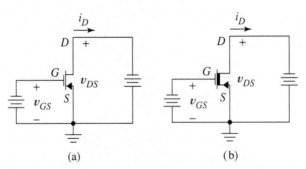

Figure 5.22   PMOS schematics, biasing circuits, and reference directions for voltage and current: (a) enhancement type; (b) depletion type.

The large-signal equivalent circuits for the three states, Fig. 5.25, have *drain currents reversed* relative to the n-channel case; however, the functions f_0 ( ) and f_a ( ) are still Eqs. (5.4) and (5.7), respectively.

The two classes of operating point analysis problems, transistor-state known, and state unknown, also apply to circuits that contain p-channel MOSFETS, and are solved in the same fashion.

EXAMPLE 5.12   Find the Q-point of the transistor in Fig. 5.26a. Assume it is active, and then verify that this assumption is correct.

Solution.   Since the transistor is assumed active,

$$I_D = \frac{8 \times 10^{-3}}{2} [V_{GS} - (-1)]^2$$

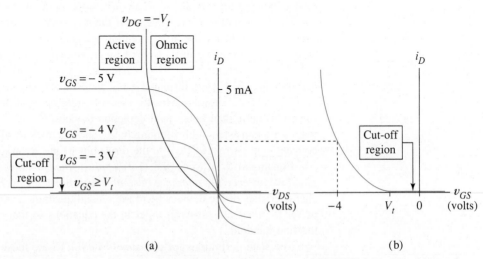

(a)   (b)

Figure 5.23   Characteristic curves for p-channel enhancement MOSFET: (a) output; (b) transfer.

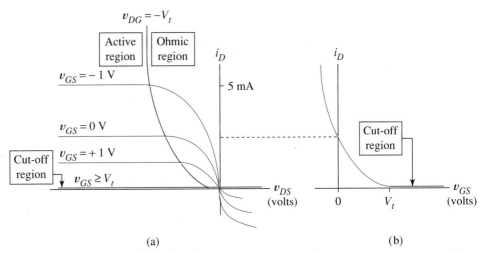

(a)   (b)

Figure 5.24   Characteristic curves for p-channel depletion MOSFET: (a) output; (b) transfer.

The equivalent circuit of Fig. 5.26b makes it obvious that

$$V_{GS} = V_G - V_S = -7.5 - (-10 \, kI_D)$$

Solving this equation for I_D and equating drain currents gives

$$\frac{V_{GS} + 7.5}{10 \, k} = \frac{8 \times 10^{-3}}{2} [V_{GS} + 1]^2$$

This is equivalent to

$$40V_{GS}^2 + 79V_{GS} + 32.5 = 0$$

which has the solutions $V_{GS} = -0.584$ V and -1.39 V, respectively. Because $V_t = -1$ V, only the latter is sufficiently negative to establish a channel. Substituting $V_{GS} = -1.39$ into either current equation gives $I_D = 0.611$ mA. Figure 5.26b shows that with this current,

$$V_D = -10 + 3.3 \, k(0.611 \text{ mA}) = -7.98 \text{ V}$$

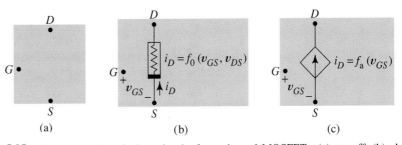

(a)   (b)   (c)

Figure 5.25   Large-signal equivalent circuits for p-channel MOSFETs: (a) cut-off; (b) ohmic; (c) active.

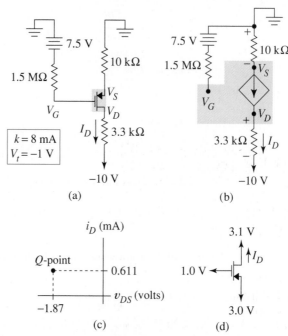

Figure 5.26 (a) Circuit schematic for Example 5.12; (b) equivalent circuit for Example 5.12; (c) transistor Q-point; (d) schematic for Exercise 5.12.

and

$$V_S = -10\,k(0.611\ \text{mA}) = -6.11\ \text{V}$$

giving $V_{DS} = -1.87$ V.

To verify active operation,

$$V_{DG} = V_D - V_G = -7.98 - (-7.50) = -0.48$$

Evaluating the inequality, $V_{DG} \le -V_t$, gives $-0.48 \le -(-1)$. Clearly, the inequality is satisfied. Figure 5.14c shows the Q-point location. ❏

Exercise 5.12 Find the drain current of the transistor in Fig. 5.26d, given that $V_t = -1.8$ V and $k = 4 \times 10^{-5}$ A/V^2.

Ans. $I_D = -1$ μA.

When we operate the p-channel MOSFET as a voltage-controlled linear resistor, our choice of reference direction for i_D gives the resistance expression

$$R_{PMOS} = \frac{1}{k|v_{GS} - V_t|} \tag{5.12}$$

which assumes the linearity condition

$$|v_{DS}| \le 0.2\ |v_{GS} - V_t|$$

MOS Integrated Circuit Families. We usually encounter MOS transistors as members of integrated circuit *families*. An IC family represents a particular class of fabrication technology, which dictates the kinds of components that can be used in a circuit and also limits fundamental attributes such as packing density, power dissipation, and speed.

Historically, the first important MOS IC family was p-channel MOS (PMOS). The key element is the transistor of Fig. 5.27a, with source and drain fabricated by diffusing acceptor impurities into an n-type substrate. The gate is aluminum deposited over a SiO₂ dielectric. The structural simplicity gave low fabrication costs, high packing densities, and high yields of usable circuits. The first LSI circuits and the first generation of microprocessors were constructed in PMOS. Diffused resistors and MOS load resistors are used in PMOS.

In the late 1970s, n-channel MOS (NMOS) technology became dominant in LSI. The NMOS transistor, Fig. 5.27b, is faster than its PMOS predecessor because electron mobility is higher than hole mobility. At at about this time it also became feasible to improve performance by constructing the transistor gate from *polysilicon,* polycrystaline silicon formed by chemical vapor deposition (CVD) over the SiO₂ insulator. During the

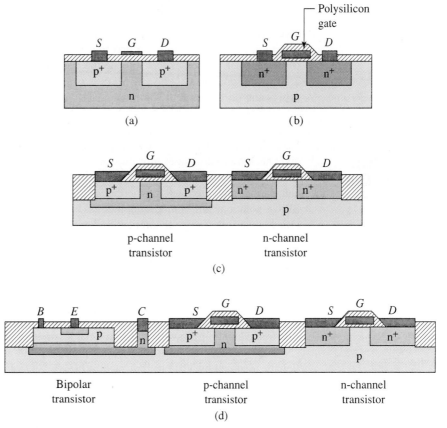

Figure 5.27   Basic devices in four IC families: (a) PMOS; (b) NMOS; (c) CMOS; (d) BiCMOS.

source and drain diffusions, the gate is also diffused with a high concentration of impurities, making it a conductor. The thin insulator under the gate is formed by selective masking, using silicon nitride to inhibit oxide growth as discussed in Chapter 4. Other advantages of NMOS include lower power dissipation, lower noise margins in digital circuits, and resistor values ranging from 50 Ω to 50 MΩ made possible by using lightly doped polysilicon. NMOS and PMOS both found their principal applications in digital logic and memory circuits.

In the late 1970s complementary symmetry MOS (CMOS) became increasingly popular in SSI and MSI, and in the 1980s became the dominant MOS technology in LSI and VLSI. The heart of CMOS is the complementary n-channel–p-channel transistor pair of Fig. 5.27c. The n-channel transistors share the same p-type substrate, whereas each p-channel transistor is fabricated within an individual n-type well, with all n-wells connected to the most positive point in the circuit to maintain reverse bias. A thick *field oxide* isolates the transistors. CMOS is comparable in density to NMOS but features lower power dissipation. It is not only suitable for digital circuits, but also is widely used in many important analog circuits such as op amps, analog multiplexers, digital to analog and analog to digital circuits, and filters.

In the 1980s yet another new technology attracted widespread attention. A combination of bipolar and CMOS, called BiCMOS, offers high density and speed and also enables greatly increased design flexibility. Figure 5.27d shows the simplest BiCMOS structure, formed by adding an npn transistor to the basic CMOS fabrication repertoire. More complex processes that include the collector buried layer and other improvements are in widespread use. With BiCMOS it is feasible to combine digital and analog signal processing on the same IC chip.

5.5
Static SPICE Model for the MOSFET

Static SPICE MOSFET models help us find operating points without having to assume and check transistor states, nor need we deal with solutions having no physical meaning. SPICE also allows us to do things that would be too tedious for hand analysis, such as calculating and plotting the nonlinear transfer characteristic of a MOSFET circuit. SPICE solutions are limited to numerical values, however, and thus provide little insight unless supported by the theoretical understanding of the user.

A built-in mathematical model determines the correct currents and voltages by iterative numerical calculations rather than by using state assumptions. Figure 5.28a shows the SPICE coding format. MOSFET names begin with M, and node numbers are listed in drain-gate-source-substrate order. In the diagram, the depletion transistor's schematic symbol does not explicitly show the substrate connection; however, the MD element line correctly indicates that its n-type substrate is connected to the most positive node in the circuit.

Immediately following the class name on the .MODEL line is the code word "NMOS" or "PMOS," which identifies the MOSFET as n- or p-channel. The threshold voltage, denoted by VTO as indicated in Fig. 5.28b, has algebraic sign appropriate to the type of transistor. Notice that the threshold voltages for the n-channel enhancement transistor and the p-channel depletion transistor are positive. For the p-channel enhancement transistor, V_t is negative.

Transconductance parameter KP (and several other parameters that might appear on the .MODEL line) applies to a device of $W/L = 1$. We can optionally append individual

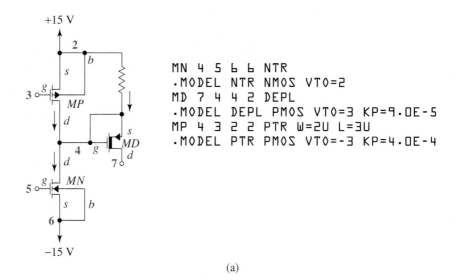

+15 V

```
MN 4 5 6 6 NTR
.MODEL NTR NMOS VTO=2
MD 7 4 4 2 DEPL
.MODEL DEPL PMOS VTO=3 KP=9.0E-5
MP 4 3 2 2 PTR W=2U L=3U
.MODEL PTR PMOS VTO=-3 KP=4.0E-4
```

−15 V

(a)

Parameter	SPICE name	Default
V_t	VTO	0.0 V
k	KP	2.0E−5 A/V^2

(b)

Figure 5.28   SPICE coding for MOSFETS: (a) examples; (b) parameter notation.

width and length dimensions at the end of the element line as in the MP description in Fig. 5.28a. This is similar to the individual area scaling available for diodes and BJTs—SPICE appropriately rescales all relevant transistor parameters in accordance with the specified W and L values.

The following example and exercise show how a SPICE simulation can improve our understanding of a complicated nonlinear circuit and even give us ideas on how to improve its performance.

EXAMPLE 5.13   Figure 5.29a shows a simple amplifier that employs a depletion MOS transistor with $k = 6 \times 10^{-4}$ A/V^2 and $V_t = -2$ V.
(a) Use SPICE transient analysis to find the shape and size of the output waveform when the input is $v_{in}(t) = 10^{-3} \sin(2\pi 1000t)$ V.
(b) From the results of part (a), determine the gain of the amplifier.
(c) Repeat part (a) using $v_{in}(t) = 2 \sin(2\pi 1000t)$ V.
(d) Examine the origin of the distortion that appears in part (c) by plotting the transfer characteristic of the amplifier using SPICE dc analysis.

Solution. (a) Figure 5.29b is the input code. The VIN line without an asterisk describes $v_{in}(t)$ for part (a). The SPICE output for the 1 mV sine wave, Fig. 5.29c, looks just like the input except that it is inverted, has greater amplitude, and has a dc offset of 2.28 volts.

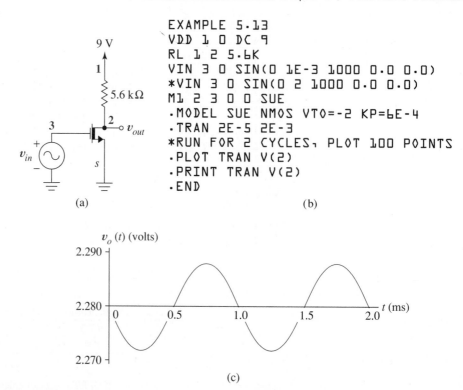

Figure 5.29   Example 5.13: (a) amplifier circuit; (b) SPICE code; (c) output waveform for small sinusoidal input; (d) output waveform for large sinusoidal input; (e) transfer characteristic showing nonlinearity.

(b) The SPICE *printout* showed the peak amplitude of the output sine wave was 7 mV. Therefore, the circuit amplified the signal by 7.

(c) Replacing the original VIN line by the one marked with an asterisk gives the output shown in Fig. 5.29d. The waveform, clearly not sinusoidal, is periodic with the same period as the input sine wave. In Sec. 1.6.2 we saw that a nonlinear transfer characteristic can produce such a result by adding harmonics of the input frequency to the amplified input. We refer to such an output as a *distorted* sine wave.

(d) Using a dc analysis to plot the transfer characteristic for the input voltage range, -2 V to $+2$ V (Fig. 3.57 shows method), gives Fig. 5.29e. The negative slope of this curve explains the inversion of the output signal. There is also an output offset voltage of 2.28 V because the transfer curve does not pass through the origin.

Imagine projecting the 1 mV amplitude input sine wave from Fig. 5.29e. The signal's small amplitude restricts it to a region where the curve approximates a straight line. In fact, the curve suggests that we can expect fairly linear operation for input sine waves with peak values as large as 100 mV, an observation easily checked with SPICE. The 2 V peak sinusoidal input swing in part (c) traverses such a wide range of the transfer curve, however, that the output voltage is severely distorted as in Fig. 5.29d. This is particularly true for positive $v_{in}(t)$ (more negative v_{out}) where a 2 V input change pro-

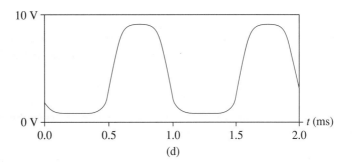

(d)

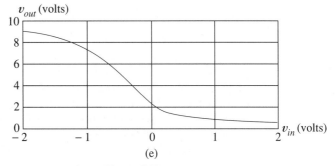

(e)

Figure 5.29   (continued)

duces an output change of less than 1 V. In Chapter 9 we learn how to quantify such distortion.   ❏

> **Exercise 5.13**   The transfer characteristic in Fig. 5.29e is relatively steep and linear for the input voltage range $-0.75 < v_{in} < 0$. This suggests that if we add a constant dc voltage (that is, a *bias* source) of $\approx -0.75/2 = -0.375$ V in series with $v_{in}(t)$, the amplifier might be able to handle input peak signal amplitudes of 350 mV or so with little distortion. Modify the input code to test this idea, and use SPICE printout to find the voltage gain.
>
> *Ans.* Voltage gain ≈ 5.4. There is some distortion, but much less than in Fig. 5.29d.

We explore biasing ideas like this at some length in Chapters 6 and 7.

MOSFETs are widely used in LSI and VLSI circuits, but are not generally available in discrete form. The transistors introduced in the next section are available as discrete devices and are used to a limited extent in ICs as well.

5.6

Junction Gate Field-Effect Transistors

The transistors we introduce here control channel conductance by gates that are in electrical contact with the channel, not insulated from it. In the metal semiconductor FET (MESFET), the channel is a compound semiconductor such as gallium arsenide, the gate a metal, and the gate-channel interface a Schottky junction. In the more traditional junction FET structures, gate and channel consist of oppositely doped silicon, and a reverse-biased pn junction forms the gate-channel interface. By common usage only the latter are called JFETs.

5.6.1 MESFET

The MESFET exploits the electron's high mobility in gallium arsenide by using a relatively new fabrication technology rather than the more mature silicon technology. The result is a device greatly superior in speed but inferior in circuit density, and presently much more expensive than silicon-based transistors. As GaAs technology improves, however, MESFETs are finding increased application, both in linear circuits that operate at microwave frequencies and in high-speed digital logic circuits.

Figure 5.30a shows the MESFET structure. Heavily doped, n-type wells formed by ion implant provide the drain and source. An n-doped GaAs channel is formed by ion implant on a substrate of intrinsic GaAs, and a metal alloy gate forms a rectifying Schottky junction with the GaAs channel. Except for its substrate material and its lack of an insulator between gate and channel, the MESFET resembles the depletion MOSFET of Fig. 5.9a. Because gallium arsenide has low conductivity by virtue of its low intrinsic carrier concentration (Table 3.1), electrical isolation between transistors is provided by the substrate itself through a micron or so of physical separation. Sometimes an ion implant of impurities between devices is used to improve the isolation.

Figure 5.30b shows the MESFET schematic, in which a dark band represents the preexisting channel; the arrow at the gate points in the forward direction of the Schottky junction. Because this device sources electrons, the reference direction for drain current is as shown. Polarities of external sources are consistent with reverse biasing the gate-source junction and with positive drain current.

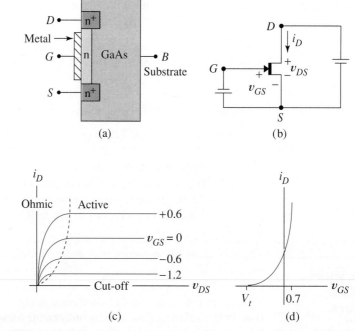

Figure 5.30 MESFET: (a) physical structure; (b) schematic symbol and bias polarities; (c) output characteristics; (d) transfer characteristic.

The MESFET output and transfer characteristics, Figs. 5.30c and d, closely resemble the depletion MOS curves of Figs. 5.9d and c. When we make gate negative relative to source, as in the depletion MOSFET, the number of channel electrons is reduced and the channel conductivity lowered. In the MESFET this occurs because the depletion region expands into the channel as indicated by black shading in Fig. 5.31a. For low v_{DS} this expansion is symmetric, resulting in a voltage-controlled linear resis-

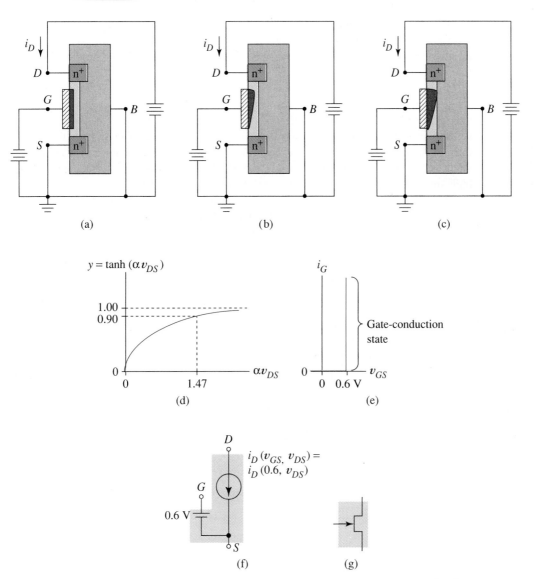

Figure 5.31 MESFET: (a)–(c) control of channel by external voltages; (d) correction for velocity saturation; (e) input characteristic; (f) model when gate conducts current; (g) schematic for enhancement device.

tance between source and drain. When v_{GS} reaches V_t, the transistor's (negative) threshold voltage, the depletion region has expanded entirely across the channel. The channel conductivity is then zero, which results in a cut-off transistor described by

$$i_D = 0 \tag{5.13}$$

Equation (5.13) holds for all v_{GS} satisfying $v_{GS} \leq V_t$.

When a channel exists, making v_{DS} increasingly positive eventually distorts the channel shape because of greater reverse bias at the drain end as in Fig. 5.31b. The increasing channel resistance results in smaller increases in drain current, Δi_D, for successive increases in drain-source voltage, Δv_{DS}, as shown in Fig. 5.30c, giving rise to a family of nonlinear ohmic-region voltage-controlled resistances. In the ohmic region, the transistor current is

$$i_D = \beta[2(v_{GS} - V_t)v_{DS} - v_{DS}^2] \tanh(\alpha v_{DS}) \tag{5.14}$$

where β is a MESFET transconductance parameter unrelated to the β of the bipolar transistor. For MESFETs V_t ranges from -3 V to -0.3 V. Equation (5.14) has the general form of the MOSFET ohmic-region equation, Eq. (5.4), except for the factor $\tanh(\alpha v_{DS})$ that corrects for *velocity saturation*. In the MESFET, electrons are accelerated to such high speeds that the straight-line proportionality of drift velocity to field intensity, indicated by μ_n, ceases to be valid. Instead, for sufficiently high field strength, the electron velocities reach terminal values of the order of 2×10^7 cm/s or so, and then no longer increase with field strength. In the velocity saturation correction, α (unrelated to the α of a BJT) is a constant determined by measurement. Typically, α falls in the range $0.3 \leq \alpha \leq 2$.

For sufficiently large v_{DS}, namely,

$$v_{DS} - v_{GS} \geq -V_t$$

or equivalently

$$v_{DG} \geq -V_t$$

the drain-end reverse bias is so great that the channel pinches off as in Fig. 5.31c. Once the channel pinches off, the drain-end depletion region expands for additional increases in v_{DS} without greatly affecting the drain current. This describes the active region where the device resembles a nonlinear VCCS described by

$$i_D = \beta[(v_{GS} - V_t)^2] \tanh(\alpha v_{DS}) \tag{5.15}$$

In Eqs. (5.14) and (5.15) it is possible to express β in the form

$$\beta = W\beta'$$

where W, the channel width, can be used for a scale factor. Because MESFETs are high-speed devices, the channel length parameter is always the smallest possible feature length (currently one-half to one micron) to minimize capacitance and maximize speed.

Figure 5.31d helps us understand the significance of velocity saturation in the MESFET. For large values of the argument, αv_{DS}, the correction factor approaches $\tanh(\infty) = 1$, and Eqs. (5.14) and (5.15) are the ordinary MOSFET equations with β

replacing $k/2$. Notice from Fig. 5.31d that the correction is close to one (within 10%) when

$$\alpha v_{DS} \leq \tanh^{-1}(0.9) = 1.47$$

For $\alpha = 2$, this means the correction is negligible for drain-source voltages in the range $v_{DS} \geq 1.47/2 = 0.735$ V; for $\alpha = 0.3$, the correction is small when $v_{DS} \geq 4.9$ V. We conclude that velocity saturation smoothly compresses the curves, that is, gives lower i_D values for low values of v_{DS} compared to values that would exist if velocity saturation were not present. The effect is most significant in the ohmic region where v_{DS} is smallest, but the active equations can also change appreciably for higher values of v_{DS}, especially when α is small. Because the velocity saturation correction is difficult to apply in hand analysis, we can *estimate* transistor Q-points by assuming the correction is one, and then rely on simulation tools that include this correction (for example, PSPICE) to determine the Q-points precisely.

Figure 5.31e shows that the MESFET has an an open-circuit input characteristic like the MOSFET only if the gate-source (Schottky) junction does not conduct. GaAs Schottky junctions have higher forward voltages (≈ 0.6 V) than silicon Schottky junctions (≈ 0.4 V). In linear applications gate conduction is avoided, and the input resistance is virtually infinite. In this case, the MESFET has three operating states with definitions identical to the MOSFET definitions of Fig. 5.6a (when we ignore velocity saturation). Equivalent circuits are those given for the MOSFET in Fig. 5.7, with $f_0(v_{GS}, v_{DS})$ given by Eq. (5.14) and $f_a(v_{GS})$ by Eq. (5.15)—both using $\alpha = \infty$ for hand analysis and initial design estimates.

Digital circuits sometimes employ intentional gate-channel conduction through the Schottky junction. The equivalent circuit of Fig. 5.31f describes this fourth MESFET state, where function $f(0.6, v_{DS})$ describes the single output curve of Fig. 5.30c labeled $v_{GS} = 0.6$ V. In the active region, this is a constant current; however, in the ohmic region it is best visualized as a nonlinear resistor.

MESFET n-channel enhancement transistors are also available. These resemble Fig. 5.30a except for the absence of the prefabricated channel under the gate. Their input, transfer, and output characteristics; equations; and properties are those of the depletion device, except for positive threshold voltages that range from 0 to 0.3 V.

5.6.2 N-CHANNEL JFET

Figure 5.32a shows the essential physical structure of an n-channel JFET. A conducting *channel* of n-type silicon is contiguous with an n-type *drain* and an n-type *source*. A *gate* of p-type material borders the channel. Figure 5.32b shows the schematic symbol; the vertical line represents the channel; an arrowhead near the source identifies gate and source, and points in the p-to n direction to indicate the JFET is n-channel. Since this *n-channel* transistor sources *electrons,* the reference direction for *conventional* drain current, i_D, is that shown. The JFET input characteristic is the volt-ampere curve of the gate-source pn junction—Fig. 5.32c. Because the gate is *never* forward biased, the input characteristic is essentially the open circuit $i_G = 0$.

Figure 5.33a shows JFET output characteristics. The n-channel JFET operates like the MESFET, with channel conduction controlled by expansion of the pn (instead of

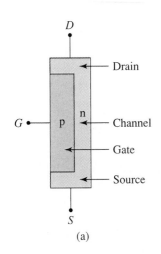

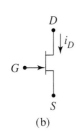

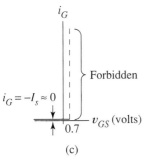

Figure 5.32 n-Channel JFET: (a) basic structure; (b) schematic symbol; (c) input characteristic.

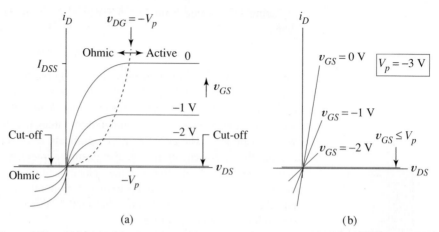

Figure 5.33   Output characteristics for JFET: (a) for all values of v_{DS}; (b) when v_{DS} is small.

Schottky) depletion region into the channel, as in Figs. 5.31a, b, and c. For the JFET, we call the negative gate-source voltage that just pinches off the channel the *pinch-off voltage*, V_P. Because operation as a voltage-controlled resistance is important for a JFET, we show the output characteristics for small v_{DS} in Fig. 5.33b.

The JFET is a three-state device with states defined in Fig. 5.34. Except for the notational change, V_P replacing V_t, the definitions are identical to those of MOSFETs (Fig. 5.6a), as are the corresponding regions of the output characteristics.

Cut-off State. In cut-off the JFET satisfies

$$i_D = 0 \tag{5.16}$$

Its equivalent circuit is that of Fig. 5.7a.

Ohmic State. In its ohmic state, drain current is given by

$$i_D = \beta \left[2(v_{GS} - V_P)v_{DS} - v_{DS}^2 \right] \tag{5.17}$$

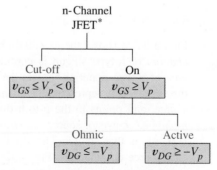

Figure 5.34   JFET state definitions.

where β is the JFET transconductance parameter. Figure 5.7b models the ohmic JFET as a nonlinear voltage controlled resistor with $f_0(v_{GS},v_{DS})$ given by Eq. (5.17). In terms of fabrication, β is given by

$$\beta = \frac{W}{L}\mu_n \frac{4\epsilon_{Si}}{3tN_D} \tag{5.18}$$

where W, L, and t are the width, length, and depth of the channel material, μ_n the electron mobility, N_D the doping concentration, and ϵ_{si} the dielectric constant of silicon.

Voltage-Controlled JFET Resistor. If v_{DS} in Eq. (5.17) is so small that the squared term is negligible, drain current and drain-source voltage are related by

$$i_D = 2\beta(v_{GS} - V_P)v_{DS} = \frac{1}{R_{N\text{-}JFET}}v_{DS}$$

It follows that the resistance of this controlled resistor is

$$R_{N\text{-}JFET} = \frac{1}{2\beta(v_{GS} - V_P)} \tag{5.19}$$

Using a factor of 10 to define small v_{DS} leads to

$$|v_{DS}| \le 0.2\,(v_{GS} - V_P) \tag{5.20}$$

as the condition to check when we use Eq. (5.19).

Active State. In its active region the JFET satisfies

$$i_D = \beta(v_{GS} - V_P)^2 \tag{5.21}$$

which also defines $f_a(v_{GS})$ in the active-region equivalent circuit of Fig. 5.7c. A plot of Eq. (5.21) is the transfer characteristic for the n-channel JFET, which looks like Fig. 5.35. Data sheets for discrete JFETs often specify β indirectly using a parameter called I_{DSS}, the drain current when the transistor is active and $v_{GS} = 0$. From Eq. (5.21) we find that the conversion equation is

$$I_{DSS} = \beta V_P^2 \tag{5.22}$$

Figures 5.35 and 5.33a show the physical meaning of I_{DSS}.

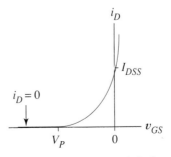

Figure 5.35   n-Channel JFET transfer characteristic for active-region operation.

EXAMPLE 5.14 Design an amplifier with voltage-controlled gain using a noninverting op amp circuit and a controlled JFET resistor. Amplifier gain must be variable and must span the range 9 to 55. JFET parameters are $\beta = 10^{-3}$ A/V^2 and $V_P = -3$ V.

Solution. The gain of the noninverting amplifier depends upon a resistor ratio. Of the two possible locations for the JFET resistor, the one in Fig. 5.36 allows the source node to be grounded. This facilitates control of the resistance, and hence gain, by a grounded control signal v_c. Amplifier input and output voltages are v_i and v_o.

To obtain a design equation, substitute Eq. (5.19) into the amplifier gain equation. Recognizing that $v_{GS} = v_c$ in Fig. 5.36 gives

$$\text{Gain} = 1 + \frac{R_F}{R_{N\text{-}JFET}} = 1 + 2 \times 10^{-3}(v_c + 3)R_F \qquad (5.23)$$

Since v_c is v_{GS} for the FET, v_c is constrained to the practical range $-3 \le v_c \le 0$. This avoids both pinching off the channel and forward biasing the gate-channel junction (see Fig. 5.35).

Since v_c in Eq. (5.23) takes on only nonpositive values, the maximum gain of 55 must occur when $v_c = 0$. Thus R_F must satisfy

$$55 = 1 + 2 \times 10^{-3}(0 + 3)\,R_F$$

giving $R_F = 9$ kΩ. With this R_F, a gain of 9 requires

$$9 = 1 + 18\,(v_c + 3)$$

or $v_c = -2.56$ V. Thus the operating range for the control voltage is $-2.56 \le v_c \le 0$.

According to Eq. (5.20), the linear resistor model is valid only for

$$|v_{DS}| = |v_i| \le 0.2\,(v_c + 3)$$

This inequality must hold even when the right side is smallest, that is, for $v_c = -2.56$ V; therefore we must restrict input signal amplitudes to

$$|v_i| \le 0.2\,(-2.56 + 3) = 88 \text{ mV}$$

For maximum input voltage of $v_i = 88$ mV, amplifier gain ranging from 9 to 55 means output voltage in the range $0.792 \le v_o \le 4.84$ V. Because the ohmic curves of Fig. 5.33b

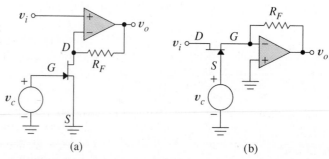

(a) (b)

Figure 5.36 (a) Noninverting amplifier with JFET resistor in feedback circuit for gain control; (b) inverting gain-controlled amplifier.

extend into the third quadrant, the circuit amplifies signals in the range -88 mV $\leq$ $v_i \leq 88$ mV. This completes the design. An example later in the chapter uses SPICE to explore this design in some detail. ❏

Exercise 5.14 Design a voltage-controlled amplifier using an *inverting* op amp circuit and a controlled JFET resistor. Amplifier gain should be variable over the range -9 to -55. JFET parameters are $\beta = 10^{-3}$ A/V^2 and $V_P = -3$ V.

Ans. The circuit is Fig. 5.36b. $R_F = 9.17$ kΩ, $|v_i| \leq 98$ mV. The control input varies over the same range as in the preceding example; that is, -2.56 V $\leq v_C \leq 0$ V.

5.6.3 p-CHANNEL JFET

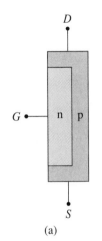

Figure 5.37a shows the p-channel JFET, a transistor with physical structure complementary to the n-channel JFET. Because of p-type channel doping, the source is a *source of holes* that flow internally to the drain. The reference direction for conventional current i_D in the schematic of Fig. 5.37b is consistent with *sourcing* of holes, and the drain-source supply is oriented to cause current flow in this direction; that is, v_{DS} is negative. The gate arrow in the schematic symbol points in the p-to-n direction, suggesting the p-type channel. As in the n-channel device, the gate to source supply polarity opposes the arrow direction to keep the gate-channel junction reverse biased.

Positive values of v_{GS} expand the depletion region into the channel, causing it to pinch off for a *positive* pinch-off voltage, V_P. Values of v_{GS} exceeding V_P give a transistor with a channel that is entirely pinched off, that is, a cut-off transistor.

When a channel exists, the JFET functions as a voltage-controlled linear resistor for small v_{DS}—for larger v_{DS} as a voltage-controlled nonlinear resistor—and for still larger values, as a voltage-controlled current source. Equations (5.16), (5.17), and (5.21), which described the n-channel JFET, also describe the p-channel JFET, provided that the reference direction for drain current is that of Fig. 5.37b. Figures 5.38a and b show the characteristic curves predicted by these equations. As for the pnp bipolar transistor, the output characteristics reside in the second quadrant. We conclude from these characteristics that the p-channel JFET differs from the n-channel JFET in the algebraic signs of V_P, v_{GS}, and v_{DS}, and in the direction of i_D. Consequently its operating mode definitions, given in Fig. 5.34, differ from the n-channel definitions only in the sense of each inequality. Equation (5.18) for β also applies after replacing subscripts n and D, respectively by p and A.

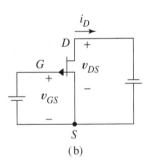

Figure 5.39 shows the large-signal circuit models for the p-channel JFET, where $i_D = f_0(v_{GS}, v_{DS})$ is Eq. (5.17) and $i_D = f_a(v_{GS})$ is Eq. (5.21).

Figure 5.37 p-Channel JFET: (a) physical structure; (b) schematic symbol.

Having described JFET physical principles, equations, large-signal models, and characteristic curves, we now examine analysis problems that arise when JFETs interact in circuits with other devices.

5.7 _____
MESFET and JFET Operating Point Analysis

JFET state definitions and equations closely resemble those of MOSFETs; consequently, we analyze JFET circuits using the same techniques we learned for MOSFETs. In hand analysis of MESFET circuits, we assume $\tanh(\alpha v_{DS}) = 1$ in Eqs. (5.14) and (5.15) and also ignore this correction in inequalities that define the transistor states. Therefore, unless there is gate conduction, MESFET hand analysis is the same as JFET hand analysis.

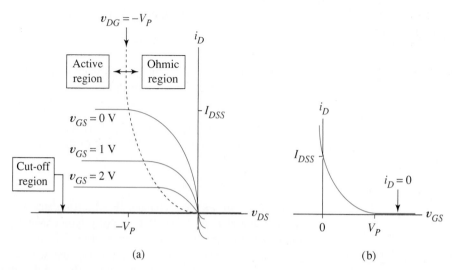

Figure 5.38 Characteristic curves for p-channel JFET: (a) output characteristics; (b) transfer characteristic.

5.7.1 ANALYSIS WHEN TRANSISTOR STATE IS KNOWN

Because JFETs are used in linear applications such as low-noise input stages for amplifiers, we use a JFET problem to illustrate analysis when we know the transistor is active.

EXAMPLE 5.15 Find the Q-point coordinates of the transistor in Fig. 5.40a.

Solution. We first draw Fig. 5.40b. Because the transistor is active,

$$I_D = 4 \times 10^{-3}(V_{GS} + 1)^2$$

For another equation involving the same unknowns, notice from the diagram that

$$V_S = 3.9 \, kI_D$$

We find V_G from the voltage divider

$$V_G = \frac{1.5}{1.5 + 3.3} \, 12 = 3.75 \text{ V}$$

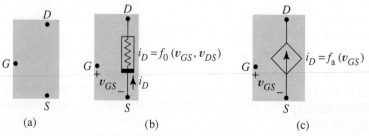

Figure 5.39 Circuit models for the p-channel JFET: (a) cut-off; (b) ohmic; (c) active.

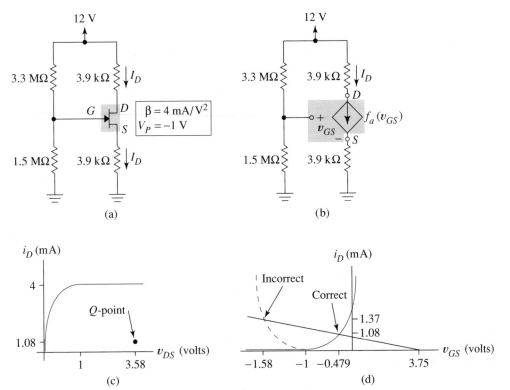

Figure 5.40 JFET amplifier circuit analysis: (a) given circuit; (b) equivalent circuit using active JFET model; (c) Q-point; (d) transfer characteristic sketch showing origin of incorrect solution.

Combining the last two equations gives

$$V_{GS} = 3.75 - 3.9\,kI_D$$

Equating the two expressions for I_D gives

$$\frac{3.75 - V_{GS}}{3.9\,k} = 4 \times 10^{-3}(V_{GS} + 1)^2$$

which is equivalent to

$$15.6\,V_{GS}^2 + 32.2V_{GS} + 11.85 = 0$$

The solutions are $V_{GS} = -1.58$ V and -0.479 V—only $V_{GS} = -0.479$ is consistent with active-state operation.

Substituting $V_{GS} = -0.479$ into either drain-current expression gives $I_D = 1.08$ mA. Subtracting the drops across the two 3.9 k resistors from the 12 V available in the drain/source circuit gives

$$V_{DS} = 12 - (2 \times 3.9\,k \times 1.08\ \text{mA}) = 3.58\ \text{V}$$

Therefore, the transistor is biased in the active region at (3.58 V, 1.08 mA), as indicated in Fig. 5.40c. In Fig. 5.40d, arrows mark the two solutions of the quadratic, showing that the incorrect solution corresponds to the left branch of the transfer function parabola that has no physical meaning.   ❏

Exercise 5.15   The p-channel transistor in Fig. 5.41a is active. Find its Q-point.

Ans. $(V_{DS}, I_D) = (-2.93 \text{ V}, 2.54 \text{ mA})$. Figure 5.41b shows where this falls on the transistor output characteristics.

5.7.2   ANALYSIS WHEN TRANSISTOR STATE IS UNKNOWN

To establish background for MESFET logic circuits in Chapter 13, the following example illustrates analysis when the transistor state is unknown,

EXAMPLE 5.16   In Fig. 5.42a, M_2 has $V_t = -1$ V and $\beta_2 = 0.8 \times 10^{-3}$ A/V². The width of M_1 is twice that of M_2, but it is otherwise identical. For simplicity, assume that $\tanh(\alpha v_{DS}) \approx 1$ for all v_{DS}. Estimate the value of V_O for
(a) $V_I = -2$ V,
(b) $V_I = 0$ V.

Solution. Because of the difference in widths, $\beta_1 = 2 \times \beta_2 = 1.6 \times 10^{-3}$ A/V². Since M_2 is a nonlinear load resistor (like the depletion MOSFET), make a rough, unscaled sketch of the load line like Fig. 5.42b.
(a) When $V_I = v_{GS1} = -2$ V, since $v_{GS1} \leq V_t$, M_1 is cut off. This places operation at point $(v_{DS1}, i_{D1}) = (2.5 \text{ V}, 0 \text{ mA})$ in Fig. 5.42b. At that point, M_2 is ohmic.
   We can formally treat ohmic operation of M_2 as an *assumption,* and then proceed mathematically. From Eq. (5.14), ignoring velocity saturation

$$i_{D2} = 0.8 \times 10^{-3}[2(0 + 1)\, v_{DS2} - v_{DS2}^2] = i_{D1} = 0$$

The obvious solution of this equation is $v_{DS2} = 0$. From Fig. 5.42a $v_{DS2} = v_{DG2}$; therefore, the inequality $v_{DG2} = 0 \leq -V_t = 1$ is satisfied. This confirms ohmic operation of M_2. We conclude that $V_O = 2.5$ V as originally suggested by the load line sketch. Notice

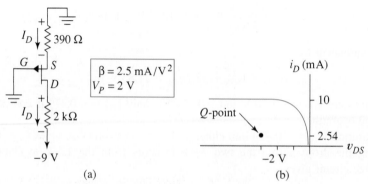

Figure 5.41   Exercise 5.15: (a) given circuit; (b) Q-point.

that the solution, $v_{DS} = 0$, that we obtained by equating drain currents is still correct when the velocity saturation factor of Eq. (5.14) is equated to zero.

(b) When $V_I = 0$, M_1 is ON, because $v_{GS1} \geq V_t$. Although Fig. 5.42b is unscaled, it suggests that for sufficiently high V_I, M_2 will be active and M_1 ohmic. Using these assumptions, we equate the active equation for i_{D2} to the ohmic equation for i_{D1}:

$$0.8 \times 10^{-3} (0 + 1)^2 = 1.6 \times 10^{-3} [2(0 + 1)V_0 - V_0^2]$$

Solving for V_O gives 0.293 V and 1.707 V. Only $V_O = 0.293$ satisfies

$$v_{DG1} = v_{DS} - v_{GS} = 0.293 - 0 \leq -V_t = 1 \text{ V}$$

which is required for M_1 to be ohmic as assumed. This value also satisfies our assumption that M_2 is active for it gives the inequality

$$v_{DG2} = (2.5 - 0.293) \geq -V_t = 1 \text{ V}$$

We conclude that $V_O = 0.293$ V. ❑

> **Exercise 5.16** Use Fig. 5.42d and a load line to find the Q-point of the transistor in Fig. 5.42c. Find the gate current. Hint: See Fig. 5.31e.
>
> *Ans.* $(V_{DS}, I_D) = (0.4 \text{ V}, 2.6 \text{ mA})$, 28 μA.

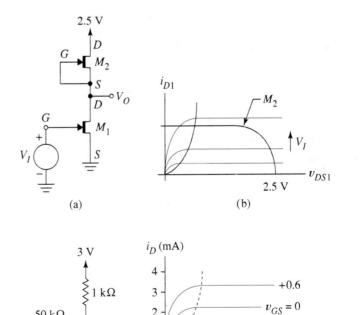

Figure 5.42 (a) Circuit for Example 5.16; (b) load line for Example 5.16; (c) circuit for Exercise 5.16; (d) transistor output characteristics for Exercise 5.16.

5.7.3 FIELD-EFFECT TRANSISTORS IN PERSPECTIVE

We have now studied seven kinds of field-effect transistors. All operate as three-state devices (the MESFET has an extra state) with similar large-signal circuit models; except for minor differences in notation, all satisfy, at least approximately, the same equations. Their state definitions are identical except for the senses of inequalities. The transfer characteristics are the key to keeping all these devices straight.

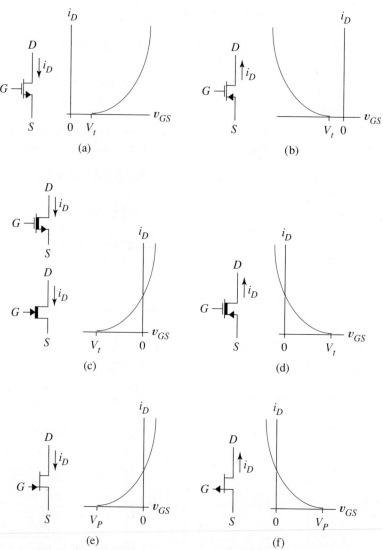

Figure 5.43   FET schematic and transfer curve: (a) n-channel enhancement MOSFET; (b) p-channel enhancement MOSFET; (c) depletion MOSFET and n-channel depletion MESFET; (d) p-channel depletion MOSFET; (e) n-channel JFET; (f) p-channel JFET.

Figure 5.43 shows the schematic symbol, current reference direction, and transfer characteristic for each FET we have studied. The n-channel devices, first column, all have *output* characteristics in the first quadrant and all satisfy the same state definitions. The p-channel devices (second column) all have second quadrant output characteristics, and their inequalities in the state definitions are reversed relative to the n-channel devices. The transfer characteristics help us visualize the sense of the ON/OFF inequality, the "v_{GS} labeling" of the output characteristics and, therefore, the output curves themselves. Transfer curves also show the range of v_{GS} for which a polynomial root has physical significance. For the reader's convenience Fig. 5.43 is reproduced at the front of the book.

5.8
Static SPICE Model for the JFET

We now introduce the SPICE JFET model, which we use in this text both for JFETs and MESFETs. Figure 5.44a shows the coding format. JFET names begin with J, and node numbers are in drain-gate-source order. An *optional* fourth node number specifies the substrate connection of an IC JFET. On the .MODEL line, the code PJF or NJF identifies the JFET as p- or n-channel.

Figure 5.44b shows SPICE notation for V_P and β. When a data sheet for a discrete JFET gives I_{DSS} we must calculate β using Eq. (5.22). SPICE assumes a value of one for W/L in Eq. (5.18). The J6 element line in Fig. 5.44a shows how to specify individual channel geometry by listing channel length and width (in meters). The basic JFET SPICE models can be used to make estimates for MESFET circuits. Newer SPICE simulators such as PSPICE employ more sophisticated MESFET models that include velocity saturation effects.

The following example demonstrates using the SPICE JFET model, and it shows how we can employ simulation to explore the simplifying assumptions we make in designing circuits.

EXAMPLE 5.17 Use SPICE to explore the linearity of the voltage-controlled amplifier designed in Example 5.14.

Solution. Figures 5.45a and b show the amplifier and SPICE code. We model the op amp as a VCVS of gain 10^5 and a 100 MΩ input resistance. SPICE steps the gain-con-

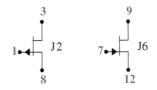

```
J2 3 1 8 ELLEN
J6 9 7 12 SID W=5U L=7U
.MODEL ELLEN PJF
.MODEL SID NJF VTO=-3 BETA=2.0E-4
```

(a)

Notation		Default value
Text	SPICE	
V_P	VTO	−2 V
β	BETA	1.0E−3 A/V^2

(b)

Figure 5.44   SPICE coding for JFETs: (a) code; (b) notation and default values.

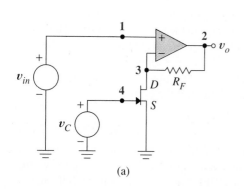

(a)

```
EXAMPLE 5.17
E1 2 0 1 3 1E5
RIN 1 3 100MEG
RF 2 3 9K
JC 3 4 0 CTRL
.MODEL CTRL NJF VTO=-3 BETA=1.0E-3
VIN 1 0 DC 0
VC 4 0 DC 0
.DC VIN -0.09 +0.09 0.018
+VC -2.5 0.0 0.5
*VC 4 0 DC -2.5
*.DC VIN -0.2 0.5 0.07
.PLOT DC V(2)
.END
```

(b)

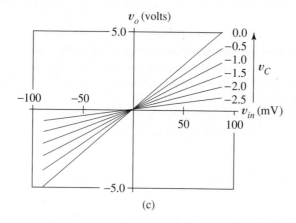

(c)

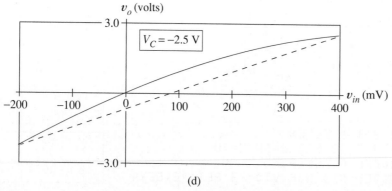

(d)

Figure 5.45 Example 5.17: (a) voltage-controlled amplifier circuit with numbered nodes; (b) SPICE code for amplifier; (c) gain curves for design values of input voltage; (d) nonlinear transfer characteristic (solid) compared with straight line.

trol source over the specified design range, -2.5 V to 0 V, in six steps of size 0.5 V, and plots the transfer characteristic for each value of VC. The VIN range of ± 90 mV for the transfer characteristics approximates the ± 88 mV range calculated in Example 5.14 for linear operation. The resulting curves of Fig. 5.45c have the specified gain variation of 9 to 55 and are quite linear. We next explore the limits of linear operation.

The inequality of Eq. (5.20) assumes the squared term in Eq. (5.17) is negligible because $v_{DS} = v_{in}$ is kept small. During the design we noticed this signal amplitude is most critical for the most negative v_{GS} values, namely, for $v_C \approx -2.5$ V, and this was the rationale for specifying the ± 88 mV range for the input signal. To explore this, the two code lines marked with asterisks in Fig.5.45b replace the two preceding lines of code for another SPICE run. This sets VC at the worst case and then allows VIN to range from -200 mV to 500 mV. The result is the solid transfer characteristic of Fig. 5.45d. Comparing it with a straight line (dashed) shows that operation is indeed nonlinear if we do not restrict input signal amplitude to values consistent with Eq. (5.20). Comparing Figs. 5.45c and d shows that the order-of-magnitude rule of thumb we used to obtain Eq. (5.20) resulted in a conservative input voltage range for linear operation. ❑

Exercise 5.17 Use SPICE to obtain a family of gain curves like Fig. 5.45c for the inverting variable-gain amplifier designed in Exercise 5.14.

Ans. See Fig. 5.46.

5.9
FET Second-Order Effects

This section discusses *second-order effects,* deviations from ideal behavior, that impose practical limitations on FET performance. Consequences of these second-order effects are that MOSFET, MESFET, and JFET characteristic curves and circuit models differ somewhat from the basic patterns studied so far.

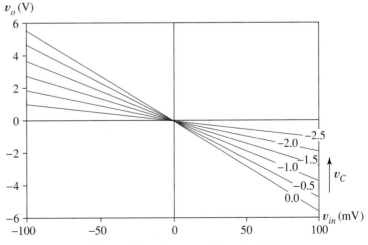

Figure 5.46 Answer to Exercise 5.17.

5.9.1 BODY EFFECT

Threshold voltage, until now treated as a constant, is actually a function of the source-to-body voltage in a MOSFET. To understand this, consider Fig. 5.47a where the n-type source is made V_{SB} volts positive relative to the p-type substrate. This reverse bias expands the depletion region around each well, uncovering additional lattice ions and lowering the electron energy of the source wells, as indicated by the dashed lines in Fig. 5.47b. Because additional energy is now required to remove an electron from the well against the forces exerted by increased numbers of ions, the energy depth of the wells increases. Threshold voltage, by definition the gate-source voltage required to reduce the surface potential under the oxide to the level of the source well (Figures 5.3a–c), clearly increases for positive V_{SB}. This increase in threshold voltage with V_{SB} is called *body effect*. By similar reasoning, the negative threshold voltage of a p-channel device becomes even more negative with negative values of V_{SB}.

In some circuits, Fig. 5.29a, for example, source and body are connected to the same point, and the threshold voltage is constant. There are many circuits, however (e.g., Figs. 5.13a, 5.15a, 5.21), in which source is not connected to body. In such cases we must include body effect to predict the circuit's performance accurately.

The theoretical expression,

$$V_t = V_{t0} + \gamma \left(\sqrt{|-2\phi_F + v_{SB}|} - \sqrt{|2\phi_F|} \right) \tag{5.24}$$

where ϕ_F is a fabrication constant and γ is called the *body-effect coefficient*, shows how threshold voltage varies with source-body voltage. Equation (5.24) expresses V_t in terms of the constant V_{t0}, threshold voltage when $v_{SB} = 0$, plus a body-effect correction term. The body-effect coefficient is positive for n-channel devices and negative for p-channel devices. A representative value for $|\gamma|$ is 0.37 (volt)$^{0.5}$.

Since threshold voltage varies as the square root of $v_{SB} = v_S - v_B$, body effect results in a circuit analysis problem that requires iterative calculations. For example, to determine V_t in Eq. (5.7), we must know v_{SB} and therefore v_S. But, as we saw in numerous examples, we need Eq. (5.7) to compute v_S! It is therefore necessary to *assume* a starting value for V_t and then repeatedly reanalyze the circuit and recalculate V_t until the process converges to a

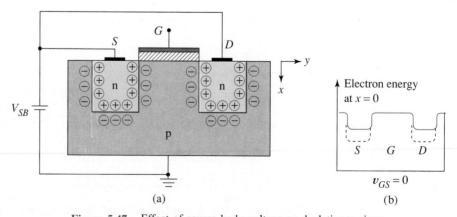

(a)   (b)

Figure 5.47 Effect of source-body voltage on depletion regions.

solution. Obviously hand calculations can be tedious when body effect is included. We therefore adopt the reasonable compromise of treating V_t as a constant in hand estimates, while developing special alertness for situations where body effect might be important. In these cases we include body effect in a computer simulation to predict actual circuit performance. In Sec. 5.11 we learn how to include body effect in SPICE models.

5.9.2 CHANNEL LENGTH MODULATION AND EARLY EFFECT

Section 4.9.3 introduced base width modulation for the BJT. The effective base width narrows with large output voltage, causing the spacing between the output characteristics to spread with v_{CE}. To describe this mathematically, an Early voltage correction factor was added to the output equation, which lead to a finite output resistance in the circuit model. In MOSFETs, MESFETs, and JFETs there is a *channel length modulation* effect, closely analogous both in concept and in overall effect to base width modulation.

Figures 5.5a and c show how reverse biasing the drain relative to the substrate causes channel pinch-off at the drain end. For large v_{DS}, the drain-end depletion region expands back into the channel as suggested by the dashed lines in Figs. 5.48a and b,

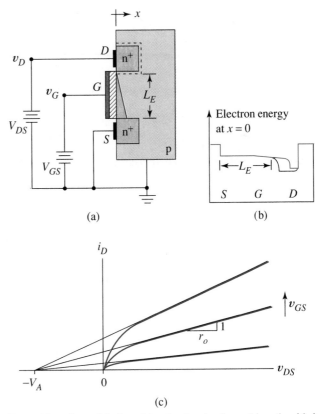

Figure 5.48   Channel length modulation: (a) reduction in channel length with increases in v_{DS}; (b) effect on electron energy diagram; (c) Early effect.

reducing the effective channel length to a lower value, L_E, than the original value L. This causes the channel resistance to decrease with increased v_{DS}. Consequently, the channel current increases, spreading the constant-v_{GS} FET curves with increased v_{DS} and adding nonzero slope as in Fig. 5.48c. A similar process occurs in MESFETs and JFETs.

The Early effect also applies to field-effect transistors. FET output curves are asymptotic to straight lines that appear to emanate from a point at $(v_{DS}, i_D) = (-V_A, 0)$, where V_A is the Early voltage indicated in Fig. 5.48c. (For FETs the Early voltage is sometimes denoted by $1/\lambda$.) The slope of the curves is the reciprocal of the output resistance, r_o, a parameter we add between drain and source in the circuit models to include Early effect in circuit analysis.

We can make Eqs. (5.4), (5.7), (5.17) and (5.21) better fit the true device curves by adding Early voltage correction factors. For example, we can describe an active MOSFET by

$$i_D = f_a(v_{GS}, v_{DS}) = \frac{k}{2}(v_{GS} - V_t)^2 \left(1 + \frac{v_{DS}}{V_A}\right) \tag{5.25}$$

which more accurately expresses the output current as a function of two variables. We then define output resistance by

$$\frac{1}{r_o} = \left.\frac{\partial i_D}{\partial v_{DS}}\right|_{Q\text{-}point}$$

A derivation like the one that produced Eq. (4.21) gives

$$r_o \approx \frac{V_A}{I_D} \tag{5.26}$$

where I_D denotes the dc drain current at the Q-point. This same expression applies to MESFETs and JFETs.

5.9.3 BREAKDOWN VOLTAGE

We saw in Sec. 4.9.2 that collector-base junctions of BJTs break down at high values of output voltage, a phenomenon that limits the useful region of operation on the output characteristics. Similar breakdown occurs in FETs. Figure 5.49a shows that in the MOSFET, breakdown occurs at the substrate-drain junction.

Figure 5.49b shows a JFET biased for active operation. For sufficiently large v_{DS} the reverse-biased gate-drain junction breaks down, and high current flows from the drain into the gate circuit. The figure also shows the effect on the output characteristics. As the node voltage on the gate is made more positive, the drain voltage required for gate-drain breakdown increases accordingly.

For both MOSFETs and JFETs, the breakdown voltage, BV, is a practical upper limit on permissible range of v_{DS} when the transistor operates as a controlled current source. Similar breakdown occurs for p-channel MOSFETs and JFETs, changing their second quadrant output characteristics accordingly.

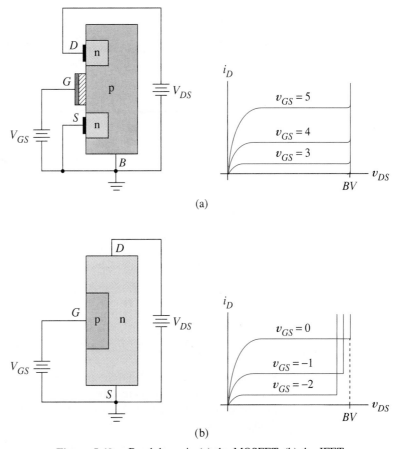

Figure 5.49   Breakdown in (a) the MOSFET, (b) the JFET.

5.9.4   VARIATIONS IN FET PARAMETERS WITH TEMPERATURE

In Sec. 3.1.2 we learned that mobility decreases with temperature because of more frequent collisions of charge carriers with the rapidly moving lattice ions. A useful empirical description of this is

$$\mu(T) = \mu(T_R)\left(\frac{T_R}{T}\right)^{1.5} \tag{5.27}$$

which relates the mobility at temperature T to its value at a reference temperature T_R, when
T and T_R are expressed in degrees Kelvin. This explains one of the temperature variations in
FETs. The mobility, μ_n, in Eqs. (5.8) and (5.18) causes k for the MOSFET and β for the
JFET to decrease with temperature; consequently, their transfer characteristics change as in
Fig. 5.50a and b, respectively. It follows that the output characteristics for both devices
crowd more closely together as in Fig. 5.50c, just the *opposite of BJT characteristics*.
Interestingly, Eq. (5.27), which affects the spacing of the MOSFET output characteris-
tics, resembles Eq. (4.19), which controls spacing of the BJT output curves.

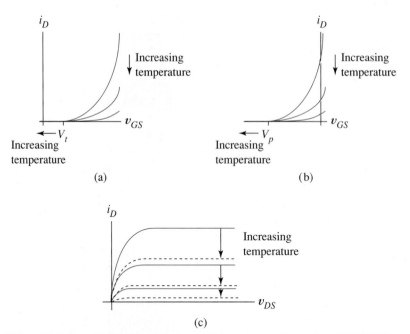

Figure 5.50   Variation of FET characteristics with temperature: (a) n-channel MOSFET transfer characteristic; (b) n-channel JFET transfer characteristic; (c) FET output characteristics.

Threshold voltages of both n- and p-channel MOSFETS *decrease* by about 2 mV/°C with increases in temperature (about the same as V_{BE} for a bipolar transistor), causing the curves of Fig. 5.50a and b to shift to the left with temperature increases as they droop downward.

Exercise 5.18   For a MOSFET, $k = 0.3 \times 10^{-4}$ A/V^2 and $V_t = 0.8$ V at 25°C. Find these parameter values at −40°C.

Ans. $k = 0.434 \times 10^{-4}$ A/V^2, 0.93 V.

EXAMPLE 5.18   The transistor in the preceding exercise is placed in the circuit of Fig. 5.51. Find its drain-source voltage at 25°C and at −40°C.

Solution. The voltage divider gives a gate voltage of 1.28 V at both temperatures. At 25°C,

$$I_D = \frac{0.3 \times 10^{-4}}{2}(1.28 - 0.8)^2 = 3.46 \ \mu A$$

giving $V_{DS} = 8 - (3.46 \ \mu A)(500 \ k) = 6.27$ V.
    At − 40°C,

$$I_D = \frac{0.43 \times 10^{-4}}{2}(1.28 - 0.93)^2 = 2.66 \ \mu A$$

making $V_{DS} = 8 - (2.66 \ \mu A)(500 \ k) = 6.67$ V.   ❑

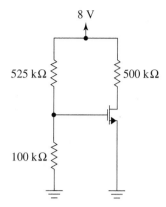

Figure 5.51   Circuit for Example 5.18

5.9.5   PARASITIC CAPACITANCES

Perhaps the most important second-order effects are those associated with the dynamic operation of FETs. Field-effect transistors, like diodes and BJTs, contain internal parasitic capacitances, which place upper limits on their operating speeds. In the next section we introduce dynamic equivalent circuits that help us understand the physical origins of these transistor capacitances, and we also describe SPICE parameters that allow us to include dynamic effects in our simulations. These very general transistor models, incidentally, include many of the other second-order effects of this section.

5.10
FET Dynamic Circuit Models

Dynamic MOSFET Model. Figure 5.52a shows a dynamic model for an n-channel MOSFET. SPICE users can implement this model using the code listed in Fig. 5.52b. The heart of the model is dependent source i_D, described by Eq. (5.3), Eq. (5.4), or Eq. (5.7), depending on the operating point. We already know that SPICE uses static parameters KP and VTO for these equations. SPICE calculations can also include corrections for body effect and Early effect through parameters GAMMA and LAMBDA, respectively. To include these corrections we simply provide nondefault values in the ".MODEL" code.

In the MOSFET of Fig. 5.52a there are depletion capacitances C_{BS} and C_{BD} between the substrate and the reverse-biased n-type wells, which we see in the circuit model of Fig. 5.52a. To describe these capacitances, SPICE uses zero-bias capacitances CBD and CBS, grading coefficient MJ and bulk junction potential PB. The SPICE model also includes the exponential volt-ampere behavior of the body-drain and body-source junctions. These junctions, characterized by reverse saturation current IS, *must be kept reverse biased in the simulation (as in the physical device) by connecting the substrate to the most negative point in the circuit.* A connection error that forward biases the body leads to excessive currents in a simulation just as it does in the physical device.

Figure 5.2a also suggests capacitance, C_{GB}, between gate and substrate, with the oxide for its dielectric. It turns out that in places where the gate slightly overlaps the drain and source materials there are additional capacitances C_{GD} and C_{GS}. The model of Fig. 5.52a includes all three. SPICE computes values for C_{GD}, C_{GS}, and C_{GB} when we

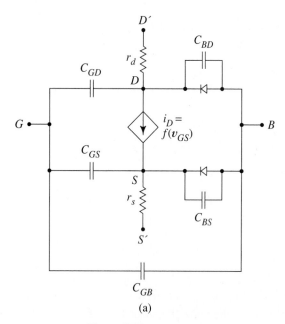

MOSFET SPICE parameters		
Parameter	SPICE name	Default
k	KP	2.0E–5 A/V^2
V_t	VTO	0.0 V
γ	GAMMA	0.0 V$^{0.5}$
$1/V_A$	LAMBDA	0.0 V^{-1}
C_{BD}	CBD	0.0 F
C_{BS}	CBS	0.0 F
m	MJ	0.5
ϕ_o	PB	0.8 V
t	TOX	∞
I_s	IS	1.0E–14 A
r_d	RD	0.0 Ω
r_s	RS	0.0 Ω

(a) (b)

Figure 5.52   n-channel MOSFET: (a) circuit model; (b) SPICE parameters.

specify a value for the oxide thickness, TOX, on the .MODEL line and when W and L are specified on the element lines of the transistors. TOX = 0.1 μ, W = 1 μ, and L = 1 μ suffice as rough estimates when actual values are unknown.

Resistors R_d and R_s represent the voltage drops created by current flowing through the wells to the *external contacts* D' and S'. SPICE includes temperature variations in threshold voltage and pn junction parameters; however variations in mobility with temperature are absent in basic versions of SPICE.

The same equivalent circuit and parameter set describes p-channel transistors if the reference direction of the current source and the diode orientations are reversed. The user needs only to specify "PMOS" on the .MODEL line, and SPICE takes care of these details.

Figure 5.53 shows a circuit model for an n-channel JFET and the corresponding SPICE parameters. Current source i_D describes the static output characteristics of the device. A nonzero value for LAMBDA introduces Early effect into the model.

The diodes add to the model the reverse saturation current I_s that flows in the input circuit of the physical device. If the gate is not reverse biased in a simulation, these diodes conduct as suggested in Fig. 5.32c, an error condition for a JFET.

Since the gate is usually reverse biased relative to the channel in the MESFET and JFET (Figs. 5.30a and, 5.32a), these devices have nonlinear depletion capacitances C_{GD} and C_{GS}. Additional depletion capacitances C_{BS} and C_{BD} connect source and drain to the substrate in IC structures. Figure 5.53 explicitly shows the depletion capacitances between gate and drain. We describe these for SPICE with the usual zero-bias values and barrier potential PB. The grading coefficient is fixed at m = 0.5. Ohmic resistances R_d and R_s complete the model.

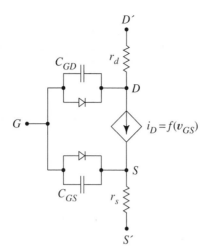

JFET SPICE parameters		
Parameter	**SPICE name**	**Default**
β	BETA	1.0E–4 A/V^2
V_P	VTO	0.0 V
$1/V_A$	LAMBDA	0.0 V^{-1}
C_{GD}	CGD	0.0 F
C_{GS}	CGS	0.0 F
ϕ_o	PB	1.0 V
I_s	IS	1.0E–14 A
r_d	RD	0.0 Ω
r_s	RS	0.0 Ω

Figure 5.53   n-Channel JFET SPICE model and SPICE parameters.

The effects of all the nonlinear transistor capacitances are very difficult to predict with hand analysis. The SPICE dynamic transistor models place accurate *numerical* results at our disposal, that show the cumulative effects of the parasitic capacitances on circuit performance. In Chapter 8 we employ linear approximations to improve our understanding, to make estimates, and to develop design equations that relate circuit performance to particular capacitances.

By including the parasitic capacitances and second-order static effects, SPICE provides a convenient tool for exploring some aspects of real device behavior otherwise too difficult to calculate. Some of the examples in the remainder of the chapter show this.

EXAMPLE 5.19   Use SPICE to see how the body effect of M_1 modifies the transfer characteristics of the circuit of Fig. 5.54a over the input range $0 \leq v_i \leq 12$ V.

Solution. Make two SPICE runs: one using the default value $\gamma = 0$, the second using $\gamma = 0.37$. Figure 5.54b shows the code, which excludes the body effect coefficient from the first run by labeling it as a comment.

Figure 5.54c compares the two transfer curves. Because $v_o = v_{SB}$ for M_1, theory predicts that body effect should be most pronounced for large values of v_o and should disappear as v_o approaches zero. The curves indeed appear to merge into a single line for low v_o. As v_o increases, the separation increases. For low v_i, however, M_2 gradually turns off, causing the curves to come together at $v_o = 12$ V in spite of body effect.

Figure 5.54c shows that when this circuit is used as an amplifier, body effect reduces the slope (gain) in the high-gain region of the curve. This same circuit is also a digital logic gate we discuss in Chapter 13. There, we learn that this gain reduction makes the gate more susceptible to noise.   ❏

Exercise 5.19   Examine the effect of channel length modulation on the circuit of Fig. 5.54a if $V_A = 50$ V. Use $\gamma = 0.37$.

Ans. Figure 5.55.

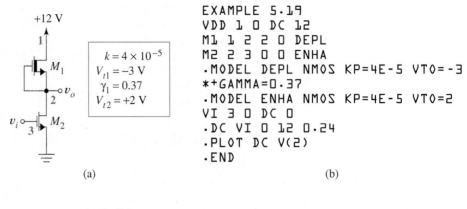

(a)

```
EXAMPLE 5.19
VDD 1 0 DC 12
M1 1 2 2 0 DEPL
M2 2 3 0 0 ENHA
.MODEL DEPL NMOS KP=4E-5 VTO=-3
*+GAMMA=0.37
.MODEL ENHA NMOS KP=4E-5 VTO=2
VI 3 0 DC 0
.DC VI 0 12 0.24
.PLOT DC V(2)
.END
```

(b)

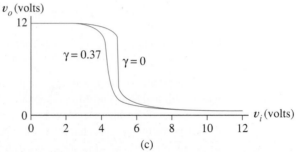

(c)

Figure 5.54   Example 5.19: (a) circuit; (b) SPICE code; (c) transfer characteristic showing consequence of body effect.

The preceding example and exercise used SPICE to explore second-order *static* FET effects. We demonstrate use of SPICE to explore *dynamic* behavior after introducing the FET-controlled switch.

5.11
FET Switch

The *FET switch* plays a central role in modern electronics. In digital circuits, its two states represent the ones and zeros required to store or transmit information encoded as binary numbers. In interface circuits, the controlled switch enables computers to control machines, lights, alarms, and other devices. In analog communications and signal proc-

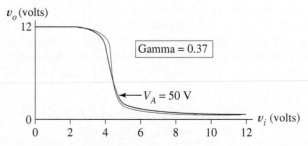

Figure 5.55   Effect of channel length modulation on circuit of Figure 5.54a.

essing, switches rapidly turn signals on and off in order to translate their information to other frequency ranges. This makes it possible to combine different signals for efficient use of communication channels and enables information to be processed by IC filters. Special switched capacitor filters, used for analog signal processing, employ switches as replacements for large resistors not available in ICs. Switches are also used in power supplies to produce drops in dc voltage without consuming large amounts of power, and to improve efficiency in power amplifiers.

5.11.1 GROUNDED SWITCH

Like the BJT switch, the FET switch exploits the similarity between two selected transistor output curves and ideal switch characteristics. MESFET and JFET switches operate in the same fashion, differing only in the required control voltages.

Figure 5.56a shows a grounded MOS transistor switch, and Fig. 5.56b the load line. Control voltage v_C takes on only two values, a subthreshold value to open the switch, and some high, superthreshold value to close the switch. Notice that the closed FET switch resembles the volt-ampere curve of a resistor, not a voltage source like the BJT. For a small voltage drop across the switch, the control voltage must be high enough that the switch resistance is much lower than R_L. FET resistor equations such as Eq. (5.5) are useful in designing switching circuits.

The transistor also switches ac voltages. Replacing V_{DD} in Fig. 5.56a by a sinusoidal source, for example, causes the load line of Fig. 5.56b to slide right and left with time, intersecting the selected output curve alternately in the first and third quadrants, thus producing a sinusoidal current through switch and load resistor when the switch is closed.

Transistor switches with one end grounded as in Fig. 5.56a are common in logic and in digital interfacing circuits. Many analog applications, such as modulators, analog multiplexers, switched-capacitor filters, and analog-to-digital conversion, require an ungrounded switch called a *bidirectional transmission gate*. Both JFETs and MOSFETs serve this purpose. We next describe the MOSFET transmission gate in some detail.

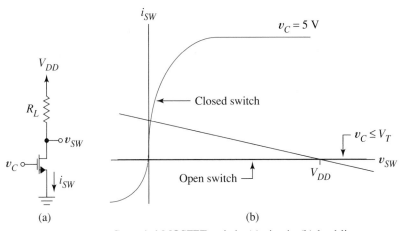

(a)   (b)

Figure 5.56   Grounded MOSFET switch: (a) circuit; (b) load line.

5.11.2 BIDIRECTIONAL TRANSMISSION GATE

The *transmission gate* is a voltage-controlled switch that has the configuration of Fig. 5.57a. For one value of v_C the switch opens, disconnecting signal v_{in} from load R_L. For a second value of v_C the switch closes, connecting v_{in} to R_L. Figure 5.57b shows how to use an NMOS (n-channel MOS) transistor as a transmission gate. Assume v_{in} is restricted to a range, $-V_{DD} \leq v_{in} \leq V_{DD}$, which is established by dual power supplies, where V_{DD} must exceed the transistor threshold voltage. As usual, the substrate is connected to the most negative point in the circuit. This FET application is especially interesting because *physical locations of the source and drain nodes change* with the direction of the switch current.

The switch in Fig. 5.57b is open when $v_C = -V_{DD}$. Although it is not yet clear whether x or y is the transistor's source node, we know the transistor is cut off because the gate cannot be more positive than *either* node when $v_C = -V_{DD}$. To close the switch, we make $v_C = +V_{DD}$. To understand the details of ON operation, we examine two cases.

Case 1. $v_C = V_{DD}$ and $v_{in} < 0$. For negative v_{in}, conventional current flows through the MOSFET from node y to node x as in Fig. 5.57c. This means node x is the *source of*

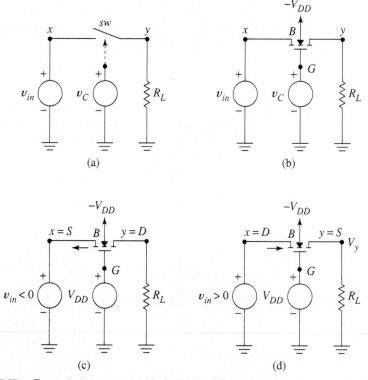

Figure 5.57 Transmission gates: (a) ideal; (b) NMOS gate circuit; (c) source and drain for negative input voltage; (d) source and drain for positive input voltage.

the *electron current* (in this n-channel device) and node y is the drain. A channel exists provided that

$$v_{GS} = V_{DD} - v_{in} \geq V_t$$

Since this inequality is satisfied *for all negative values of* v_{in}, the transistor is ON and the switch is closed. The next case brings a surprise.

Case 2. $v_C = V_{DD}$ and $v_{in} > 0$. For positive v_{in}, conventional current flows from x to y, as in Fig. 5.57d causing y to function as the transistor's source and x as its drain. For a channel to exist

$$v_{GS} = V_{DD} - V_y \geq V_t$$

where V_y takes on *positive* values. But this implies that there will be a channel *only* for $V_y \leq V_{DD} - V_t$. For a closed switch, $v_{in} \approx V_y$.

Combining Cases 1 and 2, we conclude that the NMOS switch is asymmetric; it provides a conducting channel between input and output only for v_{in} in the range

$$-V_{DD} \leq v_{in} \leq V_{DD} - V_t. \tag{5.28}$$

The NMOS transmission gate works fine provided we restrict input voltages to the proper range.

A practical consideration in using a transmission gate is the internal resistance of the switch. When conducting, the MOSFET switch is, of course, either active or ohmic, depending upon the polarity and magnitude of v_{in} for a given load resistor. We could examine every possibility by writing equations, but this would contribute little to general understanding. Sometimes switch resistance is handled by simply specifying a worst case linear resistor to approximate the switch resistance. This approach is clean and simple and leaves us with the useful intuitive idea that R_L must be large compared to this internal resistance if the transmission gate is to work properly. The following example uses SPICE to verify the asymmetric operation of the NMOS transmission gate and also helps us develop some physical feeling for the internal resistance of the MOSFET.

EXAMPLE 5.20 Use SPICE to study the transmission gate of Fig. 5.58a for $R_L = 2$ kΩ and for $R_L = 50$ kΩ. The MOSFET parameters are $V_t = 3$ V, $k = 2 \times 10^{-4}$ A/V^2, $\gamma = 0$, $V_A = \infty$, input voltages are restricted to ± 10 V.

Solution. Let v_{in} be a 5 kHz sine wave of 10 V peak value, and v_C a 10 V square wave with a 40 ms period. The sinusoidal input demonstrates switch operation for both positive and negative values of v_{in}, that is for Cases 1 *and* 2 discussed earlier. Since the period of the sine wave is 200 ms, the square wave opens and closes the transmission gate five times during each input cycle, enabling us to observe both open and closed switch conditions.

Although source and drain locations change with current direction, SPICE input code requires specific node assignments. The code of Fig. 5.58b arbitrarily specifies node 1 as drain and node 3 as source.

From the theory, we expect an asymmetrical output voltage, confined to the range -10 V $\leq V(3) \leq 7$ V. Figure 5.58c shows the SPICE output for a 50 kΩ load. The

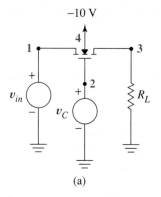

(a)

```
EXAMPLE 5.20
*OUTPUT WAVEFORMS
VDM 4 0 DC -10
MN 1 2 3 4 AL
.MODEL AL NMOS VTO=3 KP=2E-4
RL 3 0 50K
VC 2 0 PULSE(-10 10 0 0 0 0.02M 0.04M)
VIN 1 0 SIN(0 10 5000)
.TRAN 0.0003M 0.4M
.PLOT TRAN V(3)
.END
```

(b)

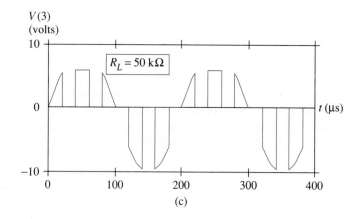

(c)

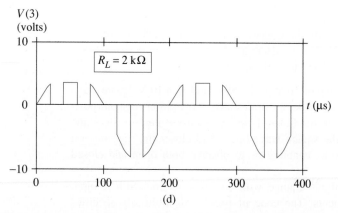

(d)

Figure 5.58 Transmission gate for Example 5.20: (a) given circuit; (b) SPICE code; (c) output for R_L = 50 kΩ; (d) output for R_L = 2 kΩ.

switch gives an excellent approximation to an open circuit whenever the control voltage swings to -10 V. As suggested by theory, the waveform has a maximum value of about 7 V and a minimum of -10 V.

Figure 5.58d shows that when we reduce R_L to 2 kΩ, the amplitude of the output waveform is smaller than expected. This suggests that 2 kΩ is not sufficiently large compared to the internal switch resistance. Intuitively speaking, there is a voltage division involving switch resistance that causes the output to have a lower amplitude than the input for the "switch-closed" case. ❏

Exercise 5.20   Use SPICE to plot the transfer function of the closed transmission gate, one plot for $R_L = 2$ kΩ and another for $R_L = 50$ kΩ.

Ans. Figure 5.59.

The curve for the 50 kΩ load in Fig. 5.59 is linear over most of the range from $v_{in} = -10$ V to $v_{in} = +7$ V, and has *slope close to one*—until the FET switch closes. This suggests that 50 kΩ is so large compared with the switch resistance that the voltage division was negligible. On the other hand, the curve for the 2 kΩ load has initial slope less than one and some significant curvature; just what we would expect from a voltage divider that includes a nonlinear switch resistor.

5.11.3   PMOS TRANSMISSION GATE

Operation of the PMOS (p-channel MOS) transmission gate of Fig. 5.60a is complementary to the n-channel gate. For an open switch, $v_C = V_{DD}$; to close the switch $v_C = -V_{DD}$. We can verify, using the kind of reasoning applied to the NMOS switch, that the p-channel FET functions as a closed switch only for

$$-V_{DD} - V_t \le v_{in} \le V_{DD} \tag{5.29}$$

where V_t is negative.

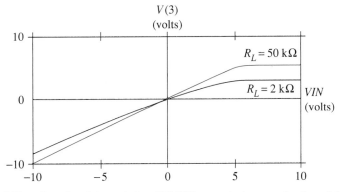

Figure 5.59   Transfer characteristic of NMOS transmission gate for $R_L = 2$ kΩ and $R_L = 50$ kΩ.

5.11.4 CMOS TRANSMISSION GATE

In some applications we need a *symmetric,* bilateral switch that will process any signal in the range $-V_{DD} \leq v_{in} \leq V_{DD}$. The complementary symmetry MOS (CMOS) gate, shown with its schematic symbol in Fig. 5.60b, satisfies this requirement. This CMOS transmission gate is simply a complementary MOS transistor pair connected in parallel. The overbar on the PMOS gate control input implies a complementary control signal. That is, when $v_C = V_{DD}$, then $\overline{v_C} = -V_{DD}$, and conversely. Thus both transistors are open circuits when $v_C = -V_{DD}$. When $v_C = V_{DD}$, we have a superposition of the n- and p-MOS switching described earlier; thus at least one transistor conducts current for either signal polarity. For most values of v_{in}, both channels conduct, providing a lower equivalent switch resistance.

The next example uses SPICE to compare the output waveform and transfer characteristics of the CMOS gate with the NMOS gate of the preceding example and exercise. It also introduces transmission gate's dynamic limitations.

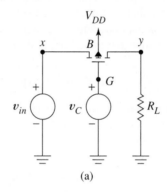

(a)

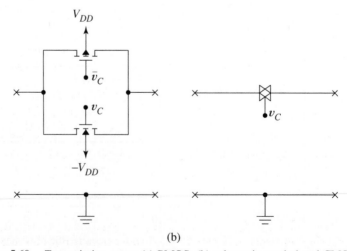

(b)

Figure 5.60 Transmission gates: (a) PMOS; (b) schematic symbol and CMOS gate.

EXAMPLE 5.21 In the NMOS gate of Fig. 5.61a, the n-channel transistor parameters are $V_t = 3$ V and $k = 2 \times 10^{-4}$ A/V², values identical to those in Example 5.20. The *complementary* p-channel transistor has $V_t = -3$ V and $k = 2 \times 10^{-4}$ A/V². The input and gate control signals are also the same as in Example 5.20.
(a) Use SPICE to find the output waveform for the given circuit, the transfer function when the gate is closed, and an estimate of the switch resistance.
(b) Determine the effect of MOSFET capacitances $CBD = CBS = 2$ FF (femtofarad) and resistances $RD = RS = 1\ \Omega$ on the output waveforms.

Solution. (a) Figure 5.61b shows the SPICE circuit and code. The VCVS, E1, with gain of -1 provides complementary excitation to the gate of the p-channel transistor. We arbitrarily define nodes 1 and 3 as drain and source for both transistors, relying on SPICE numerical algorithms to compute the details of source and drain function correctly in the simulation.

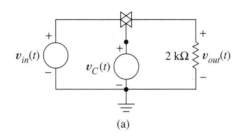

(a)

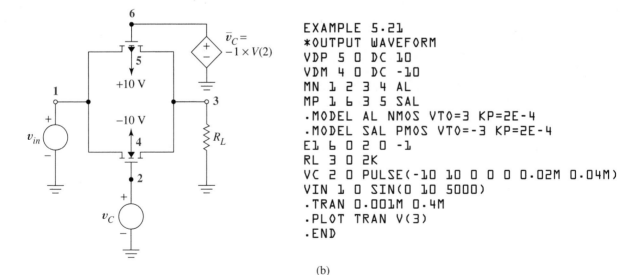

```
EXAMPLE 5.21
*OUTPUT WAVEFORM
VDP 5 0 DC 10
VDM 4 0 DC -10
MN 1 2 3 4 AL
MP 1 6 3 5 SAL
.MODEL AL NMOS VTO=3 KP=2E-4
.MODEL SAL PMOS VTO=-3 KP=2E-4
E1 6 0 2 0 -1
RL 3 0 2K
VC 2 0 PULSE(-10 10 0 0 0 0.02M 0.04M)
VIN 1 0 SIN(0 10 5000)
.TRAN 0.001M 0.4M
.PLOT TRAN V(3)
.END
```

(b)

Figure 5.61 CMOS transmission gate for Example 5.21: (a) schematic; (b) SPICE circuit and coding; (c) output waveform; (d) static transfer characteristic; (e) code for analysis of circuit dynamics; (f) output waveform for 100 MHz input. *(continued)*

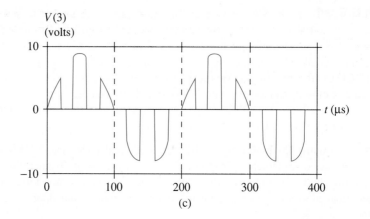

(c)

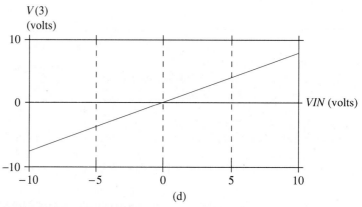

(d)

Figure 5.61 (continued)

Figure 5.61c shows an output waveform that is symmetric for the full range of input voltages, both positive and negative. Also, the transfer curve of Fig. 5.61d is symmetric and linear over the entire range of input voltage.

SPICE numerical data show that the transfer curve passes through the points (VIN, V(3)) = (−10, −8.673) and (+10, +8.673). From these values, we estimate the circuit gain to be

$$\text{Gain} = \frac{\Delta v_{out}}{\Delta v_{in}} = \frac{8.67}{10.0} = 0.867$$

Visualizing the closed switch as a linear resistor, R_{SW}, we attribute this gain to a voltage divider. That is,

$$0.867 = \frac{2\,\text{k}}{2\,\text{k} + R_{SW}}$$

Solving for the switch resistance gives $R_{SW} = 307\,\Omega$.

```
.MODEL AL NMOS VTO=3 KP=2E-4 CBD=20FF CBS=20FF RD=1 RS=1
.MODEL SAL PMOS VTO=-3 KP=2E-4 CBD=20FF CBS=20FF RD=1 RS=1
VC 2 0 PULSE(-10 10 0 0 0 1.0E-9 2.0E-9)
VIN 1 0 SIN(0 10 100MEG)
.TRAN 0.01E-8 2E-8
```

(e)

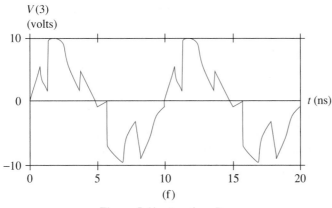

(f)

Figure 5.61 (continued)

(b) Figure 5.61e shows the changes in SPICE code used to explore switch dynamics. These mainly involve adding dynamic parameters to the transistor .MODEL statements.

In a series of SPICE runs, the frequency of the input sine wave was increased, while the period of the square wave was decreased so as to always give five switch cycles for each sine-wave cycle. Because the zero-voltage capacitances are so small, the output waveform is ideal over a wide frequency range. For a 100 MHz input, however, the output is the badly distorted waveform of Fig. 5.61f. From studying the waveform, we learn that the parasitic capacitors delay the switch opening for so long that the next "close" command arrives before the switch has fully opened, and the next "open" command comes before the switch has fully closed. Clearly, the one nanosecond intervals for open and closed switch in this simulation were too short for the transistor dynamics. The data sheet for the popular CA4096 CMOS transmission gate specifies a switch *turn off time* of 9 ns, which coincidentally appears to be a reasonable specification for the gate used in this simulation. ❏

5.12
Summary

Like bipolar transistors, field-effect transistors are three-state devices that serve as *dependent sources* in analog applications, and as *controlled switches* in digital circuits.

In one kind of FET, the control gate is insulated from the conducting channel; another FET class has a gate that is in physical contact with the channel but electrically separated by a reverse biased pn or Schottky junction.

All n-channel FETs share the same state definitions, equations, and circuit models, except for minor notational differences. Mathematically, the transfer characteristics of all n-channel FETs are the right-hand branches of parabolas, as in the figure inside the front cover; they differ only in the algebraic sign of the threshold or pinch-off voltage. Although FET equations differ greatly from BJT equations, there is one important similarity. The width-to-length ratio of an FET plays the same role as the junction area of a BJT—a current scale factor for otherwise identical transistors that can be precisely controlled during IC fabrication.

State definitions for the p-channel FETs differ from the definitions for n-channel devices only in the senses of the inequalities. Equations for p-channel devices are identical to those for n-channel devices; however, output characteristics plot in the second quadrant instead of the first. The circuit models differ only in the reference directions of the drain currents. The figure inside the front cover shows that transfer characteristics for all p-channel devices are left branches of parabolas; individual differences result from differing algebraic signs of threshold or pinch-off voltages.

In FET circuits we encounter the same general classes of problems we learned to solve for BJT circuits: finding Q-points when transistor states are known, and finding Q-points when the states are unknown. The general tools and procedures for solving these problems—load lines, equivalent circuits, and guessing and verifying states—are those we learned for BJT circuits; however, the details differ. The infinite input resistance of the FET tends to simplify circuit analysis; however, FET circuits usually require us to solve a quadratic equation and select the solution that has physical meaning.

Several novel ideas were introduced in this chapter. One is that the FET can operate as a voltage-controlled linear resistor. Another is that with gate connected to source, the depletion MOSFET and the JFET become constant-current sources, provided only that we maintain sufficient voltage across their terminals. We also learned that both depletion and enhancement MOSFETs sometimes function as nonlinear load resistors of large value that occupy minimal chip space in integrated circuits. By generalizing our linear load-line ideas, we learned how to sketch the nonlinear load lines associated with FET resistors.

FET second-order effects resemble those of bipolar transistors. For example, the field-effect transistor also has a breakdown voltage associated with a reverse-biased junction that places an upper limit on the useful active region of its output characteristics. The effect of channel length modulation on FET output characteristics is the same as that of base width modulation on BJTs—both lead to an Early effect, similar Early-effect correction factors for the output current equations, and output resistances described by analogous equations. The threshold voltage of a MOSFET changes by -2 mV/°C as does V_T for the BJT; however, the separation of the FET output characteristics decreases instead of increasing as in the bipolar transistor. Internal parasitic capacitances limit the FET's ability to operate at high frequencies. Some are depletion capacitances associated with reverse-biased junctions like those we saw in BJTs; others are "linear" capacitances associated with the insulated gates. Because there are no minority carriers, FETs lack the large diffusion capacitances associated with stored charge in BJTs. The FET has a body effect that causes threshold voltage to increase with source-body voltage. SPICE models can simulate the second-order nonlinearities of the static transistor models, and also include most second-order effects of FETs.

Because of high input impedance, and output characteristics that resemble resistor volt-ampere curves in both first and third quadrants, FETs make excellent switches. Bidirectional transmission gates constructed from FETs are widely used in both digital and linear applications.

The most important things to remember about FETs are their output and transfer characteristics, the state definitions of Figs. 5.6a and b, Equations (5.3), (5.4), (5.7), (5.8), and (5.26), and the load line concepts, including load lines for the nonlinear FET resistors sketched in Figs. 5.19d and e.

REFERENCES _____

1. ALVAREZ, A. R., ed. *BiCMOS Technology and Applications.* Boston: Kluwer Academic, 1989.
2. ANTOGNETTI, P., and G. MASSOBRIO, eds. *Semiconductor Device Modeling with SPICE.* New York: McGraw-Hill, 1988.
3. BANZHAF, W. *Computer-Aided Circuit Analysis Using SPICE.* Englewood Cliffs, NJ: Prentice Hall, 1989.
4. HODGES, D. A., and H. C. JACKSON. *Analysis and Design of Digital Integrated Circuits,* 2nd ed. New York: McGraw-Hill, 1988.

5. MEYER, J. E. *MOS Models and Circuit Simulation,* Vol. 32. RCA Rev., Somerville, NJ: (Publisher), 1971.
6. SEDRA, A. S., and K. C. SMITH. *Microelectronic Circuits,* 3rd ed., Philadelphia: Saunders College, 1993.
7. TUINENGA, P. W. *SPICE A Guide to Circuit Simulation & Analysis Using PSPICE,* Englewood Cliffs, NJ: Prentice Hall, 1988.

PROBLEMS _____

SPECIAL DIRECTIONS FOR SPICE HOMEWORK PROBLEMS: *Do not hand in lengthy SPICE printout for homework.* Instead, abstract the useful information from the SPICE output file as in the SPICE examples in the text. Include your SPICE code and a circuit diagram with nodes numbered to agree with the code. Cite relevant numerical values from the SPICE output file and discuss when appropriate. Make sketches of any relevant curves, and label appropriate points. Make small tables to present numerical data if useful for clarity.

Section 5.1

5.1 Use the MOSFET transfer and output characteristic curves of Fig. P5.1 to estimate
(a) V_t and k.
(b) i_D when $v_{GS} = 5.5$ V and $v_{DS} = 8$ V.
(c) i_D when $v_{GS} = 5.5$ V and $v_{DS} = 1$ V. (Recall that the transfer characteristic assumes *active* operation for $v_{GS} \geq V_t$.)

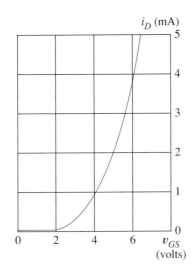
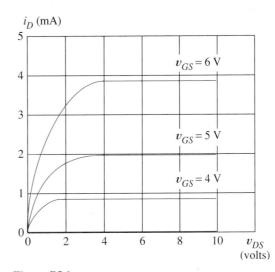

Figure P5.1

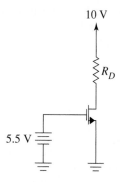

Figure P5.2

5.2 The MOSFET of Fig. P5.2 is described by the characteristic curves of Fig. P5.1. Use a load line to find (V_{DS}, I_D) when $R_D =$
(a) 5 kΩ.
(b) 2 kΩ.
(c) 1 kΩ.
Hint: What is the load line's slope?

5.3 Transfer characteristics usually apply only to an active transistor. In this problem we depart from this convention. For an FET with $k = 2 \times 10^{-2}$ A/V^2 and $V_t = 2$ V,
(a) begin with the ohmic-region equation. From this, derive a transfer characteristic (output current versus input voltage) that describes the transistor when $v_{DS} = 1$ V. Over what range of v_{GS} does this equation apply?
(b) For the same transistor, again start with the ohmic-region equation. Derive a transfer characteristic when the output is constrained by $v_{DS} = 2$ V. Over what range of v_{GS} does this equation apply?

5.4 An n-channel MOSFET has $k = 1$ mA/V^2 and $V_t = 1$ V. Find the state of the transistor and the drain current when
(a) $V_G = -2$ V, $V_D = 3$ V, $V_S = -2$ V.
(b) $V_G = -2$ V, $V_D = 2$ V, $V_S = -5$ V.
(c) $V_G = 3$ V, $V_D = 1$ V, $V_S = 1$ V.
(d) $V_G = 0$ V, $V_D = 0$ V, $V_S = -2$ V.

5.5 An n-channel MOSFET has $k = 1$ mA/V^2 and $V_t = 0.5$ V. Write the equation for i_D and sketch the circuit model when the transistor is
(a) active.
(b) in cut-off.
(c) ohmic.

5.6 (a) Sketch the transfer and output characteristics for an n-channel MOSFET having $V_t = 3$ V and $k = 1$ mA/V^2.
(b) Repeat part (a) for a MOSFET half as wide and twice as long.

5.7 A depletion MOSFET has $k = 0.05$ mA/V^2 and $V_t = -3$ V.
(a) For each state, write the drain-current equation and the inequalities that must be satisfied for its validity.
(b) Sketch the transfer and output characteristics.

5.8 Sketch i_D versus v_{DS} for the MOSFET described in Problem 5.7 when the gate is attached to the source.

5.9 An n-channel, depletion MOSFET has threshold voltage of -0.8 V. Find the state of the transistor for
(a) $V_G = -2$ V, $V_D = 3$ V, $V_S = -2$ V.
(b) $V_G = -2$ V, $V_D = -2.6$ V, $V_S = -5$ V.
(c) $V_G = 3$ V, $V_D = 3.8$ V, $V_S = 3.5$ V.
(d) $V_G = 5$ V, $V_D = 0$ V, $V_S = 6$ V.

Section 5.2

5.10 Find the transistor Q-point in P5.10.

5.11 In Fig. 5.11a, when we replace the ground symbols by -20 V and the 20 V source by ground, the transistor remains active.
(a) Redraw the circuit to fit this description.
(b) Find the transistor Q-point for this new circuit and compare it with the Q-point for the original circuit in Example 5.3.

5.12 In Fig. 5.11a, when we replace the ground symbols by -10 V and the 20 V source by 10 V, the transistor remains active.
(a) Redraw the circuit to fit this description.
(b) Find the transistor Q-point for this new circuit and compare it with the Q-point for the original circuit in Example 5.3.

5.13 Compute the Q-point for the transistor in Fig. 5.11a if the supply voltage is changed to 25 V.

5.14 (a) Show on an appropriate coordinate system the Q-point of each transistor in Fig. 5.13a, as determined in Example 5.5.

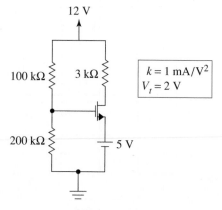

Figure P5.10

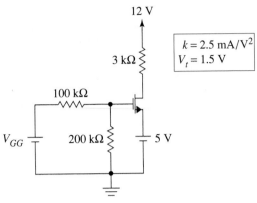

Figure P5.15

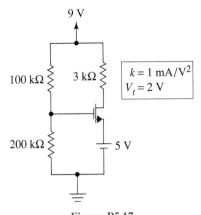

Figure P5.17

(b) Now modify the original circuit diagram by using batteries to raise each gate to a potential 2 V above ground. Find the new Q-points and show them for comparison on the diagram of part (a).
(c) Describe in words what happens to the voltage of the current source in Fig. 5.13a when the voltages of the two gates are simultaneously raised above ground potential.
(d) Describe in words what happens to the drain currents and drain-source voltages of the paired enhancement transistors when the voltages of the two gates are simultaneously raised above ground potential.

5.15 (a) Find the minimum value of V_{GG} required to turn on the transistor of Fig. P5.15.
Hint: Assume the transistor is cut off and draw the equivalent circuit.
(b) What will be the state of the transistor just as it turns on, active, or ohmic? Explain your reasoning.

5.16 Find the Q-point of the transistor of Fig. P5.16.

5.17 Find the transistor Q-point in Fig. P5.17.

5.18 The MOSFET in Fig. P5.18 is active. Its parameters are $k = 2\text{mA/V}^2$ and $V_t = 1$ V.
(a) Find the voltage across the current source.
(b) Find the numerical value of V_{DS}.

Section 5.3

5.19 Copy Fig. 5.20b. Add to the sketch two other load lines to show what happens to the original Q-points if load transistors of higher and lower W/L ratio are used. To establish a Q-point very close to the origin for given V_{GS}, should W/L be large or small?

5.20 In P5.20 both transistors have $k = 0.4 \times 10^{-3}$ A/V^2. For the load transistor $V_t = -2$ V, for M_1, $V_t = 2$ V.
(a) Sketch output characteristics of M_1 for $v_i = 0$ V, 3 V, 4 V. Your sketch need not be accurate in the ohmic region but should be correct in the active region.

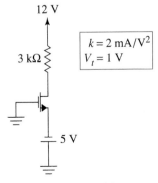

Figure P5.16

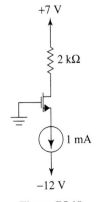

Figure P5.18

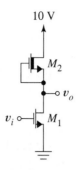

Figure P5.20

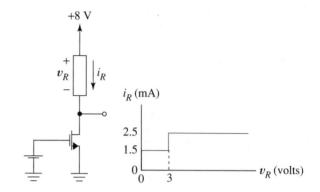

(b) Superimpose upon the characteristics of part (a) the nonlinear load line.

(c) For different values of v_i between 0 and 4 V, find on the load line the Q-point of M_1. Use this information to make a sketch of the transfer function v_o versus v_i, noting that these are the drain-source and gate-source voltages, respectively, for M_1.

5.21 Figure P5.21 shows two nonlinear MOS resistors connected in series. $V_{t1} = 1$ V, $V_{t2} = -1$ V.

(a) Use a load line to show how V_X changes as V_{BB} changes from 0 to V_{t1}.

Hint: First sketch volt-ampere curves for M_1 alone as function of V_{BB}. Then add to the diagram the load line constraint imposed by M_2.

(b) Redraw your sketch to show how the results change when W/L of M_1 is reduced.

5.22 Figure P5.22 shows a transistor constrained by a nonlinear resistor. Draw the load line on the transistor's output characteristics, and use this load line to find the transistor Q-points for $v_{GS} = 0$ V, 2 V, 4 V, and 6 V. Make a rough sketch to hand in showing how you obtained your answers.

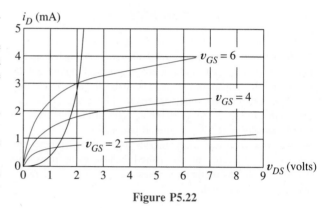

Figure P5.22

5.23 Solve Problem 5.22, but use the nonlinear resistor with volt-ampere characteristic of Fig. P5.23.

5.24 Use two n-channel MOS enhancement transistors of $V_t = 1$ V in a voltage divider design that operates between ground and $+12$ V and gives an output voltage of $+8$ V. Make the divider current 0.1 mA.

5.25 Use three n-channel MOS enhancement transistors of $V_t = 2$ V in a voltage divider design that operates between

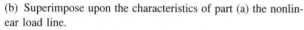

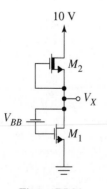

Figure P5.21

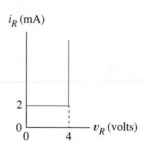

Figure P5.23

−12 V and +12 V. Provide output voltages of −4 V and +8 V. Make the divider current 0.1 mA.

Section 5.4

5.26 A p-channel MOSFET has parameters $V_t = -2$ V and $k = 3$ mA/V^2.
(a) Is this an enhancement or a depletion device? How can you tell?
(b) Sketch the input, transfer, and output characteristics.

5.27 A p-channel MOSFET has parameters $V_t = 2$ V and $k = 3$ mA/V^2.
(a) Is this an enhancement or a depletion device? How can you tell?
(b) Sketch the input, transfer, and output characteristics.

5.28 A p-channel MOSFET has $k = 2$ mA/V^2 and $V_t = -1$ V. Find the state of the transistor and the drain current when
(a) $V_G = -2$ V, $V_D = -5$ V, $V_S = -0.5$ V.
(b) $V_G = -2$ V, $V_D = 0$ V, $V_S = -3$ V.
(c) $V_G = 3$ V, $V_D = 5$ V, $V_S = 5$ V.

5.29 A p-channel MOSFET is characterized by $k = 0.5$ mA/V^2 and $V_t = -3$ V.
(a) Write the drain-current equation for each state, and list the relevant inequalities.
(b) Write an equation for i_D when $v_{GS} = -4$ V, given that the transistor is in ohmic operation.
(c) Sketch the result of part (b) as a function of v_{DS}. Show the lower limit where the equation is no longer valid.

5.30 The MOSFET in Fig. P5.30 is active. Its parameters are $k = 2$ mA/V^2 and $V_t = -2$ V.

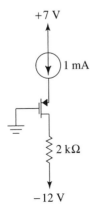

+7 V

1 mA

2 kΩ

−12 V

Figure P5.30

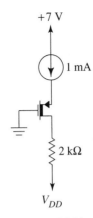

+7 V

1 mA

2 kΩ

V_{DD}

Figure P5.31

(a) Find the voltage across the current source.
(b) Find the numerical value of V_{DS} for the transistor.

5.31 In Fig. P5.31, $V_{DD} = -12$ V and the p-channel depletion MOSFET is active. Parameter values are $k = 2$ mA/V^2 and $V_t = 2$ V.
(a) Find the voltage across the current source.
(b) Find the numerical value of V_{DS} for the transistor.

5.32 The transistor in Fig. P5.31 is active when $V_{DD} = -12$ V. If we now make V_{DD} increasingly positive, at what voltage would the transistor change over to ohmic operation?

5.33 Derive an expression for a p-channel, enhancement, voltage-controlled linear resistor. Deduce the limit on the magnitude of its terminal voltage for linear operation.

5.34 Use two p-channel MOS enhancement transistors of $V_t = -1$ V in a voltage divider design that operates between ground and +12 V and gives an output voltage of +8 V. Make the divider current 0.1 mA.

5.35 Use two p-channel MOS enhancement transistors of $V_t = -1$ V in a voltage divider design that operates between ground and −12 V and gives an output voltage of −5 V. Make the divider current 0.2 mA.

5.36 The circuit of Fig. P5.36 is to give $v_o(t) = v_s(t)$ when $v_C = 0$ V. For $v_C = -5$ V the output is to be $v_o(t) = 0.02v_s(t)$.
(a) Using the transistor as a linear resistor in a voltage divider, find transistor parameters V_t and k so that the circuit satisfies specifications. Any correct answer is acceptable.
(b) If $v_s(t)$ is a sinusoid of amplitude A, find the largest A for which the linear resistance approximation used in part (a) is valid.

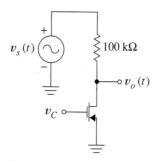

Figure P5.36

Section 5.5

5.37 Use SPICE to verify the Q-points in Example 5.5.

5.38 (a) Use a SPICE dc analysis to check the voltage divider design of Example 5.11.
(b) Add a 100 kΩ load resistor between the +7 V output of the divider and ground. Use SPICE to check for output loading by determining whether this resistor causes the node voltage drop below 7.
(c) Test loading further using successive load resistances of 10 kΩ, 1 kΩ, 100 Ω, and 1 Ω.

5.39 In Example 5.13, replace the resistor with an enhancement-type MOS load transistor with threshold voltage $V_t = +2$ and W/L one-fourth that of M_1. With SPICE obtain output waveforms for the same two sinusoidal inputs. Also plot the transfer characteristic.

5.40 Transistor parameters for Fig. P5.40 are $k_1 = k_2 = 0.9$ mA/V^2, $V_{t1} = 1$ V, $V_{t2} = -1$ V. Use SPICE dc analysis to find the transfer characteristic of the circuit.

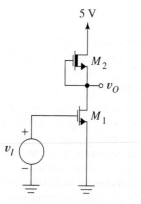

Figure P5.40

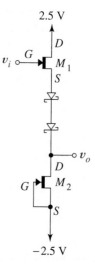

Figure P5.41

Section 5.6

5.41 In Fig. P5.41 $\beta_1 = 0.49 \times 10^{-3}$A/V^2, $\beta_2 = 0.63 \times 10^{-3}$A/V^2, and $V_t = -1.5$ V for both transistors. Forward diode drops are 0.4 V. Assume $\tanh(\alpha v_{DS}) \approx 1$. Find an expression for v_o in terms of v_i if M_1 is active.

5.42 The input characteristic of a JFET is theoretically a graph of

$$i_G = I_S(e^{v_{GS}/V_T} - 1)$$

For negative v_{GS}, we can visualize the input equivalent circuit as a dc current source and a parallel input resistance that accounts for the nonzero slope of the characteristic. For $I_S = 1.5 \times 10^{-14}$ A and $V_T = 0.025$ V, show that the input current and the input resistance are, respectively, 1.5×10^{-14} A and 3.9×10^{27} Ω when $v_{GS} = -1$ V.

5.43 Figure P5.43 shows the transfer and output characteristic curves of n-channel JFET.
(a) What are the values of I_{DSS} and V_P?
(b) Use the graphs to estimate i_D when $v_{GS} = -1.5$ V and $v_{DS} = 3$ V.
(c) Estimate i_D when $v_{GS} = -0.25$ V and $v_{DS} = 4$ V.

5.44 The JFET of Fig. P5.44 is described by the characteristic curves of Fig. P5.43. Use a load line to find (V_{DS}, I_D) when $R_D = $
(a) 1 kΩ.
(b) 2 kΩ.
(c) 5 kΩ.

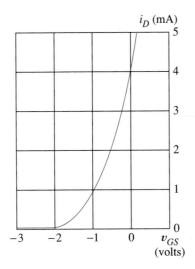

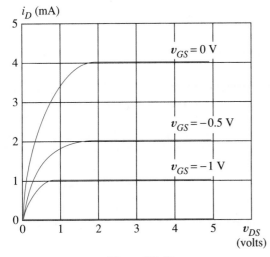

Figure P5.43

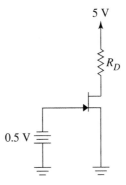

Figure P5.44

5.45 An n-channel JFET is characterized by $I_{DSS} = 1$ mA and $V_P = -0.5$ V. Find the state of the transistor for
(a) $V_G = -2$ V, $V_D = -1.75$ V, $V_S = -2$ V.
(b) $V_G = -2$ V, $V_D = -1$ V, $V_S = -2$ V.
(c) $V_G = 3$ V, $V_D = 2$ V, $V_S = 4$ V.
(d) $V_G = 3$ V, $V_D = 3$ V, $V_S = 3.25$ V.

5.46 An n-channel JFET is characterized by $I_{DSS} = 1$ mA and $V_P = -0.5$ V. Write the equation for i_D and sketch the appropriate circuit model when the transistor is
(a) active,
(b) in cut-off.
(c) ohmic.

5.47 One way to construct a constant current source is to connect gate and source nodes of a JFET (or a depletion MOSFET). Sketch the volt-ampere curve, i_D versus v_{DS}. for such a JFET for $0 \leq v_{DS}$ and state a condition necessary for proper current-source operation?

5.48 An n-channel JFET is characterized by $\beta = 0.125$ mA/V^2 and $V_P = -2$ V.
(a) Write an equation for i_D when $v_{GS} = 2$ V, given that the transistor is in ohmic operation.
(b) Sketch the result of part (a) versus v_{DS}. Show on your sketch the upper limit where the equation is no longer valid.

5.49 Sketch the input, transfer, and output characteristics for an n-channel JFET having $V_P = -3$ V and $\beta = 0.111$ mA/V^2.

5.50 An n-channel JFET with $\beta = 0.444$ mA/V^2 and $V_P = -1.5$ V is to be used as a voltage controlled resistor that spans the range

$$1 \text{ k}\Omega \leq R_{N\text{-}JFET} \leq 100 \text{ k}\Omega.$$

(a) Find the two values of v_{GS} that give the respective limiting values of resistance.
(b) Estimate the maximum permissible amplitude of v_{DS} for this resistor to operate as a linear device.

5.51 In Fig. P5.51, both JFETs have $V_P = -2$ V; however, the W/L ratio of J_1 is twice that of J_2. If $v_{DS} = 9$ V, find I_1, I_2 and V_{DS}.

5.52 Sketch the input, transfer, and output characteristics for a p-channel JFET having $V_P = 3$ V and $I_{DSS} = 4$ mA.

5.53 A p-channel JFET is characterized by $\beta = 16$ mA/V^2 and $V_P = 0.5$ V. Determine the state of the transistor for
(a) $V_G = -2$ V, $V_D = -1$ V, $V_S = -3$ V.
(b) $V_G = -1.8$ V, $V_D = -1.75$ V, $V_S = -2$ V.
(c) $V_G = 3.8$ V, $V_D = 2$ V, $V_S = 4$ V.
(d) $V_G = 3.25$ V, $V_D = 3.20$ V, $V_S = 3$ V.

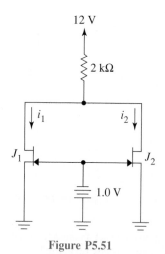

Figure P5.51

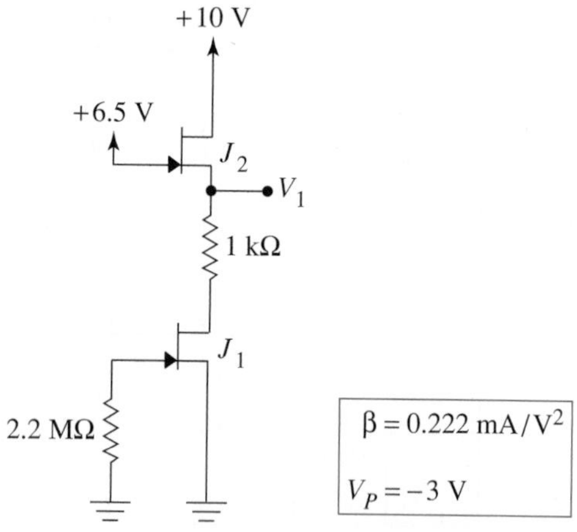

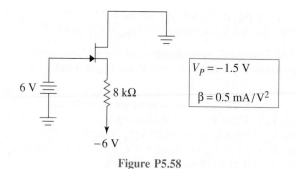

Figure P5.59

5.54 Restructure the diagram of Fig. P5.51 to one involving p-channel JFETs. This involves changing transistor schematic symbols, reference directions, and voltage-source polarities. Rewrite Problem 5.51 to make it relevant to the new diagram. Then solve the revised problem.

5.55 A p-channel JFET has $\beta = 0.0556$ mA/V^2 and $V_P = 3$ V.
(a) Write the equation for i_D for each of the three states, and write the relevant inequalities for each.
(b) Write an equation for i_D when $v_{GS} = 1$ V, given that the transistor is in ohmic operation.
(c) Sketch the result of part (b) versus v_{DS} showing the upper limit where the equation is no longer valid.

Section 5.7

5.56 (a) Compute the Q-point of the transistor of Example 5.15 if the resistor between gate and ground is 20% higher than indicated while the resistor between gate and the power supply is 20% lower.

(b) Locate the new Q-point on an i_D versus v_{DS} coordinate system.
(c) Redraw Fig. 5.40d. Sketch the new load line to compare the new solution with the original.

5.57 (a) Sketch the JFET transfer characteristic and load line for Example 5.15 showing the location of the solution.
(b) Find the new Q-point for Example 5.15 if V_P is 20% lower than the expected value of -1 V used in the original calculations.
(c) Add to the diagram of part (b) the new transfer characteristic and show that your answer to part (b) makes sense.

5.58 The transistor in Fig. P5.58 is active. Find its Q-point coordinates.

5.59 In Fig. P5.59 J_1 and J_2 are identical and active.
(a) Find the drain current of J_1.
Hint: Start with the gate current of J_1.
(b) Find the node voltage, V_1.
(c) Find the value of V_{DS} for J_1.
(d) Locate the Q-point of each transistor on output characteristic coordinates.

5.60 Find the Q-point of the JFET in Fig. P5.60. Begin by assuming active operation.

5.61 Find the Q-point of the JFET in Fig. P5.61. Begin by assuming active operation.

Section 5.8

5.62 (a) Use SPICE to find the Q-point of the JFET in Fig. 5.40a.
(b) Replace the resistor connected to the drain by a 10 kΩ re-

Figure P5.58

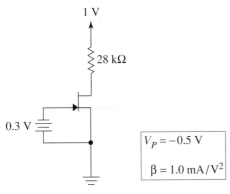

Figure P5.60

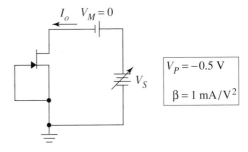

Figure P5.64

$v_C = -2.5$ V. Use $R_F = 9.17$ kΩ, $\beta = 10^{-3}$ A/V^2, and $V_P = -3$ V.

Section 5.9

5.66 Starting with Eq. (5.25) and the definition of r_o, derive Eq. (5.26). List any approximation you need to make.

5.67 Example 5.3 used the circuit model of Fig. 5.11b to compute the Q-point, $(V_{DS}, I_D) = (6$ V, 1 mA). This calculation ignored Early voltage.

To estimate the error in I_D introduced by ignoring Early effect in such an analysis, redraw Fig. 5.11b with r_o added. Use the I_D value found in the original analysis and $V_A = 120$ V to estimate the value of r_o.
(a) How much current flows through r_o if we use the original V_{DS} as an estimate of the true drain-source voltage?
(b) Express the current of part (a) as a percentage of the original drain current.
(c) Considering the additional complexity of the circuit when we include r_o *in the original analysis,* do you think r_o should be included?
(d) What conditions in a circuit analysis problem should alert you that a second analysis including r_o should be performed?

5.68 Sketch common-source output characteristics for an MOS transistor that show both Early effect and breakdown voltage.

5.69 Redraw the nonlinear load line and the $v_{GS} = 8$ V characteristic curve of Fig. 5.20b. Show on your diagram how the "$v_{GS} = 8$ V" Q-point changes as the transistor temperatures are simultaneously lowered.

Section 5.10

5.70 Redraw Fig. 5.11a, but use the MOSFET schematic symbol from Fig. 5.2b that shows the body explicitly. Add to this figure five capacitors representing the five internal capacitances of the MOS transistor. Assume that the substrate is grounded. Use the results of the dc analysis of Example 5.3 to find the dc

sistor. Then use SPICE to find the new Q-point. Use SPICE data to determine the transistor state.
Hint: .OP instruction.

5.63 Use SPICE to find currents in Problem 5.53.
Hint: .OP instruction.

5.64 The JFET in Fig. P5.64 is wired as a constant current source. V_M is an *ammeter* added to enable plotting output I_o. V_S is a voltage source of variable value.
(a) Use a SPICE dc analysis to plot I_o for $0 \le V_S \le 10$ V.
(b) Briefly discuss your simulation results relative to the theory. What operating condition must be placed upon such a current source to ensure proper operation?
(c) How would the plot change if the *W/L* ratio of the transistor were tripled during JFET fabrication?

5.65 Use SPICE dc analysis to plot v_O versus v_I twice for the amplifier of Fig. 5.36b for the input voltage range -200 mV $\le v_i \le 200$ mV. For the first plot, use $v_C = 0$ V; for the second,

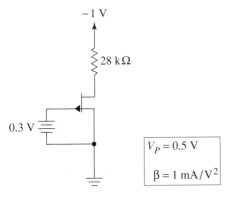

Figure P5.61

voltage across each capacitor. (Notice from Eq. (3.62) that for the depletion capacitances, knowing the dc bias voltage allows us to calculate the capacitance values.)

5.71 What effect do you expect the transistor internal capacitances to have on the operation of circuits using transistors
(a) as switches?
(b) as amplifiers?

5.72 Use SPICE to estimate the bandwidth of the amplifier of Fig. 5.29a using a rise time measurement and Eq. (1.42). Figure 5.29e shows the circuit has rather high linear gain near $v_{in} = -0.25$ V; therefore, replace v_{in} by a pulse source with initial value of -0.25 V and pulsed value of -0.24 V, that is, a 0.01 V pulse superimposed upon a dc level that places the Q-point in the high-gain region. The rise time and fall time of your pulse should be 0 s. Use transistor values from Fig. 5.29b, but add zero bias values of 0.5 pF for both CBD and CBS, and TOX $= 0.1$ μ. From the output waveform, estimate the gain and bandwidth of the amplifier.

5.73 Examine the consequence of body effect in the circuit of Fig. 5.20a.
(a) Use SPICE dc analysis to plot the drain voltage of M_1 versus V_{GS} for the circuit using default values for everything except the given transistor k and V_t values, as V_{GS} ranges from 0 to 9 V.
(b) Repeat part (a) after introducing the body effect parameter $\gamma = 0.37$ into the SPICE description of M_2.
(c) If the circuit is to be used as an amplifier by operating on the steep portion of the transfer characteristic, does body effect increase or decrease the gain?

5.74 Use SPICE to determine the consequences of body effect ($\gamma = 0.4$) on the voltage divider outputs of Example 5.11.

Section 5.11

5.75 The switch of Fig. 5.56a is to be used to switch 9 V across a load resistor of $R_L = 100$ Ω using control voltages v_C of 0 V and 9 V.
(a) Viewing the switch as a voltage-controlled linear resistance, compute the resistance of the closed switch if 99% of the available voltage is to be across the load when the switch is closed.
(b) If $V_t = 1.5$, find the value of k required of the transistor.

5.76 In Fig. 5.57b, $R_L = 1$ kΩ, $V_{DD} = v_C = 9$ V, and the transistor has parameters $V_t = 1$ V and $k = 5$ mA/V^2.
(a) For $v_{in} = -9$V, use the large signal transistor model to find i_D and v_{DS} for the transistor. What is the resistance of the switch?
(b) For $v_{in} = 8.5$ V, use the large signal transistor model to find i_D and v_{DS} for the transistor. What is the resistance of the switch?

5.77 The MOSFET switch of Fig. P5.77 operates with control voltage values of 0 V and 8 V. When the switch is closed the switch voltage is to be no more than 0.5 V.
(a) Sketch a load line that illustrates the largest permissible load resistor R_L that can be used with this switch.
(b) What is the value of this largest R_L?
(c) Suppose you construct this circuit in lab and by mistake test it with v_C pulses of 2 V instead of 8 V amplitude. Sketch the voltage waveform you would see at the drain node relative to ground.

5.78 The circuit of Fig. P5.78 puts the information, $v_{in}(t)$, onto a high-frequency carrier so that it can be radiated through space or combined with other signals on a transmission line. The

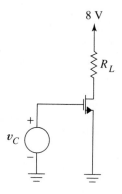

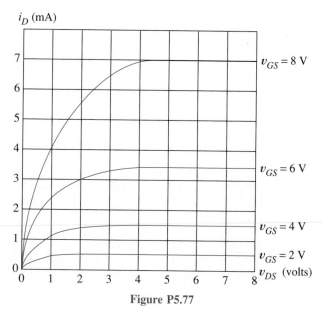

Figure P5.77

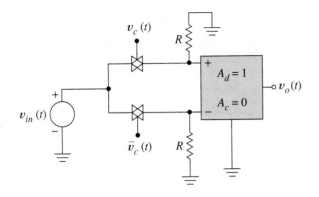

transmission gates are ideal switches opened and closed by complementary square waves. The diff amp has infinite input resistance and gains as shown on the diagram.

Sketch $v_o(t)$, *if* $v_{in}(t)$ is a sinusoid and if the square waves have period T about 1/10 that of the sine wave.

Hint: Since the amplifier has infinite input resistance, first sketch the waveforms at its two input nodes.

5.79 Use SPICE to investigate possible consequences of body effect in the low-frequency operation of the CMOS transmission gate described in Fig. 5.61b.

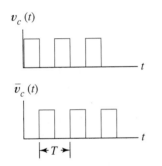

Figure P5.78

CHAPTER

6

Bias Circuits

Linear electronic circuits use transistors as dependent current sources. Because transistor output characteristics resemble dependent source curves only in the active region, we need to design circuits that set the transistor's dc operating point in this region. We call such circuits *biasing* or *bias* circuits. Although the amplifier is our central focus, the general principles of biasing apply also to other linear circuits discussed in later chapters.

6.1
Bias and Signals in Analog Circuits

A brief introduction to the transistor amplifier provides perspective and defines key concepts related to how transistors process signals. A full understanding of the notation of Fig. 1.1 is essential to what follows.

6.1.1   TRANSISTOR AMPLIFIER

Bias and Signal Components. Figure 6.1a is an amplifier biasing circuit. Gate-source voltage, drain current, and drain-source voltage have *dc bias values,*

$$v_{GS} = V_{GG}$$

$$i_D = I_D = \frac{k}{2}(V_{GG} - V_t)^2 \tag{6.1}$$

and

$$v_{DS} = V_{DS} = V_{DD} - R_L I_D$$

respectively, that are appropriate for active operation.

When we introduce a signal $v_{gs}(t)$ into the input circuit, as in Fig. 6.1b, we observe that *signal components* $i_d(t)$ and $v_{ds}(t)$ are added to the dc bias values. We now explain the origin and character of these signals, both mathematically and graphically.

With signal present, the gate-source voltage becomes

$$v_{GS}(t) = V_{GG} + v_{gs}(t) \tag{6.2}$$

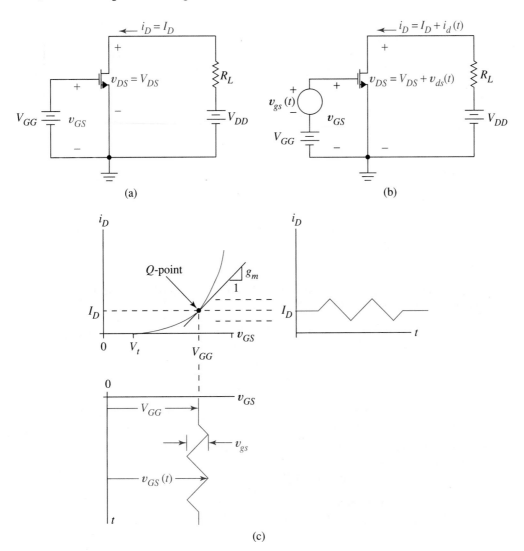

Figure 6.1 MOSFET amplifier: (a) without signal; (b) with signal; (c) transfer characteristic relating input and output ac signal waveforms.

This sum replaces V_{GG} in Eq. (6.1) giving

$$i_D(t) = \frac{k}{2}[V_{GG} + v_{gs}(t) - V_t]^2 \qquad (6.3)$$

for the drain current. Equation (6.3) suggests Fig. 6.1c, where the previously static dc operating point now moves along the transfer characteristic in response to the signal.

When we group together the constants in Eq. (6.3) and then expand the quadratic, we discover that $i_D(t)$ is actually the sum of *three* components

$$i_D(t) = \frac{k}{2}(V_{GG} - V_t)^2 + 2\frac{k}{2}(V_{GG} - V_t)v_{gs}(t) + \frac{k}{2}v_{gs}^2(t) \qquad (6.4)$$

The same result, but a different perspective, follows from expanding Eq. (6.3) as the Taylor series

$$i_D(v_{GS}) = i_D[V_{GG} + v_{gs}(t)]$$

$$= i_D(V_{GG}) + \left(\left.\frac{di_D}{dv_{GS}}\right|_{v_{GS} = V_{GG}}\right)v_{gs}(t) + \frac{1}{2!}\left(\left.\frac{d^2i_D}{dv_{GS}^2}\right|_{v_{GS} = V_{GG}}\right)v_{gs}^2(t)$$

In both expansions, the first term is Eq. (6.1) and the second is proportional to the input signal. This motivates us to express the total drain current as

$$i_D(t) = I_D + i_d(t) + d(t) \qquad (6.5)$$

where the first term is the dc bias current, the second a signal component, and the third an undesired distortion term.

Small-Signal Operation. To make the distortion negligible compared to the signal, that is, to make $d(t) << i_d(t)$, Eq. (6.4) shows we must restrict the signal amplitude so that it satisfies

$$|v_{gs}| << 2|V_{GG} - V_t|$$

In the Taylor series, this same condition ensures that the $v_{gs}^2(t)$ term makes a negligible contribution to the expansion. We call this important condition *small-signal operation.* When the signal is so small that distortion is negligible in Eqs. (6.4) and (6.5), we can represent drain current in the form

$$i_D(t) = I_D + g_m v_{gs}(t) \qquad (6.6)$$

where the proportionality constant g_m is called the *transconductance* of the transistor.

The Taylor series shows that g_m is the slope of the tangent to the transfer characteristic at the dc operating point. In terms of Fig. 6.1c, *small-signal operation means keeping the operating point excursions so small that projecting from the actual curve is equivalent to projecting from the tangent* Because the second term in Eq. (6.6) is proportional to the signal, we call it the *signal component of the drain current,*

$$i_d(t) = g_m v_{gs}(t) \qquad (6.7)$$

Amplifier Gain. We now examine the components of v_{DS}. With signal present, from Fig. 6.1b

$$v_{DS}(t) = V_{DD} - R_L[I_D + i_d(t)] = V_{DS} - R_L i_d(t) = V_{DS} + v_{ds}(t)$$

The *signal component of the drain-source voltage*

$$v_{ds}(t) = -R_L i_d(t)$$

is the output signal of the amplifier; the minus sign denotes a signal inversion. Substituting $i_d(t)$ from Eq. (6.7) and solving for the ratio of output voltage to input voltage gives the voltage gain

$$A_v = \frac{v_{ds}}{v_{gs}} = -g_m R_L \tag{6.8}$$

Obviously, the gain of the amplifier depends upon transconductance g_m. Since g_m is the coefficient of $v_{gs}(t)$ in Eq. (6.4), we can compute its numerical value using

$$g_m = k(V_{GG} - V_t)$$

A more useful expression follows from solving Eq. (6.1) for $V_{GG} - V_t$ and then substituting. This is

$$g_m = \sqrt{2kI_D} \tag{6.9}$$

Thus voltage gain in Eq. (6.8) depends upon the transistor parameter g_m, which in turn depends upon I_D, one of the coordinates of the dc operating point. Figure 6.1c shows why. A larger Q-point current I_D gives greater slope and therefore a larger projected signal current for the same signal voltage. Larger signal current flowing through R_L produces a larger output voltage.

A parallel development for a BJT amplifier shows that a small input signal $v_{be}(t)$, superimposed upon an existing dc bias voltage, causes a signal current $i_c(t)$ to be added to the dc collector current. The signal current can be written as

$$i_c(t) = g_m v_{be}(t)$$

For the BJT we calculate the transconductance using

$$g_m = \frac{I_C}{V_T} \tag{6.10}$$

where V_T is the thermal voltage. The next chapter discusses these small-signal effects in detail.

In the entire subject of electronics the concepts introduced in the foregoing paragraphs are perhaps the most difficult to master. The principal conclusions are the following. With no input signal, currents and voltages all assume dc, quiescent, or *bias* values. To avoid distortion, any applied signals must be small. Turning on such a small signal results in *small-signal components adding to* the existing bias values of voltages and currents. The various signal components are related to one another by *small-signal transistor parameters* such as g_m, which have numerical values that depend upon the dc bias values that exist in the circuit. Because parameter values depend upon transistor Q-points, *transistor biasing is closely related to the circuit's signal processing function.* Since we must start somewhere, we first master as much as we can of the biasing principles in this chapter. In the next chapter we study how circuits handle small signals and, in the process, further improve our understanding of biasing and bias circuit design.

6.1.2 BIAS CIRCUIT ANALYSIS AND DESIGN

There are two problems related to dc operating points, *analysis* and *design*. Q-point analysis means finding the operating point of each transistor in a *given circuit*. Sometimes the purpose of the analysis is to confirm that each transistor is biased in the active region—more often to obtain the information needed to calculate numerical values for the *small-signal parameters*. This is important because a subsequent *small-signal analysis* requires these parameters in order to determine how the circuit processes the signals.

We already mastered Q-point analysis in Chapters 4 and 5, so our only new analysis idea is that we must turn off signal sources before we begin a dc analysis. Chapter 1 showed that this means replacing voltage and current sources with short and open circuits, respectively.

Bias circuit *design* involves selecting a circuit structure and then finding values for resistors and dc sources so that each transistor is biased at a *prescribed Q-point;* sometimes the circuit must satisfy additional conditions as well. Unlike analysis problems, which have only one correct solution, design problems usually have many acceptable solutions. Sometimes design requires us to make a choice that, within limits, is rather arbitrary—occasionally iterative cycles of trial and error are required. As designers, our task is to find one of the better solutions from all the acceptable possibilities in a reasonable time. Because broad perspective and experience are obviously beneficial to designers, the design process can be intimidating to the beginner. It is important to develop common sense, making use of given information and relevant design guidelines but avoiding undue concern about finding the Best Possible Solution. We restrict ourselves here to reasonably well-defined design problems and active-region Q-points in order to acquire some of this experience.

6.2
Bias Circuit Design Techniques

This section introduces specific techniques for designing amplifier bias circuits. The next section follows up with theoretical principles that justify our techniques and provide additional guidelines for their implementation. While the specific circuits and techniques of Secs. 6.2–6.4 apply directly to discrete circuits, they also introduce basic principles we later employ in integrated circuit design.

6.2.1 FOUR-RESISTOR BIAS CIRCUIT

FET Circuit Design. For discrete component amplifiers, the four-resistor circuit of Fig. 6.2a, is an effective biasing structure. In a typical design task, we know the transistor parameters, the power supply voltage, and the desired Q-point coordinates; the object is to find resistor values that place the Q-point at the given location. An additional consideration is that the parallel equivalent of R_1 and R_2, usually related to the input resistance of the amplifier, should be high. It is helpful to add current and voltage values to the given diagram as they are determined, as this often reminds us of the next step in the procedure. The following example illustrates the process.

EXAMPLE 6.1 In Fig. 6.2a $V_{DD} = 12$ V, $k = 0.5 \times 10^{-3}$ A/V^2 and $V_t = 2$ V. Design the circuit to bias the transistor at $(V_{DS}, I_D) = (4$ V, 2 mA). Also make $R_P = R_1 \| R_2 = 2$ MΩ.

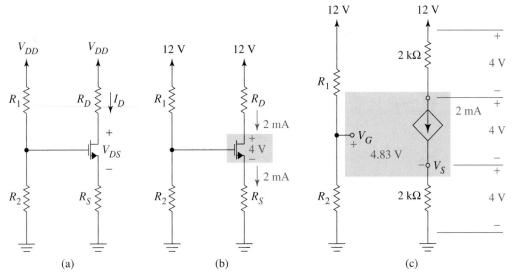

Figure 6.2 Four-resistor bias circuit for MOSFET: (a) circuit structure; (b) use of given information in a design problem; (c) information known when R_1 and R_2 values remain to be found.

Solution. First, add the given information to the diagram as in Fig. 6.2b. In the drain-source circuit, 12 V is provided by the power supply. Of this, 4 V is the required dc drop across the transistor, leaving 8 V for voltage drops across R_D and R_S. A *rule of thumb* that gives acceptable results *when no other condition is given* is to make the three drops approximately equal, 4 V each in this case. Since the given drain current must be 2 mA,

$$R_D = R_S = \frac{4\ \text{V}}{2\ \text{mA}} = 2\ \text{k}\Omega$$

Because the transistor must be active, given drain current always implies a specific gate-source voltage. In this case

$$I_D = 2 \times 10^{-3} = \frac{0.5 \times 10^{-3}}{2}(V_{GS} - 2)^2$$

It is *never* necessary to solve the quadratic formula in this situation. Instead, use

$$(V_{GS} - 2) = \pm \sqrt{\frac{2 \times 2 \times 10^{-3}}{0.5 \times 10^{-3}}}$$

which gives $V_{GS} = 4.83$ V. Figure 6.2c shows this constraint. Because the 4 V drop across R_S establishes the source voltage, $V_S = +4$ V, the node voltage at the gate is

$$V_G = V_{GS} + V_S = 4.83 + 4 = 8.83\ \text{V}$$

To establish this gate voltage, divider $R_1 - R_2$ must satisfy

$$\frac{R_2}{R_1 + R_2} 12 = 8.83$$

A good way to find specific values for R_1 and R_2 is to use the input resistance condition

$$R_P = R_1 \| R_2 = 2 \times 10^6 \, \Omega$$

Multiplying the divider equation by R_1 gives

$$\frac{R_1 R_2}{R_1 + R_2} 12 = 8.83 R_1 = R_P \times 12 = 24 \times 10^6$$

From this we find

$$R_1 = 24 \times 10^6 / 8.83 = 2.72 \, \text{M}\Omega$$

Since the parallel equivalent is 2 MΩ,

$$\frac{1}{R_2} = \frac{1}{2\text{M}} - \frac{1}{2.72\text{M}}$$

or $R_2 = 7.6$ MΩ, which completes the design.   ❏

The same design procedure applies to depletion MOSFETs and JFETs. For p-channel FETs, the power supply of Fig. 6.2a and all node voltages are negative, and all currents flow in the opposite directions, but the design procedure is the same.

Exercise 6.1   Design a circuit like Fig. 6.2 for a 12 V supply, an n-channel depletion transistor with $k = 0.5 \times 10^{-3}$ A/V^2 and $V_t = -3$ V. Make the Q-point $(V_{DS}, I_D) = (6$ V, 1 mA), with equal voltage drops across R_D and R_S. Also make $R_P = 3$ MΩ.

Ans. $R_D = R_S = 3$ kΩ, $R_1 = 18$ MΩ, $R_2 = 3.6$ MΩ.

BJT Circuit Design. The four-resistor BJT bias circuit design resembles the FET circuit design, especially when β is large; however, the procedure for finding R_1 and R_2 differs slightly because the base current loads the voltage divider. The next example illustrates the BJT procedure.

EXAMPLE 6.2   Design the bias circuit of Fig. 6.3a so that $I_C = 2$ mA, given $20 \le \beta \le 200$.

Solution. The collector-emitter voltage of the transistor is not specified. Since we have no relevant specification, we assign equal 3 V drops to V_{CE}, R_C, and R_E. Because I_C flows through R_C

$$R_C = 3/2 \times 10^{+3} = 1.5 \, \text{k}\Omega$$

Next we determine R_E. Since

$$I_E = \frac{\beta + 1}{\beta} I_C$$

the approximation, $I_E \approx I_C$, is valid for large β, say, 50 or greater. With this approximation, finding R_C and R_E is identical to finding R_D and R_S in a FET design. In this ex-

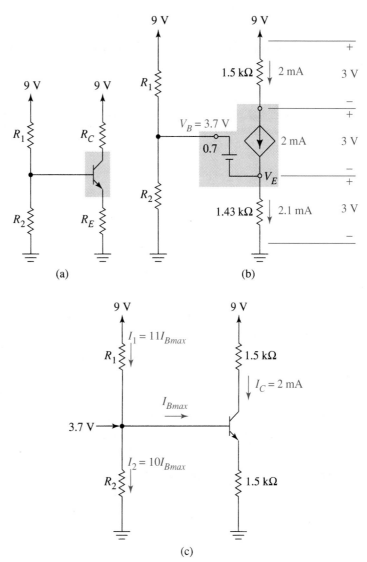

Figure 6.3 Four-resistor biasing circuit for BJT: (a) circuit structure; (b) information known when values for R_1 and R_2 are computed; (c) approximation giving near ideal voltage divider.

ample, however, β might be as low as 20. A conservative designer uses the minimum β to estimate the dc emitter current, namely,

$$I_E = \frac{\beta_{min} + 1}{\beta_{min}} I_C = \frac{21}{20} 2 \text{ mA} = 2.1 \text{ mA}$$

A 3 V drop across R_E with a 2.1 mA current requires $R_E = 1.43$ kΩ. The equivalent circuit of Fig. 6.3b shows these results.

The large signal BJT model requires that the node voltage at the base be $V_{BE} = 0.7$ V above (below) the emitter node voltage for a silicon npn (pnp) transistor. This gives $V_B = 3.7$ V as in Fig. 6.3b. This fixed dc drop replaces the calculated V_{GS} difference we need to find the gate voltage in FET designs.

If beta were infinite, base current I_C/β would be zero, and we could simply design a voltage divider to produce $V_B = 3.7$ V. Since base current is nonzero, we must include worst case loading effects when we find R_1 and R_2. Because loading of the voltage divider is worst for largest base current, worst case design uses the smallest β. In this example we need a divider that will work reasonably well for $I_{Bmax} = I_C/\beta_{min} = 2$ mA/20 $= 0.1$ mA.

Consider the diagram of Fig. 6.3c. If we assign $I_2 = 10 \times I_{Bmax}$, then loading of the divider will be negligible in the "order-of-magnitude" sense we often use in engineering approximations. This gives $I_2 = 1$ mA, so

$$R_2 = \frac{3.7}{1\text{mA}} = 3.7 \text{ k}\Omega$$

KCL requires

$$I_1 = I_2 + I_{Bmax} = 11I_{Bmax} = 1.1 \text{ mA}$$

This leaves only the computation of R_1,

$$R_1 = \frac{9 - 3.7}{1.1 \text{ mA}} = 4.8 \text{ k}\Omega$$ ❏

> **Exercise 6.2** Redesign the circuit of Example 6.2 for a Q-point of $(V_{CE}, I_C) = (2$ V, 1 mA), with the additional condition $V_E = 5$ V. Use the same transistor.
>
> **Ans.** $R_C = 2$ kΩ, $R_E = 4.76$ kΩ, $R_2 = 11.4$ kΩ, $R_1 = 6$ kΩ.

> Bias circuit design for a pnp transistor differs from preceding examples only in the polarities of the currents and voltages.

6.2.2 OTHER BIAS CIRCUITS

> There are discrete component biasing circuits that use fewer resistors and still provide adequate biasing for some purposes. The three-resistor circuits of Figs. 6.4a and c are examples. The general approach is the same, as we see in the following.

EXAMPLE 6.3 Design the circuit of Fig. 6.4a so that the Q-point is $(V_{CE}, I_C) = (5$ V, 0.5 mA) when $\beta = 40$.

Solution. Of the $12 - 5 = 7$ V yet to be assigned in the collector circuit, arbitrarily assign 3 V to R_C and 4 V to R_E. We know $I_E = (41/40) \times I_C = 0.513$ mA. Figure 6.4b shows the known information, from which

$$R_C = \frac{3}{0.5 \text{ mA}} = 6 \text{ k}\Omega$$

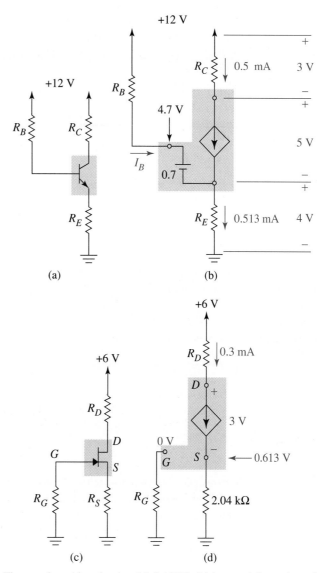

Figure 6.4 Three-resistor bias circuits: (a) for BJT; (b) known information when R_B value is determined; (c)–(d) for JFET.

and

$$R_E = \frac{4}{0.513 \text{ mA}} = 7.8 \text{ k}\Omega$$

Because the base current is $(0.5 \times 10^{-3})/40 = 12.5 \text{ }\mu\text{A}$

$$R_B = \frac{12 - 4.7}{12.5 \text{ }\mu\text{A}} = 584 \text{ k}\Omega$$

❏

Exercise 6.3 Redesign the circuit of the last example for the same Q-point and a transistor β of 400.

Ans. $R_C = 6$ kΩ, $R_E = 8$ kΩ, $R_B = 5.84$ MΩ.

EXAMPLE 6.4 The JFET in Fig. 6.4c has parameters $V_P = -1$ V and $\beta = 2$ mA/V^2. Design for the Q-point $(V_{DS}, I_D) = (3$ V, 0.3 mA$)$.

Solution. As usual, known drain current leads to V_{GS}. From Eq. (5.21)

$$0.3 \text{ mA} = 2 \times 10^{-3}[V_{GS} + 1]^2$$

which gives $V_{GS} = -0.613$ V $= 0 - V_S$ (because gate current is zero). Figure 6.4d summarizes the known numerical information and shows that

$$R_S = \frac{0.613}{0.3 \times 10^{-3}} = 2.04 \text{ k}\Omega$$

Of the 6 V provided for the drain circuit, the drain-source drop is 3 V, and the drop across R_S is 0.613 V. This leaves $6 - 3.613 = 2.39$ V for R_D; thus $R_D = 2.39/(0.3$ mA$)$ $= 7.97$ kΩ

Although gate current is zero, a physical connection between gate and ground is necessary to establish exactly zero volts at the gate node. The value of R_G is immaterial for biasing; however, in the next chapter we learn that R_G is the input resistance of the amplifier. For this reason we use a large resistor, say, $R_G = 1$ MΩ. ❑

Exercise 6.4 What changes must we make in the design of Example 6.4 if the power supply changes to 15 V but we want the same Q-point?

Ans. Change R_D to 38 kΩ.

These examples and exercises show that for simpler biasing circuits the general four-resistor design approach still works; however, we must be alert to special constraints imposed by the given information. Theory introduced next gives reasons why we might prefer one particular design over another.

6.3

Bias Circuit Design Principles

6.3.1 INTRODUCTION

The object of biasing is a transistor Q-point consistent with operation as a dependent current source; however, realistic signal processing imposes additional operating point constraints. To avoid unnecessary restrictions on output signal amplitude, we must not bias a bipolar transistor too near to either saturation or cut-off. In Fig. 6.5a the Q-point at $I_B = 25$ μA allows the instantaneous operating point to move equal distances in either direction along the load line as base current changes, the transistor remaining active all the while. Biasing at 40 μA restricts symmetric base current deviations to values of 5 μA or less. Larger base currents cause the transistor to saturate, and the output voltage *limits* at $V_{CE, sat}$. Setting the Q-point at $I_B = 5$ μA biases the transistor close to cut-off, again limiting the amplitude of the output signal. In this case, the transistor cuts off for negative base current deviations from the Q-point that exceed 5 μA.

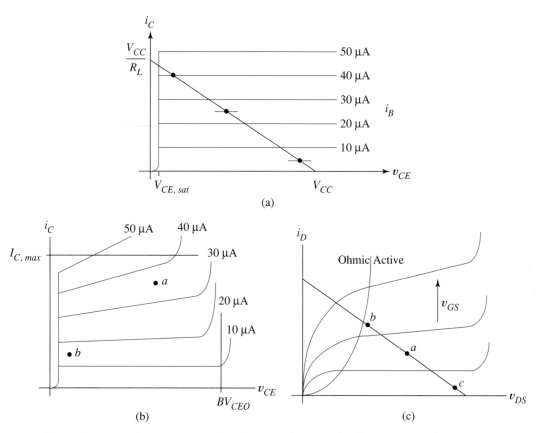

Figure 6.5 Considerations in selecting Q-point: (a) avoid proximity to cut-off and saturation; (b) point *a* has higher transconductance but is near the nonlinear region caused by breakdown, while point *b* has higher output resistance; (c) FET Q-point considerations.

Figure 6.5b shows additional concerns in choosing the Q-point. In a real transistor, curve separation, and hence transistor β, depends upon the Q-point; for example, at point *a* the transistor β is higher than at point *b*. At point *b*, however, the curves are more nearly horizontal, meaning that the transistor output resistance is higher, an advantage in some circuits. The Q-point must also be kept well away from the BV_{CEO} to avoid nonlinear distortion, and away from the maximum current $I_{C,max}$, which is specified by the manufacturer to prevent destruction of the transistor.

Similar considerations apply to FETs, as suggested in Fig. 6.5c. Q-point *a* is conservatively chosen in the active region, while points *b* and *c* might be too close to ohmic or cut-off operation, respectively. Regions of wider curve separation correspond to higher transconductance and therefore higher gain; flatter curves mean higher output resistance. There are also limits imposed by breakdown voltage and by maximum drain current.

Not only must the biasing circuit establish the Q-point at a suitable location, but it should also *keep it close* in spite of *changes in parameter values* (for example, β, V_{BE}, V_P, V_t, I_{DSS}, and k) caused by temperature changes, aging, or (in discrete circuits) component replacement. In ICs components are not replaced; however, the designer never knows in advance the *exact* values of transistor and other component parameters—only *ranges* of values. This uncertainty has the same bearing on IC design as transistor replacement has in discrete circuit design. To summarize these ideas we say the Q-point must be *stable*. We next examine biasing circuits from the viewpoint of this new requirement, *Q-point stability*.

6.3.2 *Q*-POINT STABILITY

Stability of MOSFET Biasing Circuit. Figure 6.6a shows a MOSFET amplifier circuit that employs *fixed biasing,* in which a separate supply V'_{GG} *fixes* the value of V_{GS}. The constant v_{GS} line in Fig. 6.6b shows that by proper choice of V'_{GG} we can establish any dc drain current that we wish on transfer characteristic T_1, for example, I_{D1}. Now suppose temperature decreases from T_1 to T_2, causing V_t and k to both increase; the transfer characteristic changes to the curve labeled T_2. The Q-point shifts downward from Q to Q', reducing the bias current to the lower value I_{D2}. In a particular design, bias at I_{D2} might be unacceptable. For example, it might be too close to cut-off or give voltage gain that is too high. With fixed bias, the Q-point can change a great deal if V_t and/or k change *for any reason*. We say that the fixed-bias circuit has poor Q-point stability.

The circuit of Fig. 6.6c includes a *self-biasing* resistor R_S. For this circuit, V_{GS} is not constant as in the fixed-bias circuit, but instead is given by

$$V_{GS} = V_{GG} - I_D R_S \qquad (6.11)$$

On the i_D versus v_{GS} coordinate system, Eq. (6.11) is the input circuit load line shown in Fig. 6.6d. With this circuit we are also able to establish any desired Q-point on curve T_1 by proper choice of V_{GG} and R_S, for example, the same one we used in Fig. 6.6b. With this new circuit, however, a change in transistor characteristic from T_1 to T_2, gives only a slight change in I_D. We conclude that the self-biasing resistor gives greater *Q-point stability.* An important design principle relates to our choice of R_S. Figure 6.6d shows that for a given Q-point a more horizontal load line gives better stability. Since the slope of the load line varies inversely with R_S, *larger values of R_S result in more stable Q-points.*

As a practical matter, the power supply is usually the only available dc source, so we obtain V_{GG} from a voltage divider, resulting in the four-resistor biasing circuit. Recall that in designing the four-resistor circuit, one "arbitrary" choice involves allocating fractions of the supply voltage among the three components in the drain circuit. Establishing a suitable Q-point on the output characteristics constrains the transistor's drain-source drop; R_D and R_S share the remainder of the supply voltage. We now realize that, for given I_D, a larger voltage drop $I_D R_S$ means larger R_S and therefore better stability. The dc voltage allocated to R_D must be as large as the peak value of the output signal, but this is not necessarily a large value. We therefore augment our bias circuit design strategy with an important new idea—*make the drop across R_S as large as possible for good Q-point stability.* We next show that in BJT biasing circuits we associate Q-point stability with large emitter resistors.

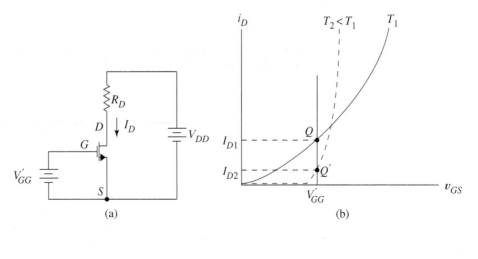

(a)

(b)

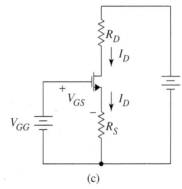

(c)

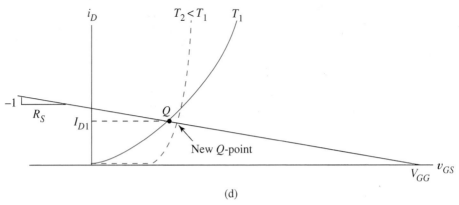

(d)

Figure 6.6 *Q*-point stability considerations: (a) for fixed biasing; (b) *Q*-point for fixed biasing at two different transistor temperatures; (c) circuit with self-biasing; (d) *Q*-point shift with temperature with self-biasing.

Stability of BJT Biasing Circuit. The four-resistor circuit of Fig. 6.7a becomes Fig. 6.7b after a Thevenin transformation. We now apply KVL to the base circuit, and, because I_C is the Q-point coordinate we wish to stabilize, write both currents in terms of I_C. The result is

$$V_{TT} = \frac{I_C}{\beta} R_T + V_{BE} + \left(\frac{I_C}{\beta} + I_C\right) R_E$$

Solving for I_C gives

$$I_C = \frac{V_{TT} - \boldsymbol{V_{BE}}}{R_T/\beta + [(\beta + 1)/\beta] R_E} \tag{6.12}$$

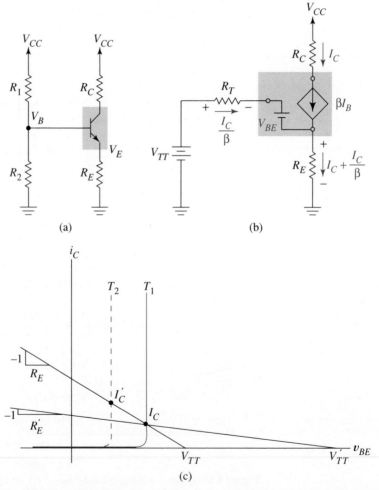

(a) (b)

(c)

Figure 6.7 Q-point stability of BJT biasing circuit: (a) actual circuit; (b) equivalent after Thevenin transformation; (c) movement of Q-point with changes in V_{BE}, and effect of larger emitter resistor.

where boldface characters denote transistor parameters that are subject to variation. By making $V_{TT} >> V_{BE}$ in the numerator, we reduce the dependence of collector current on V_{BE}. Since $V_{BE} \approx 0.7$ V for silicon at room temperature, our rule of thumb suggests we should make V_{TT} at least 7 V. Often this is not practical because of other considerations, so a lower value must suffice. Notice from Fig. 6.7b that the difference $V_{TT} - V_{BE}$ is the sum of the voltage drops across R_T and R_E. Therefore, a practical strategy for making V_{TT} large is to make this *sum* as large as possible.

In the denominator of Eq. (6.12) notice that $(\beta + 1)/\beta \approx 1$ for large β, *regardless of changes* in β. Thus if the first denominator term is much smaller than the second, variations in β have little effect on collector current. Therefore, in addition to keeping the *sum* of the drops across R_T and R_E large, we need to make R_T small. Taken together, these two conditions suggest that we make the sum large by making R_E as large as possible. We can arrive at the same conclusion about R_E by a graphical argument that closely parallels our study of MOSFET Q-point stability.

When forward active, the transistor is described by Eq. (4.5). This is the diodelike transfer characteristic

$$I_C = I_S e^{V_{BE}/V_T}$$

which looks like the curve labeled T_1 in Fig. 6.7c. When R_T is small compared to $(\beta + 1)R_E$ in Eq. (6.12), then for large β the equation reduces to

$$I_C = \frac{V_{TT} - V_{BE}}{R_E} \tag{6.13}$$

the load line equation we see in Fig. 6.7c. The intersection of the load line and the transfer characteristic establishes the Q-point coordinate $i_C = I_C$.

Now if the transistor temperature changes to a higher value, T_2, the transfer characteristic shifts to the left, and bias current increases to the new value I_C'. Notice that a larger emitter resistor, R_E, in the bias circuit reduces the slope of the load line and improves the stability of the Q-point. Comparing Figs. 6.7c and 6.6d shows that the same condition for Q-point stability applies to both BJT and FET: *make the drop across the emitter (source) resistor as large as possible for the given* $I_C(I_D)$.

The general design procedure of Sec. 6.2 easily incorporates the biasing principles we just discovered. Within practical limits we make R_E large by assigning a large fraction of the available voltage to this resistor. Then, by selecting R_1 and R_2 so that $I_2 = 10I_{Bmax}$, we ensure that the contribution of $R_T = R_1 \| R_2$ to the denominator of Eq. (6.12) is small.

This voltage divider design exemplifies *design compromises* or *trade-offs* we frequently encounter in design problems. To understand the role of R_T, visualize the divider of Fig. 6.7a in its Thevenin form, Fig. 6.7b. If actual base current turns out to be lower than the design value, the drop in R_T is less than anticipated, and base voltage is higher than planned. Higher V_B causes higher V_E, making emitter and collector currents higher than their design values. If our design had used $R_T = 0$ (this requires an impractical, ideal voltage source of *exactly* V_{TT} volts), then V_E would be unchanged by larger base current, and collector current would remain at exactly its design value. Although we cannot hope to find the ideal voltage source, we can bring R_T closer to zero as we design, by choosing I_2 to be 20 or 30 times I_{Bmax} instead of 10 times I_{Bmax}. This reduces base current loading and makes the divider more ideal. However, smaller R_1 and R_2 also give higher power dissipation—a circuit more costly to operate. When we study small signals in the next chapter, we will also learn that small values of R_1 and R_2 increase signal loading at the amplifier input. Thus the design rule $I_2 = 10I_{Bmax}$ is a *compromise* that serves as a useful starting condition. In specific designs we may sometimes wish to deviate from this rule to satisfy other conditions that might be imposed. A major task for the designer is to reconcile such conflicting requirements in the light of given specifications and common sense.

6.4
Variations on the Design Theme

6.4.1 INTRODUCTION

Practical biasing problems often include constraints in addition to given transistor Q-points, for example, prescribed values for particular node voltages. Sometimes there are both positive and negative power supplies, sometimes only a negative supply; and the use of dc current sources in biasing circuits is increasingly common The same circuit might contain two or more transistors, and these may even be of different types: npn and pnp, n-channel and p-channel FETs, or even BJTs and FETs. We now show that no new ideas are required to handle these variations. The only requirements are an understanding of principles already demonstrated and some flexibility.

6.4.2 NEGATIVE POWER SUPPLIES AND DUAL SUPPLIES

Some bias circuits that look different are actually the same because all *voltage drops and currents are identical.* For example, Fig. 6.8b differs from Fig 6.8a only in schematic notation. To change Fig. 6.8b to Fig. 6.8c, we replace the supply with series supplies that sum to V_{CC} and then move the ground reference. Neither operation changes voltage differences, branch currents, or transistor Q-point. For $\delta = 1$, the circuit is Fig. 6.8a, but for $\delta = 0$, we have the negative supply circuit of Fig. 6.8d. Figure 6.8e, with dual power supplies, represents the circuit for $0 < \delta < 1$. One case we often see is $\delta = 0.5$.

We conclude that Figs. 6.8a, d, and e are all, in fact, the *same bias circuit,* and the design techniques and theoretical principles we already know apply to all. Furthermore, since these circuit equivalences involved only changes in schematic conventions, *the conclusion also applies to FET biasing circuits.*

Our design techniques anticipated these variations by emphasizing use of node and branch voltages in a manner that *applies directly* to all cases. This means we do *not* first convert each problem to the form of Fig. 6.8a, but instead apply the principles directly to *the given circuit,* whatever its form. The following examples and exercises demonstrate this point while showing how to handle other variations in Q-point specifications.

6.4.3 DESIGN EXAMPLES

EXAMPLE 6.5 Find resistor values to establish the Q-point shown in Fig. 6.9a if $\beta = 10^{-3}$A/V^2 and $V_P = -3$ V; also, make $R_P = R_1 \| R_2 = 1$ MΩ.

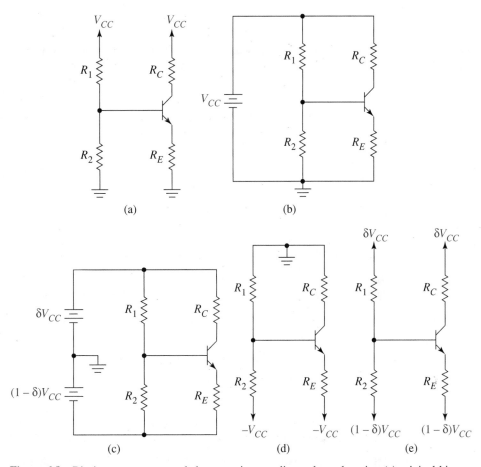

Figure 6.8 Biasing concepts extended to negative supplies and supply pairs: (a) original bias circuit; (b) alternate notation for original circuit; (c) same circuit using two supplies in series and with new ground reference; (d) negative supply circuit, an alternative notation when $\delta = 0$; (e) dual supply circuit, an alternative notation for any δ.

Solution. Of the 10 V available in the drain-source circuit, 4 V is allocated to the transistor, leaving 6 V for drops across R_D and R_S. With no specification to help us decide, we place 4 volts across R_S and the remaining 2 volts across R_D; a choice that gives better Q-point stability than assigning equal drops to R_D and R_S. With this agonizing decision behind us, $R_D = 2$ kΩ and $R_S = 4$ kΩ.

Using Eq. (5.21),

$$1 \times 10^{-3} = 1 \times 10^{-3}(V_{GS} + 3)^2$$

giving $V_{GS} = -2$ V. After placing known voltages on the diagram, Fig. 6.9b, we see that the node voltage at the gate is

$$V_G = -10 + 4 - 2 = -8 \text{ V}$$

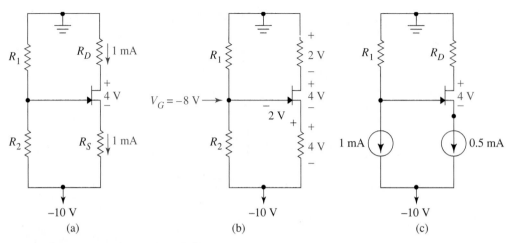

Figure 6.9 JFET bias circuit with negative supply: (a) given information; (b) information known when divider design begins; (c) with current-source biasing.

We must now design a voltage divider that divides -10 V down to -8 V using

$$\frac{R_1}{R_1 + R_2}(-10) = -8$$

To satisfy $R_P = R_1 \| R_2 = 1$ MΩ, we multiply by R_2 to give

$$R_P(-10) = -8R_2 = -10 \times 10^6$$

This leads to $R_2 = 1.25$ MΩ and $R_1 = 5.0$ MΩ. ❏

> **Exercise 6.5** In Fig. 6.9c, the transistor's drain node is to be biased at -3 V. Transistor parameters are $\beta = 2 \times 10^{-3}$A/V^2, $V_P = -1$ V. Find the necessary values for R_D, V_S, V_G, and R_1 in the order listed.
>
> *Ans.* 6 kΩ, -7 V, -7.5 V, 7.5 kΩ.

EXAMPLE 6.6 The BJT in Fig. 6.10a has β in the range $200 \le \beta \le 300$. Bias so that $I_C = 2$ mA, the *node voltage* at the collector is -5 V, and V_{CE} equals the drop across R_C.

Solution. $V_C = -5$ V implies that the drop across R_C is 5 V. Therefore, $R_C = 5/2 = 2.5$ kΩ. Allocating 5 V to V_{CE} leaves 2 V for the drop across R_E. Because the minimum β is so high, $I_E \approx 2$ mA, giving $R_E = 2/(2$ mA$) = 1$ kΩ. The emitter node voltage is

$$V_E = -12 + 2 = -10 \text{ V}$$

therefore

$$V_B = V_E + 0.7 = -9.3 \text{ V}$$

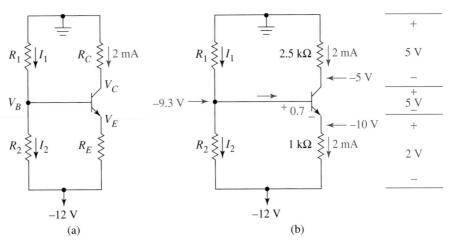

Figure 6.10 BJT biasing circuit with negative supply: (a) given Q-point information; (b) information known when divider design begins.

Figure 6.9b is the circuit diagram, updated with the information known at this point. Choosing I_2 equal to $10 \times I_{B,max}$ gives $I_2 = 0.1$ mA. Thus

$$R_2 = \frac{-9.3 - (-12)}{0.10 \text{ mA}} = 27 \text{ k}\Omega$$

and

$$R_1 = \frac{0 - (-9.3)}{0.11 \text{ mA}} = 84.5 \text{ k}\Omega$$

The point of this example is that once we learn to work comfortably with node voltages and voltage differences, we can apply the same design procedure to a variety of bias constraints. ❏

Exercise 6.6 The BJT in Fig. 6.11 has $\beta = 200$. Find I_O, R_C, and R_1 so that $I_C = 2$ mA, the *node voltage* at the collector is -5 V, and V_{CE} equals the drop across R_C.

Ans. $I_O = 2$ mA, $R_C = 2.5$ kΩ, $R_1 = 930$ kΩ.

Exercise 6.7 Find the new collector-emitter voltage for the circuit designed in Exercise 6.6 if the transistor placed in the circuit has $\beta = 300$ instead of the design value of 200.

Ans. $V_{CE} = 1.9$ V.

EXAMPLE 6.7 The pnp transistor of Fig. 6.12 has β in the range $10 \leq \beta \leq 100$. Bias for $I_C = 2$ mA and $V_C = 0$ V.

Solution. A good way to begin a pnp design is to show the current directions on the schematic as in Fig. 6.12. Since the collector is to be at ground potential, there must be 15 V across R_C. Thus $R_C = 15/(2 \text{ mA}) = 7.5$ kΩ.

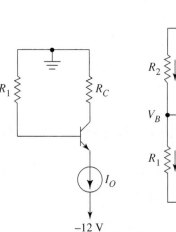

Figure 6.11 Bipolar transistor biased by a dc current source.

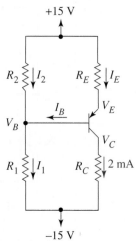

Figure 6.12 Four-resistor biasing circuit for pnp transistor using two supplies.

Because the minimum β is small, $I_E = (11/10)I_C = 2.2$ mA. The two supplies together provide the collector circuit with 30 V, of which 15 V are allocated to R_C. Anything unspecified is left to the judgment of the designer, so (somewhat arbitrarily) assign 5 V to V_{CE} and 10 V to R_E. The latter implies $R_E = 10/2.2 \times 10^{-3} = 4.55$ kΩ. With 10 V across R_E,

$$V_E = 15 - 10 = 5 \text{ V}$$

and

$$V_B = 5 - 0.7 = 4.3 \text{ V}$$

Notice from the diagram that I_2 is the smaller of the divider currents. Therefore, choose $I_2 = 10 \times I_{Bmax} = 2$ mA. Then

$$R_2 = \frac{15 - 4.3}{2 \times 10^{-3}} = 5.35 \text{ k}\Omega$$

The voltage across R_1 is $4.3 - (-15) = 19.3$ V, and its current is $(11)I_{Bmax} = 2.2$ mA; therefore,

$$R_1 = \frac{19.3}{2.2 \text{ mA}} = 8.77 \text{ k}\Omega$$

This completes the design.   ❏

The next example applies the same approach to a circuit with two transistors.

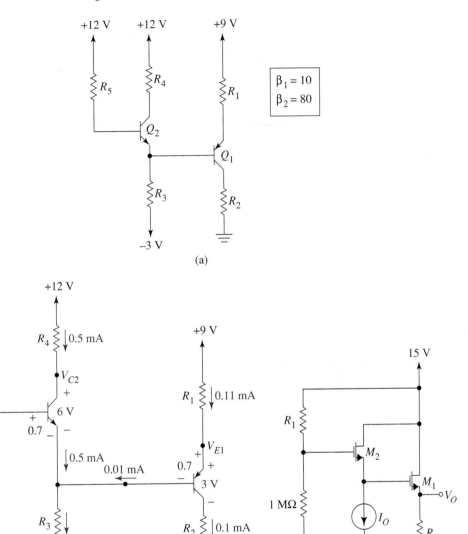

Figure 6.13 Design of direct-coupled two-transistor bias circuit: (a) bias circuit; (b) circuit illustrating computations when specified Q-points are given; (c) direct-coupled MOSFET circuit.

EXAMPLE 6.8 Design the circuit of Fig. 6.13a so that Q_1 and Q_2 are biased at $(V_{CE}, I_C) = (-3 \text{ V}, 0.1 \text{ mA})$ and $(6 \text{ V}, 0.5 \text{ mA})$, respectively.

Solution. Figure 6.13b shows the circuit diagram and the information implied by the design specifications, including base currents. Because of high β the emitter current of the npn transistor approximates the collector current.

Starting at the right, we see that of the 9 V available to the collector circuit of Q_1, 3 V is accounted for. For good Q-point stability, we assign 4 volts to the emitter resistor R_1, leaving 2 volts for R_2. This gives $R_1 = 4/0.11 = 36.4$ kΩ and $R_2 = 2/0.1 = 20$ kΩ.

Since $V_{E1} = 9 - 4 = 5$ V, $V_{B1} = 5 - 0.7 = 4.3$ V. In this circuit there is no voltage divider; however, the current in R_3 is $I_{E2} + I_{B1} = 0.51$ mA, so

$$R_3 = \frac{4.3 - (-3)}{0.51 \text{ mA}} = 14.3 \text{ k}\Omega$$

The collector voltage of Q_2 is

$$V_{C2} = V_{E2} + V_{CE2} = 4.3 + 6 = 10.3 \text{ V}$$

Therefore,

$$R_4 = \frac{12 - 10.3}{0.5 \text{ mA}} = 3.4 \text{ k}\Omega$$

Because the potential of node V_{B2} is 0.7 V above V_{B1},

$$V_{B2} = 4.3 + 0.7 = 5 \text{ V}$$

giving

$$R_5 = \frac{12 - 5}{6.25 \times 10^{-6}} = 1.12 \text{ M}\Omega$$

which completes the design. ❏

Exercise 6.8 Both transistors in Fig. 6.13c have $k = 4 \times 10^{-3}$ A/V^2 and $V_t = 0.6$ V. If V_O is to be 3 V, and $I_D = 2$ mA for both transistors, find R_S, V_{G1}, I_O, V_{G2}, and R_1. *Hint:* Calculate the quantities in the order listed.

Ans. 1.5 kΩ, 4.6 V, 2 mA, 6.2 V, $R_1 = 1.42$ MΩ.

EXAMPLE 6.9 The transistors in Fig. 6.14 are to be biased with the Q-point coordinates shown on the diagram. Find resistor values if $k = 2 \times 10^{-4}$, $V_t = -2$ V, and $\beta = 30$.

Solution. Choose 8 V for R_E and 4 V for R_C. Then

$$R_C = \frac{4}{0.2 \text{ mA}} = 20 \text{ k}\Omega$$

and

$$R_E = \frac{8}{0.21 \text{ mA}} = 38.1 \text{ k}\Omega$$

With 8 V across R_E, the node voltage at the base is

$$V_B = -15 + 8 + 0.7 = -6.3 \text{ V}$$

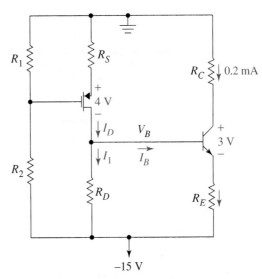

Figure 6.14 Bias circuit and Q-point information for Example 6.9.

Because I_D was not specified, we can make the current in R_D ten times the base current. This gives

$$I_1 = 10 \times \frac{0.2}{30} \text{ mA} = 0.067 \text{ mA}$$

The drop across R_D is

$$V_B - (-15) = -6.3 + 15 = 8.7 \text{ V}$$

therefore, $R_D = 8.7/(0.067 \text{ mA}) = 130 \text{ k}\Omega$.

The drain current of the FET is the sum of the base current and I_1, so $I_D = (0.2/30)$ mA $+ 0.067$ mA $= 0.074$ mA. Because drain voltage $V_D = V_B = -6.3$ V and $V_{DS} = -4$ V, the source voltage of the FET is -2.3 V. Therefore,

$$R_S = \frac{2.3}{0.074 \text{ mA}} = 31.1 \text{ k}\Omega$$

Since the FET is active,

$$I_D = 0.074 \text{ mA} = 10^{-4} (V_{GS} + 2)^2,$$

which gives $V_{GS} = -2.86$ V. With the source at -2.3 V, the gate voltage must be

$$V_G = V_{GS} + V_S = -2.86 - 2.3 = -5.16 \text{ V}$$

Designing the voltage divider requires a choice. To make $R_P = 2$ MΩ, for example, gives

$$-5.16 = \frac{R_1}{R_1 + R_2} \times (-15) \qquad \text{and} \qquad -5.16R_2 = 2 \times 10^6 \times (-15)$$

The latter gives $R_2 = 5.81$ MΩ and $R_1 = 3.05$ MΩ. This completes one of many possible designs that satisfy the given specifications.   ❏

Exercise 6.9   Modify the design of Example 6.9 by replacing R_2 with a current source, I_O, that keeps Q-points unchanged. Find I_O.

Ans. 1.69 μA.

6.5
Sensitivity

6.5.1 INTRODUCTION

The transistors and resistors we actually use to construct circuits are subject to random variations caused by manufacturing tolerances. Once the circuit is in operation, power supply voltages, environmental conditions, and component aging are additional sources of variation. In spite of all these uncertainties, the bias circuit must keep the transistor Q-points within specifications. In this situation, our understanding of Q-point stability, until now based on intuitive graphical ideas, acquires a mathematical foundation—parameter sensitivity.

The *sensitivity function* is an analysis tool we use to identify the more critical parameters in a given design, and to determine whether the design is satisfactory. We can also use sensitivity *backward* as a design tool to help us find the largest acceptable parameter variations consistent with satisfactory circuit performance.

6.5.2 SENSITIVITY DEFINITION

We define the sensitivity of a quantity Q to changes in a parameter P by the dimensionless expression

$$S_P^Q = \frac{P}{Q} \frac{dQ}{dP} \tag{6.14}$$

To understand the meaning of sensitivity, consider its incremental approximation,

$$S_P^Q \approx \frac{\Delta Q/Q}{\Delta P/P} \tag{6.15}$$

which expresses sensitivity as the ratio of the fractional increase in Q to a fractional increase in P. If an increase in P causes a decrease in Q, sensitivity is negative. Clearly, sensitivities of small magnitude are preferred.

In practice, sensitivity is simply a *number* that, once known, has important implications. For example, suppose for some bias circuit we know that the sensitivity of collector current to changes in emitter resistance is

$$S_{R_E}^{I_C} = 0.5$$

According to Eq. (6.15), if R_E increases by 1%, then there will be a resulting increase in I_C of

$$\frac{\Delta I_C}{I_C} = 0.5 \times \frac{\Delta R_E}{R_E} = 0.5 \times 0.01 = 0.005$$

Thus, when a 1% resistor is used for R_E, the uncertainty in I_C is 0.5%. This illustrates using sensitivity for *analysis*.

Turning the example around, suppose our *design specifications* require that I_C not change by more than 2%. In general terms, we can express this requirement as

$$\frac{\Delta I_C}{I_C} = S_{R_E}^{I_C} \frac{\Delta R_E}{R_E} \leq 0.02$$

If the sensitivity of our bias circuit is 0.5, this means

$$\frac{\Delta R_E}{R_E} \leq \frac{0.02}{0.50} = 0.04$$

To meet specifications, we must hold R_E variations to 4%. This shows sensitivity as a design tool.

When parameter variations are as large as 10 or 20%, sensitivity estimates require caution. Figure 6.15 suggests that the derivative in Eq. (6.14) might not always be a close approximation to the incremental ratio $\dfrac{\Delta Q}{\Delta P}$ upon which we base our conclusions. Nevertheless, sensitivity is a valuable guide, even if our predictions are sometimes only estimates. We next consider the *mathematical functions* that underlie the numerical sensitivity values.

6.5.3   THEORETICAL SENSITIVITY FUNCTIONS

One way to compute a numerical sensitivity is to evaluate a function obtained from analysis equations. For example, consider the four-resistor bias circuit of Fig. 6.7a, which is described by

$$I_C = \frac{\beta(V_{TT} - V_{BE})}{R_T + (\beta + 1)R_E} \tag{6.16}$$

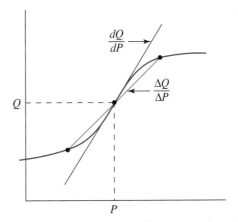

Figure 6.15   Approximating a ratio of increments by a derivative.

To find the sensitivity of quantity I_C to parameter β, we first differentiate. This gives

$$\frac{dI_C}{d\beta} = \frac{[R_T + (\beta + 1)R_E](V_{TT} - V_{BE}) - [\beta(V_{TT} - V_{BE})R_E]}{[R_T + (\beta + 1)R_E]^2}$$

The next step is to multiply by the reciprocal of the circuit quantity Q *in its original functional form.* In this case, this means

$$\frac{1}{I_C}\frac{dI_C}{d\beta} = \frac{[R_T + (\beta + 1)R_E]}{\beta(V_{TT} - V_{BE})} \frac{[R_T + (\beta + 1)R_E](V_{TT} - V_{BE}) - [\beta(V_{TT} - V_{BE})R_E]}{[R_T + (\beta + 1)R_E]^2}$$

$$= \frac{1}{\beta}\frac{[R_T + R_E]}{[R_T + (\beta + 1)R_E]}$$

This step usually leads to extensive simplification as it did here. Finally, we multiply by parameter P, β in this case, giving

$$S_\beta^{I_C} = \frac{[R_T + R_E]}{[R_T + (\beta + 1)R_E]} \tag{6.17}$$

For the same bias circuit the sensitivity to changes in V_{BE} is

$$S_{V_{BE}}^{I_C} = \frac{-V_{BE}}{V_{TT} - V_{BE}} \tag{6.18}$$

Exercise 6.10 Derive Eq. (6.18).

Equations (6.17) and (6.18) describe all the circuits of Fig. 6.8, provided that V_{TT} in Eq. (6.16) is taken to be the Thevenin equivalent voltage between the transistor base and the nonemitter side of the emitter resistor.

Once we know the sensitivity function, we substitute parameter values for particular designs to obtain numerical sensitivities.

EXAMPLE 6.10 Evaluate $S_{V_{BE}}^{I_C}$ for the bias circuit of Example 6.6.

Solution. Component values were $R_1 = 84.5 \text{ k}\Omega$ and $R_2 = 27 \text{ k}\Omega$. For this design the Thevenin voltage V_{TT} is

$$V_{TT} = \frac{27}{84.5 + 27} \times 12 = 2.91 \text{ V}$$

With 0.7 V for V_{BE}, Eq. (6.18) gives

$$S_{V_{BE}}^{I_C} = \frac{-0.7}{2.91 - 0.7} = -0.317 \qquad \square$$

Exercise 6.11 Evaluate $S_\beta^{I_C}$ for the bias circuit of Example 6.6 for both limiting values of β.

Ans. 0.0971, 0.0669.

The sensitivity equations for the FET four-resistor circuit are slightly harder to derive. Since the transistor operates in the active region, the drain current is

$$I_D = \frac{k}{2}(V_{GS} - V_t)^2$$

We saw in Fig. 6.6 that the circuit constraint, Eq. (6.11), must also be satisfied; thus we calculate sensitivities from

$$I_D = \frac{k}{2}(V_{GG} - I_D R_S - V_t)^2 \tag{6.19}$$

where I_D appears on both sides. It is hoped that the reader left calculus class with the ability to compute the derivative of I_D with respect to R_S for such an equation. Since no new electronics concepts are involved, we leave sensitivity of I_D to Problem 6.39.

Algebraic sensitivity expressions facilitate calculating numerical sensitivities, but more important, they show which circuit parameters contribute to sensitivity and in what ways. Often, examining the algebraic expression gives a clue about how to redesign a circuit for lower sensitivity. For example, rewriting Eq. (6.17) as

$$S_\beta^{I_C} = \frac{1}{1 + \beta R_E/(R_T + R_E)}$$

makes it clear that we should make R_T as small as possible to reduce sensitivity of collector current to changes in β. We previously reached the same conclusion intuitively after examining Eq. (6.12); however, the sensitivity expression further shows that (1) the best we can do is $S_\beta^{I_C} = 1/(1 + \beta)$, and (2) sensitivity is lower for higher β.

Multiparameter Sensitivity. In a given circuit there are usually many design parameters, and the total variation in a quantity like I_C depends upon the combined variations of the individual parameters. To understand this general case, suppose Q is a function of N parameters, $p_1, p_2, \ldots, p_N$; that is,

$$Q = Q(p_1, p_2, \ldots, p_N)$$

If individual parameter changes are small, the total change, ΔQ, is given by the chain rule

$$\Delta Q = \frac{\partial Q}{\partial p_1}\Delta p_1 + \frac{\partial Q}{\partial p_2}\Delta p_2 + \cdots + \frac{\partial Q}{\partial p_N}\Delta p_N$$

Dividing both sides by Q gives

$$\frac{\Delta Q}{Q} = \frac{1}{Q}\frac{\partial Q}{\partial p_1}\Delta p_1 + \frac{1}{Q}\frac{\partial Q}{\partial p_2}\Delta p_2 + \cdots + \frac{1}{Q}\frac{\partial Q}{\partial p_N}\Delta p_N$$

Multiplying each term on the right side by one, in the form p_i/p_i, gives

$$\frac{\Delta Q}{Q} \approx \frac{p_1}{Q}\frac{\partial Q}{\partial p_1}\frac{\Delta p_1}{p_1} + \frac{p_2}{Q}\frac{\partial Q}{\partial p_2}\frac{\Delta p_2}{p_2} + \cdots + \frac{p_N}{Q}\frac{\partial Q}{\partial p_N}\frac{\Delta p_N}{p_N}$$

or

$$\frac{\Delta Q}{Q} = S^Q_{p_1} \frac{\Delta p_1}{p_1} + S^Q_{p_2} \frac{\Delta p_2}{p_2} + \cdots + S^Q_{p_N} \frac{\Delta p_N}{p_N} \qquad (6.20)$$

where the ordinary derivative in Eq. (6.14) generalizes to a partial derivative for the multiparameter case. For small parameter variations, the fractional variation in Q equals the sum of the fractional parameter variations in the individual parameters, each weighted by the sensitivity of Q to that parameter.

Equation (6.20) has a number of uses. First, we can use it to estimate the expected numerical variation ΔQ for a given design when we know the expected parameter variations. Second, we can compute ΔQ for alternative circuit designs to see which gives the smallest variation. Third, by examining the relative contributions of the individual terms to $\Delta Q/Q$, we can determine which parameters are most critical to the design and conceivably assign them smaller tolerances.

The following example shows the usefulness of Eq. (6.20) when the *numerical* sensitivity values are known.

EXAMPLE 6.11 Node voltage V in a particular circuit is a function of four resistors. The sensitivities are

$$S^V_{R_1} = -0.2, S^V_{R_2} = 1.9, S^V_{R_3} = 0.2, S^V_{R_4} = 0.01$$

(a) Which resistor is the most critical in this design?
(b) Estimate the fractional variation in V if all resistors happen to be 10% above their nominal values.
(c) Estimate the *worst case* fractional variation in V if all resistors are within $\pm 10\%$ of their nominal values.
(d) Redo part (c) if the tolerance is $\pm 1\%$ for R_2 but is $\pm 10\%$ for the others.

Solution. (a) R_2 is the most critical because it contributes the highest sensitivity.
(b) Equation (6.20) gives

$$\frac{\Delta V}{V} \approx -0.2 \times 0.1 + 1.9 \times 0.1 + 0.2 \times 0.1 + 0.01 \times 0.1 = 0.191$$

making V about 19% too high.
(c) The worst case is when algebraic signs of all variations agree with signs of corresponding sensitivities, for example, R_1 is 10% low ($\Delta R_1/R_1 = -0.1$) while all other resistors are 10% high. In this unlucky circumstance,

$$\frac{\Delta V}{V} \approx -0.2 \times (-0.1) + 1.9 \times 0.1 + 0.2 \times 0.1 + 0.01 \times 0.1 = 0.231$$

(d) Holding the most critical resistor to 1% (a possibility in discrete circuits) gives a worst case variation of

$$\frac{\Delta V}{V} \approx -0.2 \times (-0.1) + 1.9 \times (0.01) + 0.2 \times 0.1 + 0.01 \times 0.1 = 0.06$$

a result that underscores the importance of identifying critical design parameters. ❑

SPICE calculates *numerical* sensitivities. The resulting ability to identify critical parameters allows us to focus our efforts on deriving sensitivity *functions* only for those parameters most likely to improve a design. Without an initial identification of critical parameters, we would be doomed to working out algebraic expressions for all the sensitivity functions. The next section includes SPICE sensitivity analysis among other topics.

6.6
SPICE Bias Circuit Analysis

6.6.1 INTRODUCTION

SPICE examples in Chapter 5 demonstrated finding transistor Q-points at various temperatures. Now that we understand bias circuit design and the concept of Q-point stability, we can better appreciate this capability. In this section we also examine other interesting uses of SPICE in bias circuit analysis.

During physical assembly of discrete circuits, each resistor is, in effect, selected from a bin labeled with some nominal value. Nominal values do not include all conceivable resistances—only decimal multiples of a few particular numbers that correspond to specified *tolerances*. Table 6.1 shows *nominal values* for resistors having 5% and 10% tolerances. (Tolerances of 1% and 20% are also available but are not included in the table.) When we use 10% resistors, we replace each resistance in the original design with the nearest nominal value from the first or third row of the table or, at additional expense, with series or parallel combinations of these values. A SPICE simulation readily shows the effects of substituting nominal values before the circuit is physically assembled.

In Chapter 4 we used SPICE to calculate temperature-induced changes in transistor parameters. With SPICE we can also specify temperature coefficients for resistors by adding the statement "$TC = value$" at the end of the resistor element line, where TC is the SPICE code for temperature coefficient α. With this information SPICE calculates resistance changes using Eq. (3.9), thus facilitating realistic simulations of circuit operation over wide ranges of temperatures.

SPICE runs using maximum, minimum, and typical values also make it easy to explore variations in parameters such as β, V_t, and k. Also, upon request, SPICE computes the numerical values of "sensitivities" during a dc analysis to help us identify critical parameters and estimate variations in important circuit quantities.

6.6.2 SPICE SIMULATION OF TEMPERATURE VARIATION

The next example illustrates the more basic SPICE capabilities related to simulating temperature variations.

TABLE 6.1 Discrete Resistor Nominal Values

Tolerance																
10% 5%	*Discrete Resistors Are Available Only in Decimal Multiples of These Values															
* *	1.0		1.5		2.2		3.3		4.7		6.8					
*		1.1	1.3	1.6	2.0	2.4	3.0	3.6	4.3	5.1	6.2	7.5	9.1			
* *		1.2		1.8		2.7		3.9		5.6		8.2				

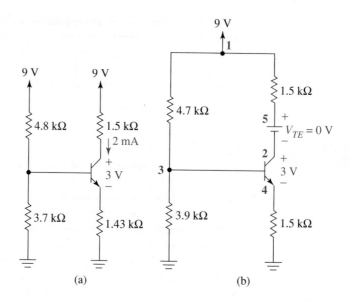

(a)   (b)

```
EXAMPLE 6.12
VCC 1 0 DC 9
VTE 5 2 DC 0
R1 1 3 4.7K TC=0.004
R2 3 0 3.9K TC=0.004
RC 1 5 1.5K TC=0.004
RE 4 0 1.5K TC=0.004
Q1 2 3 4 BIPO
.MODEL BIPO NPN BF=20 XTB=1.7
.TEMP -40 27 70
.END
```

(c)

Figure 6.16 SPICE simulation of Example 6.12: (a) original design; (b) SPICE circuit showing closest nominal 10% resistor values; (c) SPICE code for analysis of temperature-induced variations in Q-point.

EXAMPLE 6.12 Figure 6.16a shows the circuit we designed in Example 6.2. Suppose this is a design prototype for a discrete circuit that must operate from $-40°C$ to $70°C$, with collector current remaining within $\pm40\%$ of its 2 mA nominal value. Use SPICE to see if the circuit satisfies specifications with nominal 10% resistors having temperature coefficients of $\alpha = 0.004/°C$ and with $20 \leq \beta \leq 200$. Include realistic temperature variations in β as indicated by Eq. (4.19) by specifying XTB = 1.7.

Solution. In Fig. 6.16b resistor design values are replaced by nearest nominal 10% values from Table 6.1. Source VTE monitors collector current. Figure 6.16c gives the SPICE code, where the ".TEMP" line provides for analysis at the temperature extremes

TABLE 6.2 Variations of Parameters and Q-Point with Temperature

T (°C)	R1 (Ω)	R2 (Ω)	RC (Ω)	RE (Ω)	β	V_{CE} (V)	I_C (mA)	ΔI_C (%)
−40	3.44 k	2.86 k	1.10 k	1.10 k	13.0	3.40	2.46	+23
27	4.70 k	3.90 k	1.50 k	1.50 k	20.0	2.98	1.96	−2
70	5.50 k	4.57 k	1.76 k	1.76 k	25.1	2.77	1.74	−13

as well as at the 27°C nominal room temperature. For the first run, β = 20; a second run checks the case β = 200.

Table 6.2 shows selected data generated by the simulation. The middle row shows that the Q-point coordinates remain close to design values, (3 V, 2 mA), even after changing three resistances to available 10% values. The first row shows what happens when the temperature is lowered to −40°C. All resistances decrease significantly as does β. Not shown explicitly in the table are increases in V_{BE} and I_{CEO} (see Sec. 4.9.1) that were automatically included in the simulation. The net result of all these variations is an increase in both Q-point coordinates, with I_C being 23% higher than nominal. At 70°C, β and all resistances are higher than nominal, and collector current is 13% low. Notice that V_{CE} never gets close to the saturation value of 0.2 V nor does I_C approach the cut-off value of zero. ❑

After a simulation, it is a good habit to check the data to confirm our understanding of how the simulation was carried out.

Exercise 6.12 Use a calculator and Eq. (4.19) to predict, to six-digit accuracy, the 70°C value of β in Table 6.2 from the 27°C value. Use Eq. (3.9) to check the change in RE in the same fashion.

Ans. β(70) = 25.1145, RE(70) = 1.75800 kΩ.

Table 6.3 shows the Q-point variations computed during a second SPICE run using room temperature beta of 200. Again the Q-point remains well away from cut-off and saturation at all temperatures. From the two tables we conclude that the most critical circumstance is the combination of low temperature and high beta, where our original design fails to meet the specification. Although the Q-point is fairly stable, considering the extreme temperature variations, some redesign and/or closer resistor and transistor

TABLE 6.3 Q-Point Variations When β = 200

T(°C)	V_{CE} (V)	I_C (mA)	ΔI_C (%)
−40	2.70	2.86	43
27	2.49	2.17	8.5
70	2.35	1.89	−5.5

tolerances are needed before the given specifications can be satisfied. At this point a sensitivity analysis might help identify the more critical parameters for closer study.

6.6.3 SPICE SENSITIVITY ANALYSIS

SPICE does not use Eq. (6.14) to compute sensitivity. Instead, it computes related quantities called element sensitivity and normalized sensitivity. The *element sensitivity* of a quantity Q to a parameter P is defined as

$$E_P^Q = \frac{\partial Q}{\partial P}$$

From comparing definitions it is clear that

$$S_P^Q = \frac{P}{Q} E_P^Q \tag{6.21}$$

SPICE also calculates *normalized sensitivity N,* defined as

$$N_P^Q = \frac{P}{100} \frac{\partial Q}{\partial P} \approx \frac{\Delta Q}{100(\Delta P/P)}$$

It follows that

$$S_P^Q = \frac{100}{Q} N_P^Q \tag{6.22}$$

With Eqs. (6.21) and (6.22) we can convert from element or normalized sensitivity to "sensitivity."

In a dc analysis, the SPICE command ".SENS Q" gives both element and normalized sensitivities of quantity Q to variations in *all* of the circuit parameters. (This can create a rather large output file.) In SPICE, quantity Q can be any node voltage, node-voltage difference, voltage-source current, or current-source voltage.

Sometimes we must be creative to obtain the particular sensitivity we need. To find the sensitivity of a transistor output current, for example, we place an independent dc voltage source, VTE, of zero volts in series with the transistor drain or collector; then its current, I(VTE), is used for Q in the .SENS command. The following examples demonstrate these points.

EXAMPLE 6.13 Use SPICE to compute the room temperature sensitivities of I_C in the circuit of Fig. 6.16b.

Solution. Figure 6.17a gives the SPICE code. Relative to the code of Fig. 6.16c, we deleted the resistor temperature coefficients and .TEMP command, and added the .SENS command.

The last column in Fig. 6.17b lists the sensitivities. These were converted by hand from the SPICE sensitivities in the second column, using Eq. (6.22) with $Q = I_C = 1.96$ mA, the SPICE output value.

EXAMPLE 6.13
```
VCC 1 0 DC 9
VTE 5 2 DC 0
R1 1 3 4.7K
R2 3 0 3.9K
RC 1 5 1.5K
RE 4 0 1.5K
Q1 2 3 4 BIPO
.MODEL BIPO NPN
+BF=20 XTB=1.7
.SENS I(VTE)
.END
```
(a)

Element $P =$	Normalized sensitivity $N_P^{I_C}$ (amps/percent)	Sensitivity $S_P^{I_C}$
R_1	−1.465E−05	−0.733
R_2	1.450E−05	+0.725
R_C	−7.836E−14	-3.92×10^{-9}
R_E	−2.133E−05	−1.067
V_{CC}	2.667E−05	+1.334
β_F	2.569E−07	+0.013
I_S	1.690E−07	+0.008

(b)

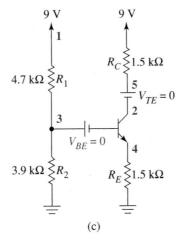

(c)

Figure 6.17 Sensitivity analysis of the circuit of Fig. 6.16: (a) SPICE code; (b) sensitivity values; (c) addition of source to obtain sensitivity to V_{BE}.

It is interesting to note that sensitivity to supply voltage V_{CC} is the highest. Of the *transistor* parameters listed, the reverse saturation current I_S is much less important than the transistor β, denoted by β_F. As we might expect from the discussion of Sec. 6.3, *RE* is the most critical of the four resistors. Notice that *RC*, isolated from the base-emitter circuit by the dependent current source in our large-signal circuit model, has little effect on I_C. ❏

Intuition suggests that sensitivity of I_C to collector resistance would increase if the dependent source were made less perfect by including transistor output resistance. The following exercise explores this conjecture.

Exercise 6.13 Use SPICE to compute $S_{RC}^{I_C}$ when the transistor in Example 6.13 has Early voltage of 95 V.

Ans. $S_{RC}^{I_C} = 0.00334$.

Conspicuously missing in Fig. 6.17b is a row for V_{BE}, which we previously identified as the most critical temperature-sensitive transistor parameter. To SPICE, V_{BE} is the voltage coordinate of the Q-point on the transistor's input characteristic, not a circuit parameter; so we must use an artifice to find this sensitivity. When we place a voltage source, V_{BE}, of zero volts in series with the base as in Fig. 6.17c, the Q-point is unchanged; however, SPICE computes sensitivities to changes in *this* V_{BE}, and these sensitivities describe the consequences of the input characteristic shifting to the right or left.

Using this extra source in the previous example gives

$$S_{V_{BE}}^{I_C} = -0.264$$

a result consistent with our previous observation that V_{BE} is the most critical of the transistor parameters in bias circuit design.

Sensitivities to changes in threshold (or pinch-off) voltage, like sensitivity to V_{BE} in the BJT, are not computed in SPICE analysis of FET circuits. We can add a dc voltage source in series with the transistor gate as in Fig. 6.18 to circumvent this omission. By differentiating I_D in the equation

$$I_D = \frac{k}{2}(v_{GS} - V_t)^2 = \frac{k}{2}(v_{NS} - VT - V_t)^2$$

we can verify that

$$E_{VT}^{I_D} = E_{V_t}^{I_D}$$

because the derivative of I_D with respect to VT is the same as the derivative with respect to V_t.

EXAMPLE 6.14 Use SPICE to compute the sensitivities of I_D and V_{DS} in the bias circuit we designed in Example 6.1. The transistor has an Early voltage of 40 V and body-effect coefficient of 0.37 volt. Use 5% resistors.

Solution. Figure 6.19a shows the circuit with resistors chosen from closest available nominal values. Source VTE makes I_D available in the sensitivity analysis, and VT in-

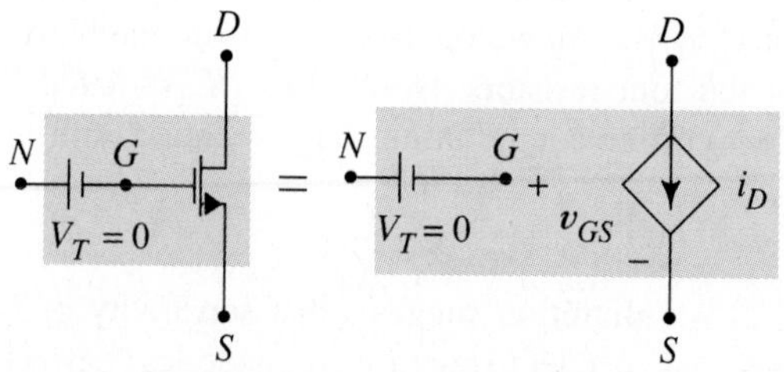

Figure 6.18 Adding a source to obtain SPICE sensitivity to threshold voltage.

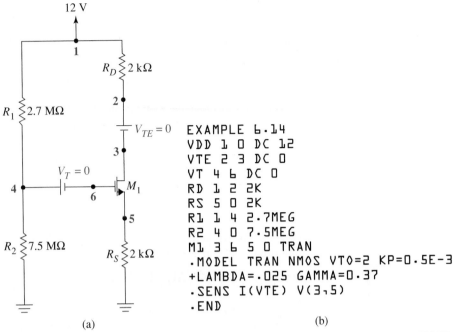

```
EXAMPLE 6.14
VDD 1 0 DC 12
VTE 2 3 DC 0
VT 4 6 DC 0
RD 1 2 2K
RS 5 0 2K
R1 1 4 2.7MEG
R2 4 0 7.5MEG
M1 3 6 5 0 TRAN
.MODEL TRAN NMOS VTO=2 KP=0.5E-3
+LAMBDA=.025 GAMMA=0.37
.SENS I(VTE) V(3,5)
.END
```

(a)   (b)

Figure 6.19   Worst case sensitivity analysis for Example 6.14: (a) biasing circuit; (b) SPICE code.

troduces sensitivity to variations in threshold voltage. Figure 6.19b shows the SPICE code. Since V(3, 5) denotes V_{DS}, the .SENS statement requests a sensitivity analysis for both Q-point coordinates.

The last column of Table 6.4 lists the drain current sensitivities calculated from the SPICE data in the preceding columns. The first five rows were computed from normalized sensitivities by using $Q = I_D = 1.87$ mA, the SPICE-calculated nominal drain current. The last entry was computed from element sensitivity using $P = V_t = 2.0$ V. The value of V_{DS} computed by SPICE was 4.51 V. Using this value for Q we compute the last column of Table 6.5.

TABLE 6.4   Sensitivities of Drain Current

Element Name P =	Element Sensitivity (amps/unit)	Normalized Sensitivity (amps/%)	Sensitivity $S_P^{I_D}$
RD	−1.827E-8	−3.654E-7	−0.020
RS	−7.007E-7	−1.402E-5	−0.749
R1	−2.896E-10	−7.820E-6	−0.418
R2	1.043E-10	7.820E-06	0.418
VDD	2.559E-04	3.071E-05	1.640
VT	3.348E-04	0.000	−0.358

TABLE 6.5   Sensitivities of Drain-Source Voltage

Element Name $P =$	Element Sensitivity (V/Unit)	Normalized Sensitivity	Sensitivity $S_P^{V_{DS}}$
RD	−1.799E-3	−3.598E-2	−0.798
RS	9.307E-4	1.861E-2	0.412
R1	1.158E-6	3.128E-2	0.694
R2	−4.17E-7	−3.128E-2	−0.694
VDD	−2.373E-2	−2.847E-3	−0.0631
VT	1.339	0.000	0.594

Exercise 6.14   Use data from Tables 6.4 and 6.5 to estimate the worst case fractional variations $\Delta I_D / I_D$ and $\Delta V_{DS} / V_{DS}$ when resistor variations are ±10%, power supply variations are ±1%, and V_t variations are 20%.

Ans. $\Delta I_D / I_D = 0.249$, $\Delta V_{DS} / V_{DS} = 0.379$.

It is prudent to use sensitivity results with care. We already noted that sensitivity calculations use derivatives to estimate ratios of incremental changes. Within this context, 10% and 20% variations are not all that small. Some cautions about worst case estimates are also appropriate. In Exercise 6.14, we assumed that *every* parameter took on its *worst possible* value so as to maximize variation in the quantity of interest—an occurrence that is, in fact, highly unlikely. In general, worst case estimates tend to be very safe but unduly pessimistic. Moreover, they can even be inconsistent. Notice that when we estimated ΔI_D, we assumed ΔR_S negative because its sensitivity in Table 6.4 is negative, whereas in estimating ΔV_{DS}, we took ΔR_S to be positive because its sensitivity in Table 6.5 is positive. Certainly the variation in R_D cannot be *both* positive and negative in the same circuit. Clearly, our worst case estimates do not take into account interrelationships between sensitivities.

Some SPICE implementations allow the user to include random selection of parameter values and large numbers of computations to provide statistical information about parameter values. Such *Monte Carlo analyses* give more realistic predictions of circuit behavior than worst case studies. Because Monte Carlo analysis involves background in statistics and is not included in all SPICE packages, we leave this useful technique for another course, self-study, or on-the-job training.

6.6.4   CURRENT-SOURCE BIASING

The biasing circuits we have studied are simple in structure, easy to design, and inexpensive for discrete realizations, and the sensitivity values demonstrated in our examples are adequate for many applications. In integrated circuit designs, however, resistors occupy considerable chip space, making them more expensive than transistors. To reduce cost and improve sensitivity, IC designers employ circuits that rely on independent current sources rather than resistors for biasing.

To ensure low sensitivity, our principal design guideline is to maximize the emitter (source) resistor within the context of a given power supply and a specified Q-point. It is clear from the examples that a design trade-off places an upper limit on the size of this resistor. With transistor voltage specified, the remainder of the supply voltage is shared by the emitter (source) resistor and collector (drain) resistor. Because these have near-identical (identical) currents, increasing emitter (source) resistance reduces collector (drain) resistance and, therefore, voltage gain.

Figure 6.20a shows an idea that might free us from this design impasse. For low sensitivity, we make the source resistor infinite; to provide specified bias current, we provide a current source. Since I_{BB} delivers the required drain current, our intuition foresees zero sensitivity. We can also achieve high gain, for we learned in Chapter 1 that the voltage across a current source can have any value; thus theoretically V_S can be small or even negative, leaving more of the supply voltage available for a larger R_D. It turns out that practical current sources require a small positive V_S and also have internal resistance; nevertheless, the idea is a good one. With the current source we can assign voltages V_{DS} and V_{RD} without the constraint of ensuring that V_S be large. This same idea works for BJT biasing, where a current source replaces the emitter resistor.

The practical current sources we have seen so far—made from depletion MOSFETs and JFETs—have volt-ampere curves like that of Fig. 6.20b. Because of Early effect there is a finite but large output resistance, r_o. When we use such a source for biasing, current-source output resistance replaces the emitter or source resistance insofar as sensitivity is concerned; however, r_o is a much larger resistance than we could use if it had to carry bias current. (I_D of 1 mA flowing through $R_S = 120$ kΩ would require a power supply higher than 120 V!) Although the current-source transistor requires a positive output voltage for active-state operation, this voltage can often be small.

To explore current source biasing, consider Fig. 6.20c, the circuit of Figure 6.19 with the 2 kΩ source resistor replaced by a current source of 2 mA and 100 kΩ output resistance. Table 6.6 compares the sensitivities computed by SPICE for the two circuits.

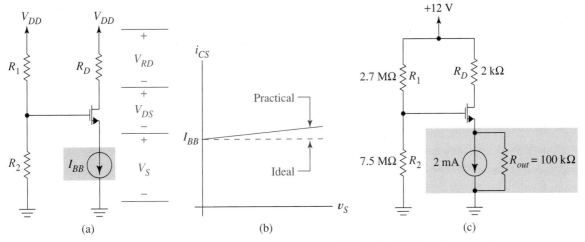

Figure 6.20 Biasing with a current source: (a) example biasing circuit; (b) volt-ampere characteristic of the current source; (c) equivalent circuit.

TABLE 6.6 Sensitivities with Current-Source Biasing

Element Name	$S_P^{I_D}$ (Biased with R_S)		$S_P^{V_{DS}}$ (Biased with R_S)	
RD	−0.000543	(−0.020)	−0.923	(−0.798)
ROUT (RS)	−0.0177	(−0.749)	0.0114	(0.412)
R1	−0.0102	(−0.418)	0.492	(0.694)
R2	−0.0102	(0.418)	−0.492	(−0.694)
VDD	0.0400	(1.640)	0.859	(−0.631)
VT	−0.00871	(−0.358)	0.422	(0.594)
IBB	0.976		0.371	

The sensitivities of the original circuit, copied from Tables 6.4 and 6.5, are listed in parentheses.

All drain-current sensitivities are reduced by one or two orders of magnitude with current-source biasing. Sensitivity to power supply, dominant in the four-resistor circuit, is greatly reduced. Sensitivities of drain-source voltage do not show the same pronounced improvements; however, drain current is more critical to small-signal parameters such as g_m.

In the next section we learn a variety of new ways to construct current sources, including some sources having exceptionally high output resistance.

6.7
Current Sources

6.7.1 INTRODUCTION

In integrated circuits, current sources usually provide the bias currents required for active-state transistor operation. They are also used as *active loads* to increase amplifier gain. Current sources and integrated circuits are a quite a good match—current sources occupy less IC chip space than biasing resistors, while most current-source designs require *matched transistors* available only in IC technology. Matched transistors have identical parameter values that track with temperature: identical parameters because they are fabricated under identical conditions, temperature tracking because transistors separated on a chip by only a micron or two have identical thermal environments.

The current sources we study next have two parts: a *reference current* that is relatively independent of temperature and parameter variations, and a *current mirror* that copies or mirrors the reference current to another circuit location.

Essential to the mirror is the *diode-connected transistor* of Fig. 6.21, a two-terminal subcircuit having a diodelike volt-ampere characteristic. In Sec. 5.3 we described the diode-connected FET as a *nonlinear resistor* in which the FET is either OFF or active. The diode-connected BJT in Fig. 6.21b is also either OFF or active by the following reasoning. Since $v_{CE} = v_{BE}$, if v_{CE} is less than cut-in, the transistor is cut off. If v_{CE} exceeds cut-in, the transistor is ON and, because

$$v_{CE} > 0.5 \text{ V} > V_{CE,sat}$$

active. The volt-ampere curve is the diodelike characteristic of Fig. 6.21b. In current sources the diode-connected transistor is always active.

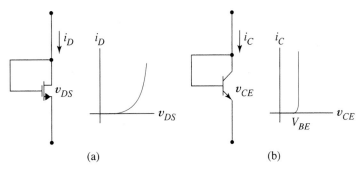

Figure 6.21 Diode-connected transistors and their volt-ampere curves: (a) MOSFET circuit; (b) BJT circuit.

The current source plays such an important role in linear ICs that its identification is often the key to understanding an unfamiliar circuit. The diode connection usually makes current sources easy to locate in a schematic.

There are several well-known current mirror circuits offering different compromises between circuit complexity, high output resistance, and insensitivity to internal parameter variations. We begin with the simplest, the *basic mirror circuit.*

6.7.2 BASIC CURRENT SOURCES

FET Current Mirror. Figure 6.22a shows the basic FET *current mirror,* a diode-connected reference transistor, M_R, and a matched mirror transistor, M_M, connected so as to have identical gate-source voltages. Output current, I_O, is a replica, possibly scaled, of an input reference current I_{REF}. We take for granted that I_{REF} is precisely determined and independent of temperature variations, and show next exactly how it is mirrored.

Active operation for M_R implies

$$I_{REF} = \frac{k_R}{2}(V_{GS} - V_{tR})^2 \tag{6.23}$$

Since I_{REF} is fixed, this implies a particular V_{GS}, say, the value we find from projecting from the blue transfer curve labeled T_1 in Fig. 6.22b. Because M_M is active and shares the *same* v_{GS} as M_M, I_O follows from projecting from the blue transfer curve for the mirror transistor in Fig. 6.22b, a plot of

$$I_O = \frac{k_M}{2}(V_{GS} - V_{tM})^2 \tag{6.24}$$

Because parameters are matched, the blue curves are identical, and $I_O = I_{REF}$. Now variations in parameters k_R and V_{tR} due to temperature changes or manufacturing tolerances cause variations in V_{GS}. For example, the "T_1" curve might turn into the "T_2" curve. Since transistors are matched, variations in k_M and V_{tM} replicate changes in k_R and V_{tR}. Thus $I_O = I_{REF}$ *in spite of manufacturing tolerances or wide variations in temperature,* as we see from the projections in Fig. 6.22b.

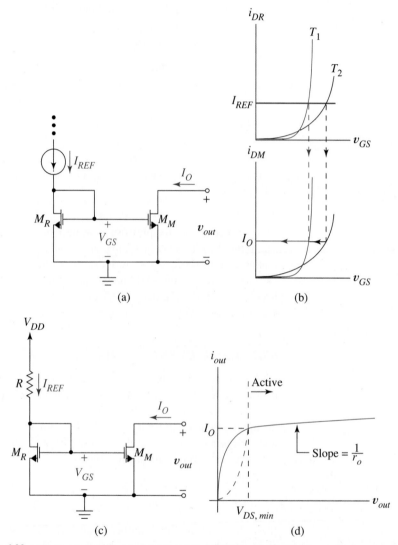

Figure 6.22 Basic current source: (a) reference current and mirror; (b) matched transistors track reference current; (c) mirror with simple reference current circuit; (d) volt-ampere curve at current-source output.

The IC fabrication process allows us to provide a temperature-independent scale factor to relate I_{REF} and I_O. Taking the ratio of Eq. (6.24) to (6.23) and recalling Eq. (5.8) gives

$$I_O = \frac{k_M}{k_R} I_{REF} = \frac{(W/L)_M}{(W/L)_R} I_{REF}$$

Thus the photographic masks that determine the transistor geometries also precisely fix the current ratio, allowing the designer to make I_O a scaled-up, scaled-down, or unscaled, temperature-independent copy of I_{REF}.

Current-Source Design. Figure 6.22c is a popular current source in which a power supply and a resistor of low temperature coefficient provide the reference current. In a typical design problem we know output current, transistor parameters, V_{DD}, and the current scale factor; design means finding R. Starting with I_O, we use the scale factor to find I_{REF}. If there is no reason to do otherwise, make $I_{REF} = I_O$. We next use Eq. (6.23) to compute V_{GS}. Then

$$I_{REF} = \frac{V_{DD} - V_{GS}}{R} \tag{6.25}$$

where only R is unknown. In a good design, V_{DD} is large compared to V_{GS}; and V_{DD} and R are both more or less independent of temperature, making I_{REF} relatively independent of temperature.

For M_M to function as a current source it must operate in the active region, as in Fig. 6.22d. This requires $V_{DG} \geq -V_t$. Since the output voltage of the *basic* current source is the drain-source voltage of the mirror transistor, we can rewrite the inequality as

$$V_{out} - V_{GS} \geq -V_t$$

which gives

$$V_{out} \geq V_{GS} - V_t$$

This sets a lower limit to the output voltage for correct operation. Notice that once we determine V_{GS} during the design, we can specify the minimum output voltage.

EXAMPLE 6.15 Design a constant current source that produces a 0.19 mA output current. Use identical transistors with $k = 0.2 \times 10^{-4}$ A/V², $V_t = 1.3$ V, and a +15 V power supply. Determine the minimum output voltage.

Solution. Solving Eq. (6.23) gives $V_{GS} = 5.66$ V. To make $I_{REF} = 0.19$ mA,

$$0.19 \times 10^{-3} = \frac{15 - 5.66}{R}$$

This gives $R = 49.2$ kΩ. Active-state operation requires that the output voltage across M_M always exceed $V_{DS,min} = V_{GS} - V_t = 5.66 - 1.30 = 4.36$ V. ❏

> **Exercise 6.15** Find values for R and $V_{DS,min}$ if the current source of Example 6.15 has the same output current and uses the same output transistor, but $I_{REF} = 3I_O$.
>
> *Ans.* 16.4 kΩ, 4.36 V.

The volt-ampere characteristic of a current source depends upon its current mirror. For the basic mirror circuit, the volt-ampere characteristic is simply the output characteristic of the mirror transistor, biased at some particular V_{GS} by the reference current in

M_R. Because of the transistor's output resistance, r_o, the output current depends slightly on output voltage as in Fig. 6.22d. As usual, we use Eq. (5.26) to compute output resistance. In Example 6.15, for $V_A = 75$ V, $r_o = 75/(0.19$ mA$) = 395$ kΩ. We usually ignore output resistances of M_R and M_M in hand-calculated designs, but include their effects in simulations. As we saw when we considered replacing the key biasing resistor with a currrent source, output resistance is a very important parameter in current-source applications.

Current-Source Analysis. Having identified the current source in a circuit diagram, often the next step is to determine its output current. More specifically, in analysis the object is to determine I_O when R, V_{DD} and the transistor parameters are given. Assuming M_R active we simply equate Eqs. (6.23) and (6.25) and solve for V_{GS}—then use Eq. (6.24) or the known scale factor to find I_O.

Exercise 6.16 The current source in Fig. 6.22c has the following parameters: $k_R = 10^{-4}$ A/V^2, $V_{tR} = V_{tM} = 0.8$ V, $V_{DD} = 10$ V, $R = 60$ kΩ, $(W/L)_M = 0.75(W/L)_R$. Find I_O.

Ans. 95 μA.

Multiple Outputs. The mirror circuit of Fig. 6.23 provides N output currents, each an individually scaled copy of I_{REF}. Since the parallel connection makes V_{GS} common to all transistors, a pair of equations like Eqs. (6.23) and (6.24) relate each output current to the reference.

BJT Current Mirror. Figure 6.24a shows the bipolar current mirror. Base currents are negligible for high-beta transistors, so for Q_R, $I_C = I_{REF}$. For active-state operation, Eq. (4.5) gives

$$I_{REF} = I_{CR} = I_{SR}e^{V_{BE}/V_T} \tag{6.26}$$

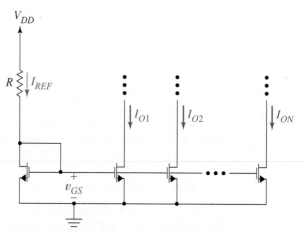

Figure 6.23 Basic current source with multiple outputs.

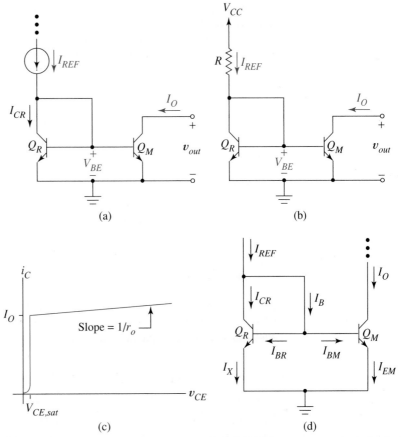

Figure 6.24 BJT constant current source: (a) basic mirror structure; (b) practical circuit for establishing I_{REF}; (c) output volt-ampere curve of current source; (d) notation for current gain calculation example.

where subscript C denotes collector current. For Q_M,

$$I_O = I_{CM} = I_{SM}e^{V_{BE}/V_T} \tag{6.27}$$

V_{BE} is identical in the two expressions because of the parallel connection of base-emitter junctions. If the transistors are near neighbors on the same IC chip, the $V_T = kT/q$ terms in the exponents are identical and track with temperature as do the saturation currents. The latter can be identical, causing I_O to mirror I_{REF} exactly, or the junction areas can be scaled to introduce a temperature-independent scale factor. The close similarity to the FET mirror is obvious.

The output resistance of the current source is the mirror-transistor output resistance r_o. Figure 6.24c shows that for active-state operation the output voltage must only be large enough to exceed $V_{CE,sat}$.

Design and Analysis of BJT Current Sources. Design and analysis are both easier for a BJT source than for a FET source because we can usually assume $V_{BE} = 0.7$ V. For example, consider designing a current source like that of Fig. 6.24b. For given $I_O = I_{REF}$ and V_{CC}, we find R from

$$I_{REF} = \frac{V_{CC} - 0.7}{R} \tag{6.28}$$

In *analysis,* we know V_{CC} and R, and solve for I_{REF}. Then output current either equals I_{REF} or is related to it by a known area ratio. Computer simulation gives more accurate results by computing the true value of V_{BE}.

EXAMPLE 6.16 Design a BJT current source to give a 1.5 mA output current using a 12 V supply. The area of the reference transistor is 1.25 times that of the mirror transistor.

Solution. In Fig. 6.24b the reference current is 1.25×1.5 mA $= 1.875$ mA. From Eq. (6.28),

$$R = \frac{12 - 0.7}{1.875 \text{ mA}} = 6.03 \text{ k}\Omega$$
❏

Current Gain of a Current Source. In some current-source applications the nonzero base currents can cause problems. When we include base currents in current-mirror analysis, the ratio of I_O to I_{REF} is a function of the transistor β, not just area ratio. That is,

$$\frac{I_O}{I_{REF}} = \frac{A_M}{A_R} f(\beta)$$

where $f(\beta)$ is a *second-order effect* in the BJT mirror called the *current gain.* The current gain is a figure of merit that helps us decide if a particular mirror circuit is suitable for a given application.

To find the current gain of any BJT mirror, first express both I_{REF} and I_O as functions of the emitter current, I_X, of the diode-connected transistor; then solve for their ratio. An efficient technique is to define I_X on the circuit diagram, and then reason systematically through the circuit, adding to the diagram each current expressed in terms of I_X and β until I_{REF} and I_O are known. For example, if we assume identical transistor areas in Fig. 6.24d, the reasoning proceeds in this order:

1. $I_{EM} = I_X$ (base-emitter junctions share the same voltage)

2. $I_{BR} = I_{BM} = \dfrac{I_X}{\beta + 1}$ (base currents of transistors with known emitter currents)

3. $I_B = I_{BR} + I_{BM} = 2\dfrac{I_X}{\beta + 1}$ (Kirchhoff's current law)

4. $I_O = I_{CR} = \dfrac{\beta}{\beta + 1} I_X$ (collector currents of transistors with known base currents)

5. $I_{REF} = I_{CR} + I_B = \dfrac{\beta + 2}{\beta + 1} I_X$ (Kirchhoff's current law)

The ratio of I_O/I_{REF} gives the current gain

$$\frac{I_O}{I_{REF}} = \frac{\beta}{\beta + 2} = f(\beta)$$

which means

$$I_O = \frac{\beta}{\beta + 2} I_{REF} \tag{6.29}$$

When β is large, the current gain is close to one. For $\beta = 399$, $f(\beta) = 0.995$; however, for $\beta = 9$, $f(\beta) = 0.818$.

An important issue related to current gain is the sensitivity of output current to uncertainties in β. Applying Eq. (6.14) to Eq. (6.29) gives

$$S_\beta^{I_o} = \frac{2}{\beta + 2} \tag{6.30}$$

For $\beta = 399$ and 9, respectively, the sensitivities are 0.005 and 0.182. Even with a very precisely fixed reference current this basic mirror can give output currents that change significantly with temperature if β is not large. We will later find that low current gain reduces performance in other ways as well. Low values of β are especially problematic in mirrors that employ pnp transistors because of their low βs.

Multiple Outputs. The BJT mirror can provide *multiple outputs* if we connect base-emitter junctions of several output transistors in parallel with the diode-connected transistor as in Fig. 6.25a. One price we pay is reduced current gain at the various outputs.

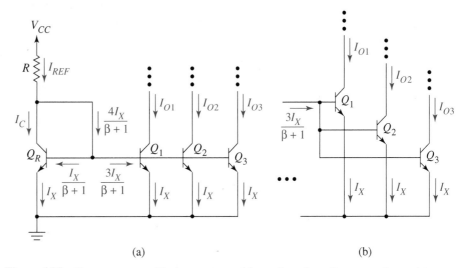

(a)   (b)

Figure 6.25 Current source with three outputs: (a) concise schematic commonly used to represent multiple outputs; (b) equivalent schematic showing the parallel connection of base-emitter junctions.

For example, with three output transistors of identical area as in Fig. 6.25a, the reference current is

$$I_{REF} = I_C + 4I_B = \left(\frac{\beta + 4}{\beta + 1}\right)I_X$$

reducing the gain at output I_{O1} to

$$\frac{I_{O1}}{I_{REF}} = \frac{\beta}{\beta + 4}$$

With N outputs, $N + 1$ replaces the "4" in this expression.

It is important to know that once we select a current-source structure of adequate current gain for an application, the initial *hand-design proceeds without further reference to current gain*. That is, we assume output currents are related to reference currents by area ratios only. Simulations, of course, include current gain.

6.7.3 HIGH-GAIN CURRENT SOURCES

Some current mirrors provide higher current gains than the basic mirror, and some have higher output resistance as well. Because infinite output resistance is the essence of a current source, high output resistance is an issue of paramount importance.

Mirror with Base Current Compensation. In the mirror of Fig. 6.26, Q_C replaces the direct "diode connection" of the reference transistor in Fig. 6.24d. The improvement in current gain occurs at the collector node of Q_R where $I_{REF} = I_{CR} + I_{BC}$ replaces I_{REF}

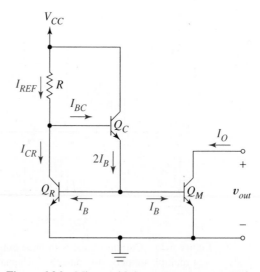

Figure 6.26 Mirror with base-current compensation.

$= I_{CR} + 2I_B$ for the basic mirror. We leave it for Problem 6.57 to show that the current gain of this mirror is

$$\frac{I_{O1}}{I_{REF}} = \frac{\beta^2 + \beta}{\beta^2 + \beta + 2} \tag{6.31}$$

For $\beta = 399$ and $\beta = 9$, this expression gives respective gains of 0.999987 and 0.9759 compared with 0.995 and 0.818 for the basic current source.

The output circuit in Fig 6.26 is that of the basic current mirror, so the output resistance is r_o. Current-source design and analysis for this circuit involve only the minor changes that result from the collector node of Q_R being two V_{BE} drops above ground instead of one.

Wilson Current Source. Another high-gain current source employs the Wilson mirror of Fig. 6.27a. It is easy to verify that its gain is

$$\frac{I_O}{I_{REF}} = \frac{\beta^2 + 2\beta}{\beta^2 + 2\beta + 2} \tag{6.32}$$

The Wilson mirror employs *current feedback,* a concept described at length in Chapter 10. An important property of current feedback is that it raises output resistance. Techniques we introduce in Chapter 7 show that the Wilson circuit output resistance is

$$R_{out} \approx \frac{\beta r_o}{2} \tag{6.33}$$

where r_o is the output resistance of Q_1.

Current gain is not an issue in MOS circuits; however, the MOS Wilson mirror of Fig. 6.27b does offer exceptionally high output resistance because of current feedback. Designing the MOS Wilson source involves computing V_{GS} for *both* M_1 and M_D, using I_O and the transistor parameters. Since both are active, this step is straightforward. The

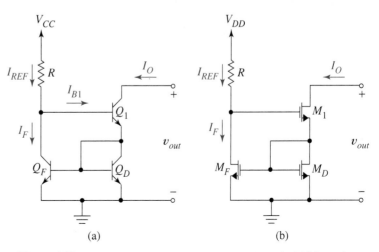

Figure 6.27 Wilson current source: (a) BJT version; (b) MOS version.

sum of these gate-source drops establishes the voltage at the drain of M_F, from which R is computed.

Analysis is essentially the same as for the basic mirror circuit. If M_1, M_D, and M_F are identical, their currents satisfy

$$I_O = I_{REF} = \frac{k}{2}(V_{GS} - V_t)^2$$

For this mirror,

$$I_{REF} = \frac{V_{DD} - 2V_{GS}}{R}$$

so we can solve these two equations for unknowns V_{GS} and I_{REF}. The minimum output voltage of the current source exceeds that of the basic mirror by the V_{GS} drop across M_D.

EXAMPLE 6.17   Design an MOS Wilson current source for 0.2 mA output current, a 9 V supply voltage, and identical transistors of $k = 10^{-4}$ A/V^2 and $V_t = 0.4$ V. Find the minimum output voltage.

Solution. Active operation requires

$$2 = \frac{1}{2}(V_{GS} - 0.4)^2$$

Thus $V_{GS} = 2.4$ V. In Fig. 6.27b, R must satisfy

$$0.2 \text{ mA} = \frac{9 - 2(2.4)}{R}$$

giving $R = 21.0$ kΩ. For M_1 to be active requires $v_{DG1} \geq -V_t$. Because $V_{GS} = 2.4$ V for M_1 and M_D, node voltages give

$$V_{DG} = v_{out} - 2(2.4) \geq -0.4 \text{ V}$$

thus $v_{out} \geq 4.4$ V.   ❏

Exercise 6.17   Find the new output current in the circuit of Example 6.17 if the supply voltage is changed to 12 V.

Ans. 0.300 mA.

Basic Current Source with Emitter Resistors. The mirror of Fig. 6.28 shows another current mirror of high output resistance. If transistor βs are high, KVL gives

$$V_{BE2} - V_{BE1} = +R_1 I_o - R_2 I_{REF}$$

If we can design so that the difference between the base-emitter drops is negligible,

$$I_o = \frac{R_2}{R_1} I_{REF} \tag{6.34}$$

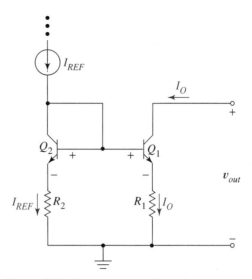

Figure 6.28 Current mirror with emitter resistors.

The small-signal techniques of the next chapter show that the output resistance of this current source is not r_o but rather

$$R_{out} \approx \left(1 + \frac{I_O R_1}{V_T}\right) r_o \tag{6.35}$$

The output resistance of Q_1 is multiplied by a factor that depends upon the ratio of the dc voltage drop, $I_O R_1$, to the 25 mV thermal voltage V_T. We conclude that even relatively small resistors introduced into the emitter circuits can significantly increase the output resistance. This particular design is also useful for *discrete* current sources when matched transistors are not available, as its operation depends primarily on the emitter resistors instead of transistor area ratio.

Cascode Current Mirror. Another current source with high output resistance is the *cascode mirror* of Figs. 6.29a and b. Small-signal analysis, introduced in the next chapter, shows that these circuits have output resistances

$$R_{out,BJT} = \frac{\beta}{2} r_o \qquad \text{and} \qquad R_{out,FET} = (1 + g_m r_o) r_o \tag{6.36}$$

where r_o is the output resistance of $Q_1(M_1)$ and g_m is given by Eq. (6.10) for the BJT circuit and Eq. (6.9) for the MOS circuit. For both BJT and FET versions, analysis and design are straightforward extensions of the corresponding problems for basic mirror circuits.

6.7.4 WIDLAR CURRENT SOURCE

The *Widlar current source* of Fig. 6.30 has two special features in addition to high output resistance: low sensitivity to supply voltage and the ability to deliver *very low output*

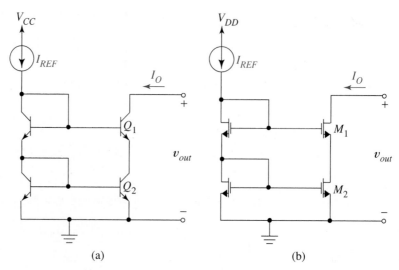

Figure 6.29 Cascode current mirrors: (a) BJT realization; (b) MOS realization.

currents even for relatively small R and R_M. In IC designs, this second feature is very useful.

In the Widlar mirror, the transistor base-emitter drops are not equal. If we ignore the base current of Q_M,

$$V_{BER} = V_{BEM} + I_O R_M \qquad (6.37)$$

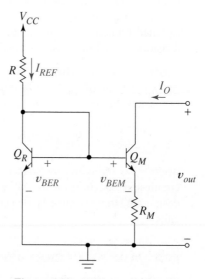

Figure 6.30 Widlar current source.

By using Eqs. (6.26) and (6.27) for matched transistors we obtain

$$V_T \ell n\left(\frac{I_{REF}}{I_O}\right) = I_O R_M \tag{6.38}$$

In analysis problems we know V_{CC} and R, so we estimate I_{REF}, using

$$I_{REF} = \frac{V_{CC} - 0.7}{R} \tag{6.39}$$

then solve Eq. (6.38) for I_O, either by iteration or by circuit simulation.

In design, I_O is known at the outset, so no iteration is necessary. We first choose a reasonable R and then compute I_{REF} from Eq. (6.39). With I_{REF} known, we then solve Eq. (6.38) for R_M. It is the log function that makes the circuit easy to design for small I_O.

The output resistance of the Widlar source is

$$R_{out} \approx \left[1 + \frac{I_O R_M}{V_T}\right] r_o \tag{6.40}$$

In Chapter 7 we will derive this result while showing exactly what conditions are necessary for this approximate equation to be valid. If R_{out} is specified as a design parameter, we can choose R_M to meet this condition at the beginning of the design.

EXAMPLE 6.18 Design a Widlar current source that produces an output of 20 μA and uses a 15 V power supply. Values of R_M and R may not exceed 10 kΩ. Calculate R_{out} for the design. Use a silicon BJT having β = 70 and Early voltage V_A = 120 V.

Solution. Arbitrarily choose R = 5 kΩ. From Eq. (6.39),

$$I_{ref} = \frac{15 - 0.7}{5\,k} = 2.86 \text{ mA}$$

Now Eq. (6.38) gives

$$0.025\, \ell n\left(\frac{2.86 \times 10^{-3}}{20 \times 10^{-6}}\right) = 20 \times 10^{-6} R_M$$

Solving gives R_M = 6.20 kΩ, which completes the design.

From Eqs. (6.40), output resistance is

$$R_{out} = \left[1 + \frac{(20 \times 10^{-6}) \times (6.2 \times 10^3)}{0.025}\right] r_o = 5.96 r_o$$

From Eq. (4.21), r_o = 120/(20 × 10⁻⁶) = 6 MΩ; therefore, R_{out} = 35.8 MΩ. ❑

Exercise 6.18 Use the transistor and power supply of Example 6.18 to design a Widlar source of 50 MΩ output resistance and 12 μA output current.

Ans. R_m = 8.33 kΩ, R = 21.9 kΩ.

6.7.5 VARIATIONS ON THE CURRENT-SOURCE THEME

After mastering current mirrors we might still fail to recognize one in an IC schematic without some additional preparation. This is because we sometimes construct mirrors with negative supplies as in Fig. 6.31a, or because they employ p-channel or pnp transistors as in Fig. 6.31b, or because they are designed for positive *and* negative supply pairs, as in Fig. 6.31c. Figure 6.31c also shows how to provide multiple dc currents that flow simultaneously from positive and negative supply mains into a complex circuit. The reader should carefully study Fig. 6.31 and convince herself or himself that *no new principles are needed to understand, analyze, or design any of these circuits.* The nearest thing to a new principle is the observation that we must include two gate-source or base-emitter drops instead of one in the analysis or design of Fig. 6.31c.

Although Fig. 6.31 serves to sensitize the reader to other possibilities, it is far from being all inclusive. The variations just illustrated for the *basic* mirror also apply to the other mirror designs such as Wilson, Widlar, and cascode. Recognizing these variations within a complex IC schematic often leads to a conceptual understanding of the circuit and quick hand estimates of the bias currents.

Figure 6.32, the schematic for the Fairchild μA733, shows how current sources are used to bias an integrated circuit. The 733 is a difference amplifier with double-ended input and output, which consists of three cascaded difference amplifier subcircuits.

Diode-connected transistor Q_R is the key to the biasing system. This transistor provides reference current, I_R, for the circuit, and transistors Q_1–Q_4 are "Fig. 6.28-type" mirrors that provide individual bias currents for the subcircuits.

When a difference-mode signal is applied to the bases of Q_5 and Q_6, the first-stage output signal appears as a difference-mode signal between the collectors, and also provides input for the second-stage difference amplifier Q_7–Q_8. The output developed between collectors of Q_7 and Q_8 provides the signal for the output stage, Q_9–Q_{10}.

Each amplifier stage has a vertical axis of symmetry and employs matched transistors with identical biasing. That is, $I_5 = I_6$, $I_7 = I_8$, $I_9 = I_{10}$. Collector-emitter voltages for each transistor pair are also identical. For signal analysis we need to know that each transistor is biased in the active region for given supply voltages V_{CC} and V_{EE}. We also need the dc collector currents in order to compute small-signal parameter values. A dc bias circuit analysis in the following example provides this information.

EXAMPLE 6.19 Use infinite beta analysis to estimate the Q-point of every transistor in the μA733 when the bases of Q_5 and Q_6 are grounded and supplies voltages are $\pm$ 12 V.

Solution. We first calculate the reference current. For $\pm$ 12 V supplies this is

$$I_R = \frac{24 - 07}{10\,k + 1.4\,k} = 2.04 \text{ mA}$$

The bias currents, according to Eq. (6.34), are

$$I_1 = I_2 = \frac{1.4\,k}{300} 2.04 \text{ mA} = 9.52 \text{ mA}$$

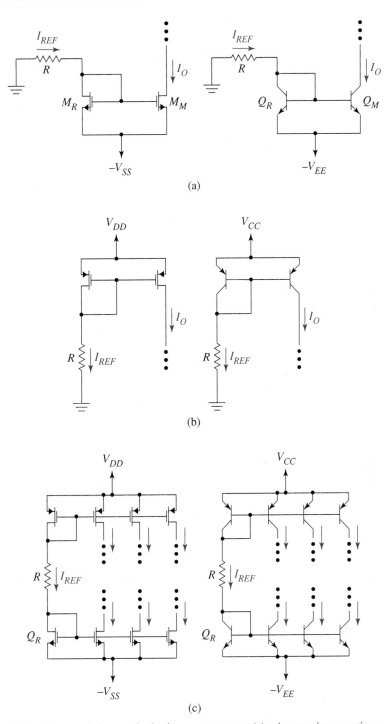

Figure 6.31 Some variation on the basic current source: (a) mirrors using negative supplies; (b) p-channel and pnp transistors; (c) multiple outputs.

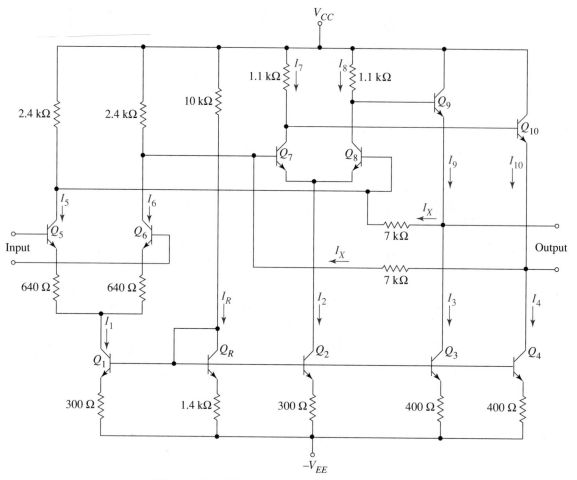

Figure 6.32 Fairchild μA733 IC amplifier schematic.

and

$$I_3 = I_4 = \frac{1.4\,\text{k}}{400}2.04\ \text{mA} = 7.14\ \text{mA}$$

Because of circuit symmetry

$$I_5 = I_6 = I_1/2 = 4.76\ \text{mA}$$

and

$$I_7 = I_8 = I_2/2 = 4.76\ \text{mA}$$

Because currents I_X are unknown, we cannot immediately find the emitter currents of Q_9 and Q_{10} or the collector voltages of Q_5 and Q_6.

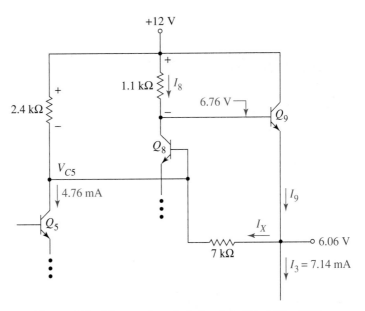

Figure 6.33   Diagram for calculating I_X in Fairchild μA733.

The collector voltages of Q_7 and Q_8 are

$$V_{C7} = V_{C8} = 12 - (1.1 \text{ k})(4.76 \text{ mA}) = 6.76 \text{ V}$$

For Q_9,

$$V_{E9} = V_{C8} - V_{BE9} = 6.76 - 0.7 = 6.06 \text{ V}$$

Figure Fig. 6.33 shows the subcircuit that is relevant to computing I_X. From KCL at the collector of Q_5,

$$\frac{12 - V_{C5}}{2.4 \text{ k}} + I_X = 4.76 \times 10^{-3}$$

Another expression for I_X is

$$I_X = \frac{6.06 - V_{C5}}{7 \text{ k}}$$

Solving gives $V_{C5} = V_{C6} = 1.98$ V and $I_X = 0.583$ mA. From the latter we find, $I_9 = I_{10} = 7.14$ mA + 0.583 mA = 7.72 mA. At this point all collector currents and voltages are known. Table 6.7 lists them.

TABLE 6.7   Estimated Transistor Q-points for μA733

	Q_1	Q_2	Q_3	Q_4	Q_5	Q_6	Q_7	Q_8	Q_9	Q_{10}	Q_R
I_C (mA)	9.52	9.52	7.14	7.14	4.76	4.76	4.76	4.76	7.72	7.72	2
V_{CE} (V)	5.39	10.3	15.2	15.2	2.68	2.68	6.48	6.48	5.93	5.93	0.7

Now we compute V_{CE} for each transistor pair. Because the emitter of Q_5 is one V_{BE} drop below its grounded base, $V_{CE5} = V_{CE6} = 1.98 - (-0.7) = +2.68$ V. Because the emitter of Q_7 is one V_{BE} drop below the collector of Q_6, $V_{E7} = V_{E8} = 1.98 - 0.7 = 1.28$ V. We previously found that the collectors of Q_7 and Q_8 were biased at 6.76 V, therefore $V_{CE7} = V_{CE8} = 6.76 - 1.28 = 6.48$ V. Finally, for Q_{10},

$$V_{E10} = V_{C7} - V_{BE} = 6.76 - 0.7 = 6.07 \text{ V.}$$

Therefore, $V_{CE10} = V_{CE9} = 12 - 6.07 = 5.93$ V. Clearly, transistors Q_5 through Q_{10} are each biased for active-state operation, well away from saturation and cut-off.

These calculations have tacitly assumed that the current sources are operating properly. Recall from Fig. 6.24c, however, that this requires that each current-source output transistor have its operating point such that $V_{CE} > V_{CE,sat} = 0.2$ V. This we investigate next.

Using node voltages and currents just computed, the collector voltage of Q_1 is

$$V_{C1} = V_{E5} - 640 \times I_5 = -0.7 - 640(4.76 \text{ mA}) = -3.75 \text{ V}$$

The emitter voltage is

$$V_{E1} = -12 + 300 I_1 = -12 + 300(9.52 \text{ mA}) = -9.14 \text{ V}$$

so

$$V_{CE1} = -3.75 - (-9.14) = 5.39 \text{ V}$$

Similar calculations for Q_2, Q_3, and Q_4, give

$$V_{CE2} = V_{C2} - V_{E2} = 1.28 - (-12 + 300 I_2) = 10.3 \text{ V}$$

and

$$V_{CE3} = V_{E9} - V_{E3} = 6.07 - (-12 + 400 \times 7.14 \text{ mA}) = 15.2 \text{ V} = V_{CE4}$$

Table 6.7 shows that all transistors, including current mirrors, are biased for active operation. ❏

Level Shifter. Most analog integrated circuits use _direct-coupled_ amplifier stages, meaning output nodes of one stage are connected directly to input nodes of the next, as in the μA733. Since all transistors must be biased for active operation, keeping collector-emitter voltages large enough to avoid saturation results in dc collector voltages that increase from one stage to the next. In the 733 our analysis showed $V_{C5} = 1.98$ V, $V_{C8} = 6.76$ V, and $V_{C9} = 12$ V. Similarly, avoiding ohmic operation in FET amplifiers leads to increasing drain voltages.

Sometimes we need to set the dc voltage of the output node(s) at a low value in spite of this trend of increasing voltage. In an op amp, for example, a bias value of zero volts at the output node causes the open-loop transfer characteristic to pass through the origin, and thereby facilitates adding external feedback components. A special circuit called a _level shifter_ serves this purpose.

The simple op amp circuit of Fig. 6.34 employs subcircuit Q_X, R_3, and R_4 as a level shifter and Q_5 as output transistor. Each of the other transistors is to be biased at (5 V, 1 mA) as indicated. With input nodes grounded, the output node is to be biased at zero.

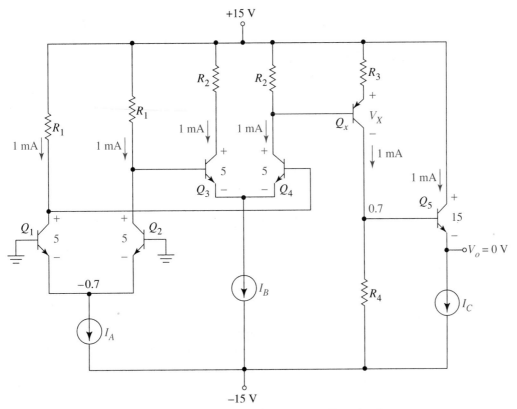

Figure 6.34 IC amplifier with level-shifting circuit.

The next exercise shows that it is easy to satisfy these conditions with current sources and a level shifter.

Exercise 6.19 In Fig. 6.34 transistor betas are so high that base currents are negligible. For the biasing conditions given on the diagram,
(a) find I_A, I_B, and I_C.
(b) find resistor values in numerical order: R_1, R_2, R_4, R_3.
(c) find V_X.

Ans. (a) $I_A = I_B = 2$ mA, $I_C = 1$ mA, (b) 10.7 kΩ, 6.4 kΩ, 15.7 kΩ, 5.7 kΩ, (c) 8.6 V.

6.7.6 CURRENT MIRRORS FOR SIGNALS

In Sec. 6.1 we learned that adding a small signal to the input of an electronic circuit results in a small signal being added to the dc bias values at the output. This idea also applies to current mirrors. In Fig. 6.35, $v_s(t)$ in the reference circuit causes a *signal component* $i_s(t)$ to be added to I_{REF}. Then *both* are mirrored to the output. This principle of *mirrored signals* is used to advantage in the *active load* circuits we study in Chapter 7.

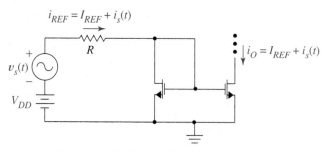

Figure 6.35 Use of mirror to copy signals.

This same principle also provides a mechanism by which noise from the power supply can enter our circuits. We address this problem next.

6.8
Special Current References

The current sources of the previous section use reference currents derived directly from the power supply. Here we explore alternative ways to establish reference currents.

6.8.1 SUPPLY-INDEPENDENT BIASING

Reference currents derived from power supplies can sometimes lead to problems. When a circuit's bias currents change with supply voltage, transistor Q-points and small-signal parameters change as well. Often this is acceptable as long as Q-points in the biasing circuit remain well within the active region; however, if specifications require small-signal operation that is more or less invariant over a range of possible power supply values, the current sources must be independent of the supply voltage.

A biasing circuit that is excessively sensitive to power supply variations can also introduce more subtle problems. First, power supplies always have a small additive noise called *ripple* that can be inadvertently mirrored into other circuits by the mechanism of Fig. 6.35. Second, a real power supply also has a small internal resistance, R_{sup}. When several circuits share the supply as in Fig. 6.36, the drop across R_{sup} caused by a signal current in one circuit can enter other circuits through their current mirrors. When the circuits are unrelated, these undesired signals are noise. If, instead, the circuits are cascaded amplifier stages, a signal from the output stage can be fed back to the first

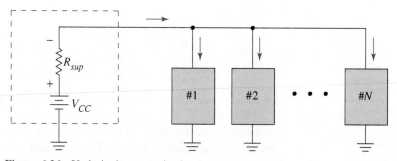

Figure 6.36 Undesired communication between circuits through a power supply.

amplifier stage through R_{sup}, adding to the external signal and creating unwanted oscillations.

We can use sensitivity to quantify the effect of supply variations on reference current. When $I_O = I_{REF}$ is derived directly from V_{CC}, as in Fig. 6.24b, we can use Eq. (6.28) to verify that the sensitivity is

$$S^{I_O}_{V_{CC}} = \frac{V_{CC}}{V_{CC} - V_{BE}} \tag{6.41}$$

This shows that sensitivity of I_O (and therefore Q-point currents) to V_{CC} is greater than one. A similar result applies to MOS designs. Thus bias current sensitivities such as

$$S^{I_D}_{V_{DD}} = 0.04$$

suggested in Table 6.6 are unrealistic unless the reference current is obtained in some other manner.

Perhaps the simplest idea for reducing supply sensitivity is to establish reference current with a simple transistor current source such as a JFET (or depletion MOSFET) with $V_{GS} = 0$, as in Fig. 6.37a. Ideally variations in V_{CC} would change only the JFET voltage without affecting I_O at all. Figure 6.37b is the large-signal equivalent of the transistor, where I_{DSS} is the "constant current" that results from $V_{GS} = 0$, and r_o comes from Early effect. With r_o present, we do not expect perfect supply independence; however, to the extent that r_o is large, we hope for an improvement over Eq. (6.41).

The output current for this circuit, its mirrored reference current, is

$$I_O = I_{DSS} + \frac{V_{CC} - V_{BE}}{r_o} \tag{6.42}$$

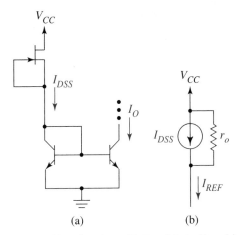

(a) (b)

Figure 6.37 A current source with reduced sensitivity of I_O to V_{CC}: (a) circuit; (b) equivalent of constant current transistor.

From this it is easy to show that the sensitivity is

$$S_{V_{CC}}^{I_O} = \frac{V_{CC}}{V_{CC} - V_{BE} + I_{DSS}r_o} = \frac{V_{CC}}{V_{CC} - V_{BE} + V_A} \tag{6.43}$$

an obvious improvement over Eq. (6.41).

Exercise 6.20   In Example 6.16 we designed a basic current source that gave a 1.5 mA output current using a 12 V supply and a 6.03 kΩ reference resistor. Suppose we use a JFET with $I_{DSS} = 1.5$ mA and $V_A = 90$ V instead of the resistor. Compare the sensitivities to V_{CC}.

Ans. 1.06, 0.118.

For the JFET to emulate a current source in Fig. 6.37 it is, of course, necessary that the JFET remain in the active state. This depends upon the threshold voltage.

EXAMPLE 6.20   Find the minimum value of V_{CC} for which the circuit of Fig. 6.37 will give reasonably constant current if $V_t = -2.1$ V.

Solution. $v_{DG} = V_{CC} - 0.7 \geq -V_t$, implies $V_{CC} \geq 2.8$ V.   ❑

Another way to reduce supply dependence is to use a Widlar source. Problem 6.75 shows that, because of the logarithmic dependence of I_O on V_{CC} in Eq. (6.38), the Widlar source has lower sensitivity than Eq. (6.41).

When low sensitivity to supply variations is important, the reference current can be derived from some voltage other than the power supply. Popular alternatives are the V_{BE} drop or thermal voltage, V_T, of a transistor, or the breakdown voltage of a zener diode. The following discussion is limited to the first two.

Reference Current Fixed by V_{BE}. Figure 6.38a shows a V_{BE} *reference* in which the output current depends upon a V_{BE} drop. To describe the circuit we ignore base currents and begin our analysis with I_{REF}. From Eq. (6.26), the value of V_{BE} required to maintain I_{REF} is

$$V_{BE} = V_T \ln\left(\frac{I_{REF}}{I_{SR}}\right) \tag{6.44}$$

The current I_O that flows through R_M must produce a drop to equal this voltage; thus

$$V_{BE} = R_M I_O \tag{6.45}$$

We establish the desired reference current by

$$I_{REF} = \frac{V_{CC} - 2V_{BE}}{R_R} \tag{6.46}$$

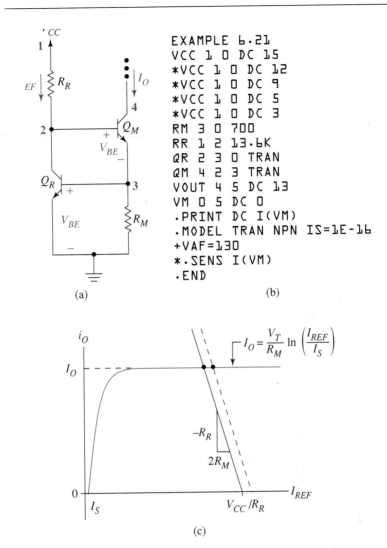

Figure 6.38 V_{BE}-based current source: (a) circuit schematic; (b) SPICE code for design of Example 6.21; (c) graphical solution of functions that relate output current to supply voltage.

We first use a design example to demonstrate the low sensitivity of this circuit; then we relate general performance to the design equations, Eqs. (6.44) through (6.46).

EXAMPLE 6.21   (a) Given $V_{CC} = 15$ V and matched transistors, design a mirror like Fig. 6.38a for $I_O = I_{REF} = 1$ mA.
(b) Use SPICE with default transistors to find values of I_O for $V_{CC} = 15$ V, 12 V, 9 V, 5 V, and 3 V. Transistor saturation current is 10^{-16} A; Early voltage is 130 V.
(c) Use SPICE to determine the sensitivity of I_O to V_{CC}, for $V_{CC} = 15$ V.

Solution. (a) For hand estimates, we find R_M from Eq. (6.45) using $V_{BE} = 0.7$ V. For 1 mA output current, this gives $R_M = 700\ \Omega$. Equation (6.46) gives

$$R_R = \frac{15 - 2(0.7)}{10^{-3}} = 13.6\ \text{k}$$

which completes the design.

(b) Figure 6.38b shows SPICE code. Five successive SPICE runs compute output current for the five listed supply voltages. A dc source, VOUT, keeps the collector of the transistor biased at 13 V so Q_M is active. An ammeter, VM, monitors the current that flows *into* the collector. SPICE results are

V_{CC}	15 V	12 V	9 V	5 V	3 V
I_O	1.105 mA	1.093 mA	1.078 mA	1.047 mA	1.013 mA

Although I_O differs slightly from the 1 mA design value at 15 V (SPICE includes base currents and computes accurate V_{BE} drops), the results demonstrate remarkable insensitivity to supply voltage.

(c) SPICE output leads to

$$S^{I_O}_{V_{CC}} = 0.047$$

This positive sensitivity is consistent with the reduction in I_O with reduced V_{CC} in the table. ❑

Exercise 6.21 By examining how the reference current is established in Fig. 6.38a, estimate a lower limit for V_{CC} for the circuit of Example 6.21. Also estimate the lowest practical value for current-source output voltage.

Ans. $V_{CC,min} = 1.4$ V, $V_{out,min} = 0.9$ V.

To understand *how* this circuit achieves low sensitivity to V_{CC}, we now substitute V_{BE} from Eq. (6.45) into Eqs. (6.44) and into (6.46), and then solve each for I_O. This gives the equations

$$I_O = \frac{V_T}{R_M} \ln\left[\frac{I_{REF}}{I_S} \right] \tag{6.47}$$

and

$$I_O = \frac{V_{CC} - R_R I_{REF}}{2R_M} \tag{6.48}$$

Solving these equations simultaneously corresponds to finding their intersection in the I_O versus I_{REF} coordinate system of Fig. 6.38c. To appreciate the true scale of the sketch, recall that I_S is very small compared to I_O. Because I_{REF} is much larger than I_S, the intersection lies on a *very flat* portion of the natural log curve. As for insensitivity to supply, notice that changes in V_{CC} cause the straight line to slide right or left as suggested by the dashed curve making little change in I_O. Another V_{BE} reference is described in Problem 6.80.

Bootstrap V_{BE} Reference Circuit. Figure 6.39a introduces an idea called *bootstrapping*, that leads to even lower supply sensitivity. This *self-biased* modification of

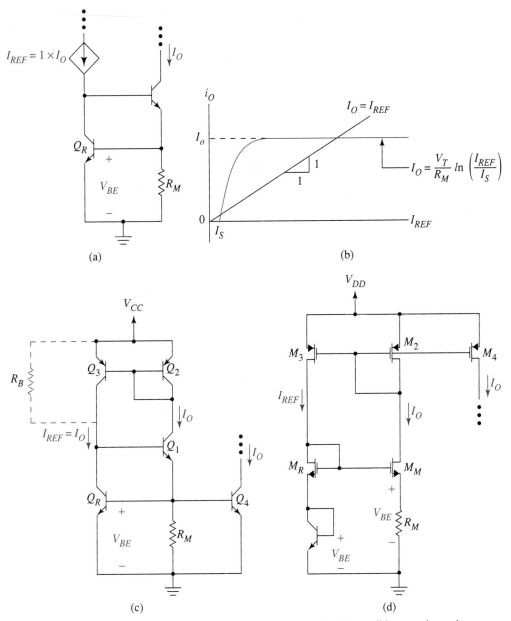

Figure 6.39 Bootstrap current reference: (a) bootstrap principle; (b) possible operating points; (c) practical circuit realization; (d) MOS bootstrap circuit.

Fig. 6.38a lifts itself by its own bootstraps by using a copy of its own output current rather than a current derived from V_{CC} for its reference. Like Fig. 6.38a, this circuit satisfies Eqs. (6.44) and (6.45); however, by virtue of the dependent source,

$$I_{REF} = I_O \qquad (6.49)$$

replaces Eq. (6.46). This means operation must be at one of the intersections shown in Fig. 6.39b. The key to low sensitivity is that Eqs. (6.47) and (6.49), which are graphed in Fig. 6.39b, do not contain V_{CC}, thus first-order operation is independent of V_{CC}.

Figure 6.39c shows a practical realization of the circuit. Transistors Q_2 and Q_3 are the unity-gain CCCS, mirroring current I_O from Q_1 into the reference circuit. Of the two possible solutions in Fig. 6.39b, the one near $i_O = I_S$ represents an output current of only a few picoamperes—too small to be useful; the other intersection corresponds to the design value. Because the undesired solution can occur in real circuits, a large resistor R_B (or a special circuit easier to realize in an IC) provides a slight current to ensure $i_O > I_S$, so the circuit operates at the desired intersection. Because the bootstrap mirror makes Q_1's collector current unavailable to the outside world, we add Q_4 to duplicate the V_{BE}-produced current I_O. A pnp transistor with base and emitter in parallel with those of Q_2 could provide an output current flowing down from V_{CC} if needed.

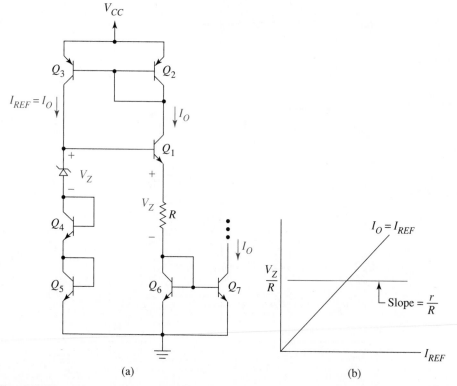

(a)

(b)

Figure 6.40 Zener current references: (a) BJT circuit; (b) graphical solution for operating point of zener reference; (c) CMOS version.

A SPICE simulation of Fig. 6.39c with transistor output resistances omitted gave identical output currents for power supply voltages of 15 V, 12 V, 9 V, 5 V, and 3 V. Sensitivity of I_O to changes in V_{CC} in this circuit was 9.23×10^{-8}! As a practical matter, the output resistances of real transistors make output current somewhat dependent upon the power supply, an idea explored in homework problems.

Figure 6.39d shows an MOS bootstrap mirror that uses a V_{BE} reference. Because the V_{GS} drops of M_R and M_M cancel, Eqs. (6.44) and (6.45) apply to this circuit, as does the bootstrap equation, Eq. (6.49).

Zener Reference Circuit. The bootstrap circuit of Fig. 6.40a is a *zener reference*. The V_{BE} drops of Q_4 and Q_5 match those of Q_1 and Q_6, making the drop across R equal the drop across the zener. Thus, in an I_O versus I_{REF} coordinate system, the operating point is the intersection of the bootstrap equation and the near-horizontal line

$$I_O = \frac{V_Z + I_{REF}r}{R} \approx \frac{V_Z}{R} \tag{6.50}$$

where r is the resistance of the zener diode. This equation applies to both design and analysis. Because this circuit does not depend upon Eq. (6.47), there is no start-up problem. Second-order dependence on V_{CC} occurs when a real zener diode is used because of the internal resistance of the zener and the dependence of I_{REF} on V_{CC} caused by transistor output resistance. Figure 6.40c shows how to modify the CMOS circuit of Fig. 6.39d to use a zener reference.

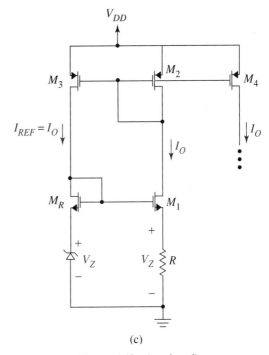

(c)

Figure 6.40 (continued)

6.8.2 TEMPERATURE-INDEPENDENT BIASING

Unless we take special design precautions, a current source can introduce temperature-induced variations into the bias circuit. After reviewing the concept of temperature coefficients, first introduced in Chapter 3, we show how to apply the idea to current-source design.

Temperature Coefficients. We define the temperature coefficient, α_P, of a parameter P by

$$\alpha_P = \frac{1}{P}\frac{dP}{dT} \tag{6.51}$$

Thus α_P denotes the fractional increase in P per unit increase in temperature T. Some of the more important parameters P related to temperature compensated circuit designs are resistances, V_{BE} drops in pn junctions, and diode breakdown voltages.

In Chapter 3 we learned that semiconductor resistors have positive temperature coefficients because charge-carrier mobilities decrease with increases in temperature. In a base-diffused resistor, α_R is about 1500 ppm/°C. Emitter-diffused resistors have lower temperature coefficients, of the order of 600 ppm/°C; however, the sheet resistance of emitter-material must be low to accommodate high-β transistors. This restricts these resistors to lower resistance values for the same geometry. Large resistors with temperature coefficients in the range of ± 200 ppm/°C or less are possible with ion implant and thin-film techniques, but these require extra fabrication steps.

The V_{BE} drop of a forward-biased, silicon, pn junction is of the order of 700 mV, and it decreases at 2 mV/°C. Thus

$$\alpha_{V_{BE}} = \frac{-2 \times 10^{-3}}{700 \times 10^{-3}} = -2860 \text{ ppm/°C}$$

The temperature coefficient of a zener diode has an algebraic sign that depends upon zener voltage, a property that makes the zener a versatile device for temperature compensation circuits. Figure 6.41 shows that a 5.7 V zener has temperature coefficient

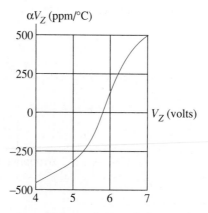

Figure 6.41 Temperature coefficients of zener diodes as a function of zener voltage.

of zero; for zener voltages close to 5.7 V, temperature coefficients range from -450 to $+500$ ppm/°C.

Temperature-Compensated Zener Reference. In the zener reference of Fig. 6.40a, the V_{BE} drops of Q_4 and Q_5 track those of Q_1 and Q_6 with temperature; therefore, the temperature variations in the output current,

$$I_O = \frac{V_Z}{R} \tag{6.52}$$

originate in V_Z and R. To find the temperature coefficient of I_O, we first differentiate with respect to temperature. The chain rule gives

$$\frac{dI_O}{dT} = \frac{\partial I_O}{\partial V_Z}\frac{dV_Z}{dT} + \frac{\partial I_O}{\partial R}\frac{dR}{dT} = \left[\frac{1}{R}\frac{dV_Z}{dT} - \frac{V_Z}{R^2}\frac{dR}{dT}\right]$$

To obtain an expression involving temperature coefficients, we divide the left side by I_O, and the right side, equivalently, by V_Z/R. In terms of the temperature coefficients this gives

$$\alpha_{I_O} = (\alpha_{V_Z} - \alpha_R) \tag{6.53}$$

This shows that a positive temperature coefficient zener reduces the temperature coefficient of the output current; furthermore, matching temperature coefficients of the zener and resistor gives a current source that is quite insensitive to temperature variations, at least in the vicinity of the design temperature.

Reviewing the temperature coefficient values discussed earlier shows that matching the coefficients can lead to difficult compromises. Positive zener temperature coefficients comparable to those of base-diffused resistors require high zener voltages, which in turn require a large supply voltage V_{CC}. Resistors with temperature coefficients as low as 500 ppm or less, which match low-voltage zeners, require extra fabrication steps. The V_{BE} multiplier, discussed next, can help in these "compromising" situations.

V_{BE} **Multiplier.** The V_{BE} multiplier circuit of Fig. 6.42a is a subcircuit that we often see in IC designs as part of current or voltage reference circuits. Its purpose is to produce a voltage, $V_{BB} = mV_{BE}$, where m is a temperature-independent scale factor. Figure 6.42b helps us see how it works. Summing voltages along the left edge of the circuit gives

$$V_{BB} = V_{BE} + R_1(I_{BB} - \beta I_B) \tag{6.54}$$

Applying KCL at the base node gives

$$I_{BB} - \beta I_B = \frac{V_{BE}}{R_2} + I_B \tag{6.55}$$

The designer chooses R_2 small enough that the first term on the right of Eq. (6.55) is much greater than I_B. Dropping I_B and then substituting the left side into Eq. (6.54) gives the design equation

$$V_{BB} = \left(1 + \frac{R_1}{R_2}\right)V_{BE} = mV_{BE} \tag{6.56}$$

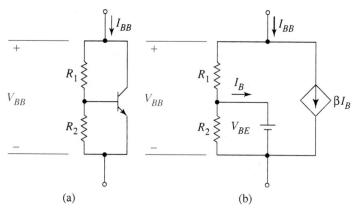

Figure 6.42 V_{BE} multiplier: (a) circuit; (b) equivalent circuit.

and the constraint

$$\frac{V_{BE}}{R_2} >> I_B \tag{6.57}$$

Because scale factor m depends on a resistor ratio, it can be precisely established in an IC design. Furthermore, first-order temperature variations in the resistors cancel, resulting in a scaled-up version of V_{BE}, V_{BB}, that tracks V_{BE} with temperature.

A second design condition ensures that most of current I_{BB} in Fig. 6.42a flows through the transistor. From Fig. 6.42b, this implies

$$\beta I_B >> \frac{V_{BE}}{R_2} + I_B$$

Equation (6.57) allows us to rewrite this as

$$\beta I_B >> \frac{V_{BE}}{R_2}$$

In fact, since most of I_{BB} is collector current, $\beta I_B \approx I_{BB}$. Substituting this approximation for the upper limit in the preceding equation, and using it again for the lower limit in Eq. (6.57), gives the design constraint

$$I_{BB} >> \frac{V_{BE}}{R_2} >> \frac{I_{BB}}{\beta} \tag{6.58}$$

which is easy to realize with a high-beta transistor.

Exercise 6.22   Design a V_{BE} multiplier for which $m = 1.13$, given that $I_{BB} = 0.89$ mA and $200 < \beta < 350$. Assume that $V_{BE} = 0.7$ V.

Ans. $R_1 = 80.8$ kΩ, $R_2 = 10.5$ kΩ is one correct answer.

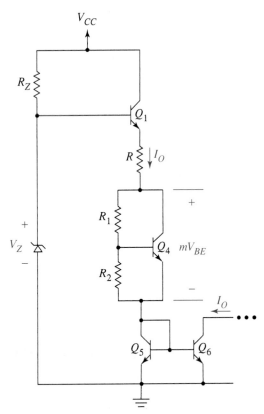

Figure 6.43 Temperature-compensated current source.

The following example shows the utility of the V_{BE} multiplier.

EXAMPLE 6.22 (a) From Fig. 6.43, find a design equation involving I_O, R, V_Z, m, and V_{BE}.
(b) From this equation, derive an expression for the temperature coefficient of I_O.
(c) Find the design condition that makes α_{I_O} zero.
(d) Design the circuit for $I_O = 0.2$ mA using a resistor, R, with temperature coefficient of 1500 ppm/°C and a 6 V zener with a 130 ppm/°C temperature coefficient. The temperature coefficient of V_{BE} is -2860 ppm and $I_S = 10^{-14}$ A.

Solution. (a) KVL gives

$$V_Z = I_O R + (m + 2)V_{BE}$$

In a design, we can solve for R, once I_O is specified and m is known. In what follows, we use temperature independence to find m.

(b) Differentiating our equation with respect to temperature, and multiplying each term by one gives

$$\frac{V_Z}{V_Z}\frac{dV_Z}{dT} = \frac{R}{R}I_O\frac{dR}{dT} + \frac{I_O}{I_O}R\frac{dI_O}{dT} + \frac{V_{BE}}{V_{BE}}(m+2)\frac{dV_{BE}}{dT}$$

Temperature coefficient definitions give

$$V_z\,\alpha_{V_Z} = R\,I_O\,\alpha_R + R\,I_O\,\alpha_{I_O} + (m+2)V_{BE}\,\alpha_{V_{BE}}$$

(c) To make I_O insensitive to temperature we set its temperature coefficient to zero. This gives a second design equation

$$V_Z\,\alpha_{V_Z} = R\,I_O\,\alpha_R + (m+2)V_{BE}\,\alpha_{V_{BE}}$$

(d) For 0.2 mA output current, the V_{BE} drops will be

$$V_{BE} = 0.025\,\ell n\left(\frac{0.2\times10^{-3}}{10^{-14}}\right) = 0.593\text{ V}$$

The nominal design equation of part (a) requires

$$6 = 0.2\times10^{-3}R + (m+2)0.593$$

The part (c) result gives

$$6\times(130) = R(0.2\times10^{-3})(1500) + (m+2)(0.593)(-2860)$$

Solving these equations gives $R = 20.4$ kΩ and $m = 1.24$. ❑

6.9
Summary

Linear electronic circuits require biasing to establish and maintain active-region operating points for the transistors. Sometimes additional considerations constrain bias circuit design, such as establishing ground potential at an output node, ensuring adequate small-signal gain, or providing high input resistance.

The four-resistor bias circuit is the circuit of choice for many discrete applications. It is important to master both analysis and design of the four-resistor circuit and its close relatives.

In analysis, the object is to find transistor Q-points from given transistor and circuit parameters and supply voltage. The dc collector and drain currents lead to numerical values for the small-signal parameters required to evaluate the circuit's signal processing capabilities. DC output voltages help us visualize how close the transistor is to cut-off or to saturation/ohmic-state operation.

In design, we are given the supply voltage and transistor parameters and Q-points; and we must calculate corresponding resistor values. There are many variations of the basic problem: positive and/or negative supplies, constraints for particular node voltages, constraints imposed by directly connected biasing circuits. We can handle all these with the same general approach. In each transistor's output circuit we apportion the existing voltage to meet the given requirements. In FET (BJT) designs a guiding principle is to make the voltage drop across the source (emitter) resistor as large as other constraints allow. Once we select voltage drops, output circuit resistances are easily calculated. The next step is to deduce input circuit requirements consistent with the output circuit design. With FETs, the key is to calculate V_{GS} from the known drain current. By adding V_{GS} to the node voltage at the source, we find the gate voltage we need to design the input circuit. For BJTs, a fixed V_{BE} difference of 0.7 V relates unknown base voltage to known emitter voltage. FET designs use a simple voltage divider at the input; for BJTs we must design the input circuit to account for worst case (minimum β) base currents.

Electronic circuits must function properly in spite of wide variations in parameter values. In this context, sensitivity theory helps us estimate what given circuits will do and design circuits with specified limits on performance variations. For such problems, we must know how to interpret numerical sensitivities, including multiparameter sensitivities, in terms of incremental approximations. SPICE sensitivity calculations are quite useful, especially in identifying the more critical parameters for more careful scrutiny. When we need a symbolic expression to better understand parameter relationships that underlie sensitivity, for example, in cases where we must modify a design to meet specifications, we need to know how to apply the definition of sensitivity to algebraic analysis equations.

In integrated circuits, where resistors are relatively expensive, dc current sources are widely used for biasing. Most often the current source consists of a reference current that flows through a diode-connected transistor, and a mirror transistor that reproduces the reference current at an output port of high internal resistance. Matched transistors result in output currents that closely duplicate the reference current over wide ranges of temperature. Scaling of the output current relative to reference current is effected by W/L ratios in FETs and by area ratios in BJTs. In both technologies, mirror transistors with parallel-connected input ports give multiple output currents with individual scaling. Basic FET and BJT current sources often use resistor-defined reference currents and simple current mirrors. It is important to master both the analysis and design of these circuits, and we must also be able to calculate their output resistances and minimum output voltages. We must also readily recognize, analyze, and design these basic circuits in their diverse forms: NMOS, PMOS, npn, and pnp and with positive, negative, or dual supplies.

There also exists a large collection of more complex current sources, more circuits than we can easily remember by name. Nevertheless, we should understand how these sources work, and be able to apply our basic analysis and design principles, including estimating minimum output voltage. Through increased complexity, these sources provide advantages over the basic circuits, such as higher output resistance, and/or reduced sensitivity to supply variations. The most prominent of these *nonbasic* circuits is the Widlar circuit with its unique logarithmic design equation. In addition to high output resistance and exceptional supply invariance, this circuit facilitates designing sources of low output current with reasonably small resistances. In bipolar current sources, there is an undesired β-dependent scale factor in addition to the desired area ratio scaling. A figure of merit called current gain describes this factor. In some applications high current gain is important, and this sometimes justifies selecting one of the complex current sources instead of a basic source.

All current sources are capable of mirroring reference current perturbations into the output. This is a useful property deliberately exploited in the active load circuits we study in the next chapter; however, it can also lead to undesired noise in bias circuits. Special supply-independent reference circuits not only reduce such noise, but also lead to bias currents that are surprisingly insensitive to the supply voltage itself, a feature suggesting the possibility of circuits with small-signal parameters relatively independent of power supply. Special design techniques can also lead to sources of dc current with temperature coefficients of zero.

REFERENCES

1. ANTOGNETTI, P., and G. MASSOBRIO, eds. *Semiconductor Device Modeling with SPICE*. New York: Masobrio, McGraw-Hill, 1988.
2. GRAY, P. R., and R. G. MEYER. *Analysis and Design of Analog Integrated Circuits,* 3rd ed. New York: John Wiley, 1993.
3. GREBENE, A. B. *Bipolar and MOS Analog Integrated Circuit Design*. New York: Wiley-Interscience, 1984.
4. TAUB, H., and D. SCHILLING. *Digital Integrated Electronics*. New York: McGraw-Hill, 1977.

PROBLEMS

SPECIAL DIRECTIONS FOR SPICE HOMEWORK PROBLEMS: *Do not hand in lengthy SPICE printout for homework.* Instead, abstract the useful information from the SPICE output file as in the SPICE examples in the text. Include your SPICE code and a circuit diagram with nodes numbered to agree with the code. Cite relevant numerical values from the SPICE output file and discuss when appropriate. Make sketches of any relevant curves, and label appropriate points. Make small tables to present numerical data if useful for clarity.

Section 6.1

6.1 (a) Replace the MOSFET in Fig. 6.1b with a bipolar transistor in the common-emitter configuration, and label your diagram with notation suitable for the bipolar circuit.
(b) Use a transfer characteristic sketch to show the output voltage waveform when the input signal is a sinusoid large enough to cause distortion.

6.2 The transistor described by Eq. (6.3) has $k = 0.2$ mA/V^2 and $V_t = 0.5$ V, and is biased at $V_{GG} = 2.5$ V.
(a) Write Eq. (6.4) using numerical values for the constant and for the coefficients of $v_{gs}(t)$ and $v_{gs}^2(t)$.
(b) Compare the magnitudes of the signal and distortion terms when
  (i) $v_{gs}(t) = 0.02 \sin 500t$.
  (ii) $v_{gs}(t) = 0.7 \sin 500t$.

6.3 In Fig. 6.1a, the transistor is biased at $V_{GG} = 2.5$ V. Parameters are $k = 0.2$ mA/V^2 and $V_t = 0.5$ V. If $R_L = 20$ kΩ, $V_{DD} = 15$ V,
(a) Find the voltage gain.

(b) Ignoring the distortion term, sketch the total node voltage waveform $v_D(t)$ in Fig. 6.1b when $v_{gs}(t) = 0.01 \sin \omega t$.

6.4 In Fig. 6.1a, $R_L = 20$ kΩ. Transistor parameters are $k = 0.2$ mA/V^2 and $V_t = 0.5$ V. Find the voltage gain if
(a) $V_{GG} = 0.75$ V.
(b) $V_{GG} = 1.5$ V.
(c) $V_{GG} = 2.0$ V.

6.5 Show that Eq. (6.9) and $g_m = k (V_{GG} - V_t)$ are equivalent.

6.6 For the amplifier of Fig. 6.1a, make a sketch of voltage gain magnitude, $|A_v|$ as a function of
(a) bias current I_D.
(b) bias voltage V_{GG}.

Section 6.2

6.7 In Fig. 6.2a, $V_{DD} = 10$ V. Design the circuit to bias the transistor at 2 V and 0.1 mA, with a 6 V dc drop across R_S. Make $R_P = 2$ MΩ. Transistor parameters are $k = 0.5 \times 10^{-3}$ A/V^2 and $V_t = 0.5$ V.

6.8 In Fig. 6.2a, $V_{DD} = 15$ V.
(a) Design the circuit to bias the transistor at 4 V and 3 mA, with a 5 V dc drop across R_D. Make $R_P = 2$ MΩ. Transistor parameters are $k = 0.5 \times 10^{-3}$ A/V^2 and $V_t = 0.5$ V.
(b) Prove that the transistor is not in the ohmic state in this circuit.

6.9 In Fig. 6.2a we wish to bias the transistor at 2 mA. Find the smallest value of V_{DS} that ensures the transistor is active if parameters are $k = 1.2 \times 10^{-3}$ A/V^2 and $V_t = 0.8$ V.
Hints: State definition, double-subscript notation.

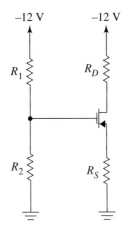

Figure P6.10

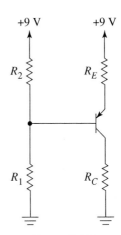

Figure P6.14

6.10 In Fig. P6.10 transistor parameters are $k = 0.5 \times 10^{-3}$ A/V^2 and $V_t = -2$ V. Design the circuit to bias the transistor at $(V_{DS}, I_D) = (-4$ V, 2 mA). Make $R_P = R_1 \| R_2 = 2$ MΩ.
Hint: First show current directions on the diagram.

6.11 In Fig. 6.2a, $V_{DD} = 9$ V.
(a) Design the circuit to bias the transistor at 1 mA. Place one-half the available voltage across R_S and divide the remaining voltage equally between the transistor and R_D. Make $R_P = 1$ MΩ. Transistor parameters are $k = 0.5 \times 10^{-3}$ A/V^2 and $V_t = 0.5$ V.
(b) Verify that your design biases the transistor in the active region.
(c) Calculate g_m for the transistor.

6.12 In Fig. P6.12, transistor parameters are $k = 0.5 \times 10^{-3}$ A/V^2 and $V_t = -0.6$ V. Make the Q-point $(V_{DS}, I_D) = (5$ V, 1 mA), with equal voltage drops across R_D and R_S. Also make $R_P = 3$ MΩ.

6.13 Design the bias circuit of Fig. 6.3a for a Q-point of (6 V, 0.5 mA) using $V_{CC} = 18$ V instead of 9 V. Make the dc drop across R_E two times the drop across R_C. Transistor β is in the range $100 \le \beta \le 300$.

6.14 (a) Design Fig. P6.14 for a Q-point of $(V_{CE}, I_C) = (-2$ V, 1 mA), with the additional condition $V_E = 4$ V. For the transistor, $20 \le \beta \le 200$.
Hint: First show directions for all currents on the diagram.
(b) Compare resistor values with those in Exercise 6.2.

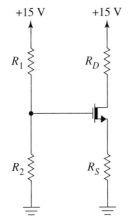

Figure P6.12

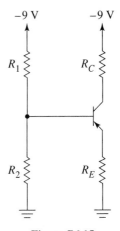

Figure P6.15

6.15 (a) Design Fig. P6.15 for a *Q*-point of $(V_{CE}, I_C) =$ $(-2 \text{ V}, 1 \text{ mA})$, with the additional condition $V_E = -4$ V. For the transistor, $20 \leq \beta \leq 200$.
Hint: First show directions for all currents on the diagram.

6.16 Design a four-resistor bias circuit for an n-channel JFET having $\beta = 1 \text{ mA/V}^2$ and $V_P = -1$ V. Use an 8 V power supply. Bias the transistor at 4 V and 0.5 mA with equal drops across drain and source resistors. R_P should be no less than 600 kΩ.

6.17 In the circuits of Fig. P6.17, the transistor has $\beta = 40$. Design each circuit so that the transistor is biased at (5 V, 0.5 mA). For circuit *c*, there is the additional requirement that 3 V be dropped across R_C.

6.18 In Fig. P6.18 each transistor has parameters $k = 0.1 \times 10^{-3} \text{ A/V}^2$ and $V_t = +2$ V. Complete the designs so that each transistor is biased at 3 V and 0.15 mA. Also,

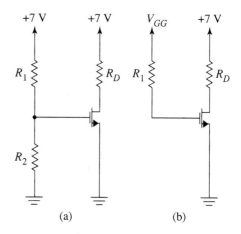

(a)   (b)

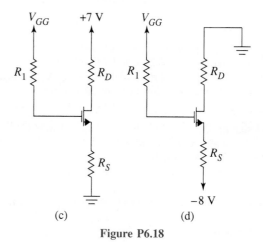

(c)   (d)

Figure P6.18

Circuit (a): In this circuit make $R_1 \| R_2 = 500$ kΩ.
Circuit (b): Does biasing the transistor place any constraint at all on R_1? Explain.
Circuit (c): Place equal dc drops across R_D and R_S.
Circuit (d): Make the drop across $R_D = 3$ V.

6.19 Redraw Fig. P6.18c with the MOSFET replaced by an n-channel JFET having parameters $V_P = -1.5$ V and $I_{DSS} = 0.5$ mA. Design the circuit for a *Q*-point of 0.2 mA and 3 V, with equal drops across R_S and R_D. Use $R_1 = 1$ MΩ.

6.20 Design a four-resistor bias circuit for a p-channel MOSFET. The available power supply is -12 V. Parameter values are $k = 3 \text{ mA/V}^2$ and $V_t = -2$ V. The *Q*-point should be at 2 mA and -5 V. The drop across R_S must be 4 V and $R_1 \| R_2 = 900$ kΩ.

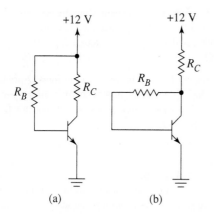

(a)   (b)

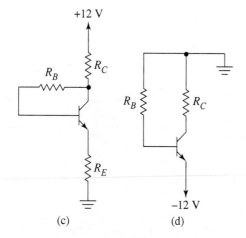

(c)   (d)

Figure P6.17

6.21 Design a four-resistor bias circuit for an n-channel deple-tion-type MOSFET. The power supply is -12 V. Parameter val-ues are $k = 3$ mA/V^2 and $V_t = -2$ V. The Q-point should be at 2 mA. Drops across R_D and R_S should be 3 V and 4 V, respec-tively. Make the parallel equivalent of the gate-biasing resistors 900 kΩ.

6.22 Choose resistor values so that the transistor of Fig. P6.22 is biased at 1 mA. Parameter values are $k = 0.5$ mA/V^2 and $V_t = 1$ V. (The answer is not unique.)

Section 6.3

6.23 Draw the n-channel JFET equivalents of Figs. 6.6a and b. Then use load line sketches to compare, qualitatively, the Q-point stabilities of the two JFET circuits as temperature de-creases. That is, redo Fig. 6.6 for the JFET case.
Hint: Temperature variation of JFET parameters was intro-duced in Sec. 5.9.4.

6.24 Figure P6.24 shows a transistor's biasing circuit and out-put characteristics.
(a) Use a load line to find the value of R_B required to bias the transistor near 2.5 V. Assume $V_{BE} = 0.7$ V.
(b) Assume that V_{BE} changes by -2 mV/°C, but the output characteristics do not change. If the temperature increases by 75°C, find the new value of base current. Could you see such a Q-point change on your load line?

6.25 (a) Use Eq. (6.12) to find I_C when $R_E = 2.5$ kΩ, $R_T = 10$ kΩ, $V_{TT} = 5$ V, $\beta = 150$, and $V_{BE} = 0.7$ V.
(b) Recalculate I_C when V_{BE} is 10% higher than the part (a) value. Find the resulting percentage change in I_C.
(c) Recalculate I_C using values of part (a) except make β 10% lower than part (a). Find the resulting percentage change in I_C.
(d) Based on parts (a) through (c), does V_{BE} or β appear to be more critical for Q-point stability?

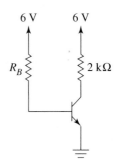

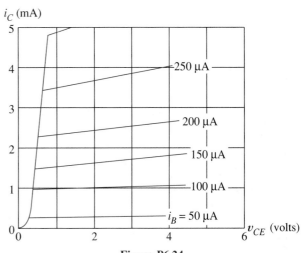

Figure P6.24

6.26 Use Eq. (6.12) to find I_C for room temperature (25°C) parameter values: $R_E = 2.5$ kΩ, $R_T = 10$ kΩ, $V_{TT} = 5$ V, $\beta = 150$, and $V_{BE} = 0.7$ V. Recalculate the value of I_C if the temperature increases to 80°C and XTB = 1.6.
Hint: Refer to Sec. 4.9.1.

Section 6.4

6.27 In Fig. 6.10a change the supply voltage to -9 V and the bias current to 0.5 mA. Then design the circuit for equal volt-ages across transistor, R_C, and R_E when the transistor has mini-mum beta of 175.

6.28 In Fig. 6.11, we want equal voltage drops across R_C, the transistor, and the current source. Find resistor and current source values if $\beta = 35$.

6.29 In Fig. 6.12, replace the $+15$ V supplies by $+12$ V sup-plies and change the bias current to 0.9 mA. Design the circuit for $100 \le \beta \le 250$.

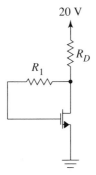

Figure P6.22

6.30 Modify the circuit of Fig. 6.13a by replacing the $+9$ and -3 V supplies with $+12$ and ground. Then design the circuit so that voltages across R_1, Q_1, and R_2 are equal, and so that the drops across R_4 and Q_2 are equal. Bias at $I_{C1} = 0.1$ mA, $I_{C2} = 0.5$ mA.

6.31 Both transistors in Fig. 6.13c have $k = 4 \times 10^{-3}$ A/V^2 and $V_t = 0.6$ V. If V_O is to be 5 V, $I_{D1} = 2$ mA and $I_{D2} = 1$ mA, find R_S, V_{G1}, I_O, V_{G2}, and R_1.

6.32 Design the bias circuits of Fig. P6.32.

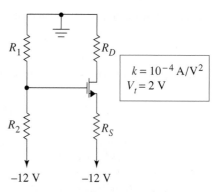

Figure P6.33

In circuit (a), bias so that $V_C = 0$ V and so that the transistor Q-point is (6 V, 100 μA), given that $40 \le \beta \le 120$.
In circuit (b), bias the transistor at (7 V, 3 mA), given that $50 \le \beta \le 100$. (Any reasonable design satisfying these conditions will do.)
In circuit (c), bias the transistor at $(-2$ V, 1 mA$)$, given that $10 \le \beta \le 500$. (Any reasonable design will do.)

6.33 Design the bias circuit of Fig. P6.33 so that the transistor is biased at (4 V, 0.4 mA), the drop across $R_D = 5$ V, and input resistance R_P is 2 MΩ.

6.34 Bias the transistor in Fig. P6.34 at $(-3$ V, 2.5 mA$)$. Also make the drop across $R_S = 3$ V.

6.35 Find resistor values for the circuit of Fig. P6.35 so that $V_G = 1$ volt, $V_D = 10$ V, $I_D = 9$ mA.

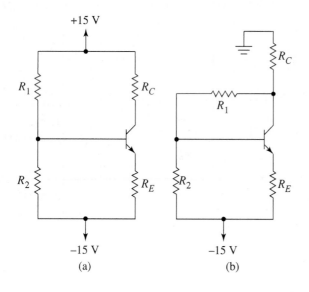

Figure P6.32

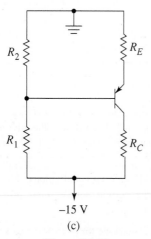

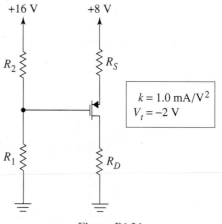

Figure P6.34

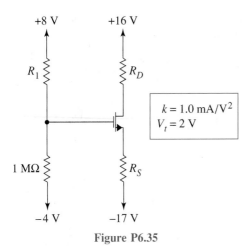

$k = 1.0\ \text{mA/V}^2$
$V_t = 2\ \text{V}$

Figure P6.35

Section 6.5

6.36 (a) Use Eqs. (6.17) and (6.18) to compute the numerical values of

$$S_\beta^{I_C} \quad \text{and} \quad S_{V_{BE}}^{I_C}$$

for the circuit of Fig. P6.36.
(b) Estimate the percentage change in I_C that would result from a 15% change in V_{BE}.
(c) In terms of sensitivity, is V_{BE} or β the more critical in maintaining the desired value of I_C? Explain.

6.37 An oscillator circuit generates a sinusoidal signal of frequency

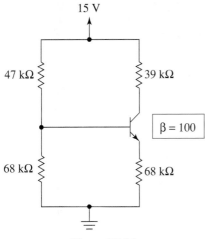

Figure P6.36

$$f_o = \frac{1}{\sqrt{LC}}$$

(a) Compute $S_C^{f_o}$.
(b) If a nominal design involves $f_o = 1$ MHz, estimate the actual frequency if C is 10% too high.
(c) If an oscillator design must hold frequency variations to within ± 100 parts per million, what would be the percentage variation allowed for C?

6.38 Starting with Eq. (6.16), show that $S_{R_T}^{I_C} = -R_T/[R_T + (\beta + 1)R_E]$. Write a verbal argument based on your expression to justify our bias circuit design practices as they relate to R_E and R_T.

6.39 Equation (6.19) implicitly defines I_D as a function of R_S for the four-resistor biasing circuit.
(a) Find an expression for the derivative of I_D with respect to R_S in terms of k, V_{GG}, I_D, R_S, and V_t.
(b) Derive an algebraic expression for $S_{R_S}^{I_D}$.
(c) Find the numerical value of $S_{R_S}^{I_D}$ for the circuit designed in Example 6.1.
(d) Use your answer from part (c) to estimate the percentage change that results in I_D if R_S increases by 10%.

6.40 (a) Starting with Eq. (6.16), show that $S_{R_E}^{I_C} = -(\beta + 1)R_E/[R_T + t(\beta + 1)R_E]$.
(b) Show that $S_{V_{TT}}^{I_C} = V_{TT}/(V_{TT} - V_{BE})$.

6.41 In a particular bias circuit, the sensitivities of I_C are
$$S_{V_{R1}}^{I_C} = -0.73,\ S_{V_{R2}}^{I_C} = +0.73,\ S_{V_{RC}}^{I_C} = -3.9 \times 10^{-3},\ S_{V_{RE}}^{I_C} = -1.07$$
$$S_{V_{CC}}^{I_C} = +1.33,\ S_\beta^{I_C} = +0.01,\ S_{V_{I_s}}^{I_C} = +0.008$$
(a) Which *transistor* parameter is most critical to maintaining I_C?
(b) Which *circuit* parameter is most critical?
(c) Estimate the percent variation in I_C if every parameter is 10% high.
(d) Estimate the worst case percent variation in I_C if every parameter varies by $\pm 10\%$.

6.42 (a) Starting with Eq. (6.16) write a general algebraic expression for $\Delta I_C/I_C$ that has the form of Eq (6.20).
(b) Evaluate the coefficients in your expression using values from Example 6.6.
Hint: See Eqs. (6.17) and (6.18) and Problems 6.38 and 6.40.
(c) Which parameter makes the greatest contribution to the uncertainty in I_C?
(d) Make a worst case estimate of $\Delta I_C/I_C$ if every parameter varies by 10%.

Section 6.6

6.43 (a) Use SPICE to check the design of Example 6.5. For a more realistic simulation, substitute nearest 10% tolerance resistor values from Table 6.1 and a transistor Early voltage of 80 V. Compare the Q-point from the simulation with the design value.
(b) Now assume each resistor has a temperature coefficient of 0.005/°C. Use SPICE to determine whether the Q-point will ever enter the ohmic region over the temperature range $-30°C \leq T \leq +50°C$?

6.44 (a) Use SPICE to plot the transfer characteristic, v_O versus v_I for the amplifier of Fig. P6.44 over the range $0 \leq v_i \leq 15$ V. Ignore body effect. From SPICE output estimate the maximum gain, and find the bias value of v_i at which it occurs?
(b) Because the source is not grounded in Fig. P6.44, body effect should be included in the analysis. Repeat part (a) after adding a body effect coefficient of 0.4 to the transistor description.

6.45 Use SPICE to compute the sensitivities of the drain current to variations in the resistors and to V_t in Fig. P6.44. Give your answers in terms of the Eq. (6.14) sensitivity definition.

6.46 Redo Problem 6.45, but find the sensitivities of V_{DS} rather than I_D.

6.47 Use SPICE to check the Q-point in the design of Example 6.6 at temperatures of $-30°C$ and $+60°C$. Use $\beta = 200$ and XTB = 1.6 for the transistor, and TC = 0.004 for the resistors.

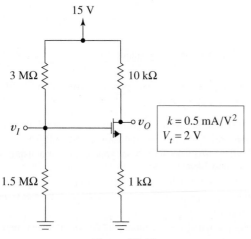

Figure P6.44

6.48 Use SPICE to compute the sensitivities of drain current for the circuit designed in Example 6.4. Convert to the Eq. (6.14) sensitivity definition.

6.49 Use SPICE to compute the collector current element sensitivities for the circuit of Example 6.2 when $\beta = 20$. Include parameters R_1, R_2, V_{CC}, and V_{BE}. See if the sensitivities improve after you replace the emitter resistor by a 2 mA current source of 65 kΩ output resistance.

Section 6.7

6.50 (a) Design an MOS current source like the one in Fig. 6.22 to give 0.2 mA output current using a 17 V supply. Transistor parameters are $k = 0.3 \times 10^{-3}$ A/V^2 and $V_t = 1.1$ V.
(b) Find the minimum output voltage for the current source.
(c) Find the output resistance if the transistor Early voltage is 110 V.
(d) Use r_o to find the value of I_O when the voltage across the current source is 12 V.
Hint: Find the additional current that flows (through r_o) when 12 V is applied.
(e) Design a BJT current source like Fig. 6.24b to give the same output current and using the same power supply as your source from part (a). For what output voltages does this current source work properly?

6.51 The diode-connected transistor of Fig. 6.22c has parameters $V_t = 1.3$ V and $k = 0.12 \times 10^{-3}$ A/V^2. Circuit parameters are $R = 34.5$ kΩ and $V_{DD} = +14$ V.
(a) Calculate the value of I_{REF} and indicate the minimum current-source voltage for proper operation.
(b) Attach a dc voltage source, V_{out}, between drain and ground, and then describe the circuit in SPICE code. Use dc analysis to obtain a plot of output current versus V_{out} for $0 \leq V_{out} \leq 14$ V. Relate this plot to a figure from the textbook.
(c) Include an Early voltage of 50 V in your SPICE description and then make a new plot to compare with the one of part (b).

6.52 Design an MOS current mirror having outputs of 1 mA, 2.5 mA, and 5 mA. The diode connected transistor has parameters $V_t = 0.5$ V and $k = 2.5$ mA/V^2. The power supply is 12 V.

6.53 (a) Design a current mirror like Fig. 6.24b to provide 2.1 mA and work with a 9 V supply.
(b) What is the minimum output voltage of the current source?
(c) If the Early voltage for the output transistor is 115 V, what is the output resistance of the current source?
(d) Model this current source with SPICE, and include a source, V_{out}, between collector and ground. Plot output current versus V_{out} for $0 \leq V_{out} \leq 9$ V.

(e) Draw a circuit diagram showing how the same mirror design of part (a) can be used with a single negative supply of -9 V and ground.
Hint: Add -9 V to power supply and ground.

6.54 When all transistor areas are identical in Fig. 6.25a, we found that $I_{O1}/I_{REF} = \beta/(\beta + 4)$. If the area of Q_3 is three times that of the other transistors, find the current gain for outputs I_{O1} and I_{O3}.

6.55 The current source of Fig. 6.25 uses $V_{CC} = 15$ V and $R = 10$ kΩ. Ignoring the current gain, estimate the value of each output current if the areas of the bases of Q_1, Q_2, and Q_3 are 0.7, 1.1, and 1.8, respectively, relative to that of Q_R.

6.56 Work out the missing details in deriving Eq. (6.30).

6.57 (a) Derive Eq. (6.31). Start by labeling the emitter current of Q_R as I_X and then add appropriate currents to the diagram. All transistor βs are identical.
(b) Compute the sensitivity of I_O to changes in β.
(c) Estimate the percentage change in I_O that would result from a 2% change in β when $\beta = 399$ and when $\beta = 9$. Compare these with the results given in the text just after Eq. (6.30).

6.58 The current source of Example 6.17 has a useful output voltage range of only $4.4 \leq v_{out} \leq 9$ V. Explain why a transistor with larger k should extend the voltage range. Then redesign the example circuit using $k = 10^{-3}$ A/V^2.

6.59 A basic BJT current mirror has two output transistors with areas of 0.6 and 1.1 relative to that of the reference transistor. Calculate the current gain at each output if $\beta = 30$.
Hint: On a diagram, start by labeling the emitter currents of the three transistors.

6.60 For the mirror of Fig. P6.60,
(a) Find R if the reference current is 2 mA.
(b) Find the highest voltage V_O consistent with operation as a current source.
(c) In what way do the emitter resistors improve the performance of this current source?
(d) Find the output resistance if the Early voltage is 110 V.

6.61 Derive Eq. (6.32).

6.62 (a) Design an MOS Wilson current source to work with a 10 V power supply. All transistors are identical with $k = 2$ mA/V^2 and $V_t = 1.1$ V. The output current is to be 1 mA.
(b) Work out a BJT Wilson current-source design to meet the specifications given in part (a) using high-beta silicon transistors. Ignore current gain.

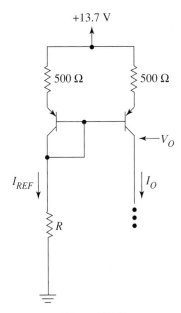

Figure P6.60

6.63 (a) Design a BJT cascode mirror that gives $I_O = 2$ mA. The reference current is to be derived from a 15 V power supply using a resistor. Specify the minimum output voltage for correct operation.
(b) Design an MOS cascode mirror that satisfies the specifications of part (a). Use identical transistors with $k = 1$ mA/V^2 and $V_t = 0.8$ V. What is the minimum current-source output voltage?

6.64 Design the Widlar current source of Fig. P6.64 for output current of 0.12 mA using resistors of 10 kΩ or less. Also compute the value of its output resistance if the BJT has an Early voltage of 100 V.

6.65 The two diode-connected transistors in Fig. P6.65 have $k = 0.8$ mA/V^2 and $|V_t| = 0.9$ V. The other transistors are identical except for W/L, which is labeled beside each transistor. Component values are $R = 212$ kΩ, $V_{DD} = +12$ V and $V_{SS} = -12$ V. Find the output currents I_1–I_4.

6.66 Design a current source of the form of Fig. 6.31c that has two output currents, 1.5 mA from a pnp transistor and 0.5 mA from an npn transistor. The transistors are silicon. Use power supplies of ± 9 V. Any of the possible correct solutions is acceptable.

6.67 The two diode-connected transistors in Fig. P6.65 have $k = 0.8$ mA/V^2 and $|V_t| = 0.9$ V and $W/L = 1$. The other transistors are identical except for W/L, which is labeled beside

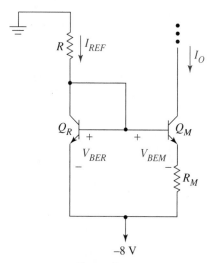

Figure P6.64

each transistor. Design the circuit so that $I_3 = 0.4$ mA for ± 15 V supplies. Find the output currents I_1–I_4.

6.68 In Example 6.19, the Fairchild µA733 was analyzed with the bases of Q_5 and Q_6 grounded. Find the Q-points of transistors Q_5–Q_8 if

(a) the bases of Q_5 and Q_6 are both raised to $+1$ V.

Hint: Imagine the four transistors in question replaced by their large signal models.

(b) Find the Q-points of the same four transistors if the bases of Q_5 and Q_6 are both lowered to -1 V.

6.69 An interesting question about an IC is whether it will work properly for power supply voltages other than those specified. Find the Q-points of transistors Q_5, Q_6, Q_7, and Q_8, in the Fairchild µA733 IC of Fig. 6.32 if the supply voltages are changed to ± 9 V.

6.70 Design a BJT current-source system, including a circuit to generate reference current, for the circuit of Exercise 6.19. Draw a complete circuit diagram showing relative transistor areas.

6.71 Design a current mirror based upon Fig. 6.28 but using pnp transistors having Early voltages of 50 V. The output current is 0.2 mA and the output resistance of the current source must be at least 25 MΩ. Use a 15 V power supply.

6.72 In Fig. P6.72 the relative areas of the transistors are given on the diagram.

(a) Find I_O.

(b) What is the purpose of Q_2?

(c) What is the advantage of connecting the resistor to ground rather than -9 V?

6.73 Use SPICE ac analysis to demonstrate the signal mirroring of Fig. 6.35. Use a 1 mV sinusoid of 1 kHz frequency in series with a 12 V supply. $R = 2.3$ kΩ, $k = 10^{-4}$ A/V^2, $V_A = 90$ V, and $V_t = 1.3$ V. Add a separate 12 V dc source at the output to keep the transistor active, and find the amplitude of the signal current that finds its way into the output current.

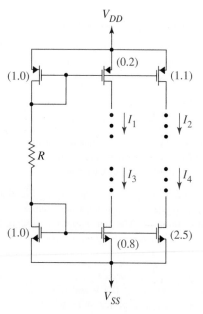

Figure P6.65

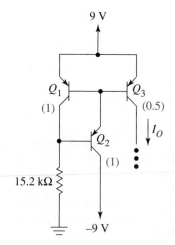

Figure P6.72

Section 6.8

6.74 A basic BJT current source uses a 15 V supply and a 47 kΩ resistor to produce a 0.304 mA output current. An alternate design uses a JFET with $I_{DSS} = 0.304$ mA and $V_A = 100$ V to replace the resistor. Compute the sensitivities of the two circuits to power supply variations.

6.75 (a) Substitute Eq. (6.39) into Eq. (6.38) to relate I_O to V_{CC} for the Widlar current source.
(b) Use your expression from part (a) to show that the sensitivity of I_O to changes in V_{CC} for the Widlar circuit is always less than Eq. (6.41).
(c) Evaluate your sensitivity expression for the Widlar design of Example 6.18.

6.76 Figure P6.76 is a circuit for measuring the sensitivity of a current source in the laboratory. The ac source simulates changes in supply voltage, and a load resistor converts corresponding changes in output current into an ac output voltage, ΔV_o.
(a) Starting with the incremental approximation to the sensitivity, relate the transconductance gain, $\Delta I_O/\Delta V_{CC}$, to $S_{V_{cc}}^{I_o}$.
(b) Use the result of part (a) to relate voltage gain, $\Delta V_O/\Delta V_{CC}$ to $S_{V_{cc}}^{I_o}$. Notice that the derivation applies to all current sources, not just the basic source shown here.
(c) For a source with nominal current of 1 mA, 12 V supply, and estimated supply sensitivity of 0.021, find R_L so overall gain will be ≈ -1.

6.77 In Fig. P6.77, the JFET parameters are $I_{DSS} = 1.5$ mA, $V_P = -1$, and $V_A = 90$ V. The bipolar parameters are $\beta = 350$ and $V_A = 180$ V. Use SPICE ac analysis to find the ac noise component of the load current if the sinusoidal noise signal has amplitude of 1 mV. Use any load voltage adequate to keep the output transistor active.

6.78 Replace the bipolar transistors in Problem 6.77 with MOSFETs having $k = 2$ mA/V^2, $V_t = 2$ V, and $V_A = 100$ V. Solve this modified problem.

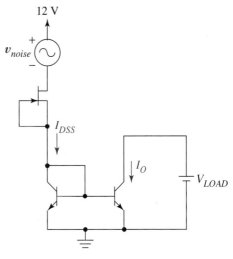

Figure P6.77

6.79 The circuit of Fig. 6.38a has $V_{CC} = 15$ V, $R_M = 748$ Ω, and $R_R = 113.5$ kΩ. Transistor parameters are $\beta = 400$ and $V_A = 100$ V. Use SPICE ac analysis to determine the effect of power supply ripple on output current. Represent the ripple voltage as a sinusoidal source of peak amplitude $1\% \times V_{CC} = 0.15$ V and 120 Hz in series with V_{CC}. Determine the ac current that appears in the output at 120 Hz. Use a separately defined voltage source of 14 V to monitor output current and keep Q_M active.

6.80 The circuit of Fig. P6.80 uses three matched transistors to produce the same result as the circuit of Fig. 6.38a. That is,

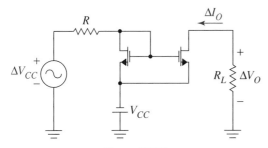

Figure P6.76

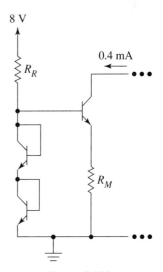

Figure P6.80

I_{REF} is given by Eq. (6.44), output current by Eq. (6.45), and the relation between I_{REF} and V_{CC} by Eq. (6.46).
(a) Design the circuit.
(b) Use SPICE to determine the sensitivity of output current to supply voltage if $\beta = 220$ and $V_A = 135$.

6.81 (a) Design a V_{BE} bootstrap reference like Fig. 6.39c for $I_O = I_{REF} = 0.4$ mA. Use $V_{CC} = 15$ V.
(b) Use SPICE to compute I_O and its sensitivity to V_{CC}. Use $\beta = 210$ for npn transistors and $\beta = 115$ for pnp transistors, but leave other parameters at their default values. Use a separate dc source of 10 V at the output to keep the output transistor from saturating and to make load current available in the SPICE output.
(c) Use SPICE to find I_O for $V_{CC} = 15$ V, 12 V, 5 V, and 3 V.
(d) Repeat parts (b) and (c) after changing the transistor Early voltages to 190 V.

6.82 Design an MOS bootstrap circuit for $V_{DD} = 15$ V and $I_O = 1$ mA using MOSFETs with $k = 0.2 \times 10^{-3}$ A/V^2 and $|V_t| = 0.9$ V; otherwise, use SPICE default values.
(a) Use SPICE to determine I_O for $V_{DD} = 15$ V and 3 V. Explain any anomaly you find.

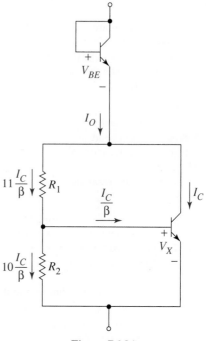

Figure P6.84

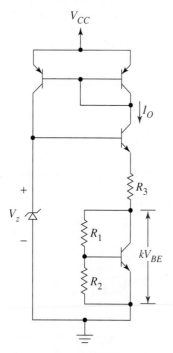

Figure P6.83

(b) Use SPICE to find the sensitivity of I_O to V_{DD} for $V_{DD} = 15$ V.
(c) Repeat part (b) after adding Early voltage of 100 V for every transistor.

6.83 Figure P6.83 is a temperature-compensated, bootstrap reference.
(a) How can you tell it is a bootstrap circuit?
(b) Derive an expression that relates the temperature coefficient of I_O to the temperature coefficients of V_z, V_{BE}, and R_3. Show mathematically what it takes to make I_O independent of temperature.
(c) Redraw the diagram showing how to obtain an externally-accessible output current I_O that does not affect the validity of your result from part (b).

6.84 The V_{BE} multiplier of Fig. P6.84 carries the same current I_O as the diode connected transistor. The multiplier is designed by choosing R_2 so that its current is 10 times the base current as shown.
(a) Assuming V_{BE} and V_X are given by

$$I_O = I_S e^{V_{BE}/V_T} \qquad \text{and} \qquad I_C = I_S e^{V_X/V_T}$$

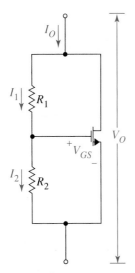

Figure P6.85

respectively, find an expression for V_X in terms of V_{BE}, I_O, and I_C.

(b) Apply KCL at the collector node of the V_{BE} multiplier to relate I_O to I_C. Substitute into the equation of part (a) so that V_X is expressed in terms of V_{BE}, V_T, and β.

(c) Find the smallest value of β such that $V_{BE} - V_X$ equals 1% of V_{BE}.

(d) Find the minimum β from part (c) for $V_{BE} = 0.6$ V.

6.85 Figure P6.85 shows a V_{GS} multiplier. Circuit design begins with a given value for output voltage V_O. Next select any value of V_{GS} that exceeds V_t and is also less than V_O. Since $V_O = (1 + R_1/R_2)V_{GS}$ the resistor ratio is fixed, and one resistor may be selected arbitrarily. The constant current I_O used to excite the circuit is determined from the given information.

(a) Show why the given equation that relates V_O to V_{GS} is correct.

(b) Using a transistor with $k = 0.3$ mA/V^2 and $V_t = 1.5$ V, design a V_{GS} multiplier to give $V_O = 3.3$ V. Include the value of I_O in your design. Any correct solution is acceptable.

CHAPTER
7

Transistor Amplifiers

In this chapter we study some transistor gain configurations that are useful both as stand-alone amplifiers and as subcircuits of larger signal processing modules. These configurations process *small signals,* so we will introduce *small-signal transistor models* and describe several new concepts associated with the meaning and use of these models. In these circuits transistors function as dependent current sources; thus we take for granted active-state operation.

7.1
Principles of Small-Signal Analysis

Figure 7.1a generalizes the FET amplifier of Fig. 6.1a to any linear electronic circuit that interacts with an FET. Each voltage and current in the circuit has a biasing component and a signal component because the circuit has dc biasing and an input signal v_i. The voltages and currents can also have distortion components that originate in the transistor. Because the transistor is biased in its active region,

$$i_D(t) = \frac{k}{2}[(V_{GS} + v_{gs}(t)) - V_t]^2$$

$$= \frac{k}{2}(V_{GS} - V_t)^2 + 2\frac{k}{2}(V_{GS} - V_t)v_{gs}(t) + \frac{k}{2}v_{gs}(t)^2 \tag{7.1}$$

where V_{GS} and $v_{gs}(t)$ are the bias and signal components of v_{GS}, respectively. To control distortion, we limit the amplitude of v_i to values that reduce Eq. (7.1) to

$$i_D(t) = \frac{k}{2}(V_{GS} - V_t)^2 + 2\frac{k}{2}(V_{GS} - V_t)v_{gs}(t) = I_D + i_d(t) \tag{7.2}$$

an equation that describes a small signal superimposed upon dc bias. This is called *small-signal operation,* and circuits operated in this fashion are called *linear* because the *signal components* of all voltages and currents follow the rules of linear circuit theory. The equivalent circuit in Fig. 7.1b models the two components of Eq. (7.2) by individual dependent sources.

474

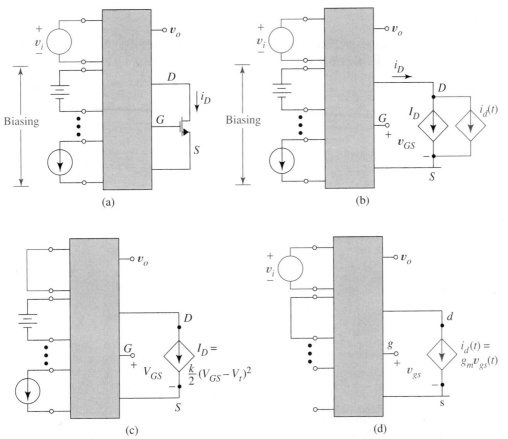

Figure 7.1 Large- and small-signal analysis: (a) amplifier with biasing and small signal; (b) transistor replaced by complete model; (c) turning-off signal gives bias circuit; (d) turning-off biasing gives small-signal equivalent circuit.

Small-Signal Equivalent Circuit. To simplify analysis, we employ a divide-and-conquer technique. To calculate I_D and the other bias values, we turn off the signal and replace the transistor by its large-signal model as in Fig. 7.1c, a technique we already mastered in Chapter 5. To compute signal components, we turn off all biasing sources and replace the transistor with its *small-signal model* as in Fig. 7.1d, where the compact notation, $i_d(t) = g_m v_{gs}(t)$, emphasizes that signal current is proportional to signal voltage. The circuit represented by Fig. 7.1d is called the *small-signal equivalent* of the original circuit, because it gives the same values for small-signal quantities as the original circuit.

The advantage of this two-step approach is that Fig. 7.1d is a linear circuit that lends itself to simple analysis even though the true circuit, Fig. 7.1a, includes a nonlinear transistor. The procedure *resembles* superposition in that we superimpose results from successive analyses to obtain the total result. It differs from true superposition, however, because the second analysis requires the results of the first. Since g_m is the coefficient of $v_{gs}(t)$ in Eq. (7.2), we need the value of V_{GS} or equivalent information from the first analysis to calculate the numerical value of g_m for the second analysis. In linear circuits, the parts of a superposition analysis are independent and can be performed in any order.

This divide-and-conquer procedure applies to circuits with more than one transistor and to circuits with all types of transistors. Before stating the procedure in general form, we first examine small-signal transistor models in more detail.

7.2
Small-Signal Transistor Models

7.2.1 FET SMALL-SIGNAL MODELS

Figure 7.2a shows the small-signal model for the MOSFET. Because small signals imply projecting from tangents that approximate the device curve near the Q-point, as in Fig. 6.1c, *transconductance* is defined by

$$g_m = \left. \frac{\partial i_D}{\partial v_{GS}} \right|_{Q\text{-}point} \tag{7.3}$$

For the n-channel MOSFET this gives

$$g_m = \left. \frac{\partial [(k/2)(v_{GS} - V_t)^2]}{\partial v_{GS}} \right|_{Q\text{-}point} = k(V_{GS} - V_t)$$

which is the coefficient of $v_{gs}(t)$ in Eq. (7.2). By substituting $(V_{GS} - V_t)$ from the Q-point current expression,

$$I_D = \frac{k}{2}(V_{GS} - V_t)^2$$

we obtain

$$g_m = \sqrt{2kI_D} \tag{7.4}$$

which we use hereafter.

In a p-channel device, *both* the gate-to-source voltage and the current that flows from drain to source are negative, thus the obvious small-signal model would be a de-

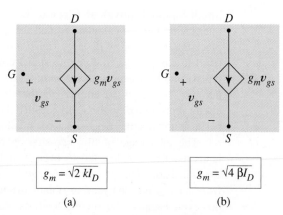

Figure 7.2 Small-signal models for n- and p-channel FETs: (a) MOSFET; (b) JFET and MESFET.

pendent current source directed from source to drain and controlled by the source-to-gate voltage. If we reverse the sense of *both* drain current and controlling voltage while keeping g_m positive, we obtain Fig. 7.2a. We conclude that the *same* small-signal model describes p- and n-channel MOSFETs.

For the n-channel JFET, the definition of g_m in Eq. (7.3) applies to Eq. (5.21). This gives

$$g_m = 2\beta(V_{GS} - V_P)$$

To express g_m in terms of bias current, we solve Eq. (5.21) for $(V_{GS} - V_P)$ and substitute. This gives

$$g_m = \sqrt{4\beta I_D} \tag{7.5}$$

Figure 7.2b is the JFET small-signal model for both n- and p-channel JFETs. (When the tanh expression approximates one in Eq. (5.15), MESFET and JFET equations are identical, so the same model is used to make estimates for MESFETs.)

FET Output Resistance. Sometimes r_o plays a vital role in small-signal analysis and should be included in the small-signal model. For example, we will find that r_o is nearly always important in output resistance calculations. To include transistor output resistance, we simply add r_o between drain and source nodes in Fig. 7.2, using $r_o = V_A/I_D$ to find its numerical value. In most other calculations, however, r_o is so large that the additional component, v_{DS}/r_o, that this parameter adds to the drain current is negligible, both compared to I_D in bias circuit analysis and compared to $g_m v_{gs}$ in small-signal analysis. For this reason, r_o is usually omitted, for it contributes little to the accuracy of the results, obscures important observations and makes the analysis more difficult.

Exercise 7.1 The transistor in Fig. 7.3a has $k = 10^{-4}$ A/V^2 and $V_A = 90$ V. Find the numerical values of g_m and r_o.

Ans. 2×10^{-4} A/V, 450 kΩ.

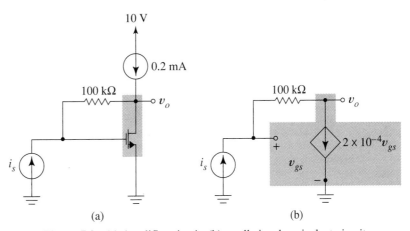

Figure 7.3 (a) Amplifier circuit; (b) small-signal equivalent circuit.

EXAMPLE 7.1   Draw the small-signal equivalent circuit for Fig. 7.3a.

Solution. Since the 0.2 mA source is a dc current used for biasing it must be turned off when drawing the small-signal equivalent; i_s is the signal source, so it must be included. Replacing the transistor with its model from Fig. 7.2a gives Fig. 7.3b.   ❏

7.2.2   BJT SMALL-SIGNAL MODEL

To derive a small-signal model for the bipolar transistor, we follow the same process we applied to MOSFETS; we linearize the device's nonlinear equations in the vicinity of the Q-point. Because the BJT input current is not zero as in the FET, its small-signal model has both input and output parameters. Voltage and current orientations are opposite in pnp and npn transistors; however, the reasoning we used for MOSFETs shows that identical small-signal models apply to npn and pnp transistors. For this reason, details are given only for the npn case.

Input Circuit. For small-signal analysis, we cannot rely on the vertical line approximation of the input characteristic as we did for large-signal analysis but must take into account its curvature. Equations (4.5) and (4.10) describe the active transistor; therefore,

$$i_B = \frac{I_S}{\beta} e^{v_{BE}/V_T} \tag{7.6}$$

describes the input characteristic of Fig. 7.4a. The biasing circuit establishes a dc quiescent bias voltage $v_{BE} = V_{BE}$ (*estimated* to be 0.7 V for room temperature silicon transistors) and a bias current I_B that are related by

$$I_B = \frac{I_S}{\beta} e^{V_{BE}/V_T}$$

When we add a small signal, v_{be}, to the bias voltage, this becomes

$$i_B = I_B e^{v_{be}/V_T} = \frac{I_S}{\beta} e^{(V_{BE} + v_{be})/V_T}$$

Taylor series expansion about the Q-point gives

$$i_B = I_B + \frac{I_B}{V_T} v_{be} + \frac{1}{2!} \frac{I_B}{V_T^2} v_{be}^2 + \cdots \tag{7.7}$$
$$= I_B + i_b + \text{distortion terms}$$

Comparing the signal terms in the two expressions, we see that $i_b = i_b(t)$ is proportional to signal voltage and can be expressed as

$$i_b(t) = \frac{1}{r_\pi} v_{be}(t) \tag{7.8}$$

where

$$r_\pi = \frac{V_T}{I_B} \tag{7.9}$$

is the *input resistance of the common-emitter transistor.*

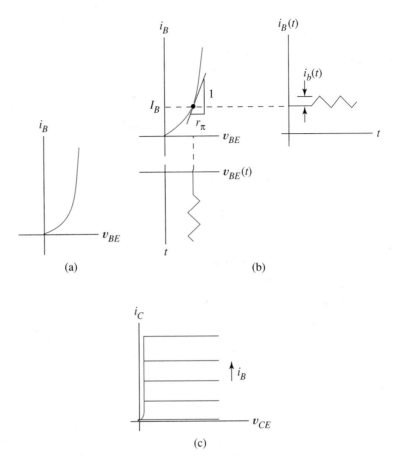

Figure 7.4 BJT characteristics for active operation: (a) input characteristic; (b) input lineariza-tion; (c) output characteristics.

Because $1/r_\pi$ in Eq. (7.8) is the coefficient of the linear term in the Taylor series expansion, it is the slope of the input characteristic at the transistor operating point as we see in Fig. 7.4b. When distortion is negligible, signals $i_b(t)$ and $v_{be}(t)$ are related by pro-jecting from the tangent to the curve. Clearly r_π is a function of Q-point, with operation at higher values of I_B giving in higher slopes and lower values of r_π, and conversely.

Output Circuit. From Fig. 7.4c the output circuit description is

$$i_C = \beta i_B$$

When signal $i_b(t)$ is added to dc bias current I_B,

$$i_C(t) = \beta[I_B + i_b(t)] = I_C + i_c(t) = I_C + \beta i_b(t)$$

Thus the small-signal equation

$$i_c(t) = \beta i_b(t) \qquad (7.10)$$

relates to operation in the vicinity of a dc operating point on the output characteristics. From Eqs. (7.8) and (7.10) we conclude that the small-signal model for the BJT looks like Fig. 7.5a.

An equivalent model that employs a voltage-controlled current source is also widely used, especially when we later extend the theory to include dynamic effects. We substitute i_b from Eq. (7.8) into Eq. (7.10) and rename v_{be} as v_π to satisfy the accepted notational convention. This gives

$$i_c = \beta \frac{1}{r_\pi} v_\pi = g_m v_\pi \qquad (7.11)$$

which describes the alternate model of Fig. 7.5b.

Two important equations relate g_m and r_π to dc Q-point values. Equating coefficients of v_π in Eq. (7.11), gives

$$g_m r_\pi = \beta \qquad (7.12)$$

a result easily remembered because g_m is a conductance, r_π a resistance, and β a dimensionless current gain. From Eqs. (7.12) and (7.9), we find

$$g_m = \frac{\beta I_B}{V_T}$$

Since I_C is central to bias circuit analysis and design, a more useful equation is

$$g_m = \frac{I_C}{V_T} \qquad (7.13)$$

where $V_T = kT/q \approx 25$ mV at room temperature. Because we rely heavily on small-signal models, Eqs. (7.12) and (7.13) are among the few equations in the text that are worth memorizing; they are given for convenience in Fig. 7.5c.

For most purposes we regard β as a constant, and r_π and g_m as functions of transistor Q-point. (According to Fig. 4.30, β in fact varies with Q-point; however, since the curve for a single temperature is rather broad and flat, we can assume it constant with little error in most cases.) The models and equations of Fig. 7.5 all apply directly to both pnp and npn transistors.

Figure 7.2 and 7.5b show that BJT and FET circuits differ only in the presence of r_π; however, for comparable bias currents BJTs have significantly higher transconductances than do FETs.

A more complete BJT model, Fig. 7.5d, includes several of the transistor's second-order effects. Since the base spreading resistance r_b is difficult to measure, small in value, and critically important only at high frequencies, we omit it in low-frequency analysis. Resistor r_μ represents the internal feedback in which base-width modulation causes the input characteristics to change slightly with output voltage as in Fig. 4.34a and 4.34b. Because r_μ is of the order of $10\beta r_o$, an open circuit is a good approximation, and omitting it usually simplifies analysis with little loss of accuracy. Base-width modulation also results in output resistance $r_o = V_A/I_C$, which is critically important in some analyses. When needed, we simply add r_o between emitter and collector nodes in the models of Figs. 7.5a or b.

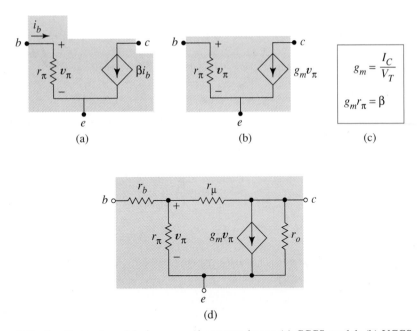

Figure 7.5 Small-signal models for npn and pnp transistors: (a) CCCS model; (b) VCCS model; (c) equations for calculating model parameters; (d) model including second-order effects.

EXAMPLE 7.2 Find the small-signal equivalent circuit of Fig. 7.6a, including numerical parameter values, if the transistors are identical and $\beta = 200$. Each transistor is biased at $I_C \approx I_E = 1$ mA.

Solution. Using $V_T = 25$ mV in Eq. (7.13) gives $g_m = 1/25$ A/V. From Eq. 7.12, $r_\pi = 200 \times 25 = 5$ kΩ. Turning off dc sources and replacing transistors with small-signal models gives Fig. 7.6b. *It is important to identify correctly the location and polarity of the controlling variable for each dependent source.* ❏

Exercise 7.2 Add r_o and r_μ to the equivalent circuit of Example 7.2, assuming $V_A = 120$ V and $r_b = 0$. Make reasonable approximations to simplify the circuit.

Ans. 120 kΩ, 240 MΩ, Fig. 7.6c.

7.2.3 OBTAINING SMALL-SIGNAL PARAMETERS FROM CHARACTERISTIC CURVES

Transistor input and output characteristics curves are sometimes included on the transistor data sheet, and they can also be displayed on an instrument called a transistor curve tracer. By viewing these plots of input and output current as functions of two variables, we can place all small-signal parameters in a broader perspective and demonstrate a practical way to obtain parameter values at any Q-point directly from the graphical information.

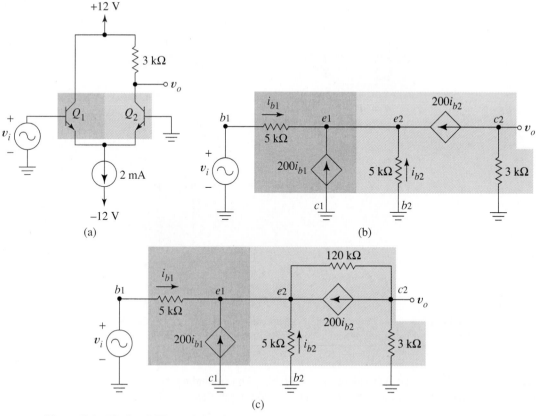

Figure 7.6 Finding BJT small-signal equivalent: (a) original circuit; (b) small-signal circuit; (c) circuit with transistor r_o and r_μ values included.

Figure 7.7a shows the common-emitter input characteristics, a *family* of curves caused by base-width modulation, as explained in Sec. 4.9.3. We view the input characteristics as plots of a mathematical function

$$i_B = f(v_{BE}, v_{CE})$$

Then we define

$$\frac{1}{r_\pi} = \frac{\partial i_B}{\partial v_{BE}}\bigg|_{Q\text{-point}} \approx \frac{\Delta i_B}{\Delta v_{BE}}\bigg|_{v = V_{CE}} \tag{7.14}$$

To compute the required differentials we draw a tangent to the $v_{CE} = V_{CE}$ curve at the Q-point and then read pairs of values for i_B and v_{BE} on this tangent; that is, $\Delta i_B = I_{B2} - I_{B1}$ and $\Delta v_{BE} = V_{BE2} - V_{BE1}$.

Figure 7.7b shows the output characteristics for a transistor biased at the Q-point (V_{CE}, I_C). Regarding collector current as a function, $i_C = g(v_{CE}, i_B)$, we define

$$\beta = \frac{\partial i_C}{\partial i_B}\bigg|_{Q\text{-point}} \approx \frac{\Delta i_C}{\Delta i_B}\bigg|_{v_{CE} = V_{CE}} \tag{7.15}$$

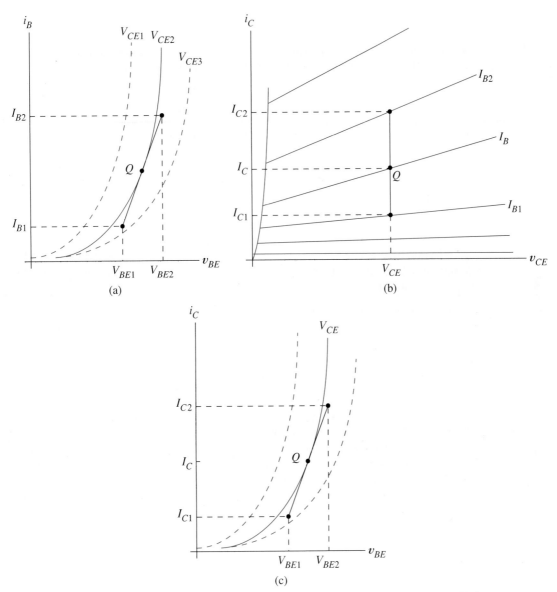

Figure 7.7 Graphical constructions for evaluating static small-signal parameters: (a) for r_π; (b) for β; (c) for g_m.

We then estimate β by measuring corresponding increments of i_C and i_B along a vertical line drawn through the Q-point as Eq. (7.15) suggests. For example, $\Delta i_C = I_{C2} - I_{C1}$ and $\Delta i_B = I_{B2} - I_{B1}$. The parameter of Eq. (7.15) is sometimes called the *ac beta* or *small-signal beta* to distinguish it from the *large-signal* or *dc beta* defined at the same Q-point by

$$\beta_{DC} = \frac{I_C}{I_B}\bigg|_{Q\text{-point}} \tag{7.16}$$

If the output characteristics were truly evenly spaced horizontal lines, as we usually assume, definitions (7.15) and (7.16) would be identical. Because real transistors include the Early effect variation of β with I_C, the distinction is sometimes important.

From the output characteristic we can also evaluate r_o using

$$\frac{1}{r_o} = \left.\frac{\partial i_C}{\partial v_{CE}}\right|_{Q\text{-point}} \approx \left.\frac{\Delta i_C}{\Delta v_{CE}}\right|_{i_B = I_B} \tag{7.17}$$

by defining increments Δi_C and Δv_{CE} along the constant I_B curve that crosses the Q-point.

To measure transconductance, we need the transfer characteristics as in Fig. 7.7c. We then use

$$g_m = \left.\frac{\partial i_C}{\partial v_{BE}}\right|_{Q\text{-point}} \approx \left.\frac{\Delta i_C}{\Delta v_{BE}}\right|_{v_{CE} = V_{CE}} \tag{7.18}$$

It should be apparent that we can use the same approach to define and measure FET parameters or, in fact, parameters for any device for which input, output, and transfer curves are available.

Small-Signal Magnitudes in BJTs and FETs. In an amplifier, the lower bound for useful signal amplitudes depends on inherent circuit noise and the upper bound on distortion. A figure of merit called *dynamic range*, defined as

$$\text{dynamic range} = 20 \log \left| \frac{\text{largest useful signal}}{\text{smallest useful signal}} \right|$$

quantifies the concept. Noise levels are beyond the scope of this text; however, we can address distortion in small-signal circuits at this time.

An input signal is a *small signal* if the distortion term produced in the transistor is at least a factor of 10 smaller than the signal term. For the MOSFET, which is described by Eq. (7.1), this gives

$$|v_{gs}| \le \frac{1}{10} \times 2|V_{GS} - V_t| = 0.2 \sqrt{\frac{2I_D}{k}} \tag{7.19}$$

For transistors biased near $I_D \approx k$, small signals are limited to peak magnitudes of about 280 mV.

For BJTs, we use Eq. (7.7) to define small signals. The error involved in truncating a Taylor series after two terms is given by the remainder R_2, where

$$R_2 = \frac{1}{2!} \frac{I_B}{V_T^2} \xi_2^2, \qquad \text{for } 0 \le \xi_2 \le |v_{be}|$$

Obviously, the largest error occurs when $\xi_2 = |v_{be}|$. For this maximum error to be negligible compared to the signal,

$$\frac{1}{2!} \frac{I_B}{V_T^2} |v_{be}|^2 \le \frac{1}{10} \frac{I_B}{V_T} |v_{be}|$$

Since $V_T = 0.025$ V at room temperature, small signals in BJT circuits satisfy

$$|v_{be}| \leq \frac{V_T}{5} = 5 \text{ mV} \tag{7.20}$$

Because its transfer characteristic is quadratic rather than exponential, the FET can process significantly larger input voltages without introducing distortion than can the BJT. In Chapter 9 we learn how to reduce distortion by using feedback; in Chapter 10 we quantify distortion so that we can measure it and meet specific design specifications. Until then, we avoid distortion by keeping signals small.

Exercise 7.3 Find the largest signal voltage that can be developed between gate and source in the circuit of Exercise 7.1 if the output signal is to be undistorted.

Ans. 400 mV.

7.2.4 DIODE-CONNECTED TRANSISTORS

We previously encountered diode-connected transistors, both FET and BJT, in constant current sources; they appear again in this chapter as load resistors and in active-load circuits. We next justify replacing any diode-connected transistor with a resistor of $1/g_m$ ohms whenever we draw a small-signal circuit.

Because of the diode connection, the transistor in Fig. 7.8a is either off or active; in all circuits of interest, it is active. But when active, the transistor's small-signal model is

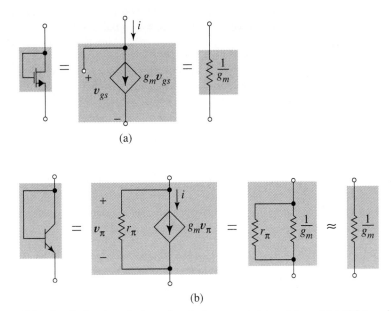

Figure 7.8 Small-signal equivalents for diode-connected transistors: (a) MOS transistor; (b) BJT.

something special—a VCCS *controlled by its own terminal voltage!* We learned in Chapter 1 that a static one-port device with current proportional to voltage is a resistor. Because the proportionality constant is $1/g_m$, the diode-connected transistor is equivalent to the resistor shown in the figure.

The same result applies to the diode-connected BJT. When we replace the active-state transistor in Fig. 7.8b by its small-signal model, we again find a current source controlled by its own terminal voltage; thus the circuit reduces to the parallel combination, $r_\pi \| 1/g_m$. From Eq. (7.12),

$$\frac{1}{g_m} = \frac{1}{\beta} r_\pi \tag{7.21}$$

This shows that $1/g_m << r_\pi$ is *always* a good approximation. (This approximation, itself, is worth noting, for it often leads to simplifications in circuit diagrams and analysis.) Finally, because small-signal models of p-channel and pnp devices are identical to those of n-channel and npn devices, respectively, the equivalence of Fig. 7.8 applies to all diode-connected transistors.

7.3
Small-Signal Equivalent Circuits

In this section we first show how to introduce signals into amplifiers without disturbing bias conditions. Then we give a general procedure for drawing the small-signal equivalent circuit.

7.3.1 DIRECT AND CAPACITIVELY COUPLED AMPLIFIERS

Having designed a biasing circuit, the next problem is to get signals into and out of the circuit without changing transistor Q-points. One way is to use direct coupling; the alternative is to use *coupling capacitors.*

Direct-Coupled Circuits. Most ICs and some discrete-component circuits are designed to be *direct coupled.* This means specially designed bias circuits keep all transistors in their active states when input signal voltage is zero. These circuits often employ dual power supplies and make extensive use of current sources for biasing, as in Fig. 6.32. With direct coupling, we simply attach the signal source directly to the input. To analyze direct-coupled circuits, we turn off the biasing sources and replace each transistor with its small-signal model as in Fig. 7.1d.

EXAMPLE 7.3 Draw the small-signal equivalent circuit for Fig. 7.9a.

Solution. Figure 7.9b shows the result of turning off dc sources V_{DD}, V_{GG}, and I_{DD} and then replacing transistors with models from Fig. 7.2.   ❏

Exercise 7.4 Draw the small-signal equivalent of Fig. 5.13a if the enhancement transistors are biased at $I_D = 1$ mA and have parameters $k = 0.5 \times 10^{-4}$ A/V^2 and $V_t = 3$ V. Transistor M_3 is wired as a constant current source.

Ans. Figure 7.9c with $g_m = 3.16 \times 10^{-4}$ A/V.

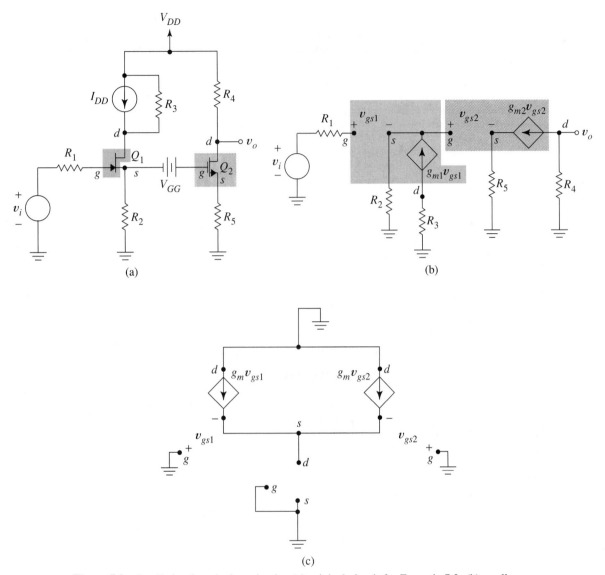

Figure 7.9 Small-signal equivalent circuits: (a) original circuit for Example 7.3; (b) small-signal equivalent for Example 7.3; (c) answer to Exercise 7.4.

An operational amplifier is the most familiar example of a direct-coupled circuit. Ideally, its transfer characteristic passes through the origin of the v_{out} versus v_{in} coordinate system, allowing us to connect sources and loads directly to input and output pins without disturbing the internal biasing. Signal processing with direct-coupled circuits is not always this easy, however. Occasionally, only a single power supply is available to power an op amp. Then the transfer characteristic no longer passes through the origin, and for correct operation, input and output dc node voltages must be one-half the supply voltage.

A signal-coupling problem can also occur when one amplifier's output is connected directly to the input of a second. A small offset voltage in the first stage can then disturb the biasing inside the second to the extent that its transistors cut off or saturate. A component called a *coupling capacitor* prevents such problems.

Capacitive Coupling. Figure 7.10a shows an amplifier and a signal source. We want to use v_i to produce slight perturbations about the *established Q-point,* and the transistor is to remain in the active state as it processes this signal. Suppose we do the obvious: connect v_i to the amplifier by closing the switch. When $v_i = 0$, the gate voltage is zero, the voltage divider that establishes bias voltage V_G is shorted out, and there is no way the gate can be more positive than the source—the transistor is cut off! This shows that we can undo the

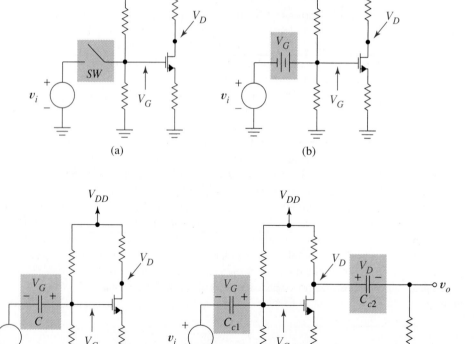

Figure 7.10 Coupling capacitor: (a) direct coupling of signal into amplifier; (b) source to maintain proper bias; (c) coupling capacitor for input signal; (d) capacitive coupling of signal at input and output.

transistor biasing by connecting a signal source directly to the amplifier if the bias circuit design did not anticipate this possibility. The same problem applies to BJT circuits.

Figure 7.10b shows a correct, though impractical, solution. A series voltage source of exactly V_G volts ensures that the node voltage at the gate is always $v_G = v_i + V_G$ as desired. This solution is impractical because batteries of all possible values are not available.

Figure 7.10c shows the solution of choice, a *coupling capacitor* C connected between the source and the amplifier. For $v_i = 0$, C *automatically* charges to V_G volts, leaving the biasing undisturbed. Now there is always a time constant RC associated with the capacitor, and by choosing large C, we make this time constant very large. Then, when we apply a *rapidly changing* signal, $v_i(t)$, the capacitor is unable to charge and discharge as the signal changes because its time constant is too large. If the capacitor voltage is $V_G = Q_C/C$ and if Q_C does not change with signal, then C acts just like the battery in Fig. 7.10b, except that it *automatically sets itself* to whatever voltage is needed. Of course, if the signal is so slow that the capacitor voltage can change appreciably during a signal cycle, the capacitor charges and discharges with changes in signal, and signal voltage develops across the capacitor. This reduces the signal voltage at the gate, causing the circuit's output signal amplitude to drop at low frequencies. (This gain change was introduced in general terms in Chapter 1 and is scheduled for further study in Chapter 8.) Another way to understand the coupling capacitor is to consider the Fourier sinusoids in a signal. If C is sufficiently large, its impedance, $1/j\omega C$, to sinusoidal signal components of all signal frequencies including the lowest, can be made negligible. In effect C acts like a short circuit to all frequencies in the input signal; however, since dc corresponds to $\omega = 0$, C is an open circuit for purposes of dc biasing. Because coupling capacitors block dc, they are always replaced by open circuits for Q-point analysis. For the same reason, we use capacitors to *couple* the signal out of the amplifier and into an external load resistor as in Fig. 7.10d.

Another use of the coupling capacitor is in *interstage coupling* of cascaded amplifiers, as illustrated by C_I Fig. 7.11. This capacitor charges to the difference in the dc levels between the interstage nodes, permitting the biasing circuits of adjacent stages to be designed individually. Without coupling capacitors, dc offsets and temperature-induced drifts are amplified by successive stages. This can make a multistage amplifier useless by driving its output into saturation, a problem illustrated in Fig. 1.36. The coupling capacitor eliminates the problem by blocking dc, but permits rapid time variations to be amplified by subsequent stages. The price we pay is that *capacitively coupled amplifiers* cannot amplify very slow signals, and so gain drops off at low frequencies.

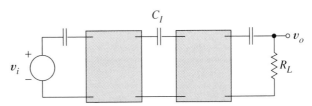

Figure 7.11 An interstage coupling capacitor C_I used to transfer signals between amplifier stages.

Coupling capacitors are used in both FET and BJT small-signal circuits. Since coupling capacitors function essentially as dc voltage sources, they are so treated in the small-signal equivalent. That is, each coupling capacitor is replaced by a short circuit, the equivalent of turning off the dc source V_G in Fig. 7.10b.

Bypass Capacitor. In discrete amplifiers there is another type of capacitor that requires special small-signal treatment: the *bypass capacitor*. We will learn to insert bypass capacitors in certain locations to increase gain, and for other reasons as well. Like coupling capacitors, bypass capacitors are large in value, function like batteries from the viewpoint of signals, and are replaced by open circuits in Q-point analysis and by short circuits in small-signal equivalent circuits. Until we study high frequency operation in the next chapter, all capacitors that appear in circuit diagrams are either coupling or bypass capacitors.

Figure 7.12a shows a two-stage amplifier with both source resistors, R_S, bypassed. Large capacitors couple the signal into the amplifier from an external source, between amplifier stages, and into an external load resistor. Figure 7.12b shows the small-signal equivalent, in which all coupling and bypass capacitors have been replaced by short circuits.

The following generalization of the ideas of Fig. 7.1 is central to much of the work in this and several successive chapters. *All transistor circuits excited by sources of small signals can be analyzed in the following way.*

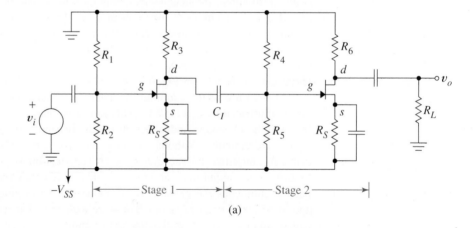

(a)

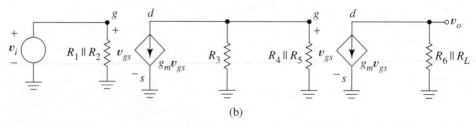

(b)

Figure 7.12 Two-stage amplifier: (a) detailed schematic of the amplifier; (b) amplifier small-signal equivalent.

Bias Circuit Analysis

1. Turn off all *signal* sources; leave bias sources on.
2. Replace each coupling and bypass capacitor with an open circuit.
3. Replace each transistor with its *large-signal model.*
4. Find the Q-point of every transistor.

Small-Signal Models

5. From the Q-point information, compute the parameter values for the small-signal model of each transistor. This requires I_C for each BJT and I_D for each FET.

Small-Signal Analysis

6. Turn off all biasing sources; leave signal sources on.
7. Replace each coupling and bypass capacitor with a short circuit.
8. Replace each transistor with its small-signal model.
9. Analyze the resulting circuit to determine the effects of the signals.

Steps 6, 7, and 8 describe exactly how to draw the small-signal equivalent circuit, extending the concept of Fig. 7.1d to circuits with more than one transistor, more than one signal source, signal currents as well as voltages, circuits with coupling and bypass capacitors, and to BJT circuits and hybrid circuits containing both BJTs and FETs.

We frequently find it useful to draw a *small-signal schematic,* defined by steps 6 and 7 only. This leaves the transistor schematic symbol in the diagram to show more clearly the location and orientation of the transistor.

In the rest of the text we use small-signal analysis to study a variety of important circuits. To explain general features and facilitate comparisons, we derive literal expressions for gains and for input and output resistances. We sometimes substitute numerical values to develop a feeling for typical magnitudes of the quantities involved; however, our object is not a collection of equations to facilitate later problem solving by plugging in numbers. Learning to apply small-signal analysis principles to these circuits and any others we may encounter in the future is itself the objective.

We begin with three basic amplifiers, named according to the orientation of the transistor relative to the input and output ports. The FET circuits are called common-source, common gate, and common drain amplifiers, respectively; the corresponding BJT circuits are named common emitter, common base, and common collector. Important for each circuit is voltage gain, current gain, and input and output resistance. Finding expressions for these provides our first examples of small-signal analysis.

7.4

Common-Emitter and Common-Source Amplifiers

7.4.1 COMMON-EMITTER AMPLIFIER

The *common-emitter amplifier,* in which the transistor is oriented with its emitter common to input and output circuits, has many applications because of its versatility. We will use small-signal analysis to demonstrate that this circuit is capable of high voltage, current, and power gain. Common-emitter input and output resistances occupy the middle ground compared with other amplifiers.

Figure 7.13a shows a discrete-component version of the common-emitter amplifier. The transistor is biased by a four-resistor circuit, coupling capacitors connect an external source and load, and a capacitor bypasses the emitter resistor. The arrows define input

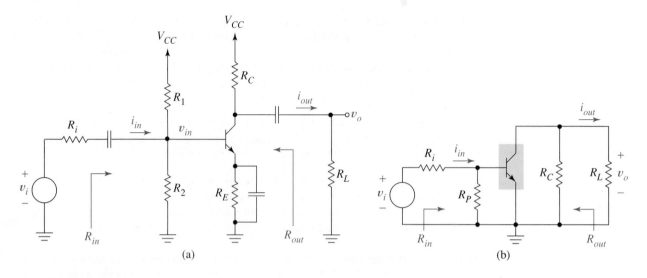

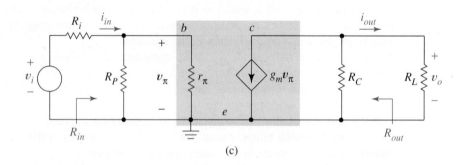

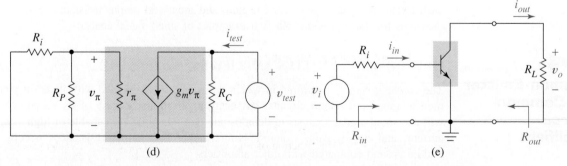

Figure 7.13 Discrete-component common-emitter amplifier: (a) original circuit; (b) small-signal schematic; (c) small-signal equivalent; (d) circuit for finding R_{out}; (e) with infinite R_P and R_C.

and output resistances seen looking into the input and output terminals of the amplifier. We will define voltage and current gains relative to the input and output nodes *of the amplifier.*

To see more clearly the roles of the various components, we use steps 6 and 7 of the general analysis procedure to draw the small-signal schematic. This gives Fig. 7.13b, where $R_P = R_1 \| R_2$. Comparing this circuit with the ideal amplifiers of Chapter 1, such as Fig. 1.19b, shows that the transistor, R_C, and R_P comprise the internal amplifier components. The small-signal schematic clearly identifies the transistor configuration, in this case showing the emitter common to the amplifier's input and output ports. For detailed analysis we usually use the small-signal equivalent, Fig. 7.13c.

To find voltage gain we first notice that v_o is produced by current $g_m v_\pi$ flowing through $R_L \| R_C$; thus

$$v_o = -g_m v_\pi \frac{R_L R_C}{R_C + R_L}$$

Because the signal at the amplifier input port is v_π,

$$A_v = (-g_m R_L) \frac{R_C}{R_C + R_L} \qquad (7.22)$$

This expresses A_v in terms of a basic gain expression, $-g_m R_L$, which includes a signal inversion, and an output loading factor introduced by biasing resistor R_C.

Now we find the current gain A_i. From the current divider in Fig. 7.13c,

$$i_{out} = \frac{R_C}{R_L + R_C}(-g_m v_\pi)$$

The amplifier's input current is

$$i_{in} = \frac{v_\pi}{R_{PP}}$$

where $R_{PP} = r_\pi \| R_P$. The ratio of these currents is

$$A_i = -g_m \frac{r_\pi R_P}{R_P + r_\pi} \frac{R_C}{R_L + R_C}$$

We obtain a more insightful result after using Eq. (7.12); namely,

$$A_i = \frac{R_P}{R_P + r_\pi}(-\beta)\frac{R_C}{R_L + R_C} \qquad (7.23)$$

Notice that input and output loading factors, current divider expressions, limit the amplifier current gain to some fraction of $-\beta$, the inherent current gain of the transistor itself.

Because the circuit of Fig. 7.13c is unilateral, to find R_{in} we visualize a test generator connected to the port identified by the R_{in} arrow. This gives

$$R_{in} = R_P \| r_\pi \qquad (7.24)$$

To find R_{out} we apply its definition as in Fig. 7.13d. Because the input circuit lacks excitation, $v_\pi = 0$, the dependent current turns off, and the test generator sees only R_C. Thus

$$R_{out} = R_C$$

With this result in mind, we notice that the output loading factor in Eq. (7.22) is exactly the kind of output loading we learned of in Chapter 1.

Ordinarily we omit r_o in common-emitter amplifier analysis because it contributes little to the results. In Fig. 7.13c, r_o would combine with R_C, usually a much smaller resistor, giving the more accurate expression,

$$R_{out} = R_C \| r_o \qquad (7.25)$$

Numerical Values. To develop a feel for what a discrete common-emitter amplifier can do, we now examine representative numerical values. Then we show how better biasing techniques could help the transistor better realize its inherent potentials, a viewpoint that relates to using common-emitter amplifiers as subcircuits in linear ICs.

In Example 4.2 we analyzed a biasing circuit like that of Fig. 7.13a using $\beta = 100$. Resistor values were $R_1 = 68$ kΩ and $R_2 = 68$ kΩ, which gives $R_P = 27.8$ kΩ, and $R_C = 39$ kΩ, and $R_E = 68$ kΩ. The transistor Q-point was (2.3 V, 0.118 mA). For this Q-point Eq. (7.13) gives $g_m = 0.118/0.025 = 4.72$ mV/A. From Eq. (7.12),

$$r_\pi = \frac{\beta}{g_m} = \frac{100}{4.72 \times 10^{-3}} = 21.2 \text{ k}\Omega$$

Assuming an Early voltage of 100 V in Eq. (4.21) gives

$$r_o = 847 \text{ k}\Omega$$

For resistances of the signal source and load let us use, respectively, 5 kΩ and 50 kΩ. Equations (7.22) and (7.23) then give voltage and current gains of

$$A_v = (-236) \times 0.438 = -103.4$$

and

$$A_i = 0.567 \times (-100) \times 0.438 = -24.8$$

Power gain is given by

$$A_P = A_v A_i = 2564$$

From Eq. (7.24) $R_{in} = 12$ kΩ, and from Eq. (7.25) $R_{out} = 37.3$ kΩ.

Now suppose that we could somehow use current sources to bias the transistor so that R_P and R_C in Fig. 7.13b become effectively infinite. Then the parasitic effects of biasing components disappear as in Fig. 7.13e, and Eqs. (7.22) through (7.25) take on forms that better demonstrate the inherent capabilities and limitations of the transistor itself, namely,

$$A_v = -g_m R_L, \quad A_i = -\beta, \quad R_{in} = r_\pi, \quad \text{and} \quad R_{out} = r_o$$

Effect of Emitter and Source Bypass Capacitors. Sometimes we willingly accept reduced gain to achieve other benefits such as reduced sensitivity to parameter variations
or improved bandwidth. Theory developed in Chapter 9 shows that we can do this by
removing the emitter (source) bypass capacitor from a common-emitter (-source) amplifier. To explore this point numerically, we find the gain of the amplifier of Fig. 7.13a
with part of R_E, R_{E1}, unbypassed as shown in Fig. 7.14a. Component values are the
same ones used for the previous numerical estimates. Biasing and Q-point stability are
unaffected because the emitter resistance in the bias circuit is still 68 kΩ. Figures 7.14b
and c show the new small-signal schematic and equivalent.

Applying KCL at the emitter node gives

$$\frac{v_\pi}{21.1 \text{ k}} + 0.00472v_\pi = \frac{v_b - v_\pi}{4 \text{ k}}$$

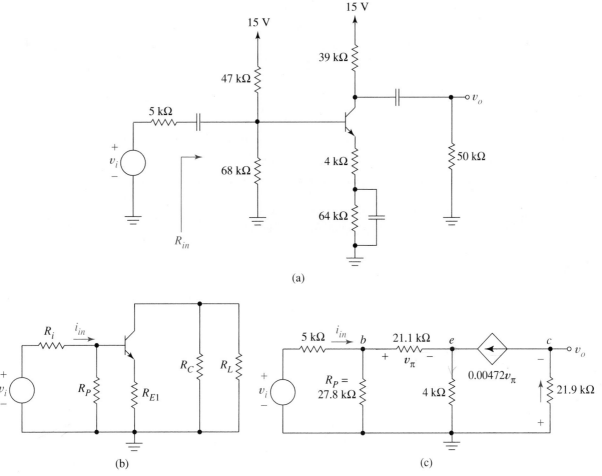

(a)

(b)   (c)

Figure 7.14   Effect of partially unbypassed emitter resistor: (a) original circuit; (b) and
(c) small-signal schematic and equivalent, respectively.

which implies that

$$v_b = 20.1 v_\pi$$

Since the dependent current develops v_o by flowing through the 21.9 k resistor,

$$v_o = (-0.00472 v_\pi) 21.9 \text{ k} = \left(-0.00472 \frac{v_b}{20.1} \right) 21.9 \text{ k}$$

The amplifier's voltage gain is now

$$A_v = -\frac{v_o}{v_b} = \frac{-0.00472}{20.1} \times 21.9 \text{ k} = -5.14$$

compared with -103.3 when the emitter resistor was completely bypassed.

Another effect of leaving emitter resistance unbypassed is to raise the input resistance. In Fig. 7.14c, R_{in} is the ratio of the voltage at node b to i_{in}. KCL at this base node gives

$$i_{in} = \frac{v_b}{27.8 \text{ k}} + \frac{v_\pi}{21.1 \text{ k}} = \frac{v_b}{27.8 \text{ k}} + \frac{v_b/20.1}{21.1 \text{ k}}$$

Solving for the ratio gives

$$R_{in} = \frac{v_b}{i_{in}} = 26.1 \text{ k}\Omega$$

compared with 12 kΩ when the entire emitter resistor was bypassed. The real change is that the equivalent resistance to the right of R_P in Fig. 7.14b has changed from $r_\pi = 21.1$ kΩ in Fig. 7.13c to 424 kΩ. This suggests that leaving emitter resistance unbypassed to raise input resistance is more effective in an environment of idealized current-source biasing where R_P is infinite.

7.4.2 COMMON-SOURCE AMPLIFIER

Like the common-emitter amplifier, the structurally related common-source circuit produces relatively high voltage and current gain, a 180° phase shift, and moderate values of input and output resistance. Figure 7.15a shows a discrete-component *common-source amplifier*. In Example 6.1 we designed the bias circuit to give a Q-point of (4 V, 2 mA) for transistor parameters $k = 0.5 \times 10^{-3}$ A/V^2 and $V_t = 2$ V. Also, we designed for $R_P = R_1 \| R_2 = 2$ MΩ.

Figures 7.15b and c show the small-signal schematic and equivalent, showing how the source is made *common to input and output* by virtue of a bypass capacitor. Comparing Figs. 7.15c and 7.13c shows that common-emitter and common-source equivalents are virtually the same, as are expressions for gain and input and output resistance. For the common-source amplifier, voltage gain and current gain are, respectively

$$A_v = (-g_m R_L) \frac{R_D}{R_D + R_L} \tag{7.26}$$

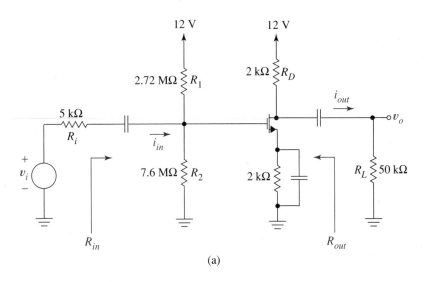

(a)

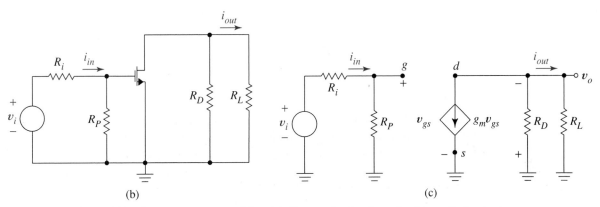

(b)   (c)

Figure 7.15 Common-source amplifier: (a) discrete-circuit schematic; (b) small-signal schematic; (c) small-signal equivalent.

and

$$A_i = (-g_m R_P)\frac{R_D}{R_D + R_L} \tag{7.27}$$

Both gain expressions include output loading factors introduced by the biasing circuit. In Eq. (7.27) the dimensionless quantity $g_m R_P$ represents current gain. Because of the FET's infinite input resistance, the common-source amplifier has inherently infinite current gain, contrasted with β for the BJT. Equation (7.27) shows that the realized current gain is limited only by R_P and thus depends upon bias circuit design.

By definition,

$$R_{in} = R_P \tag{7.28}$$

showing why we make R_P large when we design the bias circuit. The output resistance is R_D when r_o is omitted from the transistor model, or

$$R_{out} = r_o \| R_D \qquad (7.29)$$

if it is included.

Because the FET has an open circuit instead of r_π at its input, the common-source amplifier has better prospects for ending up with a high input resistance than does the common-emitter circuit. We learned in Chapter 1 that this is often desirable because it reduces loss of signal by input loading.

To determine numerical values, we first use Eq. (7.4) with $I_D = 2$ mA and $k = 0.5$ mA/V^2 to obtain $g_m = 1.41$ mA/V. For $V_A = 100$ V, $r_o = 50$ kΩ. From the diagram, $R_P = 2$ MΩ. Substituting known values into Eqs. (7.26) through (7.29) gives

$$A_v = (-70.5)(0.0385) = -2.71$$

$$A_i = (-2.82 \times 10^3)(0.0385) = -10^9$$

$$A_P = 295, \quad R_{in} = 2 \text{ M}\Omega, \quad \text{and} \quad R_{out} = 1.92 \text{ k}\Omega$$

In retrospect we notice that choosing a larger value of R_D during bias circuit design would have resulted in greater voltage and current gain. It is a good idea to examine small-signal requirements *before* designing a bias circuit.

7.4.3 LARGEST SMALL SIGNALS

Equations (7.19) and (7.20) established practical upper limits for small-signal operation. With these inequalities we can estimate the largest output and input signals we can expect in a given amplifier design without introducing distortion.

EXAMPLE 7.4 For the amplifier of Figs. 7.13a and c, we previously used the following values: $R_i = 5$ kΩ, $R_P = 27.8$ kΩ, $r_\pi = 21.2$ kΩ, $v_o/v_\pi = -103.4$.
(a) Find the largest undistorted value of v_o we can obtain.
(b) Find the largest value of v_i we can apply without having a distorted output waveform.

Solution. (a) Equation (7.20) implies that the largest permissible v_{be} in Figs. 7.13a and 7.13c is 5 mV. When gain is -103.4, $\|v_o\| \le (103.4)5 \times 10^{-3} = 0.517$ V.
(b) The input voltage divider in Fig. 7.13c attenuates v_i by

$$\frac{27.8 \text{ k} \| 21.1 \text{ k}}{5 \text{ k} + 27.8 \text{ k} \| 21.1 \text{ k}} = 0.706$$

Thus any input v_i larger than 5 mV/0.706 = 7.08 mV is likely to cause distortion. ❑

Exercise 7.5 In the amplifier of Figs. 7.15b and c, $R_i = 300$ kΩ, $R_P = 2$ MΩ, and $v_o/v_{gs} = -28.2$, (a) Find the largest undistorted value of v_o that we can expect to obtain.
Hint: Since I_D and k are not known, assume $I_D \approx k$.

(b) Find the largest value of v_i we can apply without having a distorted output waveform.

Ans. 7.98 V, 325 mV.

Estimating the largest useful output signal in this fashion helps us select a Q-point that allows signals of full amplitude without the transistor entering the cut-off or saturation (ohmic) regions.

7.4.4 APPROXIMATIONS AND EQUIVALENCES

By showing contributions of individual transistor and circuit parameters, the algebraic expressions we obtain by hand analysis help us improve our understanding and develop our intuition. However, for some circuits, straightforward analysis leads to tedious, error friendly algebra and gives expressions that are too complex to be useful. In these cases, we capitalize on approximations, equivalences, and transformations that simplify the circuit without changing its essential operation; occasionally an alternative transistor model provides an easy solution.

We already learned to ignore r_o unless we have reason to think it is important, to replace a current source controlled by its own terminal voltage by an equivalent resistor, and to replace a diode-connected transistor with a resistor. Another potential simplification exists whenever a parallel resistor combination includes a resistor of value $1/g_m$. For practical bias current values, especially for bipolar transistors, $1/g_m$ is usually much smaller than the other resistors, allowing us to replace the parallel combination by $1/g_m$. Equation (7.12) always justifies the approximation, $1/g_m << r_\pi$, when these parameters apply to bipolar transistors with identical bias currents. A *current-source transformation*, which we introduce in the next section, is another powerful analysis tool.

7.5
Common-Base and Common-Gate Amplifiers

7.5.1 COMMON-GATE AMPLIFIER

A discrete-component, common-gate amplifier design might begin with a four-resistor bias circuit as in Fig. 7.16a. A bypass capacitor connects gate to ground, signal v_i is capacitively coupled to the transistor's *source* node, and external load R_L is connected to the drain. Replacing supply voltage, bypass capacitor, and coupling capacitors by short circuits gives the small-signal schematic of Fig. 7.16b; Fig. 7.16c is the small-signal equivalent. Biasing resistors R_S and R_D are necessary evils associated with discrete-component techniques that somewhat obscure the inherent features of the common-gate circuit. For this reason we examine the more idealized version of the circuit, Fig. 7.17a, which better demonstrates the properties of IC common-gate circuits. We take for granted that the transistor is biased in its active region, perhaps by current sources in an IC amplifier. Figure 7.17b shows the small-signal equivalent.

Finding expressions for voltage and current gain and for input resistance by direct analysis of Fig. 7.17b is not difficult; however, we instead introduce a circuit equivalence that is often useful in analyzing more complex circuits. In Chapter 1 we used volt-ampere curves to examine the constraints imposed upon circuits by current sources. It follows from that discussion that we can *always* replace a current source, independent or

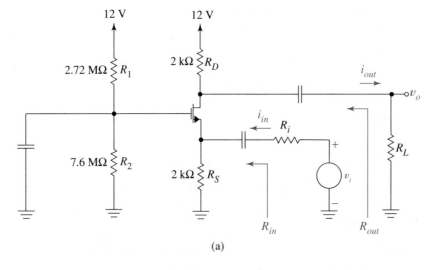

(a)

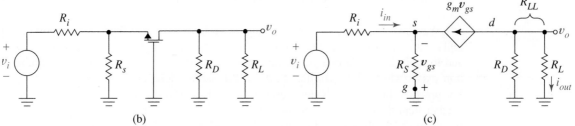

(b)   (c)

Figure 7.16 Discrete-circuit common-gate amplifier: (a) original circuit; (b) small-signal schematic; (c) small-signal equivalent.

dependent, by identical series current sources. Doing this to Fig. 7.17b gives Fig. 7.17c. This creates a new node (marked with an arrow in the diagram). Because the same current enters and leaves this node, and because the voltage across a current source is arbitrary, we can connect the newly created node to *any* point in the circuit. In this case there is an advantage to connecting the new node to ground, giving Fig. 7.17d. We call the change from Fig. 7.17b to Fig. 7.17d the *current-source transformation*. Because one current source is controlled by its own terminal voltage in Fig. 7.17d, the original circuit is equivalent to Fig. 7.17e.

From this circuit we find that the common-gate voltage gain is

$$A_b = \frac{v_o}{-v_{gs}} = \frac{-(g_m v_{gs})R_L}{-v_{gs}} = g_m R_L \tag{7.30}$$

the same as for the common-source amplifier except that there is no 180° phase shift. The current gain is

$$A_i = \frac{i_{out}}{i_{in}} = \frac{-g_m v_{gs}}{i_{in}} = \frac{-g_m[-i_{in}(1/g_m)]}{i_{in}} = 1 \tag{7.31}$$

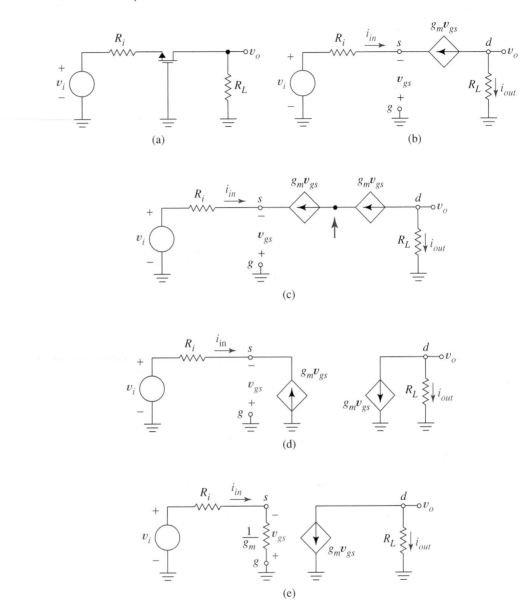

Figure 7.17 Basic common-gate amplifier: (a) small-signal schematic; (b) small-signal equivalent; (c) result of adding an identical series current source; (d) after connecting the new node to ground; (e) alternate equivalent circuit for the common-gate amplifier.

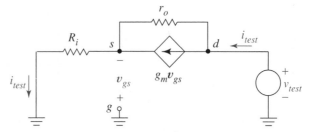

Figure 7.18 Circuit used to find common-gate output resistance.

a result quite apparent in Figs. 7.17a and 7.17b. The low input resistance,

$$R_{in} = \frac{1}{g_m}$$

(7.32)

obvious from Fig. 7.17e, is a special feature of the common-gate circuit. Low input resistance is sometimes useful, and might well be the reason for selecting the common-gate configuration.

If we apply the output resistance definition to one of the equivalent circuits of Fig. 7.17, we find R_{out} to be infinite. This is not a bad ballpark estimate; however, if we include the transistor's r_o in our analysis, we develop a better feel for the true capabilities of the common-gate circuit. Figure 7.18 shows Fig. 7.17b with r_o included, the signal generator turned off, and a test generator connected to the output. From Kirchhoff's current law at the drain node,

$$i_{test} = g_m v_{gs} + \frac{v_{test} - (-v_{gs})}{r_o}$$

Because i_{test} flows through R_i, $v_{gs} = -i_{test}R_i$. Substituting this into the foregoing equation, and solving for the ratio v_{test}/i_{test}, gives

$$R_{out} = r_o(1 + g_m R_i) + R_i \approx r_o(1 + g_m R_i)$$

(7.33)

This equation shows that common-gate output resistance is potentially much higher than r_o, the upper limit of the common-source output resistance, as indicated in Eq. (7.29).

In summary, the basic common-gate amplifier has potentially high voltage gain and output resistance, current gain of one, and low input resistance; features suggesting a CCCS with transmittance of one.

7.5.2   COMMON-BASE AMPLIFIER

The common-base amplifier resembles the common-gate circuit in both structure and properties. A discrete circuit realization might employ a four-resistor bias circuit with a bypass capacitor connecting base to ground. The signal is introduced at the emitter, and any external load is capacitively connected to the collector. The small-signal schematic for the IC version of the circuit is Fig. 7.19a; Fig. 7.19b is the small-signal equivalent. Comparing these with Figs. 7.17a and b, shows that the only substantive difference between common-gate and common-base circuits is the presence of r_π.

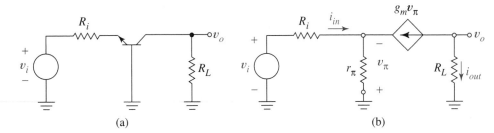

Figure 7.19 Common-base amplifier: (a) small-signal schematic; (b) small-signal equivalent circuit.

Exercise 7.6 Use the current-source transformation to find expressions for the voltage gain, input resistance, and current gain of the common-base circuit.

Ans. Equations (7.34), (7.35), and (7.36).

The common-base circuit of Fig. 7.19 has the following small-signal properties:

$$A_v = g_m R_L \tag{7.34}$$

$$R_{in} = r_e = \frac{r_\pi (1/g_m)}{r_\pi + (1/g_m)} \approx \frac{1}{g_m} \tag{7.35}$$

$$A_i = g_m \frac{r_\pi (1/g_m)}{r_\pi + (1/g_m)} = \frac{\beta}{\beta + 1} \approx 1 \tag{7.36}$$

and

$$R_{out} \approx r_o \left(1 + g_m \frac{R_i r_\pi}{R_i + r_\pi} \right) \tag{7.37}$$

EXAMPLE 7.5 Derive Eq. (7.37), and find approximations for the two cases $R_i \gg r_\pi$ and $R_i \ll r_\pi$.

Solution. Adding r_π between gate and source of Fig. 7.18 and changing the notation v_{gs} to v_π gives the test circuit for the output resistance of the common-base circuit. If we combine R_i and r_π and then repeat the common-gate analysis, Eq. (7.33) becomes Eq. (7.37). When $R_i \gg r_\pi$ the equation becomes $R_{out} \approx r_o (1 + g_m r_\pi) = r_o (1 + \beta)$; for $R_i \ll r_\pi$ we obtain

$$R_{out} \approx r_o (1 + g_m R_i)$$

the same expression we found for the common-gate amplifier, Eq. (7.33). Notice that for all R_i, common-base output resistance is higher than r_o. ❑

Like the common-gate amplifier, the common-base amplifier is a special-purpose circuit having high, noninverting voltage gain, current gain of approximately one, low input resistance, and high output resistance.

7.5.3   SPECIAL COMMON-BASE AND COMMON-GATE TRANSISTOR MODELS

We next use an example and an exercise to introduce alternative transistor models, and use the current-source transformation to relate them to the models we already know.

EXAMPLE 7.6   Use the current-source transformation to show that we can represent the BJT by the *common-base equivalent* of Fig. 7.20a, where $r_e \approx 1/g_m$.

Solution. Figure 7.20b shows the BJT model of 7.5b in a common-base orientation. Applying the transformation to the latter circuit and replacing a dependent source controlled by its own terminal voltage gives Fig. 7.20c, in which

$$r_e = r_\pi \left\| \frac{1}{g_m} \right.$$

is called the *common-base input resistance*. Resistance r_e is the common-base counterpart of r_π in Fig. 7.4b, being the slope of the tangent at the Q-point on the *common-base* input characteristic, i_E versus v_{BE}. One final step remains—replacing the VCCS in Fig. 7.20c with a CCCS. For this we recognize that

$$g_m v_\pi = g_m r_e i_e = g_m \left(\frac{r_\pi}{g_m r_\pi + 1} \right) i_e = \frac{\beta}{\beta + 1} i_e = \alpha i_e$$

The result is Fig. 7.20a.   ❏

The common-base model is especially useful for analyzing the common-base amplifier because it divides the small-signal equivalent into separate parts that are easy to analyze. Some favor the common-base over the common-emitter model as a general-purpose transistor representation because it prominently features the basic transistor mechanism, $i_c = \alpha i_e$, of Fig. 4.2. Others use the common-base model only to simplify difficult circuits, in effect skipping directly to the end result instead of applying the current transformation to the circuit. Still others ignore it entirely, relying on basics such as current transformations to simplify circuits as needed. In any event, resistance r_e tends to

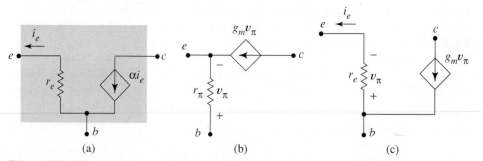

Figure 7.20   Example of current-source transformation: (a) common-base equivalent; (b) common-emitter model in common-base orientation; (c) equivalent circuit after two simplifications.

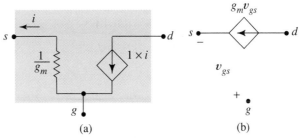

Figure 7.21 (a) Common-gate FET model; (b) starting point for model derivation.

appear from time to time and, when it does, the following alternate representations are worth knowing:

$$r_e = r_\pi \left\| \frac{1}{g_m} = \frac{r_\pi}{1 + \beta} \approx \frac{1}{g_m} \right. \tag{7.38}$$

Figure 7.21a shows a *common-gate model* for the FET. When we consider that $\alpha \approx 1$ and $r_e \approx 1/g_m$ for the BJT, there is a remarkable similarity between Figs 7.20a and 7.21a. We conclude that BJTs and FETs constrain the circuits in which they are imbedded in nearly identical fashion, an interesting fact that is not otherwise particularly obvious. Effective use of the common-gate model in analysis requires noticing that the gate current must be exactly zero (from KCL in Fig. 7.21a), a necessary property for any FET model.

Exercise 7.7 Use the current-source transformation to convert Fig. 7.21b into Fig. 7.21a.

7.6 _____

Common-Collector and Common-Drain Amplifiers

The common-base/common-gate circuits of the preceding section feature low current gain and the potential to realize low input and high output resistance. The common-collector/common-drain circuits we study next have complementary properties: low voltage gain and potentially high input and low output resistance.

7.6.1 COMMON-DRAIN (SOURCE FOLLOWER) AMPLIFIER

Figure 7.22a shows a discrete-component, *common-drain* amplifier, also called a *source follower*. The signal enters at the gate and exits at the source. A *three-resistor* bias circuit usually establishes the Q-point, for a drain resistor would contribute nothing beneficial to small-signal operation. When we turn off V_{DD} to form the small-signal schematic, we obtain Fig. 7.22b, where $R_P = R_1 \| R_2$ and where drain is common to input and output circuits. Because it better illustrates fundamental source-follower features, we now analyze the IC version, Fig. 7.22c. We assume active-state operation established by dc sources that have been turned off.

Two attributes of the ideal source follower are immediately obvious from the small-signal equivalent, Fig. 7.22d:

$$R_{in} = \infty \tag{7.39}$$

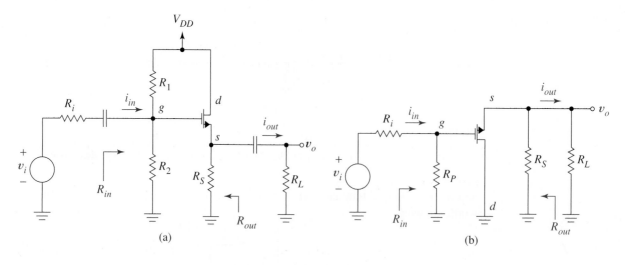

(a)

(b)

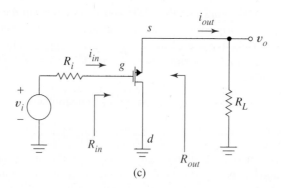

(c)

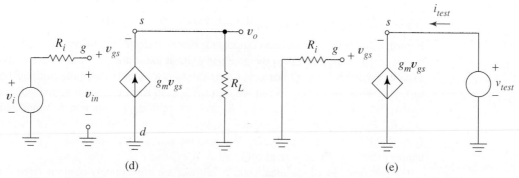

(d)

(e)

Figure 7.22 Common-drain amplifier: (a) discrete-component circuit; (b) small-signal schematic for discrete circuit; (c) small-signal schematic of IC version; (d) small-signal equivalent of IC version; (e) test circuit for finding output resistance.

and

$$A_i = \infty \tag{7.40}$$

(Input resistance and current gain are finite in the discrete-component version, because of the presence of R_P.) To find voltage gain, we observe that

$$v_o = g_m v_{gs} R_L$$

and

$$v_{gs} = v_{in} - v_o$$

Substituting for v_{gs} and solving for voltage gain gives

$$A_v = \frac{v_o}{v_{in}} = \frac{v_o}{v_{gd}} = \frac{g_m R_L}{1 + g_m R_L} \tag{7.41}$$

Notice that voltage gain is less than one as we foretold, and there is no phase shift. For output resistance, we use the test circuit of Fig. 7.22e. This gives

$$i_{test} = -g_m(-v_{test})$$

or

$$r_{out} = \frac{1}{g_m} \tag{7.42}$$

a relatively small output resistance that is independent of the source resistance R_i. It is easy to show that for Fig. 7.22b $R_{out} = R_S \| 1/g_m$.

Exercise 7.8 Assume the transistor in Fig. 7.22c is biased at 1 mA and has $k = 0.5$ m A/V². Find voltage gain and output resistance if $R_i = 5$ kΩ and $R_L = 2$ kΩ. How can the designer lower R_{out}?

Ans. 0.667, 1 kΩ. Bias at a higher drain current or specify a higher W/L for the transistor.

EXAMPLE 7.7 Assume the transistor in Fig. 7.22a and b is biased at 1 mA and has $k = 0.5$ m A/V². Find expressions for voltage gain and current gain. Evaluate voltage and current gain for $R_i = 5$ kΩ, $R_S = 5$ kΩ, $R_P = 2$ MΩ, and $R_L = 2$ kΩ.

Solution. Instead of analyzing Fig. 7.22b directly, we can exploit the similarities between Figs. 7.22 b and c. In the former, transistor current $g_m v_{gs}$ flows through $R_{LL} = R_S \| R_L$ instead of only R_L. Thus voltage gain of Eq. (7.41) becomes

$$A_v = \frac{v_o}{v_{gd}} = \frac{g_m R_{LL}}{1 + g_m R_{LL}} = \frac{10^{-3}(5 \text{ k} \| 2 \text{ k})}{1 + 10^{-3}(5 \text{ k} \| 2 \text{ k})} = 0.588$$

In Fig. 7.22b, input current is v_{gd}/R_P, output current is v_o/R_L. Thus we have

$$A_i = \frac{R_P v_o}{R_L v_{gd}} = \frac{R_P}{R_L} \frac{g_m R_{LL}}{1 + g_m R_{LL}} = \frac{2 \times 10^6}{2 \times 10^3} 0.588 = 588 \qquad \square$$

7.6.2 COMMON-COLLECTOR (EMITTER-FOLLOWER) AMPLIFIER

Like the source-follower, the *emitter-follower* or *common-collector* circuit is known for high current gain, voltage gain less than one, high input resistance, and low output resistance. In discrete-component form, it resembles Fig. 7.22a, except for the transistor. Problem 7.39 gives analysis results for the discrete-component emitter follower. Here we concentrate on the IC version, represented by the small-signal circuits of Figs. 7.23a and 7.23b.

A property that is immediately obvious from comparing these circuits is that the emitter follower current gain is

$$A_i = \frac{i_{out}}{i_{in}} = \beta + 1 \tag{7.43}$$

Because emitter current is always $(\beta + 1)$ times base current, an active-state BJT is able to scale resistances up or down by $(\beta + 1)$, a property that is useful for reducing loading at amplifier inputs and outputs. We will first show how the load resistance in the emitter-follower is scaled up to give a high input resistance. An offshoot of this development is a simple method for finding the voltage gain of the emitter follower. Then we examine how the circuit scales down resistance to give a low value at its output node.

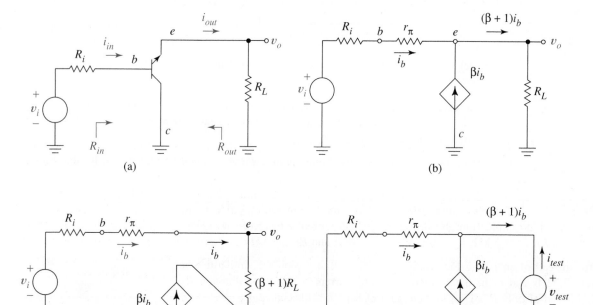

Figure 7.23 Emitter-follower amplifier: (a) small-signal schematic of basic circuit; (b) small-signal equivalent circuit; (c) after resistance scaling; (d) circuit for finding output resistance of emitter follower.

In Fig. 7.23b currents i_b and βi_b converge at the emitter node and leave together through R_L, producing a voltage drop

$$v_o = [(\beta + 1)i_b]R_L = [(\beta + 1)R_L]i_b$$

Consider replacing the circuit with the equivalent, Fig. 7.23c, in which current i_b flows through the *scaled-up* resistance, $(\beta + 1)R_L$. As we saw from the associative rule of algebra in the preceding equation, this modified circuit produces the same output voltage as the original circuit. Since Fig. 7.23c is equivalent to Figs. 7.23a and b, we see by inspection that the emitter follower's input resistance is

$$R_{in} = r_\pi + (\beta + 1)R_L \tag{7.44}$$

and its voltage gain is

$$A_v = \frac{v_o}{v_b} = \frac{(\beta + 1)R_L}{r_\pi + (\beta + 1)R_L} \tag{7.45}$$

We will call the manipulation that converted Fig. 7.23b into Fig. 7.23c the *resistance transformation*.

To find the output resistance, as defined in Fig. 7.23a, we replace R_L in Fig. 7.23b with a test generator and turn off the signal as in Fig. 7.23d. From this circuit,

$$i_{test} = -(\beta + 1)i_b = -(\beta + 1)\frac{-v_{test}}{r_\pi + R_i}$$

which means

$$R_{out} = \frac{v_{test}}{i_{test}} = \frac{r_\pi + R_i}{(\beta + 1)} \tag{7.46}$$

We conclude that in the emitter follower, Fig. 7.23a, the source resistance R_i and transistor resistance r_π are together *scaled down* to produce the low output resistance of Eq. (7.46).

We have shown that both source and emitter *followers* have voltage gain of approximately one, high input resistance, and low output resistance, attributes reminiscent of the op amp *follower* circuit we studied in Chapter 2. Like op amp followers, transistor followers are useful buffers that contribute power gain while reducing input and output loading. As we study frequency response in Chapter 8, feedback in Chapter 9, and power amplifiers in Chapter 10, our appreciation for these simple circuits will only increase.

Resistance Scaling by BJTs. The small-signal schematics of Figs. 7.24a and b show in more general form *important results about resistance scaling that we use frequently throughout the remaining chapters*. The resistance scaling derivations we used for the emitter follower apply even when there is an attachment to the collector, such as a load impedance, provided that r_o is sufficiently large. At the input port, defined by the base and node z (which need not be grounded), any resistance R between emitter and node z is scaled up by $(\beta + 1)$ before being added to r_π; at the output port, defined by emitter and node z in Fig. 7.24b, Thevenin resistance R_{TH} is added to r_π, and the two are then

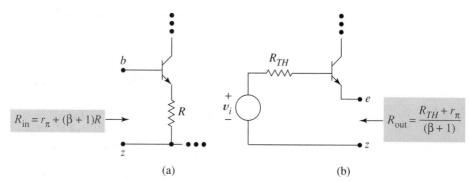

Figure 7.24   Resistance scaling property of the BJT: (a) at the base port; (b) at the emitter port.

scaled down by $(\beta + 1)$. Although the proof ultimately depends upon the transistor's small-signal equivalent circuit, we often apply the results directly upon inspection of the small-signal schematic.

EXAMPLE 7.8   Use resistance scaling rather than direct circuit analysis to find R_{in} for the circuit of Fig. 7.14a. We know that $r_\pi = 21.1$ kΩ and $\beta = 100$.

Solution. Looking directly between transistor base and ground, to the right of 47 k and 68 k, we see

$$R_{in1} = 21.1\,\text{k} + (100 + 1)4\,\text{k} = 425\,\text{k}$$

In the small-signal circuit, this resistance is in parallel with $R_P = 47\,\text{k}\|68\,\text{k}\Omega = 27.8$ kΩ. Thus

$$R_{in} = R_{in1}\|R_P = 425\,\text{k}\|(47\,\text{k}\|68\,\text{k}) = 26.1\,\text{k}$$

Notice how biasing resistance R_P reduces the effectiveness of the resistance transformation at the amplifier input. ❏

Exercise 7.9   In Fig. 7.14b, emitter resistor R_{E1} "looks back into the circuit" between emitter and ground and sees an output resistance, R_{out}. From the point of view of R_{E1}, the circuit is the same as Fig. 7.24b. Exploit this observation to find an expression for R_{out} in terms of R_i and R_P. Then evaluate your expression using numerical values from Fig. 7.14a where R_{E1} is the 4 kΩ resistor. We know that $r_\pi = 21.1$ kΩ and $\beta = 100$.

Ans. $R_{out} = 251$ Ω.

7.7 ____
SPICE Small-Signal Analysis

SPICE small-signal analysis provides accurate numerical values for quantities such as voltage and current gain, input resistance, and output resistance. These verify hand-calculated estimates, and also help us assess the importance of second-order variations such as body effect or output resistance that we usually omit to make hand analysis more tractable. SPICE first uses a dc analysis with large-signal nonlinear equations to calculate the Q-point of each transistor. Before an ac analysis it then calculates the small-signal parameter values such as g_m, r_π, β, and r_o for this Q-point. If we include the statement .OP, SPICE includes these parameters in the output file.

When we specify ac analysis, and include an ac signal source and a frequency or frequency range, SPICE constructs a linear small-signal model appropriate for the calculated Q-point and then does sinusoidal steady-state analysis to compute the requested voltages or currents in phasor form. Because SPICE linearizes the circuit model in ac analysis, the amplitude of the applied excitation is arbitrary, and all answers are proportional to this excitation. Thus to find voltage gain of an amplifier using SPICE ac analysis, we can apply a 1 V signal at the input; the resulting output phasor value at the output node is the voltage gain, even though such a value might saturate or cut off the unlinearized amplifier. We have been replacing coupling and bypass capacitors by short circuits in our small-signal analysis. In SPICE, we can assign very large values to these capacitors, such as 1000 Fd, causing them to act like very good short circuits at an analysis frequency of 1000 Hz. In the next chapter we learn to deal with realistic capacitor values.

EXAMPLE 7.9 Use SPICE to compute the voltage gain and the input and output resistances of the amplifier of Fig. 7.25a, where these quantities are defined for the amplifier proper, that is, relative to source and ground for input variables, and relative to drain and ground for output variables. Parameter values are $k = 0.5$ mA/V^2, $V_t = 2$ V, and $V_A = 100$ V.

Solution. Figure 7.25b is the SPICE code. The .AC statement asks for an ac analysis at the single frequency 1 kHz. Capacitors of 1000 Fd allow correct biasing while approximating short circuits to the 1 kHz sinusoidal test signal. Test generator IT, initially

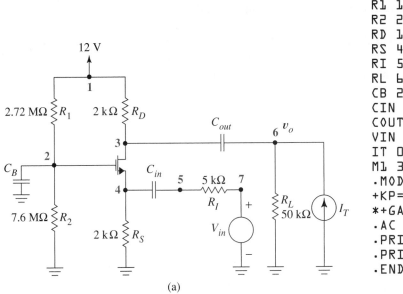

```
EXAMPLE 7.9
*COMMON GATE AMPLIFIER
VDD 1 0 DC 12
R1 1 2 2.72E6
R2 2 0 7.6E6
RD 1 3 2K
RS 4 0 2K
RI 5 7 5K
RL 6 0 50K
CB 2 0 1000
CIN 4 5 1000
COUT 3 6 1000
VIN 7 0 AC 1
IT 0 6 AC 0
M1 3 2 4 0 TRAN
.MODEL TRAN NMOS VTO=2
+KP=0.5E-3 LAMBDA=0.01
*+GAMMA=0.37
.AC LIN 1 1000 1000
.PRINT AC V(6)
.PRINT AC I(VIN)
.END
```

(a)   (b)

Figure 7.25 (a) Common-gate amplifier; (b) SPICE code for finding voltage gain and input and output resistance.

off and therefore an open circuit, is eventually turned on to measure output resistance. The first SPICE run ignores body effect because an asterisk turns the GAMMA = 0.37 statement into a comment. In a later SPICE run, the asterisk is removed.

In checking simulation results, as in laboratory, an initial examination of dc values quickly reveals connection problems and gross design errors, and should always be the first step. Here, the first SPICE run displayed reasonable dc node voltages and transistor currents.

Because gain A_v and R_{in} are defined relative to the amplifier input (node 5), the value of generator resistance RI was changed to 0.0001 Ω, effectively a short circuit. Then, with an ac voltage of VIN = 1 V applied, V(6) was numerically the amplifier gain and I(VIN) was, except for sign, the reciprocal of the input resistance. SPICE analysis gave A_v = 2.72 and R_{in} = 522 Ω. Introducing body effect alters the dc node voltages at source and drain somewhat and gives A_v = 2.85 and R_{in} = 505 Ω. For a common-gate amplifier, both changes are improvements!

To find output resistance, we turn off (set to zero) VIN, restore RI to its original 5 kΩ value and change IT to one ampere. Because R_{out} is defined to the left of RL, we also increase RL in the simulation to 1000 MΩ, effectively an open circuit. Under these conditions, V(6) is numerically the output resistance of the amplifier. Because of RD, its value was 2 kΩ, as expected. ❏

Exercise 7.10 How close to short circuits are the capacitors in Fig. 7.25b?

Ans. $|Z_C| = 1.59 \times 10^{-7}$ Ω.

7.8 _____
Multiple-Transistor Amplifiers

The common-emitter/source, common-base/gate, and common-collector/drain circuits of the preceding sections are fundamental building blocks in most linear IC designs. We now describe two ways to capitalize further on the basic properties of these elementary amplifiers. One way is to combine elementary amplifiers into multitransistor amplifiers. The other way is to use *active loads* to better exploit the *intrinsic* capabilities of the transistors, a technique we then extend to multitransistor amplifiers. We begin by reviewing and comparing properties of the elementary BJT amplifiers. Properties of FET amplifiers are quite similar.

7.8.1 SUMMARY OF ONE-TRANSISTOR AMPLIFIERS

Table 7.1 summarizes the key results of the preceding sections. The expressions for R_{out} assume an internal source resistance of R_i that is not shown in the diagrams.

Relative to the common-emitter circuit, the common-base connection offers a *scaled-down* input resistance and a *scaled-up* output resistance, both by the factor $(\beta + 1)$ if $R_i >> r_\pi$. The voltage gain is the same, except for sign, but current gain is approximately one.

The emitter follower has essentially the same current gain magnitude as the common-emitter amplifier, but its voltage gain is less than one. Because of the input resistance term, $(\beta + 1)R_L$, it is easy to achieve high input resistance with this circuit. Output resistance is much lower.

TABLE 7.1 Traditional BJT Amplifier Configurations

Name	Small-Signal Schematic	A_v	A_i	R_{in}	R_{out}
Common emitter		$-g_m R_L$	$-\beta$	r_π	r_o
Common base		$g_m R_L$	α	$\dfrac{r_\pi}{\beta + 1} = r_e$	$r_o[1 + g_m(R_i \| r_\pi)]$
Emitter follower		$\dfrac{(\beta + 1)R_L}{(\beta + 1)R_L + r_\pi}$	$\beta + 1$	$r_\pi + (\beta + 1)R_L$	$\dfrac{R_i + r_\pi}{\beta + 1}$

7.8.2 TWO-TRANSISTOR AMPLIFIERS

Figure 7.26 shows small-signal schematics for three popular two-transistor amplifiers. We tacitly assume that both transistors in each circuit have identical dc bias currents, a feature that we often see in IC biasing. To relate their properties more closely to those of the elementary amplifiers, we make extensive use of Table 7.1 results.

Cascode Amplifier. The common-emitter/common-base circuit of Fig. 7.26a is better known as the *cascode configuration*. Its input resistance is r_π, the input resistance of its common-emitter input stage. Because the load resistance of Q_1 is the low input resistance, r_e, of the common-base stage, the voltage gain of Q_1 is only

$$-g_m r_e = -g_m \frac{r_\pi}{\beta + 1} = \frac{-\beta}{\beta + 1} \approx -1$$

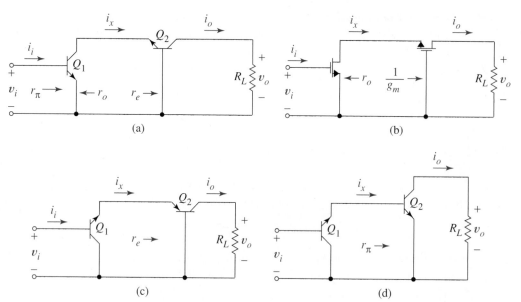

Figure 7.26 Two-transistor amplifiers: (a), (b) cascode amplifiers; (c) common-collector/common-base amplifier; (d) common-collector/common-emitter amplifier.

however, its current gain is $-\beta$. Transistor Q_2 has current gain α, giving overall cascode circuit current gain of

$$\frac{i_o}{i_i} = -\alpha\beta \approx -\beta \tag{7.47}$$

The overall cascode voltage gain is

$$A_v = \frac{v_o}{v_i} = \frac{-\alpha\beta i_i R_L}{i_i r_\pi} = -\alpha g_m R_L \approx -g_m R_L \tag{7.48}$$

which can be large because the second stage has high voltage gain. Since the common-base amplifier, Q_2, is driven by a source of resistance $R_i = r_o$, the output resistance of the cascode circuit is

$$R_{out} \approx r_o(\beta + 1) \tag{7.49}$$

These equations show that the BJT cascode amplifier has the voltage gain, current gain, and input resistance of the common-emitter amplifier, combined with the output resistance of the common-base circuit. We learn in Chapter 8 that when parasitic transistor capacitances are considered, this circuit also has excellent high-frequency response.

Exercise 7.11 For the FET cascode circuit of Fig. 7.26b, find input resistance, first-stage voltage gain, overall voltage gain, second-stage current gain, overall current gain, and output resistance.
Hint: Use Eqs. (7.30) through (7.33).

Ans. ∞, -1, $-g_m R_L$, 1, ∞, $r_o(1 + g_m r_o)$.

CC/CB Amplifier. Figure 7.26c shows the common-collector/common-base amplifier. Emitter-follower Q_1 has the input resistance r_e of the common-base amplifier, Q_2, as its load. This resistance is scaled up and added to the r_π of Q_1 to give a circuit input resistance of

$$R_{in} = r_\pi + (\beta + 1)r_e = 2r_\pi \qquad (7.50)$$

From Table 7.1, the voltage gain of Q_1 is only

$$\frac{(\beta + 1)r_e}{(\beta + 1)r_e + r_\pi} = 0.5$$

where we used Eq. (7.38); its current gain is $\alpha(\beta + 1)$. The output current is $\alpha(\beta + 1)i_i$; therefore, the overall voltage gain of the CC/CB amplifier is

$$G_v = \frac{v_o}{v_i} = \frac{\alpha(\beta + 1)i_i R_L}{i_i 2 r_\pi} = \frac{[\beta/(\beta + 1)](\beta + 1)R_L}{2r_\pi} = 0.5\, g_m R_L \qquad (7.51)$$

where, again, the voltage gain is achieved by the output stage.

CC/CE Amplifier. Figure 7.26d is the CC/CE amplifier. Because the load resistance of Q_1 is r_π,

$$R_{in} = r_\pi + (\beta + 1)r_\pi = (\beta + 2)r_\pi \qquad (7.52)$$

From Table 7.1, the first-stage voltage gain is

$$\frac{(\beta + 1)r_\pi}{(\beta + 1)r_\pi + r_\pi} = \frac{(\beta + 1)}{(\beta + 2)} \approx 1$$

The current gains of first and second stages are $(\beta + 1)$ and $-\beta$, respectively, giving overall current gain of

$$A_i = -\beta(\beta + 1) \qquad (7.53)$$

The voltage gain is

$$A_v = \frac{i_o R_L}{i_i R_{in}} = A_i \frac{R_L}{R_{in}} \qquad (7.54)$$

$$= -\beta(\beta + 1)\frac{R_L}{(\beta + 2)r_\pi} \approx -g_m R_L$$

In all three BJT two-transistor circuits, the output stage develops the voltage gain; however, first stage current gain significantly increases the power gain of the configuration. We will find in Chapter 8 that these two-transistor circuits and their FET counterparts also have advantages in frequency response over single-transistor amplifiers. Homework problems explore FET versions of Figs. 7.26c and 7.26d, called common-drain/common-gate and common-drain/common source amplifiers, respectively.

Darlington Pair. The *Darlington pair*, Fig. 7.27a, is a close relative of Fig. 7.26d that has the same high current gain; however, it is usually viewed as an *equivalent transistor*

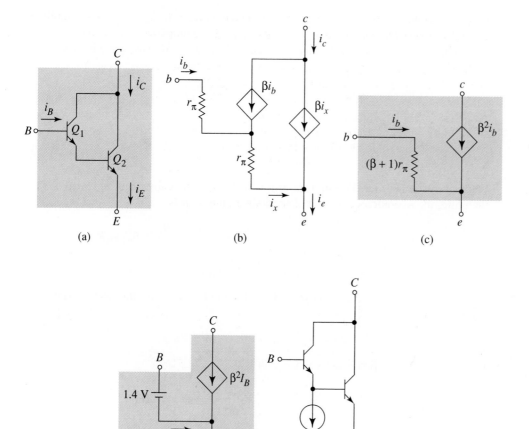

Figure 7.27 Darlington pair: (a) actual connection; (b) small-signal equivalent; (c) small-signal model; (d) large-signal model; (e) with extra biasing current.

of very high beta. The small-signal equivalent of Fig. 7.27b shows that with identical transistors

$$i_c = \beta i_b + \beta i_x = \beta i_b + \beta(\beta + 1)i_b \approx \beta^2 i_b$$

By the resistance transformation, r_π of Q_2 is scaled up and added to r_π of Q_1 to give

$$R_{in} = r_\pi + (\beta + 1)r_\pi \approx (\beta + 1)r_\pi$$

as indicated in the Darlington equivalent of Fig. 7.27c. In Problem 7.60 we derive the large-signal equivalent of Fig. 7.27d. A dc current source is sometimes added to the Darlington pair as in Fig. 7.27e so that both transistors can be biased at comparable collector currents; sometimes a resistor is used for the same purpose.

A common application of the Darlington pair is to replace the transistor in an emitter-follower output stage of an amplifier. The pair provides lower output resistance by

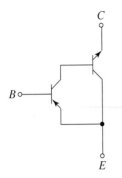

Figure 7.28 Compound pnp transistor.

scaling down the output resistance of the preceding amplifier by $(\beta^2 + 1)$ instead of $(\beta + 1)$ and ensures high power gain because its input signal current is so small.

Because IC pnp transistors have low values of β in some fabrication technologies, Darlington-connected pnp pairs are often used to provide high current gain. Figure 7.28 shows another way to obtain a high-beta equivalent for a pnp transistor. We can use small-signal models to show the equivalence of this connection to a pnp transistor having $\beta \approx \beta_p \beta_n$ and $r_\pi = r_{\pi p}$, where subscripts p and n relate parameters to the constituent transistors. As with a Darlington pair, a dc current source or biasing resistor is sometimes added to the circuit to bias the input transistor at a higher collector current than the small base current of the output transistor.

MOS Amplifier with FET Load Resistor. Figure 7.29a shows a common-source amplifier that uses a diode-connected enhancement transistor for its load resistor. Ignoring

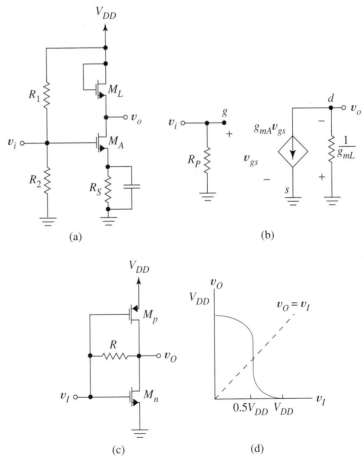

Figure 7.29 MOS two-transistor amplifiers: (a) enhancement-load amplifier; (b) small-signal equivalent of enhancement-load amplifier; (c) CMOS amplifier; (d) transfer characteristic and operating point of CMOS amplifier.

body effect in M_L, the small-signal model in Fig. 7.29b shows that the amplifier voltage gain is

$$A_v = -\frac{g_{mA}}{g_{mL}}$$

where g_{mA} and g_{mL} are the transconductances of amplifier transistor, M_A, and load transistor, M_L, respectively. Because bias current I_D is common to the devices, Eq. (7.4) implies

$$A_v = -\sqrt{\frac{k_A}{k_L}} = -\sqrt{\frac{(W/L)_A}{(W/L)_L}} \tag{7.55}$$

This shows that gain depends upon the relative width-to-length ratios of the amplifier and load transistors. An interesting feature is that this amplifier can accommodate surprisingly large signals without distortion because its nonlinear (square law) load line (see Fig. 5.20b) compensates exactly for the (square law) nonuniform spacing of the active transistor output characteristics!

Figure 7.29c shows one of the more elegant gain circuits, a complementary symmetry MOS (CMOS) amplifier. Because the basic circuit is also used as a logic inverter, we study it in some detail in Chapter 13. There we show that without resistor R the transfer characteristic has the shape of the solid curve in Fig. 7.29d, where the vertical part of the curve is a region of very high gain. Since gate currents of both transistors are zero, biasing resistor R forces the Q-point to be located on the dashed line $v_o = v_i$, biasing the amplifier in the high-gain region. Because the sources of the p- and n-channel transistors are connected, respectively, to the most positive and negative points in the circuit, neither transistor exhibits body effect. In Problem 7.63 we use a small-signal model to find the gain.

7.8.3 ACTIVE-LOAD SUBCIRCUITS

Operational amplifiers and other integrated circuits require subcircuits of high voltage gain. Since common-emitter/source and common-base/gate amplifiers as well as some of the two-transistor configurations all have voltage gains

$$|A_v| = g_m R_L$$

methods for establishing large values of R_L are of considerable interest. We already learned that we can obtain reasonable values of gain while avoiding excessive chip space or extra fabrication steps by employing nonlinear load resistors constructed from specially connected transistors. We can realize even higher gain by using the transistor's own output resistance for R_L.

Intrinsic Gain of a Transistor. Figures 7.30a and b introduce the key idea. Because current sources replace the passive biasing resistors, these are called *active load* amplifiers. Their small-signal equivalents in Figs. 7.30c and d show that both have voltage gain

$$A_v = -g_m r_o = -\mu \tag{7.56}$$

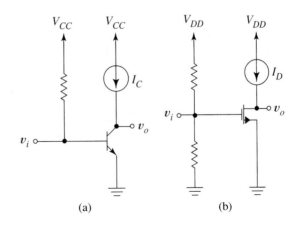

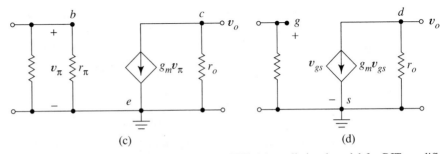

Figure 7.30 High-gain amplifiers: (a) BJT; (b) FET; (c) small-signal model for BJT amplifier; (d) small-signal model for FET amplifier.

which *depends only upon transistor parameters.* The product $\mu = g_m r_o$ is the transistor's *intrinsic gain,* a parameter indicative of the *voltage gain inherent in the transistor itself.*

For the bipolar transistor, substituting Eqs. (7.13) and (4.21) gives

$$\mu = \frac{I_C}{V_T} \frac{V_A}{I_C} = \frac{V_A}{V_T} \tag{7.57}$$

Even though g_m and r_o individually depend upon I_C, *the intrinsic gain of the BJT is independent of Q-point.* For values such as $V_A = 120$ V and $V_T = 0.025$ V, $\mu = 4800$! For the MOSFET, Eqs. (7.4) and (5.26) give

$$\mu = \sqrt{\frac{2k}{I_D}} V_A \tag{7.58}$$

which shows that the intrinsic gain of the MOSFET decreases with bias current. For $V_A = 80$ V and $k = 10^{-3}$A/V^2, an FET biased at 0.02 mA has $\mu = 800$, suggesting that large single-stage voltage gains are also possible in FET circuits. Notice that both circuits have $R_{out} = r_o$, the resistance of the transistor itself.

The particular circuits of Fig. 7.30 introduce the active-load idea in a simple way; however, the amplifiers have such high voltage gain that it is hard to bias the transistors in their active regions because resistor values are too critical. After studying difference amplifiers in the next section, we will have a much better understanding of how the biasing is actually done. In the meantime we explore active-load techniques further, using small-signal schematics that assume suitable biasing.

Common-Gate Amplifier with Active Load. The common-gate, active-load amplifiers of Figs. 7.31a and b give high open-circuit voltage gain. Applying KCL at the output node in Fig. 7.31b gives

$$\frac{v_o - v_i}{r_o} + g_m(-v_i) = 0$$

Solving for v_o/v_i gives

$$A_v = 1 + \mu \approx \mu \tag{7.59}$$

We already derived input and output resistances for this amplifier, Eqs (7.32) and (7.33), respectively; we list them again for completeness, denoting the resistance of the signal source by R_i:

$$R_{in} = \frac{1}{g_m} \tag{7.60}$$

$$R_{out} = r_o(1 + g_m R_i) \tag{7.61}$$

Problem 7.64 shows that the open-circuit gain of the common-base, active-load amplifier is also μ.

We now examine some two- and three-transistor, active-load, cascode circuits that are capable of even higher voltage gains. As before, we tacitly ignore the output resistance of the dc current sources, both to simplify the analysis and to better demonstrate the inherent capabilities of the circuits.

Figure 7.32a shows the small-signal schematic of the MOS active-load cascode amplifier. (The ideal current source load between the drain of M_2 and ground has been

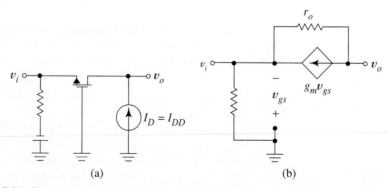

(a)   (b)

Figure 7.31 Basic common-gate active-load circuit: (a) schematic; (b) small-signal equivalent circuit.

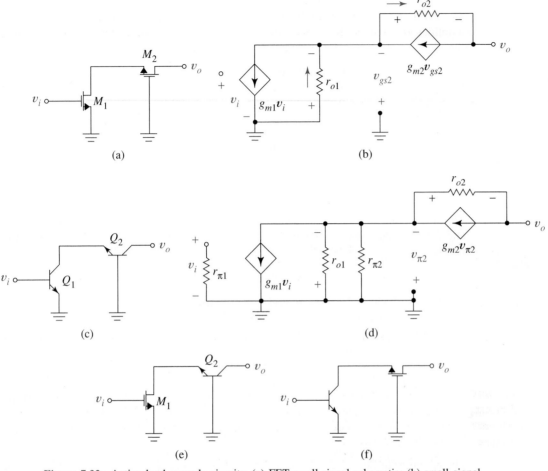

Figure 7.32 Active-load cascode circuits: (a) FET small-signal schematic; (b) small-signal equivalent of FET circuit; (c) BJT cascode circuit; (d) equivalent of BJT circuit; (e) BiCMOS circuit; (f) an alternate BiCMOS circuit.

turned off.) In Fig. 7.32b each dependent current flows only through its own output resistor; therefore,

$$v_o = -v_{gs2} - g_{m2}r_{o2}v_{gs2} = -(1 + \mu_2)v_{gs2}$$

and

$$v_{gs2} = r_{o1}g_{m1}v_i = \mu_1 v_i$$

Substituting for v_{gs2} and solving for v_o/v_i gives open-circuit gain

$$A_v = \frac{v_o}{v_i} = -\mu_1(1 + \mu_2)$$

(7.62)

$$\approx -\mu_1\mu_2$$

The cascode circuit output resistance is the output resistance of common-gate transistor M_2. This is given by Eq. (7.61), where $R_i = r_o$, the output resistance of common-emitter amplifier M_1. Thus

$$R_{out} = r_o(1 + g_m r_o) = r_o(1 + \mu) \tag{7.63}$$

(By using the same notation for the output resistances of M_1 and M_2 in this expression we assume transistors with identical Early voltages and drain currents.)

Figure 7.32c is the BJT active-load, cascode amplifier. Notice in its equivalent, Fig. 7.32d, that the approximation $r_{\pi 2} \| r_{o1} \approx r_{\pi 2}$ creates a circuit of the same structure as Fig. 7.32b where $r_{\pi 2}$ replaces r_{o1}. Thus $g_{m1} r_{\pi 2}$ replaces μ_1 in Eq. (7.62), and the gain is

$$A_v = -g_{m1} r_{\pi 2}(1 + \mu_2)$$
$$\approx -\beta_2 \mu_2 \tag{7.64}$$

The substitution $g_{m1} r_{\pi 2} = \beta_2$ assumes both transistors have the same dc collector current and therefore the same g_m, which is usually the case. Output resistance is given by Eq. (7.49),

$$R_{out} = r_o(\beta_1 + 1)$$

Figures 7.32e and f show hybrid cascode structures available in BiCMOS technology, which combines bipolar and complementary symmetry MOS. Except for the absence of $r_{\pi 1}$, the equivalent of Fig. 7.32e is the same as Fig. 7.32d. In Problem 7.67 we find that for this circuit

$$A_v \approx -g_{m1} r_{\pi 2} g_{m2} r_{o2}$$
$$= -g_{m1} \beta_2 r_{o2} \tag{7.65}$$
$$= -\frac{g_{m1}}{g_{m2}} \beta_2 \mu_2$$

The amplifier of Fig. 7.32f has the same structure as Fig. 7.32b except for $r_{\pi 1}$ across the input nodes, which does not affect the analysis. Therefore, its gain and output resistance are given by Eqs. (7.62) and (7.63). We see that all four cascode, active-load circuits have gains that greatly exceed the intrinsic gain of a single transistor.

Double-Cascode Amplifier with Active Load. Figures 7.33a and b describe the MOS-FET *double-cascode amplifier.* Again the currents are confined to individual loops, giving

$$v_x = g_{m1} r_{o1} v_i$$
$$-v_y = -v_x - g_{m2} r_{o2} v_x$$

and

$$v_o = -v_y - g_{m3} r_{o3} v_y$$

Obvious substitutions give

$$v_o = -\mu_1 v_i - \mu_2(\mu_1 v_i) - \mu_3[\mu_1 v_i + \mu_2(\mu_1 v_i)]$$

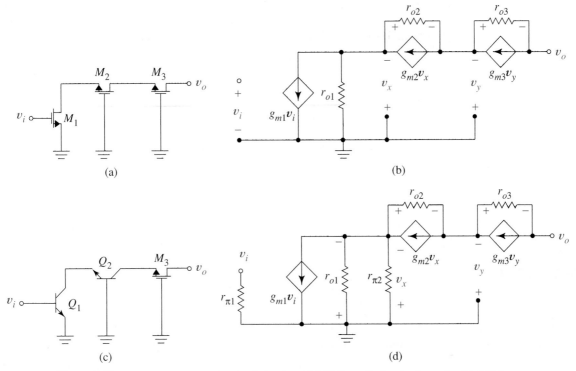

Figure 7.33 Double-cascode active-load circuits: (a) MOS small-signal schematic; (b) MOS small-signal equivalent; (c) BiCMOS small-signal schematic; (d) BiCMOS small-signal equivalent.

and the final expression

$$\frac{v_o}{v_i} = -[1 + \mu_2 + \mu_3(1 + \mu_2)]\mu_1$$

$$\approx -\mu_3\mu_2\mu_1$$

(7.66)

Exercise 7.12 Use previously derived results to find the output resistance of the double-cascode amplifier of Fig. 7.33a. The transistors and their dc drain currents are identical

Ans. $R_{out} = r_{o3}[1 + g_{m3}r_{o2}(1 + g_{m2}r_{o1})] \approx \mu^2 r_o$.

Figures 7.33c and d show a popular BiCMOS double-cascode amplifier. Because $r_{\pi2} \ll r_{o1}$, the equivalent circuit simplifies into three isolated, two-element loops through which the respective dependent currents flow. From KVL

$$v_o = -v_x - \mu_2 v_x - \mu_3 v_y$$

and

$$v_y = v_x(1 + \mu_2)$$

Substituting for v_y gives

$$v_o = -(1 + \mu_2)v_x - \mu_3(1 + \mu_2)v_x = -(1 + \mu_2 + \mu_3 + \mu_2\mu_3)v_x$$

But with $r_{o1} \| r_{\pi 2} \approx r_{\pi 2}$, we see that

$$v_x = g_{m1}r_{\pi 2}v_i$$

Thus gain is

$$A_v = \frac{v_o}{v_i} = -(1 + \mu_2 + \mu_3 + \mu_2\mu_3)g_{m1}r_{\pi 2}$$

$$\approx -\mu_2\mu_3 g_{m1}r_{\pi 2}$$

If Q_1 and Q_2 have the same bias current, $g_{m1} = g_{m2}$. From this observation we obtain the active-load, double-cascode gain expression

$$A_v = -\beta_2\mu_2\mu_3 \qquad (7.67)$$

EXAMPLE 7.10 Use previously derived results to find the output resistance of the double-cascode amplifier of Fig. 7.33c. Assume identical bipolar transistors with identical biasing.

Solution. We can use the common-gate expression for M_3, Eq. (7.61), if we use the proper expression R_i. The latter is the output resistance of the cascode amplifier, Eq. (7.49). For $\beta \gg 1$, this gives

$$R_{out} = r_{o3}(1 + g_{m3}\beta r_{o2})$$

$$\approx \mu_3 \beta r_{o2} \qquad \qquad \square$$

As a practical matter, each of these active-load circuits produces lower voltage gain than the equations predict because the output resistance of the dc current source causes output loading. That is, each small-signal schematic in Figs. 7.30 through 7.33 should have the small-signal output resistance, R_{out}, of the constant current source connected from output node to ground. Problem 7.66 examines such loading.

7.8.5 OUTPUT RESISTANCE OF A CURRENT SOURCE

Each active-load amplifier in Figs. 7.30 through 7.33 has inherently high output resistance. To realize its potential for high gain, the source of dc collector or drain current must also have high output resistance. Several current sources in Chapter 6 have high output resistance. Here we show how to use small-signal analysis to find the output resistances of a current source. We first derive a very general result on output resistance that can be applied to a variety of circuits including some current sources.

Using Emitter or Source Resistance to Increase R_{out}. A designer can usually raise the output resistance of a circuit by the simple expedient of adding resistance in series with the emitter (source) of the output transistor. We save the easier FET case for Problem 7.75, and take on the BJT case here.

Figure 7.34a shows the relevant small-signal schematic; R_{TH} is the Thevenin resistance of the input circuit; Fig. 7.34b is the equivalent circuit. Kirchhoff's current law at the collector node, using a current divider, gives

$$i_{test} = \beta\left[\left(\frac{-R_E}{R_E + R_{TH} + r_\pi}\right)i_{test}\right] + \frac{v_{test} - i_{test}R_P}{r_o}$$

$$= \left(-\beta\frac{R_E}{R_E + R_{TH} + r_\pi} - \frac{R_P}{r_o}\right)i_{test} + \frac{v_{test}}{r_o}$$

where $R_P = (R_{TH} + r_\pi)\|R_E$. Solving for the ratio, v_{test}/i_{test}, gives

$$R_{out} = R_P + \left(1 + \beta\frac{R_E}{R_E + R_{TH} + r_\pi}\right)r_o$$

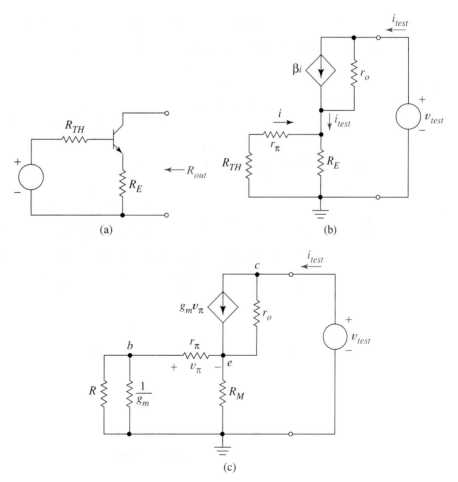

Figure 7.34 Output resistance increased by emitter resistor: (a) general small-signal schematic; (b) small-signal equivalent; (c) equivalent circuit of a Widlar source.

and from this comes the general expression

$$R_{out} \approx \left(1 + \frac{\beta R_E}{R_{TH} + r_\pi + R_E}\right) r_o \qquad (7.68)$$

Two special cases arise frequently, *both showing increases in output resistance* over r_o, the value when $R_E = 0$.

Case 1. When the sum $r_\pi + R_{TH}$ is small, specifically, $r_\pi + R_{TH} << R_E$, then Eq. (7.68) reduces to

$$R_{out} = \approx (1 + \beta) r_o \qquad (7.69)$$

Case 2. When r_π is large, specifically, when $r_\pi >> R_{TH} + R_E$, then Eq. (7.68) reduces to

$$R_{out} = \approx (1 + g_m R_E) r_o \qquad (7.70)$$

We next show that the Widlar current source is a good example of Case 2.

Widlar Current Source. Figure 7.34c is the small-signal equivalent of the Widlar circuit, Fig. 6.30. Except for notation, it is the same as Fig. 7.34b; thus we already know its output resistance. In Widlar-circuit notation, the Case 2 condition reduces to $r_\pi >> 1/g_m + R_M \approx R_M$. That is,

$$r_\pi = \beta \frac{1}{g_m} = \beta \frac{V_T}{I_O} >> R_M$$

where I_O is often very small for Widlar circuits. In terms of intrinsic gain, we can write Eq. (7.70) as

$$R_{out} \approx (1 + g_m R_M) r_o \approx \mu R_M \qquad (7.71)$$

and now we understand the origin of Eq. (6.40).

Output Resistance of Cascode Source. Cascode and double-cascode active-load circuits often employ cascode and double-cascode current mirrors to ensure current sources of high output resistance. The next example shows that the MOS cascode mirror, already familiar from Chapter 6, has output resistance

$$R_{out} \approx \mu r_o \qquad (7.72)$$

where μ is the intrinsic gain of the output transistor.

EXAMPLE 7.11 Starting with the cascode current source of Fig. 6.29b, derive Eq. (7.72).

Solution. Figure 7.35a is the small-signal equivalent. Because $v_x = 0$, the dependent source within M_2 is off as in Fig. 7.35b. Because $v_y = -i_{test} r_o$, KCL takes the form

$$i_{test} = g_m(-i_{test} r_o) + \frac{v_{test} - i_{test} r_o}{r_o}$$

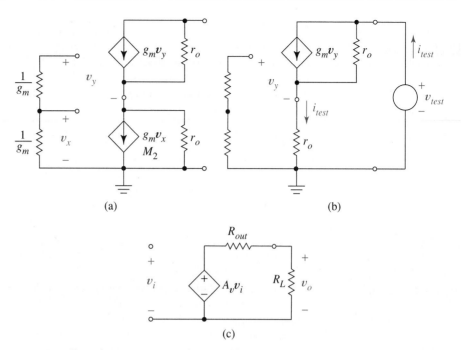

Figure 7.35 Cascode current-source output resistance: (a) small-signal equivalent; (b) simplified circuit with test generator; (c) equivalent circuit for double-cascode amplifier biased by a cascode current mirror.

which leads to

$$R_{out} = (2 + \mu)r_o \approx \mu r_o$$

Notice that if we replace the reference current source in Fig. 6.29b with a resistor, the simplest way to produce a reference current, we add a resistor to Figs. 7.35a and b, but this causes no change in the analysis or the result. ❏

By using a cascode current source to bias a double-cascode amplifier, we can achieve high voltage gain, as we see in the following.

Exercise 7.13 Figure 7.35c is the equivalent circuit of the double-cascode amplifier of Fig.7.33a, with open-circuit voltage gain given by Eq. (7.66) and output resistance calculated in Exercise 7.12. The load resistor is the output resistance of the cascode mirror circuit derived in the preceding example. Find the voltage gain, $A = v_o/v_i$, of the amplifier if all transistors are identical.

Ans. $A = \dfrac{-\mu^3}{1 + \mu} \approx -\mu^2.$

7.9 _____

Difference Amplifiers

Introduction. In Sec. 1.5 we learned that a difference amplifier not only amplifies a signal but also reduces the relative amplitudes of undesired common-mode input components that might otherwise obscure the signal. Here we investigate the internal structures of such amplifiers. A review of the definitions and concepts related to difference amplifiers is recommended at this point.

Difference amplifier *stages* based upon pairs of matched transistors are particularly well suited to IC designs. By exploiting IC transistor matching to cancel offsets and temperature-induced drift in parameter values, designers are able to create high-gain amplifiers using *direct-coupled* gain stages. Having no coupling capacitors, these circuits can amplify low frequencies, even those approaching dc. The current-source biasing employed by these stages simplifies interfacing between subcircuits of a given IC and also between the IC and user-provided external components. The basic FET circuit is called the *source-coupled pair*; the BJT version is the *emitter-coupled pair*.

Operational amplifiers exemplify circuits that employ difference amplifier stages, and we can explain many op amp parameters such as difference- and common-mode gain, input and output resistance, saturation limits, bias currents, voltage and current offsets, slew rate, and gain-bandwidth product in terms of difference-amplifier subcircuits.

For each coupled pair we first examine biasing, then use small-signal analysis to show how the circuit processes difference- and common-mode signals. Because source- and emitter-coupled pairs are more complex and versatile than single-ended amplifiers, we study them in a different manner. Instead of directly finding gain R_{in} and R_{out} by small-signal analysis, we instead construct two models *for the amplifier as a whole,* one for difference-mode signals and one for common-mode signals. These allow us to visualize the response to general excitations in terms of superposition, and also help us visualize exactly how the circuit functions in a variety of signal and loading situations. To apply small-signal techniques, we replace the amplifier as a whole with the appropriate model, rather than individual transistors, and then connect external sources and loads to the amplifier model.

7.9.1 SOURCE-COUPLED PAIR

Figure 7.36a shows the source-coupled pair; input nodes are a and b; output nodes x and y. Input voltages v_A and v_B include both bias and signal components. The dc current source, ideal in the schematic, is usually one of the mirror/reference current circuits described in Chapter 6.

Biasing. To examine biasing, we set both input nodes to ground potential as in Fig. 7.36b, a figure that includes numerical values to help clarify the biasing principles. Because of identical transistors and circuit symmetry, bias current I_O divides into equal parts, making $I_D = I_O/2 = 2$ mA for each transistor. Both drain voltages are

$$V_D = V_{DD} - R_D I_D = 10 - 5 = 5 \text{ V}$$

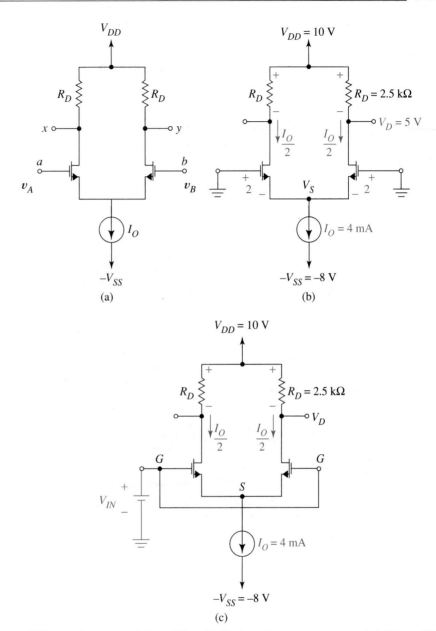

Figure 7.36 (a) Source-coupled amplifier; (b) biasing when inputs are grounded; (c) amplifier with common-mode excitation.

Because the transistors operate in their active states, the known drain currents must satisfy Eq. (5.7); and, as usual, this implies specific gate-to-source drops. For example, for $k = 4$ mA/V^2 and $V_t = 1$,

$$I_D = 2 \text{ mA} = \frac{4 \times 10^{-3}}{2}(V_{GS} - 1)^2$$

which implies $V_{GS} = 2$V. This means $V_S = -2$ V in Fig. 7.36b. From this information, we see that the voltage coordinate of each transistor Q-point is

$$V_{DS} = V_D - V_S = 5 - (-2) = 7 \text{ V}$$

The voltage across the current source,

$$V_S - (-V_{SS}) = -2 - (-8) = 6 \text{ V}$$

in this case, must exceed the minimum (for example, $V_{DS,min}$ in Fig. 6.22d).

Common-Mode Input Limits. Figure 7.36c helps us understand how a common-mode input voltage affects the biasing. Because of circuit symmetry, I_O must divide into equal parts, even in the presence of V_{IN}. Thus the drain currents and the node voltages of the drains are unchanged by V_{IN}. But because I_D does not change, neither can V_{GS}, *provided the transistors remain active*. Thus, as V_{IN} makes the gates more positive, V_S follows V_{IN}, always remaining a fixed drop ($V_{GS} = 2$ V in this example) below the gates. Because V_S increases with increasing V_{IN}, while V_D remains constant, V_{DS} diminishes until the transistor Q-points reach the ohmic region of the output characteristics. Condition

$$V_{DG} = V_D - V_{IN,max} = -V_t$$

defines this *positive common-mode limit* of the circuit. $V_{IN,max}$ is the most positive common-mode input voltage for which the transistors operate in their active states. In Fig. 7.36c

$$V_{IN,max} = V_D + V_t = 5 + 1 = 6 \text{ V}$$

The node voltage at the source also follows V_{IN} as V_{IN} goes negative, always remaining V_{GS} volts below V_{IN}. But as V_S decreases, the voltage across the current source diminishes until the minimum value for acceptable operation, $V_{CS,min}$, is reached. This condition establishes the *negative common-mode limit* of the circuit. Specifically,

$$V_S - (-V_{SS}) = (V_{IN,min} - V_{GS}) + V_{SS} \geq V_{CS,min}$$

where $V_{CS,min}$ is the voltage minimum of the current source.

EXAMPLE 7.12 In Fig. 7.37, $k = 0.2 \times 10^{-3}$ A/V^2 and $V_t = 0.5$ V.
(a) Verify that all transistors are active when $V_{IN} = 0$.
(b) Find the upper and lower common-mode limits.

Solution. We must first analyze the current source. The reference current in the 4.65 kΩ resistor is $I_O = 9.3/4.65$ k $= 2$ mA. Thus, from symmetry, $I_D = 1$ mA. The drain voltages are $V_D = 15 - 8$k$(1$ mA$) = +7$ V. For active MOSFETs,

$$1 \times 10^{-3} = 0.1 \times 10^{-3}(V_{GS} - 0.5)^2$$

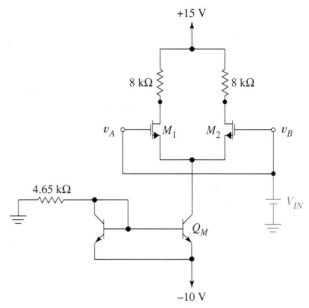

Figure 7.37   Source-coupled pair for Example 7.12.

giving $V_{GS} = 3.66$ V. With $v_G = V_{IN} = 0$, $V_S = -3.66$ V; and each MOSFET is biased at $V_{DS} = 7 - (-3.66) = 10.7$ V and $I_D = 1$ mA. With Q_M's collector at -3.66 V and its emitter at -10 V, $V_{CE} = 6.34$ V, well above the 0.2 V or so needed to avoid saturation.
(b) As V_{IN} becomes positive, the MOSFETs reach ohmic operation when

$$V_D - V_G = 7 - V_{IN,max} = -0.5 \text{ V}$$

Thus the upper common-mode limit is $V_{IN,max} = +7.5$ V. For negative input, Q_M begins to saturate when

$$V_{CE} = V_C - V_E = (V_{IN,min} - 3.66) - (-10) = 0.2 \text{ V}$$

Thus the negative common-mode limit is

$$V_{IN,min} = 0.2 - 10 + 3.66 = -6.14 \text{ V}$$

We conclude that the circuit is properly biased, provided that

$$-6.14 \text{ V} \le v_{IN} \le 7.5 \text{ V} \qquad \qquad ❏$$

Exercise 7.14   Recalculate the common-mode input range for the amplifier of Example 7.12 if the constant current source is replaced with a new source of $I_O = 1$ mA, having minimum output voltage of 2.5 V. The remainder of the circuit is unchanged.

Ans. -4.76 V $\le v_{IN} \le 11.5$ V.

Within the allowable range of common-mode input voltages, M_1 and M_2 are correctly biased for small-signal operation. Moreover, since drain currents are independent of common-mode input biasing, so are the small-signal FET models; thus the source-coupled pair automatically accommodates a wide variety of input voltages. With a supply-independent current reference for its current source, dc drain current and therefore small-signal operation would even become independent of the specific values of the power supplies. In such designs, the supplies determine the upper and lower common-mode limits but not g_m or r_o values. Although this discussion presented common-mode limits in terms of dc inputs, the same limits apply to time-varying common-mode components.

Now that we understand how the source-coupled pair is biased, we next examine its small-signal operation. Because arbitrary inputs are superpositions of difference- and common-mode components, we first derive an amplifier model that applies to difference-mode signals. Later we take on the common-mode case.

Small-Signal Difference-Mode Model. Figure 7.38a shows the source-coupled difference amplifier with a pure difference-mode input signal; Fig. 7.38b is the small-signal equivalent. The notation $g_{m1} = g_{m2} = g_m$ implies matched transistors having identical bias currents. The diagrams also include the output resistance, R_{out}, of the current source, which we soon learn could have been omitted. KCL at the source node gives

$$g_m(v_{gs1} + v_{gs2}) = \frac{v_s}{R_{out}}$$

The diagram shows that

$$\frac{v_d}{2} - v_{gs1} = v_s = -\frac{v_d}{2} - v_{gs2}$$

When we solve for v_{gs1} and for v_{gs2} and then substitute into the first equation, we obtain

$$g_m\left(\frac{v_d}{2} - v_s - \frac{v_d}{2} - v_s\right) = g_m(-2v_s) = \frac{v_s}{R_{out}}$$

which has the solution $v_s = 0$. Because of the special symmetries of circuit and excitation, there is a *virtual ground* at the connection point of the transistor sources, meaning this node remains at ground potential for arbitrary difference-mode signals. When we substitute $v_s = 0$ into the second set of equalities, we find that

$$v_{gs1} = \frac{v_d}{2}, \qquad v_{gs2} = -\frac{v_d}{2}$$

Thus exactly half of v_d always falls between gate and source of each transistor, with the polarities indicated in Fig.7.38c. Because both ends of R_{out} are at ground potential, R_{out} never carries current; so we delete it from the diagram. A further simplification in this figure involves reversing the reference directions of both the dependent current and controlling voltage for M_2.

Because the source node is *always* at ground potential, each resistor R_D in Fig. 7.38c is effectively in parallel with a dependent current source. By making source transformations and omitting the signal sources at the input, we obtain Fig. 7.38d, a model for the *entire amplifier* of Fig. 7.38a, which shows exactly how the source-coupled pair responds to pure difference-mode excitation.

This amplifier has infinite input resistance, an important feature of the source-coupled pair. Two dependent sources produce internal voltages of complementary polari-

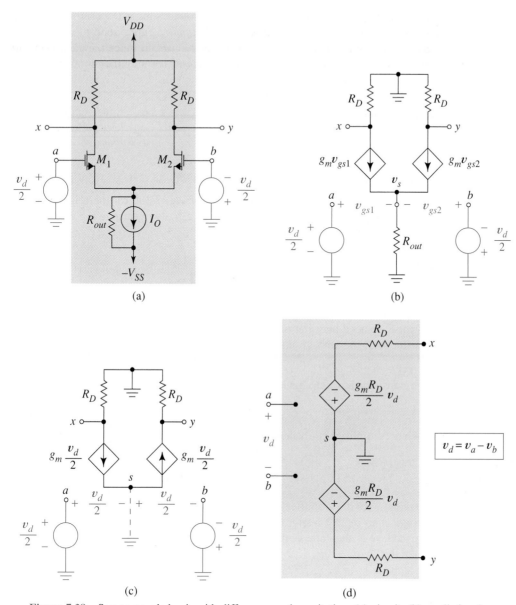

Figure 7.38 Source-coupled pair with difference-mode excitation: (a) circuit; (b) small-signal equivalent; (c) equivalent with symmetry constraints; (d) final model of source-coupled pair for difference-mode signals.

ties relative to ground, meaning that a *pure* difference-mode signal applied between nodes a and b results in a "pure" difference-mode output signal between nodes x and y. At each output node the open-circuit voltage is proportional to the difference-mode input signal, with proportionality constant, or gain, given by $g_m R_D / 2$. Between its output nodes, the circuit develops open circuit voltage, $v_x - v_y = (-g_m R_D) v_d$.

The presence of *two* amplified difference-mode voltages gives the user a number of choices. In Fig. 7.38a, we can take a *single-ended output* at either node x or node y, or we can take the difference, $v_x - v_y$, to be the output and obtain twice the gain. To summarize, with no external load the *difference-mode gain* of the amplifier is

$$A_d = -g_m R_D \tag{7.73}$$

for double-ended output, and

$$A_{d(s-e)} = \pm \frac{g_m R_D}{2} \tag{7.74}$$

for single-ended outputs.

When we substitute the small-signal equivalent of Fig. 7.38d for the original amplifier, we can see the two output options available; we can also visualize the exact nature of output loading, as we demonstrate next.

EXAMPLE 7.13 The difference amplifier of Fig. 7.39a uses transistors with $k = 4$ mA/V^2 and $V_t = 1.1$ V. Find v_o in terms of v_i if
(a) v_o is the open-circuit voltage at node x relative to node y.
(b) v_o is the open-circuit voltage at node y.
(c) v_o is the voltage across an 8 kΩ resistor R_L placed between nodes x and y.
(d) v_o is the voltage across a 10 kΩ resistor with one end grounded and the other capacitively coupled to node x.

Solution. (a) Because $I_D = 2$ mA, each transistor has transconductance

$$g_m = \sqrt{2 I_D k} = 4 \times 10^{-3} \text{ A/V}$$

therefore, the gain parameter for the difference-amplifier equivalent circuit is

$$A_{d(s-e)} = \frac{\pm (g_m R_D)}{2} = \pm 10$$

Figure 7.39b shows the small-signal equivalent with the amplifier replaced by its model. Because of the infinite input resistance, there is no input loading; therefore, $v_d = v_i$. The open-circuit voltage at node x relative to node y is the sum of the voltages across the dependent sources, or $-20v_d$. We conclude that $v_o = -20v_i$.
(b) When v_o is defined as the voltage at node y in Fig. 7.39a, from Fig. 7.39b, $v_o = +10 \, v_i$.
(c) Placing an 8 kΩ resistor between nodes x and y in Fig. 7.39a does not disturb the biasing, therefore the model of Fig. 7.39b still describes the amplifier. Connecting the 8 kΩ between nodes x and y as in Fig. 7.39c shows that there is now output loading. The internally generated sum voltage of $-20v_d$ is divided between the load resistor and the internal resistances. Only

$$v_o = \frac{8 \text{ k}}{8 \text{ k} + 5 \text{ k} + 5 \text{ k}} \times (-20v_i) = -8.89v_i$$

appears across the 8 kΩ load.

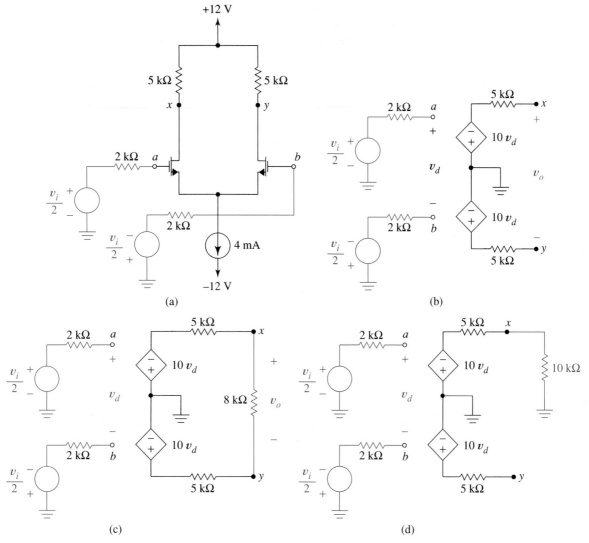

Figure 7.39 Diagrams for Example 7.13: (a) original circuit; (b) circuit with amplifier replaced by its equivalent; (c) equivalent showing output loading in part (c); (d) equivalent showing output loading in part (d).

(d) Figure 7.39d shows the small-signal circuit with load attached. Because of output loading

$$v_o = v_x = \frac{10\,\text{k}}{10\,\text{k} + 5\,\text{k}} \times (-10v_i) = -6.67v_i \qquad \square$$

> **Exercise 7.15** Find the output voltage, $v_o = v_x - v_y$ if 5 kΩ and 10 kΩ resistors are capacitively coupled from ground to nodes x and y, respectively, in Fig. 7.39a.
>
> *Ans.* $-11.7v_d$.

Small-Signal Common-Mode Model. We now explore how the common-mode component is processed by the source-coupled pair. Figure 7.40a is the small-signal equivalent of Fig. 7.36a for a common-mode signal v_c. Again we include the output resistance of the current source. At the source node, KCL requires

$$2g_m v_{gs} = \frac{v_s}{R_{out}} = \frac{v_c - v_{gs}}{R_{out}}$$

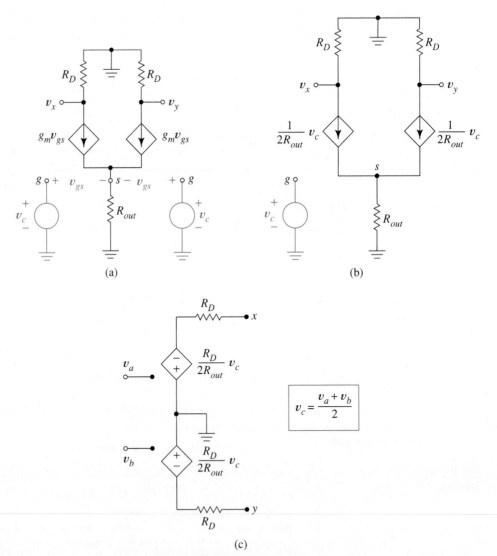

(a)

(b)

(c)

Figure 7.40 Amplifier model for common-mode inputs: (a) small-signal equivalent of Fig. 7.36a; (b) equivalent showing control of VCCS by v_c; (c) final small-signal model for pure common-mode input signal.

Solving for v_{gs} gives

$$v_{gs} = \frac{1}{1 + 2g_m R_{out}} v_c \approx \frac{1}{2g_m R_{out}} v_c$$

because R_{out} is very large. This establishes that the dependent sources produce drain currents

$$g_m v_{gs} = \frac{g_m}{2g_m R_{out}} v_c = \frac{1}{2R_{out}} v_c$$

In Fig. 7.40b we relabel the dependent sources to show this control by v_c. Next we find the Thevenin equivalent between each output node and ground. The open-circuit voltages are

$$v_x = v_y = \left(-\frac{1}{2R_{out}} v_c\right) R_D = -\frac{R_D}{2R_{out}} v_c$$

To find the Thevenin resistances, we turn off input signal v_c. But setting $v_c = 0$ turns off both dependent sources, giving $R_{THEV} = R_D$. The Thevenin transformation gives the common-mode equivalent circuit of Fig. 7.40c. We conclude that a pure common-mode input signal applied to the amplifier of Fig. 7.36a results in a pure common-mode output signal; a *pair* of voltages, identical in amplitude and phase, and proportional to v_c, at nodes x and y.

When we use the double-ended voltage, $v_x - v_y$, for our output the common-mode component is, in theory, *completely suppressed,* and the common-mode gain is

$$A_c = 0 \qquad (7.75)$$

Because of slight mismatches in component values, common-mode rejection in actual circuits falls slightly short of this ideal. When we use a single-ended output, the common-mode gain is

$$A_{c(s-e)} = -\frac{R_D}{2R_{out}} \qquad (7.76)$$

Since R_{out} is the output resistance of a current source, gain $A_{c(s-e)}$ is less than one. Indeed, for a current source of very high R_{out}, $A_{c(s-e)}$ can be much smaller than one. Thus common-mode signals are reduced in amplitude as they pass through the amplifier no matter how we define the output.

Difference Amplifier with General Excitation. Figure 7.41 provides the "big picture" of overall difference amplifier operation by combining difference- and common-mode circuits.

For arbitrary inputs, v_a and v_b, their difference-mode component, v_d, controls two dependent voltage sources of opposite phase, and produces an amplified (pure) difference-mode signal between nodes x and y. The common-mode component, v_c, simultaneously controls two dependent voltage sources of identical phase, producing an attenuated pure common-mode signal at the output nodes.

Notice that this general circuit reduces to the two special cases when *pure* difference- or common-mode signals are applied. For a pure difference-mode signal, by

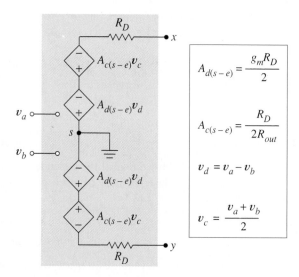

Figure 7.41 Complete equivalent circuit for source-coupled difference amplifier.

definition $v_c = 0$. This turns off both common-mode sources and reduces the circuit to Fig. 7.38d. When a pure common-mode signal is applied, by definition $v_d = 0$, and the circuit reduces to Fig. 7.40c.

The common-mode rejection ratio depends upon how we define circuit output. When output is taken between nodes x and y, the common-mode output components cancel, but the difference-mode component does not, giving infinite CMRR. When output is taken at a single node relative to ground, both common- and difference-mode components appear, and the CMRR, found from the ratio of the dependent-source transmittances in Fig. 7.41, is

$$\text{CMRR}_{(s-e)} = \frac{A_{d(s-e)}}{A_{c(s-e)}} = g_m R_{out} \tag{7.77}$$

an equation that underscores the importance of high current-source output resistance.

In FET difference amplifiers, there is no input loading whatsoever. Output loading occurs whenever there is a load, either between the output nodes or between one output node and ground. It is clear from Fig. 7.41 that CMRR is not affected by output loading because common- and difference-mode signals are subjected to identical loading factors.

Since small-signal models for n- and p-channel MOSFETs and JFETs are identical, Fig. 7.41 and all the conclusions and special cases we have discussed apply to *all* source-coupled difference amplifiers, not just the n-channel MOSFETs we selected for the derivation.

Even though Fig. 7.41 summarizes all the relevant information, to solve actual problems it is usually simplest to use Figs. 7.38d and 7.40c and superposition.

EXAMPLE 7.14 The transistors in Fig. 7.42a have parameters $k = 4 \times 10^{-3}$ A/V^2 and $V_t = 1.1$ V. The Early voltage of Q_M is 110 V. If the input excitation is

$$v_A = 0.022 \sin \omega_1 t + 0.050 \sin \omega_2 t$$

$$v_B = -0.020 \sin \omega_1 t + 0.053 \sin \omega_2 t$$

use superposition to find
(a) the single-ended output $v_x(t)$.
(b) the double-ended output $v_x(t) - v_y(t)$.
(c) the single-ended output $v_x(t)$ when an external 4 kΩ load resistor is capacitively coupled between node x and ground.

Solution. The current source of Fig. 7.42a, analyzed in Example 7.12, delivers $I_O = 2$ mA. Because $R_{out} = r_o$ for this mirror,

$$R_{out} = r_o = \frac{110}{2 \times 10^{-3}} = 55 \text{ k}\Omega$$

With each FET biased at $I_D = 1$ mA,

$$g_m = \sqrt{2kI_D} = \sqrt{2(4 \times 10^{-3})10^{-3}} = 2.83 \times 10^{-3} \text{ A/V}$$

Each dependent source in Fig. 7.38d has gain $0.5g_m R_D = 11.3$, giving Fig. 7.42b. From Fig. 7.40c, $0.5R_D/R_{out} = 0.073$, giving Fig. 7.42c.

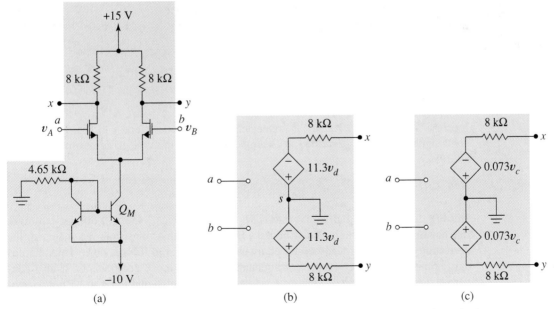

(a)   (b)   (c)

Figure 7.42 Amplifier for Example 7.14: (a) original circuit; (b) difference-mode equivalent from Fig. 7.38d; (c) common-mode equivalent from Fig. 7.40c.

The difference- and common-mode input components are

$$v_d = 0.042 \sin \omega_1 t - 0.003 \sin \omega_2 t$$

$$v_c = 0.001 \sin \omega_1 t + 0.052 \sin \omega_2 t$$

(a) From Fig. 7.42b the difference-mode component of v_x is

$$v_{xd}(t) = -11.3 \, v_d(t) = -0.475 \sin \omega_1 t + 0.034 \sin \omega_2 t$$

From Fig. 7.42c, the common-mode component is

$$v_{xc}(t) = -0.073 \, v_c(t) = -7.3 \times 10^{-5} \sin \omega_1 t - 3.80 \times 10^{-3} \sin \omega_2 t$$

By superposition,

$$v_x(t) = v_{xd}(t) + v_{xc}(t) = -0.475 \sin \omega_1 t + 0.030 \sin \omega_2 t$$

(b) From Fig. 7.42b, the difference-mode output component is

$$(v_x - v_y)_{dm} = -22.6 v_d$$

From Fig. 7.42c, the common-mode output component is

$$(v_x - v_y)_{cm} = 0$$

Therefore,

$$v_o(t) = v_x(t) - v_y(t) = -0.949 \sin \omega_1 t + 0.068 \sin \omega_2 t$$

(c) Treating the coupling capacitor as a short circuit, we attach the 4 kΩ resistor between node x and ground in Figs. 7.42b and c. This reduces the output voltage at node x in each case by the factor 4 k/(4 k + 8 k) = 1/3. Thus

$$v_o = (1/3)v_x(t)$$

Using the answer from part (a),

$$v_o = -0.158 \sin \omega_1 t + 0.010 \sin \omega_2 t \qquad \square$$

> **Exercise 7.16** A 4 kΩ resistor is connected between nodes x and y in Fig. 7.42a. If the input signal is the one described in Example 7.14, find the voltage across the 4 k resistor.
>
> *Ans.* $v_o(t) = -0.190 \sin \omega_1 t + 0.0136 \sin \omega_2 t$.

> **Double- to Single-Ended Conversion.** The circuit of Fig. 7.43a is a difference amplifier having only the single-ended output v_y. Assuming that transistors are matched, the derivation given above leads to the equivalent in Fig. 7.43b for this case. Thus the circuit amplifies difference-mode and attenuates common-mode components. It also converts a double-ended input signal to a single-ended output signal that can be further amplified by a single-ended amplifier. Since a second drain resistor would occupy chip space but have no function in this circuit, it is omitted. An analogous BJT circuit also exists.

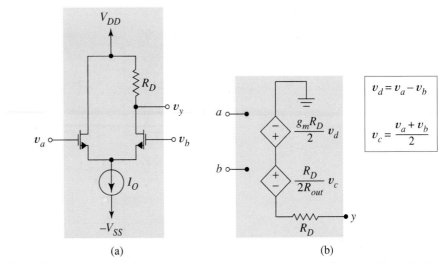

Figure 7.43 Double- to single-ended conversion circuit: (a) original circuit; (b) small-signal equivalent.

7.9.2 EMITTER-COUPLED DIFFERENCE AMPLIFIER

The emitter-coupled difference amplifier of Fig. 7.44a resembles the source-coupled circuit in topology and general operation; however, the BJTs give higher gain and finite input resistance. Before examining biasing and small-signal models, we first describe the circuit's large-signal operation. This development shows us how to avoid distortion and also provides essential background for understanding two important topics of later chapters: the emitter-coupled logic gate and the Gilbert-cell analog multiplier.

Large-Signal Operation of the Emitter-Coupled Pair. Let v_A and v_B be any voltages for which both transistors are in their active states, and, therefore, described by Eq. (4.5). In Fig. 7.44a, this means

$$i_{C1} = I_{S1}e^{(v_A - v_E)/V_{T1}} \qquad \text{and} \qquad i_{C2} = I_{S2}e^{(v_B - v_E)/V_{T2}}$$

If transistors are identical and in close proximity on an IC chip, saturation currents and thermal voltages are identical and track with temperature. Under these conditions, the ratio of collector currents is

$$\frac{i_{C1}}{i_{C2}} = e^{(v_A - v_B)/V_T} = e^{v_D/V_T} \tag{7.78}$$

where $v_D = v_A - v_B$ is the input difference voltage, *not necessarily small*. For sufficiently high β, $i_C \approx i_E$, so Kirchhoff's current law at the emitter node becomes

$$i_{C1} + i_{C2} = I_O \tag{7.79}$$

Solving for i_{C2} and substituting into Eq. (7.78) gives

$$i_{C1} = (I_O - i_{C1})e^{v_D/V_T}$$

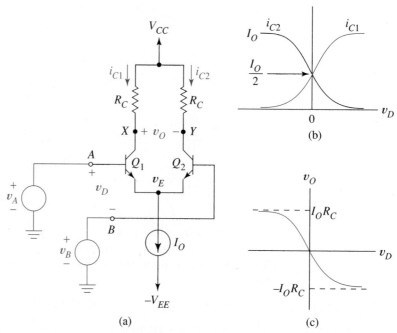

Figure 7.44 Large-signal operation of emitter-coupled pair: (a) circuit; (b) collector currents versus input voltage; (c) transfer characteristic.

When we solve for i_{C1} we obtain

$$i_{C1} = \frac{I_O e^{v_D/V_T}}{1 + e^{v_D/V_T}} = \frac{I_O}{1 + e^{-v_D/V_T}} \qquad (7.80)$$

We obtain a similar expression for i_{C2} by solving Eq. (7.79) for i_{C1} and substituting into Eq. (7.78). This is

$$i_{C2} = \frac{I_O}{1 + e^{v_D/V_T}} \qquad (7.81)$$

Figure 7.44b shows Eqs. (7.80) and (7.81) as functions of v_D. When $v_D = 0$, the transistors share I_O equally. As v_D becomes positive, Q_1 acquires a larger share of the total current at the expense of Q_2; for negative v_D, i_{C2} gains current at the expense of i_{C1}. The curves flatten at the extremes as one transistor takes all the current and the other approaches cut-off.

Now we examine the large-signal voltage transfer characteristic. In Fig. 7.44a,

$$v_O = v_X - v_Y = (V_{CC} - R_C i_{C1}) - (V_{CC} - R_C i_{C2})$$
$$= R_C(i_{C2} - i_{C1}) \qquad (7.82)$$

If we solve Eq. (7.79) for i_{C1} and substitute we obtain

$$v_o = R_C(2i_{C2} - I_O)$$

We now substitute i_{C2} from Eq. (7.81). This leads to

$$v_o = R_C \left(\frac{2I_O}{1 + e^{v_D/V_T}} - I_O \right)$$

$$= R_C I_O \left(\frac{1 - e^{v_D/V_T}}{1 + e^{v_D/V_T}} \right) \frac{e^{-v_D/2V_T}}{e^{-v_D/2V_T}}$$

$$= R_C I_O \left(\frac{e^{-v_D/2V_T} - e^{v_D/2V_T}}{e^{-v_D/2V_T} + e^{v_D/2V_T}} \right)$$

By definition of the hyperbolic tangent function,

$$v_O = -R_C I_O \tanh\left(\frac{v_D}{2V_T} \right) \tag{7.83}$$

which is Fig. 7.44c. The sketch is the transfer function of an inverting difference amplifier with zero offset voltage, which "saturates" for positive and negative values of v_D of sufficient magnitude. The slope at the origin is the small-signal voltage gain

$$\frac{v_o}{v_d} = \frac{dv_O}{dv_D} = -R_C \frac{I_O}{2V_T} = -R_C \frac{I_C}{V_T} = -g_m R_C$$

a result we will subsequently verify by small-signal analysis. The reader will find a transfer characteristic similar to Fig. 7.44c derived for source-coupled pairs in the text by Gray and Meyer cited in the references.

In the emitter-coupled logic gate and the Gilbert cell, two applications of the emitter-coupled pair we study later, we view the Q_1-Q_2 pair as a *current switch* in which control voltage v_D switches current between the two transistors. For this point of view, we substitute v_O from Eq. (7.83) into Eq. (7.82) and then solve for the current difference. This gives the *current switch equation*

$$i_{C2} - i_{C1} = -I_O \tanh\left(\frac{v_D}{2V_T} \right) \tag{7.84}$$

Biasing. Figure 7.45 shows biasing of an emitter-coupled difference amplifier. First let $V_{IN} = 0$. Because of circuit symmetry and matched transistors, each collector current is 0.5 mA. These produce 3.5 V drops across the collector resistors as in Fig. 7.45b. With V_{BE} drops of 0.7 V, the emitter voltage is $V_E = -0.7$ V. Thus

$$V_{CE} = V_C - V_E = (9 - 3.5) - (-0.7) = 6.2 \text{ V}$$

showing the transistors are biased in the active region.

The voltage across the current source is $-0.7 - (-9) = 8.3$ V. A special consideration in this bipolar circuit is the need for base currents. *Provided that there is a dc path from each base to ground*, the base currents are I_C/β. (In the absence of such paths the transistors cut off and amplification is impossible.) Some operational amplifiers have BJT difference amplifiers for input stages. The base currents necessary for active operation of the first-stage transistors are the *input bias currents*. Because of slight mismatches in the circuit halves, the bias currents differ slightly, giving rise to the *input*

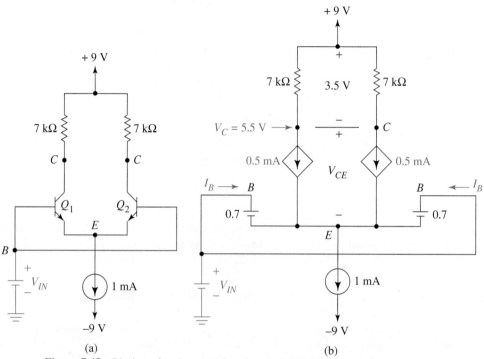

Figure 7.45 Biasing of emitter-couple pair: (a) circuit; (b) large-signal model.

offset current. Chapter 2 discussed some of the consequences of input bias and offset currents in op amps.

From Fig. 7.45b we see that as long as the transistors are active, $V_E = V_{IN} - 0.7$ V. As V_{IN} is increased, V_E also increases; however, the collector currents remain at 0.5 mA so collector voltage remains at 5.5 V. Thus V_{CE} decreases with increasing V_{IN}. When V_{CE} reaches $V_{CE,sat}$, Q_1 and Q_2 saturate. This establishes the upper limit of $V_{IN,max}$. For this example,

$$V_{IN,max} - 0.7 + 0.2 = 5.5$$

or $V_{IN,max} = 6$ V. In general, the transistors saturate when

$$V_{IN,max} - V_{BE} + V_{CE,sat} = V_{CC} - R_C \frac{I_Q}{2}$$

For negative V_{IN} the voltage across the current source decreases. If the current source in Fig. 7.45a uses a basic bipolar mirror, the minimum common-mode input, $V_{IN,min}$, for this circuit is given by

$$V_{IN,min} - 0.7 - 0.2 = -9 \text{ V}$$

or

$$V_{IN,min} = -8.1 \text{ V}$$

More generally,

$$V_{IN,min} - V_{BE} - V_{CS,min} = -V_{EE}$$

where $V_{CS,min}$ is the minimum voltage for proper operation of the current source.

Provided a dc path for base current exists and common-mode limits are not exceeded, transistors and current source operate properly over a wide range of dc input bias voltages, for which small-signal parameters are more or less invariant. Power supply values obviously affect the common-mode limits directly and, depending on the current source design, may or may not affect the values of the small-signal parameters.

The following example shows that transistor Q-point and common-mode input voltage are both related to the external resistance in the base circuit.

EXAMPLE 7.15 In Fig. 7.46, the transistors have $\beta = 80$. Find the Q-point of each transistor if $R_B = 50$ kΩ.

Solution. The base currents are $(0.5 \text{ mA})/80 = 6.25$ μA. Because base currents flow through the 50 kΩ resistors, the bases are biased at

$$V_B = -50 \times 10^3 (6.25 \times 10^{-6}) = -0.313 \text{ V}$$

The emitter voltage is $-0.313 - 0.7 = -1.01$ V. Since $V_C = 9 - 3.5 = 5.5$ V, $V_{CE} = 5.5 - (-1.01) = 6.51$ V. Thus Q_1 and Q_2 are biased at (6.51 V, 0.5 mA). ❑

Exercise 7.17 In the circuit of Example 7.15, increasing R_B eventually leads to insufficient voltage across the current source. Find the upper limit of R_B for correct biasing if the current source employs a basic BJT mirror.

Ans. $R_{B,max} = 1.30$ MΩ.

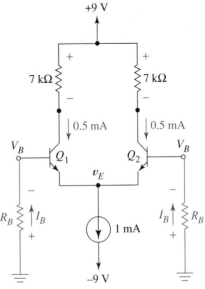

Figure 7.46 Base circuit resistors causing dc, common-mode input.

One way to reduce input bias currents is to use *superbeta transistors,* BJTs of extraordinarily narrow base width (achieved by an extra diffusion step), which have betas of 2000 to 5000. Because the narrow base widths also give small BV_{CE0} values, superbeta transistors are used only for input stages of amplifiers, where signals are small. Another way to obtain low bias current is to use a Widlar current source to provide very small collector bias currents.

Small-Signal Difference-Mode Model. Figure 7.47a shows a BJT difference amplifier with pure difference-mode excitation, and Fig. 7.47b its small-signal equivalent. As in the source-coupled pair, the special symmetry of the circuit and the excitation create a virtual ground; thus we can ignore the output resistance of the current source. Another consequence of the virtual ground is that $v_{\pi 1} = 0.5\ v_d = -v_{\pi 2}$. Figure 7.47c incorporates this observation. Figure 7.47d is equivalent because dependent currents of identical magnitude and opposite direction enter the input circuit in Fig. 7.47c. The final model, Fig. 7.47e, results from source conversions and combining r_πs.

The point of this derivation is that the amplifier of Fig 7.47a functions like the model of Fig. 7.47e when excited by a small difference-mode signal. Between the input nodes, the signal source sees an input resistance of $2r_\pi$, possibly causing input loading. The voltage v_d developed across this resistor is amplified and reproduced in two complementary parts at the output nodes just as in the FET circuit. Comparing Figs. 7.47e and 7.38d shows that the finite input resistance of the BJT circuit is the only real difference between source- and emitter-coupled pairs for difference-mode operation.

The following example demonstrates input and output loading in a two-stage, direct-coupled amplifier.

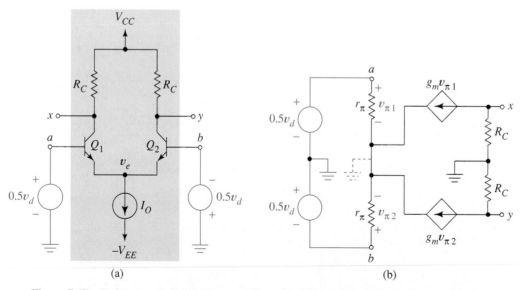

(a)   (b)

Figure 7.47 Emitter-coupled difference amplifier with difference-mode excitation: (a) actual circuit; (b) small-signal equivalent; (c) and (d) steps in the development; (e) final equivalent circuit for amplifier.

EXAMPLE 7.16 All transistors in Fig. 7.48a have $\beta = 250$. Find gain, v_o/v_i, for the two-stage amplifier.

Solution. The first stage has pure difference-mode excitation and its output produces difference-mode excitation for the second stage; therefore, the model of Fig. 7.47e is appropriate for both stages.

The first-stage transistors are biased at 0.25 mA, giving $g_m = 0.01$ A/V and $r_\pi = 25$ kΩ. The second-stage transistors are biased at 1 mA; thus $g_m = 0.04$ A/V and $r_\pi = 6.25$ kΩ. The small-signal equivalent for the entire circuit is Fig. 7.48b, where node labels relate the equivalent to the original amplifier.

Because of loading at the first stage input,

$$v_d = \frac{50\text{ k}}{50\text{ k} + 20\text{ k}}\, v_i = 0.714 v_i$$

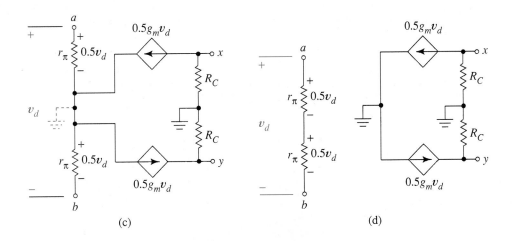

(c)

(d)

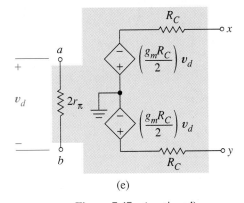

(e)

Figure 7.47 (continued)

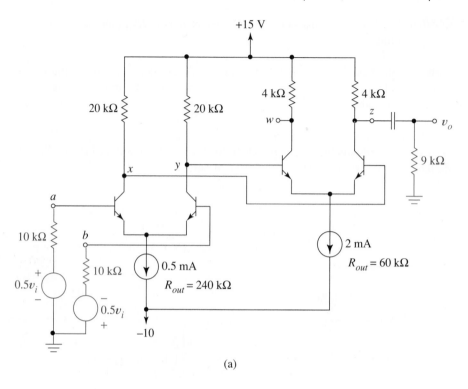

(a)

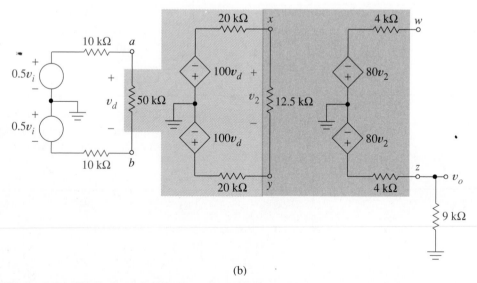

(b)

Figure 7.48 Circuit demonstrating input and interstage loading: (a) original circuit; (b) small-signal equivalent using diff amp models.

This is internally amplified by 200 and then subjected to interstage loading, giving

$$v_2 = \frac{-12.5 \text{ k}}{12.5 \text{ k} + 40 \text{ k}} \times 200v_d = -47.6v_d$$

Voltage v_2 in turn, is amplified by 80 before output loading divides it down to

$$v_o = \frac{9 \text{ k}}{9 \text{ k} + 4 \text{ k}} 80v_2 = 55.4v_2$$

Overall,

$$v_o = -55.4 \times 47.6 \times 0.714v_i = -1883v_i$$

so the gain is -1883.   ❏

Redesigning the bias circuit changes both open-circuit gain and input resistance.

Exercise 7.18 Find the new voltage gain in the preceding example if the 2 mA bias current for the second stage is changed to 1 mA.

Ans. $-1521.$

Small-Signal Common-Mode Model. Figure 7.49a shows the BJT difference ampli-fier with common-mode excitation, and Fig. 7.49b is the small-signal equivalent. Figure 7.49c is the same except that the output of the two current sources, $2g_m v_\pi$, is passed to the input circuit by a new current source. Although the purpose of this step is not yet clear, the equivalence should be. The dashed ground symbol shows where a ground connection could be made without affecting circuit operation (voltages across current sources are arbitrary).

Figure 7.49d is equivalent to 7.49c even though the two circuit parts are detached because they are excited by the same sources. Notice that the current source in the input circuit can be relabeled as

$$2g_m v_\pi = 2g_m \left(i \frac{r_\pi}{2} \right) = \beta i$$

The resistance transformation gives Fig. 7.49e. To obtain the final form of the circuit, we need to relate variable v_π, which controls the dependent sources, to the common-mode input, v_c. Figure 7.49e shows that v_π is related to v_C by the voltage divider equation

$$v_\pi = \frac{r_\pi/2}{(r_\pi/2) + (\beta + 1)R_{out}} v_c$$

Because $(\beta + 1)R_{out} >> r_\pi/2$,

$$g_m v_\pi \approx \frac{g_m r_\pi}{2(\beta + 1)R_{out}} v_c \approx \frac{1}{2R_{out}} v_c$$

where the last step assumes $\beta >> 1$. The final equivalent, Fig. 7.49f, comes from rela-beling the current sources and performing a source transformation.

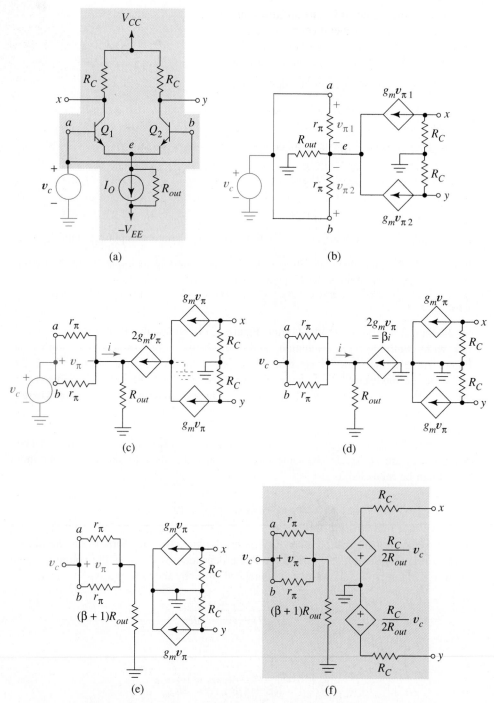

Figure 7.49 Model for common-mode input signals: (a) original circuit; (b) small-signal equivalent; (c)–(e) application of the resistance transformation; (f) final model.

Two comparisons with previous results are interesting. Notice that Fig. 7.49f has the same structure as the general common-mode model introduced as Fig. 1.29e, except that its output is double ended. Our derivation shows the origins of the common-mode input resistance components and the output resistance in Fig.1.29e, and relates the common-mode gain to specific parameters within the amplifier. Now compare Fig. 7.49f with the FET common-mode equivalent of Fig.7.40c. Except for notation, the output circuits are identical. The BJT input circuit, however, is more complex than the simple open circuit of the source-coupled pair.

Complete Equivalent for Emitter-Coupled Difference Amplifier. Figure 7.50 models the emitter-coupled differential amplifier for arbitrary input signals. Like its source-coupled counterpart, Fig. 7.41, the circuit simplifies for pure difference- and common-mode signals. With a pure difference-mode signal, $v_c = 0$. This turns off the common-mode generators and reduces the output circuit to Fig. 7.47e. It is easy to show that difference-mode symmetry in the input circuit results in exactly zero current in $(\beta + 1)R_{out}$. Thus this resistor is equivalently replaced with an open circuit, giving the input circuit of Fig. 7.47e. Common-mode excitation reduces the circuit to Fig. 7.49f.

When output is double-ended, common-mode output components cancel, giving infinite CMRR. For single-ended output, the transmittance ratio in Fig. 7.50 gives

$$\text{CMRR}_{(s-e)} = \frac{(g_m R_C)/2}{R_C/(2R_{out})} = g_m R_{out} \qquad (7.85)$$

Like Eq. (7.77) for the source-coupled pair, this equation shows that high CMRR depends upon a current source of high output resistance. For a basic current source,

$$R_{out} = r_o = \frac{V_A}{2I_C}$$

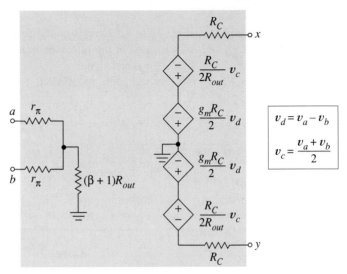

Figure 7.50 Complete model for emitter-coupled difference amplifier with arbitrary small-signal input.

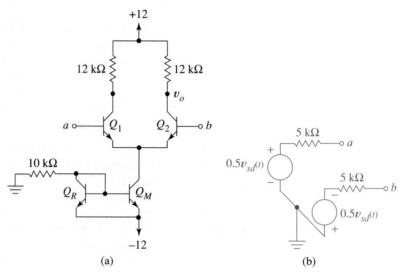

Figure 7.51 (a) Difference amplifier for Example 7.17; (b) signal source for Exercise 7.19.

where V_A is the Early voltage and I_C is the bias current of the amplifying transistors. For this special case, Eq. (7.85) implies

$$\text{CMRR}_{(s-e)} = \frac{I_C}{V_T}\frac{V_A}{2I_C} = \frac{V_A}{2V_T} = 0.5\,\mu$$

where V_A is the Early voltage and μ is the intrinsic voltage gain of the mirror transistor. For $V_A = 120$ V this gives $\approx 120/(2 \times 0.025) = 2400$, or 67.6 dB.

EXAMPLE 7.17 The transistors in Fig. 7.51a all have $\beta = 320$ and $V_A = 125$ V. The area of the mirror transistor is one-half the area of the reference transistor.
(a) Find the input resistance seen by a difference-mode signal.
(b) Find the difference-mode gain, common-mode gain, and CMRR for single-ended output.

Solution. The reference current is

$$I_{REF} = \frac{0 - (-12 + 0.7)}{10\text{ k}} = 1.13\text{ mA}$$

therefore, $I_O = 0.5 \times I_{REF} = 0.565$ mA. The collector currents of Q_1 and Q_2 are 0.283 mA, giving $g_m = 11.3 \times 10^{-3}$ A/V and $r_\pi = 28.3$ kΩ.
(a) From Fig. 7.47e, the input resistance is $2r_\pi = 56.6$ kΩ.
(b) From the same figure,

$$A_{d(s-e)} = \frac{g_m R_c}{2} = \frac{11.3 \times 10^{-3} \times 12\text{ k}}{2} = 67.8$$

For the mirror transistor,

$$R_{out} = r_o = \frac{V_A}{I_O} = \frac{125}{0.565 \times 10^{-3}} = 221 \text{ k}\Omega$$

therefore,

$$A_{c(s-e)} = \frac{R_c}{2R_{out}} = \frac{12 \text{ k}}{2 \times 221 \text{ k}} = 0.0271$$

The ratio gives $\text{CMRR}_{(s-e)} = 2502$, or 68 dB.   ❑

> **Exercise 7.19**   Find the gain, $v_o(t)/v_{sd}(t)$, if the signal source of Fig. 7.51b is connected to the amplifier of Example 7.17.
> *Hint*: Modify the answers from the preceding example to include input loading.
>
> *Ans.* 57.6.
>
> **Exercise 7.20**   Use some of the results from Example 7.17 to calculate the upper common-mode input voltage and the input bias current for the amplifier of Fig. 7.51a.
>
> *Ans.* 9.1 V, 0.884 μA.

EXAMPLE 7.18   Figure 7.52a shows the amplifier of Fig. 7.51, Example 7.17, with an external source and load attached. Find v_o/v_i.

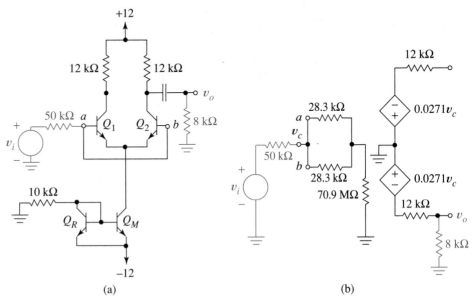

(a)   (b)

Figure 7.52   Common-mode circuit with input and output loading: (a) original circuit; (b) small-signal equivalent.

Solution. Since the excitation is common mode, we need parameter values for the model of Fig. 7.49f. In Example 7.17 we found $r_\pi = 28.3$ kΩ, $R_{out} = 221$ kΩ, and $R_C/2R_{out} = 0.0271$. Since

$$(\beta + 1)R_{out} = (321)221 \text{ k}\Omega = 70.9 \text{ M}\Omega$$

the equivalent circuit is Fig. 7.52b.

By definition, v_c is the average of the voltages at nodes a and b. From the voltage divider,

$$v_c = \frac{(28.3 \text{ k}/2) + 70.6 \text{ M}}{[(28.3 \text{ k}/2) + 70.6 \text{ M}] + 50 \text{ k}} v_i = 0.999 v_i$$

indicating minimal input loading. Because of output loading,

$$v_o = \frac{8 \text{ k}}{8 \text{ k} + 12 \text{ k}}(-0.0271)v_c = -0.0108 v_c$$

Combining these results gives the common-mode gain $v_o/v_i = -0.0108$.   ❏

EXAMPLE 7.19 Improve the CMRR of the diff amp of Example 7.17 by replacing the basic current source with a Wilson source having the same output current and three identical transistors.

Solution. Assume both output transistors in the Wilson source are identical to the mirror transistor in Fig. 7.51. From Eq. (6.33),

$$R_{out} = \frac{\beta r_o}{2} = \left(\frac{320}{2}\right)(221 \text{ k}\Omega) = 35.4 \text{ M}\Omega$$

We found in Example 7.17 that $g_m = 11.3 \times 10^{-3}$ A/V. From Eq. (7.85),

$$\text{CMRR} = 11.3 \times 10^{-3} \times (35.4 \times 10^6) = 4.00 \times 10^5$$

or 112 dB—a considerable improvement over 68 dB of the original design. Figure 7.53 shows the new current source, where the calculation

$$R = \frac{12 - 2 \times 0.7}{0.565 \text{ mA}} = 18.8 \text{ k}\Omega$$

completes the design.   ❏

> **Exercise 7.21** Compare the lower common-mode input limits of the original circuit and the circuit with the Wilson source.
>
> _Ans._ −11.1 V versus −10.4 V for the Wilson source.

Although the input resistance of an emitter-coupled difference amplifier is finite, it is not necessarily small. Since r_π is inversely proportional to I_C, we can obtain high input resistance by using a current source with very small I_O. The Widlar source is very useful for this purpose; and because the Widlar output resistance is high, CMRR is also quite high. We next describe another way to obtain high input resistance.

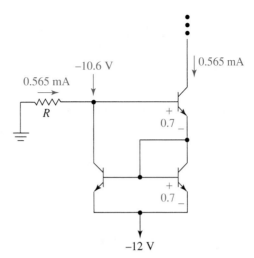

Figure 7.53   Current source of improved CMRR for Example 7.19.

7.9.3 EMITTER-COUPLED PAIR WITH UNBYPASSED EMITTER RESISTORS

Adding emitter resistors, as in Fig. 7.54a, increases the input resistance of the emitter-coupled difference amplifier. The price of this modification is a reduction in difference-mode gain. The only change in dc biasing is the addition of small dc drops $(I_O/2)R_E$. In this section we introduce *half-amplifier analysis,* a technique that exploits the special circuit and signal symmetries that occur in difference amplifiers.

Difference-Mode Model. When a pure difference-mode signal is applied to Fig. 7.54a, circuit and signal symmetry forces the emitter node to remain at ground potential. A vertical axis of symmetry through the supplies and current source divides the amplifier into identical *half-amplifiers* that differ only in the phase of the excitation voltage as in Fig. 7.54b. Clearly we can obtain all needed information by dealing with only the top half amplifier, which normally is all we bother to draw.

Figure 7.54c is the half-amplifier after resistance scaling. Control voltage v is related to v_d by

$$v = \frac{r_\pi}{r_\pi + (\beta + 1)R_E} \frac{v_d}{2}$$

therefore, the dependent current is

$$g_m v = \frac{\beta}{2[r_\pi + (\beta + 1)R_E]} v_d$$

To derive an equivalent circuit comparable to previous ones, we make a source transformation and then combine the two half-amplifier equivalents. This gives Fig. 7.54d. As a check, notice that this circuit reduces to Fig. 7.47e as R_E approaches zero. Comparing the two shows that R_E reduces gain and increases R_{in}. (Incidentally, we have now given a theoretical explanation for our numerical study of Fig. 7.14a, which showed that a par-

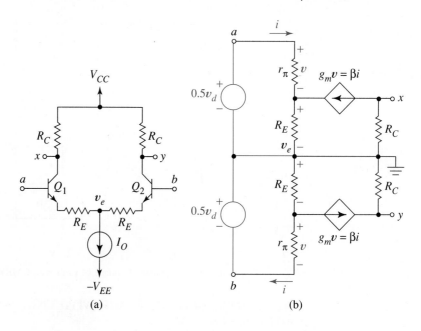

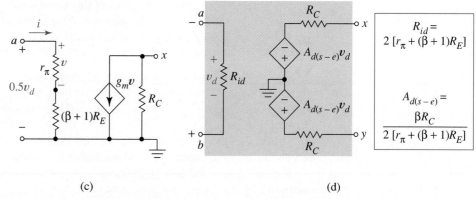

(c)　　　　　　　　　　　　　　(d)

Figure 7.54 Difference-mode model when emitter resistors are unbypassed: (a) amplifier; (b)–(c) steps in the derivation; (d) final model.

tially unbypassed emitter resistor lowers the gain and increases the input resistance of a common-emitter amplifier.)

Common-Mode Model. Figure 7.55a is a *small-signal schematic* drawn to emphasize the vertical axis of symmetry that exists when we apply a common-mode signal to a difference amplifier. Because *common-mode excitation does not create a virtual ground*, we must include the output resistance of the current source. For purposes of symmetry we show it as two parallel resistors.

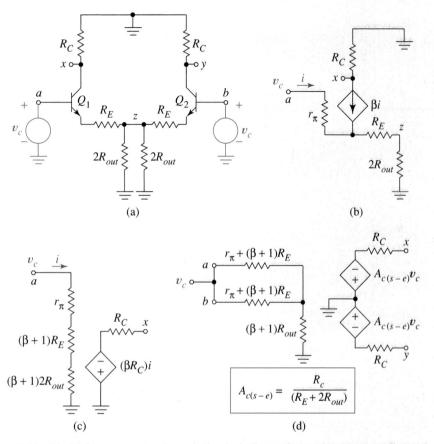

Figure 7.55 Model for common-mode excitation including emitter resistors: (a) small-signal amplifier with common-mode excitation; (b) half-amplifier; (c) half-amplifier after resistance transformation; (d) final model.

Because of symmetry in excitation and circuit, the current in the connecting wire at z is zero; thus we can cut the wire, splitting the circuit into two *half-amplifiers,* each like Fig. 7.55b. Resistance scaling and a source transformation give Fig. 7.55c. In terms of v_c the dependent source voltage is

$$\beta R_C i = \frac{\beta R_C}{r_\pi + (\beta + 1)(R_E + 2R_{out})} v_c \approx \frac{R_C}{(R_E + 2R_{out})} v_c = A_{c(s-e)} v_c$$

where $A_{c(s-e)}$ is the half-amplifier common-mode gain.

The two half-amplifiers are recombined in Fig. 7.55d to show the overall equivalent circuit. Because of the common excitation, the node voltages at the tops of the two $(\beta + 1)R_{out}$ resistors in the separate half-circuits are identical so they combine into one (reversing the manipulation to change from Fig. 7.55a to b). In the limit as R_E approaches zero, Fig. 7.55d reduces to Fig. 7.49f. We conclude that the emitter resistor decreases gain and increases input resistance for common-mode signals; both are improvements.

7.9.4 ACTIVE-LOAD AMPLIFIERS

Figures 7.56a and b show bipolar and MOSFET active-load difference amplifiers. Unlike preceding difference amplifiers, active-load circuits do not possess a vertical axis of symmetry and, therefore, do not lend themselves to half-amplifier analysis. To the extent that transistor output resistances r_o are high, however, the circuits are symmetric as

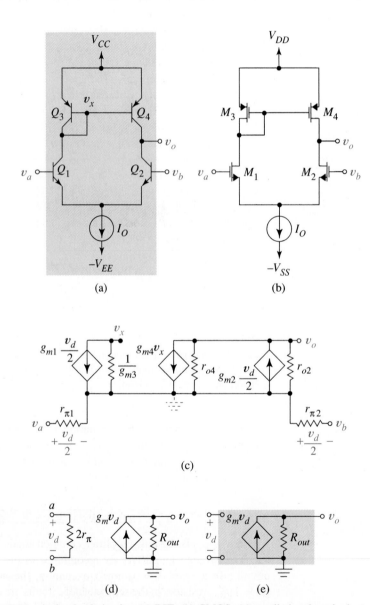

Figure 7.56 Active-load circuits: (a) BJT; (b) CMOS; (c) small-signal equivalent, assuming pure difference-mode excitation; (d) simplified equivalent for difference-mode excitation; (e) equivalent for Fig. 7.56b.

viewed from base and emitter terminals; consequently, we assume equal division of bias current and, for difference-mode analysis, the usual virtual ground at emitter or source node. Biasing and small-signal operation are nearly identical for BJT and MOSFET circuits. We select the bipolar circuit for detailed study because it has the additional complication of finite input resistance.

With inputs grounded, matched input transistors cause I_O to divide into approximately equal parts, biasing all four transistors close to $I_O/2$. After a lengthy derivation, Gray and Meyer (cited in end-of-chapter references) conclude that the Q-point voltage at the output node is one V_{BE} drop below V_{CC}.

The circuits of Figs. 7.56a and b have double-ended inputs and single-ended outputs. When a small signal is applied, the function of $Q_3(M_3)$ is to mirror the signal current of Q_1 (M_1) into the output circuit. We will find that this doubles the difference-mode gain compared to what it would otherwise be and greatly reduces the common-mode gain. Signal mirroring was mentioned in conjunction with Fig. 6.35; however, this is the first time we encounter the principle in an actual circuit. The following analysis shows exactly how it works.

Figure 7.56c is the small-signal equivalent of Fig. 7.56a, for difference-mode excitation. The diagram includes the simplification $1/g_{m3} \| r_{\pi 4} \| r_{o1} \approx 1/g_{m3}$. Identical bias currents and matched transistors make all four g_ms identical; however, the diagram includes numerical subscripts to show the origins of the various parameters.

The output voltage is

$$v_o = \left[g_{m2}\frac{v_d}{2} - g_{m4}v_x \right](r_{o4} \| r_{o2})$$

The signal at the input to Q_1 is reproduced across $1/g_{m3}$ by the dependent source giving

$$v_x = -g_{m1}\left(\frac{1}{g_{m3}}\right)\frac{v_d}{2} = -\frac{v_d}{2}$$

Substituting this v_x into the previous expression and using the common notation g_m gives

$$v_o = \left[g_m\frac{v_d}{2} + g_m\frac{v_d}{2} \right](r_{o4} \| r_{o4}) = g_m(r_{o4} \| r_{o2})v_d$$

Voltage gain is

$$A_d = \frac{v_o}{v_d} = g_m(r_{o4} \| r_{o2}) \tag{7.86}$$

Notice how the mirrored current doubles the gain by adding the second term within the brackets.

Combining current sources, output-circuit resistances, and input resistors gives the more compact equivalent of Fig. 7.56d. The open-circuit gain is potentially quite high. If we expand $r_{o4} \| r_{o2}$ in Eq. (7.86) as product over sum, and multiply numerator and denominator by g_m, Eq. (7.86) becomes

$$A_d = \frac{\mu_2\mu_4}{\mu_2 + \mu_4} \tag{7.87}$$

where μ denotes intrinsic transistor gain. The output resistance is

$$R_{out} = r_{o4} \| r_{o2} \qquad (7.88)$$

This active-load circuit is in fact a practical version of the common-emitter active-load amplifier introduced in Fig. 7.30a. The actual gain, Eq. (7.87), is the same order of magnitude as the ideal, $-\mu$, we predicted for the circuit in Eq. (7.56). Except for the absence of r_π, small-signal analysis details for the FET circuit of Fig. 7.56b are identical; the final equivalent is Fig. 7.56e, and the amplifier is described by Eqs. (7.86)–(7.88).

EXAMPLE 7.20   The active-load amplifier of Fig. 7.57a has these parameter values: $V_{AP} = 60$ V, $\beta_P = 50$, $\beta_N = 350$, and $V_{AN} = 130$ V, where subscripts P and N denote pnp and npn transistors, respectively. Use SPICE to find
(a) the Q-point of the output transistors, and the current-source voltage when both bases are grounded.
(b) the difference-mode voltage gain.
(c) the input resistance seen by a difference-mode signal source.

Solution.   Figure 7.57 shows the node numbering and SPICE code for an ac analysis. The instruction .OP requests that dc operating point information be included in the output.
(a) SPICE dc output showed Q_4 and Q_2 biased at $V_{CE} = -2.40$ V and 9.34 V, respectively, both well outside of saturation, and with magnitudes totaling about 11.7 V as expected. Bias values for Q_1 and Q_3 were 11 V and -0.72 V, respectively. Collector currents were $I_{C2} = I_{C4} = 0.297$ mA, $I_{C1} = 0.301$ mA, and $I_{C3} = 0.289$ mA, all close to the expected 0.3 mA. Because V_{CE} was greater for Q_1 than Q_2, the Early effect led to a slightly larger value for I_{C1}. We conclude that all transistors are properly biased for active operation. The node voltage difference, $V(6) - V(7) = 10.3$ V, is ample for the current source.

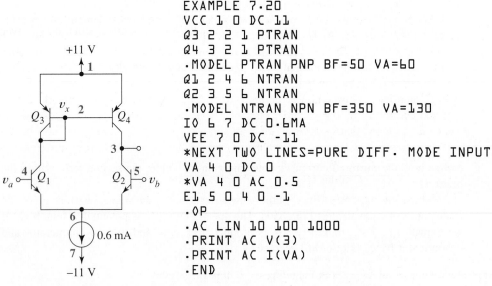

```
EXAMPLE 7.20
VCC 1 0 DC 11
Q3 2 2 1 PTRAN
Q4 3 2 1 PTRAN
.MODEL PTRAN PNP BF=50 VA=60
Q1 2 4 6 NTRAN
Q2 3 5 6 NTRAN
.MODEL NTRAN NPN BF=350 VA=130
IO 6 7 DC 0.6MA
VEE 7 0 DC -11
*NEXT TWO LINES=PURE DIFF. MODE INPUT
VA 4 0 DC 0
*VA 4 0 AC 0.5
E1 5 0 4 0 -1
.OP
.AC LIN 10 100 1000
.PRINT AC V(3)
.PRINT AC I(VA)
.END
```

Figure 7.57   SPICE node-numbered circuit and code for Example 7.20.

(b) For the ac analysis, an ac source, VA, of 0.5 V replaces the dc source at node 4. This, together with VCVS, E1, provides pure difference-mode excitation of 1 V. With this excitation the difference-mode gain is numerically equal to V(3). SPICE gave 1649 for this gain.

(c) For our one-volt input, source current I(VA) is the reciprocal of the input resistance. SPICE gave R_{in} = 94.2 kΩ. ❏

It is always a good idea to compare SPICE results with hand-analysis estimates, as the following exercise demonstrates.

Exercise 7.22 Check the SPICE results, A_d = 1649 and R_{in} = 94.2 kΩ in Example 7.20 using hand analysis.

Ans. A_d = 1645, R_{in} = 58.8 kΩ.

The voltage gain obtained in Exercise 7.22 is close to the SPICE value; however, there is a large discrepancy between the two values for input resistance. We might *surmise* that the theoretical expression $R_{in} = 2r_\pi$ is valid only when output resistances r_o of Q_1 and Q_2 are ignored, for in this case transistor current sources effectively isolate the base-emitter circuits from the Q_3–Q_4 imbalance. With resistors r_{o1} and r_{o2} in the circuit, however, the input transistors "perceive" the load unbalance, and node 6 voltage does not remain at ground potential. This conjecture is easily checked by removing the Early voltage statement from the .MODEL lines for Q_1 and Q_2. After this modification SPICE calculates R_{in} = 60.5 kΩ, a value within 4% of the theoretical prediction. Based on this exercise, prudence suggests we should not only check SPICE results against theory, but we should also augment our theoretical investigations with realistic simulations.

Because of their current mirrors, the BJT and MOS active-load amplifiers both have excellent common-mode rejection. The following comparison with passive-load circuits shows intuitively why this is so.

Figure 7.58a shows how a common-mode signal propagates to the output node in a passive-load amplifier. An increase ΔV_C above the bias value causes an increase in the voltage across R_{out}. This produces additional current ΔI_O in R_{out} that divides into equal parts. The increased drain current in M_2 flows through R_D, causing the output voltage to change for single-ended output.

Figure 7.58b shows how a current change mirrored through M_3 cancels the increased current at the output node. The small-signal equivalent of Fig. 7.58c shows the cancellation more clearly. Ignoring the output resistances of M_1 and M_2, we see that the common-mode signal produces a voltage v_x that creates identical dependent currents in M_1 and M_2. The voltage across M_3, $v_y = (-g_m v_x)(1/g_m) = -v_x$, produces a current in M_4 that is identical to the M_2 current in magnitude and direction (both flow downward when $v_x > 0$.) By KCL, the current in r_{o4} is zero, making $v_o = 0$—perfect common-mode rejection. The same ideas apply to the BJT version of the circuit.

EXAMPLE 7.21 Modify the SPICE code of Fig. 7.57 so that the input is a common-mode signal, and give the current source an output resistance of 100 kΩ. Use SPICE to find the common-mode gain with and without including the output resistances of the input transistors.

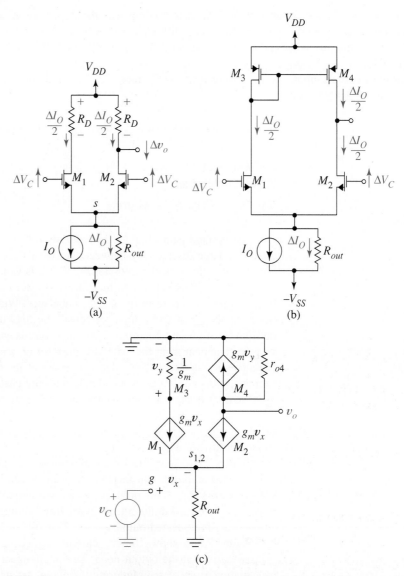

Figure 7.58 Common-mode rejection by passive and active-load circuits: (a) common-mode propagation in a passive load circuit; (b) active-load circuit showing how current change mirrors through M_3 and M_4 to oppose current change in M_2; (c) small-signal equivalent circuit.

Solution. Changes in the SPICE code are: the transmittance on the E1 line changes to $+1$, the voltage of source VA changes to 1 V, a new resistor ROUT = 100 k is added between nodes 6 and 7. Now node voltage V(3) is numerically equal to the common-mode gain.

The common-mode gains computed with and without including output resistances of Q_1 and Q_2 are, respectively, 3.28×10^{-3} and 3.68×10^{-4}. Using the difference-mode gain of 1649 from Example 7.20, these correspond to CMRR values of 114 dB and 133 dB, respectively. The output information from the simulation showed that the actual transconductance of Q_3 differed from that of Q_1, which explains the nonzero common-mode gain. ❏

Exercise 7.23 Find an expression for the common-mode gain of the circuit of Fig. 7.58c if the transconductance of M_3, g_{m3}, differs from the transconductances of the other transistors, g_m.

Ans. $A_c = \dfrac{r_{o4}}{2R_{out}}\left(\dfrac{g_m}{g_{m3}} - 1\right)$

Cascode Difference Amplifier. Figure 7.59 shows the evolution of the familiar active-load circuit into two cascode active-load circuits that are popular in MOS operational amplifiers. The circuit of Fig. 7.59a becomes Fig. 7.59b when we replace n- and p-channel transistors with complementary devices and modify biasing for proper current direction. Fig. 7.59c is simply a more conventional way to draw Fig. 7.59b. Obviously the small-signal equivalents are identical. Gain and output resistance are given by Eqs. (7.86) through (7.88), where subscripts refer to the complementary output transistors. As the first stage in a multistage amplifier, the new version of the circuit has one advantage over the original. If input signals are close to ground potential, the output voltage is below ground level, that is, close to $-V_{SS}$ instead of close to V_{DD} as in the first amplifier. Reversing transistor types has shifted the dc bias level downward, thereby allowing for the increasing dc levels of subsequent gain stages. It is now possible for the output node of the multistage amplifier to be biased at ground potential, a convenience in op amps and other circuits.

In Fig. 7.59d, the common-source input transistors of Fig. 7.59c are replaced by cascode pairs. (Visualize the small-signal schematic for difference-mode inputs, with virtual ground at the source nodes of the input transistors.) Source V_{BB} is readily established "on-chip." As in the original circuit, the output current of the leftmost cascode circuit is mirrored to the output to double the gain. We know that the cascode amplifier's transconductance is the same as the common-source transconductance, g_m; however, the cascode amplifier has greater inherent gain *potential* because of its higher output resistance. Replacing the basic mirror by a cascode mirror as in Fig. 7.59e enables the cascode amplifier to realize this potential by presenting it with a higher current-source output resistance. Of course, the cascode mirror copies signal current in the same fashion as the basic mirror. What this all means is that the cascode amplifier is also modeled by Fig. 7.56e; however R_{out} is the parallel equivalent of the output resistance of the cascode mirror, μr_o, from Eq. (7.72), and the output resistance of the cascode amplifier. From Eq. (7.63), the latter is $r_o(1 + g_m r_o)$. Thus for the cascode difference amplifier of Fig. 7.59e,

$$R_{out} = (\mu r_o) \,\|\, [(1 + g_m r_o)r_o] \approx 0.5\mu r_o \qquad (7.89)$$

and the open-circuit gain is

$$g_m R_{out} = 0.5\,\mu^2 \qquad (7.90)$$

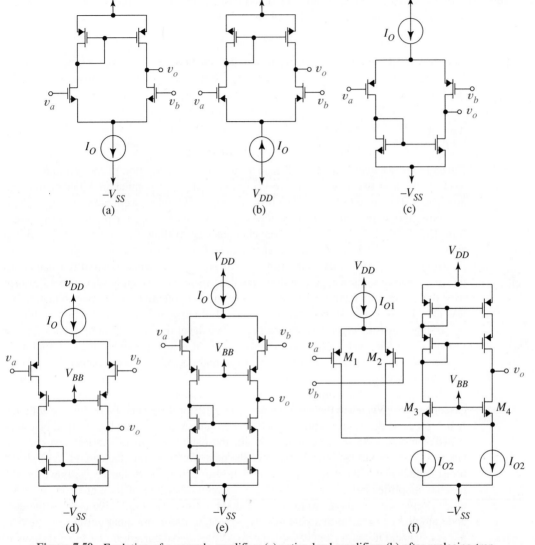

Figure 7.59 Evolution of a cascode amplifier: (a) active-load amplifier; (b) after replacing transistors with complementary types; (c) a more conventional representation of the same circuit; (d) cascode pairs replace common-source transistors; (e) cascode current mirror for higher gain and output resistance; (f) folded cascode amplifier.

Most often this circuit is used as a transconductance amplifier, and actual voltage gain is much less than the open-circuit value because of low external load resistance.

Figure 7.59f is a variation called the *folded cascode amplifier* that does develop high voltage gain and, in fact, is sometimes used as a single-stage operational amplifier. Because dc currents I_{O2}, are larger than $0.5I_{O1}$, all of the output transistors are biased properly, even though the common-gate transistors M_3 and M_4 in the input cascode circuits are n-channel devices. Since the small-signal circuit is the same as Fig. 7.59e, Fig. 7.59f is also described by Eqs. (7.89) and (7.90); however, notice that input and output nodes can all operate at ground potential.

7.10 _____
Advanced Topics in dc Design and Analysis

Now that we are familiar with the more common amplifiers, we can better understand how bias circuit design relates to small-signal and other amplifier parameters. Also related to biasing and dc analysis are input offset voltages and currents, which we also study in this section.

7.10.1 SMALL-SIGNAL DESIGN AND BIAS DESIGN

Voltage Gain, Input and Output Resistance. Because small-signal transistor parameters change with Q-point, small-signal and biasing aspects of amplifier design are closely related. Here we discuss some of the trade-offs that the designer must address.

Consider the voltage gain, input resistance, and output resistance of an amplifier. In many BJT circuits, gain involves the product

$$g_m R_L = \left(\frac{I_C}{V_T} \right) R_L$$

input resistance is related to

$$r_\pi = \frac{\beta}{g_m} = \beta \left(\frac{V_T}{I_C} \right)$$

and output resistance is proportional to

$$r_o = \frac{V_A}{I_C}$$

If a design requires high values for both gain and input resistance, because of the locations of I_C in the foregoing equations, there might be a problem in selecting a Q-point that satisfies both. Sometimes a special input stage, such as an emitter follower (for high input resistance) helps to resolve such a problem. Because of I_C, output resistance also tends to trade off against voltage gain, and a special output stage such as a common-base circuit may be needed to achieve both high gain and high output resistance.

It is also interesting to see how a specified gain, $g_m R_C$, constrains bias circuit design. Notice that a given number,

$$g_m R_C = \left(\frac{1}{V_T} \right) I_C R_C$$

implies a particular dc voltage drop, $I_C R_C$, across the collector resistor. When we first discussed four-resistor biasing circuits we lacked this background. Now we can see that, in discrete four-resistor designs, high gain trades off against Q-point stability, for portions of the available power supply voltage must be allocated to both $I_C R_C$ and $I_E R_E$. In emitter coupled pairs, assigning a larger dc bias voltage, $I_C R_C$, to achieve high gain involves reducing the upper common-mode limit. The active-load idea frees us, to some extent, from the trade-off between gain and output resistance in the BJT case. Because

$$g_m r_o = \frac{I_C}{V_T} \frac{V_A}{I_C} = \frac{V_A}{V_T}$$

gain is independent of Q-point.

In FET circuits, matters are simplified by the infinite input resistance. The equations

$$g_m = \sqrt{2kI_D}$$

for the MOSFET and

$$g_m = \sqrt{4\beta I_D}$$

for the JFET, show that high gain is associated with large bias currents. The dependence is not as strong as in BJT designs because of the square-root variation. Nonetheless, assigning a large dc drop $I_D R_S$ in the four-resistor biasing circuit for FETs to achieve Q-point stability does involve compromising with gain.

For FET active loads,

$$g_m r_o = \sqrt{\frac{2k}{I_D}} V_A$$

for MOSFETs and

$$g_m r_o = \sqrt{4\beta I_D} \frac{V_A}{I_D} = \sqrt{\frac{4\beta}{I_D}} V_A$$

for JFETs. These show that high gain and high output resistance both result from biasing at low drain currents. Other trade-offs related to biasing design come up later in the text.

Signal Amplitude and Distortion. Linear applications require active-state transistors; thus we must select Q-points so that transistors neither saturate nor cut off when signals are introduced. In this way, the required amplitude of the signal swing establishes the acceptable proximity of the Q-point to either extreme. In this regard we need to distinguish two kinds of amplifiers, small signal and large signal, or *power* amplifiers. It is only in the power amplifier that large-signal swings are necessary.

Nonlinear distortion also occurs when signals traverse appreciable regions on the transistor's transfer characteristic. Equation (7.19) shows that large bias currents (large g_m) favor low distortion for FETs; Eq. (7.20) indicates that this property is not shared by BJTs.

7.10.2 INPUT OFFSETS

Even though IC difference amplifiers employ matched components, there are inevitable small parameter differences that lead to voltage and current offsets. We now examine the factors contributing to these offsets, and relate measured offset values to manufacturing tolerances.

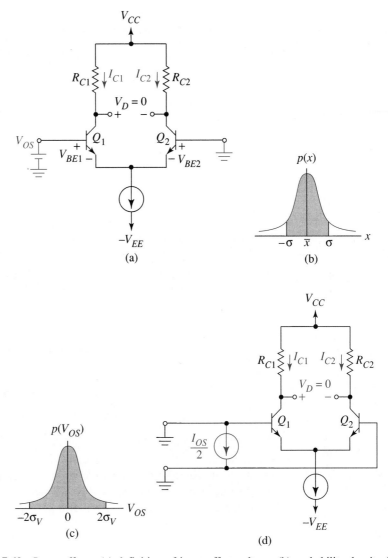

Figure 7.60 Input offsets: (a) definition of input offset voltage; (b) probability density function; (c) probability function for variations in offset voltage; (d) definition of input offset current.

Emitter-Coupled Difference Amplifier. If both bases were grounded in a real amplifier like Fig. 7.60a, output voltage, V_D, would differ from its theoretical value of zero. To counteract this offset, suppose we attach to the input an *external* source, V_{OS}, and adjust it to precisely that value that makes V_D zero. By definition, the *input offset voltage* of the difference amplifier is V_{OS}.

With both transistors active, KVL and Eq. (4.5) give

$$V_{OS} = V_{BE1} - V_{BE2} = V_T \ln\left[\frac{I_{C1}}{I_{S1}}\right] - V_T \ln\left[\frac{I_{C2}}{I_{S2}}\right] = V_T \ln\left[\frac{I_{C1}}{I_{C2}} \frac{I_{S2}}{I_{S1}}\right] \quad (7.91)$$

where the last step assumes transistor temperatures are identical. Because V_D is forced to zero by the external source,

$$V_D = I_{C2}R_{C2} - I_{C1}R_{C1} = 0$$

This implies

$$\frac{I_{C1}}{I_{C2}} = \frac{R_{C2}}{R_{C1}} \tag{7.92}$$

It is mismatches in resistors and saturation currents that cause the offset, so we want to relate V_{OS} to these parameters. Substituting Eq. (7.92) into Eq. (7.91) gives

$$V_{OS} = V_T \ell n \left(\frac{R_{C2}}{R_{C1}} \frac{I_{S2}}{I_{S1}} \right) = V_T \left[\ell n \left(\frac{R_{C2}}{R_{C1}} \right) + \ell n \left(\frac{I_{S2}}{I_{S1}} \right) \right] \tag{7.93}$$

Notice that the offset drifts with temperature because $V_T = kT/q$.

To relate V_{OS} to parameter mismatch, we define a design-center collector resistance,

$$R_C = \frac{R_{C1} + R_{C2}}{2}$$

and manufacturing variation in R_C by

$$\Delta R_C = R_{C1} - R_{C2}$$

After solving simultaneously for R_{C1} and R_{C2}, we write the ratio as

$$\frac{R_{C2}}{R_{C1}} = \frac{R_C - (\Delta R_C/2)}{R_C + (\Delta R_C/2)} = \frac{[1 - (\Delta R_C/2R_C)]}{[1 + (\Delta R_C/2R_C)]} \approx 1 - \frac{\Delta R_C}{R_C} \tag{7.94}$$

To justify the last step, use long division—then drop the other terms because $\Delta R_C/R_C$ is small. Notice that the nominal value, R_C, which varies greatly from one IC batch to the next, cancels out. The fractional variation that remains is related to the masking tolerances associated with fabricating a single chip as discussed in Sec. 3.3.6. Handling the saturation current ratio in Eq. (7.93) the same way gives

$$\frac{I_{S2}}{I_{S1}} \approx 1 - \frac{\Delta I_S}{I_S}$$

Substituting these into Eq. (7.93) gives,

$$V_{OS} = V_T \ell n \left(1 - \frac{\Delta R_C}{R_C} \right) + V_T \ell n \left(1 - \frac{\Delta I_S}{I_S} \right)$$

The small fractional variations justify the approximation $\ell n(1 + x) \approx x$. This gives the final offset expression

$$V_{OS} = -V_T \left(\frac{\Delta R_C}{R_C} + \frac{\Delta I_S}{I_S} \right) \tag{7.95}$$

We now wish to make a numerical estimate of V_{OS}.

We use a probability density curve like Fig. 7.60b to describe the random variations in parameter x that occur during IC production. The curve associates relatively high probability, p, with the more likely values of x and lower probability with less likely values. In particular, the probability that x lies between specific values, a and b, equals the area under the curve for $a < x < b$. The familiar gaussian probability curve is characterized by two parameters: a mean, $\bar{x}$, that defines the vertical axis of symmetry of the curve, and a standard deviation, σ, that describes the extent of the spreading of the curve from the mean value. The probability that $-\sigma < x < \sigma$ is about 68%; the probability that $-2\sigma < x < 2\sigma$ is 96%.

Data taken after IC production runs show variations in $x = \Delta R_C/R_C$ that are approximately gaussian with zero mean and standard deviation of 1%. Variations, $\Delta I_S/I_S$, in saturation current are also zero-mean gaussian; typical standard deviations are about 5%.

To a statistician, Eq. (7.95) describes offset voltage as $-V_T$ times the sum of two random parameters, each described by known gaussian probability curves. Because $\Delta R_C/R_C$ and $\Delta I_S/I_S$ are independent variations, probability theory predicts that the offset voltage, V_{OS}, will also be random with a gaussian probability curve like Fig. 7.60c. Furthermore, this curve theoretically has standard deviation, σ_V, that satisfies

$$\sigma_V^2 = V_T^2(\sigma_R^2 + \sigma_I^2) \tag{7.96}$$

Thus, for 1% and 5% variations in resistance and saturation current,

$$\sigma_V = \sqrt{(0.025)^2](0.01)^2 + (0.05)^2]} = 1.27 \text{ mV}$$

With this statistical information we can predict that about 96% of the difference amplifiers in a large batch will have offset voltages that fall within ± 2 standard deviations of zero; that is, virtually all offset will fall within the range

$$-2.54 \text{ mV} < V_{OS} < 2.54 \text{ mV}$$

The same kind of approach leads to a statistical description of input offset current.

Input Offset Current. A slight difference in base currents causes nonzero output voltage when both input nodes are grounded. We can force the output voltage to zero by canceling the base current difference with an external current source as in Fig. 7.60d. This takes

$$I_{OS} = I_{B1} - I_{B2}$$

and, with this source in place, Eq. (7.92) applies. Since the input offset current involves base currents instead of collector currents, we rewrite Eq. (7.92) as

$$\beta_1 I_{B1} R_{C1} = \beta_2 I_{B2} R_{C2} \tag{7.97}$$

The I_{OS} definition and Eq. (7.97) lead to

$$I_{OS} = I_{B1}\left(1 - \frac{\beta_1}{\beta_2}\frac{R_{C1}}{R_{C2}}\right)$$

Using our definitions of ΔR_C and R_C, and defining $\Delta \beta$ and β in similar fashion gives

$$I_{OS} = I_{B1}\left[1 - \left(\frac{1 + \Delta\beta/2\beta}{1 - \Delta\beta/2\beta} \right)\left(\frac{1 + \Delta R_C/2R_C}{1 - \Delta R_C/2R_C} \right) \right]$$

Expressing I_{B1} in terms of its variation, using long division, and discarding small terms inside the brackets leads to

$$I_{OS} = I_B\left(1 + \frac{\Delta I_B}{2I_B} \right)\left[1 - \left(1 + \frac{\Delta\beta}{\beta} \right)\left(1 + \frac{\Delta R_C}{R_C} \right) \right]$$

After we expand, and then discard products of small terms,

$$I_{OS} = -I_B\left[\frac{\Delta\beta}{\beta} + \frac{\Delta R_C}{R_C} \right] \tag{7.98}$$

Statistical data indicate that variations in β are gaussian with zero mean and standard deviation of about 10%. From the previous treatment of offset voltage, we conclude that input offset current is a gaussian random variable with zero mean and standard deviation

$$\sigma_I = \sqrt{I_B^2(\sigma_B^2 + \sigma_R^2)} \tag{7.99}$$

Notice from Eqs. (7.98) and (7.99) that current offset depends upon bias current and, therefore, the transistor's Q-point. In the emitter-coupled pair, biasing at high collector current to obtain high gain also increases the current offset.

Offsets in Source-Coupled Pairs. For the source-coupled pair, the offset voltage derivation closely parallels the BJT derivation. First, we note that active-state MOS transistors satisfy Eq. (5.7). Therefore, an external source attached to the input nodes must satisfy

$$V_{OS} = V_{GS1} - V_{GS2} = \left(\sqrt{\frac{2}{k_1}I_{D1}} + V_{t1} \right) - \left(\sqrt{\frac{2}{k_2}I_{D2}} + V_{t2} \right)$$

By definition, this source drives the output difference voltage V_D to zero, causing the constraint

$$I_{D2} = \frac{R_{D1}}{R_{D2}}I_{D1}$$

Including this, the offset expression becomes

$$V_{OS} = V_{GS1} - V_{GS2} = \sqrt{\frac{2}{k_1}I_{D1}} - \sqrt{\frac{2}{k_2}I_{D1}\frac{R_{D1}}{R_{D2}}} + \Delta V_t$$

$$= \Delta V_t + \sqrt{\frac{2}{\mu_n C_{ox}}I_{D1}}\left[\sqrt{\frac{1}{(W/L)_1}} - \sqrt{\frac{1}{(W/L)_2}\frac{R_{D1}}{R_{D2}}} \right]$$

where $\Delta V_t = V_{t1} - V_{t2}$, and where Eq. (5.8) was used for k.

The causes of V_{OS} are mismatches in R_D, the W/L ratios, and threshold voltages. When we express resistances and W/L ratios in terms of their means and deviations as we did for resistances in the BJT case, use long division, discard small terms, and keep only the first two terms in the power series expansion of the square root functions, we obtain

$$V_{OS} = \Delta V_t + \sqrt{\frac{2}{\mu_n C_{ox}} I_{D1}} \left[\left(\frac{-\Delta R_D}{R_D} \right) - \left(\frac{-\Delta(W/L)}{(W/L)} \right) \right] \qquad (7.100)$$

Temperature-related drift is possible through μ, and the presence of I_{D1} shows that voltage offset in the MOSFET source-coupled pair depends upon Q-point. Also, threshold voltage mismatches, ΔV_t, add directly to the input offset voltage. The offset voltage is a zero-mean random variable with standard deviation, σ_V, that is related to standard deviations of the mismatched parameters by

$$\sigma_V = \sqrt{\sigma_{V_t}^2 + \frac{2}{\mu_n C_{ox}} I_{D1}(\sigma_R^2 + \sigma_{W/L}^2)} \qquad (7.101)$$

The offset voltage of the JFET source-coupled pair is analyzed in the text by Gray and Meyer, which is listed among the references at the end of the chapter.

7.11 Summary

In this chapter we learned of the small-signal model, which describes small perturbations about the transistor's quiescent operating point. Parameter values for the model depend on both the physical structure of the transistor and its operating point; validity of the model critically depends upon keeping signals so small that distortion is negligible.

To analyze the small-signal operation of a linear electronic circuit such as an amplifier, we use a small-signal equivalent circuit. We derive this circuit from the original schematic by turning off dc biasing sources, replacing coupling and bypass capacitors by short circuits, and by replacing each transistor with its small-signal model. If we stop short of the last step, we have the small-signal schematic, in which schematic symbols still represent the transistors. This representation helps us better recognize the general role of the transistor in the circuit by showing the transistor's orientation and, with experience, can also be used to formulate circuit equations.

Small-signal analysis involves finding algebraic expressions for signal voltages and currents in a circuit. Often we employ signal ratios such as voltage, current, or power gain; input resistance; and output resistance. The algebraic expressions, though often only approximations, are especially valuable in helping us understand the underlying operating principles of our circuits—an important consideration in design. The current-source transformation and the concept of resistance scaling by BJTs are powerful tools that simplify hand analysis. Circuit simulators such as SPICE provide us with powerful numerical backup of great accuracy. Simulations are particularly useful in helping us determine the importance of second-order factors that we deliberately ignore for more insightful hand analysis.

There are three basic single-transistor amplifier configurations, common-emitter, common-base, and common-collector, each with an FET counterpart having similar properties. Common-emitter and common-source circuits are general-purpose amplifiers. Common-base and common-gate amplifiers resemble unity-gain current sources with low input resistance and high output resistance; common-collector and common-drain amplifiers resemble unity-gain voltage sources, featuring high input resistance and low output resistance. Several two-transistor configurations are also commonly used and of special importance in integrated-circuits. Properties of these two-transistor circuits are closely related to the properties of the single-transistor circuits.

In ICs, the basic one- and two-transistor circuits often occur paired in difference-amplifier structures in order to exploit the availability of matched transistors. With difference amplifiers come several important new considerations including common-mode input limits, difference- and common-mode gain, and single- and double-ended outputs. Loading at the input and output are slightly more complicated than in single-ended amplifiers. Special difference-amplifier models for difference- and for common-mode input excitation allow us to better visualize overall amplifier operation and the various options available to the user. These models also help us understand input and output loading, and are the cornerstone for analysis by superposition.

Voltage gain and output resistance of our circuits sometimes fall far short of the values inherent in the transistors themselves because of resistors we use to bias the transistors. Special active-load configurations utilize current-source biasing and signal mirroring to achieve voltage gain approaching the intrinsic gains of the transistors. Our ability to design constant current sources of very high output resistance is the key to obtaining high gain and output resistance with active-load circuits as well as for effectively rejecting common-mode signals.

There are many linkages between small-signal design and bias circuit design. With experience, a designer learns how to select an operating point and design a biasing circuit in a way that effectively compromises the various conflicting requirements of small-signal and dc operation.

Small and unavoidable mismatches in key difference amplifier parameters result in input offset voltages and currents. Although we are unable to predict the exact magnitudes and polarities of these offset quantities, product sampling and elementary statistics allow us to predict ranges of values with a high degree of certainty.

REFERENCES _____

1. ALVAREZ, A. R., ed. *BiCMOS Technology and Applications.* Boston: Kluwer Academic, 1989.
2. GRAY, P. R., and R. G. MEYER. *Analysis and Design of Analog Integrated Circuits,* 3rd ed. New York: John Wiley, 1993.
3. RANDOLPH, J. F., and M. KAC. *Analytic Geometry and Calculus.* New York: Macmillan, 1949.
4. TUINENGA, P. W. *SPICE A Guide to Circuit Simulation & Analysis Using PSpice.* Englewood Cliffs, NJ: Prentice Hall, 1988.

PROBLEMS _____

SPECIAL DIRECTIONS FOR SPICE HOMEWORK PROBLEMS: *Do not hand in lengthy SPICE printout for homework.* Instead, abstract the useful information from the SPICE output file as in the SPICE examples in the text. Include your SPICE code and a circuit diagram with nodes numbered to agree with the code. Cite relevant numerical values from the SPICE output file and discuss when appropriate. Make sketches

of any relevant curves, and label appropriate points. Make small tables to present numerical data if useful for clarity.

Section 7.2

7.1 Draw the small-signal equivalent circuit for Fig. P7.1, including transistor r_os. Simplify the circuit as much as pos-

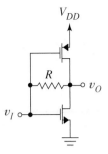

Figure P7.1

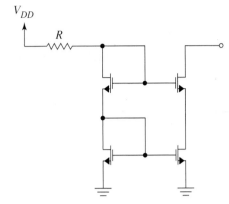

Figure P7.4

sible by assuming identical transconductances and output resistances.

7.2 Verify that the equation for g_m given in Fig. 7.2a applies to a p-channel MOSFET. Use the following procedure.
(a) Find an expression for g_m using Eq. (7.2).
(b) Substitute for $(V_{GS} - V_t)$ to express g_m in terms of I_D. You must choose the root that makes the sign of g_m agree with the slope of Fig. 5.23b.
(c) Draw a small-signal model noting that a p-channel device has its drain-current reference direction oriented from source to drain.
(d) Describe why your result is equivalent to Fig. 7.2a.

7.3 Draw the small-signal equivalent circuit for Fig. P7.3.

7.4 (a) Draw the small-signal equivalent for Fig. P7.4, including output resistances for the transistors in the mirror circuit but not for those in the reference circuit.
(b) Repeat after replacing the supply voltage and series resistor by a dc current-source reference I_{BB}.

7.5 (a) Draw the small-signal equivalent circuit for Fig. P7.5, including values for all components on the diagram. Use $\beta = 99$ to find small-signal parameters.
Hint: Ignore the base current of Q_2.
(b) The Early voltages are 110 V. Add r_o to your diagram for each transistor.

7.6 In Fig. P7.6, $\beta = 1.78$ mA/V^2 and $V_P = -1.5$ V.
(a) Find values for V_{BB} and V_{GG} consistent with the given dc current and parameter values.
Hint: v_s must be turned off for biasing calculations.
(b) Draw the small-signal equivalent circuit, giving numerical values for all parameters.

7.7 Draw the small signal equivalent of Fig. P7.7. Show values of all components including r_o. Transistor parameters are $\beta = 100$, $V_A = 100$ V.

7.8 The input and output characteristics of a transistor are given as Fig. P7.8.
(a) Estimate β, r_o, and r_π directly from the curves if the transistor is biased at $(V_{CE}, I_C) = (2\text{ V}, 2.5\text{ mA})$.

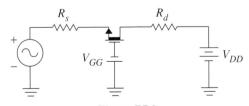

Figure P7.3

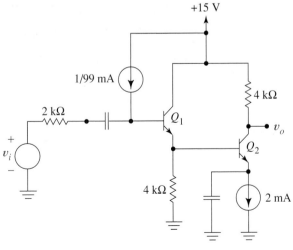

Figure P7.5

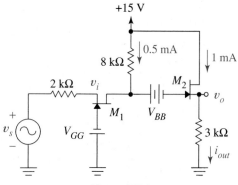

Figure P7.6

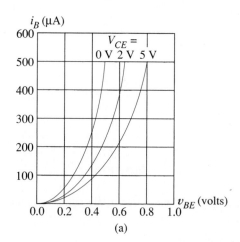

(a)

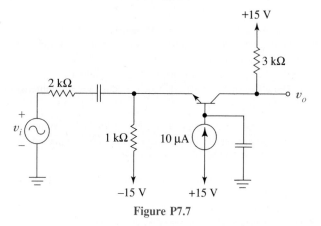

Figure P7.7

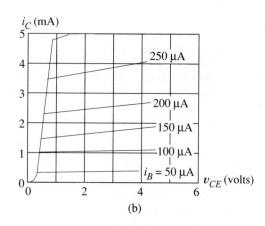

(b)

(b) Repeat part (a) if the transistor is biased at (1 V, 1 mA).

(c) Figure P5.1 shows transfer and output characteristics for a MOSFET. Define g_m as a partial derivative at a given Q-point. Then use your definition to evaluate g_m from the output characteristic at $(V_{DS}, I_D) = (6\text{ V}, 2\text{ mA})$. Make a rough sketch to show how you obtained your numerical result.

(d) Evaluate g_m at the Q-point given in part (c), except use the *transfer* characteristic. (In Fig. P5.1 the same transfer characteristic applies to *all* V_{DS} in the active region.) Make a rough sketch to show how you obtained your numerical result.

Section 7.3

7.9 In Fig. P7.9 the pnp transistors have $\beta = 40$ and $V_A = 60$, the npn transistor $\beta = 220$, $V_A = 125$. V_{BB} is an independent biasing source with value consistent with the bias current of Q_1. Draw the small-signal equivalent circuit. Include transistor output resistances, and show numerical values for all components on the diagram.

Figure P7.8

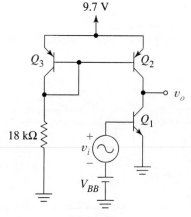

Figure P7.9

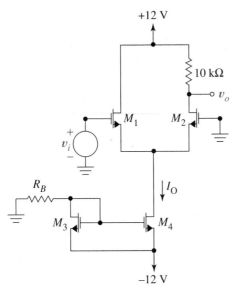

+12 V

10 kΩ

○ v_o

M_1 M_2

v_i

R_B

M_3 M_4

$\downarrow I_O$

−12 V

Figure P7.10

7.10 In Fig. P7.10, $k = 9 \times 10^{-3}$ A/V^2 and $V_t = 1.5$ V for all transistors.
(a) Find the value of R_B so that $I_O = 0.5$ mA.
(b) For the design of part (a), draw the small-signal equivalent circuit.
Hint: Bias currents of M_1 and M_2 are identical even though collector resistors differ.

(c) Add transistor output resistors to the small-signal circuit if $V_A = 100$ V for each transistor.

7.11 In Fig. P7.11, $g_m = 0.002$ A/V.
(a) Draw the small-signal equivalent.
(b) Add r_o to your circuit, assuming $V_A = 95$ V.

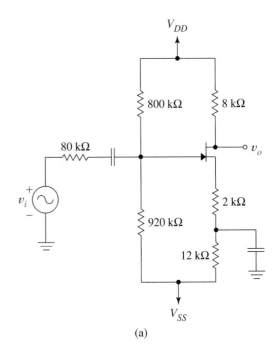

V_{DD}

800 kΩ   8 kΩ

80 kΩ

○ v_o

v_i

2 kΩ

920 kΩ

12 kΩ

V_{SS}

(a)

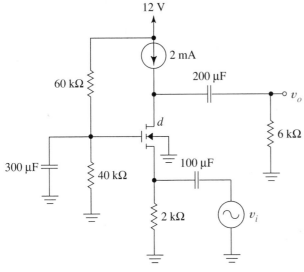

12 V

2 mA

200 μF

60 kΩ

○ v_o

d

6 kΩ

300 μF

40 kΩ

100 μF

2 kΩ v_i

Figure P7.11

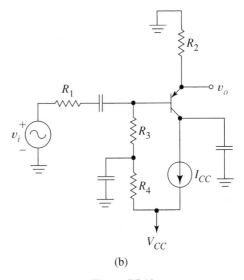

R_2

R_1 ○ v_o

v_i

R_3

R_4 I_{CC}

V_{CC}

(b)

Figure P7.12

7.12 Draw the small-signal model for each circuit of Fig. P7.12. Notice that in Fig. P7.12c the signal comes from a current source.

7.13 Draw small-signal equivalents for Figs. 6.27a and b. In the former, show the simplification that results from identical transistor biasing.

7.14 Draw the small-signal equivalent for the circuit of Fig. 6.37a, including r_o for the output transistor.
Hint: First, replace the JFET with Fig. 6.37b; then, recognize that I_{DSS} is a dc biasing current.

7.15 Draw the small-signal equivalents for both cascode mirrors of Fig. 6.29. In each circuit, include r_o for both transistors in the output circuit.

7.16 Figure P7.16, like Fig. 6.35, shows how signal or noise, $v_s(t)$, can be reproduced in the output of a mirror circuit.
(a) Draw the small-signal equivalent excluding r_o.
(b) Use the circuit from part (a) to relate $i_s(t)$ to $v_s(t)$.

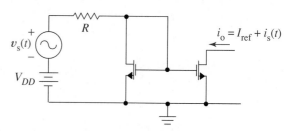

Figure P7.16

7.17 (a) Draw the small-signal equivalent for Fig. 7.29c. Simplify your circuit diagram as much as possible by assuming transistors are identical in bias currents, k, and r_o.
(b) In Fig. 7.29c, replace R by two series resistors of value $R/2$. Then add a bypass capacitor from the resistor junction point to ground.
(c) Draw the small-signal equivalent for the modified circuit using the assumptions of part (a).

7.18 Draw the small-signal equivalent for Fig. P7.18.

Section 7.4

7.19 Derive an expression for the voltage gain, $A'_v = v_o/v_i$, for the amplifier of Fig. 7.13a in terms of A_v, R_i, and R_{in}. Then find its numerical value using parameters found in the textbook analysis of the same circuit with $R_i = 5$ kΩ.

7.20 In Fig. P7.20, threshold voltages are $V_t = -3$V and $+3$ V for the depletion and enhancement devices, respectively. For both, $k = 10^{-4}$ A/V^2 and $V_A = 100$ V.
(a) Draw the small-signal equivalent circuit.

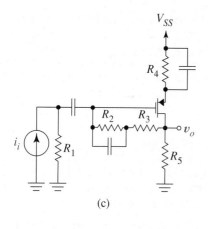

(c)

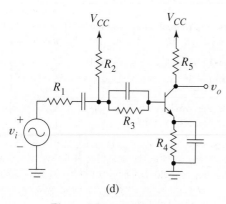

(d)

Figure P7.12 (continued)

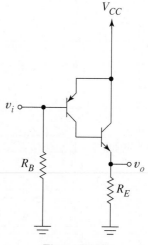

Figure P7.18

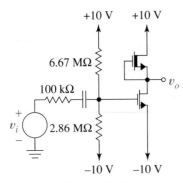

Figure P7.20

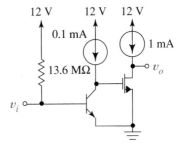

Figure P7.23

(b) Find v_o/v_{gs} and v_o/v_i.
(c) Use Eq. (7.19) to find the largest v_o consistent with undistorted small-signal operation.

7.21 Parameters for the amplifier of Fig. P7.21 are $\beta = 120$, $k = 3 \times 10^{-3}$ A/V^2, and $V_t = -1$ V. Find the gain and the largest undistorted output voltage amplitude that small-signal theory permits.

7.22 For the amplifier of Fig. P7.6,
(a) Find the voltage gain, v_o/v_i where v_i is the signal voltage at the source of M_1.
(b) Find R_{in} as seen by v_s and its 2 kΩ internal resistance.
(c) Find the current gain where input current is the current in v_s and output current is the current in the 3 kΩ load resistor.
(d) find R_{out}, as seen by the 3 kΩ load resistor looking back into the amplifier.

7.23 In the amplifier of Fig. P7.23, transistor parameters are $\beta = 120$ and $V_A = 130$ V for the BJT, and $k = 0.8 \times 10^{-4}$ A/V^2, $V_t = -0.9$ V, and $V_A = 100$ V for the MOSFET. Assume the FET's V_{DS} is appropriate for active-state operation. Use small-signal analysis, including r_o for each transistor, to

find the amplifier's voltage gain, input resistance, and output resistance. Do not be surprised by a large gain.

Section 7.5

7.24 Verify Eqs. (7.30), (7.31), and (7.32) by direct analysis of Fig. 7.17b instead of using the current-source transformation.

7.25 (a) For the discrete circuit, common-gate amplifier described by Figs. 7.16a and c, derive algebraic expressions for voltage gain, current gain, and input resistance.
(b) Rewrite the voltage gain and input resistance expressions of part (a) to show that they are Eqs. (7.30) and (7.32), but with loading factors that result from biasing.

7.26 When biased with current sources so that R_D is effectively infinite, common-source and common-gate amplifiers have output resistances given by $R_{out} = r_o$ and by Eq. (7.33), respectively. Compare numerical values of output resistances if both amplifiers use a FET biased at 2 mA with $k = 0.5$ mA/V^2 and $V_A = 100$ V. The common-gate amplifier uses a signal source of 5 kΩ internal resistance.

7.27 Show that the small-signal equivalent of the V_{GS} multiplier circuit of Fig. P6.85 is a resistor of approximately $1/g_m$ Ω. *Hint*: Use the current-source transformation and a test generator, and make an approximation involving parallel resistors.

7.28 (a) Change the amplifier of Fig. 7.13a into a common-base amplifier by moving the signal source and the bypass capacitor.
(b) Draw the small-signal schematic for this discrete-component, common-base amplifier.

7.29 Design a single-stage discrete BJT amplifier with voltage gain ≈ 500, $R_{in} \le 250$ Ω, and $R_{out} \ge 80$ kΩ. Dual supplies are acceptable but a single supply is preferred, if possible. *Hint*: Low Q-point sensitivity turns out to be a *very important issue* in meeting specifications.

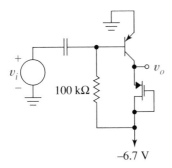

Figure P7.21

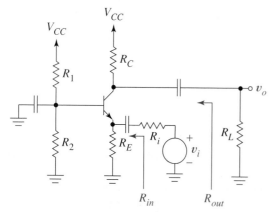

Figure P7.30

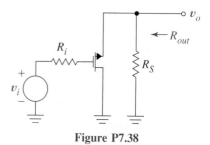

Figure P7.38

7.30 Derive expressions for voltage gain, input resistance, and current gain for Fig. P7.30, and compare them with Eqs. (7.34) through (7.36). Use the transistor model of Fig. 7.5a.

7.31 Use the common-base transistor model to find voltage gain, current gain, input resistance, and output resistance for the common-base amplifier of Fig. P7.30. Include r_o for the output resistance derivation.

7.32 Figure P7.32 is the small-signal schematic of a *cascode* amplifier. Draw the small-signal equivalent circuit using the model of Fig. 7.5a for both transistors. Use it to find voltage gain, input resistance, and current gain.

7.33 Draw an FET version of Fig. P7.32. Then find an expression for voltage gain.

7.34 In Fig. 7.19a, replace the transistor with the common-base model of Fig. 7.20a. Use this circuit to find the input resistance between emitter and ground, and the voltage gain, v_o/v_i.

7.35 Draw the small-signal equivalent circuit of Fig. P7.32 using the model of Fig. 7.5a for the common-emitter transistor and the model of Fig. 7.20a for the common-base transistor. From this circuit, find voltage gain, input resistance, and current gain.

7.36 Replace the transistor in Fig. 7.17a with the common-gate equivalent circuit of Fig. 7.21a.

(a) Then find the input resistance between source and ground, the current gain, and the voltage gain, v_o/v_s, of the amplifier.
(b) Find the voltage gain, v_o/v_i.
(c) Add r_o between source and drain in your equivalent circuit. Then find an expression for the amplifier's output resistance.

7.37 Use the common-gate FET model of Fig. 7.21a to find voltage gain, v_o/v_s, current gain, i_{out}/i_{in}, and the input resistance for the common-gate circuit of Fig. 7.16a. Notice how this model simplifies the equivalent by decoupling input and output circuits.

Section 7.6

7.38 Ignoring r_o, find an expression for R_{out} in Fig. P7.38 in two ways:
(a) Apply the definition of R_{out} and use small-signal analysis.
(b) Make use of the related output resistance, Eq. (7.42), defined in Fig. 7.22c.

7.39 Figure P7.39 shows a discrete-component emitter-follower amplifier. Use the notation $R_P = R_1 \| R_2$, $R_{II} = R_P \| R_i$, and $R_{LL} = R_E \| R_L$ to show that

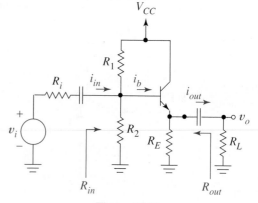

Figure P7.39

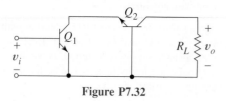

Figure P7.32

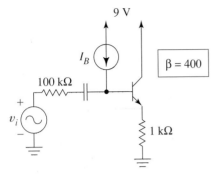

Figure P7.40

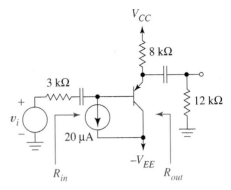

Figure P7.43

(a) $\dfrac{v_o}{v_b} = A_v = \dfrac{(\beta + 1)R_{LL}}{(\beta + 1)R_{LL} + r_\pi}$.

(b) $R_{in} = R_P \parallel \{r_\pi + (\beta + 1)R_{LL}\}$.

(c) $\dfrac{i_{out}}{i_b} = \dfrac{(\beta + 1)R_E}{R_E + R_L}$.

(d) $A_i = \dfrac{i_b}{i_{in}} = \dfrac{R_P}{R_P + r_\pi + (\beta + 1)R_{LL}}$.

(e) $R_{out} = R_E \parallel \dfrac{R_{II} + r_\pi}{\beta + 1}$.

7.40 The transistor in Fig. P7.40 is biased at $V_{CE} = 4.5$ V.
(a) Find I_B.
(b) Find the input resistance seen looking to the right between transistor base and ground.
(c) Find the fraction of the signal voltage v_i that is developed between base and ground.
(d) Find the fraction of v_i that appears at the emitter node.
(e) Find the voltage that would be developed across the 1 kΩ resistor if it were attached directly to the external source without using the amplifier.

7.41 Find the output resistance seen by the 1 kΩ resistor in Fig. P7.40 if $I_C = 2$ mA.

7.42 In Fig. P7.42 the transistor has $k = 0.5 \times 10^{-3}$ A/V² and is biased at 1 mA. Find the numerical value of R_{out}.

7.43 Find R_{in} and R_{out} in Fig. P7.43 if the transistor has $\beta = 80$. Use important previously derived results if you can. Otherwise, use small-signal analysis.

7.44 (a) Use infinite beta analysis to estimate the Q-point of each transistor in Fig. P7.44.
(b) Use the resistance scaling idea of Fig. 7.24 to find R_{in} and R_{out}. For this part, assume $\beta = 30$.

Section 7.7

7.45 Use SPICE ac analysis to find the voltage gain, input resistance, and output resistance of the amplifier in Fig. P7.5. Both transistors have $\beta = 99$ and $V_A = 130$ V. Amplifier input

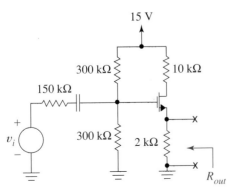

Figure P7.42

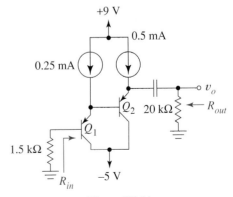

Figure P7.44

is between base of Q_1 and ground; output is between the collector of Q_2 and ground.

7.46 Use SPICE ac analysis to find the voltage gain, input resistance, and output resistance of the amplifier in Fig. P7.7. Parameters are $\beta = 100$ and $V_A = 100$ V.

7.47 In Fig. P7.7 parameters are $\beta = 100$ and $V_A = 100$ V. Add an ac voltage source between the collector and the output node. Use this as a test generator in a SPICE determination of the output resistance seen by the 3 k resistor looking back into the circuit. Compare your answer with what you expect from Eq. (7.37). What percentage error is involved in finding the common-base output resistance in this way?

7.48 Use SPICE to find the voltage gain and output resistance for the amplifier of Problem 7.20. Include a body effect coefficient of GAMMA = 0.33 in the simulation.

Section 7.8

7.49 Replace the BJT schematic symbol by its small-signal model and then verify the entries in the first row of Table 7.1. Input is an ideal voltage source. Include r_o only to find R_{out}.

7.50 Use the small-signal model of Fig. 7.20a to verify the first three entries in the second row of Table 7.1.

7.51 Construct a table like Table 7.1 for FET amplifiers. Most table entries are already worked out directly or indirectly in the text.

7.52 The transistors in Fig. P7.52 have $\beta = 99$.
(a) Draw the small-signal equivalent, showing values for all components.

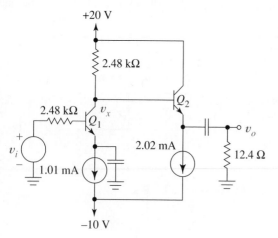

Figure P7.52

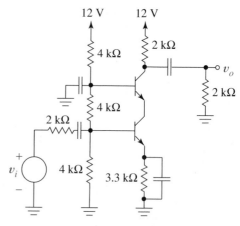

Figure P7.54

(b) Find the gain, v_x/v_i.
Hint: Use the "βi_b" model for Q_2.
(c) Find the gain, v_o/v_i.

7.53 The transistors in Fig. P7.52 have $\beta = 99$.
(a) Identify this two-transistor amplifier by name.
(b) Use SPICE ac analysis to find the gains v_x/v_i and v_o/v_i.
(c) Use SPICE to find the input resistance between base of Q_1 and ground, and also the output resistance seen by the 12.4 Ω resistor.

7.54 Both transistors in Fig. P7.54 are biased at 1 mA; $\beta = 100$. Find the voltage gain, v_o/v_i.

7.55 The transistors of Fig. P7.54 have $\beta = 275$ and $V_A = 185$ V.
(a) Identify this two-transistor amplifier by name.
(b) Use SPICE ac analysis to find v_o/v_i. Also find the input resistance to the right of the 2 k source resistor and the output resistance to the left of the 2 k load resistor.

7.56 The transistors in Fig. P7.56 are biased at 1.5 mA; parameters are $k = 10^{-4}$ A/V^2 and $|V_t| = 0.8$ V. Find v_o/v_i.

7.57 In Fig. P7.56 transistor parameters are $k = 10^{-4}$ A/V^2, $|V_t| = 0.8$ V, and $V_A = 100$ V. Use SPICE ac analysis with $f = 1000$ Hz to find
(a) v_o/v_i.
(b) the input resistance seen by the signal source.
(c) the output resistance seen by the 2.67 kΩ load resistor.

7.58 In Fig. 7.26c, replace transistors with MOSFETs, and then draw the small-signal equivalent. Assume both transistors have transconductance g_m. Ignore r_o.
(a) Use the small-signal equivalent to find voltage gain and input resistance.

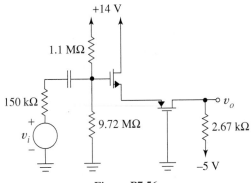

Figure P7.56

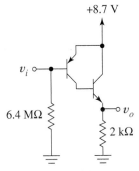

Figure P7.62

(b) Include r_o for both transistors. Then find the output resistance seen by R_L looking back into the amplifier when an ideal voltage source is connected at the input.
Hint: Use the current-source transformation.

7.59 In Fig. 7.26d, replace transistors with MOSFETs and then draw the small-signal equivalent. Assume both transistors have transconductance g_m. Include r_o only for the first-stage transistor.
(a) Use the small-signal equivalent to find voltage gain and input resistance. (Assume $g_m r_o \gg 1$.)
(b) Now include r_o for both transistors. Find the output resistance seen by R_L looking back into the amplifier when an ideal voltage source is connected at the input.

7.60 Starting with Fig. 7.27a, use large-signal transistor models and the general approach of Figs. 7.27a–c to derive Fig. 7.27d.

7.61 Figure P7.61 is the large-signal equivalent of the compound pnp transistor of Fig. 7.28.
(a) Simplify the circuit to a battery and a single CCCS.
(b) Draw the small-signal equivalent of Fig. 7.28. Simplify the circuit to a resistor and a dependent current source.

7.62 In Fig. P7.62, $\beta_P = 10$ and $\beta_N = 80$.
(a) Find the Q-point of each transistor.
(b) Find v_o/v_i.
(c) Find the input resistance between the input node and ground.

7.63 The CMOS amplifier of Fig. 7.29c uses a 5 V power supply. Parameters are $k = 0.8 \times 10^{-4}$ A/V^2, $|V_t| = 0.5$ V, and $V_A = 100$V.
(a) Find I_D. Compute numerical values for g_m and r_o assuming $v_I = 0.5V_{DD}$.
(b) Assuming R is so large it can be treated as an open circuit, draw the small-signal equivalent circuit including r_o values. Using symbols, find an expression for voltage gain.
(c) Redraw the equivalent circuit, this time including R. For this circuit find an expression for voltage gain.
(d) From the gain expression of part (c), find an inequality involving R that, when satisfied, reduces the gain to the function of part (b). Using numerical values from part (a), find the minimum value of R for which the simpler gain expression is justified.

7.64 Replace the transistor in Fig. 7.31a with a BJT. Then use the small-signal equivalent to find the voltage gain in terms of the BJT's intrinsic gain.

7.65 Find the voltage gain of the active-load circuit of Fig. P7.65.

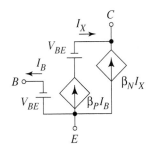

Figure P7.61

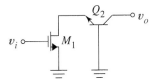

Figure P7.65

7.66 Suppose the current source used for biasing adds R_{out} between output node and ground in Figs. 7.32a and b. Use Eqs. (7.62) and (7.63) to find the voltage gain.
Hint: See Fig. 1.19.

7.67 Draw the small-signal equivalent circuit for Fig. 7.32e and write the analysis equations. Use them to verify Eq. (7.65).

7.68 (a) Draw the small-signal equivalent circuit of Fig. P7.68. Include transistor output resistances.
(b) Find v_o/v_i in terms of transistor parameters.

7.69 (a) Draw the small-signal schematic and small-signal equivalent for the common-drain, common-gate amplifier with infinite R_L.
(b) Derive amplifier voltage gain in terms of the intrinsic gains of the two transistors.
(c) Does this circuit offer an improvement in voltage gain over a single-transistor amplifier?

7.70 (a) Draw the small-signal schematic and small-signal equivalent for the common-drain, common-source amplifier with infinite R_L.
(b) Derive its voltage gain in terms of the intrinsic gains of the two transistors.
(c) Does this circuit offer an improvement in voltage gain over a single-transistor amplifier?

7.71 Figure P7.71 shows a high-gain circuit called a *double-cascode* amplifier.
(a) Find the voltage gain.
Hint: This is a cascode with a load circuit.
(b) Represent the MOSFET cascode amplifier by a VCVS and a series output resistor. Then use this model to find the voltage gain of the MOSFET double-cascode amplifier.

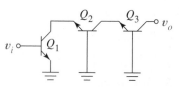

Figure P7.71

7.72 Use Table 7.1 to find the output resistance of the double-cascode amplifier of Fig. P7.71; then use approximations to simplify your answer.
Hint: The source resistance for Q_3 is the output resistance of Q_2; the source resistance for Q_2 is the output resistance of Q_1.

7.73 Use the procedure described in Problem 7.72 to find the output resistance of a double-cascode *MOSFET* amplifier.
Hint: The common-gate output resistance is given by Eq. (7.33); the common-source output resistance is r_o.

7.74 (a) Find the output resistance of Fig. 6.27b. Use reasonable approximations when appropriate.
(b) Find the numerical value for the output resistance if $R = 5$ kΩ, transistors are matched with $k = 4$ mA/V^2, $V_A = 100$ V, and $I_O = 2$ mA.

7.75 Analyze the circuit of Fig. P7.75 to show that source resistance R_S increases the output resistance over its original value of r_o much as it did for the similar BJT circuit in Fig. 7.34. Because of the FET's infinite input resistance, we obtain a single, unequivocal expression instead of different cases that depend on numerical values.

7.76 For circuits like Fig. 7.34a, the Gray and Meyer text listed among the references favors the output resistance equation

$$R_{out} \approx \left[1 + \frac{g_m R_E}{1 + (g_m R_E/\beta)} \right] r_o$$

Show that this expression follows from Eq. (7.68) if we assume only that $r_\pi \gg R_{TH}$.

7.77 Find the output resistance of the BJT cascode mirror of Fig. 6.29a.

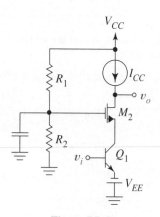

Figure P7.68

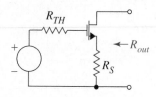

Figure P7.75

Section 7.9

7.78 In Fig. P7.78 V_{DD} = 12 V and R_D = 120 kΩ; k = 1.8 mA/V² and $V_t = 0.9$ V for all transistors. R is such that $I_{D3} = 0.1$ mA. With nodes a and b grounded,
(a) Find the Q-point of each transistor.
(b) Find the voltage across the current source.
(c) Find the common-mode limits.

7.79 Consider a cascade of two identical amplifiers, each like that of Fig. 7.36b, with gates of the second connected directly to the drains of the first. If $k = 4$ mA/V² and $V_t = 1$ V, determine whether the transistors of the second stage are biased in the active region.

7.80 (a) Draw an equivalent circuit like Fig. 7.41 for Fig. P7.78. The Early voltage of M_3 is 50 V. Problem 7.78 gives the other numerical values.
(b) Find CMRR in decibels when the output is single ended.

7.81 In Fig. P7.78 $V_{DD} = 15$ V and all transistors have $k = 1.8$ mA/V², $V_t = 0.9$ V. $V_A = 50$ V for M_3.
(a) Find I_{D3} so that the $CMRR_{(s-e)}$ is 40 dB.
(b) What is the required value of R?
(c) Select R_D so that the drop across R_D equals V_{DS} when both gates are grounded.
(d) Find the difference-mode gain of the amplifier when the output is double ended.

7.82 Find the largest small-signal, difference-mode *output* signal the amplifier of Example 7.14 can develop between its output nodes.
Hint: See Eq. (7.19).

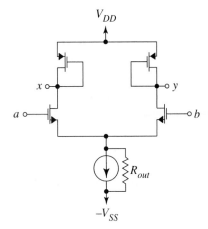

Figure P7.83

7.83 (a) For Fig. P7.83, draw a small-signal equivalent circuit like Fig. 7.41 that shows both difference- and common-mode effects. Include appropriate gain expressions.
(b) This circuit has low gain. Is it good for anything? Explain.

7.84 In Fig. 7.45a, change the supply voltages to ±5 V. Then
(a) Find the transistor Q-points and the current source voltage when bases are grounded.
(b) Find the common-mode limits assuming that the current source is a bipolar mirror circuit.

7.85 In Fig. P7.85 transistor parameters are β = 180, $V_{CE,sat} = 0.3$ V, $V_{BE} = 0.6$ V.
(a) Find the transistor Q-points and the current-source voltage when both bases are biased at +1 volt.

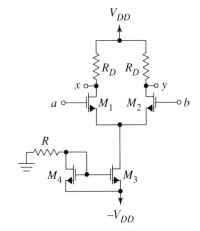

Figure P7.78

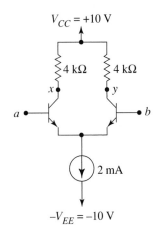

Figure P7.85

(b) Find the common-mode limits if the current source requires at least 1.2 V for proper operation.
(c) Design a basic BJT current source for this circuit using transistors that match those given. Draw the complete schematic including the current source.
(d) Do the common-mode limits change when your source replaces the one described in part (b)? Explain.
(e) Repeat part (a) after changing V_{CC} to +6 V and $-V_{EE}$ to −4 V.

7.86 In Fig. P7.85 the collectors are biased at 6 V. If the amplifier and current-source transistors require at least 0.2 V to avoid saturation, the emitters must remain between 5.8 and −9.8 V. These conditions leave transistors plus current source, together, a total *working voltage* of $5.8 − (−9.8) = 15.6$ V. Find the common-mode input bias voltage, V_{BB}, that divides this working voltage equally, that is, makes $V_{CE} = V_{CS}$, where V_{CS} is the bias voltage across the current source.

7.87 Find the transistor Q-points and the current-source voltage if a 1 MΩ resistor is connected between each input node and ground in Fig. P7.85 if $\beta = 180$.

7.88 Prove the existence of the virtual ground in Fig. 7.47b. That is, remove the dashed ground symbol and then show that $v_{\pi1} = 0.5v_d$ and $v_{\pi2} = −0.5v_d$.
Hint: Apply KCL at the emitter node. Then apply KVL to the input circuit.

7.89 A practical way to measure the difference-mode gain of a circuit is to measure v_o and v_s with one input grounded as in Fig. P7.89, and then to use

$$A_d \approx A_M = \frac{v_o}{v_s}$$

Work out the details of this theoretical justification:
(a) Find the difference- and common-mode components, v_d and v_c, of the input voltage in terms of v_s.
(b) Express v_o in terms of gains, A_d and A_c, and signals, v_d and v_c.
(c) Substitute results from parts (a) and (b) into the A_M expression.

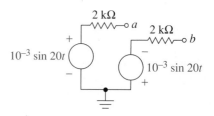

Figure P7.90

(d) The percentage error in this procedure is

$$\% \text{ error} = \frac{A_M - A_d}{A_d} \times 100$$

What error is involved if CMRR = 1000?
(e) What is the practical advantage of this method compared with applying a pure difference-mode signal?

7.90 For the difference amplifier of Fig. P7.85, $\beta = 180$.
(a) Find the CMRR in decibels if output is taken at node y. For the current source $R_{out} = 80$ kΩ.
(b) Represent this particular circuit in the form of Fig. 7.47e. Show numerical values.
(c) A 7 kΩ resistor is connected between nodes x and y. Find the signal voltage across this resistor if the input signal comes from the circuit of Fig. P7.90.
(d) Find the signal voltage developed across a grounded 3 k resistor capacitively coupled to node x if the input signal comes from Fig. P7.90.

7.91 Assume that in Fig. P7.85 the current-source output resistance is the resistance, r_o, of a bipolar transistor with $V_A = 90$ V. Compute the voltage that appears across a grounded 7 k resistor that is capacitively coupled to node x when the input is that of Fig. P7.91.

7.92 In Sec. 7.8.2 we studied small-signal schematics of several important two-transistor amplifiers without considering how they were biased. Fig. P7.92 is a practical realization of one of these amplifiers.
(a) Identify the amplifier by name.

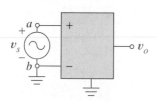

Figure P7.89

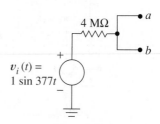

Figure P7.91

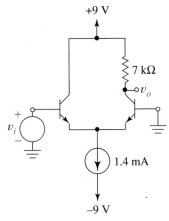

Figure P7.92

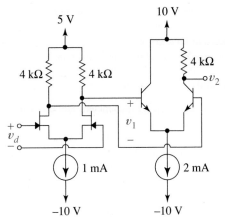

Figure P7.95

(b) Draw the small-signal equivalent showing numerical values of components if $\beta = 500$.

7.93 In Fig. P7.93, find input voltage v_d if the amplitude of the difference-mode output voltage is to be 2 V and output is
(a) $v_x - v_y$.
(b) v_y.
(c) the voltage across a 10 kΩ resistor placed between nodes x and y.

7.94 The output of the amplifier in Fig. P7.93 is to be taken at node x relative to ground.
(a) For CMRR of 70 dB, how large must be the output resistance of the current source?

(b) For the output resistance you found for part (a), is it possible to use a "basic" current mirror with an output transistor of $V_A = 90$ V? Explain.

7.95 In Fig. P7.95, JFET parameters are $\beta = 0.5$ mA/V^2 and $V_P = -2$ V; for the BJT, $\beta = 100$.
(a) Find dc values for all node voltages when both input nodes are grounded.
(b) Find small-signal gain v_1/v_d.
(c) Find small-signal gain v_2/v_d.

7.96 In Fig. P7.96 the current source mirrors a current caused by power supply ripple voltage, v_r, into the difference amplifier.

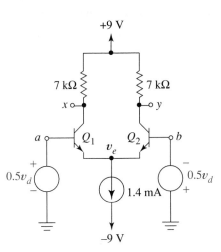

Figure P7.93

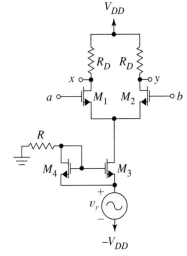

Figure P7.96

(a) Will this noise be a problem with double-ended outputs, single-ended outputs, both, or neither? Give an intuitive argument to explain your answer.

(b) Use a small-signal equivalent circuit to estimate the noise current in M_3 in terms of v_r. Ignore r_{O3}.

(c) Estimate the ripple voltage developed at node y.

7.97 In Fig. P7.97, JFET parameters are $\beta = 2$ mA/V^2 and $V_P = -1$ V.

(a) On a copy of the diagram, label the value of each dc node voltage and the dc current of each transistor when both inputs are grounded. Ignore base currents.

(b) Find the upper and lower common-mode input limits.

(c) Draw a small-signal equivalent of the entire circuit that is appropriate if a pure difference-mode signal is applied at the input. $\beta = 250$ for both second-stage transistors.

(d) Use the equivalent of part (c) to find v_o/v_d if v_d is the amplitude of a pure difference-mode input voltage.

7.98 The CMOS amplifier of Fig. 7.56b uses MOSFETs with $|V_t| = 1.2$ V, $k = 8$ mA/V^2, and $V_A = 100$ V. Supplies are ± 10 V and $I_O = 0.5$ mA. Substrates are connected in the usual fashion. Use SPICE for the following.

(a) Find the Q-point of each transistor with inputs grounded.

(b) Find the voltage gain when the input is a pure difference-mode signal.

(c) Rework part (b) after adding a body-effect coefficient of GAMMA = 0.30 to the code for the n-channel transistors.

7.99 Figure P7.99 is a simplified large-signal equivalent for the active-load amplifier of Fig. 7.56a when inputs are grounded. The circuit assumes that transistor output resistances cause collector currents to depart only slightly from $0.5I_O$.

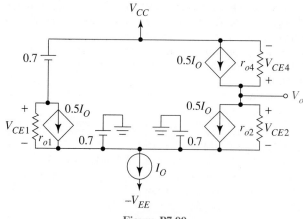

Figure P7.99

(a) Justify this estimate for the bias voltage of the output node

$$V_o = V_{CC} \frac{V_{A2}}{V_{A2} + V_{A4}} - 0.7 \frac{V_{A4}}{V_{A2} + V_{A4}}$$

where V_{Ak} is the Early voltage of Q_k.

Hint: Apply KCL at the output node to relate V_{CE2} and V_{CE4}. Find another equation to equate these same variables.

(b) Use the expression of part (a) to estimate V(3) for Example 7.20.

7.100 In Fig. P7.100, $V_{CC} = 10$ V, $R_C = 8$ kΩ, $\beta = 100$, and $I_O = 8$ μA.

(a) Find V_{CE} and I_C for the transistors when inputs are grounded.

(b) Find the upper common-mode limit if the current-source output voltage must be at least 3.3 V. Check both transistor saturation and current-source operation.

7.101 (a) Use half-amplifier analysis to find algebraic expressions for R_{in} and $A_{d(s-e)}$ for the common-base difference amplifier of Fig. P7.100 with difference-mode excitation.

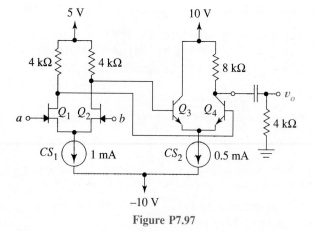

Figure P7.97

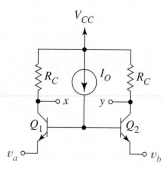

Figure P7.100

(b) Use half-amplifier analysis to find $A_{c(s-e)}$ for the amplifier of Fig. P7.100; R_{oo} is the output resistance of the current source.

(c) Find an expression for the total input resistance that the amplifier of Fig. P7.100 would present to a common-mode input signal.

(d) Combine difference- and common-mode results in the style of Fig. 7.50.

7.102 (a) Draw the small-signal equivalent of Fig. P7.102. Assume that a virtual ground exists at node 9. All pnp transistors have $\beta = 50$ and $V_A = 80$ V. npn Transistors have $\beta = 250$ and $V_A = 125$ V. Show numerical values of all circuit elements. (b) Identify or classify the amplifier, Q_3–Q_8 in terms of the subcircuits you recognize.

7.103 Use SPICE to find the difference- and common-mode gains of Fig. P7.102 using parameter values given in Problem 7.102.

7.104 Figure P7.104 is so closely related to the circuit of Fig. 7.56a that you can find its essential parameter values without doing any new circuit analysis.

(a) Find the dc collector currents of the input transistors.

(b) Find the input resistance if the input transistors have $\beta = 280$.

(c) Think about the functions of the bottom three transistors.

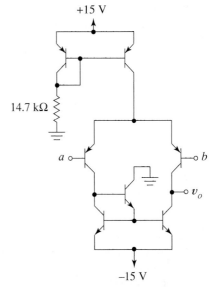

Figure P7.104

(*Hint*: Consider Fig. 6.26.) Then find the difference-mode gain if all Early voltages are 100 V.

7.105 Use SPICE to find the difference- and common-mode gains of the circuit shown in Fig. P7.104. Use parameter values from Problem 7.104.

7.106 Design a BJT difference amplifier with single-ended output that satisfies the following specifications:
1. Operates on ± 15 V power supplies
2. Has small-signal input resistance ≥ 500 kΩ
3. Uses as few resistors as possible
4. Uses no resistors of value > 100 kΩ
5. Has Early voltage of 120 V and $\beta = 200$ for all npn transistors
6. Has Early voltage of 75 V and $\beta = 50$ for all pnp transistors
For your final design, estimate the input resistance, output resistance, and difference-mode voltage gain.

Section 7.10

7.107 Use long division to verify the last step in Eq. (7.94).

7.108 Find the range of input offset voltages for an emitter-coupled pair that corresponds to three standard deviations if fractional variations in collector resistance and in saturation current have standard deviations of 2% and 6%, respectively.

7.109 Derive an expression for the input offset voltage of Fig. 7.54a using the following approach.

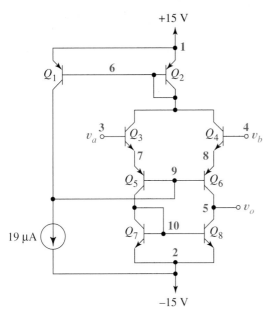

Figure P7.102

(a) Apply an input offset source, and express V_{IO} in terms of collector currents, saturation currents, and emitter resistances. Assume $I_C \approx I_E$.

(b) Use the offset definition to replace the collector current ratio by a resistor ratio, and to eliminate I_{C2} from the expression.

(c) Express saturation currents, resistor values, and I_{C1} by expressions that involve fractional variations.

(d) Use long division, discarding of small terms, and the natural logarithm approximation to reduce the expression to a simple form.

7.110 (a) Find the small-signal, difference-mode gain of the amplifier of Fig. P7.85 for double-ended output.

(b) Assume $\Delta R/R$ and $\Delta I_S/I_S$ are each three standard deviations away from the mean—that is, $\Delta R/R = 0.03$ and $\Delta I_S/I_S = 0.15$—so as to give a worst case value for the offset voltage. Calculate the offset voltage.

(c) Write SPICE code for Fig. P7.85. Make resistor and I_S values differ from the mean by three standard deviations as assumed in part (b). (Use the default value $I_S = 10^{-16}$ A for the mean.) With both input nodes grounded, use SPICE dc analysis to find the output voltage. Compare this with the input offset calculated in part (b) multiplied by the gain calculated in part (a).

7.111 An emitter-coupled pair is biased at $I_B = 0.5$ mA. Estimate a range for input offset current that statistically includes 96% of the devices if $\sigma_\beta = 0.1$ and $\sigma_R = 0.01$.

7.112 Find the design value for base current of an emitter-coupled pair if 96% of the devices are to have input offset currents of 25 μA or less. Assume $\sigma_\beta = 0.1$ and $\sigma_R = 0.01$.

7.113 (a) Write an expression for input offset voltage of Fig. P7.100 in terms of V_T, I_{C1}, I_{C2}, I_{S1}, and I_{S2}.

(b) Now make your offset expression a function of V_T, R_1, R_2, I_{S1}, and I_{S2}.

(c) Rewrite your expression in terms of V_T, $\Delta R/2R$, and $\Delta I_S/2I_S$.

(d) Simplify your part (c) expression to resemble the simple offset equations derived in the text.

CHAPTER

8

Frequency Response

In this chapter we explore the dynamic properties of small-signal electronic circuits by including capacitances in our analysis. Coupling and bypass capacitors cause gain to decrease at low frequencies; internal transistor capacitances cause gain to drop off at high frequencies. In analysis, a quantitative description of these gain limitations enables us to estimate rise time and sag of amplified pulses and predict amplitude and phase distortion of continuous time signals. In design, the ensuing equations equip us to deal with specifications related to dynamic measures of circuit performance. This chapter provides the level of detail required to better understand the general concepts of Secs. 1.6.4–1.6.9, and a review of these concepts is advisable at this time. As in previous chapters, we master simplified techniques of hand analysis to aid us in understanding and estimating while concurrently using simulations to complement our hand calculations with analyses of greater accuracy and precision.

8.1
The Wideband Amplifier

All amplifier gain functions—voltage, current, transresistance, and transconductance— change with signal frequency. For a *wideband amplifier* that includes coupling and/or bypass capacitors, the gain variation resembles Fig. 8.1. To distinguish the upper and lower regions, characterized by steep drops in gain, from the broad center region where gain is more or less constant, we use an arbitrary definition. A (dashed) line drawn 3 dB below the maximum gain value intersects the gain curve at two frequencies, ω_L and ω_H. These define the *low-, midrange-, and high-frequency regions* shown in the figure. At frequencies above ω_H, internal transistor capacitances and stray *wiring capacitance* reduce the gain; below ω_L, coupling and bypass capacitors cause the drop in gain. The response of a direct-coupled amplifier, which has no bypass or coupling capacitors, resembles Fig. 1.40b. The amplifier's bandwidth in radians per second is defined by $\omega_B = \omega_H - \omega_L \approx \omega_H$, where the approximation is usually justified because $\omega_L << \omega_H$.

Midrange-, Low-, and High-Frequency Equivalent Circuits. The circuit analysis of Chapter 7 ignored internal transistor capacitances, and treated coupling and bypass capacitors as short circuits. In the broader context of this chapter we now recognize that the static circuit models of Chapter 7 are only valid for the *midrange region* of the amplifier. Taking

589

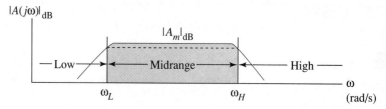

Figure 8.1 Definition of low-, midrange-, and high-frequency regions of a wideband amplifier.

small signals for granted, we now refer to the *small-signal equivalent circuits* of Chapter 7 as *midrange-equivalent circuits*. Because midrange equivalent circuits contain no capacitance, they predict (with some sacrifice in accuracy) constant midrange gain.

Low-frequency equivalent. To understand the *low-frequency region,* we use a special *low-frequency equivalent circuit.* This is constructed just like the small-signal equivalent of Chapter 7 except that coupling and bypass capacitors are included in both the circuit diagram and the equations. Since this low-frequency equivalent circuit ignores internal transistor capacitances, mathematical expressions obtained from this *low-frequency analysis* approach the midrange results in the limit as $s = j\omega \to \infty$ because in this limit the capacitors approach short circuits.

High-frequency equivalent. To understand the *high-frequency region,* we use a *high-frequency equivalent circuit.* Like the midrange equivalent, all coupling and bypass capacitors are represented by short circuits. The high-frequency equivalent includes internal capacitances of all transistors, as well as measured or estimated values of stray capacitance, when available. Because coupling and bypass capacitors are treated as short circuits, high-frequency analysis results reduce to midrange results in the limit as $s = j\omega \to 0$.

Purpose of Frequency Response Calculations. Use of three simplified equivalents rather than the complete circuit is an approximate procedure that leads to useful hand-analysis results while avoiding unnecessary complexity. Although justified only when there is a large separation between ω_H and ω_L, the technique applies to many important circuits, including audio and video amplifiers. SPICE simulations, which include *all* capacitances, give numerical frequency response curves of greater accuracy; however, unlike hand estimates, they provide no insight into the underlying physical reasons for a particular curve shape, nor do they suggest, of themselves, how a given design might be improved. By contrast, approximate hand analysis contributes to our general understanding and enables us to derive information that is useful in design. Specifications related to dynamic behavior include bandwidth, values of −3 dB frequencies, rise time, fractional sag, phase shift, and rate of *roll-off* of the frequency response asymptotes. Whenever we need numerical results to verify our work or help refine our decisions, they are readily available via simulation.

In the next section we introduce a powerful method for sketching the frequency response of any transfer function. This technique helps us understand responses of low-frequency and high-frequency equivalent circuits and gives us a general understanding of how these responses combine to form the complete frequency response. However, we almost never have occasion to apply the techniques to the complete transfer functions.

We instead rely on SPICE simulations for frequency response plots that include all the internal complexity of our circuits in the same analysis.

8.2

Analysis Tools for Dynamic Circuits

To generalize in our circuit analysis, we represent each capacitor by its complex impedance $1/sC$ and then formulate circuit equations in the usual fashion. In this notation, the expressions we obtain for gain, input impedance, and output impedance are ratios of polynomials in the complex frequency s, collectively termed *system functions*. For example, voltage gain, current gain, and input and output resistance generalize, respectively, to the polynomial ratios $A_v(s)$, $A_i(s)$, $Z_{in}(s)$, and $Z_{out}(s)$.

We frequently evaluate system functions for special values of s. To find the result of sinusoidal excitation at frequency ω, for example, we substitute $s = j\omega$. This reduces the polynomial ratio $A_v(s)$ to the complex number $A_v(j\omega)$ for each radian frequency ω. Plots of magnitude and phase of $A_v(j\omega)$ versus ω are the frequency response curves first introduced in Sec. 1.6.4. Two special values of s are useful for checking results. Since dc means zero frequency, taking the limit of a system function as $s \to 0$ gives the same end result as dc analysis, that is, of replacing capacitors by open circuits and then analyzing the circuit. Also, because $1/sC$ approaches zero as s approaches infinity, taking the limit of the system function as $s \to \infty$ gives the same result as replacing each capacitor with a short circuit before doing the circuit analysis. Recall that all the gain and input or output resistance expressions we studied in Chapter 7 assumed that coupling and bypass capacitors were large enough to treat as short circuits

Since system functions involve Laplace transform variables, we hereafter label small-signal quantities with uppercase characters whenever the context suggests system functions. Once we obtain a complex expression such as $A_v(j\omega)$, we often wish to sketch its magnitude as a function of ω. Bode plots, briefly reviewed next, give a quick and easy way to do this.

8.2.1 BODE PLOTS OF ELEMENTARY FUNCTIONS

 The *Bode plot* is an *estimate* of a frequency response curve in terms of decibel gain versus frequency. The simplicity of Bode plots arises from using asymptotic estimates to approximate certain elementary functions. We construct the plots of more complicated expressions by graphical addition of plots of *elementary functions*. A brief introduction to Bode plot sketching follows.

Amplitude Frequency Response. We can always write the gain expression of an amplifier in the form

$$A(s) = \frac{Ks^q(s + z_1)(s + z_2) \cdots (s \cdot z_M)}{(s + p_1)(s + p_2) \cdots (s + p_N)} \tag{8.1}$$

where K is a real constant; $-z_1, \ldots, -z_M$ are numbers called the *zeros* of $A(s)$, $-p_1, \ldots, -p_N$ are the *poles* of $A(s)$; and q is a positive or negative integer, or zero. For sinusoidal excitation at frequency ω, $A(s)$ becomes

$$A(j\omega) = \frac{K_p(j\omega)^q(j\omega/z_1 + 1)(j\omega/z_2 + 1) \cdots (j\omega/z_M + 1)}{(j\omega/p_1 + 1)(j\omega/p_2 + 1) \cdots (j\omega/p_N + 1)} \tag{8.2}$$

where K_P is the new constant formed after all p_is and z_is are factored out and combined with K. The amplifier's decibel gain is

$$|A(j\omega)|_{dB} = 20 \log|K_p| + 20\,q \log|j\omega| + \sum_{i=1}^{M} 20 \log \left| j\frac{\omega}{z_i} + 1 \right|$$
$$- \sum_{k=1}^{N} 20 \log \left| j\frac{\omega}{p_k} + 1 \right| \tag{8.3}$$

Each term of Eq. (8.3) is one of five elementary functions; Fig. 8.2 shows the Bode plot for each.

The first elementary function is the constant $20 \log|K_P|$, plotted in Fig. 8.2a. Each uncanceled factor s in the numerator of Eq. (8.1), called a *zero at the origin*, adds $20 \log |j\omega|$ and has a Bode plot like Fig. 8.2b—a straight line that passes through zero dB when $\omega = 1$ rad/s. An uncanceled s in the denominator, which occurs when q is negative (a *pole at the origin*), gives the elementary function $-20 \log |j\omega|$—a straight line of negative slope passing through $\omega = 1$—as in Fig. 8.2c.

In Bode plots, slopes are described in units of either dB/octave or dB/decade. An *octave* is a multiplicative factor of two and a *decade* is a factor of 10. Therefore, 400 Hz is three octaves above 50 Hz (that is, $400 = 50 \times 2^3$) and five octaves below 12,800 Hz ($400 = 12,800 \div 2^5$). Similarly, 3000 Hz is one decade above 300 Hz and three decades below 3 MHz. Since Figs. 8.2b and c are plots of $\pm 20 \log|j\omega|$, substituting $\omega = 1$ and $\omega = 10$ shows that the slopes are ± 20 dB/dec. Notice ± 20 dB/dec means that the gain

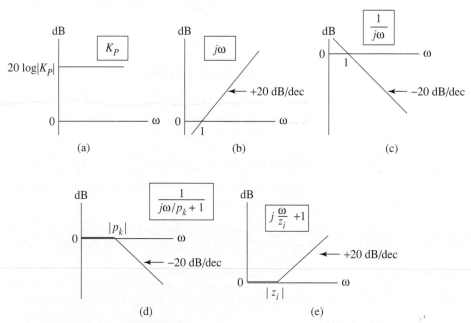

Figure 8.2 Bode plots of elementary functions: (a) constant greater than one; (b) zero at $s = 0$; (c) pole at $s = 0$; (d) pole at $s = p_k$; (e) zero at $s = z_i$.

changes by a factor of 10 (that is, ± 20 dB) as the frequency increases by a factor of 10. Likewise, substituting $\omega = 1$ and $\omega = 2$ shows that the slopes can equivalently be expressed as ± 6 dB/oc (gain doubles or halves when the frequency doubles).

Figure 8.2d is the Bode plot of a *pole term* in Eq. (8.3) of the form

$$-20 \log \left| j\frac{\omega}{p_k} + 1 \right|$$

When $\omega << |p_k|$, the function is zero dB; when $\omega >> |p_k|$,

$$-20 \log \left| j\frac{\omega}{p_k} + 1 \right| \rightarrow -20 \log \left| j\frac{\omega}{p_k} \right|$$

a straight line of -20 dB/decade slope that passes through $\omega = p_k$. The Bode plot for the term as a whole consists of these two asymptotes, each extended to the point of intersection. The original expression shows that at $\omega = p_k$, the true curve passes below the Bode approximation, with an error of $-20 \log |j1 + 1| = -20 \log \sqrt{2} = -3$ dB. The Bode plot of the zero term of Fig. 8.2e resembles the plot for the pole except that the curve *breaks upward* at 20 dB/decade. Frequencies such as z_i and p_k, where the Bode plot changes slope, are known as *break frequencies*.

Complete discussions of Bode plots also include a second-order elementary function associated with pairs of conjugate poles or zeros. Although important in electronics, second-order functions do not arise in the circuits of this chapter.

8.2.2 GRAPHICAL ADDITION OF ELEMENTARY BODE PLOTS

To sketch the amplitude frequency response curve, we first arrange the function in the form of Eq. (8.2) and then sketch the constant in decibels. Next, we graphically add to the original sketch the curve for any zeros or poles at the origin. Because this curve contributes zero dB to the original curve at $\omega = 1$ rad/s, the graphical addition is easy to do. Next we add to the existing plot the contribution of each zero and pole term, beginning with the lowest break frequency and proceeding, in order, to the highest. Each of these curves adds nothing (zero dB) until the break frequency. Above the break frequency, each either adds or subtracts from the existing curve at a rate of 20 dB/dec. It follows from the graphical addition process that poles or zeros of *mulitiplicity k* give slopes of $\pm 6k$ dB/oct ($20k$ dB/dec). Because frequency response curves typically involve large frequency intervals, Bode plots are often constructed on semilog paper.

EXAMPLE 8.1 Sketch the Bode magnitude plot for

$$A_v(s) = \frac{8 \times 10^{13} \, s(s + 20)}{(s + 30)(s + 400)(s + 10{,}000)(s + 100{,}000)}$$

Estimate the gain at 20 rad/s directly from the Bode plot, ignoring the Bode plot error.

Solution. Substituting $s = j\omega$ and factoring out pole and zero values gives

$$A_v(j\omega) = \frac{8 \times 10^{13} \times 20}{30 \times 400 \times 10^4 \times 10^5} \frac{j\omega(j\omega/20 + 1)}{(j\omega/30 + 1)(j\omega/400 + 1)(j\omega/10^4 + 1)(j\omega/10^5 + 1)}$$

where the new constant is $K_P = 133.3$ or 42.5 dB. The graphical sum of the constant and the zero at the origin, from Figs. 8.2a and b, looks like Fig. 8.3a. The next zero adds +20 dB/dec everywhere to the right of its 20 rad/s break frequency as in Fig. 8.3b. Each successive pole term leaves the curve unchanged to the left of its breakpoint, but subtracts from the slope to the right as shown in Figs. 8.3b–e. Figure 8.3 does not attempt to show correct frequency axis scaling, but only the construction procedure.

Sometimes we need to find the gain at some particular frequency such as 20 rad/s. Because 20 rad/s is one decade plus one octave above 1 rad/s, the gain is $42.5 + 20 + 6 = 68.5$ dB. (This estimate ignores the Bode error at the breakpoint.) ❏

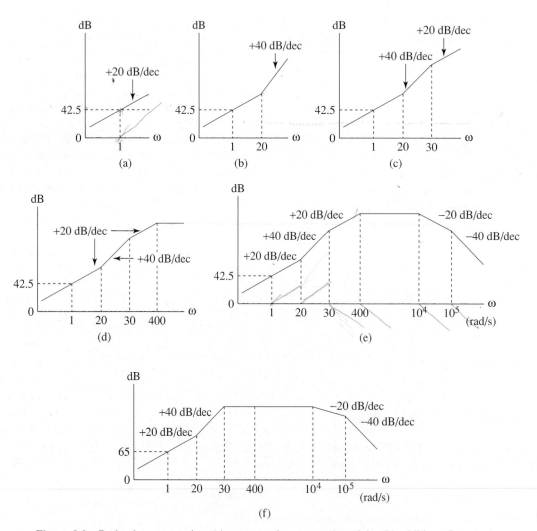

Figure 8.3 Bode plot construction: (a) constant plus zero at the origin; (b) addition of zero at 20 rad/s; (c) addition of pole at 30 rad/s; (d) addition of pole at 400 rad/s; (e) addition of poles at 10^4 and at 10^5 rad/s; (f) Bode plot for Exercise 8.1.

Exercise 8.1 In the function of Example 8.1, move the pole from 400 rad/s to 30 rad/s to create a double pole. Sketch the new Bode magnitude plot, and estimate the decibel gain at 8 rad/s.

Ans. Fig. 8.3f, 83 dB.

Bode Phase Curves. Each elementary function also has a simple phase curve. A positive or negative constant K_P contributes either zero or 180°. Zeros and poles at the origin, Figs. 8.2b and c, contribute constant phase shift of +90° and −90°, respectively, at all frequencies. A pole term like Fig. 8.2d adds a phase curve that changes smoothly from 0° to −90° with −45° phase shift at the break frequency. The straight-line approximation of Fig. 8.4, which extends one decade above and below the pole frequency, is a simple but useful estimate. The phase curve for a zero term like Fig. 8.2e changes smoothly from 0° to +90° with a +45° phase shift at the break frequency. Phase curves of elementary functions add graphically just like amplitude curves.

8.3 _____
Response to Low Frequencies

An amplifier's ability to process the low-frequency parts of a signal is limited by its bypass and coupling capacitors. Bypass capacitors, which increase gain by presenting low-impedance signal paths around certain biasing resistors, function less adequately at low frequencies, resulting in a reduction in gain. Coupling capacitors, which produce open circuits to dc bias currents, also impede slowly varying signal components. In analysis, a mathematical understanding of low-frequency response allows us to determine the lower limit for input signal frequencies for a given amplifier. In design, this knowledge allows us to select components so that our amplifier satisfies given low-frequency specifications. There are two different cases to distinguish: isolated poles and

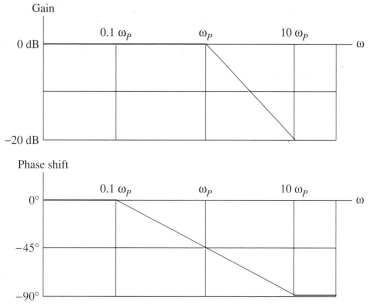

Figure 8.4 Approximate gain and phase shift contributed by a simple real pole.

interacting poles. For isolated-pole circuits we can write equations that directly relate Bode plot break frequencies to individual capacitors. For amplifiers with interacting poles, we usually settle for an estimate of the lower half-power frequency. We begin with some simple special cases that help us develop intuition and then advance to the more challenging circuits.

8.3.1 INDIVIDUAL EFFECTS OF BYPASS AND COUPLING CAPACITORS

Effect of a Bypass Capacitor. Figure 8.5a is a common-source amplifier, and Fig. 8.5b is its low-frequency equivalent circuit, where $R_P = R_1 \| R_2$ and $Z_s = R_S \| (1/sC_S)$. A current-source transformation gives Fig. 8.5c. To understand the effect of C_S, we first find the gain V_o/V_G, where V_G is the input to the amplifier proper. From Fig. 8.5c,

$$V_o = -g_m R_D V_{GS}$$

Controlling variable V_{GS} is related to node voltage V_G by the voltage divider expression

$$V_{GS} = \frac{1/g_m}{1/g_m + Z_S} V_G = \frac{1}{1 + g_m Z_S} V_G \qquad (8.4)$$

Substituting Eq. (8.4) into the previous equation leads to

$$A_v(s) = \frac{V_o}{V_G} = -g_m R_D \frac{1}{1 + g_m Z_S}$$

To show the role of C_S, we substitute

$$Z_S = R_S \| \frac{1}{sC_S} = \frac{R_S}{sC_S R_S + 1}$$

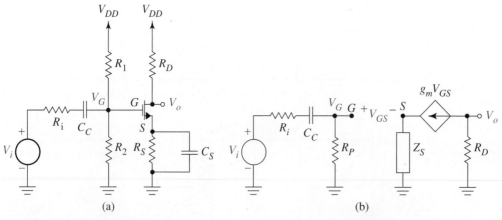

(a)   (b)

Figure 8.5 Effect of bypass capacitor on low-frequency response: (a) amplifier; (b) low-frequency equivalent circuit; (c) equivalent after source transformation; (d) effect of bypass capacitor alone; (e) effect of coupling capacitor alone; (f) addition of an external load resistor.

to obtain

$$A_v(s) = -g_m R_D \frac{1}{1 + g_m R_S/(sC_S R_S + 1)} = -g_m R_D \frac{sC_S R_S + 1}{1 + g_m R_S + sC_S R_S} \quad (8.5)$$

It is a good habit to check such a result at high and low frequencies to see if it agrees with our intuition. As $s \to \infty$, $A_V(s)$ approaches $-g_m R_D$, the gain of the common-source amplifier. This is what we expect, for it is the gain of Figs. 8.5a and b when C_S becomes

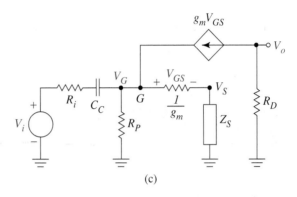

(c)

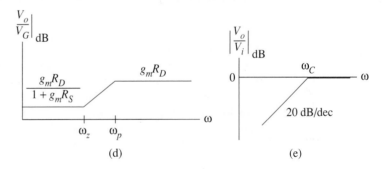

(d)   (e)

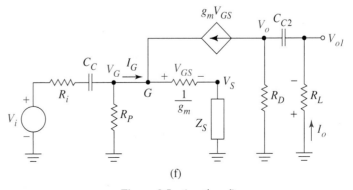

(f)

Figure 8.5   (continued)

a short circuit, as assumed in midrange analysis. When $s \to 0$, the gain approaches a lower value

$$A_v(0) = \frac{-g_m R_D}{1 + g_m R_S}$$

This is what we expect when a source resistor is left unbypassed (that is, $1/(sC_S) = \infty$); it is this reduced gain that we avoid by using the bypass capacitor.

To put Eq. (8.5) in the general form of Eq. (8.1), divide numerator and denominator by $C_S R_S$. This gives

$$\frac{V_o}{V_G} = A_v(s) = -g_m R_D \frac{s + \omega_z}{s + \omega_P} = A_{mid} \frac{s + \omega_z}{s + \omega_P} \tag{8.6}$$

where ω_Z and ω_P satisfy

$$\frac{1}{\omega_Z C_S} = R_S \tag{8.7a}$$

and

$$\frac{1}{\omega_P C_S} = \frac{R_S}{1 + g_m R_S} = \frac{(1/g_m) R_S}{(1/g_m) + R_S} = R_S \,\|\, \frac{1}{g_m} \tag{8.7b}$$

Since $R_S > R_S \,\|\, 1/g_m$, the zero frequency, ω_Z, is always lower than pole frequency ω_P.

The Bode plot of Eq. (8.6), Fig. 8.5d, shows a transition in gain from the midrange value to its "low-frequency value." We conclude that the net effect of the bypass capacitor is to add a pole and a zero to the midrange gain expression as in Eq. (8.6). The result is a Bode plot having the form of Fig. 8.5d.

In design, Eqs. (8.7) provide the information we need to *select* C_S for a prescribed lower limit to the passband; in analysis it allows us to compute this limit for a given amplifier. The equations for the break frequencies are easy to remember because they are so logical. The pole frequency is that frequency, ω_P, where the reactance of C_S equals the output resistance it sees looking back into the amplifier. The zero frequency occurs where the reactance of C_S equals R_S.

Effect of Coupling Capacitor. The overall gain V_o/V_i of the amplifier of Fig. 8.5a is actually the product of two factors: V_o/V_G, which we have already examined, and V_G/V_i, which involves the coupling capacitor. From Fig. 8.5b,

$$\frac{V_G}{V_i} = \frac{R_P}{R_P + R_i + 1/(sC_C)} = \frac{sC_C R_P}{sC_C (R_P + R_i) + 1}$$

After factoring, this becomes

$$\frac{V_G}{V_i} = \frac{R_P}{R_P + R_i} \frac{s}{s + \omega_C} \tag{8.8}$$

where ω_C satisfies

$$\frac{1}{\omega_C C_C} = R_P + R_i \tag{8.9}$$

Combining Eqs. (8.8) and (8.6) gives the overall gain

$$A_v'(s) = \frac{V_o}{V_i} = \frac{V_G}{V_i}\frac{V_o}{V_G} = \frac{s}{s + \omega_C}\left[\frac{R_P}{R_P + R_i}(-g_m R_D)\right]\frac{s + \omega_Z}{s + \omega_P} \qquad (8.10)$$

We recognize the term within brackets as the gain from V_i to V_o in Fig. 8.5b when both capacitors are short circuits, that is, the midrange gain. The first factor shows that the coupling capacitor contributes a zero at the origin and a pole at $-\omega_C$. Figure 8.5e is the Bode plot of this first factor alone. At frequencies significantly higher than ω_C, the gain is not affected by C_C; however, below ω_C there is a *roll-off* in gain of 20 dB/dec. Equation (8.9) shows that ω_C is that frequency where the reactance of C_C equals the resistance that C_C sees in the low-frequency, small-signal equivalent of 8.4b. A coupling capacitor always contributes an uncanceled numerator factor, s, causing $A(s) \to 0$ as $s \to 0$, consistent with our intuitive idea of series capacitors blocking dc.

8.3.2   MULTIPLE INDEPENDENT POLES

Shaping the Frequency Response Curve. In circuits where we can relate low-frequency poles and zeros to individual capacitors, it is easy to find the shape of the frequency response curve in *analysis,* or to structure the curve to fit given specifications in *design.* Consider the possibilities for the amplifier of Fig. 8.5a. By choosing C_C to be sufficiently large, one can make $\omega_C \ll \omega_Z$. Adding Fig. 8.5e and Fig. 8.5d then gives Fig. 8.6a. If instead we choose a smaller C_C so that ω_C lies between the pole and zero of the bypass capacitor, the frequency response resembles Fig. 8.6b. Still smaller C_C gives Fig. 8.6c, which has the same shape as Fig. 8.6b. Slopes in these curves are labeled in units of dB/decade, assuming the poles and zeros are sufficiently widely separated so that the true curve approximately follows its asymptotes.

Second Coupling Capacitor. If, in addition, we capacitively couple an external load, R_L, to the output node of Fig. 8.5a, Fig. 8.5c changes to Fig. 8.5f. This defines a new output voltage V_{o1}. Because the current source *isolates* the output RC circuit from the remainder of the amplifier, C_{C2} gives a new pole term that is *independent of the other poles and zeros.* Notice that the dependent current is controlled by V_{GS} of Eq. (8.4) for both Figs. 8.5c and 8.5f. In the latter circuit, the current divider at the output gives

$$I_o = \frac{R_D}{R_D + R_L + 1/(sC_{C2})}g_m V_{GS} = \frac{sC_{C2}R_D}{sC_{C2}(R_D + R_L) + 1}g_m V_{GS} \qquad (8.11)$$

To find the overall gain for Fig. 8.5f we use

$$\frac{V_{o1}}{V_i} = \frac{V_{o1}}{I_o}\frac{I_o}{V_{GS}}\frac{V_{GS}}{V_G}\frac{V_G}{V_i}$$

where $V_{o1}/I_o = -R_L$, and where the other factors are defined, respectively, in Eqs. (8.11), (8.4), and (8.8). Making the substitutions and simplifying shows that capacitively coupling R_L to the amplifier of Fig. 8.5a changes the gain function of Eq. (8.10) to

$$A_v'(s) = \frac{V_{o1}}{V_i} = \frac{s}{s + \omega_C}\left[\frac{R_P}{R_P + R_i}(-g_m R_D)\frac{R_L}{R_L + R_D}\right]\frac{s + \omega_Z}{s + \omega_P}\frac{s}{s + \omega_{C2}} \qquad (8.12)$$

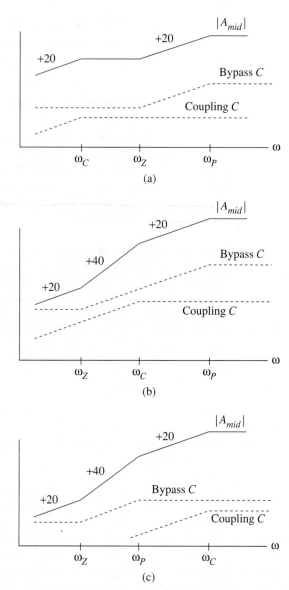

Figure 8.6 Frequency response curves for the amplifier of Fig. 8.5a: (a) $\omega_C << \omega_Z$; (b) $\omega_Z << \omega_C << \omega_P$; (c) $\omega_Z << \omega_C$.

The bracketed midrange gain term now includes the output loading factor caused by R_L. This gain term is the gain from ideal source to external load that we would obtain by assuming all capacitors are short circuits. There is also a new, standard form, coupling capacitor factor appended at the end, with a pole frequency, ω_{C2}, that satisfies

$$\frac{1}{\omega_{C2}C_{C2}} = R_D + R_L \qquad (8.13)$$

where C_{C2} is the output coupling capacitor. Just as for the input coupling capacitor, the break frequency is that frequency at which the reactance of C_{C2} equals the output resistance C_{C2} sees at its connecting nodes. In Fig. 8.6, the capacitively coupled load would add an additional 20 dB/dec to the slopes of all curves to the left of the new pole frequency ω_{C2}.

The shape of the frequency response curve at the low-frequency side is yet another factor to consider when designing the biasing circuit. In Eq. (8.12), drain resistor R_D appears, not only explicitly in the midrange gain, but also implicitly through Eq. (8.13), where a large value of R_D favors the low value of ω_{C2} usually desired. Large R_L also favors good low-frequency response. The biasing resistors that make up R_P are implicitly included in ω_C, as is the source resistance R_i. The large value for R_S associated with Q-point stability also favors good low-frequency response by leading to small values for ω_Z and ω_P.

Dominant Low-Frequency Pole. Sometimes there is one low-frequency pole that is larger than every other low-frequency pole and zero by two octaves or more. When this happens, we can use Bode plots to reason that for all signal frequencies of importance, the amplifier's frequency response resembles Fig. 1.41b, the response of the single-pole function of Eq. (1.43). This *dominant pole model* then approximates the lower half-power frequency of the amplifier and justifies using Eq. (1.44) to approximate fractional sag when the amplifier processes pulselike signals. The following design example shows how we use this dominant pole concept to select capacitor values.

EXAMPLE 8.2 The transistor in Fig. 8.7a is biased at 2 mA; parameters are $k = 0.5$ mA/V^2 and $V_t = 2$ V. (a) Choose coupling and bypass capacitors so that the lower half-power frequency is 100 Hz. (b) Sketch the Bode plot for your design, and verify it using SPICE. (c) Use SPICE to determine the percent sag in the response to a small-amplitude input pulse of 0.318 ms duration.

Solution. (a) Looking at Fig. 8.7a and visualizing the low-frequency equivalent circuit, we note that C_{C1} sees 5 k + 7.6 M$\|$2.72 M = 2.01 MΩ, and C_{C2} sees 50 k + 2.5 k = 52.5 k. At the Q-point, $1/g_m = 707.1$ Ω; therefore, C_S sees $R_{out} = (2\ k)\|(707) = 526$ Ω. Now we use the dominant pole idea. The capacitor that sees the lowest resistance is the best choice to establish the dominant pole. (This capacitor is destined to be large. Making its pole frequency lower would require it to be even larger.) We therefore place the pole of C_S at 100 Hz. According to Eq. (8.7b), this requires

$$C_S = \frac{1}{2\pi 100 \times 526} = 3.03\ \mu\text{F}$$

Also, the zero frequency contributed by C_S is the frequency f_z that satisfies

$$\frac{1}{(2\pi f_z)\ 3.03 \times 10^{-6}} = 2\ \text{k}\Omega$$

thus, $f_z = 26.3$ Hz, nearly two octaves below f_P.

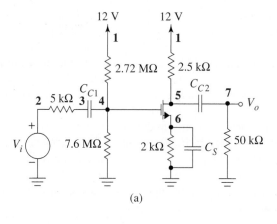

(a)

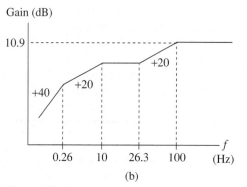

(b)

```
EXAMPLE 8.2
VDD 1 0 12
VI 2 0 AC 1
RIN 2 3 5K
CC1 3 4 0.303E-6
R1 1 4 2.72MEG
R2 4 0 7.6MEG
RD 1 5 2.5K
RS 6 0 2K
CS 6 0 3.03E-6
CC2 5 7 0.303E-6
RL 7 0 50K
M1 5 4 6 0 TRAN
.MODEL TRAN NMOS VTO=2
+ KP=0.5E-3 GAMMA=0.3
.AC DEC 20 0.02 2000
.PLOT AC VDB(7)
.PRINT AC VDB(7)
.END
```

(c)

Figure 8.7 Low-frequency design: (a) amplifier; (b) estimated Bode plot for design of Example 8.2; (c) SPICE code for frequency response; (d) Bode plot obtained using SPICE; (e) response to pulse input showing sag introduced by low-frequency limitations.

To ensure that ω_P is dominant, we place the other pole frequencies a decade or two below 100 Hz. For C_{C2}, Eq. (8.13) gives

$$\frac{1}{(2\pi\, 10)C_{C2}} = 52.5\ \text{k}$$

or $C_{C2} = 0.303\ \mu\text{F}$. Since C_{C1} faces a larger output resistance than C_{C2}, we use $C_{C1} = 0.303\ \mu\text{F}$. The break frequency of C_{C1} is then that frequency where its reactance equals 2.01 MΩ, or 0.262 Hz.

(b) Because input and output loading is small, the midrange gain should be approximately $-g_m R_D = -3.5$, or 10.9 dB. Figure 8.7b is the expected Bode plot.

Figure 8.7c is the SPICE code for computing the frequency response curve. Since signal generator VI has an amplitude of one volt, the output voltage at node 7 is numeri-

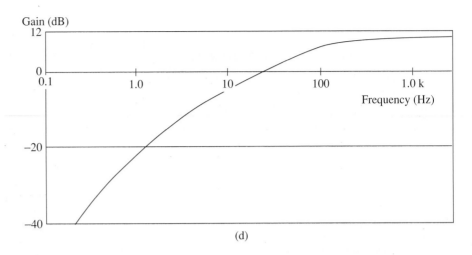

(d)

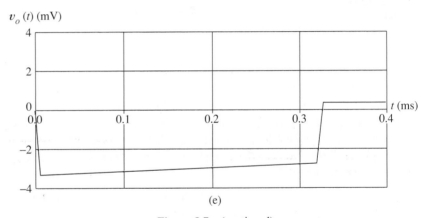

(e)

Figure 8.7 (continued)

cally equal to the amplifier gain. Because the source node of the FET is not grounded, the .MODEL line includes a body-effect factor, GAMMA = 0.3. The .AC line provides a logarithmic frequency sweep of V_i over the five-decade range, 0.02 to 2000 Hz, with 20 output points per decade. The .PLOT statement requests a plot of the voltage at node 7 (in decibel units to obtain dB gain).

Figure 8.7d is the SPICE frequency response plot. Output data show the (midrange) gain is 10.2 dB at 2000 Hz compared with 10.9 dB predicted in Fig. 8.7b. The lower gain valve might be the result of including body effect, an idea easy to test by a SPICE run with GAMMA = 0. At 100 Hz the gain is 2.98 dB below the 10.2 dB midrange value. The actual frequency response curve is smooth; however, its resemblance to the estimate of Fig. 8.7b is apparent.

(c) To find sag, we need to use a pulse source and do a transient analysis. We re-place the code for the ac signal source by

```
VI 2 0 PULSE (0 1M 0 0 0 0.318M 5M)
```

This new signal is a pulse that changes from zero to 1 mV (with zero delay time, rise time, and fall time) and lasts for 0.318 ms. (It also repeats itself every 5 ms, but this is incidental.) Unlike the unrealistically large signals allowed in SPICE ac analysis, the pulse amplitude in transient analysis must be small enough to avoid saturating the am-plifier; thus the 1 mV pulse amplitude.

When we replace

```
.AC DEC 20 0.02 2000
```

by

```
.TRAN 0.004M 0.4M
```

we specify a transient analysis, with output plotted every 0.004 ms over a computation interval of 0.4 ms. This gives 100 output points to define the waveform, and runs some-what beyond the end of the pulse.

Replacing

```
.PLOT AC VDB(7)
```

with

```
.PLOT TRAN V(7)
```

requests a plot of the output voltage as a function of time. The output waveform, Fig. 8.7e, resembles Fig. 1.41d; however, it is inverted because the amplifier has a negative midrange gain. The peak amplitude of the pulse is 3.236 mV agreeing exactly with the computed gain of 10.2 dB. The amplitude drops to 2.73 at the trailing edge of the pulse, giving a fractional sag of 15.7%.   ❏

> **Exercise 8.2**   Use the resemblance of this dominant pole design to a single-pole ampli-fier to estimate the percent sag.
>
> *Ans.* 20%.

EXAMPLE 8.3   Find the low-frequency function $V_o/V_i = A_v(s)$ for the common-emit-ter amplifier of Fig. 8.8a. Transistor parameters are $g_m = 0.056$ A/V and $r_\pi = 2411$ Ω.

Solution. The low-frequency equivalent is Fig. 8.8b. This consists of two separate subcircuits, each with its own capacitor. Each subcircuit contributes to the gain function a standard low-frequency factor of the form

$$\frac{s}{s + \omega_{C1}}$$

The input coupling capacitor has a pole frequency ω_{C1} that satisfies

$$\frac{1}{\omega_{C1}(50\ \mu F)} = 47\ k + (1M \| 2.41\ k) = 49.4\ k\Omega$$

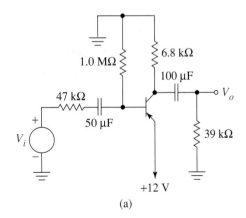

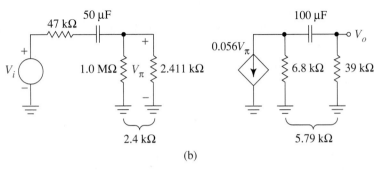

Figure 8.8 Circuits for Example 8.3: (a) original circuit; (b) low-frequency equivalent circuit.

Thus $\omega_{C1} = 0.405$ rad/s. The output capacitor contributes ω_{C2}, where

$$\frac{1}{\omega_{C2}(100 \ \mu F)} = 6.8 \ k + 39 \ k = 45.8 \ k\Omega$$

Therefore $\omega_{C2} = 0.218$ rad/s.
   The midrange gain of Fig. 8.8b is

$$A_{mid} = \frac{2.4 \ k}{2.4 \ k + 47 \ k}(-0.056 \times 5.79 \times 10^3) = -15.8$$

Putting this all together gives

$$A_v(s) = \frac{s}{s + 0.405} (-15.8) \frac{s}{s + 0.218} \qquad \square$$

> **Exercise 8.3**   We wish to use the amplifier of Fig. 8.8 for a signal containing only frequencies between 1 kHz and 100 kHz in an environment containing low-frequency noise. Find new coupling capacitors so that both low-frequency poles are located at 500 Hz.
>
> *Ans.* 6440 pF, 6950 pF.

Example 8.3 showed that low-frequency analysis of the basic common-emitter amplifier resembles that of the common-source amplifier. When there is a bypassed emitter resistor, however, the pole associated with the input capacitor is no longer isolated in a separate circuit but interacts with the bypass capacitor through r_π. We next learn how to estimate the low-frequency response in cases such as this.

8.3.3 ESTIMATING ω_L WHEN POLES AND ZEROS INTERACT

For many amplifiers, analysis of the low-frequency equivalent is much more difficult than in the preceding examples, and factoring the resulting gain expression does not give simple design equations. This is because some or all pole and zero frequencies are functions of two or more capacitors. In such cases, the best we can do is to turn to the approximate equation that we study next. This approximation does not provide us with sufficient information to sketch the Bode plot; however, it does at least allow us to estimate the lower half-power frequency, usually one important objective. Furthermore, the approximate method gives us an idea of the *relative* contributions of the individual capacitors to the lower half-power frequency, thus providing important design insight.

Method of Short-Circuit Time Constants. To estimate the lower half-power frequency ω_L of a low-frequency equivalent circuit containing N capacitors, we first compute the reciprocals

$$\omega_{Li} = \frac{1}{R_i C_i}$$

of N *short-circuit time constants,* where R_i is the output resistance seen by C_i when every other capacitor is replaced by a *short circuit.* We then estimate ω_L by

$$\omega_L \approx \sum_{i=1}^{N} \omega_{Li} \tag{8.14}$$

Equation (8.14) tacitly assumes that the gain function has a dominant low-frequency pole; however, this is a point of little practical concern. First, the method of short-circuit time constants is the only alternative at all amenable to hand analysis, so we have little choice except to assume its validity at the outset. More importantly, in design we can use the insight the technique provides to ensure that this condition is satisfied.

To demonstrate the details, we investigate the common-emitter amplifier of Fig. 8.9a and its low-frequency equivalent, Fig. 8.9b. The subcircuit containing C_2 is isolated by the current source, and so produces a simple low-frequency pole and a zero at the origin; however, the remaining two poles are both functions of C_1 and C_3 because the capacitors are not in isolated subcircuits. Therefore, we use Eq. (8.14) to estimate ω_L.

To find the resistance seen by C_1, draw Fig. 8.9c. The dependent source is active when a test generator is applied at the C_1 terminals; however, this source has no effect on R_1. Thus

$$R_1 = R_i + (R_P \| r_\pi)$$

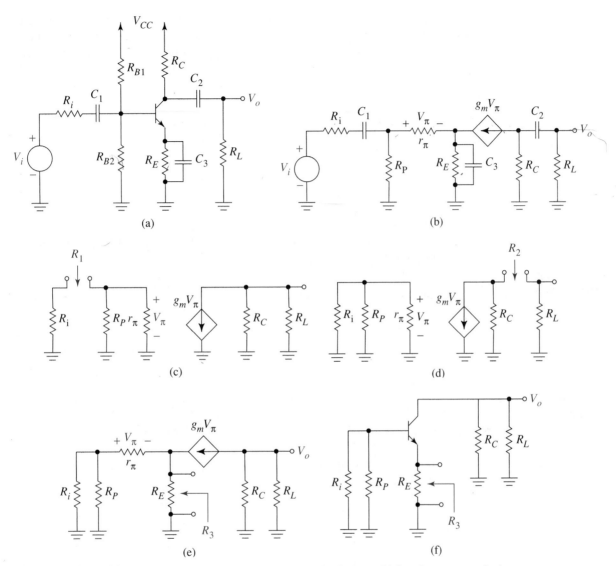

Figure 8.9 Common-emitter amplifier: (a) schematic diagram; (b) low-frequency equivalent; (c) circuit for finding R_1; (d) circuit for R_2; (e) circuit for R_3; (f) small-signal schematic for finding R_3 by inspection.

The resistance seen by C_2 comes from Fig. 8.9d. Since the definition of R_{out} requires the independent source to be turned off, the input circuit lacks excitation; thus $V_\pi = 0$, and the current source also turns off. The result is

$$R_2 = R_L + R_C$$

To find the resistance seen by C_3 we use Fig. 8.9e. At this point, most persons with experience in electronics would notice that Fig. 8.9e is the small-signal equivalent of Fig. 8.9f. Using the resistance transforming property of the BJT, we write

$$R_3 = R_E \| \left(\frac{(R_i \| R_P) + r_\pi}{\beta + 1} \right)$$

the resistance we need for the final short-circuit time constant.

By relating ω_L to individual capacitors, albeit loosely, the method of short-circuit time constants gives us insight that proves useful in design. The following exercise and example demonstrate this point.

Exercise 8.4   In Fig. 8.9a, transistor parameters are g_m = 4.72 mA/V and r_π = 21.1 kΩ. Circuit parameters are R_C = 39 kΩ, R_{B1} = 47 kΩ, R_{B2} = 68 kΩ, R_i = 60 Ω, $R_L = R_E$ = 50 kΩ, $C_1 = C_2 = C_3$ = 2 μF. The lower half-power frequency must be 50 Hz or less. Does the amplifier satisfy this specification?
Hint: Use Eq. (8.14) and the R_i expressions just derived.

Ans. No; $f_L = \omega_L / 2\pi$ = 386 Hz.

EXAMPLE 8.4   Redesign the circuit of Exercise 8.4 to bring the lower half-power frequency within the specification.

Solution. In solving Exercise 8.4, Eq. (8.14) gave

$$\omega_L \approx \omega_{L1} + \omega_{L2} + \omega_{L3} = 41.3 + 5.62 + 2381 = 2428 \text{ rad/s}$$

Notice how ω_{L3}, the term associated with the emitter bypass capacitor, dominates the expression. It makes sense to concentrate on this capacitor in our redesign. For example, if we change only C_3, we need

$$\omega_L \approx 41.3 + 5.62 + \omega_{L3} = 2\pi 50 = 314$$

which gives ω_{L3} = 267 rad/s. From the detailed analysis, R_3 = 209 Ω, therefore,

$$C_3 = \frac{1}{209 \times 267} = 17.9 \ \mu\text{F}$$

A computer simulation of the final design would be a prudent action at this point because we have not confirmed the existence of the dominant pole required for our estimate to be valid. (We only identified a term that dominated an approximation, which is quite a different thing.) In Problem 8.21 we find that the response at 50 Hz is indeed about 3.3 dB below the midrange value.

Notice that an accurate computer plot of the original frequency response would have revealed that the original design was unsatisfactory, but would have given no clue as to how to correct the problem. By contrast, our rough estimate of ω_L suggested how to *improve* the design, illustrating the conceptual power of Eq. (8.14).   ❏

From our intuitive understanding of coupling and bypass capacitors, we suspect that the true gain function in the preceding example has two zeros at the origin, a zero away from the origin and three poles; however, the *specific pole and zero locations are unknown*. Any temptation to backslide into strictly associating the individual pole frequencies of the gain function with the individual terms of Eq. (8.14) must be strongly resisted.

8.4 _____
Response to High Frequencies

In this section we show how a transistor's internal parasitic capacitances affect high-frequency response. The principles we learn here allow us to predict, for given amplifiers, practical upper limits for signal frequencies. Furthermore, they enable us to design amplifiers to meet given high-frequency specifications. There are three important cases to distinguish in high-frequency analysis. For isolated pole circuits, such as common-gate and simplified common-base amplifiers, we can write equations that directly relate Bode plot break frequencies to individual capacitors. Among the interacting pole amplifiers there are some important circuits, common-source and common-emitter amplifiers, for example, for which we can derive isolated-pole equivalents. For interacting pole circuits that have no isolated pole equivalents, we use a special expression to estimate the upper half-power frequency. Our first order of business is to introduce high-frequency transistor models.

8.4.1 HIGH-FREQUENCY TRANSISTOR MODELS

In Chapters 5 and 6 we learned that bipolar and field-effect transistors, like diodes, contain nonlinear parasitic capacitances that limit their response speeds when signals change rapidly. The high-frequency models of this section are small-signal, linearized versions of the general large-signal, nonlinear transistor models. As such, they are functions of Q-point and are valid only when signals have small amplitudes. These models allow us to understand high-frequency operation of linear electronic circuits and to make hand estimates for purposes of analysis and design.

FET Model. The model of Fig. 8.10a describes p- and n-channel JFETs and MOSFETs at high frequencies. The capacitances are only a few pF in magnitude; and measured or estimated values are often available for hand analysis. In the MOSFET, C_{gs} and C_{gd} represent the distributed capacitance spanning the oxide between gate and channel. In the JFET, they represent the depletion capacitances of Fig. 5.53. We generally ignore drain-to-body and source-to-body capacitances in hand calculations, but can easily include them in simulations to improve accuracy. When appropriate, we augment the model by adding output resistance r_o in parallel with the dependent source. At midrange and low frequencies, the capacitors act as open circuits, and the circuit reduces to the familiar midrange form of the previous chapter.

BJT Model. Figure 8.10b is the *hybrid pi* model that describes the bipolar transistor at high frequencies. Resistor r_b is the base-spreading resistance shown in Fig. 4.35. C_μ is the small-signal or incremental capacitance of the reverse-biased base-collector junction, defined as dQ_{dep}/dv_{BC} in Sec. 3.12.2. From Eq. (3.62), the value of C_μ depends upon the Q-point voltage $v_D = V_{BC}$. Capacitor C_π includes both depletion and diffusion capacitance associated with the forward-biased base-emitter junction. When required for a particular analysis, the model also includes r_o between collector and emitter nodes.

At low frequencies the hybrid pi model *almost* reduces to Fig. 7.5b. The difference is r_b, which was really present all along. Because r_b is only 10–100 Ω, critical only in

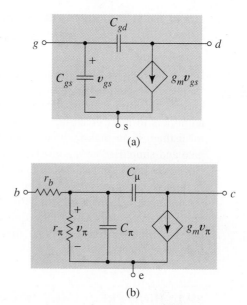

Figure 8.10   High-frequency transistor models: (a) JFET and MOSFET; (b) hybrid pi BJT model.

high-frequency analysis, and hard to measure; we customarily omit it except in high-frequency and noise analyses.

The close similarity of Figs. 8.10a and b is apparent. If both are driven by a source with internal resistance, the BJT resistors combine with the source resistance in a Thevenin transformation, giving circuits of identical form. Consequently, BJT and FET circuits are generally similar in high-frequency performance.

The BJT gives satisfactory operation over a wide range of frequencies starting with dc; however, at a certain frequency, f_β, called the *beta cut-off frequency,* transistor current gain begins to deteriorate. At its *unity-gain frequency* f_τ, current gain is reduced to unity, severely limiting its usefulness as an amplifying device. The following analysis defines and relates these important terms.

Unity-Gain Frequency. We use Fig. 8.11a, the hybrid pi model with output short-circuited, to explore inherent transistor current gain as a function of frequency. The short circuit ensures that analysis reflects the limitations of the device itself, independent of components external to the transistor.

First, notice the constraints introduced by the short circuit. No current flows in r_o because its voltage is zero. Also, because the collector node voltage is zero, the same V_π that controls the dependent source is also the voltage across C_μ. With sinusoidal excitation at frequency ω, Kirchhoff's current law gives

$$I_B = V_\pi \left[\frac{1}{r_\pi} + j\omega C_\pi + j\omega C_\mu \right]$$

Applying KCL at the collector node gives

$$I_C = (g_m - j\omega C_\mu) V_\pi$$

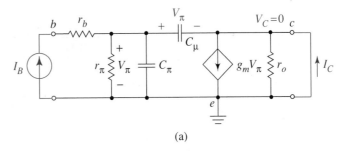

(a)

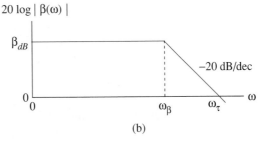

(b)

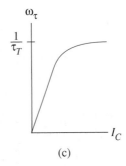

(c)

Figure 8.11 BJT high-frequency limitations: (a) circuit for computing high-frequency limitations in current gain; (b) current gain versus radian frequency; (c) unity-gain frequency versus Q-point coordinate.

The ratio of I_C to I_B gives the short-circuit current gain of the transistor as a function of ω,

$$\frac{I_C}{I_B} = \frac{g_m - j\omega C_\mu}{1/r_\pi + j\omega(C_\pi + C_\mu)}$$

To simplify, we now limit any further discussions to frequencies for which the numerator is real, that is, values of ω such that $|j\omega C_\mu| \ll g_m$ or

$$\omega \ll \frac{1}{C_\mu(1/g_m)}$$

Since $1/g_m$ and C_μ are both small, this is not very restrictive. Now we can write

$$\frac{I_C}{I_B} = \frac{g_m}{1/r_\pi + j\omega(C_\pi + C_\mu)} = \frac{\beta}{1 + j(\omega/\omega_\beta)} = \beta(\omega) \qquad (8.15)$$

where

$$\omega_\beta = \frac{1}{r_\pi(C_\pi + C_\mu)} \tag{8.16}$$

is the *beta cut-off frequency.*

Equation (8.15) generalizes our concept of transistor beta. Here we see transistor current gain as a complex function, $\beta(\omega)$, that reduces to the familiar real constant β at low frequencies, the only frequencies we have studied until now. The Bode plot of Fig. 8.11b shows what we mean by low frequencies. At ω_β the current gain is 3 dB below its dc value; for frequencies lower than this, $\beta(\omega) \approx \beta$, a real constant. At a frequency called the *unity-gain frequency,* ω_τ, $|\beta(\omega)|$ is reduced to one, or zero dB. From Eq. (8.15), by definition of ω_τ,

$$|\beta(\omega_\tau)| = 1 = \frac{\beta}{|1 + j(\omega_\tau/\omega_\beta)|} \approx \frac{\omega_\beta\beta}{\omega_\tau}$$

This relates the unity-gain and beta cut-off frequencies by

$$\omega_\tau = \beta \, \omega_\beta$$

Transistor data sheets usually give $f_\tau = \omega_\tau/2\pi$. To relate ω_τ to the circuit parameters, we substitute Eq. (8.16) to obtain

$$\omega_\tau = \beta\omega_\beta = \frac{\beta}{r_\pi(C_\pi + C_\mu)} = \frac{g_m}{C_\pi + C_\mu} \tag{8.17a}$$

or equivalently

$$f_\tau = \frac{g_m}{2\pi(C_\pi + C_\mu)} \tag{8.17b}$$

Unity-gain frequency is obviously a function of Q-point through its dependence on g_m; however C_π also depends subtly upon Q-point. This capacitance has the form

$$C_\pi = C_b + C_{je}$$

where C_b is the small-signal diffusion capacitance, and C_{je} the depletion capacitance, of the base-emitter junction. The large-signal diffusion capacitance of the active transistor is described by the Q_{FA} versus v_{BE} expression of Eq. (4.22). To obtain C_b, we differentiate Eq. (4.22) with respect to v_{BE}, and evaluate this derivative at the Q-point. Since Q_{FA} is an exponential function of V_{BE}, this gives

$$C_b = \frac{1}{V_T} \, Q_{FA}$$

Using Eq. (4.23) we obtain

$$C_b = \frac{\tau_T}{V_T} \, I_C \tag{8.18}$$

where τ_T is the transit time, the average time required for a minority carrier to drift across the base. Consequently, we can write

$$C_\pi = \frac{\tau_T}{V_T} I_C + C_{je} \tag{8.19}$$

Substituting this into Eq. (8.17a) and expressing g_m in terms of I_C gives

$$\omega_\tau = \frac{I_C/V_T}{(\tau_T/V_T) I_C + C_{je} + C_\mu} \tag{8.20}$$

For low I_C, depletion capacitances dominate the denominator and ω_τ increases in proportion to the I_C in the numerator. For higher bias current, diffusion capacitance becomes more important, causing ω_τ to approach $1/\tau_T$ asymptotically as in Fig. 8.11c.

We learned in Sec. 4.9.6 that $\tau_T \approx 1$ ns for an integrated npn transistor and 30 nsec for a lateral pnp device. Reducing base width shortens τ_T and thereby improves the fundamental high-frequency limits of the transistor. A second message from Fig. 8.11c is that here is yet another consideration in selecting a transistor Q-point. Higher I_C gives higher ω_τ.

8.4.2 COMMON-GATE AND COMMON-BASE AMPLIFIERS

The first amplifiers we study are characterized by isolated poles and straightforward design equations.

Common-Gate Amplifier. Consider the common-gate amplifier of Fig. 8.12a. We draw the high-frequency equivalent by turning off biasing sources, replacing coupling and bypass capacitors by short circuits, and replacing the transistor by its high-frequency model. The result is Fig. 8.12b where R_{LL} denotes

$$R_{LL} = R_L \| R_D \tag{8.21}$$

A current-source transformation gives Fig. 8.12c, which is easier to analyze because input and output circuits are *decoupled*. In this equivalent,

$$R_{SS} = R_S \| \frac{1}{g_m} \tag{8.22}$$

At the output node

$$V_o = -g_m V_{GS} Z_{LL} = -g_m \left(\frac{R_{LL}}{s C_{gd} R_{LL} + 1} \right) V_{GS} \tag{8.23}$$

Since V_{GS} relates to V_i through a voltage divider,

$$V_{GS} = -V_i \frac{Z_{SS}}{R_i + Z_{SS}} \tag{8.24}$$

where

$$Z_{SS} = R_{SS} \| \frac{1}{s C_{gs}} = \frac{R_{SS}}{s C_{gs} R_{SS} + 1}$$

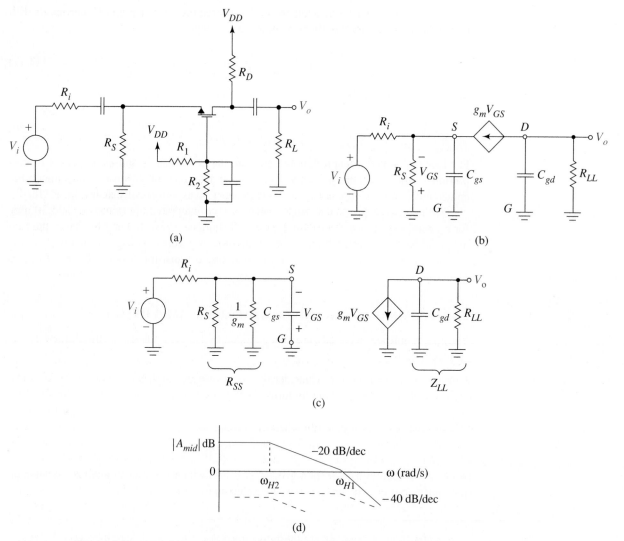

Figure 8.12 Common-gate amplifier high-frequency analysis: (a) amplifier schematic; (b) high-frequency equivalent; (c) equivalent after source transformation; (d) Bode plot and sketches of constituent curves.

Substituting Eq. (8.24) into Eq. (8.23) and solving for the voltage ratio gives

$$\frac{V_o}{V_i} = \frac{Z_{SS}}{R_i + Z_{SS}} g_m \frac{R_{LL}}{sC_{gd}R_{LL} + 1}$$

Now we substitute for Z_{SS} to obtain

$$\frac{V_o}{V_i} = \frac{R_{SS}}{sC_{gs}R_{SS}R_i + R_i + R_{SS}} g_m \frac{R_{LL}}{sC_{gd}R_{LL} + 1}$$

To obtain a form that lends itself to intuitive understanding, we divide numerator and denominator of the first factor by $R_i + R_{SS}$, rearrange, and regroup terms. This gives

$$\frac{V_o}{V_i} = \left(\frac{R_{SS}}{R_i + R_{SS}} g_m R_{LL} \right) \frac{1}{sC_{gs}[R_{SS}R_i/(R_i + R_{SS})] + 1} \frac{1}{sC_{gd}R_{LL} + 1}$$

$$= A_{mid} \frac{1}{s/\omega_{H1} + 1} \frac{1}{s/\omega_{H2} + 1} = A_{mid} \frac{\omega_{H1}}{s + \omega_{H1}} \frac{\omega_{H2}}{s + \omega_{H2}} \qquad (8.25)$$

Once we notice the logic of Eq. (8.25), we can easily reproduce the equation without repeating the derivation. First, notice that as $s \to 0$, the gain approaches the value enclosed in large parentheses. This is midrange gain, A_{mid}, which we can easily find from Fig. 8.12c by imagining the capacitors to be open circuits. Each capacitor adds its individual high-frequency pole term to the midrange expression; and the pole terms are identical in form, each approaching one as $s \to 0$. C_{gs} adds a high-frequency pole at $-\omega_{H1}$. From Eq. (8.22), we recognize ω_{H1} as that frequency where

$$\frac{1}{\omega_{H1}C_{gs}} = \frac{1}{g_m} \| R_S \| R_i$$

that is, the frequency where the reactance of C_{gs} equals the output resistance it sees in Fig. 8.12c. The contribution of C_{gd} is a pole frequency ω_{H2}, the frequency where the reactance of C_{gd} equals $R_{LL} = R_L \| R_D$, the output resistance C_{gd} sees in Fig. 8.12c. Figure 8.12d is the Bode plot, when $\omega_{H1} \gg \omega_{H2}$. Because the circuit structure of Fig. 8.12c appears often in our further studies, the memory aids for finding Eq. (8.25) directly from the circuit diagram merit special attention.

Since we can associate a specific algebraic expression with each pole frequency in this circuit, designing for a given upper half-power frequency is conceptually straightforward. Practically speaking, however, we are not free to select arbitrary values for capacitances as in low-frequency design. Transistor parameters, including internal capacitances, are functions of Q-point, and the external resistors trade off against other considerations such as gain, signal amplitude, and circuit noise. The common-gate circuit has excellent *potential* as a high-frequency amplifier because the pole frequency $\omega_{H1} \approx g_m/C_{gs}$ is often large. To achieve high bandwidth, R_D and R_L must be sufficiently small but still provide adequate midrange gain. Assuming ω_{H1} is so high that the bandwidth is determined by ω_{H2}, the gain-bandwidth product is

$$A_{mid}\omega_{H2} = \left(\frac{R_{SS}}{R_i + R_{SS}} g_m R_{LL} \right) \frac{1}{C_{gd}R_{LL}} \qquad (8.26)$$

We see from this expression that choosing R_{LL} involves a direct trade-off between gain and bandwidth, as the product is invariant with changes in R_{LL}. Large R_{LL} gives relatively high gain and low bandwidth, and conversely.

Common-Base Amplifier. Except for notational conventions, the common-gate diagram of Fig. 8.12a becomes the diagram of a common-base amplifier if the FET is replaced by a BJT. If we assume $r_b = 0$, the common-base equivalent circuit results from replacing C_{gs} and C_{gd} by C_π and C_μ, respectively, R_S by $R_E \| r_\pi$, and R_{LL} by $R_C \| R_L$ in

Fig. 8.12b. Controlling variable, V_π, for the BJT even has the same location and polarity as controlling variable V_{GS} for the FET. Thus, except for notational changes, the *common-base amplifier* has high-frequency gain described by Eq. (8.25) when $r_b = 0$. The dependent-source transformation leading to Fig. 8.12c places resistor $1/g_m$ in parallel with the input capacitor, resulting in a pole at $-\omega_{H1}$. Assuming $1/g_m$ approximates the resistance seen by C_π, we have the approximation

$$\omega_{H1} \approx \frac{g_m}{C_\pi}$$

Equation (8.17) shows that this pole frequency is comparable to the transistor's unity-gain frequency. The following example explores these conclusions as well as the error involved in ignoring r_b.

EXAMPLE 8.5   Find the complex gain function for the common-base amplifier of Fig. 8.13a when $r_b = 0$. The transistor is biased at $I_C = 0.118$ mA; the other parameters are $C_\pi = 50$ pF, $C_\mu = 5$ pF, and $\beta = 100$. Use SPICE to explore the error in ignoring r_b, if $r_b = 200$ Ω.

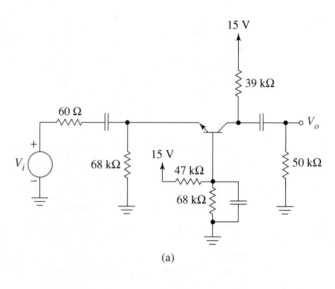

(a)

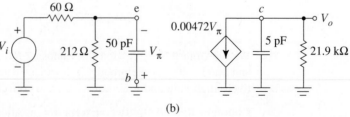

(b)

Figure 8.13 Common-base amplifier: (a) amplifier schematic; (b) high-frequency equivalent; (c) SPICE model of amplifier showing C_π, C_μ, and r_b as external components; (d) SPICE code for the amplifier; (e) frequency response with and without r_b.

Solution. At the given Q-point, $g_m = 0.118/25 = 4.72 \times 10^{-3}$ A/V. From this, $1/g_m = 212\ \Omega$ and $r_\pi = 21.2\ \text{k}\Omega$. The high-frequency equivalent circuit, after transforming the dependent source and combining parallel resistors, is Fig. 8.13b. Because this circuit is structurally identical to Fig. 8.12c, its gain function is Eq. (8.25); and we can use our memory aids to determine A_{mid}, ω_{H1}, and ω_{H2} directly.

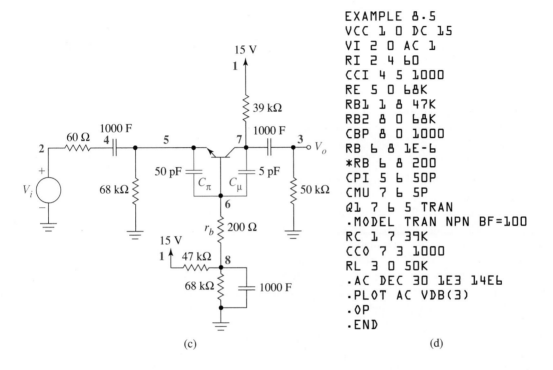

(c)

```
EXAMPLE 8.5
VCC 1 0 DC 15
VI 2 0 AC 1
RI 2 4 60
CCI 4 5 1000
RE 5 0 68K
RB1 1 8 47K
RB2 8 0 68K
CBP 8 0 1000
RB 6 8 1E-6
*RB 6 8 200
CPI 5 6 50P
CMU 7 6 5P
Q1 7 6 5 TRAN
.MODEL TRAN NPN BF=100
RC 1 7 39K
CCO 7 3 1000
RL 3 0 50K
.AC DEC 30 1E3 14E6
.PLOT AC VDB(3)
.OP
.END
```

(d)

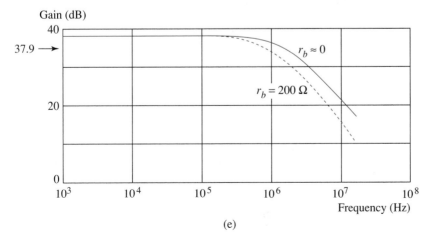

(e)

Figure 8.13 (continued)

With the capacitors open circuits, the midrange gain is

$$A_{mid} = \frac{212}{212 + 60}(0.00472 \times 21.9 \times 10^3) = 80.6$$

The pole contributed by C_π gives a break at frequency, ω_{H1}, where C_π's reactance equals $212 \| 60 = 46.8 \ \Omega$, or

$$\omega_{H1} = \frac{1}{(50 \times 10^{-12})(46.8)} = 4.27 \times 10^8 \ \text{rad/s}$$

The other pole frequency is

$$\omega_{H2} = \frac{1}{(5 \times 10^{-12})(21.9 \times 10^3)} = 9.13 \times 10^6 \ \text{rad/s}$$

Combining the gain and pole information gives

$$\frac{V_o}{V_i} = 80.6 \ \frac{1}{(s/4.28 \times 10^8) + 1} \ \frac{1}{(s/9.13 \times 10^6) + 1}$$

Because the lowest break frequency is 9.13×10^6 rad/s, the bandwidth is approximately $9.13 \times 10^6/2\pi = 1.45$ MHz.

The simulation is interesting because the small-signal capacitances we use in hand analysis, C_π and C_μ, are *not SPICE parameters*. If we knew values for the SPICE parameters, CJC, CJE, and TF, SPICE would automatically compute C_π and C_μ at the given Q-point before doing an ac analysis. In this problem, because we wish to explore the predictions of hand analysis, these parameters are unknown. To verify hand-analysis results, we can use a *static* transistor model, and augment it with *external* linear capacitors called C_π and C_μ as in Fig. 8.13c. Even though r_b is identical to SPICE parameter RB, this resistor requires special treatment here. Because we connect C_π and C_μ between the *external* BE and BC nodes, respectively in Fig. 8.13c, we must use the default SPICE value RB = 0 and then append r_b as an external component to locate it correctly relative to the capacitors.

The "low-frequency capacitors" in Fig. 8.13c have unrealistically large values of 1000 F to ensure good short circuits at all analysis frequencies. In the SPICE code of Fig. 8.13d, RB = $10^{-6} \ \Omega$ to approximate $r_b = 0$ used in the theoretical analysis. We use the replacement statement, RB = 200, in a second SPICE run.

The unity-gain frequency

$$\omega_\tau = \frac{0.00472}{55 \times 10^{-12}} \approx 2\pi \ 13.7 \times 10^6 \ \text{rad/s}$$

calculated from Eq. (8.17a), provides a practical upper limit for a useful range of frequencies for the .AC statement. One kilohertz, ordinarily well within midrange, is the lower limit, and 30 points per decade gives enough points to define the curve.

The quiescent collector current computed by SPICE was 0.118 mA, just as we assumed in the first part of the example. SPICE values for g_m and r_π were slightly lower and higher, respectively, than our example values, giving midrange gain of 78.1 in both runs compared to 80.6 in our estimate.

Figure 8.13e compares the two SPICE-computed curves. The bandwidth in the first SPICE run was 1.47 MHz, very close to our predicted value of 1.45 MHz. Changing r_b to 200 Ω gave a bandwidth of only 0.79 MHz. We conclude that unless r_b is known to have a small value, we should assume the common-base frequency response is not a simple one and estimate its bandwidth using a technique introduced in the next section. ❏

Exercise 8.5 and Problem 8.31 explore a redesign that involves sacrificing gain for increased bandwidth.

Exercise 8.5 Using SPICE data, estimate a replacement value for the 50 kΩ load resistor that will increase the bandwidth of the amplifier of Fig. 8.13a to 20 MHz, still assuming $r_b = 0$.

Hints: Equation (8.26) suggests that we can increase bandwidth at the expense of gain by reducing the size of the load resistor. (Replacing the 50 kΩ resistor does not change the biasing.)

Ans. $R_L = 1.68$ kΩ.

8.4.3 MILLER'S THEOREM

The common-gate amplifier has a simple high-frequency gain function because its high-frequency equivalent, Fig. 8.12c, consists of two disjoint subcircuits, each containing a single capacitor. In many circuits an admittance, Y, bridges between the input and output parts of an amplifier as in Fig. 8.14a, resulting in a complicated gain function that is difficult to obtain and hard to interpret. Because estimating the upper half-power frequency and understanding the factors that determine its value are the principal objectives of hand analysis, we are willing to settle for less than the exact gain function. Miller's theorem, which we describe next, gives us a related isolated-pole circuit that has *approximately* the same upper half-power frequency as the original amplifier.

In Fig. 8.14a, the node voltages at the terminals of Y are related by

$$V_o = A_M V_x$$

where A_M is the *Miller gain*. The circuit of Figs. 8.14b is equivalent to the original in the following sense. The current that leaves the input circuit of Fig. 8.14a,

$$I = Y(V_x - V_o) = [Y(1 - A_M)]V_x$$

is the same as the current that leaves the input circuit of Fig. 8.14b. Also, the current that enters the output circuit of Fig. 8.14a,

$$I = Y(V_x - V_o) = \left[Y\left(\frac{1}{A_M} - 1\right)\right]V_o = \left[Y\left(1 - \frac{1}{A_M}\right)\right](-V_o)$$

also enters the output circuit of Fig. 8.14b. Equivalence of these currents justifies *Miller's theorem,* which states that the circuits of Figs. 8.12a and b have the same voltage gain V_o/V_i and the same input impedance V_i/I_i. (Output impedance and reverse gains

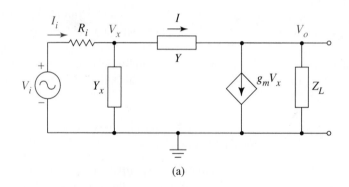

(a)

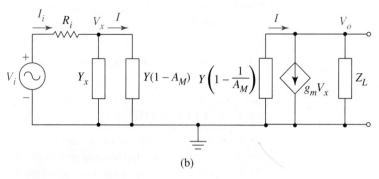

(b)

Figure 8.14 Miller effect: circuit (b) has the same gain and input impedance as circuit (a) provided that $A_M = V_o/V_x$ from circuit (a).

depend upon applying excitation to the output side and are, therefore, *not* the same for the two circuits.)

There is a catch to Miller's theorem. The only way to find the *true* A_M we need for Fig. 8.14b is to analyze the original network—exactly what we wish to avoid! A practical alternative is to make the so-called *Miller approximation*. This means we *estimate* A_M by first disconnecting Y from the output circuit of Fig. 8.14a, in effect assuming that $|I| \ll |g_m v_x|$. This gives

$$A_M = -g_m Z_L$$

never exactly true, but often a good approximation.

Special Case. When Z_L and Y are complex, the equivalent circuit can still be a challenge to analyze; however, for most circuits in this text Y is the admittance, $j\omega C$, of a capacitor, Z_L is a resistance R_L, and the approximate Miller gain is a large negative number, $-g_m R_L$. For this special case, the input circuit in Fig. 8.14b is terminated with the *Miller capacitor* $C_M = C(1 - A_M)$. For A_M large and negative, this gives a very large capacitor indeed. The scaling that turns C into the much larger capacitance C_M at the input is called *Miller multiplication* or the *Miller effect*. Because of Miller multiplication, the Miller capacitor is usually responsible for a high-frequency pole at a surprisingly low frequency. This capacitor also greatly changes the input impedance of the amplifier. Because $1/A_M \approx 0$, the admittance appended to the output circuit in Fig. 8.14b for this special case is simply C itself.

There is a simple intuitive explanation for Miller multiplication. With large *negative* A_M in Fig. 8.14a, a slight increase in V_x causes a *large* proportional *decrease* in V_o. The resulting node-voltage difference across Y causes a much larger increase in current I than if the gain were not present. From the viewpoint of the input circuit, admittance Y seems "bigger than life," as in Fig. 8.14b. There is also logic behind the Miller approximation. A consequence of high gain is that the signal current levels in the output circuit are typically much higher than current levels in the input circuit. Thus a change in I, perceived to be large by the input circuit, might well be negligible to the output circuit.

A common error in applying Miller's theorem is to confuse the internal Miller-effect gain A_M with the midrange gain, A_{mid}, of the amplifier being analyzed. The two gains are often different. Many of these statements are clarified by examples in the analysis that follows.

8.4.4 COMMON-EMITTER AND COMMON-SOURCE AMPLIFIERS

Two of the most important applications of Miller's theorem are high-frequency analysis of common-emitter and common-source amplifiers.

Common-Emitter Amplifier. The high-frequency equivalent circuit for Fig. 8.15a is Fig. 8.15b. Because a Thevenin transformation of the input circuit gives the structure of Fig. 8.14a, we can use Miller's theorem. The critical features to recognize are that C_μ bridges between input and output circuits and that V_π functions as V_x.

With C_μ detached from node c,

$$A_M = \frac{V_o}{V_\pi} = -g_m R_{LL}$$

This makes the input capacitance in Fig. 8.15c

$$C_{EQ} = C_\pi + C_M = C_\pi + C_\mu (1 + g_m R_{LL})$$

Here we see Miller multiplication, in which a small capacitor C_μ comes to have an influence far greater than its value would suggest, by virtue of its critical location in the circuit. Following Fig. 8.14b, we also place C_μ in parallel with R_{LL} tacitly assuming $|1/A_M| \ll 1$. A further simplification replaces the circuit to the left of the base by a Thevenin equivalent, giving $R_{TH} = R_i \| R_P$ and

$$V_{TH} = \frac{R_P}{R_P + R_i} V_i$$

These simplifications give Fig. 8.15c, which we can analyze by inspection. Although not identical to our original circuit, the Miller equivalent should have approximately the same bandwidth.

EXAMPLE 8.6 The common-emitter circuit of Fig. 8.15a has transistor parameters $g_m = 4.72$ mA/V, $r_\pi = 21.1$ kΩ, $r_b = 100$ Ω, $C_\pi = 50$ pF, $C_\mu = 5$ pF, and circuit pa-

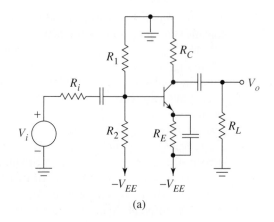

(a)

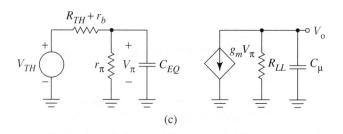

(b)

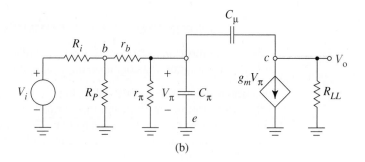

(c)

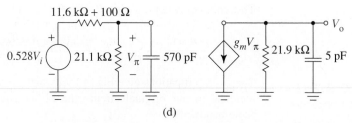

(d)

Figure 8.15 Common-emitter amplifier: (a) schematic; (b) high-frequency equivalent; (c) equivalent after Miller's theorem and Thevenin transformation; (d) numerical values for Example 8.6.

rameters $R_C = 39$ kΩ, $R_1 = 47$ kΩ, $R_2 = 68$ kΩ, $R_i = 20$ kΩ, and $R_L = 50$ kΩ. Use Miller's theorem to approximate its high-frequency gain function.

Solution. Figure 8.15d is Fig. 8.15c with numerical values. The key calculations are

$$-g_m R_{LL} = -4.72 \times 10^{-3} (21.9 \text{ k}) = -103$$

$$C_{EQ} = C_\pi + C_\mu (1 + g_m R_{LL}) = 50 + 5(1 + 103) = 50 + 520 = 570 \text{ pF}$$

$$R_P = R_1 \| R_2 = 27.8 \text{ k}\Omega$$

$$V_{TH} = \frac{27.8 \text{ k}}{27.8 \text{ k} + 20 \text{ k}} V_i = 0.582 V_i$$

$$R_{TH} = \frac{27.8 \text{ k} \times 20 \text{ k}}{27.8 \text{ k} + 20 \text{ k}} = 11.6 \text{ k}\Omega$$

The midrange gain is

$$A_{mid} = \frac{V_o}{V_i} = (0.582) \frac{21.1 \text{ k}}{21.1 \text{ k} + 11.7 \text{ k}} (-103) = -38.6$$

The 570 pF capacitor sees a resistance of $(11.7 \text{ k}) \| (21.1 \text{ k}) = 7.53$ kΩ, giving a break frequency at 2.33×10^5 rad/s. The 5 pf capacitor sees 21.9 k, giving a second pole at 9.13×10^6 rad/s. Combining these results gives the gain function

$$\frac{V_o}{V_i} = -38.6 \frac{1}{(s/2.33 \times 10^5) + 1} \frac{1}{(s/9.13 \times 10^6) + 1} \qquad \square$$

Exercise 8.6 Recalculate the two Miller equivalent pole frequencies for the amplifier of Example 8.6 if input comes from a source with resistance $R_i = 50$ Ω instead of 20 kΩ.

Ans. 11.8×10^6 and 9.13×10^6 rad/s.

Problem 8.35 uses a SPICE simulation to confirm this improvement in bandwidth.

    Gray and Meyer (listed in the references) make a theoretical comparison of the frequency responses of the true common-emitter circuit, Fig. 8.15b, and the approximate circuit, Fig. 8.15c. Their analysis shows that the true circuit has a zero and two poles, compared with only two poles for the approximate circuit. The analysis further indicates that the zero is at too high a frequency to have much effect on the bandwidth, that in both circuits one pole is often low enough to control the bandwidth, and that the two circuits give nearly the same value for this pole. The following example explores this comparison experimentally.

EXAMPLE 8.7 Use SPICE to examine the error introduced by the Miller approximation in Example 8.6.

Solution. Because the key step was replacing Fig. 8.15b with Fig. 8.15c, we use SPICE to compare the frequency responses of these circuits. Figures 8.16a and b show the two circuits with numerical values. Figure 8.16c is the SPICE code for Fig. 8.16a.

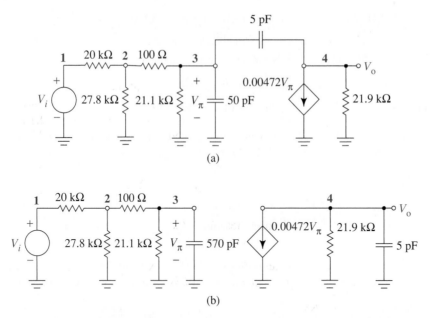

(a)

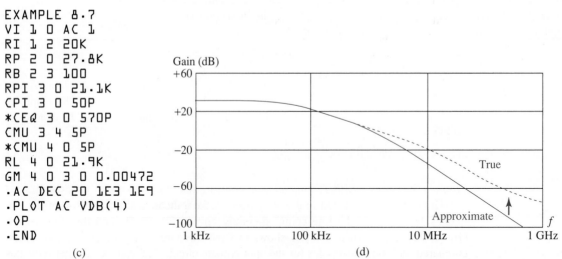

(b)

```
EXAMPLE 8.7
VI 1 0 AC 1
RI 1 2 20K
RP 2 0 27.8K
RB 2 3 100
RPI 3 0 21.1K
CPI 3 0 50P
*CEQ 3 0 570P
CMU 3 4 5P
*CMU 4 0 5P
RL 4 0 21.9K
GM 4 0 3 0 0.00472
.AC DEC 20 1E3 1E9
.PLOT AC VDB(4)
.OP
.END
```

(c)

(d)

Figure 8.16 Exploration of the Miller approximation: (a) true circuit; (b) approximate circuit; (c) SPICE code; (d) SPICE frequency response curves.

Replacing CPI and CMU by the lines that follow them for a second SPICE run gives the Miller equivalent of Fig. 8.16b.

Figure 8.16d compares the frequency responses of the two circuits. The SPICE midrange gain of 31.7 dB or $A_{mid} = -38.5$ agreed with the computed gain. The SPICE computations show that both circuits have -3dB frequencies at 2.33×10^5 rad/s, or 37.1 kHz, also agreeing with theoretical calculations. Two octaves above this half-power frequency, the curves still differed by only 0.1 dB, an error of only about 0.5%. The second pole in the approximate circuit is obviously lower than the corresponding pole, in agreement with Gray and Meyer's analysis. A noticeable slope increase in the true curve (see arrow) suggests the high-frequency zero predicted in Gray and Meyer's analysis. ❏

Dominant High-Frequency Pole. Amplifiers with two or more high-frequency poles sometimes have one pole that is at least two octaves *lower* than any other high-frequency pole or zero. This *dominant pole*, ω_D, strongly influences the amplifier's high-frequency behavior, and when it is present we can make reasonable estimates based upon the single-pole transfer function:

$$A_v(s) \approx A_{mid}\frac{\omega_D}{s + \omega_D} \tag{8.27}$$

Because Eq. (8.27) so closely resembles the gain function of the *RC* circuit of Fig. 1.40, we can use Eq. (1.42) with ω_D replacing ω_H to approximate the amplifier's rise time.

Exercise 8.7 For the amplifier analyzed in Example 8.5, (a) find the separation of the two high-frequency poles in octaves; (b) estimate the amplifier's rise time.

Ans. 5.55 octaves, 241 ns.

EXAMPLE 8.8 In the common-source amplifier of Fig. 8.17a the transistor is biased at 1 mA . Transistor parameters are $\beta = 5$ mA/V^2, $V_P = -1$ V, $C_{gs} = 2$ pF, and $C_{gd} = 3$ pF. Estimate the amplifier's midrange gain, upper -3dB frequency, and rise time.

Solution. The high-frequency model is Fig. 8.17b. From Eq. (7.5), $g_m = 4.47 \times 10^{-3}$. To find the midrange gain, analyze the circuit with capacitors replaced by open circuits. This gives

$$A_{mid} = \frac{V_o}{V_i} = \frac{245 \text{ k}}{245 \text{ k} + 50 \text{ k}}(-4.47 \times 10^{-3})(11.9 \text{ k}) = -44.2$$

Detaching the 3 pF capacitor from node d gives the Miller-effect gain

$$A_M = \frac{V_o}{V_{gs}} = -(4.47 \times 10^{-3})(11.9 \text{ k}) = -53.2$$

Figure 8.17c is the Miller equivalent circuit where

$$C_{EQ} = 2 + 3 \,(1 + 53.2) = 2 + 162.6 = 165 \text{ pF}$$

The pole frequency associated with this capacitor is

$$\omega_{H1} = \frac{1}{(165 \text{ pF})(50 \text{ k} \| 245 \text{ k})} = 1.46 \times 10^5 \text{ rad/s}$$

The second pole frequency is

$$\omega_{H2} = \frac{1}{(3 \text{ pF})(11.9 \text{ k})} = 2.8 \times 10^7 \text{ rad/s}$$

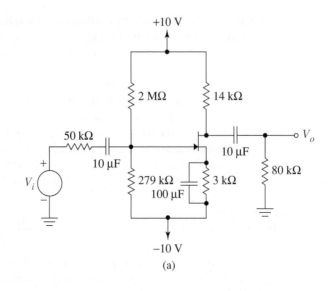

(a)

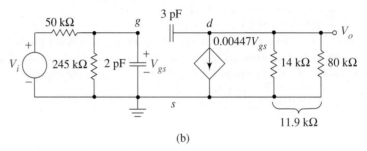

(b)

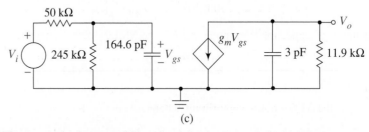

(c)

Figure 8.17 Common-source amplifier: (a) schematic diagram; (b) high-frequency equivalent; (c) equivalent after applying Miller's theorem.

which is over two octaves above ω_{H1}, specifically, $m = 7.6$ octaves, where m satisfies

$$2.8 \times 10^7 = 2^m(1.46 \times 10^5)$$

Thus we take ω_{H1} to be the dominant pole. We expect the upper half-power frequency to be close to 1.46×10^5 rad/s or 23.3 kHz. The rise time should be about $2.2/(1.46 \times 10^5) = 15.1$ μs. ❏

Exercise 8.8 One way to increase the bandwidth and reduce the rise time is to lower the source resistance. If specifications require the rise time of the amplifier in the preceding example to be 5 μs, find the replacement value of the 50 kΩ source resistance.

Ans. $R_i = 14.6$ kΩ.

Input Impedance of an Amplifier. The effect of Miller multiplication on input impedance is evident in Fig. 8.17c. At low frequencies, source V_i, with its 50 kΩ internal resistance, sees an input *resistance* of 245 kΩ; at higher frequencies the Miller capacitance causes the amplifier's input impedance to become reactive and, eventually, entirely capacitive. The amplifier has significant lowering of its input impedance at that frequency where the reactive component of Z_{in} equals the resistive component. For this amplifier, this occurs at

$$f = \frac{1}{2\pi(245\text{ k})(165\text{ pF})} = 3.94\text{ kHz}$$

Compare this with the input impedance of the common-gate amplifier of Fig. 8.12c, where there is no Miller-effect multiplication, and where the small C_{gs} contributes only its "face value" to the input reactance.

8.4.5 HIGH-FREQUENCY RESPONSE WHEN POLES AND ZEROS INTERACT

The source follower of Figs. 8.18a and b is an example of a circuit with interacting poles. Although C_{gs} bridges between input and output circuits, Miller's theorem is not very helpful because the approximate Miller gain,

$$A_M = \frac{g_m R_S}{1 + g_m R_S}$$

is positive and less than one. This leads to an "equivalent" circuit with negative output circuit capacitance. By writing the two KCL equations in matrix form and using Cramer's rule we can verify that the transfer function is actually

$$\frac{V_o}{V_{TH}} = \frac{s + g_m/C_{gs}}{R_{TH}C_{gd}(s^2 + bs + c)} \tag{8.28}$$

$$\text{where } b = \left(\frac{1}{R_S C_{gs}} + \frac{1}{R_S C_{gd}} + \frac{1}{R_{TH}C_{gd}} + \frac{g_m}{C_{gs}} \right)$$

$$\text{and } c = \frac{1 + g_m R_S}{R_S R_{TH}\, C_{gd}\, C_{gs}}$$

This shows that there is a transmission zero at $-g_m/C_{gs}$ rad/s, but the expression is otherwise of little conceptual value because the denominator cannot be factored into a product of simple one-capacitor expressions as was done in Eq. 8.25. For this, and for more complicated circuits in which we are unable to associate individual pole and zero frequencies with individual high-frequency capacitors, we fall back on an approximation that allows us to *estimate* ω_H.

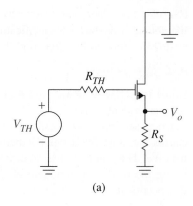

(a)

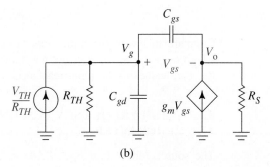

(b)

Figure 8.18   Source follower: (a) schematic; (b) high-frequency equivalent.

Method of Open-Circuit Time Constants. To estimate the upper half-power frequency, ω_H, for a high-frequency equivalent circuit with M capacitors, we first compute M time constants

$$R_j C_j = \frac{1}{\omega_{Hj}}$$

where C_j is a capacitor in the high-frequency equivalent circuit, and R_j is the resistance C_j sees when all other capacitors are replaced by *open circuits*. We then estimate ω_H using

$$\frac{1}{\omega_H} \approx \sum_{j=1}^{M} \frac{1}{\omega_{Hj}} = \sum_{j=1}^{M} R_j C_j \tag{8.29}$$

Because reciprocals of frequencies ω_{Hj} combine to form the reciprocal of the -3 dB frequency, ω_H is always *lower* than the smallest individual ω_{Hj}.

The theory behind Eq. (8.29) assumes that a dominant high-frequency pole exists; however, we generally use the equation without checking this assumption, for the purpose of the estimate is to avoid calculating the exact pole frequencies. In general, there is *no* association between the terms of Eq. (8.29) and the individual poles of the gain function.

Bandwidth of Emitter Follower. Figure 8.19a shows the small-signal schematic of an emitter follower. The BJT schematic symbol represents the high-frequency transistor equivalent, and V_{TH} and R_{TH} are the Thevenin equivalent of the signal source and the biasing resistors. Any external load resistor is lumped into R_E. The high-frequency equivalent of Fig. 8.19b shows transistor capacitances. We now use Eq. (8.29) to examine the emitter follower's capabilities as a high-frequency amplifier.

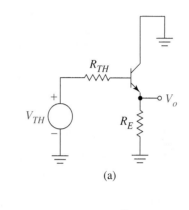

(a)

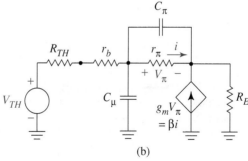

(b)

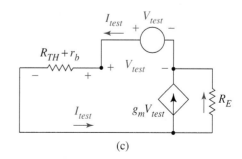

(c)

Figure 8.19 Open-circuit time constants of the emitter follower: (a) high-frequency equivalent of the follower in schematic form; (b) high-frequency equivalent; (c) circuit for finding open-circuit resistance seen by C_π.

To find the open-circuit time constant $R_\mu C_\mu$ we replace C_π with an open circuit and turn off V_{TH}. C_μ then sees $(R_{TH} + r_b)$ in parallel with the input resistance of the emitter follower; that is,

$$R_\mu = (R_{TH} + r_b) \| [r_\pi + (\beta + 1)R_E] \qquad (8.30)$$

Since the second factor is large, it is usually possible to design so that R_{TH} is small enough to satisfy $(R_{TH} + r_b) << [r_\pi + (\beta + 1)R_E]$. Then

$$R_\mu \approx R_{TH} + r_b$$

making $R_\mu C_\mu$ small.

The other open-circuit time constant, $1/\omega_\pi$, involves r_π in parallel with the input resistance defined by Fig. 8.19c. Applying KCL at the bottom node gives

$$I_{test} = g_m V_{test} + \frac{V_{test} - I_{test}(R_{TH} + r_b)}{R_E}$$

Solving for the required ratio gives

$$\frac{V_{test}}{I_{test}} = \frac{R_{TH} + r_b + R_E}{g_m R_E + 1}$$

Thus

$$R_\pi \approx r_\pi \| \left(\frac{R_{TH} + r_b + R_E}{g_m R_E + 1} \right) \qquad (8.31)$$

If specifications allow, we make $R_{TH} + r_b << R_E$. Then

$$R_\pi \approx r_\pi \| \frac{R_E}{g_m R_E + 1} \approx r_\pi \| \frac{1}{g_m} \approx \frac{1}{g_m}$$

With the approximations we have made, Eq. (8.29) gives

$$\frac{1}{\omega_H} \approx R_\mu C_\mu + R_\pi C_\pi \approx (R_{TH} + r_b)C_\mu + \frac{C_\pi}{g_m}$$

and the estimated bandwidth predicted by open-circuit time constants is

$$\omega_H \approx \frac{g_m}{C_\pi + C_\mu(R_{TH} + r_b)g_m} \qquad (8.32)$$

When we compare Eqs. (8.32) and (8.17a), we notice that the coefficient of C_μ might easily be comparable to unity. This suggests that the emitter-follower bandwidth can conceivably approach the unity-gain frequency of the transistor itself—the point of this discussion. For this reason, like the follower op amp circuit, the emitter follower is regarded as a good high-frequency configuration.

Exercise 8.9 Use open-circuit time constants to estimate the bandwidth of the source follower in Fig. 8.18.

Ans. $1/\omega_H = C_{gd}R_{TH} + C_{gs} [(R_{TH} + R_S)/(1 + g_m R_S)]$.

Problem 8.41 compares the answer to Exercise 8.9 with the Bode plot obtained from Eq. (8.28) for particular numerical values.

Capacitive Loads and Miller Effect. Miller's theorem is not useful for estimating the bandwidth of an amplifier with a highly capacitive load, because the Miller gain, A_M, is then complex. A common example is a two-stage amplifier, where the (Miller multiplied) input capacitance of the second stage loads the first stage. In such cases, we must use Eq. (8.29) to estimate the bandwidth as illustrated in the following example.

EXAMPLE 8.9 Figure 8.20a is the amplifier of Fig. 8.17a, with the input of an identical amplifier replacing the 80 kΩ load resistor. Estimate the bandwidth of the *first amplifier stage.*

Solution. The high-frequency equivalent of the first-stage amplifier, Fig. 8.20b, is identical to Fig. 8.17b to the left of the dependent source; however, the 80 kΩ load resistor is replaced by the input impedance of the second stage, a resistor of 245 kΩ in parallel with the 165 pF Miller-augmented capacitor from Fig. 8.17c. Because of this large load capacitance, *the Miller gain A_M of this circuit would be complex even for relatively low frequencies.* Therefore, we estimate using Eq. (8.29).

With 3 and 165 pF capacitors replaced by open circuits, the 2 pF capacitor sees $50\,k \| 245\,k = 41.5$ kΩ. This gives time constant $1/\omega_{H1} = 83 \times 10^{-9}$ s. With the 2 and 3 pF capacitors open, the 165 pF capacitor sees $14\,k \| 245\,k = 13.2$ kΩ; so $1/\omega_{H2} = 2.18 \times 10^{-6}$ s. The 3 pF capacitor sees the output resistance defined in Fig. 8.17c. Kirchhoff's current law gives

$$I_{test} = 0.00447(-41.5\,k)I_{test} + \frac{(-41.5\,k)I_{test} + V_{test}}{13.2\,k}$$

Solving for V_{test}/I_{test} gives 2.51 MΩ, and the time constant $1/\omega_{H3} = 7.53 \times 10^{-6}$ s. From Eq. (8.29),

$$\frac{1}{\omega_H} \approx 83 \times 10^{-9} + 2.18 \times 10^{-6} + 7.53 \times 10^{-6}$$

or $\omega_H \approx 102 \times 10^3$ rad/s, or 16.3 kHz. In Example 8.8, with a resistive load of 80 kΩ this same amplifier had a bandwidth of 23.3 kHz.   ❑

8.4.6 REDUCING BANDWIDTH

Sometimes we wish to deliberately reduce the bandwidth of an amplifier. In Sec. 1.6.6 we learned that to avoid distortion the bandwidth must accommodate the signal spectrum. However, our circuits always generate random noise, which is noise power distributed over all frequencies. Noise components within the amplifier's passband are amplified along with the signal, while those outside the passband are attenuated. Thus any bandwidth in excess of that required by our signal serves no useful purpose and, in fact, results in unnecessary output noise. In Chapter 9 we learn that we can also eliminate instability in feedback circuits by reducing bandwidth.

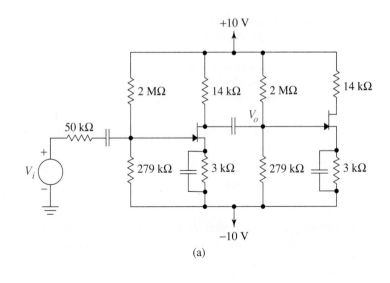

(a)

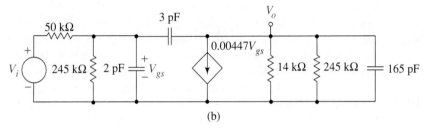

(b)

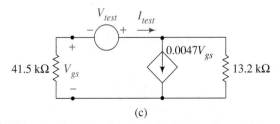

(c)

Figure 8.20 Amplifier with capacitive load: (a) two-stage common-source amplifier; (b) high-frequency equivalent of Fig. 8.15a with R_L replaced by input of an identical amplifier; (c) circuit for finding open-circuit time constant of the 3 pF capacitor.

One way to reduce bandwidth is to add an external capacitor C_{X1} between the amplifier's output node and ground, as in Fig. 8.21. Typically the required value of C_{X1} is much smaller than the coupling and bypass capacitors denoted by asterisks and is, therefore, a *high-frequency capacitor*, which we must treat as an open circuit for low-frequency and midrange calculations just like the transistor's internal capacitances. In the high-frequency equivalent, C_{X1} adds directly to the internal capacitance already present between that node and ground, increasing its open-circuit time constant and thereby reducing the bandwidth.

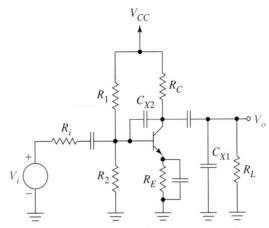

Figure 8.21   Reducing bandwidth by adding external high-frequency capacitors.

In integrated circuit designs it may be difficult to fabricate a capacitor sufficiently large to reduce bandwidth in this manner. Another approach is to place a small capacitor in a *critical location,* for example between base and collector (or gate and drain) like C_{X2}, where Miller multiplication increases its effectiveness. The next example shows this idea in a design context.

EXAMPLE 8.10   We found that the 23.3 kHz bandwidth of the amplifier in Fig. 8.17a was primarily determined by the dominant pole associated with the 165 pF capacitor of Fig. 8.17c. How large an external capacitance C_X must be attached between gate and drain to reduce the bandwidth to 5 kHz?

Solution. In Example 8.8 we found the Miller-effect gain was -53.2. Because C_X is in parallel with and thus adds directly to $C_{gd} = 3$ pF in Fig. 8.17b C_{EQ} changes from its original value of 165 pF to

$$C_{EQ} = 2 + (3 + C_X)(1 + 53.2) \text{ pF}$$

Since C_{EQ} sees $50 \text{ k} \| 245 \text{ k} = 41.5 \text{ k}\Omega$, for a 5 kHz bandwidth C_X must satisfy

$$\frac{1}{(2\pi 5 \times 10^3)(41.5 \times 10^3)} = [2 + (3 + C_X)(1 + 53.2)] \times 10^{-12}$$

or $C_X = 11.1$ pF.   ❏

In a similar way, we can use open-circuit time constants to find the capacitance needed for a particular bandwidth reduction.

Exercise 8.10   Example 8.9 showed that the amplifier of Fig. 8.20 has a bandwidth of 16.3 kHz. (a) How much capacitance should be added between drain and ground of the first transistor to reduce the bandwidth to 10 kHz?

Hint: The resistances seen at the capacitor sites do not change. (b) Repeat part (a), but instead add the capacitor between gate and drain of the first transistor.

Ans. 869 pF, 0.308 pF.

8.5

Frequency Response of Single-Stage Difference Amplifiers

We next examine the high-frequency limitations of the single-stage difference amplifier, first with difference-mode and then with common-mode excitation. Among other things, we will discover that current-source limitations cause common-mode gain to increase at higher frequencies. The emitter-coupled pair is examined in detail; however, the results also apply to the source-coupled pair.

Difference-Mode Gain. Figure 8.22a is the small-signal schematic for the BJT difference amplifier of Fig. 7.47a. Because of circuit and excitation symmetry, there is a virtual ground, indicated by the dashed ground symbol. And since all three grounds denote points of identical electrical potential, the high-frequency equivalent consists of the two half-amplifiers of Fig. 8.22b. Applying the Miller approximation to each half-amplifier gives Fig. 8.22c, where

$$C_{EQ} = C_\pi + C_\mu (1 + g_m R_C)$$

The voltage gain, $A(s)$, of the half-amplifier is

$$A(s) = \frac{V_x(s)}{0.5\, V_d(s)} = A_{mid} \frac{\omega_{H1}}{s + \omega_{H1}} \frac{\omega_{H2}}{s + \omega_{H2}}$$

For double-ended output, the voltage gain of the difference amplifier, $A_d(s)$, is

$$A_d(s) = \frac{V_o(s)}{V_d(s)} = \frac{2V_x(s)}{V_d(s)} = \frac{V_x(s)}{0.5\, V_d(s)} = A(s)$$

We conclude that the high-frequency response of the emitter-coupled pair, with difference-mode excitation and double-ended output, is identical to that of the common-emitter amplifier. The gain curve has the shape of Fig. 8.22d, with ω_{H1} usually determined by C_{EQ} and the resistance of the excitation circuit, and with ω_{H2} determined by C_μ and R_C. The half-amplifier diagram also shows that for single-ended output,

$$A_{d(s-e)}(s) = 0.5 A_d(s)$$

Exercise 8.11 For the amplifier of Fig. 7.47a, $I_O = 2$ mA and $R_C = 5$ kΩ. If $\beta = 180$, $r_b = 200\ \Omega$, $C_\pi = 20$ pF, and $C_\mu = 5$ pF, find the gain function, $A_d(s)$, for double-ended output.
Hint: Consider Fig. 8.22c.

Ans. $A_d(s) = -191 \dfrac{5.11 \times 10^6}{s + (5.11 \times 10^6)} \dfrac{4.00 \times 10^7}{s + (4.00 \times 10^6)}$

Input and Output Loading at High Frequencies. Source resistances, R_i, between each signal generator and input node in Figs. 8.22a and b would affect the resistance seen by each Miller capacitor, and thus directly affect the frequency of the dominant pole.

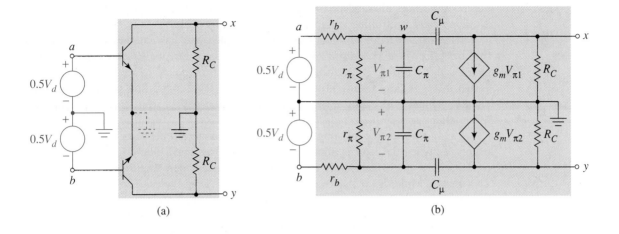

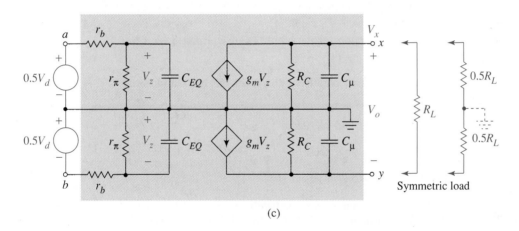

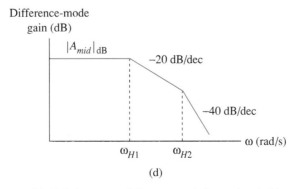

Figure 8.22 Diff amp with high-frequency difference-mode input signal: (a) small-signal schematic of high-frequency equivalent; (b) high-frequency equivalent; (c) equivalent after Miller's theorem; (d) Bode plot.

Source resistances also reduce the midrange signal amplitude, because complex voltage division makes controlling variable V_Z smaller in Fig. 8.22c.

Some care is required when the emitter coupled pair has external loads. Consider first the effect of a *symmetric resistive load* such as a resistor R_L connected between nodes x and y in Fig. 8.22a. Adding two series resistors, each of value 0.5 R_L as suggested in Fig. 8.22c, results in a virtual ground at their connection point by virtue of circuit symmetry. In effect, each $0.5R_L$ is in parallel with a resistor R_C, reducing the Miller-effect gain and also reducing C_{EQ} in Fig. 8.22c.

A symmetric load consisting of a parallel RC connected between the transistor collectors, as when the load is another diff amp, reduces the Miller-effect gain and adds capacitance to the Miller output circuit as in the single-ended amplifier of Example 8.9. This makes Miller gain A_M complex at even low frequencies. The simplest way to handle such loading is to estimate the bandwidth of the half-amplifier using open-circuit time constants. We later examine cascaded difference amplifiers to illustrate this point. Loading between one output node and ground, either resistive or RC, introduces the same cautions in applying Miller's theorem.

EXAMPLE 8.11   Find the new gain function for the amplifier of Exercise 8.11 if a 20 kΩ resistor is placed between the output nodes.

Solution. The load resistor places 10 kΩ in parallel with R_C in the half-amplifier. Thus Miller gain is

$$A_M = -g_m R_{L,EQ} = -(0.04)(5 \text{ k} \| 10 \text{ k}) = -133$$

Then $C_{EQ} = 20 \text{ pF} + 5 \text{ pF}(134) = 690$. This pole sees

$$r_b \| r_\pi = 200 \| 4.5 \text{ k} = 191 \ \Omega$$

for $\omega_{H1} = 7.59 \times 10^6$ rad/s. The other capacitor, approximately 5 pF, sees $R_{L,EQ} = 3.33$ kΩ, so $\omega_{H2} = 6.00 \times 10^7$ rad/s. Midrange gain differs from Miller gain in the r_b-r_π voltage divider, thus

$$A_{mid} = \frac{4.5}{4.7}(-133) = -127$$

Putting this all together gives

$$A_d(s) = -127 \frac{7.59 \times 10^6}{s + (7.59 \times 10^6)} \frac{6.00 \times 10^6}{s + (6.00 \times 10^6)}$$
❏

> **Exercise 8.17**   Suppose in Example 8.11, instead of placing the 10 kΩ load between collectors, we instead capacitively couple a 5 k load resistor to node x, and use the voltage at this node as output. Find $A_{d,s-e}(s)$ for this case.
> *Hint*: How would this load affect the half-amplifiers in Fig. 8.22c?
>
> *Ans.* $A_{d(s-e)}(s) = -63.5 \dfrac{5.11 \times 10^6}{s + (5.11 \times 10^6)} \dfrac{4.00 \times 10^6}{s + (4.00 \times 10^6)}$

Common-Mode Rejection. Figure 8.23a shows the difference amplifier with common-mode excitation V_C. The internal resistance and capacitance of the current source, R_{out} and C_{out}, introduce the principal high-frequency limitation for common-mode operation. Figure 8.23b shows the high-frequency, half-amplifier equivalent. Because of the symmetry, we can find the common-mode gain V_o/V_C by using only one half-circuit.

Because the resistance of the current source is large, the effect of C_{out} on high-frequency response turns out to be dominant. Because of this we can simplify the analysis by replacing the transistor with its midrange instead of its high-frequency model, giving

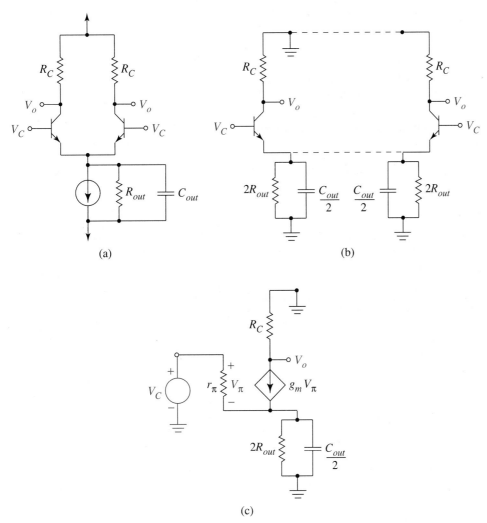

(a)

(b)

(c)

Figure 8.23 Common-mode rejection for single-ended output: (a) amplifier with common-mode excitation; (b) high-frequency model; (c) high-frequency equivalent circuit; (d) common-mode gain; (e) CMRR as function of frequency.

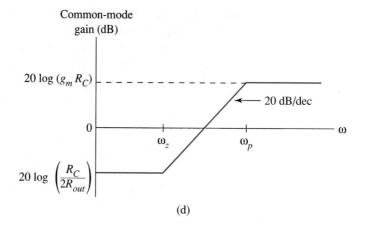

(d)

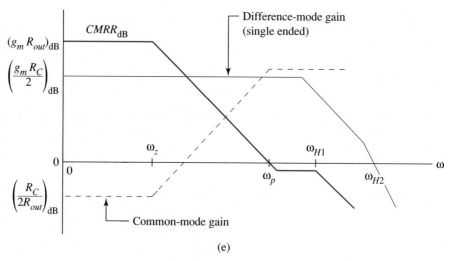

(e)

Figure 8.23 (continued)

Fig. 8.23c. In effect, we are assuming that the most important events occur at frequencies so low that the other transistor capacitances are negligible.

Now this *high-frequency* equivalent just happens to be structurally identical to the *low-frequency equivalent* of a common-emitter amplifier with unbypassed emitter resistor, $2R_{out}$! Ironically, this observation permits us to use previously established low-frequency results to predict the high-frequency behavior of Fig. 8.23a. From our (low-frequency) analysis of bypass capacitors in Sec. 8.3.1, we expect a gain function of the form

$$A_C(s) = \frac{V_o}{V_C} = A_{mid}\frac{s + \omega_Z}{s + \omega_P} \tag{8.33}$$

To find A_{mid} we assume in Fig. 8.23c that the frequency is so high that $C_{out}/2$ is a short circuit. This gives the simple common-emitter amplifier for which $A_{mid} = -g_m R_C$. Now recall that ω_z is that frequency where the reactance of $C_{out}/2$ equals $2R_{out}$, giving

$$\omega_z = \frac{1}{C_{out} R_{out}} \tag{8.34}$$

The resistance that $C_{out}/2$ sees looking back into the circuit, by resistance scaling, is

$$(2R_{out}) \parallel \frac{r_\pi}{\beta + 1} \approx \frac{r_\pi}{\beta + 1}$$

This gives pole frequency

$$\omega_p = \frac{1}{(C_{out}/2)[r_\pi/(\beta + 1)]} = \frac{2(\beta + 1)}{C_{out} r_\pi}$$

With these values Eq. (8.33) becomes

$$A_C(s) = -g_m R_C \frac{s + 1/(C_{out} R_{out})}{s + [2(\beta + 1)/(C_{out} r_\pi)]}$$

At high frequencies the gain approaches $-g_m R_C$. As $s = j\omega \rightarrow 0$,

$$A_C(s) \rightarrow \frac{-g_m R_C r_\pi}{(\beta + 1)2R_{out}} = \frac{-\beta R_C}{(\beta + 1)2R_{out}} \approx \frac{-R_C}{2R_{out}} \tag{8.35}$$

We conclude that the Bode plot of common-mode gain has the form of Fig. 8.23d.

For common-mode signals gain should be as low as possible, thus the *increase* in gain that begins in the vicinity of ω_z is *not* desired. We obtain full common-mode rejection only for common-mode signals with frequencies below ω_z. What is most important is the common-mode rejection ratio,

$$\text{CMRR}_{dB} = 20 \log|A_d| - 20 \log|A_c|$$

Figure 8.23e shows the graphical subtraction of the common-mode gain curve A_c from Fig. 8.23d (denoted by dashed lines in Fig. 8.23e) from the difference-mode curve A_d of Fig. 8.22d (thin solid lines). This gives the bold CMRR curve for one possible set of assumptions about relative values of the poles and zeros. We see that the zero frequency ω_z identifies the upper limit of the full common-mode rejection predicted by the mid-range equations of Chapter 7.

The common-mode gain (low-frequency value) and bandwidth expressions, Eqs. (8.35) and (8.34), respectively, suggest a trade-off in the current-source design. Low-common-mode gain and high-CMRR bandwidth are both desirable; making R_{out} larger to reduce common-mode gain also lowers the CMRR bandwidth.

EXAMPLE 8.12 A *basic* BJT current mirror biases a difference amplifier that has $R_C = 9$ kΩ. For the mirror transistor, $V_A = 120$ V and $C_\pi = 3$ pF. Find common-mode gain and bandwidth for current-source currents of 0.1 mA and 0.01 mA.

Solution. For the mirror transistor, $R_{out} = r_o$ and $C_{out} \approx 2C_\mu$ (see Problem 8.54). For $I_O = 0.1$ mA, $r_o = 120/0.1$ mA $= 1.2$ MΩ. From Eq. (8.35) common-mode gain is

$$A_C = \frac{-9 \text{ k}}{2 \times 1.2 \text{ M}} = -0.00375$$

and from Eq. (8.34)

$$\omega_z = \frac{1}{6 \text{ pF} \times 1.2 \text{ M}} = 1.39 \times 10^5 \text{ rad/s}$$

Reducing the current source current to 0.01 mA increases r_o to 12 MΩ, giving $A_C = 0.000375$ but reduces ω_c to 1.39×10^4 rad/s. ❑

Bandwidth of Cascaded Difference Amplifiers. When source- or emitter-coupled pairs are cascaded, the large input capacitance of the second stage appreciably loads the first stage; consequently, the simple Miller equivalent does not adequately predict bandwidth. The method of open-circuit time constants is useful in this situation, as illustrated in the following example.

EXAMPLE 8.13 Estimate the bandwidth of the two-stage amplifier of Fig. 8.24a when $R_D = 5$ kΩ and $R_i = 50$ kΩ. All four transistors have parameter values: $g_m = 6$ mA/V, $C_{gs} = 3$ pF, and $C_{gd} = 4$ pF.

Solution. Because the high-frequency equivalent for each stage has a virtual ground at the source node, each diff amp equivalent takes the form of two identical half-amplifiers that resemble the BJT circuit of Fig. 8.22b. To exploit this symmetry, we work with the half-amplifiers of Fig. 8.24b. In this circuit $C_1 = C_3 = C_{gs} = 3$ pF and $C_2 = C_4 = C_{gd} = 4$ pF.

First, we find the resistance seen by C_1. With V_i off, and C_2 through C_4 open circuits, C_1 sees $R_1 = R_i = 50$ kΩ.

Next consider the resistance seen by C_3. With C_2 an open circuit and $V_i = 0$, $g_m V_1 = 0$, that is, the dependent source is off. With C_4 also an open circuit, C_3 sees $R_3 = R_D = 5$ kΩ.

To find R_2, use Fig. 8.24c. By Kirchhoff's current law

$$I_{test} = -g_m(R_i I_{test}) + \frac{V_{test} - R_i I_{test}}{R_D}$$

Solving for the ratio gives

$$R_2 = \frac{V_{test}}{I_{test}} = R_D + (1 + g_m R_D)R_i = 1.56 \text{ M}\Omega \qquad (8.36)$$

Figure 8.24d is the circuit that gives R_4. Because this circuit differs from Fig. 8.24c only in replacement of R_i by R_D, from Eq. (8.36)

$$R_4 = \frac{V_{test}}{I_{test}} = R_D + (1 + g_m R_D)R_D = 160 \text{ k}\Omega$$

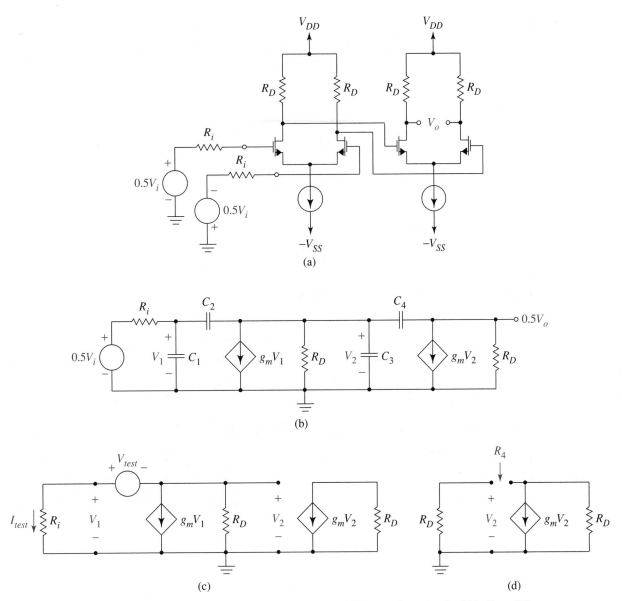

Figure 8.24 Determining bandwidth of two-stage amplifier: (a) given circuit; (b) half-amplifier for high-frequency analysis; (c) equivalent used to find resistance seen by C_2; (d) circuit used to find R_4; (e) increasing bandwidth by using a preamp.

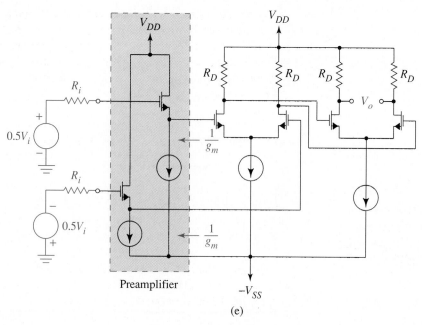

Figure 8.24 (continued)

Equation (8.29), for this circuit, is

$$\frac{1}{\omega_H} \approx R_1C_1 + R_2C_2 + R_3C_3 + R_4C_4$$

$$\approx 50\text{ k} \times 3\text{ pF} + 1.56\text{ M} \times 4\text{ pF} + 5\text{ k} \times 3\text{ pF} + 160\text{ k} \times 4\text{ pF} \quad (8.37)$$

$$\approx 150 \times 10^{-9} + 6240 \times 10^{-9} + 15.0 \times 10^{-9} + 640 \times 10^{-9}$$

or $\omega_H \approx 142 \times 10^3$ rad/s, or 22.6 kHz. ❏

Exercise 8.13 What is the midrange voltage gain of Fig. 8.24a, when output is double ended?

Ans. $(g_mR_D)^2$.

Example 8.13 motivates some important observations about resistance transformations. Notice the numerical dominance of C_2R_2 in Eq. (8.37). From Eq. (8.36) we see that reducing R_i would cause a direct increase the bandwidth. Now suppose that more bandwidth is needed but the source resistance R_i cannot be changed. We can increase bandwidth with the same R_i by using a (double-ended) source-follower-type *preamplifier* between the signal source and the diff amp input, as in Fig. 8.24e. Because the source follower has output resistance of $1/g_m$, inserting this additional stage would reduce the resistance R_2 in Eq. (8.36) to

$$R_2 = R_D + (1 + g_mR_D)\left(\frac{1}{g_m}\right)$$

Of course each source follower contributes two new high-frequency capacitances to the circuit; however, because source-follower voltage gain is less than one, these capacitors have no Miller multiplication, and the overall frequency response improves. The overall voltage gain is still $(g_m R_D)^2$ because the source followers have unity gain; however the additional stage increases the gain-bandwidth product of the amplifier. We have, in fact, *discovered* the two-transistor amplifier; for the preamplifier and first-stage difference amplifier in Fig. 8.24e combine to form a two-sided *common-drain/common-source amplifier,* the FET version of Fig. 7.26d, in differential form. The high bandwidth available from such two-transistor amplifiers is the topic of the next section.

8.6

Frequency Response of Two-Transistor Amplifiers

The two-transistor configurations of Fig. 7.26 and their FET counterparts are all good wideband amplifiers. By noting the locations of the transistor capacitances in the light of Miller multiplication we can intuitively understand their high-frequency capabilities. We explicitly discuss the BJT circuits; however, the same arguments apply to the corresponding FET circuits. Some details, such as biasing and base-spreading resistors, are omitted from the discussion to emphasize the general principles. For purposes of comparison, we begin with one-transistor configurations.

Figure 8.25a is the small-signal schematic for the common-emitter amplifier, drawn to show explicitly the locations of C_π and C_μ. Notice the strategic location of C_μ, which bridges between input and output circuits. Because this amplifier has high, inverting, midrange gain, there is significant Miller-effect multiplication, resulting in relatively low bandwidth. Compare this with the common-base circuit of Fig. 8.25b. Because neither capacitor bridges between input and output, there is no Miller multiplication, and both capacitors' contributions to the bandwidth correspond to their nominal values. This circuit has relatively high bandwidth in spite of its high midrange voltage gain, as our detailed analysis showed.

In the emitter follower of Fig. 8.25c, C_π bridges between input and output; however, because voltage gain is less than one, there is no large Miller multiplication to limit the bandwidth. As we found from our analysis, the emitter follower is a good high-frequency circuit.

Figure 8.25d shows capacitance locations in the cascode amplifier. The input stage has a strategically located capacitor; however, the load resistance of the first stage is the small input resistance, $r_e \approx 1/g_m$, of the second stage. Thus the first-stage voltage gain is approximately one, and the large Miller multiplication that usually characterizes the common-emitter amplifier is absent in this particular circuit. The common-base output stage develops voltage gain for the circuit but has no critically located capacitor to limit its bandwidth.

The common-collector/common-emitter amplifier of Fig. 8.25e has strategically located capacitors in both stages. However, the first stage is a follower with voltage gain less than one. The common-emitter output stage does have large voltage gain, which does lead to Miller multiplication; however, the Miller capacitor looks back at the low output resistance of the emitter follower. Consequently, the RC product is small and the bandwidth high *in spite of Miller multiplication.* (This was *exactly* the reasoning we used to justify adding the follower in Fig. 8.24e.)

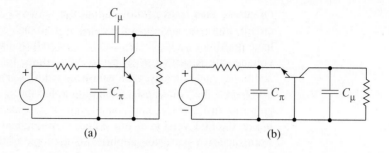

(a) (b)

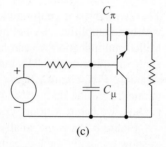

(c)

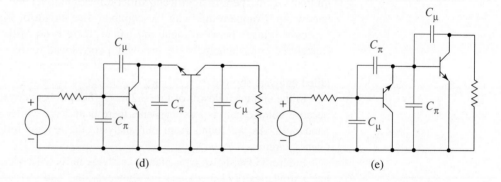

(d) (e)

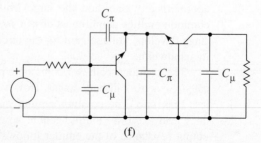

(f)

Figure 8.25 Comparison of configurations for high-frequency performance: (a) common-emitter; (b) common-base; (c) common-collector; (d) common-emitter/common-base (cascode); (e) common-collector/common-emitter; (f) common-collector/common-base.

The common-collector/common-base amplifier of Fig. 8.25f resembles the cascode circuit. The first stage develops current gain $\approx \beta + 1$ but has voltage gain less than one and, therefore, has no large Miller capacitance. The second transistor contributes the voltage gain for the cascade but has no capacitance multiplication to reduce bandwidth.

As noted in Chapter 7, current gain contributed by the first stage increases the midrange power gain of each two-transistor circuit. In the FET counterparts, power gain is not an issue, but the bandwidth improvement still justifies their use, as we saw in the discussion following Example 8.13.

The two-transistor configuration of Fig 8.25f is particularly easy to realize using IC emitter-coupled (or source-coupled) transistor pairs, and gives all the advantages of matched transistors with parameters that track with temperature. The next example demonstrates high-frequency analysis of an FET amplifier constructed in this fashion.

EXAMPLE 8.14   Estimate the bandwidth of the circuit of Fig. 8.26a if $C_{gs} = 1$ pF, $C_{gd} = 4$ pF, $g_m = 4.47$ mA/V, $R_i = 1$ kΩ, and $R_L = 1.5$ kΩ.

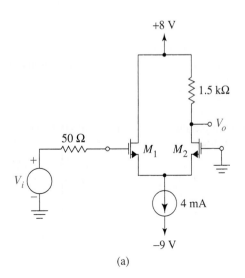

(a)

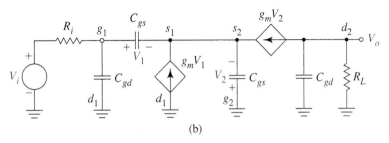

(b)

Figure 8.26   Common-drain/common-gate amplifier: (a) schematic; (b) high-frequency equivalent; (c) circuit after source transformation; (d) equivalent after applying Miller's theorem.

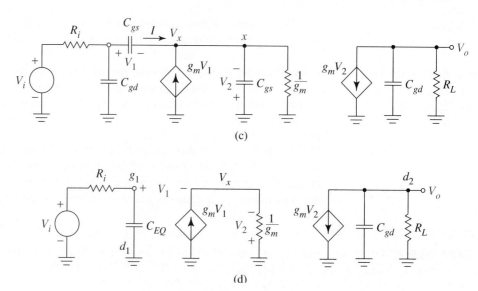

Figure 8.26 (continued)

Solution. This is a common-drain/common-gate circuit so its high-frequency struc-ture, Fig. 8.26b, resembles Fig. 8.25f. A current-source transformation applied to M_2 places a VCCS in parallel with the output impedance, and a resistor $1/g_m$ in parallel with source $g_m V_1$ as in Fig. 8.26c.

Two currents, I and $g_m V_1$, enter the RC circuit between node x and ground. The Miller approximation (ignoring current I) is justified when $|I| = \omega C_{gs} V_1 \ll g_m V_1$, that is, for frequencies that satisfy $\omega \ll g_m/C_{gs} = 4.47 \times 10^9$ rad/s. The complex imped-ance between node x and ground will be real for frequencies such that $|1/j\omega C_{gs}| \gg 1/g_m$, the same frequency range required for validity of the Miller approximation. From the biasing in Fig. 8.26a, the two g_ms are identical. We conclude that for $\omega \ll 4.47 \times 10^9$ rad/s,

$$V_x = g_m V_1 \left(\frac{1}{g_m} \right) = V_1$$

and the Miller-effect gain is

$$A_M = \frac{V_x}{V_x + V_1} = 0.5$$

Figure 8.26d shows the first stage replaced by its Miller equivalent. The capacitance C_{gs} contributed by the second stage has been canceled out by the Miller capacitance, $C_{gs} (1 - 1/A_M)$. The two capacitors C_{EQ} and C_{gd} see resistances of R_i and R_L, respec-tively. The pole frequencies of this Miller approximate circuit are therefore

$$\omega_{H1} = \frac{1}{C_{EQ} R_i} \qquad \text{and} \qquad \omega_{H3} = \frac{1}{C_{gd} R_L}$$

Since

$$C_{EQ} = C_{gd} + 0.5\, C_{gs} = 4.5 \text{ pF}$$

$\omega_{H1} = 2.2 \times 10^8$ rad/s, $\omega_{H2} = 1.67 \times 10^8$ rad/s. Since both pole frequencies are an order of magnitude less than 4.47×10^9 rad/s, the Miller approximation is justified. We use Eq. (8.29) to approximate the bandwidth. The result is $\omega_H \approx 9.49 \times 10^7$ rad/s or 15.1 MHz.

> **Exercise 8.14**   Use SPICE to find the -3 dB bandwidths for Figs. 8.26b and d.
> *Hint*:   Add a 1000 MΩ resistor between nodes s_2 and g_2 in Fig. 8.26b so that SPICE can perform the dc analysis. Values are those of Example 8.14.
>
> *Ans.* 19.4 MHz, 19.4 MHz.

EXAMPLE 8.15   Figure 8.27a is a common-collector/common-emitter amplifier. A bypass capacitor grounds Q_2's emitter at midrange and high frequencies; and both transistors have $\beta = 275$, $r_b = 50\ \Omega$, $C_\mu = 10$ pF, and $f_\tau = 100$ MHz. Compare the gain-bandwidth product of this two-transistor circuit with that of the one-transistor circuit formed when source is connected directly to the common-emitter stage without the (unity-gain) emitter follower.

Solution. First we find the midrange gain. The midrange parameters needed are $g_m \approx 2/25 = 0.080$ A/V and $r_\pi = 3.44$ kΩ for Q_1, and $g_m \approx 4/25 = 0.16$ A/V and $r_\pi = 1.72$ kΩ for Q_2. Figure 8.27b is the midrange equivalent. In Fig. 8.27a, the input resistance between the base of Q_2 and ground is $r_\pi + r_b = 1.77$ kΩ. Emitter follower Q_1 scales this resistance up by $\beta + 1$, causing $R_{in1} = (275 + 1)\,1770 = 489$ kΩ to appear in series with r_b and r_π of Q_1. The signal at the emitter of Q_1 is therefore related to V_i in Figs. 8.27a and b by the voltage division

$$\frac{V_e}{V_i} = \frac{R_{in1}}{R_{in1} + r_\pi + r_{b1} + R_i} = \frac{489\text{ k}}{489\text{ k} + 3.44\text{ k} + 50 + 200} \approx 1$$

The midrange gain of Q_2, from Fig. 8.27b, is

$$\frac{V_o}{V_e} = \frac{1.72\text{ k}}{1.72\text{ k} + 50}(-g_m R_C) = -(0.972)(160) = -156$$

The product of the two gain expressions is the voltage gain of the two-stage amplifier.
For the high-frequency transistor models, Eq. (8.17), gives

$$C_\pi = \frac{g_m}{2\pi f_\tau} - C_\mu = 117 \text{ pF}$$

for Q_1 and $C_\pi = 245$ pF for Q_2. Figure 8.27c is the high-frequency equivalent. Because of obvious interactions between the capacitors, we use open-circuit time constants to estimate the bandwidth.
Instead of hand analysis, we demonstrate use of SPICE dc analysis to find the necessary time constants. Figure 8.27d is the SPICE circuit and Fig. 8.27e the initial SPICE code. V_i is off, and each capacitor C_j in Fig. 8.27c is replaced by an independent current

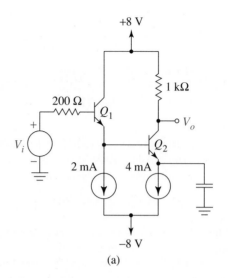

(a)

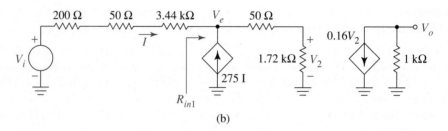

(b)

Figure 8.27 (a) Common-collector/common-emitter amplifier; (b) midrange equivalent; (c) high-frequency equivalent; (d) SPICE circuit for finding open-circuit output resistances; (e) SPICE code for finding output resistances; (f) equivalent of signal source- and common-emitter amplifier alone.

source, I_{Rj}. The independent current sources are turned on one at a time, in four separate runs, each acting as a one-ampere test generator. The resulting voltage across that current source is the output resistance at that node pair.

Table 8.1 lists the open-circuit time constants obtained using SPICE in this way. Using these data in Eq. (8.29) gives $\omega_H \approx 190 \times 10^6$ rad/s, or 30.2 MHz. The gain-bandwidth product of the two-transistor amplifier is $|-156 \times 30.2 \times 10^6| = 4.71$ GHz.

TABLE 8.1 Open-Circuit Time Constants for Fig. 8.27

j	C_j (pF)	R_j (ohms)	$1/\omega j$(ns)
1	117	12.5	1.46
2	10	6.31	0.0631
3	245	1.54	0.377
4	10	335	3.35

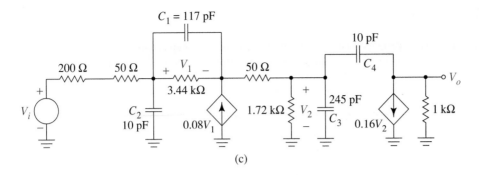

(c)

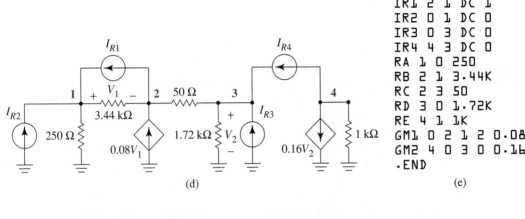

(d)

```
EXAMPLE 8.15
IR1 2 1 DC 1
IR2 0 1 DC 0
IR3 0 3 DC 0
IR4 4 3 DC 0
RA 1 0 250
RB 2 1 3.44K
RC 2 3 50
RD 3 0 1.72K
RE 4 1 1K
GM1 0 2 1 2 0.08
GM2 4 0 3 0 0.16
.END
```

(e)

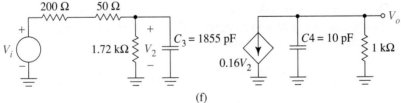

(f)

Figure 8.27   (continued)

To find the gain-bandwidth product with the first stage omitted, we use Fig. 8.27f, obtained from Fig. 8.27c by deleting the first stage and using Miller's theorem. The midrange gain is

$$\frac{V_o}{V_i} = \frac{1.72\,\mathrm{k}}{1.72\,\mathrm{k} + 50 + 200}(-0.16 \times 1\,\mathrm{k}) = -140$$

slightly lower than the two-transistor gain because of higher input loading. The open-circuit time constant associated with the 1855 pF capacitance is 218×1855 pF = 405 ns; the 10 pF capacitor has time constant of 10 ns. The two capacitors together result in an estimated bandwidth of only $f_H = 3.84 \times 10^5$ Hz compared with 30.2 Mz for the two-stage circuit.

Although the emitter follower introduced two additional high-frequency capacitors, we can see that its principal effect was to increase bandwidth. The bottom line is that the common-collector/common-emitter cascade has gain-bandwidth product of 4.9 GHz versus 53.8 MHz for the common-emitter amplifier alone. ❑

Exercise 8.15 The output of an amplifier is a succession of pulses of varying amplitude (gray level) that control the intensity of an electron beam in a display device. A complete picture must be constructed in 1/120 s or 8.33 ms so that the eye is not disturbed by flicker. To allow sufficient time for each pulse to form properly, its width, τ, must be ten times its rise time. Estimate the number, N, of distinct gray levels that can be displayed on the screen for each amplifier of Example 8.15.

Ans. 71,800 pulses for the CC/CE amplifier, 914 for the CE amplifier.

8.7
BJT Parameter Values

Until now we have emphasized two kinds of parameters for bipolar transistors, hybrid pi parameters for hand analysis and SPICE parameters for simulations. A third set of parameters, small-signal *h* parameters, are most likely to be included on data sheets of discrete transistors because they are easiest to measure with accuracy. Because *h* parameters, like hybrid pi parameters, depend upon *Q*-point, the dc conditions under which they were measured are also provided. (Sometimes the data sheet provides an alternative two-port description; for example, transistors designed for radio frequency applications are usually described by *y* parameters. With conversion tables, such as those given in the reference by Irwin, we can readily convert between the various two-port descriptions for the same operating point.)

Appendix B defines and describes the most important two-port parameters. The goal of this section is to introduce some generally accepted notational conventions for bipolar transistors, and show how to compute numerical values of hybrid pi parameters for hand analysis from data sheet information.

8.7.1 STATIC *h* PARAMETERS FOR BIPOLAR TRANSISTORS

Most often, the data sheet *h* parameters are small-signal, midrange quantities measured on a "typical" transistor that is biased at a specified *Q*-point. Small ac signals at a midrange frequency, such as one kilohertz, are applied to the ports, and current and voltage measurements are made in accordance with the definitions in Appendix B. The short- and open-circuits required by the definitions are *ac* shorts and opens, effected using capacitors and inductors so as not to disturb the *Q*-point of the biased transistor. Less frequently, the data sheet gives graphs of measured real and imaginary parts of each parameter as functions of frequency, again for a transistor biased at a specified *Q*-point.

With the transistor in the common-emitter configuration, Fig. 8.28a, the two-port notation of Appendix B is replaced by

$$v_b = h_{ie}\, i_b + h_{re}\, v_c \tag{8.38}$$

$$i_c = h_{fe}\, i_b + h_{oe}\, v_c \tag{8.39}$$

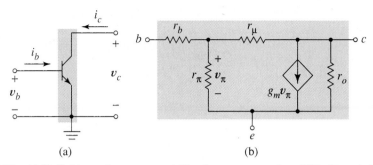

Figure 8.28 (a) Definitions of two-port variables for common-emitter BJT; (b) static hybrid pi model including parasitics.

Subscript i denotes *input* impedance; r, *reverse* voltage gain; f, *forward* current gain; and o, *output* admittance. The second subscript implies that measurements were made in the common-*emitter* configuration. (Subscripts b or c denote common-*base* and common-*collector* measurements.) Sometimes the measurement procedure is repeated over a wide range of I_C values, resulting in curves of each parameter as a function of I_C at, of course, the specified midrange frequency.

Figure 8.28b is the complete static, hybrid pi, BJT model, originally introduced as Fig. 7.5d. As the first step in relating the unknown parameters in this figure to measured h parameters, we apply the h parameter definitions to Fig. 8.28b, as explained in Appendix B; column 2 in Table 8.2 shows the results. Because $r_\mu \gg r_\pi$ and $\beta \gg 1$, the column 2 expressions reduce to column 3 expressions. As given, the table shows how to compute unknown h parameters from known hybrid pi values; however, we next show how to use these approximations in inverse fashion as part of a "recipe" to convert given first-column BJT data into unknown hybrid pi parameters.

EXAMPLE 8.16 Verify the exact equations for h_{re} and h_{oe} in Table 8.2.

Solution. From Eq. (8.38), to find h_{re} we apply an ac open circuit to the transistor input to make $i_b = 0$, and then find the reverse voltage gain. Figure 8.28 shows that un-

TABLE 8.2 Static h Parameters for Common-Emitter BJT

h Parameter	h Parameter in Terms of Hybrid Pi Parameters	Approximate Expression for h Parameter
h_{ie}	$r_b + (r_\pi \| r_\mu)$	$r_b + r_\pi$
h_{fe}	$\dfrac{r_\mu g_m r_\pi - r_\pi}{r_\mu + r_\pi}$	β
h_{re}	$\dfrac{r_\pi}{r_\pi + r_\mu}$	$\dfrac{r_\pi}{r_\mu}$
h_{oe}	$\dfrac{1}{r_o} + \dfrac{\beta + 1}{r_\pi + r_\mu}$	$\dfrac{1}{r_o} + \dfrac{\beta}{r_\mu}$

der these conditions, $h_{re} = v_\pi/v_{ce}$. This is the voltage divider expression given in the table.

While keeping $i_b = 0$, we find h_{oe} by using a test voltage, v_{ce}, that delivers test current, i_c. With this test generator in place, KCL gives

$$i_c = \frac{v_{ce}}{r_o} + g_m \frac{r_\pi}{r_\mu + r_\pi} v_{ce} + \frac{v_{ce}}{r_\mu + r_\pi}$$

Solving for the ratio, i_c/v_{ce} and substituting $g_m r_\pi = \beta$ gives the table entry.

8.7.2 FINDING HIGH-FREQUENCY PARAMETERS FROM DATA SHEET INFORMATION

In addition to numerical values for the static h parameters, the manufacturer usually provides, directly or indirectly, the unity-gain frequency f_τ and the value of C_μ, including dc conditions under which they were measured. (Alternative data sheet notations used for C_μ are C_c, C_{ob}, C_{cb}, and C_{CBO}.) From this information we wish to find numerical values for all of the hybrid pi parameters plus the value of C_π.

Data sheets sometimes give the unity-gain frequency indirectly by specifying the value of beta measured at a given radian frequency, $\omega_x = 2\pi f_x$; that is, they give $|\beta(\omega_x)|$. Because the measurement frequency ω_x is well above ω_β, the imaginary term in the denominator of Eq. (8.15) predominates, giving

$$|\beta(\omega_x)| \approx \frac{\beta \omega_\beta}{\omega_x} = \frac{\omega_\tau}{\omega_x}$$

Thus we compute ω_τ from

$$\omega_\tau = \omega_x |\beta(\omega_x)| \tag{8.40}$$

(Current gain times bandwidth is preserved.)

Starting with values for C_μ and f_τ and the static h parameters, all measured at the desired Q-point, we can now compute parameters for the hybrid pi model of Fig. 8.29 using the following sequence.

$$1.\ g_m = \frac{I_C}{V_T} \qquad\qquad 2.\ r_\pi = \frac{h_{fe}}{g_m}$$

$$3.\ r_b = h_{ie} - r_\pi \qquad\qquad 4.\ r_\mu = \frac{r_\pi}{h_{re}}$$

$$5.\ r_o = \frac{r_\mu}{r_\mu h_{oe} - h_{fe}} \qquad\qquad 6.\ C_\pi = \frac{g_m}{2\pi f_\tau} - C_\mu$$

Notice that all needed information is known at each step.

Exercise 8.16 The following BJT parameters were measured at $I_C = 1$ mA: $h_{fe} = 270$, $h_{ie} = 6.83$ kΩ, $h_{oe} = 7.9 \times 10^{-6}$ A/V, $h_{re} = 0.193 \times 10^{-3}$, and $C_{ob} = 0.9$ pF. Also, $|\beta|$ at 5.33 MHz was 75. Find the parameters for Fig. 8.29.

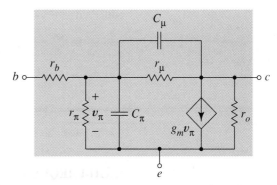

Figure 8.29 Dynamic hybrid pi model, including second-order parasitic elements.

Ans. $f_\tau = 400$ MHz, $g_m = 0.04$ A/V, $r_\pi = 6.75$ kΩ, $r_b = 80$ Ω, $r_\mu = 35$ MΩ, $r_o = 5.38$ MΩ, and $C_\pi = 15$ pF.

8.7.3 SPICE PARAMETERS

The SPICE parameters for the BJT, listed in Fig. 4.44, relate closely to the transistor's physical structure and are *the same for any operating point*. Once SPICE parameters are found, hybrid pi parameters can be calculated for any given operating point. In fact, SPICE calculates hybrid pi parameter values for each transistor before beginning an ac analysis; the SPICE command .OP causes these parameters to appear in the output file.

Variation in Frequency Response with *Q*-Point. After solving so many problems with *given* values of C_π and C_μ, it is easy to forget that the frequency response of an amplifier is highly dependent on the transistor *Q*-point. If the SPICE parameters are specified in a .MODEL statement, SPICE automatically computes the correct small-signal parameters for any *Q*-point, as illustrated in the next example. (Appendix C shows how to calculate SPICE parameters from data sheet information.)

EXAMPLE 8.17 Figure 8.30a shows a common-base amplifier. Biasing is such that I_C and V_{CB} are readily varied by changing I_B, while all nontransistor parameters remain unchanged. Use the SPICE parameters computed in Example C.1 (in Appendix C) to model the BJT in this amplifier. Plot the frequency response for $I_B = 10$ μA and for $I_B = 100$ μA to observe the effect of changing the *Q*-point.

Solution. Figure 8.30b gives the code. There are two values of I_B, 10 μA for the first run and 100 μA (initially coded as a comment) for a second run. The BJT element line has a fourth node number, an option that allows us to specify the substrate node for an IC transistor so that collector-substrate capacitance is included in the analysis. As usual, the p-type substrate is connected to the most negative node of the circuit to prevent conduction between collector and substrate. The .OP statement leads to additional output information so that we can examine the transistor biasing and the parameter values that SPICE employs for each *Q*-point.

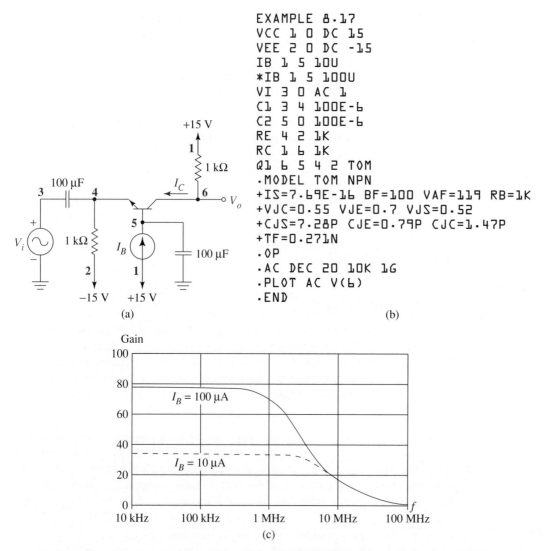

```
EXAMPLE 8.17
VCC 1 0 DC 15
VEE 2 0 DC -15
IB 1 5 10U
*IB 1 5 100U
VI 3 0 AC 1
C1 3 4 100E-6
C2 5 0 100E-6
RE 4 2 1K
RC 1 6 1K
Q1 6 5 4 2 TOM
.MODEL TOM NPN
+IS=7.69E-16 BF=100 VAF=119 RB=1K
+VJC=0.55 VJE=0.7 VJS=0.52
+CJS=7.28P CJE=0.79P CJC=1.47P
+TF=0.271N
.OP
.AC DEC 20 10K 1G
.PLOT AC V(6)
.END
```

Figure 8.30 Frequency response curves of amplifier for two different BJT Q-points: (a) circuit with Q-point determined by I_B; (b) SPICE code; (c) amplitude frequency response curves.

Figure 8.30c shows the frequency response curves. By changing transistor biasing, an amplifier with gain of 78 and bandwidth of 2.2 MHz changes to one with gain 34 and bandwidth 5.6 MHz. SPICE output data explain these differences in terms of simple models. Because of the change in I_C, static parameters g_m, r_π, and r_o each changed by an order of magnitude. As expected, β changed only slightly. Because of the differences in junction voltages and collector current, the emitter-base capacitance, CBE $= C_\pi$, changed significantly while collector-base capacitance, CBC $= C_\mu$, also changed slightly. FT $= f_\tau$ was also affected by the change in Q-point. ❑

8.8 _____
Summary

Frequency response describes how a circuit's output sinusoids relate to input sinusoids when frequency is the independent variable. Frequency response is important because it characterizes circuit dynamics in a simple mathematical form that is easy to relate to applications of the circuit and to circuit measurements.

All amplifiers have high-frequency limitations, which originate in the parasitic capacitances of the transistors, and in stray capacitances between circuit nodes and ground. Special small-signal, high-frequency-equivalent circuits describe the parasitics for various types of transistor. Circuits that employ coupling and bypass capacitors have low-frequency limitations as well.

For an amplifier, the band of useful input frequencies, called midrange, is delimited by upper and lower half-power or -3 dB frequencies; when there are no coupling or bypass capacitors, midrange extends down to dc. The amplifier's bandwidth, the upper half-power frequency for a wideband amplifier, is an important figure of merit, closely related to the amplifier's speed of response to sudden changes in its input. The amplifier's low-frequency response relates to its ability to sustain constant signal values at the output.

To better understand the frequency response of a wideband amplifier, we use distinct equivalent circuits for analysis at low, midrange, and high frequencies. In analysis, the low-frequency equivalent, which includes coupling and bypass capacitors, predicts the lower half-power frequency; in design the equivalent provides information we need to satisfy low-frequency specifications. The high-frequency equivalent, which includes transistor and wiring capacitances, is useful for predicting the upper -3dB frequency in analysis and for designing a circuit with a given bandwidth. Bode plots help us visualize and sketch the frequency response magnitude and phase curves.

Sometimes the equivalent circuits yield simple equations that directly relate individual capacitors to individual pole and zero frequencies in our Bode plots. These frequency response patterns are easily used in design and analysis, and give us a good intuitive feeling for the fundamental principles. In more complex circuits, pole and zero frequencies are related in a complicated way to two or more capacitors, and the best we can do is estimate upper and lower half-power frequencies—using short-circuit time constants for low-frequency equivalents and open-circuit time constants for high-frequency equivalents. An equivalent circuit with many poles sometimes has a dominant pole. When this occurs, the amplifier's dynamic behavior approximates that of a one-pole amplifier. In low-frequency equivalents, the pole of highest frequency is the one that might be dominant; in high-frequency equivalents, a dominant pole is the pole of lowest frequency.

When capacitance bridges between an amplifier's input and output circuits, its contribution to the circuit's dynamic behavior can be significantly multiplied through Miller effect. Miller effect is especially important in high-frequency analysis of common-emitter and common-source amplifiers because gain is large and negative in these circuits. Miller multiplication gives reduced input impedance at high frequencies and a significant reduction in bandwidth compared with what one might expect from the face value of the capacitance. The approximate Miller equivalent circuit provides a simple way to estimate the numerical values associated with Miller effect. The Miller concept is an important intuitive aid that helps us understand such things as how gain trades off against bandwidth in some circuits and why certain two-transistor circuits have such good high-frequency performance. Output capacitance of the current source causes an undesired increase in common-mode gain in emitter- and source-coupled pairs.

REFERENCES _____

1. GRAY, P. R., and R. G. MEYER. _Analysis and Design of Analog Integrated Circuits,_ 3rd ed. New York: John Wiley, 1993.
2. GRAY, P. E., and C. L. SEARLE. _Electronic Principles._ New York: John Wiley, 1969.
3. IRWIN, J. D. _Basic Engineering Circuit Analysis._ 2nd ed. New York: Macmillan, 1987.
4. MALIK, N. R. "Determining Spice Parameter Values for BJTs," _IEEE Trans. Education,_ Vol. 33, (November 1990).
5. SEDRA, A. S., and K. C. SMITH. _Microelectronic Circuits,_ 3rd ed. Philadelphia: Saunders College, 1991.

PROBLEMS _____

SPECIAL DIRECTIONS FOR SPICE HOMEWORK PROBLEMS: _Do not hand in lengthy SPICE printout for homework._ Instead, abstract the useful information from the SPICE output file as in the SPICE examples. Include your SPICE code and a circuit diagram with nodes numbered to agree with the code. Cite relevant numerical values from the SPICE output file and discuss when appropriate. Make sketches of any relevant curves, and label appropriate points. Make small tables to present numerical data if useful for clarity.

Section 8.2

8.1 Sketch the Bode amplitude plots for $A(s) =$

(a) $\dfrac{-10^7 \, s}{(s + 20)(s + 20,000)}$

(b) $\dfrac{10^4 \, (s + 10)}{(s + 500)}$

(c) $\dfrac{-7.2 \times 10^{14} \, s \, (s + 10)}{(s + 400)(s + 100,000)(s + 900,000)}$

(d) $\dfrac{1.44 \times 10^{10} \, s \, (s + 30)}{(s + 10)(s + 900)(s + 120,000)}$

8.2 Compute the magnitude and phase shift associated with $A(s)$, Problem 8.1a, at $\omega =$

(a) 10 rad/s.

(b) 1000 rad/s.

(c) 100,000 rad/s.

8.3 Find gain functions $A(s)$ that have Bode plots like those of Fig. P8.3.

8.4 Estimate the gain in dB at

(a) 10,000 rad/s for Fig. P8.3d.

(b) 4000 rad/s for Fig. P8.3d.

(c) 30,000 rad/s for Fig. P8.3a.

(d) 100,000 rad/s for Fig. P8.3d.

8.5 Estimate the radian frequency where the gain is 0 dB for

(a) Fig. P8.3a.

(b) Fig. P8.3c.

(c) Fig. P8.3d.

Section 8.3

8.6 Draw the low-frequency equivalent for Figs. P7.12a, b, and c.

8.7 In Fig. P8.7, $k = 3 \times 10^{-3}$ A/V^2.

(a) Draw the low-frequency equivalent circuit, including all component values.

(b) Draw the midrange-equivalent circuit.

(c) Add r_o to the equivalent circuits, if the Early voltage is 88 V.

8.8 Draw the low-frequency equivalent circuit for Fig. P8.8. Label each amplifier parameter with its numerical value. Use $\beta = 100$.

8.9 (a) Draw the low-frequency equivalent of Fig. P8.9. Parameter values are $\beta = 100$ and $g_m = 0.1$ A/V.

(b) Find the gain function $A_v(s) = V_o/V_i$, including all numerical values.

(c) To raise the -3dB frequency of the amplifier by about one octave, which capacitor should be changed and what should the new value be?

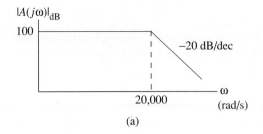

(a)

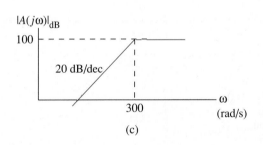

(b)

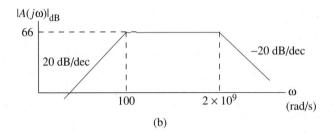

(c)

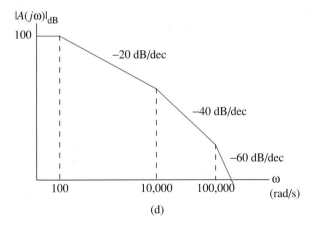

(d)

Figure P8.3

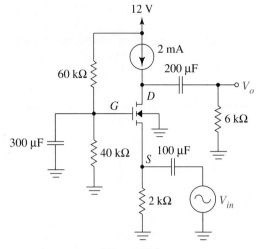

Figure P8.7

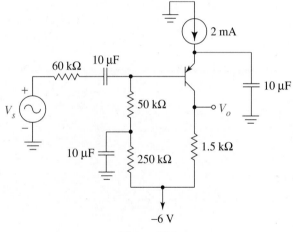

Figure P8.8

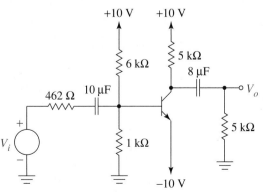

Figure P8.9

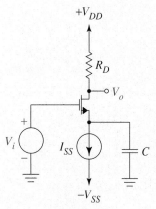

Figure P8.10

8.10 (a) Use the low-frequency equivalent circuit to derive the gain function $A_v(s) = V_o/V_i$ for Fig. P8.10.
(b) Add a capacitively coupled resistor R_L to the output of Fig. P8.10 and then find the gain function for this modified circuit.

8.11 The op amps in Fig. P8.11 have infinite gain. Find $A(s) = V_o/V_i$ for each circuit and sketch the Bode plot.

8.12 In Fig. P8.12, $I_C = 2$ mA and $\beta = 200$. Find the value of C that places the low-frequency pole of the function V_o/V_i at 20 rad/s.
Hint: Use resistance scaling.

8.13 The amplifier of Fig. P8.13 is to have a lower half-power frequency of 10^3 rad/s.
(a) Find the value of C_C if $r_\pi = 1$ kΩ and $\beta = 199$, $k = 0.5 \times 10^{-4}$ A/V^2, and $V_t = 0.5$ V for the MOSFET.
Hint: Use resistance scaling.
(b) Use SPICE to check your design.

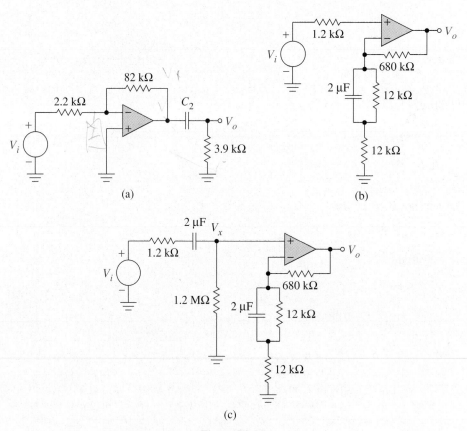

(a)

(b)

(c)

Figure P8.11

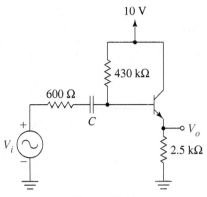

Figure P8.12

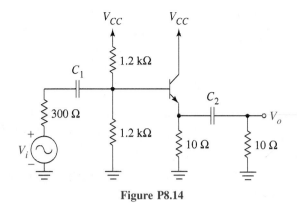

Figure P8.14

8.14 You have designed the circuit of Fig. P8.14, except for choosing values for C_1 and C_2 to make the lower -3 dB frequency about 200 rad/s. If $\beta = 99$ and $r_\pi = 1000$ Ω,
(a) Find the short-circuit time constant associated with each capacitor. (Using the resistance transforming properties of the BJT is easier than using equivalent circuits.)
(b) Select a value for one capacitor to set the lower half-power frequency at 200 rad/s and select the second capacitor so that its reciprocal time constant is two octaves lower than that of the first capacitor.

8.15 The transistor in Fig. P8.15 is biased at 4 mA and has $k = 3 \times 10^{-3}$A/V^2.
(a) Find $A(s) = V_o/V_i$.
(b) Sketch the Bode plot.

8.16 Find an expression for the lower half-power frequency of the amplifier of Fig. P8.16.

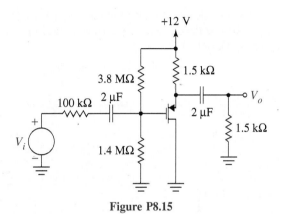

Figure P8.15

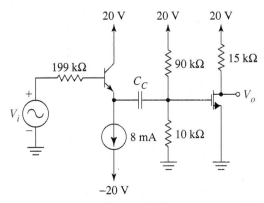

Figure P8.13

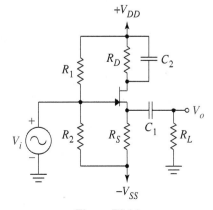

Figure P8.16

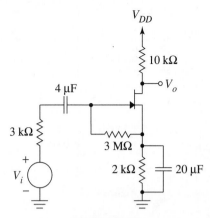

Figure P8.17

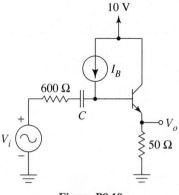

Figure P8.19

8.17 The transistor in Fig. P8.17 has $g_m = 5.6 \times 10^{-3}$ A/V. Use short-circuit time constants to estimate the lower half-power frequency.

8.18 In Fig. P8.18 $\beta = 180$ and the transistor is biased at 3 mA. For this circuit,
(a) Find the midrange gain.
(b) Find short-circuit resistances seen by C_1 and C_2.

8.19 In Fig. P8.19, $\beta = 200$.
(a) Find I_B so that the transistor is biased at $I_C = 2.5$ mA.
(b) Find the numerical value of r_π.
(c) Write an equation for C so that the low-frequency pole is located at 100 rad/s.

8.20 Use short-circuit time constants to estimate the lower half-power frequency for Fig. P8.20; $\beta = 99$, $r_\pi = 100$ Ω.

8.21 Use SPICE to check the lower half-power frequency of the redesigned amplifier of Example 8.4. Use $V_{CC} = 15$ V.

Section 8.4

8.22 In deriving Eq. (8.15) we limited our conclusions to frequencies lower than $\omega_{low} = g_m/C_\mu$, so that we could work with a simple expression.
(a) Starting with the ratio $\omega_\beta/\omega_{low}$, develop a convincing argument to show that $\omega_{low} >> \omega_\beta$.
(b) Show that $\omega_{low} >> \omega_\tau$ when $C_\pi \geq 10C_\mu$ as is often the case.

8.23 The derivation leading to Eq. (8.15) generalized the idea of transistor beta to $\beta(\omega)$, a complex function of ω.
Figure P8.23a is the common-base BJT model with parasitic capacitances added, and with terminations analogous to those of Fig. 8.11a.
(a) Derive an expression to show that α generalizes to $\alpha(\omega)$. Define an alpha cut-off frequency, ω_α, the -3dB frequency, in terms of circuit parameters.
(b) Relate ω_α to ω_τ.

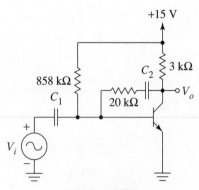

Figure P8.18

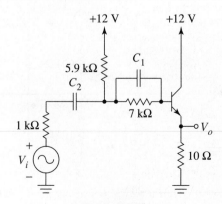

Figure P8.20

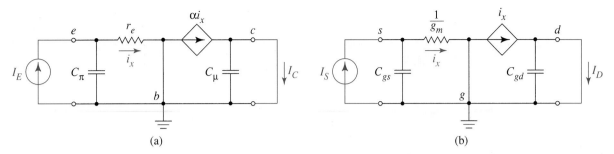

Figure P8.23

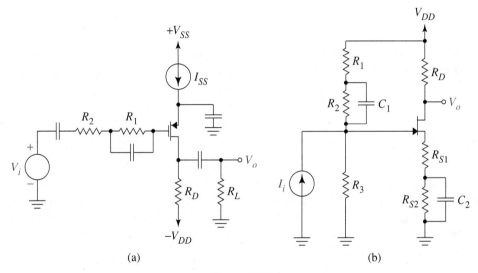

Figure P8.24

(c) Figure P8.23b is the common-gate FET model, augmented to include parasitic capacitances. Derive an expression for the short-circuit current gain of the FET and compare your answer with that for the BJT you found in part (a).

8.24 Draw the high-frequency equivalent for each circuit of Fig. P8.24.

8.25 In Fig. P8.25, transistor parameters are $k = 4 \times 10^{-3}$ A/V², $V_t = -1$ V, $C_{gs} = 2$ pF, and $C_{gd} = 3$ pF.
(a) Derive the high-frequency gain function $A_v(s) = V_o/V_i$.
(b) Sketch the Bode plot.
(c) Estimate the rise time if V_i is a small-amplitude step input.

8.26 Rework Problem 8.25 after first placing a 1 kΩ resistor in series with the signal generator and the source.

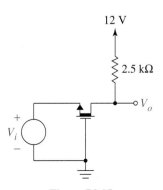

Figure P8.25

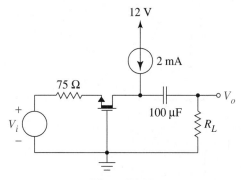

Figure P8.27

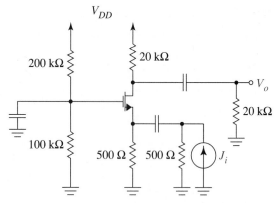

Figure P8.29

8.27 In Fig. P8.27 transistor parameters are $g_m = 0.8 \times 10^{-3}$ A/V, $C_{gs} = 2$ pF, and $C_{gd} = 3$ pF.
(a) Draw the high-frequency equivalent circuit.
(b) Use a source transformation to show that the circuit has two isolated high-frequency poles.
(c) Find the value of R_L so that there is a pole at 3.33×10^7 rad/s.
(d) Is the pole of part (c) dominant? Justify your answer.

8.28 Find the complex transresistance gain $A_R(s) = V_o/I_i$ for the circuit of Fig. P8.28.

8.29 In Fig. P8.29, transistor parameters are $C_{gs} = 3$ pF, $C_{gd} = 2$ pF, and $g_m = 4 \times 10^{-3}$ A/V. Current J_i is the input signal.
(a) Draw the high-frequency equivalent circuit.

For sinusoidal signals the transresistance gain, V_o/J_i has the general form

$$\frac{V_o}{J_i} = A_m \frac{\omega_{H1}}{j\omega + \omega_{H1}} \frac{\omega_{H2}}{j\omega + \omega_{H2}}$$

(b) From your circuit, find numerical values for ω_{H1}, ω_{H2}, and A_m.
(c) Estimate the rise time of this amplifier when a small step of current is applied.
(d) Why did part (c) specify that the current step be small?

Miller's Theorem

8.30 In Fig. P8.30, C_C and C_B are, respectively, large coupling and bypass capacitors, $g_m = 10^{-4}$ A/V, $V_A = 50$ V, $C_{gd} = 10$ pF, and $C_{gs} = 90$ pF.
(a) Draw the high-frequency equivalent circuit. Show component values.
(b) Find a *numerical* expression for the bandwidth in rad/s.
(c) The current source in the diagram is ideal. Would a real current source increase or decrease the bandwidth, compared to the ideal? Explain.

8.31 Exercise 8.5 predicted that replacing the 50 kΩ load resistor in Fig. 8.13a by 1.69 kΩ would increase the amplifier bandwidth to 20 MHz (assuming $r_b = 0$). Modify the SPICE code of Fig. 8.13d in this way and find the new bandwidth from SPICE output data. Also find the new value of gain.

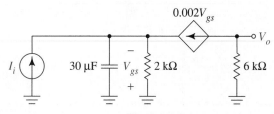

Figure P8.28

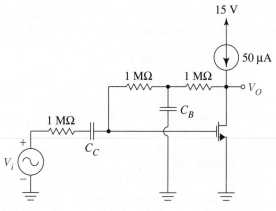

Figure P8.30

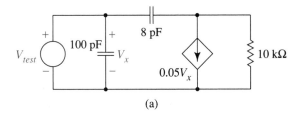

(a)

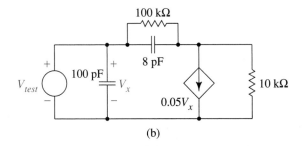

(b)

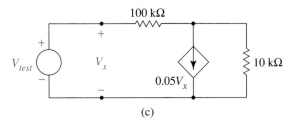

(c)

Figure P8.32

8.32 The circuits of Fig. P8.32 all have the form of Fig. 8.14a. For each, use the Miller approximation to draw the Miller equivalent, and then find the input impedance, Z_{in}, seen by V_t.

8.33 Find the exact gain function $A(s) = V_o/V_i$ for Fig. P8.33 and use the Miller approximation to find the gain function of its Miller equivalent. Compare the Bode plots of the two gain functions.

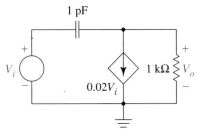

Figure P8.33

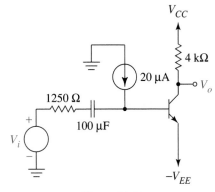

Figure P8.34

8.34 The transistor in Fig. P8.34 has $\beta = 110$, $r_b = 50\ \Omega$, $C_\mu = 50$ pF, and $C_\mu = 5$ pF.
(a) Draw the high-frequency equivalent and label all component values.
(b) Estimate the upper half-power frequency .
(c) Estimate the rise time.

8.35 Example 8.6 and Exercise 8.6 showed how the frequency response of Fig. 8.15a changes when the 20 k source resistance is reduced to 50 Ω. Use SPICE to find the bandwidth for both values of source resistance. Start with Fig. 8.16a.

8.36 Figure P8.36 is the high-frequency equivalent of an amplifier.
(a) Find the value of R_L if the upper half-power frequency is to be 5×10^6 rad/s. Assume that the Miller capacitance dominates the frequency response.
(b) Find the midrange gain, V_o/V_i for the value of R_L found in part (a).
(c) Write an expression for the voltage gain as a complex function of ω for the Miller equivalent. Include known numerical values using the answers to parts (a) and (b).

8.37 In Fig. P8.37, $g_m = 10^{-3}$ A/V, $C_{gs} = 2$ pF, and $C_{gd} = 4$ pF.
(a) Draw the Miller equivalent.

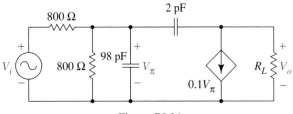

Figure P8.36

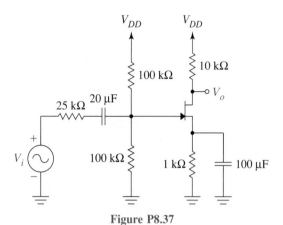

Figure P8.37

(b) The Miller circuit has two break frequencies. Calculate the-one most likely to dominate the high-frequency response.
(c) Find the midrange gain, V_o/V_i.
(d) Estimate the rise time of the output if a small step of volt-age is applied to the input.

8.38 Compute the frequency of the dominant high-frequency pole for Fig. P8.38 if parameters are $g_m = 0.2 \times 10^{-3}$ A/V, $C_{gs} = 2$ pF, and $C_{gd} = 3$ pF.

8.39 Use SPICE to find upper and lower half-power frequencies and midrange gain for the amplifier of Example 8.8.

8.40 Use SPICE to plot the magnitude of the input impedance for the amplifier of Example 8.8 for frequencies above 10^3 Hz. *Hints:* Replace the external signal source, its 50 kΩ internal resistance, and its coupling capacitor with a 1 A ac current source. Because the SPICE dynamic parameters for the transistor are

unknown, use the technique introduced in Example 8.5; that is, augment the static transistor model with linear capacitors.

Open-Circuit Time Constants

8.41 Equation (8.28) is the transfer function for the source-follower of Fig. 8.18.
(a) Find the pole and zero frequencies when $R_{TH} = 90$ kΩ, $R_S = 1$ kΩ, $C_{gs} = 2$ pF, $C_{gd} = 3$ pF, and $g_m = 0.002$ A/V.
(b) Make a rough sketch showing the general shape of the Bode plot for part (a) with break-frequencies labeled on the sketch.
(c) Use the answer to Exercise 8.9 to estimate the bandwidth using parameter values from part (a).
(d) Compare bandwidth results deduced from parts (a) and (c).

8.42 Use the open-circuit time constants of Fig. 8.15b to estimate the bandwidth of the amplifier of Example 8.6. Compare your answer with the Miller-effect prediction, $\omega_H = 2.33 \times 10^5$ rad/s.

8.43 Modify the amplifier of Fig. P8.38 by bypassing only 36 kΩ of the source resistor. Then use open-circuit time constants to estimate the upper half-power frequency.

8.44 The transistors in Fig. P8.44 have $k = 4 \times 10^{-3}$ A/V², $V_t = 1.0$ V, $C_{gs} = 2$ pF, and $C_{gd} = 4$ pF. Use open-circuit time constants to estimate the bandwidth.

8.45 In Fig. P8.45, both transistors are biased at 2 mA; both have $\beta = 120$ and $C_\pi = 100$ pF; and for both, $r_b = 0$. For Q_1, $C_\mu = 10$ pF; for Q_2, $C_\mu = 5$ pF. Estimate the rise time of the amplifier when a small current step is applied to the input node.

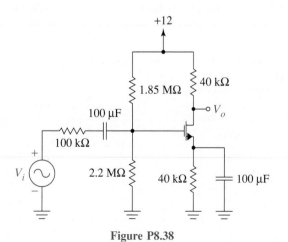

Figure P8.38

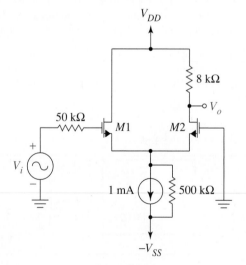

Figure P8.44

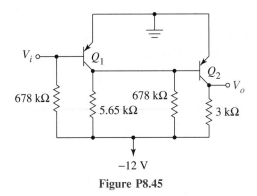

Figure P8.45

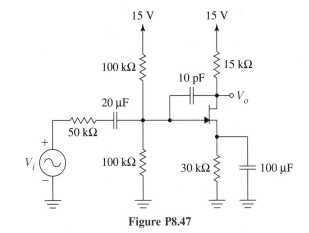

Figure P8.47

8.46 (a) Use SPICE ac analysis to find the midrange gain and bandwidth of Fig. P8.46.

(b) In your SPICE code, replace every capacitor except one with a dc current source of zero amperes. In successive SPICE *dc analyses*, use the current sources as test generators to find the resistances needed to implement Eq. (8.29).

(c) Use Eq. (8.29) to estimate the bandwidth. Also, identify the capacitor most important in determining the bandwidth, and justify your choice.

(d) Write the gain function $A(s)$ of a single-pole amplifier that approximates the frequency response of Fig. P8.46.

Reducing Bandwidth

8.47 In Fig. P8.47 the transistor has static parameters $V_P = -1$ V, $\beta = 4$ mA/V^2 and is biased at about 0.25 mA. Capacitances are $C_{gs} = 2$ pF and $C_{gd} = 4$ pF. The role of the 10 pF capacitor is to modify the high-frequency response.

(a) Draw the high-frequency-equivalent circuit.

(b) The approximate high-frequency circuit has two break fre-

quencies. Calculate the numerical value of the one most likely to dominate the high-frequency response.

(c) Find the midrange gain, V_o/V_i.

8.48 In Fig. P8.48, $g_m = 10^{-3}$ A/V, $C_{gs} = 3$ pF, and $C_{gd} = 4$ pF. Find the value of C_X that places the upper half-power frequency at 10^5 rad/s.

8.49 For Fig. P8.37,

(a) Find the capacitance that must be placed between drain and ground to reduce the upper -3dB frequency to 12 kHz. Ignore transistor capacitances.

(b) Find the capacitance required for the task if the capacitor is instead placed between drain and gate.

8.50 The transistor in Fig. P8.50 has $\beta = 210$; C_X and C_C determine the bandwidth.

(a) Design the circuit so that the transistor is biased in the ac-

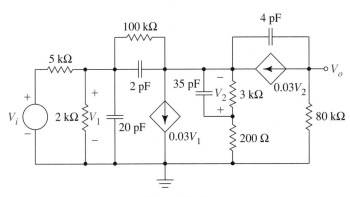

Figure P8.46

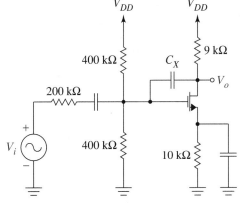

Figure P8.48

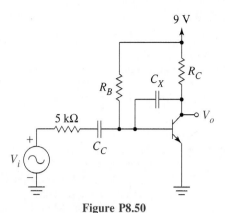

9 V

R_B

C_X

R_C

5 kΩ

C_C

V_o

V_i

Figure P8.50

tive region, the midrange gain has magnitude of at least 100, and the frequency response is within 3 dB of midrange between 600 Hz and 20 kHz. Assume the internal capacitances are negligible in this problem.
(b) Write inequalities involving C_μ and C_π that must be satisfied for the assumption of part (a) to be valid.

8.51 For the active-load amplifier of Fig. 7.56a, $I_O = 0.6$ mA, $V_{CC} = V_{EE} = 11$ V. Transistor parameters are $V_{AP} = 60$ V, $\beta_P = 50$, $C_{\mu P} = 25$ pF, $C_{\pi P} = 600$ pF, $V_{AN} = 130$ V, $\beta_N = 350$, $C_{\mu N} = 6$ pF, and $C_{\pi N} = 200$ pF, where P and N denote pnp and npn transistors, respectively. The current-source small-signal parameters are 217 kΩ and 12 pF, respectively.
(a) Apply a pure difference-mode ac input voltage; then use SPICE ac analysis to plot the frequency-response magnitude and phase curves for voltage gain.
(b) Repeat part (a) after adding a 60 pF capacitor between the base and collector of Q_2.

8.52 Replace the transistors in Fig. P8.47 with transistors having these SPICE parameters.

BETA	$1.3×10^{-3}$ **A/V²**
VTO	**−2V**
LAMBDA	$2.3×10^{-3}$
CGD	**1.5 pF**
CGS	**2.5 pF**
IS	**0.034pA**

(a) Use dc analysis to find the Q-point. The transistor in Problem 8.47 was biased at $I_D = 0.25$ mA. How much did the Q-point change from transistor replacement?
(b) Use SPICE to find the frequency response.
(c) What change in the frequency response do you expect if you change the 10 pF capacitor to 5 pF? Use SPICE to check your speculation.

Section 8.5

8.53 Find the midrange gain and the bandwidth of Fig. 7.36a when the input is a difference-mode signal v_d and the output is taken between the drains. Parameter values are $V_{DD} = 10$ V, $V_{SS} = 8$ V, $I_O = 0.5$ mA, $R_D = 10$ kΩ, $k = 4 × 10^{-3}$ A/V², $V_t = 0.9$ V, $C_{gs} = 2$ pF, $C_{gd} = 5$ pF.
Hint: Use a high-frequency half-amplifier equivalent.

8.54 Use the high-frequency BJT model to show that the output capacitance of a basic current mirror approaches $2C_\mu$ at low frequencies and $C_\mu 2C_\pi/(C_\mu + 2C_\pi)$ at high frequencies. You will need to make familiar approximations.

8.55 In Fig. P8.55, $g_m = 4.47 × 10^{-4}$ A/V, $C_{gs} = 3$ pF, and $C_{gd} = 2$ pF. Find the gain function, $A_d(s)$, for double-ended output. The signal source is balanced and has resistance $R_i = 600$ Ω.

8.56 For Fig. P8.55, $g_m = 4.47 × 10^{-4}$ A/V, $C_{gs} = 3$ pF, and $C_{gd} = 2$ pF. Find the gain function, $A_d(s)$, when a 20 kΩ resistor is placed between the output nodes and the voltage across this resistor is taken as output. The signal source is balanced and has resistance $R_i = 5$ kΩ.

8.57 For Fig. P8.55, $g_m = 4.47 × 10^{-4}$ A/V, $C_{gs} = 3$ pF, and $C_{gd} = 2$ pF. Find the gain function, $A_{d,s-e}(s)$, when a 20 kΩ resistor is capacitively coupled between each drain node and ground, and the voltage across one of these resistors is taken as output. The signal source is balanced and has resistance $R_i = 5$ kΩ.

8.58 Use SPICE to find the frequency response of the amplifier described in Problem 8.53; however, to make the simulation more realistic, also include $V_A = 105$ V. Plot dB gain versus frequency.

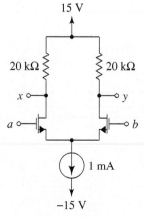

15 V

20 kΩ   20 kΩ

x y

a b

1 mA

−15 V

Figure P8.55

Hint: For a pure difference-mode signal, use an ac source of 0.5 V at one input and an appropriate VCVS of transmittance -1 at the other.

8.59 Find and label a common-mode gain curve resembling Fig. 8.23d for the amplifier of Fig. 7.38a when the input is a common-mode signal v_c and the output is taken at node y. Parameter values are $I_O = 0.5$ mA, $R_D = 10$ kΩ, $k = 4 \times 10^{-3}$ A/V^2, and $V_t = 0.9$ V. The output resistance and capacitance of the current source are 210 kΩ and 3 pF, respectively. *Hint*: Draw the half-amplifier equivalent.

8.60 Use SPICE to plot the common-mode gain for the amplifier described in Problem 8.59. For the two transistors, use $V_A = 105$ V and GAMMA = 0.32. Include transistor $C_{gs} = 2$ pF and $C_{gd} = 4$ pF in your analysis by adding external capacitors to the static model. $V_{CC} = 10$ V, $V_{SS} = 8$ V.

8.61 (a) Draw the high-frequency equivalent circuit for Fig. 6.22c with a test generator connected between the drain and source nodes of the output transistor. Include r_o in the transistor models.
(b) Find an expression for the output admittance of the current source as a function of radian frequency ω. Use simplifying assumptions to fashion a tractable expression.

8.62 Figure P8.62 is the MOSFET version of Fig. 8.23c, that is, a high-frequency, common-mode, half-amplifier model for a source-coupled pair. Find V_o/V_C as a function of complex frequency s.
Hint: Use a current-source transformation followed by a voltage divider equation. For compact notation, start by denoting the parallel RC impedance by $Z(s)$.

Section 8.6

8.63 A MOSFET version of Fig. 8.25e has transistors with $g_m = 0.0025$ A/V, $r_o = 80$ kΩ, $C_{gs} = 3$ pF, and $C_{gd} = 7$pF. The internal resistance of the signal source is 50 Ω and the load resistor is 7.5 kΩ.
(a) Draw the high-frequency equivalent circuit.
(b) Find the midrange gain.
(c) Find ω_H.

8.64 In Fig. P8.64, M_1 has capacitances $C_{gs} = 10$ pF and $C_{gd} = 11$ pF, and M_2 capacitances are $C_{gs} = 20$ pF and $C_{gd} = 21$ pF. For both, $g_m = 0.002$ A/V. Capacitors explicitly shown are large bypass and coupling capacitors.
(a) Draw the high-frequency equivalent circuit.
(b) By circuit analysis, find the midrange voltage gain.
(c) Find the pole frequencies.

8.65 The circuit of Fig. 8.25f has identical transistors biased at 0.5 mA. Source resistance is 600 Ω, and the load resistor is 15 kΩ. Parameters are $r_b = 120$ Ω, $r_\pi = 15$ kΩ, $V_A = 180$ V, $C_\pi = 28$ pF, and $C_\mu = 4$ pF.
(a) Draw the high-frequency equivalent circuit.
(b) Compute f_τ for the transistors.
(c) Use SPICE analysis to determine the upper half-power frequency of the amplifier and compare it with f_τ.

8.66 Figure P8.66 shows the small-signal schematic for an amplifier and the common-base equivalent for the transistor. Q_1 is biased at 0.5 mA and Q_2 at 0.25 mA. For both transistors, β = 300, $r_b = 0$, $C_\pi = 10$ pF, and $C_\mu = 4$ pF.
(a) Draw the high frequency equivalent circuit; for Q_2, use the

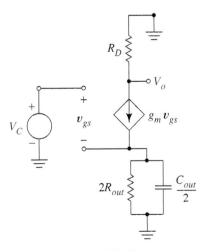

Figure P8.62

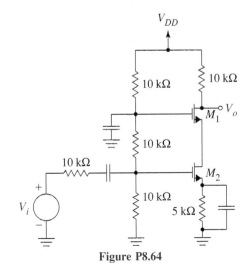

Figure P8.64

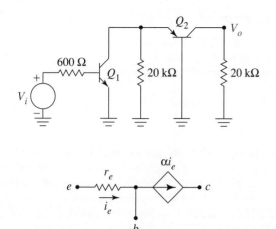

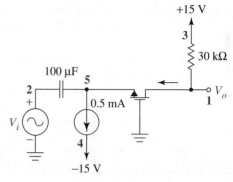

Figure P8.70

e ⚬—ⱳⱳⱳ—•——————◇—• c

_re_ ... αi_e_

b

Figure P8.66

IS = 0.014 pA   BF = 250   VAF = 70 V   RB = 10 Ω
VJC = 0.75 V   VJE = 0.75 V   CJE = 20 pF   CJC = 7 pF
TF = 0.4 ns

For each value of bias current, tabulate the following parameter values: g_m, r_π, r_o, β, C_{BE}, C_{BC}, and f_τ.

Hint: Use the .OP statement if your SPICE installation does not print out these values automatically.

8.70 Use SPICE to plot frequency response curves like those of Fig. 8.30c for the amplifier of Fig. P8.70. For the second SPICE run, change the bias current to 0.05 mA. SPICE parameters include

VTO = 3   KP = 2E-4   CBD = 20FF   CBS = 20FF
RD = 1   RS = 1
GAMMA = 0.37   LAMBDA = 0.01   TOX = 0.1U

Also add the transistor width and length parameters, $W = 1U$, $L = 1U$, to the element line that describes the transistor connections in the SPICE code.

common-base model with capacitors inserted in the proper locations. Show the numerical value of each component on the diagram.
(b) Compute the midrange gain of the amplifier.
(c) After applying Miller's theorem to Q_1, find the three pole frequencies.

8.67 (a) Use SPICE ac analysis to find the −3 dB bandwidth of the amplifier of Fig. 8.27a when both transistors are 2N2222 transistors having the following approximate SPICE parameters:
IS = 0.014 pA   BF = 250   VAF = 70 V   RB = 10 Ω
VJC = 0.75   VJE = 0.75   CJE = 20 pF   CJC = 7 pF
TF = 0.4 ns
Include the .OP code so your output file contains SPICE-calculated small-signal parameters.
(b) Draw a high-frequency equivalent circuit like Fig. 8.27 for this amplifier, using the small-signal parameter values from the SPICE output file to label transistor parameters in your circuit.
(c) Repeat part (a) after halving both of the dc bias currents.

Section 8.7

8.68 Draw the high-frequency, hybrid pi equivalent for the transistor described in Table P8.68 and label with parameter values. $I_C = 1.5$ mA.

8.69 Find the two frequency response curves, as described in Example 8.17, if the transistor is replaced with one having the following SPICE parameters:

TABLE P8.68

Parameter	Numerical Value	Measurement Condition
h_{fe}	195	$I_C = 1.5$ mA, $V_{CE} = 2$ V
h_{ie}	3.325 kΩ	$I_C = 1.5$ mA, $V_{CE} = 2$ V
h_{oe}	4.15×10^{-5} A/V	$I_C = 1.5$ mA, $V_{CE} = 2$ V
h_{re}	6.5×10^{-10}	$I_{CE} = 1.5$ mA, $V_{CE} = 2$ V
$C_{CBO} = C_\mu$	0.2 pF	$V_{CB} = 2$ V
$C_{EBO} = C_{je}$	0.40 pF	$V_{EB} = 2$ V
f_τ	1.7 GHz	$I_C = 1$ mA, $V_{CE} = 2$ V

CHAPTER 9

Feedback Circuits

Negative feedback is a general system-design strategy whereby a quantity proportional to the system's output is subtracted from a reference input, with the difference used as the actual input. The advantages of adding negative feedback to an amplifier are

1. Reduced sensitivity to parameter variations
2. Increased bandwidth
3. Reduced nonlinear distortion
4. Improved input and output resistances

One disadvantage is a reduction in gain; however, gain is easy to obtain with modern circuits, so this is a small price to pay. We simply anticipate the gain reduction and begin with a nonfeedback amplifier having excess gain. The other disadvantage, a possibility of unwanted oscillations or instability, is more serious, for oscillations make the circuit worthless as an amplifier.

Textbooks on control systems deal with general feedback theory in great detail. Here we consider only the feedback amplifier, and then consider only those aspects necessary for a basic understanding of the subject. This chapter assumes the reader has a knowledge of two-port theory at the level treated in Appendix B. Amplifiers biased for small-signal operation are assumed throughout the chapter.

9.1
Ideal Negative Feedback Theory

General Feedback Amplifier Configuration. Figure 9.1 shows the general structure of a *feedback amplifier*. A nonfeedback amplifier with gain A (the A circuit) delivers an output signal $x_o = Aw_i$ to an external load. Instead of using signal w_s as the amplifier input, a feedback circuit (β circuit) produces a feedback signal $w_f = \beta x_o$ that is subtracted from w_S to form the actual amplifier input w_i. *The feedback factor, β, used extensively in this chapter, is unrelated to the transistor beta.* To avoid confusion, in this chapter only we return to the practice of using β_F to denote the forward current gain of the BJT.

Feedback theory applies to all four amplifier types named in Fig. 1.17, but each type implies a different interpretation of Fig. 9.1. For a voltage amplifier, A denotes voltage

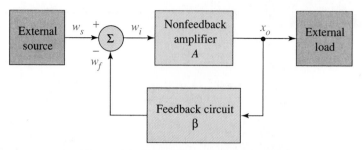

Figure 9.1 Feedback amplifier structure.

gain, and x_o and w_i are voltages. The external source must be a *source of voltage,* w_s (ideally having zero output resistance), and there is a *voltage subtraction* at the input; feedback signal w_f must also be a voltage. The feedback factor $\beta = w_f/x_o$ is a voltage gain for this case.

For a current amplifier, A denotes current gain. The input comes from a *source of current,* w_s (ideally having infinite-output resistance), and a *current subtraction* is required at the input. Here β denotes current gain of the β circuit.

In a transconductance amplifier, A is transconductance gain with dimensions of amperes/volt. Thus output x_o is a current, and the source w_s is a voltage source; the subtraction at the input is a *voltage subtraction.* For this case β has dimensions of volts/ampere, or transresistance!

Finally, the transresistance amplifier has transresistance gain A with units of volts/ampere. The output x_o is a voltage, and the external source is a current source; the subtraction is a *current subtraction.* For this case, β has dimensions of transconductance. The notation in Fig. 9.1 reminds us that the quantities involved in the subtraction at the input must have identical units, whereas the units of x_o might or might not be the same as the input units.

We next derive an *ideal feedback equation* that provides insight and leads to simple calculations and estimates. From Fig. 9.1,

$$x_o = Aw_i = A(w_s - w_f) = A(w_s - \beta x_o)$$

Solving for the gain of the feedback amplifier, $A_f = x_o/w_s$, gives

$$A_f = \frac{x_o}{w_s} = \frac{A}{1 + A\beta} \tag{9.1}$$

which applies to all four amplifier classes.

Reduction in Gain. For midrange frequencies, we will find both A and β are real quantities having the same algebraic sign. Equation (9.1) shows one of the prices we pay for negative feedback, a gain reduction. For example, suppose $A = 10^4$ and the fraction of x_o that is fed back to the input is $\beta = 0.01$. Then the resulting feedback amplifier has gain of only

$$A_f = \frac{10,000}{1 + 100} = 99$$

Usually, as in this illustration, $A\beta \gg 1$, and Eq. (9.1) gives the important approximation

$$A_f \approx \frac{1}{\beta} \tag{9.2}$$

Common practice is to obtain β from a resistor network using resistors of low-temperature coefficient. Sometimes β is determined by a voltage divider, giving

$$A_f \approx \frac{1}{\beta} = \frac{R_1 + R_2}{R_1} = 1 + \frac{R_2}{R_1}$$

In an IC realization, such a β is easily controlled to within 1% or so by masking tolerances, and temperature variations in the resistors cancel over a wide temperature range, as we saw in Sec. 3.3.6. Thus A_f is quite constant, even though A might vary greatly with temperature or other factors.

Improvement Factor. The denominator of Eq. (9.1), $1 + A\beta$, is of considerable importance in feedback theory. This factor by which the gain is reduced also turns out to be the degree of improvement effected by introducing feedback. Ideally, sensitivity, distortion, bandwidth, input resistance, and output resistance all improve by $1 + A\beta$ when we add feedback; thus we refer to the quantity as the *improvement factor.* Since $1 + A\beta$ is so important, let's examine some practical ways to determine its numerical value.

In analysis, suppose that we measure or calculate amplifier gain with and without feedback; that is, we know A and A_f. From Eq. (9.1) we can calculate the improvement factor by forming the ratio. If needed, we can also find β once $1 + A\beta$ and A are both known. In design, we typically know A and also how much improvement is required; that is, we know $1 + A\beta$. From these we can find the amount of feedback, β, we need to add. We next examine in greater detail how feedback improves the performance of an amplifier.

Let us first develop an intuitive notion of how feedback works. Suppose in Fig. 9.1, while the input w_s is held constant, output x_o decreases in amplitude *for any reason.* Being proportional to x_o, feedback signal w_f also decreases, thereby *increasing* w_i to compensate, at least partially, for the original change; a complementary scenario opposes increases in x_o. Thus feedback opposes any change in x_o caused by an event unrelated to w_s. If the cause is variation in an internal parameter p, we say that feedback has improved the sensitivity. If it is loss of gain at either low or high frequencies because of amplifier capacitances, we say feedback has improved the frequency response. If x_o changes in a manner not proportional to w_s because of nonlinearities within the amplifier, we say feedback has reduced distortion. We now examine the details to obtain a more quantitative understanding to use in analysis and design.

9.2
Effects on Sensitivity, Bandwidth, and Distortion

9.2.1 EFFECT OF FEEDBACK OF SENSITIVITY

If the sensitivity of gain A to parameter p is S_p^A, adding negative feedback results in an amplifier of gain A_f with lower sensitivity; specifically,

$$S_p^{A_f} = \frac{1}{1 + A\beta} S_p^A \tag{9.3}$$

To prove this, we need only use the definition

$$S_p^{A_f} = \frac{p}{A_f}\frac{dA_f}{dp} = \frac{p}{A_f}\frac{d[A/(1 + A\beta)]}{dp}$$

Because the circuits we design employ feedback β that is independent of p, the derivative gives

$$S_p^{A_f} = \frac{p}{A_f}\left[\frac{(1 + A\beta)(dA/dp) - A\beta(dA/dp)}{(1 + A\beta)^2}\right] = \frac{p}{A_f}\left[\frac{1}{(1 + A\beta)} - \frac{A\beta}{(1 + A\beta)^2}\right]\frac{dA}{dp}$$

Substituting for A_f from Eq. (9.1) gives

$$S_p^{A_f} = \frac{p(1 + A\beta)}{A}\left[\frac{1}{(1 + A\beta)} - \frac{A\beta}{(1 + A\beta)^2}\right]\frac{dA}{dp} = \frac{p}{A}\left[1 - \frac{A\beta}{(1 + A\beta)}\right]\frac{dA}{dp}$$

which is equivalent to Eq. (9.3).

Exercise 9.1   A nonfeedback amplifier with $A = 347$ contains a critical resistor R. By laboratory measurements we determine that $S_R^A = 1.12$. After adding negative feedback to reduce the sensitivity, we find the amplifier gain is 24. Find the sensitivity of the feedback amplifier to changes in R.

Ans. 0.077.

This demonstrates analysis. The following example uses Eq. (9.3) in a design context.

EXAMPLE 9.1   We have an amplifier with gain of 800. Sensitivity of gain to temperature change is 0.1; however, our specifications require temperature sensitivity of 0.001. Investigate the possibility of using negative feedback to bring the temperature sensitivity within specifications.

Solution. From Eq. (9.3), we need an improvement factor of

$$1 + A\beta = \frac{0.1}{0.001} = 100$$

Since $A = 800$, β must satisfy

$$1 + 800\beta = 100$$

or $\beta = 0.124$. The resulting feedback amplifier meets the temperature-sensitivity specification; however, its gain is only $800/100 = 8$.   ❏

Sensitivity of Cascaded Feedback Amplifiers. Example 9.1 showed that we can improve sensitivity by sacrificing gain. Because specifications often involve both gain and sensitivity, it is interesting to see if we can come out ahead by using a cascade of low-gain, low-sensitivity feedback amplifiers.

Consider a cascade of n feedback amplifiers, each of gain A_f. We denote the overall gain by $G = (A_f)^n$, and now need to find the sensitivity of G to parameter p. By definition,

$$S_p^G = \frac{p}{G}\frac{dG}{dp} = \frac{p}{G}n(A_f)^{n-1}\frac{dA_f}{dp}$$

Substituting $G = (A_f)^n$ gives

$$S_p^G = \frac{p}{(A_f)^n}n(A_f)^{n-1}\frac{dA_f}{dp} = n\frac{p}{A_f}\frac{dA_f}{dp}$$

We conclude that

$$S_p^G = nS_p^{A_f}$$

which shows that sensitivity increases only in proportion to n as we cascade stages, whereas G varies as the power of n. This suggests that we should be able to come out ahead using feedback. Table 9.1 illustrates the point using numbers from Example 9.1. Cascading feedback amplifiers is obviously an effective strategy, reminiscent of improving gain-bandwidth product by cascading identical stages in Example 2.8.

9.2.2 EFFECT OF FEEDBACK ON BANDWIDTH

Upper Half-Power Frequency. Consider using feedback to increase the bandwidth of an amplifier described by

$$A = A(\omega) = A_{mid}\frac{\omega_H}{j\omega + \omega_H} \tag{9.4}$$

Substituting this expression into Eq. (9.1) gives

$$A_f = \frac{A_{mid}[\omega_H/(j\omega + \omega_H)]}{1 + A_{mid}[\omega_H/(j\omega + \omega_H)]\beta}$$

$$= A_{mid}\frac{\omega_H}{j\omega + \omega_H + A_{mid}\omega_H\beta}$$

$$= A_{mid}\frac{\omega_H}{j\omega + \omega_H(1 + A_{mid}\beta)}\left(\frac{1 + A_{mid}\beta}{1 + A_{mid}\beta}\right)$$

TABLE 9.1 Gain and Sensitivity of Cascaded Feedback Amplifiers

n	G	S_p^G
1	8	0.001
2	64	0.002
3	512	0.003
4	4096	0.004

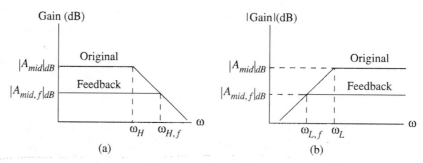

Figure 9.2 Using negative feedback to improve bandwidth: (a) high-frequency response; (b) low-frequency response.

or

$$A_f = A_{mid,f} \frac{\omega_{H,f}}{j\omega + \omega_{H,f}} \tag{9.5}$$

where

$$A_{mid,f} = \frac{A_{mid}}{1 + A_{mid}\beta} \quad \text{and} \quad \omega_{H,f} = \omega_H(1 + A_{mid}\beta)$$

Comparing Eqs. (9.5) and (9.4) shows that the feedback amplifier's gain function has the same general form as the original; however, its upper half-power frequency is higher by $1 + A_{mid}\beta$, and its midrange gain is lower by the same factor. Figure 9.2a compares the two frequency response curves. Notice the original and final gain-bandwidth products, $(A_{mid})\omega_H$ and $(A_{mid,f})\omega_{H,f}$. Since ω_H and $\omega_{H,f}$ are the bandwidths before and after adding feedback, respectively, we see that the gain-bandwidth product is preserved when we add feedback. We conclude that for a one-pole amplifier described by Eq. (9.4), negative feedback facilitates a direct trade-off of gain for bandwidth as in Fig. 9.2a. Amplifiers with more complex gain functions do not give an exact trade-off; however, we expect an amplifier with a dominant high-frequency pole to approximate such a trade-off.

Lower Half-Power Frequency. We next consider the effect of feedback on low-frequency response. Consider a nonfeedback amplifier described by

$$A = A(\omega) = A_{mid} \frac{j\omega}{j\omega + \omega_L} \tag{9.6}$$

Substituting into Eq. (9.1) gives

$$A_f = \frac{A_{mid}[\,j\omega/(j\omega + \omega_L)]}{1 + A_{mid}[\,j\omega/(j\omega + \omega_L)]\beta}$$

Algebraic manipulations similar to those for the high-frequency case lead to

$$A_f = A_{mid,f} \frac{j\omega}{j\omega + \omega_{L,f}} \tag{9.7}$$

where

$$\omega_{L,f} = \frac{\omega_L}{1 + A_{mid}\beta} \quad \text{and} \quad A_{mid,f} = \frac{A_{mid}}{1 + A_{mid}\beta}$$

From Eqs. (9.6) and (9.7), we conclude that feedback lowers the half-power frequency and midrange gain by $1 + A_{mid}\beta$; Fig. 9.2b shows how the Bode plot changes. A practical consequence is that coupling and bypass capacitors can be smaller for the same lower -3 dB frequency in a feedback amplifier. More complicated low-frequency gain functions than Eq. (9.6) are not changed in exactly this fashion; however, if there is a dominant low-frequency pole, reduction in ω_L by $1 + A_{mid}\beta$ is usually a reasonable estimate. Wideband amplifiers approximated by Eq. (9.6) at low frequencies and Eq. (9.4) at high frequencies have their passband extended at each side by $1 + A_{mid}\beta$.

Exercise 9.2 We need an amplifier with a rise time of 300 ns. The amplifier we have has rise time of 3 μs and gain of 40. Find the resulting gain and β, if we correct the rise time problem with negative feedback.

Ans. 4, 0.225.

EXAMPLE 9.2 We need an amplifier that meets the following specifications: gain ≥ 80, $f_L \leq 50$ Hz, $f_H \geq 15$ kHz, and sensitivity to power supply changes ≤ 0.2. We have an amplifier with gain $= 1000$, $f_L = 400$ Hz, and $f_H = 9$ kHz, and its sensitivity to changes in supply voltage is 1.5. Determine whether we can meet the specifications by adding feedback to the existing amplifier. If so, find an acceptable value of β, and give final specifications for the feedback amplifier.

Solution. Assume that a pair of dominant poles characterize the frequency response. Considering each specification individually, gives the following requirements:

$$\text{gain:} \quad \frac{1000}{1 + 1000\beta} \geq 80$$

$$\text{lower} - 3 \text{ dB frequency:} \quad \frac{400}{1 + 1000\beta} \leq 50 \text{ Hz}$$

$$\text{upper} - 3 \text{ dB frequency:} \quad 9000(1 + 1000\beta) \geq 15,000 \text{ Hz}$$

$$\text{sensitivity:} \quad \frac{1.5}{1 + 1000\beta} \leq 0.2$$

These lead, respectively, to the requirements $1 + 10^3\beta \leq 12.5$, $1 + 10^3\beta \geq 8$, $1 + 10^3\beta \geq 1.67$, and $1 + 10^3\beta \geq 7.5$. Sorting out these inequalities shows that any β that satisfies $8 \leq 1 + 10^3\beta \leq 12.5$ is acceptable. To leave room for error, choose $1 + 10^3\beta = 10$, which requires $\beta = 0.009$. Then the final amplifier will have gain $= 100$, $f_L = 40$ Hz, $f_H = 90$ kHz, and supply sensitivity of 0.15, all within specifications. ❏

9.2.3   EFFECT OF FEEDBACK ON NONLINEAR DISTORTION AND NOISE

Nonlinear distortion is not a problem in the small-signal amplifiers we have studied so far; however, in power amplifiers output signals must be large enough to develop a specified amount of signal power in the load. This often requires signal swings large enough to traverse a significant portion of the amplifier's nonlinear transfer characteristic. When this occurs, the output signal is the amplified input signal plus additive harmonic distortion components that arise within the amplifier itself. Amazingly, negative feedback can be used to feed these distortion components back into the input in such a way that they *subtract from themselves.* The feedback also causes some signal to subtract from itself; however, this is the gain reduction we expect with negative feedback, and we can compensate for it by adding a preamplifier. The bottom line is that negative feedback reduces internally generated distortion by $1 + A\beta$. The following quasi-numerical development shows exactly how it works.

Figure 9.3a shows how large signal swings generate distortion. To be specific suppose that when the available input signal $w_i = 0.001 \cos \omega t$ is applied, the output is the Fourier series

$$x_o(t) = 10 \cos \omega t + d(t)$$

where distortion $d(t)$ consists of Fourier harmonic terms and accounts for the distorted appearance of the output. Figure 9.3b is a linearized model of this nonlinear amplifier.

Suppose design specifications require an output signal of $10 \cos \omega t$ for the given $w_i(t)$, but also require that the distortion be much less than $d(t)$. In this type of problem we can add negative feedback as in Fig. 9.3c to cancel the distortion. For Fig. 9.3c,

$$x_o(t) = d(t) + A[w_s(t) - \beta x_o(t)]$$

or

$$x_o(t) = \frac{d(t)}{1 + A\beta} + \frac{A}{1 + A\beta} w_s(t)$$

Using $\beta = 0.01$, for example, gives

$$x_o(t) = 0.0099 d(t) + 99 w_s(t) \tag{9.8}$$

showing that distortion is reduced to 1% of its former value. Notice, however, that the new input $w_s(t)$ is multiplied by only 99 in Eq. (9.8) instead of the original 10^4. Because specifications require $10 \cos \omega t$ at the output, from Eq. (9.8) the input to the *feedback* amplifier must satisfy

$$99 w_s(t) = 10 \cos \omega t$$

or $w_s(t) = 0.101 \cos \omega t$, showing that the feedback amplifier requires a larger input voltage than the nonfeedback amplifier for the same output. The reason is evident from Fig. 9.3d. The transfer characteristic of the feedback amplifier, x_o versus w_s, is much more linear than the original curve, but has lower slope. Since only $0.001 \cos \omega t$ is available

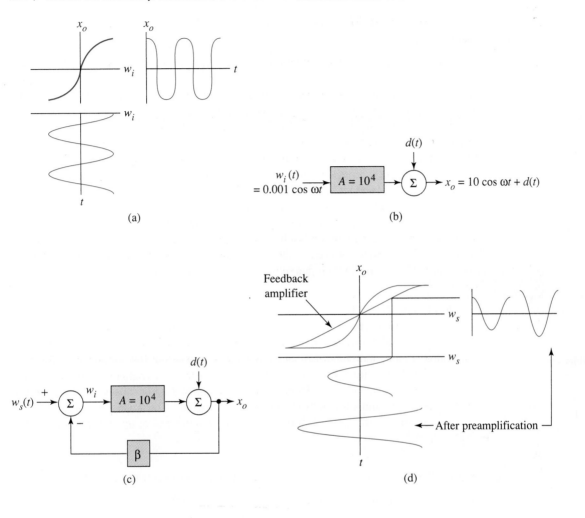

(a)

(b)

(c)

(d)

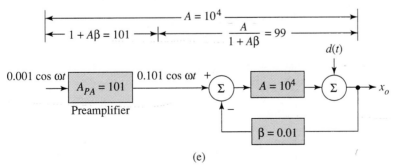

(e)

Figure 9.3 Reduction of nonlinear distortion with negative feedback: (a) origin of harmonic distortion; (b) linear model of a nonlinear amplifier; (c) canceling harmonics with feedback; (d) transfer curves of nonfeedback and feedback amplifiers; (e) feedback amplifier and preamp with example signal levels.

to us for our input signal, we must add a preamplifier of gain, A_{PA}, at the input that satisfies

$$0.001A_{PA} = 0.101$$

or $A_{PA} = 101$. Figure 9.3d shows how this larger input signal produces the required output amplitude by projection from the more linear curve of the feedback amplifier.

Figure 9.3e summarizes the design, both in general notation and in the numerical values used as examples. For this strategy to succeed, we must construct a preamp that does not produce distortion. The key observation is that the signal amplitude required at the preamp output is much smaller than the specified amplifier output voltage. That is, it is much easier to produce a distortion-free output voltage of 0.101 V than 10 V.

If $d(t)$ represents noise arising within the nonfeedback amplifier, negative feedback reduces output noise just as it reduces distortion; however, it can be shown that adding negative feedback results in no improvement in the input *signal-to-noise ratio* of an amplifier. Since this ratio is usually the critical factor, adding negative feedback is not a generally effective noise reduction strategy.

9.3
Classes of Feedback Amplifiers

Feedback amplifiers divide into four subclasses: voltage series, voltage shunt, current series, and current shunt. All have the increased bandwidth, reduced sensitivity, and lower distortion that characterize negative feedback circuits; they differ in how the feedback affects input and output resistance. We can associate the four feedback classes with the four ideal amplifiers, a logical comparison that helps us remember the differences between the feedback classes. To be concise, we call the nonfeedback amplifier of Fig. 9.1 the *A circuit* and the feedback circuit the β *circuit*. The remainder of the chapter utilizes the conventions of two-port theory, a topic described at an introductory level in Appendix B.

9.3.1   IDEAL AMPLIFIERS AND FEEDBACK CONFIGURATIONS

Our feedback amplifier design goal is one of four possibilities—a voltage, current, transresistance, or transconductance amplifier—idealized as one of the dependent sources of Figs. 9.4a–d. We add negative feedback to an existing nonfeedback amplifier in such a way that the original comes to more closely resemble our design objective. We can model our original nonfeedback amplifier or *A* circuit in any of the equivalent forms of Figs. 9.4e–h by using a source transformation and/or a change in controlling variable. Thus, the existing amplifier differs from our idealized design goal only in the presence of parasitic input and output resistance. We use feedback to modify these input and output resistances; of course, we must sacrifice gain.

If the goal is a voltage amplifier like Fig. 9.4a, we represent the *A* circuit as Fig. 9.4e. To make the given amplifier more closely resemble the ideal, negative feedback must decrease output resistance and increase input resistance. For this design goal, gain *A* in Fig. 9.1 is *voltage gain* a_v.

To design a current amplifier, we visualize the *A* circuit as Fig. 9.4f and the goal as Fig. 9.4b. Here, feedback should increase output resistance and decrease input resistance. In ideal feedback notation, $A = a_i$, *current gain*.

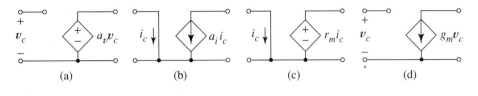

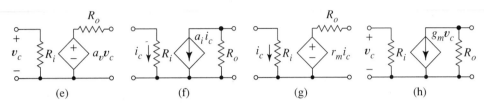

Figure 9.4 Idealized design goals and A circuit representations: (a) and (e) voltage amplifier; (b) and (f) current amplifier; (c) and (g) transresistance amplifier; (d) and (h) transconductance amplifier.

For a transresistance amplifier, feedback makes the original amplifier, Fig. 9.4g, more closely approximate Fig. 9.4c by decreasing both output resistance and input resistance. Gain A is *transresistance gain* r_m.

Finally, for a transconductance amplifier, the output and input resistances of Fig. 9.4h must both be increased to better approach the design goal of Fig. 9.4d. Here, $A = g_m$, *transconductance gain*.

Because resistances of the signal sources and loads obscure the main feedback concepts, it helps to postpone dealing with these resistances until general principles are well established. Thus we temporarily associate with each ideal amplifier an *ideal source* and an *ideal load*, the latter either a short circuit or an open circuit.

Ideal Signal Sources and Loads. The ideal amplifiers with open-circuit inputs, Figs. 9.4a and d, must be driven by voltage sources, since a current source at the input would violate KCL. The amplifiers with short-circuit inputs, Figs. 9.4b and c, must be driven by current sources, for voltage sources would violate KVL.

Open-circuit loads make sense for ideal amplifiers with voltage outputs, Figs. 9.4a and c, since a short-circuit load would violate KVL. On the other hand, the amplifiers with current outputs, Figs. 9.4b and d, use short circuits as ideal loads to avoid violating KCL.

Before adding feedback, we must make two choices: to improve output resistance, we must choose either voltage or current feedback; to improve input resistance, we must choose series or shunt feedback. As we define these feedback *classes*, we visualize both A and β circuits as two-ports.

Voltage Feedback. For voltage and transresistance amplifiers, we need to reduce output resistance so the amplifier outputs more closely resemble *voltage* sources. This takes *voltage feedback*, which employs the structure of Fig. 9.5a. The feedback network senses the A circuit's output *voltage*, and this information is fed back to the input. *Voltage feedback* opposes anything, save input signal, that attempts to change the output *voltage*; thus, voltage feedback makes the output more closely represent a *voltage*

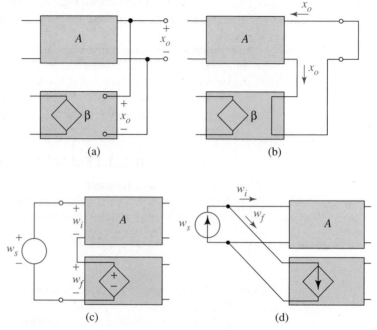

Figure 9.5 Negative feedback structures: (a) voltage feedback; (b) current feedback; (c) series feedback; (d) shunt feedback.

source by reducing output resistance. The β circuit of Fig. 9.5a has an open circuit at its input (right-hand side), thus providing an ideal load for a circuit of low output resistance.

Current Feedback. Current feedback has the structure of Fig. 9.5b, in which the feedback signal is created by sensing the amplifier's output *current. Current feedback always opposes changes in current,* making the output more closely resemble a constant *current source* (increases output resistance). Therefore, current feedback is appropriate for current and transconductance amplifiers. The current feedback configuration should present an appropriate ideal short-circuit load to the A circuit.

Once voltage or current feedback is selected, a second choice determines how the subtraction at the input is carried out.

Series Feedback. For *series feedback,* we connect the input ports of the A and β circuits in *series* as in Fig. 9.5c and use KVL to effect the voltage subtraction, $w_i = w_s - w_f$. Series feedback always increases input resistance (just as connecting resistors in series increases equivalent resistance). Because series feedback makes input resistance more closely approximate the open circuits of Figs. 9.4a and d, Fig. 9.5c uses a voltage source, w_s, for input.

Shunt Feedback. For *shunt feedback* we use the *shunt* or parallel connection of A and β circuit input ports, as in Fig. 9.5d, and KCL produces the current subtraction, $w_i = w_s - w_f$. Shunt feedback lowers the input resistance (just as connecting resistors in shunt lowers equivalent resistance). Figure 9.5d shows a current source, w_s, as the input, an appropriate choice for a connection that makes the input more closely approximate a short circuit.

In conclusion, to make a given amplifier more closely resemble the selected ideal, we apply either voltage or current feedback to lower or raise, respectively, the output resistance, and then, independent of this choice, use series or shunt feedback to increase or decrease, respectively, the input resistance. We next study, in idealized form, the four feedback topologies that result from these choices.

9.3.2 EFFECTS OF IDEAL FEEDBACK ON R_i AND R_o

Figure 9.6 shows the four feedback strategies applied to amplifiers. We examine voltage-series feedback of Fig. 9.6a and current-shunt feedback of Fig. 9.6b in detail, deriving expressions for feedback amplifier input and output resistance for each case. Because these two cases together demonstrate voltage and current feedback as well as series and shunt feedback, they include all the principles. We leave the remaining cases for home-work.

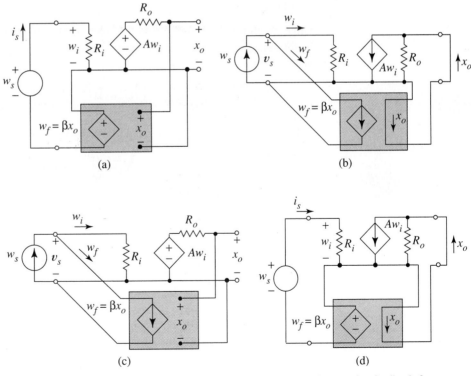

Figure 9.6 Feedback strategies for different design goals: (a) voltage-series feedback for a voltage amplifier; (b) current-shunt feedback for a current amplifier; (c) voltage-shunt feedback for a transresistance amplifier; (d) current-series feedback for a transconductance amplifier.

Voltage-Series Feedback. Figure 9.6a employs voltage feedback to reduce output resistance and series feedback to increase input resistance. We first find the input resistance seen by voltage source w_s. Using w_s as a test generator, Kirchhoff's voltage law gives

$$w_s = i_s R_i + \beta[A(i_s R_i)]$$

Dividing both sides by i_s gives the input resistance,

$$R_{if} = R_i(1 + A\beta) \tag{9.9}$$

(A similar analysis proves this same result for Fig. 9.6d.) Thus ideal series feedback multiplies the nonfeedback resistance R_i by $1 + A\beta$.

To find how voltage feedback changes output resistance, turn off w_s in Fig. 9.6a and attach a test generator to the output terminals as in Fig. 9.7a. KVL gives

$$v_{test} = R_o i_{test} + A(-\beta v_{test})$$

Solving for $R_{of} = v_{test}/i_{test}$ gives

$$R_{of} = \frac{R_o}{1 + A\beta} \tag{9.10}$$

(A similar proof gives the same result for Fig. 9.6c.) We conclude that voltage feedback ideally decreases the output resistance by $1 + A\beta$.

Current-Shunt Feedback. Now we analyze the current amplifier of Fig. 9.6b. With w_s as the test generator, because no current flows through R_o due to the short circuit, KCL at the input gives

$$w_s = \frac{v_s}{R_i} + \beta\left(A\frac{v_s}{R_i}\right)$$

After we divide both sides by v_s and take the reciprocal, we obtain

$$R_{if} = \frac{v_s}{w_s} = \frac{R_i}{1 + A\beta} \tag{9.11}$$

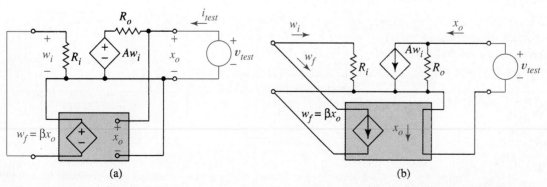

Figure 9.7 Circuits for deriving output resistance of feedback amplifiers with (a) voltage feedback; (b) current feedback.

It is easy to show that this same result applies to Fig. 9.6c. We conclude that ideal shunt feedback lowers input resistance by the improvement factor.

To find the effect of current feedback on output resistance, consider Fig. 9.7b, the test circuit for Fig. 9.6b. From KCL,

$$x_o = Aw_i + \frac{v_{test}}{R_o}$$

From the input circuit $w_i = -\beta x_o$. Substituting gives

$$x_o = -A\beta x_o + \frac{v_{test}}{R_o}$$

and

$$\frac{v_{test}}{x_o} = R_{of} = R_o(1 + A\beta) \tag{9.12}$$

The same result applies to Fig. 9.6d.

Exercise 9.3 Presently, our amplifier has $R_o = 7$ kΩ and $R_i = 11$ kΩ; however, specifications require $R_o \geq 10$ kΩ and $R_{if} \geq 190$ kΩ for the final product. What class of amplifier should serve as our ideal feedback goal, and what kind of feedback should we use to meet specifications?

Ans. transconductance amplifier, current-series feedback.

Because the improvement factor is always $1 + A\beta$, it is easy to estimate the input and output resistance of a feedback amplifier for a given amount and class of feedback, using memory aids associated with the feedback names to help us define improvement. In design, the memory aids help us determine the class and amount of feedback needed, as we see in the following example.

EXAMPLE 9.3 We have a nonfeedback amplifier with voltage gain of 920, $R_o = 3$ kΩ, and $R_i = 11$ kΩ. Determine the class of feedback and the value of β so that $R_{of} \leq 10$ Ω and $R_{if} \geq 800$k Ω. Also find A_f.

Solution. The need to lower output resistance and raise input resistance means using voltage-series feedback to approximate a voltage amplifier like Fig. 9.4a. Therefore, we visualize the given amplifier as Fig. 9.4e, where $a_v = 920 = A$.

Reducing the output resistance requires voltage feedback that satisfies

$$R_{of} = \frac{3\text{ k}}{1 + A\beta} \leq 10\ \Omega$$

or

$$1 + A\beta \geq 300$$

Increasing the input resistance requires series feedback with

$$R_{if} = (1 + A\beta)(11\text{ k}) \geq 800\text{ k}$$

or $1 + A\beta \geq 72.7$. To satisfy both conditions, select

$$1 + A\beta = 1 + 920\beta = 300$$

This means using $\beta = 0.325$. The feedback amplifier will have voltage gain $A_f = 920/300 = 3.07$, input resistance $R_{if} = 300 \times 11 \text{ k} = 3.3 \text{ M}\Omega$, and output resistance $R_{of} = 10 \ \Omega$. ❑

The next example shows the importance of first establishing the design goal.

EXAMPLE 9.4 Find the type of feedback and the value of β we must use with the nonfeedback amplifier of Example 9.3 (voltage gain = 920, $R_o = 3$ kΩ, and $R_i = 11$ kΩ) so that $R_{of} \leq 20 \ \Omega$ and $R_{if} \leq 680 \ \Omega$. Also find the gain of the feedback amplifier.

Solution. Because input and output resistance must both be lowered, the goal is a transresistance amplifier like Fig. 9.4c. This means we must visualize the given amplifier as Fig. 9.4g. In a transresistance amplifier, A is the transresistance gain, r_m. Because 920 is *voltage gain*, we must first convert the given VCVS in our A circuit to a CCVS. From Fig. 9.4g,

$$\text{voltage gain} = 920 = \frac{v_{out}}{v_{in}} = \frac{r_m i_c}{i_c R_i} = \frac{A}{R_i} = \frac{A}{11 \text{ k}}$$

Therefore, the transresistance gain of our nonfeedback amplifier is

$$A = 920 \times 11 \text{ k} = 1.01 \times 10^7 \ \Omega$$

The improvement factor must satisfy both

$$R_{of} = \frac{3 \text{ k}}{1 + A\beta} \leq 20 \ \Omega$$

and

$$R_{if} = \frac{11 \text{ k}}{1 + A\beta} \leq 680 \ \Omega$$

To satisfy both resistance specifications, $1 + A\beta$ must be the larger of $3 \text{ k}/20 = 150$ and $11 \text{ k}/680 = 16.2$, or 150. This means

$$150 = 1 + 1.01 \times 10^7 \beta$$

or $\beta = 1.48 \times 10^{-5}$ A/V. The final transresistance amplifier will then have transresistance gain of

$$A_f = \frac{1.01 \times 10^7}{150} = 6.73 \times 10^4 \ \Omega$$ ❑

Exercise 9.4 Our given amplifier is Fig. 9.4f, with $R_i = 12$ kΩ, $a_i = 280$, and $R_o = 100$ kΩ. Find the numerical value we should use for A when we apply current-series feedback. Also find the value of β for an improvement factor of 131.

Ans. $A = 0.0233$ A/V, $\beta = 5759 \ \Omega$.

The preceding exercise showed that our ideas of *reasonable* values for gain must fit the circumstances. Transconductance amplifiers develop output currents of a few milliamperes per volt of input, have feedback transresistances of a few volts per milliampere, and still give $A\beta$ much greater than one, as theory requires for a significant improvement in performance.

9.4
Feedback Theory When There Are Loading Effects

In practical feedback circuits, resistances of the signal source and the β circuit introduce loading at the input, and the β circuit and load resistor produce output loading. This section shows how to include loading effects in feedback analysis while retaining the overall conceptual simplicity of ideal feedback theory.

The first six columns of Table 9.2 summarize the patterns we have established for amplifiers with ideal terminations and β circuits. To pursue the design goal named in the first column, we apply the class of feedback given in column 2. Feedback modifies the input and output resistances of the original amplifier, as indicated in columns 3 and 4, to resemble more closely the ideal values. Consistent with each design goal is an ideal signal source and load, those in columns 5 and 6. To include circuit loading requires one additional key observation. Each β circuit of Fig. 9.6 happens to be the element subscripted as "12" for exactly one of the two-port equivalent circuits in Appendix B; the last column in Table 9.2 makes this association.

In Fig. 9.8, the ideal β circuits of Fig. 9.6 are replaced by the two-port equivalents suggested in the table; however, in all cases the element subscripted "21" is omitted, making the feedback model unilateral. This simplification involves little loss in accuracy, because the signal passing from input to output through the A circuit is generally much larger than the parallel input-to-output signal that passes through the "forward" or "21" *parameter* of the β circuit.

Figure 9.8 also shows source and load resistances. In Figs. 9.8a and c, R_L is placed in parallel with the ideal open-circuit load; in Figs. 9.8b and d, R_L is in series with the ideal short-circuit load (marked X). It is important to notice in comparing Figs. 9.8 and 9.6 that the *feedback parameter β is, in each case, identical to the two-port parameter having subscripts 12.*

TABLE 9.2 Feedback Amplifier Classifications

Design Goal	Feedback Class	Input Resistance	Output Resistance	Ideal Source	Ideal Load	β Circuit Two-Port Model
Voltage amplifier (VCVS)	Voltage series	Increases	Decreases	Voltage	Open circuit	h
Transresistance amplifier (CCVS)	Voltage shunt	Decreases	Decreases	Current	Open circuit	y
Current amplifier (CCCS)	Current shunt	Decreases	Increases	Current	Short circuit	g
Transconductance amplifier (VCCS)	Current series	Increases	Increases	Voltage	Short circuit	z

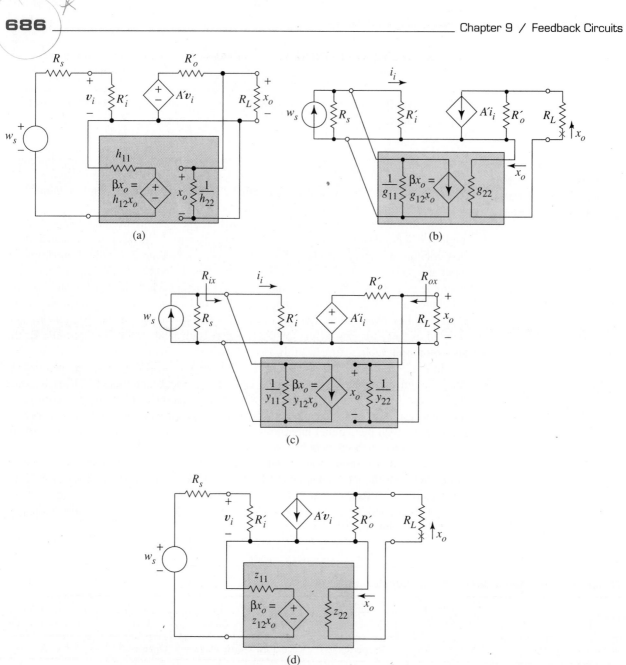

Figure 9.8 Nonideal feedback circuits: (a) voltage series; (b) current shunt; (c) voltage shunt; (d) current series.

9.4.1 ANALYSIS OF NONIDEAL FEEDBACK CIRCUITS

To include source, load, and β circuit resistances in our analysis, we first *redefine* the nonfeedback parameters A, R_i, and R_o so that they include these resistances. We then apply feedback theory *exactly as before*, using these modified parameters in the ideal equations. The following discussion of voltage-shunt feedback describes the central concepts that justify this approach for all four kinds of feedback.

Voltage-Shunt Feedback. Figure 9.8c is a transresistance-type feedback amplifier. Our object is to find its gain, A_f, output resistance, R_{of}, and input resistance, R_{if}. We first move all the resistance into the A circuit in *such a way that the resulting β circuit, the source, and the load all become ideal*. This involves "sliding" R_S, $1/y_{11}$, R_L, and $1/y_{22}$ along the existing connecting wires into the A circuit. The result, Fig. 9.9, is a transresistance amplifier with ideal current source, open-circuit load, and ideal β circuit as required for using our ideal feedback equations. Except for its more complicated A circuit, the result is Fig. 9.6c.

We must now find R_i, R_o, and A for the modified A circuit of Fig. 9.9. For the new A circuit,

$$R_i = R_s \, \| \, \left(\frac{1}{y_{11}} \right) \, \| \, R_i' \tag{9.13}$$

Looking back *from the open circuit* toward the output of the isolated A circuit, we see output resistance

$$R_o = R_o' \, \| \, R_L \, \| \, \left(\frac{1}{y_{22}} \right) \tag{9.14}$$

Since the circuit employs voltage-shunt feedback, A is the transresistance gain of the A circuit. A current divider expression from the input circuit and a voltage divider from the output give

$$A = \frac{x_o}{w_i} = \frac{(1/y_{11}) \, \| \, R_s}{[(1/y_{11}) \, \| \, R_s] + R_i} A' \frac{R_L \, \| \, (1/y_{22})}{[R_L \, \| \, (1/y_{22})] + R_o'} \tag{9.15}$$

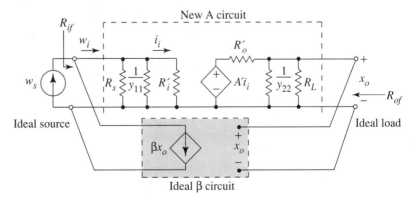

Figure 9.9 Nonideal voltage-shunt feedback circuit with R_s, $1/y_{11}$, $1/y_{22}$, and R_L associated with the A circuit.

Because shunt feedback reduces input resistance, w_s in Fig. 9.9 sees input resistance given by Eq. (9.11); however, R_i is now Eq. (9.13), A is Eq. (9.15), and β is y_{12}. Since voltage feedback reduces output resistance, the *open-circuit load* in Fig. 9.9 *sees* R_{of} given by Eq. (9.10), but with the new values for A, R_o, and β.

Input and Output Resistances of the Amplifier Proper. There is another pair of resistances that we might need to know in a practical problem, the input and output resistances of the feedback amplifier itself, *exclusive of source and load resistance.* The first is the resistance R_{ix} seen by w_s and its internal resistance R_s in Fig. 9.8c. Comparing Fig. 9.8c with Fig. 9.9, we see that R_{ix} must satisfy

$$R_{if} = R_s \| R_{ix}$$

Similarly, at the output we might need to know resistance R_{ox} seen by R_L in Fig. 9.8c when feedback is present. Comparing Figs. 9.8c and 9.9 shows that R_{ox} satisfies

$$R_{of} = R_L \| R_{ox}$$

It often happens that $R_{ix} \approx R_{if}$ and $R_{ox} \approx R_{of}$, making these extra calculations unnecessary.

9.4.2 ANALYSIS BY TURNING OFF THE FEEDBACK

Moving all resistance into the A circuit before finding A, R_i, and R_o is a key *theoretical* step that works for all four feedback classes; however, details of moving the resistors correctly depend upon whether the ideal load is a short or open circuit, and whether the ideal source is a current or a voltage. We can obtain the same results without moving resistors by simply setting $\beta = 0$ *(turning off the feedback)* in the *original circuit*. Notice that Figs. 9.8c and 9.9 are identical when $\beta = 0$. In general, we turn off the "12 source" responsible for the feedback, and then find the nonfeedback parameters including loading. This simplification does not relieve us of responsibility for remembering that R_o is the resistance seen by the ideal open-circuit or short-circuit load.

The transresistance amplifier of Fig. 9.10a introduces one additional new idea, a realistic β circuit. Here a resistive *pi circuit* (named for its resemblance to the Greek letter "π") provides the feedback. The first analysis step is to replace the β circuit with its two-port equivalent. According to Table 9.2 this takes a y-parameter equivalent, which gives Fig. 9.10b.

Exercise 9.5 Without moving resistors, find R_i, R_o, and A, for the nonfeedback amplifier of Fig. 9.10. Remember to use the correct gain definition.

Ans. $R_i = 3.19 \text{ k}\Omega$, $R_o = 2.54 \text{ k}\Omega$, and $A = -2.44 \times 10^5 \ \Omega$.

A very important insight into feedback amplifiers follows from noticing that β in Eq. (9.2) *is* the "12" two-port parameter of the β circuit; for example, in Fig. 9.8a,

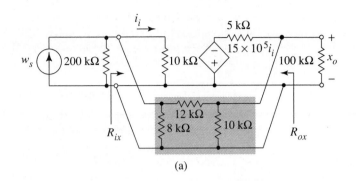

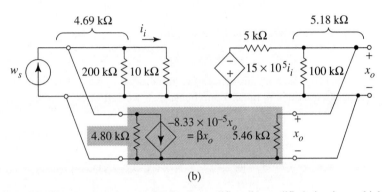

Figure 9.10 (a) Given transresistance feedback amplifier; (b) modified circuit to which ideal feedback theory applies.

$A_f \approx 1/h_{12}$. In analysis this approximation allows us to estimate A_f with little effort. In design it helps us find numerical values for our β circuit. It is clear from

$$A_f = \left(\frac{A}{1/\beta + A} \right) \frac{1}{\beta} < \frac{1}{\beta} = \frac{1}{p_{12}} \qquad (9.16)$$

that the actual A_f will always be somewhat less than the estimate we make using two-port parameter, p_{12}.

Exercise 9.6 Use Eq. (9.2) to *estimate* the transresistance gain of the feedback amplifier of Fig. 9.10. Then use the feedback equations and Exercise 9.5 answers to find A_f, R_{if}, and R_{of}.

Ans. $A_f \approx -1.2 \times 10^4 \ \Omega$, $A_f = -1.15 \times 10^4 \ \Omega$, $R_{if} = 150 \ \Omega$, and $R_{of} = 119.2 \ \Omega$.

EXAMPLE 9.5 Find numerical values for R_{ix} and R_{ox}, as defined in Fig. 9.10a.

Solution. From Exercise 9.6, ideal current source w_s in Fig. 9.10a sees $R_{if} = 150 \ \Omega$. R_{ix} in Fig. 9.10a satisfies

$$\frac{1}{150} = \frac{1}{200 \text{ k}} + \frac{1}{R_{ix}}$$

which gives $R_{ix} = 150.11$ Ω. Because ideal feedback theory gives the resistance seen looking back from the ideal open-circuit load in Fig. 9.10b, we have

$$\frac{1}{119.2} = \frac{1}{R_{ox}} + \frac{1}{100\,\text{k}}$$

so $R_{ox} \approx 119.34$ Ω.   ❏

Transconductance Amplifier Design. The following example shows how to use Eq. (9.16) in feedback amplifier design.

EXAMPLE 9.6 Find values of R_A and R_B in Fig. 9.11a so that the transconductance gain, $A_f = 1$ mA/V.

Solution. The z parameters for the β circuit are

$$z_{11} = R_A \qquad z_{22} = R_A + R_B \qquad z_{12} = \beta = R_A$$

This gives Fig. 9.11b. Since R_A influences the value of A as well as β, working directly with the feedback equations is rather complicated, so we fall back on intuition for some guidance.

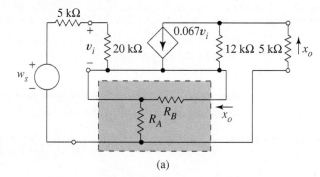

(a)

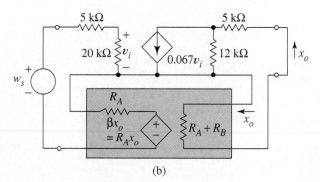

(b)

Figure 9.11 Transconductance amplifier: (a) original circuit; (b) circuit for applying ideal feedback theory.

Equations (9.2) and (9.16) cut to the heart of the design problem. They suggest that if we want $A_f = 1$ mA/V, then $1/\beta = 1/z_{12} = 1/R_A \approx 10^{-3}$, or $R_A \approx 1$ kΩ. Let us begin with this value, and then redesign, if necessary, to make things come out right at the end. Figure 9.11b shows that R_B adds to the output loading without contributing anything, so let us be brave and set $R_B = 0$ to complete the design; that is, use $R_A = 1$ kΩ and $R_B = 0$. Now we can use analysis to evaluate our amplifier.

With feedback off, the voltage division at the input and current division at the output give

$$A = \frac{x_o}{w_s} = \frac{20\,\text{k}}{20\,\text{k} + 5\,\text{k} + 1\,\text{k}}\,0.067\,\frac{12\,\text{k}}{12\,\text{k} + 5\,\text{k} + 1\,\text{k}} = 0.0344\,\text{A/V}$$

Since $\beta = 1$ kΩ,

$$1 + A\beta = 1 + (0.0344)10^3 = 35.4$$

so the transconductance gain of the feedback amplifier is

$$A_f = \frac{0.0344}{35.4} = 0.972 \times 10^{-3}$$

which is probably close enough to the specification that a redesign is not needed. If we do choose to redesign, we simply use a slightly smaller resistor for R_A. ❏

9.5 _____
FET and BJT Feedback Amplifiers

In analyzing feedback amplifiers designed by others, a critical step is to identify the class of feedback. The related design problem is to understand exactly how to implement the class of feedback suggested by the design goals. In both situations, we need to know a few *standard* feedback topologies commonly used in both FET and bipolar circuits.

9.5.1 DIFFERENCE AMPLIFIER TOPOLOGY

First, we examine the *difference amplifier topology,* a voltage-series feedback structure that employs a difference amplifier to perform the voltage subtraction at the input. A prosaic but instructive example is the noninverting op amp circuit of Fig. 9.12. To make the feedback class more obvious, dashed lines show the ground connection implicit in the op amp. The A circuit is the *open-loop* op amp with finite gain, finite input resistance, nonzero output resistance and limited bandwidth—all specified on the op amp data sheet. Divider resistors, R_1 and R_2, comprise the β circuit. The following numerical example uses negative feedback principles to justify our common assumptions of negligible input and output loading, even when the op amp parameters are quite modest.

EXAMPLE 9.7 In Fig. 9.13a, op amp parameters are $A_v' = 10^4$, $R_i' = 10$ kΩ, and $R_o' = 200$ Ω. Use feedback theory to find v_o/v_i and also the input and output resistances of the noninverting amplifier.

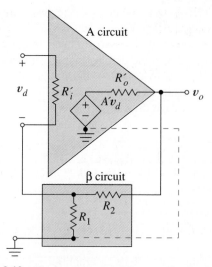

Figure 9.12   Noninverting amplifier as a feedback circuit.

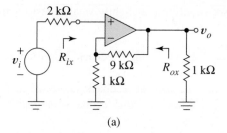

(a)

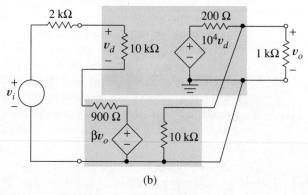

(b)

Figure 9.13   Analysis of noninverting amplifier using feedback theory: (a) amplifier; (b) feedback model.

Solution. Recognizing the voltage-series topology and consulting Table 9.2, we see that we should replace the resistive divider by an h-parameter equivalent. The h parameters are $h_{11} = 1\,\text{k} \,\|\, 9\,\text{k} = 900\,\Omega$, $1/h_{22} = 9\,\text{k} + 1\,\text{k} = 10\,\text{k}\Omega$ and

$$h_{12} = \beta = 1\,\text{k}/(1\,\text{k} + 9\,\text{k}) = 0.1$$

Figure 9.13b shows the feedback amplifier equivalent.

We next find R_i, R_o, and A from Fig. 9.13b, after setting $\beta = 0$ to turn off the feedback. Since the feedback class is voltage series, we find the voltage gain

$$A = \frac{v_o}{v_i} = \frac{10\,\text{k}}{10\,\text{k} + 2\,\text{k} + 900}\,10^4\,\frac{(1\,\text{k} \,\|\, 10\,\text{k})}{(1\,\text{k} \,\|\, 10\,\text{k}) + 200} = 6354$$

Also with $\beta = 0$, v_i sees

$$R_i = 2\,\text{k} + 10\,\text{k} + 900 = 12.9\,\text{k}\Omega$$

and with v_i turned off an ideal open-circuit load sees

$$R_o = 1\,\text{k} \,\|\, 10\,\text{k} \,\|\, 200 = 164\,\Omega$$

The improvement factor is $1 + A\beta = 1 + 6354 \times 0.1 = 636$.

The gain of the noninverting amplifier is therefore $A_f = 6354/636 = 9.99$, a value quite close to

$$(1 + R_2/R_1) = 1/\beta = 10$$

we obtain by assuming infinite gain. Since series feedback increases input resistance, $R_{if} = (12.9\,\text{k})636 = 8.2\,\text{M}\Omega \approx R_{ix}$. Because voltage feedback decreases output resistance, $R_{of} = 164/636 = 0.258\,\Omega \approx R_{ox}$.   ❏

> **Exercise 9.7**   We can turn the amplifier of Fig. 9.13a into a follower by replacing the 9 kΩ and the 1 kΩ feedback resistors by a short circuit and open circuit, respectively. This reduces the β circuit to a single pair of wires between input and output ports (one, a ground wire). Find the new h parameters for this β circuit. Then trace through the analysis steps in Example 9.7, recalculating A_f, R_{if}, and R_o as you go.
>
> **Ans.** $h_{11} = 0$, $h_{12} = 1$, $h_{22} = 0$, improvement factor = 6945, $A_f = 1$, $R_{if} = 83.3\,\text{M}\Omega$, and $R_o = 0.024\,\Omega$.

> **Exercise 9.8**   Recalculate A_f, R_{of}, and R_{if} for the noninverting amplifier of Example 9.7, but use the 741 op amp parameters: gain $= 2 \times 10^5$, input resistance $= 2\,\text{M}\Omega$, and output resistance $= 75\,\Omega$.
>
> **Ans.** $A_f = 10.0$, $R_{if} = 3.71 \times 10^{10}\,\Omega$, $R_{of} = 0.00375\,\Omega$.

EXAMPLE 9.8   Design a single-ended voltage amplifier with voltage gain of approximately 20 by adding an emitter-follower output stage and voltage-series feedback to the active-load amplifier of Fig. 7.57. SPICE analysis in Example 7.20 showed that this amplifier has voltage gain of 1649 and input resistance of 94.2 kΩ. Estimate the input and output resistance of the feedback amplifier.

Solution. First, consider the role of the emitter follower in the design. The output resistance of the active-load circuit itself is very high, namely, $r_{o4} \| r_{o2}$. The emitter follower contributes no voltage gain, but reduces output resistance by $\beta_F + 1$. This already lowered output resistance is to be further reduced by voltage feedback to a value suitable for a voltage amplifier. To bias the emitter follower at the same current as the other transistors, use a constant current source of 0.3 mA.

Since we have a difference amplifier and want voltage series feedback, we use the difference amplifier feedback topology. Because there is a 180° phase shift between node 5 and node 3 in Fig. 7.57, node 5 is the inverting input. As in the noninverting op amp circuit, a voltage divider can provide the feedback. Figure 9.14a shows the circuit configuration, where node 8 is the output and node 4 the input of the feedback amplifier. Coupling capacitor C_C ensures that the bias currents and Q-points are undisturbed by the feedback. For analysis, we replace the amplifier by a linear equivalent, and the feedback circuit by its two-port model.

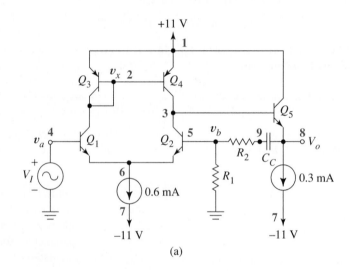

(a)

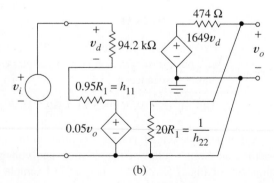

(b)

Figure 9.14 Voltage amplifier design: (a) feedback amplifier structure; (b) circuit model for aid in choosing R_1 and R_2.

From the previous SPICE analysis, $R_i' = 94.2$ kΩ is the input resistance of the A circuit. The Early voltages of 60 and 130 V and bias currents of 0.3 mA give $r_{o4} = 200$ kΩ and $r_{o2} = 433$ kΩ. Since the emitter current of Q_5 is 0.3 mA and $\beta_F = 350$, $g_{m5} = 0.3$ mA/$0.025 = 0.012$ A/V and $r_{\pi 5} = 29.2$ kΩ. The output resistance of the emitter follower is

$$R_o' \approx \frac{(r_{o4} \| r_{o2}) + r_{\pi 5}}{\beta_F + 1} = \frac{137\text{ k} + 29.2\text{ k}}{351} = 474\ \Omega$$

For $A_f \approx 1/\beta = 1/h_{12} = 20$, we have

$$h_{12} = \frac{R_1}{R_1 + R_2} = \frac{1}{20}$$

which implies $R_2 = 19R_1$. Then $h_{11} = R_1 \| R_2 = 0.95R_1$ and $1/h_{22} = R_1 + R_2 = 20R_1$, respectively. The equivalent circuit in Fig. 9.14b summarizes the known information.

To minimize loading of the feedback circuit by the amplifier input resistance, we need $h_{11} = 0.95R_1 << 94.2$ kΩ. To avoid unnecessary loading of the amplifier by the feedback network, $20R_1 >> 474\ \Omega$. A choice of standard 5% resistors that satisfies the latter conditions is $R_1 = 270\ \Omega$ and $R_2 = 5.1$ kΩ. Because of standard resistors, $R_2/R_1 = 18.9$ and make $h_{12} = \beta = 0.0503$.

Since $R_1 + R_2 = 5.31$ kΩ, we select C_C so that its reactance is 531 Ω at 200 rad/s (a reasonable small-signal short circuit down to 32 Hz). This requires $C_C = 9.4$ μF.

We now predict the performance of the circuit. Turning off the feedback in Fig. 9.14b gives

$$R_i = 94.2\text{ k} + (0.95 \times 270) = 94.5\text{ k}\Omega$$

and

$$R_o = 474 \| (20 \times 270) = 436\ \Omega$$

Including loading effects,

$$A = \frac{94.2\text{ k}}{94.2\text{ k} + 0.95 \times 270} 1649 \frac{20 \times 270}{20 \times 270 + 474} = 1512$$

For these values,

$$1 + A\beta = 1 + (1512)(0.0503) = 77.1$$

The feedback amplifier parameters are

$$R_{if} = (94.5\text{ k})(77.1) = 7.29\text{ M}\Omega$$

$$R_{of} = \frac{436}{77.1} = 5.66\ \Omega$$

and

$$A_f = \frac{1512}{77.1} = 19.6$$

EXAMPLE 9.9   Use SPICE to check the design of the preceding example.

Solution. Figure 9.15 gives the initial SPICE code for the circuit of Fig. 9.14a. The one volt ac generator VI doubles as signal source for the gain calculation and test generator for defining input resistance. For Q-point analysis, SPICE automatically turns off this source, biasing the base of Q_1 at ground potential. The .PRINT lines request the output voltage and input current needed to evaluate A_f and R_{if} at midrange frequencies. Since signal amplitude is one volt, V(8) is A_f and I(VI) is the reciprocal of R_{if}.

A zero ampere, ac current source, II, connected between node 8 and ground is the test generator for defining R_{of}. In the initial SPICE run this source is an open circuit. In a second run, we change VI to 0 and II to one ampere to make V(8) numerically the output resistance of the feedback amplifier.

The numerical values from SPICE output were $A_f = 19.6$, $R_{if} = 6.73$ MΩ, and $R_{of} = 5.85$ Ω, showing good agreement with the values from the preceding example: 19.7, 7.21 MΩ, and 5.71 Ω, respectively.   ❏

dc, ac, and Mixed Feedback. In Example 9.8 we introduced a coupling capacitor to prevent transistor Q-points from changing when feedback was added, and indicated that feedback analysis would be valid only at frequencies where this capacitor approximates a short circuit. Feedback that applies only to ac analysis is called *ac feedback*; feedback that stabilizes the dc operating points of the transistors is called *dc feedback*. Using bypass and coupling capacitors and inductors, we can design circuits that have dc but not

```
EXAMPLE 9.9
VCC 1 0 DC 11
Q3 2 2 1 PTRAN
Q4 3 2 1 PTRAN
.MODEL PTRAN PNP BF=50 VA=60
Q1 2 4 6 NTRAN
Q2 3 5 6 NTRAN
.MODEL NTRAN NPN BF=350 VA=130
I0 6 7 DC 0.6MA
I1 8 7 DC 0.3MA
VEE 7 0 DC -11
VI 4 0 AC 1
II 8 0 AC 0
Q5 1 3 8 NTRAN
R1 5 0 270
R2 5 9 5.1K
CC 9 8 9.4U
.AC LIN 1 1K 1K
.PRINT AC V(8)
.PRINT AC I(VI)
.OP
.END
```

Figure 9.15   SPICE code for simulation of Fig. 9.14a.

ac feedback, ac but not dc feedback, both in equal amounts, or both, but in differing amounts. Figure 9.16 introduces the possibilities. In each circuit, the nonfeedback amplifier is a difference amplifier with finite gain and zero offset.

Figure 9.16a has dc *and* ac feedback in equal amounts. Figure 9.16b has dc feedback but no ac feedback. The capacitor is an open circuit to dc, and the amplifier Q-point is stabilized just as in Fig. 9.16a. However, at midrange frequencies the inverting input node is shorted to ground by the capacitor, and the gain is that of the nonfeedback amplifier with R_2 as load resistor. Having no ac feedback, the amplifier does not enjoy the improvements in sensitivity, bandwidth, and input and output resistances that ac feedback would contribute. When ac plus dc is applied at the input, ac open-loop operation centers about a point on the dc transfer characteristic, which itself is stabilized by the dc feedback.

Figure 9.16c has ac feedback but no dc feedback. The Q-point established on the transfer characteristic by the dc voltage at the noninverting input is not stabilized by feedback, and thus may vary widely. For midrange frequencies, the capacitor, an ac short circuit, gives the same ac operation as Fig. 9.16a provided the Q-point is reasonable. Bandwidth is extended, midrange input resistance is raised, output resistance is lowered, and sensitivity is reduced by $1 + A\beta$, where $\beta = R_1/(R_1 + R_2)$. Figures 9.16d and e use capacitors as high-frequency short circuits to make ac and dc βs differ; for Fig. 9.16d, $\beta_{dc} > \beta_{ac}$; for Fig. 9.16e, $\beta_{dc} < \beta_{ac}$. Inductors called *choke coils* sometimes serve the same general purpose as the capacitors, by introducing selective open circuits instead of short circuits.

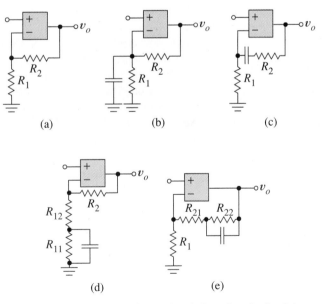

Figure 9.16 Combinations of dc and ac feedback: (a) dc and ac feedback in equal amounts; (b) dc but no ac feedback; (c) ac but no dc feedback; (d) dc feedback greater than ac feedback; (e) ac feedback greater than dc feedback.

9.5.2 SINGLE-ENDED FEEDBACK AMPLIFIERS

We next describe four commonly used strategies for adding feedback to single-ended amplifiers. To identify the four feedback classes, we must examine the origin of the signal entering the β circuit and the way in which the feedback signal returns to the A circuit. Rewards for mastering these recognition skills are a powerful insight into the designs of others and ease in adding feedback to our own amplifier designs. In these circuits, the inputs of the A and β circuits share a common ground reference, usually only implied, which adds to the challenge of identifying the feedback class.

Origin of the Feedback Signal

1. *Voltage feedback*: The output signal from the A circuit is a node voltage defined relative to the ground reference. The signal enters the β circuit through a connection between the ungrounded output nodes of the A and β circuits, as in Fig. 9.17a. The dashed lines, which make the connection easier to recognize, are usually not shown on the diagram. Figures 9.18a and c show practical implementations of voltage feedback; the β circuit is identified below each amplifier.

2. *Current feedback*: Current feedback is characterized by a kind of indirectness, which we see in Fig.9.17b. The output of the A circuit is a current, i_o; the input to the β circuit is a copy of this current. The copy is created in an unbypassed emitter or source resistor associated with the output transistor in the A circuit; however, we consider this resistor to be part of the β circuit. For an FET output transistor, the copied current exactly equals i_o; with a bipolar transistor, the copied current is $(1/\alpha)i_o \approx i_o$. See Figs. 9.18b and d for examples of current feedback.

Destination of the Feedback Signal

1. *Shunt feedback*: The essence of shunt feedback is a subtraction of feedback current from source current at an input node by means of KCL, as in Fig. 9.17c. A single connection joins the ungrounded inputs of the A and β circuits, and the dashed ground line is often only implied. Figures 9.18a and b show how the principle is implemented.

2. *Series feedback*: In series feedback, there is a KVL subtraction of β circuit feedback voltage v_f from the source voltage as in Fig. 9.17d. To implement this, the β circuit utilizes an unbypassed emitter or source resistor associated with an input transistor in the A circuit. See Figs. 9.18c and d for practical examples.

Phase Constraint. In both shunt and series feedback, the subtraction required at the input critically depends upon the relative phases of the input and feedback signals. Because resistive β circuits contribute no phase shift, the designer must ensure that a proper phase relationship exists between the A circuit input and the point where signal leaves the output side of the A circuit and enters the β circuit. There are two scenarios.

1. *Shunt feedback*: As illustrated by algebraic signs in Fig. 9.17c, there must be a 180° phase difference between the A circuit's input and output (the latter to be connected to the β circuit). This is because feedback current i_f subtracts from i_s only if node voltages within the β circuit go negative as the input node of the A circuit goes positive.

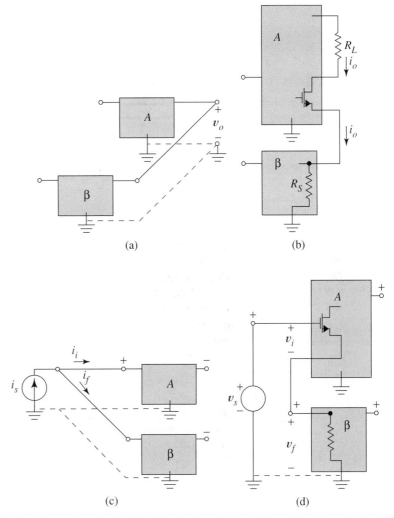

Figure 9.17 Connections for feedback in single-ended amplifiers: (a) voltage feedback; (b) current feedback; (c) shunt feedback; (d) series feedback.

2. *Series feedback*: The algebraic signs in Fig. 9.17d show the required phase relations for series feedback. There must be zero phase shift between the input node of the *A* circuit and its output (the latter to be connected to the β circuit). Only if the node voltages in the β circuit increase as v_s increases will v_f subtract from v_s.

Single-Ended Feedback Topologies. Figure 9.18 shows some single-ended feedback structures and identifies each β circuit. For clarity, biasing details are omitted. In each case, observe how the feedback follows all the connection and phase rules just discussed. Voltage feedback derives the feedback signal from the output node voltage v_o as

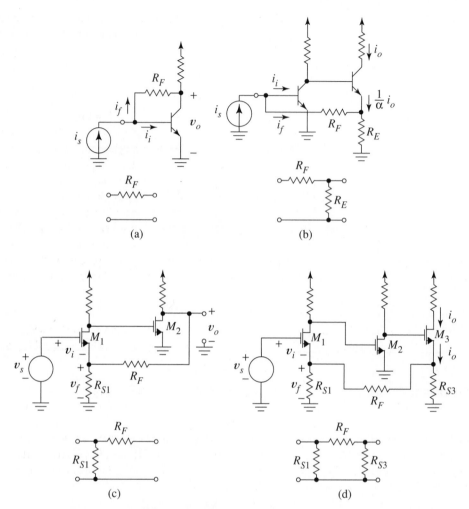

Figure 9.18 Examples of single-ended feedback structures: (a) voltage shunt; (b) current shunt; (c) voltage series; (d) current series.

in Figs. 9.18a and c. Current feedback, instead, uses a *copy* of the output current, i_o, as in Figs. 9.18b and d.

In shunt feedback, there is a current subtraction at the input node as in Figs. 9.18a and b. Because of the 180° phase shift in the common-emitter amplifier, the phases of the node voltages at the two ends of R_F differ, causing current subtraction rather than addition at the summing node. For series feedback, Figs. 9.18c and d, a feedback voltage v_f is subtracted from the signal voltage v_s using an unbypassed resistor at the amplifier input. For a voltage subtraction, the phase of the signal voltage and the voltage at the right end of R_F must be identical. In both Figs. 9.18c and d, this is accomplished by two inverting common-source stages M_1 and M_2. In Fig. 9.18d there is no phase shift between gate and source of M_3.

Exercise 9.9 In Fig. 9.19a, what is the phase shift between the gate of M_1, where the signal enters the A circuit, and the source of M_2, where signal leaves the A circuit and enters the β circuit?

Ans. 180°.

Figure 9.18 showed *examples* of each kind of single-ended feedback structure; however, we need to master the principles, for there are many variations on these basic themes. To make the subtraction mechanisms more evident, Fig. 9.18 showed ideal signal sources. In practice, there is often a capacitively coupled voltage source with internal resistance at the input, even for shunt feedback; and biasing resistors make the structures harder to recognize. Any of the feedback structures can employ either BJTs or FETs or both. Sometimes there are more amplifying stages than we showed in our examples; however, the phase constraints just given always apply.

In analysis, the first step is to use the features just described to identify the feedback class. The midrange equivalent circuit then replaces the amplifier, and the two-port equivalent replaces the feedback circuit. The next example demonstrates analysis.

EXAMPLE 9.10 Use feedback theory to find v_o/v_i and the input resistance seen by the signal source in Fig. 9.19a. Both transistors have $k = 2 \times 10^{-3}$ A/V^2 and are biased at 1 ma.

Solution. Feedback obviously involves the 7 kΩ resistor. Beginning here, we recognize ac, current-shunt feedback. This feedback class increases output resistance and lowers input resistance, so the designer's objective was a current amplifier. Table 9.2 tells us that current excitation and a short-circuit load are appropriate ideal terminations, and g parameters should describe the β circuit. Although our assignment is to find voltage gain, we first treat the circuit as a current amplifier in order to apply feedback theory. We find the voltage gain at the end of the analysis.

For the feedback circuit of Fig. 9.19b,

$$g_{11} = \left. \frac{i_1}{v_1} \right|_{i_2=0} = \frac{1}{7.5\ \text{k}} \qquad g_{12} = \left. \frac{i_1}{i_2} \right|_{v_1=0} = \frac{-500}{500 + 7\ \text{k}} = \beta = -0.067$$

$$g_{22} = \left. \frac{v_2}{i_2} \right|_{v_1=0} = 500 \| 7\ \text{k} = 467\ \Omega$$

The two-port equivalent is Fig. 9.19b with $\beta = -0.067$. We can *estimate* from Eq. (9.2) that if this feedback amplifier is well designed, its current gain is about $1/\beta = -1/0.067 = -14.9$. In some situations, this estimate might be our goal, and we need go no further. Here, we are required to examine the circuit in more detail, so we draw the midrange equivalent, convert the signal voltage source into a current source, identify the location of the short-circuit load, and replace the feedback circuit with its two-port model. The result is Fig. 9.19c, where we have replaced (5 M$\Omega \| 1.25$ MΩ) with 1 MΩ and used $g_m = 0.002$ A/V.

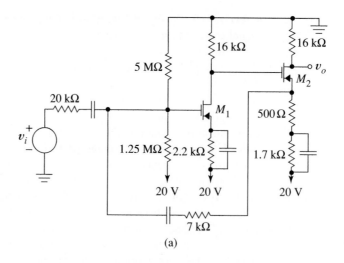

(a)

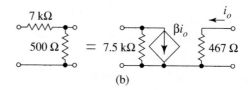

(b)

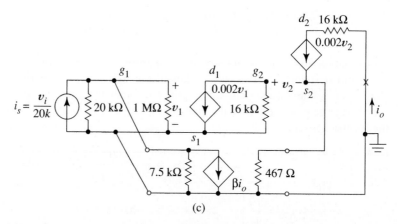

(c)

Figure 9.19   FET current-shunt feedback amplifier: (a) circuit schematic; (b) feedback network and two-port equivalent; (c) midrange circuit for applying feedback theory.

For this amplifier, A is current gain, i_o/i_s. Since $i_s = v_i/20$ k, we use the following analysis. With $\beta = 0$ we find $R_i = 20$ k$\|1M\|7.5$ k $= 5.43$ k.

$$\frac{v_1}{i_s} = 20 \text{ k}\|7.5 \text{ k}\|1M = 5.43 \text{ k}\Omega = R_i \qquad (9.17)$$

$$\frac{v_{g2}}{v_1} = -0.002 \times 16 \text{ k} = -32 \qquad (9.18)$$

$$v_{s2} = [0.002(v_{g2} - v_{s2})]467$$

which leads to

$$v_{s2} = 0.483 v_{g2} \qquad (9.19)$$

Finally

$$i_o = \frac{v_{s2}}{467}$$

Now substituting, in turn, Eqs. (9.19), (9.18), and (9.17) we obtain

$$i_o = \frac{0.484}{467}(-32)(5.43 \text{ k})i_s = -179.8 i_s$$

or

$$A = \frac{i_o}{i_s} = -179.8$$

From feedback theory, the improvement factor is

$$1 + A\beta = 1 + (-179.8)(-0.0670) = 13$$

Because of shunt feedback, $R_{if} = 5.43$ k$/13 = 417.7$ Ω. The current gain of the feedback amplifier is $A_f = -179.8/13 = -13.8$ (compared with our initial estimate of $1/\beta = -14.9$).

Now we use these results to find the voltage gain of Fig. 9.19a. In this circuit, the current that flows through the 16 kΩ drain resistor of M_2 is the *short-circuit load current* designated as i_o in Fig. 9.19c. Therefore, $v_o = (16 \text{ k})i_o$. From Fig. 9.19c, $i_s = v_i/20$ k. To find v_o/v_i, we substitute these expressions and obtain

$$\frac{v_o}{v_i} = \frac{16 \text{ k}i_o}{20 \text{ k}i_i} = \frac{16 \text{ k}}{20 \text{ k}}A_f = 11.1$$

From Fig. 9.19c, $R_{if} = 417.7$ $\Omega = 20$ k$\|R_{ix}$ where R_{ix} is the resistance seen by the external source in Fig. 9.19a. For all practical purposes, $R_{ix} = 417.7$ Ω.   ❏

There is sometimes some arbitrariness involved in identifying the β circuit. Problem 9.35 explores this point.

So far, our theory concerned midrange frequencies where $A\beta$ is large and real. Next we explore the frequency dependence of our feedback circuits.

9.5.3 FREQUENCY LIMITATIONS ON R_{if} AND R_{of}

The improvements in input and output resistance produced by negative feedback apply only to a limited band of frequencies because they depend upon $1 + A\beta$ being a large number. Recall from Chapter 8 that gain A is large and real only over the midrange frequencies of the *nonfeedback amplifier*. Above ω_H, and, if there are coupling and bypass capacitors, below ω_L, A becomes complex and decreases in magnitude. For this reason the benefits of feedback gradually disappear at high and (sometimes) at low frequencies. Also, the input and output "resistances" of a feedback amplifier become complex impedances for frequencies when A is complex. For example, $R_i(1 + A\beta)$ becomes complex when A is complex. When $|A\beta|$ becomes small compared to one, feedback amplifier input and output resistances approach their original nonfeedback values.

Figure 9.20a shows a simple feedback amplifier designed to illustrate these points. The nonfeedback amplifier has midrange voltage gain of 1000, or 60 dB, input resistance of 10 kΩ, and output resistance of 1 kΩ. The 0.0318 μF series capacitor gives the A circuit a lower half-power frequency of 500 Hz, and the 159 pF shunt capacitor gives an upper half-power frequency of 1 MHz. Ideal voltage-series feedback with $\beta = 0.1$ gives $1 + A\beta = 101$ at midrange frequencies.

Figure 9.20b, obtained by SPICE analysis, compares the open- and closed-loop gain curves, $A(\omega)$ and $A_f(\omega)$. It shows an increase in bandwidth and decrease in gain, exactly as expected. Included on the same curve are plots of input and output impedance magnitudes for the feedback amplifier.

From 500 Hz to 1 MHz (midrange), where $A(\omega)$ is uniformly large and real, we see $Z_{in} \approx 1$ MΩ and $Z_{out} \approx 10$ Ω as predicted by feedback theory. At higher frequencies, because of the reduction in $A(\omega)$, Z_{in} becomes complex and decreases in magnitude, approaching its nonfeedback value of 10 kΩ at high frequencies. (At low frequencies, Z_{in} becomes very large because of the coupling capacitor in series with the test generator.) At low frequencies, Z_{out} increases to its nonfeedback value of 1 kΩ.

In the next section we further explore the frequency dependence of feedback amplifiers and learn that this frequency dependence can lead to instability if proper precautions are not taken before introducing the feedback.

9.6
Stability of Feedback Amplifiers

Because of phase shift within the amplifier (A circuit), it is possible that a component of the feedback signal at some particular frequency ω_o is actually added rather than subtracted from the input. If this component is sufficiently large, the result is sustained oscillations at frequency ω_o—the circuit has become an *oscillator* or signal generator instead of an amplifier, and we say the circuit is *unstable*. In this section, uppercase characters denote phasors associated with sinusoidal steady-state analysis.

9.6.1 CONDITION FOR INSTABILITY

For a concrete illustration of how oscillations can occur, consider the voltage-series circuit of Fig. 9.21, where sinusoidal voltages are indicated by phasors. With the switch in position 1, we have the feedback amplifier with all resistance referred to the A circuit; position 2 gives the modified nonfeedback amplifier. At midrange frequencies, $A(\omega)$ and β are both positive and real. Therefore, with the switch in position 1, the feedback pha-

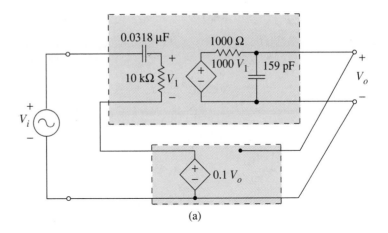

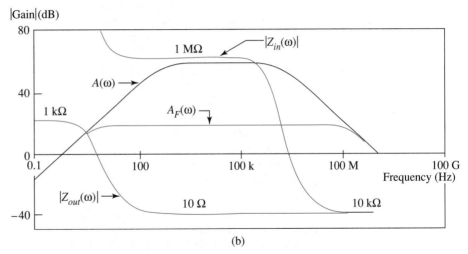

(b)

Figure 9.20 Limits of improvement in R_{in} and R_{out}: (a) equivalent of band-limited nonfeedback amplifier with ideal voltage-series feedback; (b) properties of feedback amplifier compared with frequency response of nonfeedback amplifier.

sor, $V_f = \beta A(\omega)V_i$, has the same angle as V_i. This means for sinusoidal excitation all voltages in the KVL equation, $v_s(t) = v_i(t) + v_f(t)$, have identical phase angles; hence the subtraction

$$v_i(t) = v_s(t) - v_f(t)$$

required for negative feedback.

Phase Condition. Now, with the switch in position 2, suppose signal

$$v_s(t) = V \sin \omega_o t = v_i(t)$$

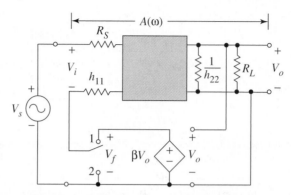

Figure 9.21 Loop gain in a feedback amplifier.

is applied to the nonfeedback amplifier. Further, suppose that ω_o is such that $A(\omega_o)$ is complex; that is,

$$A(\omega_o) = |A(\omega_o)| \angle \phi(\omega_o)$$

The output of the dependent source is then

$$v_f(t) = \beta \, |A(\omega_o)| V \sin[\omega_o t + \phi(\omega_o)] \qquad (9.20)$$

If the condition

$$\phi(\omega_o) = 180° \qquad (9.21)$$

happens to be satisfied at this excitation frequency, then according to Eq. (9.20), $v_f(t)$ is *inverted relative to* $v_s(t)$, giving

$$v_f(t) = -\beta \, |A(\omega_o)| V \sin[\omega_o t] = -\beta \, |A(\omega_o)| v_s(t)$$

Under these conditions, if the switch suddenly goes to position 1, the feedback signal reinforces or *adds to* $v_s(t)$. The KVL "subtraction" at the input now gives

$$v_i(t) = v_s(t) - v_f(t) = v_s(t) + \beta \, |A(\omega_o)| v_s(t)$$

When this occurs, we say we have *positive feedback* at frequency ω_o. It is possible for feedback to be negative for midrange components and positive for frequencies outside the -3 dB passband of $A(\omega)$, where additional phase shift occurs. For instability, a second condition is also required.

Amplitude Condition. With the switch in position 2 and excitation $V_s = V \sin \omega_o t$, suppose in Eq. (9.20) that in addition to $\phi(\omega_o) = 180°$, we also have

$$\beta \, |A(\omega_o)| \geq 1 \qquad (9.22)$$

This *amplitude condition* leads to two interesting possibilities. *Equality* in Eq. (9.22) implies

$$v_f(t) = V \sin(\omega_o t + 180°) = -V \sin \omega_o t$$

When this condition exists, we can use V_f to *replace* V_s. That is, once the circuit is running, we can flip the switch to position 1, turn off the external signal, and use V_f as the input signal. Thereafter, there would be a sustained sinusoidal output at frequency ω_o rad/s. Circuits that operate like this are called *non-self-starting oscillators.*

When the inequality condition occurs in Eq. (9.22), V_f is identical in frequency and phase to V_s but greater in amplitude. In this case, the sinusoid of frequency ω_o increases in size until circuit nonlinearities limit its amplitude. The result is a sustained oscillation with a periodic output waveform of fundamental frequency ω_o rad/s. A circuit deliberately designed to operate in this way is called a *self-starting oscillator.* If such oscillations occur inadvertently in an amplifier, we say the amplifier is *unstable.*

The amplitude and phase conditions associated with oscillations both have to do with the complex product

$$L(\omega) = A(\omega)\beta \qquad (9.23)$$

called the *loop gain,* the total gain of the feedback loop from the amplifier input back to the point of subtraction, including all loading effects at input and output. Figure 9.21 shows that $L(\omega)$ is the gain function V_f/V_s when the switch is in position 2, that is, with the feedback loop open.

We now state the *Nyquist stability criterion,* a necessary and sufficient condition for feedback amplifiers of the types we have been studying to be unstable. If there exists *any* frequency ω_o such that the *loop gain*

$$L(\omega_o) = A(\omega_o)\beta = M\underline{/180°}, \qquad \text{and } M \geq 1 \qquad (9.24)$$

then the amplifier is unstable, and there will be sustained oscillations at ω_o rad/s. Otherwise, the amplifier is stable.

The justification for such a simple stability criterion is that random noise is present in every circuit with its power distributed over *all* frequencies—an infinity of tiny uninvited generators lying in wait, each hoping to be amplified. For this reason, if there is *any* frequency whatsoever at which Eq. (9.24) is satisfied, oscillations are inevitable no matter what external voltage is attached to the input. The oscillation simply superimposes itself upon the externally applied signal and increases in amplitude of its own accord.

9.6.2 BODE PLOTS FOR UNSTABLE AND STABLE AMPLIFIERS

Among several useful tools available for studying stability are Bode plots of the amplitude and phase of $L(\omega)$. Such plots show whether or not a given feedback amplifier is stable, and, if stable, the plots tell us how stable. If a circuit is unstable, the plots provide the information we need to make it stable.

To determine if a given feedback amplifier is stable, Eq. (9.24) indicates that we need only examine the magnitude of the loop gain at that frequency, ω_o, where the phase shift is 180°. If $|L(\omega_o)|$ is greater than one, then the amplifier is unstable—if less than one, stable.

Figure 9.22a shows amplitude and phase plots of loop gain $L(\omega)$ for an unstable amplifier. We see that at frequency $\omega = \omega_o$, where the phase shift is 180°, the loop gain

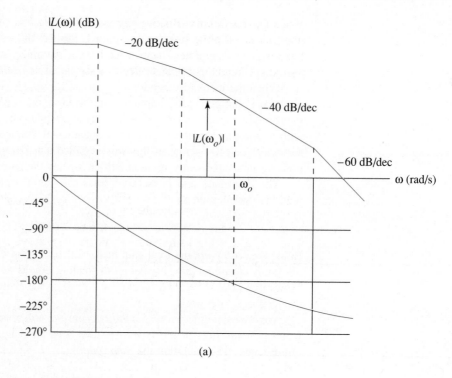

(a)

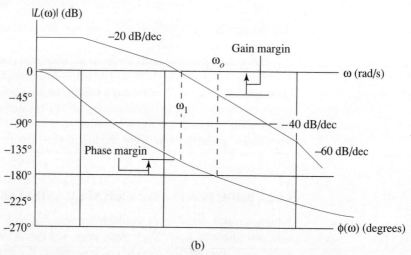

(b)

Figure 9.22 Determining stability from Bode plot of loop gain: (a) unstable amplifier; (b) stable amplifier.

$|L(\omega_o)|$ exceeds zero dB (loop gain exceeds one). Thus oscillations will occur, and the amplifier is unstable.

When an amplifier is stable, we would like to know how stable. Gain and phase margin are measures of the degree of stability.

Gain Margin. Figure 9.22b shows loop-gain plots for a stable amplifier, readily identified by the negative dB gain at ω_o. One measure of the degree of stability is the *gain margin,* defined by

$$\text{gain margin} = 0 - |L(\omega_o)|_{dB} \qquad (9.25)$$

where ω_o is the frequency at which the phase shift is 180°. Gain margin, defined to be positive for a stable amplifier, is easily read from the Bode plot, as we see in Fig. 9.22b. The larger the gain margin, the more stable the amplifier, since the amplifier is farther from satisfying the critical condition for oscillations. The unstable amplifier of Fig. 9.22a has a negative gain margin, with its magnitude indicating the extent of instability by describing the amount of excess loop gain present at ω_o.

Phase Margin. A second measure of the degree of stability is the *phase margin,* defined by

$$\text{phase margin} = 180° + \phi(\omega_1) \qquad (9.26)$$

where ω_1 is the frequency where the loop gain is zero dB, as indicated in Fig. 9.22b. Phase margin is the additional phase shift the amplifier requires at ω_1 to produce a replacement for the input signal. Phase margins of 45° to 60° are reasonable values in a conservatively designed feedback amplifier. Notice that the unstable amplifier described by Fig. 9.22a has a negative phase margin, which indicates the amount of excess phase shift present at ω_1.

9.6.3 STABILITY WHEN $A(\omega)$ IS GIVEN

Constructing Bode Plots of $L(\omega)$. The manufacturer of an off-the-shelf IC, such as an operational amplifier, provides magnitude and phase curves for the IC's open-loop gain function, $A(\omega)$. To examine the stability of a *particular* feedback circuit that employs the IC, the user must understand how to construct the Bode plot of loop gain, $L(\omega)$ for his or her particular design. The desired and known information are related in a very simple way. Expressing Eq. (9.23) in decibels gives

$$20 \log|L(\omega)| = 20 \log|A(\omega)| - 20 \log\left|\frac{1}{\beta}\right| \qquad (9.27)$$

This means the Bode plot of loop gain, $|L(\omega)|$, is related to the Bode plot of $|A(\omega)|$ by a *graphical subtraction* of the constant, $20 \log|1/\beta|$. For the amplifiers we consider in this text, β is always a real constant associated with a resistive two-port. Thus the graphical subtraction amounts to the change in coordinate system we see in Fig. 9.23a. Furthermore, the designer knows $1/\beta$. Recall from Eq. (9.2) that $1/\beta$ approximates the *desired closed-loop gain* of the feedback amplifier. Therefore, *the loop-gain magnitude in dB at each frequency is the difference between the gain of the nonfeedback amplifier*

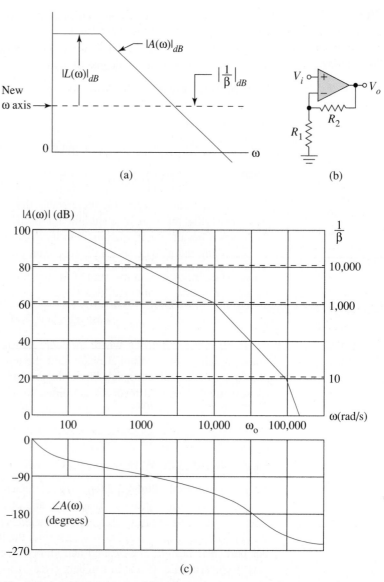

Figure 9.23 Relationship between stability and gain: (a) constructing loop-gain curve by shifting the origin of the y-axis coordinate system; (b) noninverting amplifier; (c) finding loop-gain for different values of closed-loop gain.

and the (constant) gain we wish to obtain from the feedback amplifier. As for the phase plot, when β is positive, the phase curve of $L(\omega)$ *is* the phase plot of $A(\omega)$. If β is negative, the phase curve is the phase plot of $-A(\omega)$.

Stability as a Function of β. Interestingly, it is easier to construct a stable feedback amplifier with high gain than with low gain, as we see in the following example.

EXAMPLE 9.11 Suppose the op amp of Fig. 9.23b has open-loop magnitude $|A(\omega)|_{dB}$ and phase characteristics $\angle A(\omega)$ given by the solid curves of Fig. 9.23c. Use gain margin to examine the stability of the noninverting, feedback amplifier for gains of (a) 10,000, (b) 1000, and (c) 10.

Solution. For this voltage-series feedback circuit, our design goal is a noninverting amplifier with gain of

$$\frac{1}{\beta} = 1 + \frac{R_2}{R_1}$$

Because the angle of $1/\beta$ is zero at all frequencies, the phase curve of the op amp is also the loop-gain phase curve.

(a) For gain of 10,000, $20 \log|1/\beta| = 80$ dB, so we construct the dashed line at 80 dB shown on Fig. 9.23c, and following Eq. (9.27), we regard it as the new ω axis. Relative to this new axis, the solid amplitude curve is $|L(\omega)|$. The gain margin, $0 - L(\omega_o) = 41$ dB, indicates that the amplifier is quite stable, with gain margin of 41 dB.

(b) For gain of 1000, $20 \log|1/\beta| = 60$ dB, so we construct the dashed line at 60 dB and use it as a new ω axis. The loop-gain curves show that this amplifier is also stable, but its gain margin is only 21 dB.

(c) For gain of 10 (or 20 dB), the gain margin is $0 - (39 - 20) = -19$ dB at ω_o. This amplifier is unstable. ❏

In the preceding example, the gain and phase curves suggest that for Fig. 9.23b a closed-loop gain of 39 dB, that is, $1 + R_2/R_1 = 89.1$, is the limiting case. Any amplifier with *less* gain is unstable.

Exercise 9.10 Use the graphical information to find the phase margin for each amplifier of Example 9.11.

Ans. 90°, 47°, and −50°.

Unconditionally Stable Amplifier. In general, stability depends on β; however, it is possible to construct amplifiers that are stable for all β. Figure 9.24 shows the open-loop frequency response curves for such an *unconditionally stable amplifier*. The salient feature is a dominant high-frequency pole that causes open-loop gain to drop at −20 dB/dec from its midrange value all the way to ω_1, the *unity-gain frequency*. Over this frequency range, the phase shift is *everywhere* less than 180°.

Figure 9.24 repeats the dashed-line constructions of Example 9.11 for comparison. The new dashed curve at 0 dB shows that *even a follower circuit is stable*. The price we pay for an unconditionally stable amplifier is reduced gain-bandwidth product. We soon find out why.

9.6.5 CLOSED-LOOP FREQUENCY RESPONSE

For an amplifier with a single high-frequency pole, Eq. (9.5) and Fig. 9.2a show that negative feedback implements a direct trade-off of gain for bandwidth, with no change in the general shape of the frequency response curve. Because of its dominant pole, this

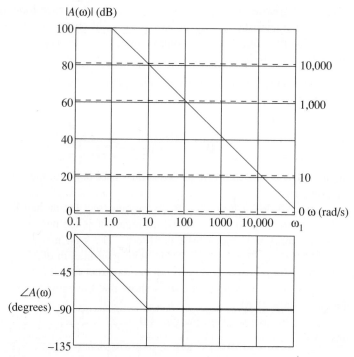

Figure 9.24 Open-loop gain curves for an unconditionally stable amplifier.

holds for an unconditionally stable amplifier. However, if $A(\omega)$ has a more general form, Fig. 9.22b, for example, then the algebra that relates $A_f(\omega)$ to $A(\omega)$ is more complicated, and so is $A_f(\omega)$.

When $A(\omega)$ describes a *two*-pole amplifier, the closed-loop frequency response, $|A_f(\omega)|_{dB}$, *relates to phase margin,* as in Fig. 9.25, and the phase-margin design specification often originates with these curves. A 45° phase margin gives a peak of about 2 dB in the closed-loop response, 60° gives a peak of about 0.5 dB, and 65.5° gives a *maximally flat response* with no peaking whatsoever. For each phase margin, the frequency response curves of Fig. 9.25 also relate closed-loop bandwidth to that frequency, ω_1, where the gain margin is 0 dB, as we see from examining the independent variable. The next example uses these curves in a design task.

EXAMPLE 9.12 Figure 9.26 shows the open-loop frequency response curves for an op amp. Design a noninverting amplifier that has a flat frequency response, and find the gain-bandwidth product of the amplifier.

Solution. The frequency response shows two break frequencies, so the nonfeedback amplifier is, at least approximately, a two-pole device. Figure 9.25 shows that flat frequency response implies a phase margin of 65.5°. From Eq. (9.26), $\phi(\omega_1) = 65.5° - 180° = -114.5°$. Therefore, we begin by finding the frequency ω_1 in Fig. 9.26 where the phase shift is about −114.5°. From Fig. 9.26, this is $\omega_1 \approx 2\pi 900$ rad/s. At this frequency we project upward to the amplitude curve to find the amount of *closed-loop*

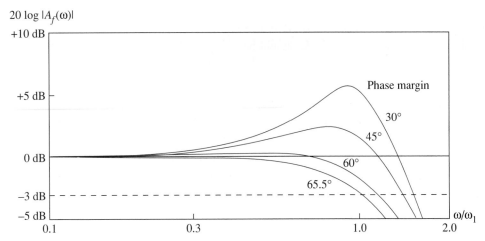

Figure 9.25 Frequency response curve for two-pole feedback amplifier as function of phase margin. Loop gain is 0 dB at $\omega = \omega_1$.

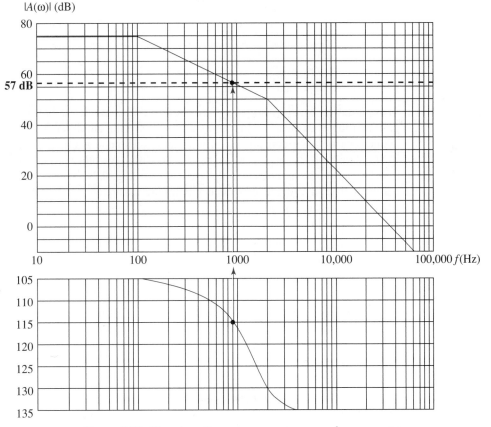

Figure 9.26 Open-loop frequency response curves for an op amp.

gain that gives this phase margin. This is 57 dB, meaning $|A(j2\pi 900)| = 708$. Thus $A_f \approx 1/\beta = 708 = 1 + R_2/R_1$. The dashed line shows $1/\beta$. We conclude that a noninverting amplifier with $R_1 = 1$ kΩ and $R_2 = 707$ kΩ should be stable with phase margin of 65.5° and have maximally flat frequency response. From Fig. 9.25, the -3dB bandwidth of the feedback amplifier is $\omega_1 = 2\pi 900$ rad/s. Since the gain is 708, gain $\times$ bandwidth = 637 kHz. ❑

Figure 9.26 shows that $A(\omega)$ has bandwidth of 100 Hz and gain of 5623 (75 dB), a gain-bandwidth product of 562 kHz. In the preceding example, gain-bandwidth product was not preserved with the addition of feedback. Instead, the product increased to 637 kHz. Figure 9.25 shows that if peaking is allowed, bandwidth can be increased further.

Exercise 9.11   For the op amp described by Fig. 9.26, find the closed-loop gain for a phase margin of 60°.

Ans. 53 dB.

9.6.6   FREQUENCY COMPENSATION

If an amplifier is unstable for given gain, $1/\beta$, we can make it stable by altering its open-loop gain curve, $A(\omega)$ (and thereby $L(\omega)$). Deliberately changing the frequency response of the nonfeedback amplifier to make a feedback amplifier stable is called *frequency compensation*. Of several available frequency compensation methods, we examine only the two most important, pole shifting and pole addition.

Compensation by Pole Shifting.   Pole shifting means moving the *lowest high-frequency pole* to a lower frequency—just enough to stabilize the amplifier for the given $1/\beta$—by adding one or more components. The specific technique we describe stabilizes with a phase margin of 45°; however, it is readily modified for other phase margins.

The solid blue amplitude and phase curves in Fig. 9.27 describe the uncompensated amplifier. The first two high-frequency poles are located at $f_a = 1$ kHz and $f_b = 10$ kHz. (A scale along the top shows symbols for critical frequencies.) We use this amplifier as an example in the general compensation procedure that follows.

1. Construct a horizontal line at $|1/\beta|_{dB}$ on the open-loop gain curve. A dashed line shows this construction in Fig. 9.27 for $|1/\beta|_{dB} = 40$ dB.
2. On this line, find the frequency of the *second-lowest high-frequency pole* of the uncompensated amplifier—this is $f_b = 10$ kHz in Fig. 9.27.
3. From this point, draw a line of -20 dB per decade slope upward to the left until it intersects the open-loop gain curve. The frequency, f_x, at this point of intersection is the new frequency of the pole. This construction, marked by arrowheads, gives $f_x = 200$ Hz.

After compensation, the two lowest high-frequency poles are at frequencies f_x and f_b. The Bode plot for the new open-loop gain characteristic follows the construction line to $|1/\beta|_{dB}$, and from there runs parallel to the original curve since locations of all poles but the first are unchanged. A dashed blue line in Fig. 9.27 shows how the phase curve

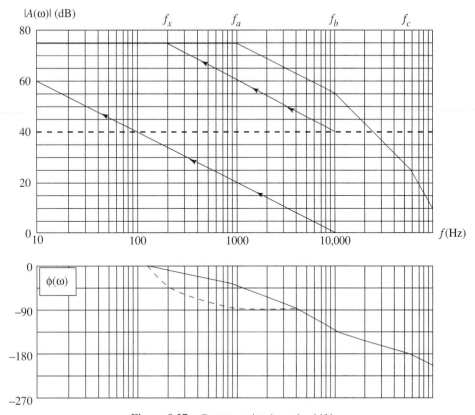

Figure 9.27 Compensation by pole shifting.

changes for the $|1/\beta|_{dB} = 40$ dB construction and illustrates why the phase margin of the compensated amplifier is 45°.

Exercise 9.12 Find the new pole location, f_x, for a feedback amplifier with 25 dB gain using the open-loop amplifier of Fig. 9.27.

Ans. ≈ 30 Hz.

If information is given mathematically, say, as a transfer function, rather than graphically, we can compute the new pole frequency f_x. Since the construction line has slope of −20 dB/dec, the number of decades, d, that separate pole frequencies, f_b and f_x, is

$$d = \frac{|A_{mid}|_{dB} - |1/\beta|_{dB}}{20 \text{ dB/dec}} \tag{9.28}$$

We can then compute f_x from

$$f_b = 10^d f_x \tag{9.29}$$

Using these equations instead of the curves gives 31.6 Hz for the answer to Exercise 9.12.

Unconditional Stability. When a feedback amplifier is likely to be used for many values of closed-loop gain—by many users as in the case of a μA741 op amp or by a single user as for a *variable-gain* amplifier (Fig. 5.36)—compensation for unconditional stability is a great convenience. No new procedure is required; we simply design the compensation for $|1/\beta|_{dB} = 0$. In Fig. 9.27, the construction for unconditional stability, at the lower left, does not fit entirely on the diagram; however, Eqs. (9.28) and (9.29) show that we must move the pole at f_a to $d = (75 - 0)/20 = 3.75$ decades below f_b, to

$$f_x = \frac{10^4}{5623} = 1.78 \text{ Hz}$$

We pay for unconditional stability in the currency of gain-bandwidth product. For example, in Fig. 9.27 the uncompensated amplifier had 1 kHz bandwidth and gain of 5623 (75 dB)—a gain-bandwidth product of 5.62 MHz. With compensation for 40 dB closed-loop gain, the bandwidth becomes 200 Hz; gain-bandwidth is 1.12 MHz. After compensated for unconditional stability, the gain-bandwidth product is only $5623 \times 1.78 = 10$ kHz. This is an important consideration, because higher gain-bandwidth in the open-loop amplifier results in superior bandwidth and other performance parameters in the closed-loop amplifier.

Pole Splitting. If the two lowest high-frequency poles (f_a and f_b in Fig. 9.27) interact, a complex and interesting phenomenon called *pole splitting* occurs. In pole splitting, not only does f_a decrease as we add compensating capacitance, but f_b simultaneously increases to a high value, usually disappearing from a Bode plot like Fig. 9.27. With pole splitting, the pole at f_b moves to the right as we add capacitance, giving us a "moving target" as we try to reference our construction line to f_b. Since the third pole, at f_c, does not move with pole splitting, we can anticipate the disappearance of f_b and instead use f_c, the third high-frequency pole, as the basis for our construction instead of f_b. Pole splitting is good news because the compensation results in a higher gain-bandwidth product than when f_a and f_b correspond to isolated poles. If we expect pole splitting we can use Eq. (9.29) with f_c substituted for f_b to estimate f_x. Gray and Meyer discuss the pole-splitting phenomenon in greater detail in one of the end-of-chapter references.

Pole Addition. *Pole addition* means adding a *new* high-frequency pole that was not originally present in the amplifier or, to the same effect, moving some very-high-frequency pole to a new location at a much lower frequency. This technique gives amplifiers of lower gain-bandwidth product than pole shifting, but is especially useful when the only information we have is the bandwidth, f_a, of the A circuit.

The graphical procedure for pole addition differs from the pole shifting only in that the construction begins at the frequency, f_a, of the lowest pole instead of at f_b. For the 40 dB amplifier in Fig. 9.27, for example, the construction would begin at 10^3 Hz. From Eq. (9.28) the new pole should be located $d = (75 - 40)/20 = 1.75$ dec below 10^3 Hz. From Eq. (9.29) this is at $f_x = 10^{-1.75} \times 10^3 = 17.8$ Hz, compared with 200 Hz from pole shifting. Pole addition is widely used to create unconditional stability, in spite of a penalty in gain-bandwidth product compared to pole shifting.

Exercise 9.13 In Fig. 9.27, we wish to add a new pole at $f = f_x$ to stabilize a feedback amplifier with gain of 50 dB. Find the bandwidth of the compensated amplifier by calculating the number of decades d that separate f_x and f_a.

Ans. $d = 1.25$, 56.2 Hz.

9.6.7 CIRCUIT ASPECTS OF FREQUENCY COMPENSATION

In Sec. 8.4.6 we learned how to reduce the bandwidth of an amplifier by adding a capacitor to the circuit. Once we find the desired pole frequency, f_x, by using graphical construction or Eqs. (9.28) and (9.29), we can use the methods of Chapter 8 to find the amount of capacitance needed for frequency compensation.

EXAMPLE 9.13 Figures 9.28a and 9.28b show the small-signal schematic of a nonfeedback, transresistance amplifier and its high-frequency equivalent. Both transistors have $\beta_F = 190$, $C_\pi = 30$ pF, $C_\mu = 5$ pF, and $g_m = 0.12$ A/V. Q_2 is represented by its common-base equivalent using $\alpha = 0.995 \approx 1$ and $r_e = 8.27$ Ω, and base-spreading resistances are ignored. Figure 9.28c shows the SPICE-generated magnitude and phase plots for transresistance gain as solid curves. From SPICE printout, the low-frequency transresistance gain is 119 dB and the -3 dB frequency, f_a, is about 2 MHz.

(a) Find a circuit location and value for a compensating capacitor that adds a new dominant pole. After feedback is added, the new pole should give stable operation for closed-loop transresistance gain of 60 dB.

(b) Use SPICE to find the frequency response of the compensated amplifier.

Solution. (a) The curves show that without compensating the amplifier, adding feedback for transresistance gain of 60 dB results in an unstable amplifier with gain margin of about -25 dB.

Equation (9.28) shows that the new pole must be

$$d = \frac{119 - 60}{20} = 2.95 \text{ dec}$$

below f_a. Thus the new pole should be located at $f_x = (2 \times 10^6) \times 10^{-2.95} = 2.24$ kHz. The circuit is a two-stage inverting amplifier, with the inversion in the common-emitter stage and voltage gain in the common-base stage. A compensating capacitor between base of Q_1 and the collector of Q_2, as in Fig. 9.28a, could use Miller multiplication to reduce the bandwidth.

From Fig. 9.28b, the midrange Miller gain is

$$\frac{V_o}{V_1} = 0.12 \frac{5 \text{ k}}{5 \text{ k} + 8.27}(-5 \text{ k}) = -599$$

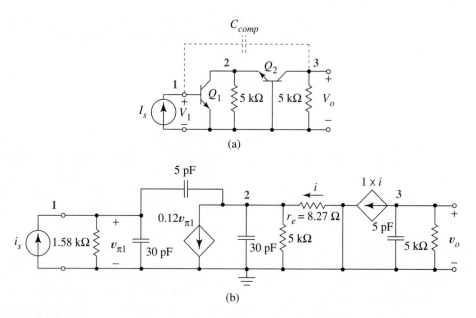

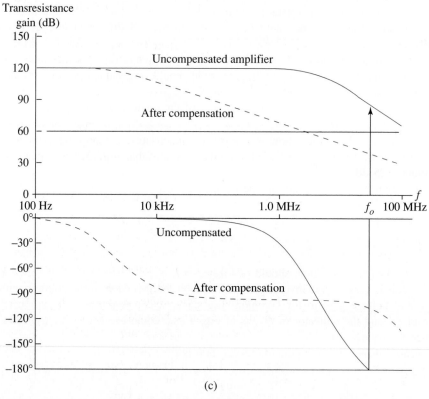

Figure 9.28 Frequency compensation: (a) uncompensated transresistance amplifier; (b) high-frequency equivalent; (c) gain and phase curves of uncompensated and compensated amplifier.

If we ignore all other capacitors, C_{comp} has an effective value of $600C_{comp}$, and would see a resistance of 1.58 kΩ. Since it must give a pole frequency at $f_x = 2.1$ kHz, C_{comp} must satisfy

$$\frac{1}{2\pi (2.1 \times 10^3)600 \, C_{comp}} = 1.58 \times 10^3$$

or $C_{comp} = 79.9$ pF.

(b) Adding this capacitor to the SPICE circuit and recomputing the frequency response curves gives the dashed magnitude and phase curves of Fig. 9.28c. It is clear that adding the proposed feedback to this compensated amplifier will result in a stable amplifier with 60 dB gain.   ❏

The preceding example featured a common-emitter/common-base amplifier with inherently high-bandwidth and pole-addition compensation. The circuit details associated with pole shifting are similar. For example, suppose f_a is determined by a Miller capacitor in a common-emitter or common-source stage. External capacitance added between base and collector (or gate and drain) then increases the Miller capacitance as in Example 8.10, shifting the pole frequency down to the desired value f_x. Use of Miller effect is particularly effective in compensating integrated circuits because the compensating capacitor need not be large.

IC Compensation When Circuit Details Are Unknown. We have taken for granted that the designer of a feedback amplifier has detailed knowledge of the *A* circuit structure and parameters, but with integrated circuits, this is not always so. Sometimes the IC manufacturer provides special external pins for frequency compensation and a *recipe* that tells exactly how to compensate the device. This information typically includes a family of open-loop gain and phase characteristics that correspond to various values of user-provided compensating components. The designer uses this information to achieve the largest bandwidth consistent with given gain and stability specifications.

9.6.8   FREQUENCY COMPENSATION AND SLEW RATE

Frequency compensation and slew rate of a feedback amplifier are closely related. This discussion amplifies upon the introductory discussion of slewing, Sec. 2.5.14. Figures 9.29a and b show an internal structure typical of many operational amplifiers. There are two gain stages: a differential-input transconductance amplifier and an inverting transresistance amplifier. A unity-gain power amplifier drives the external load. A compensating capacitor, which bridges across the transresistance amplifier, is large enough to dominate the dynamic behavior of the op amp. This capacitor is an internal component in unconditionally stable op amps; otherwise, a user-provided external component.

Figure 9.29b depicts the first stage as a differential-input VCCS and the second stage by its Miller equivalent. For the second stage, the dominant pole establishes an open-loop bandwidth of

$$\omega_H = \frac{1}{2\pi R_i C_M}, \qquad C_M = [C_C(1 + GM \, R_o)] \tag{9.30}$$

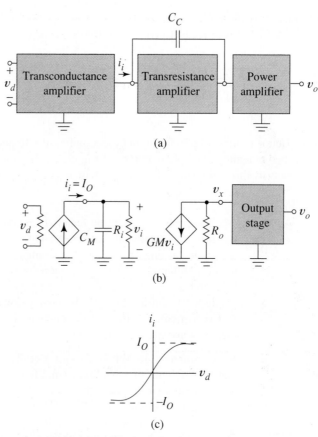

(a)

(b)

(c)

Figure 9.29 Op amp compensation: (a) op amp block diagram with compensating capacitor; (b) equivalent circuit of compensated op amp; (c) transfer characteristic of input stage.

Figure 9.29c shows the nonlinear transfer characteristic of the input stage. When input voltage v_d becomes excessive, one input transistor cuts off, forcing the output to carry bias current I_O. Figures 7.44a and b show how a cut-off input transistor causes constant output current in the other transistor. At such times, the current entering the second stage is $i_i = I_O$. From Fig. 9.29b, the differential equation for v_i is

$$C_M \frac{dv_i}{dt} = +I_O - \frac{v_i}{R_i} \approx I_O \qquad (9.31)$$

where the approximation assumes $I_O \gg v_i/R_i$. The rate of change of internal voltage v_i is, therefore,

$$\frac{dv_i}{dt} = \frac{1}{C_M} I_O$$

Since $v_o \approx v_x$, from Eq. (9.31) we conclude that the slew rate (s.r.) is

$$\text{s.r.} = \left| \frac{dv_o}{dt} \right| = \left| \frac{dv_x}{dt} \right| = GM R_o \left| \frac{dv_i}{dt} \right| = \frac{GM R_o}{C_M} I_O \qquad (9.32)$$

Substituting C_M from Eq. (9.30) gives

$$\text{s.r.} \approx \frac{I_O}{C_C} \qquad (9.33)$$

This shows that a high value for I_O, which is directly related to the dc current of the input stage, is desirable in an amplifier that is to have a high slew rate.

To relate slew rate to bandwidth, solve Eq. (9.30) for $1/C_M$ and substitute into Eq. (9.32). This gives

$$\text{s.r} = \left| \frac{dv_o}{dt} \right| = 2\pi GM R_i R_o I_O \omega_H \qquad (9.34)$$

To make slew rate large we must also make the bandwidth as large as possible; however, this must be consistent with stable closed-loop operation. We conclude that an internally compensated amplifier compromises slew rate as well as gain-bandwidth product, all in the cause of unconditional stability.

9.7
Sinusoidal Oscillators

Oscillators are intentionally unstable circuits that serve as sources of electrical waveforms. There are two broad classes of oscillators: *sinusoidal oscillators,* which produce sinusoidal waveforms, and *relaxation oscillators,* which produce triangular, or rectangular, waveforms. Both classes of oscillators are widely used for time bases in test and measurement equipment, and for signal processing in analog and digital communication systems. Here we concentrate on sinusoidal oscillators, saving relaxation oscillators for Chapter 14.

9.7.1 GENERAL THEORY OF SINUSOIDAL OSCILLATORS

A sinusoidal oscillator has three functional parts, a *phase shifter* to establish the frequency of oscillation, a *gain circuit* to compensate for energy losses in the phase shifter, and a *limiter* to control the amplitude of the oscillations. The gain circuit might be an op amp, a transistor, or a transistor amplifier; the phase shifter is typically an *RC* or *LC* circuit; the limiter might be a diode, a thermistor, or a variable-gain amplifier. In some oscillators the basic functions are combined rather than relegated to individual subcircuits. For example, the internal capacitances of the transistor that provides gain might contribute to the phase shifter, and inherent transistor nonlinearities often provide the limiting. Common to all oscillator circuits is instability, which is best understood in terms of positive feedback theory.

The voltage-shunt feedback structure of Fig. 9.30 describes many sinusoidal oscillators. A voltage amplifier with gain $A(\omega) = V_o/V_i$ provides the gain, and a feedback network described by $\beta(\omega) = V_f/V_o$ is the phase shifter. An oscillator representation, such as Fig. 9.30, differs in several ways from a feedback amplifier diagram: we regard gain $A(\omega)$ as voltage gain—not transresistance gain; $\beta(\omega)$ is defined without the notion of an

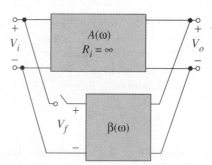

Figure 9.30 General structure of a sinusoidal oscillator.

input subtraction; the feedback network includes reactive elements to provide the phase shift required for positive feedback; and, of course, there is no external signal source. Nonlinearities that limit the signal amplitude invariably arise but do not appear in this linear model. The switch helps us examine the loop gain of the circuit.

Barkhausen Criterion. Assume that $A(\omega)$ and $\beta(\omega)$ are defined in such a way that no loading occurs when the switch is closed. The *Barkhausen criterion* states that there will be sinusoidal oscillations at frequency ω_o when the switch is closed, provided that with the switch open, the loop gain is

$$V_f/V_i = A(\omega_o)\beta(\omega_o) = M(\omega_o)\underline{/\phi(\omega_o)} = 1 \qquad (9.35)$$

When this condition is satisfied, a sinusoidal signal generator V_i attached to the input can be removed when the switch is closed, because the amplitude and phase of the signal fed back to the input are exactly those needed to replace this source.

Since the Barkhausen criterion involves a complex-valued function, it implies two conditions for oscillations, a magnitude condition and a phase condition. We use the *magnitude condition, $M(\omega_o) = 1$*, as a test to determine whether oscillations can exist in a given circuit. This condition arises from the physical requirement that the amplifier provide sufficient gain to make up exactly for energy losses in the circuit. If $M(\omega_o)$ exceeds one, the oscillator is *self-starting,* with the oscillation arising spontaneously and *increasing in amplitude* until nonlinearities cause a reduction in $M(\omega_o)$. Some oscillator circuits require a signal generator for start-up; however, self-starting circuits are the norm.

Provided that oscillations can occur, the *phase condition* $\phi(\omega_o) = 0$ determines the frequency of oscillation ω_o of the circuit. This condition physically means the signal arrives back at the input exactly in phase with itself.

9.7.2 RC OSCILLATORS

Two important oscillators use *RC* phase-shifting circuits. Both are suitable for generating oscillations at frequencies from a few hertz to hundreds of kilohertz.

Wien Bridge Oscillator. Figure 9.31a shows the *Wien bridge* oscillator with a noninverting op amp circuit for its gain element. It is also called a *bridge oscillator* because the active element is imbedded in the *RC* bridge circuit, as we see by redrawing the circuit as Fig. 9.31b.

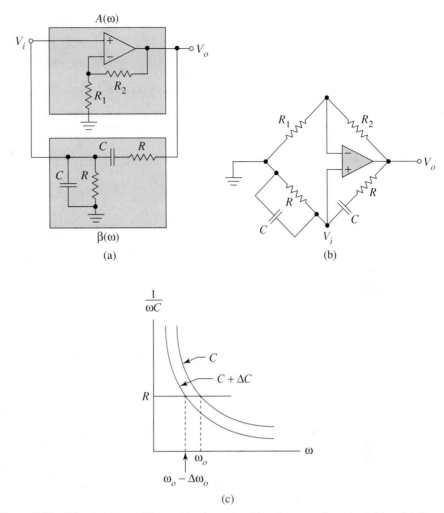

Figure 9.31 Wien bridge oscillator: (a) schematic; (b) redrawn to show the bridge; (c) illustration of frequency stability.

To find design equations for this or any oscillator, we apply the Barkhausen criterion. From Fig. 9.31a, $\beta(\omega)$ is the transfer function

$$\beta(\omega) = \frac{Z_1}{Z_1 + Z_2}$$

where Z_1 and Z_2 are the impedances of the parallel RC and series RC circuits, respectively. Since $A(\omega)$ is the gain of the noninverting amplifier, Eq. (9.35) requires

$$A(\omega_o)\beta(\omega_o) = \frac{(1 + R_2/R_1)[R/(j\omega_o CR + 1)]}{[R/(j\omega_o CR + 1)] + (R + 1/j\omega_o C)} = +1$$

where ω_o is the frequency of oscillation. Simplifying leads to the expression

$$A(\omega_o)\beta(\omega_o) = \frac{(1 + R_2/R_1)j\omega_o RC}{1 + 3j\omega_o RC - \omega_o^2 R^2 C^2} = +1 \tag{9.36}$$

For the expression to be real, the real part of the denominator must be zero. This gives the phase condition

$$\omega_o^2 = \frac{1}{R^2 C^2} \tag{9.37}$$

Solving for frequency gives

$$\omega_o = \frac{1}{RC} \tag{9.38}$$

Substituting this into Eq. (9.36) gives the amplitude condition

$$\left(1 + \frac{R_2}{R_1}\right) = 3$$

Thus if $R_2 \geq 2R_1$ the circuit will oscillate; the inequality implies oscillations of increasing amplitude.

An important attribute of an oscillator is its *frequency stability,* its ability to maintain a frequency close to the design value. From Eq. (9.38) it is easy to show that

$$S_C^{\omega_o} = S_R^{\omega_o} = -1$$

not particularly low values. To give an intuitive idea of what is involved, Fig. 9.31c shows Eq. (9.38) as the graphical solution of the equation $1/\omega C = R$. The dashed lines show how ω_o changes with variations in C. (Like the sensitivity expression, this diagram assumes that the resistor and capacitor pairs in the β circuit are matched.)

The next example demonstrates another useful way to investigate frequency stability of an oscillator—using SPICE to plot the phase of $A(\omega)\beta(\omega)$.

EXAMPLE 9.14 Plot the phase shift of the loop gain for a Wien bridge oscillator with $R = 397.9\ \Omega$ and $C = 1000$ pF. Find the change in frequency if the value of the shunt capacitor increases by 10%.

Solution. Figure 9.32a shows the phase-shifting network—Fig. 9.32b the SPICE code. An asterisk marks the statement that describes C_2 for the second run. Since Eq. (9.38) indicates an oscillation frequency of 400 kHz, the phase of $\beta(\omega)$, which is the phase of $V(1)$, is plotted from 320 to 480 kHz.

The solid curve of Fig. 9.32c shows that the phase curve of the original oscillator is roughly linear over the range of the plot and crosses zero at 400 kHz. Increasing C_2 by 10% gives the dashed curve, showing that the frequency drops to about 380 kHz, a change of 5%. Thus

$$S_{C_2}^{\omega_o} \approx -0.5$$

❏

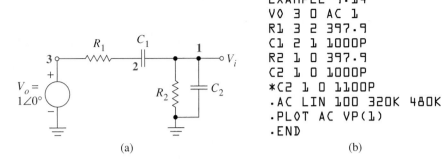

```
EXAMPLE 9.14
VO 3 0 AC 1
R1 3 2 397.9
C1 2 1 1000P
R2 1 0 397.9
C2 1 0 1000P
*C2 1 0 1100P
.AC LIN 100 320K 480K
.PLOT AC VP(1)
.END
```

(a) (b)

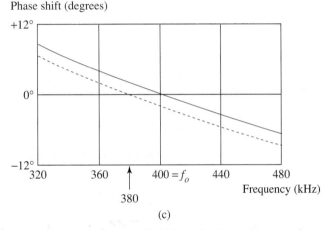

(c)

Figure 9.32 Determining frequency stability from the phase curve: (a) phase-shift circuit; (b) SPICE code for plotting phase; (c) result of change in capacitance.

From the discussion of frequency stability and the example we conclude that the components that contribute to the phase-shifting circuit should be of high quality and relatively insensitive to temperature and other environmental factors.

Phase-Shift Oscillator. Figure 9.33a shows the *phase-shift oscillator*. The phase shifter consists of three *RC* sections. The gain element is represented as an ideal inverting amplifier with voltage gain $-K$.

To find amplitude and phase conditions from the Barkhausen criterion, we analyze the phase-shifting circuit of Fig. 9.33b. The transfer characteristic of such a *ladder circuit* is most easily found by systematic analysis from output to input as suggested by the following equations.

$$I_1 = \frac{V_i}{R}, \qquad V_1 = V_i + I_1\frac{1}{j\omega C} = \left(1 + \frac{1}{j\omega RC}\right)V_i$$

$$I_2 = \frac{V_1}{R} + I_1 = \left(1 + \frac{1}{j\omega RC}\right)\frac{V_i}{R} + \frac{V_i}{R} = \left(2 + \frac{1}{j\omega RC}\right)\frac{V_i}{R}$$

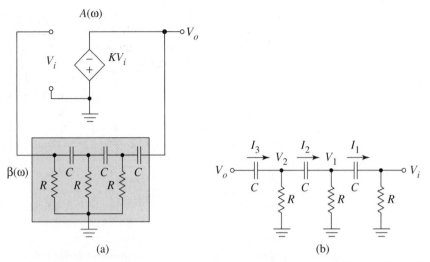

Figure 9.33   Phase-shift oscillator: (a) circuit; (b) notation for computing $\beta(\omega)$ for the oscillator.

Continuing in this fashion finally gives

$$\beta(\omega) = \frac{V_i}{V_o} = \frac{1}{1 + 6/j\omega RC - 5/\omega^2 R^2 C^2 - 1/j\omega^3 R^3 C^3}$$

Thus our design must satisfy

$$A(\omega_o)\beta(\omega_o) = \frac{-K}{1 - j(6/\omega_o RC) - 5/\omega_o^2 R^2 C^2 + j(1/\omega_o^3 R^3 C^3)} = +1$$

To find the phase condition, we set the imaginary part of the denominator to zero. This gives an oscillation frequency of

$$\omega_o = \frac{1}{\sqrt{6}RC} \tag{9.39}$$

It is easy to show that the sensitivity to changes in R and C is the same as for the Wien bridge circuit. Substituting ω_o from Eq. (9.39) into the preceding equation gives the gain condition

$$\frac{-K}{1 - 30} = +1$$

Thus the circuit oscillates for any $K \geq 29$.

Exercise 9.14   Work out the first missing step in the preceding derivation of $\beta(\omega)$; that is, find the expression for V_2 in terms of V_i.

Ans. $V_2 = \left(1 + \dfrac{3}{j\omega RC} - \dfrac{1}{(\omega RC)^2}\right)V_i$

Amplitude Limiters. To make an oscillator self-starting and to allow for uncertainties in parameter values, we usually design the circuit so that the gain condition is exceeded. The amplitude of the oscillation then increases until some nonlinearity reduces the effective loop gain. If the signal amplitude becomes too large, the signal traverses a large segment of the nonlinear transfer characteristic of the active device causing the sinusoidal output waveform to be highly distorted. There are three basic approaches to controlling the signal amplitude while keeping the waveform reasonably sinusoidal.

When the gain condition is not greatly exceeded, a design can rely on inherent transistor nonlinearities to limit the signal amplitude to a value conducive to producing a reasonably undistorted sine wave. This does not produce very *robust* designs, however, because component values are rather critically related to waveform purity.

The second approach is to insert a special nonlinear component into the loop so that loop gain begins to diminish with signal amplitude while signal amplitude is still small. A thermistor with resistance that decreases by self-heating, or a strategically placed diode are examples of components used for this purpose. An example is Fig. 9.34a, which shows the phase-shift oscillator of Fig. 9.33a with two diodes added for limiting. As long as the voltage across resistor KR has peak amplitude less than 0.5 V, the circuit is approximately the same as Fig. 9.33a; however, when output amplitude exceeds the diode cut-in voltage, the diodes introduce resistance in parallel with KR. This nonlinear resistance decreases with increasing amplitude, lowering the gain of the inverting amplifier and thereby limiting the oscillation amplitude. The following example and Problem 9.47 investigate this limiter in some detail.

EXAMPLE 9.15 Use SPICE to examine the output waveform of the phase-shift oscillator of Fig. 9.34a without and with the diode limiter. Values are $R = 10$ kΩ, $K = 50$, and $C = 3000$ pF. Use a dependent source with gain 10^5 for the op amp.

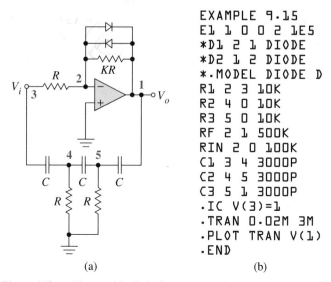

```
EXAMPLE 9.15
E1 1 0 0 2 1E5
*D1 2 1 DIODE
*D2 1 2 DIODE
*.MODEL DIODE D
R1 2 3 10K
R2 4 0 10K
R3 5 0 10K
RF 2 1 500K
RIN 2 0 100K
C1 3 4 3000P
C2 4 5 3000P
C3 5 1 3000P
.IC V(3)=1
.TRAN 0.02M 3M
.PLOT TRAN V(1)
.END
```

(a)   (b)

Figure 9.34 Phase-shift oscillator with diode limiter: (a) schematic; (b) SPICE code; (c) output with limiting diodes omitted; (d) output with limiting included.

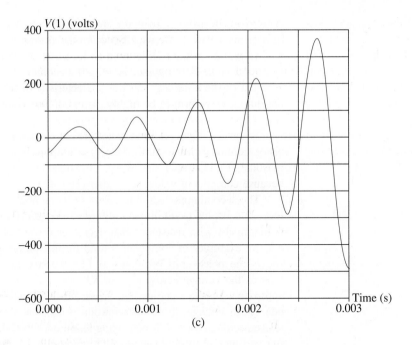

(c)

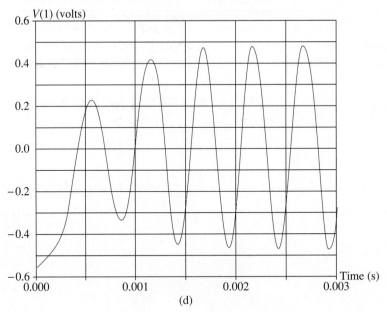

(d)

Figure 9.34 (continued)

Solution. Figure 9.34b is the SPICE code with diodes omitted. The initial condition statement, ".IC V(3) = 1," changes the initial voltage at node 3 to a nonequilibrium value to help start oscillations. Figure 9.34c shows the output without the diode statements. Since the linear circuit model has no limiting whatsoever, the oscillation amplitude quickly increases to hundreds of volts in the simulation. The output of a real op amp would be driven to its saturation limits, giving a highly distorted output voltage, a statement readily verified using the more realistic op amp model of Fig. 3.56.

Figure 9.34d shows the output with the diode statements present in the input code. The waveform quickly converges to a sinusoidal waveshape, with amplitude limited by the diodes and virtual ground to the ±0.5 V range. ❏

Exercise 9.15 Calculate the expected oscillation frequency in Example 9.15, and compute how many horizontal divisions each oscillation period should occupy in Figs. 9.34c and d.

Ans. 2.17 kHz, 0.924 division/period.

For sinusoids of very high quality, limiting can be provided by a "linear amplifier" with gain that decreases as amplitude increases. In Fig. 9.35, the voltage-controlled amplifier of Fig. 5.45 is the gain element in a Wien bridge oscillator. A half-wave rectifier with a capacitive load *(envelope detector)* smooths out small amplitude variations and produces a dc control signal, V_C, proportional to signal amplitude. The follower isolates the control circuit from the RC bridge so amplitude and phase conditions are unchanged. As oscillation amplitude increases, V_C becomes more negative, and gain automatically decreases. The detector time constant $R_D C_D$ should be several periods of the oscillator waveform.

Waveform Purity in Sinusoidal Oscillators. Because it involves nonlinear operation, limiting always introduces "impurity" into the oscillator output waveform, causing the

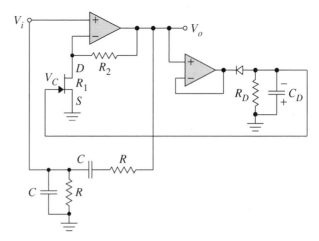

Figure 9.35 A limiter that uses signal amplitude to control loop gain.

output to differ from the desired single-frequency sinusoid. We learned in Sec. 1.6.5 that we can describe any periodic waveform as a Fourier series

$$v(t) = \sum_{n=0}^{\infty} A_n \cos(n\omega_o t + \phi_n)$$

For an oscillator, the ideal output is the $n = 1$ term alone; terms for which $n \geq 2$ are undesired *impurity* terms introduced by the limiter. The smaller these harmonic terms, the greater the *waveform purity.* A useful measure of waveform impurity is the percent *total harmonic distortion,* THD, where

$$\text{THD} = \frac{\text{rms value of harmonic components}}{\text{rms value of fundamental frequency term}} \times 100$$

If we know the fundamental frequency $f_o = 2\pi/\omega_o$ of the oscillator output, we can use SPICE to compute THD by adding a SPICE statement ".FOUR. " This asks SPICE to compute amplitudes of the first nine Fourier series components and to use them to estimate THD.

In Example 9.15, the output frequency in the simulation was actually $f_o = 2$ kHz (instead of 2.167 kHz predicted by the theory). Adding the statement

```
.FOUR 2.0E3 V(1)
```

reveals that the output voltage waveform contains 5.84% distortion. Figure 9.36 shows amplitudes of the harmonics calculated by SPICE for this example. The advantage of more sophisticated limiters like Fig. 9.35 is improved waveform purity. The oscillator designer is often required to meet a THD specification along with specifications of frequency and amplitude. Harmonic distortion is discussed in greater detail in Chapter 10 in connection with power amplifiers.

9.7.3 LC OSCILLATORS

Two important oscillators use the three-element pi structure of Fig. 9.37a for the phase shifter. In Fig. 9.37b, a transistor biased for small-signal, active operation provides the gain. From Fig. 9.37b,

$$A(\omega) = \frac{V_o}{V_i} = -g_m[R_o \| Z_1 \| (Z_2 + Z_3)]$$

and

$$\beta(\omega) = \frac{Z_3}{Z_3 + Z_2}$$

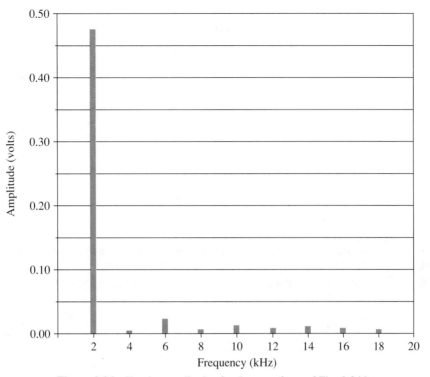

Figure 9.36 Fourier amplitudes for the waveform of Fig. 9.34d.

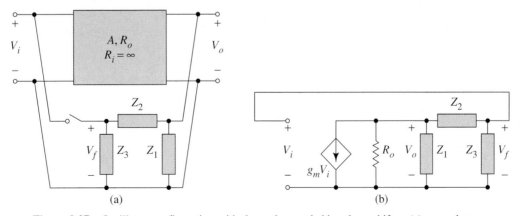

Figure 9.37 Oscillator configuration with three-element ladder phase shifter: (a) general structure; (b) oscillator with transistor as active element.

This gives the loop gain

$$A(\omega)\beta(\omega) = -g_m[R_o \| Z_1 \| (Z_3 + Z_2)]\frac{Z_3}{Z_3 + Z_2}$$

$$= -g_m \frac{1}{1/R_o + 1/Z_1 + 1/(Z_3 + Z_2)}\frac{Z_3}{Z_3 + Z_2}$$

$$= \frac{-g_m Z_3}{(Z_3 + Z_2)/R_o + (Z_3 + Z_2)/Z_1 + 1}$$

The Barkhausen criterion requires

$$M\underline{/\phi} = \frac{-g_m R_o Z_1 Z_3}{Z_1 Z_3 + Z_1 Z_2 + R_o(Z_1 + Z_2 + Z_3)} = +1 \qquad (9.40)$$

If each impedance is an LC element, then $Z_i = jX_i$, where $X_i = \omega L_i$ for an inductor and $X_i = -1/\omega C_i$ for a capacitor. Then

$$M\underline{/\phi} = \frac{g_m R_o X_1 X_3}{-(X_1 X_3 + X_1 X_2) + jR_o(X_1 + X_2 + X_3)} = 1\underline{/0°} \qquad (9.41)$$

Because the numerator is real, the phase condition is satisfied only if

$$X_1 + X_2 + X_3 = 0 \qquad (9.42)$$

When Eq. (9.42) is satisfied, the gain condition in Eq. (9.41) reduces to

$$\frac{-g_m R_o X_3}{X_3 + X_2} \geq 1 \qquad (9.43)$$

We next use these conditions to study two specific oscillators.

Colpitts Oscillator. In the Colpitts oscillator, Z_2 is an inductor and Z_1 and Z_3 are capacitors as in Fig. 9.38. Substituting $Z_2 = j\omega L_2$, $Z_1 = 1/j\omega C_1$, and $Z_3 = 1/j\omega C_3$ into Eq. (9.42) gives

$$-\frac{1}{\omega C_1} + \omega L_2 - \frac{1}{\omega C_3} = 0$$

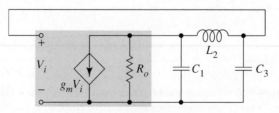

Figure 9.38 Colpitts oscillator.

Solving for ω gives the phase condition

$$\omega_o^2 = \frac{1}{L_2[C_1C_3/(C_1 + C_3)]} \tag{9.44}$$

which establishes the frequency of oscillation. A second design condition comes from Eq. (9.43),

$$\frac{g_mR_o(1/\omega_oC_3)}{-1/\omega_oC_3 + \omega_oL_2} = \frac{g_mR_o}{\omega_o^2L_2C_3 - 1} \geq 1 \tag{9.45}$$

Substituting ω_o^2 from Eq. (9.44) leads to the second design equation

$$g_mR_o \geq \frac{C_3}{C_1} \tag{9.46}$$

which specifies the gain required for sustained oscillations.

Exercise 9.16 Compute the sensitivity of the resonant frequency of a Colpitts oscillator to L_2. Use your answer to estimate the actual resonant frequency of an oscillator designed to operate at 1 MHz if L_2 is 12% high.

Ans. -0.5, 940 kHz.

EXAMPLE 9.16 The Colpitts oscillator of Fig. 9.39a has a transistor biased at 0.5 mA; $\beta_F = 120$; C_C is a large coupling capacitor. Find values for L_2, C_1, and C_3 so the circuit oscillates at $f_o = 1$ MHz. Ignore transistor capacitances.

Solution. Figure 9.39b shows the high-frequency equivalent circuit. At 0.5mA, $g_m = 0.02$ A/V and $r_\pi = 6$ kΩ. The small-signal equivalent of the oscillator is Fig. 9.39b, where

$$5.5 \text{ k}\Omega = r_\pi \| 159 \text{ k} \| 112 \text{ k}$$

From Eq. (9.46),

$$(0.02)(6 \times 10^3) = 120 \geq \frac{C_3}{C_1} \tag{9.47}$$

and from Eq. (9.44),

$$L_2C_{eq} = \frac{1}{(2\pi 10^6)^2} = 25.3 \times 10^{-15} \tag{9.48}$$

where $C_{eq} = C_1C_3/(C_1 + C_3)$.

We now have two equations involving C_1, C_3, and L_2. In an FET design we would make an arbitrary choice of one component value; however, in a BJT design we must minimize the effect of r_π, since our theory does not include a resistor in this location. Therefore, we select C_3 such that its reactance is much smaller than $r_\pi = 5.5$ kΩ. This

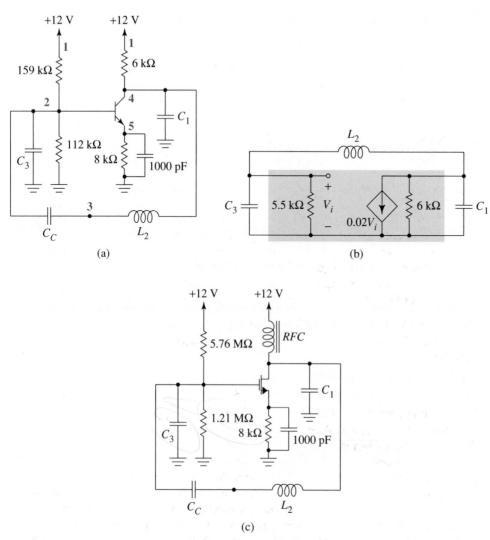

Figure 9.39 Colpitts oscillator realizations: (a) circuit diagram for a BJT oscillator; (b) small-signal equivalent for BJT circuit; (c) MOSFET oscillator.

gives the third design equation. Using two orders of magnitude to ensure r_π has negligible effect gives

$$\frac{1}{1\pi 10^6 C_3} = \frac{1}{100} 5.5 \times 10^3$$

or $C_3 = 2894$ pF.

Since our design depends upon an approximation, and since actual component values may not have exactly the values we expect, we satisfy Eq. (9.47) by using

$C_3/C_1 = 75$ instead of the limiting value of 120 to ensure self-starting oscillations. With C_3 previously selected, this gives

$$C_1 = \frac{2894}{75} = 38.6 \text{ pF}$$

and $C_{eq} = 38.1$ pF. From Eq. (9.48)

$$L_2 = \frac{25.3 \times 10^{-15}}{38.6 \times 10^{-12}} = 0.655 \text{ mH}$$

The coupling capacitor C_C must have reactance $<< \omega_o L_2$ at 1 MHz. Since the latter is 4.11 kΩ, $C_C = 0.001$ μF will do. ❏

In Colpitts oscillators, the collector or drain resistor is usually replaced by an inductor called an *rf choke*. The choke coil is a short circuit for biasing purposes but presents an open circuit at rf (radio frequency) oscillator frequencies. When a choke is used, R_o in Fig. 9.38 becomes the output resistance r_o of the transistor. The rf choke reduces power dissipation of the circuit and improves the purity of the output waveform.

Exercise 9.17 Figure 9.39c shows a MOSFET oscillator with drain biased at 1 mA through an RF choke. If transistor parameters are $k = 4\text{mA/V}^2$, $V_t = 1.2$ V, and $V_A = 70$ V, find the condition for oscillation.

Ans. $197 \geq C_3/C_1$.

We have seen that SPICE simulations are useful in verifying oscillator designs, examining distortion, and exploring sensitivity limitations. There are some practical simulation difficulties when large bypass and coupling capacitors are included in simulations along with components having short time constants. Sometimes there are convergence problems in the numerical algorithms. Sometimes a very large number of oscillator cycles are required before the circuit settles into steady-state operation. The following example shows a way to avoid time constant problems in oscillator simulations. An initial dc analysis determines the voltages across the large capacitors, which we then replace by dc voltage sources of appropriate polarity (see the discussion related to Fig. 7.10b). This substitution gives the same Q-point and the same high-frequency model without introducing large time constants.

EXAMPLE 9.17 Use SPICE to show how the oscillations arise in the circuit designed in Example 9.16.

Solution. The SPICE code of Fig. 9.40a uses values determined in the design of Fig. 9.39a. Following initial dc analysis, capacitors CE and CF were replaced, respectively, by sources VEE and VFF. Because the oscillation period is expected to be approximately 1 μs, the initial .TRAN statement runs the simulation for 30 periods to show the buildup of oscillations.

Figure 9.40b shows that the oscillation is superimposed upon a 9.2 V dc collector voltage. By measuring the average period of the last nine cycles from the output data,

```
EXAMPLE 9.17
VCC 1 0 DC 12
RB1 1 2 159K
RB2 2 0 112K
RC 1 4 6K
RE 5 0 8K
C1 4 0 38.6E-12
C3 2 0 2894E-12
L2 3 4 0.664E-3
*CE 5 0 1000P
VEE 5 0 DC 3.7456
*CF 3 2 0.001U
VFF 3 2 DC 4.714
Q1 4 2 5 TRAN
.MODEL TRAN NPN BF=120
.IC V(2)=4.5
.TRAN 2E-7 30E-6
.PRINT TRAN V(4)
.PLOT TRAN V(4)
.OP
.END
```

(a)

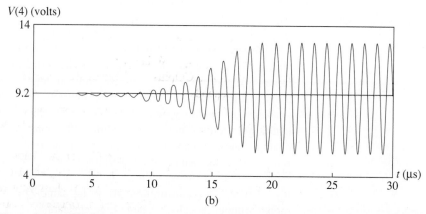

(b)

Figure 9.40 BJT Colpitts oscillator: (a) SPICE code; (b) output waveform showing initial start-up transient.

we find $f_o = 0.9414$ MHz. The oscillations reach steady-state conditions after about 20 μs.

This simulation demonstrates limiting performed by inherent transistor nonlinearities rather than by a special limiter circuit. Because we used $C_3/C_1 = 75$ in our design, the loop gain is rather high and the resulting oscillation has peak amplitude of about 3.5. Harmonic distortion is about 13%. Problem 9.54 explores how a smaller C_1 affects the buildup and purity of the waveform. ❏

Hartley Oscillator. The Hartley oscillator, Fig. 9.41, is Fig. 9.37b with inductors for Z_1 and Z_3 and a capacitor for Z_2. For this circuit, Eq. (9.42) gives

$$\omega_o L_1 - \frac{1}{\omega_o C_2} + \omega_o L_3 = 0$$

which establishes the oscillator frequency

$$\omega_o^2 = \frac{1}{C_2(L_1 + L_3)} \tag{9.49}$$

Substituting reactances into Eq. (9.43) gives

$$\frac{-g_m R_o \omega_o L_3}{\omega_o L_3 - 1/(\omega_o C_2)} = \frac{g_m R_o \omega_o^2 L_3 C_2}{1 - \omega_o^2 L_3 C_2} \geq 1$$

or

$$(g_m R_o + 1)\omega_o^2 L_3 C_2 \geq 1$$

Substituting Eq. (9.49) into this expression and simplifying gives the gain condition

$$g_m R_o \geq \frac{L_1}{L_3} \tag{9.50}$$

9.7.4 QUARTZ CRYSTAL OSCILLATORS

Frequency Stability. We know that frequency stability is an important consideration in an oscillator. The Colpitts and Hartley design equations show that in these circuits the frequency depends entirely upon the reactive components in the resonators. We are therefore concerned with variations in these components due to aging, temperature, and tolerances, especially since transistor capacitances are sometimes part of the phase shifter.

By defining $C_{eq} = C_1 C_3/(C_1 + C_3)$ for the Colpitts oscillator, we can view ω_o in Eq. (9.44) as that frequency where reactances ωL_2 and $1/\omega C_{eq}$ are equal, as illustrated by the bold curves in Fig. 9.42. The light curves show how variations in C_{eq} *and* L_2 affect the frequency of oscillation. When greater frequency stability is required than discrete capacitance and inductance can realize, quartz crystals are used in the phase shifting circuit.

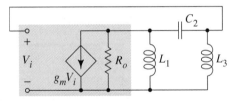

Figure 9.41   Hartley oscillator small-signal circuit.

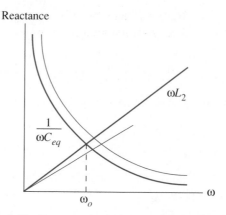

Figure 9.42 Frequency stability of a Colpitts oscillator.

Crystal Resonator. In quartz crystals, slight, reversible, physical deformations result in an electrical voltage, and, conversely, applied voltage produces physical deformations. The crystal is thus an electromechanical device in which electrical excitation and mechanical deformation are tightly coupled, a feature that makes the crystal highly useful as a transducer. Another unusual property of a quartz crystal is that, once set in motion, its energy losses per cycle are very slight. Its electrical equivalent circuit is an L-C resonant circuit like Fig. 9.43a, in which the energy losses of the crystal are embodied in the crystal parameter r, and C' is associated with the external holder that makes electrical contacts to the quartz. Representative component values for a 90 kHz crystal are $L = 137$ H, $C = 0.0235$ pF, $r = 15$ kΩ and $C' = 3.5$ pF.

We now derive an expression for the reactance of the quartz crystal that allows us to compare the crystal resonator characteristics with the L-C reactance curves of Fig 9.42.

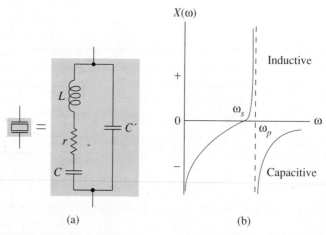

(a)   (b)

Figure 9.43 Crystal resonator: (a) schematic symbol and electrical equivalent circuit; (b) approximate reactance curve.

Near resonance where the crystal is used, $r << \omega L$. With resistance omitted to simplify the development, the impedance of the crystal in Fig. 9.43a is

$$Z = \frac{(j\omega L + 1/j\omega C)1/j\omega C'}{j\omega L + 1/(j\omega C) + 1/(j\omega C')} = \frac{-\omega^2 LC + 1}{-j\omega^3 LCC' + j\omega(C + C')}$$

where the second expression follows from multiplying numerator and denominator by $(j\omega C)(j\omega C')$. We next factor $-LC$ from the numerator and $-j\omega LCC'$ from the denominator. This gives

$$Z = jX(\omega) = \frac{-LC}{-j\omega LCC'} \frac{\omega^2 - 1/LC}{\omega^2 - (C + C')/LCC'} = \frac{-j}{\omega C'} \frac{\omega^2 - \omega_s^2}{\omega^2 - \omega_p^2} \tag{9.51}$$

where

$$\omega_s = \frac{1}{\sqrt{LC}} \quad \text{and} \quad \omega_p = \frac{1}{\sqrt{LCC'/(C + C')}}$$

define the series and parallel resonant frequencies of the crystal, respectively. Figure 9.43b is a sketch of Eq. (9.51). Intuitively, as ω approaches ω_s, L and C go into series resonance, producing zero reactance. Above this frequency, the inductive reactance dominates the LC branch, and this reactance then goes into parallel resonance with C' at ω_p. The series and parallel resonant frequencies are very close together in a quartz crystal, 88,700 and 88,998 Hz, respectively, for the nominal 90 kHz crystal described earlier. Thus the curve segment between series and parallel resonance is nearly a vertical line. Furthermore, in a *properly cut* quartz crystal, ω_p varies by only a few hundred ppm over a wide range of temperature. As we see next, this leads to a very stable oscillator frequency of frequency $\omega_o \approx \omega_p$.

Pierce Oscillator. The crystal-controlled Pierce oscillator is a Colpitts oscillator with inductor replaced by a crystal as in Fig. 9.44a. Substituting the reactance of the crystal $X(\omega)$ for the reactance of L_2 in Eq. (9.42) gives

$$X_1 + X(\omega) + X_2 = -\frac{1}{\omega C_1} + X(\omega) - \frac{1}{\omega C_2} = 0$$

The frequency of oscillation is thus the frequency ω_o that satisfies

$$X(\omega) = \frac{1}{\omega C_{eq}}$$

The graphical construction of Fig. 9.44b shows why the oscillation frequency of the Pierce oscillator is virtually independent of variations in resonator capacitance. Another popular crystal oscillator circuit is obtained by replacing one of the inductors in a Hartley oscillator with a crystal.

9.8
Summary

Adding negative feedback to an amplifier reduces sensitivity to parameter variations, increases bandwidth, and reduces nonlinear distortion. We can also use feedback to increase or decrease midrange input resistance and (independently) increase or decrease output resistance. All improvements involve multiplication or division by the improve-

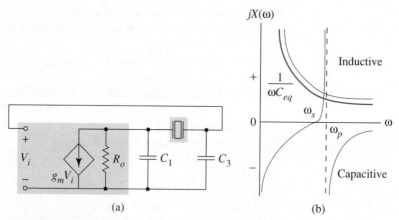

Figure 9.44 Pierce oscillator: (a) equivalent circuit; (b) sensitivity of oscillator frequency to capacitor variations.

ment factor $1 + A\beta$. The closed-loop gain is simultaneously reduced by this factor; however, this counts as an improvement once we realize that the new, lower, gain usually approximates $1/\beta$. By using resistor ratios for β we can control this closed-loop gain to $\pm 1\%$.

With four classes of feedback, we can make our nonfeedback amplifier more closely resemble an ideal voltage, current, transresistance, or transconductance amplifier at midrange frequencies. For each class of feedback, the definition of gain, A, corresponds to that of the ideal amplifier. The origin of the feedback signal defines the output circuit of the feedback amplifier—voltage (current) feedback lowers (raises) output resistance, making the original amplifier more closely resemble a dependent voltage (current) source. The way the feedback signal subtracts from the input signal determines the nature of the input circuit. Series (shunt) feedback involves a voltage (current) subtraction that increases (decreases) input resistance. This makes the amplifier better represent a voltage- (current-) controlled dependent source.

The resistive β circuits we use in practical feedback amplifiers introduce loading at input and output. To apply ideal feedback equations to this case, we represent the β circuit by a two-port equivalent—the one that corresponds to the particular kind of feedback we employ. By associating all the circuit resistance, including source and load resistance, with the nonfeedback amplifier and identifying the "1-2" parameter with β, we can apply the simple ideas of ideal feedback theory. Recognizing that the reciprocal of the "1-2" parameter approximates the closed-loop gain of the feedback amplifier leads to conceptual shortcuts in practical analysis and design.

To successfully implement feedback or evaluate the designs of others, we must be able to recognize some standard practical feedback topologies. The difference amplifier structure is a voltage-series feedback arrangement that bases the input subtraction on an amplifier's differential input. In classical voltage-feedback circuits, as in the difference amplifier topology, the output node of the nonfeedback amplifier connects directly to the β circuit. Current feedback features indirect sensing of output current using an unbypassed emitter or source resistor. Classical series feedback involves a connection from the β circuit to an unbypassed emitter or source resistor in the input circuit of the nonfeedback amplifier; shunt feedback uses a direct connection from the β circuit to an input node of the nonfeedback amplifier. The A circuit must provide a 180° phase shift to facilitate the current subtraction needed for shunt feedback; zero phase shift is required for series feedback.

To ensure that our feedback amplifier does not oscillate, we examine the magnitude and phase of its loop-gain function, the product of the A circuit and β circuit gains. If there exists no frequency where the loop gain is negative in sign and also greater than one in magnitude, the feedback amplifier will be stable. Gain and phase margins are two measures of the degree of stability that might be included in design specifications. If the feedback amplifier is destined to be unstable, we must compensate it by modifying the open-loop gain curve to produce appropriate gain and phase margins. This process, called frequency compensation, usually involves adding capacitance to reduce the bandwidth of the A circuit. This might be a custom adjustment, geared to a particular closed-loop gain, 1/β. Alternatively, we might wish our compensation to produce an unconditionally stable amplifier that will accommodate *any* closed-loop gain. Custom compensation results in amplifiers with greater gain-bandwidth product. Frequency compensation is obviously related to an amplifier's bandwidth and is also closely related to its slew rate. Larger compensating capacitors give lower slew rates—another disadvantage of unconditionally stable amplifiers.

Oscillators are circuits intentionally made unstable by positive feedback. Common to all oscillators are phase shifters, which determine the frequency of oscillation, gain circuits to make up for energy losses in the phase shifters, and limiters to control oscillation amplitude. The Barkhausen stability criterion, based upon conditions for positive feedback, gives a complex valued equation for any oscillator. The Barkhausen conditions give an equation for the frequency of oscillation and an amplitude condition that must be satisfied for oscillations to occur. In this chapter only oscillators with sinusoidal output waveforms were considered. Phase-shift and Wein bridge oscillators use *RC* circuits for phase shifters; Colpitts and Hartley oscillators employ tuned circuits.

Important considerations in sinusoidal oscillator design are the harmonic purity of the waveform and frequency stability, the relative insensitivity of the oscillation frequency to variations in circuit parameters, and environmental factors. Total harmonic distortion measures the impurity of the output waveform in terms of the amplitudes of undesired harmonics present. The sensitivity definitions introduced in Chapter 6 are useful aids in oscillator analysis and design. SPICE simulations help us address both distortion and sensitivity issues in a practical fashion. Quartz crystals introduced into oscillator phase shifters result in great improvements in frequency stability.

REFERENCES _____

1. BANZHAF, W. *Computer-Aided Circuit Analysis Using SPICE.* Englewood Cliffs NJ: Prentice Hall, 1989.
2. GRAY, P. R., and R. G. MEYER. *Analysis and Design of Analog Integrated Circuits,* 3rd ed. New York: John Wiley, 1993.
3. MILLMAN, J., and A. GRABEL. *Microelectronics.* New York: McGraw-Hill, 1987.
4. OPPENHEIM, A., and A. WILLSKY. *Signals and Systems.* Englewood Cliffs, NJ: Prentice Hall, 1983.
5. SOLOMON, J. E. "The monolithic op amp: A tutorial study, " *IEEE J. Solid-State Circuits,* Vol. SC-9 (December 1974), pp. 314–332.
6. TUINENGA, P. W. *SPICE: A Guide to Circuit Simulation & Analysis Using PSPICE.* Englewood Cliffs, NJ: Prentice Hall, 1988.

PROBLEMS

Section 9.1

9.1 An amplifier has measured gain of 800. After adding
negative feedback, measurement shows that the gain is 25. Find
the values for $1 + A\beta$ and β.

9.2 A transresistance feedback amplifier must produce an output signal voltage of 12 V when the input current is 100 mA.
The improvement factor must be 50.
(a) Find the gain of the nonfeedback amplifier.
(b) What value of β will be required?
(c) How large is w_i when the feedback amplifier develops
maximum output voltage?

9.3 The gain of a current amplifier is 300. Negative feedback
is to be added to this amplifier using improvement factor of
100.
(a) If the maximum output current of the feedback amplifier is
to be 800 mA, what is the maximum input current of the feedback amplifier?
(b) For maximum output current, find the sizes of the three currents involved in the subtraction at the input.

Section 9.2

9.4 We need an amplifier with gain of at least 100. Gain
variations over some range of temperature must be 1% or less.
We have an amplifier with gain of 200 that has 10% gain variations with temperature.
(a) What value of β should we add to the given amplifier to
reduce gain variations to 1%?
(b) Will two cascaded feedback amplifier stages like that of
part (a) have sufficient gain? Assume there are no interstage
loading effects.
(c) What will be the worst case percent variation with temperature for a two-stage amplifier consisting of a cascade of two
amplifiers like the one in part (a)?

9.5 An amplifier has gain of 100,000, and sensitivity of gain
to temperature variations is 0.20. Negative feedback is then
used to reduce the gain to 10.
(a) What is the sensitivity of the feedback amplifier to temperature variations?
(b) Find the overall gain and sensitivity to temperature variations of a cascade of five feedback amplifier stages.

9.6 Provide the algebra to show that adding negative feedback
to an amplifier described by Eq. (9.6) results in a feedback amplifier described by Eq. (9.7)

9.7 An audio amplifier with feedback needs gain of approximately 500 in a −3 dB passband extending from 60 Hz to
25 kHz. Assume this is accomplished using a feedback network
with $\beta = 0.0018$. Find the upper and lower half-power frequencies and the gain required of the original nonfeedback amplifier.
(Assume that the low- and high-frequency improvement factors
derived individually in the text apply to the more complex amplifier described here.)

9.8 A nonfeedback amplifier has $A = 800$, $\omega_L = 85$ rad/s,
$\omega_H = 9000$ rad/s, and $S_T^A = 10\%$. The required amplifier must
have $A_f \geq 75$, $\omega_{L,f} \leq 20$ rad/s, $\omega_{H,f} \geq 22,000$ rad/s, and
$S_{T'}^A \leq 20\%$.
(a) What value of β should the feedback amplifier employ?
(b) List the specifications of the resulting feedback amplifier.

9.9 Figure P9.9a shows a nonfeedback amplifier of voltage
gain A that delivers 5 W to a 5 ohm speaker when the amplifier

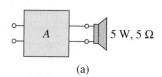

(a)

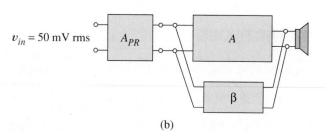

$v_{in} = 50$ mV rms

(b)

Figure P9.9

input voltage is 50 mV rms. The nonlinear distortion in the amplifier output is 1% of the output signal value.
(a) Find the numerical value of the gain A.
(b) Find the value of β required in Fig. P9.9b to reduce the distortion to 0.1%, with the same output signal amplitude.
(c) Find the value of gain A_{PR} required in the preamp.

Section 9.3

9.10 Derive input and output resistance expressions for the voltage-shunt feedback configuration of Fig. 9.6c.

9.11 Derive the input and output resistance expressions for the current-series feedback configuration of Fig. 9.6d.

9.12 Find the kind of feedback and the amount, β, we must add to a nonfeedback amplifier with voltage gain of 920, $R_i = 11$ kΩ and $R_o = 3$ kΩ, so that $R_{if} \geq 400$ kΩ and $R_{of} \leq 10$ Ω. Predict the gain of the feedback amplifier.

9.13 Find the kind of feedback and the amount, β, that we must add to a nonfeedback amplifier with voltage gain of 920, $R_i = 11$ kΩ and $R_o = 3$ kΩ, so that $R_{if} \geq 400$ kΩ and $R_{of} \geq 300$ kΩ. Predict the gain of the feedback amplifier.

9.14 For an amplifier with shunt feedback, $R_{if} = 110$ Ω, $R_o = 2$ kΩ, $R_{of} = 26$ kΩ, $A_f = 20$, $\omega_H = 10^4$ rad/s and $\omega_{L,f} = 10$ rad/s.
(a–e) Find R_i, A, β, $\omega_{H,f}$, and ω_L.
(f) Changing a resistor in amplifier A from 10 kΩ to 11 kΩ changes A_f from 20 to 21. Use sensitivity to find the new value of A.

9.15 Measurements on a nonfeedback amplifier give $A = 5000$, $\omega_H = 8$ kHz, $\omega_L = 200$ Hz, $R_{in} = 50$ kΩ, $R_{out} = 1$ kΩ, gain variation = 5%, full-signal output distortion = 2 mV. Since specifications require that the gain variation be 0.1%, ideal voltage-series feedback, just sufficient to correct the gain variation, is added.
(a) What value of β is used?
(b) Find parameter values for the feedback amplifier.

Section 9.4

9.16 (a) Redraw the voltage-series circuit Fig. 9.8a with all resistance transferred to the amplifier circuit. Your diagram should show the appropriate ideal source and load.
(b) Find A, R_{in}, and R_{out} for the A circuit you drew in part (a).
(c) Use results of part (b) to write expressions for A_f, R_{if}, and R_{of} for the voltage-series feedback circuit.

9.17 Work parts (b) and (c) of Problem 9.16 directly from the original diagram by turning off the feedback to find the A circuit parameters.

9.18 (a) Redraw the current-shunt circuit Fig. 9.8b with all resistance transferred to the amplifier circuit. Your diagram should show the appropriate ideal source and load.
(b) Find A, R_{in}, and R_{out} for the A circuit you drew in part (a).
(c) Use results of part (b) to write expressions for A_f, R_{if}, and R_{of} for the current-shunt feedback circuit.

9.19 Work parts (b) and (c) of Problem 9.18 directly from the original diagram by turning off the feedback to find the A circuit parameters.

9.20 In Example 9.5, R_{ix} and R_{ox} were nearly the same as R_{if} and R_{of} (from Exercise 9.6), respectively, because the external source and load resistances were close to ideal. In Fig. 9.10, replace the 200 kΩ source resistance with $R_s = 10$ kΩ and the 100 kΩ load resistor by $R_L = 5$ kΩ. For this new circuit, calculate R_{if}, R_{of}, R_{ix}, and R_{ox}. Notice that the β circuit does not change.

9.21 Figure P9.21 shows a feedback amplifier.
(a) Use ideal feedback theory to find gain, input resistance, and output resistance when the feedback is active.
(b) Find the input resistance seen by the signal source and its internal resistance.
(c) Find the resistance seen by the 1 kΩ load resistor looking back into the amplifier.

9.22 Figure P9.22 is a feedback amplifier. (a) Use feedback theory to find gain, input resistance, and output resistance when the feedback is active.
(b) Find the input resistance seen by the signal source and its internal resistance.

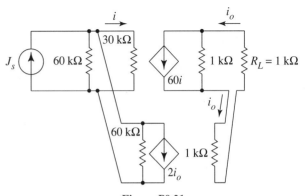

Figure P9.21

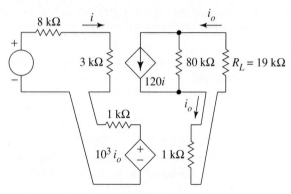

Figure P9.22

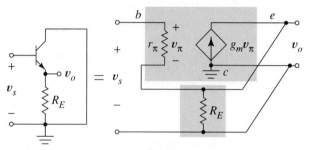

Figure P9.24

(c) Find the resistance seen by the 19 kΩ load resistor looking back into the amplifier.

9.23 Figure P9.23 shows the equivalent for a feedback amplifier. Use feedback theory to find
(a) the gain A of the amplifier without feedback.
(b) the input resistance R_i without feedback.
(c) the output resistance R_o without feedback.
(d) the input resistance seen by the signal source.

9.24 The low distortion and excellent high-frequency response of the emitter follower are credited to its feedback structure. Figure P9.24 shows the follower as a feedback amplifier.
(a) Redraw the second diagram using the proper two-port equivalent for the feedback network.
(b) From your circuit find an expression for A.
(c) From your circuit find an expression for A_f.
(d) Show that your result for A_f approximates the "usual expression"

$$A_{mid} = \frac{(\beta_F + 1)R_E}{r_\pi + (\beta_F + 1)R_E}$$

9.25 Figure P9.25 shows the source follower's small-signal equivalent, redrawn as a negative feedback circuit.
(a) Draw a new equivalent with the feedback circuit (that is, β circuit) replaced by the appropriate two-port equivalent required for feedback analysis.
(b) Find values of the two-port parameters in your circuit in terms of R_S.
(c) Find an expression for the improvement factor for this feedback circuit.

Section 9.5

9.26 The op amp in Fig. P9.26a has the following open-loop specifications:

voltage gain = 10,000   output resistance = 808 Ω
input resistance = 100 kΩ   unity-gain frequency = 100 MHz

(a) *Use ideal negative feedback theory* to determine voltage gain, output resistance, input resistance, and −3 dB frequency for the noninverting amplifier circuit.
(b) Demonstrate that you understand the meaning of input and output resistance by calculating values v_1 and v_2 for Fig. P9.26b.

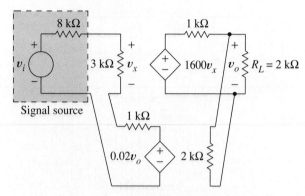

Figure P9.23

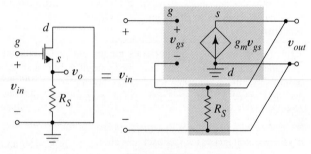

Figure P9.25

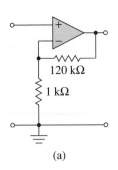

(a)

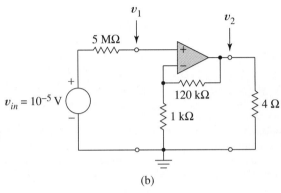

(b)

Figure P9.26

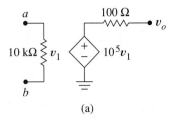

(a)

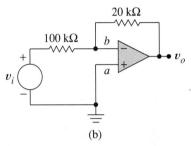

(b)

Figure P9.27

9.27 Figure P9.27a is the model for an op amp. Figure P9.27b shows voltage-shunt feedback added to this same op amp, making an inverting amplifier. Consider the 100 kΩ resistor to be part of the signal source.
(a) Draw the feedback equivalent circuit, including the correct two-port equivalent of the feedback resistor and correct signal-source and amplifier representations.
(b) From your circuit, determine A, R_i, and R_o.
(c) Use feedback theory to find A_f, R_{if}, and R_{of}.
(d) Use your feedback results to find v_o/v_i and the input resistance seen by v_i in Fig. P9.27b.
(e) Compare your answers from part (d) with expectations based upon the usual infinite-gain analysis.

9.28 Show that $\beta_{ac} > \beta_{dc}$ in Fig. 9.16e.

9.29 For each circuit of Fig. P9.29, identify the type of feedback and find the numerical value of β using the appropriate two-port parameter of the feedback circuit. State whether the feedback is dc or ac. If both, find the two-port parameter values for both.

9.30 Draw sketches corresponding to those of Fig. 9.16b and c that use choke coils rather than capacitors to produce the same kind(s) of feedback. Represent a choke coil by an inductor symbol with two parallel lines over the windings.

9.31 Measurements on the amplifier of Fig. P9.31a give the equivalent of Fig. P9.31b. You would like to add negative feedback to make the circuit resemble a current-controlled current source.
(a) What kind of feedback should you add?
(b) Assuming you could construct a β circuit having no input

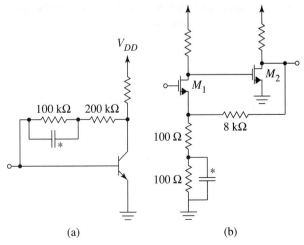

(a)　　　　　　　　　　(b)

Figure P9.29

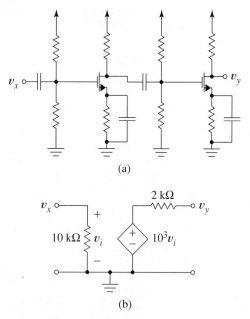

(a)

(b)

Figure P9.31

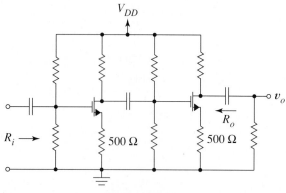

Figure P9.34

or output resistance, what numerical value should you use for
A, the gain without feedback?

(c) By adding feedback you would like to reduce the sensitivity
of gain to temperature changes from the present 10% to 0.1%.
What value of β should you use?

(d) Redraw Fig. P9.31a as given. Then show how to add components to introduce feedback without disturbing the biasing.
Do not worry about component values, only circuit structure.
Draw an X through any component that must be deleted.

(e) Derive an approximate design equation that relates the β
value from part (c) to component value(s) in your β circuit.

9.32 Use SPICE ac analysis at 1 kHz to simulate the feedback
amplifier of Example 9.10. Include an ac voltage source of 0 V
in series with the 16 kΩ load resistor, and use 1000 F to ensure
that all coupling and bypass capacitors are short circuits. For
transistors use $V_t = -0.8$ V, $k = 2 \times 10^{-3}$ A/V^2, $V_A = 120$ V.
Ignore body effect.

(a) Use SPICE to find the voltage gain of the feedback amplifier. (The example estimated 11.1.)

(b) In your SPICE code replace the voltage source and its resistance with a Norton equivalent. Then use SPICE to determine
R_{if} and R_{of}. (The example predicted $R_{if} = 417.7$ Ω.) The output
resistance should be high because of the current feedback.

(c) Open the feedback loop by changing the feedback resistor
to 1000 MΩ, and then use SPICE to find input and output resistance. (Since this open-loop simulation does not include loading

effects due to g_{11} and g_{22}, we do not expect to verify our _numerical_ predictions. We should however see input resistance increase and output resistance decrease.)

9.33 In Sec. 9.5.3 we saw that when gain A is a function of
frequency, the improvement factor, $1 + A(\omega)\beta$, becomes complex, causing R_{if} and R_{of} to become complex impedances. Describe in words the physical meaning of Eq. (9.3) when $1 + A\beta$
is complex; that is, what does complex sensitivity _mean?_

9.34 Copy Fig. P9.34. Then

(a) Add to the diagram the feedback component or components
necessary to increase the output resistance and decrease the input resistance. Add feedback to the ac circuit only.

(b) Assume the objective is a current amplifier of gain -100.
Assuming that $A_f \approx 1/\beta$ will describe the final circuit, write a
design equation that involves the parameter(s) of your feedback
circuit.

9.35 The general π- and t-shaped subcircuits of Fig. P9.35 are
often used in feedback amplifiers to contribute a specified β;

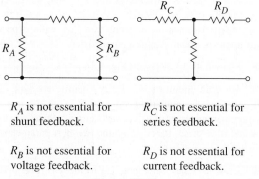

R_A is not essential for
shunt feedback.

R_C is not essential for
series feedback.

R_B is not essential for
voltage feedback.

R_D is not essential for
current feedback.

Figure P9.35

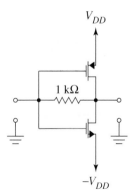

V_{DD}

1 kΩ

$-V_{DD}$

Figure P9.36

however, certain resistors can be omitted from the β circuit for certain kinds of feedback. Use appropriate "1-2 parameter" definitions to verify the statements in Fig. P9.35.
Hint: For voltage feedback, β is either y_{12} or h_{12}; for current feedback, it is either g_{12} or z_{12}; for series feedback, it is either h_{12} or z_{12}; for shunt feedback, it is either y_{12} or g_{12}.

9.36 (a) What kind of amplifier did the designer envision when designing Fig. P9.36: voltage, current, transresistance, or transconductance?
(b) Which would be more appropriate for the signal generator, a current source or a voltage source?

(c) Would a low- or high-resistance load be most appropriate?
(d) Is the feedback ac, dc, or both?
(e) Estimate the gain.

9.37 A nonfeedback amplifier has input resistance of 1 kΩ and gain given by

$$A(s) = 20,000\frac{800}{s + 800}$$

where s denotes complex frequency. A shunt feedback circuit uses β = 0.05 to lower the input resistance. Substitute the complex gain function into the "usual" expression for R_{if} to obtain impedance $Z_{if}(s)$. Make a Bode plot sketch of this function.

Section 9.6

9.38 The Bode plots of Fig. P9.38 describe the open-loop gain of an op amp. Find gain and phase margins of a noninverting feedback amplifier using this op amp when
(a) β = 0.00179.
(b) β = 0.0316.
(c) β = 0.1778.

9.39 Figure P9.39 is the loop gain of a feedback amplifier.
(a) How much must the loop gain be reduced to just make the amplifier stable?
(b) Estimate the reduction in loop gain required to make the amplifier stable with a 45° phase margin.

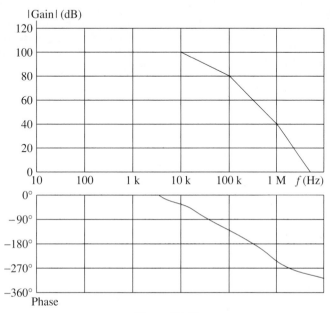

Figure P9.38

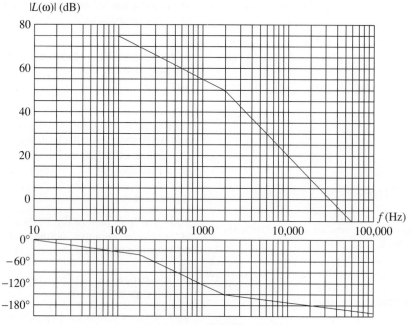

Figure P9.39

9.40 Figure P9.39 shows the frequency response curves of an op amp.
(a) Design a noninverting amplifier that has 45° phase margin, and draw its schematic diagram with resistor values.
(b) Repeat part (a) for 30° phase margin.

9.41 Figure P9.41 shows the Bode amplitude plot of an uncompensated amplifier that has pole frequencies at 0.1 kHz, 1 kHz, and 10 kHz.
(a) What is the gain-bandwidth product of the uncompensated amplifier?

(b) Make a copy of Fig. P9.41. Using the phase coordinates at the right, sketch the straight-line phase curve for each individual pole. Then use graphical addition to sketch the phase curve of the amplifier as a whole.
(c) If the amplifier is an op amp, what value of noninverting amplifier gain gives a phase margin of zero degrees?
(d) Find the gain-bandwidth product if the amplifier is compensated for unconditional stability by adding a new dominant pole.
(e) Find the gain-bandwidth product if the amplifier is compen-

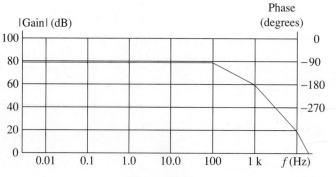

Figure P9.41

sated for closed-loop gains of 20 dB or more by adding a new dominant pole.

9.42 The lowest high-frequency pole of the amplifier described by Fig. P9.41 is to be moved to compensate the amplifier.
(a) Find the new pole frequency for this pole if the compensated amplifier is unconditionally stable.
(b) Find the new pole frequency for this pole if the compensated amplifier is stable with 45° phase margin for closed-loop gain of 50 dB.
(c) Sketch the new phase curve for the amplifier after it is compensated as in part (b).

Section 9.7

9.43 Design a Wien bridge oscillator with frequency of 500 Hz.

9.44 Repeat the SPICE run of Example 9.14 but this time increase both C_1 and C_2 by 10%. Compare the results with those of the example.

9.45 Use SPICE to find the highest and lowest oscillation frequencies of the Wien bridge oscillator of Example 9.14 over the temperature range $-40°C \le T \le +70°C$ if resistor temperature coefficients are those of Example 6.12.

9.46 (a) Design a phase-shift oscillator for a frequency of 20 kHz. Use SPICE to plot the phase curve of the phase-shifting circuit.
(b) Find the percent change in frequency if one of the resistors increases by 10%.
(c) Repeat part (b) for a 10% decrease in one capacitor.

9.47 When diodes are connected as in Fig. P9.47a to form a limiter, the nonlinear volt-ampere curve of Fig. P9.47b acts as a nonlinear resistor that decreases with increasing voltage.

(a) Show that the volt-ampere curve is described by

$$i = 2I_s \sinh\left(\frac{v}{V_T}\right)$$

(b) For large symmetrical voltage excursions of amplitude V_1, as shown in Fig. 9.34b, the *large-signal resistance* R_{LS} is defined by

$$R_{LS} = \frac{\Delta v}{\Delta i} = \frac{V_1 - (-V_1)}{i(V_1) - i(-V_1)}$$

Use the result of part (a) to find an expression for R_{LS} in terms of V_1 for the diode limiter.
(c) Evaluate R_{LS} for $V_1 = 0$, 0.25, 0.5, and 0.55 V if $I_S = 10^{-14}$ A and $V_T = 0.025$.

9.48 Repeat the SPICE simulation of Example 9.15 but include op amp limiting as suggested in Fig. 3.56. Connect the limiting diodes to ± 10 V, and use a 100 Ω output resistance. Use default diodes.

9.49 (a) Design a 2 MHz oscillator like Fig. 9.39c, using a MOSFET with $k = 4 \times 10^{-3}$A/V^2, $V_t = 1.2$ V, and $C_{gs} = C_{gd} = 2$ pF. Begin with the high-frequency equivalent circuit, assuming that your transistor is biased at 1 mA and that the Early voltage is 70 V. Include transistor capacitances as part of your design.
(b) Design a suitable biasing circuit for your oscillator transistor, including all necessary biasing and coupling capacitor values. Draw the final equivalent circuit.

9.50 Draw the complete circuit diagram for a Colpitts oscillator that uses a noninverting op amp circuit for its gain element. Use the internally compensated HA2544 op amp described in Table 2.1 to design a 10 MHz Colpitts oscillator.

9.51 Use Miller effect to redesign the oscillator of Example 9.16 if the transistor has $C_\pi = 100$ pF and $C_\mu = 5$ pF.

9.52 Use SPICE to simulate the oscillator designed in Example 9.16 using a 2N2222 transistor. The SPICE parameters for the 2N2222 are IS = 14.3 F, BF = 260, VAF = 74 V, VJC = 0.75 V, VJE=0.75 V, CJE = 22 P, CJC = 7.3 P, and TF = 411 P.

9.53 Recalculate the values of g_m and r_π in Example 9.16 if an rf choke replaces the collector resistor. Then repeat the 1 MHz oscillator design for this new biasing arrangement.

9.54 Use SPICE to determine how the frequency, amplitude and percent total harmonic distortion of the Colpitts oscillator of Example 9.17 change with C_1 by using the following procedure:
(a) For the design of Example 9.17 determine the actual frequency of oscillation by measuring the time between peaks of

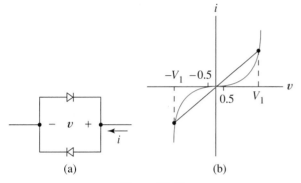

(a)   (b)

Figure P9.47

the last nine cycles in the SPICE output, dividing by 9 and taking the reciprocal. Then use a .FOUR statement with this frequency to analyze V(4). From the output note the amplitude of the fundamental frequency component and the %THD.

(b) Change the value of C_1 to 34.05 pF without changing L_2 and repeat the procedure of part (a). You may need to change the .TRAN statement so the simulation lasts for 40 μs.

(c) Explain why the results of parts (a) and (b) are expected from the theory.

9.55 A Hartley oscillator uses $L_1 + L_3 = 25$ μH and $C_2 = 40$ pF.

(a) Use SPICE to plot phase versus frequency for the phase-shifting network driven by an independent ac current source, both for the original design and when C_2 is 10% high. From Fig. 9.41, we expect the circuit to oscillate at that frequency where the phase shift is 180°. From your curves, determine the nominal design frequency and the frequency when C_2 is 10% high.

9.56 Design a Hartley oscillator using the biasing circuit of Fig. P9.56, and test it using SPICE. The frequency is to be 1.3 MHz. The transistor parameters are KP = 0.5×10^{-3}A/V^2, CBD = 1.5 pF, VTO = 2 V, and CBS = 0.5 pF. Optional: Add reasonable values for Early voltage and body-effect coefficient and see if they affect the frequency or the waveform.

9.57 Design the phase-shift oscillator of Fig. 9.34 for $\omega_o = 2\pi 90{,}000$ rad/s. Then use SPICE transient analysis to inspect the waveshape of the output voltage for two or three values of K.

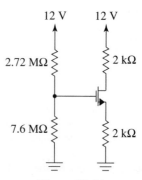

Figure P9.56

9.58 Use SPICE to plot the phase versus frequency curve for the phase-shifting circuit of Fig. P9.58. The input signal should be an ac current. Use the crystal equivalent of Fig. 9.43a. Crystal parameters are $L = 137$ H, $C = 0.0235$ pF, $r = 15$ kΩ, and $C' = 3.5$ pF. Try to determine the percent frequency change when the 32.2 pF capacitor increases by 10%.

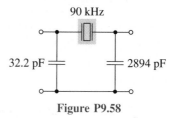

Figure P9.58

10

Power Circuits and Systems

The circuits in this chapter all develop large amounts of power. *Power amplifiers,* or *output stages,* provide the large *signal power* required by their loads. Example of loads for power amplifiers are stereo and public address speakers, deflection coils in video displays, and servo motors in *X-Y* plotters. Power supplies provide the dc power required by digital and analog electronic equipment of all kinds.

Several special concerns preoccupy the designer of an amplifier that must deliver a large amount of power to a load. The first is the amplifier's *efficiency,* defined by

$$\eta = \frac{P_{load}}{P_{in}} = \frac{\text{signal power delivered to the load}}{\text{circuit input power from the power supplies}} \qquad (10.1)$$

In circuits previously discussed, efficiency was a minor concern, for the power involved was small. With power amplifiers, however, we must consider the impact of the circuit on battery life or the utility bill, and we must also be concerned about the power supply requirements of the amplifier. Equation (10.1) shows that, for given P_{load}, low efficiency directly translates into greater requirements on the power supply. Furthermore, efficiency suggests a second issue—the ability of the circuit components to dissipate heat—for any supply power that does not leave the circuit as load power contributes to heating of transistors and resistors. More concretely, for given P_{load}, Eq. (10.1) implies

$$P_{in} - P_{load} = \left(\frac{1}{\eta} - 1\right)P_{load} = \text{circuit-component power dissipation}$$

which clearly shows how circuit heating relates to efficiency. In power circuits special efforts must be made to ensure that components are not destroyed by excessive heat. Finally, the large-signal amplitudes necessary for large output signal power make nonlinear distortion a matter of concern. Small-signal models, based upon projecting from the tangent to a transfer characteristic at some Q-point, no longer apply here, so we return to the large-signal models we used in bias circuit analysis and design.

Power amplifiers are classified according to the fraction of the time an output transistor conducts current. Class A amplifiers have output transistors in which signal current

751

flows all the time. For greater efficiency, class B amplifiers employ transistors that are active only half time—otherwise, they are cut off. Transistors in class AB amplifier;s conduct current slightly more than half time. Class AB circuits have efficiencies approximately the same as for class B circuits, but produce less distortion. *Class C amplifiers* deliver large amounts of output power at high efficiency by employing an output transistor that conducts output current for only a small fraction of a cycle. The resulting short pulses of output current excite a resonant circuit, which suppresses the distortion components that arise from the nonlinear operation of the transistor. Because applications of class C are somewhat specialized, these circuits are not discussed further in this text.

Class D amplifiers produce binary output waveforms of very high power with efficiencies approaching 100% by using transistors as switches In analog applications, the class D amplifier includes a *modulator* that first transfers information from waveform amplitude into a more suitable form such as pulse width. A lossless filter then removes undesired high-frequency terms from the amplified output.

We begin our study of power amplifiers by examining power dissipation within bipolar and field-effect transistors themselves.

10.1
Transistor Power Dissipation

Figure 10.1 shows an FET and a BJT, each biased at some dc operating point with no signal present. We can write the following equations for power dissipation P_D in these transistors. For the FET,

$$P_D = I_D V_{DS} \tag{10.2}$$

for the bipolar transistor,

$$P_D = I_C V_{CE} + I_B V_{BE} \approx I_C V_{CE} \tag{10.3}$$

From the polarities of the currents and voltages, P_D represents power being dissipated by the transistors. Next we examine exactly what dissipation means.

Elementary Heat Transfer Principles. In physics we learn that when parallel surfaces of a solid are held at uniform temperatures T_i and T_o as in Fig. 10.2a, Q joules of heat per second per unit area flow from the high-temperature side to the low-temperature side in the direction normal to the surfaces. If T_i is the larger,

$$Q = \frac{1}{K}(T_i - T_o) \text{ joules/s-m}^2$$

where K is a proportionality factor characteristic of the material. We can also use the equation in the opposite sense, thinking of the heat flow Q as the cause and the temperature difference as the effect.

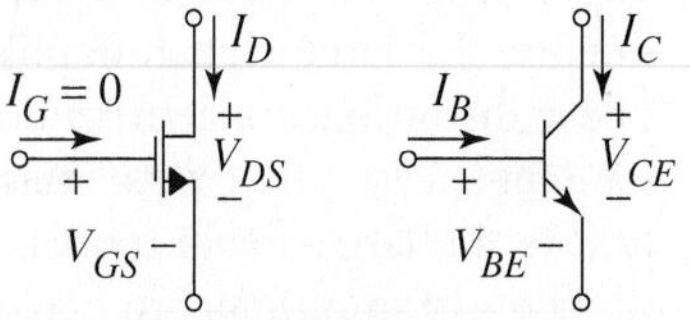

Figure 10.1 DC power dissipation in transistors.

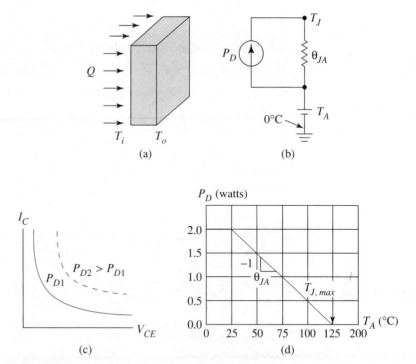

Figure 10.2 (a) Heat conduction in a solid material; (b) electrical analog of transistor power dissipation; (c) power dissipation hyperbolas for two values of average dissipation; (d) transistor power derating curve.

In the transistors of Fig. 10.1, the average dissipation P_D in joules/s is related to the temperature T_J of the drain-substrate (base-collector) junction and the temperature of the transistor's surroundings, *ambient temperature, T_A,* by

$$P_D = \frac{1}{\theta_{JA}}(T_J - T_A) \text{ joules/s} \tag{10.4}$$

where θ_{JA} is the *thermal resistance between the junction and the ambient surroundings.* The thermal resistance is a constant that is provided on the manufacturer's data sheet for a power transistor.

Figure 10.2b shows a simple circuit analog that helps us understand transistor heating described by Eq. (10.4). Node voltages represent temperatures; a constant current source represents power dissipation; a voltage source represents the fixed ambient temperature T_A relative to 0°C. When the transistor current and voltage are zero, $P_D = 0$ and the junction temperature equals the ambient value. When dc transistor current flows as in Fig. 10.1, heat is generated at the junction as charge carriers lose potential energy by flowing across the reverse biased junction. In this process electrical energy is converted to thermal energy in the crystal lattice. Junction temperature then rises above ambient, ridding the transistor of heat by causing it to flow toward the outside. In terms of Fig. 10.2b, the current source turns on, causing T_J to increase. To prevent catastrophic damage to the junction or premature aging, there is a maximum junction temperature $T_{J,max}$, that ranges from 125 to 200°C, which must not be exceeded. The designer's concern is that T_J always be less than $T_{J,max}$.

Power Dissipation Hyperbola. Figure 10.2b shows that for given temperatures T_A and $T_{J,max}$ and thermal resistance θ_{JA}, we can work backward and specify a maximum safe dissipation P_D. Once P_D is known, we can use Eqs. (10.2) or (10.3) to see how operation of the transistor is constrained, for these become equations of *power dissipation hyperbolas*. For example, in Eq. (10.3), when $P_D = P_{D1}$, a given constant, we regard V_{CE} as independent variable, I_C as dependent variable, and obtain the solid hyperbolic locus in Fig. 10.2c. Any Q-point on or beneath this hyperbola results in a safe junction temperature.

Suppose we now use a fan or air conditioner to lower the ambient temperature near the transistor. This lowers T_A in Fig. 10.2b, meaning a larger P_D, say, $P_D = P_{D2}$, is allowed for the same T_J. This implies that we can now safely operate the transistor at any Q-point on or beneath the new hyperbola labeled P_{D2} in Fig. 10.2c. Conversely, anything that raises the ambient temperature lowers P_D and reduces the permissible operating region on the transistor output characteristic.

Derating Curve. Power dissipation information is sometimes given in the alternative form of a *derating curve* like that of Fig. 10.2d. This curve tells us that at 25°C or below, this transistor can safely dissipate $P_D = 2$ W of power. If the ambient temperature exceeds 25°C, however, P_D is proportionally lower. The part of the curve with nonzero slope is Eq. (10.4), and the negative of the slope, $1/\theta_{JA}$, is called the *derating factor*. The temperature at which $P_D = 0$ on the derating curve is $T_{J,max}$. Thus from Fig. 10.2d,

$$\theta_{JA} = -\frac{125 - 25}{0 - 2} = 50°C/W$$

and $T_{J,max} = 125°C$.

Heat Sink. Figure 10.3a shows the physical structure of a bipolar power transistor. The collector is electrically and thermally attached to the case, so there are only base and emitter connecting leads. This structural geometry provides a method of lowering the inherent thermal resistance θ_{JA} by making a physical and thermal connection between transistor and a *heat sink* as in Fig. 10.3b. The heat sink is a special metal radiator designed to help rid the transistor of excess heat by convection. Heat flows from junction to transistor case, from case through a special thermal grease to the heat sink, and from the sink into the ambient atmosphere. Figure 10.3c is the electrical analog of this thermal circuit, where

$$T_J = T_A + P_D\theta_{JA,sink} = T_A + P_D(\theta_{JC} + \theta_{CS} + \theta_{SA})$$

In addition to θ_{JA}, which describes transistor heat transfer without a heat sink, the manufacturer also provides the thermal resistance θ_{JC} between junction and case. Thermal resistance θ_{SA} between sink and ambient are given by the heat-sink manufacturer or obtained from heat-sink design books. A good thermal grease designed to transfer heat from transistor to sink has $\theta_{CS} \approx 0.2°C/W$. Because $\theta_{JA,sink} < \theta_{JA}$, the analog shows that

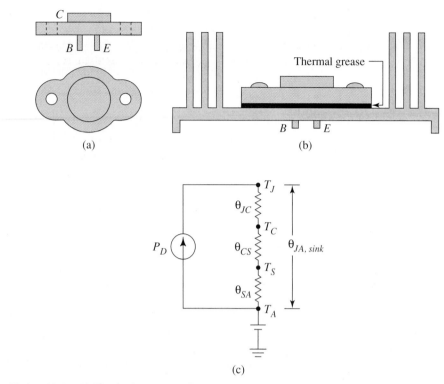

Figure 10.3 (a) Physical structure of a power transistor; (b) transistor mounted on a heat sink; (c) electrical analog of dissipation including a heat sink.

adding a heat sink raises P_D and the dissipation hyperbola to a still higher value for the same junction and ambient temperatures.

EXAMPLE 10.1 A power transistor with thermal resistance $\theta_{JA} = 12°$C/Watt is to be biased at 2 A and 10 V in a 25° C ambient temperature environment. If $T_{J,max} = 180°$C, (a) show that the transistor will overheat. (b) If a cooling system is employed, what ambient temperature is required to prevent overheating. (c) If a heat sink is used in the 25°C ambient environment, find $\theta_{JA,sink}$ required.

Solution. For the given Q-point, $P_D = 2 \times 10 = 20$ W. (a) From Eq. (10.4), $T_J = 25 + 12 \times 20 = 265°$C. Because this exceeds $T_{J,max}$, the transistor will overheat. (b) If $T_J = 180°$C, then $180 = 20 \times 12 + T_A$, or $T_A = -60°$C. (c) $180 = 20 \times \theta_{JA,sink} + 25$, giving $\theta_{JA,sink} = 7.75°$C/W. ❑

> **Exercise 10.1** For Example 10.1c, find the thermal resistance of the heat sink if the transistor manufacturer specifies $\theta_{JC} = 1.3°$C/W, and if thermal grease contributes $\theta_{CS} \approx 0.2°$C/W. Will the transistor case to be warm to the touch after the heat sink is installed?
>
> **Ans.** 6.25°C; yes—the case temperature will be 118°F.

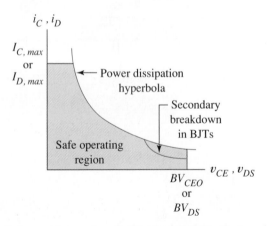

Figure 10.4   Safe operating regions for bipolar and field-effect power transistors.

Safe Operating Region. Figure 10.4 shows a *safe operating region* (SOR) for a given transistor/heat-sink arrangement. The region is bounded by the dissipation hyperbola, a maximum current rating, and the transistor's breakdown voltage. In BJTs there is also a *secondary breakdown* caused by nonuniform internal current flow that creates overheating in localized regions and further reduces the SOR as indicated. Continuous operation at a point outside the SOR is likely to damage the transistor; however, we will see that the transistor can sometimes tolerate time-*varying* excursions outside the shaded region without overheating.

Parallel Transistors. In both BJT and FET circuits, it is possible to replace a power transistor with two or more identical transistors with corresponding terminals connected. Then total transistor dissipation is divided by the number of parallel transistors. Individual heat sinks can be provided for each transistor or a common heat sink can be designed.

10.2
Power Transistors and Power Amplifiers

Power MOSFETs. In bipolar technology, the high $I_{C,max}$ required for a power transistor is achieved by making the base-emitter area large. For the MOSFET, power transistor design is not so easy because W/L rather than area scales the current (see Fig. 5.2a and Eq. (5.8)). A channel length L small enough to accommodate high current also results in a very low breakdown voltage, BV, because the depletion region expands into the channel during active operation, reducing the effective channel length L_E, as in Fig. 5.48b. For this reason, the double-diffused vertical structure of Fig. 10.5 is more successful for MOS power transistors. This transistor is called *double diffused* (sometimes called *DMOS*) because its fabrication requires both p- and n-diffusions, and it is called a *vertical* transistor for the primary direction of current flow. Like other FETs, the DMOS transistor is a three-state device.

When gate-source voltage is positive, but small, the pn junctions under the edges of the gate are reverse biased, $i_D = 0$, and the device is OFF. When v_{GS} exceeds a threshold voltage $V_t > 0$, there is an inversion in the p-material just under the gate, and the transistor turns ON. Electrons flow to the n⁻ drain through electrically parallel channels of

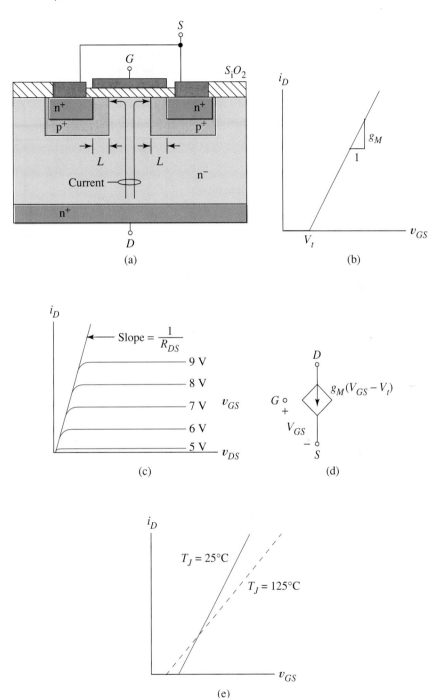

Figure 10.5 Double-diffused MOSFET power transistor: (a) physical structure; (b) transfer characteristic; (c) output characteristics; (d) active-region circuit model; (e) variation with temperature.

length L. Because the depletion region expands mostly into the n^- drain material with increased v_{DS}, the effective channel length and the real channel length L are the same. The intense electric field created in the short channel, even at rather low drain-source voltages, accelerates electrons to their maximum drift velocity, a phenomenon called *velocity saturation*. Consequences of velocity saturation are the linear transfer characteristic of Fig. 10.5b; uniformly spaced, active-region output characteristics as in Fig. 10.5c; and the *linear* active-region equation

$$i_D = g_M(v_{GS} - V_t), \qquad v_{GS} > V_t \tag{10.5}$$

where V_t is the threshold voltage and g_M is the *forward transconductance*. In terms of physical parameters,

$$g_M = 0.5\, C_{ox} W U_{max} \tag{10.6}$$

where U_{max} is the maximum drift velocity, about 10^7 cm/s for electrons in silicon, and W and C_{ox} are the channel width and oxide capacitance. Figure 10.5d shows a large-signal, static-circuit model that describes the DMOS transistor when active.

At low values of v_{DS}, the output characteristics merge into a straight line of slope $1/R_{DS}$, which is the third important static parameter of the device. For *ohmic operation*,

$$i_D = \frac{1}{R_{DS}} v_{DS} \tag{10.7}$$

Values for g_M of 1 to 10 A/V and $V_t = 4$ V are common; values of R_{DS} are of the order of 0.1 to 4 Ω, and thermal resistance θ_{JC} is of the order of 1 to 7°C/W.

As in low-power FETs, threshold voltage decreases linearly with junction temperature at -2 mV/°C, and forward transconductance also decreases. Consequently, with increasing temperature the transfer characteristic changes as in Fig. 10.5e.

Figure 10.6 defines a SPICE subcircuit for an active-state power MOSFET with $g_M = 0.2$ S and $V_t = 4$ V. The subcircuit is valid for cut-off and active but not ohmic-region operation. As in the common-gate FET model of Fig. 7.21, gate current is zero because of the unity gain CCCS. The diode and battery ensure negligible output current for $v_{GS} < V_t$ and forward current described by Eq. (10.5). In one of the references, Cor-

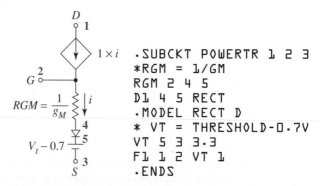

Figure 10.6 SPICE model for DMOS transistor in cut-off and active regions.

donnier and co-workers describe a SPICE model that includes dynamic effects, and tell how to measure the required parameter values.

p-Channel, double-diffused power transistors are also available; however, it is presently difficult to fabricate pairs of power transistors that are complementary in both static and dynamic parameters.

10.3
Class A Amplifier

10.3.1 CIRCUIT SCHEMATIC

Figure 10.7a is a *class A power amplifier* or output stage, a circuit that can produce peak output signal amplitudes comparable to the supply voltage with little distortion. Class A is a good choice for designs having very stringent distortion specifications; however, we will learn that its low efficiency imposes demanding requirements on the transistors and power supply.

When examining an unfamiliar circuit, it is instructive to begin with biasing—provided in this case by two power supplies and a constant current source. We have also included a dc source of $V_{BE} \approx 0.7$ V in the input to simplify the initial explanation of the circuit. In power amplifiers, we represent transistors by large-signal models; therefore, forward-active operation gives the equivalent of Fig. 10.7b. When v_I is zero, v_O is zero. Thus the transistor is biased at $(V_{CE}, I_C) \approx (V_{CC}, I_{EE})$, a Q-point easily positioned in the active region using the biasing components. Each supply delivers approximately I_{EE} amperes to the circuit, making the total circuit dissipation

$$P_{in} = 2V_{CC}I_{EE} \tag{10.8}$$

Such dissipation of a power amplifier circuit in the absence of a signal is called *standby power dissipation*.

10.3.2 MAXIMUM SIGNAL SWING

We see from Fig. 10.7b that even with a signal present

$$v_O(t) = v_I(t) \tag{10.9}$$

This linearity is not surprising, since the circuit is an emitter follower, a voltage-series feedback circuit that we expect to give low nonlinear distortion. To understand the *limits* of linear operation, we next examine the circuit's load line. Because the current-source/follower structure gives circuit constraints we have not previously seen, we go back to basics to derive an appropriate load line equation. From Fig. 10.7a, with $i_C \approx i_E$ we have

$$i_C = I_{EE} + \frac{v_o}{R_L} = I_{EE} + \frac{V_{CC} - v_{CE}}{R_L} \tag{10.10}$$

In the i_C versus v_{CE} plane, this equation describes a straight line of slope $-1/R_L$ that passes through the two points—(V_{CC}, I_{EE}) and $(V_{CC} + I_{EE}R_L, 0)$—as in Fig. 10.7c. Notice that the end points of the load line correspond (approximately) to transistor saturation and cut-off. Because signals are large in power amplifiers, we often use $V_{CE,sat} \approx 0$ to simplify the mathematics without introducing significant error.

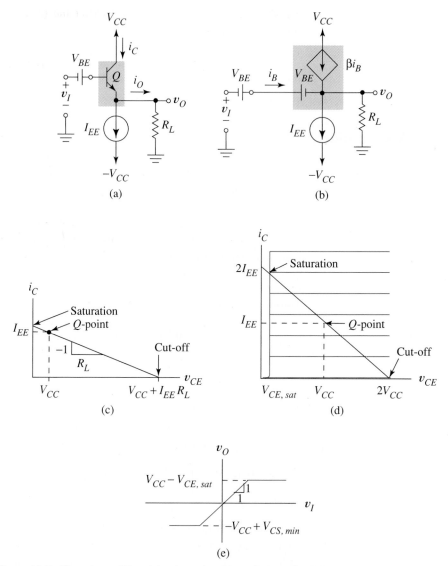

Figure 10.7 Class A amplifier: (a) schematic; (b) equivalent circuit; (c) general load line construction; (d) load line for maximum signal amplitude; (e) transfer characteristic.

In design, we select I_{EE} so that the output signal swing is as large as possible. Because

$$v_{CE}(t) = V_{CC} - v_O(t) \qquad (10.11)$$

large variations in $v_O(t)$ imply large variations in $v_{CE}(t)$—ideally ac operation over the entire load line from saturation to cut-off in Fig. 10.7c. We also need equal signal swings in either direction from the Q-point. Figure 10.7c shows that for the output signal to be both large and symmetric, we must center the Q-point on the load line. To do so we choose I_{EE} so that

$$V_{CC} = \frac{1}{2}(V_{CC} + I_{EE}R_L)$$

This gives the design equation

$$I_{EE} = \frac{V_{CC}}{R_L} \qquad (10.12)$$

which (approximately) centers the Q-point as in Fig. 10.7d. Then the transistor is active over the range

$$V_{CE,sat} \le v_{CE} \le 2V_{CC}$$

By substituting these limiting values of v_{CE} into Eq. (10.11), we tentatively conclude that the linear transfer characteristic, Eq. (10.9), should be valid for output voltages in the range

$$-V_{CC} \le v_O \le V_{CC} - V_{CE,sat}$$

We learned in Chapter 6, however, that a current source such as I_{EE} in Fig. 10.7a has a minimum output voltage, $V_{CS,min}$. We conclude that in a practical amplifier, linear operation of the circuit in Fig. 10.7a is limited to

$$-V_{CC} + V_{CS,min} \le v_O \le V_{CC} - V_{CE,sat}$$

Together with Eq. (10.9), this gives the transfer characteristic of Fig. 10.7e. Incidentally, this curve assumes a signal source with nonzero output resistance. The curve saturates because increases in source voltage v_I beyond the break points on the curve cause base current drops across the sources' output resistance rather than further increases in output voltage. If we omit the *extra* V_{BE} biasing source of Fig. 10.7a, the transfer curve shifts downward to give an output offset voltage of $-V_{BE}$ volts, however its distinguishing feature, linearity, is preserved.

10.3.3 WAVEFORMS

Figure 10.8a shows the circuit waveforms that result when we apply a sinusoidal input signal at $t = 0$. Before $t = 0$, the transistor is biased as just described. When the signal turns on, Eq. (10.9) shows that $v_O(t)$ follows $v_I(t)$ as in Fig. 10.8b. Because the current source maintains its dc value I_{EE}, the sinusoidal load current $i_O(t)$ in Fig. 10.7a adds to the dc transistor current; consequently, the collector current waveform looks like Fig. 10.8c. The phase inversion between $v_I(t)$ and $v_{CE}(t)$ in Fig. 10.8d follows from Eq. (10.11). Notice in Figs. 10.8c and d that a sufficiently large input signal drives the transistor alternately to cut-off and saturation, a result consistent with the load line of Fig. 10.7d. We next explore the efficiency of the circuit by writing equations for these waveforms.

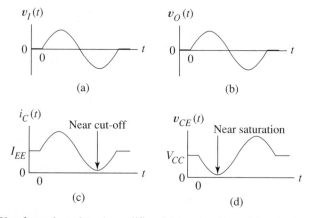

Figure 10.8 Waveforms in a class A amplifier: (a) input voltage; (b) output voltage; (c) collector current; (d) transistor voltage.

10.3.4 POWER BALANCE AND EFFICIENCY

Our objective is to account for all the power that enters and leaves the amplifier. The resulting equations reveal some important ideas related to amplifier efficiency, and show us how to make design decisions about transistor power ratings and power supply specifications.

Power Balance with No Signal. Figure 10.7a shows that when $v_I = 0$ each supply delivers $V_{CC}I_{EE}$ watts of power to the circuit. Substituting the design value of I_{EE} from Eq. (10.12) gives total circuit input power of

$$P_{in} = 2V_{CC}I_{EE} = 2\frac{V_{CC}^2}{R_L} \tag{10.13}$$

Since the output transistor is biased at (V_{CC}, I_{EE}) its power dissipation with no signal is

$$P = V_{CC}I_{EE} = \frac{V_{CC}^2}{R_L}$$

The constant current source has terminal voltage V_{CC} and current I_{EE} so it also dissipates

$$V_{CC}I_{EE} = \frac{V_{CC}^2}{R_L}$$

We conclude that, with no signal, the input power from the supply is divided equally between the transistor and the current source as we show *graphically* in the first two columns of Fig. 10.9.

Power Balance with Sinusoidal Signal. Now assume the signal in Fig. 10.8a is $v_I(t) = V_o \sin \omega t$. From Fig. 10.8c, the *average* current in the positive supply is still I_{EE}. Therefore, Eq. (10.13) gives average input power to the circuit for *both* signal and standby conditions.

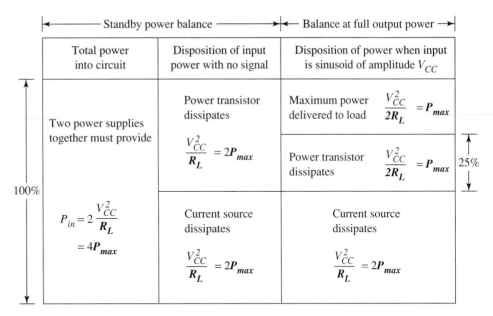

Figure 10.9 Power balance in a class A amplifier in terms of specified maximum load power P_{max} and load resistance, R_L.

With signal, the average output power is the squared rms load voltage divided by R_L,

$$P_{load} = \frac{V_o^2}{2R_L} \tag{10.14}$$

To obtain the efficiency of the class A amplifier we substitute Eqs. (10.13) and (10.14) into Eq. (10.1). This gives

$$\eta_A = 0.25\left(\frac{V_o}{V_{CC}}\right)^2 \tag{10.15}$$

The last two equations show that output power and efficiency both increase with signal amplitude, reaching practical maxima when $V_o \approx V_{CC}$.

Because maximum output power is usually a key design condition and is often given, we denote it by

$$P_{max} = \frac{V_{CC}^2}{2R_L}$$

The last column of Fig. 10.9 *graphically* shows the power balance in the circuit for the maximum amplitude case. As we turn on the signal and gradually increase its amplitude, the output transistor dissipation decreases—the power delivered to the load exactly matches this reduction in transistor dissipation. Meanwhile, average current-source dissipation stays constant. For a signal of maximum amplitude, load power equals amplifier-transistor dissipation; and efficiency is 25%. For convenience in design, Fig. 10.9 gives all power quantities in terms of P_{max}.

Class A Amplifier Design. Now consider designing a class A power amplifier. Specifications typically include *given* values, R_L and P_{max}, and we want the circuit to operate at highest efficiency. The design task involves selecting an appropriate power supply, and determining transistor and current-source parameters. For the power supply, we need dc voltage and average current ratings. For the transistor we require power dissipation, maximum current, and a lower limit for breakdown voltage. To design the current source, we need dc current and power dissipation.

First, we work out the power supply specifications. Starting with known P_{max} and R_L in the first column of Fig. 10.9, we solve for V_{CC}:

$$V_{CC} = \sqrt{2P_{max}R_L}$$

Because *each* supply requires a dc current rating I_{EE} that satisfies $V_{CC}I_{EE} = 2P_{max}$,

$$I_{EE} = \frac{2P_{max}}{V_{CC}}$$

This is the current rating of the power supply and also the dc current of the current source.

To find the transistor power rating, we note that in most applications of power amplifiers the signal is at least occasionally absent. Column 2 in Fig. 10.9 shows that the transistor must be rated for $2P_{max}$ to handle standby heating. The constant current source also requires an output transistor rated at $2P_{max}$. Figure 10.10a reminds us that the power transistor must be rated for peak current greater than $2I_{EE}$ and breakdown voltage $BV_{CEO} > 2V_{CC}$.

For special applications *in which maximum signal is always* present, column 3 in Fig. 10.9 shows that the power rating of the output transistor need only be P_{max}. For this special case, much of the design load line in Fig. 10.10b falls *above the* $P_D = P_{max}$ hyperbola. The transistor does not overheat under these conditions because *average* dissi-

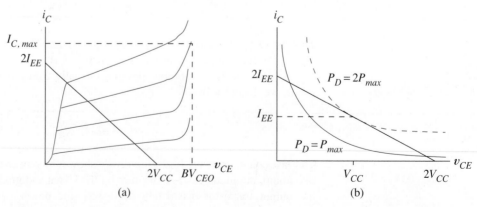

Figure 10.10 Transistor limitations in class A design: (a) breakdown and maximum current; (b) load line relative to power dissipation hyperbola.

pation is less than P_{max} for continuous ac operation. This makes sense once we notice that the high-current region of the load line corresponds to low transistor voltages and the high-voltage region to low currents. A sinusoidal signal changes slowly near its maxima and minima and more rapidly in between; thus, average power dissipation is lower than for static operation at the center of the load line.

There might be special design considerations related to the nature of the signal. If the signal is a very low frequency sinusoid with period long compared to the *thermal time constants* associated with transistor heating and cooling, the more conservative design using $P_D = 2P_{max}$ is necessary, even with signal always present, because the load line is traversed so slowly. Quasi-dc signals such as this might occur, for example, in an amplifier for a dc motor control system. Another extreme would be a large input square wave that causes the transistor to spend nearly all of its time near saturation or cut-off, where either current or voltage is low. Even choosing $P_D = P_{max}$ would be overly conservative for this case.

EXAMPLE 10.2 The class A amplifier of Fig. 10.11 must deliver $P_{max} = 4$ W of sinusoidal power to the 10 Ω load. Q_3 has one-sixth the area of Q_2. Work out the design details, including dissipation in the current source.

Solution. From Fig. 10.9, each power supply must develop at least

$$V_{CC} = \sqrt{2P_{max}R_L} = \sqrt{2 \times 4 \times 10} = 8.94 \text{ V}$$

From Eq. (10.12) the current source must provide

$$I_{EE} = \frac{8.94}{10} = 0.894 \text{ A}$$

Since there is no assurance that signal is always present, we select power ratings for Q_1 and Q_2 that will accommodate standby conditions. From Fig. 10.9, these are

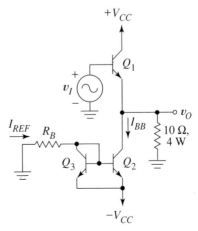

Figure 10.11 Class A amplifier.

$P_D = 2P_{max} = 8$ W. Since Q_3 has one-sixth the area of Q_2, the current in R_B should be $I_{REF} = 0.894/6 = 0.149$ A; therefore,

$$R_B = \frac{8.94 - 0.7}{0.149} = 55\ \Omega$$

The power rating of R_B must be $(0.149)^2(55) = 1.22$ W.

The dissipation required of Q_3 is 0.7×0.149 A $= 0.104$ W. To accommodate a peak output sinusoid of $V_{CC} = 8.94$ volts, breakdown voltage, BV_{CEO}, of Q_2 and of Q_1 must exceed $2V_{CC} = 17.9$ V. The positive supply provides $2P_{max} = 8$ W and its average current is $I_{EE} = 0.894$ A. The negative supply must provide $I_{EE} + I_{REF}$, so its current is $0.894 + 0.149 = 1.04$ A; therefore, the current rating of this supply must be 1.04 A.

In this design, the power supplies actually deliver $8 + 1.04 \times 8.94 = 17.3$ W to the circuit, so the circuit efficiency is $4/17.3 = 23\%$, not the theoretical maximum of 25%. ❏

The preceding example assumed an output signal amplitude of $V_o = V_{CC} = 8.94$ V peak. We know from Fig. 10.7e that the peak amplitude would actually be closer to 8.74 V to avoid transistor saturation. Sinusoidal output amplitude would be reduced still further by an output offset of 0.7 V since Fig. 10.11 does not include the input circuit source V_{BE} from Fig. 10.7a.

Exercise 10.2 Find maximum signal amplitude, V_o for the circuit designed in Example 10.2 considering current-source saturation and offset voltage. (Hint: Find the negative v_I that just saturates Q_2.) Use this amplitude and Eqs. (10.14) and (10.15) to estimate average output power and efficiency.

Ans. 8.04 V, 3.23 W, 20.2%.

Problem 10.18 establishes that MOSFET class A amplifiers operate in the same fashion and employ the same design equations as bipolar amplifiers.

10.4 Nonlinear Distortion

An important figure of merit for an amplifier is its dynamic range, defined by

$$\text{dynamic range} = 20 \log \left| \frac{\text{largest useful signal}}{\text{smallest useful signal}} \right|$$

Anyone who enjoys Beethoven symphonies appreciates electronics that can faithfully process both very small and very large signal amplitudes. For an amplifier, the smallest useful signal is marginally larger than the random noise inherent in the circuit; the largest results in barely acceptable distortion. Random noise is beyond the scope of this text; however, quantifying our intuitive notion of distortion is within our purview. Doing so enables us to write and understand specifications that describe acceptable levels of distortion, measure distortion in circuits, determine permissible signal levels for given circuits, and compare alternative designs with respect to distortion.

Nonlinear distortion occurs when signals are so large that the curvature of the static transfer characteristic affects the output waveshape. At high frequencies the nonlinear charge versus voltage curves of BJTs, MOSFETs, and diodes add to the distortion. Although we illustrate basic ideas in terms of *static* characteristics, the careful reader will notice that the distortion definition and simulation techniques can be generalized to include the nonlinear capacitances of the solid-state devices.

Harmonic Distortion. The response of any linear circuit to the sinusoid

$$v_I(t) = V_{IN} \cos(\omega_o t)$$

is itself a sinusoid of frequency ω_o, possibly differing from $v_I(t)$ in amplitude and/or phase. It is evident from graphical projections like Fig. 10.12a that the response of a nonlinear circuit is not a sinusoid but rather some periodic waveform having the same period as the input. In Sec. 1.6.5 we learned to approximate such a periodic response by a truncated Fourier series such as

$$v_O(t) = V_{DC} + V_1 \cos(\omega_o t + \phi_1) + \sum_{n=2}^{N} V_n \cos(n\omega_o t + \phi_n) \qquad (10.16)$$

where V_{DC} is the *average value* or *dc component* of $v_o(t)$. The term of frequency ω_o is the desired response; so we use V_1/V_{IN} for amplifier gain. The *components* that involve $n > 1$ are distortion terms; therefore, the distortion introduced by the nonlinearity is

$$d(t) = \sum_{n=2}^{N} V_n \cos(n\omega_o t + \phi_n) \qquad (10.17)$$

Because the frequencies of the distortion terms are *harmonics* or multiples of the excitation frequency, this is called *harmonic distortion*.

A spectral diagram, like Fig. 10.12b, helps us visualize the amplitudes of the distortion terms and the signal. Sometimes amplitudes of certain harmonics are zero. In detailed studies of Fourier series, we encounter equations that allow us to compute the values of V_n and ϕ_n from the given waveform $v_o(t)$. Simulation programs such as SPICE contain computational routines that implement these equations. In the laboratory, an in-

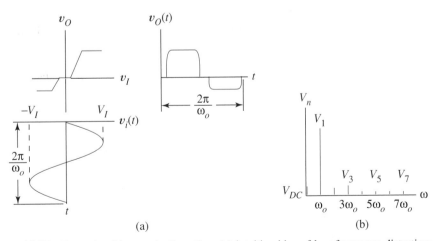

(a)   (b)

Figure 10.12 Example of harmonic distortion: (a) intuitive idea of low-frequency distortion; (b) spectral or frequency-domain view of desired output signal and undesired distortion components.

strument called a *spectrum analyzer* accepts $v_o(t)$ as input and displays the amplitudes of the fundamental and harmonics, as in Fig. 10.12b. Thus we can find the amplitudes of the fundamental and the distortion terms for any particular circuit by computation, simulation, or measurement.

To quantify distortion we use two familiar results from circuit theory: the *root mean square* or rms value of $V_1 \cos(\omega_o t + \phi)$ is $V_1/\sqrt{2}$, and the rms value of a finite sum of sinusoids of *different frequencies* is the square root of the sum of the squares of the rms values of the components. For example, for

$$d(t) = V_2 \cos(2\omega_o t + \phi_2) + V_3 \cos(3\omega_o t + \phi_3) + \cdots$$
$$+ V_N \cos(N\omega_o t + \phi_N) \tag{10.18}$$

$$V_{rms} = \left[\left(\frac{V_2}{\sqrt{2}}\right)^2 + \left(\frac{V_3}{\sqrt{2}}\right)^2 + \cdots + \left(\frac{V_N}{\sqrt{2}}\right)^2 \right]^{\frac{1}{2}} \tag{10.19}$$

The *percent total harmonic distortion, %THD*, is the ratio of rms distortion to rms signal in the output, expressed as a percent. That is

$$\%\mathrm{THD} = \frac{V_{rms}}{V_1/\sqrt{2}} \times 100 = \frac{\left[(V_2/\sqrt{2})^2 + (V_3/\sqrt{2})^2 + \cdots + (V_N/\sqrt{2})^2 \right]^{\frac{1}{2}}}{V_1/\sqrt{2}} \times 100$$

or

$$\%\mathrm{THD} = \sqrt{\left(\frac{V_2}{V_1}\right)^2 + \left(\frac{V_3}{V_1}\right)^2 + \cdots + \left(\frac{V_N}{V_1}\right)^2} \times 100 \tag{10.20}$$

Exercise 10.3 A spectrum analyzer measurement of an output voltage gives the following amplitudes for the fundamental and first four harmonics: 2.31, 0.533, 0.91, 0.21, and 0.80. Find the %THD.

Ans. 58%.

SPICE Computation of %THD. When input code includes the command ".FOUR," SPICE computes the amplitudes of the first nine harmonics of a voltage or current in a transient analysis. The command syntax is

.**FOUR** ⟨ *fundamental frequency*⟩ ⟨*output variable*⟩

where, *output variable* denotes a node voltage, node-voltage difference, or a voltage-source current. For example, given the statement

.**FOUR 100 V(3)**

SPICE analyzes the waveform of the voltage at node 3, and computes V_1 through V_9 of its Fourier series using $\omega_o = 2\pi 100$. It also evaluates Eq. (10.20).

Example 10.3 uses distortion calculations to reinforce the idea of small-signal analysis.

EXAMPLE 10.3 The amplifier of Example 5.19, with body-effect coefficient $\gamma = 0.37$ for M_1, has the nonlinear transfer characteristic of Fig. 5.54c. Find the %THD in $v_O(t)$ when (a) $v_I(t) = 4.3 + 0.001 \sin(2\pi 10^3 t)$, (b) $v_I(t) = 4.3 + 1 \sin(2\pi 10^3 t)$.

Solution. (a) This signal exemplifies small-signal analysis. The 4.3 V offset voltage is a dc input component that biases the output node at about 6.5 V. The SPICE code is Fig. 10.13a; the first column of Fig. 10.13b gives the results. The fundamental component in the output has amplitude of 14.4 mV, so small-signal analysis would show a gain of 14.4. Each harmonic is more than three orders of magnitude smaller than the fundamental, and the %THD is only twenty-seven thousandths of a percent. The computed output waveform looks like a perfect sine wave.

(b) When the amplitude of VI is changed to 1 volt and a second SPICE run is performed, the column 3 data result. The output signal now has amplitude of 6 V (gain equals 6); however, it is accompanied by large distortion terms, especially at odd harmonic frequencies. The harmonic distortion is 25.3%. Figure 10.13c shows the distorted waveform.   ❏

```
EXAMPLE 10.3
VDD 1 0 DC 12
M1 1 2 2 0 DEPL
M2 2 3 0 0 ENHA
.MODEL DEPL NMOS KP=4E-5
+VTO=-3 GAMMA=0.37
.MODEL ENHA NMOS KP=4E-5 VTO=2
VI 3 0 SIN(4.3 1E-3 1000)
*SINE WITH 4.3V OFFSET,1MV
*AMPLITUDE AND F=1000HZ
.TRAN 0.01M 1M
.FOUR 1000 V(2)
.PLOT TRAN V(2)
.END
```

(a)

		Small signal	Large signal
n		V_n	V_n
1		1.439E−02	6.011E+00
2		3.869E−06	5.729E−02
3		1.102E−07	1.385E+00
4		1.275E−07	5.728E−02
5		1.661E−07	5.630E−01
6		5.603E−08	2.435E−02
7		2.179E−08	2.374E−01
8		1.829E−07	9.337E−03
9		9.121E−08	8.388E−02
%THD		0.027	25.3

(b)

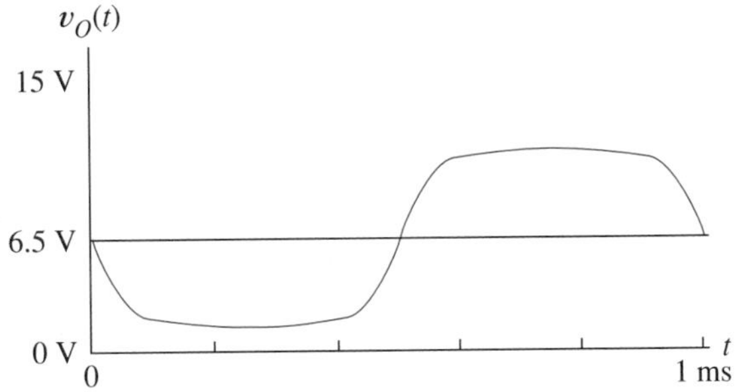

(c)

Figure 10.13 SPICE distortion analysis for the amplifier of Fig. 5.54a: (a) code; (b) SPICE output; (c) large-signal output waveform showing 25% harmonic distortion.

Intermodulation Distortion. When *two or more* sinusoids of different frequencies enter a nonlinear system, a second kind of nonlinear distortion, called *intermodulation distortion,* appears. Like harmonic distortion, intermodulation distortion produces new frequencies in the output that were not present in the input. For input frequencies ω_1 and ω_2, these new frequencies include sums and differences of the form $\omega_1 \pm k\omega_2$, where k is an integer. Notice that the frequencies of the intermodulation terms *need not be harmonically related to either input frequency.* This causes them to be exceptionally annoying to the ear in audio systems. Because realistic signals are by no means restricted to sinusoids or even to periodic waveforms, intermodulation distortion is a practical problem of no small importance. We can use a special SPICE statement called .DISTO to compute some of these distortion terms in an ac analysis. PSPICE does not support the .DISTO statement, but instead uses discrete Fourier transforms to find the amplitudes of the intermodulation frequencies contained in circuit waveforms.

In the next section we see that nonlinear distortion becomes an important consideration when we try to improve the efficiency of power amplifiers by using class B operation.

10.5
Class B Amplifier

When output power is large, the 25% efficiency of the class A amplifier translates into large supply currents, expensive transistors, large heat sinks, and high utility bills (or short battery lifetimes). The main reason for the poor efficiency of class A is its large dc bias current, I_{EE}. The class B circuit significantly improves efficiency by using transistors that are biased at cut-off.

10.5.1 BIPOLAR CLASS B AMPLIFIER

Bias and Signals. Figure 10.14a shows a class B power amplifier. The transistors are *complementary,* meaning the circuit uses an npn-pnp pair, often with matched parameter values. To examine biasing, assume that when $v_I = 0$, both transistors are cut off. This gives Fig. 10.14b, which verifies that cut-off is correct and also shows that with no signal $v_o = 0$. In fact, $v_O = 0$, for all input voltage in the range

$$-V_\gamma < v_I < V_\gamma$$

where V_γ is the cut-in voltage of the BJT. One important advantage over the class A amplifier is already obvious, *zero standby power dissipation.*

When v_I exceeds V_γ, Q_1 goes active but Q_2 remains cut off, giving the emitter follower of Fig. 10.14c. From this idealized circuit we estimate

$$v_O = v_I - V_{BE}, \qquad v_I > V_{BE} \tag{10.21}$$

(assuming $i_B \approx 0$). Equation (10.21) applies for increasing v_I until Q_1 saturates. When this happens, $v_O = V_{CC} - V_{CE,sat}$ as in Fig. 10.14d. At this point v_O clamps, and further increases in v_I drop across R_i without changing v_O.

When v_I swings negative in Fig. 10.14b, Q_2 turns ON as in Fig. 10.14e, and input and output voltages satisfy

$$v_O = v_I + V_{BE}, \qquad v_I > V_{BE} \tag{10.22}$$

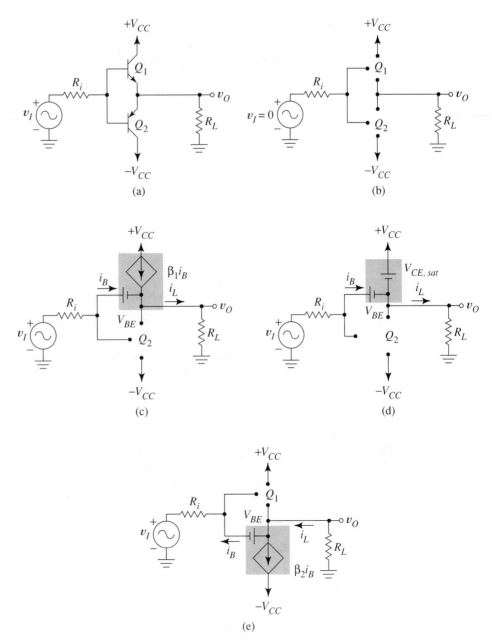

Figure 10.14 Class B amplifier: (a) schematic; (b) equivalent with no signal present; (c) equivalent for positive v_I; (d) equivalent when Q_1 saturates; (e) equivalent for negative v_I.

until Q_2 saturates at $v_O = -V_{CC} + V_{CE,sat}$. Since Q_1 *pushes* current into R_L (Fig. 10.14c) and then Q_2 *pulls* current from R_L (Fig. 10.14e), this is called *push-pull* operation.

Crossover Distortion. Figure 10.15 shows the transfer characteristic of the class B amplifier predicted from Fig. 10.14; in the true curve, smooth transitions replace the sharp corners. Because the transistors are both OFF for small v_I, the output waveform is highly distorted near the zero crossings. This *crossover distortion,* a kind of harmonic distortion, is a serious problem that is especially severe for small-signal amplitudes. Crossover distortion also occurs in the FET class B amplifier we discuss next.

10.5.2 MOS CLASS B AMPLIFIER

Figure 10.16a is an MOS class B power amplifier, and Fig. 10.16b the equivalent circuit for $v_I = 0$. Again both transistors are cut off, there is no standby power dissipation, and $v_O = 0$. When v_I exceeds the threshold voltage of M_1, this transistor goes active, giving the source follower of Fig. 10.16c. For double-diffused MOSFETs described by Eq. (10.5), the output is

$$v_O = i_D R_L = g_M(v_I - v_O - V_t)R_L \tag{10.23}$$

Solving for v_O gives

$$v_O = \frac{g_M R_L}{1 + g_M R_L}(v_I - V_t), \qquad v_I > V_t \tag{10.24}$$

which closely resembles the BJT expression, Eq. (10.21), when $g_M R_L \gg 1$. This approximation is usually valid because g_M is large for power MOSFETs. If M_2 is comple-

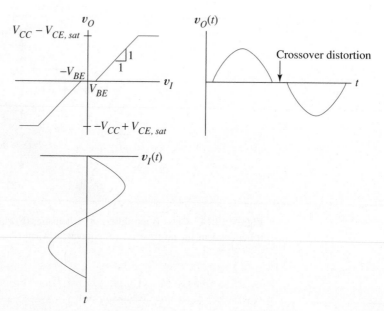

Figure 10.15 Transfer characteristic of class B amplifier creating crossover distortion.

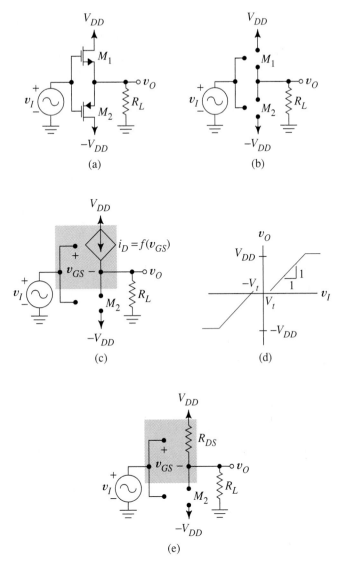

Figure 10.16 Class B FET amplifier: (a) schematic; (b) with $v_I = 0$; (c) with $v_I > V_t$; (d) transfer characteristic; (e) mechanism for saturation of transfer curve.

mentary, it has the same g_M, and also $V_{t2} = -V_{t1} = -V_t$. When M_2 conducts, a similar analysis gives

$$v_O = \frac{g_M R_L}{1 + g_M R_L}(v_I + V_t), \qquad v_I < V_t \qquad (10.25)$$

which resembles Eq. (10.22) for the BJT circuit. Figure 10.16d shows the resulting transfer characteristic. When v_O swings close to V_{DD} or $-V_{DD}$, the conducting MOSFET enters its ohmic state and becomes a resistor, R_{DS}, as suggested in Fig. 10.5c. This resistor combines with R_L to give a voltage divider as in Fig. 10.16e. For $R_{DS} << R_L$, the transfer characteristic saturates at voltages close to the supply values.

When the required output current is not too large, we can use square-law MOSFETS for output transistors. Because of the negative feedback of the source follower, the transfer characteristic of the class B amplifier still closely resembles Fig. 10.16d, a result easily demonstrated by simulation.

There are two ways to reduce harmonic distortion in a class B amplifier—applying negative feedback around the output stage or changing the biasing to create a class AB amplifier—both are widely used.

10.5.3   REDUCING DISTORTION WITH FEEDBACK

To reduce distortion with feedback, we include the power amplifier within the feedback loop of a high-gain preamplifier, a procedure we demonstrate using SPICE in the following examples and exercises.

EXAMPLE 10.4   Figure 10.17a shows a class B amplifier driven by

$$v_I(t) = 12 \sin 2\pi 1000t$$

Use SPICE to plot the transfer characteristic and output waveform, if the output transistors are square-law devices with $k = 0.2$ A/V^2 and $|V_t| = 3$ V.

Solution.   Figure 10.17b gives the SPICE code for computing the transfer characteristic. To produce the output waveform, each comment replaces the preceding statement.

The SPICE output of Fig. 10.17c confirms that the transfer characteristic resembles Fig. 10.16d, featuring a *dead zone* for $-V_t < v_I < V_t$. Voltage-series feedback linearizes the transfer characteristic in the intervals when the transistors are active, but there is no

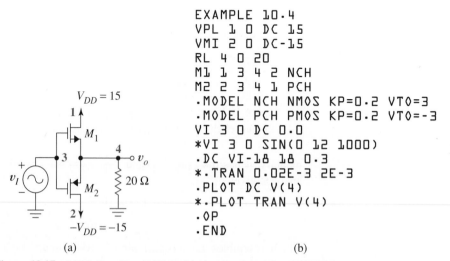

```
EXAMPLE 10.4
VPL 1 0 DC 15
VMI 2 0 DC-15
RL 4 0 20
M1 1 3 4 2 NCH
M2 2 3 4 1 PCH
.MODEL NCH NMOS KP=0.2 VTO=3
.MODEL PCH PMOS KP=0.2 VTO=-3
VI 3 0 DC 0.0
*VI 3 0 SIN(0 12 1000)
.DC VI-18 18 0.3
*.TRAN 0.02E-3 2E-3
.PLOT DC V(4)
*.PLOT TRAN V(4)
.OP
.END
```

$V_{DD} = 15$

M_1

v_o

v_I

M_2

$20\ \Omega$

$-V_{DD} = -15$

(a)   (b)

Figure 10.17   MOS class B amplifier: (a) circuit schematic; (b) SPICE code; (c) transfer characteristic; (d) output waveform showing crossover distortion.

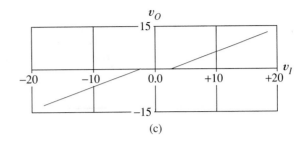

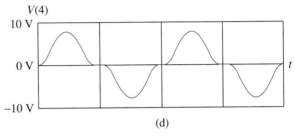

Figure 10.17 (continued)

feedback when the transistors are off. Saturation of the transfer characteristic is not apparent in the output even though the transistor is driven into ohmic operation. The transient response waveform of Fig. 10.17d shows crossover distortion. Determining the consequences of body effect in the transistors, if any, is Problem 10.28. ❏

> **Exercise 10.4** Use SPICE to analyze the crossover distortion and compute the %THD of the circuit of Fig. 10.17a.
>
> *Ans.* 18.4 %, Figure 10.18.

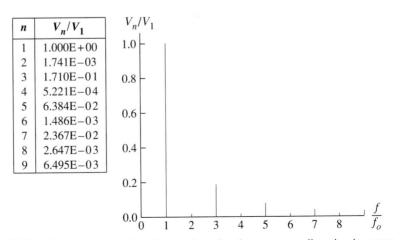

n	V_n/V_1
1	1.000E+00
2	1.741E−03
3	1.710E−01
4	5.221E−04
5	6.384E−02
6	1.486E−03
7	2.367E−02
8	2.647E−03
9	6.495E−03

Figure 10.18 Line spectrum and tabulated values showing crossover distortion in output of Figure 10.17a.

EXAMPLE 10.5   Figure 10.19a shows the amplifier of Fig. 10.17a enclosed within the feedback loop of a follower circuit. Use SPICE to verify that the negative feedback reduces crossover distortion. Also compare the %THD for the feedback circuit with that for the nonfeedback circuit of the preceding example.

Solution. The SPICE code of Fig. 10.19b defines a simple, high-gain, static op amp model—subcircuit OPAMP. With an infinite gain op amp in Fig. 10.19a, we would expect the voltage at node 4 to follow the voltage of node 5—with a real op amp we hope to approximate this. SPICE output in Fig. 10.19c confirms that this occurs, by showing that V(4) has no visible crossover distortion. The figure also shows the mechanism that corrects the output waveshape. The op amp output, V(3), becomes exactly that input waveform needed to produce a sinusoidal output when projected from Fig. 10.17c! Adding feedback reduces the %THD from 18.4% to 0.0131%.

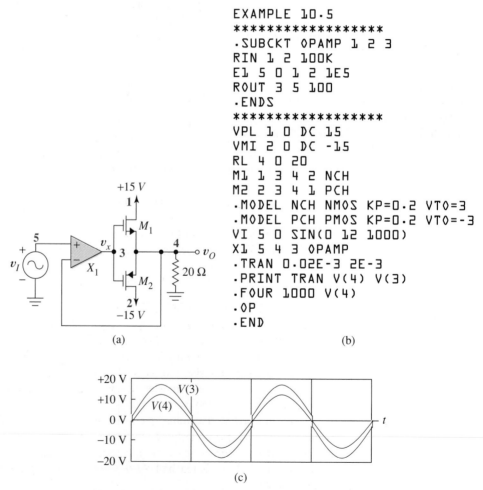

```
EXAMPLE 10.5
****************
.SUBCKT OPAMP 1 2 3
RIN 1 2 100K
E1 5 0 1 2 1E5
ROUT 3 5 100
.ENDS
****************
VPL 1 0 DC 15
VMI 2 0 DC -15
RL 4 0 20
M1 1 3 4 2 NCH
M2 2 3 4 1 PCH
.MODEL NCH NMOS KP=0.2 VTO=3
.MODEL PCH PMOS KP=0.2 VTO=-3
VI 5 0 SIN(0 12 1000)
X1 5 4 3 OPAMP
.TRAN 0.02E-3 2E-3
.PRINT TRAN V(4) V(3)
.FOUR 1000 V(4)
.OP
.END
```

(a) (b)

(c)

Figure 10.19 Class B amplifier with feedback to reduce crossover distortion: (a) schematic; (b) SPICE code; (c) circuit and op amp output waveforms.

Notice in Fig. 10.19c that V(3) has peak value greater than the ± 15 V power supply voltages. The output of a real op amp would saturate unless it employed separate supply voltages higher than the $\pm$ 15 V used for the power amplifier. ❏

> **Exercise 10.5** Example 10.5 showed that enclosing the class B amplifier within the feedback loop of a high-gain op amp significantly reduces distortion. Like many op amp concepts, the idea also applies to amplifiers of more modest design. To appreciate this, find the new %THD if the open-loop op amp gain is only 100.
>
> ***Ans.*** %THD = 0.169%. The waveform still looks sinusoidal to the eye.

For voltage gain greater than one, the class B circuit can be enclosed in the feedback loop of an inverting or noninverting amplifier instead of a follower.

The class AB circuit discussed next uses a different method to reduce crossover distortion. Power dissipation, efficiency, and design of class B amplifiers are all practically the same as for the class AB amplifiers, so we discuss these in the next section.

10.6
Class AB Amplifier

10.6.1 OPERATING PRINCIPLES

The class AB power amplifier eliminates crossover distortion by biasing both output transistors just above cut-in so that they conduct slight currents when $v_I = 0$. In Figs. 10.20a and b, diode-connected transistors biased with current I_{BB} develop V_{BB} volts between the transistor gates (bases). For the FET, $V_{BB} > 2|V_t|$. For the BJT, $V_{BB} > 2|V_\gamma|$ where V_γ is the cut-in voltage. Alternatively, a V_{GS} or V_{BE} multiplier circuit can produce V_{BB}. Because I_{BB} and the bias currents in the output transistors are small, standby power dissipation is still negligible.

When v_I becomes sufficiently positive in Fig. 10.20b, Q_2 cuts off while Q_1 remains active as in Figure 10.20c. For large v_I, the models suggest

$$v_O = v_I + 0.5\, V_{BB} - V_{BE} \approx v_I$$

For negative v_I, Q_1 cuts off and Q_2 goes active; and again $v_O \approx v_I$ is a reasonable approximation. The linearized models of Fig. 10.20c help us understand large-signal operation, but they are inaccurate when currents are small (near $v_I = 0$), for they show neither the simultaneous operation of the transistors nor the gradual transition of the base-emitter voltage as the transistor turns on. The dashed blue curves of Fig. 10.20d suggest how the individual nonlinear transfer characteristics combine to produce the solid transfer characteristic described by $v_O = v_I$.

As in class B, the transfer curve saturates when the signal source contains resistance; however, the underlying mechanism is different. When v_I swings positive in Fig. 10.20c, the voltage across current source Q_5 diminishes until the current-source minimum voltage, V_{min}, is reached. At this point, assuming a saturated transistor in the current source,

$$v_O = V_{CC} - V_{min} - V_{BE} \approx V_{CC} - 0.2 - 0.7 = V_{CC} - 0.9$$

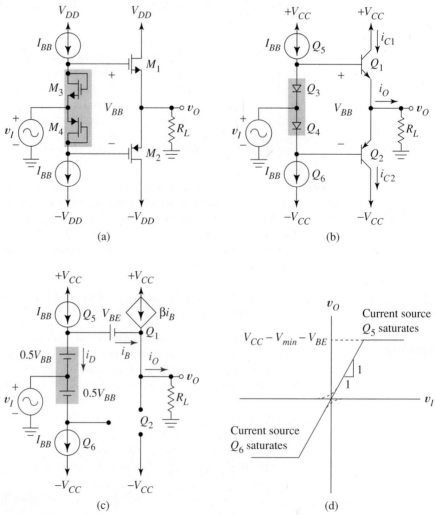

Figure 10.20 Class AB amplifiers: (a) MOSFET circuit; (b) BJT circuit; (c) large-signal equivalent for (b) when $v_I > 0$; (d) transfer characteristic of BJT class AB amplifier.

At this voltage Q_1 is still active because

$$v_{CE1} = V_{CC} - v_O \approx 0.9 \text{ V}$$

Similarly, the lower saturation limit of the transfer characteristic depends on current source Q_6. Figure 10.20d shows the complete transfer characteristic. As we pointed out for class B, the drop across the internal resistance of the preamp (not shown in the diagram) actually produces the saturation of the transfer characteristic for the BJT amplifier. For MOSFET circuits the nonlinear ohmic resistance of the output transistor forms a divider with R_L.

10.6.2 POWER BALANCE AND EFFICIENCY FOR CLASS B AND AB AMPLIFIERS

Our next goal is to account for all the power that enters and leaves class AB (and class B) amplifiers. We first develop a load line to help us visualize overall class AB operation. Using the load line, we derive design equations, and then graphically summarize the overall power relationships in a scaled diagram.

Because the class AB amplifier employs alternate operation of two transistors, the most useful load line shows output characteristics of both transistors plotted in an output current versus output voltage coordinate system. We relate the transistors to the new coordinate system by the following observations. When Q_1 conducts and Q_2 is off, Fig. 10.20b gives

$$i_O \approx i_{C1} \quad \text{and} \quad v_O = V_{CC} - v_{CE1}$$

a simple change in variables. New variables, i_O and v_O, are related to the i_{C1} versus v_{CE1} characteristic of Q_1 by reflecting the output characteristic about its current axis ($v_O = -v_{CE1}$) and then translating it to the right by V_{CC} volts. This gives the curves labeled Q_1 in Fig. 10.21a.

From Fig. 10.20b, when Q_2 conducts and Q_1 cuts off the variable changes are

$$i_O \approx -i_{C2} \quad \text{and} \quad v_O = -V_{CC} - v_{CE2}$$

These imply that the second quadrant characteristics of *pnp transistor* Q_2 are first reflected about the voltage axis ($i_O = -i_{C2}$), then reflected about the current axis ($v_O = -v_{CE2}$), and finally translated to the left by V_{CC}. This gives the curves labeled Q_2 in Fig. 10.21a.

In the new coordinate system, the load line is $i_o = v_o/R_L$, shown in Fig. 10.21a. With no signal, both transistors are biased *close to* cut-off, point 1. For positive v_I, Q_1 conducts, Q_2 cuts off, and operation moves up the load line toward saturation of Q_1 at point 2 . For negative v_I, Q_2 conducts, and operation moves down the load line toward saturation of Q_2 at point 3 (producing negative values for both v_O and i_O). In explaining Fig. 10.20d, we noted that the output transistors never quite reach saturation. In the interest of simplicity, because V_{CC} and the output signal amplitude are ordinarily large compared to $|V_{CE,sat}|$, we ignore the fine points and simply assume the Q-point moves from $-V_{CC}$ to $+V_{CC}$.

Before we continue, notice that we could use Fig. 10.20a to derive changes in variables *exactly* like those just mentioned (except for notational differences). This gives a load line diagram just like Fig. 10.21a for the MOSFET class AB amplifier, where MOSFET output characteristics replace the corresponding BJT characteristics. This means that we can visualize the MOSFET and BJT circuits in the same way. Furthermore, all of the power and efficiency equations we develop next apply to both forms of amplifier.

To derive power relations, we assume $v_I(t) = v_O(t) = V_o \sin \omega t$ where, for generality, V_o can take on any value in the range $0 \le V_o \le V_{CC}$. From Fig. 10.21a we deduce that the transistor currents are the half-rectified waves of Fig. 10.21b. We previously en-

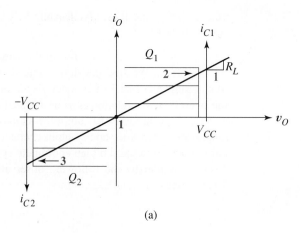

(a)

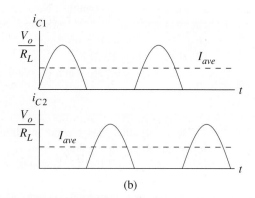

(b)

Figure 10.21 (a) Load line for class AB amplifier; (b) transistor and power supply current waveforms.

countered a half-rectified voltage waveform in Fig. 3.40b and Eq. (3.46). For the case at hand,

$$I_{ave} = \frac{V_o}{\pi R_L}$$

We next calculate the average input power to the circuit. Each power supply maintains V_{CC} volts while delivering one of the currents of Fig. 10.21b to the circuit. Therefore, the supplies together deliver to the circuit average power of

$$P_{in} = 2V_{CC}\frac{V_o}{\pi R_L} \tag{10.26}$$

Thus P_{in} is proportional to signal amplitude, V_o, in class B, not constant as in class A. Because $v_O(t) = V_o \sin \omega t$, the average load power is

$$P_{load}(V_o) = \frac{V_o^2}{2R_L} \tag{10.27}$$

The ratio P_{load}/P_{in} gives the efficiency of the class AB (and class B) amplifier:

$$\eta_{AB} = \frac{\pi}{4} \frac{V_o}{V_{CC}} \qquad (10.28)$$

As in the class A circuit, efficiency improves with signal amplitude. If we ignore the small transistor saturation voltages, then maximum efficiency corresponds to signal amplitude $V_o \approx V_{CC}$, giving $\pi/4 = 78.5\%$, a considerable improvement over 25% for class A.

Because complementary transistors alternately work half time and idle half time, their average dissipations are equal, and their collective power dissipation accounts for the difference between P_{in} and P_{load}. The average dissipation of *each transistor* is, therefore, $P_D = 0.5(P_{in} - P_{load})$, which, from Eqs. (10.26) and (10.27) is

$$P_D(V_o) = P_{in} - P_{load} = \frac{V_{CC}V_o}{\pi R_L} - \frac{V_o^2}{4R_L} \qquad (10.29)$$

We see that transistor dissipation is a parabolic function of signal amplitude. Figure 10.22 shows Eqs. (10.28) and (10.29) plotted together as functions of signal amplitude. Transistor dissipation has a relative maximum for $V_o = 2V_{CC}/\pi$, which from Eq. (10.29) is

$$P_{D,max} = \frac{V_{CC}^2}{\pi^2 R_L} \qquad (10.30)$$

From Eq. (10.28), this occurs when output circuit efficiency is 50%.

10.6.3   CLASS B AND AB DESIGN

To understand how this all fits together, we orient the equations around a design assignment in which a *given* maximum output power $P_{load} = \boldsymbol{P_{max}}$ is required for a *given* load resistance $\boldsymbol{R_L}$.

Our first concern is that the amplifier be capable of delivering rated power $\boldsymbol{P_{max}}$ to the specified load resistor, and that it do so at maximum efficiency. From Eq. (10.28), maximum efficiency of 78.5% implies $V_o \approx V_{CC}$. Since $\boldsymbol{P_{max}} = 0.785\ P_{in}$ for this case, we can work backward and calculate the required supply power, $P_{in} = 1.27\boldsymbol{P_{max}}$. The height of the first column of Fig. 10.22b represents this total average input power. The second column is scaled to show where this input power goes: $\boldsymbol{P_{max}}$ into the load and the remaining $0.27\ \boldsymbol{P_{max}}$ into transistor heating. By specifying $\boldsymbol{P_{max}}$ and $\boldsymbol{R_L}$, and simultaneously requiring maximum efficiency, we indirectly specify both the total amount of power required of the two power supplies, P_{in}, and the supply voltage V_{CC}. From this information we can also find the average current required of each supply.

Our next concern is to select transistors with appropriate power ratings and to provide necessary heat sinking. If we could be certain of maximum signal amplitude at all times, that is, $V_o = V_{CC}$, then from the second column of Fig. 10.22b each transistor would need to dissipate only $0.135\boldsymbol{P_{max}}$. A more conservative and usually more appropriate design provides for the possibility that $V_o < V_{CC}$. Figure 10.22a shows that making the signal smaller than V_{CC} can actually increase transistor dissipation. Thus *it is the maximum dissipation condition that determines the dissipation requirements of the transistors.*

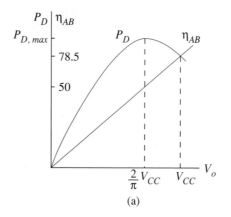

(a)

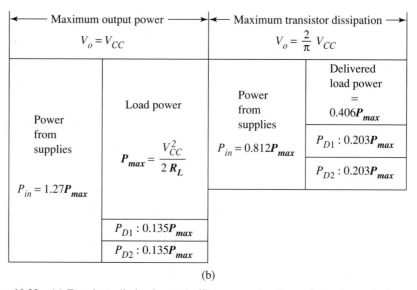

(b)

Figure 10.22 (a) Transistor dissipation and efficiency as functions of signal amplitude; (b) power balance in the class AB amplifier—scaled graphically and expressed in terms of maximum load power P_{max}—for maximum output and maximum dissipation cases.

According to Eq. (10.27), reducing V_o to $(2/\pi V_{CC})$ reduces the load power to

$$P_{load} = \left(\frac{2}{\pi}\right)^2 \frac{V_{CC}^2}{2R_L} = \left(\frac{2}{\pi}\right)^2 P_{max} = 0.405\, P_{max}$$

From Eq. (10.26), this reduced signal amplitude also reduces P_{in} from the power supplies to only $0.811P_{max}$ instead of $1.27P_{max}$, leaving the remaining

$$0.5(0.811 - 0.405)P_{max} = 0.203P_{max}$$

to be dissipated by each transistor. The last two columns of Fig. 10.22b show graphically the changes caused by reducing signal amplitude. The conclusion is obvious. We use

P_{max} and the maximum output power case to find the power supply specifications, but to protect the transistors from overheating in the event of a signal reduction, we provide for individual transistor dissipation of $0.203P_{max}$. Class B and class AB FET and BJT designs both use the power relations in Fig. 10.22b, so the only extra task for class AB is designing the biasing circuit.

EXAMPLE 10.6 Design a class B power amplifier with the same specifications as the class A design of Example 10.2, that is, 4 W into a 10 Ω load.

Solution. From column 2, Fig. 10.22b, for maximum output power, the supply voltage is

$$V_{CC} = \sqrt{2 \times 10 \times 4} = 8.94 \text{ V}$$

The supplies together must provide the circuit with $1.27 \times 4 = 5.08$ W, 2.54 W each. Thus each must be rated for average current of $I = 2.54/8.94 = 0.284$ A. From column 4, each power transistor must be able to dissipate $0.203 \, P_{max} = 0.812$ W. ❑

Table 10.1, which compares the class A and class B designs of Examples 10.6 and 10.2, respectively, shows that dramatically reduced requirements on both supply currents and output transistors result from using a circuit of greater efficiency. Once we understand the basic design procedure, we can usually handle any small changes needed to make the procedure more realistic.

Exercise 10.6 Redesign the amplifier of Example 10.6 by increasing V_{CC} in such a way that the original specifications are satisfied with load line excursions within the practical limits $-V_{CC} + 0.9 \leq V_o \leq V_{CC} - 0.9$. Find the supply voltage, required average supply current, and required dissipation of each transistor.
Hint: On Fig. 10.22a, V_o that gives 4 W of power is smaller than our new V_{CC}.

Ans. 9.84 V, 0.284 A, $P_D = 0.983$ W.

For class AB operation, I_{BB} in Fig. 10.20b must be large enough to keep the biasing diodes (or device) ON, even for peak signal current. When i_O reaches its instantaneous maximum in Fig. 10.20c, diode current i_D must still exceed some minimum, I_{min}, or the diode turns off and disconnects v_I from the circuit. By KCL, this requires

$$i_D = I_{BB} - \frac{i_{O,max}}{\beta + 1} \geq I_{min} \qquad (10.31)$$

(When the output transistors are MOSFETs, this problem does not occur.)

TABLE 10.1 Class A and B Power Amplifiers Delivering 4 W to a 10 Ω Load

	Dissipation (W)	Power Supply Ratings		Average Supply Power (W)	
	Each Transistor	Voltage (V)	Current (A)	No Signal	With Signal
Class A	8	8.94	0.895, 1.04	17.3	17.3
Class B	0.81	8.94	0.284 A	0	5.08

EXAMPLE 10.7 (a) Convert the class B circuit of Example 10.6 to a class AB amplifier by adding current sources and diodes. Output transistors are complementary, with $\beta = 39$; $I_s = 10^{-14}$ A for diodes.
(b) Estimate the quiescent collector currents of the output transistors, if each is described by the transfer characteristic

$$i_C = 6.0 \times 10^{-14} \exp\left(\frac{v_{BE}}{V_T}\right)$$

(The power transistors have base-emitter junctions with six times the areas of the biasing diodes.)

Solution. (a) In Example 10.6 we found the peak output voltage was 8.9 V, thus $i_{o,max} = 8.94/10 = 894$ mA. To keep the diodes forward biased just above cut-in at 0.55 V (a somewhat arbitrary choice), we need

$$I_{min} = 10^{-14} \exp\left(\frac{0.55}{0.025}\right) = 35.8 \ \mu A$$

From Eq. (10.31),

$$I_{BB} \geq \left(\frac{894}{40}\right) mA + 0.036 \ mA = 22.3 \ mA$$

(b) Under standby conditions $I_{BB} = 22.3$ mA flows through each diode. Since the active transistor area is six times the diode area, the standby collector current is $6 \times 22.3 = 134$ mA. ❏

Because of the high standby collector currents in the preceding example the supplies must deliver standby power of $2 \times (8.94 \times 0.134) = 2.40$ W, an unacceptable value for 4 W load power. However, re-examining Eq. (10.31) shows an obvious way to reduce standby power—use output transistors of higher β. For this and other reasons we discuss later, Darlington pairs often replace the power transistors in practical circuits.

Exercise 10.7 Compute the standby dissipation for the class AB circuit of Example 10.7 when Darlington pairs replace the output transistors (each of the four transistors have $\beta = 39$).

Ans. 63.8 mW.

10.6.4 OTHER POWER AMPLIFIER CONSIDERATIONS

Instantaneous and Average Dissipation. As we apply the theory to specific design tasks, we must take care that the assumptions upon which the theory is based apply to the case at hand. For example, the design equations for class AB operation assume the signal frequency is much higher than the *thermal time constants* of the transistors. That is, that motion along the load line is so rapid that the junction does not have time to heat and cool during each cycle. Under these conditions the instantaneous operating point can pass above the dissipation hyperbola, without average dissipation exceeding the provisions of the heat sink. For very slow signals, however, when junction temperature can change during a signal cycle, *instantaneous dissipation* rather than average dissipation becomes the critical factor. For an extreme example, notice that an amplifier designed to control the speed of a dc motor might operate for a prolonged interval at point A in Fig. 10.23. Maximum instantaneous dissipation occurs at the center of the load line where it is

$$P_D = \frac{I_{max}}{2}\frac{V_{CC}}{2} = \frac{V_{CC}}{2R_L}\frac{V_{CC}}{2} = \frac{V_{CC}^2}{4R_L} = 0.5P_{max} \tag{10.32}$$

instead of $0.135P_{max}$ or $0.203P_{max}$ that apply to "fast" signals.

For another application that differs widely from our assumptions, consider what happens if a square wave rather than a sine wave is developed across the load resistor in a class B amplifier. This means the Q-point jumps back and forth between the extremities of the load line of Fig. 10.21a. Since both transistors deliver high current at low voltage under these conditions, we intuitively expect much higher efficiency than 78%. Problem 10.38 explores this conjecture, and Sec. 10.8 expands the idea into the class D amplifier, a general-purpose amplifier having higher efficiency than class B.

Power Gain. A power amplifier must have power gain greater than one; otherwise, the preceding *driver* stage would also have to be a power amplifier. FET circuits have infinite power gain; bipolar circuits do not. Because voltage gain is only one in a BJT output stage, current gain must exceed one, making a high value of dc beta (h_{FE} on the manufacturer's data sheet) important for output transistors. Figure 10.14c reminds us that high input resistance accompanies high power gain in class B and AB amplifiers, because the driver stage sees $R_{in} = (\beta + 1)R_L$. Consequently, even though the driver circuit must develop voltage of full output amplitude, this voltage is developed at a low current level, a current lower than the output current by a factor of $(\beta + 1)$. Power transistors sometimes have high β; however, because of manufacturing tolerances minimum values can be as low as 5 or 10, especially for integrated pnp transistors. For these rea-

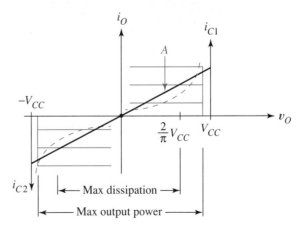

Figure 10.23 Load line, dissipation hyperbola for average power, and point of maximum instantaneous power for class AB amplifier.

sons the Darlington connection of Fig. 10.24, which increases the effective β to β², is popular in power amplifiers. Output stages sometimes employ the compound pnp transistor of Fig. 7.28 for the same reason. Figure 10.24 also shows that if bias current is allowed to flow through source v_I, only a single bias source I_{BB} is needed for class AB operation. This circuit lacks the classical symmetry of Fig. 10.20b and even has an output offset; however, the dissipation $I_{BB}V_{CC}$ of the second current source is eliminated, as is its hardware.

Also associated with the power amplifier input resistance is a special distortion problem. This occurs when the transistor betas are not matched and when the resistance R_i of the driver is not small compared to R_{in} of the power amplifier. Figures 10.14c and e show that R_{in} takes on different values when $\beta_1 \neq \beta_2$, and this results in different voltage gains for positive and negative voltage swings. From Fig. 10.14c,

$$v_O = R_L i_L = R_L(\beta_1 + 1)i_B$$

where i_B is given by

$$i_B = \frac{v_I - (v_O + V_{BE})}{R_i}$$

Substituting i_B and solving for v_o gives

$$v_O = \frac{(\beta_1 + 1)R_L}{(\beta_1 + 1)R_L + R_i}(v_I - V_{BE})$$

A similar derivation for Fig. 10.14e gives

$$v_O = \frac{(\beta_2 + 1)R_L}{(\beta_2 + 1)R_L + R_i}(V_{BE} + v_I)$$

If R_i is negligible, these expressions approximate Eqs. (10.21) and (10.22); however, if R_i is large, the voltage gain differs for the two signal polarities, giving a transfer charac-

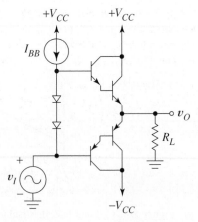

Figure 10.24 Output stage employing Darlington connected transistors.

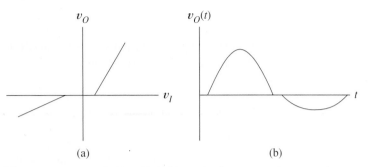

Figure 10.25 Consequences of either high driver circuit output resistance or low-power amplifier input resistance, or both: (a) unsymmetric transfer characteristic; (b) unsymmetric output waveform.

teristic resembling Fig. 10.25a and a distorted output waveform like Fig. 10.25b. Output transistors of high beta help us avoid this problem in addition to giving high power gain and lower standby power.

Thermal Runaway. Even though we select transistors with adequate power ratings, it is still possible for them to overheat in class AB circuits. Biasing voltage V_{BB} in Figs. 10.20a and b must have a negative temperature coefficient to prevent a transistor self-destruct mechanism called *thermal runaway.* To understand runaway, suppose that V_{BB} does not vary at all with temperature. Then in the FET class AB circuit, at temperature T_1 each transistor is biased at $v_{GS} = 0.5V_{BB}$, and $i_D = I_{D1}$ as in Fig. 10.26a. Because the transistor dissipates $\approx V_{DD}I_{D1}$ watts, its junction temperature increases to T_2, causing its transfer characteristic to shift as in the figure. But this increases the current to I_{D2}, increasing the dissipation and leading to further increases in junction temperature. In runaway, this cycle continues until the transistor overheats. Figure 10.26b shows that the same mechanism occurs in the BJT.

When diodes, diode-connected transistors, or V_{GS}, or V_{BE} multipliers produce V_{BB}, the vertical lines in Fig. 10.26 *automatically* shift to the left with increasing temperature,

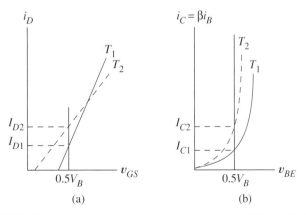

Figure 10.26 Basis for thermal runaway: (a) in the power FET; (b) in the BJT.

compensating for the changing transistor characteristic by keeping bias current approximately constant or even reducing it with increases in temperature. In IC designs, the components that establish V_{BB} are physically located so as to share the same thermal environment as the output transistors.

Short-Circuit Protection. Since large amounts of power are involved in an output stage, the designer needs to be alert to other issues that might not be important in low-power circuits. For example, for a class A amplifier an accidental open-circuit load produces the horizontal load line shown in Fig. 10.27a. Consequently, instantaneous dissipation in the power transistor approaches $2V_{CC}I_B = 4P_L$, which can damage the transistor if the signal frequency is low.

   An accidental short circuit from output node to ground presents a potentially serious problem in both class A and class B and AB amplifiers. Figure 10.27a shows the vertical load line that results from a short circuit in a class A amplifier. With the emitter voltage held to 0 V by the short circuit and with v_I limited only by the internal resistance of the signal source, there is a large increase in base current and, consequently, collector current, forcing operation high above the dissipation hyperbola, as in Fig 10.27a. Figure 10.27b shows the same danger exists in class B and AB amplifiers. To prevent these problems, power amplifiers often include short-circuit protection. In the class AB circuit of Fig. 10.28, Q_3, Q_4, R_{P3}, and R_{P4} provide short-circuit protection. Normally cut off, Q_3 turns on only when the output current produces a drop across R_{P3} that exceeds the cut-in value, that is, for

$$i_{O,max}R_{P3} > 0.5 \text{ V}$$

Once on, Q_3 diverts input current from the base of Q_1 directly into R_L, thereby limiting the collector current and power dissipation of Q_1 to safe values. Because Q_3 is not across the power supply like Q_1, the power it dissipates is not destructive. For large negative i_O, Q_4 and R_{P4} protect Q_2 in the same way. Problem 10.40 outlines a SPICE exploration of this interesting circuit.

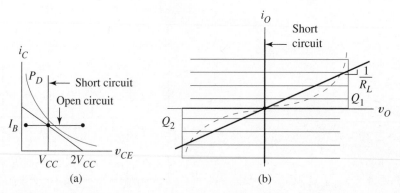

Figure 10.27 Modified load lines presented by accidental open- and short-circuit faults: (a) class A; (b) class B and AB.

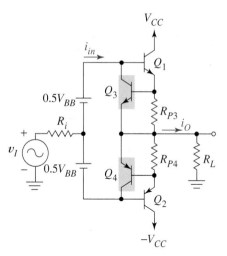

Figure 10.28 Short-circuit protection for class AB amplifier.

Foldover Current Limiting. When there is an accidental short circuit between the output pin and the negative or positive supply, the load line is replaced by the vertical line

$$v_O = V_{CC} \quad \text{or} \quad v_O = -V_{CC}$$

as in Fig. 10.29a. Short-circuit protection, which sets safe current limits at $\pm i_{O,max}$ for $v_O = 0$ (dashed lines), still permits excessive dissipation for shorts to the power supplies through operation into the shaded areas above and below the dissipation hyperbolas. We could set lower short-circuit current limits, but this would further limit the maximum normal output current of the amplifier. Figure 10.29b shows a modification of the current-limiting circuit called a *foldover circuit* that gives more effective short-circuit protection by using two additional resistors R_1 and R_2. The base and emitter voltages of Q_3 are

$$v_B = (v_O + i_O R_{P3})\frac{R_2}{R_2 + R_1}, \qquad v_E = v_O$$

These give

$$v_{BE} = v_B - v_E = \frac{-R_1}{R_2 + R_1}v_O + \frac{R_2 R_{P3}}{R_2 + R_1}i_O \tag{10.33}$$

Setting v_{BE} to the cut-in value relates the current $i_O = i_{O,max}$ to v_O by

$$i_{O,max} = \frac{R_1}{R_2 R_{P3}}v_O + \frac{(R_2 + R_1)}{R_2 R_{P3}}V_\gamma \tag{10.34}$$

Figure 10.29c shows with dashed curves that resistor values can be chosen for this equation that protect the transistor from short circuits and shorts to the supplies while allowing greater load currents than the "horizontal-line" short-circuit protection of Fig. 10.29a. Problem 10.41 gives some assistance with the details.

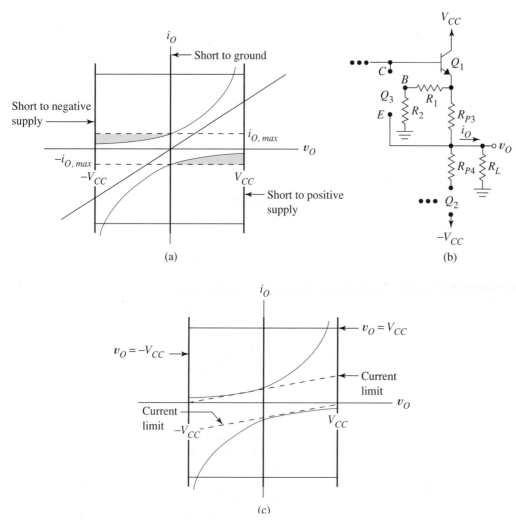

Figure 10.29 Foldover protection circuit: (a) the problem; (b) foldover circuit; (c) current limits possible with foldover protection.

Power Supply Decoupling. In high-gain amplifiers, positive feedback is sometimes introduced accidentally through the power supply connections. The origin of this subtle problem is impedance Z_{PS} originating in the power supply and in the conductors that carry bias current I_B to the various amplifier stages. As suggested by Fig.10.30a, these conductors also carry *signal* currents such as I_{sig} that originate in the individual gain stages. In the output stage, phasor I_{sig} might describe a large current. If Z_{PS} were zero there would be no problem; however, compared to the small signal levels in the first stage, the feedback voltage $V_f = I_{sig} Z_{PS}$ can be appreciable. As we learned in Chapter 9, instability can result if V_f happens to arrive in phase with the input signal.

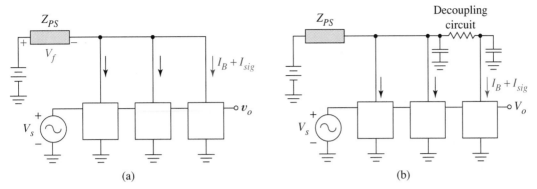

Figure 10.30 Preventing feedback through the power supply: (a) the problem; (b) solution in the form of a decoupling circuit.

A practical solution to this problem is to add a large, low-inductance *decoupling capacitor* to the circuit at the *physical point* where the power supply lead enters the output stage, to short out I_{sig}. This reduces the signal in the power supply line without affecting the biasing. When the problem is especially severe, a series resistance and another capacitor are added, forming a *decoupling circuit* as in Fig. 10.30b. Signal must then pass from the output through a two-stage *RC* filter before developing voltage across Z_{PS}. Of course, the decoupling resistor adds an undesired dc drop in the supply line and dissipates power.

10.7
Power Op Amps

Nowadays the easiest way to provide a large amount of signal power to a load is to use a *power op amp*. These have all the usual features we associate with op amps but also possess an output stage, usually class AB or B, that is capable of delivering large amounts of power. Used in conventional feedback configurations with moderate amounts of closed-loop gain, they offer low distortion and design simplicity. Data sheets for power op amps contain thermal resistances and %THD ratings in addition to the "usual" parameters such as open loop gain, offset voltage and current, and CMRR. The latter have valves comparable to those of low-power op amps.

For loads requiring large amounts of power, two or more op amps can operate in parallel—as in Fig. 10.31a, for example. Here a low-power op amp OA_1 in a non-inverting amplifier configuration, uses multiple power op amps configured as followers to deliver the load power. Because of unequal offsets in OA_2–OA_N, small equalization resistors of 0.1 Ω or so are placed between each follower output node and the circuit output node. Alternately, we could use the inverting amplifier configuration for OA_1. Figure 10.31b shows another arrangement, in which power amplifier OA_1 develops circuit gain and also contributes output power.

Op Amp Dissipation Calculations. Like power transistors, power op amps must sometimes be provided with heat sinks to allow them to dissipate power without exceeding critical transistor junction temperatures. Because most of this dissipation is in the output transistors, the class B theory we learned is easily applied to op amp circuits. The only

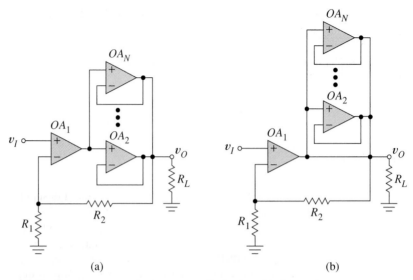

Figure 10.31 Power op amps sharing power dissipation: (a) driver plus power output stages; (b) all power op amp circuit.

real novelty is that the op amp thermal resistances are defined for *pairs* of output transistors, and we must design the heat sinks accordingly.

EXAMPLE 10.8 A single power op amp with a class B output stage must deliver 95 W to a 4 Ω resistive load. Find the thermal resistance required of the heat sink θ_{SA}. The available op amp has $\theta_{JC} = 2.3°C/W$ and maximum junction temperature of 200°C. Assume that a thermal grease of $\theta_{CS} = 0.2°C/W$ conducts heat from the op amp case to the heat sink.

Solution. Driving the 95 W, 4 Ω load at maximum efficiency, from Fig. 10.22 requires

$$V_{CC} = \sqrt{2 \times 4 \times 95} = 27.6 \text{ V}$$

From the same figure, for a conservative design, the *two output transistors together* must dissipate $2 \times 0.203 \times 95 = 38.6$ W, which is the dissipation required of the op amp. Assuming ambient temperature is 25°C, the dissipation requirement gives

$$38.6 \, (2.3 + 0.2 + \theta_{SA}) + 25 = 200$$

or $\theta_{SA} = 2.03°C/W$. ❑

Exercise 10.8 We wish to explore an alternative solution to the design task of Example 10.8, using N of the available op amps in parallel *without heat sinks*. How much power must each op amp dissipate?

Ans. 0.406(95/N) watts.

EXAMPLE 10.9   If the op amp thermal resistance in Exercise 10.8 is $\theta_{JA} = 10°C/W$, find the number N of op amps needed.

Solution. Assuming each op amp dissipates its power independently in an ambient environment of 25°C,

$$(0.406) \times \left(\frac{95}{N}\right)\theta_{JA} + 25 = 200$$

or $N = 2.20$. Since N must be an integer, use $N = 3$.   ❏

In Example 10.9 and Exercise 10.8, the supply voltages required for parallel operation are 27.6 V, the same as calculated in Example 10.8. This is because the peak-to-peak output voltage of each op amp is still approximately the voltage difference between positive and negative supplies. (Visualize the three class B circuits with output nodes all connected to the same R_L.) Individual op amps provide one-third the output power when connected in parallel because each op amp provides only one-third the current in the load resistor.

The *bridge amplifier* of Fig. 10.32a demonstrates another way for two op amps to share dissipation. Because of the phase opposition of the follower and inverter outputs, an input sinusoid of amplitude V_{CC} gives an output sinusoid of amplitude $2V_{CC}$ and maximum output power of

$$P_{max} = \frac{(2V_{CC})^2}{2R_L} \tag{10.35}$$

Figures 10.32b and c show how the four class B output transistors alternately conduct and cut off for positive and negative input voltage and, in the process, clarify the term "bridge amplifier."

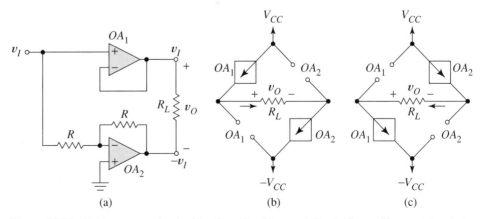

Figure 10.32   Bridge power circuit: (a) schematic; (b) output circuits for positive v_i; (c) output circuits for negative v_i.

We now examine the power balance for this circuit. For sinusoidal input

$$v_I(t) = \frac{V_o}{2} \sin \omega_o t$$

the current delivered by *each* power supply is a full-rectified sine wave of amplitude V_o/R_L, with average value $I_{ave} = (2/\pi)(V_o/R_L)$. The supplies together therefore deliver

$$P_{in} = 2V_{CC}I_{ave} = \frac{4V_{CC}V_o}{\pi R_L} \tag{10.36}$$

watts, of which

$$P_{load} = \frac{V_O^2}{2R_L} \tag{10.37}$$

goes to the load. This leaves $P_{OA} = 0.5(P_{in} - P_{load})$ to be dissipated by *each* op amp, where

$$P_{OA} = \frac{2V_{CC}V_o}{\pi R_L} - \frac{V_O^2}{4R_L} \tag{10.38}$$

The V_o that maximizes P_{OA} is $V_o = (4/\pi)V_{CC}$. Substituting this into Eq. (10.38) and then using Eq. (10.35) to express the result in terms of maximum output power gives

$$P_{OA,max} = \frac{2}{\pi^2}\boldsymbol{P_{max}} = 0.203\boldsymbol{P_{max}} \tag{10.39}$$

Referring to Fig. 10.22b shows that in this circuit the maximum dissipation of *each op amp* is the same as the dissipation of *each transistor* in a single class B stage delivering the same output power.

Exercise 10.9 In Examples 10.8 and 10.9, we explored two op amp configurations that might supply 95 W to a 4 Ω load. Determine whether we can use two op amps with $\theta_{JA} = 10°C/W$ to deliver the same power in a bridge configuration without a heat sink. Maximum junction temperature is 200°C.

Ans. No. $T_J = 218°C$.

10.8
Class D Amplifier

By using transistor switches to alternately connect the load to positive and negative supplies, the *class D amplifier* achieves nearly 100% efficiency. There is little transistor dissipation because high current is delivered through a closed switch ($v_{DS} \approx 0$), and high voltage develops across an open switch ($i_D \approx 0$). Since this amplifying scheme involves binary-valued input and output waveforms, it would seem to be too specialized for general analog signals; however, preprocessing with a *pulse-width modulator* provides appropriate signal input for any time function. The high efficiency of the total amplifying system justifies the added complexity of a low-power modulator circuit.

10.8.1 PULSE-WIDTH MODULATION (PWM)

The circuit of Fig. 10.33a causes the pulse width of its binary-valued output signal $v_O(t)$ to follow the amplitude of analog signal $v_I(t)$. A high-frequency pulse train $v_K(t)$ (Fig. 10.33b) produces the sawtooth waveform $v_S(t)$ of Fig. 10.33c by closing the BJT switch for a brief interval every T second to discharge the capacitor. When the transistor is cut off, the capacitor charges linearly at a rate determined by the capacitance and the current source.

The comparator circuit continually compares the analog signal $v_I(t)$ with $v_S(t)$ as in Fig. 10.33c. It produces high output, $v_O = V_M$, whenever $v_I(t) > v_S(t)$ and low output, $v_O = V_M$, for $v_I(t) < v_S(t)$. The result is the PWM waveform $v_O(t)$ of Fig. 10.33d, where shading helps the eye recognize the pattern. This binary-valued waveform has a low-frequency component that contains exactly the information of $v_I(t)$. Thus $v_O(t)$ is the sum of an amplitude-scaled input signal and some high-frequency components that eventually must be removed by filtering, a process that is simplified by operating at a much higher pulse frequency than our illustrative diagrams suggest. To amplify a PWM signal, we need only reproduce it with a circuit that has high power-delivering capabilities, that is, at higher voltages and/or currents.

The overall PWM system is described by a simple static characteristic. If we define the *duty cycle* δ as the fraction of each period T that $v_O(t)$ is positive, we see from Fig. 10.33c and d that the circuit is described by

$$\delta = \frac{1}{V_P} v_I \qquad (10.40)$$

where V_P is the peak value of the sawtooth input waveform. This is Fig. 10.33e.

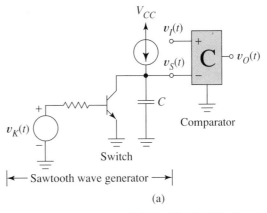

(a)

Figure 10.33 Pulse-width modulation: (a) modulating circuit; (b) pulse train of period T; (c) comparator input signals; (d) comparator output waveform and its low-frequency information bearing component; (e) duty cycle of output waveform versus amplitude of input.

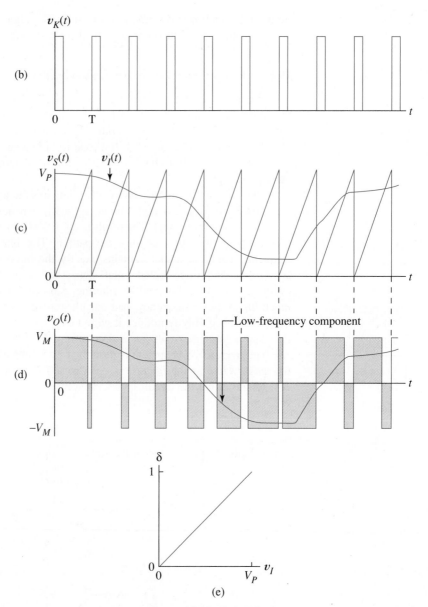

Figure 10.33 (continued)

10.8.2 CLASS D CIRCUIT

Figure 10.34a shows the complete class D amplifier, including a CMOS output circuit. First, consider the biasing. When v_X is zero, M_1 and M_2 are both cut off, and coupling capacitors C_C charge to V_{DD} volts. During operation, v_X is a PWM wave like Fig. 10.33d with peak value V_M. The capacitors function as dc voltage sources, and v_X is superimposed upon the dc bias values of the gates. This means, for example, that the gate of M_1 is driven between peak values of $V_{DD} + V_M$ and $V_{DD} - V_M$. For sufficiently high V_M, M_1 and M_2 alternate between cut-off and ohmic operation in complementary fashion. A *lossless, LC,* lowpass filter transfers the information to R_L while removing the undesired high-frequency components. Because the filter is lossless, the signal power in R_L equals the total waveform power developed at the amplifier output.

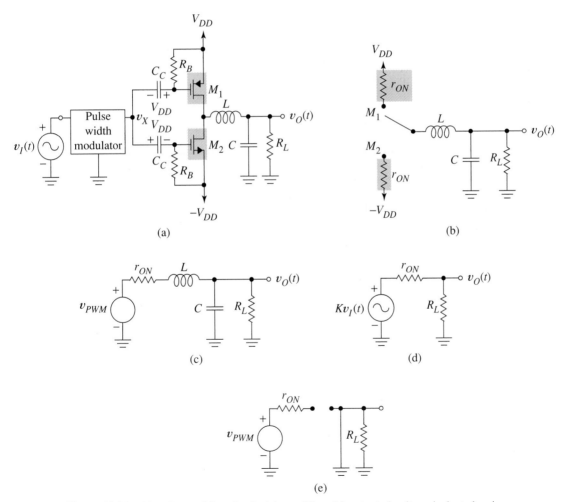

Figure 10.34 Class D amplifier circuit: (a) amplifier; (b) output circuit equivalent showing "ON" resistance of ohmic transistors and complete PWM signal; (c) simplified equivalent; (d) output equivalent for signal frequencies only; (e) output equivalent for high frequencies.

Figure 10.34b represents the output circuit. The transistors are shown as lossy switches by using their ohmic-state equivalents. The filter and load in Fig. 10.34c see the same signal as in Fig. 10.34b where $v_{PWM}(t)$ denotes a binary-valued PWM output waveform like Fig. 10.33d, but of peak value V_{DD} (and also inverted since positive v_X turns on M_2). The information components of this waveform see $\omega L \approx 0$ and $1/\omega C \approx \infty$, giving Fig. 10.34d, where K denotes the voltage gain of the amplifier. The high-frequency components see $\omega L \approx \infty$ and $1/\omega C \approx 0$ as in Fig. 10.34e, a lossless load, and contribute nothing to the system's power dissipation. Since Fig. 10.34d accounts for all of the power delivered by the power supply and developed in the load, instantaneous load and input power for

$$v_I(t) = 0.5V_P + V_I \cos(\omega t)$$

are

$$P_{LOAD} = \frac{[KV_I R_L/(R_L + r_{on})]^2}{2R_L} \quad \text{and} \quad P_{IN} = \frac{(KV_I)^2}{2(R_L + r_{on})}$$

giving an output circuit efficiency of

$$\eta = \frac{P_{LOAD}}{P_{IN}} = \frac{1}{1 + r_{ON}/R_L} \times 100 \tag{10.41}$$

which approaches 100% when M_1 and M_2 are driven to ohmic values, r_{ON}, much smaller than R_L.

EXAMPLE 10.10 In Fig. 10.34a M_1 and M_2 are complementary MOSFETS with parameters $k = 0.06$ A/V^2, $|V_t| = 1.2$ V, and $R_L = 10$ Ω. The output stage uses $V_{DD} = 20$ V. Find the minimum voltage, V_M, of the square wave at v_X for 80% efficiency.

Solution. From Eq. (10.41), 80% efficiency requires $r_{ON} = 2.5$ Ω when $R_L = 10\Omega$. When the MOSFET is used as a linear resistor, r_{ON} is given by Eq. (5.5). For this resistance, the gate voltage that turns the transistor on must satisfy

$$2.5 = \frac{1}{0.06 \, (V_{GS} - 1.2)}$$

or $V_{GS} = 7.87$ V, provided that

$$|v_{DS}| \leq 0.2 \, |(v_{GS} - V_t)| \leq 0.2 \times 6.67 = 1.33 \text{ V}$$

Because coupling capacitors are charged to $|V_{DD}| = 20$ volts, for M_2

$$v_{GS}(t) = v_G(t) - v_S(t) = [v_X(t) - V_{DD}] - (-V_{DD}) = v_X(t)$$

Since the peak value of v_X is V_M, $V_M = 7.87$ V. ❏

The frequency, $\omega_s = 2\pi/T$, of the sawtooth generator of Fig. 10.33a should be low enough that the parasitics of the transistors are negligible, but high enough that the output filtering task is not too difficult. Before filtering, the frequency spectrum of the

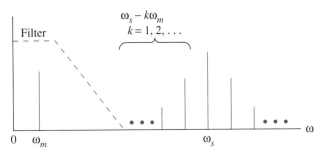

Figure 10.35   Spectrum of class D output before lowpass filtering.

PWM waveform has the general form of Fig. 10.35 when the input signal is a sinusoid of frequency ω_m. The high-frequency noise consists of a sinusoid at the switching fre-quency ω_s plus additional *sideband* components at frequencies $\omega_s \pm k\omega_m$, where k takes on integer values. The output filter, shown as a second-order lowpass LC filter in Fig. 10.34, must pass ω_m but attenuate ω_s and its more important sidebands. We learn about designing such filters in Chapter 12. For a rule of thumb, ω_s is at least four times the highest frequency ω_m contained in the input signal.

Exercise 10.10   Instead of using the rule of thumb for filtering, let us use our wits. Suppose in Fig. 10.35 the Bode plot of the filter is flat up to ω_m and then drops off at -40 dB/dec. To ensure good attenuation of high-frequency noise, we decide that the noise at $\omega = \omega_s - 7\omega_m$ should be reduced by at least 20 dB as it passes through the filter. Find the sampling frequency ω_s in terms of ω_m.

Ans. $\omega_s = 10.2\ \omega_m$.

Often class D amplifiers operate in a closed-loop feedback configuration like Fig. 10.36. The input to the pulse-width modulator is

$$v_X(t) = \frac{V_P}{2} + v_I(t) - v_O(t) = \frac{V_P}{2} + \epsilon(t)$$

where $V_P/2$ biases the input at the half-way level in Fig. 10.33c. Error, $\epsilon(t)$, the differ-ence between the desired output and the actual output, controls the switches so as to reduce the error. This causes the output to follow the input.

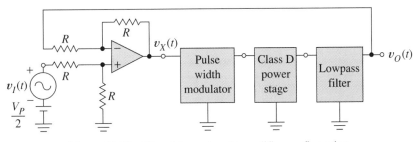

Figure 10.36   Closed-loop class D amplifier configuration.

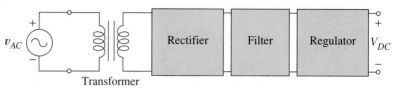

Transformer

Figure 10.37 Subcircuits of a power supply.

10.9

Power Supplies

The diagram of Fig. 10.37 shows basic functions common to many power supplies. A transformer provides electrical isolation between the ac line voltage and the output and converts the line voltage to a more suitable amplitude. A half- or full-wave rectifier converts the transformed ac voltage into pulsating dc, and a filter consisting of one or more lossless L or C elements smooths out most of the voltage variation. The filter output is *unregulated output voltage,* dc with a small additive ac ripple. An unregulated output is suitable when the current drawn from the power supply changes only slightly, and where small perturbations in output voltage caused by line voltage variations are tolerable. Most electronic circuits require a regulator circuit to keep the output constant in spite of load variations or changes in unregulated voltage. In many cases, the regulator also provides protection against an accidental short circuit across the output terminals. Since we are already familiar with the transformer and rectifier from Chapter 3, we begin with the power supply filter.

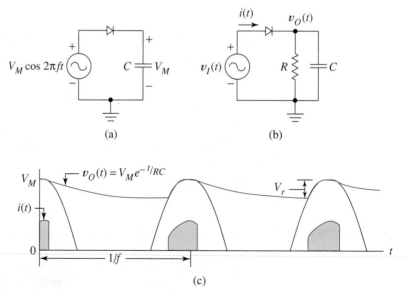

Figure 10.38 Half-wave rectifier and filter: (a) basic dc supply with no load; (b) rectifier-filter schematic; (c) rectifier-filter voltage and current waveforms.

10.9.1 RECTIFIERS WITH FILTERS

Half-Wave Rectifier/Filter. If the diode in Fig. 10.38a is ideal, during an initial transient, capacitor C charges through the diode to V_M volts, and remains charged to this constant value. For the circuit to be useful, however, it must supply dc power to an external load R as in Fig. 10.38b. It is useful to view Fig. 10.38b as the ideal half-wave rectifier with a filter capacitor C in parallel with the load. As we see in the steady-state curves of Fig. 10.38c, the rectifier operation is greatly modified by the capacitor. Because C is chosen so that the RC time constant is quite long compared to the period $1/f$ of the input sinusoid, node voltage $v_O(t)$ cannot drop as fast as $v_I(t)$. Since v_O and v_I are the two diode node voltages, the diode turns off shortly after the input voltage reaches its peak, leaving the load voltage to decay exponentially. If $RC >> 1/f$, v_O changes only slightly before the input voltage again turns on the diode. This makes the *ripple voltage* V_r very small. The diode current consists of a sequence of short bursts, highlighted by shading in the diagram. As RC is increased, these current bursts become more narrow and higher: more narrow because the diode conducts for shorter intervals, higher because the average diode current must always equal the dc current in R. The average current approaches V_M/R as the time constant becomes large.

In a typical design problem, we know the desired output voltage $V_{DC} \approx V_M$ and the maximum dc output current I_{DC}; thus, we can determine the smallest value for load resistor R. The frequency, f, and the largest permissible value of peak-to-peak ripple, V_r, are also specified. We now derive a design equation that relates the acceptable ripple amplitude to the size of the filter capacitor.

The charge ΔQ that leaves the capacitor during each cycle is related to V_r in Fig. 10.38c by

$$\Delta Q = C\Delta V = CV_r$$

This charge is replaced by the average current flowing through the diode. For maximum output current this is

$$I_{DC} = \frac{\Delta Q}{\Delta t} = \frac{CV_r}{1/f} = fCV_r \tag{10.42}$$

If the ripple is small, $V_{DC} \approx V_M$, and average diode current in Eq. (10.42) must equal the output current; that is, $I_{DC} = V_M/R$. Substituting this into Eq. (10.42) and solving for C gives the design equation

$$C = \frac{1}{fR}\frac{V_M}{V_r} \tag{10.43}$$

Thus, reducing the permissible fractional ripple V_r/V_M means increasing the capacitor size.

Once the design is completed, C, f, and V_M in Eq. (10.43) are constants. As R grows larger (less current is drawn from the supply), V_r becomes smaller. Since C always remains charged to nearly V_M volts, the diode must be rated for PIV $\geq 2V_M$ volts. The following example illustrates using Eq. (10.43) in a design.

EXAMPLE 10.11 Design a power supply using a half-wave rectifier and capacitor filter that gives 12 V and 25 mA with a ripple of 2%.

Solution. Assume $f = 60$ Hz. Select the transformer turns ratio so that $V_M = 12.7$ V instead of 12 V to allow for diode offset. For maximum current, $R = 12$ V/0.025 A = 480 Ω. From Eq. (10.43)

$$C = \frac{1}{60 \times 480 \times 0.02} = 1736 \ \mu F$$

The PIV rating for the diode must be at least 25.4 V.   ❏

> **Exercise 10.11** Estimate the percentage ripple in the power supply of Example 10.11 when the supply delivers 5 mA.
>
> *Ans.* 0.4%.
>
> **Full-Wave Rectifier/Filter.** Figure 10.39 shows how a capacitor filters the output of a full-wave rectifier. For the same RC, the ripple is smaller than for the half-wave case because the capacitor has only half the time to discharge. To obtain the new design equation for C, we replace the half-wave period, $1/f$, in Eq. (10.43) by the full-wave period $0.5/f$. This gives
>
> $$C = \frac{1}{2fR} \frac{V_M}{V_r} \tag{10.44}$$
>
> To find peak inverse voltages for the diodes for the various rectifier designs we simply note that C is always charged to approximately V_M. Because the filter capacitor is bulky and expensive, a benefit of the full-wave circuit is a smaller C for the same ripple. Diode offsets reduce the output voltage to a value somewhat smaller than the value V_M that we predict with ideal diodes. This effect is readily observed in simulations and is easily anticipated by the designer.
>
> ### 10.9.2 VOLTAGE REGULATORS
>
> A power supply sometimes requires a voltage regulator to maintain constant output voltage in spite of variations in dc loading, unregulated voltage, or both. Two figures of merit describe voltage regulators: voltage regulation (VR) and line regulation (LR).

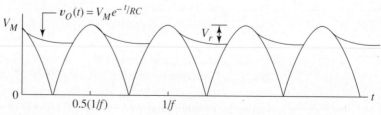

Figure 10.39 Output waveforms for full-wave rectifier, with and without capacitor filter.

To describe variations in output voltage due to changes in load resistance we have

$$\text{VR} = \frac{\text{no-load voltage} - \text{full-load voltage}}{\text{full-load voltage}} \times 100 = \left(\frac{V_{NL}}{V_{FL}} - 1 \right) \times 100 \quad (10.45)$$

where V_{NL} and V_{FL} denote, respectively, the no-load voltage (open-circuit output) and full-load voltage, which corresponds to maximum rated load current.

To describe variations in output caused by changes in the unregulated voltage, we define

$$\text{LR} = \frac{\text{change in regulated voltage}}{\text{change in unregulated voltage}} \times 100 = \frac{\Delta V_R}{\Delta V_U} \times 100 \quad (10.46)$$

We now examine two classical types of voltage regulator, shunt and series. Shunt regulators operate in parallel with the external load, diverting current in such a way that the output voltage remains constant. Series regulators use a transistor in series with the load, while a feedback system adjusts the transistor current to whatever value is required to maintain constant output voltage.

Shunt Regulator. The zener voltage regulator of Fig. 3.30 and Example 3.10 is a linear shunt regulator: linear because it operates where the zener acts as a linear voltage source, shunt because it is connected in parallel with the external load. The following example shows that a low incremental zener resistance is the key to good voltage and line regulation.

EXAMPLE 10.12 (a) Find expressions for voltage and line regulation for the zener regulator of Fig. 10.40.
(b) Evaluate voltage and line regulation using $V_U = 15$ V, $R_S = 10$ Ω, $V_Z = 5$ V, $r = 0.1$ Ω, and $R_L = 25.6$ Ω.

Solution. (a) Figure 10.40b shows the Thevenin equivalent of the regulator. From Eq. (10.45), since $r \ll R_S$

$$\frac{\text{VR}}{100} = \frac{V_{TH1}}{[R_L/(R_L + R_{TH1})]V_{TH1}} - 1 = \frac{R_{TH1}}{R_L} \approx \frac{r}{R_L}$$

To find the line regulation, first replace the zener model and R_L, Fig. 10.40a, by a Thevenin equivalent. This leaves V_R unchanged and keeps V_U in the diagram.

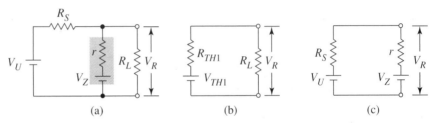

$$\text{(a)} \qquad\qquad\qquad \text{(b)} \qquad\qquad\qquad \text{(c)}$$

Figure 10.40 Shunt regulator: (a) equivalent circuit; (b) circuit used to find voltage regulation; (c) circuit used to find line regulation.

Figure 10.40c is a good approximation, since $r \ll R_L$. By superposition, V_R has components due to both V_Z and V_U; in LR we are interested only in the latter. This is

$$V_{R,U} = \frac{r}{r + R_S} V_U$$

We conclude from Eq. (10.46) that

$$LR = \frac{\Delta V_R}{\Delta V_U} 100 = \frac{r}{r + R_S} 100$$

(b) Using the expressions just derived, $VR = [(0.1)/25.6] \times 100 = 0.391\%$ and $LR = 0.99\%$. ❏

Exercise 10.12 If the unregulated supply voltage in Example 10.12 has a ripple voltage of 5 mV, how large is the ripple that appears at the output?

Ans. 049.5 μV.

Series Regulator. Figure 10.41a shows a series regulator. A fraction of V_R is continuously compared with a zener reference voltage V_Z, and the difference used to control the current of the series transistor. If V_R drops for any reason, transistor current increases, increasing V_R and decreasing the drop V_S across the transistor; an increase in V_R decreases the emitter current and V_S increases.

Figure 10.41b is the same circuit redrawn to show the op amp/transistor pair as the A circuit in a voltage-series feedback amplifier—the β circuit consists of R_F and R_1. The low output resistance of the emitter follower,

$$R_{out} = \frac{R_o + R_\pi}{\beta_F + 1}$$

where R_o is the op amp output resistance, is further reduced by the voltage feedback. Consequently, V_R looks back at a nearly constant source of voltage

$$V_R = \left(1 + \frac{R_F}{R_1}\right) V_Z$$

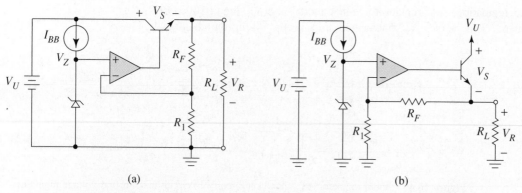

(a) (b)

Figure 10.41 Series regulator: (a) traditional diagram; (b) viewed as a feedback amplifier.

The zener Q-point, biased by I_{BB}, does not change with loading at the regulator output. This circuit obviously lends itself nicely to IC design, since the resistor ratio and zener voltage have very low temperature variations. In Problem 10.54 we learn that voltage regulation depends on the output resistance of the circuit and line regulation depends upon the zener and current-source resistances.

There are several commonly used variations of this same basic circuit. To give lower output resistance, and therefore better voltage regulation, "the" transistor is sometimes a Darlington pair. Current source I_{BB} can be a sophisticated IC current source of very high output resistance to give the best possible line regulation. A simpler current source might be an n-channel JFET or depletion MOSFET with gate connected to source. Bias current I_{BB} can even be provided by a simple resistor rather than a current source, with some sacrifice of line regulation.

Protection Circuits. Voltage regulators for power supplies often employ the same short-circuit protection schemes introduced for power amplifiers in Figs. 10.28 and 10.29.

Operational Power Supply. If the op amp of Fig. 10.41b is a power op amp, the BJT is not required, and the op amp output can be directly connected to the load. Such a circuit is called an *operational power supply.* Then Fig. 10.41b reduces to a simple noninverting amplifier that scales the zener voltage to the desired value at the output and delivers whatever current is demanded by the load. Inverting configurations are also used in operational power supplies.

EXAMPLE 10.13 The operational power supply in Fig. 10.42 is to develop a regulated voltage of 12 V to load R_L at $I_L \leq 3$ A. JFET parameters are $\beta = 8.89 \times 10^{-4}$ A/V^2 and $V_P = -1.5$ V. The zener voltage is $V_Z = 6.7$ V at $I_Z = 1$ mA. Complete the design by finding values for R_B, R_1, and R_F, and compute the worst case dissipation of the op amp.

Solution. To bias the zener at 1 mA, V_{GS} of the JFET must satisfy

$$1 \text{ mA} = 8.89 \times 10^{-4}(V_{GS} + 1.5)^2$$

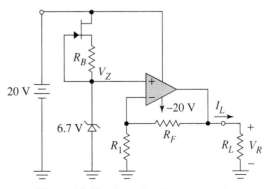

Figure 10.42 Operational power supply.

or $V_{GS} = -0.44$ V. Since this is the voltage across R_B, we need $R_B = 0.44/1$ mA $= 440\ \Omega$. R_1 and R_F must satisfy

$$12 = \left(1 + \frac{R_F}{R_1}\right)6.7$$

which gives $R_F/R_1 = 0.791$. For $R_1 = 10$ kΩ, $R_F = 7.91$ kΩ.

When the op amp delivers 3 A at 12 V, the voltage across the conducting op amp output transistor (output stage assumed class AB) is $20 - 12 = 8$ V; thus $8 \times 3 = 24$ W is the internal dissipation. An op amp and appropriate heat sink must be provided to dissipate this 24 W on a steady-state basis. ❏

As for power amplifiers, we can define the efficiency of a regulator by

$$\eta = \frac{\text{regulated output power}}{\text{input power to regulator circuit}}$$

In the operational power supply and the regulators of Fig. 10.41, most of the power loss is in the output transistor.

Exercise 10.13 Estimate the efficiency of the regulator of Example 10.13. Also, write expressions for the maximum dissipation of the series transistor and efficiency of the series regulator of Fig. 10.41a.

Ans. 60%, $P_D = (V_U - V_R)I_{R,MAX}$, $\eta = V_R/V_U$.

The preceding example and exercise showed that internal dissipation results in low efficiency when large voltages are dropped across conventional regulators. The regulator class discussed next has lower internal dissipation and, therefore, higher efficiency for the same output specifications.

Switching Regulators. Transistor switching schemes that give high efficiency in class D power amplifiers provide the same benefit in voltage regulators; thus power supplies that employ *switching regulators* can be lighter and more compact than those using linear regulators. Switching regulators also offer greater versatility, as they can be designed to give output voltage that is higher than the unregulated voltage or even of different polarity. Disadvantages include added complexity in control circuitry, filtering required to remove switching noise from the output, and electromagnetic interference (emi) that can introduce noise into nearby equipment.

A switching regulator has two parts: a *converter* circuit to perform the switching and a feedback system to control the switching rate. There are three basic converter structures: *buck, boost,* and *buck-boost.* Figure 10.43a shows a buck-type converter. The total output voltage of the supply is $v_R = V_R + v_r$, where V_R is the regulated dc output voltage and v_r a small ac ripple. Transistor S_1 is periodically switched between cut-off and ohmic states by source v_C at frequency $f_s = 1/T_s$.

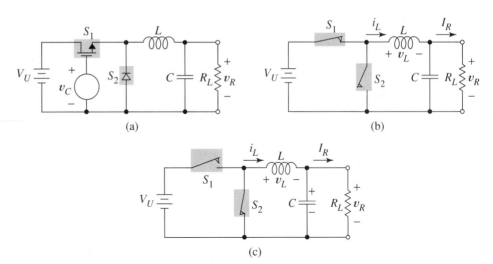

Figure 10.43 Buck-type switching converter: (a) schematic; (b) state 1 equivalent; (c) state 2 equivalent.

Figure 10.43b shows the transistor and diode replaced by ideal switches. When v_C causes S_1 to close, diode S_2 simultaneously opens, giving the circuit's first switch state, state 1. The voltage across L is then

$$v_L \approx V_U - V_R$$

This expression assumes C previously charged to V_R and ignores the small ripple v_r.

Let δ denote the control-voltage *duty cycle,* the fraction of the switching period for which S_1 is closed, where $0 < \delta < 1$. After δT_s seconds in state 1, v_C then opens S_1. Now v_L becomes negative, as the inductor tries to maintain constant current. This causes the diode to conduct as in Fig. 10.43c, the circuit's switch state 2. Now $v_L \approx -V_R$. At the end of the switching period, v_C again changes the state of S_1, and a new switching cycle begins.

Figure 10.44a shows the inductor voltage waveform v_L. Just as the average current in a capacitor must be zero, so also must the average voltage across an inductor. Thus

$$\frac{(V_U - V_R)\delta T_s + (-V_R)(1 - \delta)T_s}{T_s} = 0$$

which gives the important result

$$V_R = \delta V_U \qquad (10.47)$$

To obtain a regulated voltage V_R from an unregulated voltage V_U, v_C must provide the required duty cycle δ. More to the point, Eq. (10.47) is independent of load resistance R_L, suggesting perfect voltage regulation from this idealized circuit. To make V_R impervious to changes in V_U, that is, provide good line regulation, we need a feedback circuit to monitor the output continuously and adjust the duty cycle as required.

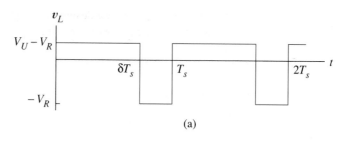

(a)

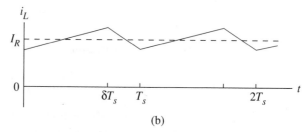

(b)

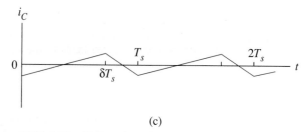

(c)

Figure 10.44 Key waveforms for buck-type switching converter.

Now we focus on the ripple, v_r. The inductor current is

$$i_L = \frac{1}{L}\int v_L(t)dt$$

therefore, the inductor current is that of Fig. 10.44b. The average inductor current is I_R, the dc output current of the power supply. Because the capacitor cannot carry dc current, it follows from KCL that the time-varying part of i_L must flow through C, giving Fig. 10.44c. Since

$$i_C = C\frac{dv_R}{dt}$$

there is a time-varying ripple voltage v_r across R_L. The designer makes this ripple small compared to V_R by selecting large C. This is further explored in Problem 10.59.

We still need to derive a design equation to guide us in selecting L and C. Figures 10.44a and b show that, between δT_s and T_s, $i_L(t)$ has constant slope that relates to $v_L(t)$ through

$$L\frac{di_L}{dt} = -V_R$$

Therefore, the peak-to-peak value, $i_{L,p\text{-}p}$, of the time-varying part of $i_L(t)$ gives the approximation

$$L\frac{\Delta i_L}{\Delta t} = L\frac{-i_{L,p\text{-}p}}{(1-\delta)T_s} = -V_R$$

Solving gives the expression

$$i_{L,p\text{-}p} = \frac{V_R}{L}(1-\delta)T_s = \frac{V_R(1-\delta)}{Lf_s} \tag{10.48}$$

To prevent the ideal diode from conducting during the second part of the switching cycle, we see from Fig. 10.44b that the design must satisfy the condition

$$I_R - \frac{1}{2}i_{L,p\text{-}p} \geq 0$$

Substituting Eq. (10.48) leads to the design equation

$$Lf_s \geq \frac{V_R(1-\delta)}{2I_R} \tag{10.49}$$

With regulated output voltage V_R specified, and δ previously selected using Eq. (10.48), this equation allows us to choose appropriate values for f_s and L. For Eq. (10.49) to apply under all loading conditions, we must choose f_s and L using the *minimum* value expected for I_R in Eq. (10.49), not rated or maximum value. Finally, to obtain effective filtering of the high-frequency switching components we select C to satisfy $1/(\omega_s C) >> \omega_s L$, or

$$C \geq \frac{10}{(2\pi f_s)^2 L} \tag{10.50}$$

If the design allows the diode to open during the last $(1-\delta)T_s$ second of the switch cycle, the circuit has a third state—both switches open. Designing the circuit so that the diode opens between δT_s and T_s leads to more complicated operation, but has the advantage of using a smaller inductor. Mitchell describes this operating mode in one of the references cited at the end of the chapter.

Figures 10.45a and b, respectively, show the boost and buck-boost converters. We make the same assumptions for these circuits as we did for the buck converter, namely,

1. The ripple components in the output can initially be ignored.
2. S_1 closes for the first δT_s second of every T_s second switching cycle.
3. S_2 automatically closes when S_1 is opened.

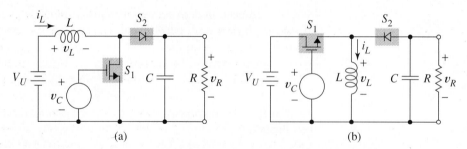

Figure 10.45   (a) Boost converter; (b) buck-boost converter.

For the boost converter, this leads to

$$v_{L,AVE} = \frac{V_U(\delta T_s) + (V_U - V_R)(1 - \delta)T_s}{T_s} = 0$$

and the design equation

$$V_R = \frac{1}{1 - \delta} V_U \tag{10.51}$$

Since $\delta < 1$, this converter *boosts* the unregulated voltage to a higher value at the regulated output.

For the buck-boost converter, similar analysis leads to

$$v_{L,AVE} = \frac{V_U(\delta T_s) + V_R(1 - \delta)T_s}{T_s} = 0$$

and the buck-boost design equation

$$V_R = \frac{-\delta}{1 - \delta} V_U \tag{10.52}$$

Because $0 < \delta < 1$, the polarities of V_R and V_U are *opposite*. The buck-boost name follows from noticing that $|V_R| > V_U$ for $\delta > 0.5$ and $|V_R| < V_U$ for $\delta < 0.5$.

Figure 10.46 is a transformer-coupled version of the buck-boost converter, called a *flyback converter*. When S_1 closes, V_U is applied directly to L just as in Fig. 10.45b.

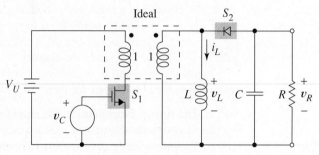

Figure 10.46   Flyback converter.

When S_1 opens, the transformer secondary current goes to zero and S_2 closes. We can produce V_U directly from the ac line voltage with a rectifier circuit, because we obtain electrical isolation from the converter transformer. Since the latter operates at the high switching frequency instead of line frequency, it is much lighter and more compact than the usual power transformer. A turns ratio of $1:n$ instead of $1:1$ changes V_U to nV_U in Eq. (10.52).

The complete switching power supply is closely related to the class D amplifier circuit of Fig. 10.36. The converter circuit, consisting of switches and filter, replace the class D power stage and filter. A dc zener reference voltage replaces the time-varying input signal, $v_I(t) + V_P$, in Fig. 10.36. A pulse width modulator converts the difference between desired output and filtered output into a signal that controls the switches.

10.10 ____
Summary

Because large amounts of power are involved in power amplifiers and voltage regulators, designers of these circuits must anticipate and provide for the power dissipation of transistors and resistors. Power resistors have simple maximum power ratings; transistors are described in terms of thermal resistances that relate power dissipation to internal junction temperatures. These thermal resistances allow the designer to provide a suitable thermal environment by adding either cooling or heat sinking, or both, as needed to maintain safe junction temperatures.

To deliver large amounts of output power, signal voltages must be large. This means instantaneous operating points traverse large ranges on the transfer characteristics of a power amplifier, making nonlinear distortion a potential problem. Harmonic distortion is readily quantified in a simple figure of merit called total harmonic distortion, usually expressed as a percentage of the desired output signal. Total harmonic distortion is used as a specification to guide power amplifier design and is readily determined during SPICE simulation. Feedback circuits like emitter followers are commonly used in power amplifiers to help keep distortion small.

The class A power amplifier produces output voltage having low distortion; however, its efficiency upper limit is only 25%. This leads to high transistor dissipation and high supply current requirements per watt of output power. Class A amplifiers also consume large amounts of standby power when there is no signal present.

The class B amplifier has no standby dissipation and, per watt of output power, gives lower power supply currents and transistor dissipation than class A because of its high efficiency—78%. However, the class B transfer characteristic has a dead zone that causes crossover distortion. To retain the benefits of high-efficiency class B operation and also achieve low harmonic distortion, designers sometimes use the class B design for the output stage of a feedback amplifier. Another way to reduce distortion is to add biasing to keep the output transistors turned on slightly when there is no signal, a technique called class AB operation. The biasing voltage used for this purpose must have a negative temperature coefficient to avoid thermal runaway. Power op amps, which usually include a class B or AB output stage, provide a simple way to provide high power to a load. Power and heat-sinking requirements of power op amps are computed by the same methods used for other class B amplifiers.

Still higher efficiencies are possible with class D amplifiers, in which the output transistors operate as switches that connect the power supplies to the load. This amplifying scheme requires some overhead in signal processing circuitry: a pulse-width modulator to convert the signal voltage to a pulse sequence, and a filter to remove high-frequency noise added by the modulation process. An overall feedback loop further improves class D amplifier performance.

Basic power supplies consist of half- or full-wave rectifiers followed by filters to reduce output ripple. More sophisticated applications additionally require a special circuit to reduce voltage regulation (drop off of output voltage with load current) and line regulation (changes in output voltage caused by undesired changes in supply voltage). Simple shunt regulators employ a zener diode of low internal resistance in parallel with the load to automatically compensate for changes in load current or in unregulated voltage. Series regulators employ a power transistor in series with the load to change load current whenever load voltage tends to change for any reason. In series regulators a zener reference provides the standard with which output voltage is compared.

Switching regulators use switches and lossless LC elements to regulate output voltage in an efficient fashion. These versatile regulating schemes can also provide output of different polarity or larger amplitude than the unregulated supply.

REFERENCES _____

1. APEX MICROTECH. *High Performance Amplifier Handbook.* Tucson, AZ: Apex Microtech, 1987.
2. CORDONNIER, C., R. MAIMOUNI, H. TRANDUC, P. ROSSEL, D. ALLAIN, M. NAPIERALSKA. "Spice Model for TMOS Power Mosfets," *Application Note AN1043/D.* Motorola, Inc., 1989.
3. GAUEN, K. "Designing with TMOS Power MOSFETs," *Application Note AN913.* Motorola, Inc. 1989.
4. MAURO, R. *Engineering Electronics: A Practical Approach.* Englewood Cliffs, NJ: Prentice Hall, 1989.
5. MITCHELL, D. M. *Switching Converter Analysis.* New York: McGraw-Hill, 1988.
6. NATIONAL SEMICONDUCTOR CORP. *Linear Databook 1.* 1989.
7. PAULY, D. "High Fidelity Switching Audio Amplifiers Using TMOS Power MOSFETs," *Application Note AN1042.* Motorola, Inc., 1989.
8. SCHILLING, D., and C. BELOVE. *Electronic Circuits, Discrete and Integrated.* New York: McGraw-Hill, 1989.
9. STREETMAN, B. *Solid State Electronic Devices.* Englewood Cliffs, NJ: Prentice-Hall, 1980.

PROBLEMS _____

SPECIAL DIRECTIONS FOR SPICE HOMEWORK PROBLEMS: *Do not hand in lengthy SPICE printout for homework.* Instead, abstract the useful information from the SPICE output file as in the SPICE examples. Include your SPICE code and a circuit diagram with nodes numbered to agree with the code. Cite relevant numerical values from the SPICE output file and discuss when appropriate. Make sketches of any relevant curves, and label appropriate points. Make small tables to present numerical data if useful for clarity.

Section 10.1

10.1 The transistor in Fig. P10.1 is forward active.
(a) Compute the transistor's power dissipation when $v_I = 0$. Assume base current is negligible and the signal is turned off.

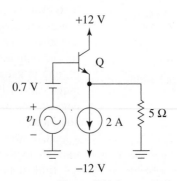

Figure P10.1

(b) If $\theta_{JA} = 15°C/W$, compute the junction temperature if the transistor is in an ambient environment of 30°C.

10.2 The transistor in Fig. P10.1, biased at (12 V, 2 A), is equipped with a heat sink. Thermal resistances are $\theta_{JC} = 2.1°C/W$, $\theta_{CS} = 0.9°C/W$, and $\theta_{SA} = 5°C/W$.
(a) Calculate the junction temperature when the ambient temperature is 5°C.
(b) What are the temperatures of the heat sink and the transistor case when ambient temperature is 12°C?

10.3 Two transistors, each with $\theta_{JC} = 1.4°C/W$, $\theta_{CS} = 0.3°C/W$, and dissipation of 12 W, share the same heat sink. If the junction temperatures are 200°C and the ambient temperature is 10°C, use a thermal equivalent circuit to find the thermal resistance of the heat sink.

10.4 A power transistor with maximum junction temperature of 180°C must dissipate 80 W. If $\theta_{JC} = 1.2°C/W$ and $\theta_{CS} = 0.3°C/W$, determine the thermal resistance of the heat sink required
(a) if the sink operates in a 25°C environment.
(b) if the sink operates in a cooled atmosphere of $-4°C$.

10.5 An amplifier design requires a power transistor to dissipate 19 W in an ambient temperature of 35°C.
(a) An available transistor has $\theta_{JA} = 15°C/W$ and $T_{J,max} = 195°C$. Show that this transistor would overheat.
(b) Investigate the feasibility of using two parallel transistors instead.

10.6 A power transistor has the following specifications: $I_{D,max} = 20$ A, $BV_{DSO} = 40$ V, $\theta_{JA} = 8°C/W$, and $T_{JMAX} = 200°C$. On the same I_D versus V_{DS} coordinate system, sketch the safe operating region for 25 and 40°C ambient temperatures.

10.7 Draw a sketch of the safe operating region for a MOSFET, and locate three different Q-points *on* the dissipation hyperbola. Draw a tangent load line through each. Comment on the relative merits of the Q-points relative to signal amplitude if symmetric signals cause the instantaneous operating point to move along the load line in both directions.

Section 10.2

10.8 The double-diffused power transistor in Fig. P10.8 is described by

$$i_D = 0.05(v_{GS} - 2)$$

(a) Find the transistor's power dissipation.
(b) The junction temperature must not exceed 150°C. Will the transistor overheat if ambient temperature is 50°C and $\theta_{JA} = 200°C/W$? Explain.

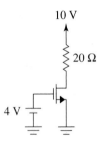

Figure P10.8

(c) Name two ways to reduce the junction temperature of the transistor without changing the operating point.

10.9 (a) Use SPICE dc analysis to verify that the model of Fig. 10.6 has a transfer characteristic like Fig. 10.5b.
(b) Sketch the i_D versus v_{DS} (output) characteristics of the model directly from the circuit diagram of Fig. 10.6. Your curves should differ from Fig. 10.5c in one important way.

10.10 Draw a three-terminal equivalent circuit that describes the transistor of Fig. 10.5 in ohmic operation.

10.11 Sketch transfer and output characteristics for a p-channel DMOS transistor. Then devise a SPICE equivalent circuit analogous to the one in Fig. 10.6 for the p-channel device.

10.12 Draw a flow chart that summarizes the operating regions for the power transistor of Fig. 10.5 as Fig. 5.6a does for the square-law MOSFET. Label your flow chart. with appropriate inequalities.

10.13 The double-diffused transistor in Fig. P10.13 has $V_t = 2$ V, $g_M = 0.3$ A/V, and $R_{DS} = 0.5$ Ω. Use your understanding of transistor models to find the values of v_O and v_I at the point when the transistor just changes to ohmic operation. *Hint:* Read Problem 10.10.

10.14 Assume the DMOS transistor in Fig. P10.14 is active. Its parameters are $g_M = 3$ A/V, $V_t = 2$ V.

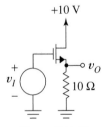

Figure P10.13

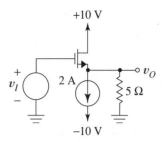

Figure P10.14

(a) Write the equation that relates v_O to v_I when the transistor is active.
(b) Find the value of v_I that just cuts off the transistor.

Section 10.3

10.15 Re-sketch (and label numerical values) the transfer characteristic of Fig. 10.7e for the case when the battery labeled V_{BE} in Fig. 10.7a is missing, that is, zero volts. Assume that I_{EE} is a *basic* BJT mirror circuit. Use $V_{CC} = 15$ V, $V_{CE,sat} = 0.2$ V, and $V_{BE} = 0.7$ V.

10.16 (a) Design a class A amplifier that will deliver 200 mW to a 9 Ω load. Assume $V_{CE,sat} = 0$ V.
(b) Compute the power rating of the output transistor if
  (i) the signal is always present and is very fast compared to the thermal time constants.
  (ii) the signal is always present but is sometimes very slow compared to thermal time constants.
  (iii) the signal is sometimes absent.
(c) Find the minimum allowable breakdown voltage for the current-source transistor.
(d) Find the power and dc current rating required for each power supply.

10.17 Design a BJT current source for the output stage specified in Problem 10.16 using a basic current-source circuit.
(a.) Draw the complete circuit diagram of the power amplifier and current source.
(b.) Find the power rating required of the current-source transistor.

10.18 (a) Draw the circuit schematic for a MOSFET version of Fig. 10.7a. Use an external biasing battery of V_{GS} volts instead of V_{BE}.
(b) Draw the large-signal equivalent for a double-diffused MOSFET version of Fig. 10.7b.
(c) Find an expression that V_{GS} of part (a) must satisfy if its

role is comparable to V_{BE} in Fig. 10.7a. This expression should involve the current-source current and DMOS parameters.
(d) From your equivalent circuit, derive the transfer function equation that relates v_O to v_I.
(e) Sketch the transistor output characteristic and load line for the MOSFET circuit and indicate the limits of linear operation for the transfer function.
(f) Sketch waveforms corresponding to those of Fig. 10.8 for the MOSFET amplifier.

10.19 (a) Draw the large-signal equivalent of the circuit of Fig. 10.7a if the transistor is a square-law MOSFET. Replace the V_{BE} source with a battery of V_{GS} volts.
(b) From your equivalent circuit, derive an equation that relates v_O to v_I when the transistor is active.
(c) Write an equation for v_I at the point where the transistor enters the ohmic region.
(d) This circuit has a surprisingly linear transfer characteristic considering the transistor is a square law device. Give an intuitive justification.

10.20 Make a SPICE model of the amplifier designed in Example 10.2. Include a signal source resistance of 100 Ω.
(a) In a dc analysis, plot the transfer characteristic for $-12 \le v_I \le +12$ V. Examine your curve for things such as gain, offsets, and saturation; and write a short paragraph to enumerate and explain your observations.
(b) Use a transient analysis to find the output waveform for a 1000 Hz sinusoidal input signal of 8.94 V peak value. Reconcile the waveform with the transfer characteristic of part (a).
(c) Using part (b) results, reduce the amplitude of the input signal to the largest value consistent with a fairly undistorted output waveform. Estimate the average ac load power for this input amplitude.

10.21 In Fig. P10.21, the transistor is a double-diffused MOSFET with $g_M = 0.5$ A/V, $V_t = 2$ V, and $R_{DS} = 0.1$ Ω.
(a) If $v_O = 0$ and the transistor active, find v_I.

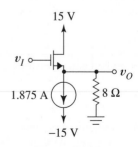

Figure P10.21

(b) Draw a large-signal equivalent circuit assuming the transistor is active.
(c) Use your equivalent circuit to find an equation that expresses v_O as a function of v_I.
(d) Estimate the most positive value of v_O that this amplifier can develop without distortion.

Section 10.4

10.22 A spectrum analyzer measurement of the output of a power amplifier with sinusoidal excitation gives these relative amplitudes:

> fundamental : 1.0000   third harmonic : 0.0050
> fifth harmonic : 0.0001

Compute the percent harmonic distortion assuming all other harmonics are negligible.

10.23 Figure P10.23 models an amplifier that saturates when v_O reaches approximately ± 5.7 V.
(a) Use SPICE to find its transfer characteristic for -10 V $\le v_S \le +10$ V.
(b) Find the %THD when v_S is a 100 Hz sine wave with peak value of 20 mV, 30 mV, and 50 mV.

10.24 In the large-signal data of Fig. 10.13b the third harmonic ($n = 3$) distortion component is by far the largest.
(a) Compute the ratio of the rms third harmonic distortion to the total rms distortion.
(b) Suppose the third harmonic distortion could be completely removed by a filter without affecting the fundamental or any other distortion amplitudes. What would be the %THD after filtering?

10.25 Use SPICE to compute the percent total harmonic distortion for the amplifier of Example 10.3 when the excitation is $v_I(t) = 4.3 + 0.2 \sin(2\pi 10^3 t)$.

Section 10.5

10.26 Simulate the amplifier of Fig. 10.14a with SPICE. Use $R_L = 10\ \Omega$, $R_I = 0.1\ \Omega$, $V_{CC} = 12$ V, and default transistors.
(a) Plot the transfer characteristic v_O versus v_I.
(b) Plot the output waveform for $v_I = 8 \sin 377t$.
(c) Repeat parts (a) and (b) after making the following changes. Make $\beta = 30$ for the npn transistor, 10 for the pnp transistor, and $R_I = 1$ kΩ. (This demonstrates different input loading for the two emitter followers when source resistance is high and transistors mismatched, a result discussed in connection with Fig. 10.25.)

10.27 Find an equation that relates v_I and v_O for the class B DMOS power amplifier at the point where the n-channel device changes to ohmic operation.

10.28 Use SPICE to repeat the simulation of Example 10.4. From output data, determine the slope of the linear region of the transfer characteristic. Then repeat the simulation with body-effect coefficient of GAMMA = 0.4 and determine the new transfer characteristic's slope. Can you detect a difference in the transfer characteristic or output waveform attributable to body effect?

10.29 Rework Example 10.4 using the transistor model of Fig. 10.6 instead of the built-in SPICE MOSFET model. This requires you to construct a SPICE model for the complementary p-channel device.

10.30 In Fig. 10.14a, $R_L = 3\ \Omega$, load power is 40 watts when v_O is a sinusoid.
(a) What is the peak value of the current that flows through each supply?
(b) If Q_1 has $\beta = 100$ and $R_i = 500\ \Omega$, find the value of v_I when it reaches its positive peak.
(c) If Q_2 has $\beta = 10$ and $R_i = 500\ \Omega$, find the value of v_I when it reaches its negative peak.
(d) Use your results from parts (b) and (c) to sketch the input waveform v_I it takes to produce the sinusoidal output voltage we assume. Does the circuit have this problem if $R_i = 0$? Explain.

Section 10.6

10.31 (a) Find the voltage and current rating of each power supply in Fig. P10.31 if the amplifier must deliver a maximum of 25 W of power to the load resistor. Assume sinusoidal excitation.
(b) Find the power rating required for each output transistor, assuming the signal is faster than the thermal time constants. The circuit must operate with less than maximum output power.

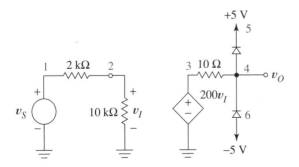

Figure P10.23

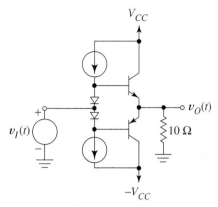

Figure P10.31

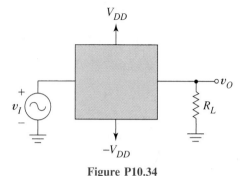

Figure P10.34

(c) What thermal resistance is required to keep junction temperatures below 180°C when the transistors operate without heat sinks in an ambient temperature of 14°C?

10.32 An MOS class B amplifier uses ±16 V supplies. The output transistors are double-diffused MOSFETS with $g_M = 1$ A/V, $R_{DS} = 0.6$ Ω, and $|V_t| = 2$V. $R_L = 5$ Ω.
(a) Sketch the transfer characteristic of the amplifier, including expected saturation of the curve. Find the amplitude of the input sine wave that drives the output to the edge of saturation.
(b) Convert your amplifier to class AB by adding two diode-connected lateral (ordinary) NMOS transistors for which $k = 10$ mA/V^2 and $V_t = 1$ V. Bias these with dc currents just sufficient to hold the *output* transistors at cut-in.

10.33 In the class B circuit of Fig. 10.14a, $R_L = 5$ Ω and $R_i = 0$. When a sinusoidal input voltage is applied, 25 watts of output power are delivered to R_L. Ignore crossover distortion.
(a) Find the peak value of v_O.
(b) What value is required for V_{CC} if the circuit operates near its maximum efficiency of 78.6%?
(c) What power rating is required for the individual power transistors? Allow for the possibility of lower peak value than you calculated in part (a).
(d) Find the peak value of the input current if $\beta = 80$.

10.34 The class AB power amplifier of Fig. P10.34 is *now* operating at 70% efficiency *as it delivers* 140 watts to R_L.
(a) How much average power is each supply delivering to the circuit?
(b) What is the average power dissipation of *each* of the two output transistors as the circuit delivers the 140 watts?
(c) If the ambient temperature is 10°C, find the junction temperature of the output transistors. Individual thermal

resistance values are $\Theta_{JC} = 4.0$°C/W, $\Theta_{CS} = 0.2$°C/W, and $\Theta_{JC} = 1.8$°C/W.
(d) What is the temperature of the transistor case?

10.35 Design a bipolar class B output stage using two 9 V batteries to drive an 8 Ω speaker. Ignore crossover distortion.
(a) What is the average output power when the input is a sinusoid of maximum amplitude?
(b) What power rating is required of each output transistor?
(c) Find the average power and peak instantaneous power required of the batteries.
(d) For a 7 V peak sinusoidal output, compute the average output power and the efficiency of the circuit.
(e) Under estimated conditions of usage, each battery can deliver 12 ampere-hours of charge. Estimate the time between battery replacements for continuous operation at maximum efficiency.

10.36 A bipolar class B amplifier is to deliver $P_{max} = 40$ watts into a resistive load when the sinusoidal signal is of maximum amplitude.
(a) Find the power dissipation rating of each power transistor if the signal does not always have maximum amplitude.
(b) In Fig. P10.36, find the minimum acceptable value for $I_{C,max}$ if $R_L = 5$ Ω.
Hint: Visualize how the circuit works.
(c) In Fig. P10.36, find the minimum allowable value of breakdown voltage, BV_{CEO}.
Hint: Visualize how the circuit works.
(d) Name one advantage of using a Darlington pair to replace each output transistor.

10.37 Make a SPICE model for the amplifier completed in Example 10.7.
(a) Find the %THD in the output waveform when the input is a 1000 Hz sine wave of 8.9 V amplitude and 0.1 Ω output resistance.

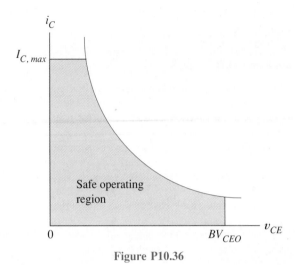

Figure P10.36

with the maximum output current if protection circuitry is not provided.
(b) Use SPICE dc analysis to plot collector currents of Q_1 and Q_3 as functions of v_I for -35 V $\leq v_I \leq +35$ V.
(c) Repeat part (b) after replacing R_L by a short circuit. Explain your results in terms of the operating principles of the circuit.
(d) Change R_{P3} and R_{P4} so that output transistors are nominally protected for 0.6 A. Make another SPICE run to check the new design with $R_L = 10$ Ω.

10.41 With the help of Fig. 10.29c, design a foldover protection circuit like Fig. 10.29b for an amplifier with $V_{CC} = 15$ V and short-circuit current $i_{o,max} = 5$ A.
(a) Evaluate Eq. (10.34) for the short-circuit condition to obtain the first design equation.
(b) Use the slope of the current limit line in Fig. 10.29c to obtain a second design equation.
(c) Choose one resistor value and then solve for the others using your design equations. Any of the possible correct solutions will do.

10.42 Even when $Z_{PS} = 0$ in Fig. 10.30a, power supply hum can enter an output stage as noise. Negative feedback can reduce this hum in the same way it reduces distortion. Suppose the *output* waveform of a two-stage power amplifier with voltage gain of 10 has two components, 2 volts of signal and 2 volts of power supply hum.
(a) How large is the input signal? (This is *not* a trick question.) Negative feedback with $\beta = 1$ is now introduced.
(b) How large must the input signal be (after adding feedback) if the output signal of the feedback amplifier is still 2 volts?
(c) What is the amplitude of the output hum of the feedback circuit?
(d) Draw an op amp–based preamp that will make up for the loss in gain introduced by adding feedback.

10.43 Figure P10.43 shows Fig. 10.20a, modified to use a V_{GS} multiplier. Assume that M_1 and M_2 each have $|V_t| = 2$ V. Select V_B so that each transistor is biased 5% beyond its threshold voltage. If the parameters of M_3 are $k = 2$ mA/V^2 and $V_t = 1$ V, use the Problem 6.85 design procedure to find I_{BB}, R_1, and R_2.

10.44 Work Problems 10.30a, b, and 10.30c as given. Then reanalyze after replacing each transistor with a Darlington pair. Each transistor of each pair has $\beta = 10$.

(b) Remove the current sources and diodes making the circuit a simple class B amplifier. Find %THD using the same signal.

10.38 In Fig. 10.23, $V_{CC} = 10$ V and $R_L = 2$ Ω. The output voltage of this class B amplifier is a square wave that alternately drives each output transistor to the edge of saturation.
(a) Sketch output voltage and the current that flows in each power supply.
(b) Compute the average load power.
(c) Compute the average input power delivered to the circuit by the two supplies together.
(d) Compute average dissipation of each output transistor.
(e) Compute the efficiency of the circuit.

10.39 Make a SPICE model of the class AB amplifier designed in Examples 10.6 and 10.7. Drive the amplifier with an ac signal source with 5 Ω internal resistance.
(a) Examine the quiescent biasing conditions and compare them with the values assumed in the design. Write a sentence or two summarizing the results.
(b) Plot the transfer characteristics for -9.0 V $\leq v_I \leq 9.0$ V. Examine the curve for linearity, offset, and permissible signal amplitude. Summarize your observations and how they relate to the theory.
(c) Repeat the simulation of part (b) after reducing both biasing currents to $I_{BB} = 5$ mA. Describe and explain the result.

10.40 In Fig. 10.28, $V_{CC} = 35$ V, $R_L = 10$ Ω, $R_{P3} = R_{P4} = 0.143$ Ω, $V_B = 1.04$ V, and $R_i = 4$ Ω. For all transistors, $\beta = 80$. For Q_1 and Q_2, $I_S = 10^{-14}$ A. For Q_3 and Q_4, 10^{-13} A.
(a) If the protection transistors have cut-in voltages of 0.5 V, estimate the maximum output current allowed. Compare this

Section 10.7

10.45 The identical power op amps in Fig. P10.45 have class B output stages. Maximum load power is to be 75 W.
(a) What voltage is required of the power supplies that supply the op amps?

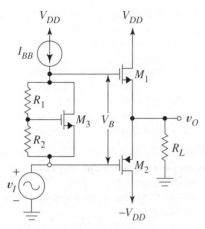

Figure P10.43

(b) If each power op amp is in a separate case with θ_{JC} = 2.3°C/W and if θ_{CS} = 0.2°C/W, find the thermal resistance required of the heat sink θ_{SA} for each op amp to keep the junction temperature at 180°C or less. (The signal will always be much faster than the thermal time constants.)

10.46 Repeat Problem 10.45 but use four parallel op amps.

10.47 Explore the average power dissipation ratings required for the input and feedback resistors in the bridge amplifier of Fig. 10.32 if the output voltage is a sinusoid of maximum amplitude. The power supply voltage used for the op amps is V_{CC}. How can the designer keep the dissipation requirements of these resistors small?

10.48 Using op amps, design a bridge amplifier with overall voltage gain of 4 that will deliver 115 W to an 8 Ω load. No heat sink is to be used.
(a) Provide a circuit diagram with resistor values.
(b) Find the thermal resistance θ_{JA} required of each op amp if maximum junction temperature is 190°C.
(c) Draw the diagram of a modified circuit that uses two parallel op amps in place of each op amp in your original circuit, and

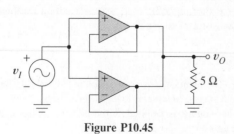

Figure P10.45

calculate the value of θ_{JA} required in this circuit. Assume individual heat sinks for op amps.

Section 10.8

10.49 A pulse-width modulator operates as follows. "The input to an integrator (with a resistor between op amp output and capacitor) is a constant $-V_M$ volts. A bipolar transistor switch, driven by a train of short pulses, shorts the integrator output. An op amp, operating as an open-loop device, is the comparator."
(a) Find an upper limit for the period, T, of the pulse train in terms of V_M, R, C, and the saturation voltage, V_{SAT}, of the integrator op amp. Assume the pulses have negligible width.
(b) What determines the peak-to-peak amplitude of the PWM output?
(c) Is the slew rate of the comparator important if the PWM circuit is part of a class D amplifier? Explain.

10.50 In Fig. 10.34a, M_1 and M_2 are complementary MOS-FETS with parameters $k = 0.06$ A/V², $|V_t| = 1.2$ V, and $R_L = 5$ Ω. The output stage uses $V_{DD} = \pm 15$ V. Find the minimum voltage, V_M, of the square wave at v_X for 90% efficiency.

10.51 The class D amplifier of Fig. 10.34 is to amplify a signal with spectral content in the range $\omega \leq \omega_m$. The modulator switching waveform has fundamental frequency ω_s. To pass the signal, filter capacitor C must have reactance much greater than R_L at ω_m. To eliminate switching noise, the same reactance must be much less than R_L at $\omega_s - 3\omega_m$.
(a) Show that these conditions and usual engineering approximations imply the design equation.

$$\frac{10}{\omega_s - 3\omega_m} \leq R_L C \leq \frac{1}{10\omega_m}$$

(b) Use the same kind of reasoning on the role of the inductor to derive

$$\frac{10}{\omega_s - 3\omega_m} \leq \frac{L}{R_L} \leq \frac{1}{10\omega_m}$$

Section 10.9

10.52 A shunt regulator like that of Fig. 10.40a is to deliver 5 volts at 1.2 amperes from an unregulated 9 V source with voltage regulation of 2% and line regulation of 5%.
(a) Find specifications for the zener diode (voltage, maximum current, and internal resistance) and the value of the series resistor.
(b) Find the minimum power rating for R_S.

10.53 The specifications in Problem 10.52 result in a large zener current. Replace the line regulation specification with the requirement that the maximum zener current be 600 mA. Work the problem for this maximum zener current, and then solve for the line regulation.

10.54 Find expressions for the voltage regulation and line regulation for the series regulator of Fig. 10.41.
Hints: Draw the Thevenin equivalent seen by R_L. Use ideal feedback theory. For line regulation, use small-signal analysis and include the internal resistances of the current source and zener diode. The op amp has infinite gain.

10.55 Design a power supply like Fig. 10.42 that gives 800 mA of output current at $V_R = 5$ V. The JFET has $\beta = 2.08 \times 10^{-3}$ A/V^2 and $V_P = -1.2$ V. A replacement zener provides 3.2 V when biased at 0.8 mA. Compute the dissipation required of the output transistor.

10.56 (a) Design a regulated power supply that operates from 120 V, 60 Hz line voltage and delivers 1 A from a regulated 9 V output. Use a full-wave rectifier with capacitor filter and a series regulator like that of Fig. 10.41a. The zener voltage is 6.7 V at 0.9 mA.
(b) Find the power dissipation required of the transistor, the diode, the current source, and the resistors at full load.
(c) Simulate your power supply using SPICE, including current source resistance of 100 kΩ for I_{BB}. Use your judgment to study the supply (by simulation) and briefly report on the results you thought were important and/or interesting. To stimulate your imagination, how about output ripple voltage, sensitivities, variations in zener current with load, diode currents in the rectifier, and transformer input current?

10.57 Make a SPICE model of the supply designed in Example 10.13. The JFET has an Early voltage of 50 V.
(a) Use SPICE data to plot a voltage regulation curve, that is, V_R versus R_L, for $4 \ \Omega \le R_L \le 100 \ \Omega$.
(b) Use SPICE to compute the sensitivities of the output voltage to changes in unregulated voltage, R_B and R_1.
(c) Add an ac voltage source in series with the 20 V source to simulate a 1% ripple at 120 Hz. Use SPICE ac analysis to find the percent ripple that appears at the output when $R_L = 4 \ \Omega$.

10.58 A switching converter like that of Fig. 10.43a is to produce a regulated output of 10 V from an unregulated 17 V supply.
(a) Find values for duty cycle δ and capacitance C. Select L and f_s so that the inductor current is nonnegative for output current as low as 1 A. Any of the possible correct designs are acceptable.
(b) Sketch the i_L waveform for $I_R = 1$ A and for 10 A, labeling all critical amplitudes in your sketch.
(c) Redo the i_L sketch for $I_R = 10$ A if the switching frequency f_s is doubled relative to your design value.

10.59 The capacitor current in Fig. 10.44c approximately equals the time-varying component of the inductor current.
(a) Recognizing that

$$\frac{dv_R}{dt} = \frac{1}{C} i_C$$

sketch the ripple component of v_R.
(b) Derive an expression for the ripple v_r in the output voltage that allows one to select C based on a specification for fractional ripple v_r / V_R.

10.60 Sketch the inductor current waveform of Fig. 10.44 for the case where i_L reaches zero at some time t_1, where $\delta T < t_1 < T$. Also sketch the diode voltage waveform.

10.61 For the boost-type converter of Fig. 10.45a, sketch $v_L(t)$ and $i_L(t)$ assuming that $V_U < V_R$. Then find expressions for $i_{L,p-p}$ and Lf_s that correspond to Eqs. (10.48) and (10.49). In finding the expression for Lf_s, assume the circuit has only two switch states as in the buck-circuit analysis.

10.62 For the buck-boost-type converter of Fig. 10.45b, sketch $v_L(t)$ and $i_L(t)$ assuming that $V_U > V_R$. Then find expressions for $i_{L,p-p}$ and Lf_s that correspond to Eqs. (10.48) and (10.49). Assume the circuit has only two switch states as we assumed in the buck-circuit analysis.

10.63 Figure P10.63 shows a current converter, the dual of the buck voltage converter.
(a) Describe in words how the converter works if v_C closes S_1 for the first δT_s seconds of each cycle and opens it during the remainder of the cycle.
Hint: The capacitor current is analogous to the inductor voltage in the dual circuit.
(b) Sketch $i_C(t)$. From your sketch derive an equation that relates I_R, the dc component of i_R, to I_U.
(c) Find an expression for the peak-to-peak capacitor voltage.
(d) Find the inequality that must be satisfied for two-state operation of the circuit.

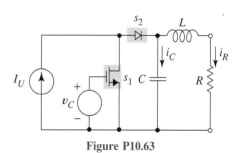

Figure P10.63

CHAPTER
11

Analog Integrated Circuits

Integrated circuits fit into two broad application categories, digital and nondigital. Nondigital or *analog* circuits are often called *linear ICs,* even though many involve nonlinear operation. This chapter introduces some well-known analog circuits and uses the analysis techniques of previous chapters to describe them. We investigate digital circuits in Chapters 13 and 14.

11.1
Operational Transconductance Amplifier

The *operational transconductance amplifier* (OTA) is an off-the-shelf IC that performs the VCCS function suggested by its name. A differential input facilitates its use in negative feedback configurations, many reminiscent of op amp circuits. The OTA's unique feature, a transconductance controlled by an *external* bias current, makes the transconductance amplifier useful in many applications, linear and nonlinear, that require variable gain.

In the OTA schematic of Fig. 11.1a, diode symbols denote diode-connected transistors, a common convention in IC diagrams. Thus each dashed box contains a Wilson mirror like Fig. 6.27a or its pnp equivalent. Also, the pnp output transistors in the mirror circuits represent Darlington pairs, equivalent transistors of high current gain.

dc Analysis. Figure 11.1b, which shows OTA biasing for dual 15 V supplies, assumes active-state operation of all transistors. An external resistor, R_B, establishes the desired current I_X in mirror Q_3, and Wilson circuits mirror bias currents throughout the circuit. All node voltages are fixed by the supplies and base-emitter drops except for the voltage at the output node, which depends upon the external load circuit.

Increasing V_I eventually saturates Q_1 and Q_2. This takes

$$v_{CE1} = v_{CE2} = 13.6 - (V_I - 0.7) = 0.2 \text{ V}$$

giving an upper common-mode input limit of $V_{I,max} = 14.1$ V. For negative common-mode input, current source Q_3 saturates when

$$v_{CE3} = (V_I - 0.7) - (-15) = 0.2 \text{ V}$$

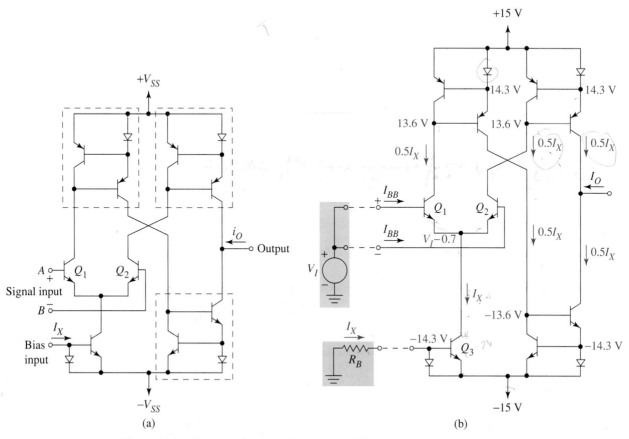

Figure 11.1 Operational transconductance amplifier: (a) schematic; (b) biasing.

This gives $V_{I,min} = -14.1$ V. The OTA data sheet specifies that the circuit operates for supply voltages from ± 2 V to ± 18 V. A knowledgeable user can easily redo the dc analysis for those supply voltages required by a particular application.

The two-port arrangement of Fig. 11.2a suggests that the Wilson mirror is a current-shunt, feedback circuit. Mirror Q_5–Q_6 senses $i_x \approx i_{out}$ and creates a proportional feedback current, i_f, that subtracts from i_{in} at a summing node. Current-shunt feedback, of course, gives a device of low input resistance and high output resistance, ideal attributes for a circuit that must function as a CCCS for both bias and signals as it does in the OTA. Figure 11.2b is an approximate small-signal equivalent that contributes little error to the small-signal OTA analysis that follows.

Difference-Mode Signal. Figure 11.3a shows how the OTA responds to small-signal, difference-mode excitation. Collector currents are replicated in the output circuit with the relative phases indicated on the diagram. The OTA output resistance (parallel equivalent of two Wilson mirror output resistances) is so much larger than any external load

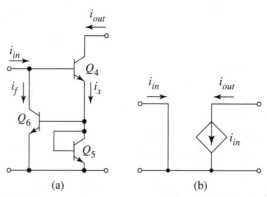

Figure 11.2 Wilson current mirror: (a) as a current-shunt feedback circuit; (b) closed-loop equivalent.

resistance that the overall circuit is best visualized as Fig. 11.3b. The input resistance, from Q_1 and Q_2, is

$$R_{in} = 2r_\pi = \frac{2\beta}{g_m} = \frac{4\beta V_T}{I_X} \tag{11.1}$$

To find GM

$$i_o = i_{c1} - i_{c2} = \left(g_m \frac{v_d}{2}\right) - \left(g_m \frac{-v_d}{2}\right) = 2\left(g_m \frac{v_d}{2}\right) = GM v_d \tag{11.2}$$

Therefore, in Fig. 11.3b,

$$GM = g_m = \frac{I_X}{2V_T} \tag{11.3}$$

In the OTA schematic representation of Fig. 11.3c, the linked-circle current-source symbol suggests a transconductance device, and explicitly shows the input current that controls GM. The OTA data sheet states that the circuit operates over the three-decade range

$$1 \ \mu\text{A} < I_X < 1000 \ \mu\text{A}$$

For this range, Eq. (11.3) predicts

$$20 \times 10^{-6} \ \text{A/V} < GM < 20 \times 10^{-3} \ \text{A/V}$$

and, using $\beta = 150$, Eq. (11.1) gives

$$15 \ \text{M}\Omega \le R_{in} \le 15 \ \text{k}\Omega$$

Although R_{in} is not high for large I_X, external series feedback around the OTA can give application circuits of high input resistance.

Common-Mode Gain. Figure 11.4 shows the OTA response to a common-mode signal. Because $i_{c1} = i_{c2}$, $i_o = 0$, thus nonzero common-mode gain is a second-order effect. Common-mode rejection ratios of 110 dB are typical for OTAs.

The following example illustrates using the OTA as a variable-gain amplifier.

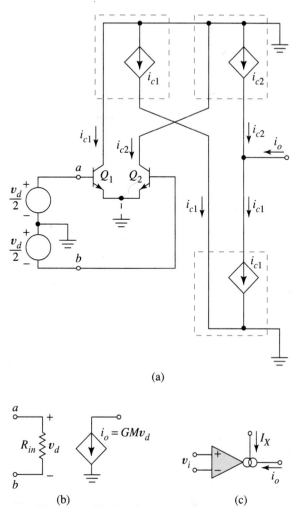

(a)

(b)   (c)

Figure 11.3 Operational transconductance amplifier: (a) simplified equivalent circuit; (b) final conceptual OTA model; (c) schematic symbol showing the circuit's gain control input.

EXAMPLE 11.1 In Fig. 11.5a, variable resistor R_B controls GM and therefore voltage gain, and the follower gives low output resistance. Find the voltage gain and required R_B values for I_X of 1 μA and 1000 μA. The OTA uses ±5 V supplies.

Solution. From the small-signal equivalent of Fig. 11.3b and Eq. (11.3),

$$\frac{v_o}{v_i} = -GM\,R_L = \frac{-I_X}{2(0.025)}(50\ \text{k})$$

Thus gain varies from 1 to 1000.

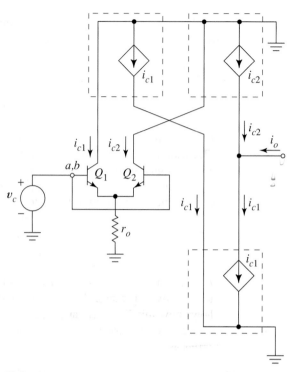

Figure 11.4 Common-mode operation of the transconductance amplifier.

From Figs. 11.1b and 11.5a we see that for ±5 V supplies,

$$I_X = \frac{5 - 0.7}{R_B}$$

Thus R_B must vary from 4.3 MΩ to 4.3 kΩ for I_X to vary between 1 and 1000 μA. ❑

Exercise 11.1 The signal source of Fig. 11.5b is to provide signal input to the amplifier of Example 11.1. Find the value of R_B required to make the input resistance of the OTA

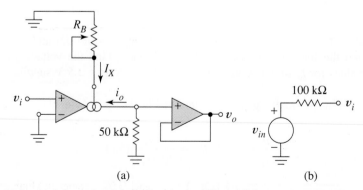

(a)   (b)

Figure 11.5 (a) Schematic of variable-gain amplifier; (b) signal source for Exercise 11.1.

equal 100 kΩ if β = 150 for the OTA transistor. Find the peak amplitude of the amplifier output signal for this R_B if the peak amplitude of v_{in} is 20 mV.

Ans. 28.7 kΩ, 1.5 V.

A more esoteric application of the OTA is the voltage-controlled filter of Problem 11.3.

Input Limits for Large Signals. Some applications require intentional operation of the OTA with large input signals. To describe these circuits and better understand the limits of small-signal theory, we now examine large-signal OTA operation.

The development that led to Eqs. (7.80) and (7.81) showed that the collector current of an emitter-coupled input transistor like Q_1 is

$$i_{C1} = \frac{I_X}{1 + e^{-v_D/V_T}} \tag{11.4}$$

where the input difference voltage v_D is not necessarily small. Because this equation assumes active operation, however, it tacitly implies positive base current. Figure 11.6a shows how this limits the amplitude of $v_D(t)$.

Because Wilson mirrors work for large as well as small signals, the OTA output current is

$$i_O = i_{C1} - i_{C2} = i_{C1} - (I_X - i_{C1}) = 2i_{C1} - I_X$$

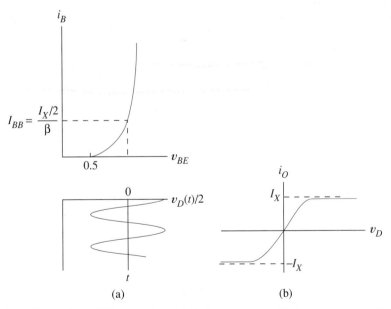

(a)   (b)

Figure 11.6 Large-signal OTA operation: (a) limit on input amplitude imposed by active operation; (b) large-signal transfer characteristic.

Substituting i_{C1} from Eq. (11.4) gives

$$i_O = I_X\left(\frac{2}{1 + e^{-v_D/V_T}} - 1\right) = I_X\left(\frac{1 - e^{-v_D/V_T}}{1 + e^{-v_D/V_T}}\right)\frac{e^{+v_D/2V_T}}{e^{+v_D/2V_T}}$$

or

$$i_O = I_X\left(\frac{e^{+v_D/2V_T} - e^{-v_D/2V_T}}{e^{+v_D/2V_T} + e^{-v_D/2V_T}}\right) = I_X \tanh(v_D/2V_T) \qquad (11.5)$$

which looks like Fig. 11.6b.

In linear, open-loop applications, v_D must be small enough that the hyperbolic tangent in Eq. (11.5) is well approximated by the first term in its power series. This approximation, which gives the small-signal results of Eqs. (11.2) and (11.3), generally requires $v_D \leq 40$ mV or so. In linear applications using negative feedback, larger v_D is possible because feedback reduces the distortion. In some open-loop applications such as the Schmitt trigger circuit, the OTA operates well outside the linear region, where output current takes on one of the saturation values $\pm I_X$. The next example illustrates large-signal operation of a transconductance amplifier.

EXAMPLE 11.2 (a) Find the hysteresis curve for the OTA Schmitt trigger of Fig. 11.7a when supply voltages are ± 12 V. Assume infinite input resistance for the OTA.
(b) Find hysteresis width for $R_B = 230$ kΩ and $R = 5$ kΩ.
(c) Check to make sure the output transistors do not saturate for these values.

Solution. (a) We expect steady-state operation on the transfer curve of Fig. 11.6b at either I_X or $-I_X$. To get started, assume v_I is so large and positive that $v_D \gg 0$. Then $i_O = I_X$, and the voltage at the inverting input node in Fig. 11.7a is

$$v_O = -RI_X = -V_S$$

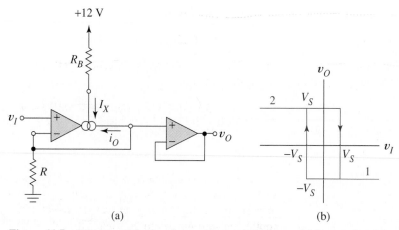

(a)   (b)

Figure 11.7 OTA as a Schmitt trigger: (a) circuit diagram; (b) hysteresis curve.

The assumption, $v_D > 0$, remains true for all v_I such that

$$v_D = v_I - (-RI_X) > 0$$

that is, for $v_I > -RI_X = -V_S$. We conclude that we have described operation in region 1, Fig. 11.7b.

Now assume that v_I is so negative that $v_D << 0$. Then $v_O = RI_X = V_S$, which holds for

$$v_D = v_I - (RI_X) < 0$$

that is, for $v_I < RI_X = V_S$. This gives region 2 in Fig.11.6b.

(b) Now we must relate V_S to specific design values. Figure 11.1a shows the biasing details. The bias current for the OTA is

$$I_X = \frac{12 - 0.7 - (-12)}{R_B} = \frac{23.3}{R_B} \tag{11.6}$$

Thus,

$$V_S = RI_X = 23.3 \frac{R}{R_B}$$

The width of the hysteresis region is $2V_S$. For $R_B = 230$ k and $R = 5$ kΩ,

$$2V_S = 2(23.3)\left(\frac{5}{230}\right) = 1.01 \text{ V}$$

We see that hysteresis width can be controlled by I_X, a special feature of the OTA circuit.

(c) Because our equations assumed active operation of all transistors, it is prudent to check for saturation of the output transistors. Figure 11.1a shows that for 12 V supplies, the output transistors saturate when the output node reaches

$$\pm(V_{SS} - 0.7 - 0.2) = \pm 11.1 \text{ V}$$

Since the Schmitt trigger output voltage varies between

$$\pm V_S = \pm 0.505 \text{ V}$$

the output transistors never saturate. If i_O were forced to flow through a sufficiently large R, however, the transistors would saturate. ❏

> **Exercise 11.2** For the Schmitt trigger of Example 11.2, find the smallest R for which the output transistors would saturate.
>
> *Ans.* $R_{min} = 110$ kΩ.

The *amplitude modulator* of Fig. 11.8 is another OTA application. Problem 11.6 shows that an amplitude-modulated wave appears at the output when $v_c(t)$ and $v_m(t)$ are, respectively, a high-frequency carrier signal and a lower-frequency modulating signal containing information to be transmitted.

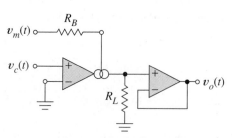

Figure 11.8 An amplitude modulator using an operational transconductance amplifier.

11.2

The 741 Operational Amplifier: A Case Study

The following detailed analysis reviews and reinforces key principles of previous chapters, as we apply the principles to a single complex IC, the 741 operational amplifier. We are able to verify published parameter values using simple hand analysis, and these same techniques enable us to estimate parameters for conditions *not* described on the data sheet, for example, supplies of $+5$ V and -6 V instead of $\pm$ 15 V, should our application require it.

11.2.1 OVERVIEW OF THE 741

Figure 11.9a is a block diagram of the μA741. Transconductance and voltage amplifiers, together, provide the op amp's high voltage gain, and a class AB output stage efficiently delivers power to an external load. The second stage is bridged by a compensating capacitor that establishes the unity gain frequency, ensures unconditional stability, and sets the amplifier's slew rate.

The simplified schematic of Fig. 11.9b shows the more important details. Recognizing the virtual ground induced by difference-mode excitation, we identify Q_1–Q_3 and Q_2–Q_4 as paired cascode amplifiers biased with a current source. Active load, Q_5–Q_6, provides the high output resistance required of a transconductance amplifier, and also mirrors signal i_{c3} into the output circuit, where it adds to i_{c4} to form output current i_o.

Q_{16} and Q_{17} are a common-collector/common-emitter amplifier that uses the output resistance of a current source for its load. An unbypassed emitter resistor provides a high load resistance for the transconductance amplifier, to improve first-stage voltage gain.

In the output stage, V_{BB} biases Q_{14} and Q_{20} slightly above cut-in. Emitter follower, Q_{23}, lowers the output resistance presented to the class AB circuit and increases the resistance that loads the voltage amplifier.

The dashed lines in the complete 741 schematic, Fig. 11.10, enclose the actual three-stage amplifier. As in the mirror of Fig. 6.28, active load Q_5–Q_6 uses emitter resistors R_1 and R_2 to give higher output resistance. Transistor Q_7 (with bias current provided by R_3) improves the current gain of the mirror (see Fig. 6.26). The second stage differs from Fig. 11.9b only in the addition of biasing resistor R_9.

Subcircuit Q_{18}, Q_{19}, and R_{10} in Fig. 11.10 provides the DC voltage V_{BB} of Fig. 11.9b. Transistors Q_{15}, Q_{21}, Q_{24}, and Q_{22} are cut off except when load current becomes excessive. Protection for Q_{14} reduces the base current of Q_{14} when activated by a shorted output node. The protection for Q_{20} uses a mirror, Q_{22}, to reduce the base of current of Q_{16}.

R = 200(250+1) = 500000 Ω

$h_{fe} = 250$
$R = 200$

$\dfrac{100000}{50} = 2000$

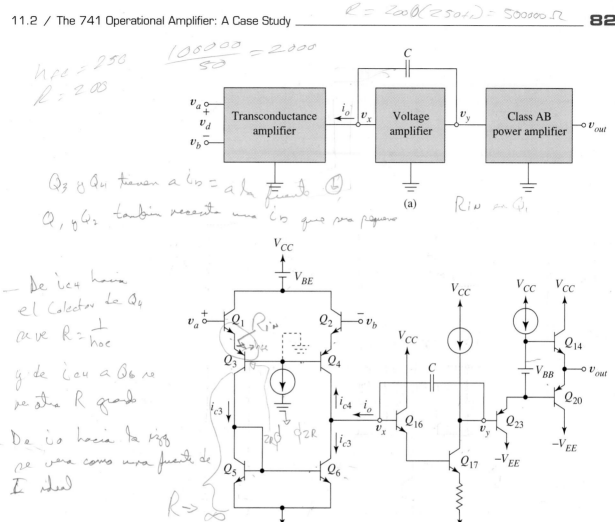

Q_3 y Q_4 tienen a $i_b =$ a la fuente ①
Q_1 y Q_2 también necesita una i_b que sea pequeña

R_{in} en Q_1

— De i_{c4} hacia
el Colector de Q_4
se ve $R = \dfrac{1}{h_{oe}}$

y de i_{c4} a Q_6 se
ve otra R grande

— De i_o hacia la I_{O1}
se vera como una fuente de
I ideal

$R \to \infty$

Figure 11.9 741 Operational amplifier: (a) block diagram; (b) simplified schematic.

The components outside the dashed box in Fig. 11.10 establish the bias currents. Q_{11}, Q_{12}, and R_5 establish the reference current; Q_{13A} and Q_{13B} mirror scaled copies into the last two stages. Widlar mirror, Q_{10}–Q_{11}–R_4, produces a bias current, I_W, that is low enough to give the op amp a very high input resistance.

A current-shunt feedback mechanism stabilizes the bias current in the first amplifier stage. Base currents from Q_3 and Q_4, combined as $2I_B$, subtract from I_W to form the difference, I_O, which is then mirrored into bias current, I_{O1}, by Q_8. We will find that the feedback equations of Chapter 9 describe this system, but because loop gain is less than one, we refer to the circuit's function as *stabilization* rather than classical negative feedback.

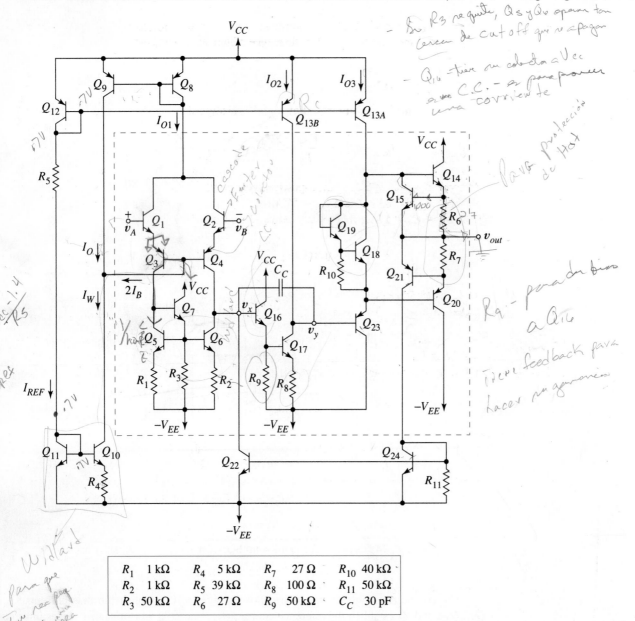

Figure 11.10 Schematic diagram of 741 operational amplifier.

R_1	1 kΩ	R_4	5 kΩ	R_7	27 Ω	R_{10}	40 kΩ
R_2	1 kΩ	R_5	39 kΩ	R_8	100 Ω	R_{11}	50 kΩ
R_3	50 kΩ	R_6	27 Ω	R_9	50 kΩ	C_C	30 pF

Because the 741 does not employ supply-independent biasing, Q-points and small-signal parameters depend on the supply voltages. The dc and small-signal analyses that follow assume $V_{CC} = V_{EE} = 15$ V. They also assume that all npn transistors have $\beta_n = 250$ and $V_A = 130$ V, and pnp transistors have $\beta_p = 50$ and $V_A = 52$ V. In dc analysis we use $I_S = 10^{-14}$ ampere for all transistors except Q_{14} and Q_{20}. Wherever reasonable, we ignore base currents of npn transistors compared to collector currents. Resistor values are tabulated in Fig. 11.10 to reduce clutter in the diagram.

11.2.2 dc ANALYSIS

Reference and Bias Currents. For ± 15 V supplies, the reference current is

$$I_{REF} = \frac{V_{CC} + V_{EE} - 2V_{BE}}{R_5} = \frac{30 - 2 \times 0.7}{39 \text{ k}} = 0.733 \text{ mA}$$

Mirrors, Q_{13A} and Q_{13B}, with respective areas of 0.25 and 0.75 relative to Q_{12}, produce currents $I_{O3} = 183$ μA and $I_{O2} = 550$ μA.

Since $R_4 = 5$ kΩ, the Widlar source satisfies Eq. (6.38); that is,

$$0.025 \ \ell n \left(\frac{0.733 \text{ mA}}{I_W} \right) = I_W (5 \text{ k})$$

so $I_W = 18.5$ μA. The first and second columns of Table 11.1 list the dc currents computed so far.

Stabilization of Input-Stage Biasing. We next examine the feedback mechanism that stabilizes I_{O1}. Mirror Q_8–Q_9 has current gain that satisfies Eq. (6.29), where I_{O1} is the reference current and I_O the output current. Therefore,

$$I_{O1} = \frac{\beta_P + 2}{\beta_P} I_O$$

where β_P denotes beta for the pnp transistors. Without feedback stabilization (and the subtraction of $2I_B$), I_O would be the Widlar current I_W, and I_{O1}, which biases all the first-stage transistors, would be highly sensitive to β_P.

If we regard I_{O1} as the output variable and I_W as the input variable of a feedback system, we can write

$$I_{O1} = \frac{\beta_P + 2}{\beta_P} I_O = \frac{\beta_P + 2}{\beta_P} (I_W - 2I_B) = A(I_W - 2I_B) \tag{11.7}$$

where the nonfeedback gain is

$$A = \frac{\beta_P + 2}{\beta_P}$$

Assume input voltages v_A and v_B have equal dc components consistent with active operation of Q_1 and Q_2, and collector currents equal emitter currents for these transistors. Then the feedback current provided by the bases of Q_3 and Q_4 is

$$2I_B = \frac{I_{O1}}{\beta_P + 1} \tag{11.8}$$

Handwritten margin notes:
$y_1 = I_W R$
$y_2 = .025 \ln \left(\frac{.733 \text{ mA}}{I_W} \right)$
ecu. Trascendental
forma de resolver ecu. Trascendental

which identifies the *feedback beta* as $\beta = (\beta_P + 1)^{-1}$. Equation (9.1) then gives

$$I_{O1} = \frac{A}{1 + A\beta} I_W = \frac{(\beta_P + 2)/\beta_P}{1 + [(\beta_P + 2)/\beta_P] [1/(\beta_P + 1)]} I_W$$

$$= \frac{\beta_P^2 + 3\beta_P + 1}{\beta_P^2 + 2\beta_P + 1} I_W \approx I_W = 18.5 \ \mu A$$

The feedback approximation, $A\beta \gg 1$, is not appropriate here. Indeed, if it were valid we see that I_{O1} and I_W would be related by the transistor-dependent parameter $(\beta_P + 1)$. In this system, if $\beta_P^2 \gg 3\beta_P$, the squared terms dominate both numerator and denominator. This means for *any* $\beta_P > 30$, the approximation $I_{O1} \approx I_W$ is a good one. This development justifies the column 3 entries in Table 11.1.

Special Bias Currents. The currents listed in the last column of Table 11.1 are all special, Q_{14} and Q_{20} because they are standby currents of power transistors, the remaining ones because their low values justify special analysis methods. Whenever bias currents are low we forgo the approximation, $V_{BE} = 0.7$ and, instead, calculate V_{BE}.

From $I_{C6} \approx 9.3 \ \mu A$ we use Eq. (4.5) to show that

$$V_{BE5} = V_{BE6} = 0.025 \ \ell n \left(\frac{9.3 \ \mu A}{10^{-14} \ A} \right) = 0.516 \ V$$

The voltage drop across R_3 is

$$V_{50 \, k} = I_{ES} R_1 + V_{BE5} = (9.3 \ \mu A) 1 \ k + 0.516 = 0.525 \ V$$

The emitter current of Q_7 is $V_{50 \, k}/50 \ k = 10.5 \ \mu A$.

In the voltage amplifier, the 550 μA current in Q_{17} produces a drop of 55 mV across R_8. The emitter current of Q_{16} is the current in R_9,

$$I_{E16} = \frac{0.055 + 0.7}{50 \ k} = 15.1 \ \mu A$$

Because the emitter current of Q_{16} is much less than that of Q_{17}, the "usual" assumption $I_{E16} \gg I_{B17}$ is questionable. Using $\beta_n = 250$ gives $I_{B17} = 550/250 = 2.2 \ \mu A$. From this comes a refined estimate, $I_{E16} \approx 15.1 + 2.2 = 17.3 \ \mu A$.

TABLE 11.1 **μA741 Collector Currents for ±15 V Supplies**

Q_{11}	0.733 mA	Q_{13B}	550 μA	Q_1	9.3 μA	Q_{16}	17 μA
Q_{12}	0.733 mA	Q_{17}	550 μA	Q_2	9.3 μA	Q_7	11 μA
Q_{10}	18.5 μA	Q_{13A}	183 μA	Q_3	9.3 μA	Q_{19}	15 μA
Q_9	18.5 μA	Q_{23}	183 μA	Q_4	9.3 μA	Q_{18}	168 μA
Q_8	18.5 μA			Q_6	9.3 μA	Q_{14}	160 μA
				Q_5	9.3 μA	Q_{20}	160 μA

We next analyze the circuit that produces V_{BB} to reduce crossover distortion. Mirror Q_{13A} provides 183 μA to the joint Q_{18}–Q_{19} collector node. Assuming $I_{C18} \gg I_{C19}$, we initially estimate

$$V_{BE18} \approx 0.025 \, \ell n\left(\frac{183 \text{ μA}}{10^{-14}}\right) = 0.591 \text{ V}$$

R_{10} then carries

$$\frac{0.591}{40 \text{ k}} = 14.8 \text{ μA} \approx I_{E19}$$

From this we find

$$V_{BE19} = 0.025 \, \ell n\left(\frac{14.8 \text{ μA}}{10^{-14}}\right) = 0.528 \text{ V}$$

Knowing $I_{C19} = 14.8$ μA, we now refine our estimate of I_{C18} to $183 - 14.8 = 168$ μA. Using this to recalculate V_{BE18} gives

$$V_{BE18} = 0.025 \, \ell n\left(\frac{168 \text{ μA}}{10^{-14}}\right) = 0.589 \text{ V}$$

so $V_{BB} = V_{BE18} + V_{BE19} = 0.589 + 0.528 = 1.12$ V. If we ignore small drops in R_6 and R_7 and assume that the high-current output transistors have areas three times those of the other transistors, then $V_{BE14} = V_{EB20} = V_{BB}/2 = 0.56$ V. This biases the output transistors just above cut-in at

$$I_{C14} = I_{C20} = 3 \times 10^{-14} e^{0.56/0.025} = 160 \text{ μA}$$

This completes bias circuit analysis of the 741 and Table 11.1.

Op Amp Power Dissipation and Input Bias Currents. The sum of the dc currents leaving the positive supply in Fig. 11.10 equals the sum of currents entering the negative supply. This is

$$I_{C12} + I_{C9} + I_{C8} + I_{C13B} + I_{C13A} + I_{C14} + I_{C16} + I_{C7} = 1.69 \text{ mA}$$

We estimate the 741 standby dissipation as 1.69 mA × 30 V = 50.7 mW. The 741 data sheet lists 50 mW as typical dissipation for ±15 V supplies. From the collector currents of Q_1 and Q_2, we estimate input bias currents of 9.3 μA/250 = 37 nA. The data sheet lists 80 nA.

Common-Mode Limits. We use Fig. 11.11 to estimate limiting values for common-mode inputs. The figure shows V_{BE} drops for Q_8, Q_7, and Q_{16} in addition to previously calculated bias currents and voltages. The collectors of Q_1 and Q_2 are biased at

$$V_{C1} = 15 - 0.534 = 14.5 \text{ V}$$

the collector voltage of Q_5 is

$$V_{C5} = -15 + 0.527 + 0.519 = -14 \text{ V}$$

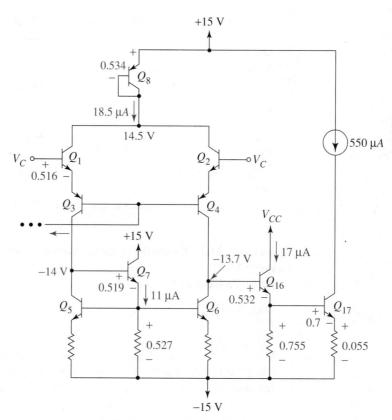

Figure 11.11   Circuit for finding 741 input voltage limits.

and the collector voltage of Q_6 is

$$V_{C6} = -15 + 0.755 + 0.532 = -13.7V$$

For collector currents of 9.25 μA the V_{BE} drops of Q_1 and Q_2 are 0.516 V as indicated. As V_C increases, the emitter voltages of Q_1 and Q_2 follow the base voltage up until Q_1 and Q_2 saturate. Saturation occurs when

$$V_C - 0.516 + 0.2 = 14.5$$

Thus the upper common-mode limit is $V_{CMAX} = 14.8$ V.

As V_C decreases, the emitter voltages of Q_3 and Q_4 follow V_C down, simultaneously reducing the emitter-collector drops across Q_3 and Q_4. The lower common mode limit occurs when Q_4 saturates, since $V_{C4} > V_{C3}$. This happens when

$$V_C - 0.516 - 0.2 = -13.7$$

giving $V_{CMIN} = -13$ V. The 741 data sheet specifies a typical input voltage range of $\pm$ 13 V with 15 V supplies.

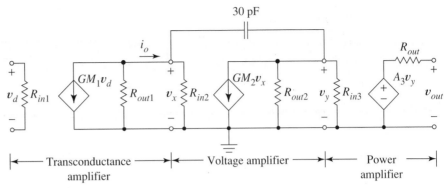

Figure 11.12 General equivalent circuit form for 741 operational amplifier.

Exercise 11.3 Estimate the upper common-mode limit for the μA741 when supplies are ±5 V.

Hint: Problem 11.9 shows that for ±5V, $I_W = 13.8 \ \mu A = I_{C8}$.

Ans. 4.78 V.

11.2.3   SMALL-SIGNAL ANALYSIS

The object of our small-signal analysis is to replace the block diagram of Fig. 11.9a with the equivalent circuit of Fig. 11.12, including parameter values. As in the preceding bias study, this analysis assumes ±15 V supplies.

Transconductance Amplifier. The simplified circuit of Fig. 11.13a gives correct expressions for GM_1 and R_{in1}. Because Table 11.1 gives 9.3 μA for the collector currents,

$$g_m = \frac{9.3 \times 10^{-6}}{0.025} = 0.372 \text{ mA/V}$$

for Q_1 through Q_6. Since difference-mode signals create a virtual ground at the bases of Q_3 and Q_4, we draw the half-amplifier, small-signal schematic of Fig. 11.13b. Clearly the 741 input circuit is a differential, common-collector/common-base amplifier, which has input resistance given by Eq. (7.50). Therefore, the input resistance offered by the 741 to a difference-mode source is

$$R_{in1} = 4r_{\pi 1} \tag{11.9}$$

which has numerical value

$$R_{in1} = 4\frac{250}{0.372 \times 10^{-3}} = 2.69 \text{ M}\Omega$$

This estimate agrees well with the typical value listed on the 741 data sheet, 2 MΩ.

Now we need to find the first-stage transconductance, GM_1. Our strategy stems from comparing the schematic, Fig. 11.13c, with the equivalent circuit, Fig. 11.13d. We

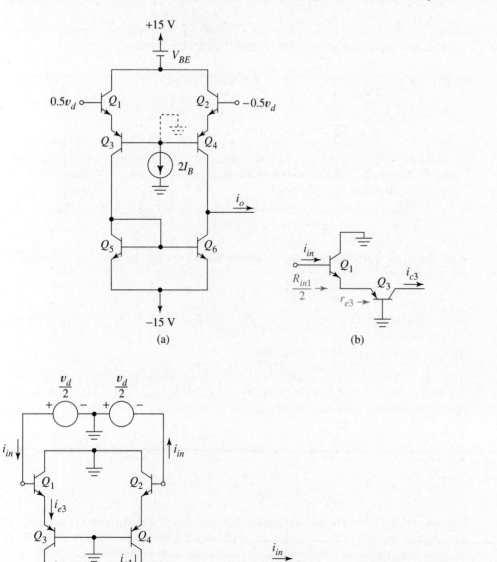

Figure 11.13 Circuits used to find R_{in1} and GM_1 for the 741: (a) approximate amplifier schematic; (b) half-amplifier small-signal schematic; (c) circuit for finding transconductance; (d) circuit that defines GM_1.

simply equate the short-circuit current, i_{sc}, that results from difference-mode excitation v_d in the two representations. In Fig. 11.13c, i_{in} is converted into

$$i_{e3} = (\beta_1 + 1)i_{in}$$

by Q_1, and then into

$$i_{c3} = \alpha_3(\beta_1 + 1)i_{in} \approx \beta_1 i_{in}$$

by Q_3. The active-load mirrors i_{c3} into the collector of Q_6. The other half-amplifier, physically symmetric but excited out of phase, gives $i_{c4} \approx -\beta_2 i_{in}$. Applying KCL at the output node and using $\beta_2 = \beta_1$ gives

$$i_{sc} = i_{c4} - i_{c3} = -2\beta_1 i_{in} = -2\frac{\beta_1}{R_{in1}}v_d$$

Substituting Eq. (11.9) gives

$$i_{sc} = -2\frac{\beta_1}{4r_{\pi1}}v_d = -\frac{g_{m1}}{2}v_d = -GM_1 v_d$$

We conclude that

$$GM_1 = 0.5g_{m1} \qquad (11.10)$$

For ±15 V supplies this gives the numerical value $GM_1 = 0.186$ mA/V.

All that remains for first-stage analysis is finding the output resistance, R_{out1}. Because resistors R_1–R_3 in Fig. 11.10 affect R_{out1}, we must include them in this analysis. The most straightforward approach is to apply the R_{out} definition to the small-signal schematic of Fig. 11.14. We obtain the same result with less work, however, by reducing the task to two subtasks, and using a previous result.

Looking back from the output node with signal generators off in Fig. 11.14, we see resistance R_{o4} between the collector of Q_4 and ground, and resistance R_{o6} between the collector of Q_6 and ground. We first find these two output resistances, then use $R_{out1} = R_{o4} \| R_{o6}$.

For R_{o6} we can use a result previously derived for Fig. 7.34a, here using $R_E = 1$ kΩ and $R_{TH} = R_X$, where R_X is the resistance looking back from the base of Q_6. Obviously $R_X < 50$ kΩ. Since $I_{C6} = 9.3$ μA and $\beta = 250$, $r_{\pi6} = 672$ kΩ. We conclude that the "Case 2" condition and Eq. (7.70) apply here. This gives

$$R_{o6} = \left(1 + \frac{9.3 \times 10^{-6}}{0.025}10^3\right)r_{o6} = 1.37 r_{o6} \qquad (11.11)$$

Using $V_A = 130$ V gives $r_{o6} = 14$ MΩ and $R_{o6} = 19.2$ MΩ.

Figure 7.34a also defines R_{o4}, provided that

1. the transistor parameters are those of Q_4.
2. $R_{TH} = 0$ (the base of Q_4 is grounded).
3. $R_E = r_{e2}$, the common-base input resistance of Q_2.

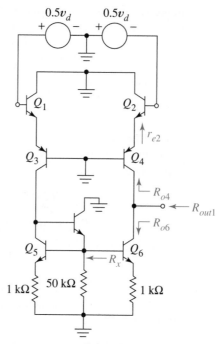

Figure 11.14 Small-signal schematic for the 741's transconductance amplifier.

Since $r_{\pi 4} = 50/0.38 \times 10^{-3} = 132$ kΩ and $R_X = 0$, the Case 2 assumption, $r_\pi \gg R_{TH} + R_E$ also applies here. Using Eq. (7.70), with $R_E = r_{e2} \approx 1/g_m = 2.63$ kΩ, we find

$$R_{o4} = \left(1 + \frac{9.3 \times 10^{-6}}{0.025}2.63 \times 10^3\right)r_{o4} = 1.98 r_{o4}$$

For $V_A = 52$ V and $I_{C4} = 9.3$ μA, $r_{o4} = 5.59$ MΩ. Thus, $R_{o4} = 11.1$ MΩ. Now we combine the two output resistance results to obtain

$$R_{out1} = 19.2 \text{ M} \| 11.1 \text{ M} = 7.03 \text{ M}\Omega$$

The first-stage amplifier in Fig. 11.17 summarizes the results we have just found: $GM_1 = 0.186$ mA/V, $R_{in1} = 2.67$ MΩ, and $R_{out1} = 7.03$ MΩ.

Voltage Amplifier Analysis. Figure 11.15a is the small-signal schematic for the voltage amplifier; r_{o13B} is the resistance of active load, Q_{13B}. Resistance scaling gives the circuit's input resistance

$$R_{in2} = r_{\pi 16} + (\beta_{16} + 1)\{50 \text{ k} \| [r_{\pi 17} + (\beta_{17} + 1)100]\}$$

A numerical evaluation of this expression in Figure 11.15b uses Table 11.1 values for I_{C16} and I_{C17}. The result is

$$R_{in2} = 5.67 \text{ M}\Omega$$

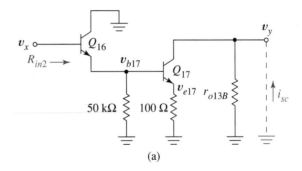

(a)

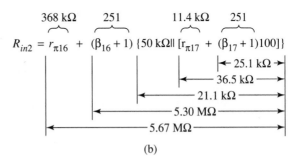

(b)

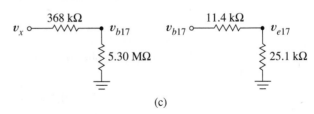

(c)

Figure 11.15 Voltage amplifier: (a) small-signal schematic; (b) components of input resistance; (c) voltage divisions that establish gain.

To find GM_2 for Fig. 11.15a, we find the current that would flow in response to v_x through a short circuit (dashed line) between collector of Q_{17} and ground. The voltage dividers of Fig. 11.15c help us through the three-step analysis. The first divider gives

$$v_{b17} = \frac{5.30\text{ M}}{5.30\text{ M} + 368\text{ k}}v_x = 0.935\,v_x$$

where the resistances came from Fig. 11.15b. From the second divider

$$v_{e17} = \frac{25.1\text{ k}}{25.1\text{ k} + 11.4\text{ k}}v_{b17} = 0.688\,v_{b17}$$

Finally, the short-circuit current from Fig. 11.15a is

$$i_{sc} = i_{c17} \approx \frac{v_{e17}}{100} = 0.01 \times 0.689 \times 0.935\,v_x = GM_2 v_x$$

Thus the transconductance for the voltage amplifier of Fig. 11.15a is $GM_2 = 6.43 \times 10^{-3}$ A/V.

From Fig.11.15a, the amplifier's output resistance is

$$R_{out2} = r_{o13B} \| R_{o17}$$

where R_{o17} is the output resistance of Q_{17} and

$$r_{o13B} = \frac{52}{550 \ \mu A} = 94.5 \ k\Omega$$

We cannot use Eq. (7.70) to find R_{o17} as we previously did for R_{o6} and R_{o4} because the key assumption does not hold. Problem 11.15 shows that $R_{o17} = 197$ kΩ. We conclude that $R_{out2} = (94.5 \ k) \| (197 \ k) = 63.9$ kΩ. Our three key results—input resistance, transconductance, and output resistance for the voltage amplifier—are summarized in the voltage amplifier equivalent circuit of Fig. 11.17.

Power Amplifier Analysis. Small-signal analysis of a class AB amplifier is suspect, since the circuit's purpose is to develop a *large signal*. Nonetheless, we now perform such an analysis, emphasizing the large-signal parameter β (h_{FE}) rather than g_m, making approximations for r_π that seem relevant, and regarding our results only as useful rough estimates.

When Q_{14} conducts, the power amplifier/driver circuit is Fig. 11.16a. The external 2 kΩ load was arbitrarily chosen so we can make numerical estimates. The 163 Ω resistor shows the effects of the biasing circuit, Q_{18}–Q_{19}–R_{10}, on signals. (See Problem 11.17.) Short-circuit protection resistor R_6 is omitted, as it complicates the analysis and contributes little to the result.

As output current increases with signal, $r_{\pi14}$, the reciprocal of the slope of the input characteristic of Q_{14}, decreases. For a rough estimate, we select the instant when $v_{out} = 7.5$ V. Then Q_{14}'s collector current is 7.5/2 k = 3.75 mA, and

$$r_{\pi14} = \frac{250 \times 0.025}{3.75 \ mA} = 1.67 \ k\Omega$$

By working backward from the load with resistance scaling, as in Figure 11.16b, we find the input resistance of Q_{23} at this instant is $R_{in3} = 9.29$ MΩ. Notice that neither our rough estimate of $r_{\pi14}$ nor the 163 Ω equivalent had much effect on the result.

When Q_{20} conducts, we again estimate r_π for the output transistor at half of peak output current. This is

$$r_{\pi20} = \frac{50 \times 0.025}{3.75 \ mA} = 334 \ \Omega$$

Figure 11.16c shows the input resistance calculation. The result is $\overline{R_{in3}} = 3.84$ MΩ. In Fig. 11.17 we show this smaller input resistance value, as it will lead to the more pessimistic (lower) estimate of second-stage voltage gain.

Next we examine voltage gain for the power amplifier. From Fig. 11.16a, the gain is

$$\frac{v_{out}}{v_y} = \left(\frac{v_{out}}{v_{e23}}\right)\left(\frac{v_{e23}}{v_y}\right) \tag{11.12}$$

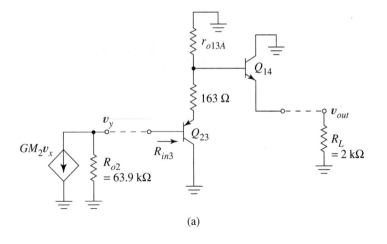

(a)

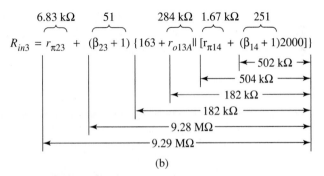

(b)

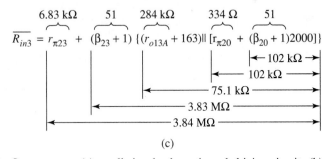

(c)

Figure 11.16 Output stage: (a) small-signal schematic and driving circuit; (b) components of R_{in3} when Q_{14} conducts; (c) components of R_{in3} when Q_{20} conducts.

Because $163\ \Omega \ll 182\ \text{k}\Omega$ in Fig. 11.16b, $v_{out}/v_{e23} = v_{out}/v_{b14}$. With this observation, we can express both ratios in Eq. (11.12) as voltage divider expressions, using Fig. 11.16b. The result is

$$A_{v3} = \frac{v_{out}}{v_{b14}} \times \frac{v_{c23}}{v_y} = \frac{502\ \text{k}}{504\ \text{k}} \times \frac{9.28\ \text{M}}{9.29\ \text{M}} = 0.995 \approx 1$$

When Q_{20} conducts, the approach is the same, but data come from Fig. 11.16c. The result is

$$\overline{A_{v3}} = \frac{v_{out}}{v_{b20}} \times \frac{v_{c23}}{v_y} = \frac{102 \text{ k}}{102 \text{ k}} \times \frac{3.83 \text{ M}}{3.84 \text{ M}} = 0.997 \approx 1$$

In spite of a 5-to-1 ratio in the output transistor betas, the voltage gain is close to one for both positive and negative voltage swings. Notice that this result is largely due to using emitter followers in the driver and output stages.

Finally, we estimate the 741's output resistance . When Q_{14} conducts, R_{out3} is the resistance seen by R_L in Fig. 11.16a plus protection resistance R_6; that is,

$$R_{out3} = \frac{[(R_{out2} + r_{\pi 23})/(\beta_{23} + 1) + 163]\|r_{o13A} + r_{\pi 14}}{\beta_{14} + 1} + R_6$$

$$= \frac{[(63.9 \text{ k} + 6.83 \text{ k})/51 + 163]\|284 \text{ k} + 1.67 \text{ k}}{251} + 26 = 38.2 \ \Omega$$

Similarly, when Q_{20} conducts,

$$\overline{R_{out3}} = \frac{[(R_{out2} + r_{\pi 23})/(\beta_{23} + 1)]\|(163 + r_{o13A}) + r_{\pi 20}}{\beta_{20} + 1} + R_7$$

$$= \frac{[(63.9 \text{ k} + 6.83 \text{ k})/51]\|(163 + 284 \text{ k}) + 667}{51} + 26 = 66.1 \ \Omega$$

In Fig. 11.17 we show the higher of our estimates. The data sheet for the 741 lists $R_{out} = 75 \ \Omega$.

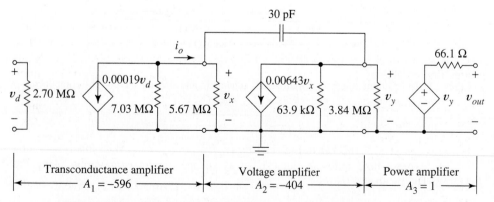

Figure 11.17 Completed small-signal model for 741 operational amplifier.

Gain, Bandwidth, and Slew Rate. Figure 11.17 is the small-signal equivalent of the 741. The diagram shows the voltage gains of each stage, including interstage loading. From these we predict

$$A = A_1 A_2 A_3 = 2.41 \times 10^5$$

for the open-loop gain of the 741, compared with the published value of 2×10^5.
The Miller capacitance,

$$C_M = (1 + 404)30 \text{ pF} = 12{,}150 \text{ pF}$$

suggests a bandwidth of

$$\omega_H = \frac{1}{(7.03 \text{ M} \| 5.67 \text{ M})1.22 \times 10^{-8}} = 26.1 \text{ rad/s}$$

or 4.15 Hz. Our equivalent circuit has gain-bandwidth product or unity-gain frequency of

$$f_\tau = 2.43 \times 10^5 \times 4.15 = 1.01 \text{ MHz}$$

The published value is 1 MHz.
Equation (9.33) predicts the slew rate, where I_O is the dc bias current of the first stage. For the 741, $I_O = I_{C8}$, 18.5 μA in Table 11.1. Since $C_C = 30$ pF, our estimate is

$$\text{s.r.} = \frac{18.5 \text{ μA}}{30 \times 10^{-12} \text{ F}} = 0.617 \text{ V/μs}$$

compared to the published value of 0.5 V/μs.

11.3
Analog Multiplier

A key operation in communications systems is a process called *mixing,* deliberately combining input signals of different frequencies in a time-varying or nonlinear device to produce an output signal that contains new frequencies. The *analog multiplier* mixes by producing an output that is the instantaneous product of signals applied to its two inputs.

To radiate information from an antenna, it is first necessary to combine the information with a high-frequency *carrier* signal, a mixing process called *modulation.* Modulation also allows us to transmit multiple messages over a single communication channel after first combining each message with an individual carrier frequency. Analog multipliers sometimes serve as *modulators* in such applications. Communication receivers employ multipliers as *demodulators* to reverse the modulation process, that is, to remove the original information from the carrier. Signal multiplication can be used to facilitate high-quality filtering in radio receivers and compensate for signal fading in automatic gain control (AGC) circuits.

The simplest multiplier, conceptually, is the large-signal, *four-quadrant multiplier,* a circuit whose output is the product of its two inputs regardless of their algebraic signs. We first introduce a simpler *two-quadrant multiplier* and then extend the concepts to more complex circuits.

11.3.1 TWO-QUADRANT MULTIPLIER

Figure 11.18a shows a basic two-quadrant multiplier, an emitter-coupled pair biased by current I_O and common-mode input sources V_{BB}. One input is a difference-mode signal, v_D, applied to the bases of Q_1 and Q_2. The other input signal, v_C, enters the circuit through the current source. The development leading to Eq. (11.5) showed that for active-state operation of Q_1 and Q_2,

$$i_{C1} - i_{C2} = I_O \tanh\left(\frac{v_D}{2V_T}\right) \tag{11.13}$$

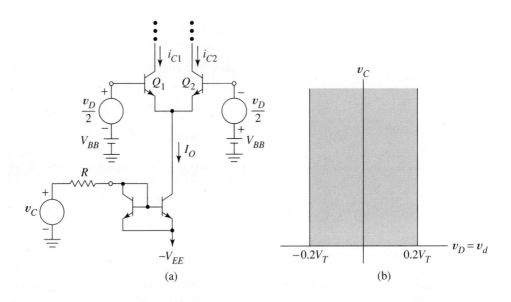

(a)   (b)

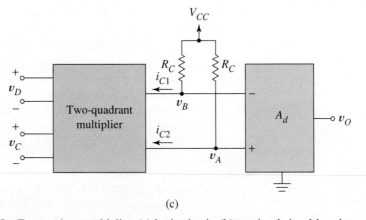

(c)

Figure 11.18 Two-quadrant multiplier: (a) basic circuit; (b) restricted signal locations required for validity of Eq. (11.16); (c) multiplier with single-ended output.

where i_{C1}-i_{C2} is the output. From the current mirror,

$$I_O = \frac{v_C - V_{BE} + V_{EE}}{R}, \qquad v_C > -V_{EE} + B_{BE} \tag{11.14}$$

where the inequality states a necessary condition for active operation of Q_1 and Q_2. Substituting into Eq. (11.13) gives

$$i_{C1} - i_{C2} = \frac{v_C - V_{BE} + V_{EE}}{R} \tanh\left(\frac{v_D}{2V_T}\right) \tag{11.15}$$

a general, large-signal result.

If v_D is a small signal, v_d, then the hyperbolic tangent is approximated by its argument, giving

$$i_{C1} - i_{C2} = \frac{1}{2RV_T} v_d v_C + \frac{V_{EE} - V_{BE}}{2RV_T} v_d \tag{11.16}$$

Our critical assumptions, active operation and small-signal input, respectively, require

$$v_C > -V_{EE} + V_{BE} \quad \text{and} \quad |v_d| < 0.2 \, V_T \tag{11.17}$$

Under these conditions, the circuit generates the desired product $v_d v_C$; however, Eq. (11.16) contains an error term in addition to the signal product. We can eliminate the error by using the bias condition $V_{EE} = V_{BE} \approx 0.7$ V. Equation (11.17) then requires us to restrict signals to

$$v_C > 0 \quad \text{and} \quad |v_d| < 0.2 \, V_T \tag{11.18}$$

These inequalities define an acceptable band in the first and second signal quadrants of input signal space as sketched in Fig. 11.18b—hence the term *two-quadrant multiplier*.

Figure 11.18c shows how a difference amplifier of gain A_d transforms the current difference into a single-ended output voltage. That is,

$$v_o = A_d(v_A - v_B) = A_d[(V_{CC} - R_C i_{C2}) - (V_{CC} - R_C i_{C1})] = A_d R_C(i_{C1} - i_{C2})$$

With the error term eliminated, Eq. (11.16) shows that the overall multiplier system satisfies

$$v_O = \frac{A_d R_C}{2RV_T} v_d v_C \tag{11.19}$$

EXAMPLE 11.3 A transducer produces a slowly varying voltage between 0 and 5 V to denote pressure in a flow process. Design a multiplier like Fig. 11.18c to convert the pressure information into the amplitude of a 1 MHz sinusoidal carrier waveform. A 5 V "pressure" input should correspond to an output sinusoid of 10 V amplitude. Use supplies $V_{CC} = +15$ V and $V_{EE} = 0.7$ V.

Solution. Since the pressure signal is nonnegative, introduce it as v_C. We arbitrarily choose to use the maximum possible input amplitude for the sinusoidal carrier signal, which, from Fig. 11.18b, gives

$$v_d(t) = 0.2V_T \sin \omega_c t = 0.005 \sin 2\pi 10^6 t$$

To give a 10 V output sine wave when $v_C = 5$ V, Eq. (11.19) requires that A_d, R, and R_C satisfy

$$10 = \frac{A_d R_C}{0.05 R} 0.005 \times 5$$

This means

$$\frac{A_d R_C}{R} = 20$$

so we arbitrarily select $R = R_C = 1$ kΩ and use a diff amp of gain $A_d = 20$.

Now we work backward to establish suitable bias voltage V_{BB}. With the carrier absent and $v_C = 5$ V, Eq. (11.14) shows that Q_1 and Q_2 in Fig. 11.19a have dc collector currents of $I_O/2 = 0.5(5/R) = 2.5$ mA. This biases the collectors at

$$V_{C1} = V_{C2} = V_{CC} - R_C\frac{I_O}{2} = 15 - 1 \text{ k}(2.5 \text{ mA}) = 12.5 \text{ V}$$

2.5 V below the 15 V supplies.

When $v_C = 5$ V in Fig. 11.18c, we expect the diff amp input, $v_a - v_b$, to be a sinusoid of amplitude $|v_o|/A_d = 10/20 = 0.5$ V; thus each collector voltage in the multiplier has a signal swing of 0.25 V. An input bias voltage of $V_{BB} = 10$ V puts the emitters at $10 - 0.7 = 9.3$ V and thereby biases Q_1 and Q_2 at $V_{CE} = 12.5 - 9.3 = 3.2$ V, a value that should easily accommodate the 0.25 V signal swing. Therefore, the complete multiplier is Fig. 11.19a. ❏

> **Exercise 11.4** Sketch $v_o(t)$ for the multiplier designed in Example 11.3 when $v_C = 2.1$ V.
>
> *Ans.* Sinusoid of 10^6 Hz and amplitude of 4.2 V.

EXAMPLE 11.4 Use SPICE with 5 V pulse and ramp signals for $v_C(t)$ to test the design of Example 11.3.

Solution. Figure 11.19b shows the SPICE code. Dependent sources simulate the difference amplifier and provide a pure difference-mode input. Source VC provides pulse and ramp test inputs.

Figure 11.19c shows the response to 5 V pulses that turn the difference amplifier on and off to create sinusoidal output bursts. The output amplitude is slightly lower than expected, perhaps because of the more accurate V_T calculated by SPICE.

The ramp test results of Figure 11.19d verify that the multiplier circuit has the linear relationship between output voltage and input pressure that such an application requires.

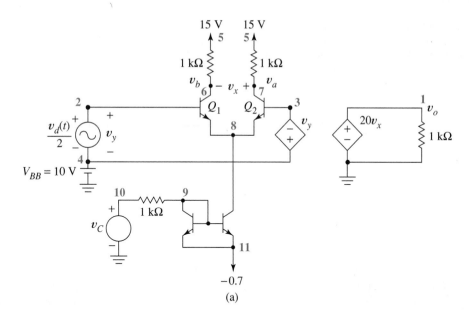

(a)

```
EXAMPLE 11.4
VCC 5 0 DC 15
VEE 11 0 DC -0.7
R 10 9 1K
RC1 5 6 1K
RC2 5 7 1K
RL 1 0 1K
QM1 9 9 11 NDEV
QM2 8 9 11 NDEV
Q1 6 2 8 NDEV
Q2 7 3 8 NDEV
.MODEL NDEV NPN BF=100
E1 4 3 2 4 1
E2 1 0 7 6 20
VBB 4 0 DC 10
VC 10 0 PULSE(0 5 5U 0 0 5U 10U)
*VC 10 0 PULSE(0 5 5U 5U 5U 0.1U 10U)
VD2 2 4 SIN(0 2.5M 1E6)
.TRAN 0.025U 25U
.OP
.PLOT TRAN V(1)
.END
```

(b)

Figure 11.19 Example 11.4: (a) circuit design; (b) SPICE code; (c) response to pulses; (d) response to ramps.

$v_o(t)$ (volts)

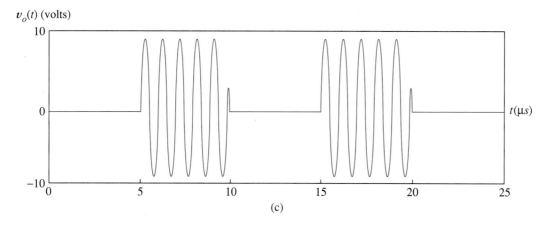

(c)

10 V

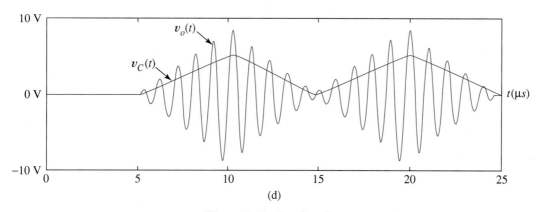

(d)

Figure 11.19 (continued)

Another practical detail in our design is worth exploring. The idea of using $V_{EE} = V_{BE}$ to eliminate the error term originated with Eq. (11.16). Since our design used the *approximation*, $V_{EE} = 0.7$ V, of the unknown V_{BE}, we suspect that the actual circuit might produce a small error term. Indeed, upon examining $v_o(t)$ in the SPICE amplitude listings, we discover that there is a sinusoidal output of 42 mV amplitude between the pulses in Fig. 11.19c, an output too small to see in the scale of the diagram.   ❏

Exercise 11.5   Because the circuit of Fig. 11.19a erroneously produces a sinusoidal output of 42 mV when $v_C = 0$, prudent usage of the circuit requires that we restrict variations at the control input to a practical range

$$v_{C,min} \le v_C \le 5 \text{ V}$$

Find $v_{C,min}$

Ans. $v_{C,min} = 210$ mV.

11.3.2 SMALL-SIGNAL FOUR-QUADRANT MULTIPLIER

Gilbert Cell. The *Gilbert cell*, Figure 11.20a, is a *four-quadrant multiplier* provided that both inputs are small signals. Useful in its own right, the Gilbert cell is also an important building block for other circuits. Inputs are v_{D1} and v_{D2}, and output is $i_x - i_y$. With appropriate notational changes, Eq. (11.13) describes this circuit by

$$i_{C3} - i_{C4} = i_{C1} \tanh\left(\frac{v_{D2}}{2V_T}\right) \tag{11.20}$$

and

$$i_{C6} - i_{C5} = i_{C2} \tanh\left(\frac{v_{D2}}{2V_T}\right) \tag{11.21}$$

equations that assume all collector currents are nonnegative. The output is

$$i_X - i_Y = (i_{C3} + i_{C5}) - (i_{C4} + i_{C6}) = (i_{C3} - i_{C4}) - (i_{C6} - i_{C5})$$

or

$$i_X - i_Y = (i_{C1} - i_{C2}) \tanh\left(\frac{v_{D2}}{2V_T}\right) \tag{11.22}$$

A final substitution from Fig. 11.20a gives the large-signal Gilbert cell equation

$$i_X - i_Y = I_O \tanh\left(\frac{v_{D1}}{2V_T}\right) \tanh\left(\frac{v_{D2}}{2V_T}\right) \tag{11.23}$$

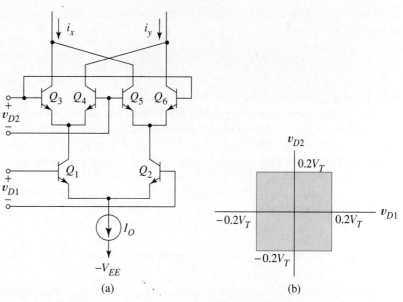

(a)   (b)

Figure 11.20 Gilbert multiplier cell: (a) circuit schematic; (b) restriction on amplitudes of inputs for validity of Eq. (11.24).

When $v_{D1} = v_{d1}$ and $v_{D2} = v_{d2}$ satisfy the *small-signal conditions*, $|v_{d1}/2V_T| \ll 1$ and $|v_{d2}/2V_T| \ll 1$, that is, for the order-of-magnitude approximation

$$|v_{d1}| \leq 0.2V_T \quad \text{and} \quad |v_{d2}| \leq 0.2V_T$$

then Eq. (11.23) becomes

$$i_x - i_y = \frac{I_O}{4V_T^2} v_{d1}v_{d2} \tag{11.24}$$

Figure 11.19b shows the four-quadrant input space for this multiplier. For the Gilbert cell there is no error term that requires critical biasing as in Eq. (11.16).

EXAMPLE 11.5 (a) Find the frequencies that appear in the output of Fig. 11.20a for small signals

$$v_{D1} = v_{d1}(t) = A \sin \omega_c t$$

and

$$v_{D2} = v_{d2}(t) = B \sin \omega_m t$$

when $\omega_c \gg \omega_m$.
(b) Select signal amplitudes, establish bias levels, and add any necessary components so the Gilbert cell output is converted to a difference voltage.

Solution. (a) From Eq. (11.24),

$$i_x - i_y = \frac{I_O}{4V_T^2} AB \sin \omega_m t \sin \omega_c t$$

To find the individual frequencies contained in the trigonometric product, we employ a familiar identity that gives

$$i_x - i_y = \frac{A B I_O}{8V_T^2}[\cos(\omega_c - \omega_m)t - \cos(\omega_c + \omega_m)t]$$

Thus the output spectrum consists of the difference and sum frequencies, $\omega_c - \omega_m$ and $\omega_c + \omega_m$. In communication theory these are called upper and lower *sideband* frequencies, and $i_x - i_y$ is called a double-sideband, suppressed-carrier wave, an important class of signals in long-distance communication systems. A modulator like this, which produces sidebands but not the carrier frequency in the output, is called a *balanced modulator*.
(b) The largest signal amplitudes consistent with Eq. (11.24) are $A = B = 0.2 V_T = 0.005$ V. If we arbitrarily choose $I_O = 4$ mA, the peak output signal amplitude is

$$|i_x - i_y|_{peak} = \frac{4 \text{ mA}}{4(0.025)^2}(0.005)^2 = 40 \text{ μA}$$

Causing i_x and i_y to flow through collector resistors of $R_C = 4$ kΩ (another somewhat arbitrary choice) produces a difference-mode output of amplitude $(4 \text{ kΩ})(40 \text{ μA}) = 0.16$ V. Figure 11.21 shows the complete circuit.

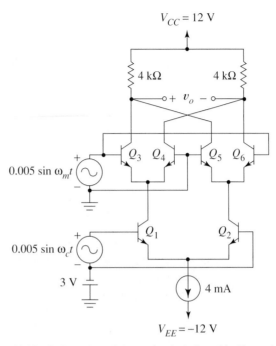

Figure 11.21 Balanced modulator circuit designed in Example 11.5.

Now let us examine the biasing. The collector currents in Q_1 and Q_2 and the resistor currents are 2 mA. If we select $V_{EE} = V_{CC} = 12$ V, as indicated, then the collector voltages of Q_3, Q_4, Q_5, and Q_6 become

$$12 - (4 \text{ k}\Omega)(2 \text{ mA}) = 4 \text{ V}$$

Biasing the bases of Q_3, Q_4, Q_5, and Q_6 at 0 V while biasing the bases of Q_1 and Q_2 at -3 V should ensure active operation with signals. ❏

> **Exercise 11.6** For the circuit of Fig. 11.21,
> (a) What goes wrong if the 3 V bias source is replaced by zero V?
> (b) To increase output signal amplitude, the resistors in Fig. 11.21 are increased to 6 kΩ. Suggest changes in biasing which would make this possible. What would be the new output signal amplitude?
>
> **Ans.** Q_1 and Q_2 saturate. Change the -3 V source to -6 V and bias the sin $\omega_m t$ input at -3 V, 0.24 V.

11.3.3 LARGE-SIGNAL FOUR-QUADRANT MULTIPLIER

We now describe two special input circuits that together extend the Gilbert cell's four-quadrant operation to large signals.

Voltage-to-Current Circuit. We first show that the emitter-coupled pair of Fig. 11.22a converts a *large* input voltage into a proportional output current difference. This justifies

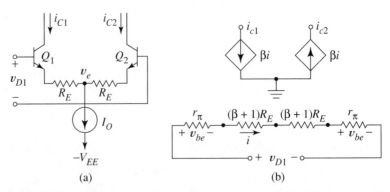

Figure 11.22 Voltage-to-current circuit: (a) schematic; (b) small-signal model.

adding resistances to the emitters of Q_1 and Q_2 in Fig. 11.20a to remove the small-signal restriction at the "lower" Gilbert cell input port.

When bases are grounded in Fig. 11.22a, we assume that transistors are in active-state operation, and each collector current is $I_O/2$. Now consider applying a differential input voltage v_{D1}, *not necessarily small*. Taking for granted for the time being that we can use small-signal analysis, we draw Fig. 11.22b. The derivation of Eq. (7.20) showed that the BJT small-signal model is valid provided that v_{be} is 5 mV or less. Figures 11.22a and b show that v_{be} is only that part of v_{D1} that drops across r_π; specifically,

$$v_{be} = \frac{r_\pi}{2r_\pi + 2(\beta + 1)R_E} v_{D1} \approx \frac{1}{2g_m R_E} v_{D1}$$

when $(\beta + 1)R_E \gg r_\pi$. By substituting $v_{be} = 5$ mV into this equation, we find that our small-signal model is valid for *input* signals that satisfy

$$|v_{D1}| < \frac{g_m R_E}{100} = \frac{[(I_O/2)/V_T]R_E}{100} = \frac{I_O R_E}{5} \tag{11.25}$$

Combining bias and signal analysis results gives

$$i_{C1} = \frac{I_O}{2} + \beta \frac{v_{D1}}{R_{in}}$$

where R_{in} is the circuit's input resistance, shown explicitly in Fig. 11.22b. Substituting for R_{in} and making the usual approximations for a high-beta transistor gives

$$i_{C1} = \frac{I_O}{2} + \frac{1}{2R_E} v_{D1} \tag{11.26}$$

and

$$i_{C2} = \frac{I_O}{2} - \frac{1}{2R_E} v_{D1} \tag{11.27}$$

Therefore,

$$i_{C1} - i_{C2} = \frac{1}{R_E} v_{D1} \tag{11.28}$$

We conclude that adding resistors R_E to the emitters of Q_1 and Q_2 in Fig. 11.20a expands the admissible signal band for v_{D1} in Fig. 11.20b to that described by Eq. (11.25). The Gilbert circuit, so modified, combines Eq. (11.28) with (11.22) to give

$$i_X - i_Y = \frac{1}{R_E} v_{D1} \tanh\left(\frac{v_{D2}}{2V_T}\right) \tag{11.29}$$

We next describe a special *predistortion circuit* that compensates for the nonlinear hyperbolic tangent function in this equation.

Predistortion Circuit. Figure 11.23a shows a *predistortion circuit*, a voltage-to-current circuit that uses diode-connected load transistors to produce output voltage v_X. Since the output is the difference of the V_{BE} drops of matched, diode-connected output transistors

$$v_X = V_T \ln\left(\frac{i_{C1}}{I_S}\right) - V_T \ln\left(\frac{i_{C2}}{I_S}\right) = V_T \ln\left(\frac{i_{C1}}{i_{C2}}\right) \tag{11.30}$$

Because bias and signal components of the collector currents are those of Eqs. (11.26) and (11.27),

$$v_X = V_T \ln\left(\frac{I_O/2 + v_{D2}/2R_E}{I_O/2 - v_{D2}/2R_E}\right) = V_T \ln\left(\frac{1 + v_{D2}/I_O R_E}{1 - v_{D2}/I_O R_E}\right)$$

The identity

$$\ln\left(\frac{1+x}{1-x}\right) = 2 \tanh^{-1}x, \qquad |x| < 1$$

(true because both sides have the same Taylor series) gives the equivalent equation

$$v_X = 2V_T \tanh^{-1}\left(\frac{v_{D2}}{I_O R_E}\right), \qquad \frac{v_{D2}}{I_O R_E} < 1 \tag{11.31}$$

When this *predistorted* signal, v_X, provides the "v_{D2} input" for the Gilbert multiplier equation (11.29), we have

$$i_X - i_Y = \frac{1}{R_E} v_{D1} \tanh\left[\tanh^{-1}\left(\frac{1}{I_O R_E} v_{D2}\right)\right] = \frac{1}{I_O R_E^2} v_{D1} v_{D2} \tag{11.32}$$

The large-signal multiplier circuit of Fig. 11.23b incorporates both the voltage-to-current circuit for signal v_{D1} and the predistortion circuit for signal v_{D2}. Equations (11.25) and (11.31) limit input signal amplitudes to

$$|v_{D1}| \le \frac{I_O R_E}{5} \quad \text{and} \quad |v_{D2}| \le I_O R_E \tag{11.33}$$

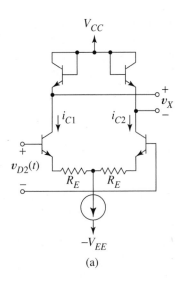

(a)

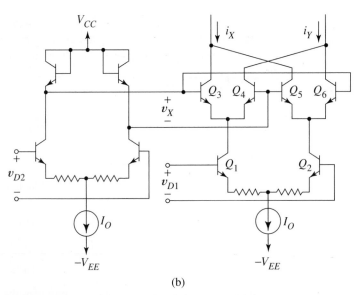

(b)

Figure 11.23 Large-signal four-quadrant multiplier: (a) predistortion circuit; (b) complete multiplier.

EXAMPLE 11.6 Design a large-signal, four-quadrant multiplier for maximum-amplitude input signals of $|v_{D1}| = 1$ V and $|v_{D2}| = 5$ V. Use ± 15 V supplies.

Solution. Selecting $|v_{D1}| = 1$ V in Eq. (11.33) gives $I_O R_E = 5$ V. If we choose $I_O = 1$ mA, then $R_E = 5$ kΩ, and Eq. (11.32) becomes

$$i_X - i_Y = 4 \times 10^{-5} v_{D1} v_{D2}$$

Figure 11.24 shows the known biasing information; collector resistors R_C are added to produce output voltage v_o. Because each carries 0.5 mA of bias current, resistors R_C must be small enough to avoid saturating Q_3–Q_6. If we select bias voltage $V_{CE} = 0.6$ V to give a relatively large output voltage but small signal swing, then R_C must satisfy

$$15 - R_C (0.5 \text{ mA}) = 13.6 + 0.6 = 14.2 \text{ V}$$

or $R_C = 1.6$ kΩ. The output signal voltage satisfies

$$v_o = (i_X - i_Y)1.6 \text{ k} = 0.064 v_{D1} v_{D2}$$

The maximum instantaneous output voltage would correspond to $v_{D1}v_{D2} = 5$ V; thus $|v_{o,max}| = 0.32$ V. The signal swings at the collectors of Q_3–Q_6, of amplitude $0.32/2 = 0.16$ V, are accommodated by the selected bias voltage $V_{CE} = 0.6$ V.   ❏

Lowering the positive supply voltage of the predistortion circuit to a value less than 15 V in the preceding example would enable us to use larger resistors R_C to obtain higher output voltage.

Exercise 11.7   Find the new value of R_C that biases Q_3–Q_6 at $V_{CE} = 3.4$ V if we replace the 15 V supply biasing the predistortion circuit by a 5 V supply. Determine the maximum instantaneous output signal voltage for this case.

Ans. $R_C = 16$ kΩ, $|v_{o,max}| = 3.2$ V.

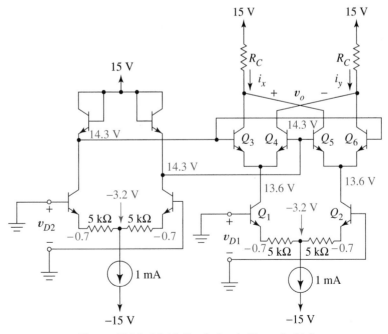

Figure 11.24 Multiplier design in Example 11.6.

11.3.4 BALANCED MODULATOR

The balanced modulator of Fig. 11.20 has another mode of operation, which involves driving Q_1 and Q_2 alternately between cut-off and active states with a carrier-frequency signal of large amplitude. This gives the equivalent circuits of Fig. 11.25, as the removal of bias current alternately cuts off different pairs of output transistors. Figure 11.25a is an emitter-coupled amplifier described by

$$v_o = -g_m R_C v_{d2} \qquad (11.34)$$

Figure 11.25b gives

$$v_o = g_m R_C v_{d2} \qquad (11.35)$$

To establish minimum input amplitude for this mode of operation, notice from Equation (11.13) that v_{D_1} diverts virtually all current through a single input transistor when $|i_{C1} - i_{C2}| > 0.99 I_O$, that is, when the input satisfies

$$|v_{D1}| > 2V_T \, \tanh^{-1} 0.99 = 132 \text{ mV} \qquad (11.36)$$

We next show that when the circuit operates in this fashion it is a balanced modulator. The output predicted by switching between Eqs. (11.34) and (11.35) at carrier frequency is equivalent to

$$v_o(t) = [-g_m R_C v_{d2}(t)] v_{D3}(t) \qquad (11.37)$$

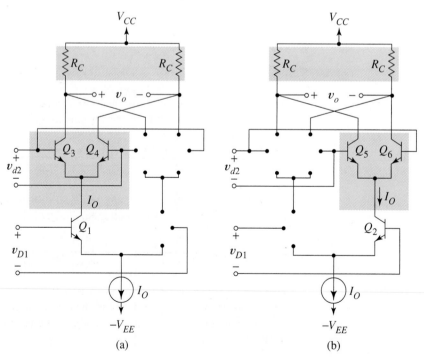

Figure 11.25 Balanced modulator with large carrier input: (a) $v_{D1} > 0$; (b) $v_{D1} < 0$.

where $v_{D3}(t)$ is a *fictitious* square wave of unit amplitude and carrier frequency that provides the changes in algebraic sign. For example, let $v_{D2}(t) = V_M \cos \omega_m t$, a low-frequency, modulating wave and let $v_{D1}(t)$ be a square wave of frequency $\omega_c \gg \omega_m$ having amplitude that satisfies Eq. (11.36). Then Fig. 11.26a shows the resulting waveforms.

The Fourier series of $v_{D3}(t)$ is

$$v_{D3}(t) = \sum_{n=1,3,5,\ldots}^{\infty} \frac{4}{n\pi} \sin n\omega_c t \tag{11.38}$$

For these waveforms, Eq. (11.37) gives

$$v_o(t) = -g_m R_C V_M \sum_{n=1,3,5,\ldots}^{\infty} \frac{4}{n\pi} \sin n\omega_c t \cos \omega_m t$$

A well-known identity shows this equivalent to

$$v_o(t) = \sum_{n=1,3,5,\ldots}^{\infty} \frac{-2g_m R_C V_M}{n\pi} [\sin(n\omega_c - \omega_m)t + \sin(n\omega_c + \omega_m)t] \tag{11.39}$$

The balanced modulator output waveform results from removing the terms of index $n > 1$ with a filter as in Fig. 11.26b.

11.3.5 DUAL-GATE MOSFET

A special transistor, the dual-gate MOSFET of Fig. 11.27a, is suitable for mixer, multiplier, and AGC applications. In the figure, a bias voltage connects the second gate to the source; Fig. 11.27b shows its schematic. When $v_{G2,S} = 4$ V, the output characteristics, Fig. 11.27c, resemble depletion MOSFET characteristics; however, the channel that exists for $v_{G1} = 0$ is created by biasing gate 2—not by fabrication. With the second gate, the channel conductivity can be dynamically controlled. We see this in Fig. 11.27d, which shows the output characteristics for $v_{G2,S} = 3$ V. The transfer characteristic family, Fig. 11.27e, depicts the drain current as a nonlinear function of the two gate-source voltages.

After establishing a static Q-point with a bias circuit, we apply small signals $v_{g1,s}$ and $v_{g2,s}$ at the gates. The resulting small-signal drain current is

$$i_d = \frac{\partial G}{\partial v_{G1,S}} v_{g1,s} + \frac{\partial G}{\partial v_{G2,S}} v_{g2,s} + \frac{\partial^2 G}{\partial v_{G2,S}\, \partial v_{G1,S}} v_{g1,s} v_{g2,s} \tag{11.40}$$

where each partial derivative is evaluated at the Q-point. This expression assumes signals so small that all higher-order terms are small enough to be ignored.

To construct a small-signal equivalent , we rewrite this equation in the small-signal notation

$$i_d = g_{m1} v_{g1,s} + g_{m2} v_{g2,s} + g_{m1,2} v_{g1,s} v_{g2,s} \tag{11.41}$$

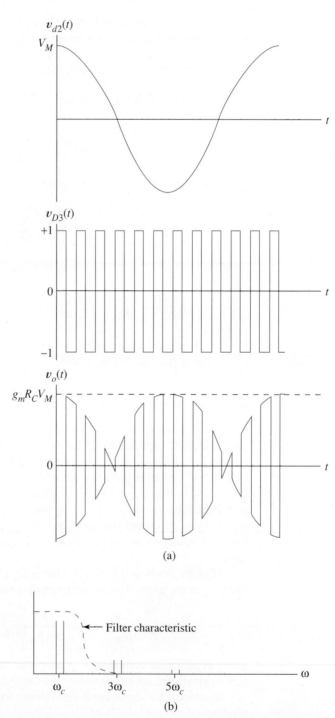

Figure 11.26 Balanced modulator: (a) equivalent waveforms for large-signal carrier and small-signal modulating sinusoid; (b) filter operation required to remove unwanted frequencies.

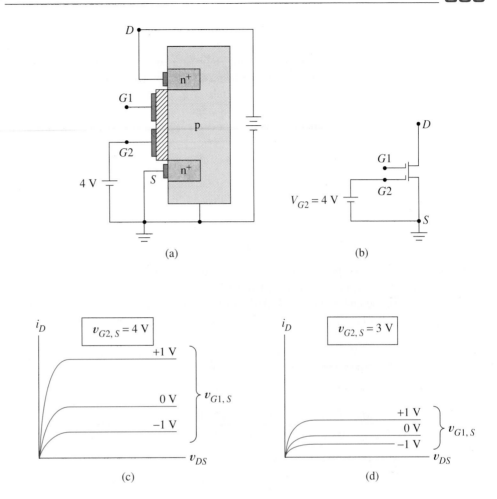

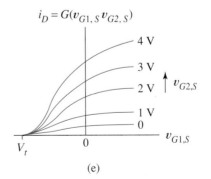

Figure 11.27 Dual-gate MOSFET: (a) physical structure; (b) schematic representation; (c) output characteristics for $v_{G2,S} = 4$ V; (d) output curves for $v_{G2,S} = 3$ V; (e) drain current as function of two variables.

which suggests the equivalent circuit of Fig. 11.28a. For the device to multiply, either the last term in Eq. (11.41) must be dominant or the other terms removed by filtering.

Sometimes the transistor data sheet lists small-signal parameters for a given Q-point. If this information is not given, or if we choose to use a different Q-point, we can estimate the values of g_{m1} and g_{m2} from a family of transfer characteristics by the methods introduced in Sec. 7.2.3. In addition to transfer characteristics, the data sheet sometimes includes a curve of g_{m1} versus $v_{G2,S}$ like that of Fig. 11.28b. Because

$$g_{m1,2} = \frac{\partial(\partial G/\partial v_{g1,s})}{\partial v_{G2,S}} = \frac{\partial g_{m1}}{\partial v_{G2,S}} \tag{11.42}$$

$g_{m1,2}$ is the slope of this g_{m1} curve at the Q-point.

EXAMPLE 11.7   Figures 11.29a and b describe a dual-gate MOSFET. Find g_{m1}, g_{m2}, and $g_{m1,2}$ graphically when the operating point is $(V_{G1,S}, I_D) = (2.5\ \text{V}, 23\ \text{mA})$.

Solution. Figure 11.29a shows that $V_{G2,S} = 4\ \text{V}$ at the designated Q-point. An asterisk marks the Q-point on each curve, and the two figures show the constructions.

In Fig. 11.29a, the end points of the dashed vertical (constant v_{G1}) line through the Q-point give g_{m2}; that is,

$$g_{m2} \approx \left. \frac{\Delta i_D}{\Delta v_{G2,S}} \right|_{V_{G1,S} = 2.5} = \frac{(30 - 15) \times 10^{-3}}{5 - 3} = 0.0075\ \text{A/V}$$

The slope of the tangent to the curve at $V_{G2,S} = 4\ \text{V}$ is

$$g_{m1} \approx \left. \frac{\Delta i_D}{\Delta v_{G1,S}} \right|_{V_{G2,S} = 4} = \frac{(26 - 20.5) \times 10^{-3}}{3.5 - 1.5} = 0.00275\ \text{A/V}$$

In Fig. 11.29b the slope of the tangent is

$$g_{m1,2} \approx \left. \frac{\Delta g_{m1}}{\Delta v_{G2,S}} \right|_{V_{G1,S} = 2.5} = \frac{(4 - 1.5) \times 10^{-3}}{5 - 3} = 0.00125\ \text{A/V}^2 \qquad \square$$

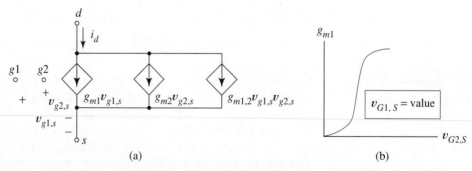

(a)   (b)

Figure 11.28   Dual-gate MOSFET: (a) small-signal equivalent circuit; (b) g_{m1} as a function of $v_{G2,S}$ evaluated at the Q-point value of $V_{G1,S}$.

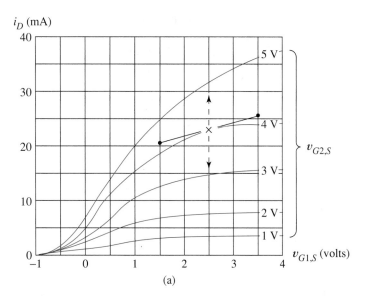

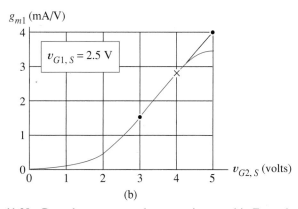

Figure 11.29 Data sheet curves and constructions used in Example 11.7.

Exercise 11.8 Find the coefficients in Eq. (11.40) when the Q-point in Fig. 11.29 is $(V_{G1,S}, I_D) = (2.5 \text{ V}, 14.5 \text{ mA})$

Ans. $g_{m2} \approx 8 \text{ mA/V}$, $g_{m1} \approx 1.5 \text{ mA/V}$, and $g_{m1,2} \approx 1.25 \text{ mA/V}^2$.

11.4
Phase-Locked Loop

A *phase-locked loop* is a negative-feedback circuit that accepts a frequency for its input, and presents either frequency or voltage as its output. Applications include demodulators for amplitude and frequency modulation, frequency multipliers, and frequency synthesizers. Phase-locked loops can be constructed from subcircuits or purchased as MSI devices. This section emphasizes basic operating principles common to all phase-locked loops, and hardware used in some IC implementations. We first discuss the phase-locked loop in terms of idealized components. Subsequently, we describe some practical components and relate their characteristics to loop operation.

11.4.1 LOOP COMPONENTS

Figure 11.30 shows the general phase-locked loop structure. The input signal, a sine (or square) wave of radian frequency ω_i and phase θ_i, is applied to a device called a *phase detector*. Phase-detector output, filtered and amplified, controls the frequency of a voltage-controlled oscillator (VCO). The VCO output, a sine or square wave of frequency ω_o, provides a second input for the phase detector. The output of the phase-locked loop, depending upon the particular application, is either the voltage v_3 that controls the VCO, or the frequency, ω_o, of the VCO output. Instantaneous phases and angular frequencies are related in the usual fashion; that is,

$$\omega_i = \frac{d\theta_i}{dt} \quad \text{and} \quad \omega_o = \frac{d\theta_o}{dt} \tag{11.43}$$

Figure 11.31 shows a simplified characteristic for each loop component. The filter and amplifier are familiar. The phase detector, mathematically described by

$$v_1 = K_P(\theta_i - \theta_o) \tag{11.44}$$

where K_P is a constant, produces output v_1 whenever the input and VCO signals differ in phase. The voltage-controlled oscillator, characterized by Fig. 11.31d, operates at its *free-running frequency*, ω_{FR}, when v_3 is zero. Positive or negative values of v_3 cause its frequency to increase or decrease relative to ω_{FR} according to

$$\omega_o = \omega_{FR} + K_o v_3 \tag{11.45}$$

where K_o is a constant. We next show how the phase detector and VCO interact during normal operation.

11.4.2 PHASE-LOCKED OPERATION

Static Lock Condition. To understand phase lock, assume for a starting point that the phase detector inputs are both sinusoids of frequency ω_{FR}, and assume both have identical phase. In Fig. 11.32a we represent this condition by rotating vectors whose instantaneous vertical projections are the two phase-detector inputs. Both vectors rotate at the constant angular velocity ω_{FR} rad/s, and there is no phase difference. This means that in Figs. 11.30 and 11.31, v_1, v_2, and v_3 are all zero. But according to Fig. 11.31d, $v_3 = 0$ is

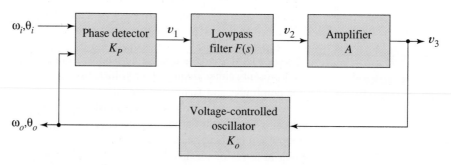

Figure 11.30 Phase-locked loop block diagram.

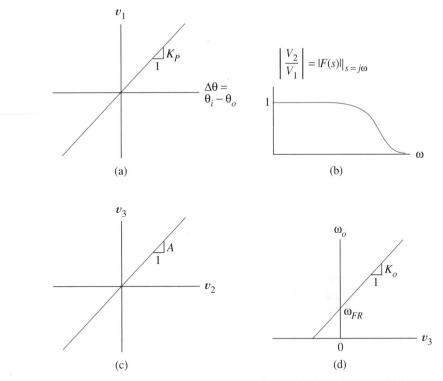

Figure 11.31   Idealized loop component characteristics: (a) phase-detector; (b) lowpass filter; (c) amplifier; (d) voltage-controlled oscillator.

exactly the VCO input voltage required to *keep* the local oscillator at frequency ω_{FR}, that is, matched to input frequency ω_i. Thus the loop is in equilibrium.

Now suppose ω_i suddenly takes on a new, higher value. Then, in Fig. 11.32a the ω_i vector, rotating faster, begins to pull ahead of the ω_o vector. This creates an increasing phase difference, $\Delta\theta = \theta_i - \theta_o$, as in Fig. 11.32b. The resulting phase-detector output voltage, v_1, is filtered and amplified in the loop of Fig. 11.30. In Fig. 11.31d, VCO frequency ω_o increases with the amplified error voltage, v_3, until it matches the new, higher, value of ω_i. Once again both vectors rotate at the same (higher) angular velocity, but now with a constant positive phase difference, $\Delta\theta$, representing a new state of equilibrium for the loop. A similar scenario ensues if ω_i decreases, with a negative phase difference leading to a new equilibrium through a negative v_3 and a lower VCO output frequency. Insofar as the loop is able to adjust the VCO frequency in response to changes in the input frequency in this fashion, we have a *phase-lock* condition, and we say the loop is *in lock*.

When the loop is in lock, v_3 is proportional to the VCO frequency, itself identical to the input frequency. If we substitute $\omega_i = \omega_o$ in Eq. (11.45) we find

$$v_3 = \frac{\omega_i - \omega_{FR}}{K_o}$$

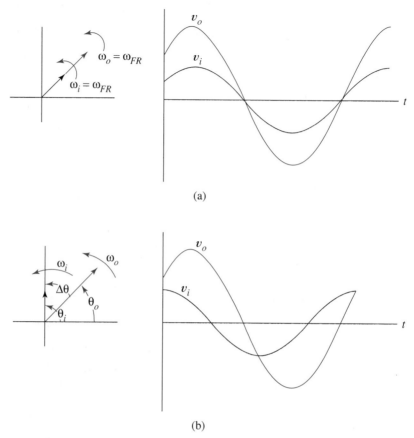

(a)

(b)

Figure 11.32 Creation of the loop error signal: (a) when inputs are identical in frequency and phase, phase difference is the constant, 0; (b) an incrase in input frequency causes, a positive phase error $\Delta\theta$.

We conclude that v_3 can be used as the loop output in applications where the function of the loop is to determine unknown input frequency ω_i.

Phase-locked loops often employ square waves for both input and VCO output. For square waves, the rotating vector concept of Fig. 11.32 still applies; however, the angular velocities represent the fundamental frequencies of the square waves. Binary-valued waveforms extend the use of phase-locked loops to a variety of interesting applications. For example, Fig. 11.33 shows a loop functioning as a *frequency multiplier*. A digital frequency divider produces a square wave output of frequency ω_o/N, which locks onto a stable input reference frequency, ω_{REF}. When the loop is in lock, the VCO operates at $\omega_o = N\omega_{REF}$, with frequency stability the same as that of the reference frequency. The frequency multiplier is transformed into a versatile waveform synthesizer by using a programmable divider to give alternate choices of N, and by providing a selection of alternate crystal-controlled reference frequencies at its input.

Linear circuit analysis applies to a loop in lock. We now exploit this observation by using linear analysis to describe the dynamic behavior of the loop in lock.

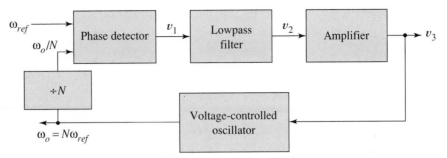

Figure 11.33 Frequency multiplier configuration.

Mathematical Description of a Loop in Phase Lock. We first describe each device in Fig. 11.30a by a transfer function, starting with the phase detector. The Laplace transform of Eq. (11.44) is

$$V_1(s) = K_P[\theta_i(s) - \theta_o(s)]$$

Since angular velocity, not phase angle, is the input to the loop, we convert the phase functions into angular velocities using Eq. (11.43). For zero initial conditions this gives

$$sV_1(s) = K_P[s\theta_i(s) - s\theta_o(s)] = K_P[\Omega_i(s) - \Omega_o(s)] \qquad (11.46)$$

where $\Omega(s)$ denotes the transform of angular velocity. To simplify the mathematics we choose to keep track of *deviations from the free-running frequency* rather than instantaneous frequencies themselves. Thus, we substitute

$$\Omega_i(s) = \frac{\omega_{FR}}{s} + \Delta\Omega_i(s) \quad \text{and} \quad \Omega_o(s) = \frac{\omega_{FR}}{s} + \Delta\Omega_o(s)$$

The result is a phase-detector equation relating output voltage to frequency deviations:

$$V_1(s) = \frac{K_P}{s}[\Delta\Omega_i(s) - \Delta\Omega_o(s)] \qquad (11.47)$$

From the filter function, $F(s)$, and the amplifier in Figs. 11.31b and c we obtain,

$$V_3(s) = AF(s)V_1(s)$$

and

$$V_3(s) = AF(s)\frac{K_P}{s}[\Delta\Omega_i(s) - \Delta\Omega_o(s)] \qquad (11.48)$$

The voltage-controlled oscillator of Fig. 11.31d satisfies

$$\Delta\Omega_o(s) = K_o V_3(s) \qquad (11.49)$$

Substituting $V_3(s)$ from Eq. (11.48) into Eq. (11.49) gives

$$\Delta\Omega_o(s) = K_o F(s)A\frac{K_P}{s}[\Delta\Omega_i(s) - \Delta\Omega_o(s)] \qquad (11.50)$$

Solving for the ratio of output variable $\Delta\Omega_o(s)$ to input variable $\Delta\Omega_i(s)$ gives the transfer function

$$\frac{\Delta\Omega_o(s)}{\Delta\Omega_i(s)} = \frac{K_P A\, K_o F(s)}{s + K_P A\, K_o F(s)}$$

when ω_o is the output. We can relate the phase-locked loop to negative feedback theory by interpreting this equation in terms of Fig. 9.1. Output variable x_o is $\Delta\Omega_o(s)$, input variable w_s is $\Delta\Omega_i(s)$, and $\beta = 1$. The phase detector provides the subtraction. If loop gain, $K_o F(s)A\, K_P$, is sufficiently high, we expect the overall gain of the phase-locked loop to be $A_f \approx 1/\beta = 1$; that is, the output frequency should be locked to the input frequency. This does indeed describe the steady-state condition when the loop is in lock, as we already explained using Fig. 11.32. By using Laplace transforms, however, we can also study the transient behavior of the loop. If we use v_3 for the output, substituting Eq. (11.49) leads to

$$\frac{V_3(s)}{\Delta\Omega_i(s)} = \frac{K_P A\, F(s)}{s + K_P A\, K_o F(s)} \tag{11.51}$$

Dynamics of a First-Order Loop. When there is no loop filter, that is, when $F(s) = 1$ in Eq. (11.51), the dynamics are those of the first-order lowpass system

$$\frac{V_3(s)}{\Delta\Omega_i(s)} = \frac{K_P A}{s + K_P A\, K_o} \tag{11.52}$$

characterized by time constant $\tau = 1/(K_P A\, K_o)$.

Figure 11.34 shows the response of the system when input frequency, initially $\omega_x < \omega_{FR}$, abruptly changes to a value $\omega_y > \omega_{FR}$ for T seconds and then returns to its original value. Figure 11.34a shows the frequency variation, Fig. 11.34b the input voltage waveform, and Fig. 11.34c the output voltage $v_3(t)$. Because $\omega_o(t)$ is directly proportional to $v_3(t)$, we see that the VCO *frequency* increases exponentially to its new value rather than changing instantaneously as does the input. Unless the loop is designed to have a short time constant (high loop gain) compared to T, the circuit distorts the frequency pulse just as an RC lowpass filter distorts a pulse by adding rise and fall times to the response. Although frequency might seem a little odd as a bearer of information, we see that any information encoded as changes in frequency is filtered by the lowpass system of Eq. (11.52) just as voltage information was filtered in previous lowpass circuits in this text. From Eq. (11.52), we conclude that there is a *filter bandwidth*, in this case, $\omega_B = K_P A\, K_o$ rad/s.

Dynamics of a Second-Order Loop. For reasons that become clear when we study practical phase detectors, phase-locked loops require a filter function $F(s) \neq 1$. When $F(s)$ is a first- or higher-order filter, then the phase-locked loop exhibits second- or higher-order dynamic behavior. We now briefly examine the important second-order case.

If we substitute the lowpass filter function

$$F(s) = \frac{\omega_H}{s + \omega_H} \tag{11.53}$$

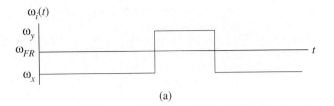

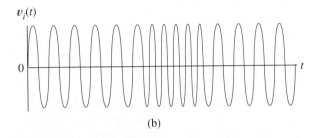

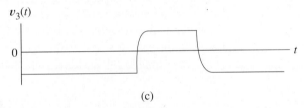

Figure 11.34 Transient response of phase-locked loop when $F(s) = 1$: (a) input frequency change; (b) input voltage waveform; (c) output voltage.

into Eq. (11.51), the result is

$$\frac{V_3(s)}{\Delta\Omega_i(s)} = \frac{(K_P A\,\omega_H)/(s + \omega_H)}{s + (K_P A\,K_o\,\omega_H)/(s + \omega_H)} = \frac{K_P A\,\omega_H}{s^2 + \omega_H s + K_P A\,K_o\,\omega_H}$$

In terms of loop gain,

$$K_L = K_P A\,K_o \tag{11.54}$$

the transfer function is

$$\frac{V_3(s)}{\Delta\Omega_i(s)} = \frac{(K_L/K_o)\omega_H}{s^2 + \omega_H s + K_L\,\omega_H}$$

In standard second-order system notation, this equation takes the form

$$\frac{V_3(s)}{\Delta\Omega_i(s)} = G(s) = \frac{1}{K_o}\,\frac{\omega_N^2}{s^2 + 2\zeta\omega_N s + \omega_N^2} \tag{11.55}$$

where the *undamped natural frequency* ω_N and the *damping ratio* ζ are given in terms of loop parameters by

$$\omega_N = \sqrt{K_L \omega_H} \tag{11.56}$$

and

$$\zeta = \frac{1}{2} \sqrt{\frac{\omega_H}{K_L}} \tag{11.57}$$

Thus, the information that enters the loop encoded as frequency changes appears as output $v_3(t)$, but only after being filtered by the second-order lowpass filter function $G(s)$.

Figure 11.35a shows the frequency response, Eq. (11.55), in terms of system parameters. We must distinguish between two different bandwidths, the *filter bandwidth*, ω_H, and the *loop bandwidth*, which defines the frequency response of the closed-loop system. The dashed line shows that when the damping ratio is less than 0.707, the -3 dB loop bandwidth is larger than the undamped natural frequency ω_N. Table 11.2 shows how loop bandwidth relates to damping ratio and ω_N. Equations (11.56) and (11.57) show that high-loop gain, K_L, gives large signal bandwidth and short response time to transient changes in signal frequency. Excessively high values of loop gain, however, lower the damping ratio, introducing overshoot or even ringing in the system step response as we see in Fig. 11.35b.

EXAMPLE 11.8   A phase-locked loop employs a first-order, lowpass loop filter $F(s)$ with unity gain and bandwidth ω_H. The overall filter response $G(s)$ should have -3 dB bandwidth of 720 Hz and the slightly peaked frequency response that corresponds to $\zeta = 0.5$. Find values for ω_H and K_L.

Solution. From Table 11.2, the normalized bandwidth is

$$\frac{2\pi 720}{\omega_N} \approx 1.27$$

TABLE 11.2
Normalized Loop
Bandwidth for Various
Damping Ratios

ζ	Bandwidth/ω_N
0.1	1.545
0.25	1.485
0.5	1.27
0.707	1.0
1.0	0.645

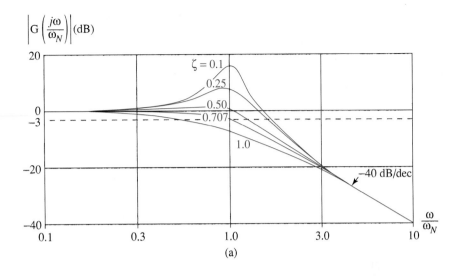

$$\left|G\left(\frac{j\omega}{\omega_N}\right)\right|\text{(dB)}$$

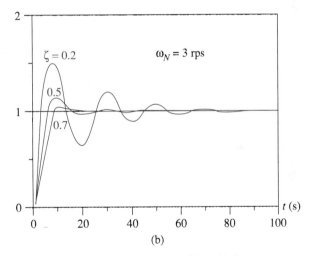

Figure 11.35 Phase-locked loop with first-order filter: (a) frequency response; (b) response to a step change in input frequency.

or $\omega_N = 3562$ rad/s. We have two design equations: Eq. (11.56), which gives

$$\sqrt{K_L\omega_H} = 3562 \text{ rad/s}$$

and Eq. (11.57), which for $\zeta = 0.5$ is

$$1 = \sqrt{\frac{\omega_H}{K_L}}$$

From the product of the equations,

$$\omega_H = 3562 \text{ rad/s}$$

Thus the bandwidth of $F(s)$ should be $\omega_H = 3562$ rad/s. Substituting this into the first equation gives

$$3562 = \sqrt{K_L\, 3562}$$

The loop gain should be $K_L = K_P A\, K_o = 3562\ \text{s}^{-1}$. ❏

Exercise 11.9 Find ω_H and loop gain K_L for a second-order loop with bandwidth of 10 rad/s using $\zeta = 1$.

Ans. $\omega_H = 31$ rad/s, $K_L = 7.75\ \text{s}^{-1}$.

11.4.3 PHASE DETECTOR

We next discuss three different kinds of phase detector. The first is a digital exclusive OR circuit, an important phase detector in its own right and a useful starting point for studying other phase detectors.

Digital Phase Detector. In the *EOR phase detector* of Fig. 11.36a, the input voltages, $v_i(t)$ and $v_o(t)$, are square waves with fundamental frequencies ω_i and ω_o. All voltages alternate between the logic one and zero levels of the EOR. For convenience, we assume an ideal EOR gate that operates at logic one and logic zero levels of 5 V and -5 V, respectively. The truth table of Fig. 11.36b describes the EOR function in terms of voltage levels. The output is -5 V when inputs are alike and 5 V when they differ.

Figure 11.36c shows input waveforms and the corresponding EOR output waveform and defines phase in terms of the period of the square wave. The output, related to inputs at each instant by the truth table, is periodic when input frequencies are identical. It is the average or dc component of this output waveform that actually characterizes the phase detector. A lowpass filter like that in Fig. 11.30 removes undesired frequencies from the phase detector output but passes the dc component.

To find the phase-detector characteristic, first visualize $v_o(t)$ in Fig. 11.36c shifted to the left so that the phase difference is zero. Then EOR inputs are always identical, so $v_1(t)$ is constant at -5 V. If $v_o(t)$ is shifted to the right so that it lags $v_i(t)$ by 90°, the inputs are the same half the time and differ half the time, and the output is a 5 V square wave with average value of zero. For 180° phase difference, the inputs *always differ*, so the output is constant at $+5$ V. If we shift the phase of $v_o(t)$ continuously relative to $v_i(t)$ from 0° to 180°, the average of $v_1(t)$ increases in proportion to the overlap in the waveforms as in Fig. 11.36d.

Since a positive shift in $v_o(t)$ results in the same output waveform as a negative shift of equal amount, the phase characteristic is an even function of phase difference $\Delta\theta$. Since a shift $\Delta\theta + n360°$, for any integer n, produces the same output waveform as a shift of $\Delta\theta$, the phase detector characteristic is also a periodic function of $\Delta\theta$.

For the EOR phase detector, the rotating vectors in Fig. 11.32 represent fundamental frequencies of the input square waves. Because the characteristic of Fig. 11.36d gives zero output voltage for 90° instead of for zero, our explanation of phase-locked loop operation must be modified slightly. With the EOR phase detector, a

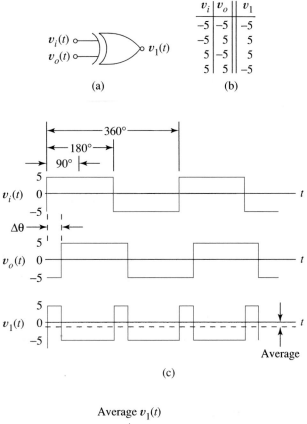

v_i	v_o		v_1
----	----		----
-5	-5		-5
-5	5		5
5	-5		5
5	5		-5

(a)　　　　　　　　　　　　(b)

(c)

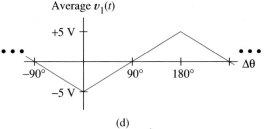

(d)

Figure 11.36 EOR phase detector: (a) circuit; (b) truth table using logical one and zero voltages; (c) input and output waveforms; (d) phase-detector characteristic.

90° phase difference corresponds to the free-running frequency. When ω_i pulls ahead, the phase difference advances beyond 90° toward 180°; when ω_i slips behind in phase, the phase difference decreases from 90° toward zero. The slope of the phase detector characteristic of Fig. 11.36d is

$$K_P = \frac{10}{\pi} \text{ V/rad} \tag{11.58}$$

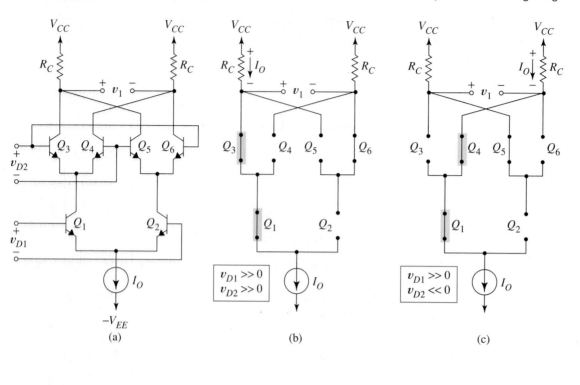

(a)

(b)

(c)

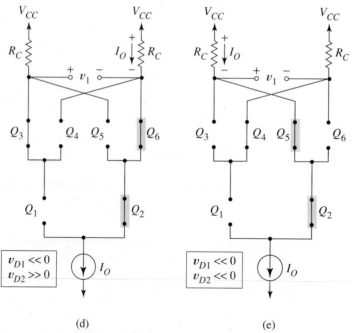

(d)

(e)

Figure 11.37 Large-signal Gilbert cell phase detector: (a) complete circuit; (b)–(e) equivalents showing how bias current is routed to output for different excitation combinations.

EXAMPLE 11.9 A phase detector characterized by Fig. 11.36d has a dc output voltage of -4 V. If the phase detector is part of a phase-locked loop, in lock, with parameters $A = 2$, $\omega_{FR} = 10^7$ rad/s, and $K_o = 10^5$ (volt-second)$^{-1}$,
(a) Find the input frequency.
(b) Sketch the rotating phasors.

Solution. (a) The VCO input voltage is $V_3 = -4 \times 2 = -8$ V. From Eq. (11.45), $\omega_o = 10^7 + 10^5 x(-8) = 9.2 \times 10^6$ rad/s. Since the loop is in lock, this is the input frequency.
(b) From the slope of Fig. 11.36d,

$$\frac{5 - (-4)}{180° - \Delta\theta} = \frac{10}{180°}$$

which gives $\Delta\theta = 18°$. The sketch resembles Fig. 11.32b but with $\Delta\theta = 18°$. ❑

Exercise 11.10 A phase-locked loop uses the phase detector of Fig. 11.36d, and the loop is in lock.
(a) If $A = 2$, $\omega_{FR} = 10^7$ rad/s and $K_o = 10^5$ V/s, find the input frequency for which the phase detector output voltage is 4.8 V.
(b) What is the phase difference between input and output when $V_1 = 4.8$ V?

Ans. 1.096×10^7 rad/s, $176.4°$.

Large-Signal Gilbert Cell Phase Detector. When *both* inputs to the Gilbert cell circuit of Fig. 11.37a are large, *all* transistors operate between cut-off and active states, diverting the bias current along various pathways controlled by the two inputs. Signals are square waves or sine waves of amplitude sufficient to cut off one transistor in each emitter-coupled pair. Figures 11.37b–e show the four possibilities by representing each transistor as a switch. The notation $>> 0$ means signal amplitude is sufficient to cut off transistors as required. (Equation (11.36) gives the required amplitude.)

The diagrams show that the output voltage relates to input as in Fig. 11.38a. Since this is the EOR function of Fig. 11.36b, input waveforms are processed as described in Fig. 11.36c, and we conclude that the Gilbert cell phase detector is characterized by

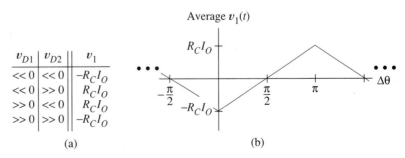

v_{D1}	v_{D2}	v_1
$<< 0$	$<< 0$	$-R_C I_O$
$<< 0$	$>> 0$	$R_C I_O$
$>> 0$	$<< 0$	$R_C I_O$
$>> 0$	$>> 0$	$-R_C I_O$

(a)   (b)

Figure 11.38 Gilbert cell with large signals: (a) truth table; (b) phase-detector characteristic.

Fig. 11.38b. When used in a phase-locked loop, 90° phase difference corresponds to operation at the free-running frequency. The phase detector constant is

$$K_P = \frac{2R_C I_O}{\pi} \text{ V/rad} \tag{11.59}$$

Small-Signal Gilbert Cell Phase Detector. In Sec. 11.3.4 we learned that when a large signal is used for $v_{D1}(t)$ and a small signal for $v_{D2}(t)$ in Fig. 11.37a, the Gilbert cell output is given by Eqs. (11.34) and (11.35). In small-signal notation, this means

$$v_1(t) = -g_m R_C v_{d2}(t), \qquad v_{D1}(t) >> 0 \tag{11.60}$$

and

$$v_1(t) = g_m R_C v_{d2}(t), \qquad v_{D1}(t) << 0 \tag{11.61}$$

These describe a small-signal amplifier with binary-valued phase shift (zero or 180°) that is controlled by $v_{D2}(t)$. Less obvious is that with a sinusoid for $v_{d2}(t)$ and a square wave for $v_{D1}(t)$ the circuit detects phase. We next show that this is so.

Figure 11.39 shows input and output waveforms when the phase difference between sine and square waves is $\Delta\theta$ rad/s and

$$v_{d2}(t) = V_I \sin\theta_i, \qquad \theta_i = \omega_i t$$

The phase-detector output is the average value of $v_1(t)$. Therefore,

$$
\begin{aligned}
v_1(t)_{AVE} &= \frac{1}{\pi} \int_0^{\Delta\theta} g_m R_C V_I \sin\theta_i \, d\theta_i - \frac{1}{\pi} \int_{\Delta\theta}^{\pi} g_m R_C V_I \sin\theta_i \, d\theta_i \\
&= -\frac{g_m R_C V_I}{\pi}(\cos\Delta\theta - 1) + \frac{g_m R_C V_I}{\pi}(\cos\pi - \cos\Delta\theta_i) \tag{11.62} \\
&= -\frac{2\,g_m R_C V_I}{\pi}\cos\Delta\theta
\end{aligned}
$$

Figure 11.39b is a sketch of Eq. (11.62), the phase-detector characteristic. Like the other phase detectors we have examined, there is a 90° phase difference at the free-running frequency. The novelty of Fig. 11.39b is the nonlinearity of the curve; nevertheless, this phase detector maintains phase lock in the ordinary fashion. For linear analysis at frequencies near ω_{FR}, the slope of the curve is

$$K_P = \frac{2\,g_m R_C V_I}{\pi} \tag{11.63}$$

It is important to note that K_P is a function of signal amplitude, a feature that complicates loop operation.

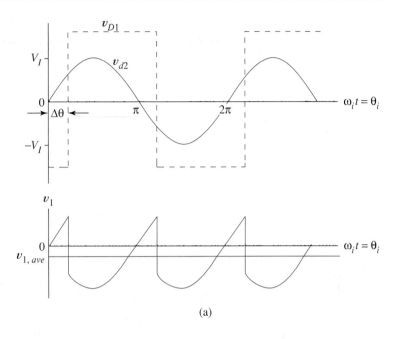

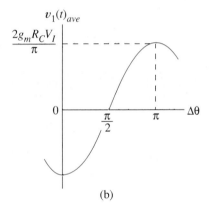

Figure 11.39 Gilbert cell phase detector with one small input: (a) input and outwave waveforms; (b) phase-detector characteristic.

11.4.4 LOCK AND CAPTURE RANGES

Now we have the background to understand two additional phase-locked loop parameters, lock range and capture range. Once in lock, the loop remains in lock only for input frequencies, ω_i, that satisfy

$$\omega_{FR} - \frac{\Delta\omega_L}{2} < \omega_i < \omega_{FR} + \frac{\Delta\omega_L}{2}$$

where $\Delta\omega_L$ is the *lock range*. Capture range has to do with a loop that is not in lock. The loop is able to *capture* (lock onto) the input frequency ω_i only if ω_i comes within the interval

$$\omega_{FR} - \frac{\Delta\omega_C}{2} < \omega_i < \omega_{FR} + \frac{\Delta\omega_C}{2}$$

where $\Delta\omega_C$ is the *capture range*. We next relate lock and capture ranges to the other loop parameters.

Lock Range. We understand from Figs. 11.30 and 11.31 that whenever a change in ω_i alters the phase difference $\Delta\theta$, the phase detector initiates a correction of the VCO frequency so the loop can remain in lock, and it is the amplified change in v_1 that adjusts the VCO to the new value. We now know that there are limits to the amplitude of v_1 and, consequently, to the amount of frequency correction possible. When the phase difference becomes so great that further frequency correction becomes impossible, the loop slips out of lock. Starting with Fig. 11.40 and working backward through the loop we find

$$\frac{\Delta\omega_L}{2} = K_o\Delta v_3 = K_o A \Delta v_{1,max} \tag{11.64}$$

For the exclusive OR phase detector of Fig. 11.36d, $\Delta v_{1,max} = 5$ volts, giving

$$\Delta\omega_L = 10\,K_o A$$

The large-signal Gilbert cell detector has lock range of

$$\Delta\omega_L = 2\,K_o A\,R_C I_O \tag{11.65}$$

which includes design parameters R_C and I_O. When one Gilbert cell input is a small signal of amplitude V_I, from Fig. 11.39b,

$$\Delta\omega_L = K_o A\frac{4g_m R_C}{\pi}V_I \tag{11.66}$$

In this case, lock range depends on signal amplitude V_I.

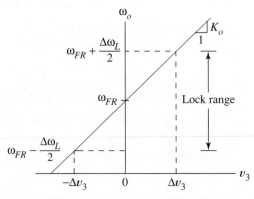

Figure 11.40 VCO characteristic showing how lock range is determined by maximum correction voltages $\pm\Delta v_3$.

Capture Range. A conceptual experiment introduces the idea of capture range. Imagine a loop with its VCO operating at free-running frequency ω_{FR} when input frequency is $\omega_i \ll \omega_{FR}$. The loop is not in lock. The arrowheads in Fig. 11.41a show what happens as ω_i slowly increases from its original value at the extreme left side of the diagram to a value $\omega_i \gg \omega_{FR}$. The VCO operates at ω_{FR} until the input reaches that frequency, $\omega_{FR} - 0.5\,\Delta\omega_C$, which defines the lower edge of the *capture range*. Here the input signal is *captured*, meaning the VCO frequency locks onto ω_i. Once in lock, ω_o follows ω_i to the upper edge of the *lock range*. As ω_i increases beyond $\omega_{FR} + 0.5\,\Delta\omega_L$, the loop slips out of lock, and ω_o reverts to ω_{FR}. As ω_i is now slowly reduced, operation follows a different path. Capture occurs at $\omega_{FR} + 0.5\,\Delta\omega_C$, and the loop slips out of lock only when the input drops below $\omega_{FR} - 0.5\,\Delta\omega_L$. An intuitive description of the capture mechanism follows.

Our derivations of phase-detector characteristics assumed a locked loop with two inputs of identical frequency. When the loop is not in lock, the phase-detector output contains sum and difference frequencies $(\omega_{FR} + \omega_o)$ and $(\omega_{FR} - \omega_o)$. We know this is true for the small-signal Gilbert cell phase detector, because in Sec. 11.3.4 we showed that the circuit functions as a balanced modulator when input frequencies differ. We need this idea (which holds for the other phase detectors as well) to understand the capture mechanism.

Capture Transient. We now rethink the experiment of Fig. 11.41a, this time including the nonlinearity of the phase detector. Because the capture process is nonlinear, our explanation is highly intuitive and approximate.

Figure 11.41b shows input and VCO frequencies, ω_i and ω_o, and how their difference relates to the filter passband when $\omega_i \ll \omega_{FR}$. Because the difference frequency, $\omega_o - \omega_i$, in the phase detector output greatly exceeds the filter bandwidth, filter output $v_2(t)$ is zero and the VCO frequency is constant at $\omega_o = \omega_{FR}$.

As ω_i approaches ω_{FR} in Fig. 11.41c, a difference-frequency voltage v_1 enters the filter and leaves as a low amplitude voltage $v_2(t)$. This produces a difference-frequency VCO input voltage, $v_3(t) = Av_2(t)$, which pulls ω_o back and forth about ω_{FR} at a rate of $\omega_o - \omega_i$. Because of the small amplitude of $v_3(t)$, this involves only small frequency changes. (The difference frequency in the filter responds to changes in ω_o, but for simplicity we ignore this.)

As ω_i approaches ω_{FR} more closely, the frequency difference grows smaller and the filter output amplitude grows larger. Consequently, the excursions of ω_o become larger and slower, and ω_o swings nearer to ω_i as in Fig. 11.41d. Capture occurs when the maximum negative excursion of the ω_o line in Fig. 11.41d reaches ω_i, that is, when $\omega_{FR} - 0.5\,\Delta\omega_C = \omega_i$. When this occurs, the difference-frequency component of voltage $v_2(t)$ passing through the filter is $\omega_{FR} - \omega_i = 0.5\,\Delta\omega_C$. Figure 11.41e suggests how the VCO voltage might actually look during capture.

At the instant the amplified filter output, $v_2(t)$, pulls the VCO frequency by $0.5\,\Delta\omega_C$, we have

$$\frac{\Delta\omega_C}{2} = K_o A |v_2|_{max} \qquad (11.67)$$

Now let us examine the difference-frequency signal that *enters* the filter at this capture instant. Until the loop locks, the vectors in Fig. 11.32b rotate at different speeds, causing

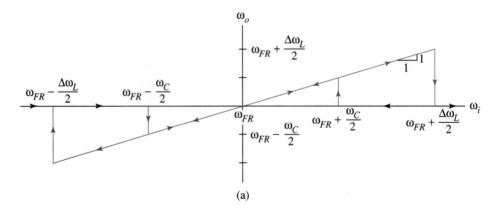

(a)

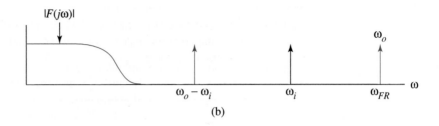

(b)

(c)

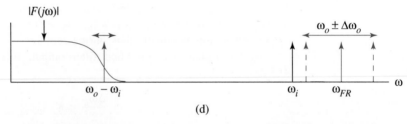

(d)

Figure 11.41 Capture: (a) steady-state frequency diagram; (b) loop frequencies when ω_i $\ll \omega_o$; (c) loop frequencies as ω_i approaches ω_o; (d) loop frequencies close to lock; (e) capture transient waveform.

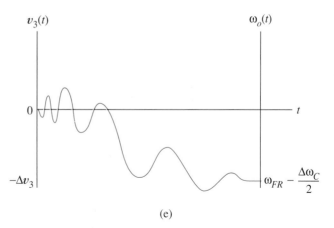

(e)

Figure 11.41 (continued)

the phase difference to change continually, and suggesting that phase-detector output $v_1(t)$ has a peak amplitude of $|v_1|_{max}$. At the instant of capture the difference frequency is $0.5\,\Delta\omega_C$. Therefore, filter output and input satisfy

$$|v_2|_{max} = \left| F\left(j\frac{\Delta\omega_C}{2} \right) \right| |v_1|_{max}$$

Since the peak amplitude of the phase detector output, $|v_1|_{max}$, is related to the lock range by Eq. (11.64), we have

$$|v_2|_{max} = \left| F\left(j\frac{\Delta\omega_C}{2} \right) \right| \frac{\Delta\omega_L}{2K_oA} \qquad (11.68)$$

Combining this with Eq. (11.67) gives

$$\left(\frac{2}{\Delta\omega_L} \right) \frac{\Delta\omega_C}{2} = \left| F\left(j\frac{\Delta\omega_C}{2} \right) \right| \qquad (11.69)$$

which can be solved by iteration for a given filter function $F(s)$. To see this, replace $0.5\,\Delta\omega_c$ by the variable u and notice that the iteration corresponds to locating the intersection of the straight line and the filter characteristic in Fig. 11.42 for a given constant, $\Delta\omega_L$.

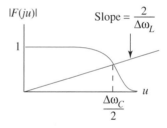

Figure 11.42 Graphical solution of Eq. (11.69) to estimate capture range.

11.4.5 VOLTAGE-CONTROLLED OSCILLATOR

The voltage-controlled oscillator we study here is an extension of the current-controlled relaxation oscillator of Fig. 11.43a, a circuit useful in its own right for generating triangular and square waves. The oscillator is an interconnection of familiar subcircuits: two constant current sources, a mirror that doubles its reference current, a timing capacitor whose voltage provides input for a Schmitt trigger, and a switch, Q_s. Figure 11.43b shows the Schmitt trigger characteristic, where V_L might be negative in some implementations.

When v_O is high, Q_s is closed, and reference current is diverted from the mirror, causing I_x to be zero. Then C charges by virtue of constant current I_O, causing v_X to increase linearly until it reaches V_2, the upper Schmitt trigger input threshold. At this

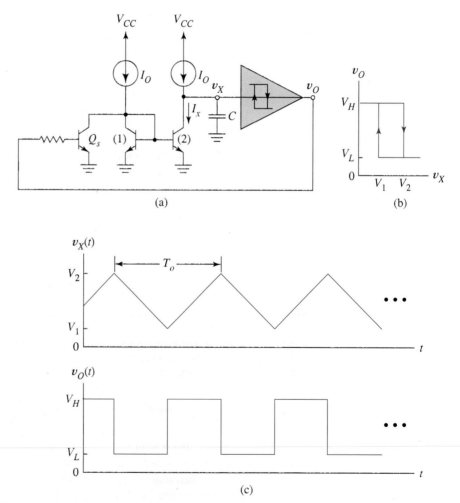

(a)

(b)

(c)

Figure 11.43 Current-controlled oscillator: (a) circuit; (b) Schmitt trigger characteristic; (c) output waveforms.

point, v_O drops to V_L, cutting off Q_s and adding discharge current $I_X = 2I_O$ to the capacitor circuit by virtue of the double-area mirror transistor. The timing capacitor, now with net *discharge* current of I_O, discharges linearly until $v_X = V_1$. The circuit then reverts to its original state and repeats its previous behavior.

Figure 11.43c shows the steady-state waveforms. Since the capacitor charges and discharges at the rate

$$\left| \frac{dv_X}{dt} \right| = \frac{I_O}{C} \text{ V/s}$$

and since it takes $(T_o/2)$ s to charge from V_1 to V_2, the slope is

$$\frac{V_2 - V_1}{0.5\,T_o} = \frac{I_O}{C} \text{ V/s}$$

Therefore, the frequency is

$$f_o = \frac{1}{T_o} = \frac{I_O}{2C(V_2 - V_1)} \text{ Hz} \tag{11.70}$$

hence the name *current-controlled oscillator*. To make a voltage-controlled oscillator, we replace each current source with a *voltage-controlled current source*, our next circuit.

Voltage-Controlled Current Source. Figure 11.44a is a voltage-controlled current source, with bias current I_{BB} provided to ensure that both transistors operate in their active regions. Since the base-emitter drops of the transistors approximately cancel, the voltage across R is $V_{CC} - v_C$. Assuming high β for the pnp transistor gives

$$I_O = \frac{V_{CC} - v_C}{R}, \qquad V_{min} \leq v_C \leq V_{CC} \tag{11.71}$$

where V_{min} depends upon the design of biasing source I_{BB}.

Figure 11.44b shows the complete voltage-controlled oscillator. Comparing this circuit with Fig. 11.43a shows the following changes: identical voltage-controlled sources with compound transistors for high beta (see Fig. 7.28) replace the current-source transistors, a Wilson mirror replaces the basic mirror for improved current gain, and biasing current I_{BB} comes from a separate mirror circuit. If we substitute the current of Eq. (11.71) into Eq. (11.70), we obtain the VCO characteristic

$$\omega_o = 2\pi f_o = 2\pi \frac{V_{CC} - v_C}{2\,CR(V_2 - V_1)} \text{ rad/s} \tag{11.72}$$

This describes an oscillator with frequency linearly related to v_c as in Fig. 11.44c.

Interconnecting subcircuits like phase detectors and VCOs to form systems like phase-locked loops often requires *signal conditioning*. For example, there are two important differences between the VCO characteristics of Figs. 11.44c and 11.31d. The former has negative, not positive slope. Also, it does not allow one to pull frequency in a direction that depends upon the polarity of the input signal as required in the phase-locked

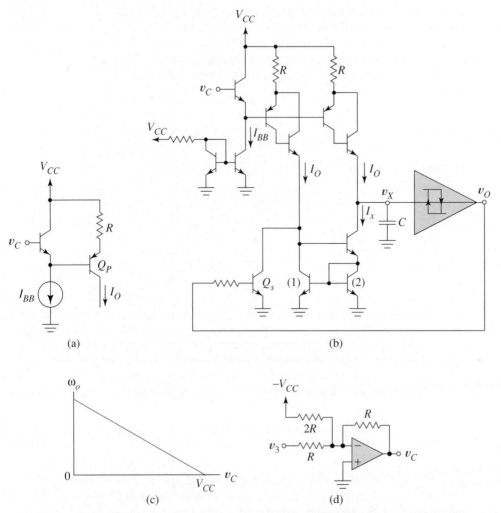

Figure 11.44   VCO circuit: (a) basic voltage-controlled current source; (b) complete voltage-controlled oscillator; (c) VCO characteristic; (d) circuit implementing variable transformation.

loop. In this case, solving the conditioning problem means devising a circuit that performs the change in variables

$$v_C = -v_3 + \frac{V_{CC}}{2} \tag{11.73}$$

The operating range of v_c in Eq. (11.71), $0 < v_C < V_{CC}$, requires that the conditioning circuit work over the range

$$-\frac{V_{CC}}{2} < v_3 < +\frac{V_{CC}}{2} - V_{min} \tag{11.74}$$

Substituting v_C from Eq. (11.73) into Eq. (11.72) gives the VCO equation

$$\omega_o = \pi \frac{V_{CC}/2 + v_3}{RC(V_2 - V_1)} \qquad (11.75)$$

which assumes that v_3 satisfies Eq. (11.74). Notice that this equation does have the form of Fig. 11.31d. The circuit of Fig. 11.44d provides a convenient IC realization of Eq. (11.73) for a system that includes a negative supply. Problem 2.82 analyzes a Schmitt trigger circuit with characteristic like that of Fig. 11.43b. The Schmitt trigger circuit might also require signal conditioning to adjust its input or output levels, or both, to match the requirements of this application.

11.5
Summary

The operational transconductance amplifier is an off-the-shelf IC realization of a voltage-controlled current source, with transconductance controlled by an external current I_X. Circuits such as voltage-controlled filters and amplifiers use this current-control feature. When a second signal causes I_X, and therefore the circuit transconductance, to change with time, the circuit becomes time variant. Then it can produce sum and difference frequencies at its output, leading to frequency mixing applications.

The 741 operational amplifier is the cascade of a transconductance amplifier, an inverting voltage amplifier, and a class AB output stage. A compensating capacitor straddles the voltage amplifier, ensuring unconditional stability while limiting bandwidth and slew rate. Biasing employs current sources, with dc negative feedback used to stabilize the Q-points of the first-stage transistors. DC analysis is routine except that V_{BE} drops must be computed for those transistors biased at low collector currents. Once dc collector currents are estimated, small-signal analysis techniques predict with reasonable accuracy the data sheet specifications.

This chapter discussed three circuits that perform analog multiplication. The first, an emitter-coupled pair with a signal introduced through the current source, is capable of two-quadrant multiplication, in which one multiplicand must be positive. An additional constraint is a critical biasing condition required to eliminate an extra, undesired output component. Because the output of the basic circuit is the difference of collector currents, external resistors are necessary to produce a double-ended output voltage.

The second multiplier circuit was the Gilbert cell, an interconnection of three emitter-coupled pairs. When both differential input voltages are small, the Gilbert cell is a four-quadrant multiplier, meaning both inputs can be signed variables. Modifications to the Gilbert cell result in a large-signal, four-quadrant, multiplier. In applications, it is necessary to add output resistors and sometimes input biasing to ensure that all transistors remain active.

The basic Gilbert cell has a special operating mode in which one input is large enough to drive a pair of input transistors between active and cut-off states. The result is a small-signal amplifier in which the algebraic sign of the gain is controlled by a second signal. In his mode, the Gilbert cell circuit can function as a balanced modulator or a phase detector.

The third circuit capable of analog multiplication is actually a discrete device rather than an IC, the dual-gate MOSFET. A signal applied to the extra gate modulates the channel conductivity and thus controls the transconductance, much as the current-source input controls the gain in the OTA. Very compact in size, these devices perform small-signal analog multiplication at higher frequencies than the more complex multiplier ICs.

The phase-locked loop is a negative feedback circuit in which the input variable is frequency, and the output is either frequency or voltage. A phase detector keeps the VCO frequency locked onto the input frequency by sensing the phase difference between the input and the VCO output and by generating a correction voltage for the VCO. The closed-loop dynamics of the loop depend upon the characteristics of the loop filter and upon loop gain, the product of the static gains of the loop components.

Two kinds of phase detectors were discussed at the transistor level. The EOR-type detector, realizable by a Gilbert cell with large input signals, gives a linear relation between output voltage and phase difference. When one Gilbert cell input is a small signal, the phase-detector characteristic takes the form of a cosine function with amplitude proportional to signal amplitude. The phase-detector characteristic is important because the maximum amplitude of its output voltage determines the lock range of the loop. The loop also has a capture range that depends on the lock range and the characteristics of the loop filter.

The VCO circuit we studied consists of a Schmitt trigger, a timing capacitor, a voltage-controlled constant current source, and a current-switching circuit that alternately diverts the controlled current into or out of the timing capacitor. Triangular and square-wave outputs are available with frequency controlled by input voltage.

REFERENCES _____

1. GHAUSI, M. S. *Electronic Devices and Circuits: Discrete and Integrated.* New York: Holt, Rinehart and Winston, 1985.
2. GRAY, P. R., and R. G. MEYER. *Analysis and Design of Analog Integrated Circuits,* 3rd ed. New York: John Wiley, 1993.
3. GREBENE, A. B. *Bipolar and MOS Analog Integrated Circuit Design.* New York: Wiley-Interscience, 1984.
4. NILSSON, J. W. *Electric Circuits,* 2nd ed. Reading, MA: Addison-Wesley, 1985.
5. OPPENHEIM, A. V., and A. S. WILLSKY. *Signals and Systems.* Englewood Cliffs, NJ: Prentice Hall, 1983.

PROBLEMS _____

SPECIAL DIRECTIONS FOR SPICE HOMEWORK PROBLEMS: *Do not hand in lengthy SPICE printout for homework.* Instead, abstract the useful information from the SPICE output file as in the SPICE examples in the text. Include your SPICE code and a circuit diagram with nodes numbered to agree with the code. Cite relevant numerical values from the SPICE output file and discuss when appropriate. Make sketches of any relevant curves, and label appropriate points. Make small tables to present numerical data if useful for clarity.

Section 11.1

11.1 Redraw Fig. 11.1b with the supply voltages changed to 5 V and -3 V. On the diagram, show values for node voltages and currents when $R_B = 2$ kΩ.

11.2 (a) Find the power dissipation of Fig. 11.1b when $R_B = 20$ kΩ.

(b) Write a general expression for the dissipation of the IC of Fig. 11.1a for bias current I_X when the supplies are $\pm V_{SS}$.

11.3 Figure P11.3 is a voltage-controlled filter, a first-order lowpass filter with cut-off frequency controlled by V_C.

(a) Use sinusoidal steady-state analysis to find the complex gain function V_o/V_{in} in terms of OTA transconductance gain GM. Assume the OTA input resistance in Fig. 11.3b is infinite.

(b) Design a filter with cut-off frequency programmable from 100 to 5000 rad/s. Use supplies of ± 15 V and I_X values that fall within the 1–1000 μA range of the OTA. Any correct solution is acceptable.

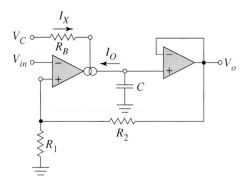

Figure P11.3

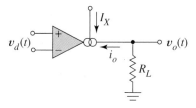

Figure P11.7

$$v_d(t) = \sum_{n=1,3,5,...} \frac{4A}{n\pi} \cos(n\omega_o t)$$

Find an expression for $v_o(t)$ consisting of cosines and products of cosines with numerical coefficients for each odd value of n.

11.4 Change the bias current in Example 11.2 so that the hysteresis width doubles.

11.5 Design an OTA-based Schmitt trigger like that of Example 11.2 with $V_S = 8$ V. Any correct solution is acceptable.

11.6 An *amplitude-modulated wave* of the form

$$v(t) = K(1 + m \cos \omega_s t) \cos \omega_c t$$

results when a sinusoidal *modulating signal* of frequency ω_s is placed upon a carrier of frequency ω_c. K is the amplitude of the modulated wave, and m is the *modulation index*, a number between 0 and 1.
(a) Show that $v_o(t)$ in Fig. 11.8 can be written in this form when $v_s(t) = V_M \cos \omega_s t$ and $v_c(t) = V_C \cos \omega_c t$.
Hint: The circuit is a small-signal amplifier with signal input $v_c(t)$. Gain changes with time when I_X has both dc and ac components. Use power supplies of $\pm V_{SS}$ volts.
(b) What condition on signal amplitude V_M ensures that $1 > m > 0$?

11.7 In Fig. P11.7, $v_d(t)$ is a *small-signal* square wave of peak amplitude A volts. I_X is a nonzero signal, as yet undetermined, but consistent with active operation of the OTA transistors.
(a) Find a practical maximum value for A.
Hints: Use Eq. (7.20) and Fig. 11.1a.
(b) Starting with Eqs. (11.2) and (11.3), find an expression for the small-signal output voltage $v_o(t)$ in terms of $v_d(t)$, I_X, V_T, and R_L.
(c) If I_X can have a maximum instantaneous value of 200 μA, find the value of R_L that makes the maximum instantaneous value of $v_o(t) = 13$ V when the square wave of part (a) is applied.
(d) If I_X is to have the form

$$I_X(t) = I_{X,ave} + B \cos \omega_i t$$

find the value of B so that 0.1 μA $\le I_X(t) \le$ 200 μA.
(e) The Fourier series for the square wave $v_d(t)$ is

11.8 Make a SPICE model of an operational transconductance amplifier like Fig. 11.1 but use MOS transistors instead of BJTs (including diode-connected devices). Transistor parameters are $k = 2 \times 10^{-4}$ A/V² and $|V_t| = 1$ V, and supplies are ± 15 V. Connect an external load resistor of 100 kΩ between output node and ground.
(a) Find the ac voltage gain for bias resistors R_B of 12.9 kΩ and 1.29 kΩ.
(b) Repeat part (a) after adding LAMBDA = 0.01 to the .MODEL lines.

Section 11.2

11.9 Recalculate the collector currents in the first two columns of Table 11.1 for ±5 V supplies.

11.10 When ±5V supplies are used in Fig. 11.10, $I_{C13A} = 55$ μA.
(a) Compute the new V_{BB} developed by Q_{19}–Q_{18}–R_{10} for this case.
(b) Compute the new bias currents for Q_{14} and Q_{20}.

11.11 (a) Do μA741 input bias currents increase or decrease when the supply voltage is lowered? Explain.
(b) For ±5 V supplies, $I_W = 13.8$ μA. Estimate the op amp's input bias currents.

11.12 Figure P11.12 shows the second-stage biasing when ±5 V power supplies are used with the 741.
(a) Find I_{C16}. Since I_{C17} is small, calculate V_{BE17} rather than assuming it is 0.7 V.
(b) Find the dc node voltage at the collector of Q_6.

11.13 When ±5 V supplies are used with the 741, $I_W = 13.8$ μA, and $V_{C6} = -3.87$V. Recalculate the upper and lower common-mode limits for this case.

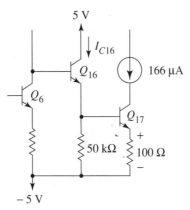

Figure P11.12

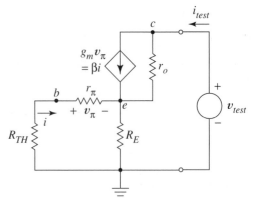

Figure P11.16

11.14 If ± 5V supplies reduce I_W to 13.8 μA, find the corresponding values for R_{in} and G_{M1} for the transconductance amplifier of Fig. 11.17.

11.15 (a) Show that the output resistance defined by Fig. P11.15 is

$$R_{out} = \left(1 + \frac{\beta R_E}{R_E + R_B}\right) r_o$$

(b) Use part (a) to show that $R_{o17} = 433$ kΩ in Fig. 11.15a. *Hints*: We found the output resistance of the transconductance amplifier (741's first stage) was $R_{out1} = 7.03$ MΩ. Also, $r_{\pi 16} = 368$ kΩ, $r_{\pi 17} = 11.4$ kΩ, and $r_{o17} = 236$ kΩ.

11.16 Several output resistance developments employing different approximations and different notations involved the same circuit, Fig. P11.16.

(a) Translate the expression for R_{out} in Problem 11.15 into the notation of Fig. P11.16.
(b) Derive the Widlar result, Eq. (7.71), from your part (a) answer by assuming $r_\pi \gg R_{TH} + R_E$.
(c) Show that $R_{out} = r_o(1 + \beta)$ when $r_\pi + R_{TH} \ll R_E$.

11.17 (a) Show that Fig. P11.17 is equivalent to the resistor

$$R_{eq} = \frac{1 + g_{m19}R}{1 + g_{m18}R} \frac{1}{g_{m19}}, R = 40 \text{ k} \| r_{\pi 18}$$

Hint: Replace the diode-connected transistor with its resistor equivalent, and use the small-signal hybrid pi model for Q_{18}.
(b) Show that for $\beta = 250$ and the ± 15 V supplies assumed in Table 11.1, $R_{eq} = 160$ Ω.

11.18 Use the general result of Problem 11.17a to find the equivalent resistance of Fig. P11.17 when ± 5 V supplies make $I_{C13A} = 55$ μA in Fig. 11.10.

11.19 When the 741 is biased using ± 5 V supplies, I_{C8} is 13.8 μA. Estimate the slew rate.

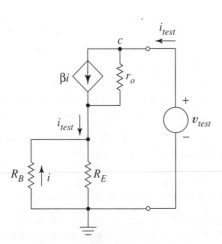

Figure P11.15

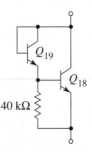

Figure P11.17

Section 11.3

11.20 In Eq. (11.19), v_C is the output of a summing amplifier, which applies weights R_F/R_1 to a dc input, $-V_{DC}$, and R_F/R_2 to a modulating signal, $-V_s \cos \omega_s t$. A carrier signal $v_d = V_c \cos \omega_c t$ is the small-signal input. Show that $v_o(t)$ can be written in the form of the amplitude-modulated wave:

$$v(t) = K(1 + m \cos \omega_s t) \cos \omega_c t$$

(a) Find design expressions for K and m in terms of circuit and system parameters.
(b) The multiplier design of Example 11.3 used $A_d = 20$ and $R_C = R = 1$ kΩ. To convert the circuit into a modulator, find values of R_1, R_2, R_F, and V_{DC} so that $K = 2$ and $m = 0.9$ when signals are $V_c = 0.005$ V and $V_s = 0.1$ V. Any correct answer is acceptable.
(c) Draw a diagram of the summing amplifier.

11.21 Problem 11.20 shows that if v_C in Fig. 11.19a comes from the circuit of Fig. P11.21, and if

$$v_d(t) = 0.005 \cos 2\pi 10^4 t$$

the output should be the amplitude-modulated wave

$$v(t) = 2(1 + 0.9 \cos 2\pi 1000t) \cos 2\pi 10,000t$$

Modify the SPICE code of Fig. 11.19b to verify this by simulation.

11.22 In Examples 11.3 and 11.4, we used $A_d = 20$ and $R_C = R = 1$ kΩ in Eq. (11.19). Try to eliminate the difference amplifier in a redesign by making $A_d = 1$ and then compensating by making $R_C = 20$ kΩ. Output should be the difference v_x in Fig. 11.19a. What goes wrong?
Hint: Check the circuit for maximum input voltage.

11.23 To examine the linearity of the Gilbert cell multiplier, use SPICE to produce a family of curves. Plot $(i_x - i_y)$ versus v_{D1} for $v_{D2} = -5$ mV, -3 mV, -1 mV, 1 mV, 3 mV, 5 mV.
(a) Restrict input values to $|v_{Di}| \leq 0.2\ V_T$.
(b) Extend the range of v_{D1} to $|v_{D1}| \leq 2\ V_T$.

11.24 We want to operate the circuit in Fig. P11.24 as a small-signal, four-quadrant multiplier.

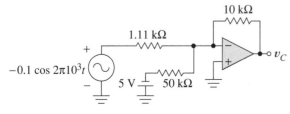

Figure P11.21

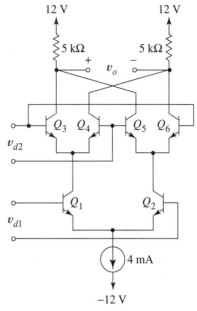

Figure P11.24

(a) Write a numerical expression for the output voltage in terms of the input voltages.
(b) Assume that the current source requires at least 0.2 V to operate properly. Add signal sources and dc bias batteries, V_{XX} and V_{YY}, to the diagram. Find values for V_{XX} and V_{YY} so that, with signals turned off, the available dc voltage is evenly partitioned between top and bottom-row transistors and current source; that is, $V_{CE3} = V_{CE1} = V_{CS}$, where V_{CS} is the dc drop across the current source.

11.25 Use SPICE to find the double-sideband, suppressed carrier output waveform of the circuit designed in Example 11.5. Sketch the waveform showing its shape and amplitude, paying particular attention in the vicinity of the zero crossing.

11.26 Find suitable values for R_C, R_E, and I_O so that Fig. P11.26 is a voltage-controlled difference amplifier. The gain should be continuously variable from -20 to $+20$ as control variable v_{D2} changes from -1 to $+1$ V. Bias the output transistors at $V_{CE} = 1.4$ V. Specify the maximum allowable output and input signals.

11.27 In Example 11.6 and Exercise 11.7 we saw that a larger collector resistor R_C gives increased voltage gain, provided that the quiescent value of V_{CE} allows for this increased signal without saturating the output transistors. In Fig. 11.24, assume that the node voltages at the emitters of Q_3–Q_6 do not change appreciably with signal. Then changes in v_{CE} are caused only by

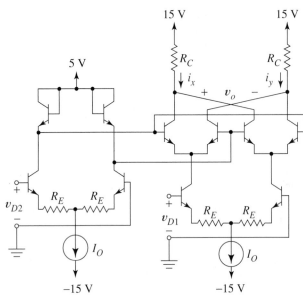

Figure P11.26

changes in collector voltage v_C in response to signals. The extreme case is when the collectors approach to within 0.2 V of the emitters at the peak of the input signal.

(a) For Fig. 11.24, write an equation for v_{CE} at the instant the output transistor saturates. This equation should include both biasing and signal, and should involve R_C and the signal drop across R_C, $v_o/2$, as unknowns.

(b) Write an equation for the value of v_o in terms of R_C when $|v_{D1}v_{D2}| = 10 \text{ V}^2$.

(c) Use the results of parts (a) and (b) to find the maximum value of R_C.

(d) For the R_C of part (c), find the maximum instantaneous value of v_o.

11.28 On a frequency axis, draw a line for each frequency that appears in the expansion Eq. (11.39) for $n = 1, 3$, and 5 when $\omega_c = 1.02 \times 10^8$ rps, $\omega_m = 0.01 \times 10^8$ rps, and

$$\frac{-g_m R_c V_M 4}{2\pi} = -1$$

Show the amplitude of each line.

11.29 (a) Design a balanced modulator that operates in the binary fashion of Fig. 11.25. Your circuit should use ± 9 V supplies and produce a peak output voltage of 3 V. Do not design the lowpass filter.

(b) Check your design by a SPICE simulation that uses a square-wave carrier with amplitude that satisfies by Eq. (11.36).

Use a transient analysis to verify that the output waveform is correct.

11.30 Express the drain current of the dual-gate MOSFET in Example 11.7 as a sum of simple sinusoidal terms when

$$v_{g1,s}(t) = 0.002 \cos(2\pi 10^8 t)$$

and

$$v_{g2,s}(t) = 0.003 \cos(2\pi 1.3 \times 10^6 t).$$

Show the numerical value for each term.

11.31 Use Fig. 11.29a to estimate values of g_{m1} at $v_{G1,S} = 1$ V for $v_{G2,S} = 1, 2, 3, 4, 5$ V. Use this data to sketch a g_{m1} versus $v_{G2,S}$ curve like Fig. 11.29b.

Section 11.4

11.32 The input voltage to a phase-locked loop is a sine wave of frequency ω_{FR}. Suddenly at $t = 0$ the frequency jumps to $\omega_{FR} + \Delta\omega$ and then decreases exponentially back to ω_{FR} with time constant of 1 ms.

(a) Sketch instantaneous frequency versus time.

(b) Sketch instantaneous voltage versus time.

(c) Find the Laplace transform expression $\Delta\Omega(s)$.

11.33 A phase-locked loop has idealized components like those in Fig. 11.31 except that $F(s) = 1$. Numerical values are $K_P = 2$ V, $A = 10$, and $K_o = 10^3 \text{ s}^{-1}\text{V}^{-1}$.

(a) What is the numerical value of loop gain?

(b) Sketch the impulse response of the phase-locked loop when the output variable is frequency.

(c) Sketch the impulse response of the phase-locked loop when the output variable is voltage.

11.34 The phase-locked loop described in Problem 11.33 is in lock at the free-running frequency. Find $v_3(t)$ if input frequency is given by

$$\Delta\omega_i(t) = \Delta\omega e^{-1000t} u(t)$$

where $u(t)$ denotes the unit step function and $\Delta\omega$ is the deviation from the free-running frequency.

11.35 A phase-locked loop has step response like that shown in Fig. 11.35b for $\zeta = 0.5$. The input frequency changes abruptly to a higher value for an interval of time and then returns to its original value. Assuming that the loop operates with square waves and remains in lock throughout the transient, show that you understand the meaning of Fig. 11.35b by sketching input and output voltage waveforms.

11.36 A second-order, phase-locked loop has loop gain $K_L = 1.4 \times 10^8 \text{ s}^{-1}$ and a filter bandwidth of $\omega_H = 1.26 \times 10^7$ rad/s. Find the damping ratio ζ and the loop bandwidth.

11.37 A phase-locked loop employs a phase detector described by Fig. 11.36d. The other loop constants are $A = 2$, $\omega_{FR} = 60 \times 10^3$ rad/s, and $K_o = 1.53 \times 10^3$ s^{-1}V^{-1}. The low-frequency gain of the filter is one. Find the lock range of the loop.

11.38 A phase-locked loop uses the phase detector of Fig. 11.38b. If $A = 4$, $K_o = 2 \times 10^4$ s^{-1}V^{-1}, and filter gain $F(0) = 1$, find values for I_O and R_C so that the loop gain is 4×10^5s^{-1}. Find the lock range for the loop.

11.38 A phase-locked loop uses the phase detector of Fig. 11.39b.
(a) If $A = 1$, $K_o = 2 \times 10^3$ s^{-1}V^{-1}, and filter gain $F(s) = 1$, find an expression for loop gain (when $\Delta\theta \approx \pi/2$) in terms of design parameters g_m, R_C, and V_I.
(b) Find values for design parameters so that the loop gain is 1000 s^{-1}.
(c) Determine the lock range for the values of part (b).

11.39 Fig. 11.41 suggests that the capture range for a PLL is smaller than the lock range. Use the estimate of Eq. (11.69) to show that this is so.

11.40 The filter in a phase-locked loop has the transfer characteristic

$$F(s) = \frac{5 \times 10^4}{s + 5 \times 10^4}$$

Estimate the capture range if the lock range is 160×10^3 rad/s.

11.41 A current-controlled oscillator like Fig. 11.43a employs the Schmitt trigger circuit of Fig. P11.41.
(a) Why does the circuit include a follower?
(b) If the op amps are HA2544 devices using ±15 V supplies (Table 2.1), find values for V_H, V_L, V_1, and V_2.
(c) In Fig. 11.43c, the Schmitt trigger output pulses change instantaneously; however, the op amps of Fig. P11.41 will have gradual transitions because of slewing. Find the highest oscillation frequency f_o so that slewing of the output-circuit HA2544 occupies only one-tenth of a period. Assume the follower does not slew.
(d) Select a timing capacitor C. Then find the range of values of I_O required so that the oscillator operates over a two-decade

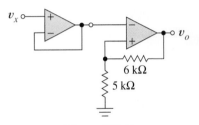

Figure P11.41

range of frequencies with the frequency of part (c) as the upper limit.

11.42 Work Problem 11.41 using the 741 op amp instead of the HA2544.

11.43 The voltage-controlled current source of Fig. P11.43 employs npn transistors with $\beta = 220$ and $V_A = 100$ V; the pnp transistor has $\beta = 60$ and $V_A = 75$ V.
(a) Use SPICE to plot I_O versus v_C for $0 \le v_C \le 12$ V. Explain any deviations from the straight-line behavior described by Eq. (11.71).
(b) Use SPICE to plot I_O versus v_3 after adding the transformation circuit of Fig. 11.44d at the input to produce v_C. Use a high-gain VCVS for the op amp.

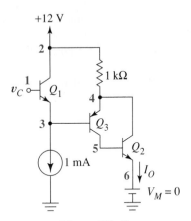

Figure P11.43

CHAPTER
12

Filters

12.1
Basic Filter Concepts

We can visualize a signal as a superposition of sinusoidal components that occupy a band of frequencies. A filter separates this signal from additive noise or competing signals by selectively passing the desired frequencies while attenuating undesired frequencies. Sometimes a filter intentionally modifies the signal itself by emphasizing or weighting certain frequencies relative to others or by changing the relative phases of the various signal components. In this chapter we learn how to specify various kinds of filtering tasks and how to design filters to satisfy such specifications. There is a vast literature that spans all aspects of filter theory. The introductory concepts of this chapter will, it is hoped, provide a sound foundation for exploring and comprehending this literature.

The *ideal filter* characteristic of Fig. 12.1 is central to filter theory. Because it passes all frequencies lower than one rad/s with its gain of one and removes all frequencies higher than one rad/s with its gain of zero, it is called a *lowpass filter with cut-off frequency* of one rad/s. The frequency band where gain is one is the *passband,* and the band where gain is zero is the *stopband*. The phase shift of the ideal filter is zero at all frequencies.

In a filter, it is *relative,* not absolute gain, that is important, so any scale factor that affects all frequencies equally contributes nothing to the filtering. For example, a filter

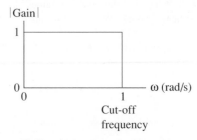

Figure 12.1 Ideal lowpass filter.

with gain of -50 for $\omega \le 1$ and 0 for $\omega > 1$ is equivalent to the ideal lowpass filter cascaded with an amplifier of gain -50. If specifications require some particular scale factor, we can always add an amplifier or voltage divider at the filter output.

Unfortunately the ideal lowpass filter cannot be *realized* (achieved) by a circuit with a finite number of elements, so we next examine a practical circuit that approximates it.

12.1.1 FIRST-ORDER ACTIVE FILTER

Figure 12.2a shows an active filter in which an inverting preamplifier of gain minus one compensates for the incidental signal inversion in the filtering stage that follows. In terms of Laplace transform voltages, the transfer function $T(s)$ is

$$T(s) = \frac{V_o(s)}{V_i(s)} = \frac{[(1/s)1]/[(1/s) + 1]}{1} = \frac{1}{s + 1} \tag{12.1}$$

which has the pole-zero plot of Fig. 12.2b, a simple pole on the negative real axis of the s plane. Because its denominator is a polynomial in s of degree one, $T(s)$ is a *first-order filter.*

To find the frequency response of the first order filter, we set $s = j\omega$ in Eq. (12.1). In polar notation the complex-valued frequency function is

$$T(j\omega) = |T(j\omega)|\,\angle \phi(\omega) = \frac{1}{\sqrt{\omega^2 + 1}}\,\angle{-\tan^{-1}(\omega)} \tag{12.2}$$

The resulting frequency response curves in Fig. 12.2c show a -3 dB bandwidth of 1 rad/s. The amplitude curve approximates the ideal of Fig. 12.1 in the sense that the gain is within 3 dB of the ideal unity gain (zero dB) for $\omega \le 1$ rad/s. In the stopband, $\omega > 1$, the approximation to the ideal of zero gain improves with increasing ω.

We are conditioned by amplifier studies to think of transfer functions in terms of gain. In a filter, however, we have seen that the measure of success is often the extent to which stopband signals are reduced or *attenuated* relative to passband signals. For this reason we often visualize filter functions in terms of *normalized gain,* that is, circuit gain divided by maximum passband gain. This gives frequency response curves with gain ≤ 1. Once gain is normalized in this fashion, we sometimes flaunt our success by referring to the reciprocal of the gain (a number greater than one) as the *attenuation.* In decibels, $(\text{attenuation})_{dB} = -(\text{gain})_{dB}$. Some filter references emphasize this viewpoint by showing plots of attenuation (instead of gain) versus frequency. We consistently use gain plots in this text, although we sometimes find it useful to refer to attenuation in discussing filter specifications.

Specifications often state the required gain or phase shift of the filter at a particular radian frequency ω. For the first-order filter we can compute gain from the design equation

$$|T(j\omega)| = \frac{1}{\sqrt{\omega^2 + 1}} \tag{12.3}$$

In decibels this gives

$$20 \log |T(j\omega)| = 20 \log |(\omega^2 + 1)^{-0.5}|$$

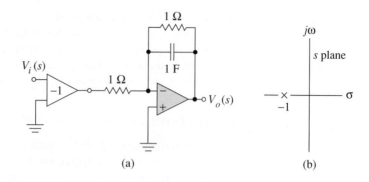

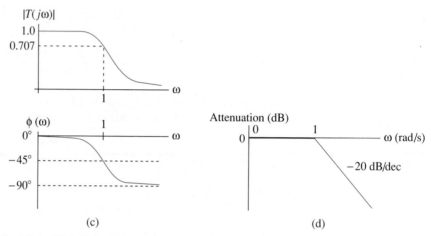

Figure 12.2 First-order active lowpass filter: (a) schematic; (b) pole-zero plot of transfer function; (c) frequency response curves; (d) Bode plot.

We see that for $\omega \gg 1$, the gain drops at -20 dB/dec (or -6 dB/oct) as in Fig. 12.2d. For any value of ω the phase shift of this filter is given by

$$\phi(\omega) = -\tan^{-1}\omega \qquad (12.4)$$

This shows that for $\omega \ll 1$ phase shift is negligible, and for $\omega = 1$ the phase shift is $-45°$. Maximum phase shift, $-90°$, is closely approached at high frequencies.

12.1.2 FREQUENCY-SCALING

Frequency-scaling is a procedure that translates the frequency and phase curves of *any linear filter* to new frequencies. Scaling is central to filter design, for it allows us to orient *all* designs around the same starting point, some kind of *prototype filter* with a specified *reference frequency* of one rad/s. For first-order lowpass filters, the reference frequency is the cut-off frequency.

Frequency-Scaling a Transfer Function. Let p denote the complex frequency variable before frequency scaling and $s = \alpha p$ the scaled frequency, for positive, real scale factor α. Then for *any* transfer function, $T_o(p)$, the frequency-scaled transfer function $T(s)$ is

$$T(s) = T_o(p)|_{p = s/\alpha} = T_o\left(\frac{s}{\alpha}\right) \tag{12.5}$$

For the first-order function, for example, frequency scaling gives the new function

$$T(s) = \left.\frac{1}{p + 1}\right|_{p = s/\alpha} = \frac{\alpha}{s + \alpha} \tag{12.6}$$

Notice that scaling by α moves the pole in Fig. 12.2b from -1 to $-\alpha$.

To see exactly how scaling works, notice that at 17 rad/s the unscaled lowpass filter of Eq. (12.1) has magnitude and phase shift given by

$$T_o(j17) = \frac{1}{j17 + 1} = 0.059\underline{/-86.6°}$$

Scaling this filter by $\alpha = 9800$, according to Eq. (12.6), gives the new function

$$T(j\omega) = \frac{9800}{j\omega + 9800}$$

At $\omega = 9800 \times 17$ the magnitude and phase of the new, scaled filter is

$$T(j9800 \times 17) = \frac{9800}{j9800 \times 17 + 9800} = \frac{1}{j17 + 1} = 0.059\underline{/-86.6°}$$

To demonstrate the generality of frequency scaling, notice that for any filter function $T_o(p)$,

$$T_o(p)|_{p = s/\alpha} = \left.\frac{K(p - z_1)(p - z_2)\cdots(p - z_M)}{(p - p_1)(p - p_2)\cdots(p - p_N)}\right|_{p = s/\alpha}$$

frequency-scaling by α moves every pole p_i to αp_i and every zero z_j to αz_j. (The constant also changes from K to $\alpha^{N/M}K$, which is immaterial insofar as filtering is concerned.) Equation (12.5) shows exactly how the magnitude and phase curves of any $T_o(p)$ are affected by such scaling. Evaluating the original function at arbitrary frequency $p = j\lambda$ gives the complex number $T_o(j\lambda)$. As in the example, the new function takes on the identical complex value when *evaluated at the scaled frequency $j\omega = j\alpha\lambda$* as we see from

$$T(j\omega) = T_o\left(j\frac{\omega}{\alpha}\right) \tag{12.7}$$

Because Eq. (12.5) was not restricted to lowpass filter functions or even to functions of any particular order, our frequency scaling conclusion applies to *any linear filter function whatsoever.*

Frequency Scaling a Circuit. Frequency scaling also works on *circuits.* Since the complex frequency variable p in a filter transfer function $T_o(p)$ can arise in circuit analysis only from impedance terms, $1/pC$ and pL, the frequency scaling $s = \alpha p$ can be carried out in any given circuit by scaling the values of all energy storage elements directly. Specifically, under frequency scaling the impedance of each capacitor C and inductor L changes to C/α and L/α, respectively, because

$$\frac{1}{pC} = \frac{1}{(s/\alpha)C} = \frac{1}{s(C/\alpha)}$$

and

$$pL = \left(\frac{s}{\alpha}\right)L = s\left(\frac{L}{\alpha}\right)$$

The conclusion is easy to remember. *To scale a given circuit's magnitude and phase curves to higher frequencies by a factor $\alpha > 1$, reduce the size of all inductors and capacitors by α. To scale the frequency response to lower frequencies, increase each L and C value by the appropriate scale factor.* This means we can rescale the first-order filter of Fig. 12.2a so that it has any cut-off frequency we wish. To obtain a -3 dB bandwidth of 10 kHz, for example, we simply change C from 1 F to $1/(2\pi 10,000) = 15.9$ μFd.

Scaling Filter Specifications. In a filter *design,* we begin with specifications, and from them determine the required frequency-scale factor, α. For example, suppose we require a first-order lowpass filter that produces $-20°$ phase shift at 800 Hz. We first use Eq. (12.4) to find the frequency for which the unscaled filter gives $-20°$ phase shift,

$$\omega = \tan[-(-20°)] = 0.364 \text{ rad/s}$$

To obtain this same phase shift at 800 Hz, we must frequency-scale by α that satisfies

$$\alpha\, 0.364 = 2\pi 800 \text{ rad/s}$$

or $\alpha = 13.8 \times 10^3$. To frequency-scale the physical filter, we simply change the 1 F capacitor (Fig. 12.2a) to $1/\alpha = 72.5$ μF.

Because frequency scaling leaves resistors unchanged, our circuit diagram often contains impractical component values even after frequency scaling. A second scaling procedure allows us to change component values *without changing the frequency response!*

12.1.3 IMPEDANCE SCALING

Impedance-scaling a circuit by a positive real scale factor β means replacing every resistor R by a scaled resistor βR, every inductive impedance sL by scaled impedance $\beta sL = s(\beta L)$, and every capacitive impedance $1/sC$ by scaled impedance $\beta/sC = 1/(sC/\beta)$. This is sometimes called magnitude scaling; however, the term *impedance scaling* helps us remember how to do it. When impedances are all scaled up to higher values, circuit currents are reduced, but voltage ratios such as $T(s)$ are unchanged. The *frequency response magnitude and phase curves are unchanged by impedance scaling.*

To demonstrate the usefulness of impedance scaling, recall that frequency-scaling the circuit of Fig. 12.2a by $\alpha = 2\pi 10^4$ increased the bandwidth to 10 kHz by changing the capacitor from 1 F to 15.9 μF. If we now impedance-scale by $\beta = 10,000$, the result is the circuit of Fig. 12.3a, which has the frequency response of Fig. 12.3b.

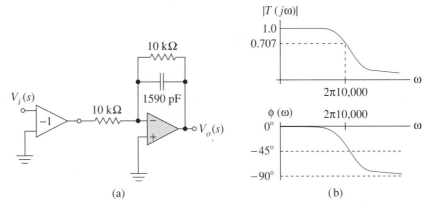

Figure 12.3 (a) Filter of Fig. 12.2a after frequency-scaling by $\alpha = 2\pi \times 10^4$ and impedance-scaling by 10^4; (b) magnitude and phase responses of the scaled filter.

12.1.4 FREQUENCY-SCALING RC-ACTIVE CIRCUITS BY SCALING RESISTORS

Many filters include only resistors, capacitors and active elements. *For these RC-active circuits,* it is possible to frequency-scale by changing resistors instead of capacitors. To prove this, suppose we first frequency-scale an *RC*-active filter by $\alpha > 1$ by reducing the value of each capacitor C_i to C_i/α. The resulting filter function $T(s)$ now has the desired frequency response. We next scale down each impedance by $\beta = 1/\alpha$. Each resistor R_j takes on a new value R_j/α, while each capacitor, previously scaled to C_i/α, is now impedance scaled to $\alpha(C_i/\alpha) = C_i$, its original value! Because impedance scaling does not change the frequency response, we conclude that we can *frequency-scale RC-active filters to higher frequencies by reducing resistances instead of capacitances.* This principle is sometimes useful in finding practical component values, and is also useful in tuning, that is, adjusting the frequency response to fit specifications, because widely varying component values are easier to find in resistors than in capacitors. Thus if we need to design a filter that tunes to a number of different cut-off frequencies by means of a selector switch, it is easier to change frequency ranges by replacing resistors rather than capacitors.

12.1.5 HIGHPASS FILTER

Sometimes we wish to use a filter to pass high frequencies and attenuate low frequencies. For this we use a *highpass* filter, shown in ideal form in Fig. 12.4. Frequency- and impedance-scaling concepts apply to highpass filters without modification.

RC-CR **Transformation.** Once we know how to design an *RC*-active circuit to realize a lowpass filter with one rad/s cut-off frequency, we can always find a corresponding *RC* active circuit to realize a highpass filter with a one rad/s cut-off frequency. The key principle, called the *RC-CR transformation,* is described by Mitra in one of the references.

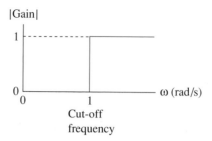

Figure 12.4 Ideal highpass filter with 1 rad/s cut-off frequency.

To transform a lowpass *RC*-active circuit with one rad/s cut-off frequency into a highpass *RC*-active circuit with 1 rad/s cut-off frequency, replace every resistor of R_i ohms by a capacitor of $C_i = 1/R_i$ farad, and every capacitor of C_j farad by a resistor of $R_j = 1/C_j$ ohms. Gains of active elements are not changed.

We now use this transformation to "discover" a first-order highpass filter. (Problem 12.17 demonstrates that the transformation also works for more complex filters.)

12.1.6   FIRST-ORDER HIGHPASS FILTER

Applying the *RC-CR* transformation to the circuit of Fig. 12.2a gives Fig. 12.5a. Analysis techniques from Chapter 2 show that its voltage gain function is

$$T(s) = \frac{s}{s + 1} \tag{12.8}$$

Figure 12.5b is the pole-zero plot, and the frequency response curves are Fig. 12.5c. Substituting $s = j\omega$ into Eq. (12.8) gives

$$|T(j\omega)| = \frac{\omega}{\sqrt{\omega^2 + 1}} \tag{12.9}$$

and

$$\phi(\omega) = \frac{\pi}{2} - \tan^{-1}\omega \tag{12.10}$$

Expressing the gain of Eq. (12.9) in decibels gives the Bode plot of Fig. 12.5d. The following example illustrates first-order, highpass filter design.

EXAMPLE 12.1   Design a first-order highpass filter with 0 dB gain at high frequencies and attenuation of 30 dB at 40 Hz. Use only 5000 pF capacitors.

Solution. From Eq. (12.9), a highpass *prototype* filter with cut-off at one rad/s has 30 dB attenuation at a frequency λ_x given by

$$10^{-30/20} = 0.03162 = \frac{\lambda_x}{\sqrt{\lambda_x^2 + 1}}$$

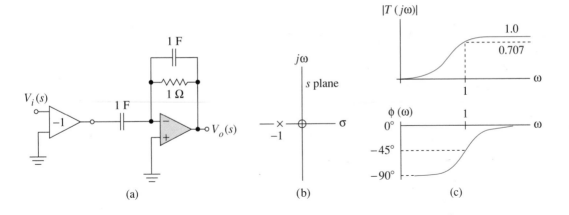

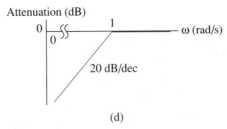

Figure 12.5 First-order RC-active highpass filter: (a) schematic; (b) pole-zero plot; (c) frequency response curves; (d) Bode plot.

or

$$9.998 \times 10^{-4}(\lambda_x^2 + 1) = \lambda_x^2$$

Solving gives $\lambda_x = 0.03164$ rad/s. Therefore, the scale factor is

$$\alpha = \frac{2\pi40}{\lambda_x} = \frac{2\pi40}{0.03164} = 7.943 \times 10^3$$

Frequency-scaling the capacitors in Fig. 12.5a by α gives $C = 126.3$ μF. To use 5000 pF capacitors, we must also impedance-scale by

$$\beta = \frac{126.3 \times 10^{-6}}{5000 \times 10^{-12}} = 25.3 \times 10^3$$

We conclude that the specified filtering task can be performed by the circuit of Fig. 12.6b. The resistor at the noninverting input was inserted to reduce offset due to input bias current. An inverting amplifier can be added if required. ❏

Exercise 12.1 Suppose that specifications in Example 12.1 called for 200 kΩ resistors *instead of* 5000 pF capacitors, but the same frequency response. Find the capacitor values for this case.

Ans. 631.5 pF.

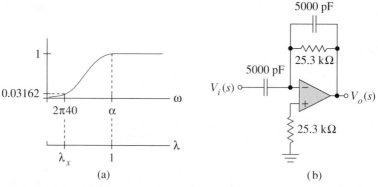

Figure 12.6 Highpass filter design for Example 12.1: (a) specifications related to one rad/s prototype filter; (b) final design.

12.2
Second-Order Active Filters

12.2.1 LOWPASS FILTER FUNCTION

Another filter function that approximates the ideal lowpass filter is the second-order lowpass function

$$\frac{V_o(s)}{V_i(s)} = T(s) = \frac{1}{s^2 + \left(\dfrac{1}{Q}\right)s + 1} \tag{12.11}$$

where Q is a design parameter. Factoring the denominator shows that for $Q \geq 0.5$, $T(s)$ has a pair of conjugate poles located at

$$\sigma \pm j\omega = -\frac{1}{2Q} \pm j\sqrt{1 - \frac{1}{4Q^2}} \tag{12.12}$$

From Eq. (12.12), the distance of these poles from the origin of the s plane is

$$\sqrt{\sigma^2 + \omega^2} = 1$$

independent of Q. The pole-zero plot, Fig. 12.7a, shows that as Q increases from 0.5 toward infinity, the conjugate poles move along the unit circle toward the $j\omega$ axis.

Substituting $s = j\omega$ into Eq. (12.11) gives the frequency response

$$T(j\omega) = \frac{1}{-\omega^2 + j(1/Q)\omega + 1} = |T(\omega)| \underline{/\phi(\omega)}$$

where

$$|T(\omega)| = \frac{1}{\sqrt{(1 - \omega^2)^2 + \left(\dfrac{\omega}{Q}\right)^2}} \tag{12.13}$$

is the magnitude and

$$\phi(\omega) = -\tan^{-1}\left[\frac{\omega}{Q(1 - \omega^2)}\right] \qquad (12.14)$$

is the phase shift.

Figures 12.7b and 12.2b show that for $Q = 1/\sqrt{2} = 0.707$, the second-order magnitude response superficially resembles the first-order curve. Actually, the second-order function approximates the ideal lowpass filter more closely, by giving higher passband gain and increased stopband attenuation. In the stopband, gain rolls off at -40 dB/decade instead of -20 dB/decade.

Ordinary and dB plots of $|T(\omega)|$, Figs. 12.7c and Fig. 12.7d, respectively, show that $T(s)$ describes a *family* of frequency response functions, distinguished by means of parameter Q. All are lowpass filter functions in the sense of giving higher gain at low frequencies and diminished gain at high frequencies. For the higher values of Q, the proximity of the poles to the imaginary axis causes a peaking in the frequency response curve. Although some curves are poor approximations to the ideal lowpass filter function, they describe useful subcircuits for more complex filters we study later. For this reason it is important to be able to design a second-order lowpass circuit for *any* specified value of Q.

Figure 12.7e shows the phase curves. Higher Q gives progressively less phase shift for $\omega < 1$, with the greatest change in phase occurring near $\omega = 1$. Compared to the first-order filter, this function produces a $-90°$ phase shift at $\omega = 1$ instead of $-45°$ and introduces a maximum phase shift of $-180°$ instead of $-90°$.

One rad/s is the reference frequency for second-order lowpass filters. Although one rad/s is sometimes called the cut-off frequency, this can lead to confusion since the -3dB bandwidth is greater than one rad/s for some values of Q. For any Q, frequency

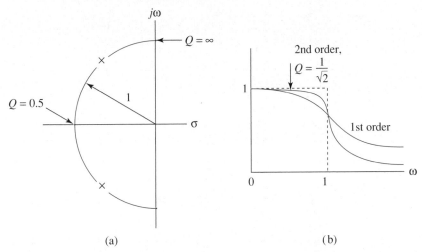

(a)   (b)

Figure 12.7 Second-order lowpass filter with one rad/s reference frequency: (a) pole-zero plot; (b) comparison with first-order filter; (c) magnitude response; (d) magnitude response in decibels; (e) phase response.

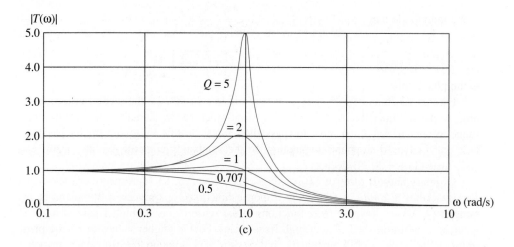

(c)

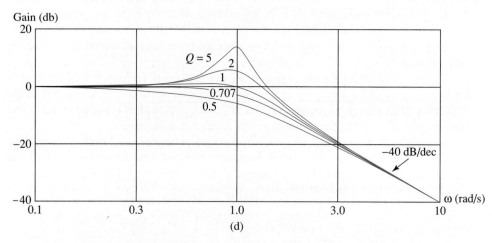

(d)

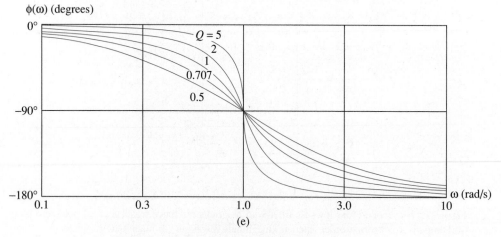

(e)

Figure 12.7   (continued)

scaling shifts the magnitude and phase curves to new frequencies in the usual way. Any peak in the frequency response curve is translated to a new frequency with amplitude unchanged. Frequency scaling by α changes the unit circle pole locus in Fig. 12.7a to a circle of radius α.

12.2.2 SALLEN–KEY LOWPASS FILTER

The *Sallen–Key circuit* of Fig. 2.15a, redrawn in Fig. 12.8a with unit R and C values, realizes the second-order filter function. If we repeat the derivation of Eq. (2.26) but employ s instead of $j\omega$, we find for $R = C = 1$ that

$$\frac{V_o}{V_i} = T(s) = \frac{A}{s^2 + (3 - A)s + 1} \tag{12.15}$$

To realize the filter function of Eq. (12.11) for some particular Q, we can use a Sallen–Key circuit, provided that A satisfies

$$Q = \frac{1}{3 - A} \tag{12.16}$$

Equation (12.15) includes a gain factor, A, that Eq. (12.11) lacks; however, this is inconsequential in most filtering problems. Although precise adjustment of A becomes difficult for high Q because the difference, $3 - A$, in the denominator becomes small, the Sallen–Key filter is adequate for most of the filters we study in this chapter. (Problems 12.14 and 12.15 explore this sensitivity problem.)

12.2.3 LOWPASS FILTER DESIGN

Second-order filter design involves two fundamental parameters, the circuit Q and the frequency-scale factor. In addition there might be a component constraint that necessitates impedance scaling. Filter Q is often specified by indicating in some fashion the degree of peaking desired in the frequency response. The frequency-scale factor is indicated by either specifying the -3 dB frequency or, alternately, by requiring a particular gain or phase at some specified frequency.

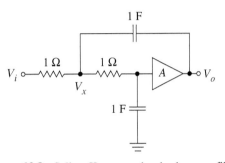

Figure 12.8 Sallen–Key second-order lowpass filter.

EXAMPLE 12.2   Design a second-order Sallen–Key filter with $Q = \sqrt{2}$ that has attenuation of 18.36 dB at 60 kHz. Use only 2000 pF capacitances in the design.

Solution. It is a good idea to sketch the desired characteristic as in Fig. 12.9. Two frequency axes show how the specifications for the frequency-scaled filter (with unknown α) relate to the unity reference frequency filter for which we have design equations. We first find the radian frequency λ_x at which the attenuation is 18.36 dB for the one rad/s filter. Since gain of -18.36 dB corresponds to $|T(\lambda_x)| = 0.1208$, Eq. (12.13) gives

$$T(\lambda_x) = \frac{1}{\sqrt{(1 - \lambda_x^2)^2 + (\lambda_x/\sqrt{2})^2}} = 0.1208$$

Solving gives $\lambda_x = 3$. From Fig. 12.9, $\alpha\lambda_x = 2\pi 60 \times 10^3$, so we must frequency-scale by $\alpha = 2\pi 20 \times 10^3$ rad/s. To do this, we change the capacitors in Fig. 12.8 to $1/(2\pi 2 \times 10^4) = 7.958$ μF.

To reduce the capacitances to 2000 pF, as specifications require, the circuit impedance level must be increased by $\beta = 7.958 \times 10^{-6}/2 \times 10^{-9} = 3.979 \times 10^3$. The final circuit looks like Fig. 12.8, but with resistors of 3.979 kΩ and capacitors of 2000 pF. From Eq. (12.16), a noninverting amplifier with gain $A = 3 - 0.707 = 2.293$ gives the required Q.   ❏

> **Exercise 12.2**   Calculate the phase shift in degrees at 60 kHz for the filter designed in the preceding example.
>
> *Ans.* $-165.1°$

12.2.4 HIGHPASS FILTER

Lowpass to Highpass Transformation. If $T_{LP}(p)$ is any lowpass filter function with one rad/s reference frequency, then the *lowpass to highpass transformation,*

$$p = \frac{1}{s} \tag{12.17}$$

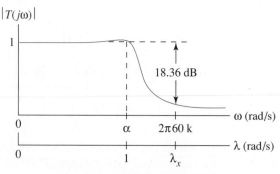

Figure 12.9 Filter specifications for Example 12.2.

gives a highpass filter function $T_{HP}(s)$ with 1 rad/s reference frequency; that is,

$$T_{HP}(s) = T_{LP}(p)|_{p = 1/s} \tag{12.18}$$

Like the frequency-scaling relation of Eq. (12.5), the lowpass to highpass transformation is merely a change in the independent variable. This transformation maps the *magnitude and phase* of the original filter at any complex frequency p to a new frequency s, and Eq. (12.17) relates the source and destination frequencies.

It is easy to verify that this transformation changes the first-order lowpass filter of Eq. (12.1) into the highpass filter of Eq. (12.8). For the second-order function of Eq. (12.11), the transformation gives

$$T(s) = \frac{1}{(1/s)^2 + (1/Q)(1/s) + 1}$$

which we usually write as

$$T(s) = \frac{s^2}{s^2 + (1/Q)s + 1} \tag{12.19}$$

For each lowpass filter function defined by a particular value of Q in Eq. (12.11), Eq. (12.19) gives a corresponding highpass filter of the same Q. Figures 12.10a and b show the frequency response curves of the second-order highpass filter. Comparing these with the lowpass curves of Fig. 12.7 shows how the magnitude and phase of the lowpass filter at frequency λ map to the new frequency $\omega = 1/\lambda$ for the highpass filter.

It is relatively easy to see specifically how a general lowpass function changes under the transformation of Eq. (12.17). The transformation gives

$$T_{HP}(s) = T_{LP}(p)|_{p = 1/s} = \frac{K(1/s - z_1)(1/s - z_2) \cdots (1/s - z_M)}{(1/s - p_1)(1/s - p_2) \cdots (1/s - p_N)}$$

Now doing some algebra to convert the function to the ordinary form of a ratio of factored polynomials,

$$T_{HP}(s) = \left[\frac{K(1/s - z_1)(1/s - z_2) \cdots (1/s - z_M)}{(1/s - p_1)(1/s - p_2) \cdots (1/s - p_N)} \frac{s^M}{s^N} \right] \frac{s^N}{s^M}$$

$$= \left\{ \frac{K(1 - sz_1)(1 - sz_2) \cdots (1 - sz_M)}{(1 - sp_1)(1 - sp_2) \cdots (1 - sp_N)} \right\} \frac{s^N}{s^M}$$

Finally, we obtain the algebraic form

$$T_{HP}(s) = (-1)^{N-M} \hat{K} s^{N-M} \frac{(s - 1/z_1)(s - 1/z_2) \cdots (s - 1/z_M)}{(s - 1/p_1)(s - 1/p_2) \cdots (s - 1/p_N)} \tag{12.20}$$

in which a new constant $\hat{K}$ is defined because of the factoring of poles and zeros. We see that each finite pole and zero of the lowpass filter maps into a new location given by Eq. (12.17). If $N > M$, a zero at the origin of order N-M also results.

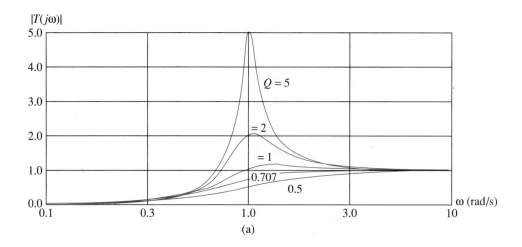

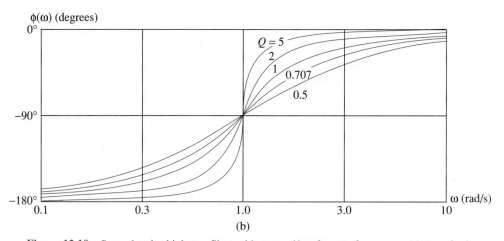

Figure 12.10 Second-order highpass filter with one rad/s reference frequency: (a) magnitude response; (b) angle response.

Sallen–Key Highpass Filter. To transform the Sallen–Key lowpass circuit of Fig. 12.8 into the Sallen–Key highpass circuit of Fig. 12.11 we use the *RC-CR* transformation. We can verify by direct analysis that the Sallen–Key highpass filter function is

$$T(s) = \frac{As^2}{s^2 + (3 - A)s + 1} \tag{12.21}$$

Comparing Eq. (12.21) with Eq. (12.19) shows that, except for gain A, the functions are identical provided that $A = 3 - 1/Q$.

For first- and second-order highpass filters, it would be easy to relate specifications directly to filter transfer functions, as we did for lowpass filters. However, more complex highpass designs require that we first convert highpass specifications into lowpass specifications, so we might as well learn the technique now. Starting with a sketch, we

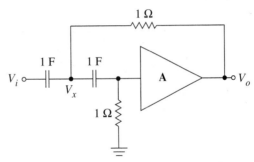

Figure 12.11   Sallen–Key highpass filter having reference frequency of one rad/s.

simply anticipate using both the lowpass to highpass transformation and frequency scaling, as in the following example.

EXAMPLE 12.3   Design a second-order, highpass filter that has phase shift of $-120°$ at 80 Hz and $Q = 1$.

Solution.   Figure 12.12 shows the desired and prototype highpass responses for $Q = 1$. Once we know the frequency, λ_x, where $-120°$ phase shift occurs, we can find α and then frequency-scale the filter of Fig. 12.11. In a highpass design there is generally one additional step. We first find the frequency $\lambda_y = 1/\lambda_x$, where the one rad/s *lowpass* filter with $Q = 1$ has phase shift of $-120°$. From Eq. (12.14), we find

$$\phi(\lambda_y) = -\tan^{-1}\left[\frac{\lambda_y}{1(1 - \lambda_y^2)}\right] = -120°$$

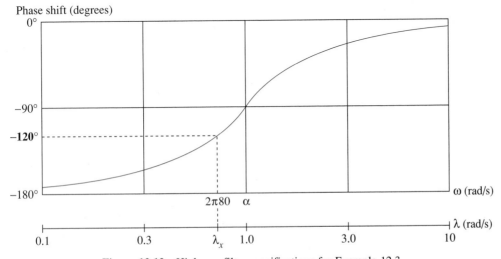

Figure 12.12   Highpass filter specifications for Example 12.3.

This gives

$$-1.732(1 - \lambda_y^2) = \lambda_y$$

for which $\lambda_y = 1.330$ rad/s. The highpass filter with unity reference frequency gives the same phase shift at $\lambda_x = 1/1.330 = 0.7519$ rad/s. From our sketch $\alpha\lambda_x = 2\pi 80$. Therefore, the filter of Fig. 12.11 must be frequency-scaled by $\alpha = 668.5$. This requires replacing the 1 F capacitors by $1/668.5 = 1496$ μF capacitors. Impedance scaling would lead to more practical R and C values. Selecting $A = 2$ to give $Q = 1$ completes the design.   ❏

> **Exercise 12.3** Design a second-order, Sallen–Key, highpass filter that has gain of 0.01 at 80 Hz and $Q = 1$. Do not impedance-scale.
> *Hint*: When the time comes, use $u = \lambda_y^2$.
>
> *Ans.* $C = 198$ μF, $R = 1$ Ω, and $A = 2$.

12.2.5 BIQUADRATIC FILTER CIRCUIT

So far we have realized second-order lowpass and highpass filter functions by means of individual Sallen–Key lowpass and highpass circuits. A circuit called the *biquadratic filter* or *biquad* simultaneously performs both lowpass and highpass filtering of an input signal and provides the results at two different output nodes. Furthermore, we introduce a new kind of filter function to describe the signal at the biquad's third output node.

To introduce the biquad we begin with the lowpass filter function

$$\frac{V_{LP}}{V_i} = \frac{K}{s^2 + (1/Q)s + 1} \tag{12.22}$$

which is Eq. (12.11) with gain K. Cross-multiplying and solving for $s^2 V_{LP}$ gives

$$s^2 V_{LP} = -\frac{1}{Q} s V_{LP} - V_{LP} + K V_i \tag{12.23}$$

Suppose we could produce $s^2 V_{LP}$ by using some circuit. According to Fig. 12.13a, we could then obtain the desired output, V_{LP}, using two inverting integrators, since frequency-domain division by s corresponds to time-domain integration.

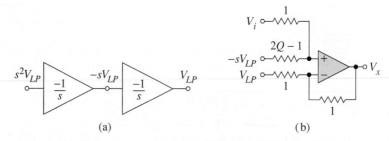

Figure 12.13 Biquad derivation: (a) circuit to obtain V_{LP} by successive integrations; (b) summing circuit to produce $s^2 V_{LP}$.

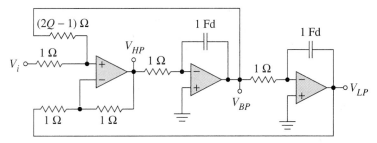

Figure 12.14 Biquad filter circuit with unity reference frequency.

Now Eq. (12.23) shows that *we can obtain* s^2V_{LP} by adding to a scaled input $V_i(s)$ a weighted sum of outputs *from* Fig. 12.13a. The circuit of Fig. 12.13b, practical for $Q > 0.5$, produces this weighted sum, as we now show. Because input V_{LP} sees an inverting amplifier while the other inputs each see a voltage divider followed by a noninverting amplifier of gain 2, we find by superposition

$$V_x = -V_{LP} + \left[\frac{2Q-1}{(2Q-1)+1}V_i + \frac{1}{1+(2Q-1)}(-sV_{LP}) \right]\left(1 + \frac{1}{1}\right) \quad (12.24)$$

$$= -\frac{1}{Q}sV_{LP} - V_{LP} + \left(\frac{2Q-1}{Q}\right)V_i$$

which is the same as the right side of Eq. (12.23), provided we take K to be

$$K = \frac{2Q-1}{Q} \quad (12.25)$$

Figure 12.14 shows Fig. 12.13b as part of the complete biquad circuit, where V_{LP} identifies the *lowpass output node* for the transfer function described by Eqs. (12.22) and (12.25). Because every coefficient in Eq. (12.24) depends upon a resistor *ratio*, both Q and K are invariant under impedance scaling. For positive resistors, this particular biquad circuit requires $Q > 0.5$. (The references describe other biquad circuits that do not have this restriction.)

Since the biquad produces V_{LP} from V_{HP} by two inverting integrations, we deduce from Eq. (12.22) that

$$\frac{V_{HP}}{V_i} = \frac{Ks^2}{s^2 + (1/Q)s + 1} \quad (12.26)$$

Therefore, we can realize any second-order highpass filter by using this highpass output. After introducing the bandpass filter, we will discuss the third biquad output, labeled V_{BP}.

12.2.6 BANDPASS FILTER

Ideal Bandpass Filter. Another filter class is the *bandpass filter,* idealized as in Fig. 12.15a. The reference frequency is the *band-center* frequency, $u_C = 1$ rad/s. This filter

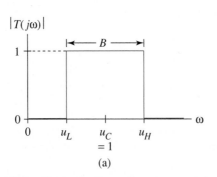
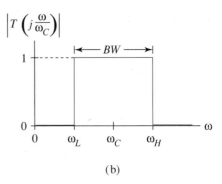

Figure 12.15   Bandpass filter characteristic: (a) with one rad/s reference frequency; (b) after frequency scaling.

passes all frequencies within a *passband* defined by upper and lower *cut-off frequencies,* u_H and u_L, and rejects all others. The width of the passband is the *bandwidth B*, given in rad/s by

$$B = u_H - u_L \qquad (12.27)$$

The band-center frequency of any bandpass filter is defined as the *geometric mean* of the cut-off frequencies. For Fig. 12.15a this means the filter's one rad/s band-center frequency is given by

$$u_C = 1 = \sqrt{u_H u_L} \qquad (12.28)$$

Frequency-Scaled Bandpass Filter. Consider how the ideal characteristic of Fig. 12.15a changes when we frequency-scale with a nonnegative scale factor, $\alpha = \omega_C$. The band-center frequency moves to ω_C rad/s; the upper and lower edges of the passband move to $\omega_H = \omega_C u_H$ and $\omega_L = \omega_C u_L$, respectively, as in Fig. 12.15b. After scaling, the band-center frequency is the geometric mean

$$\sqrt{\omega_H \omega_L} = \sqrt{(\omega_C u_H)(\omega_C u_L)} = \omega_C \sqrt{u_H u_L} = \omega_C \qquad (12.29)$$

Taking the log of both sides gives

$$\log(\omega_C) = \frac{\log(\omega_H) + \log(\omega_L)}{2}$$

showing that "band-center frequency" means the average in terms of a logarithmic frequency scale. Hereafter for brevity we refer to the band-center frequency as the *center frequency,* with this interpretation implied.

When we frequency-scale, bandwidth B is also scaled to a new value BW, as we see from

$$BW = \omega_H - \omega_L = \omega_C(u_H - u_L) = \omega_C B \qquad (12.30)$$

Because center frequency and bandwidth both frequency-scale by the same factor, their ratio remains constant, or invariant under frequency scaling. The Q or quality factor *of a bandpass filter* is defined as the ratio of the band-center frequency to the bandwidth, that is.

$$Q = \frac{\omega_C}{BW} = \frac{1}{B} \tag{12.31}$$

Second-Order Bandpass Function. The ideal bandpass filter is not physically realizable; however, a realizable second-order transfer function that *approximates* the ideal is

$$T(s) = \frac{s/Q}{s^2 + s/Q + 1} \tag{12.32}$$

where Q is a parameter. Figure 12.16a compares the amplitude response, $|T(j\omega)|$, from Eq. (12.32) with the ideal bandpass characteristic. By substituting $s = j1$ into Eq. (12.32) we readily verify unity gain at one rad/s. By checking the limits as $s = j\omega$ approaches zero and infinity, we see that the gain approaches zero at both extremities. By solving for the frequencies that satisfy

$$|T(j\omega)| = \frac{1}{\sqrt{2}}$$

it is possible (after some challenging algebra!) to compute expressions for the upper and lower -3 dB frequencies. In the notation of Fig. 12.15a, the results are

$$u_H, u_L = \sqrt{1 + \left(\frac{1}{2Q}\right)^2} \pm \frac{1}{2Q} \tag{12.33}$$

These upper and lower half-power frequencies define the passband for the second-order filter. An important result follows from Eq. (12.33). This is

$$B = u_H - u_L = \frac{1}{Q} \tag{12.34}$$

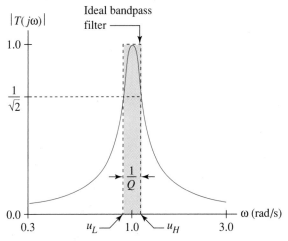

Figure 12.16 Second-order bandpass characteristic compared with ideal bandpass curve.

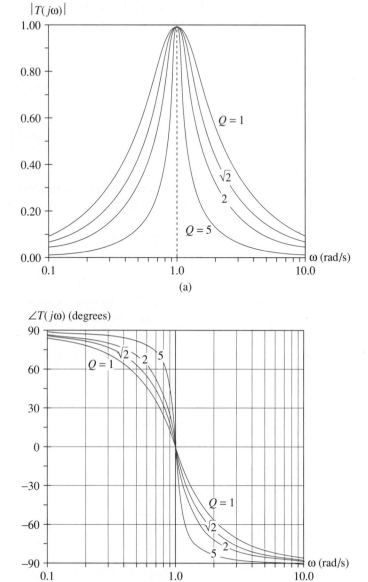

Figure 12.17 Second-order bandpass filter characteristics: (a) magnitude response; (b) phase response.

In other words, *parameter Q* of the second-order filter function is the ratio of center frequency to -3 dB bandwidth, the quality factor of the bandpass filter. In Fig. 12.16, the passband is symmetric relative to the center frequency because frequency is plotted on a logarithmic scale. Figure 12.17 compares the magnitude and phase curves for several values of Q.

Biquad Bandpass Output. The third output of the biquad of Fig. 12.14, labeled V_{BP}, is the *bandpass output*. Because output V_{BP} is output V_{HP} processed by an inverting integrator, we divide Eq. (12.26) by $-s$ to find the transfer function from input to V_{BP}. This gives

$$\frac{V_{BP}}{V_i} = \frac{-Ks}{s^2 + (1/Q)s + 1} = \frac{-[(2Q-1)/Q]s}{s^2 + (1/Q)s + 1} = \frac{-Hs/Q}{s^2 + (1/Q)s + 1} \qquad (12.35)$$

Comparing this equation with Eq. (12.32) we see that, except for the inconsequential Q-dependent voltage gain, $-H$, output V_{BP} gives second-order bandpass filtering of the circuit's input signal.

In a bandpass design, the required filter bandwidth determines Q, a value also dependent on the band-center frequency, which itself is fixed by frequency scaling. We see this in the following example.

EXAMPLE 12.4 Design a second-order bandpass filter with -3 dB bandwidth extending from 800 Hz to 925 Hz.

Solution. A sketch like Fig. 12.18a helps us translate given specifications into specifications for the unscaled filter. Final bandwidth is $BW = 2\pi(925 - 800) = 785.4$ rad/s. The final center frequency is the geometric mean

$$\omega_C = 2\pi\sqrt{800 \times 925} = 5.405 \times 10^3 \text{ rad/s} = \alpha$$

From Eq. (12.31), the required Q is

$$Q = \frac{5.405 \times 10^3}{785.4} = 6.882$$

We start with the biquad of Fig. 12.14. One resistor is $2Q - 1 = 12.764$ Ω; the others are 1 Ω resistors. At output V_{BP}, this circuit would have a bandpass characteristic with correct Q but center frequency of one rad/s. Therefore, we frequency-scale by $\alpha = 5.405 \times 10^3$ rad/s, reducing all capacitor values to $1/\alpha = 185$ μF. Impedance-scaling by $\beta = 10^3$ to increase resistor size gives the circuit of Fig. 12.18b. Incidentally, from Eq. (12.35), this circuit has voltage gain of $-H = -(2Q - 1) = -12.8$ at the center frequency. If this is a problem we can use an inverting *attenuator* like Fig. 12.18c at input or output. ❑

Exercise 12.4 Design a second-order, biquad, bandpass filter like Fig. 12.14, but with center frequency of 10 kHz and bandwidth of 5 kHz.

Ans. Capacitors are 15.9 μF. Five 1 Ω resistors, one 3 Ω resistor.

12.2.7 BANDSTOP AND ALL-PASS FILTERS

The *simultaneous* availability of lowpass, highpass, and bandpass functions makes the biquad a very versatile circuit. We next show that summing amplifiers can combine biquad outputs to realize two additional filter functions, the bandstop and all-pass functions.

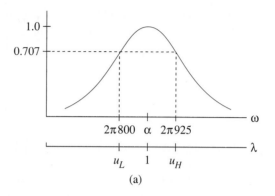

$$2\pi 800 \quad \alpha \quad 2\pi 925$$

$$u_L \quad 1 \quad u_H$$

(a)

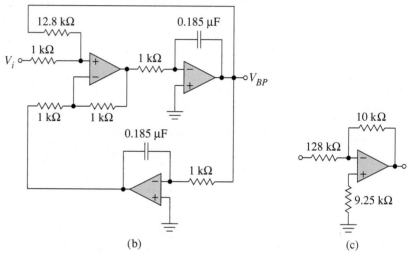

(b)   (c)

Figure 12.18 Bandpass filter design: (a) specifications; (b) bandpass filter that satisfies specifications of Example 12.4; (c) circuit to correct sign and gain of the bandpass filter.

Bandstop (Notch) Filter. First, consider using a summing amplifier to combine the lowpass and highpass outputs of a biquad by forming the weighted sum

$$V_{BS} = -\frac{1}{K}(V_{HP} + V_{LP}), \qquad K = \frac{2Q - 1}{Q} \qquad (12.36)$$

where K is the lowpass and highpass gain of the biquad. From Eqs. (12.22) and (12.26) the new output gives the *bandstop function*

$$\frac{V_{BS}}{V_i} = T(s) = \frac{-(s^2 + 1)}{s^2 + (1/Q)s + 1} \qquad (12.37)$$

which has the pole-zero plot of Fig. 12.19a. In addition to conjugate poles on the unit circle, $T(s)$ has zeros on the imaginary axis at $\pm j1$. Since $T(j1) = 0$, this function completely rejects a signal component of $\omega = 1$ rad/s. Also, as $s = j\omega \to \infty$ and $s = j\omega \to 0$, $|T(j\omega)| \to 1$.

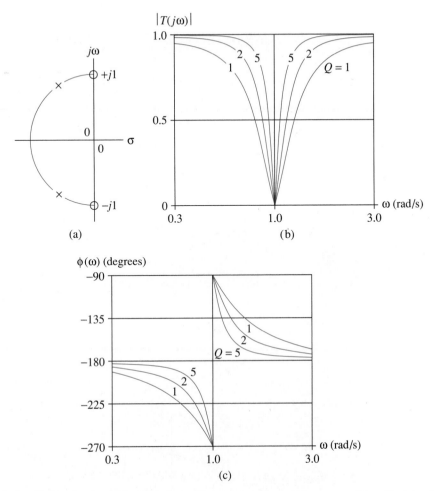

Figure 12.19 Bandstop filter function with notch or reference frequency of 1 rad/s: (a) pole-zero plot; (b) magnitude response; (c) phase response.

The reference frequency for a notch filter is the notch frequency. Figure 12.19b shows the amplitude frequency response curve for several values of Q. As Q increases, the poles move closer to the zeros, giving a more narrow stopband. Figure 12.19c shows corresponding phase curves. The relationship between the pole locations and the frequency response is better understood by examining the complex number $|T(j\omega)|$ at any frequency ω.

Factoring Eq. (12.37) into pole and zero terms and evaluating the expression at $s = j\omega$ gives a ratio of products of complex numbers for any selected frequency ω; that is,

$$T(j\omega) = \left. \frac{-(s + j1)(s - j1)}{(s - p)(s - \hat{p})} \right|_{s = j\omega} = \frac{-(j\omega + j1)(j\omega - j1)}{(j\omega - p)(j\omega - \hat{p})} = -\frac{M_1\angle\theta_1 \, M_2\angle\theta_2}{M_3\angle\theta_3 \, M_4\angle\theta_4} \quad (12.38)$$

where p denotes a pole frequency and $\hat{p}$ its complex conjugate.

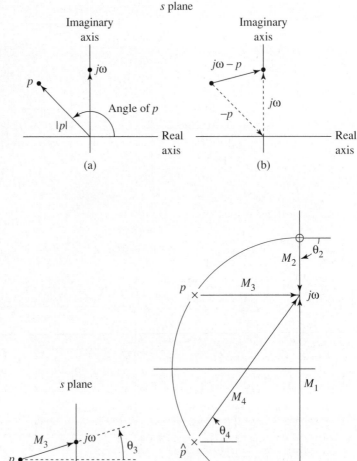

Figure 12.20 Visualizing a frequency response curve from its pole-zero plot: (a) vectors representing pole p and variable $j\omega$; (b) the vector that represents a pole term; (c) vector representation of $M_3 \underline{/\theta_3}$; (d) four vectors representing bandstop filter response as function of ω.

Since p is complex, we can show its magnitude and angle by a vector directed from the origin to point p in the complex frequency plane, as in Fig. 12.20a. The figure also shows *variable* $j\omega$ as a vector. Figure 12.20b represents the pole term, $(j\omega - p)$ from Eq. (12.38) as the vector sum, $(-p) + j\omega$. Figure 12.20c represents the term $(j\omega - p)$ as the complex vector, $M_3 \underline{/\theta_3}$, directed from pole p to point $j\omega$. Both M_3 and θ_3 change as frequency variable ω changes. By the same reasoning, the zero terms in Eq. (12.38) are vectors directed from the zeros to point $j\omega$.

It follows that the magnitude frequency response curve is a plot of

$$|T(j\omega)| = \frac{M_1(\omega)\,M_2(\omega)}{M_3(\omega)\,M_4(\omega)}$$

where the magnitudes for one value of ω are shown in Fig. 12.20d. As $j\omega$ increases from the position shown in the figure, M_2 decreases and the other vector lengths increase until $j\omega = j1$ is reached, where M_2 has zero length. For $\omega > 1$ all vectors increase in length with further increases in ω. As ω approaches infinity, the vector lengths become equal and $|T|$ approaches one.

Since poles are on the unit circle, at $\omega = 0$ the gain magnitude is also one. For high Q, pole p is positioned close to the zero at $j1$; thus, vectors $M_3\angle\theta_3$ and $M_2\angle\theta_2$ approximately cancel each other until ω is very close to 1, as do vectors $M_4\angle\theta_4$ and $M_1\angle\theta_1$. This explains the variations in the magnitude curves of Fig. 12.19b with increasing Q.

We next use the vectors to explain the phase curves of Fig. 12.19c. The minus sign in Eq. (12.38) adds a constant $180°$ phase shift to the phase curves. The variation in the phase plot comes from the poles and zeros. From Eq. (12.38), the phase $\phi(\omega)$ of $-T(j\omega)$ is

$$\phi(\omega) = \theta_1(\omega) + \theta_2(\omega) - \theta_3(\omega) - \theta_4(\omega)$$

From the vectors in Fig. 12.20d, since $\theta_3(0) = -\theta_4(0)$, when $j\omega \to 0$,

$$\phi(j0) = 90° + (-90°) - [-\theta_4(0)] - \theta_4(0) = 0°$$

We can also see that as $\omega \to \infty$, $\phi(\omega) \to 90° + 90° - 90° - 90° = 0°$. The phase of the vector $M_2\angle\theta_2$ changes abruptly from $-90°$ to $+90°$ as ω passes through zero. When p is closer to $j0$ (for high Q), the angle of vector $M_3\angle\theta_3$ changes quite rapidly with ω, explaining the variation of the curves of Fig. 12.19c with Q.

There are many ways to construct second-order bandstop filters in addition to the biquad method. Section 12.4 suggests two other bandstop filter structures.

Allpass Function. Now consider using the summing amplifier of Fig. 12.21a to combine the three biquad outputs into

$$V_{AP} = -\frac{1}{K}\left(V_{HP} + \frac{1}{Q}V_{BP} + V_{LP}\right) \tag{12.39}$$

where

$$K = \frac{2Q - 1}{Q}, \qquad Q > 0.5$$

From Equations (12.26), (12.35), and (12.22) the result is

$$\frac{V_{AP}}{V_i} = -T(s)$$

where

$$T(s) = \frac{s^2 - s/Q + 1}{s^2 + s/Q + 1} \tag{12.40}$$

is known as the second-order *allpass function*.

The pole-zero plot of Fig. 12.21b shows a special pole-zero pattern that characterizes all pass transfer functions in general; the negative of every pole is a zero, and the negative of every zero is a pole. Because of this, the complex number

$$T(j\omega) = \frac{M_1 \angle \theta_1 \, M_2 \angle \theta_2}{M_3 \angle \theta_3 \, M_4 \angle \theta_4}$$

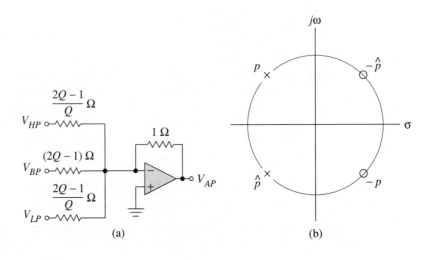

(a)   (b)

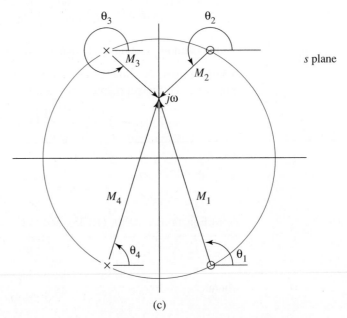

(c)

Figure 12.21 Allpass filter function: (a) summing amplifier that creates allpass filter from biquad outputs; (b) allpass pole-zero plot; (c) vectors showing how allpass frequency response is developed; (d) phase response of allpass function.

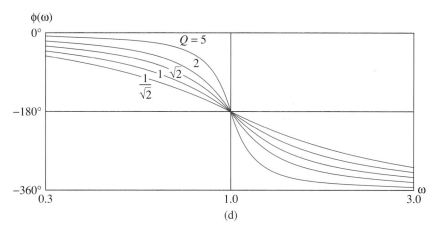

Figure 12.21 (continued)

that results from evaluating Eq. (12.40) at any point $s = j\omega$ has magnitude independent of frequency. Figure 12.21c shows that *for any* ω, $M_2 = M_3$ and $M_1 = M_4$. We conclude that $|T(j\omega)| = 1$ for $0 \le \omega \le \infty$, so this circuit has no effect whatsoever on the relative amplitudes of the signal frequencies in the output. For this reason, $T(s)$ is called an *allpass* function. $T(s)$ *does*, however, alter the phase of the signal component.

In Sec.1.6.7 we learned that a circuit produces no phase distortion *provided that the phase curve has constant slope*. Also, we learned that a circuit with a linear phase curve produces a delay equal to the slope of the phase curve. When we design a filter to process signal amplitudes, the filter sometimes introduces *phase distortion* because its phase curve is not linear. Since the phase shifts of cascaded networks add while their amplitude curves multiply, we can sometimes process the filter output with an allpass *phase equalization* circuit and, by proper design, correct the phase distortion introduced by the filter. Stand-alone, allpass circuits are useful for adding desired time delay to a signal. We see from Fig. 12.21d that higher Q gives higher slopes and thus greater delays near $\omega = 1$ rad/s; however, the frequency band of signals for which the slope, and therefore the delay, is constant decreases with Q.

The *group delay*, or simply *delay D* in seconds, introduced by a circuit having phase shift $\phi(\omega)$ rad/s is

$$D(\omega) = \frac{d\phi(\omega)}{d\omega} \ \text{s}$$

From Eq. (12.40) we see that the phase shift contributed by the second-order, allpass filter is

$$\phi(\omega) = \arg\left[\frac{1 - \omega^2 - j(\omega/Q)}{1 - \omega^2 + j(\omega/Q)}\right] = -2 \tan^{-1}\left[\frac{\omega}{Q(1 - \omega^2)}\right] \qquad (12.41)$$

where we have made use of the odd-function property of the inverse tangent function. Van Valkenburg's design text, included in the references, provides a good introduction to delay equalization design.

12.3 _____
Lowpass Active Filters of Order *n*

In this section we learn how to apply filter design concepts to specifications too exacting to be satisfied by first- or second-order filters. We study two special approximations of the ideal lowpass filter that lead to filter *families*. Filter families give us the flexibility to approximate the ideal as closely as we wish by using filters of higher order.

12.3.1   FILTER SPECIFICATIONS

We first introduce the general idea of approximating the ideal lowpass filter with a continuous function of ω. With three parameters, we are able to define specifications that control the critical features of the approximation. Figure 12.22 shows the ideal lowpass characteristic and parameters: δ, γ, and ω_s. We use δ to denote a minimum passband gain and γ to specify a maximum stopband gain. A *selectivity factor,* ω_s defines a *transition band,* where the approximating function changes from satisfying the passband specification to satisfying the stopband specification. The solid curve shows how a lowpass filter function is constrained by the specifications. Allowing *ripple* in the passband, the stopband, or both admits into consideration approximations with narrower transition bands than would otherwise be possible. Many different approximation functions have proven useful in filter design, and entire books on the subject are readily available. We limit our studies to the two simplest and best-known approximation functions, Butterworth and Chebyshev.

12.3.2   BUTTERWORTH APPROXIMATION

Figure 12.23 shows the magnitude frequency response curves for a few members of the Butterworth filter *family.* For increasing order *n* the filters more closely approximate the

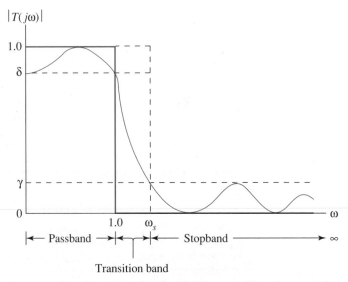

Figure 12.22   Approximating the ideal lowpass filter characteristic.

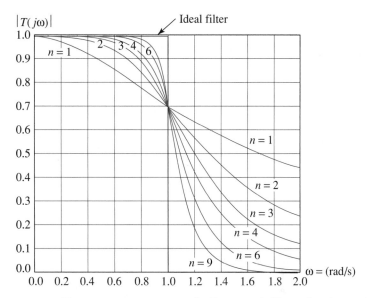

Figure 12.23 Frequency response curves for Butterworth filters of various orders.

ideal. Since all curves have gain of $1/\sqrt{2}$ at the one rad/s reference frequency, $\delta = 1/\sqrt{2}$ is often used for these filters; however, other values of δ simply require special frequency scaling. Butterworth filters are known as *maximally flat* filters because the first $2n - 1$ derivatives of $|T(j\omega)|$ at $\omega = 0$ are zero, making its response very flat at low frequencies.

We introduce the Butterworth filter function $T(s)$ in a roundabout way because the product $T(s)T(-s)$ is much easier to describe mathematically. This product can be written as

$$T(s)T(-s) = \frac{1}{1 + (-1)^n s^{2n}} = \frac{1}{B(s)B(-s)} \tag{12.42}$$

where n is the *order* of the filter and $B(s)$ is the *Butterworth polynomial* of order n.

By definition, the poles of $T(s)T(-s)$ are the $2n$ complex numbers s that make the denominator zero, that is, values of s that satisfy

$$s^{2n} = -(-1)^n \tag{12.43}$$

When n is odd, the poles satisfy

$$s^{2n} = 1 = 1\ e^{jm360°} = 1\ \underline{/m\ 360°}, \qquad m = 0, 1, 2, \ldots$$

Here we recall that we can represent the complex number "1" in either exponential or polar notation, or as a vector in the complex plane, using any integer multiple of 360° for its angle. Taking the "$2n$"th root gives the distinct poles

$$s = 1\ e^{m360°/(2n)} = 1\ e^{m180°/n}, \qquad m = 0, 1, \ldots, 2n - 1 \quad \text{for } n \text{ odd} \tag{12.44}$$

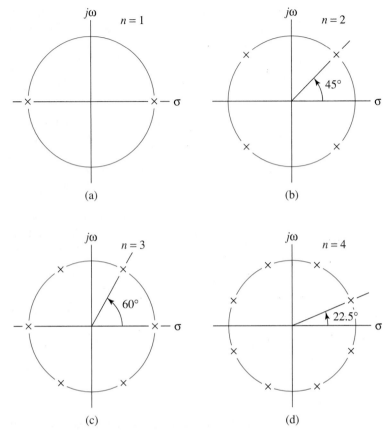

Figure 12.24 Locations of complex poles for $T(s)T(-s)$: (a) $n = 1$; (b) $n = 2$; (c) $n = 3$; (d) $n = 4$.

Figures 12.24a and c show the poles for $n = 1$ and $n = 3$. In general, for odd n the poles lie on the unit circle, uniformly spaced at angular intervals of $180°/n$, with the pole for $m = 0$ located at $s = +1$.

When n is even, the poles satisfy

$$s^{2n} = -1 = 1\, e^{180° + m360°}, \qquad m = 0, 1, 2, \ldots$$

which leads to the poles

$$s = \left(1\, e^{180° + m360°}\right)^{1/(2n)} = 1\, e^{(90° + m180°)/n}$$

$$\text{for m} = 0, 1, \ldots, 2n - 1, n \text{ even} \tag{12.45}$$

illustrated in Figs. 12.24b and d for $n = 2$ and 4. For even n, the $2n$ poles are evenly spaced around the unit circle, but the first pole ($m = 0$) is located at $90°/n$.

To be stable, our filters may have no poles in the right half of the complex plane. Therefore, starting with the denominator polynomial of Eq. (12.42) in factored form, we partition it into the product of two polynomials: $B(s)$, which consists of only the left-half plane pole factors, and $B(-s)$, which contains the remaining factors. The result, $B(s)$, is the *Butterworth polynomial* of degree n. The Butterworth filter function of order n is the function

$$T(s) = \frac{1}{B(s)}$$

which has n left-half poles, all on the unit circle.

Although $B(s)$ cannot be expressed in a simple closed form, its poles are easily computed for any n using either Eq. (12.44) or (12.45), and from the roots of $B(s) = 0$ we can easily construct $B(s)$. A convenience in practical filter design is the general availability of tabulated results such as Table 12.1, which show $B(s)$ factored into first- and second-order polynomials. Such tables save us the work of computing left-half plane pole values and then combining conjugate pole terms into second-order terms.

Once we decide that we need a Butterworth filter of a particular order, for example, $n = 5$, we can use a table such as Table 12.1 to design a cascade of active filters that together give the desired Butterworth response. For example, for $n = 5$,

$$\begin{aligned} T(s) &= \frac{1}{(s+1)(s^2+0.6180s+1)(s^2+1.6180s+1)} \\ &= T_1(s)T_2(s)T_3(s) \\ &= \frac{1}{(s+1)}\frac{1}{(s^2+0.6180s+1)}\frac{1}{(s^2+1.6180s+1)} \end{aligned}$$

(12.46)

where $T_1(s)$ is a first-order filter and $T_2(s)$ and $T_3(s)$ are second-order filters. From the preceding sections we already know how to design circuits that realize first- and second-order filter functions. To design Butterworth filters, we now need only to learn how to derive the required filter order from given specifications. We address this next.

Butterworth Filter Frequency Response. It follows from Eq. (12.42) that the squared-magnitude frequency response curve of the Butterworth filter is

$$|T(j\omega)|^2 = \frac{1}{1+\omega^{2n}}$$

TABLE 12.1 Butterworth Filter Denominators

n	Factors of $B(s)$
1	$(s+1)$
2	$(s^2+1.4142s+1)$
3	$(s+1)(s^2+s+1)$
4	$(s^2+0.7654s+1)(s^2+1.8478s+1)$
5	$(s+1)(s^2+0.6180s+1)(s^2+1.6180s+1)$
6	$(s^2+0.5176s+1)(s^2+1.4142s+1)(s^2+1.9318s+1)$

therefore

$$|T(j\omega)| = \frac{1}{\sqrt{1 + \omega^{2n}}} \qquad (12.47)$$

Figure 12.23 is, in fact, Eq. (12.47) plotted for selected values of n. The curves for $n = 1$ and 2 describe the first- and second-order filters we learned to design in preceding sections. For higher n the filters become more complex and expensive; however, their filter functions more closely approach the ideal. From Eq. (12.47) we see that for $\omega \gg 1$, $20 \log|T(j\omega)| \Rightarrow -20\ n \log \omega$; that is, the frequency response *rolls off* at $-20n$ dB/decade.

The following exemplifies Butterworth filter design, including the new idea of selecting n.

EXAMPLE 12.5 Design a maximally flat, lowpass, active filter with -3 dB bandwidth of 3 kHz and attenuation of at least 60 dB at 15 kHz

Solution. Figure 12.25a shows how specifications relate to the unscaled filter characteristic. The unscaled filter must have 60 dB attenuation at $\lambda_x = 15$ k/3 k = 5 rad/s. Since -60 dB is gain of 0.001, we have from Eq. (12.47)

$$10^{-3} = \frac{1}{\sqrt{1 + 5^{2n}}}$$

giving $n = 4.29$. Because the filter order must be an integer, we use $n = 5$, which gives attenuation *exceeding* 60 dB at 15 kHz.

From Table 12.1 we find the denominator of the unscaled filter function and Eq. (12.46). We can realize the first-order factor by frequency-scaling the circuit of Fig. 12.2a by $\alpha = 2\pi3000 = 1.8850 \times 10^4$ rad/s. Each second-order factor can be realized using either a Sallen–Key or a biquadratic filter. From Eqs. (12.46) and (12.11) we see that one filter needs to have $Q = 1/0.6180 = 1.618$ and the other $Q = 1/1.6180 = 0.618$. For a Sallen–Key design, Eq. (12.16) shows that these filters require amplifiers of gain $A = 3 - 1/Q = 2.382$ and 1.3819, respectively. After frequency-scaling by $\alpha = 2\pi3000$, we obtain Fig. 12.25b. All that remains is a final impedance-scaling, for example, by 10^4, to obtain more reasonable values. ❏

Exercise 12.5 Compute the decibel attenuation at 15 kHz for the filter designed in the preceding example.

Ans. -69.9 dB.

Example 12.5 showed how to design a Butterworth filter when the -3 dB bandwidth and the stopband attenuation at some particular frequency are specified. If the cut-off frequency is specified in some other way, for example, by requiring a -1 dB passband, a somewhat different approach is required. The following example shows this.

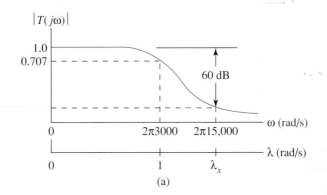

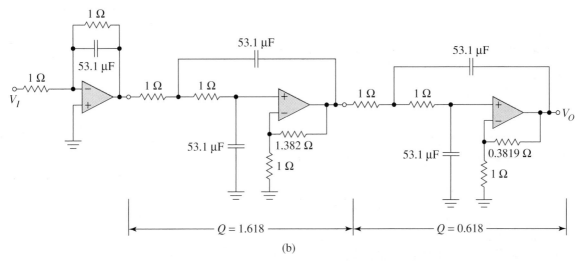

Figure 12.25 Butterworth filter of Example 12.5: (a) specifications; (b) filter design after frequency scaling.

EXAMPLE 12.6 Design a Butterworth lowpass filter with a -1 dB passband of 5 kHz that has attenuation of at least -20 dB for $f > 12$ kHz.

Solution. We begin by sketching Fig. 12.26. The frequency scale factor, $\alpha = 2\pi f_c$, is initially unknown; however, because we will eventually frequency-scale the *basic* Butterworth filter, we see from the diagram that

$$\frac{\lambda_x}{\lambda_y} = \frac{\alpha 2\pi \, 12{,}000}{\alpha 2\pi \, 5{,}000} = \frac{12}{5}$$

Attenuation of 1 dB at λ_y means the gain is 0.891. From Eq. (12.47)

$$|T(j\lambda_y)| = 0.891 = \frac{1}{\sqrt{1 + (\lambda_y)^{2n}}} \qquad (12.48)$$

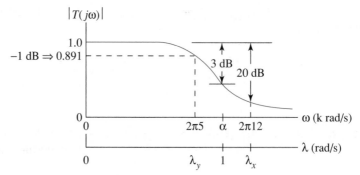

Figure 12.26 Specifications for filter of Example 12.6.

For attenuation of 20 dB at λ_x,

$$|T(j\lambda_x)| = 0.1 = \frac{1}{\sqrt{1 + (\lambda_x)^{2n}}} = \frac{1}{\sqrt{1 + [(12/5)\lambda_y]^{2n}}}$$

Squaring and taking the reciprocals gives

$$1.2601 = 1 + (\lambda_y)^{2n} \quad \text{and} \quad 100 = 1 + \left(\frac{12}{5}\lambda_y\right)^{2n}$$

Substituting $(\lambda_y)^{2n} = 0.2601$ from the first into the second gives

$$100 = 1 + 0.2601\left(\frac{12}{5}\right)^{2n}$$

When we solve for n we obtain

$$2n \log\left(\frac{12}{5}\right) = \log\left(\frac{100 - 1}{0.2601}\right)$$

which gives $n = 3.394$. If n could take on this noninteger value, we could solve for both λ_x and λ_y; these would be in the exact 12/5 ratio shown on the diagram, and we would obtain the exact attenuation specified at both λ_x and λ_y. However, since n must be an integer, we use the next higher integer value, $n = 4$, in Eq. (12.48). This gives

$$\lambda_y = (1.2601 - 1)^{1/8} = 0.8451 \text{ rad/s}$$

This fixes the -1 dB bandwidth to be exactly 5 kHz. The consequence of using $n = 4$ instead of 3.394 is that the filter will have 20 dB attenuation at a frequency slightly lower than λ_x. After frequency-scaling, we will obtain attenuation greater than 20 dB at 12 kHz. Figure 12.26 shows that the -3 dB cut-off frequency must be scaled from one rad/s to

$$\frac{\alpha}{1} = \frac{2\pi 5000}{\lambda_y} = \frac{2\pi 5000}{0.8451} = 2\pi 5.916 \text{ kHz}$$

giving $\alpha = 3.717 \times 10^4$ rad/s.

The remainder of the design is straightforward. From Table 12.1 the unscaled filter has the transfer function

$$T(s) = \frac{1}{(s^2 + 0.7654s + 1)} \frac{1}{(s^2 + 1.8478s + 1)}$$

$$= \frac{1}{(s^2 + s/1.307 + 1)} \frac{1}{(s^2 + s/0.5412 + 1)}$$

Each second-order section is realized by one of the active circuits we have studied, frequency-scaled to meet specifications, and then impedance-scaled to practical component values.   ❏

Exercise 12.6   Find the attenuation at 12 kHz for the filter designed in the preceding example.

Ans. 24.6 dB.

12.3.3   CHEBYSHEV LOWPASS APPROXIMATION

Figure 12.27 shows frequency responses for the first three members of the Chebyshev *filter family*. As with Butterworth filters, parameter n specifies the filter order. For $n > 1$, Chebyshev characteristics are not monotonic in the passband like the Butterworth functions, but exhibit a *ripple* or variation in gain. The designer controls the minimum passband gain, δ, by means of a Chebyshev parameter, ε. By allowing passband ripple, the Chebyshev filter achieves greater stopband attenuation than does a Butterworth filter of the same order.

Designing Chebyshev *equal ripple filters* resembles designing Butterworth filters. Widely available tables provide functions $T(s)$ that describe a lowpass filter of one rad/s cut-off frequency for any desired values of n and ε. We initially design this prototype

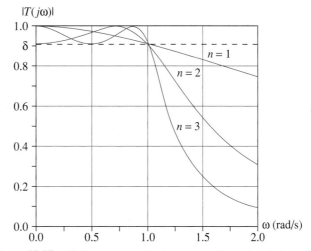

Figure 12.27   Chebyshev magnitude response for orders 1 through 3.

filter as a cascade of first- and second-order filters and then frequency-scale to fit the specifications. Because the Chebyshev poles are not located on the unit circle, however, formulating the one rad/s prototype filter requires an additional operation, frequency *prescaling*. The following details of Chebyshev filter theory explain the frequency response curves and show how we can translate given specifications into required values for n and ϵ.

Chebyshev Polynomials. The Chebyshev filter function $|T(j\omega)|$ is described indirectly by its magnitude-squared function

$$|T(j\omega)|^2 = \frac{1}{1 + \epsilon^2 C_n^2(\omega)} \tag{12.49}$$

where the *Chebyshev polynomial of degree n*, $C_n(\omega)$, is defined by

$$C_n(\omega) = \cos(n \cos^{-1}\omega) \tag{12.50}$$

At this point it is not obvious that $C_n(\omega)$ is a polynomial, much less that it is good for anything. We address both points next.

Problem 12.33 shows how to prove that $C_n(\omega)$ satisfies the *recursion formula*

$$C_{n+1}(\omega) = 2\omega C_n(\omega) - C_{n-1}(\omega) \tag{12.51}$$

This means that we can generate a Chebyshev polynomial once we find the two consecutive polynomials of next-lower order. Fortunately, the first two polynomials are obvious. For $n = 0$ and $n = 1$, Eq. (12.50) gives, respectively, the polynomials

$$C_0(\omega) = \cos(0 \cos^{-1}\omega) = 1$$

and

$$C_1(\omega) = \cos(\cos^{-1}\omega) = \omega$$

Using Eq. (12.51) we now generate successive higher-order polynomials such as

$$C_2(\omega) = 2\omega^2 - 1$$
$$C_3(\omega) = 2\omega(2\omega^2 - 1) - \omega = 4\omega^3 - 3\omega$$
$$C_4(\omega) = 8\omega^4 - 8\omega^2 + 1$$

and so on

showing that $C_n(\omega)$ indeed describes a polynomial in ω of order n. Figure 12.28 shows the polynomials for $n = 1$, 2, and 3. Resisting any compulsion to generate more polynomials, we instead step back and note that the polynomials $C_n(\omega)$ have some remarkable properties. Although we do not present a proof, the following properties apply to all Chebyshev polynomials.

1. $C_n(1) = 1$ for all n.
2. $C_n(-1) = 1$ for even n; $C_n(-1) = -1$ for n odd.
3. All roots of $C_n(\omega) = 0$ are real and in the range $|\omega| < 1$.
4. $|C_n(\omega)| \to \infty$ for $|\omega| \to \infty$.
5. $C_n(\omega)$ oscillates with *equal ripple* between -1 and $+1$ when $|\omega| \le 1$.

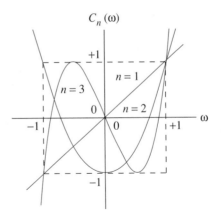

Figure 12.28 Chebyshev polynomials of degrees 1, 2, and 3.

Chebyshev Filter Function. Knowing something of Chebyshev polynomials, we next examine how they are used to approximate the ideal lowpass filter, using the third-order polynomial to be specific. The function of Eq. (12.49) defines the Chebyshev magnitude-squared frequency response curve. Figure 12.29a illustrates the third-order Chebyshev polynomial; Fig. 12.29b shows $\epsilon C_3(\omega)$ for $\epsilon = 0.45$. Notice how ϵ reduces the

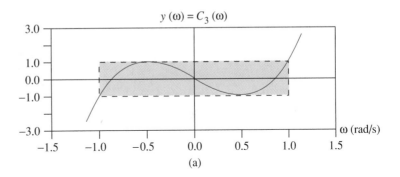

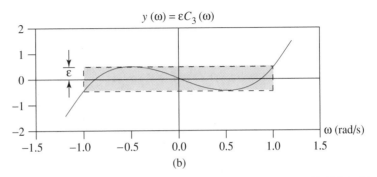

Figure 12.29 Chebyshev magnitude-squared frequency response curve: (a) $C_3(\omega)$; (b) $\epsilon C_3(\omega)$ for $\epsilon = 0.45$; (c) $\epsilon^2 C_3^2(\omega)$; (d) $1 + \epsilon^2 C_3^2(\omega)$; (e) $|T(j\omega)|^2$.

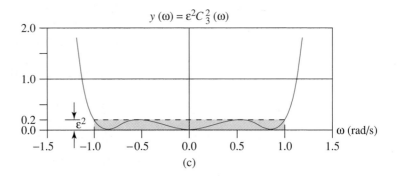

(c)

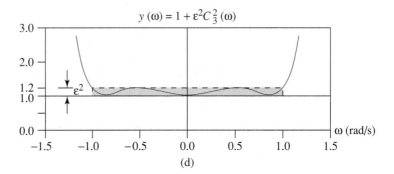

(d)

$$|T(j\omega)|^2 = \frac{1}{1 + \epsilon^2 C_3^2(\omega)}$$

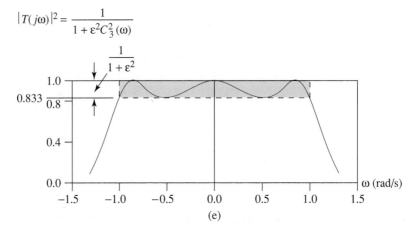

(e)

Figure 12.29 (continued)

ripple to ± 0.45 V for $-1 \le \omega \le +1$. Squaring this product in Fig. 12.29c gives a ripple with three minima (for $n = 3$) over the range $-1 \le \omega \le +1$. The peak-to-peak amplitude of the ripple becomes $\epsilon^2 = 0.2$. Squaring also produces an even function of ω and large functional values for large $|\omega|$. Adding 1 raises the curve as in Fig. 12.29d, making each relative minimum tangent to $y = +1$ with ripple amplitude still ϵ^2. The final squared-magnitude function, Fig. 12.29e and Eq. (12.49), is the reciprocal of the function of Fig. 12.29d for every ω. The points of tangency to the $y = +1$ line are the same after taking the reciprocal while all other points are folded down to values less than one, giving an approximation to the ideal filter for $\omega \geq 0$. Notice that the final peak-to-peak ripple amplitude within the passband is $1/(1+\epsilon^2)$.

Figure 12.30 compares Chebyshev filter, squared transfer characteristics for $n = 1$, 2, and 3. Within the passband, each curve has ripple bounded by 1.0 and $1/(1+\epsilon^2)$. Odd-order filter curves all pass through the point $(0, 1)$, whereas even-order filters have dc attenuation at $\omega = 0$. Higher-order filters have more ripples in the passband and greater attenuation in the stopband. Passband ripples are of equal amplitude. Taking the square root at each frequency gives Fig. 12.27.

Chebyshev Filter Design. From Eq. (12.49), the transfer function of the Chebyshev filter is

$$|T(j\omega)| = \frac{1}{\sqrt{1 + \epsilon^2 C_n^2(\omega)}} \tag{12.52}$$

In a filter design, we begin with a specification for minimum passband gain δ or, equivalently, *passband ripple* $r_{dB} = -20 \log \delta$. Because the passband minima occur at those frequencies where $C_n(\omega) = 1$ (see Fig. 12.30), it follows from Eq. (12.52) that

$$r_{dB} = -20 \log \delta = -20 \log\left(\frac{1}{1 + \epsilon^2}\right)^{0.5}$$

Thus

$$\epsilon = \sqrt{10^{r_{dB}/10} - 1} \tag{12.53}$$

which allows us to compute ϵ when r_{dB} is specified.

In a typical design problem, there is also a specification for maximum stopband gain that creates the need to evaluate Eq. (12.52) for $\omega > 1$. Because the inverse cosine in Eq. (12.50) is imaginary for such frequencies, we use an alternative form for the Chebyshev polynomial,

$$C_n(\omega) = \cosh(n \cosh^{-1} \omega), \qquad |\omega| > 1 \tag{12.54}$$

Because the proof of this equation is somewhat distracting, we outline it in Problem 12.35 and continue our design discussion.

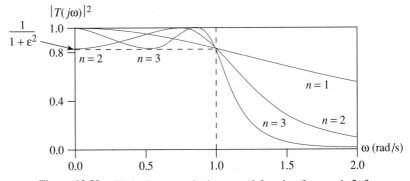

Figure 12.30 Chebyshev magnitude-squared function for $n = 1, 2, 3$.

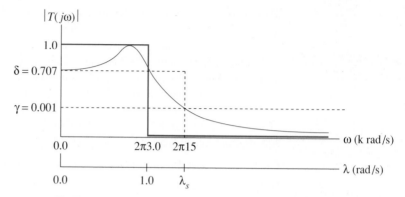

Figure 12.31 Specifications for filter design of Exercise 12.7.

To design a Chebyshev filter with gain γ at stopband frequency ω_s, we start with Eq. (12.52) written in the form

$$|T(j\omega_s)| = \gamma = \frac{1}{\sqrt{1 + \epsilon^2[\cosh(n \cosh^{-1} \omega_s)]^2}} \tag{12.55}$$

Solving for n gives the design equation

$$n = \frac{\cosh^{-1}\left[(1/\gamma^2 - 1)^{0.5}/\epsilon\right]}{\cosh^{-1} \omega_s} \tag{12.56}$$

With ω_s and γ known and ϵ found from Eq. (12.53), we can for the filter order n.

Exercise 12.7 Figure 12.31 shows specifications for a Chebyshev lowpass filter with -3 dB bandwidth of 3 kHz and at least 60 dB attenuation at 15 kHz. Use Eqs. (12.53) and (12.56) to find ϵ and n.

Ans. $\epsilon = 0.998$, $n = 3.317$ (use $n = 4$).

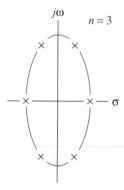

Figure 12.32 Locations of poles of $T(s)T(-s)$ for a Chebyshev third-order filter.

The filter orders in Exercise 12.7 and Example 12.5 (identical specifications) show that by allowing passband ripple, the Chebyshev design is of lower order ($n = 4$) than the Butterworth filter ($n = 5$).

Chebyshev Pole Locations. Although we leave the proof for books on filter theory, such as those in the references, $T(s)T(-s)$ has its $2n$ poles positioned in an elliptical pattern in the complex s plane as in Fig. 12.32. For the Chebyshev filter function $T(s)$, we select only those denominator factors that correspond to left-half plane poles. The final Chebyshev function then has the form

$$T(s) = \frac{K}{s^n + a_{n-1}s^{n-1} + \cdots + a_o} \tag{12.57}$$

where $K = a_o$ for n odd and $K = a_o/(1 + \epsilon^2)^{0.5}$ for n even. Table 12.2 gives the denominator of $T(s)$ in factored form for two different values of passband ripple. Several filter

TABLE 12.2 Chebyshev Denominator Factors for 0.5 and 3.0 dB Ripple

Passband Ripple Value (dB)	n	Factors
0.5	1	$(s + 2.863)$
	2	$(s^2 + 1.426s + 1.5164)$
	3	$(s + 0.626)(s^2 + 0.626s + 1.142453)$
	4	$(s^2 + 0.350s + 1.062881)(s^2 + 0.846s + 0.35617)$
	5	$(s + 0.362)(s^2 + 0.224s + 0.012665)(s^2 + 0.586s + 0.476474)$
	6	$(s^2 + 0.156s + 1.022148)(s^2 + 0.424s + 0.589588)(s^2 + 0.58s + 0.157)$
3.0	1	$(s + 1.002)$
	2	$(s^2 + 0.2986s + 0.83950649)$
	3	$(s + 0.299)(s^2 + 0.2986s + 0.83950649)$
	4	$(s^2 + 0.17s + 0.902141)(s^2 + 0.412s + 0.1961)$
	5	$(s + 0.177)(s^2 + 0.11s + 0.936181)(s^2 + 0.288s + 0.377145)$
	6	$(s^2 + 0.076s + 0.95402)(s^2 + 0.208s + 0.522041)(s^2 + 0.286s + 0.089093)$

books listed in the references include more extensive tables. If available tables are inadequate, we can always compute the pole values using

$$s_k = \sigma_k + j\omega_k$$

$$= -\sinh\left[\frac{1}{n}\sinh^{-1}\left(\frac{1}{\epsilon}\right)\right]\sin\left(\frac{2k-1}{n}\frac{\pi}{2}\right) \quad (12.58)$$

$$+ j\cosh\left[\frac{1}{n}\sinh^{-1}\left(\frac{1}{\epsilon}\right)\right]\cos\left(\frac{2k-1}{n}\frac{\pi}{2}\right), \quad k = 1, 2, \ldots, n$$

and from these construct first- and second-order factors as in the table.

Notice that the first- and second-order factors in Table 12.2 have constant terms that are not 1 like the constants in the Butterworth functions of Table 12.1. Because of this, Chebyshev filters require an additional design step, frequency prescaling, which we explain next.

Frequency Prescaling. Table 12.2 suggests that if we ignore the overall gain factor K in Eq. (12.57), cascaded filters described by first- and second-order functions have the forms

$$T_k(s) = \frac{1}{s + a_k}$$

and

$$T_k(s) = \frac{1}{s^2 + a_{1k}s + a_{0k}} = \frac{1}{s^2 + (\omega_k/Q_k)s + \omega_k^2} \quad (12.59)$$

Because the poles are not on the unit circle, constants a_k and a_{0k} are not unity as for the Butterworth factors in Table 12.1. Dividing numerator and denominator of Eq. (12.59) by ω_k^2 gives

$$T_k(s) = \frac{1/\omega_k^2}{(s/\omega_k)^2 + \dfrac{1}{Q_k}(s/\omega_k) + 1}$$

where

$$\omega_k = \sqrt{a_{0k}} \quad \text{and} \quad Q_k = \frac{\omega_k}{a_{1k}} = \frac{\sqrt{a_{0k}}}{a_{1k}} \tag{12.60}$$

This shows that to realize a term like Eq. (12.59) we can start with a second-order filter of unit reference frequency and quality factor Q_k and then frequency-*prescale* using scale factor $\alpha = \omega_k$. For active filters such as those of Figs. 12.8 and 12.14, this means we use capacitors of $1/\omega_k$ farad instead of 1 farad *in the prototype filter design*. We call this prescaling, because a *subsequent* frequency scaling of *all* stages is necessary to obtain the specified cut-off frequency of the Chebyshev filter as a whole. Strictly speaking, we should also provide gain $1/\omega_k^2$ for each filter stage; however, we ignore this detail just as we ignore K in Eq. (12.57). When n is odd, the first-order filter section must be prescaled by $\alpha = a_k$.

EXAMPLE 12.7 Complete the design of the fourth-order, lowpass filter of Exercise 12.7 with the additional constraint that all resistors be at least 5 kΩ.

Solution. Since we need $n = 4$ for a filter with 3 dB passband ripple, Table 12.2 shows that $T(s)$ for the initial filter can be realized by the cascade

$$T(s) = \frac{K_1}{s^2 + 0.17s + 0.902141} \, \frac{K_2}{s^2 + 0.412s + 0.1961}$$

For the first filter, Eq. (12.60) gives

$$\omega_{k1} = 0.9498, \qquad Q_1 = \frac{0.9498}{0.17} = 5.587$$

For the second,

$$\omega_{k2} = 0.44283, \qquad Q_2 = \frac{0.44283}{0.412} = 1.0748$$

For a Sallen–Key realization, we ignore K_1 and K_2, frequency-prescale Fig. 12.28 and calculate gain A for each filter section. This gives Fig. 12.33a. This is a fourth-order Chebyshev filter with 3 dB bandwidth of one rad/s. Notice that the capacitor values in farads, from prescaling, are $1/\omega_{k1}$ and $1/\omega_{k2}$ in the respective filters.

We must now frequency-scale by $\alpha = 2\pi 3000$ to obtain the desired final bandwidth, a process that changes capacitor values to 55.9 μF and 119.8 μF, respectively, in the two filters. Subsequent impedance scaling by 5000 gives the final circuit of Fig. 12.33b.

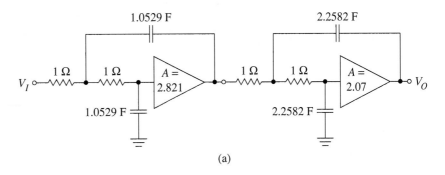

(a)

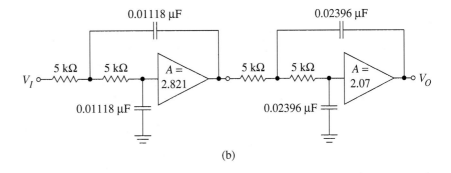

(b)

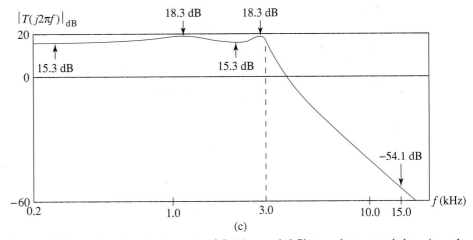

(c)

Figure 12.33 Active filter for Example 12.7: (a) cascaded filter sections prescaled to give a 1 rad/s bandwidth; (b) filter circuit after final frequency and impedance scaling; (c) SPICE computed frequency response of scaled filter.

Figure 12.33c shows the SPICE-computed frequency response of this filter. The passband peaks show gain of 18.3 dB instead of 0 dB that would have resulted had we corrected the amplitude scale factors during our design. The ripple amplitude and bandwidth are correct, however, and the attenuation at 15 kHz is $18.3 - (-54.1) = 72.4$ dB, which exceeds the minimum requirement of 60 dB. ❏

12.3.4 IMPEDANCE SCALING OF INDIVIDUAL SECTIONS

All our filter designs conceptually begin with a prototype filter with given R and C values, often unity. We then frequency- and impedance-scale *the entire circuit* to fit specifications. However, it is obviously possible to *individually* impedance-scale the various subcircuits, and we can use this extra degree of freedom to reduce the spread of component values in a design. For example, impedance-scaling the second subcircuit in Fig. 12.33b by

$$\beta = \frac{0.02396}{0.01118} = 2.1431$$

results in identical capacitances *for the entire circuit* while increasing second-stage resistor values to $2.1431(5 \text{ k}\Omega) = 10.716 \text{ k}\Omega$. Obviously other compromises between spreads of capacitor and resistor values can be achieved by $1 < \beta < 2.1431$. This idea obviously applies to all multistage filter circuits.

12.4
Active Extensions of Passive Filter Theory

The literature contains abundant information on filters constructed from resistors, capacitors, and inductors. Some *RLC* designs, especially *ladder structures,* have low sensitivity to parameter variations. Passive filter theory is *directly* applicable to high-frequency circuits, for which inductors are easy to fabricate, but not to audio or subaudio frequencies or to ICs where high-quality inductors are unavailable. We first review some basic passive filters and then show how a *synthetic inductor* can extend the advantages of passive designs to some IC- and low-frequency discrete circuits.

12.4.1 SECOND-ORDER PASSIVE FILTERS

It is easy to verify that the filters of Fig. 12.34a and b have second-order lowpass and highpass transfer characteristics described by Eqs. (12.11) and (12.19), respectively. Equally easy to analyze are the bandpass circuit of Fig. 12.34c, described by Eq. (12.32), and the bandstop filter of Fig. 12.34d, which, except for algebraic sign, is described by Eq. (12.37). We can frequency-scale these circuits to make their frequency response curves satisfy given specifications and we can impedance-scale to find practical realizations. In particular, each can be impedance-scaled to a specified load resistance value. For complex filters, such as Butterworth and Chebyshev filters of arbitrary order, we can cascade passive filters (using a first-order *RC* or *RL* circuit for odd *n*), using follower circuits for interstage isolation.

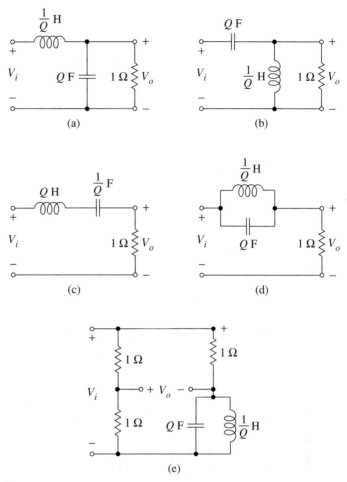

Figure 12.34 Second-order passive filters with reference frequencies of 1 rad/s: (a) lowpass; (b) highpass; (c) bandpass; (d) band-reject or notch filter; (e) allpass circuit.

Figure 12.34e shows a second-order allpass circuit, analyzed as follows. First notice that the impedance of the parallel LC subcircuit is

$$Z = \frac{(s/Q)[1/(sQ)]}{s/Q + 1/(sQ)} = \frac{1}{Qs + Q/s} = \frac{s}{Q(s^2 + 1)}$$

Voltage dividers give

$$V_o = 0.5\, V_i - \frac{s/[Q(s^2 + 1)]}{s/[Q(s^2 + 1)] + 1}\, V_i = \left[0.5 - \frac{s}{s + Q(s^2 + 1)}\right] V_i$$

Dividing numerator and denominator of the second term by Q gives the second-order allpass function

$$T(s) = \frac{V_o}{V_i} = \left(0.5 - \frac{s/Q}{s^2 + s/Q + 1}\right) = 0.5\left(\frac{s^2 - s/Q + 1}{s^2 + s/Q + 1}\right)$$

which is Eq. (12.40) except for gain of 0.5.

Exercise 12.8   Design a bandpass filter like Fig. 12.34d with $Q = 85$, band-center frequency of 455 kHz, and a 5 kΩ load resistance.

Ans. $L = 20.57\ \mu\text{H}$, $C = 594.6\ \text{pF}$ is one correct answer.

The synthetic inductor of the next section, and similar electronic circuits, help us extend the benefits of passive filter theory to *RC*-active circuits.

12.4.2   SYNTHETIC INDUCTOR

The *synthetic inductor,* Figs. 12.35a and b, is an *RC-active* building block that is practical for some discrete and integrated designs. The following analysis proves that the circuit has the input impedance of an inductor. It also establishes that we can frequency- and impedance-scale this circuit in the usual fashion.

First, observe that I_{in} flows through R_1 because of the infinite gain of *OA*#1. Infinite gain also forces

$$\frac{I_x}{sC} + I_{in}R_1 = 0$$

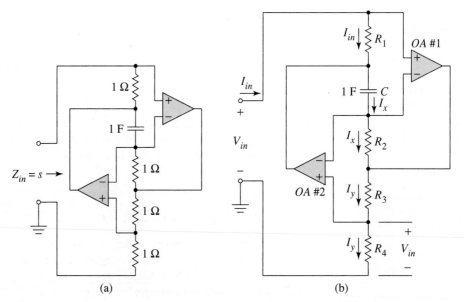

(a)   (b)

Figure 12.35   Synthetic inductor: (a) basic circuit; (b) circuit used to justify scaling procedures.

Because of the infinite gain of *OA*#2,

$$I_y R_3 + I_x R_2 = 0$$

Finally, because the sum of the input voltages of *OA*#1 and *OA*#2 are zero, KVL gives

$$V_{in} = I_y R_4$$

Starting with the last equation and substituting gives

$$V_{in} = I_y R_4 = \frac{-R_2}{R_3} I_x R_4 = \frac{-R_2}{R_3}(-sCR_1)I_{in}R_4$$

Thus the input impedance is

$$Z_{in} = \frac{V_{in}}{I_{in}} = \frac{sCR_1R_2R_4}{R_3} = sL \qquad (12.61)$$

the design equation for the synthetic inductor. In principle we can realize any of the circuits of Fig. 12.34 in *RC*-active form by replacing the inductor by this active circuit and choosing appropriate values. Notice from Eq. (12.61) that impedance-scaling *C* and all resistors by β gives *s*(β*L*), and frequency-scaling the capacitor by α gives *s*(1/α)*L*. Therefore, we can use Fig. 12.35a in our designs, and then scale the entire filter circuit in the usual way. For proper operation R_4 in Fig. 12.35b must be grounded. Interchanging *C* and R_3 gives another realization that is more practical for real op amps.

EXAMPLE 12.8 Design a second-order *RC*-active notch filter, based on Fig. 12.34d, with *Q* = 5 to reject a noise frequency of 1 kHz. The circuit must have a 10 kΩ load resistance.

Solution. Figure 12.36a is a passive circuit that has *Q* = 5 and a notch at 1 rad/s. In Fig. 12.36b we replace the inductor by the circuit of Fig. 12.35a after impedance-prescaling to give an inductance of 0.2 H. We next frequency-scale the circuit by α = $2\pi 10^3$ to position the notch, and impedance-scale by 10^4 to arrive at the final design, Fig. 12.36c. Interchanging source and load would allow the source to be grounded. ❑

12.4.3 DOUBLY TERMINATED *LC* LADDER FILTERS

Lowpass Filters. Our present understanding of filters opens the door to many other results from filter theory. As but one example, consider the doubly terminated, lowpass ladders of Fig. 12.37. Both circuits use *n* energy storage elements to realize the same lowpass filter function of order *n*. The difference is that for odd *n*, Fig. 12.37a has fewer inductors and Fig. 12.37b fewer capacitors. For a Chebyshev or Butterworth ladder filter

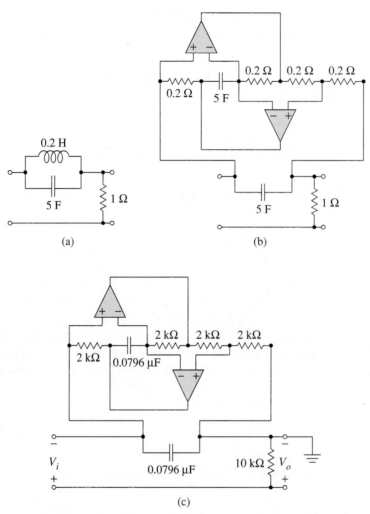

Figure 12.36 Schematics for filter design of Example 12.8: (a) passive filter with $Q = 5$ and notch at 1 rad/s; (b) RC-active realization of Fig. 12.36a; (c) final design having notch at 1 kHz.

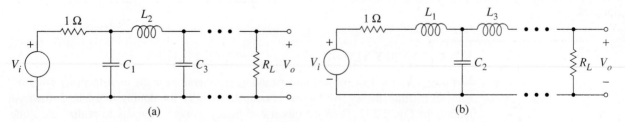

Figure 12.37 Doubly terminated ladder structure: (a) minimum inductance circuit; (b) minimum capacitance circuit.

of order n we need only consult precalculated tables such as Tables 12.3 or 12.4 for component values for the filter of 1 rad/s reference frequency. (Some of the references give more complete tables.)

Highpass Ladder Filters. The lowpass to highpass transformation of Eq. (12.17) has a useful corollary for passive filters with reference or cut-off frequencies of 1 rad/s. In analysis, all appearances of complex variable p result from terms pL_{LP} or $1/pC_{LP}$, where L_{LP} and C_{LP} are inductance or capacitance values in the lowpass circuit. The lowpass to highpass transformation changes these analysis terms in the following ways:

$$pL_{LP} \Rightarrow \frac{L_{LP}}{s} = \frac{1}{sC_{HP}}$$

and

$$\frac{1}{pC_{LP}} \Rightarrow \frac{s}{C_{LP}} = sL_{HP}$$

Thus by replacing each inductor of L_{LP} H by a capacitor of $1/L_{LP}$ F and each capacitor of C_{LP} F with a $1/C_{LP}$ H inductor we can transform the lowpass circuit into a highpass filter with 1 rad/s reference frequency. For example, notice that the highpass filter of Fig. 12.34b is the lowpass filter of Fig. 12.34a after a lowpass to highpass transforma-

TABLE 12.3 Butterworth Lowpass Minimum Inductance (Minimum Capacitance) Ladder, 1 rad/s Bandwidth

n	$C_1(L_1)$	$L_2(C_2)$	$C_3(L_3)$	$L_4(C_4)$	$C_5(L_5)$	$L_6(C_6)$	$C_7(L_7)$	R_L
2	1.414	1.414						1.000
3	1.000	2.000	1.000					1.000
4	0.7654	1.848	1.848	0.7654				1.000
5	0.6180	1.618	2.000	1.618	0.6180			1.000
6	0.5176	1.414	1.932	1.932	1.414	0.5176		1.000
7	0.4450	1.247	1.802	2.000	1.802	1.247	0.4450	1.000

TABLE 12.4 Chebyshev Lowpass Minimum Inductance (Minimum Capacitance) Ladder, 1 rad/s bandwidth

Passband Ripple Value (dB)	n	$C_1(L_1)$	$L_2(C_2)$	$C_3(L_3)$	$L_4(C_4)$	$C_5(L_5)$	$L_6(C_6)$	$C_7(L_7)$	R_L
0.1	3	1.0316	1.1474	1.0316					1.0000
	5	1.1468	1.3712	1.9750	1.3712	1.1468			1.0000
	7	1.1812	1.4228	2.0967	1.5734	2.0967	1.4228	1.1812	1.0000
1.0	3	2.0236	0.9941	2.0236					1.0000
	5	2.1349	1.0911	3.0009	1.0911	2.1349			1.0000
	7	2.1666	1.1115	3.0936	1.1735	3.0936	1.1115	2.1666	1.0000

tion. For an *RC*-active realization, we replace every inductor with an impedance-prescaled, synthetic inductor.

Exercise 12.9   Design a third-order, Chebyshev, highpass, *RC*-active filter based upon a doubly terminated passive ladder. Cut-off frequency is 900 Hz and passband ripple is 1 dB. Use the minimum number of synthetic inductors.

Ans. Figure 12.38.

EXAMPLE 12.9   Use SPICE to plot the magnitude frequency response of Fig. 12.38 if the op amp has gain of 2×10^5 but is otherwise ideal. Then repeat for a 741 op amp.

Solution. Figure 12.39a shows the SPICE code. The code includes two op amp subcircuits: a high-gain VCVS called GENERIC, and M741, the 741 op amp subcircuit introduced in Fig. 2.34a that includes the internal resistances of the 741 as well as its open-loop frequency response.

Figure 12.39b and the solid curve of Fig. 12.39c show the frequency response using op amp GENERIC. Both agree exactly with original specifications over a wide frequency range. The 60 dB/dec slope expected of a third-order filter is clearly evident.

The dashed curve in Fig. 12.39c, which was computed after replacing the X1 and X2 statements with the 741 references that follow them, shows the frequency response we would obtain if we used 741 op amps. Because the 741's open-loop bandwidth is only 5 Hz, the high gain we count on for operation of the synthetic inductor begins to diminish at low frequencies, and the passband ripple is much higher than specifications

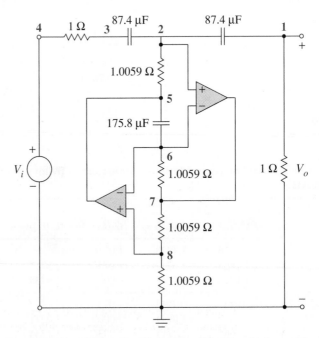

Figure 12.38   Third-order, Chebyshev, *RC*-active, highpass filter.

```
EXAMPLE 12.9            RL1 2 5 1.0059
.SUBCKT GENERIC 1 2 6   RL2 6 7 1.0059
E1 6 0 1 2 2E5          RL3 7 8 1.0059
.ENDS                   RL4 8 0 1.0059
.SUBCKT M741 1 2 6      CL 5 6 175.8U
RIN 1 2 2MEG            X1 2 6 7 GENERIC
ROUT 5 6 75             *X1 2 6 7 M741
EI 3 0 1 2 2E5          X2 8 6 5 GENERIC
R 3 4 1                 *X2 8 6 5 M741
C 4 0 0.0318            VI 4 0 AC 1
E2 5 0 4 0 1            .AC DEC 30 1000 100E6
.ENDS                   .PRINT AC VDB(1)
ROUT 1 0 1              .END
C1 2 1 87.4U
C2 3 2 87.4U
RI 4 3 1
```

(a)

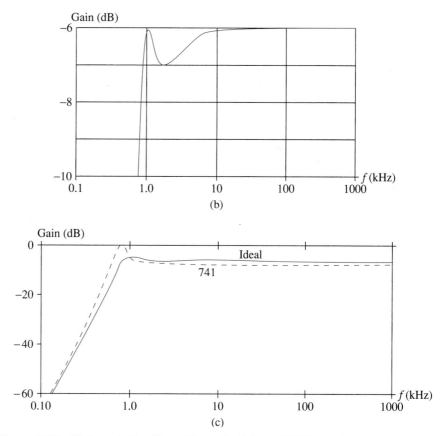

(b)

(c)

Figure 12.39 Highpass ladder filter of Example 12.9: (a) SPICE code for Fig. 12.38; (b) ideal response of Fig. 12.38; (c) responses using ideal op amps and μA741s; (d) responses of 741 realization and passive asymptotic circuit; (e) responses using ideal op amps and HA2544s.

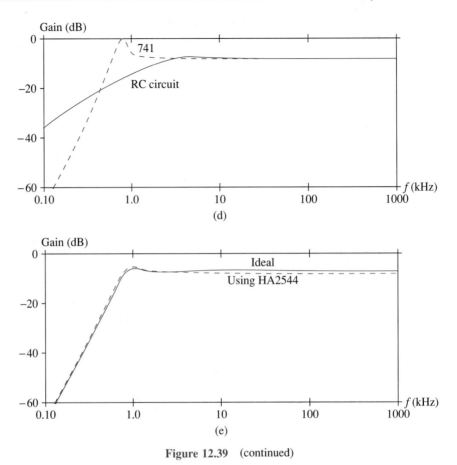

Figure 12.39 (continued)

allow. At still higher frequencies, the op amp gains further degenerate until the op amps become, for all practical purposes, passive circuit elements.

Figure 12.39d shows the response of the circuit if the 741s are simply replaced by their 2 MΩ input and 75 Ω output resistances. At high frequencies, this passive circuit predicts exactly the response of the 741 circuit in Fig. 12.39c as the gain of the op amps becomes negligible. Figure 12.39e compares the ideal filter with a realization using the high-frequency HA2544 op amp.   ❏

12.5
Switched-Capacitor Circuits

After scaling IC filters to realistic capacitance values, the resulting resistances are sometimes too large for IC fabrication. In this section we examine a solution to this problem, replacing the resistors with resistor equivalents consisting of capacitors and switches. Because capacitors and switches are lossless, power dissipation is negligible in the resulting *switched-capacitor circuits*. Furthermore, the critical features of these circuits depend upon capacitor ratios and a switching frequency, both of which can be controlled to within 1%.

12.5.1 BASIC CONCEPTS

Figure 12.40a shows a resistor and one idealized switched-capacitor equivalent. In practical circuits, the switches might be the MOS transmission gates of Sec. 5.11. A two-phase clock, consisting of nonoverlapping pulse trains ϕ_1 and ϕ_2, alternately opens and closes the switches at a frequency f_s high enough that v_1 and v_2 cannot change appreciably during a single switching period. When the ϕ_1 switch closes, C_R charges to v_1 volts by storing charge

$$q_1 = C_R v_1$$

The ϕ_1 switch then opens and the ϕ_2 switch closes. The capacitor now charges to v_2 volts by readjusting its stored charge to

$$q_2 = C_R v_2$$

When the first switch recloses to begin the next cycle, there has been a net movement of

$$\Delta q = q_1 - q_2 = C_R (v_1 - v_2)$$

coulombs of charge from the node at v_1 volts to the node at v_2 volts. Since this charge transfer takes $1/f_s$ seconds, the average current is

$$i = \frac{\Delta q}{\Delta t} = C_R \frac{(v_1 - v_2)}{1/f_s}$$

Ignoring the detailed pulsating nature of the current, we have

$$R = \frac{v_1 - v_2}{i} = \frac{1}{C_R f_s} \tag{12.62}$$

Using $f_s = 100$ kHz and $C_R = 20$ pF, for example, gives a 500 kΩ resistor.

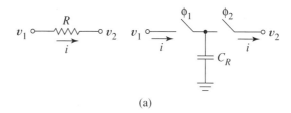

(a)

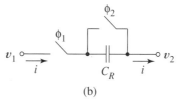

(b)

Figure 12.40 Switched-capacitor resistors.

Figure 12.40b is also a switched-capacitor resistor. When the ϕ_1 switch closes, $C_R (v_1 - v_2)$ coulombs of charge move as C_R charges. When the ϕ_2 switch closes, the capacitor discharges. Again Eq. (12.62) describes the average operation.

Equation (12.62) shows a unique feature of switched-capacitor circuits; resistance is inversely proportional to f_s. We learned in Sec. 12.1.4 that we can frequency-scale RC-active circuits by changing only resistors; therefore, if f_s is common to all resistors in a switched-capacitor filter, we can use the clock to frequency-scale. This feature facilitates precise tuning by adjusting f_s and also allows us to change the cut-off or reference frequency to adapt the same filter to different filtering tasks.

12.5.2 SWITCHED-CAPACITOR BUILDING BLOCKS

Integrators. The switched-capacitor circuits discussed next provide useful stand-alone functions and also constitute building blocks for more complex circuits. Figure 12.41a is an integrator with a switched-capacitor input resistor. When the ϕ_2 switch closes, the virtual ground causes C_R to discharge to zero volts, transferring all the stored charge to C. Replacing the resistance in the integrator equation by Eq. (12.62) gives

$$v_o = -f_s \frac{C_R}{C} \int v_i(t)dt \qquad (12.63)$$

showing how the capacitor ratio and f_s define the integrator coefficient. When C_R is small, the stray circuit capacitances that exist between each node and ground contribute

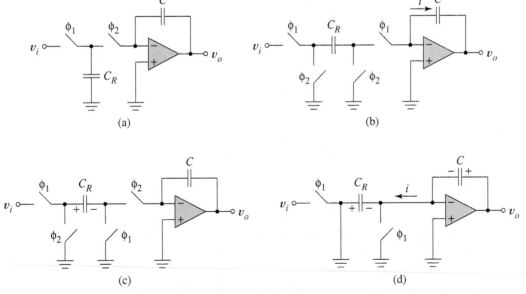

Figure 12.41 Switched-capacitor integrators: (a) inverting integrator; (b) stray capacitance insensitive inverting integrator; (c) noninverting integrator; (d) charge transfer mechanism of noninverting integrator.

to uncertainties in C_R. The integrator of Fig. 12.41b uses additional switches to avoid this problem. During the ϕ_1 cycle, C_R charges to v_i volts, as charging current flows through C. During the ϕ_2 cycle, C_R discharges. The net current flow is from v_i onto C as in the unswitched integrator. Overall operation is described by Eq. (12.63).

By reassigning the clock phases of two switches in Fig. 12.41b, we obtain the *non-inverting integrator* of Fig. 12.41c. During ϕ_1, C_R charges to v_i with the indicated polarity. During the ϕ_2 cycle, the virtual ground forces C_R to discharge. This is accomplished by current i in Fig. 12.41d, which transfers the charge to C with the polarity opposite that of the inverting integrator.

Amplifiers. Figure 12.42a shows a switched-capacitor, inverting amplifier. During ϕ_1, $q_i = C_{R1}v_i$ coulomb of charge is stored on C_{R1}. During ϕ_2, this charge moves to C_{R2} as current i, making v_o negative for positive v_i. With switch ϕ_1 open,

$$v_o = -\frac{q_i}{C_{R2}} = -\frac{C_{R1}}{C_{R2}}\, v_i$$

The output voltage appears as samples of the output of an amplifier with voltage gain

$$\frac{v_o}{v_i} = -\frac{C_{R1}}{C_{R2}} \tag{12.64}$$

alternating with output values of zero (when switches ϕ_1 are closed.)

Figure 12.42b is a noninverting amplifier that combines the gain mechanism of the inverting amplifier with the switching of the noninverting integrator, to give gain that is the negative of Eq. (12.64).

First- and Second-Order Filters. The switched-capacitor building blocks can be used to construct lowpass filters like those studied in Secs. 12.1 and 12.2, opening the way to switched-capacitor realizations of the multistage Butterworth and Chebyshev filters of arbitrary order.

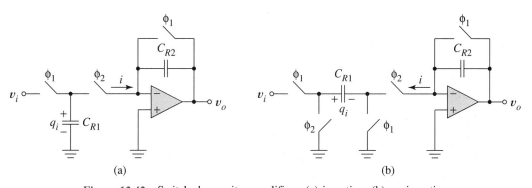

(a)   (b)

Figure 12.42 Switched-capacitor amplifiers: (a) inverting; (b) noninverting.

Figure 12.43 is the switched-capacitor equivalent of the first-order lowpass filter of Fig. 2.14a. Replacing the resistances in Eq. (2.24) with switched values from Eq. (12.62) gives

$$\frac{V_o}{V_i} = \frac{-C_{R1}/C_{R2}}{1 + j(\omega/\omega_H)}, \qquad \omega_H = \frac{C_{R2}f_s}{C}$$

The voltage gain is established by a capacitor ratio, and cut-off frequency by f_s and a different capacitor ratio. For a general second-order filter section, we can replace the resistors and the noninverting amplifier of Fig. 12.8 with switched-capacitor equivalents.

A switched-capacitor biquadratic circuit exploits the availability of noninverting integrators. Beginning with the inverting lowpass function

$$\frac{V_{LP}}{V_i} = \frac{-1}{s^2 + (1/Q)s + 1} \tag{12.65}$$

we cross-multiply giving

$$s^2 V_{LP} = -\frac{1}{Q} sV_{LP} - V_{LP} - V_i$$

We now implement this equation with noninverting integrators as in Fig. 12.44a. The minus sign in Eq. (12.65) enables us to use a simple summing amplifier at the input as in Fig. 12.44b. With a clock operating at $f_s = 1$ Hz this circuit simultaneously gives the lowpass transfer function of Eq. (12.65), the bandpass function with gain $-Q$

$$\frac{V_{BP}}{V_i} = \frac{-Q(s/Q)}{s^2 + (1/Q)s + 1} \tag{12.66}$$

and the highpass function

$$\frac{V_{HP}}{V_i} = \frac{-s^2}{s^2 + (1/Q)s + 1} \tag{12.67}$$

The four resistors in the diagram can be realized by switched-capacitor circuits or, for increased versatility, left to the IC user to provide as external components. In a commer-

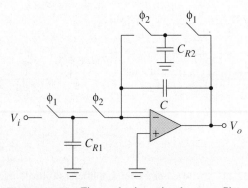

Figure 12.43 First-order inverting lowpass filter.

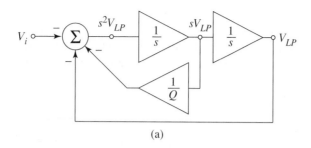

(a)

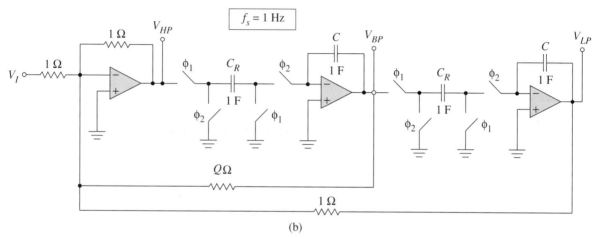

(b)

Figure 12.44 Biquadratic circuit: (a) block diagram; (b) filter schematic for reference frequency of 1 rad/s showing resistance and capacitor values for $f_s = 1$ Hz.

cial switched-capacitor filter IC, an input pin is provided on the IC for an external clock f_s, with the two nonoverlapping switching signals created on the chip itself.

Scaling Switched-Capacitor Circuits. Switched-capacitor filters can be frequency- and impedance-scaled in the usual way; however, in scaling we must differentiate between the two kinds of capacitors in these circuits. In Figs. 12.41 through 12.44, capacitors C_R contribute to resistances; capacitors C function as true capacitors. Frequency scaling, therefore, involves scaling only capacitors C (or instead, scaling resistors—including resistors $1/f_s C_R$). To impedance-scale we multiply impedances of both kinds of capacitor by the impedance scale factor β.

12.5.3 PRACTICAL CONSIDERATIONS

Because switched-capacitor circuits are sampled data systems, there are some special constraints on the clock frequency and on the waveforms that control the switches.

The *Nyquist sampling theorem* requires f_s to be at least two times the highest input frequency being processed by the circuit. Intuitively, this ensures that v_1 and v_2 in

Fig. 12.40 do not change appreciably during a switching cycle. If a lowpass filter is to separate a signal of 600 Hz bandwidth from 1800 Hz noise, for example, then it is necessary that $f_s \geq 3600$ Hz. Failure to satisfy this criterion results in *aliasing distortion* in which each input frequency $f_k > 0.5 f_s$ has its energy transferred to $f_s - f_k$ in the output spectrum, constituting an irreversible information loss.

Circuit diagrams clearly indicate that correct operation requires the ϕ_1 switches to open *before* ϕ_2 switches close and ϕ_2 switches to open *before* ϕ_1 switches close. Although each switch operates at frequency f_s with period $T_s = 1/f_s$, this means that the duration of switch closure, T_c, must be a limited fraction κ of T_s; that is,

$$T_c = \kappa T_s, \qquad 0 < \kappa < 0.5 \qquad (12.68)$$

as in Fig. 12.45a. Theory beyond the scope of the text suggests that for ideal operation T_c should approach zero, giving instantaneous sampling. However, this is not practical in

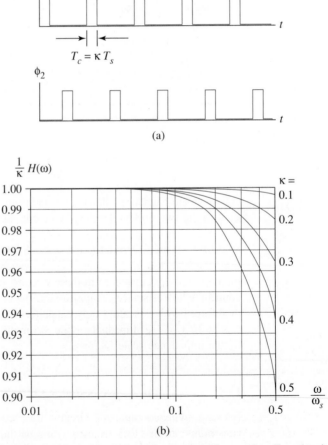

Figure 12.45 Aperture effect: (a) comparison of switch closure time T_c and switch period T_s; (b) amplitude distortion introduced by nonzero switch closure times.

a real circuit because the energy in the individual signal pulses would be zero. Consequently, practical sampled data systems introduce a special frequency distortion called *aperture distortion* that is increasingly severe for larger values of κ. This *aperture effect* is equivalent to cascading the filter transfer function, $T(j\omega)$, with the aperture function

$$H(\omega) = \kappa \frac{\sin[\pi \ \kappa(\omega/\omega_s)]}{[\pi \ \kappa(\omega/\omega_s)]} \qquad (12.69)$$

where ω_s is the switch frequency in rad/s. Notice the two factors of $H(\omega)$: uniform attenuation of all frequencies by κ, and filtering by a $\sin x/x$ function. The latter introduces amplitude distortion as in Fig. 12.45b, but no phase distortion. For a given duty cycle, κ, the amplitude distortion becomes negligible at sufficiently high sampling frequencies.

Our discussions have tacitly assumed that switches and signal sources have no internal resistance. We saw in Sec. 5.11.2, however, that MOSFET switches do have internal resistance, and we know this is also true for practical signal sources. These resistances cause exponential charging and discharging of the switch capacitors in practical circuits. To approach the ideal of instantaneous charging and discharging, we need to ensure that all time constants τ_k are short compared with the switching period. Practically speaking,

$$4\tau_k \leq 0.1 T_c \qquad (12.70)$$

This might involve using follower circuits or parallel switches to reduce resistance.

In some circuits, such as the amplifiers of Fig. 12.42, op amp slewing can introduce problems. Since the op amp outputs in these circuits must rise from zero to maximum signal value in a time $\ll T_c$, we have

$$\frac{V_{out,max}}{s.r.} \leq 0.1 T_c \qquad (12.71)$$

EXAMPLE 12.10 Design a fourth-order, lowpass, switched-capacitor, Butterworth filter with 3 dB bandwidth of 8000 rad/s to filter input frequencies as high as 32,000 rad/s. Use cascaded biquad sections.

Solution. From Table 12.1,

$$T(s) = T_1(s)T_2(s) = \frac{1}{(s^2 + 0.7654s + 1)} \frac{1}{(s^2 + 1.8478s + 1)}$$

Each section of this prototype filter has the general form of Fig. 12.44b with bandwidth of 1 rad/s. Columns 2 and 3 of Table 12.5 show component values after frequency-scaling capacitors C by $\alpha = 8000$. The row labeled "R" describes all resistors with Fig. 12.44b values of 1 Ω, and row "$1/Q$" describes the remaining resistors. The last row gives the switch frequency.

For discrete resistors to have reasonable values, let us impedance-scale by $\beta = 10^5$. This gives the values in the last two columns of Table 12.5.

Since $\omega_s = 1$ rad/s does not satisfy the sampling theorem, we must now select the final clock frequency. Equation (12.62) shows we must simultaneously increase f_s and reduce C_R in such a way that the switched-capacitor resistances do not change. To filter

TABLE 12.5   Effects of Scaling on Filter Parameters

Parameter	Filter Parameters After Frequency Scaling		Filter Parameters After Impedance Scaling	
	$T_1(s)$	$T_2(s)$	$T_1(s)$	$T_2(s)$
C	125 μF	125 μF	1250 pF	1250 pF
C_R	1 F	1 F	10 μF	10 μF
R	1 Ω	1 Ω	100 kΩ	100 kΩ
$1/Q$	0.7654 Ω	1.8478 Ω	76.5 kΩ	184.8 kΩ
ω_s	1 rad/s	1 rad/s	1 rad/s	1 rad/s

**TABLE 12.6   Design
That Satisfies the
Sampling Theorem**

	$T_1(s)$	$T_2(s)$
C	1250 pF	1250 pF
C_R	785 pF	785 pF
R	100 kΩ	100 kΩ
$1/Q$	76.5 kΩ	184.8 kΩ
ω_s	80 k rad/s	80 k rad/s

input frequencies as high as 32,000 rad/s, we need $\omega_s > 64{,}000$ rad/s. To leave room for error, select $\omega_s = 80{,}000$ rad/s. Then capacitors C_R must satisfy

$$R = 10^5 = \frac{1}{C_R[80 \times 10^3/(2\pi)]}$$

This gives C_R of 785 pF, and the final design of Table 12.6.   ❏

Exercise 12.10   In the filter of Example 12.11, the final spread in capacitor values was $1250 : 785 = 1.59 : 1$. Find the minimum spread that can result from reducing f_s.

Ans. 1.27.

EXAMPLE 12.11   Select a duty cycle, κ, for the filter switches in Example 12.10 to ensure aperture distortion is negligible.

Solution. Figure 12.45b shows that the aperture effect improves stopband attenuation in a lowpass filter; however, its contribution to the passband attenuation must not be excessive. Since the filter of Example 12.10 has 3 dB attenuation at 8000 rad/s, let us make sure that the attenuation contributed by aperture effect at this frequency is no more than 0.3 dB. This means the aperture-effect gain must exceed

$$\frac{1}{\kappa}H(8000) = 10^{-0.3/20} = 0.966$$

From Fig. 12.45c we see that for

$$\frac{\omega}{\omega_s} = \frac{8,000}{80,000} = 0.1$$

gain from aperture effect exceeds 0.995, even for $\kappa \Rightarrow 0.5$; so we may use any duty cycle compatible with nonoverlapping switch operation. ❏

EXAMPLE 12.12 Design a switched-capacitor, fourth-order, Butterworth, highpass filter with -3 dB cut-off frequency of 8000 rad/s that must process signal frequencies as high as 20,000 rad/s.

Solution. (a) Because the filter order and cut-off frequency are the same as for Example 12.10, we begin with the cascaded biquads described by Table 12.6, but use the highpass outputs.

To process signals of 20,000 rad/s, select $\omega_s = 40,000$ rad/s. Now capacitors C_R must satisfy

$$R = 10^5 = \frac{1}{C_R[40 \times 10^3/(2\pi)]}$$

or $C_R = 1571$ pF.

Notice that the lowpass nature of aperture distortion can create a severe problem in a highpass filter. To successfully process signals up to 20,000 rad/s, the aperture distortion must be negligible at

$$\frac{\omega}{\omega_s} = \frac{20,000}{40,000} = 0.5$$

Figure 12.45 shows that a switch duty cycle of $\tau \leq 0.3$ ensures aperture gain of at least 0.965 at all frequencies of interest. ❏

12.6 _____
Summary

A filter processes the frequency components of signals applied to its input by changing their relative amplitudes, phases, or both. Any gain or phase change that is uniform for all frequencies is immaterial to the filtering process. The major filter classes are lowpass, highpass, bandpass, band reject, and allpass. Each of the first four classes has an idealized frequency response that involves unity gain for frequencies in some passband and zero gain for the other frequencies, the stopband. Idealized allpass filters have uniform gain and linear phase shift—pure time-delay circuits.

Filters cannot be realized in ideal forms with finite numbers of components, so we must work with functions and circuits that only approximate ideal filters. Lowpass and highpass filters can be approximated by simple first-order transfer functions and related circuits. All five filter types can be approximated by second- and higher-order transfer functions and circuits.

Filter design focuses on simple transfer functions and circuit building-blocks that have unit reference frequencies. Sallen–Key, biquad, and passive *RLC* circuits demonstrate the basic ideas in this text, but the literature describes many others. These circuits employ components, such as one farad capacitors and one ohm resistors, that are physically impractical to use. Frequency scaling, which involves changing all capacitor and inductor values by the same scale factor, α, translates the magnitude and phase of the prototype filter to a new frequency, α times the original frequency, for any positive, real α. In *RC*-active circuits, an alternative is to frequency-scale by changing only resistors. Impedance scaling, which involves multiplying every impedance by the same scale factor, β, leaves the magnitude and phase functions unchanged, but changes all *R*, *L*, and *C* component values. Impedance scaling is an important tool for obtaining a final design with realistic components. A lowpass-to-highpass transformation, $p = 1/s$, applies to both functions and filter circuits. With this tool we can transform *RLC* lowpass filters into related highpass filters, and conversely. An *RC-CR* transformation allows us to do the same with *RC*-active filters. (Similar transformations that relate lowpass to bandpass and band-reject functions are described in the literature but omitted here.) Equipped only with these basic scaling and transformation ideas, we can sometimes rescale and transform existing filters to satisfy given specifications.

In filter design, we use a sketch to relate our specifications to a lowpass, prototype filter with unit reference frequency. We then use the known magnitude and phase functions of the prototype filter to find unknown parameters such as filter *Q* and filter order. Once this prototype filter is conceptualized, it takes only frequency and impedance scaling and, perhaps, a lowpass-to-highpass transformation to convert it into a finished design. SPICE simulations provide valuable checks on the basic design and help us explore the consequences of second-order effects on the overall response.

With Butterworth, Chebyshev, and approximation functions not covered in the text, we can construct filters that approximate ideal filter functions with increasing accuracy as the filter order (as well as complexity and expense) increases. Using filter tables or calculations, we can relate Butterworth or Chebyshev filters of any order to prototype filters consisting of cascaded first- and second-order *RC*-active filter sections or double-terminated ladder circuits. It then takes only frequency and impedance scaling to obtain the final design. Synthetic inductors exemplify modern *RC*-active circuits and techniques that sometimes allow us to make use of classical filter theory in ICs where it is impractical to fabricate inductors.

Switched-capacitor filters replace large resistors in IC filter realizations with lossless networks of switches and capacitors. Most of the structures and concepts of conventional circuit theory apply to switched capacitor circuits, including frequency and impedance scaling. Because they are sampled-data systems, however, the two-phase, nonoverlapping switches are constrained to operate at twice the frequency of the highest signal component to be processed. There is also the possibility of aperture distortion, which constrains the frequency and duty cycles of the switches.

REFERENCES _____

1. Huelsman, L. P. *Theory and Design of Active RC Circuits.* New York: McGraw-Hill, 1968.
2. Karni, S. *Network Theory: Analysis and Synthesis.* Boston: Allyn & Bacon, 1966.
3. Mitra, S. K. "A Network Transformation for Active RC Networks," *Proc. IEEE,* Vol. 55, (1967), pp. 2021-2022.
4. Sedra A., and K. Smith. *Microelectronic Circuits,* 3rd ed. Philadelphia: Saunders College, 1991.
5. Van Valkenburg, M. E. *Introduction to Modern Network Synthesis.* New York: John Wiley, 1967.
6. Van Valkenburg, M. E. *Analog Filter Design.* New York: Holt, Rinehart and Winston, 1982.

PROBLEMS _____

Section 12.1

12.1 Find the gain and phase of the filter of Fig. 12.2 at $\omega = 3.8$ rad/s.

12.2 Find the radian frequency at which the filter of Fig. 12.2 gives 40° phase shift. What is the attenuation in decibels at that frequency?

12.3 The first-order filter of Fig. 2.14a has cut-off frequency given by $\omega_H = 1/R_2 C_2$. Find the sensitivities of ω_H to R_2 and to C_2.

12.4 (a) Sketch the magnitude frequency response curve for Fig. P12.4, showing gain and bandwidth values on the sketch.
(b) Rescale the filter so that its cut-off frequency is 4.2 kHz. Draw the diagram of the new filter and sketch its magnitude frequency response curve.
(c) Frequency-scale the filter so that its cut-off frequency is 160 Hz. Draw the diagram of the new filter, and sketch its frequency response magnitude curve.

12.5 A filter has the transfer function

$$F(s) = \frac{-1}{s^2 + 3s + 1}$$

Find the new filter function $F_N(s)$ if this filter is frequency-scaled by $\alpha = \omega_N$.

12.6 Starting with Fig. 12.2a, design a first-order lowpass filter with cut-off frequency of 2 kHz.

12.7 What frequency-scale factor α would be necessary to scale the filter of Fig. 12.2a so that filter gain is 0.1 at $\omega = 2000$ rad/s?

12.8 A first-order filter has cut-off frequency $\omega_H = 1/RC$. Show that the sensitivity of the cut-off frequency is not affected by the scaling; that is, show that sensitivity of new cut-off frequency ω_{HN} to new filter capacitance C_N is the same as the original sensitivity.

12.9 Frequency-scale the circuit of Fig. P12.9 by $\alpha = 10^3$, and then impedance-scale to make the resistance 2 kΩ. Sketch the circuit showing values after each step.

12.10 Do the frequency scaling of Problem 12.4b in two ways, with capacitance and with resistance, and draw the two circuit diagrams. Compare the gains of the circuits at $\omega = 0$.

12.11 Design a first-order, highpass filter that has attenuation of 30 dB at 300 Hz. Use capacitor values of 2000 pF.

Section 12.2

12.12 Design a second-order, lowpass Sallen–Key filter with $Q = 1$ and cut-off frequency of 3.8 kHz. Use only 10 kΩ resistors in the design. Draw the circuit diagram of your final circuit.

12.13 Design a second-order lowpass Sallen–Key filter with $Q = 1$ and cut-off frequency of 3.8 kHz. Use resistors for the initial frequency scaling. Use only 0.01 μF capacitors in the final design. Draw the circuit diagram of your final circuit.

12.14 A Sallen–Key design uses a noninverting amplifier of gain $A = 1 + R_2/R_1$ to achieve $Q = 5$. Determine the largest and smallest values of Q if the design uses 1% resistors, and find the percent error in Q for each case.

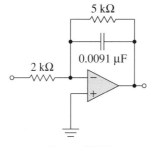

Figure P12.4

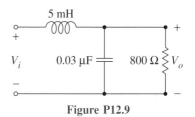

Figure P12.9

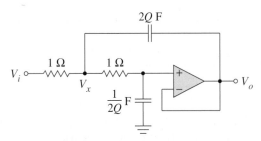

Figure P12.16

12.15 Suppose we wish to design a Sallen–Key second-order filter with $Q = 50$.
(a) Find the value of A.
(b) Compute S_A^Q, and evaluate it for this design.
(c) Show that the sensitivity expression of part (b) can be written as $S_A^Q = 3Q - 1$.

12.16 Show that Eq. (12.11) describes Fig. P12.16.

12.17 Figure P12.17 is described by the lowpass function

$$F(s) = \frac{V_O}{V_I} = \frac{-1}{s^2 + 3s + 1}$$

At its 1 rad/s reference frequency, gain is 1/3.
(a) Draw the circuit that results from applying the *RC-CR* transformation.
(b) Use circuit analysis to find the transfer function of the new circuit. Compare the result with $F(1/s)$, the result of applying the lowpass-to-highpass transformation to the lowpass function.
(c) Scale the new circuit so that it has reference frequency of 420 Hz and uses 20 kΩ resistors.
(d) Keep the same reference frequency as in part (c) but rescale so that each $C = 0.01$ μF.

12.18 Design a biquad lowpass filter that satisfies the specifications given in Example 12.2.

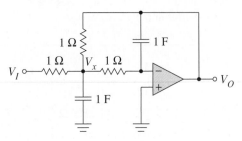

Figure P12.17

12.19 A second-order bandpass filter has a −3 dB bandwidth of 0.95 kHz and center frequency of 1.9 kHz.
(a) Find its upper and lower half-power frequencies.
(b) Design such a filter using a biquad circuit and 1000 pF capacitors. Draw a diagram of the final circuit.

12.20 In Example 12.4 we designed the filter of Fig. 12.18b to have −3 dB bandwidth extending from 800 Hz to 925 Hz.
(a) Use SPICE to verify the correct magnitude frequency response. Model op amps using a *subcircuit* VCVS device with gain of 10^5.
(b) Find how the results change if the resistor that determines the circuit Q is 10% high.
(c) Repeat part (a) after rescaling the capacitors by $\alpha = 10^3$.
(d) Replace the subcircuit with the μA741 subcircuit of Fig. 2.34a, and repeat the simulations of parts (a) and (c). Compare the simulations, and state what you learned from the comparisons.

12.21 Design a second-order biquad circuit that will reject 1.82 kHz but give gain of no more than 0.8 for frequencies higher than 5 kHz.
Hint: Start with a frequency response sketch. Use Eq. (12.37).

12.22 Starting with Eq. (12.41), find the group delay function $D(\omega)$.

12.23 Figure 12.34e shows an allpass filter that has transfer function of $0.5T(s)$, where $T(s)$ is given by Eq. (12.40). Use SPICE transient analysis to compare input and output waveforms when the input is a 10 V square wave with fundamental frequency of 1.3 rad/s and
(a) $Q = 5$.
(b) $Q = 1$.

Section 12.3

12.24 Use Eq. (12.44) to calculate the poles of the fifth-order Butterworth filter function. Then verify the entry for $n = 5$ in Table 12.1.
Hint: $(s - p)(s - p^*) = s^2 + 2\,\text{Re}[p] + |p|^2$, where * denotes the conjugate, Re[ ] denotes the real part, and | | denotes magnitude.

12.25 Write $T(s)$ in factored form for a third-order Butterworth lowpass filter that has −3 dB cut-off frequency-scaled by $\alpha = 10^3$ rad/s. Simplify your expression so that the coefficients of s and s^2 are one.

12.26 Find a Butterworth lowpass filter function that has a bandwidth of 2 kHz and at least 43 dB attenuation at 7 kHz.

12.27 Draw the diagram of a filter that satisfies the specifications of Example 12.5 using biquadratic filter sections. Show all component values.

12.28 Draw the circuit diagram of a lowpass Butterworth filter that has bandwidth of 2 kHz and at least 43 dB attenuation at 7 kHz. All resistor values must be 10 kΩ. Use Sallen–Key filter sections.

12.29 Draw the circuit diagram of a lowpass Butterworth filter that has bandwidth of 2 kHz and at least 43 dB attenuation at 7 kHz. All resistor values must be 10 kΩ. Use the passive first- and second-order filter sections of Figs. P12.29 and 12.34a with op amp followers for interstage isolation.

12.30 (a) Rescale the filter designed in Example 12.5 so that resistors are all 1 kΩ.
(b) Use SPICE to determine the frequency response of the filter of part (a) when infinite gain op amps are used.
(c) Repeat part (b) except use the 741 op amp subcircuit of Fig. 2.34a.

12.31 (a) Complete the filter design of Example 12.6 using sections like Fig. P12.16 with 4.7 kΩ resistors. Draw the circuit diagram showing component values.
(b) Rescale the circuit of part (a) so that its 1 dB bandwidth is changed to 2.5 kHz. Show the circuit diagram and sketch the frequency response curve, showing on the diagram the frequencies where gain is −3 dB and where gain is −20 dB.

12.32 (a) Complete the filter design of Example 12.6 using Sallen–Key sections with 4.7 kΩ resistors. Draw the circuit diagram showing component values.
(b) Use SPICE to check the frequency response of the circuit, using high-gain dependent sources for the op amps.

12.33 Prove that the $C_n(\omega)$ defined by Eq. (12.50) satisfies the recursion equation (12.51).
Hint: Write an expression for $C_{n+1}(\omega)$ using Eq. (12.50). Then use a trigonometric identity that involves the sum of two angles. Starting with $C_{n-1}(\omega)$, use the same approach.

12.34 Find the order of a Chebyshev lowpass filter that has 2 dB bandwidth of 3 kHz and at least 65 dB attenuation at 15 kHz.

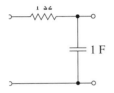

Figure P12.29

12.35 Prove Eq. (12.54) using the following procedure.
(a) Use the exponential definitions of cosine and hyperbolic cosine functions to prove the identity $\cos(jy) = \cosh(y)$.
(b) Because the inverse cosine in Eq. (12.50) is imaginary for $|\omega| > 1$, define variable y by

$$\cos^{-1}\omega = jy \qquad (P.1)$$

Substitute expression (P.1) into Eq. (12.50), and then use the identity from part (a) to simplify.
(c) Solve (P.1) for ω, and then use the identity of part (a) to simplify.
(d) Solve the expression of part (c) for y and then substitute into the C_n expression from part (b).

12.36 (a) Use Eq. (12.58) to compute the poles for a third-order Chebyshev filter with 1 dB passband attenuation.
(b) Write $T(s)$ for the filter, with the denominator written as the product of first- and second-order polynomials.
(c) Write $T(s)$ for the filter, if the 1 dB passband is to be 120 MHz.

12.37 Design a third-order Chebyshev lowpass filter with a 0.5 dB bandwidth of 900 rad/s using a biquad section. All resistors should be 4.7 kΩ if possible. A signal inversion is acceptable, and the absolute gain is not important.

12.38 Realize the filter specifications of Example 12.7 using second-order filter sections like Fig. P12.16.

12.39 Figure 12.25b is a Butterworth filter with 3 kHz bandwidth and attenuation of at least 60 dB at 15 kHz. Impedance-scale the individual sections by 5000, 10,000, and 20,000, respectively. Then use SPICE with infinite-gain op amps to verify that the frequency response still satisfies specifications.

12.40 Individually impedance-scale the sections of Fig. 12.33b so that all capacitors have values of 5000 pF.

Section 12.4

12.41 Verify by circuit analysis that the circuit of Fig. 12.34b is described by Eq. (12.19).

12.42 We want to replace the fourth-order, 3 kHz filter of Fig. 12.33b with two passive filters, using a follower circuit for isolation.
(a) Ignoring constants, frequency-prescale two passive filters like Fig. 12.34a to give the $T(s)$ used in Example 12.7. Draw the diagram for the prototype filter.
(b) Frequency-scale by 2π3000 to obtain the proper bandwidth, and then impedance-scale to give 500 Ω resistor values. Draw the diagram for this filter.

(c) Use SPICE to verify that the final filter has frequency response like Fig. 12.33c. Use a VCVS with gain of 10^6 for the op amp.

12.43 Use the circuit of Fig. 12.34c to realize a bandpass filter with center frequency of 2.8 kHz and $Q = 7$. The final filter should use a 500 Ω resistor.

12.44 Use the circuit of Fig. 12.34d to realize a *trap* to remove 400 Hz from a signal. Use $Q = 10$ and $R = 10$ kΩ.

12.45 (a) Design a seventh-order, lowpass, Chebyshev passive ladder filter using 500 Ω resistors that has a 0.1 dB passband of 100 MHz.
(b) Use SPICE to verify the frequency response of the design.

12.46 Work out the design for Problem 12.44. Then replace the inductor with a synthetic inductor that employs a capacitor of 0.001 F. Use an ungrounded output resistor.

12.47 (a) Use SPICE to repeat the HA2544 results of Fig. 12.39e.
(b) Once the inductor is replaced by a synthetic inductor, a realization of this type becomes an *RC*-active circuit. We learned in Sec. 12.1.4 that an *RC*-active circuit can be frequency-scaled by changing resistors instead of capacitors. In your SPICE code for part (a) with infinite-gain op amp, make every resistor larger by a factor of 10 and see if the frequency response curve is rescaled accordingly.

12.48 Design a third-order, Butterworth, highpass, active filter based upon a doubly terminated passive ladder structure with bandwidth of 5 kHz. Use a minimum number of active elements. *Hint*: See Exercise 12.9.

12.49 (a) Frequency-scale the allpass circuit of Fig. 12.34e to a reference frequency of 800 Hz. Use $Q = 5$. Use SPICE to check the magnitude and phase characteristics.
(b) Replace the inductor with a synthetic inductor, and verify this design using SPICE.

Section 12.5

12.50 Replacing the resistors and noninverting amplifier A of Fig.12.8 with switched-capacitor equivalents gives a general second-order switched-capacitor filter section with four parameters, C_{R1}, C_{R2}, C_{R3}, *and* C_{R4}.
(a) Draw a diagram of this filter and label all the capacitors.
(b) Write a design equation that relates the capacitors in the amplifier realization to the circuit Q.

12.51 (a) If $v_i = 2 \sin 20t$ and $C_{R1} = 2C_{R2}$ in Fig. 12.42, carefully sketch $v_o(t)$ for each amplifier. Label values of critical amplitudes.

(b) Sketch the output of Fig. 12.42b if the switch across C_{R2} is erroneously operated by the ϕ_2 clock instead of the ϕ_1 clock.

12.52 Draw the diagram for a switched-capacitor version of the highpass filter of Fig. 12.5a. Include a switched-capacitor inverting amplifier to provide the gain of -1 indicated in the diagram. Label the diagram with component values appropriate for a cut-off frequency of 20 Hz and filtering of frequencies up to 2 kHz.

12.53 Design a second-order lowpass biquad filter with bandwidth of 400 Hz and Q of 5 that processes frequencies up to 4 kHz. Use no resistors. Draw the diagram showing reasonable capacitance values and giving an appropriate sampling frequency f_s. If switch closure is $0.25T_s$, find the gain caused by aperture effect at 4 kHz. Any correct answer is acceptable.

12.54 Design a third-order, switched-capacitor, lowpass, Chebyshev filter with 0.5 dB passband ripple and cut-off frequency of 600 Hz. The circuit must be capable of processing input frequencies as high as 2400 Hz. Use a first-order section like Fig. 12.43 and a biquad section.

12.55 Figure P12.55 shows a switched-capacitor, inverting amplifier and one of the clock waveforms. The clock frequency is 50 kHz, and $v_i(t) = 0.02 \sin(\omega_i t)$.
(a) What is the highest frequency permissible for $v_i(t)$, if the output waveform is to be accurate?
(b) The CMOS switches must be connected to *both* positive and negative power supplies. Why?

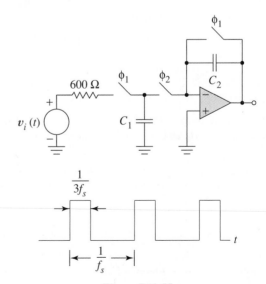

Figure P12.55

(c) Sketch the *transfer function*, v_{out} versus v_i for a CMOS inverter that operates between $\pm$ 5 volt power supplies.

(d) What output would you expect to see when $v_i(t) = $ V sin ωt if the switch above C_2 were incorrectly wired to the ϕ_2 clock? Explain.

12.56 Figure P12.56 shows the switched-capacitor version of the summing amplifier in Fig. 12.44b. The two-phase clock operates at 20 kHz with a duty cycle of 0.25%. Find the value of R_s so that the source can fully charge (four time constants) the capacitor during the time the switch is closed. The resistance of the ϕ_1 switch is 270 Ω.

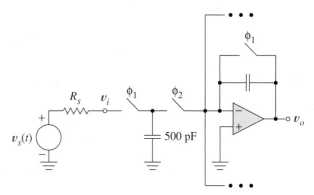

Figure P12.56

Digital Logic Circuits

The principles involved in processing signals with digital electronic circuits were well known in the 1940s and 1950s; however, since digital techniques require larger circuits with many more components than analog methods, they were used sparingly. Rapid progress in integrated circuit fabrication in the 1960s eventually made digital circuits so cost-effective that they came to rival and eventually surpass analog circuits in most applications. The inherent advantages of digital circuits are reliability and flexibility.

Digital circuits are more reliable than analog circuits because they employ transistors as simple binary switches that are either cut off or saturated (ohmic), and use signals that take on only two distinct values. In designing such circuits, it is relatively easy to address uncertainties in component values, component aging, and environmental factors such as temperature. Compare, for example, the design of a transistor switch, which must be either cut off or saturated when its input is zero or five volts, with designing a circuit with a biased, active-state transistor that must process, without introducing distortion, an analog signal that can take on *any* values.

The flexibility of digital design arises from the fact that all circuits are constructed of basic switching-circuit building blocks called logic gates, with many identical gates interconnected in various ways to perform various signal-processing functions. Interconnection schemes are logically organized around a hierarchy of functional units such as counters, registers, coders, decoders, and memory, where all or nearly all are constructed from logic gates. Increasing signal-processing precision simply involves adding additional, identical hardware to carry out computations with more bits. The flexibility of digital hardware is perhaps most obvious when the processing circuit takes the form of a computer, where the same hardware can perform a wide variety of different tasks at different times and where throughput can be enhanced by parallel processing.

This chapter assumes that the reader is familiar with the basic principles of digital logic such as AND, OR, NOR, and NOT gates and their uses, which are covered in many books on digital logic. It also assumes familiarity with transistor state definitions, analysis problems involving unknown transistor states, and nonlinear load lines, all treated in detail in Chapters 4 and 5.

In this chapter, our object is to develop a familiarity with *circuit-level implementa-tions* of logic gates. This improves our understanding of practical matters associated with design and operation of gate circuits such as propagation delays, fan-out and fan-in limitations, and noise margins. Circuit-level understanding is especially useful in inter-facing between digital and analog devices and between different digital logic *families*.

We begin with an abstract concept, the ideal logic gate. This allows us to make some key definitions and provides a standard of comparison for the practical circuits that follow. The latter are organized into *logic families,* collections of logic gates based upon particular transistor technologies. Each family embodies unique strengths and weak-nesses, and is characterized by its own power supply requirements, voltage or current levels used to represent logic values, and a particular power/delay trade-off.

We begin the study of each family with a detailed examination of its basic inverter. After deducing the inverter's circuit-level static and dynamic properties, we then show how we can augment the circuit to perform other logic functions such as AND, OR, NAND, and NOR. The basic tools for studying static properties of logic circuits are *large-signal* transistor models. The dynamic properties of diodes, BJTs and FETs, which we studied in earlier chapters, give insight into switching delays and transient behavior.

13.1
Ideal Logic Gates and Practical Approximations

The ideal logic gate is a conceptual standard with which we compare real gates to assess their performance. The ideal gate has zero power dissipation, both during level transi-tions and while preserving a fixed logic level. It also has unlimited fan-out; that is, any number of load gates may be attached at the output without a degradation in perfor-mance. The number of inputs, its fan-in, is also unlimited. In ideal gates, output transi-tions instantaneously follow input level changes, and the transitions themselves are in-stantaneous.

13.1.1 STATIC TRANSFER CHARACTERISTICS

Physical gate circuits use distinct voltage levels to represent binary logic values. In a positive logic system, the only systems we consider in this text, the more positive level represents logic 1. An ideal positive logic inverter has the transfer characteristic of Fig. 13.1a, where V_H represents logic 1 and V_L represents logic 0. The characteristic shows that an input $v_I = V_L$ gives $v_O = V_H$, and input $v_I = V_H$ gives $v_O = V_L$, as re-quired for an inverter. The origin of the coordinate system in Fig. 13.1a is immaterial; that is, a logic circuit can be designed so that V_H and V_L are both positive, so V_H is positive and V_L is negative, or so both levels are negative.

The transfer characteristic implies an automatic *regeneration* of the correct logic level each time digital information is processed by the gate. Notice on the v_I axis of Fig. 13.1a that an input logic 1, ideally V_H volts, could in fact be nearly as low as V_M, and the inverter would still recognize it as a logic 1 and produce a correct, full-value, output of V_L volts. Similarly an input zero could be almost V_M volts and still be correctly rec-ognized as a zero by the inverter, and would still result in a full-scale logic 1 output. This regeneration feature makes the gate's operation robust, that is, forgiving of small errors in input logic levels. It also helps the circuit operate in the presence of additive noise.

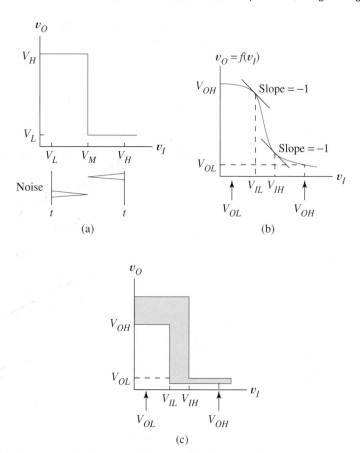

Figure 13.1 Inverter logic levels: (a) ideal transfer characteristic; (b) representative transfer characteristic for a real gate; (c) range of inverter characteristics reflecting parameter variations.

In an ideal world, V_H and V_L would not need to differ by much, so long as they are distinct. In practical applications, however, where the gate operates in the presence of noise, a large *logic swing*, $V_H - V_L$, contributes to reliability. The construction at the bottom of Fig. 13.1a shows how noise spikes, if sufficiently small in amplitude, fail to cause an error in the gate output. A measure of the gate's ability to operate correctly in the presence of additive noise is its *noise immunity*. We quantify this idea by defining the *noise margins* of a gate. The *high noise margin, NMH* $= V_H - V_M$, is the amplitude of the largest negative noise that, superimposed upon an input logic 1, would not cause a logic error in the output. Similarly, the *low noise margin NML* $= V_M - V_L$ defines the largest positive noise spike that could be added to a logic 0 input without causing error.

Figure 13.1(b) shows a transfer characteristic more typical of a real inverter,

$$v_O = f(v_I)$$

where $f(\ )$ is some monotonically decreasing function of v_I. Because of the gradual transition between levels, we use different voltages to describe input logic levels. The following definitions are applicable to most gate circuits:

$$V_{OH} = f(0), \qquad V_{OL} = f(V_{OH})$$

Input low and high values, V_{IL} and V_{IH}, respectively, are defined as those v_I values where the slope of the function is minus one, as indicated in Fig. 13.1b. It is necessary that $V_{OL} < V_{IL}$ and $V_{IH} < V_{OH}$. These inequalities ensure that the output level of one gate is an appropriate input level for a second gate. Most of our work in this chapter involves individual gate characteristics like this one.

When we also consider practical manufacturing tolerances and loading effects, we visualize each gate *family* in terms of a *transfer region* as depicted in Fig. 13.1c, where worst-case performance defines the shaded area. The manufacturer guarantees that every transfer characteristic will fall within the shaded region. Provided the user follows certain operational guidelines, this means that each gate in the logic family will correctly interpret any input, $v_I \leq V_{IL}$, as a logic 0, and produce a corresponding output $v_O \geq V_{OH}$. Furthermore, a gate will take any input, $v_I \geq V_{IH}$, to be a logic 1, and produce a corresponding output voltage, $v_O \leq V_{OL}$. The idea is easily visualized using Fig. 13.2a, which shows a logic inverter with the regions defined in Fig. 13.1c illustrated at input and output. Notice that the overlap of input and output ranges for logic 1 and for logic 0 ensures the desired *regeneration* of the logic levels.

For a practical gate, we need to modify our noise immunity concept slightly. Figure 13.2b shows the output of gate #1 providing input for gate #2. Using worst-case output and input values, we define

$$NML = V_{IL} - V_{OL} \quad \text{and} \quad NMH = V_{OH} - V_{IH} \tag{13.1}$$

where we use Fig. 13.1b values for an individual gate, Fig. 13.1c values for a gate family. As the following development shows, logic swing is closely related to noise margins.

$$\text{logic swing} = V_{OH} - V_{OL} = V_{OH} - V_{OL} - V_{IH} + V_{IH} - V_{IL} + V_{IL}$$
$$= NHM + NML + \Delta V \tag{13.2}$$

where

$$\Delta V = V_{IH} - V_{IL}$$

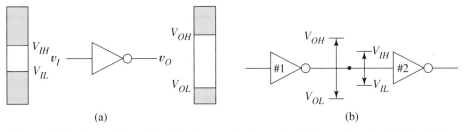

(a)   (b)

Figure 13.2 Comparison of input and output logic levels: (a) for an inverter; (b) for driving and load inverters.

For the ideal of Fig. 13.1a, the logic swing is the sum of the noise margins. For the imperfect gates of Figs. 13.1b and c, part of the logic swing is *overhead*, ΔV, associated with the transition region between logic 1 and logic 0.

13.1.2 DYNAMIC CHARACTERISTICS

Figure 13.3 defines the most important dynamic limitations of a logic gate. When the inverter input is the rectangular test signal, $v_I(t)$, the output waveform is $v_O(t)$. Because of parasitic capacitances, output transitions between logic levels are gradual rather than instantaneous. We characterize them as a 90 to 10% fall time t_f and a 10 to 90% rise time t_r. The *leading edge* of the output pulse is delayed relative to the leading edge of the input by t_{PHL}, the *high-to-low propagation delay*, defined as the time measured between the *midpoints* of the input and output transitions as the output changes from its high-to-low value. There is also a *low-to-high propagation delay*, t_{PLH}, between the midpoints of the *trailing edges* of input and output waveforms. Most often gate delays are characterized by a single number, the *propagation delay* t_P defined by

$$t_P = \frac{t_{PHL} + t_{PLH}}{2} \tag{13.3}$$

Figure 13.4 shows the SPICE pulse waveform, which is useful in transient simulations of gate operation. After source name and node numbers, the element line for an independent pulse source follows the syntax

```
PULSE( VALUE1 VALUE2 DELAY RISE_TIME FALL_TIME WIDTH PE-
RIOD)
```

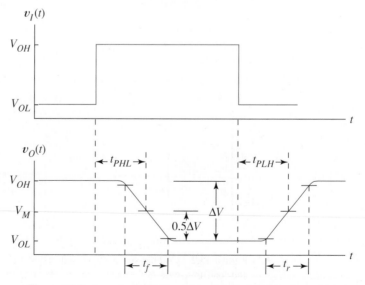

Figure 13.3 Inverter definitions associated with gate dynamics.

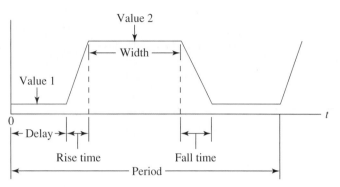

Figure 13.4 SPICE pulse waveform and parameter definitions.

In SPICE code, rise and fall times are the total transition times indicated in the diagram rather than 10 to 90% values. Zero rise and fall times approximate the test signal of Fig. 13.3.

13.2
NMOS and PMOS Logic Circuits

We learned in Chapter 5 that NMOS and PMOS are two similar all-transistor circuit families featuring high circuit density. Except for transistor type, NMOS employs the same circuit structures as PMOS; it receives greater emphasis here because of its superior speed.

13.2.1 STATIC TRANSFER CHARACTERISTIC

Figure 13.5a shows an NMOS inverter that uses an enhancement-load transistor. Inverter input and output are, respectively, the gate-source voltage and drain-source voltage of M_1. Figure 13.5b shows characteristic curves of M_1 for $v_{GS} = V_{OH}$ and $v_{GS} = V_{OL}$, and also the nonlinear load line contributed by M_2 and the supply. To emphasize gate variables, the horizontal axis and characteristic curves are, respectively, labeled v_O and v_I. To simplify, we initially ignore body effect in M_2.

To derive the transfer characteristic, we begin with $v_I \leq V_{t1}$. For this low input, M_1 is cut off, giving Fig. 13.5c. M_2 must be either OFF or active because it is diode-connected, as we showed in Sec. 5.3.1. With the slightest nonzero trickle of current in M_2, this transistor is active. Since $v_{G2} = V_{DD}$ and $v_{S2} = v_O$, M_2 then satisfies

$$i_{D2} = \frac{k_2}{2} [(V_{DD} - v_O) - V_{t2}]^2 = 0 \tag{13.4}$$

Solving for v_O gives

$$v_O = V_{DD} - V_{t2} \tag{13.5}$$

and we see that operation is at point 1 on the load line.

Now for *any* input in the range $v_I \leq V_{t1}$, M_1 remains cut off with output given by Eq. (13.5). This gives the horizontal line on the transfer characteristic, Fig. 13.5d, which

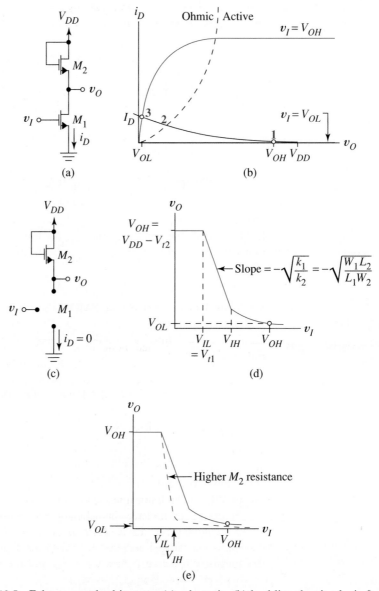

Figure 13.5 Enhancement-load inverter: (a) schematic; (b) load line showing logic 0 and 1 values; (c) when M_1 is cut off to give logic 1; (d) transfer characteristic; (e) effect of higher M_2 resistance on transfer characteristic.

defines both V_{OH} and the logic 0 input range. Because of the sharp break point and steep slope, $V_{IL} = V_{t1}$.

As v_I in Fig. 13.5c is further increased, it eventually exceeds V_{t1}, and M_1 turns on. From the load line of Fig. 13.5b we observe that when M_1 changes to active operation, M_2 remains active. Thus there is a range of input values for which both M_1 and M_2 are active. Equating the active equations for M_1 and M_2 and expressing all voltages in terms of the node voltages v_I, v_O, and V_{DD} of Fig. 13.5a gives

$$\frac{k_1}{2}(v_I - V_{t1})^2 = \frac{k_2}{2}(V_{DD} - v_O - V_{t2})^2 \tag{13.6}$$

which relates v_O to v_I on the transfer curve of Fig. 13.5d when both transistors are active. Taking the square root of both sides and solving for v_O gives

$$v_O = (V_{DD} - V_{t2}) - \sqrt{\frac{k_1}{k_2}} \, (v_I - V_{t1}) \tag{13.7}$$

the equation of a straight line of negative slope that passes through

$$(v_I, v_O) = (V_{t1}, V_{DD} - V_{t2})$$

From Fig. 13.5b, increasing v_I further causes M_1 to change to ohmic operation at point 2. Equating equations for drain currents with M_1 ohmic and M_2 active gives the equation for the quadratic curve segment in Fig. 13.5d for larger values of v_I. V_{OL} is found by substituting $v_I = V_{OH}$ into this equation as suggested in Fig. 13.5d. Also, we can compute the value of V_{IH} by finding the point where the slope of this equation equals minus one. These details are left for Problem 13.4.

Design Trade-offs Involving the Transfer Characteristic. Since slope, $-(k_1/k_2)^{0.5}$ in Fig. 13.5d, is controlled by the relative width-to-length ratios of M_1 and M_2, the shape of the transfer characteristic is closely related to fabrication geometry. Figure 13.5e shows that a transfer characteristic with steeper slope more closely approaches the ideal of Fig. 13.1a by giving a slightly higher logic swing and higher noise margins. The designer must weigh this benefit against the possibility of lower circuit density that might result from widely differing width-to-length ratios.

A different design trade-off makes it possible to increase V_{OH}. We saw in Figs. 5.17c and d, that returning the gate of M_2 to a separate supply of $V_{DD} + V_{t2}$ volts translates the nonlinear load line to the right in Fig. 13.5b by V_{t2} volts. This makes $V_{OH} = V_{DD}$.

Body Effect. An interesting detail, explored in Problem 13.5, is the consequence of including the body effect of M_2 when calculating the transfer characteristic. It is when v_O is close to V_{DD} that the source voltage of M_2 differs most from the substrate voltage (zero). Only as gate output voltage approaches zero does V_{t2} approach the assumed value of V_{t20}. Not only does the slope of the transfer characteristic change because of body effect, but so does $V_{OH} = V_{DD} - V_{t2}$.

13.2.2 AVERAGE POWER DISSIPATION

Figure 13.5b shows that power dissipation is zero when gate output is logic 1 because the inverter's current is zero. At logic 0, point 3, the average dissipation in the gate equals the power from the supply $V_{DD}I_D$, where I_D is the gate current when $v_O = V_{OL}$. Because M_2 is active for logic 0 output, power dissipation in the inverter is

$$V_{DD}I_D = V_{DD}\frac{k_2}{2}[(V_{DD} - V_{OL}) - V_{t2}]^2 = \frac{V_{DD}k_2}{2}[(V_{DD} - V_{t2}) - V_{OL}]^2$$

By assuming the gate spends half its time with output at logic 1 and half at logic 0 we estimate the average gate dissipation to be

$$P_{D,ave} = \frac{V_{DD}k_2}{4}(V_{OH} - V_{OL})^2 \qquad (13.8)$$

Thus average dissipation is proportional to the square of the *logic swing*, $V_{OH} - V_{OL}$. According to Eqs. (13.8) and (13.2), higher dissipation relates to higher noise margins, another design compromise.

13.2.3 PROPAGATION DELAY

To estimate the inverter's propagation delay, we lump internal transistor capacitances and load-gate input capacitances into the single load capacitor C_L of Fig. 13.6a, a procedure that sacrifices accuracy for design insight. (For accurate numerical calculations of gate delay, we simulate the actual circuit dynamics.) Figure 13.6b shows the form of the output when a pulse is applied to the input. We now examine how the circuit produces this output.

For $t < 0$, M_1 is cut off as in Fig. 13.6c, C_L is charged to V_{OH}, and the circuit is in equilibrium at point a in Fig. 13.6d. At $t = 0$, v_I changes from V_{OL} to V_{OH}, *instantly* creating a channel in M_1 for, by hypothesis, all circuit dynamics are embodied in C_L. The characteristic curve of M_1 instantly changes to the one labeled $v_I = V_{OH}$. The Q-point cannot move directly to the logic 0 final equilibrium at point c because the voltage across C_L, v_O, cannot change instantaneously. With the capacitor in the circuit, however, operation is not constrained to the load line, for Fig. 13.6e shows that capacitor current (not included in the "static" load-line derivation) can flow *during transient conditions*. In this circuit, when

$$\frac{dv_O}{dt} \neq 0$$

instantaneous operation *can* leave the load line. Therefore, operation jumps instantaneously from point a to point b in Fig. 13.6d, and M_1 switches to active-state operation. M_1 *becomes* constant current source I_X, discharging C_L even as resistor M_2 simultaneously tries to charge C_L from the power supply. Because I_X greatly exceeds i_{D2}, C_L discharges, and v_O decreases as indicated by arrowheads in Fig. 13.6d. For each value of v_O in this figure, the $v_I = V_{OH}$ curve shows the discharge current provided by M_1, and the load line curve shows the charging current from M_2.

To estimate the high-to-low propagation delay defined in Fig. 13.3, we need to estimate the time, t_{PHL}, it takes for C_L to discharge from its initial value of V_{OH} to V_M, the mean output voltage shown in Fig. 13.3 and also in Figs. 13.6b and d. Let I_{HL} denote the *average* discharge current as v_O drops from V_{OH} to V_M. Then from

$$-C_L\frac{dv_O}{dt} = I_{HL}$$

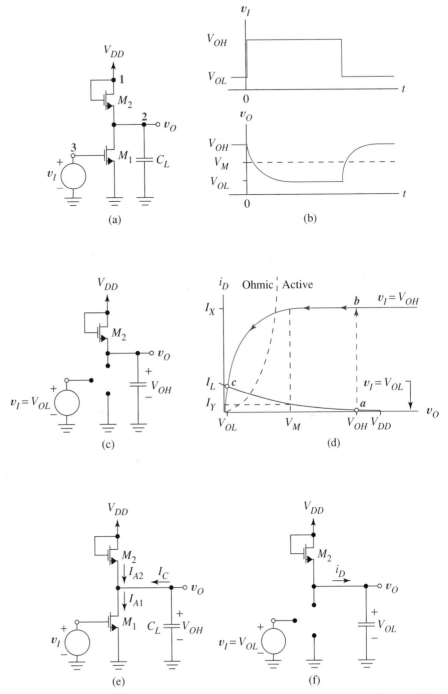

Figure 13.6 NMOS switching transients: (a) inverter with capacitive load; (b) input and output waveforms; (c) equivalent circuit for zero-to-one output transient; (d) characteristic curves and load line; (e) circuit during one-to-zero transient; (f) circuit beginning the zero-to-one transient.

we estimate

$$-C_L \frac{\Delta v_O}{\Delta t} = C_L \frac{(V_{OH} - V_M)}{t_{PHL}} = C_L \frac{[V_{OH} - (V_{OH} + V_{OL})/2]}{t_{PHL}} = I_{HL}$$

Solving for the unknown delay gives

$$t_{PHL} = \frac{C_L(V_{OH} - V_{OL})}{2I_{HL}} \tag{13.9}$$

From Fig. 13.6d, the average discharge current during the interval of interest is

$$I_{HL} = \frac{I_X + (I_X - I_Y)}{2} \approx I_X$$

where the approximation is valid if the resistance of M_2 is sufficiently high. Since M_1 is active during this interval, with $v_{GS} = v_I = V_{OH}$,

$$I_{HL} \approx I_X = \frac{k_1}{2}(V_{OH} - V_{t1})^2$$

Substituting this into Eq. (13.9) gives

$$t_{PHL} = \frac{C_L(V_{OH} - V_{OL})}{k_1(V_{OH} - V_{t1})^2}$$

Since the voltage intervals in numerator and denominator are comparable, we sacrifice accuracy for insight by assuming $V_{t1} \approx V_{OL}$. This gives

$$t_{PHL} = \frac{C_L}{k_1(V_{OH} - V_{OL})} \tag{13.10}$$

Eventually v_O becomes so low that M_1 changes to ohmic operation and the circuit settles into its logic 0 equilibrium, point c in Fig. 13.6d.

At the end of the input pulse, v_I returns to subthreshold value, V_{OL}, and M_1 cuts off, giving Fig. 13.6f. In Fig. 13.6d, the initial charging current is I_L at point c, and after the interval t_{PLH} the charging current is reduced to I_Y. Problem 13.6 verifies that the average charging current in this interval is

$$I_{LH} = \frac{I_L + I_Y}{2} = \frac{5}{16} k_2(V_{OH} - V_{OL})^2 \tag{13.11}$$

To estimate t_{PLH}, we use the estimate

$$C_L \frac{\Delta v_O}{t_{PLH}} = C_L \frac{V_M - V_{OL}}{t_{PLH}} = I_{LH} = \frac{5}{16} k_2(V_{OH} - V_{OL})^2$$

which gives

$$t_{PLH} = \frac{16C_L}{5k_2} \frac{[(V_{OH} + V_{OL})/2 - V_{OL}]}{(V_{OH} - V_{OL})^2} = \frac{1.6C_L}{k_2(V_{OH} - V_{OL})} \tag{13.12}$$

From Eqs. (13.3), (13.10), and (13.12) we conclude that for the NMOS logic family the propagation delay is

$$t_P = \left(\frac{0.5}{k_1} + \frac{0.8}{k_2}\right) \frac{C_L}{(V_{OH} - V_{OL})} = K \frac{C_L}{(V_{OH} - V_{OL})} \qquad (13.13)$$

Trade-Offs Involving Dynamic Behavior. Equation (13.13) suggests that small internal and load capacitances favor short delays. Because every load gate connected between the inverter output and ground adds its input capacitance to C_L, the fan-out limit for NMOS is determined by the propagation delay that can be tolerated in the particular application.

The appearance of *logic swing* $(V_{OH} - V_{OL})$ in the denominator means that better noise immunity tends to accompany shorter delays in NMOS. This might seem counter-intuitive, since a larger logic swing means more charge must be moved to or from C_L to effect a change in output state. Certainly a constant charging current would have to flow longer for a larger swing. However, in NMOS the average currents depend on the *square* of $(V_{OH} - V_{OL})$; thus shorter delays accompany larger logic swings and noise margins.

Notice that K in Eq. (13.13) depends on k_1 and k_2. Increasing these constants provides higher charging currents for the same logic swing, thus reducing propagation delay. But larger W/L ratios generally increase the physical size of the transistor on the chip and impact negatively on circuit density.

13.2.4 POWER DELAY PRODUCT

An important figure of merit for any logic family is its *power-delay product (PDP)*, the product of average power dissipation and propagation delay, which has units of energy. If a gate is changing logic levels at maximum speed, its *PDP* represents the energy required per change in logic transition or per logic decision! Obviously a low *PDP* is best. From Eqs. (13.8) and (13.13) we estimate

$$PDP \approx K_1 C_L V_{DD}(V_{OH} - V_{OL}) \text{ joules} \qquad (13.14)$$

where K_1 is a constant. Low *PDP* accompanies small supply voltage, small logic swing, and physically small gate structures that minimize parasitic capacitance.

EXAMPLE 13.1 Use SPICE to plot the transient response of an NMOS gate with enhancement load when there are no load gates. Find the *PDP* from simulation data.

Solution. Figure 13.7a shows the code for the inverter of Fig. 13.6a with $C_L = \infty$. Transistor capacitances are specified indirectly through W/L ratios and oxide capacitances, as in the discussion of Fig. 5.52. To ensure high gain and low V_{OL}, the load transistor has W/L of 1:3, the inverter 2:1. Body effect is included. A pulse of 5 V amplitude and 10 ns duration is applied at $t = 2$ ns to test the circuit.

Figure 13.7b shows the output waveform. From examining the numerical values, we find $V_{OL} \approx 0.251$ and $V_{OH} \approx 3.08$ V. The latter reduced below 3.5 V by body effect in M_2, as expected. Because of the high-resistance load transistor, $t_{PLH} = 4.71$ ns, whereas the capacitance discharged rapidly under the influence of the current source, giving $t_{PHL} = 0.683$ ns, for average of $t_P = 2.70$ ns.

```
EXAMPLE 13.1
VDD 1 0 DC 5
M2 1 1 2 0 TRAN W=5U L=1.5U
M1 2 3 0 0 TRAN W=10U L=5U
.MODEL TRAN NMOS VTO=1.5
+KP=7E-6 GAMMA=0.37 TOX=1.0E-7
VS 3 0 PULSE(0 5 2N 0 0 10N 50N)
.OP
.TRAN 0.19N 40N
.PLOT TRAN V(2)
.END
```

(a)

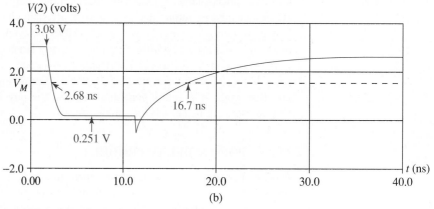

(b)

Figure 13.7 Transient response of an NMOS inverter with no load: (a) SPICE code; (b) output voltage.

Separate dc runs at both logic levels established average dissipation of 29.8 μW , so
$PDP = (2.70 \text{ ns})(29.8 \text{ μW}) = 80.5 \text{ Fj}$. ❏

Exercise 13.1 Find the equivalent load capacitance, C_L, so that Eq. (13.10) predicts the value of t_{PHL} found in Example 13.1.

Ans. 27.1 FF.

13.2.5 OTHER LOGIC FUNCTIONS

We transform the NMOS inverter to a multi-input NOR gate by adding parallel input transistors, as in Fig. 13.8a. To understand the configuration, notice that output voltage v_Z can be low only if there is a path from the output node to ground, that is, only if v_A OR v_B OR v_C is high enough to create a channel. Viewing the parallel transistors as switches, this means

$$\overline{Z} = A + B + C \quad \text{or} \quad Z = \overline{(A + B + C)}$$

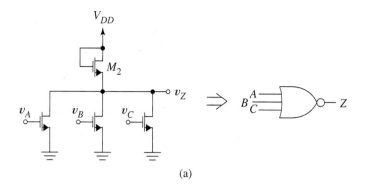

(a)

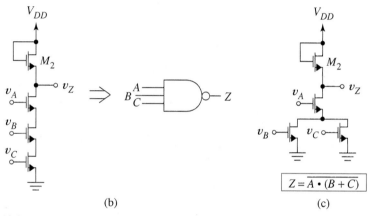

$$Z = \overline{A \cdot (B + C)}$$

(b)   (c)

Figure 13.8 Extension of the NMOS inverter to other logic functions: (a) three-input NOR; (b) three-input NAND; (c) three-input AND-OR function.

where "+" denotes the logic OR operation and the overbar denotes complement. The additional input transistors that increase gate fan-in also increase the propagation delay by adding to the circuit capacitance.

Figure 13.8b shows that series input transistors result in a NAND function, since the output can be low only if *all* inputs are simultaneously high. Since current from M_2 must flow through the combined lengths of the series transistors, a higher W/L must be used for M_2 or lower W/L for the input transistors to give the same V_{OL} as the inverter discussed earlier. Thus the number of inputs, or fan-in, trades off with physical size constraints associated with circuit density and low capacitance.

Figure 13.8c shows how series-parallel transistor groups give complex AND-OR logic functions.

13.2.6 NMOS GATE WITH DEPLETION LOAD

Figure 13.9a shows an NMOS inverter with depletion transistor M_2 for its load; Fig. 13.9b shows the nonlinear load line with body effect in M_2 ignored. For $v_I < V_{t1}$; M_1 is cut off, M_2 is ohmic but conducting no current, and operation is at the logic 1 static

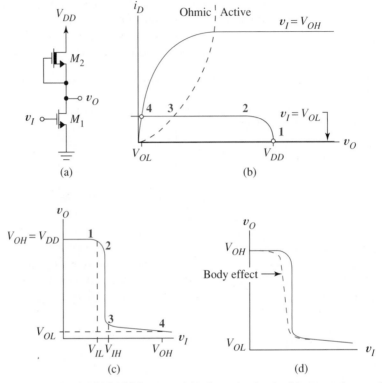

Figure 13.9 Depletion-load NMOS inverter: (a) schematic circuit; (b) output characteristics and load line; (c) transfer characteristic; (d) change in transfer characteristic caused by body effect.

equilibrium value, point 1 in Fig. 13.9b. For this circuit, we see that $V_{OH} = V_{DD}$. When v_I exceeds V_{t1}, M_1 changes to active operation near point 1, but M_2 initially stays ohmic. Additional increases in v_I move the operating point up the load line to point 2, where M_2 also becomes active. If both Early voltages are high, the transistor output characteristics are nearly horizontal, causing the slightest increase in v_I to move operation to point 3, where M_1 enters ohmic operation. Between points 3 and 4, the quadratic ohmic characteristic of M_1 causes the transfer curve to trail off along a curved locus as in Fig. 13.9c. Because of the high output resistance of M_1, Fig. 13.9c closely resembles the ideal of Fig. 13.1a and offers larger noise margins than the enhancement-load inverter. Another advantage is that the very steep transfer characteristic is independent of the relative physical dimensions of M_1 and M_2, depending mainly on the transistor output resistances.

Unfortunately, body effect in M_2 makes the transfer characteristic less ideal than it would otherwise be, as Fig. 13.9d shows. Because the substrate of M_2 is connected to ground, body effect is most pronounced at high output voltages. Early effect in M_1 and M_2 also reduce the slope of the transfer characteristic because load line and output characteristics are not horizontal.

Like the enhancement-load inverter, power dissipation is negligible when output is logic 1 (point 1, Fig. 13.9b), but nonzero for logic 0 (point 4). To estimate average power, we assume that the gate spends half of its time in each logic state. We make hand estimates of propagation delays from a lumped-capacitor model, using information obtained from the transistor equations (as previously illustrated for the enhancement-load inverter) or graphically as in the following example.

EXAMPLE 13.2   Estimate the PDP for the inverter of Fig. 13.10.

Solution. The power dissipation with logic 1 output (point a) is zero. At logic 0 (point c) the drain currents are 1 mA. Thus gate dissipation is

$$P_{AVE} = \frac{0 + (12\ \text{V})(1\ \text{mA})}{2} = 6\ \text{mW}$$

If v_I switches suddenly from V_{OL} to V_{OH}, M_1 turns active at $t = 0^+$ and discharges the capacitor with a current of 3 mA (point b). By the time v_O is reduced to

$$V_M = 0.5\ (12 + 0.3) = 6.15\ \text{V}$$

M_2 is providing a charging current of 1 mA, giving a net discharge current of 2 mA. The average discharge current during the interval $0 \le t \le t_{PHL}$ is $0.5(3 + 2) = 2.5$ mA. We estimate t_{PHL} using

$$C_L \frac{dv_O}{dt} \approx (20 \times 10^{-12}) \frac{6.16 - 12}{t_{PHL}} = -2.5 \times 10^{-3}$$

which gives $t_{PHL} = 46.8$ ns.

For the low-to-high transition, with initial equilibrium at point c, v_I suddenly changes to V_{OL} and M_1 cuts off. The characteristic curves show that M_2 provides a constant charging current of 1 mA throughout the interval $0 \le t \le t_{PLH}$. Therefore, our estimate is

$$C_L \frac{dv_O}{dt} \approx (20 \times 10^{-12}) \frac{6.15 - 0.3}{t_{PLH}} = 1.0 \times 10^{-3}$$

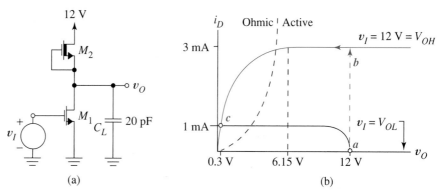

(a) (b)

Figure 13.10   Switching transients in depletion-load inverter: (a) circuit schematic; (b) output characteristics and load line.

giving $t_{PLH} = 117$ ns. Averaging the two delays gives

$$t_P = 0.5\,(46.8 + 117) = 81.9 \text{ ns}$$

The *PDP* is

$$PDP = (6 \times 10^{-3})(81.9 \times 10^{-9}) = 491 \text{ pJ} \qquad \square$$

> **Exercise 13.2** Recalculate t_p for Example 13.2 if M_2 is replaced by a transistor with $k_2 = 0.3 \times 10^{-4}$ A/V^2 and $V_t = -0.9$ V. Assume that V_{OL} does not change significantly.
>
> *Ans.* 4.81 μs.

13.3 _____
CMOS Logic Circuits

Gates in the complementary symmetry MOS (CMOS) logic family have static properties that closely approximate the ideal. Static power dissipation is typically less than 10 nW/gate; noise margins, together, are as high as 90% of the logic swing; and propagation delays are of the order of 25 to 50 ns. CMOS is a very versatile family that operates with power supply voltages from 4.5 to 15 V.

The CMOS inverter, Fig. 13.11a, consists of a p-channel transistor, M_P, and an n-channel transistor, M_N. Since the substrates of M_P and M_N are tied, respectively, to the most positive and negative points in the circuit, there is no body effect, and both threshold voltages are constants.

13.3.1 STATIC INVERTER CHARACTERISTICS

We can predict the shape of the inverter's transfer characteristic directly from the equations for M_N and M_P. To do this, we first use the transistor state definitions to relate each state of M_N to a region in the v_O versus v_I plane. From Fig. 13.11a, we adopt the following notation for M_N: $v_S = 0$, $v_G = v_I$, and $v_D = v_O$. In this notation, the state definitions are

$$\text{cut-off} \Rightarrow v_I \leq V_{tN}$$

$$\text{active} \Rightarrow v_O - v_I \geq -V_{tN}$$

$$\text{ohmic} \Rightarrow v_O - v_I \leq -V_{tN}$$

In Fig. 13.11b these regions correspond to "left of the vertical line through V_{tN}," "above the line of unity slope through V_{tN}," and "below the line through V_{tN}." For M_P, Fig. 13.11a implies $v_G = v_I$, $v_S = V_{DD}$, and $v_D = v_O$. Since M_P is a p-channel transistor, Fig. 5.6a gives

$$\text{cut-off} \Rightarrow v_I - V_{DD} \geq V_{tP}$$

$$\text{active} \Rightarrow v_O - v_I \leq -V_{tP}$$

$$\text{ohmic} \Rightarrow v_O - v_I \geq -V_{tP}$$

The complete set of state inequalities thus partition the first quadrant into eight regions, as indicated in Fig. 13.11b. The CMOS transfer characteristic crosses through five of them.

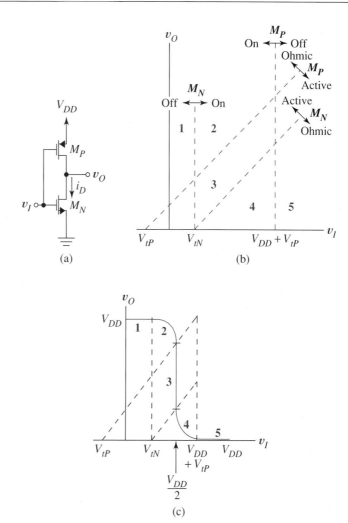

Figure 13.11 CMOS: (a) inverter schematic; (b) regions of operation defined by states of M_P and M_N; (c) transfer characteristic when transistors have complementary parameter values.

To find the transfer characteristic, begin with $v_I = 0$. From Fig. 13.11b, M_N is off and M_P is either ohmic or active. Since M_N is off, its current and that of M_P are zero. In Fig. 13.11a, $v_I = 0$ implies that the gate of M_P is very negative relative to its source. If M_P were active, its current would not be zero for large negative v_{GS}; therefore, M_P is ohmic. Because M_N and M_P are in series, currents are identical, so we equate the cut-off current of M_N to the ohmic current for M_P. This gives

$$\frac{k_P}{2}\left[2(v_I - V_{DD} - V_{tP})(v_O - V_{DD}) - (v_O - V_{DD})^2\right] = 0 \qquad (13.15)$$

This equation has the solution $v_O = V_{DD}$, which describes the transfer characteristic, not only for $v_I = 0$, but for all $v_I \leq V_{tN}$. Region 1 of Fig. 13.11c is a graph of this solution.

As v_I increases beyond V_{tN}, Fig. 13.11b shows that M_N changes to active operation. Equating drain currents, active region for M_N and ohmic for M_P, gives the equation for region 2 in Fig. 13.11c:

$$\frac{k_P}{2}\left[2(v_I - V_{DD} - V_{tP})(v_O - V_{DD}) - (v_O - V_{DD})^2\right] = \frac{k_N}{2}(v_I - V_{tN})^2 \qquad (13.16)$$

Though formidable in appearance, Eq. (13.16) is simply a second-order equation that relates v_I and v_O. It is easy to verify that the equation is satisfied by the point $(v_I, v_O) = (V_{tN}, V_{DD})$, where it joins with the region 1 curve.

As v_I increases still further, Fig. 13.11b suggests that M_P next changes to active operation, region 3. Equating currents when both transistors are active gives

$$\frac{k_N}{2}(v_I - V_{tN})^2 = \frac{k_P}{2}(v_I - V_{DD} - V_{tP})^2 \qquad (13.17)$$

which is *independent of* v_O. This is consistent with the vertical line labeled region 3 in Fig. 13.11c. A particularly useful inverter design results when transistors have complementary parameter values. This means $-V_{tP} = V_{tN} = V_t$ and $k_P = k_N = k$. Substituting these special values into Eq. (13.17) and taking the square root of both sides (positive root for left side and negative root for right side as indicated by Figs. 5.43a and b) gives

$$v_I - V_t = -(v_I - V_{DD} + V_t)$$

Solving for v_I gives $v_I = 0.5V_{DD}$ for region 3 in Fig. 13.11c. The rest of the curve is symmetric. In region 4 we equate the active equation for M_P to the ohmic equation for M_N. In region 5 we equate the OFF equation of M_P to the ohmic equation for M_N.

Real transfer characteristics deviate slightly from Fig. 13.11c because of channel-length modulation. That is, the region 3 curve is not vertical if the output resistances of the two transistors are included in the analysis, a thesis easily explored by simulation. Because M_N and M_P have matched parameters, the transfer characteristic is relatively independent of temperature.

One unique feature of CMOS is already apparent. At both static logic levels, regions 1 and 5, a transistor is cut off, giving average power dissipation of zero, a useful feature fully exploited in calculator memories and many other applications. For CMOS, the only appreciable power dissipation occurs during *changes* in logic level. This is explored after transient effects are examined.

Notice that Eq. (13.16) and the corresponding equation for region 4 allow us to determine V_{IH} and V_{IL} analytically. For each equation, we equate the derivative dv_O/dv_I to -1. Each resulting equation, which involves v_O and v_I, must be solved simultaneously with the original equation to determine the logic levels. Problems 13.16 and 13.17 provide guidance with the details. Here we simply state that for the inverter with the symmetric transfer characteristic of Fig. 13.11c,

$$V_{IL} = \frac{1}{8}(3V_{DD} + 2V_t) \qquad (13.18)$$

and

$$V_{IH} = \frac{1}{8}(5V_{DD} - 2V_t) \tag{13.19}$$

Since $V_{OH} = V_{DD}$ and $V_{OL} = 0$, it follows that

$$NML = NMH = \frac{1}{8}(3V_{DD} + 2V_t) \tag{13.20}$$

CMOS is highly valued for its versatility. This equation shows that noise margins are readily adapted to particular noise environments by selection of V_{DD}.

13.3.2 PROPAGATION DELAY AND POWER DISSIPATION

Switching Delays. To explore CMOS switching transients, we represent internal and load capacitances by a single equivalent capacitor, C_L, as in Fig. 13.12a. The initial con-

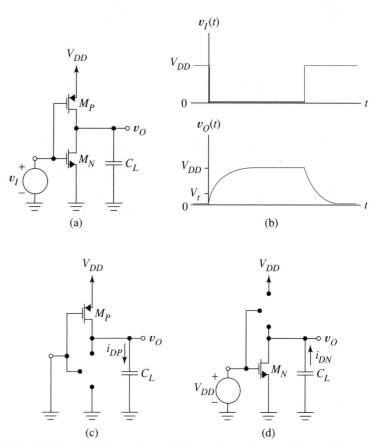

Figure 13.12 CMOS switching transients: (a) lumped-capacitor equivalent; (b) input and output waveforms; (c) circuit for zero-to-one transient; (d) circuit for one-to-zero transient.

ditions for the zero-to-one output transient are $v_I = V_{OH}$ (logic 1) and C_L discharged to $V_{OL} = 0$. At $t = 0$, v_I changes to $V_{OL} = 0$ volt as in Fig. 13.12b. Since we ignore internal capacitances, M_N instantly cuts off, and M_P, with $v_{DS} = -V_{DD}$ volts, changes to active operation. The result is Fig. 13.12c, where output voltage satisfies

$$C_L \frac{dv_O}{dt} = \frac{k_P}{2}(0 - V_{DD} + V_t)^2 \tag{13.21}$$

where $V_{tP} = -V_t$. Because the charging current is constant at first, the output voltage increases linearly in accordance with Eq. (13.21); however, as v_O increases, the magnitude of the drain-source voltage decreases; consequently, M_P changes over to ohmic operation when $v_{DG} = v_O - 0 = -V_{tP} = V_t$. Thereafter, Eq. (13.21) is replaced by

$$C_L \frac{dv_O}{dt} = \frac{k_P}{2}\left[2(-V_{DD} + V_t)(v_O - V_{DD}) - (v_O - V_{DD})^2\right] \tag{13.22}$$

By equating dv_O/dt to zero in this equation, we learn that the circuit's final equilibrium is at $v_O = V_{DD}$. The solutions to Eqs. (13.21) and (13.22) describe the zero-to-one transient of $v_O(t)$ in Fig. 13.12b.

The one-to-zero transient is explored in the same way using Fig. 13.12d. Here M_N discharges C_L from its initial value of V_{DD} volts to its final equilibrium, linearly at first while M_N is active, and then nonlinearly when M_N is ohmic, with differential equations very similar to Eqs. (13.21) and (13.22). Solving these equations gives the trailing edge of the output pulse of Fig. 13.12b. Equations (13.21) and (13.22) both show that the rate of change of output voltage is inversely proportional to C_L. As in NMOS, fan-out, which increases C_L, must be traded against switching speed

Equation (13.21) is linear and (13.22) separable, so closed-form solutions exist. Therefore, it is possible to find analytical expressions for propagation delay of the CMOS gate (see Problem 13.22); however, the procedure and result contribute little to our basic understanding. Problem 13.23 uses SPICE to compute numerical solutions to the more complex differential equations that describe the CMOS inverter when internal capacitances are associated with individual transistors and C_L is a true load capacitance. With such simulations, we can determine more accurate values for propagation delays.

Power Dissipation. Although CMOS dissipates no power at either logic 0 or logic 1, dissipation does occur during transitions when both transistors are on. Surprisingly, we can predict the correct average dissipation from our lumped-capacitor model, even though this model incorrectly predicts that the transistors never conduct simultaneously.

It is evident from Fig. 13.12c that current flows from the power supply during the zero-to-one transient, dissipating power in M_P and storing energy in C_L. In Fig. 13.13a, we generalize Fig. 13.12c by replacing M_P by a general nonlinear resistor, and use this circuit to calculate the energy dissipated when a capacitor is charged through any nonlinear resistor, $i = f(v)$. The energy dw dissipated during interval dt is

$$dw = p\,dt = vi\,dt = (V_{DD} - v_O)\left(C_L \frac{dv_O}{dt}\right)dt$$

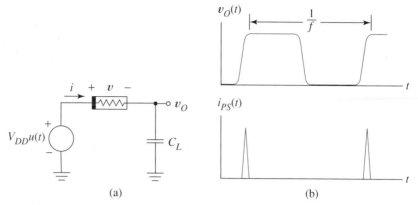

Figure 13.13 CMOS power dissipation: (a) charging a capacitor through a nonlinear resistor; (b) output voltage and *predicted* supply current for CMOS.

where p denotes instantaneous power dissipation in the nonlinear resistor. If C_L is initially uncharged, the energy, W, dissipated during the charging interval is

$$\int_0^W dw = \int_0^\infty (V_{DD} - v_O)\left(C_L \frac{dv_O}{dt}\right) dt = \int_0^{V_{DD}} (V_{DD} - v_O)C_L \, dv_O$$

The integration gives

$$W = \frac{1}{2}C_L V_{DD}^2 \text{ joule} \qquad (13.23)$$

which is the energy dissipated in M_P in Fig. 13.12c as C_L is charged. Figure 13.12d suggests that the stored energy from C_L is dissipated in M_N during the one-to-zero transient.

Figure 13.13b shows the power-supply current that flows into the *idealized* CMOS gate when the input is a periodic pulse train with zero rise and fall times. From Figs. 13.12c and d, current flows only during the zero-to-one output transitions. Each current spike delivers $(1/2)C_L V_{DD}^2$ joules of energy to be dissipated in M_P and an additional $(1/2)C_L V_{DD}^2$ joules to be stored on C_L. The latter is then dissipated in M_N during the second half-cycle, when no supply current flows. The average power dissipation of the idealized inverter is, therefore,

$$P_{AVE} = 2\left(\frac{1}{2}C_L V_{DD}^2\right)\frac{\text{joules}}{\text{cycle}} f \frac{\text{cycles}}{\text{s}} = C_L V_{DD}^2 f \text{ watts} \qquad (13.24)$$

As in NMOS, average dissipation is directly proportional to the square of the logic swing, $V_{OH} - V_{OL} = V_{DD}$.

Although Eq. (13.24) gives correct average power dissipation in a CMOS gate, the true circuit details are much more complex than our derivation suggests because of the internal capacitances of the transistors and the finite rise and fall times of the input waveform. The main difference is that both transistors conduct simultaneously during each logic transition. The resulting current spikes can be troublesome, since large di/dt values can magnetically couple noise into nearby circuits. Noise spikes originating in

CMOS transient currents can also pass into other circuits through the internal resistance of a shared power supply and by capacitive coupling between nodes.

EXAMPLE 13.3   A CMOS inverter uses a 12 V power supply and transistors with $V_{tP} = -1.3$ V, $V_{tN} = 1.3$ V, $k_P = k_N = 2 \times 10^{-4}$ A/V^2. Both have zero voltage capacitances $CBS = CBD$ of 20 femtofarad and $RD = RS = 1$ Ω.
(a) Use SPICE to find the static transfer characteristic.
(b) Use SPICE to find the power supply current when the input is a 12 V input pulse of 30 ns duration with 1 ns rise and fall times.
(c) Repeat part (b) after adding an external load of 40 pF at the inverter output.

Solution. (a) Figure 13.14a shows the circuit and initial SPICE code. The resulting transfer characteristic, Fig. 13.14b, verifies that the curve has the general shape predicted in Fig. 13.11c.

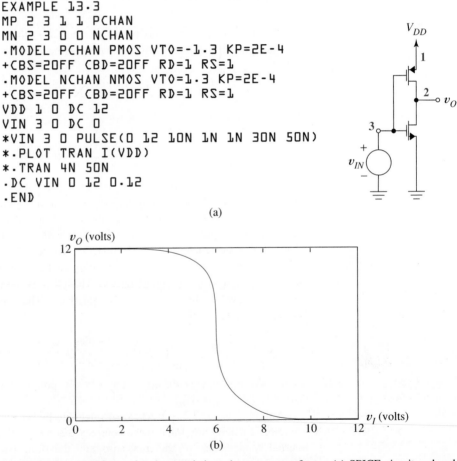

```
EXAMPLE 13.3
MP 2 3 1 1 PCHAN
MN 2 3 0 0 NCHAN
.MODEL PCHAN PMOS VTO=-1.3 KP=2E-4
+CBS=20FF CBD=20FF RD=1 RS=1
.MODEL NCHAN NMOS VTO=1.3 KP=2E-4
+CBS=20FF CBD=20FF RD=1 RS=1
VDD 1 0 DC 12
VIN 3 0 DC 0
*VIN 3 0 PULSE(0 12 10N 1N 1N 30N 50N)
*.PLOT TRAN I(VDD)
*.TRAN 4N 50N
.DC VIN 0 12 0.12
.END
```

(a)

(b)

Figure 13.14   CMOS transfer characteristic and current waveforms: (a) SPICE circuit and code for transfer characteristic; (b) computed transfer characteristic; (c) supply current for unloaded gate; (d) transistor drain currents with 40 pF load capacitance.

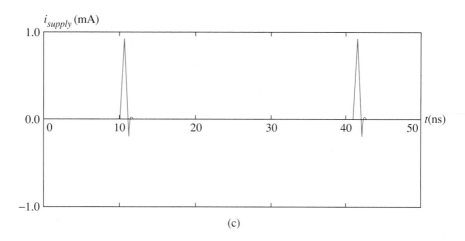

(c)

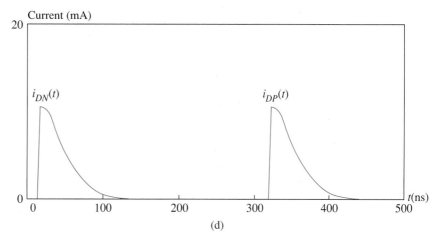

(d)

Figure 13.14 (continued)

(b) The input pulse, delayed by 10 ns, is marked by an asterisk in the SPICE listing, as is the .TRAN statement that replaced the .DC statement for the second SPICE run. The power supply current waveform computed by SPICE is Fig. 13.14c. With no load, the supply current consists of very narrow spikes with amplitude of 1.3 mA *for both logic transitions*. The rate of change of these spikes is very large. Bypass capacitors are often added to the circuit between the power supply lines and ground to minimize noise coupling to other circuits through the supply lines.

(c) After adding the 40 pF load capacitor, the original pulse width was too short for the circuit time constants, so the pulse was widened to 300 ns. Figure 13.14d shows the resulting power supply current waveform. The peak value is 11.4 mA, much larger than the current of Fig. 13.14c for the gate without load capacitance. ❑

The CMOS power-delay product is a function of frequency and therefore not directly comparable to the *PDP* in other logic families. However, in the sense that the

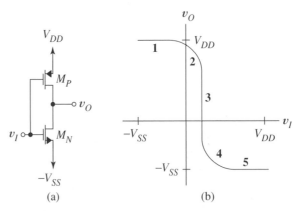

Figure 13.15 CMOS with two supplies: (a) schematic; (b) transfer characteristic.

PDP denotes the energy required to change the state of the inverter, Eq. (13.23) can be used as the *PDP* for CMOS. Comparing Eq. (13.23) with Eq. (13.14) shows a close similarity. This is logical because the same basic process is involved in changing logic levels in both families: charging and discharging capacitance through nonlinear resistors.

13.3.3 OPERATION WITH TWO POWER SUPPLIES

The versatility of CMOS extends to its power supplies. For simplicity, we described the gate in terms of a single positive supply; however, CMOS works the same with a negative supply, or with two supplies as in Fig. 13.15a. For logic-one input, $v_I \approx V_{DD}$ so M_P cuts off, leaving the output node connected through a low-impedance path to $-V_{SS}$. When v_I is close to $-V_{SS}$, M_N cuts off, and $v_O = V_{DD}$. Thus the transfer characteristic has the form of Fig. 13.15b. The transistor states in each numbered region are exactly those described for Figs. 13.11b and c. Notice that Fig. 13.15b reduces to Fig. 13.11c for $V_{SS} = 0$. In fact Fig. 13.15b applies to any pair of supply voltage values that satisfy $V_{DD} > -V_{SS}$, provided their separation is large enough to permit the transistor state changes that define regions 1–5 (assuming transistor current, breakdown, and dissipation ratings are not exceeded). Furthermore, the essence of dynamic operation is unchanged by the alternate supply configurations. We conclude that Eqs. (13.20), (13.23), and (13.24) describe the circuit for two supplies if V_{DD} is replaced by $V_{DD} + V_{SS}$. An important consequence of this power supply versatility is ease in interfacing between CMOS and the various other logic families.

13.3.4 OTHER LOGIC FUNCTIONS

Figure 13.16 shows how the CMOS inverter is augmented to perform other logic functions. As with NMOS, we can understand the logic functions by visualizing the transistors as ideal switches. In CMOS, input transistors are paired with pull-up transistors so that a high input voltage not only creates a path from the output node to ground but also breaks the path from output node to supply.

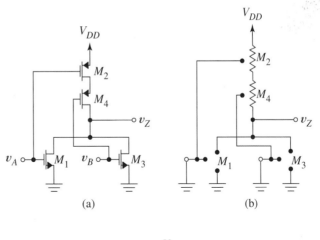

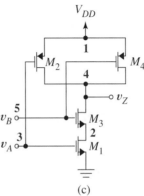

Figure 13.16 CMOS logic gates: (a) two-input NOR; (b) equivalent of (a) when both inputs are low; (c) two-input NAND.

Figure 13.16a is a two-input NOR, because v_Z takes on a LOW value only if v_A OR v_B OR both are HIGH enough to turn on their respective transistors to form a path to ground. Figure 13.16b shows an equivalent when both inputs are low. Since output current is zero, M_2 and M_4 are both ohmic, providing a low resistance path to V_{DD}. Figure 13.16c shows a two-input, CMOS NAND. For this circuit, the output voltage can be low only when both inputs are high; therefore, the circuit realizes

$$Z = \overline{A \cdot B}$$

Extending CMOS beyond the simple inverter to a circuit of two or more inputs in this fashion is conceptually straightforward, but introduces some new complexities for the designer. These include body effect (see M_4 in Fig. 13.16a), transfer characteristics of differing shape depending upon how many parallel inputs are used, and more complicated gate dynamics.

BiCMOS Inverter. The BiCMOS inverter of Fig. 13.17 is a modification of the basic CMOS inverter that facilitates more rapid charge and discharge of load capacitance

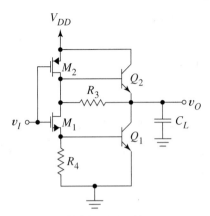

Figure 13.17 BiCMOS inverter.

while preserving the property of nearly zero standby power dissipation. When v_I is logic 0, M_1 and Q_1 cut off, and M_2 supplies base current for Q_2, which charges the load capacitance.

When v_I is logic 1, M_2 and Q_2 cut off. If C_L has been charged to logic 1 before the high input, Q_1 goes into its active state, with base current routed from C_L through R_3 and M_1. Q_1 thus helps discharge the load capacitance and then cuts off as final equilibrium is approached. Meanwhile, M_1, R_3, and R_4 provide a dc path to ground ensuring that $v_O = 0$ V. A detailed examination involving circuit models is guided by Problem 13.29.

13.4

Gallium Arsenide Logic Circuits

We introduced the compound semiconductor, gallium arsenide, in Chapter 3, and we described the gallium arsenide MESFET in Sec. 5.6.1. Because GaAs technology is still new, standards for GaAs gates do not yet exist as they do for silicon gates. Here we briefly explore several of the MESFET logic families that have been proposed. In general MESFET gates have low propagation delays, high power dissipation, and small noise margins. In addition to introducing digital GaAs technology, this section shows how to apply our general modeling concepts in a developing field and introduces some new ideas such as level shifting in digital circuits.

13.4.1 SCHOTTKY DIODE FET LOGIC

Figure 13.18a shows the basic inverter in a logic family called Schottky diode FET logic (SDFL). Although the gate requires positive and negative supplies, inputs and outputs take on values between 0 and 2.5 V. Current source B_1 keeps S_1 and S_2 ON. The inverting pair B_2 and B_3 function much like the NMOS inverter with depletion load. All transistors have lengths of one micron, and channel widths in microns are shown on the diagram. B_3 is wider than B_2 to give low logic 0 output voltages. As we will presently see, the relative widths of B_2 and B_1 affect the fan-out.

We now explore the circuit by hand analysis, using $V_{ON} = 0.6$ V for Schottky diodes and $V_t = -0.5$ V for the depletion transistors. We assume a unit width transistor has $\beta = 0.07$ mA/V^2. For the geometries indicated on the diagram, then, $\beta_1 = 0.35$ mA/V^2,

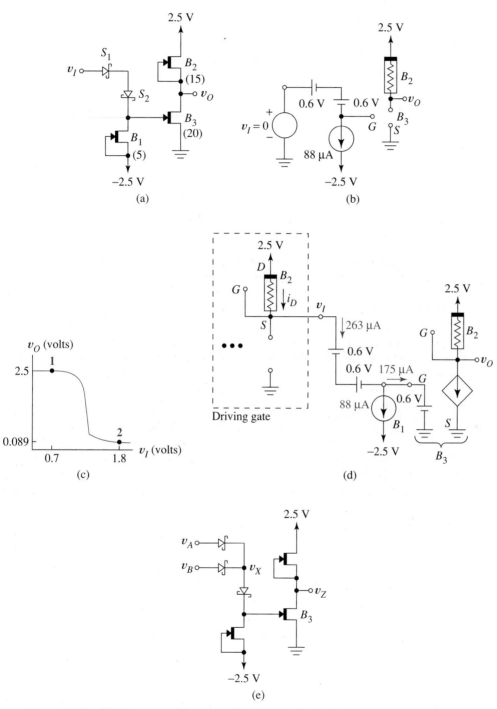

Figure 13.18 SDFL inverter: (a) circuit; (b) equivalent for logic 0 input; (c) hand-estimated transfer characteristic; (d) equivalent for logic 1 input; (e) SDFL two-input NOR gate.

$\beta_2 = 1.05 \text{ mA/V}^2$, and $\beta_3 = 1.40 \text{ mA/V}^2$. For simple hand analysis, we ignore velocity saturation and Early effect.

Logic 1 Output. When the input is logic 0, say, 0 V, we assume that B_3 is cut off and B_2 is ohmic, as for the NMOS inverter at logic 1 output. B_1 is a current source of

$$I_{D1} = 0.35 \times 10^{-3}(0 + 0.5)^2 = 88 \ \mu\text{A}$$

Figure 13.18b is the equivalent circuit. Because the gate voltage of B_3, -1.2 V, is less than $V_t = -0.5$ V, B_3 is cut off as assumed. B_2 is ohmic but delivering zero current, so $v_O = 2.5$ V.

Now as V_I increases, the gate voltage of B_3 increases. B_3 begins to turn on when

$$v_I - 1.2 = -0.5 \text{ V}$$

that is, when $v_I = +0.7$ V. This is point 1 on the transfer characteristic of Fig. 13.18c.

Logic 0 Output. Now we investigate the circuit when it has a logic 1 input provided by the output of an identical gate, as in Fig. 13.18d. Because there is a current path from the positive supply through the diodes to the gate of B_3, the Schottky gate-source junction of B_3 conducts, so we model the transistor as in Fig. 13.18d. For the load transistor of the driving gate,

$$v_{DG} = 2.5 - 1.8 = 0.7 \text{ V}$$

therefore, B_2 in the preceding stage is active and supplies input current

$$i_D = 1.05 \times 10^{-3}(0 + 0.5)^2 = 263 \ \mu\text{A}$$

Of this, 88 μA goes through current source B_1 and the remainder through the gate-source junction of B_3. The excess current, $263 - 88 = 175 \ \mu\text{A}$, is available to drive other load gates. Each such gate would absorb at least 88 μA in its current source, so $175/88 = 2$ *additional* load gates could be held at logic 1 input for a total fan-out of 3 for this gate design.

To find the logic 0 output voltage, we equate the output circuit currents of B_3 and B_2, assuming the former ohmic and the latter active. This gives

$$1.4 \times 10^{-3}\left[2(0.6 + 0.5)(v_O) - (v_O)^2\right] = 1.05 \times 10^{-3}(0.50)^2$$

which has the solution $v_O = 0.0891$ V and is consistent with our assumptions about the states of the output transistors. This is the logic 0 value at point 2 in Fig. 13.18c.

Notice from Fig. 13.18d that if v_I is reduced below 1.8 V, the gate-source voltage of B_3 will begin to decrease, always staying two Schottky drops below v_I. This decreases the current in B_3 and raises v_O as indicated in Fig. 13.18c. The transfer characteristic we have derived is the *open-circuit* characteristic of the gate. Unlike NMOS and CMOS gates that draw no static input current, SDFL gates have different transfer characteristics under load.

Figure 13.18e shows how additional Schottky diodes at the input change the inverter to a multiple-input NOR gate. If v_A or v_B is high, the gate of B_3 conducts, and v_Z is low. If only one input is low, its input diode will be OFF. When both inputs are low, B_3 cuts off, and the output is high.

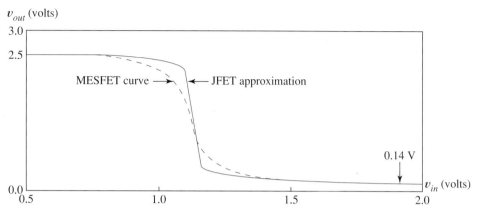

Figure 13.19 SPICE estimated SDFL transfer characteristics obtained from JFET and MESFET models.

Because gate-source conduction is normal in some GaAs logic families such as SDFL, an n-channel JFET is a more appropriate SPICE model than a depletion MOSFET when a true MESFET model is not available. Figure 13.19 compares the results of JFET and PSPICE MESFET (using $\alpha = 2$) modeling of the open-circuit gate we have just analyzed. Problem 13.37 provides additional guidance with SPICE simulations.

Noise margins and propagation delays can be estimated by applying the hand-analysis principles used for NMOS gates. These ideas are pursued in Problems 13.35 and 13.36. Because of velocity saturation, simulation with realistic models is perhaps more appropriate. This is illustrated for the next GaAs logic family.

13.4.2 BUFFERED FET LOGIC

Figure 13.20a shows the buffered FET logic (BFL) inverter. B_1 and B_2 are a depletion-load inverter like Fig. 13.9a; however, because MESFETs have negative V_t, v_I must go negative to turn off B_1. This means the simple two-MESFET circuit cannot function as a stand-alone logic gate, because the output range of v_X (0 to 2.5 V) is never sufficiently negative to provide input for an identical gate. To create compatible output and input levels, it is necessary to add a level shifter consisting of B_3, B_4, and the diodes.

We first consider the operation of the inverter alone, assuming no loading by the level shifter. All MESFETs have $V_t = -0.5$ V and $\beta = 0.7 \times 10^{-3}$ A/V² except for B_1, for which $\beta_1 = 1.4 \times 10^{-3}$ A/V². This approximate analysis ignores velocity saturation and Early effect.

The output characteristic of B_1 and load line for B_2 resemble Fig. 13.9b. Therefore, for $v_I \leq V_t$, B_1 is cut off, B_2 is ohmic, and the output is $v_X = 2.5$ V. This gives segment **1** in Fig. 13.20b.

When we make $v_I \geq V_t$ B_1 turns on. B_2 remains ohmic as long as

$$v_{DG2} = 2.5 - v_X \leq 0.5 = -V_t$$

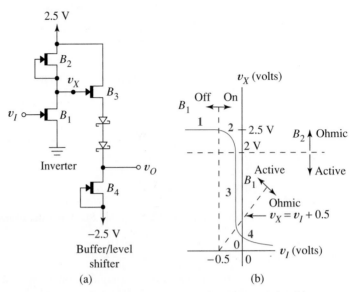

Figure 13.20 BFL inverting gate: (a) schematic; (b) characteristic of inverter subcircuit.

that is, for $v_X \geq 2$ V. As v_I is increased, B_1 changes to active operation when

$$v_{DG1} = v_X - v_I \geq 0.5 = -V_t$$

or for $v_X \geq v_I + 0.5$. Figure 13.20b shows these inequalities in the $v_X - v_I$ plane. In segment **2** the drain current of active B_1 equals the drain current of (ohmic state) B_2, so the transfer characteristic is described by

$$1.4 \times 10^{-3}(v_I + 0.5)^2 = 0.7 \times 10^{-3}\left[2(0 + 0.5)(2.5 - v_X) - (2.5 - v_X)^2\right] \quad (13.25)$$

This quadratic function is the equation for segment **2** in Fig. 13.20b.

Once the curve crosses the $v_X = 2$ V line, both MESFETs are active, and we have segment **3**, which satisfies

$$1.4 \times 10^{-3}(v_I + 0.5)^2 = 0.7 \times 10^{-3}(0 + 0.5)^2 = 175 \ \mu\text{A} \quad (13.26)$$

the equation for a vertical line at $v_I = -0.146$ V.

Finally, for sufficiently large v_I, B_1 also becomes ohmic and

$$i_{D1} = 1.4 \times 10^{-3}\left[2(v_I + 0.5)v_X - v_X^2\right] = i_{D2} = 175 \ \mu\text{A} \quad (13.27)$$

describes segment **4**.

Now we study the level shifter. Because B_4 is configured as a current source, we assume B_4 is active, and supplying current

$$i_{D4} = 0.7 \times 10^{-3}(0 + 0.5)^2 = 175 \ \mu\text{A} \quad (13.28)$$

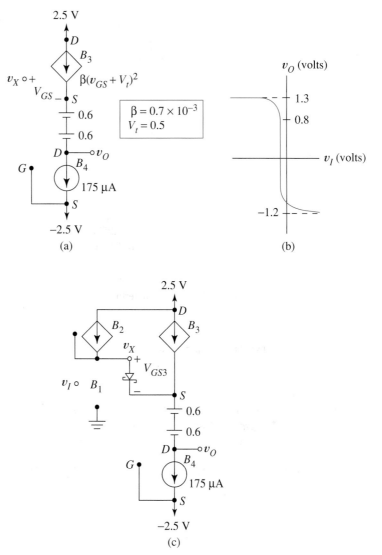

Figure 13.21 Level shifter: (a) equivalent with both transistors active; (b) transfer characteristic after ideal level shifting; (c) equivalent when B_3 is ohmic with gate current supplied from B_2.

to the Schottky diodes. Assuming B_3 is also active, we draw Fig. 13.21a. Since B_3 and B_4 are identical, active, and carry the same current, v_{GS} of B_3 must match $v_{GS} = 0$ of B_4. From the circuit, this implies the level shifter equation

$$v_O = v_X - 1.2 \qquad (13.29)$$

As its name suggests, the circuit ideally shifts the characteristic of Fig. 13.20b downward by 1.2 volts giving the overall inverter characteristic of Fig. 13.21b.

Actually, when v_X exceeds 2 V in Fig. 13.20a, B_3 goes into its ohmic state. Once this happens, it is possible for the drain currents of B_3 and B_4 to differ, as gate current begins to flow from B_2 into B_3. This current is so small that the v_X versus v_I characteristic is unchanged; however, the effect of the current is to shift v_O downward *slightly* more than Eq. (13.29) predicts because of the slight diode drop, V_{GS3}, shown in Fig. 13.21c. Exploring this by simulation shows that near $v_X = 2$ V, Eq. (13.29) merges into the new equation

$$v_O = v_X - 1.45 \tag{13.30}$$

which causes the $v_X = 2.5$ V level of Fig. 13.20b to map into $2.5 - 1.45 = 1.05$ V instead of 1.3 V. Thus we expect the transfer characteristic to have the shape of Fig. 13.22a. Figure 13.22b is the transfer characteristic obtained from a PSPICE simulation. The dashed curve shows how the transfer characteristic changes when there is a load gate. This is because the gate-source junction of the input transistor conducts, clamping the output of the preceding gate to 0.56 V.

Estimating the noise margins from the SPICE transfer curve output data, which includes an output load gate, gives

$$NMH = 0.56 - (-0.07) = 0.63 \text{ V}$$

$$NML = -0.215 - (-1.154) = 0.939 \text{ V}$$

The average power dissipation from the simulation was 1.1 mW, a result easily checked by hand calculation. For logic 0 input, Figs. 13.21a and b show that 175 μA flows through supplies totaling 5 V, for 0.875 mW. When input is logic 1, an additional 175 μA flows from the +2.5 V supply to ground through B_1 and B_2; thus the total logic 1 power is $(0.875 + 0.438) = 1.313$ mW. The average is 1.094 mW.

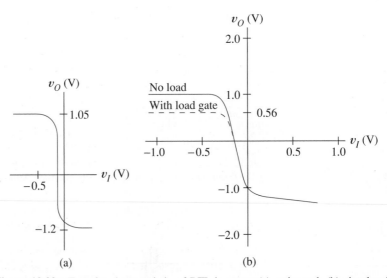

Figure 13.22 Transfer characteristic of BFL inverter: (a) estimated; (b) simulated.

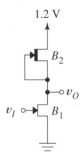

Figure 13.23 Direct-coupled FET logic inverter.

A PSPICE simulation of the logic gate used an input pulse of 1.4 V superimposed upon a dc voltage of -1 V to study the gate's dynamic response. One load gate was attached to the output of the test gate. Zero-bias capacitance values of 0.5 FF were used for the Schottky diodes and also for C_{GS} and C_{GD} of each transistor. The simulation gave a propagation delay of $t_P = 300$ ps. Combining this with the 1.1 mW dissipation gave a *PDP* of 0.33 pJ for the test gate.

As with NMOS, we can obtain the NOR function by adding input transistors in parallel with B_1 in Fig. 13.20a. Presently, fan-in and fan-out are limited to two or three.

13.4.3 DIRECT-COUPLED FET LOGIC

A simpler gate structure and a single supply are advantages of the *direct-coupled FET logic* gate (DCFL) of Fig. 13.23. Input and output levels are compatible in this gate because V_t is positive for enhancement transistor B_1. Adding input transistors in parallel with B_1 gives the NOR function. Since the circuit is conceptually identical to the NMOS gate of Fig. 13.9a, except for clamping of the output voltage when gate current flows into the Schottky gate-source junction of the load gate, we leave its exploration for Problem 13.40.

13.5 _____
TTL Logic
Circuits

Transistor-transistor logic (TTL) is the longest-lived and one of the most successful of logic families. Contributing to its initial success were its use of a single 5 V supply and a very workable compromise between speed and power dissipation. Its ability to drive off-chip loads enabled it to dominate discrete logic applications involving SSI and MSI, and it is still the de facto standard for interfacing between computers and other devices. Another reason for TTL's longevity has been its ability to evolve into newer forms according to application needs and improvements in technology. Early-on, low-power and high-speed subclasses of TTL allowed designers to apply TTL to special classes problems without having to master hardware design details for new families. A more recent evolution in TTL involved replacing certain transistors with *Schottky clamped transistors,* a change that led to a significant improvement in speed while retaining the same general circuit structures and design rules. Classical TTL gates established robust logic 0 levels using saturated transistors as closed switches; however, minority charge stored in the bases of these transistors resulted in zero-to-one propagation delays much larger than the one-to-zero delays. As we will see, Schottky clamping gives a nonsaturating transistor switch that resembles the classical switch in every respect but charge storage delay. Because classical TTL has today been largely supplanted by this *Schottky TTL* (STTL), we introduce TTL principles using Schottky transistors. A glance at Fig. 13.37e, which compares the responses of classical and Schottky TTL inverters to the same input pulse, should provide adequate motivation for the following study of Schottky clamping.

13.5.1 SCHOTTKY CLAMPED TRANSISTOR

Figure 13.24a shows a Schottky clamped transistor and Fig. 13.24b its schematic representation. We learned in Chapter 3 that the silicon Schottky diode has two special features: an ON voltage of 0.4 V and rapid switching by virtue of employing no minority

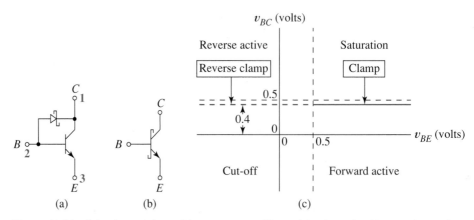

Figure 13.24 Schottky transistor: (a) components; (b) special schematic; (c) operating regions compared to ordinary bipolar transistor.

charge carriers. We now show that the Schottky transistor is a four-state device, having two familiar and two unfamiliar states.

Figure 13.24c shows how the clamping diode modifies the transistor's *ordinary* four-state behavior, as originally defined in Fig. 4.4. When $v_{BC} < 0.4$ V, the diode is open, and the transistor operates in its usual forward active or cut-off states. However, because of the Schottky diode, v_{BC} cannot exceed the 0.4 V level indicated by the bold horizontal line. Therefore, the transistor can never saturate or operate in its reverse active mode. Instead, we have two new transistor states defined along the bold line where the Schottky diode conducts. One state, called the *clamp state*, obtains when $v_{BE} \geq 0.5$ V; the other, *reverse clamp*, applies when $v_{BE} \leq 0.5$ V. We now explore these new transistor states.

Clamp Operation. We can model the clamped transistor as in Fig. 13.25a. Writing KCL at base and collector nodes gives the matrix equation

$$\begin{bmatrix} 1 & 1 \\ -1 & \beta_F \end{bmatrix} \begin{bmatrix} i_D \\ i_{BI} \end{bmatrix} = \begin{bmatrix} i_B \\ i_C \end{bmatrix}$$

which has the solution

$$i_D = \frac{\beta_F i_B - i_C}{\beta_F + 1} \tag{13.31}$$

In addition to $v_{BE} \geq 0$, a necessary condition for clamp is that the Schottky diode conduct; that is, $i_D \geq 0$. From Eq. (13.31), this means

$$\beta_F i_B \geq i_C \tag{13.32}$$

The *clamp* state thus *resembles* transistor saturation; however, the definition of i_B in Fig. 13.25a shows that it *applies to the Schottky transistor as a whole.*

There is another similarity between a clamped Schottky transistor and an ordinary saturated BJT. Notice from Fig. 13.25a that for the clamped transistor, $V_{CE,clamp} = 0.7 - 0.4 = 0.3$ V, a constant voltage that is independent of i_C. This means we can use

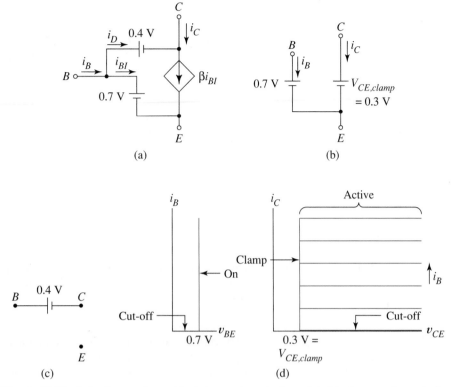

Figure 13.25 Schottky transistor: (a) clamp-mode analysis model; (b) clamp-mode equivalent; (c) reverse clamp equivalent circuit; (d) input and output characteristics.

the simpler equivalent of Fig. 13.25b to represent the clamped Schottky transistor, provided that Eq. (13.32) is satisfied. After comparing this model with Fig. 4.10, we conclude that clamp in a Schottky transistor is nearly identical to saturation in an ordinary transistor. The main difference is that the Schottky transistor can switch out of clamp very quickly because diffusion (charge storage) capacitance is not involved in switching.

Incidentally, it is possible for a clamped transistor to have a negative collector current. If the base-collector junction is very heavily forward biased, i_D in Fig. 13.25a can exceed βi_B. Figure 4.8 showed that a saturated transistor can also have negative collector current if base-emitter forward bias is sufficiently high.

Reverse Clamp. Figure 13.24c shows that if the clamping diode is ON and the base-emitter junction is reverse biased, another state is possible. Since the underlying BJT is in cut-off, the circuit model for reverse clamp is that of Fig. 13.25c.

Active and Cut-Off States. If Eq. (13.32) is not satisfied, the clamping diode is OFF, and we have the BJT alone, which can be either active or cut off depending upon whether or not $i_B \geq 0$. In these regions, we model the Schottky transistor by the usual active and cut-off BJT models, respectively.

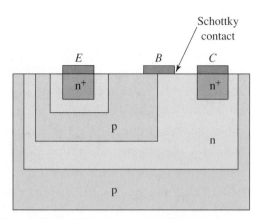

Figure 13.26 Physical structure of Schottky transistor.

From this development we conclude that we can visualize the Schottky transistor of Fig. 13.24b as a device with input and output characteristic curves like those of Fig. 13.25d. Compared with an unclamped transistor, the only difference is that $V_{CE,sat} \approx 0.2$ V is replaced by $V_{CE,clamp} \approx 0.3$ V.

Schottky Transistor Physical Structure. Figure 13.26 shows that no special steps are required to fabricate a Schottky transistor. The aluminum or platinum base contact is simply allowed to overlap base and collector regions during metalization. The contact with the base silicon is ohmic; however, since the collector is lightly doped in the overlap region, the contact to the collector forms a rectifying Schottky junction, making the device functionally equivalent to Fig. 13.24a.

13.5.2 TRANSFER CHARACTERISTIC OF BASIC TTL INVERTER

Figure 13.27a shows a simplified TTL inverter. The circuit develops logic 1 output by causing Q_2 to cut off and logic 0 by clamping Q_2. The analysis that follows assumes transistors and diode parameters are

$$\beta_F = 30, \quad V_{CE,clamp} = 0.3 \text{ V}, \quad V_{BE,on} = 0.7 \text{ V}, \quad \text{and } V_D = 0.4 \text{ V}$$

Logic 1 Output. Assume v_I is a logic-zero voltage produced by another TTL gate; then $v_I = V_{CE,clamp} \approx 0.3$ V. Because the circuit is an inverter, *assume* Q_2 is cut off. This gives Fig. 13.27b, where we assume the diode is OFF because the base current of Q_2 is negligible.

In Fig. 13.27b, the circuit appears to forward-bias the base-emitter junction of Q_1, so we reason that this transistor is ON. Then

$$i_{B1} = \frac{5 - (0.3 + 0.7)}{2.8 \text{ k}} = 1.43 \text{ mA} \tag{13.33}$$

Because the leakage current from the diode and cut-off transistor ($i_{C1} \approx 10^{-16}$ A) ensures that $30i_{B1} > i_{C1}$ we conclude that Q_1 is clamped.

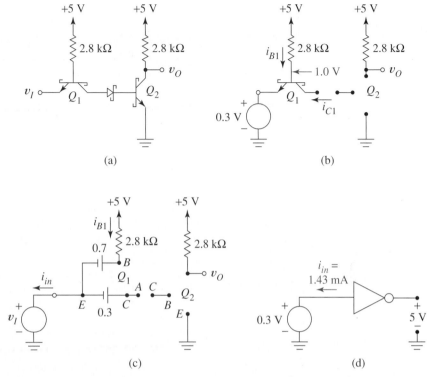

Figure 13.27 Simplified TTL inverter: (a) schematic; (b) with Q_2 cut off; (c) with Q_1 clamped and Q_2 cut off; (d) input current when input is logic 0.

Because Q_1 clamps, we draw Fig. 13.27c. At the input we see that v_I must be able to *sink* (that is, absorb) a current $i_{IN} = i_{B1}$, which from Eq. (13.33) is

$$i_{IN} \approx 1.43 \text{ mA} \tag{13.34}$$

Figure 13.27d shows this important result from the perspective of the gate user.

We can use Fig. 13.27c to verify that Q_2 is cut off, because when $v_I = 0.3$ V, the resulting anode voltage of 0.6 V is insufficient to turn on both the diode and the transistor. The equivalent circuit also shows that as we increase v_I, the voltage at the diode's anode increases and eventually reaches a value sufficient to turn on both diode and transistor. If the cut-in voltage of Q_2 is 0.5 V, this takes

$$v_I + 0.3 = 0.4 + 0.5 \text{ V} \quad \text{or} \quad v_I \approx 0.6 \text{ V}$$

We have just explained segment 1 of the gate's static transfer characteristic, Fig. 13.28. Because $v_O = v_{CE2}$ is large at cut-in, Q_2 changes to active operation for $v_I = 0.6^+$ volts, giving segment 2. Instead of examining this region in detail, it is simpler to skip to segment 3.

Logic 0 Output. For a logic 1 input such as 5 V, Q_2 should be clamped, giving Fig. 13.29a. With Q_2 clamped, i_{B2} must be positive and i_{C1} negative. Since the Schottky

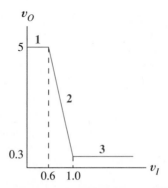

Figure 13.28 Transfer characteristic for simplified inverter when there is no output loading.

diode conducts when $v_{BC1} = 0.4$ V and the (parallel) base-collector junction only when $v_{BC1} = 0.5$ V, we reason that Q_1's Schottky diode is conducting. This makes

$$V_{B1} = 0.7 + 0.4 + 0.4 = 1.5 \text{ V}$$

If this is true,

$$v_{BE1} = 1.5 - 5 = -3.5 \text{ V}$$

and the base-emitter junction is reverse biased. We therefore assume reverse clamp for Q_1 and use the model of Fig. 13.25c to draw Fig. 13.29b. The currents and voltages in this circuit confirm the reverse clamp assumption. The base current of the output transistor is

$$i_{B2} = \frac{5 - 1.5}{2.8 \text{ k}} = 1.25 \text{ mA} \tag{13.35}$$

and

$$i_{C2} = \frac{5 - 0.3}{2.8 \text{ k}} = 1.68 \text{ mA} \tag{13.36}$$

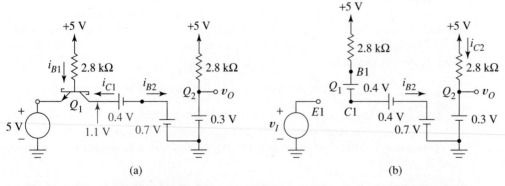

(a)   (b)

Figure 13.29 Equivalent circuit for logic 0 output: (a) assuming Q_2 is clamped; (b) assuming also that Q_1 is reverse clamped.

Because

$$\beta_F i_{B2} = 30(1.25 \text{ mA}) = 37.5 \text{ mA} > i_{C2}$$

Q_2 is indeed in clamp. As v_I in Fig. 13.29b decreases, Q_1 remains in region 3 of Fig. 13.28; however, we have yet to justify region 3's lower limit, $v_I = 1.0$ V.

As v_I in Fig. 13.29b is reduced from 5 V to lower values, the reverse bias of the base-emitter junction of Q_1 diminishes. Eventually the cut-in value is reached, and Q_1 begins to change state. This happens when

$$v_{B1} - v_I = 1.5 - v_I = 0.5 \text{ V}$$

that is, when $v_I = 1.0$ V. This completes the explanation of segment 3 in Fig. 13.28. During the transition between logic 1 and logic 0 output, Q_2 goes through forward active operation, and the circuit functions as an inverting amplifier, giving segment 2.

13.5.3 STATIC LOADING EFFECTS

The transfer characteristic of Fig. 13.28 is a *no-load* characteristic, because the inverter in our analysis had no external load gates. We now examine the effects of loading at the inverter output.

In TTL logic, *fan-out* is limited by the amount of current the output transistor of the source gate can sink (from external load gates at logic 0) without leaving clamp. Figure 13.30a shows gate S trying to maintain logic 0 while sinking currents from N load gates. Figure 13.30b shows the output equivalent circuit of S. Because loading does not affect the input of S, the output transistor has base current of 1.25 mA as given by Eq. (13.35). To remain in clamp it is necessary that $\beta i_B \geq i_C$, *even with the external load.* From Fig. 13.30b this means

$$30(1.25 \times 10^{-3}) \geq N \, 1.43 \times 10^{-3} + \frac{5 - 0.3}{2.8 \text{ k}}$$

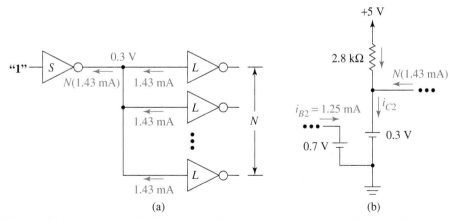

Figure 13.30 Fan-out: (a) loading at logic 0 output; (b) equivalent circuit for calculating fan-out.

which gives fan-out of $N \leq 25.05$ load gates, or an integer maximum of $N = 25$ gates. A lower value of N results for the more practical case when uncertainties in component values are considered.

13.5.4 TTL OUTPUT CIRCUITS

TTL gates employ two kinds of output circuit, *passive pull-up* and *active pull-up*. Passive pull-up is simpler; active is more common. The main functional difference is the speed with which the gate charges capacitances associated with external loads. As in preceding studies of gate dynamics, we simplify the TTL circuit by lumping all internal and load capacitances into a single lumped-equivalent, C_L.

Passive Pull-Up. Passive pull-up employs an output transistor and a *pull-up resistor,* most often an external component attached by the user between collector and supply, as in Fig. 13.31a. An input circuit, not shown in the figure, translates input logic levels into appropriate base currents to drive the output transistor. Points *a* and *c* in Fig. 13.31b show the equilibrium logic 1 and 0 operating points, respectively.

   Assume that the inverter is initially in equilibrium with output at logic 1. Q_2 is cut off and C_L charged to 5 V. If a change in the gate's input voltage causes base current to increase suddenly to $i_B = I_1$, v_O cannot change instantaneously to its logic 0 value at point *c* because v_O is the voltage across a capacitor. As in the NMOS gate described by Fig. 13.6d, the output transistor temporarily goes active, and operation jumps to point *b* in Fig. 13.31b. Once the transistor is active, C_L discharges quickly according to

$$-C_L \frac{dv_O}{dt} = \beta I_1 - \frac{5 - v_O}{R_{PU}} \approx \beta I_1$$

to point *c*. The resulting propagation delay, t_{PHL}, is small, and there is no problem with this transition.

   The serious limitation of passive pull-up shows up during the zero-to-one output transient. When Q_2, initially clamped, suddenly cuts off, C_L simply charges through R_{PU} to point *a* with time constant $R_{PU}C_L$. The only way to make the time constant, and,

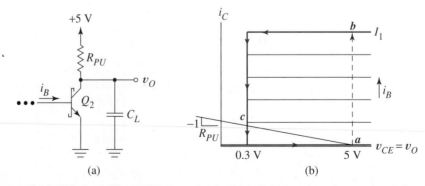

Figure 13.31 Passive pull-up: (a) output circuit; (b) switching trajectories on output characteristics.

therefore, t_{PLH} small, is to make R_{PU} very small; however, this increases the current co-ordinate of point *c*, raising the power dissipation at logic 0 to a high level. In interfacing circuits, passive pull-up is useful because a slow zero-to-one transition or high power dissipation can often be justified—a slow transition, by slow peripheral equipment—high power by the rarity of the logic 0 event. We examine interfacing applications in Chapter 14 that further clarify these ideas. For ordinary logic decisions within a digital circuit, however, hundreds or even thousands of gates might be involved, so high dissipation per gate and slow decision making are both prohibitive. For these more routine decision-making applications, the active pull-up circuit examined next gives small propagation delays for both transients, and low power dissipation as well.

Active Pull-Up. Figure 13.32a is a simplified schematic for an active pull-up, TTL output circuit. Darlington pair Q_4–Q_3 stacked above output transistor Q_2 is called a *totem pole* output stage. Transistor switch Q_5, controlled by the gate input signal, opens to produce a logic 1 output and closes for logic 0.

Output Zero-to-One Transient. Assume that C_L has previously been discharged to 0.3 V by a clamped transistor Q_2. Suddenly a logic 0 gate input causes Q_5 to open. The base current of Q_2 disappears, and Q_2 cuts off as in Fig. 13.32b. Current i_{B4} allows Darlington-pair emitter follower, Q_4–Q_3, to charge the load capacitance. With v_O initially held to 0.3 V by C_L, Q_4 and Q_3 turn on, with base-emitter voltages as shown on the diagram. The base current of Q_4 is

$$i_{B4} = \frac{5 - (1.4 + v_O)}{900} = \frac{2.6 - v_O}{900} \tag{13.37}$$

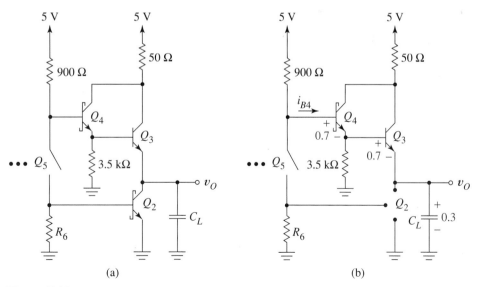

(a) (b)

Figure 13.32 Totem pole output circuit with driver: (a) simplified schematic; (b) equivalent for logic 1 output; (c) for logic 1 output showing Q_4 clamped; (d) Thevenin equivalent of (c); (e) circuit for logic 1 output when Q_4 is active; (f) Thevenin equivalent of (e).

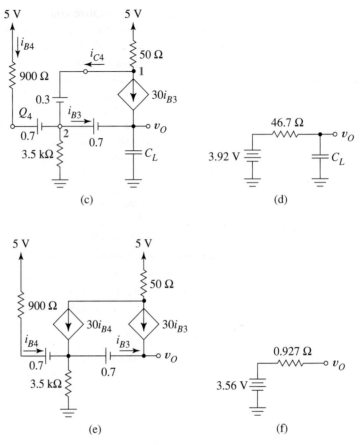

Figure 13.32 (continued)

When the transient begins, $v_O = 0.3$ V; therefore, $i_{B4} = 3.67$ mA. With such a high base current, we should suspect Q_4 is clamped. If so, Q_3 is active because $v_{CE3} = 0.7 + 0.3 = 1.0$ V. (Q_3 is not a Schottky transistor because it cannot saturate in this circuit.) These assumptions give the equivalent of Fig. 13.32c.

To verify that Q_4 is clamped, we apply KCL at nodes 1 and 2, giving

$$\frac{5 - (v_O + 0.7 + 0.3)}{50} = i_{C4} + 30i_{B3} \tag{13.38}$$

and

$$i_{B4} = \frac{5 - (v_O + 0.7 + 0.7)}{900} = i_{B3} - i_{C4} + \frac{v_O + 0.7}{3.5\text{ k}} \tag{13.39}$$

Evaluating both expressions when $v_O = 0.3$ V and solving for i_{C4} gives

$$i_{C4} = -0.9\text{ mA}$$

We previously noted that negative collector current is possible for a clamped transistor. Because $\beta i_{B4} = 30(0.367 \text{ mA}) > i_{C4}$, we conclude that the transistor states are correct. From Eqs. (13.38) and (13.39) we find that $i_{B3} = 2.49$ mA when $v_O = 0.3$ V. At this initial instant, the charging current for C_L is $31 i_{B3} = 77.2$ mA! In Problem 13.44 we discover that the Thevenin equivalent charging circuit is Fig. 13.32d.

Equation (13.37) indicates that base current i_{B4} diminishes as C_L charges and v_O rises, suggesting that Q_4 might come out of clamp before the transient ends—in fact this happens. In Fig. 13.32e we examine the remainder of the transient by replacing the clamp model for Q_4 with the active model. This leads to the final Thevenin circuit of Fig. 13.32f. The capacitance charges to a final value of 3.6 V, and Q_4 remains on the borderline between forward active and cut-off states, prepared to deliver current if v_O should decrease slightly, say, through leakage from the load capacitance.

To appreciate the totem pole's ability to charge a load capacitance quickly, compare Figs. 13.32d and f with the passive pull-up circuit of Fig. 13.27b. The totem pole time constants are two orders of magnitude smaller for the same C_L. Changing the 2.8 k resistor to a value such as 30 Ω is not feasible because this would make the gate's power dissipation excessive at logic 0.

Exercise 13.3 Find the logic 0 output circuit dissipation in Fig. 13.29b, both with the 2.8 k resistor and with a 30 Ω collector-resistor replacement.

Ans. 8.39 mW, 783 mW.

Output One-to-Zero Transient. With C_L initially charged to a logic 1 value of 3.6 V in Fig. 13.32a, a logic 1 applied to the gate input closes Q_5, which in turn supplies base current to Q_2. Because Q_2 is ON with $v_{CE} = 3.6$ V, it is active, so we visualize the circuit as in Fig. 13.33a. Notice that the base of Q_4 is at 0.7 V and, since the 3.5 k resistor

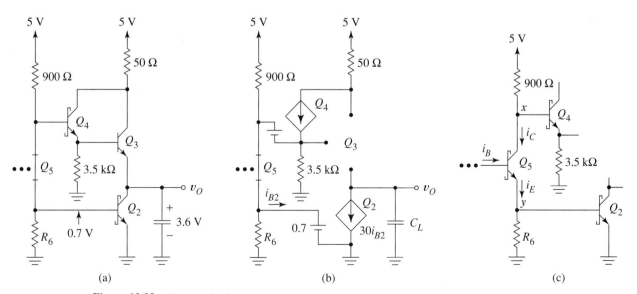

Figure 13.33 Totem pole during one-to-zero output transient: (a) initial condition; (b) equivalent circuit during capacitor discharge; (c) phase inverter driving circuit.

provides a path from its emitter to ground, Q_4 is ON. However, this makes the base of Q_3 negative relative to its emitter, so Q_3 cuts off. The resulting equivalent circuit, Fig. 13.33b, shows that there is no load resistor R_{PU} simultaneously trying to charge C_L from the power supply as in the passive pull-up circuit. Consequently, there is a rapid discharge of C_L at constant current down to a logic 0 of 0.3 V or so, at which point Q_2 clamps.

In Fig. 13.33c, transistor switch Q_5, together with its collector and emitter resistors, forms a circuit called a *phase inverter*. During transitions between clamp and cut off, Q_5 operates in its forward active mode as an amplifier with two outputs that act in phase opposition. As its base current increases, i_C and i_E both increase, causing the voltage at node x to decrease while the voltage at y increases. This tends to simultaneously turn Q_2 on and Q_4 off. Reducing base current of Q_5 has the opposite dual effect. The phase inverter gives a static transfer characteristic featuring a very abrupt transition from high to low output values, a feature that increases the fraction of the logic swing that pertains to noise margins in Eq. (13.2), and reduces the overhead ΔV.

13.5.5 TTL INVERTER WITH TOTEM POLE OUTPUT

Static Transfer Characteristic. Figure 13.34a shows a complete TTL inverter with totem pole output. Input transistor Q_1 operates between clamp and reverse clamp just as it did in the resistive pull-up circuit of Fig. 13.27a. The emitter resistor in the phase inverter is replaced by a *squaring circuit* consisting of Q_6 and its two resistors for reasons that we explain presently. With a steady-state logic 0 input, Q_1 clamps, diverting most of Q_1's base current to the input node; consequently, Q_5, Q_6, and Q_2 all cut off, giving Fig. 13.34b. Notice that v_I must sink 1.43 mA, the same current we predicted from the introductory circuit of Fig. 13.27b.

As v_I is increased, the voltage at node 6 increases until it equals the sum of the cut-in voltages of Q_2 and Q_5. When v_I satisfies

$$v_I + 0.3 \approx 0.5 + 0.5$$

that is, when $v_I \approx 0.7$ V, Q_2 and Q_5 begin to turn on. The *squaring circuit* consisting of Q_6 and its associated resistors also turns on at the same time. (If there were a simple resistor between node 5 and ground instead of the squaring circuit, Q_5 would turn on for $v_I = 0.2$, but Q_2 would not cut in until the input increased by another 0.5 V. This was the case in the classical TTL circuit.) When v_I reaches $2(0.7) - 0.3 = 1.1$ V, Q_2, Q_5, and Q_6 are all active and constitute an inverting amplifier between node 6 and the output. With further increases of v_I, the internal behavior is complex, with various transistors changing states as the output voltage drops. The circuit eventually reaches the final equilibrium of Fig. 13.34c, with the clamped output transistor producing the logic 0 output level. Notice that Q_3 is cut off and that the emitter current of Q_4 is 0.3/3.5 k = 86 μA, very small as we assumed earlier using simpler models.

When v_I decreases in Fig. 13.34c, there can be no change in the output until Q_1 changes state. The base-emitter voltage of Q_1 reaches cut-in when v_I satisfies

$$v_{B1} - v_{E1} = [2(0.7) + 0.4] - v_I = 0.5 \text{ V}$$

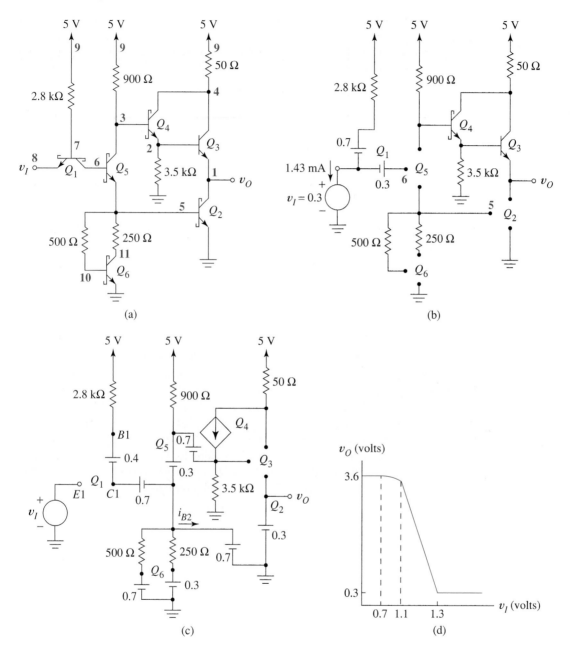

Figure 13.34 TTL inverter with active pull-up: (a) schematic; (b) equivalent when input = "0"; (c) equivalent when input = "1"; (d) transfer characteristic; (e) equivalent with N load gates.

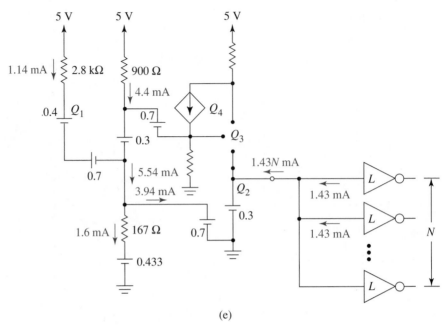

(e)

Figure 13.34 (continued)

or $v_I = 1.3$ V. To summarize, we have now deduced that the transfer characteristic resembles Fig. 13.34d.

Using Fig. 13.34e we estimate the fan-out. This figure is Fig. 13.34c with the squaring circuit replaced by its Thevenin equivalent, and N load gates attached to the output. With totem pole output, no current flows directly from the 5 V power supply into the collector of Q_2. Thus the entire quantity, βi_{B2} is available for sinking load current. Since node voltages are known, by ignoring the small base current of Q_4 it is easy to verify the currents in Fig. 3.34e. We conclude that the output will remain in clamp for fan-out N that satisfies

$$30(3.94) \geq 1.43\,N$$

or $N = 82$ load gates. Considering worst case variations in β and other components, variations over temperature, and more accurate nonlinear analysis, the practical fan-out reduces to about 10.

The TTL gate of Fig. 13.34a was a significant improvement over the simpler classical TTL gate of Fig. 13.35 that used saturating transistors. Structures and operating principles of the two circuits are very similar; however, to avoid the long propagation delay associated with minority charge storage, clamped transistors in the Schottky circuit replace saturated TTL transistors. The squaring circuit involving Q_6 eliminated a TTL transfer characteristic problem in which Q_5 turned on at a lower input voltage than Q_2. Examples that follow compare the two circuits.

EXAMPLE 13.4 (a) Construct a SPICE subcircuit for a Schottky clamped transistor. The npn transistor has nondefault parameters: BF = 30, TF = 0.2 ns, TR = 20 ns, RC = 20 Ω, CJE = 0.3 pF, and CJC = 0.15 pF.

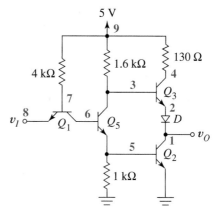

Figure 13.35 Classical TTL inverting gate.

(b) Use this subcircuit to find the static transfer characteristic of Fig. 13.34a.
(c) Compute the transfer characteristic of the classical TTL gate of Fig. 13.35, and compare the two characteristics.

Solution. (a) Figure 13.36a describes the Schottky transistor with node numbering as in Fig. 13.24a. There are two special items in the .MODEL statement for SDIODE. The large reverse saturation current typical of a Schottky diode reduces the diode's static forward voltage. The explicit statement TT = 0 underscores that the diode has no diffusion capacitance. The model does include depletion capacitance by virtue of CJO. Notice that the model for TRAN includes both diffusion and depletion capacitance.
(b) Figure 13.36b shows selected parts of the code used to find the transfer characteristic of Fig.13.34a. The non-Schottky transistor has the parameters specified in the problem statement and includes depletion capacitance. The solid curve of Fig. 13.36c shows the result, a curve similar to our hand-analysis prediction, Fig. 13.34d.
(c) The dashed curve of Fig. 13.36c is the computed characteristic of the classical TTL gate of Fig. 13.35 using identical BJT .MODEL statements but no diode clamps. Two distinct curve segments correspond to successive turn-on of Q_5 and Q_2, creating a broad transition region between logic 1 and logic 0 output. By contrast, the squaring circuit causes a single region of very high gain, more like the ideal of Fig. 13.1a. V_{OL} is close to zero for the TTL gate instead of the 0.2 V that we would have estimated using a saturation model, because with no load the collector current went to zero even though the base current was high, bringing the Q-point of the output transistor to $(v_{CE}, i_C) = (0, 0)$. ❑

Power-Delay Product. Because the transient behavior of Fig. 13.34a is too complex for realistic hand analysis, we explore gate dynamics by simulation.

EXAMPLE 13.5 (a) Use SPICE to find the power-delay product for the TTL inverter of Fig. 13.34a. Have v_I switch between 0.3 and 3.6 V for the propagation delay measurement.
(b) Use SPICE to determine the power-delay product for the classical TTL inverter of Fig. 13.35.

⋮

```
*******
·SUBCKT SCHOT 1 2 3
DS 2 1 SDIODE
·MODEL SDIODE D IS=2E-11
+CJO=1PF TT=0
Q 1 2 3 TRAN
·MODEL TRAN NPN BF=30
+TF=0·2N TR=20N RC=20
+CJE=0·3P CJC=0·15P
·ENDS
*******
```

(a)

```
XS1 6 7 8 SCHOT
XS5 3 6 5 SCHOT
XS6 11 10 0 SCHOT
XS4 4 3 2 SCHOT
XS2 1 5 0 SCHOT
Q3 4 2 1 NTRAN
·MODEL NTRAN NPN BF=30
+RC=20 TF=0·2N TR=20N
+CJE=0·3P CJC=0·15P
VI 8 0 DC 0
·DC VI 0 3 0·05
·PLOT DC V(1)
·END
```

(b)

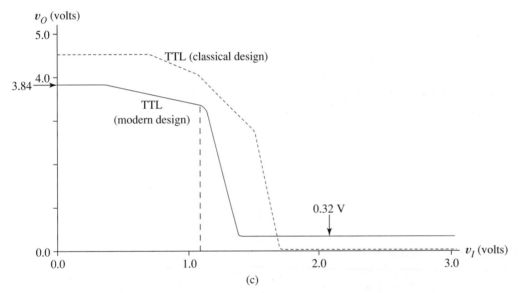

(c)

Figure 13.36 TTL simulation: (a) subcircuit description of a Schottky transistor; (b) partial SPICE code for Fig. 13.34a; (c) transfer characteristics of modern and classical TTL inverters.

Solution. (a) Successive dc analyses with VI set to logic 0 and then to logic 1 values in Fig. 13.36b gave dissipations values of 12.6 mW and 27.1 mW, respectively, for average dissipation of 19.9 mW. For the transient investigation, the VI, .DC, and .PLOT statements in Fig. 13.36b were replaced by

```
VI 8 0 PULSE(0.3 3.6 10N 1N 1N 30N 210N)

.TRAN 0.4N 80N

.PLOT TRAN V(1)
```

A trial run or two, to ensure that dynamic equilibrium was achieved during the pulse, led to the choice of a 30 ns pulse width. A delay of 10 ns was added to make the leading edge easy to see. The 80 ns duration of the transient analysis was selected to ensure that the trailing edge of the output could be observed, and the step size of 0.4 ns was selected to give 200 output points.

The transient analysis gave the solid curve of Fig. 13.37. From the data we estimate that the midpoint of the leading edge of the output pulse occurred at about 11.7 ns, the midpoint of the trailing edge at 42.2 ns. Thus we estimate $t_{PHL} \approx 1.7$ ns and $t_{PLH} \approx 2.2$ ns for a propagation delay of $t_P = 2$ ns. Combining this with the dissipation of 19.9 mW gives $PDP = 39.8$ pJ.

(b) For classical TTL, the average dissipation was 4.81 mW with output at "1" and 18.5 mW with output at "0," for average dissipation of 11.7 mW.

For the same input pulse, the classical TTL output is the dashed curve in Fig. 13.37. The large storage time associated with the "0" to "1" transition was the major problem that Schottky clamping was designed to correct. The first part of the storage time is the delay associated with bringing Q_5 in Fig. 13.35 out of saturation. This is fairly short because Q_1 spends a brief interval in the forward active state, providing a current source to remove base current from Q_5 at a high rate. Once Q_5 is out of saturation, a second and longer plateau results as stored charge leaves the base of Q_2. Even though the npn transistors in both simulations have identical SPICE descriptions, we see that Schottky clamping removes both of these storage delays.

Data in the output file show the delays for the classical circuit were $t_{PHL} \approx 0.16$ ns and $t_{PLH} \approx 23.4$ ns, for $t_P = 11.8$ ns compared to 2.2 for TTL. For classical TTL,

$$PDP = (11.7 \times 10^{-3})(23.4 \times 10^{-9}) = 138 \text{ pJ}$$

Although not required by the problem statement, Fig. 13.38 shows the power supply current of the simulated TTL gate during the output pulse. TTL users customarily employ special power supply decoupling capacitors to prevent the abrupt transitions from being inadvertently coupled into other circuits by the mechanism of Fig. 10.30. We must

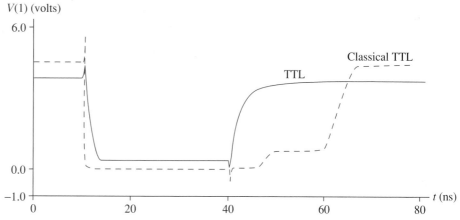

Figure 13.37 Comparison of TTL and classical TTL responses to the same input pulse.

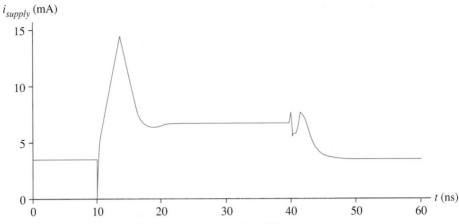

Figure 13.38   Power supply current for a TTL gate with a pulse input.

also be concerned about the derivative of the current waveform, because voltages magnetically induced in adjacent circuits (Fig. 1.28a) are proportional to the rate of change of current. From data in the output file we find that the leading edge of the large current spike in Fig. 13.38 increases 8.26 mA in 2.3 ns, giving a rate of change of 3.58×10^6 A/s! Problem 13.46 suggests exploring the supply current when the gate has an external load capacitance.   ❏

13.5.6 MULTIPLE-INPUT NAND STRUCTURE

The inverter of Fig. 13.34a becomes a NAND, when input transistor Q_1 is replaced by the multiple-emitter Schottky transistor of Fig. 13.39a. This structure is equivalent to connecting the base-collector junctions of three Schottky transistors in parallel as in

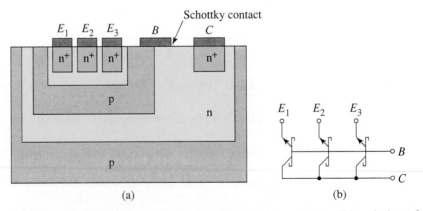

Figure 13.39   TTL NAND gate: (a) multiple-emitter transistor structure; (b) equivalent of a three-emitter transistor; (c) TTL NAND; (d) input circuit with all inputs at "1"; (e) equivalent input circuit when one input is low.

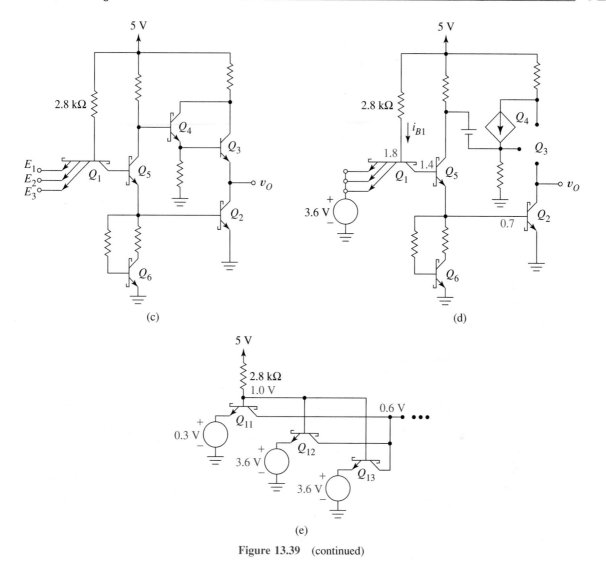

Figure 13.39 (continued)

Fig. 13.39b. In fact, this is how to model a multiple-emitter transistor in SPICE simulations. Figure 13.39c shows the complete three-input NAND.

To understand how the multiple-input NAND works, consider Fig. 13.39d where all inputs are logic 1. With Q_3 in cut-off; Q_4 nearly so; and Q_2, Q_5, and Q_6 clamped to provide the logic 0 output, Q_1 provides the input currents for the clamped transistors. Its base voltage is 1.8 V, and all three emitter-base junctions of Q_1 are reverse biased, placing the device in the reverse clamp mode. Base current i_{B1} is the same as for the circuit of Fig. 13.34c, which had a single input.

When at least one input is low, the input circuit is related to Fig. 13.39e. Q_{11} (and any other input transistor with low emitter voltage) has its base-emitter junction forward biased, bringing the bases of all three transistors to 1 V. Clamping of the transistor with low input causes the (joint) collector voltage to approximate 0.6 V. Any input transistor with high emitter voltage then has its base-emitter junction reverse biased, and its base-collector junction clamped by the Schottky diode. This implies reverse clamp operation. The low collector voltage ensures that Q_2 and the transistors in its driving circuit in Fig. 13.39c are cut off, and the output is logic 1. In summary, the circuit output is low only if all inputs are high, giving the NAND function.

Exercise 13.4   (a) In Fig. 13.39e, find the emitter current of Q_{11}.
Hints: Q_5, which uses the collector currents for its base current, is cut off. Draw the equivalent circuit assuming correct states for the input transistors. (b) What would be the value of the emitter current of Q_{11} if emitters of Q_{11} and Q_{12} are both low? (c) What would be the emitter current of Q_{11} if emitters of Q_{11} and Q_{12} and Q_{13} are all low?

Ans. 1.43 mA, 0.715 mA, 0.477 mA.

13.6
Emitter-Coupled Logic

Emitter-coupled logic (ECL) is a bipolar logic family that features high speed at the expense of power dissipation and noise margins. Transistors are restricted to active or cut-off states by circuit design (instead of Schottky clamping); therefore, as in modern TTL, ECL avoids the long propagation delays introduced by saturated transistors. Small resistors allow large currents to charge and discharge transistor capacitances quickly. Because of these large currents, we change our modeling strategy slightly. For ECL we assume that the base-emitter drop of an ON transistor is 0.9 V and cut-in is 0.7 V instead of the 0.7 V and 0.5 V, respectively, we use when currents are smaller.

Basic Operation. Figure 13.40 shows a basic ECL noninverting logic *buffer.* Emitter-coupled transistors, Q_1 and Q_2, operate between cut-off and active states. Although structurally the same as a difference amplifier, this circuit is sometimes called a *current switch* to emphasize its large-signal operating mode. Q_3 is an active-state, common-collector buffer that provides low output resistance, and V_R is a temperature-compensated, on-chip reference voltage. The 50 Ω resistor and -2 V source represent the load seen by the ECL gate in a practical circuit environment, often a matched transmission line with a 50 Ω characteristic resistance. ECL employs a negative power supply to reduce noise, an interesting feature explored in Problem 13.52.

The gate works as follows. When v_I is low, Q_1 cuts off and Q_2 is active with a relatively large collector current. This current flows through the 245 Ω resistor, dropping the collector of Q_2 below ground. Since Q_3 is always active, v_O is one V_{BE} drop below Q_2's collector. When v_I is a high voltage, Q_1 is active, and Q_2 cuts off. Q_2's collector current goes to zero, and v_O is approximately one V_{BE} drop below ground. We now use circuit analysis to confirm this general description and discover some additional features.

Static Transfer Characteristic. To develop the static transfer characteristic of the inverter, first consider the transistor states when v_I is much more negative than V_R, say, $v_I = -5$ V. Because voltages v_I and V_R suggest that the base current of Q_2 is larger than that of Q_1, we guess that Q_1 is OFF and Q_2 active. This gives Fig. 13.40b. The node voltage at the emitter of Q_1 is

$$V_E = -1.29 - 0.90 = -2.19 \text{ V}$$

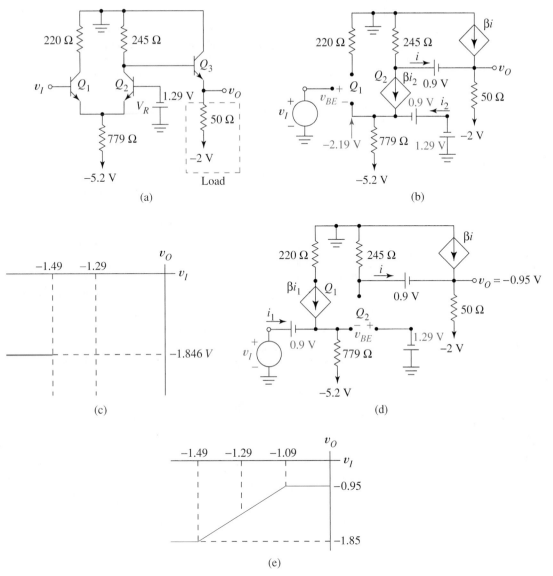

Figure 13.40 Emitter-coupled logic: (a) basic circuit; (b) equivalent when input is low; (c) transfer characteristic for low input; (d) equivalent when input is high; (e) complete transfer characteristic.

This verifies that Q_1 is cut off when $v_I = v_B = -5$ V. Furthermore, the circuit shows that Q_1 *remains* cut off until

$$v_{BE} = v_I - (-2.19) = V_\gamma$$

the cut-in voltage. Since we use $V_\gamma = 0.7$ for ECL, Q_1 remains in cut-off for

$$v_I \leq -1.49 \text{ V} \tag{13.40}$$

We next find the emitter current of Q_2. For high β we can ignore base current i_2. Then the emitter current, which flows through the 779 Ω resistor, is

$$i_{E2} = \frac{-2.19 - (-5.2)}{779} = 3.86 \text{ mA}$$

High beta also implies

$$i_{C2} \approx i_{E2} = 3.86 \text{ mA}$$

Ignoring the base current i of Q_3 gives

$$v_O + 0.9 + 245 \,(3.86 \text{ mA}) = 0$$

or $v_O = -1.846$ V. To verify the assumption that Q_2 is active, we note that

$$v_{CE2} = v_C - v_E = (-1.846 + 0.9) - (-2.18) = 1.24 \text{ V}$$

which exceeds $V_{CE,sat} = 0.2$ V. To summarize, the transfer characteristic is $v_O = -1.846$ V for $v_I \le -1.49$ V as in Fig. 13.40c.

We next assume that v_I is so positive compared to V_R that Q_1 is active and Q_2 is cut off, that is, Fig. 13.40d. Notice how the output is independent of v_I. Summing voltages from the -2 V source and using $\beta = 100$ gives

$$-2 + 50(101)\,i + 0.9 + 245i = 0 \qquad (13.41)$$

or $i = 0.208$ mA. This means the output voltage is

$$v_O = -2 + 50(101)\,0.208 \text{ mA} = -0.95 \text{ V}$$

Now v_O stays at -0.95 V as we make v_I more negative by pulling down the voltage at the emitter of Q_2. The circuit does not change until the base-emitter voltage of Q_2 reaches cut-in at $v_{BE} = 0.7$ V. That is,

$$v_{BE} = -1.29 - (v_I - 0.9) \ge +0.7$$

or $v_I \ge -1.09$ V. To summarize, $v_O = -0.95$ V for $v_I \ge -1.09$ V. Figure 13.40e shows this line segment added to the partial curve of Fig. 13.40c. We then connect the two horizontal lines by a third line that approximates the transition between states when both transistors are active. This completes the static transfer characteristic.

Exercise 13.5 Find the input current i_1 that must be "sourced" at the gate input in Fig. 13.40a when $v_I = -1.2$ V.
Hint: Use the appropriate equivalent circuit.

Ans. 39.4 μA.

Noise Margins. From Fig. 13.40e, $V_{OH} = -0.95$ V, $V_{OL} = -1.85$ V, $V_{IH} = -1.09$ V, and $V_{IL} = -1.49$ V, giving noise margins

$$NMH = -0.95 - (-1.09) = 0.14 \text{ V}$$

and

$$NML = -1.49 - (-1.85) = 0.36 \text{ V}$$

These small noise margins are one of the limitations of ECL; however, the small logic swing contributes to switching speed by allowing parasitic capacitances to be charged and discharged quickly.

The noise margins change with fan-out or output loading as we see in the following example.

EXAMPLE 13.6 Recalculate the transfer characteristic and fan-out for the gate of Fig. 13.40 when the 50 Ω load resistance is replaced by 50 kΩ.

Solution. Replacing the 50 Ω load resistance by 50 kΩ in Fig. 13.40b changes i slightly; however, as in the first analysis, the effect of this current is negligible compared to the collector current of Q_2. We conclude that the transfer characteristic is essentially unchanged for logic 0 input.

Replacing 50 Ω with 50 kΩ in Fig. 13.40d changes Eq. (13.41) to

$$-2 + 50 \text{ k}(101)i + 0.9 + 245i = 0$$

giving $i = 0.218$ μA. Thus the output voltage for logic 1 input is

$$v_O = -2 + 50 \text{ k}[(101)0.218 \text{ } \mu\text{A}] = -0.9\text{V}$$

compared to -0.95 V for the 50 Ω load. It is obvious from the circuit that the value of v_I that turns on Q_1 is independent of the load resistor. Thus, with the 50 k load, the transfer characteristic of Fig. 13.40e has a transition region that terminates at the level -0.90 instead of -0.95. Only V_{OH} changes with loading. The new noise margin value is $NMH = -0.90 - (-1.09) = 0.19$, showing that reduced loading increases the noise margin. ❏

Second ECL Output. The complete ECL gate has two outputs: an OR output, which we have just examined, and a NOR output produced by the inverting subcircuit of Fig. 13.41a, which we explore now. When v_I has a low logic value, Q_1 cuts off, giving Fig. 13.41b. Because the base current of Q_4 is small, the drop across the 220 Ω resistor is negligible, and $v_O = -0.9$ V. Q_2 is active but has no effect on v_O. Since this equivalent holds until Q_1 turns on at

$$v_I = -2.19 + 0.7 = -1.49 \text{ V}$$

the first transfer characteristic segment is that of Fig. 13.41c.

An increase in v_I causes a transition during which both Q_1 and Q_2 are active. Eventually, v_I becomes so large that Q_2 cuts off, giving the Fig. 13.41d. From KVL,

$$v_I = -5.2 + i_{E1} (779) + 0.9 \tag{13.42}$$

Ignoring the base current of Q_4 and assuming $i_{C1} \approx i_{E1}$, gives

$$v_O = 0 - 220 \text{ } i_{E1} - 0.9$$

Solving Eq. (13.42) for i_{E1} and substituting gives

$$v_O = -220 \left(\frac{v_I + 4.3}{779} \right) -0.9 = -0.282v_I -2.11 \tag{13.43}$$

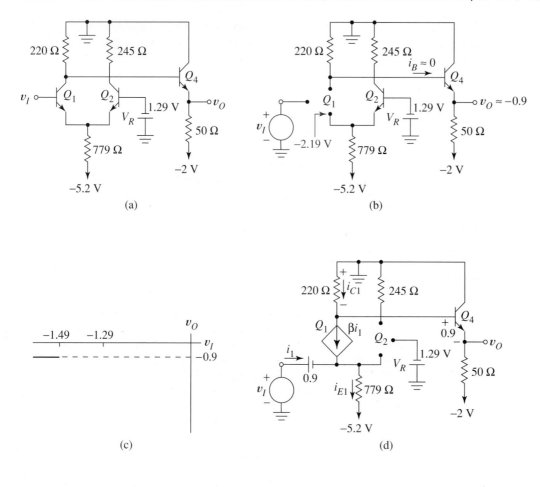

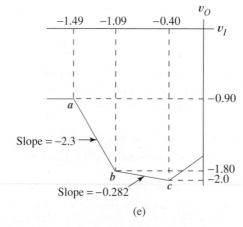

Figure 13.41 ECL NOR output: (a) schematic; (b) equivalent when input is low; (c) transfer characteristic for low input; (d) equivalent when input is high; (e) transfer characteristic.

the equation of a straight line of negative slope that relates v_O to v_I—segment b–c in Fig. 13.41e. To establish the lower limit of v_I to which Eq. (13.43) applies, we see from Fig. 13.41d that reducing v_I to the point where Q_2 cuts in takes

$$v_I = -1.29 - 0.7 + 0.9 = -1.09 \text{ V}$$

This gives point b in Fig. 13.41e. Substituting $v_I = -1.09$ V into Eq. (13.43) gives $v_O = -1.80$ V, the other coordinate of point b. Segment b–c, when both transistors are active, is also sketched. As v_I increases beyond -1.09 V in Fig. 13.41d, the collector voltage of Q_1 decreases as its emitter voltage increases until Q_1 saturates at $v_{CE2} = 0.2$ V. Once this happens, input and output voltages are related by

$$v_O + 0.9 - 0.2 + 0.9 = v_I$$

That is,

$$v_O = v_I - 1.6 \text{ V} \qquad (13.44)$$

Equations (13.43) and (13.44) are both true at the point where Q_1 saturates. Simultaneous solution shows that Q_1 saturates when $v_I = -0.40$ V. Once Q_1 saturates, the transfer characteristic changes to Eq. (13.44), a straight line of slope $+1$, also shown in Fig.13.41e. This region where Q_1 is saturated is of theoretical and practical interest and is even shown on some ECL data sheets. Operation in this region is not permitted because the diffusion capacitance of Q_1 would greatly reduce switching speed. Thus we must limit input voltages to $v_I \leq -0.4$ V.

Exercise 13.6   Use Fig. 13.41e to estimate ECL noise margins.

Ans. NMH = 0.19 V, *NML* = 0.31 V.

The gate we have been studying is known as the 10K series ECL gate. The more complete circuit diagram of Fig. 13.42a shows additional features. First, we see that voltage reference V_R consists of two diodes, two resistors, and an active-state output transistor Q_5. The 907 and 4.98 kΩ resistors, in conjunction with the power supply and diodes, establish the dc base voltage of Q_5. Current supplied by Q_5 is available to maintain the emitter voltage of Q_5 at the the desired -1.29 V in spite of state changes in Q_2 and the input transistors. The dc drops in D_1 and D_2 compensate for temperature changes in V_{BE5} and V_{BE2}. Problem 13.51 further explores this *voltage reference*.

Figure 13.42a shows how fan-in is expanded by adding input transistors in parallel with Q_1. The result is a three-input logic circuit with two outputs, depicted schematically by Fig. 13.42b. Gate operation is deduced from the previous development in this way. If inputs A, B, and C are all well below V_R, all three input transistors cut off, and Q_2 is active. With no current in the 220 Ω resistor, the NOR output is high and the OR output is low. In fact, only if A OR B OR C is high does current flow through the 220 Ω resistor making the NOR output low, and only under this condition does Q_2 cut off making the OR output high.

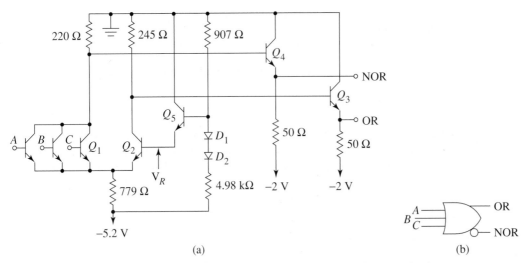

Figure 13.42 Three-input, 10K series ECL gate: (a) equivalent; (b) logic diagram.

EXAMPLE 13.7   For the inverter of Fig. 13.43a, use SPICE to
(a) Check the static NOR characteristic predicted in Fig. 13.41e.
(b) Find the static power dissipation for logic 0 and logic 1 inputs.
(c) Find the propagation delay.

Solution. (a) Figure 13.43b is partial SPICE code for the transfer function plot. Transistor parameters are the same as those previously used to investigate TTL gates.

The SPICE plot of Fig. 13.43c shows all four regions predicted by hand analysis and sketched in Fig. 13.41e. The numerical predictions using $V_{BE} = 0.9$ and $V_\gamma = 0.7$ for ECL were fairly good. Notice that computing the SPICE transfer plot without hand analysis gives no insight into the reasons for the shape of the curve. In particular we would have no idea that the last region is to be avoided because it involves saturation of the input transistor.

(b) Since Fig. 13.43c shows our estimates of $V_{OH} = -0.9$ V and $V_{OL} = -1.81$ V were reasonable, we use these as input values for our SPICE computation of power dissipation. With logic 1 input of -0.9 V, dissipation was 78.6 mW; with input of -1.81 V, dissipation was 76.6 mW.

(c) To find propagation delay, the input voltage was changed to the pulse:

```
VIN 2 0 PULSE(-1.81 -0.9 10N 0 0 30N 210N)
```

The initial inverter output, shown in Fig. 13.43d, followed the dashed transition from logic 1 to logic 0. Examining the internal waveforms of the gate showed that the -1.29 V reference voltage, supposedly constant, changed with the input pulse as in Fig. 13.43e. To increase the current-delivering capability of the source, the junction area of Q_5 was increased relative to the other transistors by using

```
Q5 0 8 7 TR 15
```

This reduced the variation at Q_5's emitter voltage and gave the solid-curve output waveform of Fig. 13.43d. Output data showed that for this modified gate, $t_{PHL} = 1.3$ ns and

t_{PLH} = 1.54 ns, giving propagation delay t_P = 1.42 ns. No change in the logic 1 or logic 0 power dissipation resulted from changing the area of Q_5. ❏

Exercise 13.7 The transfer characteristic of Fig.13.43c shows that when $v_I = 0$, v_O takes on some value more negative than -1.5 V. What value does our theoretical development predict?

Ans. -1.6 V.

Wired OR Function. ECL has a special feature that is useful to the logic designer, the *wired OR function*. This means we can obtain a *free* OR function by direct wiring ECL outputs together, thereby saving a logic gate. Figure 13.44 helps to show how the ECL's

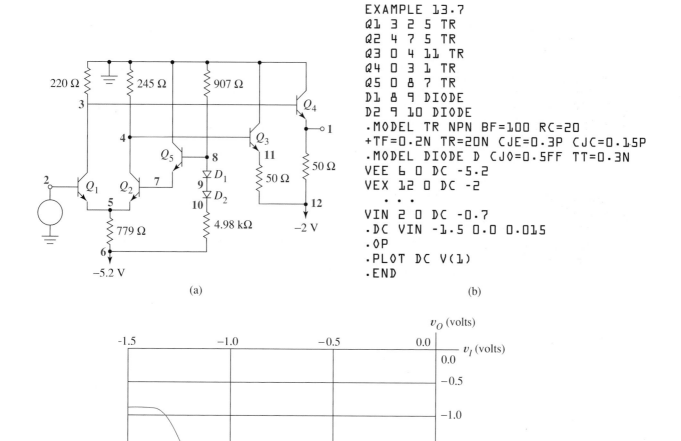

```
EXAMPLE 13.7
Q1 3 2 5 TR
Q2 4 7 5 TR
Q3 0 4 11 TR
Q4 0 3 1 TR
Q5 0 8 7 TR
D1 8 9 DIODE
D2 9 10 DIODE
.MODEL TR NPN BF=100 RC=20
+TF=0.2N TR=20N CJE=0.3P CJC=0.15P
.MODEL DIODE D CJO=0.5FF TT=0.3N
VEE 6 0 DC -5.2
VEX 12 0 DC -2
    ...
VIN 2 0 DC -0.7
.DC VIN -1.5 0.0 0.015
.OP
.PLOT DC V(1)
.END
```

(a)

(b)

(c)

Figure 13.43 ECL inverter simulation: (a) schematic; (b) partial SPICE code; (c) static transfer characteristic; (d) pulse response for transistor Q_5 of two areas; (e) variation in internal reference voltage that caused dashed transition in (d).

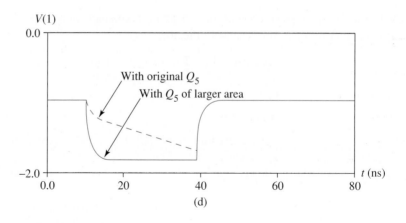

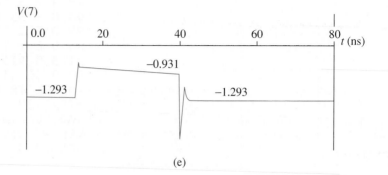

Figure 13.43 (continued)

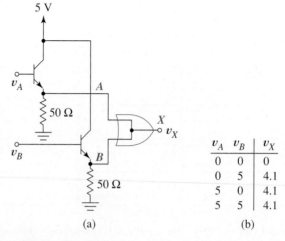

Figure 13.44 Wired OR function: (a) implemented by connecting two buffer outputs; (b) input and output voltage combinations.

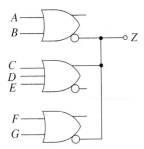

Figure 13.45 Example of wired OR logic function.

emitter-follower output circuits lend themselves to the wired OR. Suppose both bases are at 0 volts, representing logic 0 inputs for this circuit. Then both transistors cut off and $v_X = 0$, thus row 1 in the "voltage truth table" of Fig. 13.44b. Now if both bases are raised to 5 volts, both transistors are on, and v_X is one diode drop below the base, as in the last row of the truth table. Finally if one base is grounded and the other held at 5 V, only one transistor conducts, giving a high or logic 1 output. In terms of logic variables, $X = A + B$. Thus, connecting the outputs is equivalent to adding the OR function suggested by Fig. 13.44a. The actual ECL OR and NOR outputs combine in exactly the same way, the only difference being the voltage levels that represent the logic values.

EXAMPLE 13.8 Find logic function Z if gates are direct wired as in Fig. 13.45.

Solution. First find the logic function at the output node of each gate if the gates were not wired together. Then OR these outputs. This gives

$$Z = \overline{(A + B)} + (C + D + E) + \overline{(F + G)}$$

where "+" denotes OR and the overbar denotes the complement. ❏

13.7
Summary

Logic gates represent and process digital data by using transistors as highly reliable switches that produce high and low voltages to represent logic 1 and 0. Various logic circuit families are available to perform the same basic functions; however, each implements the logic using different solid-state technologies, and thus has characteristic strengths and weaknesses that result from different design constraints and compromises.

The basic gate in each family is the logic inverter, which is described by a static transfer characteristic. This characteristic defines logic 1 and 0 values for input and output, which, in turn, define the gate's noise margins. The ideal inverter has a very steep and narrow transition between output levels that requires the inverter to have high gain. This allows most of the logic swing to be productively utilized in the form of noise margins. Inverter dynamics are characterized by output rise and fall times and by propagation delays that are defined relative to the leading and trailing edges of an ideal input pulse. In each family the basic inverter is augmented in some fashion to produce other logic functions such as OR, NOR, AND, and NAND.

NMOS and PMOS are compact IC logic families that use enhancement- or depletion-transistors as nonlinear load resistors. Enhancement load resistors must be relatively long and narrow compared to the transistor switch for low logic 0 output voltages; depletion loads do not have this *ratio problem*. NMOS and PMOS dissipate no power at logic 1 output and low power at logic 0. Because of the MOSFET's high input impedance there are no static loading effects; additional load gates add only load capacitance. Fan-out is limited by capacitive loading and the resulting propagation delays. The inverter is expanded into a multiple-input NOR by adding transistor switches in parallel with the switch in the inverter. Because currents available for external loads are small, NMOS and PMOS are not available as LSI and MSI devices.

CMOS uses complementary n- and p-channel transistors as a switch pair in the basic inverter, with a logic 1 input connecting the output node to ground and a logic 0 input connecting output to the supply. The static transfer characteristic of a CMOS gate has high gain and a narrow transition region centered at half the supply voltage, resulting in high noise margins for a given logic swing. Static dissipation is zero, and average dissipation is proportional to switching frequency. As in NMOS and PMOS, CMOS fanout is limited by propagation delay that increases with capacitive loading. CMOS has long been popular for discrete logic in SSI and MSI, and is one of the most important families for VLSI technology as well. Much current development is centered around BiCMOS, a hybrid technology that augments the low dissipation and high noise margins of CMOS with the current-drive capabilities of bipolar transistor output circuits.

The fastest gates currently available are discrete structures using the relatively new technology of gallium arsenide. Since this is an area characterized by rapid advancements, no single gallium arsenide logic family has yet emerged to claim a position of dominance. To develop sufficient circuit-level familiarity with GaAs gates to be able to follow the progress in this emerging area in depth, we studied three families representative of current technology: Schottky diode FET logic, buffered FET logic, and direct-coupled FET logic. One novelty was having to account for gate conduction in our analysis of FET circuits; another was use of a level shifter to make output logic levels compatible with input logic levels in BFL.

TTL is the dominant bipolar logic technology. To eliminate the long propagation delays caused by minority charge storage in saturated transistors, TTL now makes extensive use of the Schottky transistor, a BJT with a Schottky clamping diode connecting base and collector. Schottky transistors have four operating states: cut-off, active, clamp, and reverse clamp; however, only clamp and reverse clamp are unique to the Schottky transistor, because the diode conducts only for these states. Clamp differs from saturation in two ways: a clamp voltage $V_{CE,clamp} \approx 0.3$ V replaces $V_{CE,sat} \approx 0.2$ V, and charge storage delay is absent. Reverse clamp involves conduction of the Schottky diode with the underlying BJT in cut-off. Both new states are easy to understand and model using equivalent circuits.

TTL gates dissipate power for both logic 1 and 0 output voltages and require that the source of an input logic 0 be capable of sinking a significant amount of current. Fan-out is limited by dc loading considerations, specifically, the number of input currents from load gates that the output transistor is able to sink with output at logic 0 while still remaining in the clamp state. The principal TTL output circuit is the active totem pole circuit, which provides high currents for quickly charging and discharging capacitive loads. Passive pull-up, provided by a user-supplied, externally connected pull-up resistor in the open collector TTL gate, provides flexibility for interfacing between TTL and nonlogic devices. Fan-in provided by multiple-emitter transistors result in a multi-input NAND circuit. Because of large currents and small internal capacitances, TTL gives shorter propagation delays than FET gates.

Emitter-coupled logic is the fastest of the bipolar logic families. The basic gate has both OR and NOR outputs, each buffered by an emitter follower. An emitter-coupled current switch provides the internal logic, with input transistors operating between cut-off and active states. ECL uses a negative supply of -5.2 V. Optimized for high speed, ECL has noise margins of only a fraction of a volt, relatively high currents, and high power dissipation in both logic states.

REFERENCES _____

1. ALVAREZ, A. R., ed. *BiCMOS Technology and Applications.* Boston: Kluwer Academic, 1989.
2. BANZHAF, W. *Computer-Aided Circuit Analysis Using SPICE.* Englewood Cliffs, NJ: Prentice Hall, 1989.
3. BUCHANNAN, J. E. *BiCMOS/CMOS Systems Design.* New York: McGraw-Hill, 1991.
4. HODGES, D. A., and H. C. JACKSON. *Analysis and Design of Digital Integrated Circuits.* New York: McGraw-Hill, 1988.
5. KIAEI, S., S.-H. CHEE, and D. ALLSTOT. "CMOS Source-Coupled Logic for Mixed-Mode VLSI," *Proc. IEEE Inter-*

nat. Symp. on Circuits and Systems. New Orleans: 1990, pp. 1608–1611.
6. MANO, M. M. *Computer Engineering Hardware Design.* Englewood Cliffs, NJ: Prentice Hall, 1988.
7. NATIONAL SEMICONDUCTOR. *CMOS Logic Databook.* Santa Clara, CA: National Semiconductor, 1988.
8. SEDRA, A. S., and K. C. SMITH. *Microelectronic Circuits,* 3rd ed. Philadelphia: Saunders College, 1991.
9. TAUB, H., and D. SCHILLING. *Digital Integrated Electronics.* New York: McGraw-Hill, 1977.

PROBLEMS _____

SPECIAL DIRECTIONS FOR SPICE HOMEWORK PROBLEMS: *Do not hand in lengthy SPICE printout for homework.* Instead, abstract the useful information from the SPICE output file as in the SPICE examples. Include your SPICE code and a circuit diagram with nodes numbered to agree with the code. Cite relevant numerical values from the SPICE output file and discuss when appropriate. Make sketches of any relevant curves, and label appropriate points. Make small tables to present numerical data if useful for clarity.

Section 13.1

13.1 Estimate V_{OH}, V_{OL}, V_{IH}, V_{IL}, *NMH*, and *NML* as well as you can directly from (a) Fig. P13.1a, (b) Fig. P13.1b, (c) Fig. P13.1c.

13.2 The inverter transfer characteristic of Fig. P13.2 is described by

$$v_O = -\frac{1}{6} v_I^2 + 10 \qquad \text{for } v_I \leq 4 \text{ V}$$

and

$$v_O = 0.115 \, v_I^2 - 2.38 \, v_I + 15 \qquad \text{for } v_I \geq 4 \text{ V}$$

(a) Find V_{OH} and V_{OL}.
(b) Find V_{IH} and V_{IL}.
(c) Find the noise margins.

13.3 Figure P13.3 shows the output waveform of an inverter when the input is an ideal positive pulse of 10 ns duration, applied at $t = 14$ ns. From the diagram estimate
(a) fall time, (b) rise time, (c) t_{PHL}, (d) t_{PLH}, (e) V_{OH}, (f) V_{OL}.

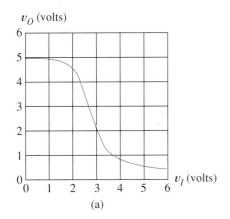

(a)

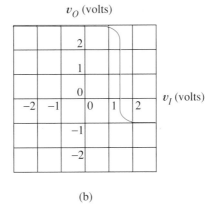

(b)

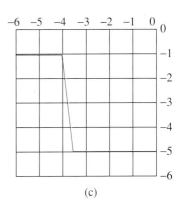

(c)

Figure P13.1

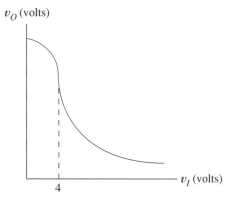

Figure P13.2

Section 13.2

13.4 (a) For Fig. 13.5a write the equation that relates v_O to v_I when M_1 is ohmic and M_2 active. Then find V_{OL} by substituting $v_I = V_{OH}$ into this equation as suggested in Fig. 13.5d. *Hint:* $V_{OL} \ll V_{DD}$. (b) Find conditions V_{IH} must satisfy by finding the point where the slope of the equation of part (a) is minus one.

13.5 Use SPICE to plot the transfer characteristic of Fig. 13.5a for GAMMA = 0 and for GAMMA = 0.4. Use $k = 0.3 \times 10^{-3}$ A/V^2 and $V_t = +1.3$ V for M_1. M_2 is identical except its width is one-fourth that of M_1. Does body effect improve the static operation of the gate? Explain.

13.6 In Fig. 13.6f, active-state transistor M_2 charges C_L from V_{OL} to V_{OH}.
(a) Write an expression for the initial drain current in terms of V_{OH} and V_{OL}.
(b) Write an expression for the drain current when $v_O = V_M$. Substitute for V_M so that your expression involves V_{OH} and V_{OL}.
(c) Use results from (a) and (b) to show that the average charging current is given by Eq. (13.11).

13.7 Draw an NMOS circuit that realizes the function
(a) $\bar{Z} = (A \cdot B + C) \cdot D$,
(b) $\bar{Z} = (A + B) \cdot D \cdot E$,
(c) $\bar{Z} = (A \cdot B + C \cdot D) \cdot E$.

13.8 Use SPICE to find the new values for t_{PLH} and V_{OL} when W/L for M_2 is changed to 5U/5U in the gate of Example 13.1.

13.9 Use SPICE to find t_P when the gate of Example 13.1 has an external load capacitance of 1 pF.

13.10 Ignoring body effect,
(a) Find a mathematical expression to relate v_O and v_I in Fig. 13.9a when
(i) M_1 is off and M_2 is ohmic.
(ii) M_1 is active and M_2 is ohmic.
(iii) M_1 and M_2 are both active.
(iv) M_1 is ohmic and M_2 is active.
(b) Draw a sketch of the transfer characteristic and label regions a–d to agree with the equations you derived.

13.11 Instead of a 20 pF load capacitance as in Example 13.2, suppose the gate of Fig. 13.10a is loaded with the input capacitances of N load gates, each input capacitance being 1.8 pF. If C_L consists only of input capacitances, what is the fan-out N if $t_P \leq 50$ ns?

13.12 Figure P13.12 shows a depletion-load, NMOS inverter with dynamics lumped into a single output capacitance.
(a) From the graphical information, estimate V_{OH} and V_{OL}.
(b) Find the power dissipation of the inverter when the output is at logic 0.
(c) With the output at logic 1, the input instantly changes from 0 V to 10 V. Estimate dv_O/dt at $t = 0^+$, the instant after switching.
(d) Rework part (c) when the gate is driving 10 load gates, each with input capacitance of 2 pF, in addition to the original load capacitance.

13.13 The transistor in Fig. P13.13 has $V_t = 2$ V, $k = 2$ mA/V^2, and no internal capacitance.
(a) Input v_I is gradually increased until the transistor turns on. What is the value of v_I when the transistor begins to conduct?

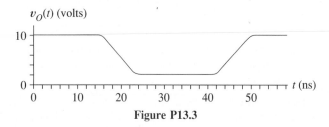

Figure P13.3

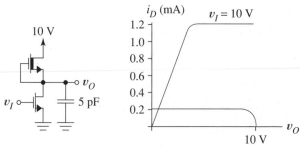

Figure P13.12

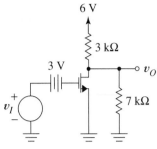

Figure P13.13

(b) What is the initial state of the transistor as it turns on? Guess and verify.

(c) Write an equation relating v_O to v_I when the transistor is in the state you found for part (b).

(d) Find an inequality that must hold for parts (b) and (c) to be valid.

(e) Write an equation that relates v_O to v_I for large values of v_I.

(f) After reviewing the preceding parts of the problem, sketch the transfer characteristic. Label values at key points on the curve.

13.14 For Fig. P13.14,

(a) What are the steady-state logic 1 and logic 0 output voltage values?

(b) The input voltage of the inverter instantaneously increases from 2 V to 12 V. Use the given graphical information to estimate the propagation delay. Assume transistor capacitances are included in the 4 pF load capacitor.

Section 13.3

13.15 Figure P13.15 is the transfer characteristic of a CMOS inverter. Use this graphical information to estimate the noise margins.

13.16 Show that V_{IL} for the CMOS gate is given by Eq. (13.18).

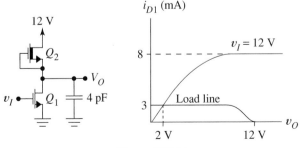

Figure P13.14

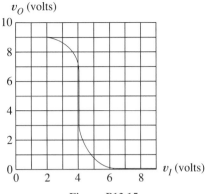

Figure P13.15

Hint: In region 2, the transfer characteristic is described by Eq. (13.16). First substitute $k_P = k_N = k$ and $-V_{tP} = V_{tN} = V_t$. Then show that

$$\frac{dv_O}{dv_I} = -1 \quad \text{implies} \quad v_O = v_I + \frac{V_{DD}}{2}$$

Use this to eliminate v_O in Eq. (13.16). Solve for $v_I = V_{IL}$. (The algebra requires a measure of faith.)

13.17 Show that V_{IH} for the CMOS gate is given by Eq. (13.19).

Hint: Instead of using the definition, use Eq. (13.18) and the symmetry of the characteristic curve.

13.18 What are the noise margins for a CMOS gate with a single 12 V supply, if the transistor threshold voltages are 0.8 V in magnitude?

13.19 A CMOS inverter uses a power supply of +12 V. The transistors have $k_P = k_N = 3 \times 10^{-5}$ A/V^2, $|V_t| = 1.0$ V, and Early voltages of 90 V.

(a) Use SPICE dc analysis to plot the transfer characteristic.

(b) Repeat part (a) using Early voltage of 50 V.

13.20 In Fig. P13.20 the transistors are matched, having k for their transconductance parameters and $V_{tN} = -V_{tP} = V_t$ for threshold voltages. Find the value of v_O when the input is logic 0.

Hint: Assume the value of R_L is so large (5.12) applies.

13.21 In Fig. P13.20 the input is zero volts. Find the value of R_L (in terms of MOSFET parameter k) that just brings the p-channel transistor to active operation.

13.22 A CMOS inverter uses a power supply of +12 V. The transistors have $k_P = k_N = 3 \times 10^{-4}$ A/V^2 and $|V_t| = 1.0$ V.

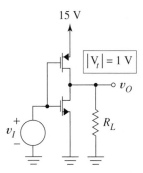

Figure P13.20

With the load capacitor $C = 40$ pF initially uncharged, the input suddenly goes to logic 0 as in Fig. 13.12b.
(a) Write and then solve the differential equation that applies at $t = 0^+$.
(b) Find the largest value of t, t_1, for which your solution in part (a) is valid. Explain.
(c) Find the new differential equation that applies for $t \geq t_1$. _Separate_ the differential equation into parts, one that leads to integration with dv_o and the other an integration involving dt.

13.23 Use SPICE to find t_{PLH} and t_{PHL} for a CMOS gate with a 2 pF load capacitance. For the n-channel transistor, parameters are VTO = 1.5, KP = 7E − 6, GAMMA = 0.37, TOX = 1.0E − 7, W = 10U, and L = 5U. The p-channel parameters _that differ from n-channel parameters_ are VTO = −1.5, W = 25U, and L = 5U. $V_{DD} = 5$ V.

13.24 Write the two differential equations that describe the discharge transient of Fig. 13.12d and indicate the range of v_O for which each is valid.

13.25 (a) Use SPICE to find $v_O(t)$ for the inverter of Fig. 13.14a when input is the pulse included in the code in Fig. 13.14a.
(b) Repeat part (a) after adding a 10 pF load capacitor between output and ground.

13.26 The SPICE code of Fig. 13.14a predicts the transfer characteristic of Fig. 13.14b. Use SPICE to find the curve when both transistors have Early voltage of 40 V. Compare your curve with the original.

13.27 Sketch the transfer characteristic for the CMOS inverter ($V_{tP} = -V_{tN}$ and $k_P = k_n$) of Fig. 13.15a for
(a) $V_{DD} = 0$, $V_{SS} = 5$ V.
(b) $V_{DD} = V_{SS} = 4$ V.
(c) $V_{DD} = +3$ V, $V_{SS} = 8$ V.
(d) Use SPICE to verify the result of part (c).

13.28 Find the noise margins for a CMOS gate that uses ±4 V supplies if the transistors have threshold-voltage magnitudes of 1.1 V.

13.29 (a) Sketch an equivalent circuit for Fig. 13.17 when the circuit is in equilibrium with $v_I = 0$ V. Give the probable state of each transistor
(b) C_L is initially discharged to zero volts. Sketch an equivalent circuit for Fig. 13.17 at the instant after v_I has switched from V_{DD} to zero volts. Give the probable state of each transistor.
(c) Sketch an equivalent circuit for Fig. 13.17 when the circuit is in equilibrium with $v_I = V_{DD}$ volts. Give the probable state of each transistor
(d) C_L is initially charged to V_{DD} volts. Sketch an equivalent circuit for Fig. 13.17 at the instant after v_I has switched from zero to V_{DD} volts. Give the probable state of each transistor.

13.30 Figure P13.30 shows a proposed BiCMOS inverting circuit. MOSFET parameters are those of Fig. 13.14a. BJT parameters are those of Fig. 4.45b.
(a) Use SPICE to find the static transfer characteristic and the logic 0 and one power dissipation.
(b) Use SPICE transient analysis to plot the output voltage when the input is a voltage pulse of 5 V amplitude.

13.31 Draw a diagram showing how to convert the BiCMOS inverter of Fig. 13.17 into a two-input NAND.

13.32 Write SPICE code for the two-input NAND of Fig. 13.16. Use the same .MODEL statements as in Fig. 13.14a, except add body-effect coefficient $\gamma = 0.37$. Modify the transistor element lines by adding W = 25U, L = 5U for M_2 and M_4, and W = 20U and L = 5U for M_3 and M_1. $V_{DD} = 12$ V.
(a) Find the static transfer function of the gate, V(4) versus V(5), when V(3) = 1.
(b) Find the response of the two-input gate to the input pulse described in the code of Fig. 13.14a, when V(3) is held at 12 V.

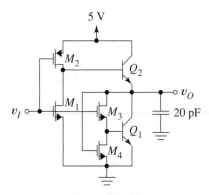

Figure P13.30

Section 13.4

13.33 (a) For Fig. 13.18a, sketch the nonlinear load line formed by B_2 on the output characteristics of B_3. Your sketch should show the correct numerical values of the active-region current of B_2 and the B_3 curve corresponding to gate-source conduction. A unit-width transistor has $\beta = 0.07$ mA/V^2 and $V_t = -0.5$ V.
(b) In Fig. 13.18b, when $v_I - 1.2 \geq V_t = -0.5$ V, B_3 first enters the active state while B_2 remains ohmic. Ignoring velocity saturation, write the equation that relates v_O to v_I for operation in this region.
(c) For what value of v_I does B_2 become active?
(d) Write the equation that relates v_O and v_I when both are active.
(e) For what value of v_O does B_3 become ohmic?

13.34 Figure P13.34 shows an SDFL inverter with transistor widths (in microns) given in parentheses. For a MESFET of one micron width, $\beta = 0.07$ mA/V^2. Threshold voltage is $V_t = -0.5$ V. Schottky diodes conduct for forward bias of 0.6 V.
(a) Find the fan-out.
(b) Find v_O when input comes from an identical gate with logic 1 output.

13.35 Make a SPICE subcircuit model of the inverter described in Problem 13.34, using JFETs for transistors. Model the diodes with default valves, except use EG = 0.69.
(a) Write SPICE code for a two-inverter cascade, with output of one inverter connected to the input of a second. Use dc analysis to plot the transfer function of the first inverter in the cascade, for $0 \leq v_I \leq 2.5$ V.
(b) From SPICE data *for the transfer characteristic of the loaded inverter,* use standard definitions to find the noise margins.
(c) Modify the circuit of part (a) by adding a second load inverter. Find the transfer function of the first inverter for this case.

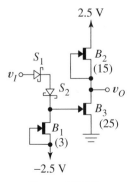

Figure P13.34

(d) Repeat part (b) using the transfer characteristic from part (c).

13.36 Here we estimate t_{PLH} for the SDFL gate of Fig. 13.18a by hand analysis, ignoring velocity saturation.
(a) If a load capacitor, $C_L = 5$ pF is connected between output and ground, what are the initial capacitor voltage, the final voltage, and the average V_M for the gate described in the text?
Hint: See Fig. 13.18c.
(b) Find the charging current at the instant following the input's change to logic 0.
Hint: See Fig. 13.18b.
(c) Find the charging current when the capacitor has charged to V_M volts.
Hint: See Fig. 13.18b.
(d) Use your results to estimate t_{PLH}.

13.37 (a) Construct a SPICE model for the SDFL inverter of Fig. 13.18a. Use n-channel JFETs to approximate the MESFETs. Parameter values for a $1 \mu \times 1 \mu$ device are $\beta = 0.07 \times 10^{-3}$ A/V^2, VTO $= -0.5$ V, LAMBDA = 0.05, CGS = CGD = 0.5 FF, RD = RS = 1 Ω, PB = 1, and FC = 0.5. The zero-bias capacitances and βs must be scaled according to channel width. This is done automatically by using the area parameter on the JFET element line. To model the Schottky diodes, use default diodes except for CJO = 1.5 FF, EG = 0.69. Of course the diode transit time stays at its default value of zero.
(b) Use your model to verify the general shape of the static transfer characteristic of Fig. 13.18c.
(c) Determine the waveform at the output, and use data from the output file to estimate the propagation delay when the input is the pulse

> **PULSE(0.89 1.8 10N 0 0 30N 210N)**

13.38 Use Eq. (13.27) to find V_{IH} for the BFL gate of Fig. 13.20a.
Hint: First differentiate Eq. (13.29) with respect to v_I to see how to use Eq. (13.27).

13.39 Write a linear equation that relates v_O to v_X in Fig. P13.39, and specify the range of v_O over which it is valid. Parameter values are $V_{t1} = V_{t2} = -0.5$ V, $\beta_1 = 11.2 \times 10^{-4}$ A/V^2, and $\beta_2 = 0.7 \times 10^{-4}$ A/V^2. The Schottky diode ON voltage is 0.7 V.

13.40 In Fig. 13.23 the depletion and enhancement MESFETs have threshold voltages of -0.5 V and $+0.1$ V and channel widths of 5 μ and 50 μ, respectively. For both transistors $\beta = 7 \times 10^{-5}$ A/V^2 for a device of $W = 1\mu$, $L = 1\mu$, and $\alpha = 0$.
(a) Sketch the output characteristics of B_1 for $v_I = 0$ and for $v_I = 0.6$ V (the ON voltage of the gate-source junction.) Add to your sketch, the nonlinear load line of B_2.

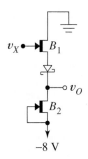

Figure P13.39

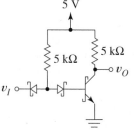

Figure P13.42

(b) Write the equation that describes the gate when both transistors are active.
(c) Sketch the transfer characteristic of the gate.

Section 13.5

13.41 Figure P13.41 is a classical TTL inverter that uses a saturating output transistor. For the transistor, $V_{CE,sat} = 0.2$ V and $\beta = 30$.
(a) Assume a logic 0 of $v_I = 0.2$ V is applied. Find the state of Q_1 and find the value of the input current.
(b) If v_I is now increased, for what value of v_I will the state of a transistor change? Assume cut-in occurs at 0.5 V.
(c) Assume v_I is 5 V. Find the states of Q_1 and Q_2 and the base current of Q_2. *Hint:* $\beta_R = 0.01$.
(d) Find the fan-out of the inverter.
(e) Starting with $v_I = 5$ V as in part (c), find the value of v_I for which Q_1 begins to change state.
(f) Sketch the no-load transfer characteristic for the inverter. Assume a straight-line change between the two extreme conditions studied in (a)–(e).

13.42 Figure P13.42 shows a Schottky clamped *diode transistor logic* gate. Parameter values are $V_D = 0.3$ V, $V_{CE,clamp} = 0.4$ V, and $\beta = 20$.
(a) For an applied voltage of 5 V, assume that the input diode is OFF, the internal diode ON, and the transistor in clamp. Draw a

model of the circuit and verify the assumptions about diode and transistor states.
(b) Compute the value of the diode currents and the base current. What is the output voltage?
(c) Now assume the input voltage applied to the gate is obtained from another gate of this family that itself is developing a logic 0 at its output. Assume that the input diode is ON, the other diode is OFF, and the transistor is cut off. Draw a model of the circuit under these conditions and verify these assumptions.
(d) Determine the direction and value of the gate input current for the condition described in part (c).
(e) Find the fan-out of this gate family.

13.43 Assume the capacitor in Fig. P13.43 was previously discharged to 0.3 V. This circuit is to charge it for $t \geq 0$. $V_{CE,clamp} = 0.3$ V, $V_{BE,on} = 0.7$ V, and $\beta = 30$. Assume Q_D is in clamp and Q_P active at $t = 0$. Verify the assumptions at $t = 0$.

13.44 Starting with Fig. 13.32c, verify the Thevenin equivalent of Fig. 13.32d.

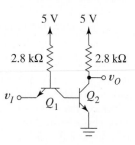

Figure P13.41

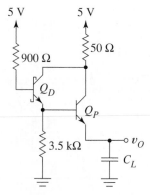

Figure P13.43

Hints: Replace the 3.5 k resistor and components to its left by a Thevenin equivalent. Using KCL at the output node is a good start in finding the Thevenin voltage.

13.45 Starting with Fig. 13.32e, verify the Thevenin charging circuit of Fig. 13.32f. See the hints in the preceding problem.

13.46 Use SPICE to find the supply current waveform of Fig. 13.35 and its maximum rate of change if a 100 pF capacitor is added between the output node and ground.

13.47 Use SPICE to plot the transfer characteristic for the gate of Fig. 13.34a for $-35°C$ and for $+65°C$. To the transistor .MODEL lines in Example 13.4 add XTB = 1.8, and give each resistor a first-order temperature coefficient of 0.004/°C.

13.48 Redraw Fig. 13.39e with each transistor replaced by the appropriate equivalent circuit.

13.49 Figure P13.49 shows a BiCMOS (bipolar CMOS hybrid) logic inverter. The output is at its logic 0 value of 0 V. At $t = 0$, v_I changes instantly to zero volts, upsetting the equilibrium.
(a) Assume that M_1 and Q_1 are cut off and M_2 and Q_2 are active at $t = 0^+$. Draw the equivalent circuit and use it to compute the charging current of C_L at $t = 0^+$ if $C_L = 10$ pF.
(b) Verify that the state assumed for M_2 is correct.
(c) Find the value of v_O for which M_2 will change state.

Section 13.6

13.50 Figure 13.40d shows that an ECL gate output must be able to source a current i_1 for each load gate it tries to hold at logic 1.
(a) From the circuit, relate i_1 to v_I.

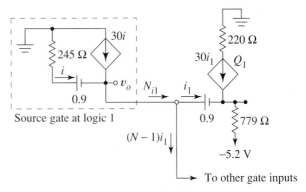

Figure P13.50

(b) Figure P13.50 shows the logic 1 output loading of one ECL gate by N identical gates. (Figure P13.50 shows direct gate connections rather than the 50 Ω transmission line connections of the other diagrams.) Find an equation that relates the logic 1 output voltage v_O to N. From this equation find v_O values for $N = 1, 2, 3$.

13.51 Use circuit analysis to estimate V_R in Fig. 13.42a. Assume Q_5 is active. Ignore I_{B2}.

13.52 ECL uses a negative supply voltage to reduce output noise. To see this, consider Fig. P13.52, the output circuit of Fig. 13.40a when v_I is logic 1. Source v_n represents noise entering the circuit through the power supply. If we ground point a, we have a positive supply and need to deal with the noise voltage at node c relative to a. If we ground point b, creating a negative supply, we deal with noise at c relative to b.
(a) Replace the transistor with its large signal model using $\beta = 100$, and then turn off both dc sources so that v_n is the only independent source.
(b) Place a reference ground at node b and calculate the noise voltage at nodes c and a.
(c) Compare the noises at c relative to a and at c relative to b.

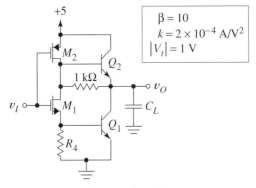

Figure P13.49

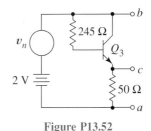

Figure P13.52

CHAPTER 14

Digital Memory, Interfacing, and Timing Principles

Introductory digital logic books describe digital circuits in terms of highly idealized logic elements, waveforms, and delays. This chapter describes the operating principles of circuits used for digital memory, pulse shaping, and timing more realistically by emphasizing circuit-level principles. It also addresses numerous topics related to digital interfacing. These interfacing concepts add a global perspective to our digital circuit studies by refining our understanding of IC design; of interconnections between logic families; and of interconnections between digital circuits and external devices such as switches, relays, and LEDs. Some interfacing topics, such as wired logic, differentiators, and level shifters, are already familiar. Others such as tristate gates and transmission line effects are new. Transmission line theory explains the complex waveforms that often occur when high-speed gates are interconnected. We end the chapter by discussing basic principles of IC memories.

Multivibrators. *Multivibrators* are two-state, digital circuits that employ positive feedback during state changes. There are three classes of multivibrators: bistable, monostable, and astable.

The *bistable multivibrator,* also known as a *flip-flop, latch,* or *binary,* is characterized by two *stable states,* states that can persist indefinitely unless deliberately changed. The binary is a basic memory element capable of storing a single bit of information.

In the *monostable multivibrator,* only one state is stable. The other is a *quasi-stable state,* which is transient in nature. The monostable, also known as a one-shot, accepts a triggering signal as input and produces an output pulse of specified duration for its output.

Both states of the *astable multivibrator* are quasi-stable, making this circuit a free-running oscillator used for waveform generation or as a *clock* to control circuit timing.

14.1
Bistable Circuits

The essence of a bistable is the cross-coupled inverter pair of Fig. 14.1a. The individual transfer characteristics, v_X versus v_I and v_O versus v_X, each have the general shape of Fig. 13.1b. With static loading effects (if any) included, the two-stage circuit has a transfer characteristic resembling the solid curve of Fig. 14.1b. An external conductor (dashed lines) completes the positive feedback loop and adds the additional straight-line

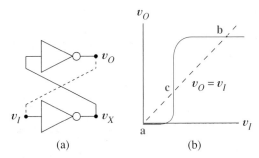

Figure 14.1 Bistable circuit: (a) basic configuration; (b) operating point constraints.

constraint, $v_O = v_I$. Points **a** and **b** represent stable operating points or states; point **c** is an *unstable operating point,* which we subsequently explain. When used as a digital memory, one stable operating point corresponds to logic 1 and the other to logic 0. We can construct a bistable using inverters from any logic family.

Stable States. To explore bistable operation, consider the CMOS implementation in Fig. 14.2a. The logic value, Q, denotes the state of the latch. Input capacitances C_1 and C_2 are lumped equivalents that represent the parasitic transistor capacitances.

When $Q = 1$, inverter M_2-M_4 has a logic 1 output and M_1-M_3 a logic 0 output, giving Fig. 14.2b. This state corresponds to point **b** in Fig. 14.1b. Figure 14.2c corresponds to point **a.** To be sure **a** state is *stable,* we must verify that it persists in spite of small transient noise perturbations.

Figure 14.2d shows the $Q = 0$ equivalent in the presence of noise v_n. With no noise, C_2 is charged to V_{DD} volts and keeps Q_2 in ohmic operation. A noise pulse with the polarity indicated in the diagram causes C_2 to discharge with time constant $R_3 C_2$, reducing the gate voltage of M_2 and bringing it closer to cut-off. If v_n is not *both* large enough and long enough to turn off M_2, however, C_2 simply recharges to V_{DD} when the noise ends. A similar argument shows that the $Q = 1$ state is stable. We show later that when a state is unstable the slightest noise voltage triggers a chain of events that leads to a state change. But first we examine ways to cause the circuit to change from one stable state to the other.

Changing the State. Figure 14.3 shows the latch connected by a switch, SW, to an external source with internal resistance R_I. To store a bit, we first set v_I to the desired logic 1 or 0 voltage and then briefly close SW. Let us examine the timing of this operation.

Figure 14.4a shows the latch in the $Q = 1$ state. Because the equilibrium voltage of C_1 is V_{DD} volts, closing SW with $v_I = V_{DD}$ volts causes no change. When $v_I = 0$ as in the diagram, however, capacitor C_1 discharges through the Thevenin resistance $R_I \| R_4$, and v_X drops exponentially from its initial value of V_{DD} toward a final (Thevenin) voltage of

$$V_F = \frac{R_I}{R_I + R_4} V_{DD} \tag{14.1}$$

Because of C_2's isolation, there can be no change in v_Y until M_3 turns on. This happens when

$$v_X = V_{DD} - V_t \tag{14.2}$$

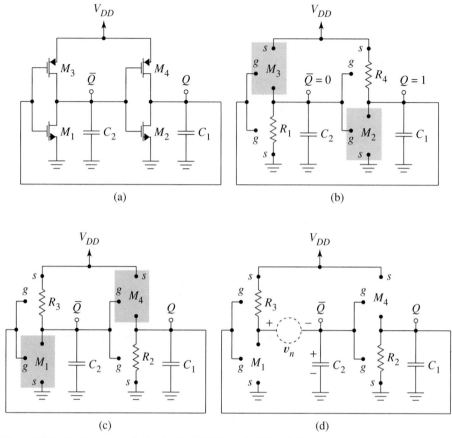

Figure 14.2 Stable states of a latch: (a) CMOS latch; (b) logic 1 state; (c) logic 0 state; (d) circuit for examining stability.

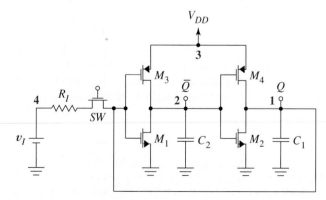

Figure 14.3 Latch with provision for changing the stored value using an external source.

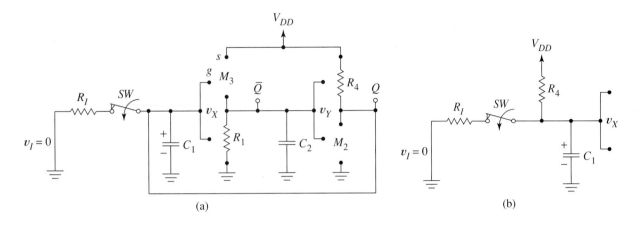

(a)   (b)

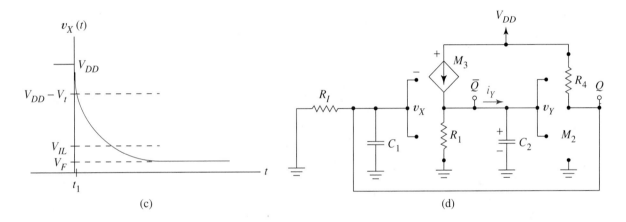

(c)   (d)

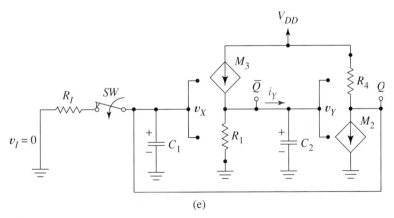

(e)

Figure 14.4 Latch dynamics: (a) $Q = 1$ equilibrium state; (b) input circuit that discharges C_1; (c) discharge waveform; (d) M_3 begins to change the state of M_2; (e) regenerative action as a two-stage amplifier.

where $-V_t$ is the threshold voltage of M_3. There is a delay, t_1, as C_1 discharges suffi-ciently to turn on M_3, which we now examine.

For *any* linear RC circuit with constant excitation, the difference between initial and final equilibrium values vanishes exponentially; thus the following generic solution de-scribes the capacitor voltage.

$$\text{response} = \text{final value} + (\text{initial value} - \text{final value})e^{-t/(RC)} \qquad (14.3)$$

For the input circuit, Fig. 14.4b, t_1 must satisfy

$$v_X(t_1) = V_F + (V_{DD} - V_F)e^{-t_1/(R_E C_1)} = V_{DD} - V_t \qquad (14.4)$$

where V_F is given by Eq. (14.1) and $R_E = R_I \| R_4$. Figure 14.4c shows the waveform. Notice that if *SW* reopens before $t = t_1$, C_1 recharges to V_{DD} and the latch does not change state. Thus switch closure of insufficient duration causes only a transient voltage change at Q and no change at $\overline{Q}$.

After t_1 seconds, M_3 turns on, giving Fig. 14.4d. Only now does C_2 begin to charge. The switching process is still reversible at this point, however, for if *SW* is reopened before M_2 turns on, v_X returns to V_{DD}, M_3 turns off, C_2 discharges to zero, and the circuit reverts to its original state. Obviously, a necessary condition for switching is that v_Y reach the threshold, V_t, of M_2. Once M_2 is on, both inverters conduct simultaneously, giving the two-stage, positive feedback amplifier of Fig. 14.4e. As v_Y increases, v_X de-creases, providing higher charging current from M_3. When v_X becomes so low that loop gain exceeds one, positive feedback accelerates the simultaneous charging of C_2 and dis-charging of C_1. The circuit continues to "flip" regeneratively, with the next critical event being cut-off of M_4. Once this happens, the state change is irreversible. Opening *SW* at this point disconnects the external source, but C_1 continues to discharge through M_2. Both capacitances reach their final $Q = 0$ values, as in Fig. 14.5, and the transition is complete.

If a new input is applied after M_4 cuts off, but *before* capacitor voltages settle to their final values, the timing and waveforms differ from those based upon initial equilib-rium. To avoid this, we must observe a *settling time*, four or five $R_2 C_1$ time constants, before trying to change the state again.

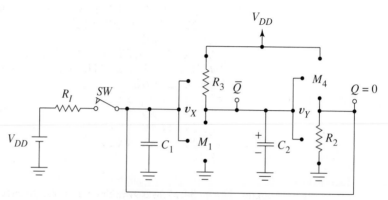

Figure 14.5 Latch in its $Q = 0$ equilibrium state.

To store a one with the latch in its $Q = 0$ state as in Fig. 14.5, we bring v_I to V_{DD} and close *SW*. C_1 charges through Thevenin resistance $R_I \| R_2$, turning on M_1, which then discharges C_2. When v_y is sufficiently low, M_4 turns on and assists V_I in charging C_1. When M_2 cuts off, the change continues even if the source is disconnected by opening the switch.

EXAMPLE 14.1 The latch of Fig. 14.6a has values $R_{in} = 200\ \Omega$, $C_1 = C_2 = 0.4$ pF, $k_p = k_n = 2 \times 10^{-4}\ \mathrm{A^2/V}$, and $|V_t| = 1$ V. It presently stores a logic 1.

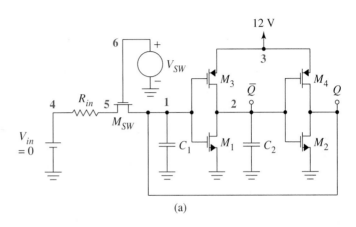

(a)

```
EXAMPLE 14.1
VDD 3 0 DC 12
M1 2 1 0 0 NFET
M2 1 2 0 0 NFET
M3 2 1 3 3 PFET
M4 1 2 3 3 PFET
MSW 1 6 5 0 NSW
.MODEL NSW NMOS VTO=1 KP=1
C1 1 0 0.4PF
C2 2 0 0.4PF
.MODEL NFET NMOS VTO=1 KP=2E-4
*+CBD=200FF CBS=200FF
.MODEL PFET PMOS VTO=-1 KP=2E-4
*+CBD=200FF CBS=200FF
VIN 4 0 DC 0
VSW 6 0 PULSE (0 13 1N 0.01N 0.01N 2N 40N)
*VSW 6 0 PULSE (0 13 1N 0.01N 0.01N 0.2N 40N)
RIN 4 5 200
.IC V(1) = 12
.TRAN 0.1N 4N
.PLOT TRAN V(1)
.PLOT TRAN V(2)
.END
```

(b)

Figure 14.6 CMOS latch switching transients: (a) circuit schematic; (b) SPICE code; (c) waveforms for 2 ns switch closure; (d) waveforms for 0.2 ns closure.

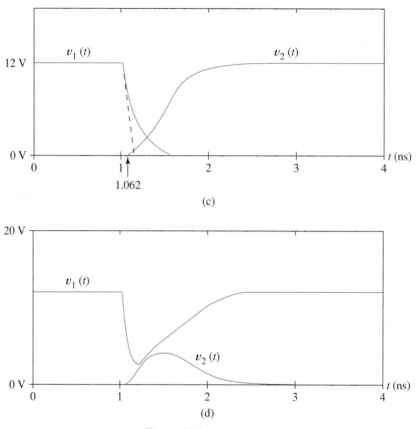

Figure 14.6 (continued)

(a) Use SPICE to find the "state voltages" at nodes 1 and 2 when the switch closes for 2 ns. Find the value of t_1 from the output data.

(b) Find the waveforms if the switch is closed for only 0.2 ns.

Solution. (a) Fig. 14.6b shows the code. M_{SW} is a MOSFET switch with large KP to ensure negligible switch resistance. Further idealizations are that the switch is closed by a pulse of amplitude $V_{DD} + V_t$ and that switch dynamics are negligible. Switch closure is delayed until 1.0 ns to show initial conditions more clearly. Transistors M_1–M_4 are, by default, static models with circuit dynamics introduced by external capacitors. (Zero voltage capacitances are given in comment statements for use in a homework problem.) The .IC statement sets the initial FF state to logic 1.

Figure 14.6c shows the output waveforms. As expected, $v_1(t)$ begins its exponential drop immediately upon switch closure. Expanding the time scale in vicinity of 1 ns shows that $v_2(t)$ begins to change only at 1.062 ns; therefore, $t_1 \approx 0.062$ ns. When the switch reopens at 2 ns, the latch continues to settle into its new logic 0 state because the critical internal switching has occurred. (Problem 14.6 examines the zero-to-one state change.)

(b) Replacing the first *VSW* statement by the second, changes the pulse duration to 0.2 ns for the second simulation. Figure 14.6d shows that at $t = 1.2$ ns $v_2(t)$ has not yet reached the input switching threshold of the second inverter. Because switch closure was too short, the FF state does not change state. ❑

Exercise 14.1 (a) Compute R_4, the ohmic resistance of M_4, at the beginning of the transient in Example 14.1.
Hint: Eq. (5.5).
(b) In Fig. 14.6a, if the voltage at node 1 is pulled down to the point where the two-stage amplifier has maximum gain, we expect regeneration to continue the state-change process. Use Eq. (14.1) to estimate the largest $R_I = R_{in}$ for which this can occur in Example 14.1.
Hint: For what input voltage is the gain of a CMOS inverter maximum?

Ans. 455 Ω, 455 Ω.

Sometimes the designer provides special inputs to control the latch state. Figure 14.7a shows an NMOS latch with control inputs S and R, and with dynamics embodied in

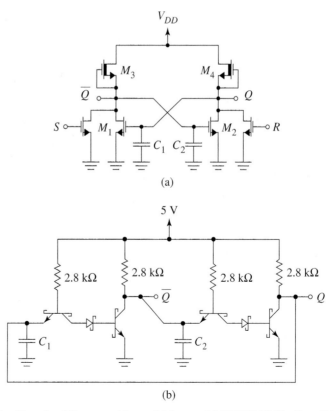

(a)

(b)

Figure 14.7 Changing FF states with special inputs: (a) NMOS *SR* flip-flop; (b) TTL latch; (c) TTL flip-flop using NANDs; (d) equivalent circuit just after $\overline{S}$ goes low.

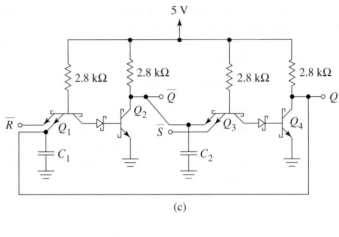

(c)

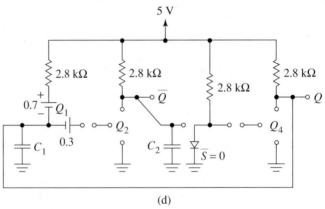

(d)

Figure 14.7 (continued)

parasitic input capacitances. Both inputs are normally low to keep input transistors in cut-off. To store a 1, we hold *set input S* high long enough to initiate regenerative action by discharging C_2; to store a 0 we use the *reset R* input. The switching dynamics and timing constraints resemble those described for the CMOS latch; however, there are two advantages to this switching method over direct switching with an external source. First, no current is required of the external signal source, nor is its internal resistance critical. Second, the input transistor briefly goes into active operation, providing high current to more quickly discharge the input capacitance. To see this, notice that C_2 is initially charged to V_{DD} volts when S goes high.

Unlike CMOS, in this latch there is always one transistor path from power supply to ground, causing power dissipation in both static states.

Figure 14.7b shows a basic TTL latch circuit. In TTL, it is convenient to incorporate control inputs into multiple-emitter transistors as in 14.7c. Both inputs have rest values of 5 V. To store a 1, input $\overline{S}$ must be held low long enough to initiate regenerative action. First Q_4 cuts off, giving Fig. 14.7d. Now C_1 must charge to the point where Q_1 changes to reverse clamp and turns on Q_2. Similar operation holds for the TTL gates with active pull-up. Since Fig. 14.7a represents cross-coupled NORs and Fig. 14.7c cross-coupled NANDs, we see that different logic families lend themselves naturally to different latch representations.

Unstable Equilibrium. We can learn more about the unstable latch state, point c in Fig. 14.1b, by analyzing the equivalent circuit of Fig. 14.8, where both inverter transistors are in their active states. The notational change, replacing v_O by v_Y, emphasizes the symmetry of the circuit for this development. For MOS flip-flops, R_o is the inverter small-signal output resistance. (For a bipolar circuit, the input resistance has been absorbed into the output resistance of the preceding amplifier by a Thevenin transformation.) At the instant depicted in the diagram, the past history of the device is embodied in initial capacitor voltages $V_X(0)$ and $V_Y(0)$.

The capacitor currents are

$$C_{in}\frac{dv_X}{dt} = \frac{-A\,v_Y - v_X}{R_o}, \qquad C_{in}\frac{dv_Y}{dt} = \frac{-A\,v_X - v_Y}{R_o}$$

which differ only in subscript locations. Dividing by C_{in}, replacing $C_{in}R_o$ by τ, and taking the Laplace transform of the first gives

$$sV_X(s) - V_X(0) = -\frac{A}{\tau}V_Y(s) - \frac{1}{\tau}V_X(s)$$

The second equation follows by interchanging X and Y subscripts. In matrix form the two equations are

$$\begin{bmatrix} \left(s + \frac{1}{\tau}\right) & \frac{A}{\tau} \\ \frac{A}{\tau} & \left(s + \frac{1}{\tau}\right) \end{bmatrix} \begin{bmatrix} V_X(s) \\ V_Y(s) \end{bmatrix} = \begin{bmatrix} V_X(0) \\ V_Y(0) \end{bmatrix}$$

We next solve for transforms of $v_X(t)$ and $v_Y(t)$ in terms of initial conditions. The determinant is

$$\Delta = \left(s + \frac{1}{\tau}\right)^2 - \left(\frac{A}{\tau}\right)^2 = s^2 + \frac{2}{\tau}s + \frac{(1 - A^2)}{\tau^2} = \left(s + \frac{1 - A}{\tau}\right)\left(s + \frac{1 + A}{\tau}\right)$$

From the circuit diagram, A^2 is the loop gain. The last expression shows that when loop gain exceeds one (positive feedback), the circuit has a right half-plane pole and is therefore unstable.

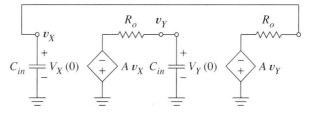

Figure 14.8 Flip-flop active-state equivalent circuit with arbitrary capacitor voltages.

From the poles, we know that $v_X(t)$ and $v_Y(t)$ are both sums of exponential terms. The term associated with the negative real pole in each expression vanishes quickly, leaving the dominant term from the right-half plane. By Cramer's rule the transform of $v_X(t)$ is

$$V_X(s) = \frac{(s + 1/\tau)V_X(0) - (A/\tau)V_Y(0)}{[s + (1 - A)/\tau][s + (1 + A)/\tau]} = F(s)$$

Therefore, the coefficient of the dominant part of $v_x(t)$ is

$$K = F(s)\left(s + \frac{1 - A}{\tau}\right)\Bigg|_{s = -\frac{1-A}{\tau}} = \frac{1}{2}[V_X(0) - V_Y(0)]$$

Using the same procedure for the left-half plane pole gives

$$v_X(t) = \frac{1}{2}[V_X(0) - V_Y(0)]e^{+[(A-1)/\tau]t} + \frac{1}{2}[V_X(0) + V_Y(0)]e^{-[(A+1)/\tau]t}$$

$$\approx \frac{1}{2}[V_X(0) - V_Y(0)]e^{+[(A-1)/\tau]t} \tag{14.5}$$

Interchanging the x and y subscripts in Eq. (14.5) gives

$$v_Y(t) \approx \frac{1}{2}[V_Y(0) - V_X(0)]e^{+[(A-1)/\tau]t} \tag{14.6}$$

Now suppose that the circuit finds itself at *exactly* the unstable equilibrium point of Fig. 14.1b. Then $V_X(0)$ *exactly* equals $V_Y(0)$ in Eqs. (14.5) and (14.6), and dc conditions prevail. However, if a noise perturbation causes $V_X(0)$ to differ even slightly from $V_X(0)$, Eqs. (14.5) and (14.6) predict a quick transition to one of the stable states. For example, if $A = 20$ and $V_X(0) > V_Y(0)$, then

$$v_X(t) \approx \frac{1}{2}[V_X(0) - V_Y(0)]e^{19t/(R_o C_{in})}$$

and

$$v_Y(t) \approx \frac{1}{2}[V_Y(0) - V_X(0)]e^{19t/(R_o C_{in})}$$

We see that $v_X(t)$ quickly becomes large and the transistors limit, with v_X reaching its logic 1 value; $v_Y(t)$ simultaneously drops quickly to logic 0. If $V_X(0) < V_Y(0)$ the flip-flop goes to the opposite state. The equations also show that time constant τ, which comes from R_o and C_{in}, is effectively reduced by $A - 1$, justifying earlier references to the fast response times associated with positive feedback circuits.

Schmitt Trigger. Chapters 2 and 11 described Schmitt trigger circuits and showed that they had two stable states; thus Schmitt triggers are bistable multivibrators. In Chapter 2 we learned how the circuit can extract binary information from large amounts of additive noise. Because of its regenerative switching, the Schmitt trigger is also useful in interfaces where fast logic transitions are required.

Figure 14.9a shows yet another Schmitt trigger, called an *emitter-coupled multivibrator*. For one stable state Q_1 is cut off and Q_2 in clamp; in the other state, the operating modes are reversed.

Figure 14.9b shows the stable state that exists when v_I is a very low voltage. Applying KCL at the emitter gives

$$\frac{v_{X1}}{R} = \frac{[5 - (v_{X1} + 0.7)] + [5 - (v_{X1} + 0.3)]}{qR}$$

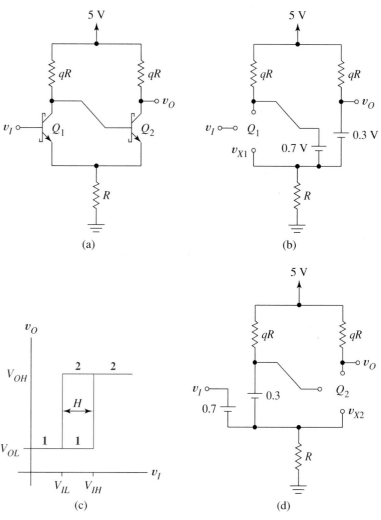

(a)

(b)

(c)

(d)

Figure 14.9 Emitter-coupled Schmitt trigger: (a) schematic; (b) region 1 stable state; (c) transfer characteristic; (d) region 2 stable state; (e) hysteresis width as a function of parameter q.

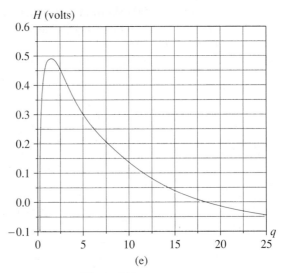

Figure 14.9 (continued)

where q is a design parameter. Multiplying by qR gives

$$qv_{X1} = 9 - 2v_{X1}$$

and

$$v_{X1} = \frac{9}{2 + q} \qquad (14.7)$$

The output voltage for this state is

$$V_{OL} = \frac{9}{2 + q} + 0.3 \qquad (14.8)$$

This operating mode remains valid for increasing v_I in Fig. 14.9b until

$$v_I \approx v_{X1} + 0.5 = V_{IH}$$

at which point Q_1 begins to turn on. This establishes region 1 in Fig. 14.9c. Substituting Eq. (14.7) gives

$$V_{IH} = \frac{9}{2 + q} + 0.5 \qquad (14.9)$$

Once v_I passes V_{IH}, regenerative feedback occurs. The collector current of Q_1 increases, pulling the base of Q_2 downward. As Q_2 comes out of clamp, its previously large current contribution to R decreases, reducing the voltage at the emitter node of Q_1 as its base voltage is increased. Consequently, operation flips quickly to region 2. This corresponds to the stable equivalent circuit, Fig. 14.9d, from which

$$v_O = V_{OH} = 5 \text{ V} \qquad (14.10)$$

As v_I is lowered from a very high value, operation remains in region 2 in Fig. 14.9c until $v_I = V_{IL}$. At this *point*, where Q_1 comes out of clamp and changes to active operation in Fig. 14.9d,

$$i_E = \frac{\beta + 1}{\beta} i_C \approx i_C$$

This means v_{X2} satisfies

$$\frac{v_{X2}}{R} \approx i_C = \frac{[5 - (v_{X2} + 0.3)]}{qR}$$

or

$$v_{X2} = \frac{4.7}{1 + q}$$

Thus regenerative switching begins at $v_I = V_{IL}$, where

$$V_{IL} = v_{X2} + 0.7 = \frac{4.7}{1 + q} + 0.7 \qquad (14.11)$$

Once active, Q_1 functions as a phase inverter. The 180° phase shift between base and collector of Q_1 cause the base voltage of Q_2 to increase with decreases in v_I, while emitters of Q_1 and Q_2 are forced down by v_I. Thus the base-emitter voltage of Q_2 increases rapidly, quickly reaching cut-in. This regeneration ensures a rapid transition to region 1 operation. From Eqs. (14.9) and (14.11), the hysteresis of the transfer characteristic for this circuit is

$$H = V_{IH} - V_{IL} = \left(0.5 + \frac{9}{2 + q}\right) - \left(0.7 + \frac{4.7}{1 + q}\right)$$

$$= \frac{4.3q - 0.4}{(1 + q)(2 + q)} - 0.2 \qquad (14.12)$$

the function sketched in Fig. 14.9e. Problem 14.9 finds that maximum hysteresis, $H_{max} = 0.493$, occurs when $q = 1.59$. Allowing the collector-resistor values to differ from one another, and using a power supply of other than 5 V, gives increased design latitude, but more complex equations.

EXAMPLE 14.2 Design an emitter-coupled Schmitt trigger using a 5 V supply that gives $V_{OL} = 2$ V. Find the hysteresis width.

Solution. From Eq. (14.8) we need $q = 3.3$. We can use $R = 1$ kΩ and $qR = 3.3$ kΩ for the circuit. From Eqs. (14.9) and (14.11), respectively, we have $V_{IH} = 2.20$ V and $V_{IL} = 1.79$. From Eq. (14.12), $H = 0.41$ V. ❑

Exercise 14.2 Design an emitter-coupled Schmitt trigger using a 5 V supply that gives maximum hysteresis. Find V_{OH}, V_{OL}, V_{IH}, and V_{IL}.

Ans. 5 V, 2.81 V, 3.01 V, and 2.51 V.

14.2
Monostable
Circuits

The monostable multivibrator, also called a *pulse stretcher,* or a *one shot,* is "normally" in its stable state. When *triggered* by an input pulse, the circuit switches to its *quasi-stable* state for a time interval, τ, determined by its design parameters, and then spontaneously returns to its stable state. The monostable output is a voltage indicative of its state; thus the overall circuit function is to produce an output pulse of predetermined duration when triggered by a rather arbitrary input pulse. A common application is to establish a time delay of fixed length between two events.

One monostable structure consists of two capacitively coupled NOR gates from any logic family, with feedback from output to input. For simplicity, we use the idealized CMOS gate defined in Fig. 14.10 to introduce the key ideas. Idealizations include zero output resistance, and a transfer characteristic with sharp break points instead of a gradual transition from 1 to 0 values. Figure 14.11a shows a monostable constructed from idealized NORs. R and C control circuit timing, and we assume that transistor capacitances have negligible effects on circuit operation compared to C.

Stable State. The stable state exists when the monostable is in dynamic equilibrium, and $v_I = 0$. In Fig. 14.11a, because C is an open circuit and $R_{in} = \infty$, $v_X = V_{DD}$. With this voltage at the gate #2 input, $v_O = 0$ V. With v_I and v_O both zero, $\overline{v_O}$ has the logic 1 value of V_{DD}. Figure 14.11b uses equivalent circuits to illustrate this stable state. Notice that C is uncharged.

Astable State. Figure 14.11c shows the stable-state conditions we have described for $t < 0$. At $t = 0$, a *triggering signal* $v_I(t)$ disturbs the equilibrium. For example, this might be the exponential at the top of Fig. 14.11c. When $v_I(t)$ first crosses the NOR #1

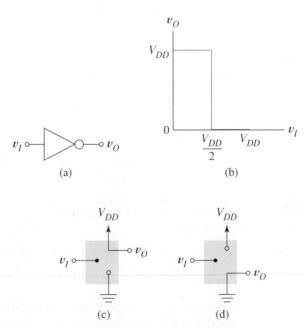

Figure 14.10 Idealized CMOS inverter: (a) schematic; (b) transfer characteristic; (c) equivalent circuit for $v_I < 0.5V_{DD}$; (d) equivalent circuit for $v_I > 0.5V_{DD}$.

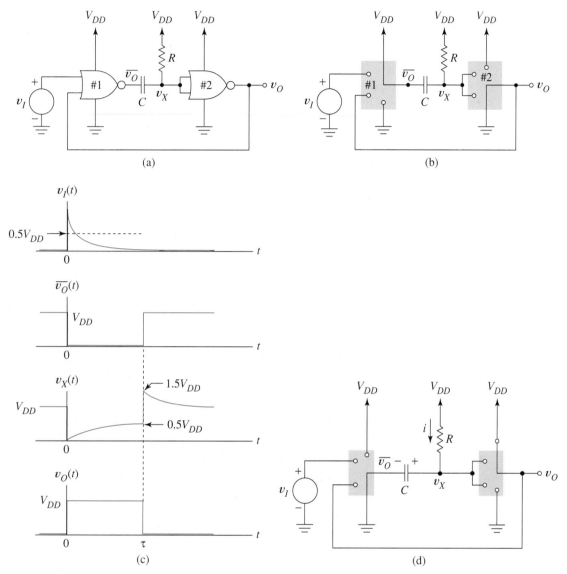

Figure 14.11 Monostable circuit: (a) CMOS realization; (b) equivalent circuit for stable state; (c) key circuit waveforms; (d) equivalent circuit for quasi-stable state.

threshold, $0.5 V_{DD}$, $\overline{v_O}$ drops to zero. Since the voltage across C cannot change instantly, v_X also drops to zero, making the inputs of NOR #2 zero, and forcing v_O to logic 1 or V_{DD} volts. Because v_O provides a second input to NOR #1, once v_O changes to logic 1 the logic value $v_I(t)$ ceases to be of immediate importance. Even if $v_I(t)$ now goes to zero, the output of NOR #1 stays at zero, for the circuit has entered the quasi-stable state, represented by Fig. 14.11d. (In a practical gate, $v_I(t)$ must remain above the $0.5 V_{DD}$ threshold for at least two propagation delays to produce the quasi-stable state.)

Once the circuit enters the quasi-stable state, $v_X(t)$ increases exponentially from its initial value of zero toward a final value of V_{DD}. The capacitor charges according to

$$V_X(t) = V_{DD} + (0 - V_{DD})e^{-t/(RC)} \tag{14.13}$$

with waveform shown in Fig. 14.11c. This equivalent circuit applies until $v_X(t)$ reaches $0.5 \, V_{DD}$, the input threshold of NOR #2. From Eq. (14.13), this takes τ second, where

$$\tau = -RC \, \ell n \, 0.5 = 0.693 RC \tag{14.14}$$

Now the NOR #2 output changes to zero, making the second input to NOR #1 logic 0. The NOR#1 output then switches to logic 1, reestablishing the equivalent circuit of Fig. 14.11b. Figure 14.11c shows that the output waveform is a pulse of duration τ. A complementary output pulse $\overline{v_O}$ is available at the NOR #1 output.

Recovery Time. At the instant that $v_O(t)$ again goes low in Fig. 14.11c, $v_X = 0.5 V_{DD}$ volts. It takes about four time constants for C to discharge to $v_X = V_{DD}$, our original equilibrium value. If a new triggering pulse arrives before this *recovery time* is over, the timing of the circuit will be influenced by this nonstandard initial condition, and the output pulse duration will not be given by Eq. (14.14).

Figure 14.12 shows a monostable constructed from simple Schottky inverters. The stable state consists of Q_1 and Q_3 in cut-off and Q_2 in clamp. For this state v_X is lower than V_{CC}, for R_B must carry sufficient current to keep Q_2 clamped. Thus C is charged to $V_{CC} - v_X$ with polarity as shown.

A positive pulse at v_I sufficient to drive Q_1 into clamp causes v_X to drop suddenly to zero or below, cutting off Q_2 and raising v_O high enough to clamp Q_3. Now the input pulse can end, because the circuit is in its quasi-stable state. *Timing capacitor C eventu- ally discharges and then charges with opposite polarity until v_X reaches the cut-in volt- age of Q_2. Then the circuit reverts to its original stable state after a suitable recovery time. Operation is essentially the same as that of the CMOS circuit above, but the charg-

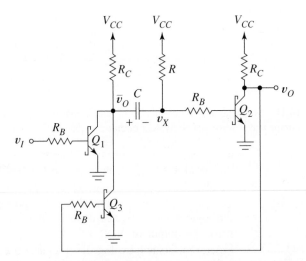

Figure 14.12   Monostable circuit using Schottky transistors.

ing equations and thresholds are somewhat different. Problem 14.13 is a detailed examination of this circuit.

14.3
Astable Circuits

In the astable multivibrator, operation alternates between the two quasi-stable states at a rate that depends upon design parameters. The circuit has no input. An output continuously indicates the state of the circuit by its logic level. The circuit is sometimes called a *relaxation oscillator* to distinguish it from the sinusoidal oscillators of Chapter 12.

One way to make an astable circuit is to connect two inverters with an *RC* feedback circuit as in Fig. 14.13a. Circuit analysis is particularly easy if the inverters are the idealized CMOS devices defined by Fig. 14.10. Analysis begins by assuming inverter logic states and a capacitor initial condition consistent with these assumptions. To begin, let us assume that the output of inverter #1 is logic 0, the output of inverter #2 is logic 1, and the capacitor is uncharged. This gives Fig. 14.13b, which represents one quasi-stable state. The waveform values for $t = 0$ in Fig.14.13c are consistent with these initial conditions.

In Fig. 14.13b the capacitor charges toward $v_X = V_{DD}$ according to

$$v_X(t) = V_{DD}(1 - e^{-t/(RC)})$$

Because v_X is the input to inverter #2, the circuit changes to its other quasi-stable state at t_1, the instant when $v_X(t)$ reaches the switching threshold, $0.5V_{DD}$. We see this in the $v_X(t)$ waveform of Fig. 14.13c. At $t = t_1$, the voltage across C is $0.5V_{DD}$. Although t_1 is readily calculated from the exponential function given earlier, it is of minor importance because we have been tracing start-up from an arbitrarily chosen initial condition.

Figure 14.13d shows the second state of the circuit at $t = t_1^+$. Because of the charge stored on the capacitor, the circuit and the waveform sketch show that

$$v_X(t_1^+) = 1.5V_{DD}$$

Now $v_X(t)$ decreases exponentially from this value toward zero. We deduce the latter by noting that when we replace the capacitor in Fig. 14.13d with an open circuit (equilibrium model), $v_X = 0$. From the initial and final values and Eq. (14.3),

$$v_X(t) = 1.5V_{DD}e^{-(t-t_1)/(RC)}$$

which is sketched in Fig. 14.13c. We see from the waveform that at $t = t_2$, when v_X has decreased to the threshold of inverter #2, the circuit will again change states. Solving for $0.5T = t_2 - t_1$, the width of the first representative output pulse, gives

$$0.5T = t_2 - t_1 = -RC \, \ell n \left(\frac{0.5V_{DD}}{1.5V_{DD}} \right) = 0.405 \, RC \qquad (14.15)$$

Figure 14.13e shows that the original state has been reestablished, and also shows the capacitor voltage at $t = t_2^+$. Thereafter, $v_X(t)$ changes periodically between 1.5 V_{DD} and

$-0.5V_{DD}$, with a state change occurring every T seconds. Complementary output waveforms are available at the two inverter output nodes. It follows from Eq. (14.15) that the oscillation frequency, $f = 1/T$, of the astable in Hz is

$$f = \frac{1.23}{RC} \tag{11.16}$$

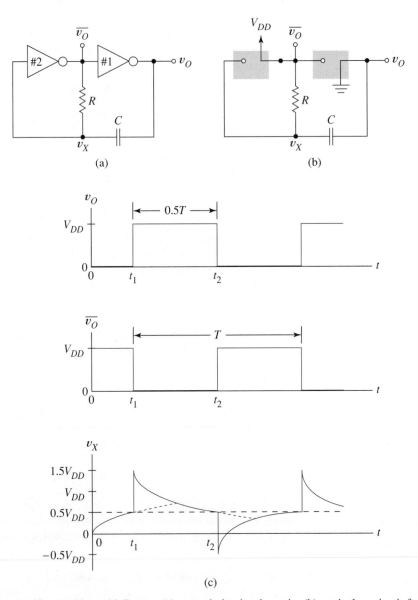

Figure 14.13 Astable multivibrator: (a) general circuit schematic; (b) equivalent circuit for one quasi-stable state; (c) voltage waveforms; (d) second quasi-stable state just after switching at $t = t_1$; (e) circuit equivalent just after switching at $t = t_2$.

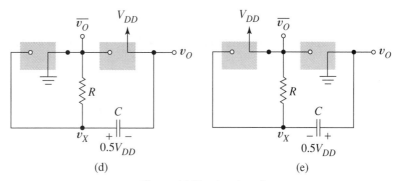

Figure 14.13   (continued)

Figure 14.14a shows another astable multivibrator, an integrator/Schmitt trigger cascade with feedback from output to input. The Schmitt trigger is described by Fig. 14.14b. In Sec. 2.6.2 we found the hysteresis limits for this Schmitt trigger to be

$$v_X = \left(\frac{R_1}{R_2}\right)V_M \quad \text{and} \quad v_Y = \left(\frac{R_1}{R_2}\right)V_P \tag{14.17}$$

where $-V_M$ and V_P are the negative and positive saturation limits of the op amp.

For the initial condition in our analysis, we assume at $t = 0$ operation is at the point

$$(v_{O2}, v_{O1}) = (0 , V_P)$$

on the transfer characteristic. This starting point implies C is uncharged and $i = V_P/R$. The two output voltage functions are

$$v_{O1}(t) = V_P \quad \text{and} \quad v_{O2}(t) = -\frac{V_P}{RC}t$$

as in Fig. 14.14c. From the second equation, $v_{O2}(t)$ decreases linearly in Fig. 14.14b, until that instant, t_1, when $v_{O2} = -v_Y$. Then v_{O1} switches to $-V_M$, and the circuit enters its other quasi-stable state with C still charged to $v_C = -v_Y$ volts.

Since $v_O = -V_M$, now $i = -V_M/R$, and the integrator output is

$$v_{O2}(t) = -v_y + \frac{V_M}{RC}(t - t_1) \tag{14.18}$$

In Fig. 14.14b, operation moves to the right along the bottom branch of the hysteresis curve until $v_I = v_X$ at $t = t_2$, where the state again changes, as indicated in Fig. 14.14c. From Eq. (14.18), the time spent in the $-V_M$ state is

$$t_2 - t_1 = \frac{RC(v_X + v_Y)}{V_M} \tag{14.19}$$

With the circuit again in its V_P state, the $v_{O2}(t)$ decreases linearly to $-v_Y$ at t_3, a process that takes

$$t_3 - t_2 = \frac{RC}{V_P}(v_X + v_Y) \tag{14.20}$$

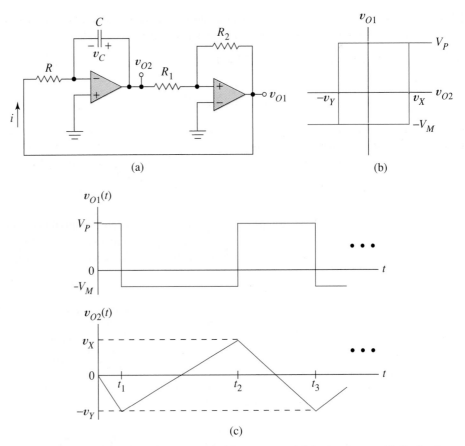

Figure 14.14 Astable circuit: (a) cascaded integrator and Schmitt trigger with feedback; (b) hysteresis curve of Schmitt trigger; (c) output waveforms.

The circuit continues in this fashion, generating the periodic waveforms of Fig. 14.14c. By adding Eqs. (14.19) and (14.20) we find that the period of the waveform, $T = t_3 - t_1$, is

$$T = RC(v_X + v_Y)\left(\frac{1}{V_P} + \frac{1}{V_M}\right)$$

Substituting Eq. (14.17) gives

$$T = RC\frac{R_1}{R_2}(V_P + V_M)\left(\frac{1}{V_P} + \frac{1}{V_M}\right) \qquad (14.21)$$

Sometimes the sawtooth waveform at v_{O2} is the desired output, as in the following example.

EXAMPLE 14.3   Design a circuit to generate the sawtooth waveform with period of $T = 10$ ms. The duration of the falling part of the waveform should be 1 ms.

Solution. From Eqs. (14.19) and Eq. (14.20),

$$\frac{t_2 - t_1}{t_3 - t_2} = \frac{V_P}{V_M}$$

therefore this design requires $V_P = 9V_M$. In the circuit of Fig. 14.14a, the saturation limits V_P and V_M are given parameters of the op amp and power supplies, and their ratio is often close to one. Adding a 6.3 V zener limiter to the Schmitt trigger circuit, as in Fig. 14.15b, gives the new saturation limits $V_P = V_Z = 6.3$ and $V_M = 0.7$ V.

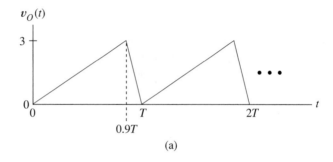

(a)

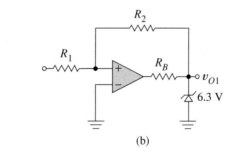

(b)

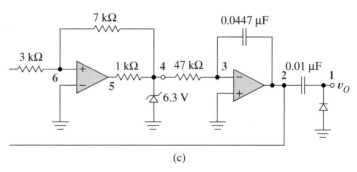

(c)

Figure 14.15 Sawtooth waveform generator: (a) waveform; (b) Schmitt trigger with limiter to control saturation limits; (c) sawtooth generator circuit.

To obtain the required 3 V peak-to-peak amplitude, from Fig. 14.14c and Eq. (14.17) we need

$$v_X + v_Y = \frac{R_1}{R_2}(V_M + V_P) = \frac{R_1}{R_2}(0.7 + 6.3) = 3 \text{ V}$$

Thus $R_1/R_2 = 3/7$.

For the proper period, Eq. (14.21) requires

$$0.01 = RC\,\frac{3}{7}\,(0.7 + 6.3)\left(\frac{1}{0.7} + \frac{1}{6.3}\right)$$

or $RC = 2.1 \times 10^{-3}$ s. To satisfy this condition we choose $R = 47$ kΩ (arbitrarily) and $C = 0.0447$ μF. To adjust the dc level, we clamp the negative peak of the output to zero with a clamping circuit. The final design is Fig. 14.15c. ❏

Because of the nonideal diode in the clamping circuit, the output waveform would actually oscillate between limits of -0.7 V and 2.3 V instead of 0 and 3. If this is a problem, the anode of the diode can be connected to a voltage divider having output of $+0.7$ V.

Slewing and delays introduced by the op amp's frequency response can cause astable circuits to malfunction. The period of 0.01 s in the example is so low that even a rather slow op amp such as a 741 can be used in the realization; however, if a faster sawtooth were needed, op amps with high bandwidths and slew rates might be required.

Because the period T varies directly with C in Eq. (14.21), we can rescale the astable of Example 14.3 to have a higher oscillation frequency by using a smaller capacitor. The next exercise explores intuitively the upper frequency limit imposed by slewing.

Exercise 14.3 Find the frequency of oscillation and required C that make the falling part of the waveform in Fig. 14.15a 10 times as long as it would take the op amp to slew between the same voltages. (We then assume slewing is negligible.) Our op amp slews at 150 V/μs.

Ans. $f = 500$ kHz, $C = 8.94$ pF.

14.4
555 Timer Circuit

The 555 timer is an MSI device available in both TTL and CMOS logic that provides a convenient way to construct high-performance monostables and astables. The dashed box of Fig. 14.16 encloses the internal elements of the 555 timer. Three matched, integrated resistors R form a temperature-independent voltage divider. An *SR* flip-flop provides internal memory, and a high-current buffer circuit provides digital output levels corresponding to the IC family. Comparators #1 and #2, each with one input set by the divider, provide steering logic for the flip-flop, and a switch controlled by $\overline{Q}$ provides an open or short circuit at an external output pin.

14.4.1   555 MONOSTABLE

To realize a monostable with the 555 timer, it is necessary only to connect external components C_X and R_X, as in the diagram.

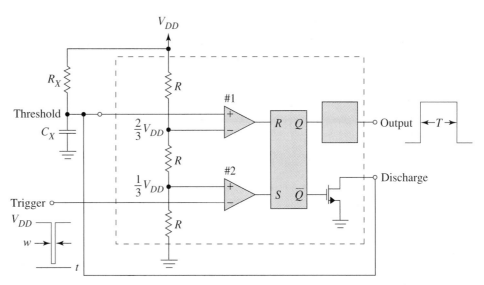

Figure 14.16   555 timing circuit configured with external components for monostable operation.

Stable State. The stable state is characterized by the internal flip-flop state $Q = 0$. We see from Fig. 14.16 that, for this state, the 555 output is zero, the discharge switch is closed, and C_X is discharged. As long as the trigger input stays at the logic 1 level, these assumptions imply $(S, R) = (0, 0)$; thus the assumed state will persist indefinitely.

Quasi-Stable State. A negative pulse that brings the trigger input below $V_{DD}/3$ changes the comparator #2 output to 1, changing flip-flop input to $(S, R) = (1, 0)$. Now Q goes to 1, as does the output voltage. The discharge switch opens, allowing C_X to charge through R_X toward V_{DD} volts. The circuit is now in its quasi-stable state. When the input pulse ends, S returns to zero; however, the circuit remains in its quasi-stable state, because returning to $(S, R) = (0, 0)$ does not change the FF state.

The R input stays at zero until C_X charges past $2V_{DD}/3$. Thus after T seconds, where T is given by

$$\frac{2}{3}V_{DD} = V_{DD}(1 - e^{-T/(R_X C_X)})$$

the output of comparator #1 changes to one, making the excitation $(S, R) = (0, 1)$. Now the internal circuit returns to its original state, the output pulse ends, and the discharge switch closes, quickly discharging C_X. This brings the R input back to zero and re-creates the original initial conditions. Solving the preceding equation gives the output pulse duration

$$T = R_X C_X \ln 3 = 1.1 R_X C_X \tag{14.22}$$

Notice in the derivation that the coefficient of $R_X C_X$ is *independent of V_{DD}* and determined by the ratio of IC resistors. For a 555 monostable, pulse width T can vary from microseconds to hours, and the temperature stability is better than 50 ppm/°C.

14.4.2 ASTABLE MULTIVIBRATOR

Figure 14.17a shows the 555 configured for astable operation. Two external resistors and a capacitor define the timing. Flip-flop state $Q = 1$ denotes one quasi-stable state, $Q = 0$ denotes the other.

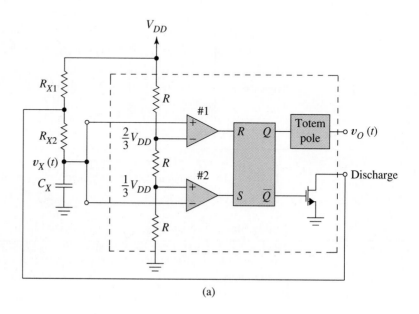

(a)

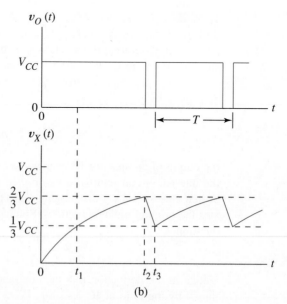

(b)

Figure 14.17 555 TTL timing circuit configured with external components for astable operation.

For purposes of analysis, we use the initial conditions $Q = 1$ and C_X uncharged. Because $v_X(0) = 0$, the flip-flop inputs are initially $(S, R) = (1, 0)$, which is consistent with our state assumption, and the discharge switch is open. As C_X charges toward V_{DD} with time constant

$$\tau_1 = (R_{X1} + R_{X2})C_X \tag{14.23}$$

v_X increases exponentially according to

$$v_X(t) = V_{DD}(1 - e^{-t/\tau_1})$$

The output of comparator #2 changes at $t = t_1$, the time defined by

$$\frac{V_{DD}}{3} = V_{DD}(1 - e^{-t_1/\tau_1}) \tag{14.24}$$

making $(S, R) = (0, 0)$. The flip-flop state is unchanged, and charging continues. At $t = t_2$, defined by

$$\frac{2 V_{DD}}{3} = V_{DD}(1 - e^{-t_2/\tau_1}) \tag{14.25}$$

the output of comparator #1 changes, and the flip-flop input becomes $(S, R) = (0, 1)$. Now the flip-flop changes state. Figure 14.17b shows the waveforms $v_X(t)$ and $v_O(t)$.

When the flip-flop state changes, the discharge switch closes, and C_X discharges with time constant $\tau_2 = R_{X2}C_X$. When $v_X(t)$ drops below $2V_{DD}/3$, the flip-flop input becomes $(S, R) = (0, 0)$. When $v_X(t)$ drops to $V_{DD}/3$, at $t = t_3$ defined by

$$\frac{2}{3}V_{DD}e^{-(t_3 - t_2)/\tau_2} = \frac{1}{3}V_{DD} \tag{14.26}$$

then the output of comparator #2 changes, giving $(S, R) = (1, 0)$. We see that after an initial transient, the circuit settles down to periodic operation as indicated in Fig. 14.17b.

To find the period, we note that $T = (t_3 - t_2) + (t_2 - t_1)$. From Eq. (14.26),

$$t_3 - t_2 = \tau_2 \ln 2$$

From Eqs. (14.25) and (14.24),

$$t_2 = \tau_1 \ln 3 \quad \text{and} \quad t_1 = \tau_1 \ln \left(\frac{3}{2} \right)$$

thus

$$t_2 - t_1 = \tau_1 \left[\ln 3 - \ln \left(\frac{3}{2} \right) \right] = \tau_1 \ln 2$$

Adding the first and third equations gives the period

$$T = t_3 - t_1 = (\tau_1 + \tau_2) \ln 2$$

Substituting for the two time constants leads to the design equation

$$T = 0.693(R_{X1} + 2R_{X2})C \tag{14.27}$$

The period is independent of V_{DD}, and the numerical coefficient depends upon IC resistor ratios.

In some applications, the *duty cycle* δ, defined here as the fraction of the period T for which the output is high, is important. From the preceding expressions, we find

$$\delta = \frac{t_2 - t_1}{T} = \frac{\tau_1 \ell n\, 2}{(\tau_1 + \tau_2)\, \ell n\, 2} = \frac{R_{X1} + R_{X2}}{R_{X1} + 2\,R_{X2}} = \frac{1}{1 + R_{X2}/(R_{X1} + R_{X2})} \quad (14.28)$$

which restricts the duty cycle to $0.5 < \delta < 1.0$. From the design viewpoint, Eqs. (14.25) and (14.28) show that we can use a resistor *ratio* to fix the duty cycle, and then independently select one resistor to satisfy some other constraint such as power dissipation in the charging circuit, and finally select a capacitor to set the frequency.

The CMOS version of the 555 timer is capable of astable operation at frequencies up to 3 MHz, has less than 1 mW typical power dissipation with a 5 V power supply, and is rated for maximum output current of 100 mA.

14.5
Interfacing Problems and Principles

Interface designs usually involve applying principles covered earlier in the text plus common sense. In this section we identify some generic problems associated with interfaces and suggest how they might be solved. We also describe two specialized interface circuits: A/D and D/A converters.

One class of interface problem involves interconnecting logic and analog circuits, for example, analog circuits that provide inputs and outputs for digital systems. Representative input sources are keyboards, telephone lines, flow meters, pressure transducers, photodetectors, clock circuits, and A/D converters. Examples of output devices are displays, printers, telephone lines, relays, and D/A converters.

Another interfacing problem involves interfacing different logic families. One motivation is to establish communication between independently designed systems For example, a digital radio receiver implemented in one logic family might require an interface so it can provide data to a computer realized in another. Interfacing between logic families might also be appropriate when some function, memory, for example, is unavailable or more expensive in the primary logic family.

14.5.1 LOGIC SWING TOO SMALL

For our first generic problem, consider the case where a circuit's logic swing is too small for its digital load gate. Figure 14.18a shows an example. A symmetric CMOS gate with transistors of $|V_t| = 1$ V and input logic limits calculated from Eqs. (13.18) and (13.19) receives input from a source of 2 V clock pulses. If the clock voltage were translated upward it would still be inadequate to drive the gate, since the 2 V clock amplitude is less than $V_{IH} - V_{IL}$. Clearly amplification is required; however, since the signal information is binary, preserving the exact input waveshape is less important than reliability. Therefore, the amplifying circuit should be digital in the sense of employing transistors as switches, *not* a linear amplifier from Chapters 7 and 8.

Figure 14.18b shows how a BJT switch with pull-up and base resistors can make the clock compatible with the input requirements of the gate. The transistor switches between saturation and cut-off, producing an input signal for the inverter that changes from

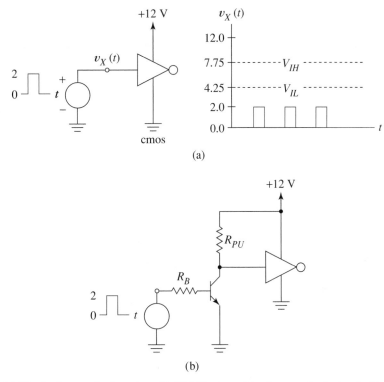

Figure 14.18 Inadequate logic swing: (a) CMOS view of inadequate clock signal; (b) interface circuit to amplify the digital signal.

0.2 to 12 V. This logic swing, 11.8 V in amplitude, exceeds $V_{IH} - V_{IL} = 3.5$ V and also spans the input logic thresholds.

The interface design is straightforward. Because the static input current for the CMOS circuit is zero, the BJT saturates provided that

$$\beta I_B = \beta \frac{2 - 0.7}{R_B} \geq \frac{12 - 0.2}{R_{PU}} = I_C \qquad (14.29)$$

where β means the *dc beta* denoted by h_{FE}. For reliable interface operation, R_B should be small enough to satisfy the inequality in spite of expected parameter variations. The inversion (or logic complement) inherent in the interface circuit is usually acceptable (or easily corrected using a logic inverter). An MOS switch can perform the same interfacing function, as we show in the following example.

EXAMPLE 14.4 Figure 14.19 shows an NMOS interface circuit intended to solve the problem of Fig. 14.18a.

Solution. The first requirement is that the threshold voltage of the transistor be well below 2 V so that the switch opens and closes in response to the signal. When the input pulse is high, the MOSFET should be driven into ohmic operation and act as a resis-

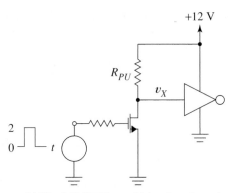

Figure 14.19   MOSFET amplifying interface circuit.

tor—a resistor small enough to produce an output voltage well below $V_{IL} = 4.25$ V. We satisfy this condition somewhat arbitrarily by selecting $V_{X,low} = 0.2$ V.

Approximating the resistance R_I of the ohmic NMOS transistor by Eq. (5.5), we have

$$R_I = \frac{1}{k(2 - V_t)} \qquad (14.30)$$

subject to the condition

$$|v_{DS}| \le 0.2\ |(2 - V_t)|$$

Because we have selected $V_{X,low} = v_{DS,low} = 0.2$ V, Eq. (14.30) is valid for $1 \le 2 - V_t$, so we select $V_t = 1.0$ V to satisfy the necessary condition. With this threshold voltage, the transistor turns on and off at the midpoint of the 2 V clock signal.

The low value of V_X is provided by the voltage division

$$\frac{R_I}{R_I + R_{PU}}\ 12 = 0.2 \qquad (14.31)$$

Solving for the divider ratio in Eq. (14.31) and inverting gives

$$1 + \frac{R_{PU}}{R_I} = 60$$

Substituting Eq. (14.30) with $V_t = 1$ gives

$$1 + R_{PU}k(2 - 1) = 60 \qquad (14.32)$$

We have one final design choice. Using a MOSFET with $k = 4 \times 10^{-4}$ A/V², for example, requires $R_{PU} = (59/4) \times 10^4 \approx 148$ kΩ.   ❏

Exercise 14.4   The interface of Fig. 14.18b is to drive the gate of Fig. 14.18a. Find R_B and R_{PU} so that the interface operates for $20 \le \beta \le 100$, even if R_B and R_{PU} independently vary by as much as 20%.

Ans. $R_B = 10$ kΩ (arbitrary), $R_{PU} = 6.81$ kΩ.

Noise Margins for Interfaces. The interface example of Fig. 14.18b suggests the idea of noise margins for interfaces. Between the transistor and the gate we have $NMH = 12 - 7.75 = 4.25$ V and $NML = 4.25 - 0.2 = 4.05$ V. For the design of Exercise 14.4, the minimal clock voltage for saturating the transistor is $0.7 + R_{Bmin} I_B = 0.7 + 10$ k$(8.6 \ \mu A) = 1.57$ V. Thus between clock and interfacing transistor we have $NMH = 2 - 1.57 = 0.43$ V and $NML = 0.7 - 0 = 0.7$ V.

14.5.2 OUTPUT LEVELS DO NOT OVERLAP INPUT LEVELS

If the logic swing has sufficient amplitude, it might still be incorrectly aligned relative to the input requirements as in Fig. 14.20a. This problem requires a *level shift*, which, in general, might need to be either positive or negative. Figure 15.20b shows idealized circuits that shift $v_I(t)$ upward or downward by ΔV volts.

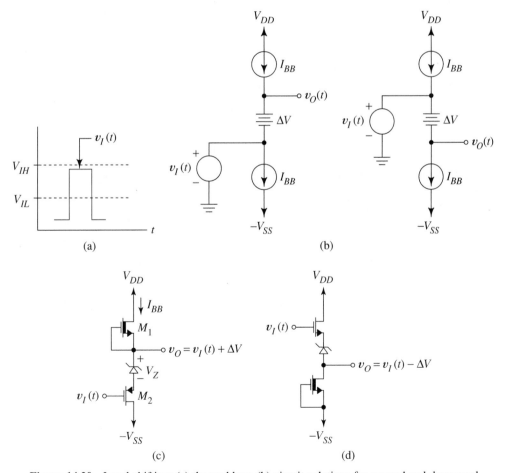

Figure 14.20 Level shifting: (a) the problem; (b) circuit solutions for upward and downward shifts; (c), (d) transistor-level shifter implementations.

Figure 15.20c shows one possible level shifter implementation. A zener diode biased at I_{BB} provides part of the level shift. Depletion and enhancement MOSFETs in active-state operation are the current sources. Parameters for M_1 must satisfy

$$I_{BB} = \frac{k_1}{2}(0 - V_{t1})^2 \tag{14.33}$$

where I_{BB} must be suitable for maintaining zener operation. With its current established at I_{BB}, M_2 has a constant negative V_{GS} that satisfies

$$I_{BB} = \frac{k_2}{2}(V_{GS} - V_{t2})^2$$

The positive level shift is then

$$\Delta V = V_Z - V_{GS}$$

or, more specifically

$$\Delta V = V_Z - V_{t2} + \sqrt{\frac{2I_{BB}}{k_2}}, \text{ where } V_{t2} < 0 \tag{14.34}$$

Active-state operation of M_1 requires

$$V_{DG1} = V_{DD} - v_O \geq -V_{t1}$$

or

$$V_{DD} \geq v_O - V_{t1} \tag{14.35}$$

which provides a lower limit for V_{DD} in a given design. Similarly, active-state operation of M_2 requires

$$V_{DG2} = -V_{SS} - v_I(t) = -V_{SS} - v_O + \Delta V \leq -V_{t2}$$

showing that V_{SS} must satisfy

$$V_{SS} \geq \Delta V + V_{t2} - v_{O,min} \tag{14.36}$$

A design typically begins with known values for the required shift ΔV and the output levels $v_{O,max}$ and $v_{O,min}$ that will result from the shift. The designer then uses the foregoing equations to select M_1, M_2, and the diode, and to ensure that supply voltages are adequate. Figure 14.20d shows a similar circuit that downshifts by ΔV. The same design principles apply although the equation details differ. A number of similar circuits utilizing JFETs, BJTs, or current sources employing mirror circuits can implement the same general level-shifting strategy.

Level Shifting with Capacitors. The level-shifting strategy of Fig. 14.20 is _dc coupled_ and, therefore, works for dc inputs and time-varying signals. When the input is a periodic waveform such as a clock signal, simpler circuits such as coupling capacitors or clamping circuits sometimes suffice.

Figure 14.21a shows a 5 V square wave superimposed upon a 2.5 V dc level. To make this signal compatible with the input logic levels indicated in the figure, we need

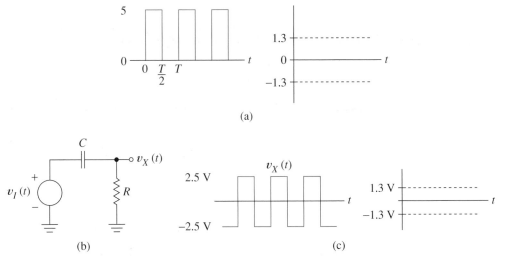

Figure 14.21 Shifting levels of periodic signals: (a) illustration of the level mismatch; (b) *RC* highpass circuit; (c) output waveform.

only an *RC* highpass circuit like Fig. 14.21b. The capacitor charges to the average of the input waveform, 2.5 V in this case, and produces a zero-average output of original amplitude as in Fig. 14.21c. From Chapter 1, the fractional sag in the output waveform is

$$\text{f.s.} \approx \frac{T}{2\,RC}$$

where *T* is the period. For the sag to be less than 1%, for example, we need $RC \geq 50T$.

Figures 14.22a through c show the steady-state capacitor voltages and the output waveforms produced by clamping circuits with ideal diodes, when the input waveform is that of Fig. 14.21a. The basic clamping circuit was previously described in the discussion of Fig. 3.47a, and Problem 3.52 explored some of its variations. Figure 14.22d serves to remind us that if the diode never conducts, the capacitor cannot charge and the waveform is unchanged. As explained in Chapter 3, we can easily accommodate real diodes by appropriate changes in the dc source voltages.

The circuits of Figs. 14.18b, and 14.19 sometimes offer a simple solution to a level-shifting problem, because they shift the dc level as they amplify logic swing.

14.5.3 GATE OUTPUT CANNOT SOURCE OR SINK REQUIRED LOAD CURRENT

At an interface with a peripheral device, a gate might be required to either source or to sink load current. For example, in Fig. 14.23a a logic inverter is to turn on a load device when v_I is logic 0 and turn it off when v_I is logic 1. For loads such as LEDs, relays, piezoelectric alarm circuits, and laser diodes we usually specify their ON operation by a rated voltage V_L and a minimum dc current I_L. Because the output resistance, R_{high}, of

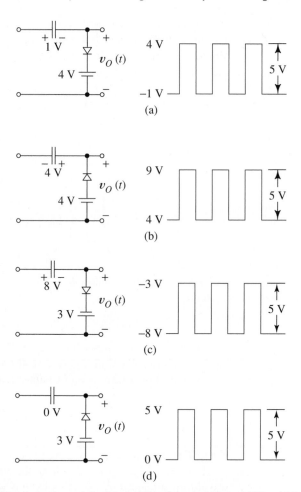

Figure 14.22 Clamp circuit used for level shifting the waveform of Fig. 14.21a: (a) positive peak clamped to +4 V; (b) negative peak clamped to +4 V; (c) positive peak clamped to −3 V; (d) circuit fails to clamp negative peak to −3 V because diode never conducts.

the gate is fixed, the interface usually requires an additional resistor R to prevent excessive current.

An alternate arrangement, Fig. 14.23b, activates the load for v_I = logic 1 and requires the gate to *sink* load current I_L through the gate resistance R_{low}. This value of v_O need not conform to the conventions of the logic family, ($v_O \geq V_{IH}$ or $v_O < V_{IL}$), because no logic gates are driven by v_O. The principal interface consideration is that the gate is able to supply minimum current I_L without exceeding the specified power dissipation of the (gate) IC chip.

Choosing between Figs. 14.23a and b depends upon whether or not the load needs to be grounded, and also upon the logic family. NMOS with enhancement or depletion load employs a high-resistance pull-up transistor and a low resistance inverting transistor; therefore, this gate can sink load current more easily than it can source it. TTL also favors such an arrangement because gates are designed to sink large currents to provide fan-out. CMOS, because of its symmetric structure, is equally capable of sourcing and sinking current. A simple expedient for providing increased output current is to connect gates in parallel as in Fig. 14.23c.

If the logic gates themselves are incapable of providing needed current, special interface *buffers* like those of Fig. 14.23d give greater design freedom. Notice that voltage

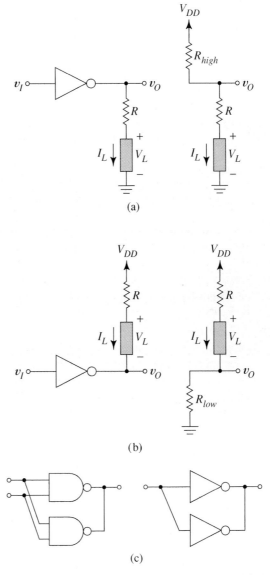

(a)

(b)

(c)

Figure 14.23 Using logic gates to control nonlogic loads: (a) by sourcing current; (b) by sinking current; (c) increased current through parallel operation; (d) using an interface buffer to provide current; (e) BiCMOS scheme using BJT output buffer for increased current.

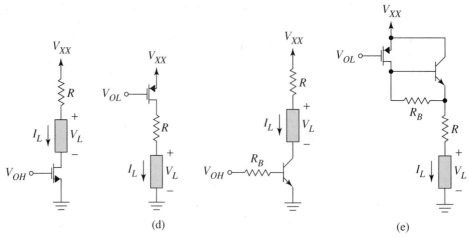

Figure 14.23 (continued)

V_{XX} need not be the supply voltage of the logic family. It is only necessary that the output logic level, V_{OH} or V_{OL}, be adequate to operate the interface transistor. Special interface transistors can be designed for the greater W/L or A required by the current requirements of the load device. In BiCMOS gates, bipolar transistors sometimes augment the output current of a MOSFET as in Fig. 14.23e.

EXAMPLE 14.5   Design an interface like the second circuit of Fig. 14.23d to interface 5 V CMOS logic to an optical communication line using a diode laser rated at 1.75 V and 50 mA. Specify R, k, and V_t for the transistor, and compute the transistor power dissipation when the laser is on.

Solution. It is more economical to use the existing power supply when possible, so we start with $V_{XX} = 5$ V. Figure 14.24 results when the CMOS circuit applies 0 V to the gate of the interface transistor, where R_M is the interface transistor's ohmic-state resistance, Eq. (5.12). Using known values, we identify these design equations:

$$R_M + R = \frac{5 - 1.75}{50 \times 10^{-3}} = 65 \ \Omega \tag{14.37}$$

and

$$R_M = \frac{1}{k|-5 - V_t|}, \qquad V_t < 0 \tag{14.38}$$

which is subject to the condition

$$|v_{DS}| \leq 0.2 \ |-5 - V_t| \tag{14.39}$$

and

$$v_{DS} = -R_M \ (50 \times 10^{-3}) \tag{14.40}$$

Since there are five unknowns (R, k, V_t, v_{DS}, R) and four equations, we begin with a choice. In Eq. (14.39) choose $V_t = -0.8$ V. There is no assurance that this is the best choice, a wise choice, or even an acceptable choice, but it gets us started. (In design, second guessing is acceptable, and usually our grasp of the problem improves as we progress.) From Eq. (14.39) our design now requires

$$|v_{DS}| \le 0.2 \ |-5 + 0.8| = 0.84$$

Next we select $v_{DS} = -0.5$ V to satisfy this inequality with a little room to spare. From Eqs. (14.40) and (14.38), we now obtain

$$R_M = \frac{0.5 \text{ V}}{50 \text{ mA}} = 10 \ \Omega = \frac{1}{k|-5 + 0.8|}$$

R dissipates 138 mW.

This requires $k = 1/42 = 23.8 \times 10^{-3}$ A/V^2. Also, since $R_M = 10$ Ω, Eq. (14.37) implies $R = 55$ Ω. When the laser is on, the power dissipation in the transistor is

$$P_D = (50 \times 10^{-3})^2 \ 10 = 25 \text{ mW}$$

The value of k is somewhat high. We can reduce it in a redesign by selecting larger V_{XX}, a smaller V_t, or both. ❏

> **Exercise 14.5** Rework the interfacing problem of Example 14.5, changing as little as possible, but using $V_{XX} = 10$ V to reduce k.
>
> **Ans.** $|v_{DS}| = 1.54$ V (somewhat arbitrary), $k = 3.5 \times 10^{-3}$ A/V^2, $R = 134$ Ω, $P_D = 77$ mW.

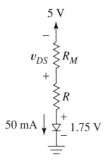

Figure 14.24 Equivalent circuit for the design in Example 14.5.

14.5.4 INTERFACES BETWEEN LOGIC FAMILIES

Interfaces between logic families involve problems already introduced, such as mismatches in logic swing, incompatible logic levels, and requirements for sourcing and sinking currents. Interface circuits generally are designed for fan-out of one and operate at the speed of the slower of the families. *Interface noise margins,* which sometimes require attention, use output values for the driving gate and input values for the driven gate.

CMOS/TTL Interface. Figure 14.25a shows an interface from CMOS to TTL. The lowest logic 1 input voltage for the TTL gate is $v_O \ge 2$ V at $i_{IN} \approx 0$. For a satisfactory logic 0, TTL requires $v_O \le 0.8$ V and $i_{IN} \approx -2$ mA. This means with M_2 off, M_1 must present a sufficiently low ohmic-state resistance. Specifically, the logic 0 output resistance of M_1 must not exceed $R_1 = 0.8/2\text{mA} = 400$ Ω. From Eq. (5.5), parameters for M_1 and V_{DD} must satisfy

$$\frac{1}{k(V_{DD} - V_t)} \le 400$$

For example, for $V_{DD} = 5$ V and $V_t = 1$ V, we need $k \ge 6.25 \times 10^{-4}$A/V^2. For larger V_{DD}, k could be correspondingly reduced.

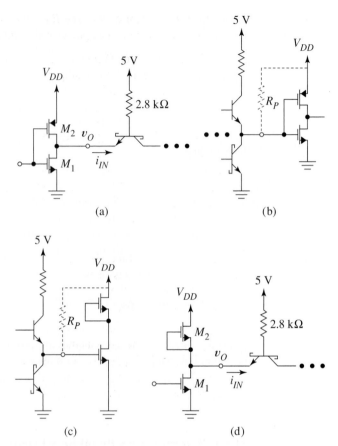

Figure 14.25   Interface circuits: (a) CMOS drives TTL; (b) TTL drives CMOS; (c) TTL drives NMOS; (d) NMOS drives TTL.

In Fig. 14.25b, TTL drives CMOS. In Chapter 13 we estimated that TTL active pull-up produces a logic 1 output voltage of about 3.6 V. For CMOS, according to Eq. (13.19), logic 1 input must be at least

$$V_{IH} = \frac{1}{8}(5V_{DD} - 2V_t)$$

For $V_{DD} = 5$ V and $V_t = 1$ V, $V_{IH} = 2.88$ giving

$$NMH = 3.6 - 2.88 = 0.720 \text{ V}$$

a rather small noise margin. For $V_{DD} \geq 6.16$ V, the V_{IH} equation gives an unacceptable *negative noise margin*.

Pull-up resistor R_P in Fig. 14.25b is a simple expedient that improves the noise margin for low values of V_{DD} and allows the interface to work for $V_{DD} > 6.16$. This resistor simply *pulls up* the logic 1 interface voltage to V_{DD} volts. At logic 0, the clamped TTL output transistor gives 0.3 V, which is easily less than V_{IL} for CMOS; the clamped transistor sinks the current of R_P. The R_P value is not critical to the operation of the interface. A large resistor reduces power dissipation in R_P for logic 0 interface voltage.

NMOS/TTL Interface. Figure 14.25c shows a TTL-to-NMOS interface. When the interface voltage is logic 0, the clamp voltage of the output transistor easily holds the gate node of the load gate below $V_{IL} = V_t$ as required. For low V_{DD}, the 3.6 V output of the totem pole circuit barely exceeds V_{IH} for the NMOS gate, and a pull-up resistor is generally required. If the NMOS gate has a depletion instead of enhancement load, the steep curve and lower V_{IH} of Fig.13.9c suggest that the low TTL logic 1 value is less likely to be a problem.

When NMOS drives TTL as in Fig. 14.25d, the current sinking problem at logic 0 is similar to that of the CMOS-to-TTL interface; however, ohmic-state transistor M_1 must also sink the current of M_2. Because of the low input current demand of the TTL gate at logic 1, the NMOS gate produces the required voltage provided that V_{DD} is greater than $V_{IH} = 2$ V.

NMOS-CMOS Interface. Direct connections between NMOS and CMOS are possible when the power supplies of the two families do not differ greatly. If the supply of the driving gate is too small, a pull-up resistor might be needed. Problem 14.41 explores this interface more carefully.

Interfaces With ECL. With the aid of Figs. 13.40e and 13.41e we recall that ECL input and output logic levels are negative, and the voltage swing is small; consequently, interfaces with other logic families inevitably involve level shifting, and either amplification or compression of the logic swing. The commercially available level translators described next provide a ready solution for these more difficult interfacing problems.

An ECL-to-TTL *translator* circuit operates between the two given supplies as in Fig. 14.26a, accepting ECL levels for input and providing TTL output levels. Figure 14.26b shows the important features of the translator. The input circuit resembles an ECL gate, with Q_1 and Q_6 switching the bias current to either the 790 Ω resistor or the 188 Ω resistor, respectively, for v_I logic 1 or 0. Transistors Q_2, Q_3, and Q_4 are the TTL totem pole output circuit of Fig.13.33a. When v_I is logic 1, Q_3 cuts off and Q_2 clamps, to produce the usual TTL logic 0 level at v_O. When v_I is logic 0, Q_4 clamps, while Q_2 cuts off because of the increased current through the 188 Ω resistor. A similar translator emulates the TTL input and the ECL output.

14.5.5 PROVIDING PULSES IN RESPONSE TO EVENTS

Some interfacing problems require us to provide a narrow input pulse for a counter or other digital circuit in response to some event. The event might be a level change in a digital circuit, or it could be a threshold crossing in an analog circuit, say, temperature passing a critical value. Especially useful for such problems are the monostable, the comparator, and the *RC differentiator.*

The differentiator of Fig. 14.27a produces a narrow output pulse for every input-level transition. Unlike the *RC* level shifter of Fig. 14.21, the differentiator has a very short time constant. Figure 14.27b shows the circuit's output when the input is a pulse train of period T when $RC \ll T$. A positive output spike of amplitude E accompanies

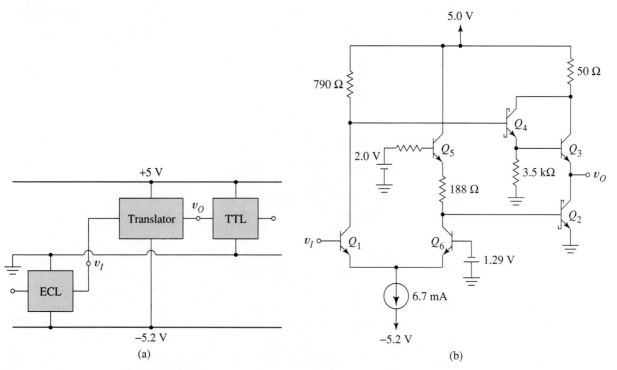

Figure 14.26 ECL-to-TTL translator: (a) connection to supplies and logic circuits; (b) translator circuit.

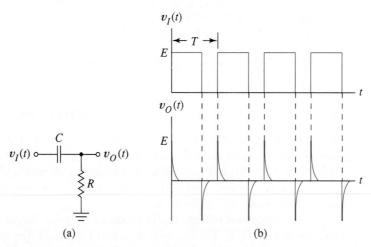

Figure 14.27 *RC* differentiator: (a) circuit; (b) input and output waveforms; (c) use of diode limiter to remove negative output pulses.

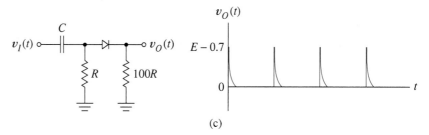

Figure 14.27   (continued)

each positive input transition; a negative spike of the same amplitude accompanies each negative transition. The average value of the output is, of course, zero because the capacitor cannot pass dc. If needed, a diode circuit readily eliminates the negative spikes as in Fig. 14.27c—opposite diode polarity eliminates the positive spikes instead. If we need well-shaped pulses, we can use the differentiator output to trigger a monostable.

Figure 14.28a shows the differentiator used in an interface between a flow rate transducer and a flip-flop. Normally flow rate, $v_T(t)$, is greater than 2.20 V, the Schmitt trigger output is 5 V, and the flip-flop is reset, with input $T = 0$. If $v_T(t)$ ever drops below 1.79 V, however, the flip-flop is to change state. The T flip-flop is a CMOS device with input switching threshold near $V_{DD}/2 = 6$ V.

The differentiator/switch circuit must convert the 5- to 2-volt-level change of the Schmitt trigger into a positive input pulse large enough to trigger the flip-flop. The transistor switch, with gate biased at 2 V by R_1 and R_2, is normally ohmic. When the Schmitt trigger changes state, the 3 V drop at its output is instantaneously transmitted to the transistor gate by C. The MOSFET briefly cuts off, producing a positive 12 V pulse for the flip-flop's T input. As C charges to its new equilibrium value, the gate voltage exponentially returns to 2 V. The following example explores some details.

EXAMPLE 14.6   (a) If the transistor threshold voltage is 0.5 V, find k so that the steady-state drain voltage is 0.4 V.
(b) Find values for R_1, R_2, and C so that the transistor gate is biased at 2 V and so that the differentiator time constant is 1 μs.
(c) Estimate the width of the FF input pulse.

Solution. (a) The ohmic resistance of the MOSFET, R_{out}, must satisfy

$$\frac{R_{out}}{R_{out} + 50 \text{ k}} 12 = 0.4$$

This gives $R_{out} = 1.72$ kΩ. From Eq. (5.5),

$$1.72 \text{ k}\Omega = \frac{1}{k(2 - 0.5)}$$

which is satisfied by $k = 3.88 \times 10^{-4}$ A/V².

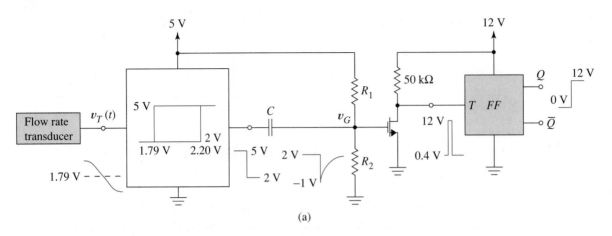

(a)

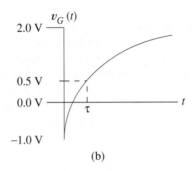

(b)

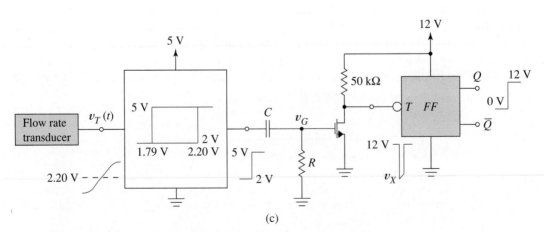

(c)

Figure 14.28 Interface between transducer output and a flip-flop: (a) interfacing circuit; (b) differentiator output waveform; (c) redesigned interface for Exercise 14.6.

(b) For a 2 V quiescent gate voltage, R_1 and R_2 must satisfy

$$\frac{R_2}{R_2 + R_1} 5 = 2$$

A 1 μs time constant requires

$$\frac{R_2 R_1}{R_2 + R_1} C = 1 \ \mu s$$

If we select $C = 100$ pF, the parallel equivalent resistance must be 10 kΩ. Multiplying the first resistance equation by R_1 gives

$$10\ k(5) = 2R_1$$

Thus $R_1 = 25$ kΩ and $R_2 = 16.7$ kΩ.

(c) Figure 14.28c shows that the transistor cuts off for τ s, where, according to Eq. (14.3), τ satisfies

$$0.5 = 2 + (-1 - 2)e^{-\tau/10^{-6}}$$

This gives $\tau = 0.693$ μs. ❏

> **Exercise 14.6** Figure 14.28c outlines a redesign of the interface of Example 14.6 to provide for setting the flip-flop when flow transducer output *exceeds* 2.2 V. Find values for R and C so that the interface transistor is ohmic for 0.7 μs. Also check the minimum value, v_X of the flip-flop input pulse to ensure that it is lower than the 6 V threshold. Use an interface transistor with $k = 3.88 \times 10^{-4}$ A/V² and $V_t = 0.5$ V.
>
> *Ans.* $C = 100$ pF (arbitrary), $R = 3.91$ kΩ, $v_{X,min} = 0.242$ V.

14.5.6   TRANSIENT VOLTAGE LEVELS ENDANGER INPUT TRANSISTORS

The base-emitter junction of a high-β transistor has a low reverse breakdown voltage because of the narrow base width; the thin gate oxide of a MOSFET can be damaged by excessive voltage. For these reasons most logic gates have built-in *protective diodes* to shelter input transistors from excessive voltages. Figures 14.29a and b show the diodes that limit input voltages of CMOS and TTL inverters. Normally open circuits, these diodes only conduct when input signals exceed normal logic levels. The diode limiters introduced in Sec. 3.9 are useful for this purpose. Figures 14.29c and d show how a limiter protects an interface CMOS gate. Notice in Fig. 14.29d that, after limiting, the gate input voltage, $v_X(t)$, is suitable for CMOS input.

Schottky diodes are preferable to junction diodes in limiters because they clamp at lower voltages and do not introduce charge storage delays. In Sec. 14.6 we will see how transmission line effects can result in dangerously high transient voltages if diode protection is not present.

Figure 14.29e shows another situation that requires a protective diode, an inductive load circuit such as a motor or relay. When the transistor switch is closed, the diode is

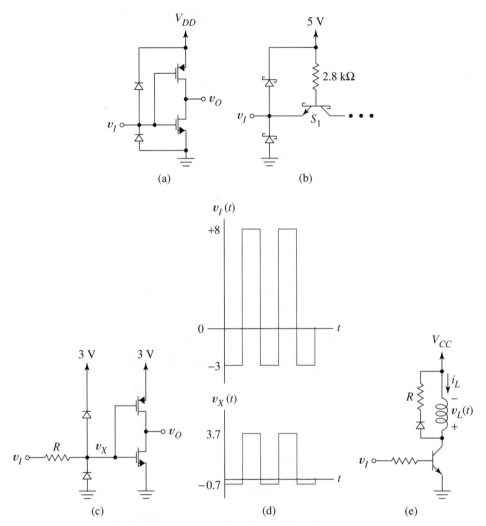

Figure 14.29 Input overvoltage protection: (a) for CMOS gate; (b) for TTL gate; (c) for CMOS interfacing circuit; (d) input and output waveforms for the limiter of (c); (e) protection in relay or other switched magnetic circuit.

open and dc current flows through the inductor. When the switch suddenly opens, however, there is an induced inductor voltage

$$v_L(t) = L\frac{di_L}{dt}$$

with the polarity indicated that acts to prevent change in the coil current. Without diode protection, $v_L(t)$ added to V_{CC} can easily destroy the transistor because i_L decreases so rapidly. When the circuit contains the diode, the induced voltage causes the diode to conduct, and the energy stored in the coil is dissipated in R.

14.5.7 MULTIPLE GATES SHARE A LOAD

Special gate designs are required when several logic gates share a common output node. In Fig. 14.30a, for example, several gates that communicate with independent peripheral devices share a computer interrupt input pin. In this arrangement, any individual gate must be able to pull the logic level of the pin to zero, no matter what the inputs to the other gates. We learned in Chapter 13 that logic gates generally provide one transistor to pull the output to logic 1 and another to pull it to logic 0. Figure 14.30b shows the fault situation that arises when different gates try to impose different logic values on the interrupt pin. A *pull-up transistor* in one gate becomes connected in series with the *pull-down transistor* in another gate. Because the circuit resistance is so low, destructively large currents can destroy one or both gates.

A special logic circuit called an *open drain gate (open collector gate)* solves this class of interfacing problems. In such a gate, a pull-down transistor pulls the logic level to zero by ohmic or clamp operation, whereas a user-supplied *external pull-up resistor* shared by all the connected gates, causes pull-up to logic 1. Figure 14.31 shows the outputs of three such gates sharing pull-up resistor R_P. The internal logic that controls the pull-down transistor is omitted. What is important to notice is that *any* gate that so wishes can pull v_O down by turning on its output transistor; however, v_O goes to logic 1 only if all gates want logic 1.

The parameters of the output transistors of the driving gates are generally given. The interfacing task involves selecting V_{DD} and R_P such that v_O is below V_{IL} (for the load circuit) for the worst case logic 0, and above V_{IH} for logic 1. Design considerations include any current in the interfacing circuit that originates or terminates in the load. The following example shows several kinds of constraints that can arise in such a problem.

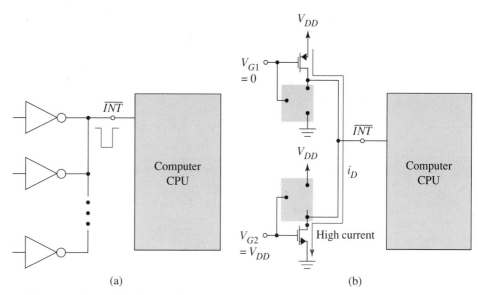

Figure 14.30 Interfacing multiple gates to a single load: (a) computer interrupt example; (b) origin of high currents when gates have active pull-up.

EXAMPLE 14.7 In Fig. 14.31a, three open-drain logic gates are interfaced with a single-load circuit. The load requirements are $V_{IH} = 1.5$ V and $V_{IL} = 0.5$ V. At logic 1, the input resistance of the load circuit is 40 kΩ to ground. At logic 0 the interfacing gate must sink 1.4 mA of current. The logic gates use an 8 V supply, and the gate terminals of the output transistors operate between 0 and 8 V. Find suitable values for V_{DD}, R_P, k, and V_t.

Solution. A logic 1 occurs only when all output transistors are cut off. This gives the design equation

$$v_O = \frac{40\,\text{k}}{40\,\text{k} + R_P} V_{DD} \geq 1.5 \tag{14.41}$$

Once we select V_{DD}, this equation will give an upper limit for R_P. We defer this decision until we examine the logic 0 constraints.

Figure 14.31b shows the constraint for logic 0 input, where R_Z is the ohmic resistance of one output transistor. A logic 0 can be established by $N = 1$, 2, or 3 output transistors in parallel; however, $N = 1$ gives the highest v_O. For a logic 0 input, then, $N = 1$ is the worst case. KCL gives

$$\frac{V_{DD} - v_O}{R_P} + 1.4\,\text{mA} = \frac{v_O}{R_Z}$$

Solving for v_O gives

$$v_O = \frac{(V_{DD} + 1.4\,\text{mA}\,R_P)R_Z}{R_P + R_Z}$$

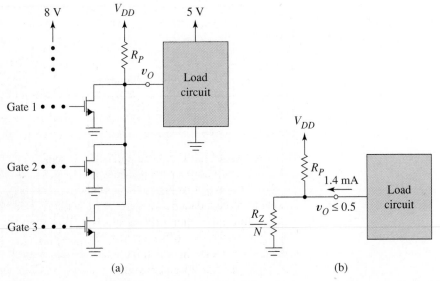

Figure 14.31 Open collector/open drain gates: (a) connected to share a common pull-up resistor; (b) equivalent circuit for calculating R_P when N gates have low outputs.

The design requirement $v_O \leq 0.5$ V now becomes

$$(V_{DD} + 1.4 \times 10^{-3} R_P)R_Z \leq 0.5(R_P + R_Z) \tag{14.42}$$

Equation (14.42) shows that low V_{DD} helps satisfy the inequality. It is usually best to use an available supply, so we use $V_{DD} = 5$ V. Now Eq. (14.41) gives $R_P \leq 93.3$ kΩ. Substituting $V_{DD} = 5$ V into Eq. (14.42) gives

$$R_Z \leq \frac{0.5 R_P}{4.5 + 1.4 \times 10^{-3} R_P}$$

Selecting $R_P = 10$ kΩ satisfies $R_P \leq 93.3$ kΩ, gives a pull-up resistor of fairly low power dissipation, and leads to $R_Z \leq 270$ Ω. This choice is somewhat arbitrary and can be reconsidered later if necessary.

For the output transistors in Fig. 14.31a to be ohmic when $v_O = 0.5$ V,

$$v_{DG} = 0.5 - 8 \leq -V_t$$

We conclude that we need $V_t \leq 7.5$ V, which is not very restrictive. Let us simply select a reasonable value such as $V_t = 0.8$ V. From our condition $R_Z \leq 270$ and Eq. (5.5) we have

$$270 \leq \frac{1}{k(8 - 0.8)}$$

which gives $k \leq 5.14 \times 10^{-4}$ A/V^2 subject to the condition

$$|v_{DS}| \leq 0.2|(v_{GS} - V_t)|$$

For this design, this condition is

$$|v_{DS}| \leq 1.44 \text{ V}$$

which is satisfied for the logic 0 condition $v_O \leq 0.5$. The final design is $V_{DD} = 5$ V, $R_P = 10$ kΩ, $V_t = 0.8$ V, and $k = 5.14 \times 10^{-4}$ A/V^2. ❏

Exercise 14.7 The design of Example 14.7 produced a worst-case input logic 0 of 0.5 V, making no allowance for noise margin. Keeping $V_{DD} = 5$ V, $R_P = 10$ kΩ, and $V_t = 0.8$ V, find the factor by which W/L of the output transistors must be increased to give worst case NML of 0.3 V.

Ans. New $W/L = 2.55 \times$ (old W/L).

Tristate Logic Gates. Sometimes multiple gates time-share a single output node, that is, use the same node but at different times. This occurs in addressing and data communication structures like Fig. 14.32a, where each of three separate memory modules must be able to place a data bit on a common data line under the direction of control circuitry. This kind of application utilizes special *tristate* logic gates shown schematically in the diagram. In addition to the ordinary logic 1 and logic 0 outputs, tristate gates have a *floating output* state in which the internal gate circuitry is effectively disconnected from the output node. A special *tristate* input, v_{TR} on the diagram, is available to the control circuits that operate the interface. When v_{TR} is logic 0, the gate output is disconnected from the data line; when logic 1, the gate is connected.

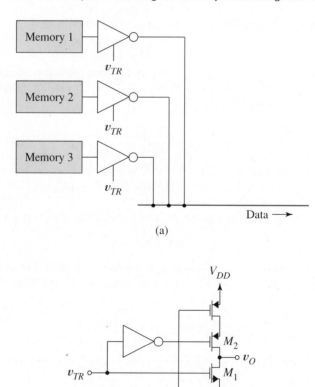

Figure 14.32 Tristate gates: (a) an application where several gates time-share a common node; (b) CMOS tristate gate structure.

Figure 14.32b shows a CMOS tristate inverter. When v_{TR} is zero, M_1 and M_2 cut off, and the output node floats. When v_{TR} is one, both transistors turn on, connecting the inverter transistors to the output node, and the inverter operates in its usual fashion. Similar tristate gates are available in other logic families.

14.5.8 CONVERSIONS BETWEEN ANALOG AND DIGITAL

For reasons of precision, accuracy, convenience, versatility, reliability, and cost, most modern data processing installations use digital techniques. These require that data functions be represented by time sequences of binary numbers, in which binary ones and zeros are represented by voltage or current levels. Many original sources of information—temperature, pressure, and sound transducers, for example—are inherently analog, meaning their outputs take the form of continuously varying time waveforms of voltage or current. At the interface between such an analog data source and the digital processing hardware is an *analog-to-digital converter* (A/D), a circuit that accepts an analog signal waveform as input and produces a corresponding binary number sequence as output. The A/D converter, operating at some constant *sampling rate* f_s, produces as output f_s evenly spaced binary numbers per second, each the binary number equivalent of the signal amplitude at the sampling instant. Representative values for the number of binary digits or *bits* developed by the A/D converter are 8, 12, and 16.

Often, after data processing is completed, the results must be employed in some physical way, for example, to drive a motor, a speaker, or a display system. When this load is an analog device that requires input in the form of a continuously varying time waveform, we use a special electronic circuit, the *digital-to-analog converter* (D/A), also commonly operating with 8, 12, or 16 bits.

Figure 14.33 is a block diagram of a digital processing system in which both input and output are inherently analog. An analog *antialiasing* filter, designed according to principles described in Chapter 12, limits the bandwidth of the input to that strictly necessary to pass the information. Such filtering eliminates the special *aliasing distortion* that occurs when signal or noise frequencies in $v_I(t)$ are more than half the sampling frequency f_s. The sample-and-hold circuit, introduced in Sec. 2.2.7, presents the A/D converter with an input signal that does not change during conversion. The A/D converter converts analog signal $v_A(t)$ into a sequence of *n*-bit binary numbers, a_o–a_{n-1} that provide input for the digital data processing hardware. Often the binary bits appear in parallel as shown in the diagram; however, some A/Ds produce the *n* output bits as a serial time sequence on a single pair of wires. The *n*-bit binary number output, b_o–b_{n-1}, is converted into the analog waveform $v_B(t)$ by a D/A converter. Although continuously varying, $v_B(t)$ contains sharp corners and nearly vertical edges that are consequences of digital processing. From the spectral point of view, the corners and edges represent high-frequency additive noise that usually must be removed by an analog lowpass filter. Because the D/A converter is the simpler of the two new circuits we discuss it first.

D/A Converter. The *R/2R ladder* circuit of Fig. 14.34a is central to one D/A circuit. Two important features of this resistive ladder circuit are that the resistance looking to the right from any node is $2R$ and that the input resistance of the circuit is R, regardless of the length of the ladder. To verify these assertions, start at the rightmost node and work toward the left. This resistive ladder is fabricated as an IC, using either diffused or thin film resistors of precisely controlled ratios that track with temperature.

In the D/A circuit, the ladder is connected by *n* voltage-controlled switches to a current-to-voltage converter or ground, as in Fig. 14.34b. The switches are often MOSFETs or MOSFET pairs, operated between ohmic and cut-off regions. Because every switch

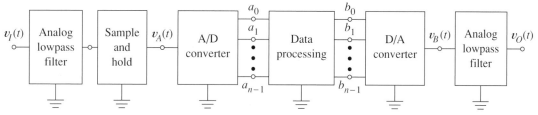

Figure 14.33 System to produce analog output $v_O(t)$ by digitally processing analog input $v_I(t)$.

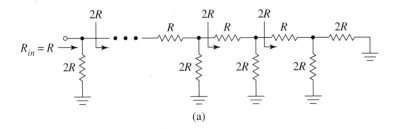

(a)

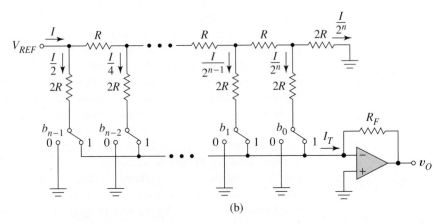

(b)

Figure 14.34 *R/2R* type D/A: (a) *R/2R* ladder; (b) ladder-based converter.

connects a resistor to either ground or virtual ground, the resistances seen at each node are the same as for Fig. 14.34a *for all possible combinations of switch positions.*

Reference voltage V_{REF} initiates input current

$$I = V_{REF}/R \qquad (14.43)$$

which then divides into equal parts at the input node. Because the resistance to ground equals the resistance looking to the right, a division into two equal currents occurs *at every node of the ladder.*

Now let

$$b^{n-1}b^{n-2} \ldots b^2b^1b^0$$

represent a binary number, currently inside a digital processor, which is to be converted into an analog value. Each b^k is either 1 or 0, and the binary digits are listed in positional notation in order of decreasing significance. The voltage-controlled switches of Fig. 14.34b interface the processor to the A/D, where a binary bit $b^k = 1$ closes the switch to the noninverting input and $b^k = 0$ diverts the branch current to ground. From KCL, current I_T that enters the current to voltage converter is

$$I_T = \left(\sum_{k=0}^{n-1} b_k \frac{I}{2^{n-k}} \right) = \frac{I}{2^n} \left(\sum_{k=0}^{n-1} b_k 2^k \right)$$

and the circuit output is

$$v_O = -I_T R_F = -\frac{R_F I}{2^n}\left(\sum_{k=0}^{n-1} b_k 2^k\right)$$

Substituting from Eq. (14.43) gives

$$v_O = -\frac{V_{REF}}{2^n}\frac{R_F}{R}\left(\sum_{k=0}^{n-1} b_k 2^k\right) \qquad (14.44)$$

The expression within parentheses is the decimal equivalent of the binary number we describe in positional notation as $b^{n-1}b^{n-2} \ldots b^1 b^0$; the term outside parentheses is a scale factor. When all b_ks are zero except b_o, Eq. (14.44) gives the minimum nonzero output voltage the D/A is capable of producing, the *resolution* of the D/A. The example that follows reveals some of the details of operation.

EXAMPLE 14.8 The A/D converter of Fig. 14.34b uses $R_F = R$, $V_{REF} = -5$ V, and $n = 12$ bits.
(a) Find the value of v_O that represents the binary number 000000000001.
(b) Find the maximum v_O.
(c) Suppose the binary input that controls the switches is obtained from the 12-bit counter of Fig. 14.35a. Find the waveform $v_O(t)$ that results as the counter counts up from 0 to $2^{12} - 1$, clears to zero, and then repeats itself in response to a constant frequency clock input.
(d) Estimate the minimum slew rate for the op amp if the clock frequency is 50 kHz.

Solution. (a) From Eq. (14.44), $v_O = -(-5/2^{12})2^0 = 1.221$ mV. This is the D/A's resolution.
(b) The largest output value occurs when all bits are one. Since the sum S of the first 12 terms in the power series 2^k is

$$S = \sum_{k=0}^{12-1} 2^k = 2^{12} - 1 = 4095$$

from Eq. (14.4) we obtain

$$v_{O,MAX} = -\left(\frac{-5}{4096}\right)(4095) \approx 5 \text{ V}$$

(c) The output is the stairstep function suggested by Fig. 14.35b, with each step 1.221 mV in height.
(d) With a 50 kHz clock, each *stairstep* duration is a 20 μs clock period. When the counter clears, the D/A output must drop from 5 V to zero in a negligible part of a clock period, say, 2 μs. Thus the op amp needs a slew rate of at least $5/2 = 2.5$ V/μs. ❏

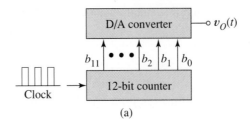

(a)

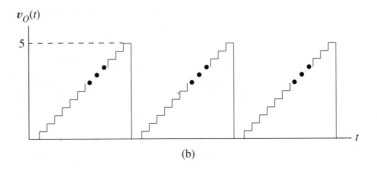

(b)

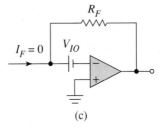

(c)

Figure 14.35 D/A driven by a digital counter: (a) circuit; (b) D/A output waveform; (c) circuit for finding output component due to offset.

Equation (14.44) shows that for an n-bit D/A, the smallest nonzero output, expressed as a percentage of the largest output, is

$$\frac{\Delta v}{v_{max}} = \frac{1}{2^n - 1} \times 100 \qquad (11.45)$$

For $n = 8$, 12, and 16 bits, respectively, we see the precision is 0.39%, 0.024%, and 0.0015%.

The input offset voltage of the op amp can adversely affect the performance of the D/A. Since the D/A reduces to Fig. 14.35c when all switches are open, we can use superposition to find the effect of offset voltage.

The following exercise critically explores two features overlooked in the preceding D/A design.

Exercise 14.8 (a) Use Equation (14.44) to compute the sensitivity of v_O to R_F.
(b) Find the contribution of the op amp's input offset voltage to v_O in Fig.14.34b.
(c) Is the 741 of Table 2.1 a suitable choice for the design of Example 14.8? Explain.

Ans. (a) $S^{v_O}_{R_f} = 1$. (b) $v_O = \pm V_{IO}$. (c) For a 741, $V_{IO} = 5$ mV, an excessive value since the height of each step is only 1.221 mV. The 741's slew rate of 0.5 V/μs is also inadequate.

In an all-IC design, the ratio R_F/R can be precisely controlled. If an IC resistive ladder is combined with a discrete resistor, the voltage range and the minimum voltage are both scaled by R_F; however, this has no effect on the *uniformity* of step size. If R_F is a *trim pot,* we can adjust the circuit for proper operation, at least at one temperature.

Obviously, the importance of offset on D/A operation relates to the value of V_{REF}, the number of bits, and the op amp, so op amps specially designed for small offsets are desirable. Because offset voltage changes with temperature, offset compensation is of limited usefulness, and in critical applications might have to be redone before each use of the D/A.

Using a computer algorithm to generate a ramp waveform like Fig. 14.35b provides a simple way to check a D/A interface for correct and complete connection of D/A inputs, proper operation of D/A hardware, and correct dynamic behavior. This idea is also the basis for a versatile digital waveform generator. We need only replace the counter output by a more general periodic number sequence generated by a special-purpose sequential circuit or a computer.

Problem 14.44 explores another D/A circuit, one that uses *binary weighted resistors* and an operational amplifier.

A/D Converter. Figure 14.36a shows the external terminals of an *analog-to-digital converter.* There is an analog input, an *n*-bit digital output, and a pair of nodes used for interfacing. In response to a pulse at its *convert input,* the A/D, by some process, produces the *n*-bit digital code that represents the analog input voltage. The conversion process requires an interval t_c called the *conversion time,* during which the digital output bits are being computed and therefore may be invalid. As a warning, a busy/done output

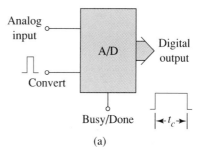

(a)

Figure 14.36 Analog-to-digital converter: (a) schematic representation showing inputs and outputs; (b) one possible realization of the converter; (c) waveforms.

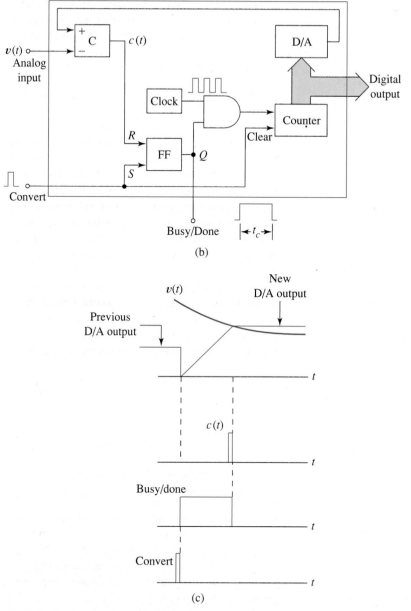

(b)

(c)

Figure 14.36 (continued)

goes high during the conversion and returns to zero when the conversion is complete. At this time the correct n-bit number is available at the output and remains latched until the next convert signal.

A/D converters differ in a number of ways. The digital output might be in straight binary, 2's complement, or some other code. There are also variations in the permissible range of input voltage, 0 to +15 V or −5 to + 5 V, for two examples. Sometimes tristate gates at the output must be enabled by a control signal before the digital output is available at the output pins. This allows the A/D to share data lines with other devices. A/D converters also differ in the internal circuit mechanism used for the conversion. We next discuss two of these mechanisms.

Counting-Type A/D. Figure 14.36b shows a *counting-type A/D* that generates simple binary output. In its inactive state, the previously calculated digital output is latched in the counter, and the flip-flop is reset, holding one AND input low to prevent the internal clock pulses from reaching the counter. The analog signal provides continuous input to comparator C; however, since the flip-flop is reset, $c(t)$ does not affect the flip-flop state.

A convert pulse simultaneously clears the counter and sets the flip-flop. This brings the busy/done line high and allows clock pulses to reach the counter through the AND. The D/A output is a ramp that starts at zero as in Fig. 14.36c. So long as the D/A output is less than the analog signal, the comparator output is zero; however, at the instant the ramp voltage exceeds $v(t)$, the comparator output changes to one and resets the flip-flop. This returns the system to its original state, except for the new value stored in the counter. The busy/done line returns to zero, indicating that the conversion is complete.

In this A/D the conversion time is variable. The worst case value for t_c is the clock period times 2^{n-1}, the time required for the ramp to rise from zero to its maximum value. Analog input voltages for this circuit must be in the range $0 < v_{in} < V_M$, where V_M is the maximum value of the A/D output. The analog sample is represented by its digital equivalent, with precision expressed as a percent by

$$\text{precision} = \frac{\text{smallest D/A output}}{\text{range of analog input}} \times 100 = \frac{V_M/(2^n - 1)}{V_M} \times 100 = \frac{1}{2^n - 1}\%$$

the same expression we found for the D/A in Eq. (14.45).

To be of negligible importance, the slew rate of the D/A subcircuit must accommodate worst case changes from maximum to minimum analog voltage in a small fraction of a clock cycle. For example, if the last sample in Fig. 14.36b was the maximum, V_M, the D/A output must drop within a negligible portion of a clock cycle.

Successive Approximation A/D. Another A/D circuit mirrors an elegant solution to a well-known child's game: guessing an unknown number. The game begins with one child visualizing a number between 0 and 100. The solution strategy begins with a standard guess of 50. If this is too high, the second guess is 25, otherwise 75. By using each guess to halve the range of uncertainty, the guessing procedure converges to the unknown number with surprising quickness.

Figure 14.37a shows a *successive approximation A/D* that applies the same principle. The unknown number is the current n-bit digital equivalent of $v(t)$, a binary number between 0 and 2^{n-1}. When the circuit receives a convert pulse, successive approximation logic (special hardware or a computer algorithm) clears the n D/A inputs to zero and then makes its first "guess" by setting the most significant bit to 1. Figure 14.37b shows for the 4-bit case that this is equivalent to guessing half of 2^{4-1}. In Fig. 14.37a, the internal D/A presents the analog equivalent, $v_X(t)$, of the current approximation to the comparator. If $v_X(t)$ is lower than $v(t)$, as in the example, the comparator output remains at zero, and

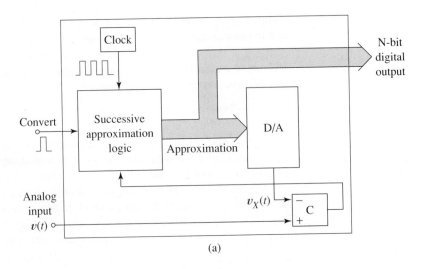

(a)

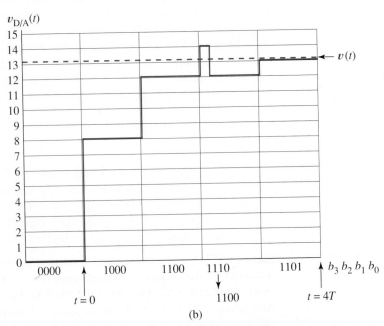

(b)

Figure 14.37   Successive approximation-type A/D converter; (a) block diagram; (b) 4-bit example.

the logic refines the approximation by changing the next bit to 1. If the comparator output changes to 1 because the first guess is too high, the "guessed bit" is returned to zero before proceeding. The logic continues in this fashion until n such guesses have been made (and possibly corrected). The result is the best n-bit approximation to $v(t)$. The example shows how a 4-bit system converges to 13 when $v(t) = 13.2$ volts. Notice that the conversion is *always* complete after n clock periods. The internal D/A output must change at most by only half the range of the analog input voltage in one cycle; thus slew rate need not be so high as in the counting-type D/A.

A busy/done output is easily added to give a package like Fig. 14.36a. Sometimes A/D chips also provide a separate *n*-bit output latch so that transient bits generated during conversions never appear at the output, and some include tristate gates between the D/A outputs and the external pins to facilitate time sharing of data buses.

14.6
Transmission Line Effects

Until now we tacitly ignored delays introduced by signal travel between gates, assuming these delay are small compared to the gate propagation delays. The delays associated with interconnection wiring become increasingly troublesome, however, when gates become faster or the interconnecting wires become longer. We rely heavily upon the theory of ideal lossless transmission lines for an understanding of such problems. In practical terms, a connection becomes a transmission line when the time the signal takes to travel from driver to receiver is more than twice the rise (or fall) time.

The development that follows differs from most transmission line discussions in emphasizing graphical techniques. Graphical interpretations of the line equations enable us to extend the theory to a new class of problems: linear lines terminated by nonlinear logic circuits. Our object is a quality of knowledge that enables us to both understand the causes of problems and formulate solutions. Before we can handle the nonlinearities, however, we must first review the fundamental principles and timing of reflections from linear terminations.

14.6.1 IDEAL LOSSLESS LINE

When closely examined, connections such as twisted pairs, coaxial cables, wire over ground, and striplines all exhibit distributed capacitance and inductance; that is, the line actually consists of an infinity of infinitesimally small capacitances and inductances. Consequently, we visualize the connection schematic of Fig. 14.38a as Fig. 14.38b, where *L* and *C* denote the distributed inductance in henries per meter and the distributed capacitance in farads/meter. Because of distributed capacitance, we cannot instantly change the voltage across the line as we could for ideal connecting wires, nor can we instantly change the current because of the distributed inductance.

Suppose the line in Figs. 14.38b and c is in *equilibrium* with $v_s(t) = 0$ for $t < 0$. If at least one resistor is finite and nonzero, equilibrium implies that all differential capacitances are uncharged and inductor currents are zero; that is, there is no energy stored in the line. Suddenly changing $v_s(t)$ to *E* volts at $t = 0$ upsets this equilibrium at the *sending end*, $x = 0$. The linear partial differential equation that describes the line, the *wave equation*, predicts that a voltage *disturbance* of v^+ volts and a simultaneous current disturbance of i^+ amperes then begin to travel down the line at a constant *propagation velocity u*, charging the differential capacitors and creating currents in the differential inductors as they go. The wave equation also predicts that forward-moving disturbances such as v^+ and i^+ are constrained by

$$\frac{v^+}{i^+} = R_o \tag{14.46}$$

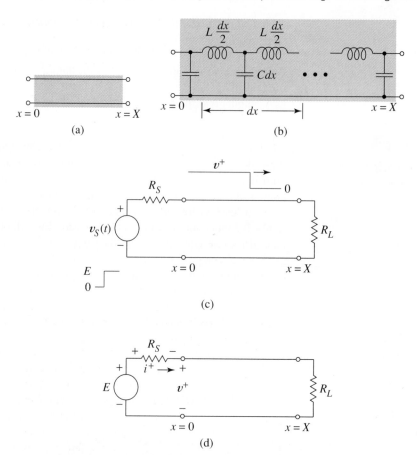

Figure 14.38 Ideal lossless transmission line: (a) schematic representation; (b) conceptual model using capacitance and inductance per unit length; (c) propagation of a voltage disturbance down the line; (d) sending-end constraint at $t = 0^+$.

where R_o is a constant called the *characteristic resistance* of the line. An observer at the *receiving end*, $x = X$, sees no change until $t = T = X/u$, when the disturbances reach the load end. At this instant, depending upon the value of R_L, the line either achieves a new equilibrium or additional disturbances begin.

Books in electromagnetic theory derive expressions for L and C for various kinds of lines based upon the physical dimensions and separation of the conductors and the dielectric constant of the material between the conductors. They also prove that for the ideal lossless line,

$$u = \sqrt{\frac{1}{LC}} \quad \text{and} \quad R_o = \sqrt{\frac{L}{C}}$$

where $u \approx 3 \times 10^8$ m/s, the velocity of light. Typically, $30 \leq R_o \leq 300$ Ω.

14.6.2   DISTURBANCES ON IDEAL LINES

Initial Disturbances. We now examine more closely the formation of line disturbances in response to a step input. At the sending end, voltage and current disturbances must satisfy Eq. (14.46). For zero initial conditions and voltage step E, Fig. 14.38d, shows that v^+ and i^+, with reference directions as defined in the diagram, must also satisfy the *sending-end constraint*

$$v^+ = E - R_S i^+ \qquad (14.47)$$

Using Eq. (14.46) for i^+ gives

$$v^+ = E - R_S \frac{v^+}{R_o}$$

which leads to the voltage divider expression

$$v^+ = \frac{R_o}{R_o + R_S} E \qquad (14.48)$$

We can use Eqs. (14.48) and (14.46) to find v^+ and i^+ for a given line and given terminations.

These disturbances usually mark the beginnings of transients that lead eventually to new equilibrium values of line voltage and current that are consistent with the newly applied input voltage. If neither line termination is a finite, nonzero resistor, however, sustained oscillations that resemble oscillations in an LC circuit can occur, since we have introduced energy into a lossless circuit with no possibility for dissipation.

Now that we know how a line transient begins, we examine the events that follow, and their timing. There are three different cases, with critical differences that depend upon how resistors R_S and R_L individually compare with R_o. When $R_L = R_o$, we say that the line is *matched* at the load end. When $R_S = R_o$, we say the line is matched at the sending end. If the line is not matched at either end, it is *mismatched.*

Case 1: $R_L = R_o$. In Fig. 14.39a, a switch provides step-function excitation by switching from position 1 to position 2 at $t = 0$. With switch in position 1, we have the initial equilibrium. Because there is no excitation for $t < 0$, the line voltage and current, $v(x)$ and $i(x)$, are both zero for all x, and all differential capacitances and inductances have zero initial conditions.

At $t = 0$, the switch changes to position 2, and disturbances v^+ and i^+ travel down the line at velocity u, taking $T = X/u$ second to reach the load. Consequently, line voltage and current are functions of both distance x and time t, expressed using the notation $v(x,t)$ and $i(x,t)$. Figure 14.39b shows the voltage and current at selected times. At $t = T$ the disturbances impinge upon the load resistor. Because of the special termination and Eq. (14.46), Ohm's law is satisfied and the transient ends. Notice that for $t > T$ the line capacitance has everywhere been charged to

$$v(x) = v^+ = \frac{R_o}{R_o + R_S} E$$

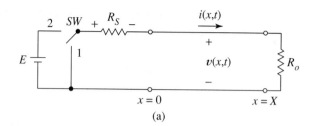

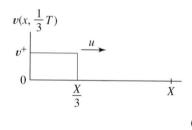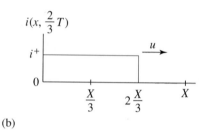

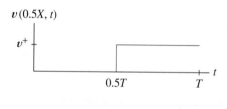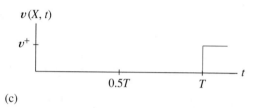

Figure 14.39 Step voltage input on ideal line with matched load: (a) diagram; (b) voltage and current disturbances at fixed times; (c) voltage waveforms at center and at end of the line.

and the line inductance has current

$$i(x) = i^+ = \frac{E}{R_o + R_S}$$

These final equilibrium values are exactly what we would expect from the dc circuit of Fig. 14.39a. Figure 14.39c shows the voltage waveforms that we observe at $x = X/2$ and at $x = X$. We conclude that when a line is matched at the load end, the line introduces a simple delay of T seconds in changing from the original to the new equilibrium values.

14.6.3 REFLECTIONS FROM THE LOAD AND THE SOURCE

Case 2: $R_L \neq R_o$, $R_S = R_o$. For this case, the line is matched at the sending end but mismatched at the load. Until $t = T$, the step-induced transient is just like Case 1 with $v^+ = E/2$ because of the matched source, but an interesting thing happens at $t = T$. Because the disturbance ratio that impinges upon the load is

$$\frac{v^+}{i^+} = R_o \neq R_L$$

it appears that Ohm's law is to be violated! The wave equation, however, shows that line disturbances, v^- and i^-, that travel in the negative direction are also possible and that whenever such *negative-going* waves exist, they must be in the ratio

$$\frac{v^-}{i^-} = -R_o \tag{14.49}$$

Load-end reflection coefficient. To maintain the credibility of Ohm's law, new *additive* disturbances, called *reflections*, begin at the load end at precisely $t = T$ and travel toward the sending end. The reference directions for v^- and i^- are the same as for v^+ and i^+ even though they travel in the opposite direction. These reflections assume exactly those values required to satisfy Ohm's law at $t = T$. That is,

$$\frac{v^+ + v^-}{i^+ + i^-} = R_L = \frac{v^+ + v^-}{v^+/R_o - v^-/R_o}$$

Factoring out R_o and dividing by v^- gives

$$R_L = R_o \frac{v^+/v^- + 1}{v^+/v^- - 1}$$

and solving for v^-/v^+ leads to

$$v^- = \rho_L v^+ \tag{14.50}$$

where ρ_L is the *load-end reflection coefficient*, defined by

$$\rho_L = \frac{R_L - R_o}{R_L + R_o} \tag{14.51}$$

The reflection coefficient is simply a real number that we can compute for given R_L and R_o. The importance of ρ_L is that we can use it to calculate the unknown reflected voltage v^- from the known disturbance v^+. Including the possibilities of open- and short-circuit loads,

$$-1 \le \rho_L \le +1$$

Using Eq. (14.46) and Eq. (14.49) to replace voltages by currents in Eq. (14.50) gives the companion expression

$$i^- = -\rho_L i^+$$

Once launched, the voltage and current reflections travel in the negative x direction at velocity u, *adding to the values already established by the initial disturbances.* Figures 14.40a and b show the line voltages and currents at several instants when $\rho_L > 0$. Because the positive reflection coefficient resulted in positive v^-, according to Eq. (14.49) i^- must be negative. When ρ_L is negative, a negative voltage reflection sweeps back down the line, partially discharging the distributed capacitance. Meanwhile, the current wave adds to the current previously established in the line inductance.

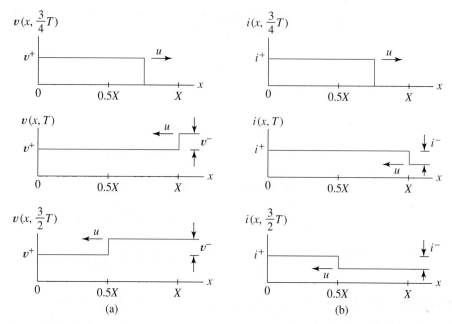

Figure 14.40 Reflections when load is mismatched: (a) line voltage at different times; (b) line current at different times.

At $t = 2T$ seconds, the reflected waves arrive at the sending end, where the total wave, initial plus reflected, must satisfy the sending-end constraint. Figure 14.41 shows that this is

$$v^+ + v^- = E - R_o(i^+ + i^-) \tag{14.52}$$

Because the line is matched at the sending end for Case 2, Eq. (14.47), with $R_S = R_o$, gives

$$v^+ = E - R_o i^+$$

Substituting this for v^+ in Eq. (14.52) reduces the sending-end constraint to $v^- = -R_o i^-$. Because this is satisfied by the reflected current and voltage, there are no further reflec-

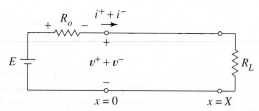

Figure 14.41 Reflection constraints when sending end is matched.

tions; and we conclude that after $2T$ seconds the transient ends with line capacitance charged to

$$v^+ + v^- = \frac{E}{2} + \rho_L \frac{E}{2} = \frac{E}{2}\left[1 + \frac{R_L - R_o}{R_L + R_o}\right] = \frac{R_L}{R_L + R_o} E$$

The final line current is

$$i^+ + i^- = \frac{E}{2R_o} - \rho_L \frac{E}{2R_o} = \frac{E}{2R_o}\left(1 - \frac{R_L - R_o}{R_L + R_o}\right) = \frac{E}{R_L + R_o}$$

Notice that these final values are exactly what we predict when we treat the line of Fig. 14.41 as a pair of ideal conductors; however, they occur only after the two-pass charging process. Thus final equilibrium values are not established until $t = 2T$ when the reflections reach the sending end.

The next example shows how we can combine the reflection and timing ideas to predict the waveforms at the sending and receiving ends.

EXAMPLE 14.9 The ideal lossless line of Fig. 14.42a is in equilibrium for $t < 0$.
(a) Find the equilibrium values of line current and line voltage for $t < 0$ and also for $t = \infty$, that is, long after the line transient has ended.
(b) Sketch $v(x, 0.5T)$, $i(x, 0.5T)$, $v(x, 1.5T)$, and $i(x, 1.5T)$.
(c) Sketch the voltage and current waveforms that appear at the sending and receiving ends of the line; that is, sketch $v(0, t)$, $i(0, t)$, $v(X, t)$, and $i(X, t)$.

Solution. (a) To find the equilibrium values, we treat the line as a pair of ideal connecting wires. For $t < 0$, $v = i = 0$ because $v_i(t) = 0$. After the transient has ended, $v_i(t) = 7$ V; therefore, everywhere on the line,

$$v = \frac{50}{50 + 300}\, 7 = 1 \text{ V}, \qquad i = \frac{7}{50 + 300} = 20 \text{ mA}$$

The rest of the problem shows how the transition from initial to final values takes place.
(b) The initial disturbances are

$$v^+ = \frac{300}{300 + 300}\, 7 = 3.5 \text{ V}, \qquad i^+ = \frac{v^+}{R_o} = 11.6 \text{ mA}$$

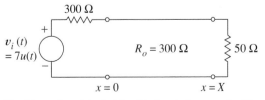

Figure 14.42 Line mismatched at load but not at source for Example 14.9.

These travel the length of the line in T s and then generate reflections that we compute using

$$\rho_L = \frac{50 - 300}{50 + 300} = -0.714$$

Thus $v^- = (-0.714)\, 3.5 = -2.5$ V and $i^- = -(-0.714)\, 11.6$ mA $= 8.28$ mA. Figures 14.43a and b show the voltage and current along the line at $0.5T$ and at $1.5T$. Because of

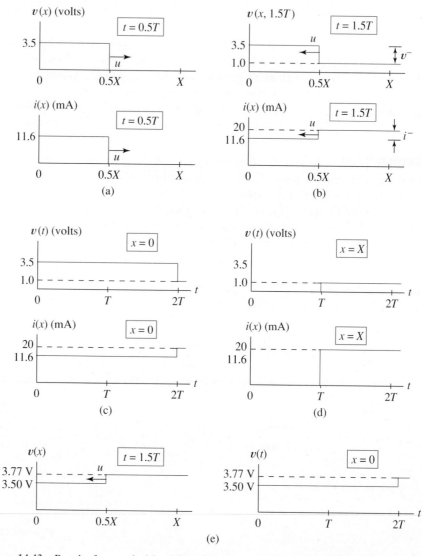

(a)

(b)

(c)

(d)

(e)

Figure 14.43 Results for matched load line of Example 14.9: (a) line voltage and current at $t = 0.5T$; (b) line voltage and current at $1.5T$; (c) sending-end voltage and current; (d) receiving-end voltage and current; (e) answers to Exercise 14.9.

the negative reflection coefficient, the reflected voltage differs in sign from v^+. The first voltage wave charges the line capacitance to 3.5 V, a value higher than the equilibrium voltage. The reflection removes the excess charge, giving a line charged to the expected equilibrium value.

The initial current waveform is less than the equilibrium value, but the reflection adds on exactly what is needed to establish the correct final equilibrium. Because the line is matched at the source end, there are no further reflections.

(c) By visualizing the delays and how reflections are superimposed upon the original values, we sketch Figs. 14.43c and d. In Fig. 14.43c, observing at the sending end, we see the initial voltage and current transitions caused by the initial disturbances. Values then remain constant for T s as these travel to the load end, and then for another T s as the reflections come back from the load. At $t = 2T$ the reflections superimpose themselves upon the original values to give the final equilibrium values.

To sketch Fig. 14.43d we imagine ourselves observing voltage and current at the load. For the first T s, we see only the initial values. At the same instant the initial disturbances arrive, the additive reflections start down the line. At $x = X$ we see only the superimposed values, that is, $(v^+ + v^-)$ and $(i^+ + i^-)$, which are consistent with Ohm's law at the load. In summary, at the load we observe delayed jumps from initial to final equilibrium values. ❏

Exercise 14.9 Find the initial $(t < 0)$ and final $(t = \infty)$ values of line voltage and current if the load resistor in Fig. 14.42 is changed to 350 Ω. Also, sketch $v(x)$ at $t = 1.5T$ and $v(t)$ at $x = 0$ for this case.

Ans. $v(x, 0^-) = 0$, $i(x, 0^-) = 0$, $v(x, \infty) = 3.77$ V, $i(x, \infty) = 10.8$ mA. See Fig. 14.43e.

Case 3: $R_L \neq R_o$, $R_S \neq R_o$. In this case the line is mismatched at both source and load. Until $t = 2T$ the scenario is like Case 2; however, when the reflection arrives at the sending end, there is a problem with the sending-end constraint. From Fig. 14.44 the additive disturbances at $t = 2T$ must satisfy

$$v^+ + v^- = E - R_S(i^+ + i^-) \tag{14.53}$$

Because the initial disturbances satisfied $v^+ = E - R_S i^+$, these terms cancel in Eq. (14.53), leaving the constraint

$$v^- = -R_S i^- \tag{14.54}$$

However, v^- and i^- are in the ratio of $-R_o$. Therefore new, additive, sending-end reflections v^{2+} and i^{2+} start toward the load from the sending end at $t = 2T$. These add,

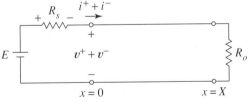

Figure 14.44 Reflection constraints when line is mismatched at sending end.

respectively, to v^- and i^- so as to satisfy Eq. (14.54). Because they travel in the $+x$ direction they must also satisfy

$$\frac{v^{2+}}{i^{2+}} = R_o$$

To satisfy Ohm's law,

$$\frac{v^{2+} + v^-}{i^{2+} + i^-} = -R_S = \frac{v^{2+} + v^-}{v^{2+}/R_o - v^-/R_o} = \frac{v^{2+}/v^- + 1}{v^{2+}/v^- - 1} R_o$$

Solving for v^- in terms of v^{2+} leads us to the *sending-end reflection coefficient* ρ_S defined by

$$\rho_S = \frac{R_S - R_o}{R_S + R_o} \tag{14.55}$$

and the associated equations:

$$v^{2+} = \rho_S v^- \tag{14.56}$$

and

$$i^{2+} = -\rho_S i^- \tag{14.57}$$

14.6.4 THE BOUNCE DIAGRAM

When a line is mismatched at both source and load, new reflections arise every T seconds. The character of the reflections depends upon the sizes of R_S and R_L relative to R_o. Sometimes successive voltages at a point on the line resemble stairstep samples of an exponential waveform from an RC or RL circuit. Alternately, the waveform might resemble samples of a damped sinusoid that suggest the kind of *ringing* we expect in LC circuits. If at least one termination contains resistance, voltage and current converge to the values we expect for a simple pair of wires. If neither termination contains resistance, there are sustained oscillations as in an ideal LC circuit. This comes as no surprise since the ideal transmission line has no mechanism for energy losses.

We can predict transmission line waveforms using a time and space sketch of the line activity called the *bounce diagram*. This diagram gives a systematic way to apply the reflection equations when there are multiple reflections and gives an intuitive feeling for the timing of the reflections. When we progress to graphical methods for studying nonlinear terminations, our experience with bounce diagrams will help us visualize the timing.

Figure 14.45b is the bounce diagram for the line of Fig. 14.45a. The x axis represents distance, and the vertical axis shows time. The lines with arrowheads represent the reflections that travel at constant velocity u.

For the given R_S and R_L, Eqs. (14.51) and (14.55) give $\rho_L = +3/5$ and $\rho_S = +1/3$. At $t = 0$ and $x = 0$, the initial voltage disturbance of $v^+ = 10$ V starts down the line, arriving at the load after $t = T$ s. The first reflection, $v^- = \rho_L v^+ = 6$ V starts back toward the source at this instant. At $t = 2T$, the reflected voltage arrives at the source,

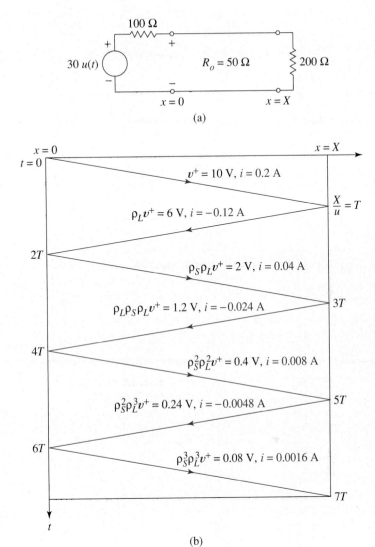

Figure 14.45 Finding reflections from a bounce diagram: (a) doubly mismatched line; (b) bounce diagram showing reflections.

generating a second reflection $v^{2+} = \rho_S(\rho_L v^+)$, and so forth. Notice how the bounce diagram in Fig. 14.45b depicts these events. The diagram also shows the same ideas for currents.

We now show how to use information from the bounce diagram to sketch waveforms. By treating the line in Fig. 14.45a as a simple pair of wires, we see that the equilibrium values for line voltage are $v_{initial} = 0$ and

$$v_{final} = \frac{200}{200 + 100} 30 = 20 \text{ V}$$

Figure 14.46a shows the load-end voltage. All voltage values for this sketch come from the $x = X$ border of the bounce diagram. After a delay of T s, the first disturbance arrives. At this instant the reflected voltage *adds to* the incident voltage. From the bounce diagram at $x = X$ and $t = T$ we form the sum

$$v(X, T) = 10 + 6 = 16 \text{ V}$$

No further changes occur at the load until another $2T$ seconds have elapsed, time for the first reflection to reach the source and a second reflection of 2 V to arrive at the load. This second reflection plus a new reflection of 1.2 V *add to the voltage already present*, giving $v(X, 3T) = 16 + 2 + 1.2 = 19.2$ V. Similarly, at $t = 5T$ the voltage increases by $0.4 + 0.24$ to 19.84. An observer at the load sees a stairstep voltage sequence that converges to the expected final equilibrium value of 20 V.

Figure 14.46b shows the sending-end voltage. At $t = 0$ this jumps from its initial value of zero to 10 V as the initial disturbance starts down the line. At $t = 2T$, reflections of 6 V and 2 V add another 8 V. The observer sees a sequence that converges to the expected equilibrium value; however, the step sizes and timing differ from those at the load. We can also sketch current waveforms using current information from Fig. 14.45b.

14.6.5 GRAPHICAL SOLUTION OF LINEAR LINE EQUATIONS

To extend the theory to lines with nonlinear terminations such as logic gates, we first apply the volt-ampere concepts of Chapter 1 to lines with *linear* terminations, using our

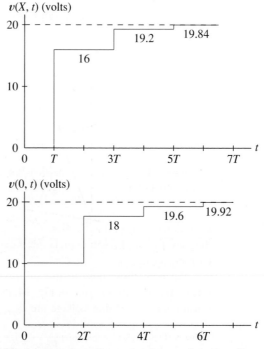

Figure 14.46 Load- and sending-end voltages when $R_S > R_o$ and $R_L > R_o$.

bounce diagram experience to visualize the timing. Once we can solve linear problems graphically, curved lines for nonlinear terminations replace the straight lines used for linear terminations, but techniques remain unchanged.

Figure 14.47a shows a transmission line with linear terminations. At the source, i and v must satisfy

$$i = \frac{E\,u(t) - v}{R_S} \tag{14.58}$$

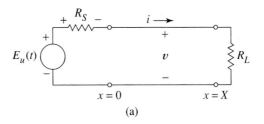

(a)

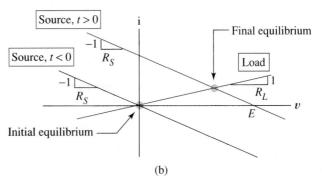

(b)

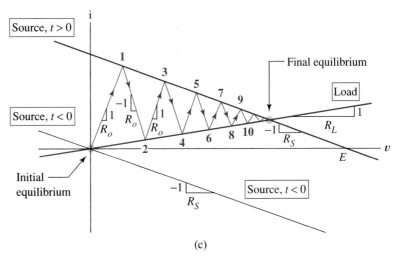

(c)

Figure 14.47 Reflection pattern for $R_L > R_o$ and $R_S > R_o$: (a) terminated line; (b) initial and final equilibrium locations for step of E volts; (c) reflection diagram.

where $u(t)$ is the unit step function. For the two values of $u(t)$, Eq. (14.58) gives the two straight-line equations of negative slope in Fig. 14.47b, the *sending-end constraints* imposed upon the line by the source-end termination. One is a straight line through the origin, the other a straight line through $(v, i) = (E, 0)$, both with slope $-1/R_S$. At $x = 0$, sending-end voltage and current must always be a point, (v, i), on one of these lines.

At $x = X$, i and v must always satisfy the *load constraint,*

$$i = \frac{v}{R_L}$$

a straight line through the origin with slope $1/R_L$, also sketched in Fig. 14.47b.

Now we examine the initial and final equilibria. For $t < 0$, steady-state operation must simultaneously satisfy the $t < 0$ source constraint and the load constraint. The intersection of the corresponding curves gives the *initial equilibrium* at $(v, i) = (0, 0)$.

As $t \to \infty$ voltage and current converge to the intersection of the load constraint, and the $t > 0$ source constraint, that is, to the point marked *final equilibrium* in Fig. 14.47b. The horizontal coordinate of each equilibrium point is the equilibrium voltage and the vertical coordinate the equilibrium current. An advantage of graphical solutions is that we can represent current and voltage on the same diagram.

We next determine exactly how v and i change during their transition from initial to final values. At $t = 0$, voltage step E disturbs the initial equilibrium. Starting at the initial equilibrium points in Figs. 14.47b and c, disturbances v^+ and i^+ arise. From Eq. (14.46), these must be in the ratio $i^+/v^+ = 1/R_o$. They must also satisfy the sending-end constraint. To find their values, we construct a line of slope $1/R_o$ from the initial equilibrium point to the $t > 0$ sending-end constraint curve. This line ends at Point **1** in Fig. 14.47c. The voltage coordinate of this point is the graphical solution of Eqs. (14.46) and (14.47). Since point **1** is on the *sending-end* line (as are all odd-numbered points in the diagram), its coordinates are the voltage and current at $x = 0$, in this case when $t = 0$.

Next we find (graphically) and add (graphically) to (v^+, i^+) a new *vector* (v^-, i^-) such that Ohm's law is satisfied at the load end and such that v^- and i^- satisfy Eq. (14.49). To do this we construct a line of slope $-1/R_o$ from point **1** to the load-end constraint curve, the line from point **1** to point **2**. Since the construction *adds the vectors* (v^+, i^+) and (v^-, i^-), the coordinates of point **2** are the voltage and current that appear at the load end at $t = T$. We continue the construction in this fashion as in Fig. 14.47c, alternately drawing lines of slope $1/R_o$ and $-1/R_o$ between the $t > 0$ source and load constraint curves. The voltage coordinates of successive odd- (even-) numbered points, are the sequence of voltages that appear at the source (load) end. The current coordinates of odd- (even-) numbered points are the sequence of current values at source (load). Notice that each voltage and current sequence converges to the expected final equilibrium value. Hereafter we refer to diagrams like Fig. 14.47c as *reflection diagrams.*

We can sketch voltage and current waveforms using number sequences obtained from the reflection diagram and timing visualized from bounce diagrams, as in Fig. 14.48. In Fig. 14.47c, points 1, 3, 5, ... represent events at the sending end, so they are

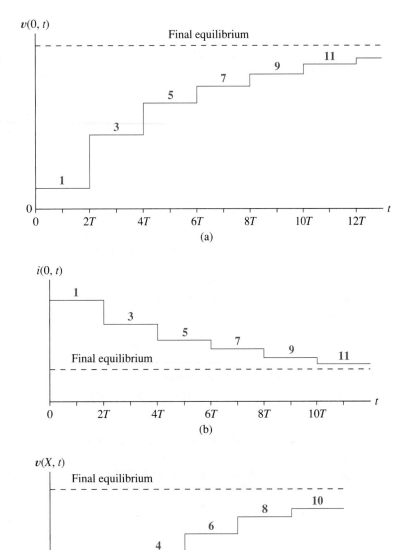

Figure 14.48 Voltage and current sketches for the line of Fig. 14.47a using the construction of Fig. 14.47c: (a) sending-end voltage waveforms with values corresponding to odd-numbered points on the input curve; (b) sending-end current waveform; (c) load-end voltage waveform taken from points on the load curve.

associated, respectively, with times $t = 0$, $2T$, $4T$, ... in Figs. 14.48a and b. Points 2, 4, 6, ... represent events at the load end; thus they are associated with times $t = T$, $3T$, $5T$, ... in Fig. 14.48c.

As we decrease R_L in Figs. 14.47a and c, the point of final equilibrium changes and the number of reflections required to reach it decreases. The load constraint line becomes steeper until it becomes coincident with the first construction line of slope $1/R_o$. This graphically describes Case 1, where the line is matched at the load end and there are no reflections.

General Construction Procedure. We summarize (and slightly generalize) the graphical construction by the following algorithmic procedure:

Reflection Diagram Construction
1. On an i versus v coordinate system, sketch the sending-end constraints for $t < 0$ and $t > 0$. Also sketch the load constraint. Use the current and voltage reference directions defined in Fig. 14.47a.
2. From the initial equilibrium point, construct a line of slope $1/R_o$ to the $t > 0$ source constraint curve. This is point **1**.
3. From point **1** construct a line of slope $-1/R_o$ to the load constraint curve. This is point **2**.
4. From point **2** construct a line of slope $1/R_o$ to the $t > 0$ source constraint curve. This is point **3**.
5. Continue constructing lines that alternate in slope and destination in this fashion until the construction either converges to the final equilibrium point or repeats itself.

The diagrams of Figs. 14.47 and 14.48 show the kind of waveforms generated when $R_L > R_o$ and $R_S > R_o$. Three other general scenarios are possible. Figure 14.49a shows the construction for $R_L > R_o$ and $R_S < R_o$. The result, Fig. 14.49b, is an oscillatory load-end voltage, quite unlike the waveform of Fig. 14.48c. Problems 14.62 and 14.63 explore the remaining cases.

An interesting feature of Figs. 14.49a and b is that the load voltage at point **2** exceeds the input voltage E, an event we might not anticipate from static circuit analysis. In the presence of line inductance and capacitance, voltages higher than the input voltage are possible in transmission lines just as in LC resonant circuits. Because excessive voltage can damage components, it is valuable to be able to anticipate such events. Notice that reducing R_L to a lower value in Fig. 14.49a would reduce the peak load voltage because point **2** would move to the left. The first two construction lines, based upon slopes of $\pm 1/R_o$, would not change. This shows how the reflection diagram aids intuition, not only suggesting a potential problem, but also providing the means to deal with it. Sometimes a reflection diagram reveals a high transient current that can pose a threat to a component.

Reflections When Source Is Turned Off. Reflections also occur when a source is turned *off*. Although the waveforms differ from the turn-on waveforms, the only change

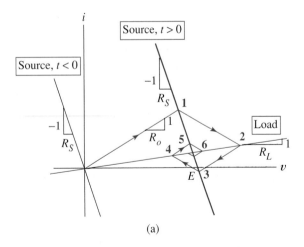

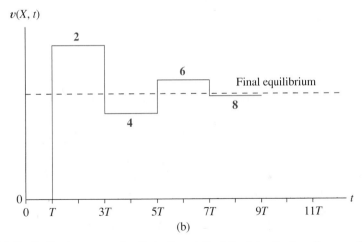

(a)

(b)

Figure 14.49 Reflection pattern for $R_L > R_o$ and $R_S < R_o$: (a) reflection diagram; (b) voltage waveform at the load end.

in the graphical construction is that roles of initial and final equilibrium points are inter changed. Since we are interested in 1 to 0 logic changes as well as 0 to 1 changes, the turn-off transients are also important. Figure 14.50a has the same volt-ampere constraint curves as Fig. 14.49a; however, now we use the "source ON" curve for the initial condition and the "source OFF" curve for the final equilibrium. The waveforms of Figs. 14.50b and 14.49b both describe transitions from initial to final values with ringing and overshoot. Figure 14.50b reveals that the load-end voltage becomes negative during the turn-off transient. Sometimes the reflection diagram shows that a current or voltage exceeds a maximum value for the terminating device. An appropriate response might be to add protective diodes to the circuit.

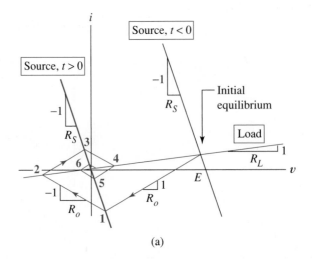

(a)

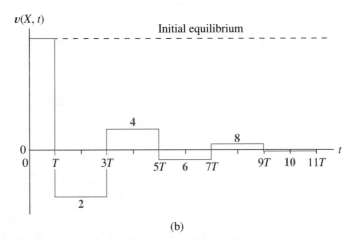

(b)

Figure 14.50   Reflections caused by turning a source off: (a) reflection diagram; (b) load voltage waveform.

14.6.6 REFLECTION PROBLEMS IN LOGIC CIRCUITS

We now have the background to understand the problems reflections can create in a circuit like Fig. 14.51a. Figure 14.51b shows one problem waveform that might appear at the input of the load gate. Although the sending-end output changes instantly from 0 to 1, multiple reflections cause the receiving-end voltage to converge to V_{OH} in a long sequence of small steps, causing a delay of many propagation times (seven in this figure) before the input voltage crosses V_{IH}. And the delay is not the only problem. If the load gate has high gain, the constant input at a level just below V_{IH} holds the gate in the high-gain region of its transfer characteristic. This can cause the load gate to oscillate because gate designs do not anticipate prolonged operation in the high-gain region. Figure 14.51c shows a different problem. The 0 to 1 change is correctly detected after a

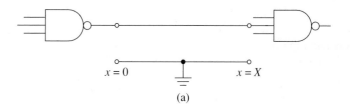

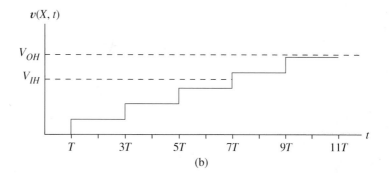

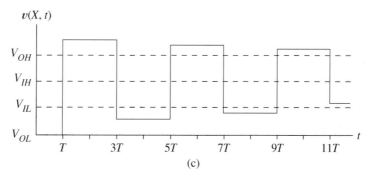

Figure 14.51 Problems caused by reflections: (a) circuit; (b) unnecessary delay; (c) false input logic values.

minimal line delay T; however, because of the oscillatory nature of the reflections, the load gate perceives a false sequence of transitions as the waveform alternates between V_{IL} and V_{IH}. To explore these matters realistically, we must extend our reflection diagram techniques to include the nonlinear volt-ampere characteristics.

14.6.7   NONLINEAR LINE CONSTRAINTS

The introductory examples all used a voltage source with internal resistance at the sending end and a resistor for the load. We next apply the reflection diagram to other terminations including nonlinear devices. The next example introduces several new ideas.

EXAMPLE 14.10   In Fig. 14.52a, a 3-meter transmission line connects a current source to a diode load. The volt-ampere curve shows the diode has forward voltage of

0.5 V and a 3 V breakdown voltage. The initial dc current of -2.5 mA changes suddenly to $+10$ mA at $t = 0$. Line parameters are $u = 3 \times 10^8$ m/s and $R_o = 200\ \Omega$. (a) Find the initial and final equilibrium conditions graphically. (b) Construct the reflection diagram. (c) Sketch $i(3, t)$ and $v(0, t)$.

Solution. (a) First draw the volt-ampere characteristics for the source and load *using the coordinate system defined by the v and i reference directions of the transmission line*. In Fig. 14.52b constant current lines, -2.5 mA for $t < 0$ and $+10$ mA for $t > 0$, describe the source. The diode characteristic, constructed on the same coordinate system, describes the load. Circles show the initial and final equilibria.

(b) At $t = 0$, operation instantaneously changes to a point on the $i = 10$ mA line. To find this point, construct a line of slope

$$\frac{\Delta i}{\Delta v} = \frac{1}{R_o} = \frac{1}{200} = \frac{5\ \text{mA}}{1\ \text{V}}$$

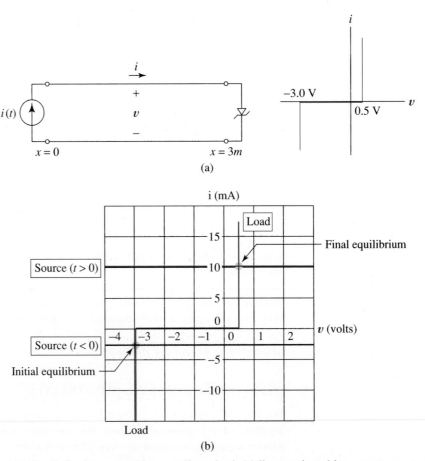

(a)

(b)

Figure 14.52 Reflections caused by a nonlinear load: (a) line terminated by current source and diode; (b) graphical source and load constraints; (c) reflection diagram; (d) waveforms at source and load.

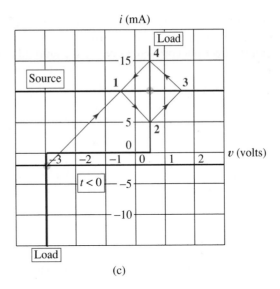

(c)

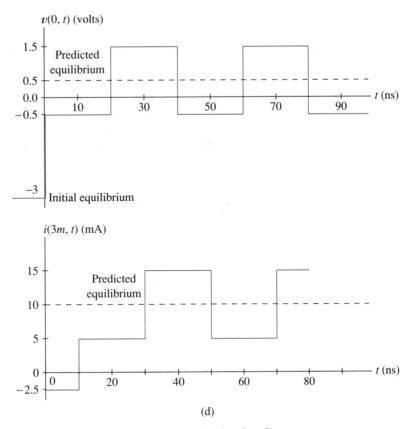

(d)

Figure 14.52 (continued)

from the initial equilibrium point to the $t > 0$ source constraint. This gives point **1** in Fig. 14.52c. Following the usual procedure, we construct lines to points **2, 3, 4,** and **1.** In this example, the sequence does not converge to the final equilibrium because the circuit contains no resistance to dissipate energy and make oscillations die out. Instead, periodic operation ensues, with point **5** coincident with point **1,** point **6** coincident with point **2,** and so forth. This circuit never reaches equilibrium but instead continues to oscillate.
(c) For this line, $T = 3/(3 \times 10^8) = 10$ ns. Using information from Fig. 14.52c while mentally visualizing the reflections to deduce the timing, we draw the waveforms of Fig. 14.52d. ❏

> **Exercise 14.10** On Fig. 14.52b use graphical construction to find the maximum and minimum values for sending-end voltage and load-end current if the zener diode in the preceding example is replaced by a short circuit. Use $R_o = 200 \ \Omega$.
>
> **Ans.** -2.5 to 2.5 V (at sending end), -2.5 to 25 mA (at load end).

14.6.8 REFLECTIONS IN LOGIC CIRCUITS

We now examine reflections associated with several logic families. Hand estimates of inverter delays were simplified by attributing all dynamics to a load capacitor and consciously ignoring internal gate capacitances. In the same fashion, we here attribute all dynamics to the ideal line in order to isolate and better understand reflections. Once we grasp the basic concepts, we can use simulations for more complete analyses.

The *load gate* at $x = X$ in Fig. 14.53 provides a load constraint, in general, a non-linear input characteristic. As with the diode in the preceding example, we simply draw the input characteristic for the load gate in terms of our i versus v coordinate system. At $x = 0$, the *driving gate* presents the line with two constraints, one for steady-state logic 0 and the other for steady-state logic 1. In general these are two nonlinear volt-ampere curves, i_{out} versus v_{out}. Before plotting them, we must first convert each into i and v transmission line coordinates by the change in variables:

$$i = -i_{out}, \qquad v = v_{out}$$

CMOS Terminations. When the load is CMOS, the *load constraint* is the open-circuit equation, $i = 0$. Figure 14.54a shows this, as well as the two sending-end constraints that we explain next.

Figure 14.54b shows how the line is constrained by the source gate for logic 0 output. The constraint is the single MOSFET output curve corresponding to logic 1 input,

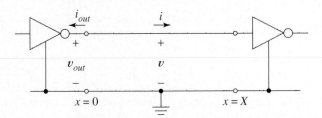

Figure 14.53 Line terminated by logic gates.

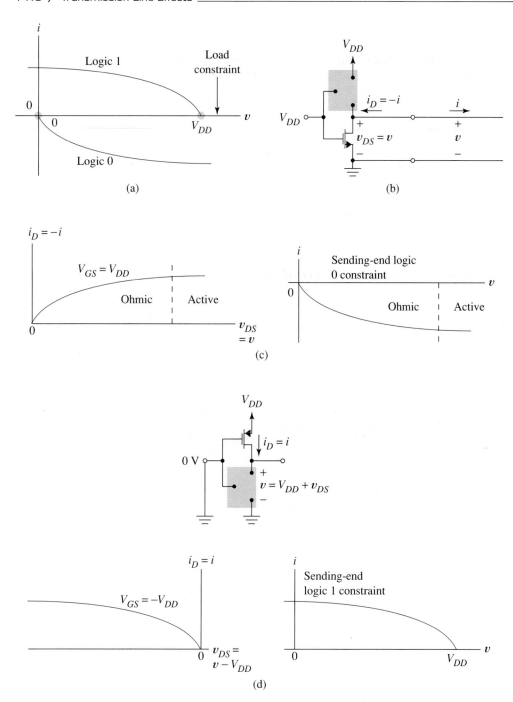

Figure 14.54 CMOS reflection diagram construction: (a) sending and load constraints; (b) variable transformation for logic 0; (c) logic 0 sending-end constraint; (d) logic 1 variables and sending-end constraint.

$v_{GS} = V_{DD}$; however, the change in variables $i = -i_D$ requires that the curve first be reflected about the voltage axis as we show in Fig. 14.54c.

For logic 1 output, Fig. 14.54d shows that the sending-end constraint is the output characteristic of the p-channel transistor translated to the right by V_{DD} volts.

We conclude that for CMOS terminations the complete set of constraints are those of Fig. 14.54a, which uses circles to identify the logic 0 and 1 equilibrium conditions.

CMOS Reflections. The kind of reflections that occur with CMOS or any logic family depends upon how R_o is related to the circuit parameters of the gate. When R_o is relatively small, the reflection diagram for 0 to 1 transitions in CMOS resembles Fig. 14.55a. The load voltage converges to its final value in the sequence indicated by points **2, 4,** and **6,** a response resembling Fig. 14.51b. Clearly, smaller R_o results in longer delays while larger R_o (within limits) reduces the delay. It is also easy to see that a 1 to 0 transition would give similar results involving negative instead of positive line currents.

Figure 14.55b shows the construction when R_o is very large. The voltage at the load converges to its final value in an oscillatory fashion, which can give the false zero problem of Fig. 14.51c. Actually, our CMOS load gate construction is unrealistic in omitting

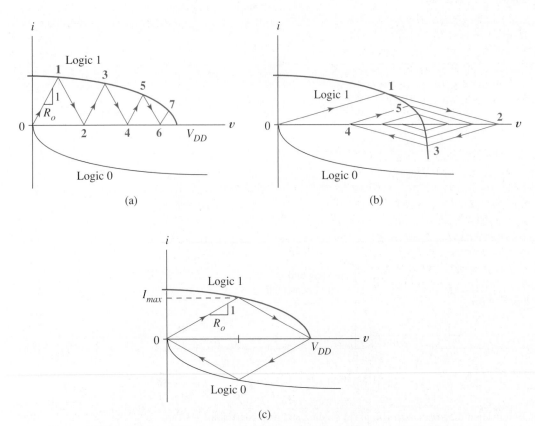

(a)

(b)

(c)

Figure 14.55   CMOS reflection patterns: (a) low R_o; (b) high R_o; (c) special R_o for no reflections.

the protective diodes mentioned in Sec. 14.5.5. These prevent voltages greater than $V_{DD} + 0.7$ V, a situation explored in Problem 14.68.

Figure 14.55c shows that for CMOS there exists an *optimal* R_o for which there are no reflections, and for which transitions from 0 to 1 and 1 to 0 both occur in minimal time of T seconds. For the special CMOS design $k_n = k_p = k$ and $V_{tn} = -V_{tp} = V_t$, this R_o is readily calculated. Since R_o is the reciprocal of the slope of the construction lines, Fig. 14.55c gives

$$\frac{1}{R_o} = \frac{\Delta i}{\Delta v} = \frac{I_{max}}{0.5V_{DD}}$$

In Problem 14.65, we find

$$\frac{1}{R_o} = k\left(\frac{3}{4}V_{DD} + V_{tp}\right) \tag{14.59}$$

where $V_{tp} = -V_t$ is negative. Since R_o is ordinarily in the range of 30–300 Ω, Eq. (14.59) suggests that optimizing line interfaces impinges on gate design through choice of k and V_t, and also might be a factor in selecting V_{DD}. Sometimes specially designed gates called *line drivers* interface between "ordinary" logic and long connecting lines.

EXAMPLE 14.11 A CMOS inverter has parameters $k_n = k_P = 2 \times 10^{-4}$ A/V², $|V_t| = 1.3$ V, $CBS = CBD = 20$ FF (femtofarad), and $RD = RS = 1$ Ω. Use SPICE to find load-end voltage waveforms for $R_o = 50$ Ω, 500 Ω, and 5000 Ω. For the transmission line, $T = 10$ ns. The inverter operates between +12 V and ground.

Solution. Figure 14.56b lists the SPICE code for Fig. 14.56a. The "T1" statement describes the transmission line, where first and second node pairs identify, respectively, the sending and receiving ends of the line. Z0 and TD denote the characteristic resistance R_o and line delay T. To generate reflections associated with 0 to 1 changes at the sending end, v_{in} changes from 12 V to 0 at $t = 0$ as described by the SPICE pulse function.

Three separate runs give the SPICE outputs of Figs. 14.56c and d. Notice how these waveforms agree with Figs. 14.55a and b. Parasitic capacitances introduced nonzero rise times and rounding of the corners of the steps; however, the general nature of the reflections was predicted by the reflection diagrams. We see that the 50 Ω line causes a delay of nearly 90 ns before the input to the load gate reaches 6 V. For this CMOS family, it takes very large R_o such as $R_o = 5$ kΩ to produce false logic 0 inputs like those we use in Fig. 14.56d. ❏

Exercise 14.11 Use Eq. (14.59) to calculate the optimal value of R_o for the CMOS gate of the preceding example. Use SPICE to find the voltage at the load end of the line for this R_o.

Ans. $R_o = 649$ Ω, Fig. 14.56e.

The gradual transition in Fig. 14.56e is the result of a sending-end waveform that does not change instantly as assumed in the reflection diagrams because of internal capacitances, an effect that is difficult to account for in hand analysis.

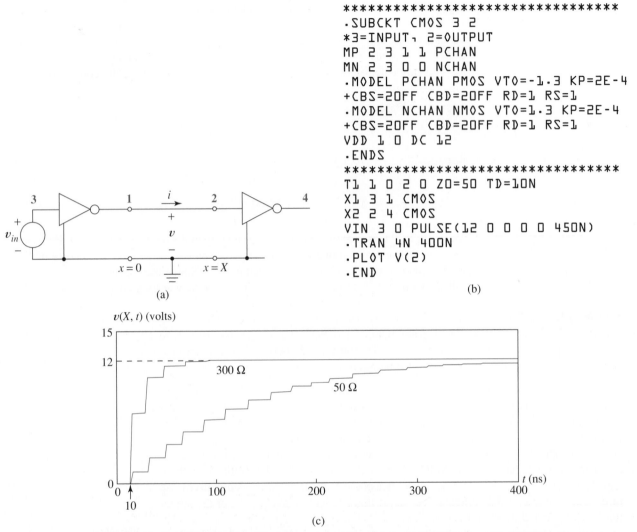

```
EXAMPLE 14.11
*********************************
.SUBCKT CMOS 3 2
*3=INPUT, 2=OUTPUT
MP 2 3 1 1 PCHAN
MN 2 3 0 0 NCHAN
.MODEL PCHAN PMOS VTO=-1.3 KP=2E-4
+CBS=20FF CBD=20FF RD=1 RS=1
.MODEL NCHAN NMOS VTO=1.3 KP=2E-4
+CBS=20FF CBD=20FF RD=1 RS=1
VDD 1 0 DC 12
.ENDS
*********************************
T1 1 0 2 0 Z0=50 TD=10N
X1 3 1 CMOS
X2 2 4 CMOS
VIN 3 0 PULSE(12 0 0 0 0 450N)
.TRAN 4N 400N
.PLOT V(2)
.END
```

(a)   (b)

(c)

Figure 14.56 SPICE simulation of line reflections: (a) given circuit with CMOS gates; (b) SPICE code; (c) load waveforms for low R_o; (d) load waveform for high R_o; (e) load waveforms for $R_o = 649 \ \Omega$.

NMOS Reflections. The NMOS inverter of Fig. 14.57a, like CMOS, presents an open-circuit load constraint. Figure 14.57b shows that the logic 1 sending-end constraint is the nonlinear load line like the one in Fig. 5.20b. Figure 14.57c shows construction of the logic 0 sending-end constraint. Curve i_1 is the output characteristic of M_1 when $v_{GS} = V_{DD}$, and i_2 is the same as Fig. 14.57b. Line current i is the difference, $i_2 - i_1$, easily visualized by graphical construction. We conclude that the reflection diagram constraint curves take the form of Fig. 14.57d; then it is easy to deduce the nature of the reflections for various R_o values. It is also easy to use SPICE simulations to find the reflection waveforms. We leave these explorations for homework problems and go on.

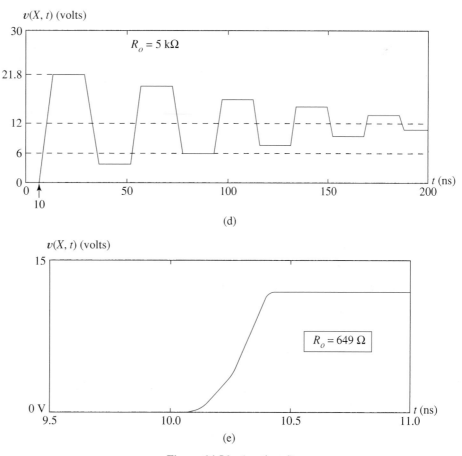

$v(X, t)$ (volts)

$R_o = 5\ k\Omega$

(d)

$v(X, t)$ (volts)

$R_o = 649\ \Omega$

(e)

Figure 14.56 (continued)

An obvious idea for eliminating line reflections when the load gate presents an open circuit would be to terminate the line in its characteristic resistance as in Fig. 14.58a. This indeed eliminates reflections; however, it also changes the logic 0 and 1 equilibrium values as we see in Fig. 14.58b. Excessive power dissipation in the resistor and the possibility of input logic 1 voltages being too low must both be closely examined for such a solution.

The following example shows that we can use SPICE dc analysis to examine more realistic versions of the reflection diagram; for example, we can include body and Early effects in the source constraints.

EXAMPLE 14.12 Use SPICE to find the constraint curves for Fig. 14.57a when $V_{DD} = 10$ V and transistor parameters are $k_1 = 2 \times 10^{-4}$ A/V^2, $k_2 = 2 \times 10^{-5}$ A/V^2, $V_{t1} = V_{t2} = 1.1$ V, $V_A = 60$ V, GAMMA = 0.3, and LAMBDA = 1/60 volt^{-1}.

Solution. Figure 14.59a is the SPICE test circuit and Fig. 14.59b the code. A single .MODEL statement describes both transistors, with individual W/L statements on the el-

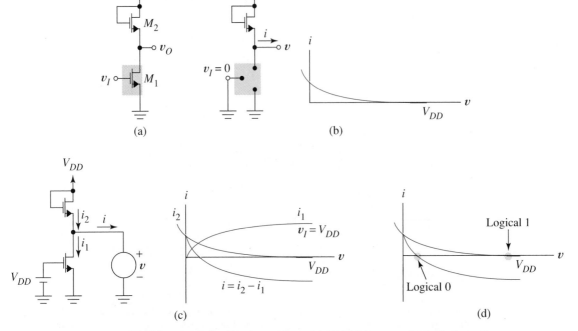

Figure 14.57 NMOS transmission line interactions: (a) NMOS inverter; (b) logic 1 sending-end circuit and volt-ampere constraint; (c) logic 0 sending-end circuit constraint and construction of its volt-ampere characteristic; (d) final sending-end constraints and equilibrium values.

ement lines leading to different k values. The two VIN statements allow us to find the required curves in two separate runs. Notice that because the SPICE current-reference direction is into the positive voltage terminal as shown in the figure, the SPICE currents satisfy our transmission line reference conventions as well. The solid curve of Fig. 14.59c is the requested logic 1 constraint. The dashed curve that follows from replacing the continuation code + with the comment code * ignores body and Early effects. The solid curves of Fig. 14.58d, together with the $i = 0$ line, are the reflection diagram constraints, with equilibrium points encircled. The dashed curve shows how the diagram changes when we ignore body effect and Early effect. Unlike for CMOS, the reflection construc-tions for 1 to 0 and 0 to 1 transitions will be quite dissimilar, with high output currents resulting from 1 to 0 transitions when R_o is small. ❑

Exercise 14.12 Figure 14.59d shows the logic 0 equilibrium is about 0.5 V and the logic 1 equilibrium is near 8.9 V. Assume the characteristic resistance of the line is such that there is a direct transition from logic 1 to logic 0 with no load-end reflection when $\gamma = 0.32$ and $\lambda = 0.017$.
(a) Estimate v^+.
(b) Estimate the value of R_o required.

Ans. -4.2 V, about 700 Ω.

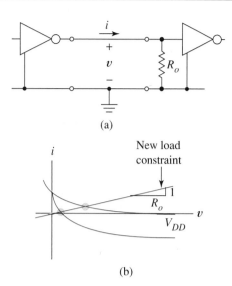

(a)

(b)

Figure 14.58 NMOS line interaction when load resistor matched to the line is added: (a) schematic; (b) new equilibrium values.

The analysis and simulation techniques described for CMOS and NMOS apply to the various bipolar logic families. Because these gates do not have infinite input resistance, however, there is the additional nontrivial task of finding the input volt ampere characteristic of the load gate. A simple analytical approach is to use piecewise linear analysis as we did in Example 14.10. An alternative is to use SPICE as in the preceding example to compute the graphical input and load constraint curves.

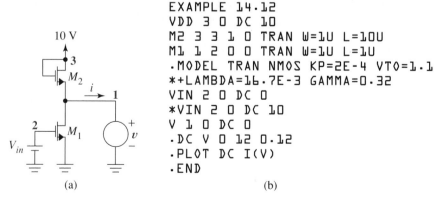

```
EXAMPLE 14.12
VDD 3 0 DC 10
M2 3 3 1 0 TRAN W=1U L=10U
M1 1 2 0 0 TRAN W=1U L=1U
.MODEL TRAN NMOS KP=2E-4 VTO=1.1
*+LAMBDA=16.7E-3 GAMMA=0.32
VIN 2 0 DC 0
*VIN 2 0 DC 10
V 1 0 DC 0
.DC V 0 12 0.12
.PLOT DC I(V)
.END
```

(a)   (b)

Figure 14.59 Using SPICE to find line constraints imposed by logic gates: (a) circuit for finding sending-end constraint; (b) SPICE code for sending-end volt-ampere curve when input is logic 0; (c) SPICE output comparing the logic 1 sending-end constraint with and without body effect; (d) complete set of transmission line constraints determined using SPICE.

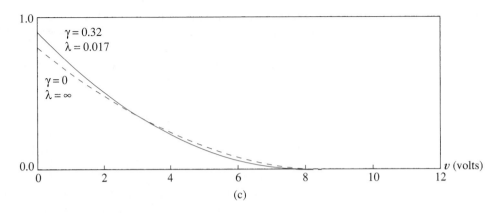

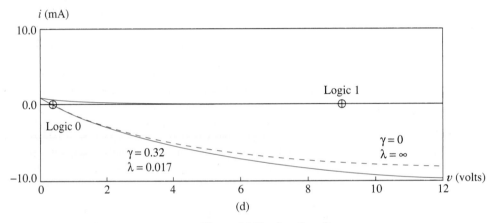

Figure 14.59 (continued)

14.7
Solid-State Memory

Introduction. Solid-state memory is widely used in general- and special-purpose computers and in custom-designed application hardware. Organized physically and logically into arrays of replications of one basic memory *cell* that stores a single bit of binary information, memories lend themselves to economical IC realizations. VLSI technology enables construction of large memory structures having small physical size and weight at very low unit cost. In addition to packing density, speed and power dissipation are important considerations for solid-state memories.

The simplest kind of memory is the *read-only memory* or ROM, in which one can read the stored bits but cannot change them. ROMs are widely used in computers to store microprograms and programs, and in special-purpose hardware. ROM is *nonvolatile*; that is, it retains its contents when power is off. This feature makes ROM programs useful for establishing initial conditions in digital systems upon power up. In many modern instruments and devices, this resident *firmware* can be updated by simply changing ROMs. Read only memory is also used extensively for lookup tables, a fast alternative to software-based computations.

Contrasting with ROM is *read/write memory* used for temporary storage of programs and data. Unlike ROM, the content of a read/write memory cell is easily changed;

read/write memory also differs from ROM in being *volatile*. Therefore, memory contents must be reloaded whenever the system is powered up. It is common usage to refer to read/write memory as *random access memory* or RAM. Strictly speaking, *random access* implies that individual cells or groups of cells can be accessed in any order, a trait that applies to ROM as well as read/write memory.

14.7.1 READ-ONLY MEMORY

In isolation, the term ROM usually denotes *mask programmed memory*, in which memory cell contents are irrevocably stored during memory fabrication. Because of setup costs, such memories are expensive in small quantities, but have low unit costs in large quantities, attributes that lend themselves to high-volume applications. Prototypes or products of limited production volume use other kinds of ROMs we discuss later.

Figure 14.60a shows the block diagram of a basic ROM cell. Cell contents are accessible by means of *word* and *bit* nodes, also called *row* and *column* nodes, respectively. The simplest ROM cell is that of Fig. 14.60b. A "1" is stored by fabricating a diode from the word to the bit line, a "0" by leaving an open circuit, as in Fig. 14.60b. To read the cell contents, we apply a logic 1 voltage to the word line and observe the voltage that appears across a *pull-down* resistor that is connected between bit node and ground. Thus a "1" voltage is the word-line voltage less a diode drop; a "0" gives zero volts.

The MOS ROM cell of Fig. 14.60c is programmed to one by forming the connection to the gate during metalization, to zero by omitting the connection. Bringing the word line high closes the transistor switch, placing V_{CC} on the bit line for programmed ones but leaving the infinite resistance of the nonexistent channel for zeros.

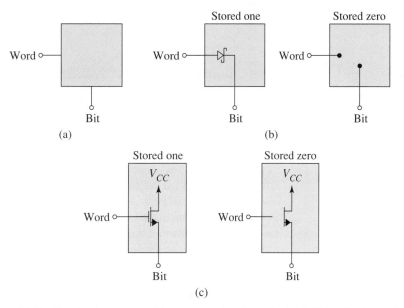

Figure 14.60 Read-only memory: (a) memory cell schematic; (b) ROM cell using a diode; (c) cell using a MOSFET; (d) 16-bit ROM memory; (e) memory array schematic.

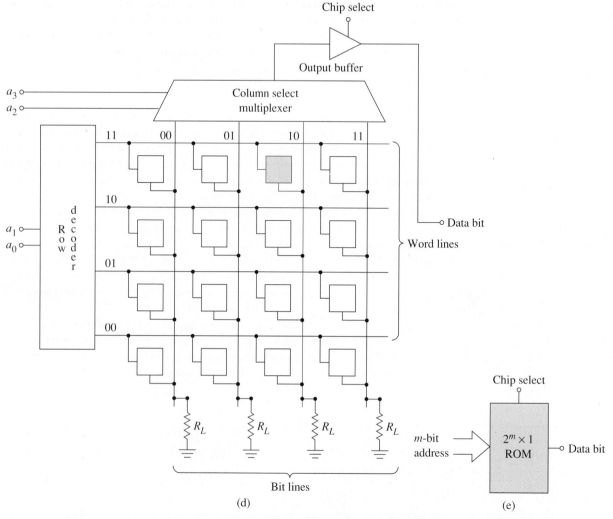

Figure 14.60 (continued)

The cells in an IC memory chip are organized into a regular array of rows and columns, exemplified by the 16×1 ROM of Fig. 14.60d, where rows of cells share *word lines* and columns of cells share *bit lines*. Each cell has a unique *address* determined by its word- and bit-line connections. A row decoder and column-select multiplexer allow the user to access any of the 16 cells using only 4 external pins. When we apply inputs to the two least significant address digits, a_0 and a_1, a high voltage appears on exactly one of the word lines. Contents of all word-addressed cells are then simultaneously accessible on the 4-bit lines. The highest two address digits, a_2 and a_3, however, connect only one of these cells to the output buffer. For example, the shaded cell is located at address $a_3 a_2 a_1 a_0 = 1011$.

In the memory of Fig. 14.60d, the stored information appears at the output node only if the *chip select*, a tristate control input, is high. The usefulness of the chip select for static interconnections and for dynamics is discussed presently.

Often memory arrays have many more rows and columns than Fig. 14.60d. More generally, external pins provided for m address bits are used to select the contents of one out of 2^m data cells. Figure 14.60e schematically represents the ROM memory as a whole.

Figure 14.61a shows how to use parallel address and chip select connections to four ROM chips to organize a memory of 4-bit words. Figure 14.61b shows how the tristate outputs permit parallel output connections from two or more memories like Fig. 14.61a to share a data bus, a concept we saw in Fig. 14.32. The 4-bit addressed word is placed on the bus only when that memory's chip select is asserted, thereby avoiding contention with other memory outputs.

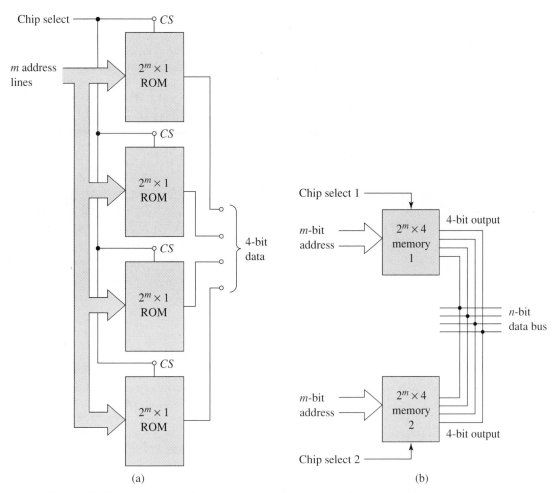

(a)   (b)

Figure 14.61   ROM hierarchical memory organization: (a) chips organized into a memory containing 4-bit words; (b) 4-bit memory outputs sharing a data bus.

PROM and EPROM. For prototype development, where code modifications are frequent, or for applications involving small numbers of final products, mask-programmed ROMs are too expensive. For such applications, *programmable* ROMS are available. The term PROM, an acronym for programmable ROM, usually means a *field-programmable* ROM device having cells similar to the one in Fig. 14.62a. The PROM originally contains ones at every location. The user changes the stored one to zero whenever desired using a special programmer, often computer controlled, which opens the fuse by addressing the cell and then passing high current through the transistor. A logic 1 voltage on the word line closes the transistor switch in the unmodified locations and a stored 1 appears as a high emitter voltage. At each location where the fuse is open, the bit line is pulled to ground through a resistor. Since such programming is irreversible, subsequent design changes require programming a new PROM.

The EPROM (erasable ROM) might employ the dual-gate MOS structure of Fig. 14.62b, in which the second gate is electrically isolated. When the chip is purchased, the second gate is uncharged in all memory cells; the transistor is an *ordinary* MOS switch. A programming device, under computer control, addresses each cell in which a zero is desired and uses a special high programming voltage to store electrons on the second gate. This increases the threshold voltage to such a high value that a logic 1 voltage on the word line no longer turns on the transistor. Since there is never a channel, the bit node of the addressed transistor is pulled to logic 0 by a pull-down resistor. The stored charge remains on the gates for many years unless deliberately altered. The EPROM chip can be entirely deprogrammed (erased) by illuminating a special window with ultraviolet light of sufficient intensity. This removes the electrons from the second gates and restores each cell to its original state.

A dual-gate transistor is also the basic EAROM (electrically alterable ROM) cell. EAROMs are designed so that the charge on individually addressed gates can be changed from one to zero under program control instead of having all set to one by ultraviolet radiation as in the EPROM. EAROMs differ from EROMs, then, in having a memory that can be selectively edited. Also called E^2ROMs (electrically erasable ROMs), these devices resemble the read/write memories described later; however, because they require a long time to write bits into memory compared to their short read time, they are used in ROM applications.

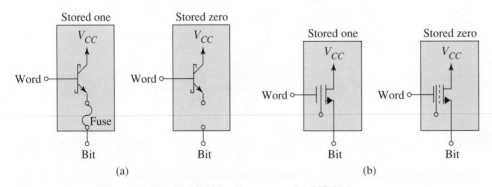

Figure 14.62 (a) PROM cell structure; (b) EPROM structure.

ROM Dynamics. Like other circuits, ROMs have parasitic capacitances that introduce delays. From Fig. 14.60d we might suspect that memory cells attached in parallel to each word and bit line might give rise to rather large capacitances for large memories, and indeed this is the case. Figure 14.63 shows how these capacitances affect the timing of a read operation. Once a valid address is established at the decoder inputs in Fig. 14.60d, there is delay introduced by the capacitances within the decoder before the word-line voltage $v_A(t)$ in Fig. 14.63a reaches its final value. Further delay arises from charging the word-line capacitance C_W and the bit-line capacitance C_B. Notice that C_W, the lumped equivalent of the line capacitance, increases with the number of memory cells connected to that particular word line, the number of columns in the memory. C_B increases with the number of rows. Because we measure delays from the midpoints of the waveform transitions, we can associate with a memory read instruction an overall worst case delay t_{AA} called the *access time*. The timing diagram of Fig. 14.63b, with

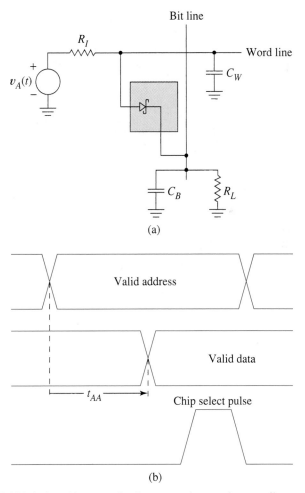

Figure 14.63 ROM timing: (a) row and column capacitances that contribute to delay; (b) timing diagram.

TABLE 14.1 Typical PROM Memory Information

Device	Family	Organization	t_{AA}	Power
PROM—PL87X288B	TTL	$32 \times 8 = 256$ bits	15 ns	0.7 W
EPROM—NMC27C1024	CMOS	$64k \times 16 = 1$ M bit	170 ns	110 mW
EEPROM—NMC9346	NMOS	1024×1 bit	2 ns	60 mW

worst case t_{AA}, represents the memory as a whole. Since new cells can be addressed as frequently as every t_{AA} seconds, t_{AA} is also called the _read cycle time_ of the ROM.

In some systems, the chip select is held high and output data are simply ignored until t_{AA} s after the chip is addressed. In other systems, there are possibilities of reading invalid data. In these cases, the chip-select pulse is applied only after waiting for at least t_{AA} s. Enabling the tristate output in this way guarantees that invalid data never appears at the memory output pins. Table 14.1 gives information for three off-the-shelf PROMs.

14.7.2 RANDOM ACCESS MEMORY

Introduction. Random access memory chips, like ROM chips, are organized as arrays of individually addressable memory cells. RAMs are more complex than ROMs, however, for there must be a provision for writing data into the cells as well as for reading cell contents.

There are two classes of RAMs, _static_ and _dynamic_. In static RAMs (SRAMs), binary digits are stored in flip-flops; dynamic RAMs (DRAMs) store information in IC capacitors, which occupy much less chip space. Since capacitors are lossless, dynamic RAMs also dissipate less power per bit than do static RAMs. Because of their high packing density and low dissipation, dynamic RAM chips as large as 4 megabits per chip are feasible. A disadvantage of dynamic RAM is its complicated control circuitry, which must include a provision for periodically refreshing the stored information. We now examine each type of RAM.

SRAM. Figure 14.64a shows external and internal designs for a static RAM cell. Logic 1 voltage applied to the word line closes switches M_1 and M_2, making the stored bit observable at the output pins. A high word line also provides the external circuit access to the flip-flop for purposes of changing its contents, as discussed in Sec. 14.1. Control circuitry determines whether a read or a write operation takes place, as we see next.

Figure 14.64b shows the general organization of a 16×1 RAM chip. The least significant address bits activate the word lines in some row, connecting each cell flip-flop to a different _pair_ of column lines. The most significant address bits use a decoder to connect exactly one column-pair to an _internal data bus_. In the illustration, address $a_3 a_2 a_1 a_0 = 1011$ activates the shaded cell and simultaneously connects its output to the bus by closing switches M_3 and M_4. (Three unselected MOS _column-connect switch pairs_ are omitted to simplify the diagram.)

The internal data bus is a pair of conductors that facilitates two-way communication between addressed cells and the external pins of the RAM. For a read operation, the read/write input, $R/\overline{W}$, is held at logic 1, closing switch M_5 that connects the addressed

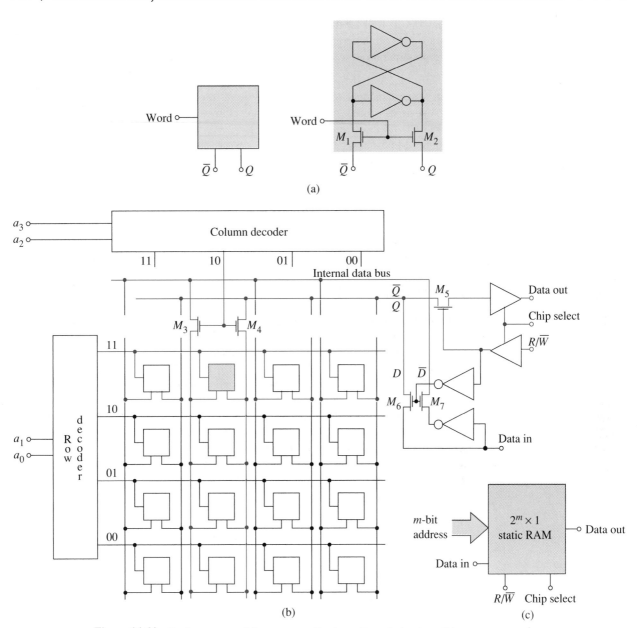

Figure 14.64 Static memory: (a) memory cell schematic and structure; (b) memory organization; (c) schematic.

cell to the *data-out* pin. To write, $R/\overline{W}$ is brought to logic 0, closing M_6 and M_7. Now the logic value available at the *data-in* pin and its complement are simultaneously applied to the addressed flip-flop, forcing it to the proper state.

Sometimes RAMs have a chip select input, as shown in the figure, that isolates the data-out pin from M_5 and also disconnects the internal $R/\overline{W}$ function from the external

$R/\overline{W}$ pin. As in ROMs, the chip select facilitates interfacing memory units with other memories and with external data and address buses. Figure 14.64c is a convenient schematic representation of a RAM chip.

RAM Timing. A RAM read operation involves an access time, t_{AA}, as defined in Fig. 14.63b. Since the input $R/\overline{W}$ is normally high, the principal difference between ROM and RAM access times are the additional delays associated with switches inside the memory cell and in column selection.

Figure 14.65 shows the timing for a write operation. By visualizing the internal events, it is easy to understand why the delays defined on the diagram are necessary in the light of variations in component and parasitic values. First there is the address setup time, t_{AS}, the minimum time that valid address information must be present before $R/\overline{W}$ is brought low. This delay ensures that the *slowest* cell on the IC chip will be connected to data bus before data are written. Thus t_{AS} allows time for worst case delays within the decoders, for charging the word- and bit-line capacitances, and for complete closure of all connecting switches. A minimum width, t_W, for the write pulse ensures sufficient time for the worst case cell to change to the proper stable state by regenerative action. Data setup time, t_{DS}, is the minimum time data must be valid before the end of the write pulse. This allows for worst case delays in closing necessary switches and charging the capacitances of the internal data bus. The address hold time, t_{AH}, is the minimum time during which the address must remain valid after write pulse ends. Were the address to change too soon, the data could also be inadvertently written to an incorrect address. This RAM essentially uses *S-R* excitation as in Fig. 14.7a (in addition to some series switches) to write into memory. A minimum pulse width, t_W, avoids the flip-flop problem that was illustrated in Fig. 14.6d, that is, the FF reverting to its original state because its input was not stable for a sufficient time to initiate irreversible regenerative action.

The data hold time, t_{DH}, is the minimum time data must be valid after the write pulse ends. This gives the capacitances of the circuits associated with the write pulse

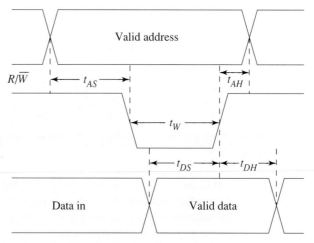

Figure 14.65 RAM timing diagram for a write operation.

sufficient time to change to their final read values. If data change before this happens, incorrect data might be inadvertently written to the addressed location, since the cell "memorizes" the *last* value present on the internal data bus before the cell is disconnected.

An example of a modern SRAM is the 256 k × 1 NM5100 BiCMOS memory, which is designed for compatibility with 100 K ECL. Internally organized as a 256-row by 1024-column cell array, the device has read access time of 15 ns. With reference to Fig. 14.65, write timing is specified by $t_{AS} = 2$ ns, $t_W = 12$ ns, $t_{DS} = 14$ ns, $t_{AH} = 3$ ns, and $t_{DH} = 3$ ns. Power dissipation of the chip is 1.1 W, or 4.2 μW per cell. Some SRAMs, including the NM5100, have a special feature called a *sense amplifier*, not illustrated in Fig. 14.64b, which reduces power dissipation and increases speed. Sense amplifiers are described in the following discussion of DRAMs.

Dynamic RAM. The simple capacitor/switch cell of Fig. 14.66a is the basis for memory chips of very low power dissipation and high density: low dissipation because of lossless components, high density because the compact IC element of Fig. 14.66b replaces the six-transistor cell of Fig. 14.64a. The capacitance between the grounded conductor and the source stores the information, with a charged capacitor indicating a stored one. Creating a channel with the word line closes the switch. Because capacitors slowly lose stored charge, it is necessary to read and rewrite, that is, *refresh* stored data every 4 ms or so. This refresh operation involves a complex configuration of external components and makes the cell temporarily unavailable for ordinary read/write operations; nev-

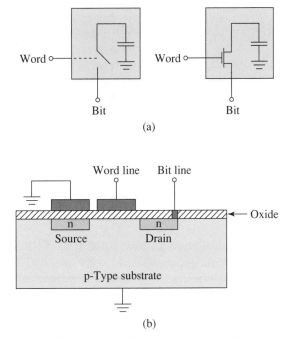

Figure 14.66 Memory cell: (a) circuit; (b) physical structure.

ertheless the advantages of the simple cell persist, since many cells share the same control components, and since refresh makes the cells inaccessible only about 2% of the time. In spite of the refresh hardware, DRAMs occupy only about one-sixth the chip space of SRAMs of comparable memory.

Memory Refresh. To refresh the memory, control circuitry periodically brings each word line high. The contents of all cells in the addressed row are simultaneously read and then rewritten, using parallel hardware units called *sense amplifiers*, one for each column.

To understand the refresh system we begin with Fig. 14.67a, where we wish to read the contents of the addressed memory cell. Because each bit line interfaces with many cells, it is physically long, and consequently has parasitic capacitance much larger than the cell capacitance, for example, 1 pF versus 0.05 pF as in the diagram. Suppose the cell capacitance happens to be charged to a logic "one" value of 5 V, and the bit-line capacitance is uncharged. When the word line goes high, the charge stored in the cell is shared with the parallel bit-line capacitance. The resulting voltage, V_1, of the parallel capacitors satisfies

$$Q = CV = 0.05 \times 10^{-12}(5) = 1.05 \times 10^{-12}V_1$$

Thus a voltage V_1 of only 0.24 V represents a logic 1, whereas 0 V represents logic 0. This difference is too small to allow for practical considerations such as residual charge left on the bit-line capacitance before the refresh begins. We next learn how additional hardware corrects this problem.

In Fig. 14.67b, the memory cell environment is modified through addition of a second bit line, a *dummy cell*, and some switches and associated control logic. Activating a chip-select pin (not shown) initiates refresh by producing a pulse ϕ_1. This pulse closes switches that *precharge both* bit-line capacitances and also the dummy cell capacitance to 2.5 V. Because A is low, dummy cell contents are isolated from the bit line. After the ϕ_1 switches open, the word line and the dummy cell are *simultaneously* addressed, the latter by bringing A high.

Addressing the dummy cell *always* keeps the voltage of bit line 2 at its precharge value of 2.5 V. If the memory cell holds a zero because its capacitance is uncharged, then conservation of charge requires the voltage V_0 of the parallel equivalent capacitor of value 1.05 pF in Fig. 14.67c to satisfy

$$Q = CV = 0.05 \times 10^{-12}(0) + 1.00 \times 10^{-12}(2.5) = 1.05 \times 10^{-12} \, V_0$$

or $V_0 = 2.38$ V. For a stored one the voltage V_1 satisfies

$$Q = CV = 0.05 \times 10^{-12}(5.0) + 1.00 \times 10^{-12}(2.5) = 1.05 \times 10^{-12} \, V_1$$

which gives $V_1 = 2.62$ V. It may seem that we have made little progress because the difference between V_1 and V_0, 0.24 V, is the same as we found for Fig. 14.67a. However, comparing the memory and dummy cell outputs with a difference amplifier gives a reliable method for reading the state of the memory cell. For the difference amplifier (sense amplifier) we use a flip-flop as in Fig. 14.67c. Large resistors, R, formed by doped polysilicon, contribute little to the power dissipation of the chip. Once the two bit-line capacitances have charged, a pulse ϕ_2 activates the flip-flop. From our analysis of unstable

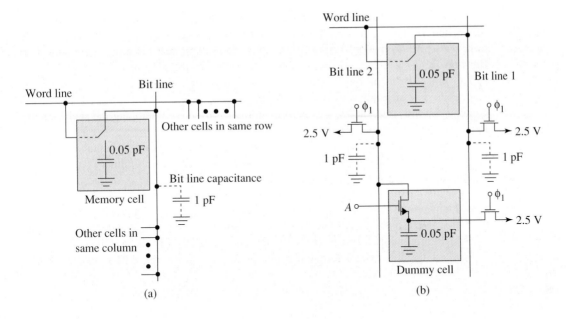

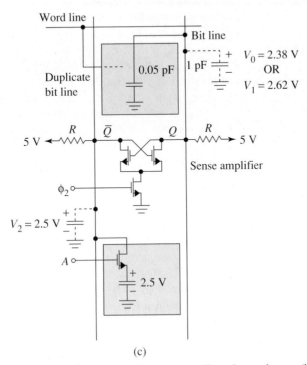

Figure 14.67 Refreshing dynamic memory: (a) memory cells and capacitance; (b) switches and capacitances associated with refresh; (c) sense amplifier.

equilibrium and Fig. 14.8 we know that the flip-flop will quickly go to the $Q = 1$ state if the bit line is charged to 2.62 V, but will go to the $Q = 0$ state for 2.38 V. We see that the events prior to the ϕ_2 pulse establish initial conditions that force the sense-amplifier flip-flop to $Q =$ "1" if a "1" was previously stored in the cell, or a "0" if a "0" was stored. Since the memory cell is still connected when this happens, the cell capacitance charges to 5 V if a one was previously stored or to 0 if a zero was stored. Once cell contents are thus refreshed, ϕ_2 goes to its normal zero value, removing the sense amplifier from the circuit.

Since all columns have sense amplifiers like this that operate simultaneously, the entire chip is refreshed in the time it takes to address all the rows. Ordinary read or write operations also result in refreshing. The sense-amplifier hardware is shared by all cells in the same column; thus its addition has only a minor impact on packing density and chip dissipation. The logic required to interface the bit lines to an internal data bus, and the latter to the data-in and data-out pins, are similar to that already described in Fig. 14.64b. Representative of DRAM units currently available is the 41C1000-08-ND, a 1 Meg × 1 memory with internal structure of 2048 columns and 512 rows and access time of 80 ns.

14.8 _____
Summary

This chapter introduced a number of new principles, most associated with internal and external dynamics of digital circuits. Our first topic was the multivibrator family, a class of two-state circuits that employ positive feedback.

Bistable multivibrators called flip-flops employ their two stable states to store a binary digit: one state represents a stored logic one, the other a stored zero. The most common bistable is a pair of inverters from some logic family, cross coupled to form a binary memory unit called a latch. Cross-coupling two-input gates provides an elegant way to control the flip-flop states and gives rise to _SR_ flip-flops. Schmitt triggers, also bistables, are effective in removing noise from digital data and are used when very rapid logic transitions are needed. All bistables have an unstable equilibrium point. To change the state of the bistable, it is necessary to charge the internal capacitances just past this unstable balance point. Once this is done, the circuit completes the state change on its own at high speed and without further assistance from an outside source.

The monostable multivibrator has a stable rest state and a quasi-stable transient state. An external triggering source upsets the rest-state equilibrium and forces the monostable into its quasi-stable state. The circuit remains in its quasi-stable state for a predetermined time and then spontaneously returns to its stable state. An output variable displays a different logic level for each internal state, thus providing an output pulse of predetermined length after each triggering event, the characteristic response of the monostable to an input pulse. The triggering pulse has amplitude and duration requirements that are dictated by the need to force the monostable into regenerative action. Once this positive feedback mechanism is invoked, the triggering pulse can be removed.

The astable multivibrator is a signal generator or relaxation oscillator characterized by two quasi-stable states, spontaneous rapid transitions between these states caused by positive feedback, and output logic levels indicative of the current internal state. The frequency of the astable, like the delay of the monostable, is determined by the designer. The 555 timer circuit is a popular MSI device often used to realize both monostable and astable multivibrators with performance parameters that are relatively independent of temperature and power supply voltage.

A number of common interfacing problems were defined, and examples of solutions for each were given. These included logic swing too small, output logic levels that do not overlap with input levels required by gates, and inadequate gate output current for sourcing or sinking load current. We found simple transistor switches, level-shifting circuits, and clamping circuits to be useful for correcting output levels, and parallel gates or additional output transistors were used to augment available output current.

Some logic families have output circuits compatible with logic input requirements of others and require no interfacing. In other cases, simple pull-up resistors or transistor switches solve the interface problem. Occasionally, off-the-shelf ICs specially designed for a particular interface are needed to satisfy demanding requirements. An interface between families needs to operate only at the speed of the slower family, and the interface design is further simplified by using interfacing components with fan-out of one. Interface noise margins, defined using output values from the interfacing circuit and input values from the driven gate family, are conceptually useful.

Most logic families have special gates that are designed to solve certain interfacing problems. These include "open-collector"-type circuits that can cooperatively manage the logic level at a shared node by sharing a pull-up resistor, and tristate gates that can be individually disconnected from a node they share with other gates.

A digital-to-analog converter transforms a binary number sequence into a continuous analog waveform, usually voltage. The D/A converter is a fast operational amplifier interfaced with a reference voltage through an input network of precision resistors and switches. The number of switches determines the precision with which the binary-number input can be translated into voltage.

The A/D converter performs the inverse operation of the D/A; it converts a periodically sampled analog voltage into a binary number sequence. To facilitate interfacing, the A/D has an input that receives a pulse when a conversion is desired, and an output that signals when a conversion is complete and the output data are valid. A number of different A/D designs exist, which differ in internal details as well as in conversion times. As in the D/A, resolution and precision depend upon the number of bits.

Transmission line models for intergate connections become important when the delays caused by travel between gates becomes comparable to the propagation delay of the gate. This means we need to worry about long connecting lines for any logic family, and short connections for the faster families.

The ideal, lossless transmission line model, which characterizes the line by its characteristic resistance, R_o, and propagation velocity, u, provides us with considerable insight into the kinds of problems that can arise in connections. Reflection coefficients, real numbers that we can easily compute, tell us in the case of linear terminations, whether or not there will be sending- and receiving-end reflections. The bounce diagram helps us visualize the timing of multiple reflections and helps us sketch sending- and receiving-end waveforms. With step inputs of current or voltage, there is usually a transition from an initial equilibrium to a final equilibrium. The transient waveshapes, monotonic or oscillatory, depend on the magnitudes and signs of the reflection coefficients. For lossless terminations, oscillations occur, and dynamic equilibrium is never reestablished. From studying linear line terminations, we learn that the connecting wires can introduce problems such as long delays and false logic transitions at the receiving end. They can also lead to large currents and/or voltages at either end of the line that might otherwise be overlooked.

Because logic gates provide nonlinear terminations to the transmission lines, a graphical construction provides the generality we need to study more complex reflections. In general, three nonlinear volt-ampere curves, two from the sending-end gate and one from the receiving-end gate, provide the constraints for the line equations. A graphical construction, derived from the constraints placed upon the line disturbances by linear transmission line theory, helps us visualize the kinds of reflections that can occur for various kinds of logic gates. Furthermore, the construction allows us to evaluate the use of modified gate/line interfaces including nonlinear elements such as diodes. SPICE simulations, reinforced by user understanding, permit us to include gate delays and other second-order effects in our interface explorations.

Various kinds of memory—ROM, PROM, EAROM, DRAM, and SRAM—were discussed in terms of their internal and external structures. Memories are organized as arrays of basic memory units called cells, each characterized by a row address and a column address. A memory cell can be as simple as a diode or a capacitor, or as complex as a flip-flop. In the case of RAMs, the simplest cell structures lead to high packing density and low power dissipation; however, control and interfacing circuitry is more complex, for it must provide for periodically refreshing the cell contents. There are timing constraints that memory control signals must satisfy. These are dictated by delays associated with the internal capacitances of the memory cells and the gates that interface the cells with the outside world.

REFERENCES _____

1. HAZNEDAR, H. _Digital Microelectronics._ Redwood City, CA: Benjamin/Cummings, 1991.
2. HIROSE, A., and K. LONNGREN. _Introduction to Wave Phenomena._ New York: John Wiley, 1985.
3. HODGES, D. A., and H. C. JACKSON. _Analysis and Design of Digital Integrated Circuits._ New York: McGraw-Hill, 1988.
4. MAURO, R. _Engineering Electronics._ Englewood Cliffs, NJ: Prentice Hall, 1989.
5. MITCHELL, F. H., JR., and F. H. MITCHELL, SR. _Introduction to Electronics Design._ Englewood Cliffs, NJ: Prentice Hall, 1988.
6. NATIONAL SEMICONDUCTOR CORP. _Memory Databook._ Santa Clara, CA: National Semiconductor Corp., 1990.
7. SEDRA, A., and K. SMITH. _Microelectronic Circuits,_ 3rd ed. Philadelphia: Saunders College, 1991.

PROBLEMS _____

SPECIAL DIRECTIONS FOR SPICE HOMEWORK PROBLEMS: _Do Not Hand in Lengthy SPICE Printout for Homework._ Instead, abstract the useful information from the SPICE output file as in the SPICE examples. Include your SPICE code and a circuit diagram with nodes numbered to agree with the code. Cite relevant numerical values from the SPICE output file and discuss when appropriate. Make sketches of any relevant curves, and label appropriate points. Make small tables to present numerical data if useful for clarity.

Section 14.1

14.1 In Fig. 14.2a, $V_{DD} = 9$ V and transistor parameters are $k = 0.8 \times 10^{-3}$ A/V^2 and $|V_t| = 0.5$ V. Find the resistances in Figs. 14.2b and c.

14.2 Sketch the output voltage waveforms, $v_X(t)$ and $v_Y(t)$, for the latch of Fig. 14.4 as the stored logic 1 is changed to 0 by closing the switch.

14.3 In Fig. 14.4a, $C_1 = C_2 = 4$ pF, $V_{DD} = 9$ V, $R_1 = R_4 \approx 150$ Ω, and $R_I = 15$ Ω. For the transistors $k = 0.8 \times 10^{-3}$ A/V^2 and $|V_t| = 0.5$ V.
(a) Write a numerical equation for $v_X(t)$ that holds until M_3 turns on.
(b) Calculate the delay t_1.
(c) Calculate the actual value of B$_1$ at the instant M_3 begins to conduct. *Hint*: Assume M_1 is a voltage-controlled resistor.

14.4 Find an inequality that R_I and R_2 must satisfy if v_X in Fig. 14.5 is ever to drive the inverter pair into the maximum gain region.

14.5 Example 14.1 simplified the flip-flop dynamics by lumping transistor capacitances into equivalent input capacitances. Modify the code of Fig. 14.6b by eliminating input capacitances and including zero voltage capacitances in the .MODEL statements.
(a) Examine the node voltages $v_1(t)$ and $v_2(t)$ generated by this dynamic model. From your results would you say the original model gives a reasonable idea of the internal operation of the FF? Explain.
(b) Explore more realistic values of KP for the transistor switch. What goes wrong if KP is too small?

14.6 Starting with the code of Fig. 14.6b,
(a) use SPICE to explore the effect on gate switching of increasing RIN.
(b) Reformulate the code so that the initial state is logic 0, RIN = 10 Ω, and the input voltage is logic 1. Use SPICE to plot the waveforms for a 2 ns switch closure.

14.7 The flip-flop in Fig. 14.7a is in the $Q = 0$ state with $S = R = 0$. At $t = 0$, the S input suddenly goes to V_{DD} volts while the R input remains at 0 V.
(a) Draw the equivalent circuit that applies at the instant after $t = 0$.
(b) Write the differential equation that applies at this instant in terms of input capacitance C_2, the gate voltage v_G of M_2, resistance R_3 of M_3 and the parameters k_s and V_{ts} of the S-input transistor.

14.8 Estimate the power dissipation for the FF in Fig. 14.7b.

14.9 (a) Use Eq. (14.12) to find the range of q for which the circuit of Fig. 14.9a has $H > 0$.

(b) Sketch the hysteresis curve and label critical values when $q = 3.25$.
(c) Find the value of q that maximizes hysteresis H, and also the maximum value of H.

14.10 Show that for sufficiently high q, it is possible to design an emitter-coupled Schmitt trigger like Fig. 14.9a such that $V_{OL} < V_{IL}$.

Section 14.2

14.11 In Fig. P14.11 the inverter has infinite input resistance and the given transfer characteristic. Source v_s has the value +10 volts and the circuit has attained equilibrium.
(a) What are the values of v_X and v_O?
(b) What is V_C?
(c) At $t = 0$, v_S suddenly changes to zero. Write the equation that describes $v_X(t)$ for $t > 0$.
(d) Sketch $v_X(t)$ and $v_O(t)$ on the same coordinate system.
(e) Compute the width of the output pulse.

14.12 In Fig. 14.11b, $V_{DD} = 5$ V and $v_I = 0$. C is charged, with v_X 0.2 V more positive than $\overline{v}_O$ because of a previous pulse. That is, the recovery time is not yet over.
(a) Find the voltages v_O and v_X at this instant.
(b) If the circuit is now triggered by an input pulse from v_I, how long will the monostable output pulse last?

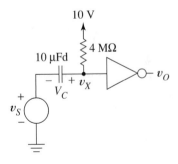

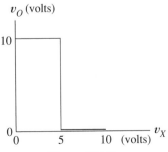

Figure P14.11

14.13 In Fig. 14.12, $V_{CC} = 5$ V, $R_C = R_B = 1$ kΩ, $R = 10$ kΩ, and $C = 0.005$ μF. All transistors have $V_{CE,clamp} = 0.3$ V, $V_{BE,clamp} = 0.7$ V, and $\beta = 20$. The cut-in voltage of Q_2 is 0.5 V.

(a) Draw the equivalent circuit for the stable state assuming $v_I = 0$.

(b) Label all node voltages and all current values on the diagram. Also show the capacitor voltage.

(c) An input pulse v_I now instantly drives Q_1 into clamp, which, in turn, cuts off Q_2 and drives Q_3 into clamp. Draw the equivalent circuit at this instant and compute the minimum amplitude of the pulse at v_I.

Hints: Consider gate fan-out and the current in C. Assume Q_1 and Q_3 have equal collector currents when in clamp.

(d) Write the differential equation for $v_X(t)$ for this quasi-stable circuit.

(e) Find the duration τ of the output pulse.

Section 14.3

14.14 The op amp in Fig. P14.14 saturates at ± 15 V. Use $v_O = +15$ V and $v_B = [1 \text{ k}/(1 \text{ k} + 3.3 \text{ k})](-15)$ V for initial conditions.

(a) Use an equivalent circuit to calculate $v_B(t)$. Find the time when the op amp changes state. Then change the circuit model and calculate $v_B(t)$ for the next time interval

(b) Sketch the waveforms of $v_O(t)$ and $v_B(t)$.

(c) Find the period of the waveform.

14.15 Sketch the waveforms that appear at nodes 4, 2, and 1 in Fig. P14.15. Use the following procedure.

(a) Assume the voltage at node 4 is positive. Find its value. Find the integrator input current. Sketch the voltage at node 2 versus time while these conditions hold. Find the slope of this voltage-time curve.

(b) Find the input threshold voltages for the Schmitt trigger. On the diagram from part (a) mark the maximum and minimum integrator output voltages.

(c) Determine the integrator input current when the Schmitt trigger is in its second state. Add to the sketch of part (a) the next segment of node voltage 2. Complete the sketches of the node voltages at nodes 2 and 4.

(d) Sketch the output voltage of the clamping circuit at node 1.

14.16 Draw a sketch to show how the waveforms of Fig. 14.14c would change if the op amp had significant slewing.

Hint: Sketch $v_{O1}(t)$ first.

14.17 Use SPICE to simulate the astable of Fig. P14.15 for three or four cycles. Omit the clamping circuit from the simulation. Use the 741 subcircuit of Fig. 2.34. Use .IC V(2) = 0.1 to set the initial condition for the integrator capacitor. Sketch the SPICE waveforms for nodes 4 and 2.

Section 14.4

14.18 Design two 555-based monostables that use a 9 V supply and produce output pulses of width 2.3 μs and 12 hours, respectively. For each, give your values for R_X and C_X and sketch the capacitor and output voltages. Any correct solution is acceptable.

14.19 Trigger pulses of 1 μs width and fundamental frequency 16 kHz provide input for a monostable designed to give a pulse width of 2.2×10^{-4} s. Use a sketch of input and output

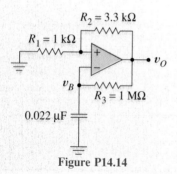

Figure P14.14

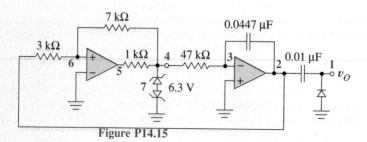

Figure P14.15

pulses to show that this circuit divides the input frequency by four.

Hint: Your sketch should locate the ends of the output pulses relative to the input pulses.

14.20 Find values for the external components so that the astable of Fig. 14.17a has frequency of 800 kHz and duty cycle of 0.8. Your design should also satisfy the condition that for a supply of 5 V, the maximum supply current delivered to the external components does not exceed 0.1 mA.

Hint: Find the maximum current with (ideal) switch open and with switch closed. Then examine the ratio to see if one must be larger than the other.

Section 14.5

14.21 Use SPICE to verify the design of Example 14.4. Use transistors of $k = 0.2 \times 10^{-4}$ A/V^2 and $|V_t| = 1$ V for the inverter and the interface. Print or plot SPICE transient analysis output from 0 to 200 ns for an input pulse of 100 ns width. Verify correct waveforms V_X at the inverter output.

14.22 Design a level shifter like Fig. 14.20c for a square wave input. Specifications are $\Delta V = 7$ V, $k_1 = 4 \times 10^{-4}$ A/V^2, $k_2 = 10^{-4}$ A/V^2, $V_{t1} = V_{t2} = -1.5$ V, $v_{IN,max} = 3$ V, $v_{IN,min} = -2$ V.
(a) Using the given specifications, sketch the input voltage.
(b) Using the given specifications, sketch the output voltage the circuit is to produce.
(c) Determine the zener voltage and power supplies required.

14.23 Design a level shifter like Fig. 14.20d. Specifications are $k_E = k_D = 10^{-4}$ A/V^2, $V_{tE} = -V_{tD} = 1.5$ V. The input is a square wave with offset characterized by $v_{IN,max} = 3$ V, $v_{IN,min} = -2$ V, $V_Z = 2$ V.
(a) Find the bias current.
(b) Find the value of ΔV.
(c) Sketch the output waveform if the input is a square wave plus offset with values given.

14.24 The pulse levels of Fig. P14.24 must be shifted upward so that the mean of the input signal is aligned with the mean of the dashed-line input gate levels.
(a) Find the required ΔV, and show how the shifted signal lines up relative to the input levels.
(b) For part (a) find the interface noise margins, *NMH* and *NML*. What additional operation on the signal would improve the noise margins?
(c) Specify suitable zener voltage and transistor parameters k_1, k_2, V_{t1}, and V_{t2} for a level-shifting circuit like Fig. 14.20c to implement the shift of part (a). Any reasonable values that satisfy all these conditions is acceptable.

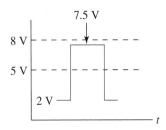

Figure P14.24

(d) Find minimum V_{DD} and V_{SS} so that the circuit works properly.

14.25 (a) Design an interface circuit that will convert the TTL-level pulses of Fig. 14.21a into ECL levels, -1.52 V and -1.06 V, when the waveform frequency is 1 MHz.

Hint: Attenuate the amplitude, and then use a clamping circuit to shift the dc level. Adjust the dc source in the level shifter to allow for a diode drop of about 0.5 V. Make the time constant long to avoid sag in the output pulses.
(b) Use SPICE to verify your design.

14.26 In Fig. P14.26, the input is a pulse sequence with minimum and maximum amplitudes of -1.52 V and -1.06 V.
(a) Find the voltage V_{BB} that centers the gate voltage pulses about the MOSFET threshold voltage of $V_t = 0.7$ V.
(b) Would gain added between the clamp and the MOSFET improve the reliability of the circuit? Explain.

14.27 The triangular waveform in Fig. P14.27 must be transformed into a 0 to 10 volt waveform. The figure shows a circuit proposed to perform this function.
(a) What goes wrong if the triangular wave is connected directly (or through a follower) to the clamping circuit without using the inverter? Assume the capacitor is initially uncharged.
(b) Sketch the waveform v_A in the given circuit.

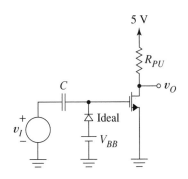

Figure P14.26

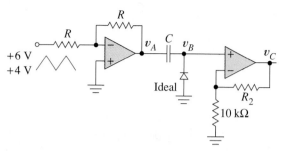

Figure P14.27

(c) What is the minimum value of R that can be used if the op amp can deliver no more than ± 25 mA without distorting the signal?
Hint: The loading from the clamp is negligible after the initial transient.
(d) Sketch the waveform v_B in the given circuit.
(e) Find R_2.
(f) Will the maximum common-mode input rating of ± 13 V be exceeded for either op amp? Explain.

14.28 Figure P14.28 shows a 5 V CMOS inverter interfaced with a 12 V CMOS inverter directly and through an interface circuit. All internal transistors have $|V_t| = 1$ V.
(a) Compute and compare the noise margins at nodes 1 and 2.
(b) Does the interface circuit perform amplification, level shifting, or both? Explain your answer.
(c) Using the threshold voltage of the interface transistor, compute noise margins for node 3.

14.29 Design a circuit like the first of Fig. 14.23d to solve the interfacing problem of Example 14.5.

14.30 CMOS logic using a 9 V supply is to be interfaced to an LED rated at 4 V and 50 mA. Design an interface like the first circuit of Fig. 14.23d. Specify the value of R, k, and V_t for the transistor, and the transistor power dissipation when the diode conducts.

14.31 Figure P14.31 shows the interface between an NMOS logic inverter and a piezoelectric buzzer with internal oscillator, which is rated at 5 V and 15 mA. The depletion transistor has $k = 10^{-4}$ A/V² and $V_t = -0.7$ V.
(a) Find the resistance, R_M, of the depletion transistor assuming ohmic-region operation and linear resistance. Also write the inequality that must be satisfied for this expression to be valid.
(b) Determine the maximum base current if the condition of part (a) is to be satisfied.
(c) Assuming the BJT saturates to turn on the buzzer, find the value of R.
(d) Find R_B consistent with parts (a) and (b). Would a smaller value of R_B give acceptable operation of the input circuit? Explain.
(e) Find the minimum value of β to satisfy the conditions established so far.
(f) If a Darlington pair replaces the BJT, estimate the new value of R_B and the β of the transistors in the identical pair.

14.32 Figure P14.32 shows an inverter driving an optical isolation circuit. The forward voltage of the LED is 3 V.
(a) If the gate is TTL with the totem pole output circuit of Fig. 13.32, find R for a 5 mA "on" current for the optical isolator.
Hint: Assume transistor states are the same as for zero output current.
(b) Assume the gate is delivering 5 mA with the resistor of part (a) in place. Check the forward active assumption you made in part (a).

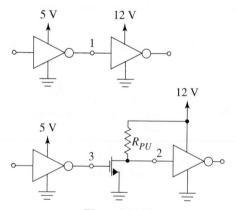

Figure P14.28

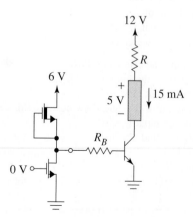

Figure P14.31

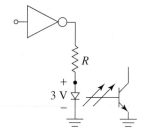

Figure P14.32

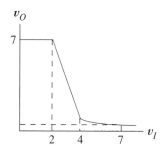

Figure P14.37

14.33 Figure P14.33 shows a BJT used to enhance the load current delivered by an NMOS inverter when output is logic 1. (M_2 is cut off by a logic 0 input.) Parameters are $k_1 = 2 \times 10^{-4}$ A/V², $V_{t1} = 1.2$ V, $\beta = 90$. Assume Q is active.
(a) Find the drain current of M_1.
(b) Find the value of R consistent with the voltage and current indicated on the diagram.
Hint: A diode-connected transistor is either off or active.

14.34 In Fig. 14.23e, $V_{XX} = 8$ V, $k = 10^{-4}$ A/V², $V_t = -1.2$ V, $\beta = 70$, $V_L = 2$ V, and $I_L = 30$ mA. The MOSFET is the p-channel transistor in a CMOS gate with the other transistor cut off. Assume the MOSFET satisfies Eq. (5.12). Draw a circuit diagram and show numerical values on your diagram as you find them.
(a) Find the resistance of the MOSFET.
(b) Assume the drop across the MOSFET is the largest possible for the assumption of a linear resistor. Find V_{SD} for this case and the corresponding drain current.
(c) Find the value of R so that the voltages across the various components add up to 8 V.
(d) Find R_B.

14.35 Find the value of k for the interface of Fig. 14.25a if $|V_t| = 0.9$ V and $V_{DD} = 9$ V.

14.36 In Fig. 14.25b, $V_{DD} = 10$ V, C_{in} for the CMOS gate is 2.5 pF, and $R_P = 50$ kΩ.
(a) Find the power dissipation in R_P when the interface logic value is zero and when it is one.
(b) The input to the CMOS gate has been logic 1 for a long time. Find the rate of change of input voltage at the instant the TTL output goes to logic 0.

14.37 When $V_{DD} = 7$ V in Fig. 14.25c, the NMOS transfer curve is that of Fig. P14.37. Estimate *NMH* and *NML* at the interface (NMOS input)
(a) when R_P is infinite.
(b) when $R_P = 20$ kΩ.

14.38 In Fig. 14.25d, the driving gate uses $V_{DD} = 5$ V. Transistor parameters are $V_t = 1.1$ V, $k_1 = 2 \times 10^{-3}$ A/V², and $k_2 = 0.4 \times 10^{-3}$ A/V². Determine whether this driving gate is able to sink the required TTL input current of 1.43 mA when its input voltage is 5 V.

14.39 In Fig. P14.39, both transistors in the NMOS gate have $V_t = 0.5$ V; the lower transistor has $k = 0.2 \times 10^{-4}$ A/V². Width-to-length ratios are shown in parentheses. The CMOS input parameters are $V_{IL} = 2.125$ V, $V_{IH} = 2.875$ V.
(a) Write a quadratic equation for V_{OL} for the NMOS gate when $v_I = 5$ V. Assume the lower transistor satisfies Eq.

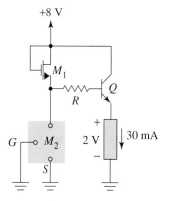

Figure P14.33

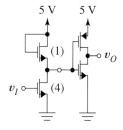

Figure P14.39

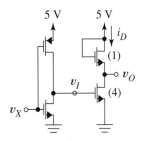

Figure P14.40

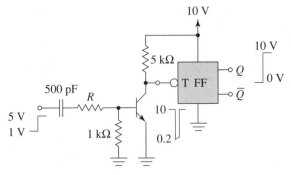

Figure P14.42

(5.5), and the load is an active-state transistor. Find V_{OL}.
(b) Calculate *NML* for the interface.
(c) Calculate *NMH* for the interface.

14.40 All transistors in Fig. P14.40 have $|V_t| = 1.1$ V.
(a) Compute *NML* for the interface.
Hint: $v_X = 5$ V for this case.
(b) To find V_{IH} for the NMOS gate we must find the point on the transfer curve where $dv_O/dv_I = -1$. To simplify the problem, assume that

$$v_O \approx i_D R_{NMOS}$$

where i_D comes from assuming the load resistor is an active transistor and where R_{NMOS} comes from Eq. (5.5). (Width-to-length ratios of the NMOS transistors are shown on the diagram in parentheses.) The result should be an equation relating v_O and v_I. Assume $v_O^2 = 0$.
(c) Differentiate the equation of part (b) and determine V_{IH} by equating the derivative to -1.
(d) Find *NMH* for the interface.

14.41 In Fig. P14.41 the transistors in the NMOS gate both have $k = 0.2 \times 10^{-4}$ A/V². For the depletion device, $V_t = -0.5$ V; the enhancement transistor has $V_t = +0.5$ V. The CMOS input parameters are $V_{IL} = 3.25$ V, $V_{IH} = 4.75$ V.
(a) Estimate V_{OL} for the NMOS gate by treating the depletion gate as a constant current source and using ohmic resistance as in Eq. (5.5) to describe the enhancement device when $v_I = 5$ V.

Then compute *NML* for the interface.
(b) Estimate *NMH* for the interface.
(c) Add a 100 kΩ pull-up resistor from the interface node to the 8 V supply. Rework parts (a) and (b) including this resistor.

14.42 The differentiator/switch interface in Fig. P14.42 is to provide a pulse to set the CMOS flip-flop when there is a sudden input level-shift as indicated.
(a) Ignoring internal transistor dynamics, use a large-signal model to find an expression for base current, $i_B(0^+)$, in terms of R at the instant after the level change.
(b) Select R so that the forced β is 2. Then find $i_B(0^+)$.
(c) Use the circuit model of part (a) to predict the final value of $i_B(t)$, assuming (incorrectly) that the transistor will stay in its ON state forever.
(d) Use the initial and final values and R value just computed, to write an exponential equation for $i_B(t)$.
(e) Estimate the pulse width by finding how long it takes the base current to reach zero.

14.43 In Fig. P14.43, $\beta = 20$ and the diode has an offset of 0.7 V.
(a) What is the transistor state when $v_I = 5$ V? What is the value of inductor current?

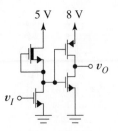

Figure P14.41

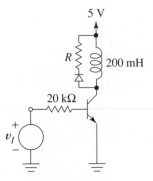

Figure P14.43

(b) Suppose the transistor cuts off instantaneously when v_I suddenly changes from the 5 V of part (a) to zero. Draw the equivalent circuit. Take into account that the inductor current cannot change instantly.
(c) Find the largest R so that v_{CE} of the transistor does not exceed $BV_{CE0} = 20$ V.
(d) For the R of part (c), estimate how long it takes the inductor current to reach 10% of its initial value.

14.44 Figure P14.44 shows a *binary weighted resistor* D/A converter.
(a) Assuming infinite gain, find a series expression for I in terms of V_{REF}, R, and the binary digit values b_i.
(b) Starting with your answer to part (a), show that output voltage can be written in the form

$$v_O = \frac{R_F}{R} V_{REF} \left(\sum_{k=0}^{n-1} b_k 2^k \right)$$

14.45 The number of bits in the D/A of Example 14.8 is reduced to $n = 8$, but the design otherwise remains the same.
(a) Investigate the importance of input offset voltage if a 741 op amp is used in this 8-bit design.
(b) Find V_{REF} so that the offset is only one-tenth of the minimum nonzero D/A output voltage.

14.46 We are designing an $R/2R$-type D/A with resolution of 4 mV that uses a reference voltage of 15 V. Find the minimum integer n that gives this resolution when $R_F/R = 1$.

14.47 A controller requires analog voltages between 0 and 8 V with precision of 0.4% for its input. Design an $R/2R$ analog-to-digital converter that meets these requirements. Our parts catalog shows that IC ladder circuits are readily available with $R = 25$ kΩ, 50 kΩ, and 100 kΩ.

14.48 The conversion time of an A/D must be 10 μs or less. Find the minimum clock frequencies required for $n = 8$, 12, and 16 bit precision for

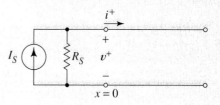

Figure P14.50

(a) a counting-type A/D circuit
(b) a successive-approximation circuit.

14.49 (a) A counting-type A/D converts analog input voltages of 0 to 5 V into 8-bit digital numbers. If the conversion time is 100 μs, find the minimum slew rate for the D/A output if worst case slewing takes only 0.1 of a clock cycle.
(b) Work out the slew rate for a successive approximation A/D with the same specifications as part (a).

Section 14.6

14.50 Use direct analysis of Fig. P14.50 and Eq. (14.46) to show that when a sending-end disturbance is launched by a current step, the sending-end constraint is the current divider equation

$$i^+ = \frac{R_S}{R_S + R_o} I_S$$

14.51 In Fig. P14.51, the switch upsets an equilibrium by switching from position 1 to position 2 at $t = 0$. If $R_o = 50$ Ω, $u = 3 \times 10^8$ m/s, and T is the time it takes a wave to travel to the end of the line, sketch $v(0.1\text{m}, t)$ and $i(x, T/2)$, when
(a) $R_S = 100$ Ω.
(b) $R_S = 200$ Ω.

14.52 The terminations of a line with characteristic resistance R_o are a load resistance R_L and a sending-end voltage source of V_1 volts and R_S Ω.
(a) Find the equilibrium values, V_E and I_E, for line voltage and current in terms of the given quantities.
(b) At $t = 0$ the source voltage suddenly changes from V_1 to V_2 volts, upsetting the equilibrium. Use basic equations to find

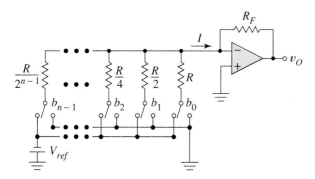

Figure P14.44

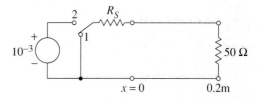

Figure P14.51

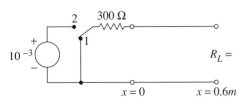

Figure P14.53

the disturbance v^+, given that it adds to the initial conditions. That is, the sending-end constraint relates $(I_E + i^+)$ to $(V_E + v^+)$. *Hint*: Draw sketches of the line for $t < 0$ and $t > 0$.

14.53 The ideal lossless line of Fig. P14.53 is in equilibrium for $t < 0$ with the switch in position 1. At $t = 0$ the switch changes to position 2. If the propagation velocity is 3.0×10^8 m/s and $R_o = 300$ Ω,
(a) Sketch v $(x, 0.5T)$, i $(x, 0.5T)$, v $(x, 1.5T)$, and i $(x, 1.5T)$.
(b) Sketch the voltage and current waveforms that appear at the sending and receiving ends of the line; that is, sketch v $(0, t)$, i $(0, t)$, v $(0.6m, t)$, and i $(0.6m, t)$.

14.54 Sketch i $(0.6m, t)$ and v $(x, 1.5T)$ if R_L in Fig. P 14.53 is changed to 0. $R_o = 300$ Ω, $\mu = 3 \times 10^8$m/s.

14.55 In Fig. P14.50, $R_s = R_o = 200\Omega$, and the line is in equilibrium with $I_S = 0$. At $t = 0$, I_S suddenly changes to 1 mA. Sketch v $(X/2, t)$ and v (X, t).
Hint: Read the statement of Problem 14.50.

14.56 Use the bounce diagram of Fig. 14.45b to sketch v $(X/2, t)$. Take care to visualize the timing correctly.

14.57 From Fig. 14.45a, find the final equilibrium value of the line current. Use Fig. 14.45b to sketch the current, i $(X/2, t)$. Take care to depict the timing correctly.

14.58 (a) Draw the bounce diagram that describes Fig. 14.45a if R_L is changed from 200 Ω to 50/3 Ω.
(b) Find the final equilibrium value for the line voltage.
(c) Sketch v $(0, t)$ and v (X, t).

14.59 For each problem described in the table, sketch v $(0, t)$ and v (X, t). Show all important numerical values in each

sketch. In each case, $u = 3 \times 10^8$ m/s and the line has zero initial conditions.

Problem No.	R_S (Ω)	R_o (Ω)	X (m)	R_L (Ω)	Voltage Step
14.59(a)	50	50	10	0	0 to 5 V
14.59(b)	50	50	0.1	20	0 to 5 V
14.59(c)	50	50	5	70	0 to 10 V
14.59(d)	0	50	10	0	0 to 10 V

14.60 Use bounce diagrams to find v $(X/2, t)$, v (X, t), i $(X/2, t)$, and i (X, t) for the following line problems. Compare your solutions with expected steady-state values.
(a) $R_o = 100$ Ω, $R_S = 50$ Ω, $R_L = \infty$, and input is the voltage step $30u(t)$.
(b) $R_o = 50$ Ω, $R_S = 0$ Ω, $R_L = \infty$, and input is the voltage step $50u(t)$.

14.61 Sketch input voltage and input current for the reflection diagram of Fig. P14.61.

14.62 On ruled paper construct the reflection diagram for the line problem described by $X = 0.6$ m, $R_L = 50$ Ω, $R_S = 400$ Ω, and $R_o = 100$ Ω. Use information from your diagram to sketch v $(0, t)$, v $(0.6m, t)$ and i $(0, t)$, assuming the line is initially in equilibrium and the input is a 2 volt step at $t = 0$. Propagation velocity is $u = 2.8 \times 10^8$ m/s.

14.63 Repeat Problem 14.62 using $X = 0.6$ m, $R_L = 50$ Ω, $R_S = 50$ Ω, and $R_o = 100$ Ω.

14.64 Sketch the reflection constructions for *1 to 0 transitions* for Figs. 14.55a and b. Number the points on the curves, odd numbers for sending-end points and even numbers for receiving-end points. Use the same R_o values as those on the figures.

14.65 Show that the characteristic resistance that produces no reflections in CMOS for $k_p = k_n = k$ and $V_{tn} = -V_{tp} = V_t$ is given by Eq. (14.59).
Hint: Carefully relate the quantities in Fig. 14.55c to the PMOS transistor in Fig. 14.54d. Notice that I_{max} is the current when transistor voltage is $V_{DD}/2$.

14.66 Figure P14.66 shows logic 0 and logic 1 source constraints and the load constraint for a logic gate connected to a nonlinear load through an ideal, lossless transmission line. The line is 0.03 m long and the propagation velocity is 3×10^8 m/s.
(a) Show the reflection construction for a "0" to "1" transition when $R_o = 100$ Ω (10 mA/volt).

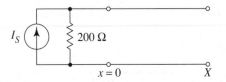

Figure P14.55

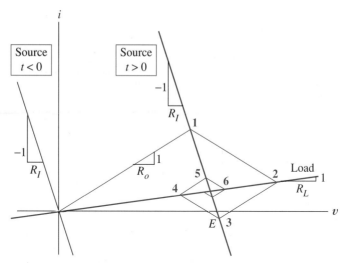

Figure P14.61

(b) From the construction of part (a) sketch the current waveform at the *load end*. Show numerical values of current and time.

(c) What are the maximum values of *sending-end* voltage and current for the transition of part (a)?

(d) If the line resistance is 50 Ω (20 mA/volt), estimate how long a "1" to "0" transient will take.

Hint: Draw a small circle surrounding the final equilibrium on your diagram to define "close to equilibrium."

14.67 We found that the CMOS inverter of Example 14.11 gives no reflections when terminated with a line of $R_o = 649$ Ω.
(a) See if you can find a new value of V_{DD} so that the same inverter gives no reflections when $R_o = 300$ Ω.
(b) Repeat part (a) for $R_o = 50$ Ω.
(c) Using $V_{DD} = 12$ V as in the example, find a new value of transistor parameter k so that there are no reflections for $R_o = 50$ Ω.

14.68 Redo the SPICE simulation that gave Fig. 14.56d after first adding protection diodes (with default parameters) at the input of the load gate (node 2) as suggested in Fig. 14.29a.

14.69 Two NMOS gates like Fig. 14.57a are connected by a 0.03 m length of lossless transmission line characterized by $R_o = 300$ Ω, $u = 2.5 \times 10^8$ m/sec. Transistor parameters are $k_1 = 2 \times 10^{-4}$ A/V^2, $k_2 = 2 \times 10^{-5}$ A/V^2, $V_{t1} = V_{t2} = 1.1$ V, $V_A = 60$ V, GAMMA = 0.3, LAMBDA = 1/60 volt^{-1}. Use SPICE to find the sending- and load-end voltages for a zero-to-one logic transition of the sending gate. $V_{cc} = 12$ V.

14.70 Work Problem 14.69 using a one-to-zero input logic transition.

14.71 Figure P14.71 shows a resistor-transistor logic (RTL) inverter. (a) Use piecewise linear analysis to find i_{IN} versus v_I.

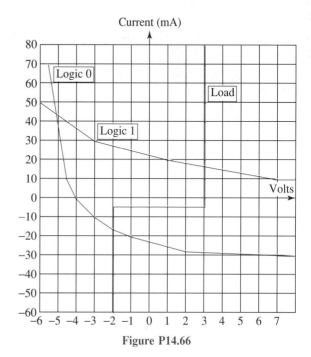

Figure P14.66

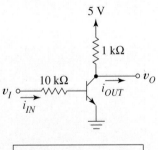

$$V_{BE,on} = V_{BE,sat} = 0.7 \text{ V}$$
$$\beta = 40 \quad V_{CE,sat} = 0.1$$

Figure P14.71

Figure P14.72

(b) Use piecewise linear analysis to find i_{OUT} versus v_O when $V_{IN} = 5$ V.

(c) Use piecewise linear analysis to find i_{OUT} versus v_O when $V_{IN} = 0.1$ V.

(d) Plot all sending and load constraints.

14.72 Figure P14.72 shows a diode-transistor logic (DTL) inverter. For SPICE simulations use default diodes and $\beta = 30$.

(a) Use dc analysis to plot i_{IN} versus v_I for $0 \le v_I \le +4$ V.

(b) Plot i_{OUT} versus v_O when $v_I = 0.2$ V using dc analysis.

(c) Plot i_{OUT} versus v_O when $v_I = +4$ V using dc analysis.

(d) Use the SPICE results to make a rough sketch showing the general appearance of the reflection diagram for DTL.

14.73 Figure P14.73a shows two CMOS gates connected by a transmission line of $R_o = 200 \ \Omega$ (1 V per 5 mA). A 1 kΩ resistor parallels the load-gate input. Figure P14.73b shows the CMOS output constraints.

(a) Construct the load constraint on the diagram.

(b) Sketch $v(X, t)$ for a "1" to "0" input transition.

(c) If $T = 4.2$ ns, estimate how long it takes the load gate to perceive a change from logic "0" to logic "1" when $R_o = 100$

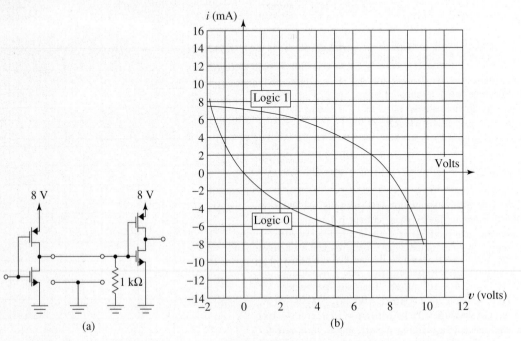

Figure P14.73

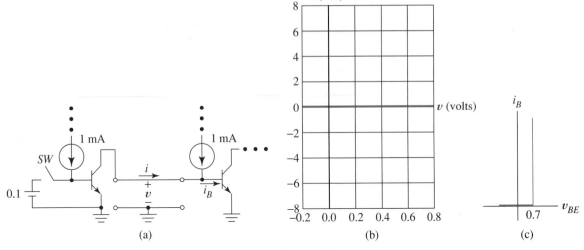

(a) (b) (c)

Figure P14.74

Ω (1 V per 10 mA). Use $V_{IH} \approx V_{IL} \approx 0.5\ V_{DD}$ for your estimate.

14.74 Figure P14.74 shows two *current injection logic* (I^2L) inverters connected by a transmission line. For the transistors, $\beta = 6$ and $V_{CE,sat} = 0.1$ V. When input is a "0" (closed switch), the transistor is cut off. When the input is "1" (open switch), the collector current is $\beta i_B = 6$ mA, except for when the transistor saturates, giving $v_{CE} = V_{CE,sat}$ and $i_C \le \beta i_B$.
(a) On a coordinate system like Fig. P14.74b, sketch the sending-end constraint that applies when the switch is closed.
(b) Add a sketch of the sending-end constraint when the switch is open.
(c) Now add the receiving end constraint.
Hint: Transistor input characteristic is Fig. P14.74c.

14.75 Use SPICE to model the circuit consisting of two inverters like Fig. P14.71 connected at the source and load ends of an ideal, lossless, transmission line. A voltage source at the sending-end inverter input changes from 0.2 V to 5 V at $t = 0$. Line parameters are Z0 = 50Ω and TD = 20 ns
(a) Use a SPICE *static* transistor model with BF = 40 and transient analysis to find the input waveform of the load inverter for the first 120 ns.
(b) Now repeat the simulation after adding dynamic parameters to the transistors. Add CJE = 0.15PF, CJC = 0.15PF, TF = 0.2N, TR = 20N.

14.76 Figure P14.76a shows a line with ECL buffers at input and output, showing one way to interface ECL circuits to give low reflections. Figure 13.40a is the diagram for the input buffer. Notice that the components in the dashed

box are actually connected at the load end of the line in Fig. P14.76a.
(a) Use SPICE to find $v_o(t)$ for the input transition shown. Transistor parameters are
  BF = 30
  CJE = 0.15PF
  CJC = 0.15PF
  TF = 0.2N
Use .IC V(1) = -1.85.

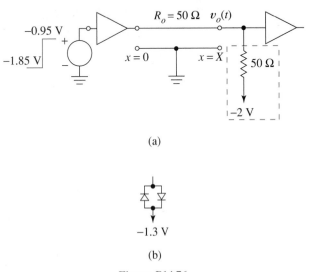

(a)

(b)

Figure P14.76

Line delay is TD = 5 ns. Let the simulation run for 20 ns.
(b) Repeat the simulation after replacing the circuit within the dashed box of Fig. P14.76a with the circuit of Fig. P14.76b. This is another interfacing method used for ECL.

(c) Repeat the simulation a third time with the components in the dashed box simply deleted, that is, for a direct connection of gate output to input by the line, with no attention paid to interfacing.

SPICE Analysis

This appendix uses examples to introduce beginners to fundamentals of SPICE simulation. Additional examples integrated throughout the text introduce many more techniques of SPICE programming, each as its need arises, and a table inside the back cover helps the reader quickly return to key examples as needed. Although textbook examples are self-contained, a reference that discusses SPICE parameter and syntax options in greater depth is sometimes useful; several such references are listed in this appendix. The examples in the text all run on student PSPICE; however, to make the code as universally applicable as possible, the examples deliberately eschew special features not available in all versions of SPICE.

SPICE performs three kinds of circuit analysis: dc analysis, ac analysis, and transient analysis. In this introduction we use dc and ac analysis to demonstrate SPICE fundamentals. In transient analysis, time is the independent variable, and the circuit excitation is a time function such as a step, a ramp, an exponential function, a triangular wave, or a sinusoid. We encounter many examples of transient analysis in the text.

A.1
dc Analysis

Figure A.1a shows a dc circuit containing resistors, an independent current source, and two independent voltage sources. The first step is to give SPICE the circuit diagram. We begin by numbering the nodes of our circuit. Any node numbers will do, provided that we designate the ground reference as node 0. Next, we use lines of SPICE code to describe the circuit. Figure A.1b shows the code for Fig. A.1a.

SPICE code satisfies a few simple rules. The first line of code must be a title; the last must be the control statement ".END". Between these, *element lines* describe the individual circuit elements. Special control statements, which always begin with ".", specify the kind of analysis and describe the output desired by the user.

Every resistor requires a name that begins with R, and this name is the first entry on its element line. Then come the resistor's node numbers and its resistance in ohms. The resistor descriptions of Fig. A.1b show four acceptable number formats. The information fields on the element line are separated by one or more spaces. Any line that begins with an asterisk is a comment, which is ignored by SPICE. Comments afford a convenient

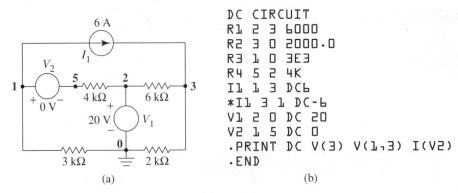

```
DC CIRCUIT
R1 2 3 6000
R2 3 0 2000.0
R3 1 0 3E3
R4 5 2 4K
I1 1 3 DC6
*I1 3 1 DC-6
V1 2 0 DC 20
V2 1 5 DC 0
.PRINT DC V(3) V(1,3) I(V2)
.END
```

(a)   (b)

Figure A.1

way to explain sections of code, and any number of comments may be placed anywhere in the code.

Names for independent current sources begin with I. Node numbers indicate the current reference direction by listing the "from node" first, and the "to node" second. The letters DC indicate a source of direct current, and the current is given in amperes. The comment that follows the current-source element line shows an equivalent description of the same current source.

Independent voltage sources require names that begin with V. The voltage in volts on the element line is the voltage at the first-listed node relative to the second.

The ".PRINT DC" line is a control statement that asks SPICE to place in the output file certain values computed during the dc analysis. In this example, these are: the voltage at node 3, V(3); the voltage at node 1 relative to node 3, V(1, 3); and the current that flows into the positive terminal and out of the negative terminal of V2, I(V2). Source V2 contributes nothing to the circuit, but it serves as an "ammeter," providing a way to find the current in the 4 kΩ resistor, since SPICE does not include resistor currents in its output.

Given the code of Fig. A.1b, SPICE computes and places in the output file all dc node voltages and all voltage-source currents, and also lists numerical values for the quantities listed in the ".PRINT" statement.

Figures A.2 describes a dc circuit that includes a voltage-controlled current source and a current-controlled voltage source. Each current-controlled voltage source, such as the one that produces voltage at node 3 relative to node 4, *requires* a name that begins with H. Next come its nodes, using the same reference-direction convention as an independent voltage source, followed by the name of the voltage source whose current controls the dependent source. Finally, we list the transmittance (ratio of dependent voltage to controlling current) in ohms.

A voltage-controlled current source, like the one that directs current from node 2 to 3, requires a name that begins with G. After its node numbers come the nodes that define the controlling voltage. In this case, the voltage at node 2 relative to 0 controls the source. Finally, we give the transmittance in amperes per volt. Table A.1 lists SPICE abbreviations for various power of 10 multipliers. Notice that M is the SPICE abbrevia-

TABLE A.1   SPICE Abbreviations

Abbreviation	Name	Multiplier
F	femto	10^{-15}
P	pico	10^{-12}
N	nano	10^{-9}
U	micro	10^{-6}
M	milli	10^{-3}
K	kilo	10^{+3}
MEG	mega	10^{+6}
G	giga	10^{+9}
T	tera	10^{12}

tion for 10^{-3}, *not* 10^6. We encounter the SPICE code for the other kinds of dependent sources elsewhere in the text. These examples required no special code line to direct SPICE to perform a dc analysis because SPICE *always* performs a dc analysis.

A.2
AC Analysis

Figure A.3 shows SPICE code for an ac circuit. Inductors and capacitors require names that begin, respectively, with L and C. In this circuit, V1 is an ac voltage source, and I1 is an ac current source. SPICE expects magnitude listed first and phase (in degrees) second, and if there is no value for phase, SPICE uses zero degrees.

The .AC line asks SPICE to do an ac analysis from 1000 Hz to 1000 Hz in 1 linearly spaced step, that is, to analyze the circuit at the single frequency of 10^3 Hz. The comment that follows shows how to obtain a frequency response involving 100 points linearly spaced from 0.1 Hz to 1000 Hz. The .PRINT line asks SPICE to produce in the output file a table that lists, for each analysis frequency, the magnitude of the voltage at node 2 relative to node 4, VM(2, 4), and the phase of this same voltage, VP(2, 4). The same line also requests tabulated values for the magnitude of the current that flows from

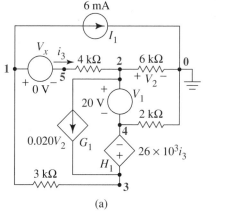

```
DEPENDENT SOURCES
R1 5 2 4K
R2 2 0 6K
R3 4 0 2K
R4 1 3 3K
I1 1 0 DC 6M
V1 2 4 DC 20
VX 1 5 DC 0
H1 3 4 VX 26K
*H1 4 3 VX -26K
G1 2 3 2 0 20M
.WIDTH OUT=80
.END
```

(a)   (b)

Figure A.2

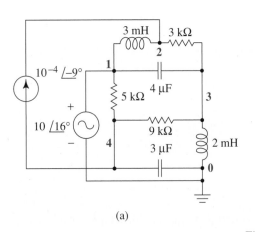

```
SINGLE-FREQUENCY ANALYSIS
R1 2 3 3K
R2 4 3 9K
R3 1 4 5K
L1 1 2 3M
L2 3 0 2M
C1 4 0 3U
C2 1 3 4U
V1 1 0 AC 10 16
I1 4 2 AC 0.1M -9
.AC LIN 1 1E3 1E3
*.AC LIN 100 0.1 1000
.PRINT AC VM(2,4) VP(2,4) IM(V1)
.END
```

(a)   (b)

Figure A.3

node 1 to ground through voltage source V1 at each analysis frequency. Since the .AC line requested analysis at only one frequency, the table has only one row.

REFERENCES _____

1. ANTOGNETTI, P., and G. MASOBRIO, eds. *Semiconductor Device Modeling with SPICE*. New York: McGraw-Hill, 1988.
2. BANZHAF, W. *Computer-Aided Circuit Analysis Using SPICE*. Englewood Cliffs, NJ: Prentice Hall, 1989.
3. TUINENGA, P. W. *SPICE A Guide to Circuit Simulation & Analysis Using PSPICE*. Englewood Cliffs, NJ: Prentice Hall, 1988.

Two-Port Networks

The concepts we explore here help us analyze interconnected devices and subcircuits. The theory is especially germane to feedback amplifiers and is essential background when we want to extract information from transistor data sheets.

B.1
Definition of a Two-Port

A two-port is a device or circuit that communicates with the "outside world" only through input and output *ports*, as in Fig. B.1. Each port is an external terminal pair with port currents and voltages defined by the *standard reference convention* of Fig. B.1. Essential to the definition of a port is that the current entering one port *terminal* also exits from the other, for example i_1. This discussion is strictly limited to *linear* two-ports.

Most two-ports in this text happen to include a direct connection between an input node and an output node (dashed line in Fig. B.1); however, this connection is not essential for two-port theory to apply. A transformer is a two-port that lacks this connection. The two-ports of greatest interest in this course represent transistors biased to operate with small-signals, simple resistor circuits, and interconnections of *RLC* elements and dependent sources.

Two-port theory plays the same role for two-ports that Thevenin's and Norton's theorems play for one-ports. Two-port theory:

- ❏ leads to simple, standard-format, equivalent circuits that can be obtained either by measurement or by circuit analysis
- ❏ applies to arbitrarily complex as well as simple circuits
- ❏ facilitates suppression of unneeded details
- ❏ provides simplified equations for devices and circuits
- ❏ applies to either dc or sinusoidal steady-state ac descriptions of devices or circuits
- ❏ correctly describes only those variables *external* to the two-port
- ❏ simplifies handling of interconnections of circuits

1143

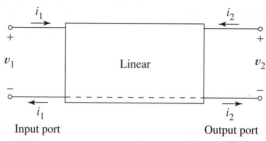

Figure B.1 Notational conventions for two-port theory.

Unlike Thevenin's and Norton's theorems, two-port theory does not apply if the two-port contains independent sources; however, dependent sources are permissible.

We examine four of the possible two-port representations here. For a particular device or circuit, some or all of the representations might exist; and the intended application helps us decide which to use. Sometimes the choice depends upon which parameters can be measured or calculated with the greatest ease or precision, or both. Other two-port descriptions exist, such as ABC and scattering parameters, but their applications lie beyond the scope of our studies.

B.2
z Parameters

Equations. In terms of z parameters, the two-port of Fig. B.1 is described by :

$$v_1 = z_{11}i_1 + z_{12}i_2$$
$$v_2 = z_{21}i_1 + z_{22}i_2$$

(B.1)

Figure B.2a helps us understand the meaning of these equations. Since the circuit is *linear*, Eqs. (B.1) simply describe each port voltage as a *superposition* of responses to the current sources attached to the ports. By viewing Eqs. B.1 as basic expressions of the linearity of circuit within the two-port, we conclude that the equations are valid for *arbitrary* external terminations, not just current sources.

z-Parameter Definitions. Equations (B.1) lead to definitions that give a meaning to each z parameter. In the first of Eqs. (B.1), setting $i_2 = 0$ and solving for z_{11} gives

$$z_{11} = \left. \frac{v_1}{i_1} \right|_{i_2 = 0}$$

(B.2)

To make i_2 equal zero, we open-circuit the output port as in Fig. B.2b. This diagram shows why z_{11} is called the *open-circuit input impedance* of the two-port. It is the impedance that we see at the input port when the output port is terminated in an open circuit.

From the second equation, with $i_2 = 0$,

$$z_{21} = \left. \frac{v_2}{i_1} \right|_{i_2 = 0}$$

(B.3)

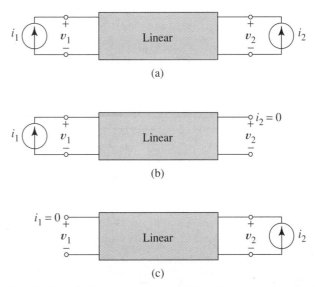

Figure B.2 (a) Terminations for interpreting z parameters; (b) terminations for defining z_{11} and z_{21}; (c) terminations for defining z_{12} and z_{22}.

From Fig. B.2b, z_{21} is named the *open-circuit, forward transfer impedance: transfer* impedance because it is voltage at one port divided by current at the *other* port, *forward* because it is the ratio of an *output* variable to an *input* variable.

Setting $i_1 = 0$ in Eqs. (B.1) leads to Fig. B.2c and definitions of *open-circuit, reverse transfer impedance* and *open-circuit, output impedance,*

$$z_{12} = \left. \frac{v_1}{i_2} \right|_{i_1 = 0} \quad \text{and} \quad z_{22} = \left. \frac{v_2}{i_2} \right|_{i_1 = 0} \tag{B.4}$$

respectively.

Finding z Parameters by Measurement. To measure z_{11} and z_{21}, we open-circuit the output port and apply a source to the input port as in Fig. B.2b. We then measure the numerator and denominator quantities of Eqs. (B.2) and (B.3) and compute the ratios. To measure z_{22} and z_{12}, we open-circuit the input port and connect a source to the output, as in Fig. B.2c. After measuring the required variables, we compute the ratios of Eq. B.4. Because the two-port equations are valid for any terminations, we can use either a current source or a voltage source for these measurements. Sometimes one type of source is physically more convenient than the other or leads to more precise measurements. If sinusoidal steady-state excitation is appropriate, the voltage and current variables and the z parameters all involve both magnitude and phase.

Finding Parameters by Circuit Analysis. To find z parameters by circuit analysis, we apply the terminations of Figs. B.2b and c to the circuit diagram and then solve for the ratios using linear circuit analysis. For sinusoidal steady state, this analysis involves the usual complex algebra. Sometimes the analysis is carried out using Laplace transform

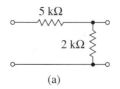

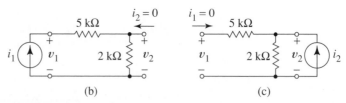

Figure B.3 Circuits for z-parameter calculations: (a) original circuit; (b) circuit for finding z_{11} and z_{21}; (c) circuit for finding z_{12} and z_{22}.

variables. As in measurements, the source may be either voltage or current. The following example demonstrates finding z parameters by circuit analysis.

EXAMPLE B.1   Find the z parameters for the circuit of Fig. B.3a

Solution. This circuit contains no capacitance or inductance, so dc analysis is appropriate. The open-circuit input impedance, z_{11}, is the resistance at the input port when the output port is open-circuited, as in Fig. B.3b. Following the letter of the definition, we would find $v_1 = i_1 \times 7 \text{ k}\Omega$. Then, from the definition, find that $z_{11} = 7 \text{ k}\Omega$. A more intuitive approach is to simply observe that the source i_1 "sees" an input impedance of 7 kΩ when it "looks into" the circuit. Also from Fig. B.3b, $v_2 = (2 \text{ k})i_1$. Using Eq. (B.3), $z_{21} = 2 \text{ k}\Omega$.

The open-circuit output impedance, z_{22}, is simply the 2 kΩ seen at the output nodes in Fig. B.3c. Also from Fig. B.3c, $v_1 = (2 \text{ k})i_2$ (because the voltage drop caused by i_1 in the 5 kΩ resistor $= 0$). From the first definition in Eq. (B.4), $z_{12} = 2 \text{ k}.\Omega$ ❏

z-**Parameter Circuit Model.** From Eqs. (B.1) it is easy to draw a simple equivalent circuit for any linear two-port in terms of z parameters. From the first equation, v_1 is the sum of two voltages: one the drop caused by i_1 as it flows through an impedance z_{11}, the other a voltage that *depends on* i_2. To model the latter, we need a CCVS. Figure B.4a

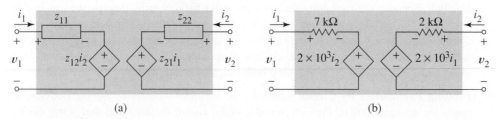

Figure B.4   (a) Equivalent circuit using z parameters; (b) z-parameter equivalent for voltage divider of Example B.1.

shows how these observations translate into a circuit diagram. This figure also shows how a simple two-component output circuit realizes the second of Eqs. (B.1). No matter how large or complex the actual circuit within the two-port "box," once the z parameters are known, we can always replace it with Fig. B.4a.

Figure B.4 shows the z-parameter equivalent of the voltage divider of Example B.1 and Fig. B.3a. Although the equivalent is more complex than the original divider, such circuit models are conceptually very important in unifying feedback amplifier theory.

Similar equivalent circuits exist for each of the other two-port parameter descriptions we will study. In each case the "11" and "22" parameters physically represent loading effects at the input and output ports, respectively. The "21" parameter always describes forward signal flow through the two-port and the "12" parameter represents reverse" signal flow or *internal feedback* of the signal from the output back to the input. Two-ports for which the "12" parameter is zero are appropriately called *unilateral* two-ports because signals flow only in the forward direction. Examining Figs. B.3a and B.4b suggests that an internal feedback mechanism exists even within the prosaic voltage divider, a small gem of understanding that does not occur to most of us when we look at the divider circuit itself.

B.3
y Parameters

The y-parameter, two-port equations are

$$i_1 = y_{11}v_1 + y_{12}v_2$$
$$i_2 = y_{21}v_1 + y_{22}v_2$$

(B.5)

Figure B.5a depicts these equations as expressions of superposition; therefore, they obtain for *any* terminations, not just for voltage sources.

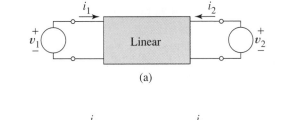

(a)

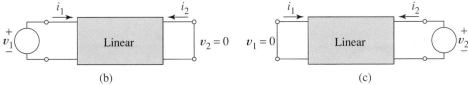

(b)   (c)

Figure B.5 Terminations for y parameters: (a) terminations to justify superposition interpretation; (b) terminations for finding y_{11} and y_{21}; (c) terminations for finding y_{12} and y_{22}.

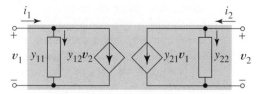

Figure B.6 Equivalent circuit for y parameters.

y-Parameter Definitions. Isolating the individual terms in Eq. (B.5) by turning off the sources one at a time leads to the definitions

$$y_{11} = \left. \frac{i_1}{v_1} \right|_{v_2 = 0,} \qquad y_{21} = \left. \frac{i_2}{v_1} \right|_{v_2 = 0}$$

$$y_{12} = \left. \frac{i_1}{v_2} \right|_{v_1 = 0,} \qquad y_{22} = \left. \frac{i_2}{v_2} \right|_{v_1 = 0} \tag{B.6}$$

the measurement/analysis circuits of Figs. B.5b and B.5c, and the descriptive names

y_{11}: short-circuit input admittance

y_{21}: short-circuit, forward transfer admittance

y_{12}: short-circuit, reverse transfer admittance

y_{22}: short-circuit output admittance

It is worth noting that y_{11} (y_{22}) is *not* the reciprocal of z_{11} (z_{22}) because the parameters are defined for different terminating conditions.

y-Parameter Equivalent Circuit. The defining equations, (B.5), have the equivalent circuit interpretation of Fig. B.6, where we take each equation to represent a current summation at an internal node. Again, we need dependent sources to represent the transfer parameters.

EXAMPLE B.2 Find the y parameters of the two-port of Fig. B.7a by circuit analysis.

Solution. Since the diagram contains LC elements, interpret Eqs. (B.5) as *phasor equations* that relate voltage and current phasors at some angular frequency ω. Then the desired y parameters are complex numbers. Figure B.7b shows how to terminate the two-port for finding y_{11} and y_{21}. Complex impedances represent the reactive elements, so we use uppercase characters for phasor variables.

Because of the short circuit, the voltage across L_2 and, therefore, its current are zero. Notice also that the short-circuit places V_1 directly in parallel with C_3. Since y_{11} is the admittance that V_1 "sees" looking into the circuit, y_{11} is the admittance of L_1 in parallel with the admittance of C_3. Thus

$$y_{11} = \frac{1}{j\omega L_1} + j\omega C_3$$

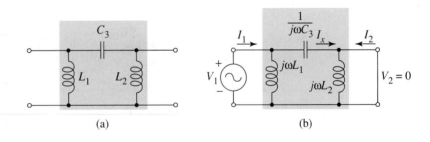

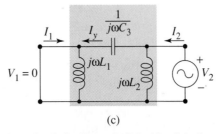

Figure B.7 Equivalent circuit for Example B.2: (a) original two-port; (b) two-port terminated for finding y_{11} and y_{21}; (c) two-port terminated for finding y_{22} and y_{12}.

To find y_{21}, first notice that $I_2 = -I_x$. But since C_3 is in parallel with V_1, $I_x = j\omega C_3 V_1$. Substituting $-j\omega C_3 V_1$ for I_2 in the y_{21} definition in Eq. (B.6) gives

$$y_{21} = -j\omega C_3$$

From Fig. B.7c, the same kinds of analysis gives

$$y_{22} = j\omega C_3 + \frac{1}{j\omega L_2}$$

and

$$y_{12} = -j\omega C_3 \qquad \square$$

B.4

h Parameters

Equations and Equivalent Circuit. Figure B.8a shows the terminations which justify the h-parameter equations, Eqs. (B.7), as statements of superposition. For this *hybrid* description, the input termination is that used for z parameters, while the output termination is that used for y parameters.

$$v_1 = h_{11}i_1 + h_{12}v_2$$
$$i_2 = h_{21}i_1 + h_{22}v_2$$
(B.7)

h-Parameter Definitions. Employing the two-port procedures and terminology that, by now, are becoming quite familiar, we use Eqs. (B.8) to define the short-circuit input impedance, h_{11}; the open-circuit, reverse voltage gain, h_{12}; the short-circuit, forward cur-

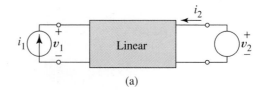

(a)

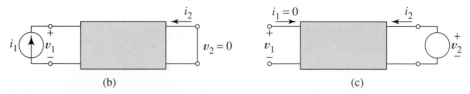

(b)   (c)

Figure B.8 Two-port terminations for *h* parameters: (a) for understanding definitions; (b) for finding h_{11} and h_{21}; (c) for finding h_{12} and h_{22}.

rent gain, h_{21}; and the open-circuit, output admittance, h_{22}. Figures B.8b and B.8c show the corresponding terminations used for measurement or analysis.

$$h_{11} = \left.\frac{v_1}{i_1}\right|_{v_2=0,} \qquad h_{21} = \left.\frac{i_2}{i_1}\right|_{v_2=0}$$

$$h_{12} = \left.\frac{v_1}{v_2}\right|_{i_1=0,} \qquad h_{22} = \left.\frac{i_2}{v_2}\right|_{i_1=0}$$

***h*-Parameter Equivalent Circuit.** Interpreting the first line of Eq. (B.7) as KVL applied to an input circuit loop and the second line as KCL applied at an output node leads to the circuit model of Fig. B.9.

EXAMPLE B.3 Find the *h* parameters for the circuit of Fig. B.10a.

Solution. Since the circuit contains no *LC* elements, use dc analysis. For finding h_{11} and h_{21}, we place a short circuit across the output port and attach an independent current source to the input. This gives Fig. B.10b. To simplify the analysis, notice that r_o is shorted out and that the voltage across r_μ is v_1 (alias v_π).

Kirchhoff's current law at the upper input node gives

$$i_1 = \left(\frac{1}{r_\pi} + \frac{1}{r_\mu}\right)v_1 \qquad\qquad (B.9)$$

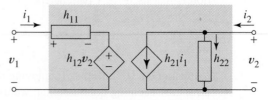

Figure B.9 *h*-Parameter equivalent circuit.

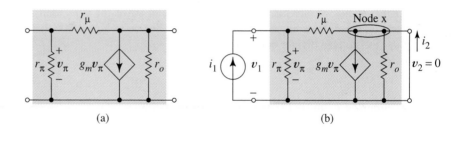

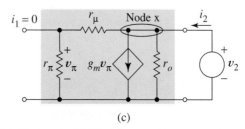

(c)

Figure B.10 (a) Linear two-port for Example B.3; (b) termination for finding h_{11} and h_{21}; (c) termination for finding h_{12} and h_{22}.

It follows from the definition that $h_{11} = r_\pi \parallel r_\mu$.

Kirchhoff's current law at node x gives

$$\frac{v_1}{r_\mu} = g_m v_1 - i_2$$

Thus

$$i_2 = \left(g_m - \frac{1}{r_\mu}\right) v_1 \qquad\qquad (\text{B.10})$$

Solving Eq. (B.9) for v_1 and substituting into Eq. (B.10) gives

$$i_2 = \frac{(g_m - 1/r_\mu)}{(1/r_\pi + 1/r_\mu)} i_1 \qquad\qquad (\text{B.11})$$

The ratio, i_2/i_1, is

$$h_{21} = \frac{(g_m - 1/r_\mu)}{(1/r_\pi + 1/r_\mu)}$$

Figure B.10c shows the terminations to use for finding the other h parameters. Applying KCL at node x gives

$$i_2 = \frac{v_2}{r_o} + g_m v_\pi + \frac{v_2}{r_\mu + r_\pi} \qquad\qquad (\text{B.12})$$

Examining (B.12) shows we are not quite done. We cannot solve for i_2/v_2 until v_π, in the second term on the right, is expressed in terms of v_2. Notice in Fig. B.10c that v_π is related to v_2 by the voltage divider equation

$$v_\pi = \frac{r_\pi}{r_\pi + r_\mu} v_2 \qquad (B.13)$$

Using this to eliminate v_π in (B.12) and solving for the defining ratio gives

$$h_{22} = \frac{i_2}{v_2} = \frac{1}{r_o} + \frac{g_m r_\pi}{r_\pi + r_\mu} + \frac{1}{r_\pi + r_\mu} \qquad (B.14)$$

In applying the definition of h_{12} we notice that the input voltage, v_1, in Fig. B.10c is the same as v_π; thus from (B.13),

$$h_{12} = \frac{v_1}{v_2} = \frac{r_\pi}{r_\pi + r_\mu} \qquad (B.15)$$

❑

The preceding example shows that while the details of applying the definitions can be challenging, the principles are simple. For very complex numerical problems, we can use SPICE to work out the details as long as we understand how to set up the circuits, provide the proper terminations, and interpret the results. After defining g parameters we will demonstrate this application of SPICE.

B.5
g Parameters

The final two-port parameter set is another *hybrid* description. Since h is already taken, we call these g parameters.

g-Parameter Equivalent Circuit. Figure B.11a shows terminations that suggest the following two-port description:

$$i_1 = g_{11}v_1 + g_{12}i_2$$
$$v_2 = g_{21}v_1 + g_{22}i_2 \qquad (B.16)$$

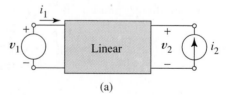

(a)

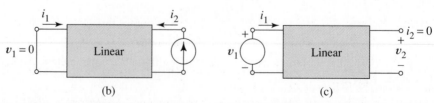

(b)   (c)

Figure B.11 Terminations for g parameters: (a) for understanding definitions; (b) for finding g_{21} and g_{22}; (c) for finding g_{11} and g_{21}.

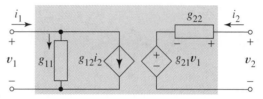

Figure B.12 Equivalent circuit for g parameters.

g-Parameter Definitions. From Eq. (B.16), we derive the following g-parameter definitions:

$$g_{11} = \frac{i_1}{v_1}\bigg|_{i_2 = 0,} \qquad g_{21} = \frac{v_2}{v_1}\bigg|_{i_2 = 0}$$

$$g_{12} = \frac{i_1}{i_2}\bigg|_{v_1 = 0,} \qquad g_{22} = \frac{v_2}{i_2}\bigg|_{v_1 = 0}$$

These define, respectively, *the open-circuit input admittance*, the *open-circuit forward voltage gain*, the *short-circuit reverse current gain*, and the *short-circuit output impedance*.

g-Parameter Equations and Equivalent Circuit. Figure B.12 shows the equivalent circuit that corresponds to Eqs. (B.16).

SPICE Analysis of Two-Ports. An easy way to find *numerical values* for dc two-port parameters is to use the transfer function statement, .TF, in SPICE. We write the code for the two-port circuit in the usual fashion and terminate the two-port with independent sources of the types used in the superposition representation of the two-port equations. That is, we use current sources for z parameters, voltage sources for y parameters, and so forth. We then carry out the SPICE analysis in two runs. In the first run, the input source is given unit value and the output source is turned off. SPICE then gives input resistance, output resistance, and the "21" parameter. The input or output resistance calculated by SPICE will be either the desired parameter or its reciprocal. We use a second SPICE run to find the "12" parameter, this time with the input source turned off and the output source assigned unit value. Some care is necessary to correctly interpret the algebraic signs of the results, since SPICE follows the passive sign convention for sources. The following examples demonstrate the procedure.

EXAMPLE B.4 Use SPICE to find h parameters for the circuit of Fig. B.10a when $r_\pi = r_\mu = r_o = 1$ kΩ and $g_m = 0.1$ A/V.

Solution. Figure B.13a is Fig. B.10a, with input and output terminations appropriate for finding h parameters, and Fig. B.13 is the code. Source V2 is set to zero to produce the short circuit at the output required for two of the h parameters. The .TF line requests that SPICE compute the ratio of the short-circuit output current to input current by defining these for SPICE as I(V2) and I1, respectively.

Figure B.13c shows the SPICE results and their two-port interpretation. SPICE has computed the output resistance under proper terminating conditions; however, h_{22} is de-

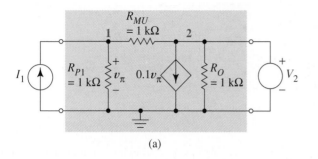

```
*H-PARAMETER CIRCUIT
RPI 1 0 1K
RMU 1 2 1K
RO 2 0 1K
G 2 0 1 0 0.1
I1 0 1 DC 1
V2 2 0 DC 0
.TF I(V2) I1
.END
```

(a)   (b)

Spice Output	Interpretation
Input resistance = 500 Ω	h_{11}
Output resistance = 19.4 Ω	$1/h_{22}$
I(V2)/I1 = -49.5	$-h_{21}$

(c)

Figure B.13 SPICE model: (a) circuit for finding h parameters for Fig. B.10a; (b) SPICE code; (c) SPICE output and interpretation of results.

fined as the output admittance under these conditions. Also, SPICE defines the current in V2 as the current flowing into the "+" terminal. Since this is opposite to the two-port sign convention, the transfer function found by SPICE was the negative of h_{21}. We can verify these values by substituting numerical values into the algebraic expressions derived in Example B.3.

To find the remaining h parameter, we change the three lines of code preceding .END in Fig. B.13c to

```
I1 01 DC 0
V2 2 0 DC 1
.TF V(1) V2
```

These changes turn off the input source, turn on the output source, and ask SPICE to calculate the reverse open-circuit voltage gain as transfer function. SPICE again calculates input and output resistances, but also produces the new result V(1)/V2 = −0.500, which is h_{12}. ❑

EXAMPLE B.5 Use SPICE to find z parameters for Fig. B.10a when parameter values are $r_\pi = r_\mu = r_o = 1$ kΩ and $g_m = 0.1$ A/V.

Solution. Figure B.14a shows the two-port, now terminated with current sources to find z parameters, and Fig. B.14b shows the code. Figure B.14c shows the results and

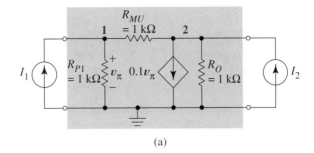

(a)

```
*Z-PARAMETER CIRCUIT
RPI 1 0 1K
RMU 1 2 1K
RO 2 0 1K
G 2 0 1 0 0.1
I1 0 1 DC 1
I2 0 2 DC 0
.TF V(2) I1
.END
```

(b)

Spice Output	Interpretation
Input resistance = 19.4 Ω	z_{11}
Output resistance = 19.4 Ω	z_{22}
V(2)/I1 = -961.2	z_{21}

(c)

Figure B.14   (a) SPICE model for finding z parameters for Fig. B.10a; (b) SPICE code; (c) interpretation of SPICE output.

their interpretation in terms of z parameters. To find the remaining parameter, we change the three lines of code preceding .END to

```
I1 01 DC 0

I2 0 2 DC 1

.TF V(1) I2
```

These changes turn off the input source, turn on the output source, and ask for the reverse, open-circuit transfer impedance. Included in the SPICE output is the new result $V(1)/I2 = -961.2\ \Omega$, which is z_{12}.   ❏

With these concepts, we can also use the SPICE .TF command and dc analysis to find the y and g parameters for any circuit.

.TF is not available for ac analysis; however, it is easy to apply independent sources of unit current and voltage as required by the definitions to find any two-port parameters, as we show in the next example.

EXAMPLE B.6   Use SPICE to find the y parameters of the passive two-port of Fig. B.15a at 1.59 MHz.

Solution.   The voltage sources in Fig. B.15a are not part of the two-port under test, but are external terminations provided for test purposes. Thus the code of Fig. B.15b applies

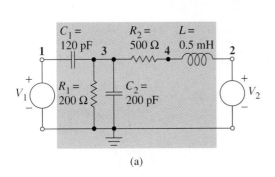

```
Y-PARAMETERS
C1 1 3 120P
C2 3 0 200P
R1 3 0 200
R2 3 4 500
L 4 2 .5E-3
V1 1 0 AC 1
V2 2 0 AC 0
.AC LIN 1 1.59E6 1.59E6
.PRINT AC IM(V2) IP(V2)
+ IM(V1) IP(V1)
.END
```

(a)   (b)

IM(V1) = 1.09E-3, IP(V1) = -101° y_{11} = 1.09E-3 /79°	IM(V2) = 2.02E-4, IP(V2) = 97.5° y_{22} = 2.02E-4 /-82.5°
IM(V2) = 4.08E-5, IP(V1) = -25° y_{21} = 4.08E-5 /155°	IM(V1) = 4.08E-5, IP(V1) = -25° y_{12} = y_{21} = 4.08E-5 /155°

(c)

Figure B.15 Circuit for Example B.6: (a) circuit schematic; (b) SPICE code for finding y_{11} and y_{21}; (c) SPICE results.

a voltage phasor of unit amplitude to the input and a short circuit to the output. The .PRINT line requests magnitudes and phases of input and output currents. Since the input voltage has unit amplitude and zero phase, the output current magnitudes and phases would be those of y_{11} and y_{21} if the two-port current convention were the same as the SPICE sign convention; however, we need a 180° phase correction of the output. That is, the two-port currents are $I_2 = -\text{I(V2)}$ and $I_1 = -\text{I(V1)}$. Figure B.15c shows the results, including those obtained by running the program a second time with V2 ON and V1 OFF. ❑

Our closing observations are

1. The "two-source" procedure introduced in the last example is very straightforward compared to using .TF because it is related so closely to the definitions. This procedure applies to both dc and ac problems, avoiding .TF entirely and making less to remember.
2. Although SPICE analysis is easily applied to very complex circuits, it gives only numerical solutions that fall far short of symbolic expressions in providing design insight.
3. A good understanding of both the theory and the SPICE conventions is needed to make effective use of SPICE results.

PROBLEMS _____

B1 Find z parameters for the nonideal amplifier of Fig. P-B1.

B2 Find the z parameters at $\omega = 10^4$ rad/s for Fig. P-B2.

B3 Draw a z-parameter equivalent circuit for Fig. P-B3. Show numerical values.

B4 Find the y parameters for Fig. P-B4 as a function of ω.

B5 For Fig. P-B5, draw the equivalent circuit, including values, using (a) z parameters, (b) y parameters, (c) h parameters, (d) g parameters.

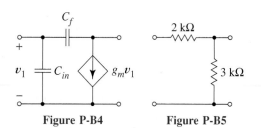

Figure P-B4   **Figure P-B5**

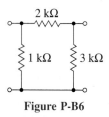

Figure P-B6

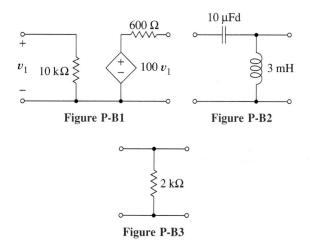

Figure P-B1   **Figure P-B2**

B6 For Fig. P-B6, draw the equivalent circuit, including values, using (a) z parameters, (b) y parameters, (c) h parameters, (d) g parameters.

B7 Find the h parameters for Fig. B.4a. Then answer the question: "Is it possible to convert from one set of two-port parameters to another?"

Figure P-B3

C

SPICE Parameters for Bipolar Transistors

Figures 4.40 and 4.44 show, respectively, the dynamic SPICE transistor model and the parameters required to use it in simulations. For simplicity, the calculations that follow are limited to finding those parameters necessary for forward active operation. Once these are found, we can accurately simulate forward active operation of the BJT *at any Q-point.*

From the manufacturer's data sheet we usually begin with the following information:

1. Coordinates of some particular point (V_{BE}, I_E) on the transistor input characteristic
2. Measured values of h_{fe}, h_{ie}, h_{oe}, and h_{re}, and the value of I_C where the measurements were made
3. Measured values of the depletion capacitances, specifically collector-base capacitance C_μ, emitter-base capacitance C_{je}, collector-substrate capacitance C_{CIO}, and the junction voltages at which the measurements were made (We sometimes encounter alternative notations such as $C_\mu = C_c = C_{ob} = C_{cb} = C_{CBO}$, and $C_{je} = C_{EB0}$.)
4. The unity-gain frequency, f_τ, and the coordinates, (V_{CE}, I_C), at which the frequency measurement was made.

Calculation Procedure. We now list a calculation sequence that leads to the static and dynamic SPICE parameters. I_C is the current at which the h parameters were measured. The SPICE parameters we seek are highlighted in the equations.

Static Parameters

1. Compute the saturation current from measured values of I_E and V_{BE}; that is,

$$\mathbf{IS} = I_E e^{-V_{BE}/V_T} \tag{C.1}$$

using $V_T = 0.025$ V at room temperature.

2. Find forward beta from

$$\mathbf{BF} = \beta = h_{fe} \qquad \text{(C.2)}$$

3. Compute g_m using

$$g_m = \frac{I_C}{V_T} \qquad \text{(C.3)}$$

Then

$$r_\pi = \frac{\beta}{g_m} \qquad \text{(C.4)}$$

$$r_\mu = \frac{r_\pi}{h_{re}} \qquad \text{(C.5)}$$

$$r_o = \frac{r_\mu}{r_\mu h_{oe} - \beta} \qquad \text{(C.6)}$$

4. $$\mathbf{VAF} = r_o I_C \qquad \text{(C.7)}$$

5. Find the "zero-bias base resistance" from

$$\mathbf{RB} = r_X = h_{ie} - r_\pi \qquad \text{(C.8)}$$

DYNAMIC MODEL PARAMETERS

The following calculations give the SPICE parameters that model dynamics in a forward active transistor.

Depletion Capacitances. First, calculate the zero-voltage values of the collector-base, emitter-base, and collector-substrate depletion capacitances, using measured values of each capacitance at a known voltage. Unless information is given to the contrary, assume the collector-substrate and collector-emitter junctions are abrupt and use

$$\mathbf{MJC} = \mathbf{MJS} = 0.5$$

A graded junction is a good model for a diffused collector-emitter interface; therefore,

$$\mathbf{MJE} = 0.33$$

Reasonable values for the built-in barrier potentials are

$$\mathbf{VJC} = 0.55 \text{ V}$$

$$\mathbf{VJE} = 0.7 \text{ V}$$

$$\mathbf{VJS} = 0.52 \text{ V}$$

6. Knowing C_μ at a particular voltage V_{CB}, calculate **CJC** from

$$C_\mu = \frac{\textbf{CJC}}{(1 + V_{CB}/\textbf{VJC})^{\textbf{MJC}}} \tag{C.9}$$

7. Find **CJE,** from C_{je} and V_{EB} using

$$C_{je} = \frac{\textbf{CJE}}{(1 + V_{EB}/\textbf{VJE})^{\textbf{MJE}}} \tag{C.10}$$

8. Compute **CJS** from

$$C_{CI0} = \frac{\textbf{CJS}}{(1 + V_{C1}/\textbf{VJS})^{\textbf{MJS}}} \tag{C.11}$$

(Omit this step for discrete transistors.)

Forward Transit Time. Some final calculations give the forward transit time **TF**. If f_τ was measured at a different I_C than the h parameters and the capacitances, capacitances and h parameters must first be recomputed.

9. i. Use Eq. (C.3) to compute g_m at the I_C value where f_τ was measured.
   ii. From Eq. (C.9), recompute C_μ using the value of V_{CB} at which f_τ was measured. If V_{CE} is given for the f_τ measurement, use $V_{CB} = V_{CE} - 0.7$ V. Now **CJC** is known.
   iii. With the new values of g_m and C_μ, compute

$$C_\pi = \frac{g_m}{2\pi f_\tau} - C_\mu \tag{C.12}$$

   iv. Estimate C_{je}, the part of C_π attributed to depletion capacitance, *at the Q-point where f_τ was measured.* For strongly forward-biased pn junctions Eq. (C.10) does not give correct results, and a suggested estimate is

$$C_{je} = 2 \times \textbf{CJE} \tag{C.13}$$

   v. Compute the charge-storage part, C_b, of C_π from

$$C_b = C_\pi - C_{je} \tag{C.14}$$

   vi. To find the forward transit time, τ_T, we use Eq. (8.18). In SPICE notation,

$$\textbf{TF} = \frac{V_T}{I_C} C_b \tag{C.15}$$

EXAMPLE C.1 For the CA3086, a general-purpose npn transistor array, the manufacturer provides curves of V_{BE} versus temperature for several values of I_E. A room temperature value from one of these curves is $(V_{BE}, I_E) = (0.68$ V, 0.5 mA$)$. Table C.1 lists other data sheet information. Find the SPICE parameters for the BJTs.

TABLE C.1 Measured Values for CA3086 npn Transistor Array

Parameter	Numerical Value	Measurement Conditions
h_{fe}	100	$I_C = 1$ mA, $V_{CE} = 3$ V
h_{ie}	3.5 kΩ	$I_C = 1$ mA, $V_{CE} = 3$ V
h_{oe}	15.6×10^{-6} A/V	$I_C = 1$ mA, $V_{CE} = 3$ V
h_{re}	1.8×10^{-4}	$I_C = 1$ mA, $V_{CE} = 3$ V
$C_{CBO} = C_\mu$	0.58 pF	$V_{CB} = 3$ V
$C_{EBO} = C_{je}$	0.60 pF	$V_{EB} = 3$ V
$C_{CI} = C_{CI0}$	2.8 pF	$V_{CI} = 3$ V
f_τ	550 MHz	$I_C = 3$ mA, $V_{CE} = 3$ V

Solution. We find the static parameters using Eqs. (C.1) through (C.8):

$$\mathbf{IS} = \frac{0.5 \times 10^{-3}}{e^{0.68/0.025}} = 7.69 \times 10^{-16} \text{ A}$$

$$\mathbf{BF} = h_{fe} = 100$$

$$g_m = 0.001/0.025 = 0.04 \text{ A/V}$$

$$r_\pi = 100/0.04 = 2.5 \text{ k}\Omega$$

$$r_\mu = 2.5 \text{ k}/1.8 \times 0^{-4} = 1.39 \times 10^7 \Omega$$

$$r_o = \frac{1.39 \times 10^7}{1.39 \times 10^7 (15.6 \times 10^{-6}) - 100} = 199 \text{ k}\Omega$$

$$\mathbf{VAF} = 119 \times 10^3 \times 10^{-3} = 119 \text{ V}$$

$$\mathbf{RB} = 3.5 \text{ k}\Omega - 2.5 \text{ k}\Omega = 1 \text{ k}\Omega$$

From Eqs. (C.9) through (C.11),

$$\mathbf{CJC} = 0.58 \times 10^{-12}(1 + 3/0.55)^{0.5} = 1.47 \text{ pF}$$

$$\mathbf{CJE} = 0.60 \times 10^{-12}(1 + 3/0.7)^{0.33} = 1.04 \text{ pF}$$

$$\mathbf{CJS} = 2.8 \times 10^{-12}(1 + 3/0.52)^{0.5} = 7.28 \text{ pF.}$$

To find **TF,** we first compute C_μ at $V_{CB} = 3 - 0.7 = 2.3$ V, the bias voltage when f_τ was measured.

$$C_\mu = \frac{1.47 \times 10^{-12}}{(1 + 2.3/0.55)^{0.5}} = 0.648 \text{ pF}$$

Next, we compute C_π at this same Q-point. For Eq. (C.12) we require g_m at $I_C = 3$ mA or $g_m = 0.003/0.025 = 0.12$ A/V. Then

$$C_\pi = \left(\frac{0.120}{2\pi 550 \times 10^6}\right) - 0.648 \times 10^{-12} = 34.1 \text{ pF}$$

From the estimate of Eq. (C.13),

$$C_{je} = 2 \times \textbf{CJE} = 1.60 \text{ pF}$$

and from Eq. (C.14),

$$C_b = 34.1 - 1.60 = 32.5 \text{ pF}$$

Finally, Eq. (C.15) gives

$$\textbf{TF} = \frac{0.025}{0.003}32.5 \times 10^{-12} = 0.271 \text{ ns}$$

❏

Index

A

Abrupt junction, 209
Absolute value, 190
ac analysis with SPICE, 601–3, 1141–42
ac beta, 483
Acceptor impurities, 147
Access time, 1117, 1120
ac feedback, 696–97
A circuit (in feedback amplifier), 669, 687, 698–99
Active filter, 890, 892
Active load, 428, 518–24, 828, 837
  amplifier, 558–65
  circuit, 447, 461
Active pull-up, 998–99, 1037
Active region:
  of transistor, 273
  of transistor characteristics, 390
Active state, 316
  BJT, 235–36, 241
  depletion MOSFET, 319
  enhancement MOSFET, 313–14, 319
  FET model, 315, 337
  JFET, 349
  MESFET, 347
ac to dc converter, 195–96
A/D (analog-to-digital) converter, 1075, 1079, 1125
  counting-type, 1081
  successive approximation, 1081–83
Address, 1114
  bits, 1115

bus, 1120
  hold time, 1120
  setup time, 1120
AGC (automatic gain control), 857
Aging of components, 401, 414
Alarm circuit, 1059
Aliasing distortion, 948
All-pass filter, 911
  circuit, 915–16
  function, 915–17
Alpha:
  forward, 233
  reverse, 234
Ambient temperature, 753
Amplifier:
  audio, 742
  biasing circuit, 254
  buffer, 72, 509
  capacitances, 46
  capacitively coupled, 489
  cascaded, 21, 55, 498
  cascode, 513–14, 644, 828
  class A, 751, 759–66
  class AB, 752, 777–91
  class B, 752, 770–77
  class C, 752
  class D, 752, 794–800
  CMOS, 518
  common-base, 491, 502, 615–19
  common-collector, 508–9
  common-collector/common-base, 514, 515
  common-collector/common-emitter, 514, 515
  common-drain, 491, 505–7

  common-emitter, 491–96
    at high-frequencies, 621–23
  common-gate, 491, 499–502
    at high frequencies, 613–15
  common-source, 491, 496–99
    at high frequencies, 625–27
  current, 16, 21, 670
  difference, 23, 74–77, 528–57, 566–71
    emitter-coupled, 541–57
    source-coupled, 528–41
  digital-signal, 1054
  direct-coupled, 446
  double-cascode, 582
  dynamics, 46–55
  feedback, 669–721
  frequency response, 48, 49, 589–91
  half-power frequency, 47, 601, 606, 628 (*see also* −3 dB frequency)
  ideal, 12
  input impedance, 627
  instrumentation, 74
  inverting, 17, 68
  multiple-transistor, 512–18
  multistage, 23 (*see also* Cascaded amplifiers)
  need for, 14
  nonideal, 18
  noninverting, 72, 125–26
  power, 751–800
  pulse response, 51–54
  sense, 1122–23
  stable, 707–17, 741
  summing, 70, 125–26

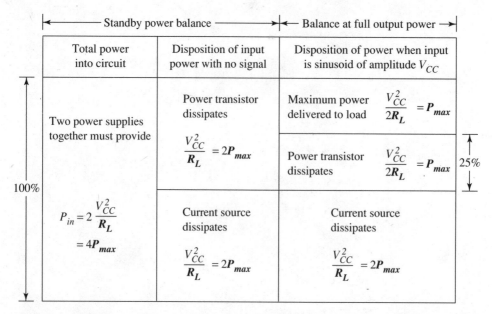

	Standby power balance		Balance at full output power
	Total power into circuit	Disposition of input power with no signal	Disposition of power when input is sinusoid of amplitude V_{CC}
100%	Two power supplies together must provide $$P_{in} = 2\,\frac{V_{CC}^2}{R_L}$$ $$= 4P_{max}$$	Power transistor dissipates $$\frac{V_{CC}^2}{R_L} = 2P_{max}$$	Maximum power delivered to load $\dfrac{V_{CC}^2}{2R_L} = P_{max}$ Power transistor dissipates $\dfrac{V_{CC}^2}{2R_L} = P_{max}$ (25%)
		Current source dissipates $$\frac{V_{CC}^2}{R_L} = 2P_{max}$$	Current source dissipates $$\frac{V_{CC}^2}{R_L} = 2P_{max}$$

Class A Amplifier Power Balance in Terms of Specified Maximum Load Power, P_{max}, and Load Resistance, R_L

Maximum output power $V_o = V_{CC}$		Maximum transistor dissipation $V_o = \dfrac{2}{\pi} V_{CC}$	
Power from supplies $$P_{in} = 1.27P_{max}$$	Load power $$P_{max} = \frac{V_{CC}^2}{2\,R_L}$$	Power from supplies $$P_{in} = 0.812P_{max}$$	Delivered load power $= 0.406P_{max}$
			$P_{D1} : 0.203P_{max}$
			$P_{D2} : 0.203P_{max}$
	$P_{D1} : 0.135P_{max}$		
	$P_{D2} : 0.135P_{max}$		

Class B Amplifier Power Balance in Terms of Specified Maximum Load Power, P_{max}, and Load Resistance, R_L